T0181108

Lecture Notes in Computer Science 12829

More information about this subseries at http://www.springer.com/series/7412

Frank Nielsen · Frédéric Barbaresco (Eds.)

Geometric Science of Information

5th International Conference, GSI 2021
Paris, France, July 21–23, 2021
Proceedings

Editors
Frank Nielsen
Sony Computer Science Laboratories Inc.
Tokyo, Japan

Frédéric Barbaresco
THALES Land & Air Systems
Meudon, France

ISSN 0302-9743 ISSN 1611-3349 (electronic)
Lecture Notes in Computer Science
ISBN 978-3-030-80208-0 ISBN 978-3-030-80209-7 (eBook)
https://doi.org/10.1007/978-3-030-80209-7

LNCS Sublibrary: SL6 – Image Processing, Computer Vision, Pattern Recognition, and Graphics

Cover page painting: Woman teaching Geometry, from French medieval edition of Euclid's Elements (14th century) © The British Library, used with granted permission.

This Springer imprint is published by the registered company Springer Nature Switzerland AG
The registered company address is: Gewerbestrasse 11, 6330 Cham, Switzerland

Preface

At the turn of the twenty-first century, new and fruitful interactions were discovered between several branches of science: information sciences (information theory, digital communications, statistical signal processing, etc.), mathematics (group theory, geometry and topology, probability, statistics, sheaves theory, etc.), and physics (geometric mechanics, thermodynamics, statistical physics, quantum mechanics, etc.). The aim of the Geometric Science of Information (GSI) biannual international conference is to discover mathematical structures common to all these disciplines by elaborating a "General Theory of Information" embracing physics, information science, and cognitive science in a global scheme.

As for GSI 2013, GSI 2015, GSI 2017, and GSI 2019 (https://franknielsen.github.io/GSI/), the objective of the 5th International SEE Conference on Geometric Science of Information (GSI 2021), hosted in Paris, was to bring together pure and applied mathematicians and engineers with a common interest in geometric tools and their applications for information analysis. GSI emphasizes the active participation of young researchers to discuss emerging areas of collaborative research on the topic of "Geometric Science of Information and their Applications". In 2021, the main theme was "LEARNING GEOMETRIC STRUCTURES", and the conference took place at Sorbonne University, France, during July 21–23, 2021, both in-person and virtually via video conferencing.

The GSI conference cycle was initiated by the Léon Brillouin Seminar team as early as 2009 (http://repmus.ircam.fr/brillouin/home). The GSI 2021 event was motivated in the continuity of the first initiative launched in 2013 (https://www.see.asso.fr/gsi2013) at Mines ParisTech, consolidated in 2015 (https://www.see.asso.fr/gsi2015) at Ecole Polytechnique, and opened to new communities in 2017 (https://www.see.asso.fr/gsi2017) at Mines ParisTech, and 2019 (https://www.see.asso.fr/gsi2019) at ENAC. In 2011, we organized an Indo-French workshop on the topic of "Matrix Information Geometry" that yielded an edited book in 2013, and in 2017, we collaborated with the CIRM seminar in Luminy, TGSI 2017 "Topological & Geometrical Structures of Information" (http://forum.cs-dc.org/category/94/tgsi2017).

GSI satellites events were organized in 2019 and 2020 as FGSI 2019 "Foundation of Geometric Science of Information" in Montpellier, France (https://fgsi2019.sciencesconf.org/) and SPIGL 2020 "Joint Structures and Common Foundations of Statistical Physics, Information Geometry and Inference for Learning" in Les Houches, France (https://www.springer.com/jp/book/9783030779566), respectively.

The technical program of GSI 2021 covered all the main topics and highlights in the domain of the "Geometric Science of Information" including information geometry manifolds of structured data/information and their advanced applications. This Springer LNCS proceedings consists solely of original research papers that have been carefully peer-reviewed by at least two or three experts. Accepted contributions were revised before acceptance.

As with the previous GSI conferences, GSI 2021 addressed interrelations between different mathematical domains like shape spaces (geometric statistics on manifolds and Lie groups, deformations in shape space, etc.), probability/optimization and algorithms on manifolds (structured matrix manifold, structured data/information, etc.), relational and discrete metric spaces (graph metrics, distance geometry, relational analysis, etc.), computational and Hessian information geometry, geometric structures in thermodynamics and statistical physics, algebraic/infinite dimensional/Banach information manifolds, divergence geometry, tensor-valued morphology, optimal transport theory, manifold and topology learning, and applications like geometries of audio-processing, inverse problems, and signal/image processing. GSI 2021 topics were enriched with contributions from Lie group machine learning, harmonic analysis on Lie groups, geometric deep learning, geometry of Hamiltonian Monte Carlo, geometric and (poly)symplectic integrators, contact geometry and Hamiltonian control, geometric and structure preserving discretizations, probability density estimation and sampling in high dimension, geometry of graphs and networks, and geometry in neuroscience and cognitive sciences.

The GSI 2021 conference was structured in 22 sessions as follows:

- **Probability and Statistics on Riemannian Manifolds** - Chairs: Xavier Pennec, Cyrus Mostajeran
- **Shapes Spaces** - Chairs: Salem Said, Joan Glaunès
- **Geometric and Structure Preserving Discretizations** - Chairs: Alessandro Bravetti, Manuel de Leon
- **Lie Group Machine Learning** - Chairs: Frédéric Barbaresco, Gery de Saxcé
- **Harmonic Analysis on Lie Groups** - Chairs: Jean-Pierre Gazeau, Frédéric Barbaresco
- **Geometric Mechanics** - Chairs: Gery de Saxcé, Frédéric Barbaresco
- **Sub-Riemannian Geometry and Neuromathematics** - Chairs: Alessandro Sarti, Dario Prandi
- **Statistical Manifold and Hessian Information Geometry** - Chairs: Noemie Combe, Michel Nguiffo Boyom
- **Information Geometry in Physics** - Chairs: Geert Verdoolaege, Jun Zhang
- **Geometric and Symplectic Methods for Hydrodynamical Models** - Chairs: Cesare Tronci, François Gay-Balmaz
- **Geometry of Quantum States** - Chairs: Florio Maria Ciaglia, Michel Berthier
- **Deformed Entropy, Cross-Entropy, and Relative entropy** - Chairs: Ting-Kam Leonard Wong, Léonard Monsaingeon
- **Geometric Structures in Thermodynamics and Statistical Physics** - Chairs: Hiroaki Yoshimura, François Gay-Balmaz
- **Geometric Deep Learning** - Chairs: Gabriel Peyré, Erik J. Bekkers
- **Computational Information Geometry 1** - Chairs: Frank Nielsen, Clément Gauchy
- **Computational Information Geometry 2** - Chairs: Giovanni Pistone, Goffredo Chirco
- **Optimal Transport and Learning** - Chairs: Yaël Frégier, Nicolas Garcia Trillos

- **Statistics, Information, and Topology** - Chairs: Pierre Baudot, Michel Nguiffo Boyom
- **Topological and Geometrical Structures in Neurosciences** - Chairs: Pierre Baudot, Giovani Petri
- **Manifolds and Optimization** - Chairs: Stéphanie Jehan-Besson, Bin Gao
- **Divergence Statistics** - Chairs: Michel Broniatowski, Wolfgang Stummer
- **Transport information geometry** - Chairs: Wuchen Li, Philippe Jacquet

In addition, GSI 2021 has hosted 6 keynote speakers:

- Yvette Kosmann-Schwarzbach on "Structures of Poisson Geometry: old and new"
- Max Welling on "Exploring Quantum Statistics for Machine Learning"
- Michel Broniatowski on "Some insights on statistical divergences and choice of models"
- Maurice de Gosson on "Gaussian states from a symplectic geometry point of view"
- Jean Petitot on "The primary visual cortex as a Cartan engine"
- Giuseppe Longo on "Use and abuse of "digital information" in life sciences, is Geometry of Information a way out?"

We would like to thank everyone involved in GSI 2021 for helping to make it such an engaging and successful event.

June 2021

Frank Nielsen
Frédéric Barbaresco

Organization

Conference Co-chairs

Frank Nielsen	Sony Computer Science Laboratories Inc., Japan
Frédéric Barbaresco	Thales Land and Air Systems, France

Local Organizing Committee

Gérard Biau	Sorbonne Center for AI, France
Xavier Fresquet	Sorbonne Center for AI, France
Michel Broniatowski	Sorbonne University, France
Jean-Pierre Françoise	Sorbonne University, France
Olivier Schwander	Sorbonne University, France
Gabriel Peyré	ELLIS Paris Unit and ENS Paris, France

Secretariat

Marianne Emorine	SEE, France

Scientific Committee

Bijan Afsari	Johns Hopkins University, USA
Pierre-Antoine Absil	Université Catholique de Louvain, Belgium
Stephanie Allasonnière	University of Paris, France
Jesus Angulo	Mines ParisTech, France
Marc Arnaudon	Bordeaux University, France
John Armstrong	King's College London, UK
Anne Auger	Ecole Polytechnique, France
Nihat Ay	Max Planck Institute, Germany
Roger Balian	CEA, France
Frédéric Barbaresco	Thales Land and Air Systems, France
Pierre Baudot	Median Technologies, France
Daniel Bennequin	University of Paris, France
Yannick Berthoumieu	Bordeaux University, France
Jeremie Bigot	Bordeaux University, France
Silvere Bonnabel	Mines ParisTech, France
Michel Boyom	Montpellier University, France
Marius Buliga	Simion Stoilow Institute of Mathematics of the Romanian Academy, Romania
Giovanna Citi	Bologna University, Italy
Laurent Cohen	Paris Dauphine University, France
Nicolas Couellan	ENAC, France

Ana Bela Cruzeiro	Universidade de Lisboa, Portugal
Remco Duits	Eindhoven University of Technology, Netherlands
Stanley Durrleman	Inria, France
Stephane Puechmorel	ENAC, France
Fabrice Gamboa	Institut Mathématiques de Toulouse, France
Jean-Pierre Gazeau	University of Paris, France
François Gay-Balmaz	Ecole Normale Supérieure, France
Mark Girolami	Imperial College London, UK
Audrey Giremus	Bordeaux University, France
Hatem Hajri	IRT SystemX, France
Susan Holmes	Stanford University, USA
Stephan Huckeman	University of Göttingen, Germany
Jean Lerbet	University of Évry Val d'Essonne, France
Nicolas Le Bihan	Grenoble Alpes University, France
Alice Le Brigant	Pantheon-Sorbonne University, France
Luigi Malago	Romanian Institute of Science and Technology, Romania
Jonathan Manton	University of Melbourne, Australia
Gaetan Marceau-Caron	Mila, Canada
Matilde Marcolli	CALTECH, USA
Jean-François Marcotorchino	Sorbonne University, France
Charles-Michel Marle	Sorbonne University, France
Hiroshi Matsuzoe	Nagoya Institute of Technology, Japan
Nina Miolane	Stanford University, USA
Jean-Marie Mirebeau	Paris-Sud University, France
Ali Mohammad-Djafari	Centrale Supelec, France
Antonio Mucherino	IRISA, University of Rennes 1, France
Florence Nicol	ENAC, France
Frank Nielsen	Sony Computer Science Laboratories Inc., Japan
Richard Nock	Data61, Australia
Yann Ollivier	Facebook, France
Steve Oudot	Inria, France
Pierre Pansu	Paris-Saclay University, France
Xavier Pennec	Inria, France
Giovanni Pistone	Collegio Carlo Alberto, Italy
Olivier Rioul	Telecom ParisTech, France
Gery de Saxcé	Lille University, France
Salem Said	Bordeaux University, France
Alessandro Sarti	EHESS, Paris, France
Rodolphe Sepulchre	Liège University, Belgium
Olivier Schwander	Sorbonne University, France
Stefan Sommer	Copenhagen University, Denmark
Dominique Spehner	Grenoble Alpes University, France
Wolfgang Stummer	Friedrich-Alexander-Universität Erlangen, Germany
Alain Trouvé	Ecole Normale Supérieure Paris-Saclay, France

Geert Verdoolaege	Ghent University, Belgium
Rene Vidal	Johns Hopkins University, USA
Jun Zhang	University of Michigan, USA
Pierre-Antoine Absil	Université Catholique de Louvain, Belgium
Florio M. Ciaglia	Max Planck Institute, Germany
Manuel de Leon	Real Academia de Ciencias, Spain
Goffredo Chirco	INFN, Naples, Italy
Hông Vân Lê	Institute of Mathematics of Czech Academy of Sciences, Czech Republic
Bruno Iannazzo	Università degli Studi di Perugia, Italy
Hideyuki Ishi	Osaka City University, Japan
Koichi Tojo	RIKEN, Tokyo, Japan
Ana Bela Ferreira Cruzeiro	Universidade de Lisboa, Portugal
Nina Miolane	Stanford University, USA
Cyrus Mostajeran	University of Cambridge, UK
Rodolphe Sepulchre	University of Cambridge, UK
Stefan Horst Sommer	University of Copenhagen, Denmark
Wolfgang Stummer	University of Erlangen-Nürnberg, Germany
Zdravko Terze	University of Zagreb, Croatia
Geert Verdoolaege	Ghent University, Belgium
Wuchen Li	UCLA, USA
Hiroaki Yoshimura	Waseda University, Japan
Jean Claude Zambrini	Universidade de Lisboa, Portugal
Jun Zhang	University of Michigan, USA

Abstracts of Keynotes

Structures of Poisson Geometry: Old and New

Yvette Kosmann-Schwarzbach

Professeur des Universités honoraire, CNRS, France
http://www.cmls.polytechnique.fr/perso/kosmann/

Abstract: How did the brackets that Siméon-Denis Poisson introduce in 1809 evolve into the Poisson geometry of the 1970's? What are Poisson groups and, more generally, Poisson groupoids? In what sense does Dirac geometry generalize Poisson geometry and why is it relevant for applications? I shall sketch the definition of these structures and try to answer these questions.

References

Libermann, P., Marle, C.-M.: Symplectic Geometry and Analytical Mechanics. D. Reidel Publishing Company (1987)

Marsden, J.E., Ratiu, T.S.: Introduction to Mechanics and Symmetry. Texts in Applied Mathematics, vol. 17, 2nd edn. Springer, New York. https://doi.org/10.1007/978-0-387-21792-5 (1998)

Laurent-Gengoux, C., Pichereau, and Vanhaecke, P.: Poisson Structures, Grundlehren der mathematischen Wissenschaften 347, Springer, Heidelberg (2013). https://doi.org/10.1007/978-3-642-31090

Kosmann-Schwarzbach, Y.: Multiplicativity from Lie groups to generalized geometry. In: Grabowska, K., et al., (eds.) Geometry of Jets and Fields, vol. 110. Banach Center Publications (2016)

Alekseev, A., Cattaneo, A., Kosmann-Schwarzbach, Y., Ratiu, T.: Lett. Math. Phys. **90** (2009). Special volume of LMP on Poisson Geometry, guest editors

Kosmann-Schwarzbach, Y., (éd.) Siméon-Denis Poisson : les Mathématiques au service de la science, Editions de l'Ecole Polytechnique (2013)

Kosmann-Schwarzbach, Y.: The Noether Theorems: Invariance and Conservation Laws in the Twentieth Century, Sources and Studies in the History of Mathematics and Physical Sciences. Springer, New York (2011). https://doi.org/10.1007/978-0-387-87868-3. translated by B. E. Schwarzbach

Exploring Quantum Statistics for Machine Learning

Max Welling

Informatics Institute, University of Amsterdam and Qualcomm Technologies
https://staff.fnwi.uva.nl/m.welling/
ELLIS Board Member (European Laboratory for Learning
and Intelligent Systems)
https://ellis.eu/

Abstract: Quantum mechanics represents a rather bizarre theory of statistics that is very different from the ordinary classical statistics that we are used to. In this talk I will explore if there are ways that we can leverage this theory in developing new machine learning tools: can we design better neural networks by thinking about entangled variables? Can we come up with better samplers by viewing them as observations in a quantum system? Can we generalize probability distributions? We hope to develop better algorithms that can be simulated efficiently on classical computers, but we will naturally also consider the possibility of much faster implementations on future quantum computers. Finally, I hope to discuss the role of symmetries in quantum theories.

Reference

Bondesan, R., Welling, M.: Quantum Deformed Neural Networks, arXiv:2010.11189v1 [quant-ph], 21 October 2020. https://arxiv.org/abs/2010.11189

Some Insights on Statistical Divergences and Choice of Models

Michel Broniatowski

Sorbonne Université, Paris

Abstract: Divergences between probability laws or more generally between measures define inferential criteria, or risk functions. Their estimation makes it possible to deal with the questions of model choice and statistical inference, in connection with the regularity of the models considered; depending on the nature of these models (parametric or semi-parametric), the nature of the criteria and their estimation methods vary. Representations of these divergences as large deviation rates for specific empirical measures allow their estimation in non-parametric or semi parametric models, by making use of information theory results (Sanov's theorem and Gibbs principles), by Monte Carlo methods. The question of the choice of divergence is wide open; an approach linking non-parametric Bayesian statistics and MAP estimators provides elements of understanding of the specificities of the various divergences in the Ali-Silvey-Csiszar-Arimoto class in relation to the specific choices of the prior distributions.

References

Broniatowski, M., Stummer, W.: Some universal insights on divergences for statistics, machine learning and artificial intelligence. In: Nielsen, F., (eds.) Geometric Structures of Information. Signals and Communication Technology, pp. 149–211 (2019). Springer, Cham. https://doi.org/10.1007/978-3-030-02520-5_8

Broniatowski, M.: Minimum divergence estimators, Maximum Likelihood and the generalized bootstrap, to appear in "Divergence Measures: Mathematical Foundations and Applications in Information-Theoretic and Statistical Problems" Entropy (2020)

Csiszár, I., Gassiat, E.: MEM pixel correlated solutions for generalized moment and interpolation problems. IEEE Trans. Inf. Theory **45** (7), 2253–2270 (1999)

Liese, F., Vajda, I.: On divergences and informations in statistics and information theory. IEEE Trans. Inf. Theory **52**(10), 4394–4412 (2006)

Gaussian States from a Symplectic Geometry Point of View

Maurice de Gosson

Faculty of Mathematics, NuHAG group, Senior Researcher at the University of Vienna
https://homepage.univie.ac.at/maurice.de.gosson

Abstract: Gaussian states play a ubiquitous role in quantum information theory and in quantum optics because they are easy to manufacture in the laboratory, and have in addition important extremality properties. Of particular interest are their separability properties. Even if major advances have been made in their study in recent years, the topic is still largely open. In this talk we will discuss separability questions for Gaussian states from a rigorous point of view using symplectic geometry, and present some new results and properties.

References

de Gosson, M.: On the Disentanglement of Gaussian quantum states by symplectic rotations **358** (4), 459–462 (2020). C.R. Acad. Sci. Paris

de Gosson, M.A.: On Density operators with Gaussian weyl symbols. In: Boggiatto, P., et al. (eds.) Advances in Microlocal and Time-Frequency Analysis. Applied and Numerical Harmonic Analysis, pp. 191–206. Birkhäuser, Cham (2020). https://doi.org/10.1007/978-3-030-36138-9_12

de Gosson, M.: Symplectic coarse-grained classical and semiclassical evolution of subsystems: new theoretical aspects. J. Math. Phys. **9**, 092102 (2020)

Cordero, E., de Gosson, M., Nicola, F.: On the positivity of trace class operators. in Advances in Theoretical and Mathematical Physics **23**(8), 2061–2091 (2019). to appear

Cordero, E., de Gosson, M., Nicola, F.: A characterization of modulation spaces by symplectic rotations. J. Funct. Anal. **278**(11), 108474 (2020). to appear in

The Primary Visual Cortex as a Cartan Engine

Jean Petitot

CAMS, École des Hautes Études en Sciences Sociales, France
http://jean.petitot.pagesperso-orange.fr/

Abstract: Cortical visual neurons detect very local geometric cues as retinal positions, local contrasts, local orientations of boundaries, etc. One of the main theoretical problem of low level vision is to understand how these local cues can be integrated so as to generate the global geometry of the images perceived, with all the well-known phenomena studied since Gestalt theory. It is an empirical evidence that the visual brain is able to perform a lot of routines belonging to differential geometry. But how such routines can be neurally implemented? Neurons are «point-like» processors and it seems impossible to do differential geometry with them. Since the 1990s, methods of "in vivo optical imaging based on activity-dependent intrinsic signals" have made possible to visualize the extremely special connectivity of the primary visual areas, their "functional architectures." What we called «Neurogeometry» is based on the discovery that these functional architectures implement structures such as the contact structure and the sub-Riemannian geometry of jet spaces of plane curves. For reasons of principle, it is the geometrical reformulation of differential calculus from Pfaff to Lie, Darboux, Frobenius, Cartan and Goursat which turns out to be suitable for neurogeometry.

References

Agrachev, A., Barilari, D., Boscain, U.: A Comprehensive Introduction to Sub-Riemannian Geometry, Cambridge University Press (2020)

Citti, G., Sarti, A.: A cortical based model of perceptual completion in the roto-translation space. J. Math. Imaging Vis. **24**(3), 307–326 (2006). https://doi.org/10.1007/s10851-005-3630-2

Petitot, J.: Neurogéométrie de la vision. Modèles mathématiques et physiques des architectures fonctionnelles, Les Éditions de l'École Polytechnique, Distribution Ellipses, Paris (2008)

Petitot J.: Landmarks for neurogeometry. In: Citti, G., Sarti, A., (eds.) Neuromathematics of Vision. LNM, pp. 1–85. Springer, Heidelberg (2014). https://doi.org/10.1007/978-3-642-34444-2_1

Petitot, J.: Elements of Neurogeometry. Functional Architectures of Vision. LNM, Springer, 2017. https://doi.org/10.1007/978-3-319-65591-8

Prandi, D., Gauthier, J.-P.: A Semidiscrete Version of the Petitot Model as a Plausible Model for Anthropomorphic Image Reconstruction and Pattern Recognition, https://arxiv.org/abs/1704.03069v1 (2017)

Use and Abuse of "Digital Information" in Life Sciences, is Geometry of Information a Way Out?

Giuseppe Longo

Centre Cavaillès, CNRS & Ens Paris and School of Medicine, Tufts University, Boston
http://www.di.ens.fr/users/longo/

Abstract: Since WWII, the war of coding, and the understanding of the structure of the DNA (1953), the latter has been considered as the digital encoding of the Aristotelian Homunculus. Till now DNA is viewed as the "information carrier" of ontogenesis, the main or unique player and pilot of phylogenesis. This heavily affected our understanding of life and reinforced a mechanistic view of organisms and ecosystems, a component of our disruptive attitude towards ecosystemic dynamics. A different insight into DNA as a major constraint to morphogenetic processes brings in a possible "geometry of information" for biology, yet to be invented. One of the challenges is in the need to move from a classical analysis of morphogenesis, in physical terms, to a "heterogenesis" more proper to the historicity of biology.

References

Islami, A., Longo, G.: Marriages of mathematics and physics: a challenge for biology, invited paper. In: Matsuno, K., et al. (eds.) The Necessary Western Conjunction to the Eastern Philosophy of Exploring the Nature of Mind and Life, vol. 131, pp. 179¬192, December 2017. SpaceTimeIslamiLongo.pdf. Special Issue of Progress in Biophysics and Molecular Biology

Longo, G.: How Future Depends on Past Histories and Rare Events in Systems of Life, Foundations of Science, (DOI), (2017). biolog-observ-history-future.pdf

Longo, G.: Information and causality: mathematical reflections on cancer biology. In Organisms. Journal of Biological Sciences, vo. 2, n. 1, 2018. BiologicalConseq-ofCompute.pdf

Longo, G.: Information at the threshold of interpretation, science as human construction of sense. In: Bertolaso, M., Sterpetti, F. (eds.) A Critical Reflection on Automated Science Will Science Remain Human? Springer, Dordrecht (2019). Information-Interpretation.pdf

Longo, G., Mossio, M.: Geocentrism vs genocentrism: theories without metaphors, metaphors without theories. Interdisciplinary Sci. Rev. **45** (3), 380–405 (2020). Metaphors-geo-genocentrism.pdf

Contents

Shapes Spaces

Geometry of Quantum States

Geometric and Structure Preserving Discretizations

Geometric Mechanics

Deformed Entropy, Cross-Entropy, and Relative Entropy

Transport Information Geometry

Manifolds and Optimization

Divergence Statistics

Optimal Transport and Learning

Probability and Statistics on Riemannian Manifolds

From Bayesian Inference to MCMC and Convex Optimisation in Hadamard Manifolds

Salem Said[1(✉)], Nicolas Le Bihan[2], and Jonathan H. Manton[3]

[1] CNRS, Laboratoire IMS, Talence, France
salem.said@u-bordeaux.fr
[2] CNRS, GIPSA-lab, Talence, France
nicolas.le-bihan@gipsa-lab.grenoble-inp.fr
[3] The University of Melbourne, Melbourne, Australia
jmanton@unimelb.edu.au

Abstract. The present work is motivated by the problem of Bayesian inference for Gaussian distributions in symmetric Hadamard spaces (that is, Hadamard manifolds which are also symmetric spaces). To investigate this problem, it introduces new tools for Markov Chain Monte Carlo, and convex optimisation: (1) it provides easy-to-verify sufficient conditions for the geometric ergodicity of an isotropic Metropolis-Hastings Markov chain, in a symmetric Hadamard space, (2) it shows how the Riemannian gradient descent method can achieve an exponential rate of convergence, when applied to a strongly convex function, on a Hadamard manifold. Using these tools, two Bayesian estimators, maximum-*a-posteriori* and minimum-mean-squares, are compared. When the underlying Hadamard manifold is a space of constant negative curvature, they are found to be surprisingly close to each other. This leads to an open problem: are these two estimators, in fact, equal (assuming constant negative curvature)?

Keywords: Hadamard manifold · Gaussian distribution · Bayesian inference · MCMC · Convex optimisation

1 MAP Versus MMS

In the following, M is a symmetric Hadamard space: a Hadamard manifold, which is also a symmetric space (for a background discussion, see Appendix A). Recall the definition of a Gaussian distribution on M [9]. This is $P(x,\sigma)$ whose probability density function, with respect to Riemannian volume, reads

$$p(y|x,\sigma) = (Z(\sigma))^{-1}\exp\left[-\frac{d^2(y,x)}{2\sigma^2}\right] \tag{1}$$

where $Z(\sigma)$ is a normalising constant, and $d(\cdot,\cdot)$ is the distance in M.

The maximum-likelihood estimator of the parameter x, based on independent samples $(y_1,\dots,y_N)$, is equal to the Riemannian barycentre of these samples.

F. Nielsen and F. Barbaresco (Eds.): GSI 2021, LNCS 12829, pp. 3–11, 2021.
https://doi.org/10.1007/978-3-030-80209-7_1

The one-sample maximum-likelihood estimator, given a single observation y, is therefore $\hat{x}_{ML} = y$.

Instead of maximum-likelihood, consider a Bayesian approach to estimating x, based on the observation y. Assign to x a prior density, which is also Gaussian,

$$p(x|z,\tau) = (Z(\tau))^{-1} \exp\left[-\frac{d^2(x,z)}{2\tau^2}\right] \quad (2)$$

Then, Bayesian inference concerning x is carried out using the posterior density

$$\pi(x) \propto \exp\left[-\frac{d^2(y,x)}{2\sigma^2} - \frac{d^2(x,z)}{2\tau^2}\right] \quad (3)$$

where $\propto$ indicates a missing (unknown) normalising factor.

The maximum-*a-posteriori* estimator $\hat{x}_{MAP}$ of x is equal to the mode of the posterior density $\pi(x)$. From (3), $\hat{x}_{MAP}$ minimises the weighted sum of squared distances $d^2(y,x)/\sigma^2 + d^2(x,z)/\tau^2$. This is expressed in the following notation,

$$\hat{x}_{MAP} = z \,\#_\rho\, y \quad \text{where } \rho = \frac{\tau^2}{\sigma^2 + \tau^2} \quad (4)$$

which means $\hat{x}_{MAP}$ is a geodesic convex combination of the prior barycentre z and the observation y, with respective weights $\sigma^2/(\sigma^2+\tau^2)$ and $\tau^2/(\sigma^2+\tau^2)$.

The minimum-mean-squares estimator $\hat{x}_{MMS}$ of x is the barycentre of the posterior density $\pi(x)$. That is, $\hat{x}_{MMS}$ is the global minimiser of the variance function (in the following, vol denotes the Riemannian volume)

$$\mathcal{E}_\pi(y) = \frac{1}{2}\int_M d^2(y,x)\pi(x)\text{vol}(dx) \quad (5)$$

whose existence and uniqueness are established in [7] (Chapter 4). While it is easy to compute $\hat{x}_{MAP}$ from (4), it is much harder to find $\hat{x}_{MMS}$, as this requires minimising the integral (5), where $\pi(x)$ is known only up to normalisation. Still, there is one special case where these two estimators are equal.

Proposition 1. *If $\sigma^2 = \tau^2$ (that is, if $\rho = 1/2$), then $\hat{x}_{MMS} = \hat{x}_{MAP}$.*

Remark: the proof of this proposition, and of all the propositions introduced in the following, can be found in [7] (Chapter 4).

2 Are They Equal?

Proposition 1 states that $\hat{x}_{MMS} = \hat{x}_{MAP}$, if $\rho = 1/2$. When M is a Euclidean space, it is famously known that $\hat{x}_{MMS} = \hat{x}_{MAP}$ for any value of ρ. In general, one expects these two estimators to be different from one another, if $\rho \neq 1/2$.

However, when M is a space of constant negative curvature, numerical experiments show that $\hat{x}_{MMS}$ and $\hat{x}_{MAP}$ lie surprisingly close to each other, and that they even appear to be equal.

It is possible to bound the distance between $\hat{x}_{MMS}$ and $\hat{x}_{MAP}$, using the so-called fundamental contraction property [10] (Theorem 6.3).

$$d(\hat{x}_{MMS}, \hat{x}_{MAP}) \leq W(\pi, \delta_{\hat{x}_{MAP}}) \tag{6}$$

where W denotes the Kantorovich distance, and $\delta_{\hat{x}_{MAP}}$ is the Dirac distribution at $\hat{x}_{MAP}$. Now, the right-hand side of (6) is equal to the first-order moment

$$m_1(\hat{x}_{MAP}) = \int_M d(\hat{x}_{MAP}, x)\pi(x)\text{vol}(dx) \tag{7}$$

Of course, the upper bound in (6) is not tight, since it is strictly positive, even when $\rho = 1/2$, as one may see from (7).

It will be seen below that a Metropolis-Hastings algorithm can be used to generate geometrically ergodic samples $(x_1, x_2, \ldots)$, from the posterior density π. Using these samples, it is possible to approximate (7), by an empirical average,

$$\bar{m}_1(\hat{x}_{MAP}) = \frac{1}{N}\sum_{n=1}^{N} d(\hat{x}_{MAP}, x_n) \tag{8}$$

In addition, these samples can be used to compute a convergent approximation of $\hat{x}_{MMS}$. Precisely, the empirical barycentre $\bar{x}_{MMS}$ of the samples $(x_1, \ldots, x_N)$ converges almost-surely to $\hat{x}_{MMS}$ (see the discussion after Proposition 3, below).

Numerical experiments were conducted in the case when M is a space of constant curvature, equal to -1, and of dimension n. The following table was obtained for $\sigma^2 = \tau^2 = 0.1$, using samples $(x_1, \ldots, x_N)$ where $N = 2 \times 10^5$.

dimension n	2	3	4	5	6	7	8	9	10
$\bar{m}_1(\hat{x}_{MAP})$	0.28	0.35	0.41	0.47	0.50	0.57	0.60	0.66	0.70
$d(\bar{x}_{MMS}, \hat{x}_{MAP})$	0.00	0.00	0.00	0.01	0.01	0.02	0.02	0.02	0.03

and the following table for $\sigma^2 = 1$ and $\tau^2 = 0.5$, again using $N = 2 \times 10^5$.

dimension n	2	3	4	5	6	7	8	9	10
$\bar{m}_1(\hat{x}_{MAP})$	0.75	1.00	1.12	1.44	1.73	1.97	2.15	2.54	2.91
$d(\bar{x}_{MMS}, \hat{x}_{MAP})$	0.00	0.00	0.03	0.02	0.02	0.03	0.04	0.03	0.12

The first table confirms Proposition 1. The second table, more surprisingly, shows that $\hat{x}_{MMS}$ and $\hat{x}_{MAP}$ can be quite close to each other, even when $\rho \neq 1/2$. Other values of σ^2 and τ^2 lead to similar orders of magnitude for $\bar{m}_1(\hat{x}_{MAP})$ and $d(\bar{x}_{MMS}, \hat{x}_{MAP})$. While $\bar{m}_1(\hat{x}_{MAP})$ increases with the dimension n, $d(\bar{x}_{MMS}, \hat{x}_{MAP})$ does not appear sensitive to increasing dimension.

Remark: from numerical experiments, it appears that $\hat{x}_{MMS}$ and $\hat{x}_{MAP}$ may be equal to one another. It is an open problem to mathematically prove or disprove this equality ($\hat{x}_{MMS} = \hat{x}_{MAP}$).

3 Metropolis-Hastings

Often, the first step in Bayesian inference is sampling from the posterior density. Here, this is $\pi(x)$, given by (3). Since $\pi(x)$ is known only up to normalisation, a suitable sampling method is afforded by the Metropolis-Hastings algorithm. This algorithm generates a Markov chain $(x_n\,;n \geq 1)$, with transition kernel [6]

$$Pf(x) = \int_M \alpha(x,y)q(x,y)f(y)\text{vol}(dy) + \rho(x)f(x) \tag{9}$$

for any bounded measurable $f : M \to \mathbb{R}$, where $\alpha(x,y)$ is the probability of accepting a transition from x to dy, and $\rho(x)$ is the probability of staying at x, and where $q(x,y)$ is the proposed transition density

$$q(x,y) \geq 0 \,\text{and} \int_M q(x,y)\text{vol}(dy) = 1 \quad \text{for } x \in M \tag{10}$$

In the following, (x_n) will be an isotropic Metropolis-Hastings chain, in the sense that $q(x,y) = q(d(x,y))$ only depends on the distance $d(x,y)$. In this case, the acceptance probability $\alpha(x,y)$ is given by $\alpha(x,y) = \min\{1, \pi(y)/\pi(x)\}$.

The aim of the Metropolis-Hastings algorithm is to produce a Markov chain (x_n) which is geometrically ergodic. Geometric ergodicity means the distribution π_n of x_n converges to π, at a geometric rate (that is, exponentially fast – See [3], Theorem 15.0.2). If the chain (x_n) is geometrically ergodic, then it satisfies the strong law of large numbers (for any suitable function $f : M \to \mathbb{R}$) [3]

$$\frac{1}{N}\sum_{n=1}^{N} f(x_n) \longrightarrow \int_M f(x)\pi(dx) \text{ (almost-surely)} \tag{11}$$

as well as a corresponding central limit theorem (see [3], Theorem 17.0.1). Then, in practice, the Metropolis-Hastings algorithm generates samples $(x_1, x_2, \ldots)$ from the posterior density $\pi(x)$.

In [7] (Chapter 4), the following general statement is proved, concerning the geometric ergodicity of isotropic Metropolis-Hastings chains.

Proposition 2. *Let M be a symmetric Hadamard space. Assume $(x_n\,;n \geq 1)$ is a Markov chain in M, with transition kernel given by (9), with proposed transition density $q(x,y) = q(d(x,y))$, and with strictly positive invariant density π. The chain (x_n) is geometrically ergodic if the following assumptions hold,*

(a1) there exists $x^* \in M$, such that $r(x) = d(x^*,x)$ and $\ell(x) = \log \pi(x)$ satisfy

$$\limsup_{r(x)\to\infty} \frac{\langle \text{grad}\, r, \text{grad}\, \ell\rangle_x}{r(x)} < 0$$

(a2) if $n(x) = \text{grad}\, \ell(x)\|\text{grad}\, \ell(x)\|$, then $n(x)$ satisfies

$$\limsup_{r(x)\to\infty} \langle \text{grad}\, r, n\rangle_x < 0$$

(a3) there exist $\delta_q > 0$ and $\varepsilon_q > 0$ such that $d(x,y) < \delta_q$ implies $q(x,y) > \varepsilon_q$

Remark: the posterior density π in (3) verifies the Assumptions (a1) and (a2). To see this, let $x^* = z$, and note that,

$$\text{grad}\, \ell(x) = -\frac{1}{\tau^2} r(x) \text{grad}\, r(x) - \frac{1}{\sigma^2} \text{grad}\, f_y(x)$$

where $f_y(x) = d^2(x, y)/2$. Then, taking the scalar product with grad r,

$$\langle \text{grad}\, r, \text{grad}\, \ell \rangle_x = -\frac{1}{\tau^2} r(x) - \frac{1}{\sigma^2} \langle \text{grad}\, r, \text{grad}\, f_y \rangle_x \tag{12}$$

since grad $r(x)$ is a unit vector [5] (see Page 41). Now, grad $f_y(x) = -\text{Exp}_x^{-1}(y)$ (Exp is the Riemannian exponential map). Since $r(x)$ is a convex function of x,

$$r(y) - r(x) \geq \langle \text{grad}\, r, \text{Exp}_x^{-1}(y) \rangle$$

for any $y \in M$. Thus, the right-hand side of (12) is strictly negative, as soon as $r(x) > r(y)$, and Assumption (a1) is indeed verified. That Assumption (a2) is also verified can be proved by a similar reasoning.

Remark: on the other hand, Assumption (a3) holds, if the proposed transition density $q(x, y)$ is a Gaussian density, $q(x, y) = p(y|x, \tau_q^2)$. In this way, all the assumptions of Proposition 2 are verified, for the posterior density π in (3). Proposition 2 therefore implies that the Metropolis-Hastings algorithm generates geometrically ergodic samples from this posterior density.

4 The Empirical Barycentre

Let $(x_n\,; n \geq 1)$ be a Metropolis-Hastings Markov chain in M, with transition kernel (9), and invariant density π. Assume that the chain (x_n) is geometrically ergodic, so it satisfies the strong law of large numbers (11).

Let $\bar{x}_N$ denote the empirical barycentre of the first N samples $(x_1, \ldots, x_N)$. This is the unique global minimum of the variance function

$$\mathcal{E}_N(y) = \frac{1}{2N} \sum_{n=1}^{N} d^2(y, x_n) \tag{13}$$

Assuming it is well-defined, let $\hat{x}$ denote the barycentre of the invariant density π. It turns out that $\bar{x}_N$ converges almost-surely to $\hat{x}$.

Proposition 3. *Let (x_n) be any Markov chain in a Hadamard manifold M, with invariant distribution π. Denote $\bar{x}_N$ the empirical barycentre of $(x_1, \ldots, x_N)$, and $\hat{x}$ the Riemannian barycentre of π (assuming it is well-defined). If (x_n) satisfies the strong law of large numbers (11), then $\bar{x}_N$ converges to $\hat{x}$, almost-surely.*

According to the remarks after Proposition 2, the Metropolis-Hastings Markov chain (x_n), whose invariant density is the posterior density $\pi(x)$, given by (3), is geometrically ergodic. Then, by Proposition 3, the empirical barycentre $\bar{x}_{MMS}$, of the samples $(x_1, \ldots, x_N)$, converges almost-surely to the minimum mean square error estimator $\hat{x}_{MMS}$ (since this is the barycentre of the posterior density π). This provides a practical means of approximating $\hat{x}_{MMS}$. Indeed, $\bar{x}_{MMS}$ can be computed using the Riemannian gradient descent method, discussed in the following section.

Remark: the proof of Proposition 3 is nearly a word-for-word repetition of the proof in [1] (Theorem 2.3), which first established the strong consistency of empirical barycentres (in [1], these are called Fréchet sample means). In [1], the focus is on i.i.d. samples $(x_1, x_2, \ldots)$, but the proof only uses the strong law of large numbers (11), and nowhere requires the samples to be independent.

5 Riemannian Gradient Descent

Since the minimum mean square error estimator $\hat{x}_{MMS}$ could not be computed directly, it was approximated by $\bar{x}_{MMS}$, the global minimum of the variance function $\mathcal{E}_N$, defined as in (13). This function $\mathcal{E}_N$ being (1/2)-strongly convex, its global minimum can be computed using the Riemannian gradient descent method, which even guarantees an exponential rate of convergence.

This method is here studied from a general point of view. The aim is to minimise a function $f : M \to \mathbb{R}$, where M is a Hadamard manifold, with sectional curvatures in the interval $[-c^2, 0]$, and f is an $(\alpha/2)$-strongly convex function.

This means f is $(\alpha/2)$-strongly convex along any geodesic in M. In particular,

$$f(y) - f(x) \geq \langle \mathrm{Exp}_x^{-1}(y), \mathrm{grad}\, f(x) \rangle_x + (\alpha/2) d^2(x, y) \tag{14}$$

for $x, y \in M$.

This implies that f has compact sublevel sets. Indeed, let x^* be the global minimum of f, so $\mathrm{grad}\, f(x^*) = 0$. Putting $x = x^*$ and $y = x$ in (14), it follows that

$$f(x) - f(x^*) \geq (\alpha/2) d^2(x^*, x) \tag{15}$$

Accordingly, if $S(y)$ is the sublevel set of y, then $S(y)$ is contained in the closed ball $\bar{B}(x^*, R_y)$, where $R_y = (2/\alpha)(f(y) - f(x^*))$. Therefore, $S(y)$ is compact, since it is closed and bounded [5].

The Riemannian gradient descent method is based on the iterative scheme

$$x^{t+1} = \mathrm{Exp}_{x^t}(-\mu \mathrm{grad}\, f(x^t)) \tag{16}$$

where μ is a positive step-size, $\mu \leq 1$. If this is chosen sufficiently small, then the iterates x^t remain within the sublevel set $S(x^0)$. In fact, let $\bar{B}_0 = \bar{B}(x^*, R_{x^0})$ and $\bar{B}_0' = \bar{B}(x^*, R_{x^0} + G)$, where G denotes the supremum of the norm of $\mathrm{grad}\, f(x)$, taken over $x \in \bar{B}_0$. Then, let H_0' denote the supremum of the operator norm of $\mathrm{Hess}\, f(x)$, taken over $x \in \bar{B}_0'$.

Lemma 1. *For the Riemannian gradient descent method (16), if $\mu \leq 2/H_0'$, then the iterates x^t remain within the sublevel set $S(x^0)$.*

Once it has been ensured that the iterates x^t remain within $S(x^0)$, it is even possible to choose μ in such a way that these iterates achieve an exponential rate of convergence towards x^*. This relies on the fact that x^* is a "strongly attractive" critical point of the vector field grad f. Specifically, putting $y = x^*$ in (14), it follows that

$$\langle \mathrm{Exp}_x^{-1}(x^*), \mathrm{grad}\, f(x)\rangle_x \leq -(\alpha/2)d^2(x,x^*) + (f(x^*) - f(x)) \tag{17}$$

Now, let $C_0 = cR_{x^0}\coth(cR_{x^0})$.

Proposition 4. *Let $\bar{H}'_0 = \max\{H'_0, 1\}$. If $\mu \leq 1/(\bar{H}'_0 C'_0)$ (this implies $\mu \leq 2/H'_0$) and $\mu \leq 1/\alpha$, then the following exponential rate of convergence is obtained*

$$d^2(x^t, x^*) \leq (1 - \mu\alpha)^t d^2(x^0, x^*) \tag{18}$$

A Symmetric Hadamard Spaces

The present work focuses on symmetric Hadamard spaces. These are Hadamard manifolds which are also symmetric spaces. This class of spaces includes: spaces of constant negative curvature (precisely, hyperbolic spaces), cones of covariance matrices (real, complex, or quaternion), as well as Siegel domaines (recently used in describing spaces of block-Toeplitz covariance matrices). These spaces play a prominent role in several applications [2,4,8].

A.1 Hadamard Manifolds

Definition
A Hadamard manifold is a Riemannian manifold which is complete, simply connected, and has negative sectional curvatures [5]. Completeness means that geodesics extend indefinitely in both directions, and is equivalent to the property that closed and bounded sets are compact (by the Hopf-Rinow theorem [5]). The simple connectedness and negative curvature properties combine to ensure that the Riemannian exponential map is a diffeomorphism. In particular, any two points x, y in a Hadamard manifold M are connected by a unique length-minimising geodesic $c : [0,1] \to M$, with $c(0) = x$, and $dc/dt|_{t=0} = \mathrm{Exp}_x^{-1}(y)$.

Convexity
Now, this geodesic $c : [0,1] \to M$ is called the segment between x and y. For $t \in [0,1]$, the point $x \,\#_t\, y = c(t)$ is said to be a geodesic convex combination of x and y, with respective weights $(1-t)$ and t. In a Hadamard manifold [5], any open geodesic ball $B(x^*, R)$ (or any closed geodesic ball $\bar{B}(x^*, R)$) is convex: if x and y belong to this ball, so does $x \,\#_t\, y$ for all $t \in (0,1)$.

A function $f : M \to \mathbb{R}$ is called $(\alpha/2)$-strongly convex, if $(f \circ c)(t)$ is an $(\alpha/2)$-strongly convex function of t, for any geodesic $c : [0,1] \to M$. That is, if there exists $\alpha > 0$, such that for any x and y in M, and $t \in [0,1]$,

$$f(x \,\#_t\, y) \leq (1-t)f(x) + tf(y) - (\alpha/2)t(1-t)d^2(x,y) \tag{19}$$

When f is C^2-smooth, this is equivalent to (14).

Squared Distance Function
For $y \in M$, consider the function $f_y : M \to \mathbb{R}$, where $f_y(x) = d^2(x,y)/2$. If M is a Hadamard manifold, then this function has several remarkable properties.

Specifically, it is both smooth and (1/2)-strongly convex (irrespective of y). It's gradient is given by $\operatorname{grad} f_y(x) = -\operatorname{Exp}_x^{-1}(y)$, in terms of the Riemannian exponential map. On the other hand, on any geodesic ball $B(y,R)$, the operator norm of its Hessian is bounded by $cR\coth(cR)$, where $c > 0$ is such that the sectional curvatures of M lie in the interval $[-c^2, 0]$ (see [5], Theorem 27).

A result of the strong convexity of these functions f_y is the uniqueness of Riemannian barycentres. If π is a probability distribution on M, its Riemannian barycentre $\hat{x}_\pi$ is the unique global minimiser of the so-called variance function $\mathcal{E}_\pi(x) = \int_M f_y(x)\pi(dy)$. If M is a Hadamard manifold, each function f_y is (1/2)-strongly convex, and therefore $\mathcal{E}_\pi$ is (1/2)-strongly convex. Thus, $\mathcal{E}_\pi$ has a unique global minimiser.

A.2 Symmetric Spaces

Definition
A Riemannian symmetric space is a Riemannian manifold M, such that for each $x \in M$ there exists an isometry $s_x : M \to M$, which fixes x and reverses geodesics passing through x: if c is a geodesic curve with $c(0) = x$, then $s_x(c(t)) = c(-t)$. As a consequence of this definition, a Riemannian symmetric space is always complete and homogeneous. Completeness was discussed in Appendix A.1. On the other hand, homogeneity means that for each x and y in M, there exists an isometry $g : M \to M$, with $g(x) = y$. A symmetric space M may always be expressed as a quotient manifold $M = G/K$, where G is a certain Lie group of isometries of M, and K is the subgroup of elements of G which fix some point $x \in M$.

A Familiar Example
A familiar example of a symmetric Hadamard manifold is the cone $\mathcal{P}_d$ of $d \times d$ symmetric positive-definite matrices, which was discussed in some length in [8,9]. A geodesic curve through $x \in \mathcal{P}_d$ is of the form $c(t) = x\exp(tx^{-1}u)$ where exp is the matrix exponential, and u a symmetric matrix, $dc/dt|_{t=0} = u$. The isometry s_x is given by $s_x(y) = xy^{-1}x$ for $y \in \mathcal{P}_d$, and one easily checks $s_x(c(t)) = c(-t)$.

Let $G = \mathrm{GL}(d,\mathbb{R})$ denote the group of invertible $d \times d$ real matrices. Then, $g \in G$ defines an isometry $\tau_g : \mathcal{P}_d \to \mathcal{P}_d$, given by $\tau_g(x) = gxg^\dagger$ ($\dagger$ the transpose). Typically, one writes $\tau_g(x) = g(x)$. Note that $x = \mathrm{I}_d$ (the identity matrix) belongs to $\mathcal{P}_d$, and then $g(x) = x$ if and only if $g \in K$, where $K = O(d)$ is the orthogonal group. In fact, one has the quotient manifold structure $\mathcal{P}_d = \mathrm{GL}(d,\mathbb{R})/O(d)$.

References

1. Bhattacharya, R., Patrangenaru, V.: Large sample theory of instrinsic and extrinsic sample means on manifolds I. Ann. Stat. **31**(1), 1–29 (2003)
2. Jeuris, B., Vandebril, R.: The Kähler mean of block-Toeplitz matrices with Toeplitz structured blocks. SIAM J. Matrix Anal. Appl. (2016). https://lirias.kuleuven.be/handle/123456789/497758
3. Meyn, S., Tweedie, R.L.: Markov Chains and Stochastic Stability. Cambridge University Press, Cambridge (2008)
4. Nielsen, F., Nock, R.: Hyperbolic Voronoi diagrams made easy (2009). https://arxiv.org/abs/0903.3287
5. Petersen, P.: Riemannian Geometry, 2nd edn. Springer, New York (2006). https://doi.org/10.1007/978-0-387-29403-2
6. Roberts, R.O., Rosenthal, J.S.: General state-space Markov chains and MCMC algorithms. Probab. Surv. **1**, 20–71 (2004)
7. Said, S.: Statistical models and probabilistic methods on Riemannian manifolds. Technical report, Université de Bordeaux (2021). http://arxiv.org/abs/2101.10855
8. Said, S., Bombrun, L., Berthoumieu, Y., Manton, J.H.: Riemannian Gaussian distributions on the space of symmetric positive definite matrices. IEEE Trans. Inf. Theory (2016). http://arxiv.org/abs/1507.01760
9. Said, S., Hajri, H., Bombrun, L., Vemuri, B.C.: Gaussian distributions on Riemannian symmetric spaces: statistical learning with structured covariance matrices. IEEE Trans. Inf. Theory **64**(2), 752–772 (2018)
10. Sturm, K.T.: Probability measures on metric spaces of nonpositive curvature. Contemp. Math. **338**, 1–34 (2003)

Finite Sample Smeariness on Spheres

Benjamin Eltzner[1(✉)], Shayan Hundrieser[2], and Stephan Huckemann[1]

[1] Felix-Bernstein-Institute for Mathematical Statistics in the Biosciences, Georg-August-Universität Göttingen, Göttingen, Germany
benjamin.eltzner@mpibpc.mpg.de

[2] Institute for Mathematical Statistics, Georg-August-Universität Göttingen, Göttingen, Germany

Abstract. *Finite Sample Smeariness* (FSS) has been recently discovered. It means that the distribution of sample Fréchet means of underlying rather unsuspicious random variables can behave as if it were smeary for quite large regimes of finite sample sizes. In effect classical quantile-based statistical testing procedures do not preserve nominal size, they reject too often under the null hypothesis. Suitably designed bootstrap tests, however, amend for FSS. On the circle it has been known that arbitrarily sized FSS is possible, and that all distributions with a nonvanishing density feature FSS. These results are extended to spheres of arbitrary dimension. In particular all rotationally symmetric distributions, not necessarily supported on the entire sphere feature FSS of Type I. While on the circle there is also FSS of Type II it is conjectured that this is not possible on higher-dimensional spheres.

1 Introduction

In non-Euclidean statistics, the Fréchet mean [6] takes the role of the expected value of a random vector in Euclidean statistics. Thus an enormous body of literature has been devoted to the study of Fréchet means and its exploitation for descriptive and inferential statistics [1,2,7,9,12]. For the latter, it was only recently discovered that the asymptotics of Fréchet means may differ substantially from that of its Euclidean kin [5,8]. They are called *smeary*, if the rate of convergence of sample means to a limiting distribution is slower than the usual Euclidean $1/\sqrt{n}$ where n denotes sample size. Initially, such examples were rather exotic. More recently, however, it has been discovered by [11] that also for a large class of classical distributions (e.g. all with nonvanishing densities on the circle, like, e.g. all von-Mises-Fisher distributions) Fréchet means behave in a regime up to considerable sample sizes as if they were smeary. We call this effect *finite sample smeariness* (FSS), also the term *lethargic means* has been suggested. Among others, this effect is highly relevant for asymptotic one- and two-sample tests for equality of means. In this contribution, after making

Acknowledging DFG HU 1575/7, DFG GK 2088, DFG EXC 2067, DFG CRC 803, meteoblue AG, and the Niedersachsen Vorab of the Volkswagen Foundation.

F. Nielsen and F. Barbaresco (Eds.): GSI 2021, LNCS 12829, pp. 12–19, 2021.
https://doi.org/10.1007/978-3-030-80209-7_2

the new terminology precise, we illustrate the effect of FSS on statistical tests concerning the change of wind directions in the larger picture of climate change.

Furthermore, while we have shown in [11] that FSS of any size can be present on the circle and the torus, here we show in Theorem 6, using results from [3], that FSS of arbitrary size is also present on spheres of arbitrary dimension, at least for local Fréchet means. For such, on high dimensional spheres, distributions supported by barely more than a geodesic half ball may feature arbitrary high FSS. Moreover, we show that the large class of rotationally symmetric distributions on spheres of arbitrary dimension, e.g. all Fisher distributions, feature FSS. This means not only that the finite sample rate may be wrong, also the rescaled asymptotic variance of Fréchet means may be considerably different from the sample variance in tangent space. Although the modulation of FSS is small for concentrated distributions, one has to take care to avoid these pitfalls.

2 Finite Sample Smeariness on Spheres

In this section, we introduce the concepts smeariness and finite sample smeariness. While they extend at once to general manifolds, here, we restrict ourselves to spheres. Let $\mathbb{S}^m$ be the unit sphere in $\mathbb{R}^{m+1}$ for $m > 1$ and $\mathbb{S}^1 = [-\pi, \pi)/\sim$ with $-\pi$ and π identified be the unit circle, with the distance

$$d(x, y) = \begin{cases} \arccos(x^T y) & \text{for } x, y \in \mathbb{S}^m, \\ \min\{|y - x|, 2\pi - |y - x|\} & \text{for } x, y \in \mathbb{S}^1. \end{cases}$$

For random variables $X_1, \ldots, X_n \overset{\text{i.i.d.}}{\sim} X$ on $\mathbb{S}^m$, $m \geq 1$, with silently underlying probability space $(\Omega, \mathbb{P})$ we have the *Fréchet functions*

$$F(p) = \mathbb{E}[d(X, p)^2] \quad \text{and} \quad F_n(p) = \frac{1}{n}\sum_{j=1}^{n} d(X_j, p)^2 \text{ for } p \in \mathbb{S}^m . \tag{1}$$

We work under the following assumptions. In particular, the third Assumption below is justified by [15, Lemma 1].

Assumption 1. *Assume that*

1. *X is not a.s. a single point,*
2. *there is a unique minimizer* $\mu = \mathrm{argmin}_{p \in \mathbb{S}^m} F(p)$, *called the* Fréchet population mean,
3. *for $m > 1$, μ is the north pole $(1, 0, \ldots, 0)$ and $\mu = 0$ on $\mathbb{S}^1$,*
4. *$\widehat{\mu}_n \in \mathrm{argmin}_{p \in \mathbb{S}^m} F_n(p)$ is a selection from the set of minimizers uniform with respect to the Riemannian volume, called a* Fréchet sample mean,

Note that $\mathbb{P}\{X = -\mu\} = 0$ for $m > 1$ and $\mathbb{P}\{X = -\pi\} = 0$ on $\mathbb{S}^1$ due to [8,12].

Definition 2. *We have the* population variance

$$V := F(\mu) = \mathbb{E}[d(X,\mu)^2]\,,$$

which on $\mathbb{S}^1$ *is just the classical variance* $\mathbb{V}[X]$, *and the* Fréchet sample mean variance

$$V_n := \mathbb{E}[d(\widehat{\mu}_n,\mu)^2]$$

giving rise to the modulation

$$\mathfrak{m}_n := \frac{nV_n}{V}\,.$$

Conjecture 3. *Consider* $X_1,\ldots,X_n \overset{\text{i.i.d.}}{\sim} X$ *on* $\mathbb{S}^1$ *and suppose that* $J \subseteq \mathbb{S}^1$ *is the support of* X. *Assume Assumption 1 and let* $n > 1$.

Then $\mathfrak{m}_n = 1$ *under any of the two following conditions*

(i) J *is strictly contained in a closed half circle,*
(ii) J *is a closed half circle and one of its end points is assumed by* X *with zero probability.*

Further, $\mathfrak{m}_n > 1$ *under any of the two following conditions*

(iii) the interior of J *contains a closed half circle,*
(iv) J *contains two antipodal points, each of which is assumed by* X *with positive probability.*

Finally, suppose that X *has near* $-\pi$ *a continuous density* f.

(v) If $f(-\pi) = 0$ *then* $\lim_{n\to\infty}\mathfrak{m}_n = 1$,
(vi) if $0 < f(-\pi) < \frac{1}{2\pi}$ *then* $\lim_{n\to\infty}\mathfrak{m}_n = \frac{1}{(1-2\pi f(-\pi))^2} > 1$.

For $\mu = 0$, claims (i) and (ii) follow from Equation (10) in [8] which yields $\widehat{\mu}_n = n^{-1}\sum_{i=1}^n X_i$, thus $nV_n = \mathbb{E}[X^2] = V$. The rest of the conjecture will be proved in a future version of [11]. In [8] it has been shown that $2\pi f(-\pi)$ can be arbitrary close to 1, i.e. that $\lim_{n\to\infty}\mathfrak{m}_n$ can be arbitrary large. In fact, whenever $2\pi f(-\pi) = 1$, then $\lim_{n\to\infty}\mathfrak{m}_n = \infty$. These findings give rise to the following.

Definition 4. *We say that* X *is*

(i) Euclidean *if* $\mathfrak{m}_n = 1$ *for all* $n \in \mathbb{N}$,
(ii) finite sample smeary *if* $1 < \sup_{n\in\mathbb{N}}\mathfrak{m}_n < \infty$,
(ii$_1$) Type I finite sample smeary *if* $\lim_{n\to\infty}\mathfrak{m}_n > 1$,
(ii$_2$) Type II finite sample smeary *if* $\lim_{n\to\infty}\mathfrak{m}_n = 1$,
(iii) smeary *if* $\sup_{n\in\mathbb{N}}\mathfrak{m}_n = \infty$.

3 Why is Finite Sample Smeariness Called Finite Sample Smeariness?

Under FSS on the circle in simulations we see typical shapes of modulation curves in Fig. 1. For statistical testing, usually building on smaller sample sizes, as detailed further in Sect. 4, the initial regime is decisive, cf. Fig. 2:

There are constants $C_+, C_-, K > 0$, $0 < \alpha_- < \alpha_+ < 1$ and integers $1 < n_- < n_+ < n_0$ satisfying $C_+ n_-^{\alpha_+} \leq C_- n_+^{\alpha_-}$, such that

(a) $\forall n \in [n_-, n_+] \cap \mathbb{N}: \quad 1 < C_- n^{\alpha_-} \leq \mathfrak{m}_n \leq C_+ n^{\alpha_+}$.
(b) $\forall n \in [n_0, \infty) \cap \mathbb{N}: \quad \mathfrak{m}_n \leq K$.

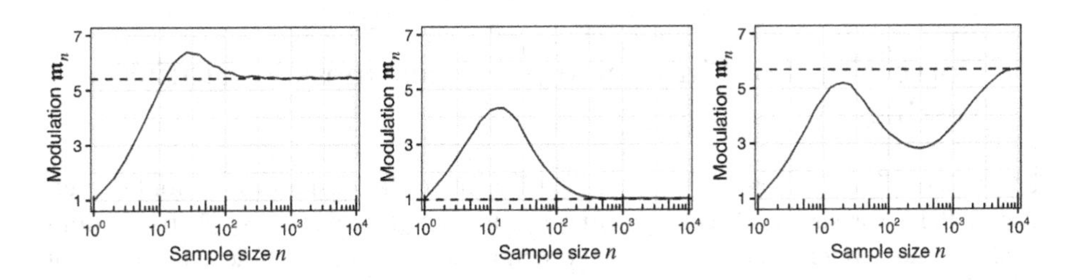

Fig. 1. Modulation $\mathfrak{m}_n$ for von Mises distribution (Mardia & Jupp, 2000) with mean $\mu = 0$ and concentration $\kappa = 1/2$ (left), conditioned on $[-\pi + 0.2, \pi - 0.2]$ (center), and conditioned on $[-\pi, -\pi + 0.1] \cup [-\pi + 0.2, \pi + 0.2] \cup [\pi - 0.1, \pi)$ (right). The dashed lines represent the respective limits of $\mathfrak{m}_n$ obtained by Conjecture 3 (v), (vi).

Although under FSS, $\mathfrak{m}_n$ is eventually constant, i.e. the asymptotic rate of $\widehat{\mu}_n$ is the classical $n^{-1/2}$, for nonvanishing intervals of sample sizes $[n_-, n_+]$, the "finite sample" rate is (in expectation) between

$$\left(n^{-\frac{1}{2}} <\right) \quad n^{-\frac{1-\alpha_-}{2}} \text{ and } n^{-\frac{1-\alpha_+}{2}},$$

i.e. like a smeary rate, cf. [11].

Of course, as illustrated in Fig. 1, the modulation curve can be subject to different regimes of α_- and α_+, in applications, typically the first regime is of interest, cf. Sect. 4.

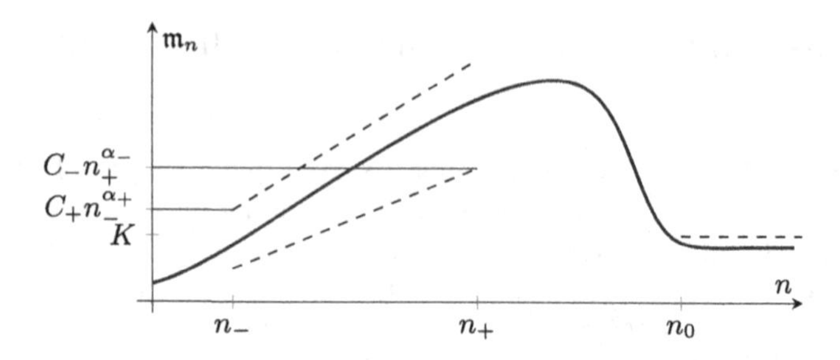

Fig. 2. Schematically illustrating the modulation curve $n \mapsto \mathfrak{m}_n$ for FSS on the circle. Along $[n_-, n_+]$ the curve is between the lower ($C_- n^{\alpha_-}$) and upper ($C_+ n^{\alpha_+}$) bounds (dashed), satisfying the condition $C_+ n_-^{\alpha_+} \leq C_- n_+^{\alpha_-}$, and for $n \geq n_0$ it is below the horizontal upper bound (dashed).

4 Correcting for Finite Sample Smeariness in Statistical Testing

The central limit theorem by [2] and [7] for an m-dimensional manifold M, cf. also [1,9,10] for sample Fréchet means $\widehat{\mu}_n$, has been extended by [5] to random variables no longer avoiding arbitrary neighborhoods of possible cut points of the Fréchet mean μ. Under nonsmeariness it has the following form:

$$\sqrt{n}\,\phi(\widehat{\mu}_n) \xrightarrow{\mathcal{D}} \mathcal{N}\left(0, 4\,H^{-1}\Sigma H^{-1}\right) .$$

Here $\phi : M \to \mathbb{R}^m$ is a local chart mapping μ to the origin of $\mathbb{R}^m$, H is the expected value of the Hessian of the Fréchet function F from (1) in that chart at μ and Σ is the covariance of $\phi(X)$. In practical applications, H is usually ignored, as it has got no straightforward plugin estimators, and $4H^{-1}\Sigma H^{-1}$ is simply estimated by the empirical covariance $\widehat{\Sigma}_n$ of $\phi(X_1), \ldots, \phi(X_n)$ giving rise to the approximation

$$n\phi(\widehat{\mu}_n)^T \widehat{\Sigma}_n^{-1} \phi(\widehat{\mu}_n) \xrightarrow{\mathcal{D}} \chi^2_m , \tag{2}$$

e.g. [1,2]. For finite samples sizes, this approximation depends crucially on

$$\mathfrak{m}_n = \frac{\mathbb{E}[n\|\phi(\widehat{\mu}_n)\|^2]}{\mathbb{E}[\text{trace}(\widehat{\Sigma}_n)]} = 1 ,$$

and it is bad in regimes whenever $\mathfrak{m}_n \gg 1$.

This is illustrated in Fig. 3 where two samples from von Mises distributions with concentration $\kappa = 1/2$ are tested for equality of Fréchet means. Indeed the quantile based test does not keep the nominal level, whereas the bootstrap based test, see [4], keeps the level fairly well and is shown to be consistent under FSS on $\mathbb{S}^1$, cf. [11].

Moreover, Table 1 shows a comparison of p-values of the quantile test based on (2) and the suitably designed bootstrap test for daily wind directions taken at

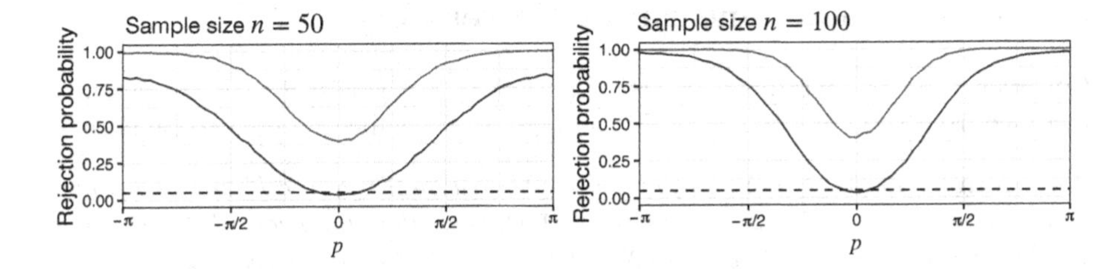

Fig. 3. Empirical rejection probabilities of quantile based tests (red) and bootstrap based tests (blue) to test for significance 95% if two samples of size $n = 50$ (left) and $n = 100$ (right) have identical Fréchet means. The two samples are taken independently from a von Mises distribution with mean $\mu = 0$ and $\mu = p$, respectively, and concentration $\kappa = 1/2$. The dashed line represents 5%.

Table 1. Comparing p-values of the quantile based test for equality of means of yearly wind data from Basel (Fig. 4), based on (2) with the bootstrap test amending for FSS proposed in [11] for $B = 10.000$ bootstrap realizations.

p-value	2018 vs. 2019	2019 vs. 2020	2018 vs. 2020
Quantile based test	0.00071	0.27	0.019
Bootstrap based test	0.047	0.59	0.21

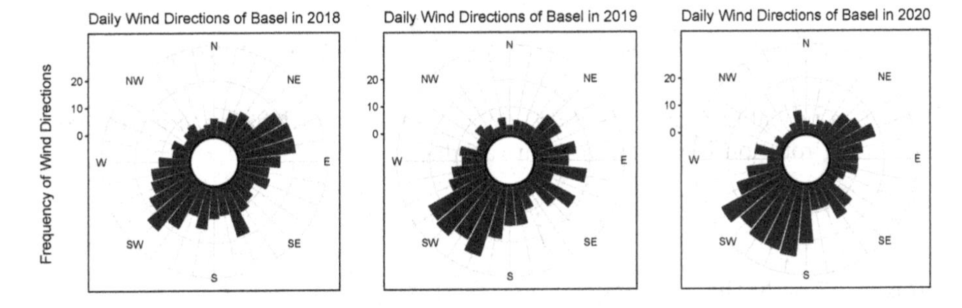

Fig. 4. Histograms of daily wind directions for Basel (provided by meteoblue AG) for 2018 (left), 2019 (center), and 2020 (right).

Basel for the years 2018, 2019, and 2020, cf. Fig. 4. While the quantile based test asserts that the year 2018 is highly significantly different from 2019 and 2020, the bootstrap based test shows that a significant difference can be asserted at most for the comparison between 2018 and 2019. The reason for the difference in p-values between quantile and bootstrap based test is the presence of FSS in the data, i.e. $\mathfrak{m}_n \gg 1$. Indeed, estimating for $n = 365$ the modulation $\mathfrak{m}_n$ of the yearly data using $B = 10.000$ bootstrap repetitions, as further detailed in [11], yields $\mathfrak{m}_n^{2018} = 2.99$, $\mathfrak{m}_n^{2019} = 2.97$, and $\mathfrak{m}_n^{2020} = 4.08$.

5 Finite Sample Smeariness Universality

Consider $p \in \mathbb{S}^m$ parametrized as $(\theta, \sin\theta q) \in \mathbb{S}^m$ where $\theta \in [0, \pi]$ denotes distance from the north pole $\mu \in \mathbb{S}^m$ and $q \in \mathbb{S}^{m-1}$, which is rescaled by $\sin\theta$.

Theorem 5. *Let $m \geq 4$, Y uniformly distributed on $\mathbb{S}^{m-1}$ and $K > 1$ arbitrary. Then there are $\theta^* \in (\pi/2, \pi)$ and $\alpha \in (0,1)$ such that for every $\theta \in (\theta^*, \pi)$ a random variable X on $\mathbb{S}^m$ with $\mathbb{P}\{X = (\theta, \sin\theta Y)\} = \alpha$ and $\mathbb{P}\{X = \mu\} = 1 - \alpha$ features $\sup_{n\in\mathbb{N}} \mathfrak{m}_n \geq \lim_{n\to\infty} \mathfrak{m}_n > K$. In particular, $\theta^* = \frac{\pi}{2} + \mathcal{O}(m^{-1})$.*

Proof. The first assertion follows from [3, Theorem 4.3] and its proof in Appendix A.5 there. Notably $\theta^* = \theta_{m,4}$ there. The second assertion has been shown in [3, Lemma A.5].

Theorem 6. *Let X be a random variable on $\mathbb{S}^m$ with $m \geq 2$ with unique non-smeary mean μ, whose distribution measure is invariant under rotation around μ and is not a point mass at μ. Then μ is Type I finite sample smeary.*

Proof. From [3], page 17, we see that the Fréchet function $F^{(\theta)}$ for a uniform distribution on the $\mathbb{S}^{m-1}$ at polar angle θ evaluated at a point with polar angle ψ from the north pole is

$$a(\psi, \theta, \phi) := \arccos\left(\cos\psi\cos\theta + \sin\psi\sin\theta\cos\phi\right)$$
$$F^{(\theta)}(\psi) = \left(\int_0^{2\pi} \sin^{m-2}\phi\, d\phi\right)^{-1} \int_0^{2\pi} \sin^{m-2}\phi\, a^2(\psi, \theta, \phi)\, d\phi\,.$$

Defining a probability measure $d\mathbb{P}(\theta)$ on $[0, \pi]$, the Fréchet function for the corresponding rotation invariant random variable is

$$F(\psi) = \int_0^{\pi} F^{(\theta)}(\psi)\, d\mathbb{P}(\theta)\,.$$

On page 17 of [3] the function

$$f_2(\theta, \psi) := \frac{1}{2}\sin^{m-1}\theta \int_0^{2\pi} \sin^{m-2}\phi\, d\phi\, \frac{d^2}{d\psi^2} F^{(\theta)}(\psi)$$

is defined and from Eq. (5) on page 19 we can calculate

$$\begin{aligned} f_2(\theta, 0) &= \sin^{m-2}\theta\left(\frac{1}{m-1}\sin\theta + \theta\cos\theta\right)\int_0^{2\pi}\sin^m\phi\, d\phi \\ &= \sin^{m-1}\theta\int_0^{2\pi}\sin^{m-2}\phi\, d\phi\left(\frac{1}{m} + \frac{m-1}{m}\theta\cot\theta\right), \end{aligned}$$

which yields the Hessian of the Fréchet function for $d\mathbb{P}(\theta)$ as

$$\mathrm{Hess}F(0) = 2\mathrm{Id}_m \int_0^{\pi}\left(\frac{1}{m} + \frac{m-1}{m}\theta\cot\theta\right) d\mathbb{P}(\theta)\,.$$

One sees that $\theta \cot \theta \leq 0$ for $\theta \geq \pi/2$. For $\theta \in (0, \pi/2)$ we have

$$\tan \theta > \theta \quad \Leftrightarrow \quad \theta \cot \theta < 1 \quad \Leftrightarrow \quad \left(\frac{1}{m} + \frac{m-1}{m} \theta \cot \theta \right) < 1 \,.$$

Using $\Sigma[\mu]$ to denote the CLT limit $n\mathrm{Cov}[\widehat{\mu}_n] \to \Sigma[\mu]$, cf. [2], we get the result

$$\mathrm{Hess}F(\mu) < 2\mathrm{Id}_m \quad \Rightarrow \quad \Sigma[\mu] > \mathrm{Cov}\left[\log_\mu X\right] \quad \Rightarrow \quad \mathrm{trace}\left(\Sigma[\mu]\right) > \mathrm{Var}[X] \,.$$

The claim follows at once.

Conjecture 7. *Let X be a random variable supported on a set $A \subset \mathbb{S}^m$ whose convex closure has nonzero volume and which has a unique mean μ. Then μ is Type I finite sample smeary.*

References

1. Bhattacharya, R., Lin, L.: Omnibus CLTs for Fréchet means and nonparametric inference on non-Euclidean spaces. Proc. Am, Math. Soc. **145**(1), 413–428 (2017)
2. Bhattacharya, R.N., Patrangenaru, V.: Large sample theory of intrinsic and extrinsic sample means on manifolds II. Ann. Stat. **33**(3), 1225–1259 (2005)
3. Eltzner, B.: Geometrical smeariness - a new phenomenon of Fréchet means (2020). arXiv:1908.04233v3
4. Eltzner, B., Huckemann, S.: Bootstrapping descriptors for non-Euclidean data. In: Nielsen, F., Barbaresco, F. (eds.) GSI 2017. LNCS, vol. 10589, pp. 12–19. Springer, Cham (2017). https://doi.org/10.1007/978-3-319-68445-1_2
5. Eltzner, B., Huckemann, S.F.: A smeary central limit theorem for manifolds with application to high-dimensional spheres. Ann. Stat. **47**(6), 3360–3381 (2019)
6. Fréchet, M.: Les éléments aléatoires de nature quelconque dans un espace distancié. Annales de l'Institut de Henri Poincaré **10**(4), 215–310 (1948)
7. Hendriks, H., Landsman, Z.: Mean location and sample mean location on manifolds: asymptotics, tests, confidence regions. J. Multivari. Anal. **67**, 227–243 (1998)
8. Hotz, T., Huckemann, S.: Intrinsic means on the circle: uniqueness, locus and asymptotics. Ann. Inst. Stat. Math. **67**(1), 177–193 (2015)
9. Huckemann, S.: Inference on 3D procrustes means: tree boles growth, rank-deficient diffusion tensors and perturbation models. Scand. J. Stat. **38**(3), 424–446 (2011a)
10. Huckemann, S.: Intrinsic inference on the mean geodesic of planar shapes and tree discrimination by leaf growth. Ann. Stat. **39**(2), 1098–1124 (2011b)
11. Hundrieser, S., Eltzner, B., Huckemann, S.F.: Finite sample smeariness of Fréchet means and application to climate (2020). arXiv:2005.02321
12. Le, H., Barden, D.: On the measure of the cut locus of a Fréchet mean. Bull. Lond. Math. Soc. **46**(4), 698–708 (2014)
13. Mardia, K.V., Jupp, P.E.: Directional Statistics. Wiley, New York (2000)
14. Meteoblue, A.G.: History+ platform (2021). https://www.meteoblue.com/en/weather/archive/export/basel_switzerland_2661604. Accessed 09 Feb 2021
15. Tran, D., Eltzner, B., Huckemann, S.F.: Smeariness begets finite sample smeariness. In: Geometric Science of Information 2021 Proceedings. Springer, Cham (2021)

Gaussian Distributions on Riemannian Symmetric Spaces in the Large N Limit

Simon Heuveline[1,2], Salem Said[3], and Cyrus Mostajeran[4(✉)]

[1] Centre for Mathematical Sciences, University of Cambridge, Cambridge, UK
[2] Mathematical Institute, University of Heidelberg, Heidelberg, Germany
[3] CNRS, University of Bordeaux, Bordeaux, France
[4] Department of Engineering, University of Cambridge, Cambridge, UK
csm54@cam.ac.uk

Abstract. We consider the challenging problem of computing normalization factors of Gaussian distributions on certain Riemannian symmetric spaces. In some cases, such as the space of Hermitian positive definite matrices or hyperbolic space, it is possible to compute them exactly using techniques from random matrix theory. However, in most cases which are important to applications, such as the space of symmetric positive definite (SPD) matrices or the Siegel domain, this is only possible numerically. Moreover, when we consider, for instance, high-dimensional SPD matrices, the known algorithms can become exceedingly slow. Motivated by notions from theoretical physics, we will discuss how to approximate these normalization factors in the large N limit: an approximation that gets increasingly better as the dimension of the underlying symmetric space (more precisely, its rank) gets larger. We will give formulas for leading order terms in the case of SPD matrices and related spaces. Furthermore, we will characterize the large N limit of the Siegel domain through a singular integral equation arising as a saddle-point equation.

Keywords: Gaussian distributions · Riemannian symmetric spaces · Random matrix theory · SPD matrices · High-dimensional data

1 Gaussian Distributions on Riemannian Symmetric Spaces

It is widely known that an isotropic Gaussian distribution on $\mathbb{R}^N$ takes the form $p_{\mathbb{R}^N}(x;\bar{x},\sigma) = (2\pi\sigma)^{-n/2}\exp(-\frac{1}{2\sigma^2}(x-\bar{x})^2)$, where $\bar{x}$ and σ denote the mean and standard deviation, respectively. Gaussian distributions can naturally be generalized to Riemannian manifolds (M, g) with the property

This work benefited from partial support by the European Research Council under the Advanced ERC Grant Agreement Switchlet n.670645. S.H. also received partial funding from the Cambridge Mathematics Placement (CMP) Programme. C.M. was supported by Fitzwilliam College and a Henslow Fellowship from the Cambridge Philosophical Society.

F. Nielsen and F. Barbaresco (Eds.): GSI 2021, LNCS 12829, pp. 20–28, 2021.
https://doi.org/10.1007/978-3-030-80209-7_3

$$Z_M(\bar{x},\sigma) := \int_M \exp(-\frac{1}{2\sigma^2} d_g(x,\bar{x})^2)\, dvol_g(x) < \infty, \tag{1}$$

where $dvol_g(x)$ denotes the Riemannian volume measure on M and $d_g : M \times M \to \mathbb{R}$ denotes the induced Riemannian distance function. Z_M will be referred to as the **partition function**. In this case,

$$p_M(x;\bar{x},\sigma) := \frac{1}{Z_M(\bar{x},\sigma)} \exp(-\frac{1}{2\sigma^2} d_g(x,\bar{x})^2), \tag{2}$$

is a well-defined probability distribution on M.

In numerous applications requiring the statistical analysis of manifold-valued data, it is important to be able to compute the partition function Z_M. A key difference between $p_{\mathbb{R}^N}$ and p_M in general is that Z_M is typically a complicated integral while $Z_{\mathbb{R}^N}(\bar{x},\sigma) = (2\pi\sigma)^{n/2}$ is readily available. On a general Riemannian manifold, Z_M will depend in a highly non-linear and intractable way on the mean $\bar{x}$ as the dominant contribution to Z_M will come from local data such as the curvature at $\bar{x}$. Hence, to have a chance of actually computing the full partition function analytically, we should restrict to spaces that look the same locally at any point. Riemannian symmetric spaces formalize this intuition and we are fortunate that precisely such spaces appear in most applications of interest (see, for example, [2,4,6,10–12,18]).

Definition 1. *A* ***Riemannian symmetric space*** *is a Riemannian manifold (M,g) such that for each point $p \in M$ there is a global isometry $s_p : M \to M$ such that $s_p(p) = p$ and $d_p s_p = -id_{T_pM}$, where $d_p s_p$ is the differential of s_p at p and id_{T_pM} denotes the identity operator on the tangent space T_pM.*

In the general theory of symmetric spaces [3,5], it can be shown that any symmetric space is isomorphic to a quotient of Lie groups $M \cong G/H$, where H is a compact Lie subgroup of G, and that integrals over M can be reduced to integrals over the respective Lie groups and their Lie algebras (more precisely, certain subspaces thereof) by the following proposition [12].

Proposition 2. *Given an integrable function on a non-compact symmetric space $(M = G/H, g)$, we can integrate in the following way:*

$$\int_M f(x) dvol_g(x) = C \int_H \int_{\mathfrak{a}} f(a,h) D(a) da dh \tag{3}$$

where dh is the normalized Haar measure on H and da is the Lebesgue measure on a certain subspace $\mathfrak{a} \subset \mathfrak{g} = Lie(G)$ [5,12]. The function $D : \mathfrak{a} \to \mathbb{R}^+$ is given by the following product over roots $\lambda : \mathfrak{a} \to \mathbb{R}^+$ (with m_λ-dimensional root space) $D(a) = \prod_{\lambda>0} \sinh^{m_\lambda}(|\lambda(a)|)$.

Proposition 2 and the fact that its isometry group acts transitively on a symmetric space allows us to prove that the partition function of a symmetric space does not depend on $\bar{x}$: $Z_{G/H}(\bar{x},\sigma) = Z_{G/H}(\sigma)$ (see [12]). Moreover, with Proposition 2 at hand, the problem of determining $Z_{G/H}$ reduces to calculating an integral over $\mathfrak{a}$, which is a linear space. Computing the resulting integrals can now be achieved numerically using a specifically designed Monte-Carlo algorithm [11]. Even though this works well for small dimensions, $\mathfrak{a} \cong \mathbb{R}^N$ is typically still a high-dimensional vector space when G/H is high-dimensional. The known algorithms start to break down for $N \approx 40$ [12], even though cases involving $N > 250$ are of relevance to applications such as electroencephalogram (EEG) based brain-computer interfaces [2,15]. It should be noted as an aside, however, that for some spaces, N does not depend on the dimension of the underlying symmetric space. For instance, in hyperbolic d-space, $N = 1$ independently of d [6].

Fortunately, for the space of positive definite Hermitian matrices, the resulting integrals have previously been studied in the context of Chern-Simons theory [8,17] and, in fact, they fit within the general theory of random matrices [9]. In this paper, we will extend such approaches to study the corresponding integrals for a broader range of symmetric spaces of interest, including the space of symmetric positive definite (SPD) matrices and the Siegel domain. We should like to note that we have recently discovered that ideas similar to the ones explored in this paper were developed independently by [14] from a slightly different perspective, which we will draw attention to throughout this work. However, we expect that our work will complement [14] by providing a stronger emphasis on the large N limit, including an analysis of the saddle-point equation.

2 Partition Functions at Finite N and Random Matrices

Motivated by applications, we will study the following spaces:

1. The spaces of symmetric, Hermitian and quaternionic Hermitian positive definite matrices will be denoted by $\mathcal{P}_{\mathbb{F}}(N) \cong GL(N,\mathbb{F})/K$ where $K \in \{U(N), \mathrm{O}(N), Sp(N)\}$ for $\mathbb{F} \in \{\mathbb{R}, \mathbb{C}, \mathbb{H}\}$, respectively. In each case, $\mathfrak{a} \subset Lie(GL(N,\mathbb{F})) \cong End(N,\mathbb{F})$ is the N-dimensional space of diagonal matrices. We will denote the partition functions by $Z_\beta(\sigma)$ where $\beta \in \{1,2,4\}$, respectively. $\mathcal{P}(N) := \mathcal{P}_{\mathbb{R}}(N)$ is commonly referred to as the **space of SPD matrices**.
2. The **Siegel domain** $\mathbb{D}(N) \cong Sp(2N,\mathbb{R})/U(N)$ is the space of complex symmetric $N \times N$ matrices z such that the imaginary part $\mathrm{Im}(z)$ is a positive definite matrix. We will denote its partition function by Z_S.

Using Proposition 2, it can be seen [6] that their respective partition functions are given by

$$Z_\beta(\sigma) = \frac{C_{N,\beta}(\sigma)}{(2\pi)^N N!} \int_{\mathbb{R}_+^N} \prod_{i=1}^N \Big(\exp\big(- \frac{\log^2(u_i)}{2\sigma^2} \big) \Big) |\Delta(u)|^\beta \prod_{i=1}^N du_i \tag{4}$$

where $\Delta(u) := \prod_{i<j}(u_i - u_j)$ is the **Vandermonde determinant**, $C_{N,\beta}(\sigma) := \frac{\omega_\beta(N)(2\pi)^N}{2^{NN_\beta}} \exp\Big(-NN_\beta^2 \frac{\sigma^2}{2} \Big)$ and $N_\beta := \frac{\beta}{2}(N-1)+1$. This is in the well-known form of a random matrix partition function:

$$Z(\sigma) = \int_{(a,b)^N} \prod_{i=1}^N \Big(\exp\big(- V(u_i;\sigma) \big) \Big) |\Delta(u)|^\beta \prod_{i=1}^N du_i, \tag{5}$$

even though the potential takes the non-standard form $V_{SW}(x;\sigma) := \frac{1}{2\sigma^2}\log^2(x)$ in our case. For instance, with the quadratic potential $V_Q(x;\sigma) := \frac{1}{2\sigma^2}x^2$, the random matrix partition function from Equation (5) corresponds to the famous orthogonal, unitary and symplectic random matrix ensembles for $\beta \in \{1,2,4\}$, respectively [9], on which we will comment further in Example 8. In fact, matrix models with the potential V_{SW} also appear in the physics literature as the partition functions of $U(N)$ Chern-Simons theory on S^3 ([8,17]). Moreover, the large N limit of $U(N)$ Chern-Simons theory is of physical interest and has been well studied in the theoretical and mathematical physics literature such as [7] Chap. 36.2 and [1], which motivates Sect. 3. Z_S also turns out to be of the form (5) with $\beta = 1$ and the potential $V_S(u;\sigma) := \frac{\log^2(u+\sqrt{u^2-1})}{8\sigma^2}$:

$$Z_S(\sigma) = vol(U(N))2^{\frac{N(N+1)}{2}} N! \int_{(1,\infty)^N} \prod_{i=1}^{N} \exp\big(-V_S(u_i;\sigma)\big)\Delta(u)\prod_{i=1}^{N} du_i. \quad (6)$$

Integrals of the form (5) can generally be calculated, by bringing the Vandermonde determinant to a suitable form involving orthogonal or skew-orthogonal polynomials.

Definition 3. *Let $V : (a,b) \to \mathbb{R}$ be a given potential. A set of polynomials $\{R_i : i = 1,\dots,N\}$, with $R_j(x) = a_j x^j + \dots$ being of degree j is called*

1. ***orthogonal with potential** V if they form an orthonormal basis for the space of degree N polynomials with respect to the inner product*

$$\langle f,g\rangle_2 = \int_a^b \exp\big(-V(x)\big)f(x)g(x)dx. \quad (7)$$

2. ***skew-orthogonal with potential** V if they bring the skew-symmetric product*

$$\frac{1}{2}\langle f,g\rangle_1 = \int_a^b\int_a^b f(x)g(y)\,\mathrm{sign}(x-y)\exp(-V(x))\exp(-V(y))dxdy \quad (8)$$

to the standard form, meaning

$$\begin{cases} \langle R_{2k}, R_{2l}\rangle_1 &= \langle R_{2k+1}, R_{2l+1}\rangle_1 = 0 \\ \langle R_{2k}, R_{2l+1}\rangle_1 &= -\langle R_{2l+1}, R_{2k}\rangle_1 = \delta_{kl}. \end{cases} \quad (9)$$

If we are given orthogonal or skew-orthogonal polynomials for a given potential V, then its corresponding partition function with $\beta = 1$ or $\beta = 2$, respectively, can be calculated in terms of the polynomials' leading order coefficients [9]. A similar story, which we will not go into also works for the quaternionic case $\beta = 4$. Fortunately, orthogonal polynomials for the potential V_{SW} have previously appeared in the literature as **Stieltjes-Wigert polynomials** [16], which can be used to arrive at the following result [6].

Proposition 4. *The partition function Z_2 for $\mathcal{P}_{\mathbb{C}}(N)$ is given by*

$$Z_2(\sigma) = \frac{\omega_2(N)}{2^{N^2}}(2\pi\sigma^2)^{\frac{N}{2}} \exp\left((N^3 - N)\frac{\sigma^2}{6}\right) \prod_{k=1}^{N-1} (1 - e^{-k\sigma^2})^{N-k}.$$

Unfortunately, even though [14] provides explicit calculations for the first few skew-orthogonal polynomials of V_{SW}, general expressions for such polynomials have yet to be found as noted in [6,14]. However, if we let a_i and b_i denote the leading order coefficients of skew-orthogonal polynomials $\{P_i = a_i x^i + \ldots | i = 1, \ldots, N\}$ for V_{SW} and skew-orthogonal polynomials $\{Q_i = b_i x^i + \ldots | i = 1, \ldots, N\}$ for V_S, respectively, then we can obtain exact formulas for Z_1 and Z_S at finite N in terms of a_i and b_i [6]. For technical convenience, we will focus on the even-dimensional cases ($N = 2m$) in this work. The odd-dimensional cases can be treated in a similar manner subject to a number of technical modifications. See [14] for details on the treatment of the odd-dimensional case.

Proposition 5. *If $N = 2m$, the partition functions Z_1 and Z_S are given by*

$$Z_1(\sigma) = \frac{\omega_1(N)}{2^{Nm}} \exp\left(- N((N-1)/2 + 1)^2(\sigma^2/2)\right)\left(\prod_{l=1}^{N} a_l\right)^{-1} \tag{10}$$

$$Z_S(\sigma) = vol(U(N))2^{m^2(N+1)}\left(\prod_{l=1}^{N} b_l\right)^{-1}. \tag{11}$$

The coefficients a_i and b_i can be found by bringing the inner product (8) into the standard form (9), which can be achieved numerically via a symplectic Gram-Schmidt algorithm [13]. Nonetheless, such a computation becomes exceedingly challenging for large N, which motivates the consideration of the large N limit provided in the following section.

3 Partition Functions in the Large N Limit and the Saddle-Point Equation

We will now change our point of view and compute the large N limit of the partition function Z, rather than try to compute it at finite N. Here, the **large N limit** specifically refers to the limit where $N \to \infty$ while the **'t Hooft parameter** $t := \sigma^2 N$ is kept fixed. In this limit, Z has an asymptotic expansion in N^{-2}, known as the **genus expansion** [6,8]: $\log(Z(\sigma)) \sim \sum_{g=0}^{\infty} f_g(t)N^{2-2g}$. The first term in this series becomes an increasingly good approximation as N gets larger, which is very useful since we are able to compute it.

Since fixing t implies that $\sigma^2 \to 0$, the large N limit is also referred to as the **double scaling limit** [14]. Interestingly, one may also consider other limits such as $\sigma^2 \to 0$ and $\sigma^2 \to \infty$ while N is fixed, which we will not consider in this work (see [14] (II, 3. a, b) for further details). In the case of Z_2, the large N limit can be directly computed from Proposition 4 and is found to be

$$\frac{1}{N^2}\log(Z_2(\sigma)) \sim -\frac{1}{2}\log\left(\frac{2N}{\pi}\right) + \frac{3}{4} + \frac{t}{6} - \frac{Li_3(e^{-t}) - \zeta(3)}{t^2} \tag{12}$$

where $Li_3(x) := \sum_{k=1}^{\infty} \frac{x^k}{k^3}$ (for $|x| < 1$) is the trilogarithm.

Equation (12) crucially relies on having a closed form expression for the partition function Z_2 for any finite N in the first place (Proposition 4). This is not the case for Z_1 and Z_S, since we do not know the asymptotics of a_i and b_i from Proposition 5. Therefore, we will now discuss a different, more powerful approach to the large N limit inspired by ideas from theoretical physics: a saddle-point approximation. This will enable us to directly obtain the large N limit of Z_2 in Proposition 9 by solving a certain singular integral equation, the **saddle-point equation**.

As above, we split $Z_\beta(\sigma) = C_{N,\beta}(\sigma)\tilde{Z}_\beta(\sigma)$, where $\tilde{Z}_\beta$ is in an appropriate form for the use of saddle-point methods as discussed in [8] (Equation (1.46)):

$$\tilde{Z}_\beta(\sigma) = \frac{1}{N!}\int_{\mathbb{R}_+^N} \exp\left(N^2 \tilde{V}_{SW}(\boldsymbol{\lambda}, \beta; t)\right) \prod_{k=1}^{N} \frac{d\lambda_k}{2\pi}, \tag{13}$$

where we have introduced the **effective potential**:

$$\tilde{V}_{SW}(\boldsymbol{\lambda}, \beta; t) := -\frac{1}{2tN}\sum_{i=1}^{N}\log^2(\lambda_i) + \frac{\beta}{N^2}\sum_{i<j}\log|\lambda_i - \lambda_j|. \tag{14}$$

In the large N limit, the dominant contributions to $\tilde{Z}_\beta$ will come from the saddle points of $\tilde{V}_{SW}$. Note that this is analogous to a **semiclassical limit**, which is further discussed in [6]. For the Siegel domain, we have a similar effective potential, which takes the form

$$\tilde{V}_S(\boldsymbol{\lambda}; t) := -\frac{1}{8tN}\sum_{i=1}^{N}\log^2\left(\lambda_i + \sqrt{\lambda_i^2 - 1}\right) + \frac{1}{N^2}\sum_{i<j}\log|\lambda_i - \lambda_j|. \tag{15}$$

Remark 6. *The effective potential $\tilde{V}_{SW}$ gives rise to a physical interpretation of the theory and its large N limit: the N eigenvalues can be seen as static particles in the potential V_{SW} interacting through a logarithmic Coulomb repulsion (the second term of the effective potential). As $\sigma^2 = \frac{t}{N}$ decreases, the repulsion becomes weak and all particles can sit next to each other close to the minimum of the potential ($\lambda = 1$), while the particles tend to spread out for large σ^2 as observed in Fig. 1. Now, the large N limit can be seen to correspond to the addition of more and more particles (i.e. $N \to \infty$) while letting their repulsion become increasingly weak (fixing $t = N\sigma^2$). Finding the limiting distribution ρ_t (that still depends on t) turns out to characterize the large N limit of the partition function. It can be obtained by solving the saddle-point equation for $\tilde{V}$ in a* *continuum limit as discussed below. Motivated by the physics literature, we will refer to this* ρ_t *as the* **master field**. *In the random matrix theory literature, this approach is known as the* **Coulomb gas method** *[9].*

Remark 7. *A further observation, from the effective action* (14) *is that up to an overall factor of* $\tilde{V}$ *(which leaves the saddle-point equation invariant), we can absorb the* β *by rescaling* $t \mapsto t/\beta$. *So, in the large* N *limit,* $\tilde{Z}_\beta(\sigma) \sim \tilde{Z}_2(\sqrt{\beta/2}\,\sigma)$ *irrespective of the choice of* $\beta \in \{1,2,4\}$, *which is remarkable considering the quite distinct geometric origins associated with the different values of* β. *We refer to this phenomenon as* **universality**: *the three cases* $\mathbb{K} \in \{\mathbb{R}, \mathbb{C}, \mathbb{H}\}$ *start out differently, but "flow" to a universal limit (characterized by the master field) as* $N \to \infty$. *This is specially useful for the particularly important case of* $\beta = 1$, *which can now be related to the large* N *limit of* Z_2 *derived in Equation* (12)*:*

$$\frac{1}{N^2}\log(Z_\beta(\sigma)) \sim \frac{1}{N^2}\left[\log\left(Z_2\left(\sqrt{\tfrac{\beta}{2}}\sigma\right)\right) - \log\left(\frac{C_{\infty,\beta}(\sigma)}{C_{\infty,2}(\sqrt{\tfrac{\beta}{2}}\sigma)}\right)\right] \tag{16}$$

where $C_{\infty,\beta}$ *are the large* N *limits of the prefactors, which are discussed in [6]. Below, we will give another formula for the large* N *limit of* Z_β, *by solving the saddle-point equation explicitly.*

Example 8 (Wigner's semicircle law). *For simplicity, since* $V_Q(\lambda;\sigma) = \frac{1}{2\sigma^2}\lambda^2$ *has a simpler form than* V_{SW} *or* V_S *yet illustrates all the necessary ideas, we will begin by considering its saddle-point equation. As motivated in Remark 6 and following [8] ((1.47)–(1.53)), we are interested in the saddle points of the effective potential, which in this case reads* $\tilde{V}_Q(\boldsymbol{\lambda};t) := -\frac{1}{tN}\sum_i \lambda_i^2 + \frac{\beta}{N^2}\sum_{i<j}\log|\lambda_i - \lambda_j|$ *in analogy with equations* (14) *and* (15). *The saddle points are characterized by* $\frac{d}{d\lambda_k}\tilde{V}_Q(\boldsymbol{\lambda};t) = 0$ *for all* $k = 1,\dots,N$, *which is simply*

$$\frac{1}{\beta t}\lambda_k = \frac{1}{N}\sum_{j\neq k}\frac{1}{\lambda_k - \lambda_j} =: P\left(\int \frac{\rho_{t,N}(\lambda')}{\lambda' - \lambda_k}d\lambda'\right) \tag{17}$$

where $\rho_{t,N}$ *is formally given by* $\rho_{t,N}(\lambda) = \frac{1}{N}\sum_j \delta(\lambda - \lambda_j)$ *and* P *is a discrete Cauchy principal value. The large* N *limit can now be regarded as a continuum limit, in which* $\rho_{t,N}$ *becomes a continuous function* ρ_t. *Equation* (17) *becomes*

$$\frac{1}{\beta t}\lambda = P\left(\int_{-\infty}^{\infty}\frac{\rho_t(\lambda')d\lambda'}{\lambda - \lambda'}\right) \iff \frac{1}{2t}\lambda = P\left(\int_{-\infty}^{\infty}\frac{\rho_{\frac{2}{\beta}t}(\lambda')d\lambda'}{\lambda - \lambda'}\right) \tag{18}$$

where now P *is the actual Cauchy principal value. The saddle-point equation* (18) *can be solved using resolvent methods and in [8] (Equation (1.82)) it is shown that the solution in this case is simply a semicircle of radius* $2\sqrt{t}$: $\rho_t^Q(\lambda) = \frac{1}{2\pi t}\sqrt{4t - \lambda^2}\chi_{C(t)}(\lambda)$, *where* $\chi_{C(t)}$ *denotes the characteristic function supported on the interval* $C(t) = [-2\sqrt{t}, 2\sqrt{t}]$. *This celebrated semicircle law was first derived by Wigner in 1955 [19].*

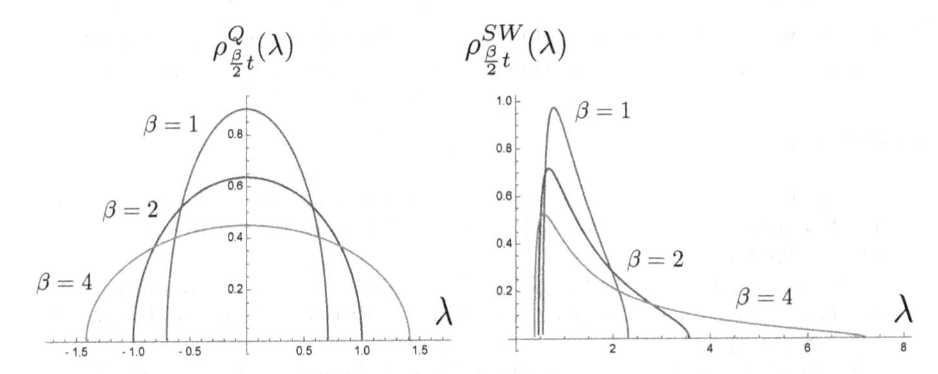

Fig. 1. The master field $\rho^Q_{\beta t/2}$ of Wigner's semicircle law (left) and $\rho^{SW}_{\beta t/2}$ (right) for fixed $t = 1/4$ and different choices of β. The red, blue and green curves correspond to the real, complex and quaternionic cases, respectively ($\beta = 1, 2, 4$).

In fact, the steps of Example 8 that lead to the saddle-point equation (18) can be performed analogously for the potentials V_{SW} and V_S. Moreover, the saddle-point equation for V_{SW} can even be solved analytically using resolvent methods [6,8], which leads to the following result.

Proposition 9. *Let $\beta > 0$. If $N \to \infty$ while the 't Hooft parameter $t = N\sigma^2$ is fixed, we get*

$$\frac{1}{N^2}\log(\tilde{Z}_\beta(\sigma)) \sim F_{uni}\left(\frac{\beta}{2}t\right) + \mathcal{O}(N^{-2}) \tag{19}$$

where

$$F_{uni}(t) = -\frac{1}{2t}\int_{\mathcal{C}(t)} \rho^{SW}_t(\lambda)\log^2\lambda\, d\lambda + \int_{\mathcal{C}(t)^2} \rho^{SW}_t(\lambda)\rho^{SW}_t(\lambda')\log(|\lambda - \lambda'|)d\lambda d\lambda'$$

and the master field ρ^{SW}_t is given by

$$\rho^{SW}_t(\lambda) = \frac{1}{\pi t\lambda}\tan^{-1}\left[\frac{\sqrt{4\lambda - (1 + e^{-t}\lambda)^2}}{1 + e^{-t}\lambda}\right]\chi_{\mathcal{C}(t)} \tag{20}$$

where $\mathcal{C}(t) = [2e^{2t} - e^t + 2e^{\frac{3t}{2}}\sqrt{e^t - 1}, 2e^{2t} - e^t - 2e^{\frac{3t}{2}}\sqrt{e^t - 1}]$.

In Fig. 1, Wigner's semicircle distribution ρ^Q_t is plotted alongside ρ^{SW}_t for different choices of $\beta \in \{1, 2, 4\}$. It can be observed that as t decreases (or equivalently, β decreases), the distibutions tend to concentrate around the classical minima $\lambda = 0$ (for V_Q) and $\lambda = 1$ (for V_{SW}). Conversely, they tend to spread out as t increases.

Finally, we note a similar equation that characterizes the master field in the case of the Siegel domain:

$$\frac{\log\left(\lambda + \sqrt{\lambda^2 - 1}\right)}{4t\sqrt{\lambda^2 - 1}} = P\left(\int_1^\infty \frac{\rho^S_t(\lambda')d\lambda'}{\lambda - \lambda'}\right). \tag{21}$$

At present, we are unaware of a solution to this equation and finding one either numerically or using resolvent methods is to be carried out in future work.

References

1. Auckly, D., Koshkin, S.: Introduction to the Gopakumar-Vafa large N duality. The Interaction of Finite-Type and Gromov-Witten Invariants (BIRS 2003),vol. 8, 195–456 (2006)
2. Barachant, A., Bonnet, S., Congedo, M., Jutten, C.: Multiclass brain-computer interface classification by Riemannian geometry. IEEE Trans. Biomed. Eng. **59**(4), 920–928 (2012)
3. Eberlein, P.: Geometry of Nonpositively Curved Manifolds. University of Chicago Press, Chicago(1996)
4. Fletcher, P.T., Joshi, S.: Principal geodesic analysis on symmetric spaces: statistics of diffusion tensors. In: Sonka, M., Kakadiaris, I.A., Kybic, J. (eds.) CVAMIA/MMBIA -2004. LNCS, vol. 3117, pp. 87–98. Springer, Heidelberg (2004). https://doi.org/10.1007/978-3-540-27816-0_8
5. Helgason, S.: Differential Geometry, Lie Groups, and Symmetric Spaces. Academic Press, New York (1979)
6. Heuveline, S., Said, S., Mostajeran, C.: Gaussian distributions on Riemannian symmetric spaces, random matrices and planar Feynman diagrams. arXiv preprint arXiv:2106.08953 (2021)
7. Hori, K., et al. : Mirror Symmetry, vol. 1. American Mathematical Society, Providence (2003)
8. Marino, M., et al.: Chern-Simons Theory, Matrix Models, and Topological Strings. No. 131, Oxford University Press, Oxford (2005)
9. Mehta, M.L.: Random matrices. Elsevier (2004)
10. Pennec, X., Sommer, S., Fletcher, T.: Riemannian Geometric Statistics in Medical Image Analysis. Academic Press, San Diego (2019)
11. Said, S., Bombrun, L., Berthoumieu, Y., Manton, J.H.: Riemannian Gaussian distributions on the space of symmetric positive definite matrices. IEEE Trans. Inf. Theor. **63**(4), 2153–2170 (2017)
12. Said, S., Hajri, H., Bombrun, L., Vemuri, B.C.: Gaussian distributions on Riemannian symmetric spaces: statistical learning with structured covariance matrices. IEEE Trans. Inf. Theor **64**(2), 752–772 (2018)
13. Salam, A.: On theoretical and numerical aspects of symplectic Gram-Schmidt-like algorithms. Num. Algorith. **39**(4), 437–462 (2005)
14. Santilli, L., Tierz, M.: Riemannian Gaussian distributions, random matrix ensembles and diffusion kernels. arXiv preprint arXiv:2011.13680 (2020)
15. Seeck, M., Koessler, L., Bast, T., Leijten, F., Michel, C., Baumgartner, C., He, B., Beniczky, S.: The standardized EEG electrode array of the IFCN. Clin. Neurophysiol. **128**(10), 2070–2077 (2017)
16. Szegő, G.: Orthogonal Polynomials, vol. 23. American Mathematical Society, Providence (1939)
17. Tierz, M.: Soft matrix models and Chern-Simons partition functions. Mod. Phys Lett. A **19**(18), 1365–1378 (2004)
18. Tupker, Q., Said, S., Mostajeran, C.: Online learning of Riemannian hidden markov models in homogeneous Hadamard spaces. arXiv preprint arXiv:2102.07771 (2021)
19. Wigner, E.P.: Characteristic vectors of bordered matrices with infinite dimensions. Ann. Math. **62**(3), 548–564 (1955)

Smeariness Begets Finite Sample Smeariness

Do Tran(✉), Benjamin Eltzner, and Stephan Huckemann

Felix-Bernstein-Institute for Mathematical Statistics in the Biosciences, Georg-August-Universität at Göttingen, Göttingen, Germany
do.tranvan@uni-goettingen.de

Abstract. Fréchet means are indispensable for nonparametric statistics on non-Euclidean spaces. For suitable random variables, in some sense, they "sense" topological and geometric structure. In particular, smeariness seems to indicate the presence of positive curvature. While smeariness may be considered more as an academical curiosity, occurring rarely, it has been recently demonstrated that *finite sample smeariness* (FSS) occurs regularly on circles, tori and spheres and affects a large class of typical probability distributions. FSS can be well described by the *modulation* measuring the quotient of rescaled expected sample mean variance and population variance. Under FSS it is larger than one – that is its value on Euclidean spaces – and this makes quantile based tests using tangent space approximations inapplicable. We show here that near smeary probability distributions there are always FSS probability distributions and as a first step towards the conjecture that all compact spaces feature smeary distributions, we establish directional smeariness under curvature bounds.

1 Introduction

For nonparametric statistics of manifold data, the Fréchet mean plays a central role, both in descriptive and inferential statistics. For quite some while it was assumed that its asymptotics can be approximated under very general conditions by that of means of data projected to a suitable tangent space, e.g. [1,2,7,9,10]. Under existence of second moments, these follow a classical central limit theorem. In the last decade, however, other asymptotic regimes have been discovered, yielding so called *smeary* limiting rates, limiting rates that are slower than the classical $n^{-1/2}$, where n denotes sample size, e.g. [5,8]. While such smeary distributions are rather exceptional, more recently, it was discovered that these exceptional distributions affect the asymptotics of a large class of otherwise unsuspicious distributions, for instance all Fisher-von-Mises distributions on the circle, cf. [11]: for rather high sample sizes the rates are slower than $n^{-1/2}$ and eventually an asymptotic variance can be reached that is higher than that of

Acknowledging DFG HU 1575/7, DFG GK 2088, SFB CRC 803 and the Niedersachsen Vorab of the Volkswagen Foundation.

F. Nielsen and F. Barbaresco (Eds.): GSI 2021, LNCS 12829, pp. 29–36, 2021.
https://doi.org/10.1007/978-3-030-80209-7_4

tangent space data. While this effect on the circle and the sphere is explored in more detail by [6], here we concentrate on rather general manifolds and discuss recent findings concerning two conjectures.

Conjecture 1

(a) *Whenever there is a random variable featuring smeariness, there are nearby random variables featuring finite sample smeariness.*
(b) *All compact Riemannian manifolds feature smeariness.*

Here, we prove Conjecture (a) under the rather general concept of *power smeariness* and Conjecture (b) for *directional smeariness* under curvature bounds. We also provide for simulations, showing that classical quantile based tests fail under the presence of finite sample smeariness, suitably designed bootstrap tests, however, amend for it.

2 Assumptions, Notation and Definitions

Let M be a complete Riemannian manifold of dimension $m \in \mathbb{N}$ with induced distance d on M. Random variables $X_1, \dots, X_n \overset{\text{i.i.d.}}{\sim} X$ on M with silently underlying probability space $(\Omega, \mathbb{P})$ induce *Fréchet functions*

$$F(p) = \mathbb{E}[d(X,p)^2] \text{ and } F_n(p) = \frac{1}{n}\sum_{j=1}^{n} d(X_j,p)^2 \text{ for } p \in M.$$

We also write F^X and F_n^X to refer to the underlying X.

Lemma 2. *If* $F(p) < \infty$ *for some* $p \in M$*, the set of minimizers* $\operatorname{argmin}_{p\in M} F(p)$ *is not void and compact. In particular,* $\operatorname{argmin}_{p\in M} F_n(p)$ *admits a probability measure.*

Proof. If $F(p) < \infty$ for some $p \in M$, then due to the triangle inequality $F(p) < \infty$ for all $p \in M$. Further, by the completeness of M a minimizer of the Fréchet function is assumed, by continuity the set of minimizers is closed and due to

$$d(\mu_1, \mu_2) \le \mathbb{E}[d(\mu_1, X)] + \mathbb{E}[d(\mu_2, X)] \le \sqrt{\mathbb{E}[d(\mu_1, X)^2]} + \sqrt{\mathbb{E}[d(\mu_2, X)^2]}$$

it is bounded. Due to [13], M can be isometrically embedded in a finite dimensional Euclidean space, hence the set of minimizers is compact. Thus one can define probability measures on the set of minimizers on $F(p)$, and as well on $F_n(p)$.

We work under the following additional assumptions.

Assumption 3. *Assume*

1. *X is not a.s. a single point,*
2. *$F(p) < \infty$ for some $p \in M$,*
3. *there is a unique minimizer* $\mu = \operatorname{argmin}_{p\in M} F(p)$*, called the* Fréchet population mean,

4. $\widehat{\mu}_n \in \operatorname{argmin}_{p\in M} F_n(p)$ *is a measurable selection from the empirical mean set, called a* Fréchet empirical mean,
5. *and that the cut locus* $\mathrm{Cut}(\mu)$ *of* μ *is either void or can be reached by two different geodesics from* μ.

The last point ensures that $\mathbb{P}\{X \in \mathrm{Cut}(\mu)\} = 0$ *due to [12].*

Definition 4. *With the Riemannian exponential* $\exp_\mu$, *well defined on the tangent space* $T_\mu M$, *let*

$$\rho(X,x) := d(X, \exp_\mu x)^2,$$
$$f(x) := F(\exp_\mu(x)),$$
$$f_n(x) := F_n(\exp_\mu(x)).$$

We also write f^X *and* f_n^X *to refer to the underlying* X. *Further, with the Riemannian logarithm* $\log_\mu p = (\exp_\mu)^{-1}(p)$, *well defined outside of* $\mathrm{Cut}(\mu)$, *we have, for* $x \in T_\mu M$ *close to* 0,

$$\rho(X,x) = \| \log_\mu X - x\|^2 + \mathcal{O}(\|x\|^2). \tag{1}$$

We define

i. the population variance

$$V := F(\mu) = f(0) = \mathbb{E}[d(X,\mu)^2] = \mathrm{trace}\left(\mathrm{Cov}[\log_\mu X]\right);$$

ii. the Fréchet sample mean variance

$$V_n := \mathbb{E}[d(\widehat{\mu}_n, \mu)^2]; \text{ and}$$

iii. the modulation

$$\mathfrak{m}_n := \frac{nV_n}{V}.$$

We shall also write $\widehat{x}_n := \log_\mu \widehat{\mu}_n$ *for the image of the empirical Fréchet mean* $\hat{\mu}_n$ *in the tangent space at* μ. *Again, if necessary, we write* $V^X, V_n^X, \widehat{\mu}_n^X, \widehat{x}_n^X$ *and* $\mathfrak{m}_n^X$ *to refer to the underlying* X.

Remark 5. *The* modulation factor $\mathfrak{m}_n$ *was introduced in [15] to indicate the influence of curvature on the rate of convergence of Fréchet empirical mean.*

Assumption 6. *In order to reduce notational complexity, we also assume that*

$$f(x) = \sum_{j=1}^{m} T_j |(Rx)_j|^{r+2} + o(\|x\|^{r+2}) \tag{2}$$

with some $r \geq 0$, *where* $(Rx)_j$ *is the* j*-th component after multiplication with an orthogonal matrix* R *and* $T_1, \ldots, T_m$ *are positive.*

With these definitions, we can define various asymptotic regimes.

Definition 7. *We say that X is*

(i) Euclidean *if* $\mathfrak{m}_n = 1$ *for all* $n \in \mathbb{N}$,
(ii) finite sample smeary *if* $1 < \sup_{n\in\mathbb{N}} \mathfrak{m}_n < \infty$,
(iii) smeary *if* $\sup_{n\in\mathbb{N}} \mathfrak{m}_n = \infty$,
(iv) r-power smeary *if (2) holds with* $r > 0$.

If the manifold M is a Euclidean space and if second moments of X exist, then Assumptions 3 hold and due to the classical central limit theorem, X is then Euclidean (cf. Definition 7). In case of M being a circle or a torus, as shown in [11], X is Euclidean only if it is sufficiently concentrated. As further shown there, if X is spread beyond a geodesic half ball on the circle or the torus, it features finite sample smeariness, which, on the circle and the Torus manifests, among others, in two specific subtypes, cf. [6]

In consequence of the *general central limit theorem* (GCLT) from [5] under Assumptions 3 and 6,

$$n^{\frac{1}{2r+2}}(R^t\widehat{x}_n)_j \xrightarrow{\mathcal{D}} \mathcal{H}_j \text{ for all } 1 \le j \le m \tag{3}$$

where R^t is the transpose of $R, T = \text{diag}(T_1, \ldots .T_m)$ and $(\mathcal{H}_1|\mathcal{H}_1|^r, \ldots, \mathcal{H}_m|\mathcal{H}_m|^r)$ is multivariate Gaussian with zero mean and covariance

$$\frac{4}{(r+2)^2} T^{-1} \operatorname{Cov}[\log_\mu X]\, T^{-1},$$

we have at once that r-power-smeary for $r > 0$ implies smeariness.

3 Smeariness Begets Finite Sample Smeariness

Theorem 8. *Under Assumptions 3 and 6, if there is a random variable on M that is r-power smeary, r > 0, then there is one that is finite sample smeary. More precisely, for every K > 1 there is a random variable Y with* $\sup_{n\in\mathbb{N}} \mathfrak{m}_n^Y \ge K$.

Proof. Suppose that X is r-power smeary, $r > 0$ on M. For given $K > 1$ let $0 < \kappa < 1$ such that $\kappa^{-2} = K$ and define the random variable X_κ via

$$\mathbb{P}\{X_\kappa = \mu\} = \kappa \text{ and } \mathbb{P}\{X_\kappa = X\} = 1 - \kappa.$$

With the sets $A = \{X_\kappa = \mu\}$ and $B = \{X_\kappa = X\}$, the Fréchet function of X_κ is given by

$$\begin{aligned} F^{X_\kappa}(p) &= \int_A d(p,\mu)^2\, d\mathbb{P}^{X_\kappa} + \int_B d(X,\mu)^2\, d\mathbb{P}^{X_\kappa} \\ &= \kappa d(p,\mu)^2 + (1-\kappa) F^X(p) = \kappa \|\log_\mu(p)\|^2 + o(\|\log_\mu(p)\|^2), \end{aligned}$$

where we use Eq. (1) for the last equality. Thus, Assumption 6 holds with $T_j = \kappa$, $r = 0$ and $R = \text{id}$, therefore X_κ has the unique mean μ and population variance

$$V^{X_\kappa} = F^{X_\kappa}(\mu) = (1-\kappa)F^X(\mu) = (1-\kappa)V^X \tag{4}$$

by hypothesis. Since $\text{Cov}[\log_\mu X_\kappa] = (1-\kappa)\text{Cov}[\log_\mu X]$, we have thus with the GCLT (3) for $T_j = \kappa$, $r = 0$ and $R = \text{id}$,

$$\sqrt{n}\widehat{x}_n^{X_\kappa} \xrightarrow{\mathcal{D}} \mathcal{N}\left(0, \frac{1-\kappa}{\kappa^2}\text{Cov}[\log_\mu X]\right).$$

This yields $nV_n^{X_\kappa} = \frac{1-\kappa}{\kappa^2}V^X$ and in conjunction with (4) we obtain

$$\mathfrak{m}_n^{X_\kappa} = \frac{1}{\kappa^2}.$$

Thus, $Y = X_{K^{-1/2}}$ has the asserted property.

4 Directional Smeariness

Definition 9 (Directional Smeariness). *We say that X is directional smeary if (2) holds for $r = 2$ with some of the $T_1, \ldots, T_m$ there equal to zero.*

Theorem 10. *Suppose that M is a Riemannian manifold with the sectional curvature bounded from above by $\mathbf{K} > 0$ such that there exists a simply connected geodesic submanifold of constant sectional curvature $\mathbf{K}$. Then M features a random variable that is directional smeary.*

Proof. Suppose $N \subset M$ is a simply connected, totally geodesic submanifold of M with constant sectional curvature $\mathbf{K}$. Let $\mu \in N$ and consider orthogonal unit vectors $V, W \in T_\mu N$. Let us consider a point mass random variable X with $P(\{X = \mu\}) = 1$, a geodesic $\gamma(t) = \exp(tV)$, and a family of random variables X_t defined as

$$P\{X_t = \delta_{\gamma(t)}\} = P\{X_t = \delta_{\gamma(-t)}\} = \frac{1}{2}.$$

We shall show that we can choose t close to $\pi/\sqrt{\mathbf{K}}$ and $\epsilon > 0$ sufficiently small such that the random variable $Y_{t,\epsilon}$, which is defined as

$$P\{Y_{t,\epsilon} = X_t\} = \epsilon,\ P\{Y_{t,\epsilon} = X\} = 1-\epsilon,$$

is directional smeary.

Let us write $F^{t,\epsilon}(p)$ for the Fréchet function of $Y_{t,\epsilon}$. Suppose for the moment that μ is the unique Fréchet mean of $Y_{t,\epsilon}$, we shall show that for sufficient small ϵ and t close to $\pi/\sqrt{K}$, the Hessian at μ of $F^{t,\epsilon}$ vanishes in some directions, which will imply that $Y_{t,\epsilon}$ is directional smeary as desired.

We claim that $\nabla^2 F^{t,\epsilon}(\mu)[W,W] = 0$, which will fulfill the proof. Indeed, it follows from [16, Appendix B.2] that

$$\nabla^2 F^{t,\epsilon}(\mu)[W,W] = (1-\epsilon) + 2\epsilon(t\sqrt{\mathbf{K}})\cot(t\sqrt{\mathbf{K}}).$$

Hence, if we choose t close to $\pi/\sqrt{\mathbf{K}}$ such that

$$t\sqrt{\mathbf{K}}\cot(t\sqrt{\mathbf{K}}) = -\frac{1-\epsilon}{2\epsilon} \tag{5}$$

then $\nabla^2 F^{t,\epsilon}(\mu)[W,W] = 0$ as claimed.

It remains to show that μ is the unique Fréchet mean of $Y_{t,\epsilon}$ for t close to $\pi/\sqrt{\mathbf{K}}$ and ϵ satisfies Eq. 5. Indeed, for any $p \in M$ we have

$$F^{t,\epsilon}(p) = \frac{1}{2}\epsilon\mathbf{d}^2(p,\gamma(t)) + \frac{1}{2}\epsilon\mathbf{d}^2(p,\gamma(-t)) + (1-\epsilon)\mathbf{d}^2(p,\mu).$$

For small ϵ then $F^{t,\epsilon}(\mu) \leq F^{t,\epsilon}(p)$ if $\mathbf{d}(p,\mu) \geq \pi/\sqrt{\mathbf{K}}$. Thus, it suffices to show that μ is the unique minimizer of the restriction of $F^{t,\epsilon}$ on the open ball $B(\mu,\pi/\sqrt{\mathbf{K}})$. Let us consider a model in the two dimensional sphere $S^2_{\mathbf{K}}$ of curvature $\mathbf{K}$ with geodesic distance $\mathbf{d}_s$. Let $\tilde{\mu}$ be the South Pole, $\tilde{V} \in T_{\tilde{\mu}}S^2_{\mathbf{K}}$ be a unit vector and $\tilde{\gamma}(t) = \exp_{\tilde{\mu}} t\tilde{V}$. Let t and ϵ satisfy Eq. (5) and consider the following measure on $S^2_{\mathbf{K}}$

$$\zeta = (1-\epsilon)\delta_{\tilde{\mu}} + \frac{\epsilon}{2}(\delta_{\tilde{\gamma}(t)} + \delta_{\tilde{\gamma}(-t)}).$$

Write F_ζ for the Fréchet function of ζ, then for any $q \in S^2_{\mathbf{K}}$,

$$F_\zeta(q) = \frac{1}{2}\epsilon\mathbf{d}_s^2(q,\tilde{\gamma}(t)) + \frac{1}{2}\epsilon\mathbf{d}_s^2(q,\tilde{\gamma}(-t)) + (1-\epsilon)\mathbf{d}_s^2(q,\tilde{\mu}).$$

It follows from the definition of ζ that $F_\zeta(\tilde{\mu}) = F^{t,\epsilon}(\mu)$. Direct computation of F_ζ on the sphere $S^2_{\mathbf{K}}$ verifies that $\tilde{\mu}$ is the unique Fréchet mean of ζ.

On the other hand, suppose that $\tilde{p} \in S^2_{\mathbf{K}}$ with $\mathbf{d}_s(\tilde{p},\tilde{\mu}) = \mathbf{d}(p,\mu) < \pi/\sqrt{\mathbf{K}}$ and $\angle(\log_{\tilde{\mu}}\tilde{p},\tilde{V}) = \angle(\log_\mu p, V)$. Because $\mathbf{K}$ is the maximum sectional curvature of M, Toponogov's theorem, c.f. [3, Theorem 2.2] implies that

$$\mathbf{d}_s(\tilde{p},\tilde{\gamma}(-t)) \leq \mathbf{d}(p,\gamma(-t)) \text{ and } \mathbf{d}_s(\tilde{p},\tilde{\gamma}(t)) \leq \mathbf{d}(p,\gamma(t)).$$

Thus $F_\zeta(\tilde{p}) \leq F^{t,\epsilon}(p)$. Because $\tilde{\mu}$ is the unique minimizer of F_ζ and $F_\zeta(\tilde{\mu}) = F^{t,\epsilon}(\mu)$ it follows that μ is the unique minimizer of $F^{t,\epsilon}|_{B(\mu,\pi/\sqrt{\mathbf{K}})}$ as needed.

Remark 11. *Examples for spaces that fulfill conditions in Theorem 10 include spheres, the Grassmanians, compact simple Lie groups, or in general globally symmetric spaces of compact type.*

5 Simulations

In the analysis of biological cells' filament structures, buckles of *microtubules* play an important role, e.g. [14]. For illustration of the effect of FSS in Kendall's shape spaces Σ_m^k of k landmark configurations in the m-dimensional Euclidean space, e.g. [4], we have simulated two groups of 20 planar buckle structures each without and with the presence of *intermediate vimentin* filaments (generating stiffness) and placed 5 mathematically defined landmarks on them, leading to two groups in Σ_2^5 as detailed in [17]. Figure 1 shows typical buckle structures.

We compare the two-sample test based on suitable χ^2-quantiles in tangent space with the test based on a suitable bootstrap procedure amending for FSS, cf. [6,11]. In order to assess the effective level of the test we have generated a control sample of another 20 buckles in the presence of vimentin filaments. As clearly visible in Table 1, the presence of FSS results in a higher level of the quantile-based test and a reduced power, thus making it useless for further evaluation. In contrast the bootstrap-based test keeps the level, making its rejection of equality of buckling with and without vimentin credible.

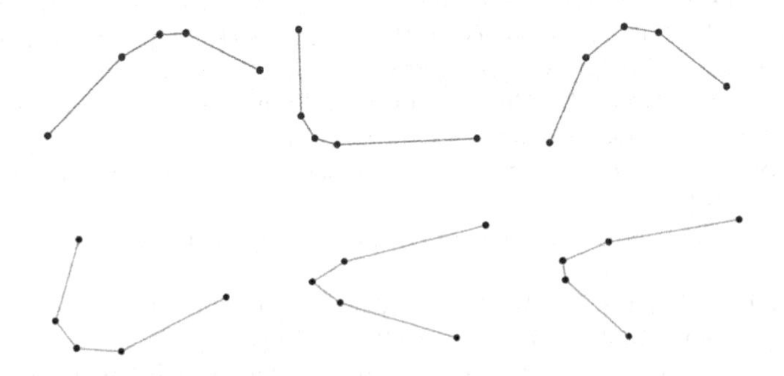

Fig. 1. Microtubule buckle structures with 5 landmarks. Upper row: without vimentin filaments. Lower row: in the presence of vimentin filaments (generating stiffness).

Table 1. Reporting fraction of rejected hypothesis of equality of means using 100 simulations of two-sample test at nominal level $\alpha = 0.05$ based on quantiles (top row) and a suitable bootstrap procedure (bottom row) under equality (left column) and under inequality (right column).

	Both with vimentin	One with and the other without vimentin
Quantile based	0.11	0.37
Bootstrap based	0.03	0.84

References

1. Bhattacharya, R., Lin, L.: Omnibus CLTs for Fréchet means and nonparametric inference on non-Euclidean spaces. Proc. Am. Math. Soc. **145**(1), 413–428 (2017)
2. Bhattacharya, R.N., Patrangenaru, V.: Large sample theory of intrinsic and extrinsic sample means on manifolds II. Ann. Stat. **33**(3), 1225–1259 (2005)

3. Cheeger, J., Ebin, D.G.: Comparison Theorems in Riemannian Geometry, vol. 365. American Mathematical Society, Providence (2008)
4. Dryden, I.L., Mardia, K.V.: Statistical Shape Analysis, 2nd edn. Wiley, Chichester (2016)
5. Eltzner, B., Huckemann, S.F.: A smeary central limit theorem for manifolds with application to high-dimensional spheres. Ann. Stat. **47**(6), 3360–3381 (2019)
6. Eltzner, B., Hundrieser, S., Huckemann, S.F.: Finite sample smeariness on spheres. In: Nielsen, F., Barbaresco, F. (Eds.) GSI 2021, LNCS 12829, pp. 12–19. Springer, Heidelberg (2021). https://doi.org/10.1007/978-3-030-80209-7_2
7. Hendriks, H., Landsman, Z.: Mean location and sample mean location on manifolds: asymptotics, tests, confidence regions. J. Multivar. Anal. **67**, 227–243 (1998)
8. Hotz, T., Huckemann, S.: Intrinsic means on the circle: uniqueness, locus and asymptotics. Ann. Inst. Stat. Math. **67**(1), 177–193 (2015). https://doi.org/10.1007/s10463-013-0444-7
9. Huckemann, S.: Inference on 3D procrustes means: tree boles growth, rank-deficient diffusion tensors and perturbation models. Scand. J. Stat. **38**(3), 424–446 (2011)
10. Huckemann, S.: Intrinsic inference on the mean geodesic of planar shapes and tree discrimination by leaf growth. Ann. Stat. **39**(2), 1098–1124 (2011)
11. Hundrieser, S., Eltzner, B., Huckemann, S.F.: Finite sample smeariness of Fréchet means and application to climate. arXiv preprint arXiv:2005.02321 (2020)
12. Le, H., Barden, D.: On the measure of the cut locus of a Fréchet mean. Bull. Lond. Math. Soc. **46**(4), 698–708 (2014)
13. Nash, J.: The imbedding problem for Riemannian manifolds. Ann. Math. **63**, 20–63 (1956)
14. Nolting, J.-F., Möbius, W., Köster, S.: Mechanics of individual keratin bundles in living cells. Biophys. J . **107**(11), 2693–2699 (2014)
15. Pennec, X.: Curvature effects on the empirical mean in Riemannian and affine Manifolds: a non-asymptotic high concentration expansion in the small-sample regime. arXiv preprint arXiv:1906.07418 (2019)
16. Tran, D.: Behavior of Fréchet mean and central limit theorems on spheres. Bras. J. Prob. Stat. arXiv preprint arXiv:1911.01985 (2019)
17. Tran, D., Eltzner, B., Huckemann, S.F.: Reflection and reverse labeling shape spaces and analysis of microtubules buckling. Manuscript (2021)

Online Learning of Riemannian Hidden Markov Models in Homogeneous Hadamard Spaces

Quinten Tupker[1], Salem Said[2], and Cyrus Mostajeran[3](✉)

[1] Centre for Mathematical Sciences, Wilberforce Road, Cambridge CB3 0WA, UK
[2] CNRS, University of Bordeaux, Bordeaux, France
[3] Department of Engineering, University of Cambridge, Cambridge CP2 1PZ, UK
csm54@cam.ac.uk

Abstract. Hidden Markov models with observations in a Euclidean space play an important role in signal and image processing. Previous work extending to models where observations lie in Riemannian manifolds based on the Baum-Welch algorithm suffered from high memory usage and slow speed. Here we present an algorithm that is online, more accurate, and offers dramatic improvements in speed and efficiency.

Keywords: Hidden Markov model · Riemannian manifold · Gaussian distribution · Expectation-maximization · k-means clustering · Stochastic gradient descent

1 Introduction

Hidden Markov chains have played a major role in signal and image processing with applications to image restoration, speech recognition, and protein sequencing. The extensive and well-developed literature on hidden Markov chains is almost exclusively concerned with models involving observations in a Euclidean space. This paper builds on the recent work of [10] in developing statistical tools for extending current methods of studying hidden Markov models with observation in Euclidean space to models in which observations take place on Riemannian manifolds. Specifically, [10] introduces a general formulation of hidden Markov chain models with Riemannian manifold-valued observations and derives an expectation-maximization (EM) algorithm for estimating the parameters of such models. However, the use of manifolds often entails high memory usage. As such, here we investigate an "online" low memory alternative for fitting a hidden Markov process. Since this approach only ever "sees" a limited subset of the data, it

This work benefited from partial support by the European Research Council under the Advanced ERC Grant Agreement Switchlet n. 670645. Q.T. also received partial funding from the Cambridge Mathematics Placement (CMP) Programme. C.M. was supported by Fitzwilliam College and a Henslow Fellowship from the Cambridge Philosophical Society.

F. Nielsen and F. Barbaresco (Eds.): GSI 2021, LNCS 12829, pp. 37–44, 2021.
https://doi.org/10.1007/978-3-030-80209-7_5

may be expected to inherently sacrifice some accuracy. Nonetheless, the dramatic increase in speed and efficiency often justifies such online algorithms.

The motivation for the development of hidden Markov models on manifolds and associated algorithms is derived from the wide variety of data encoded as points on various manifolds, which fits within the broader context of intense and growing interest in the processing of manifold-valued data in information science. In particular, covariance matrices are used as descriptors in an enormous range of applications including radar data processing [5], medical imaging [7], and brain-computer interfaces (BCI) [1]. In the context of BCI, where the objective is to enable users to interact with computers via brain activity alone (e.g. to enable communication for severely paralysed users), the time-correlation of electroencephalogram (EEG) signals are encoded by symmetric positive definite (SPD) matrices [1]. Since SPD matrices do not form a Euclidean space, standard linear analysis techniques applied directly to such data are often inappropriate and result in poor performance. In response, various Riemannian geometries on SPD matrices have been proposed and used effectively in a variety of applications in computer vision, medical data analysis, and machine learning. In particular, the affine-invariant Riemannian metric has received considerable attention in recent years and applied successfully to problems such as EEG signal processing in BCI where it has been shown to be superior to classical techniques based on feature vector classification [1]. The affine-invariant Riemannian metric endows the space of SPD matrices of a given dimension with the structure of a Hadamard manifold, i.e. a Riemannian manifold that is complete, simply connected, and has non-positive sectional curvature everywhere. In this work, we will restrict our attention to Hadamard manifolds, which encompass SPD manifolds as well as examples such as hyperbolic d-space.

2 Hidden Markov Chains

In a hidden Markov model, one is interested in studying a hidden, time-varying, finitely-valued Markov process $(s_t; t = 1, 2, \dots)$ that takes values in a set S. When $s_t = i$ for $i \in S$, we say that the process is in state i at time t. We assume that the process is time-stationary, so that there exists a transition matrix A_{ij}, which specifies the conditional probabilities $A_{ij} = \mathbb{P}(s_{t+1} = j | s_t = i)$. If $\pi_i(t) = \mathbb{P}(s_t = i)$ is the distribution of s_t, then $\pi_j(t+1) = \sum_{i \in S} \pi_i(t) A_{ij}$ describes the transition from time t to $t+1$. The states s_t are "hidden", meaning that we can never know their true values and can only observe them through random outputs y_t that take their values in a Riemannian manifold M and are assumed to be generated independently from each other.

We assume that M is a homogeneous Riemannian manifold that is also a Hadamard space, and that y_t is distributed according to a Riemannian Gaussian distribution [9] with probability density

$$p(y; c, \sigma) = \frac{1}{Z(\sigma)} \exp\left(-\frac{1}{2\sigma^2} d^2(y, c)\right), \tag{1}$$

where $c \in M$ and $\sigma > 0$ denote the mean and standard deviation of the distribution, respectively, and $Z(\sigma)$ is the normalization factor. In particular, on the space of $d \times d$ symmetric positive definite matrices $M = \mathcal{P}_d$, we have $\mathcal{P}_d = \mathrm{GL}(d)/\mathrm{O}(d)$, where $\mathrm{GL}(d)$ is the general linear group of invertible $d \times d$ matrices, which acts transitively on $\mathcal{P}_d$ by $g \cdot y = gyg^\dagger$, where $g^\dagger$ denotes the transpose of g. The isotropy group is the space of $d \times d$ orthogonal matrices $\mathrm{O}(d)$. The affine-invariant Riemannian distance takes the form

$$d^2(y, z) = \mathrm{tr}\left[\left(\log(y^{-1}z)\right)^2\right], \tag{2}$$

where tr denotes the trace operator and log the principal matrix logarithm. It is easy to see that this distance indeed satisfies the group invariance property $d(y, z) = d(g \cdot y, g \cdot z)$ for all $g \in G$, $y, z \in \mathcal{P}_d$.

Our assumption that y_t is distributed according to a Riemannian Gaussian distribution is expressed as

$$p(y_t|s_t = i) = p(y_t; c_i, \sigma_i), \tag{3}$$

where $p(y_t|s_t = i)$ denotes the conditional density with respect to the Riemannian volume measure. It follows that the probability density of y_t is a time-varying mixture density

$$p(y_t) = \sum_{i \in S} \pi_i(t) p(y_t; c_i, \sigma_i). \tag{4}$$

The objective is to estimate the transition matrix (A_{ij}) and the parameters (c_i, σ_i) given access only to the observations $(y_t; t = 1, 2, \ldots)$. This problem is addressed in [10] via an expectation-maximization (EM) algorithm using an extension of Levinson's forward-backward algorithm [3,8].

3 Online Estimation of Hidden Markov Models

The algorithm that we propose consists of a combination of an initialization phase and a fine-tuning phase. The initialization phase consists of running Riemannian k-means on a limited subset of the data, while the fine-tuning phase is based on an algorithm described by Krishnamurthy and Moore [6]. Specifically, [6] describes a hidden Markov model as a length n chain $s_1, \ldots s_n$ that switches between N different states, according to a transition matrix A, so that $A_{ij} = \mathbb{P}(s_{k+1} = j|s_k = i)$, and starts in an initial state $\pi = \pi(1) \in \mathbb{R}^N$ given by $\pi_i = \mathbb{P}(s_1 = i)$. As before, we assume that this Markov chain is "hidden", meaning that we never know the true state s_k, but instead only see a representative y_k of s_k. In our case specifically, we assume that y_k is a Riemannian Gaussian random variable $y_k \sim N(c_i, \sigma_i^2)$ with mean c_i and standard deviation σ_i.

Initializing the algorithm using k-means on a limited subset of the data is straightforward. After k-means has been completed using the Riemannian center of mass [2], we count transitions between clusters to estimate the transition matrix, and estimate the means c_i of the Gaussian distributions as the means of the clusters. Estimating the standard deviation of these Gaussian distributions is more tricky. Here we introduce

$$\delta = \frac{\partial}{\partial \eta} \log(Z(\eta)), \tag{5}$$

where η is the natural parameter $\eta = \frac{-1}{2\sigma^2}$. In the Gaussian case, where $Z(\eta)$ is the normalization constant of the distribution, it is not hard to see that $\delta = \mathbb{E}(d^2(y, c))$, the expected value of the square Riemannian distance from the Gaussian mean c to an observation y. In general, it can be quite challenging to compute $Z(\eta)$. Fortunately, recent work by Santilli et al. [11] outlines a method for calculating $Z(\eta)$ for SPD matrices in arbitrary dimension using orthogonal polynomials. In particular, [11] provides explicit formulas for dimensions 2, 3, and 4, which could be used to establish a relationship between δ and σ, allowing us to estimate σ. While this completes the initialization phase of the algorithm, we will continue to use this conversion frequently during the fine-tuning section of the algorithm as well.

For the fine-tuning step, we use a stochastic approximation method derived in [6] based on the Kullback-Leibler information measure, which leads to a stochastic gradient descent algorithm based on

$$\lambda^{(k+1)} = \lambda^{(k)} + \mathcal{J}^{-1} \partial_{\lambda^{(k)}} \log f(y_1, \ldots, y_{k+1} | \lambda^{(k)}) \tag{6}$$

where $\lambda^{(k)} = (A^{(k)}, c_i^{(k)}, \sigma_i^{(k)}, \pi_i^{(k)})$ is the kth estimate of the model parameters, y_k is the kth observation, $\mathcal{J}$ is the Fisher information matrix, and the last derivative term can be called the score vector. Using superscript k to denote the kth approximation of each quantity (so, for instance, $A^{(k)}$ is the kth approximation to the transition matrix), we collect definitions of the relevant conditional probabilities in Table 1, where $f(x|y)$ denotes the conditional probability density function on the distribution of the observation and $p_i(y)$ denotes the probability density function of the ith Gaussian distribution. Here Δ is the size of the "minibatch" of observations that the algorithm can see (so stores in memory) at any given moment, and the formulas given are the approximations used given the limited size of the minibatch.

To continue working out what Eq. (6) means in this case, we find that for the Fisher information matrix of the transition matrix, it is helpful to define

$$\mu_j^{(i)} = \frac{\sum_{t=1}^{k+1} \zeta_{t|k+1}(i,j)}{(A_{ij}^{(k+1)})^2}, \tag{7}$$

which is truncated to the size of the minibatch in practice. Similarly, for the score vector of the transition matrix we find it useful to define

$$g_j^{(i)} = \frac{\zeta_{k+1|k+1}(i,j)}{A_{ij}^{(k+1)}}, \tag{8}$$

from which we can express our transition matrix update rule as

$$A_{ij}^{(k+1)} = A_{ij}^{(k)} + \frac{1}{\mu_j^{(i)}} \left(g_j^{(i)} - \frac{\sum_{h=1}^{N} g_h^{(i)} / \mu_h^{(i)}}{\sum_{h=1}^{N} 1/\mu_h^{(i)}} \right). \tag{9}$$

Table 1. Conditional probabilities

Symbol	Definition	Calculation
$\alpha_t(i)$	$f(y_1,\dots,y_t,s_t=i\vert\lambda^{(t-1)})$	$\alpha_t(j)=\sum_{i=1}^N \alpha_{t-1}(i)A_{ij}^{(t-1)}p_j^{(t-1)}(y_t)$, $\alpha_1(i)=\pi_i(i)p_i^{(1)}(y_1)$
$\beta_{t\vert k}(i)$	$f(y_{t+1},\dots,y_k\vert s_t=i,\lambda^{(k-1)})$	$\beta_{k\vert k+\Delta}=A^{(k-1)}P^{(k-1)}(k+1)\dots A^{(k-1)}P^{(k-1)}(k+\Delta)\mathbf{1}$ where $P^{(k)}(s)=\mathrm{diag}(p_1^{(k)}(y_s),\dots,p_N^{(k)}(y_s))$ and $\mathbf{1}$ is a vector of ones
$\gamma_{t\vert k}(i)$	$f(s_t=i\vert y_1,\dots,y_k,\lambda^{(k-1)})$	$\gamma_{t\vert k}(i)=\frac{\alpha_t(i)\beta_{t\vert k}(i)}{\sum_{j=1}^N \alpha_t(j)\beta_{t\vert k}(j)}$
$\zeta_{t\vert k}(i,j)$	$f(s_t=i,s_{t+1}=j\vert y_1,\dots,y_k,\lambda^{(k-1)})$	$\frac{\alpha_t(i)A_{ij}^{(t-1)}\beta_{t+1\vert k}(j)p_j^{(t-1)}(y_{t+1})}{\sum_i\sum_j \alpha_t(i)A_{ij}^{(t-1)}\beta_{t+1\vert k}(j)p_j^{(t-1)}(y_{t+1})}$

Similar work on the means and standard deviations yields the following update rules when working with Gaussian distributions on the real line [6]:

$$c_i^{(k+1)} = c_i^{(k)} + \frac{\gamma_{k+1|k+1}(i)(y_{k+1}-c_i^{(k)})}{\sum_{t=1}^{k+1}\gamma_{t|k+1}(i)} \tag{10}$$

$$(\sigma_i^2)^{(k+1)} = (\sigma_i^2)^{(k)} + \frac{\gamma_{k+1|k+1}(i)((y_{k+1}-c_i^{(k)})^2-(\sigma_i^2)^{(k)})}{k+1}, \tag{11}$$

which, after adjusting the step sizes used, can be converted to update rules on a Riemannian manifold as

$$c_i^{(k+1)} = c_i^{(k)}\#_\tau y_{k+1}, \qquad \tau = \frac{\gamma_{k+1|k+1}(i)}{\sum_{t=1}^{k+1}\gamma_{t|k+1}(i)}, \tag{12}$$

$$\delta_i^{(k+1)} = \delta_i^{(k)} + \frac{\gamma_{k+1|k+1}(i)\left(d\left(y_{k+1},c_i^{(k)}\right)^2-\delta_i^{(k)}\right)}{\sqrt{k}}, \tag{13}$$

where $x\#_\tau z$ denotes the unique point on the Riemannian geodesic from x to z that satisfies $d(x,x\#_\tau z)=\tau d(x,z)$ for $\tau\in[0,1]$. In particular, in the case of $\mathcal{P}_d$ equipped with the Riemannian distance given in Eq. (2), we have

$$x\#_\tau z = x^{1/2}(x^{-1/2}zx^{-1/2})^\tau x^{1/2}. \tag{14}$$

Finally, we may note that since these intermediate calculations, particularly the conditional probabilities mentioned, function as the "memory" of the algorithm, it is not sufficient to initialize the algorithm by transferring only the values of A, c_i, δ_i. Instead, one must calculate values of $\alpha,\beta,\gamma,\zeta$ from the k-means data as well. Fortunately, using the clusters obtained, one can do this easily by using the final estimates obtained for A, c_i, δ_i from the k-means algorithm.

4 Computational Experiment

We consider a computational experiment comparing the EM algorithm from [10] with our algorithm. For that, we generate a chain of length 10000 with values taken in a three-element set $S = \{1, 2, 3\}$, an initial distribution $\pi = \begin{pmatrix}1 & 0 & 0\end{pmatrix}$ (i.e. certainly starting in state $i = 1$), and transition matrix

$$A = (A_{ij}) = \begin{pmatrix} 0.4 & 0.3 & 0.3 \\ 0.2 & 0.6 & 0.2 \\ 0.1 & 0.1 & 0.8 \end{pmatrix} \tag{15}$$

with means and standard deviations given by $(c_1, \sigma_1) = (0, 0.2)$, $(c_2, \sigma_2) = (0.29 + 0.82i, 1)$, and $(c_3, \sigma_3) = (-0.29 + 0.82i, 1)$ as in [10]. Here, the outputs y_t are generated from a Riemannian Gaussian model in the Poincaré disk model of hyperbolic 2-space. That is, each y_t takes values in $M = \{z \in \mathbb{C} : |z| < 1\}$, and

$$d(y, z) = \operatorname{acosh}\left(1 + \frac{2|y - z|^2}{(1 - |y|^2)(1 - |z|^2)}\right) \tag{16}$$

$$Z(\sigma) = (2\pi)^{3/2} \sigma e^{\frac{\sigma^2}{2}} \operatorname{erf}\left(\frac{\sigma}{\sqrt{2}}\right) \tag{17}$$

where erf denotes the error function. We use the Poincaré disk here rather than $\mathcal{P}_2$ simply for ease of visualization. Moreover, the Poincaré disk with distance (16) is isometric to the space of 2×2 symmetric positive definite matrices of unit determinant equipped with the affine-invariant Riemannian distance (2).

Applying our algorithm to this example, we obtain the results in Table 2. We begin by comparing the speed and accuracy of our algorithm with the EM algorithm. Here, the online algorithm is the clear winner, since it matches or exceeds the EM algorithm on accuracy, and for $\Delta = 200$ is around 450 times faster (all tests performed on a standard personal laptop computer)—a remarkable improvement. We observe a pattern that accuracy decreases rapidly for very small Δ and is roughly stable for $\Delta \geq 200$. Also, if we only use k-means without any fine-tuning, we see accuracy is slightly lower (0.90), although it is quite fast (5 s runtime). Finally, we note that the runtimes scale sublinearly with Δ, and in particular, the scaling is very linear for small Δ.

Regarding the estimates of the transition matrix, we measure the error in the transition matrix as the root mean squared error (RMSE) in the Frobenius norm of the difference between the estimated and true transition matrices $\|A - A'\|_{Frob}$. We consequently see that the EM algorithm underperforms all forms of the online algorithm for $\Delta < 200$. However, we also observe that the full online algorithm rarely outperforms pure k-means significantly. In fact, when data clusters are spread out as they are in this example, we find that the fine-tuning step does not improve on k-means. However, for data clusters that do overlap significantly, we have observed instances where the fine-tuning step improves significantly on k-means.

Figure 1 depicts results on mean estimation for minibatch sizes of $\Delta = 100$ and $\Delta = 200$. Note that the clustering of estimated means is more focused and closer to the true mean for $\Delta = 200$ than for $\Delta = 100$, as expected. We also observe that the EM mean appears to consistently struggle to accurately estimate the mean at the center of the disk.

Table 2. Online algorithm accuracy, runtime, and estimates for selected transition matrix elements for different minibatch sizes Δ.

Minibatch size, Δ	Accuracy	Runtime/s	A_{11}	A_{22}	A_{33}	Transition RMSE
True values	–	–	0.4	0.6	0.8	0
40	0.48	1.37	0.40	0.30	0.40	1.49
60	0.50	1.93	0.31	0.34	0.28	1.39
80	0.76	2.54	0.36	0.49	0.45	1.26
100	0.86	3.21	0.48	0.63	0.59	1.13
200	0.98	5.81	0.46	0.60	0.77	1.12
300	0.94	8.39	0.57	0.63	0.75	0.95
1000	0.95	28.58	0.51	0.64	0.76	0.94
5000	0.95	70.69	0.41	0.58	0.70	0.91
k-means only	0.90	4.99	0.53	0.65	0.56	0.93
EM	0.90	2623.69	0.31	0.88	0.96	1.29

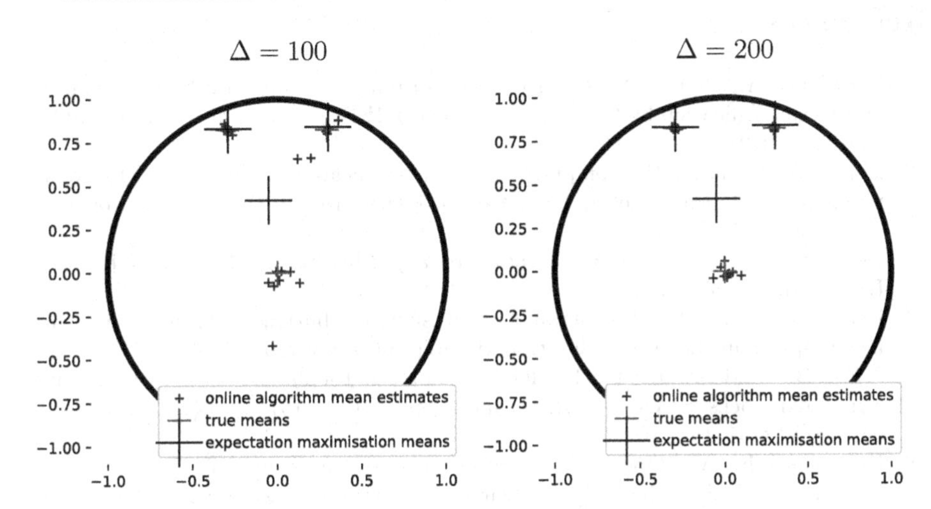

Fig. 1. Mean estimates for multiple runs of the online algorithm for the example in the Poincaré disk for minibatch sizes of $\Delta = 100$ and $\Delta = 200$ compared with the EM algorithm estimates. As expected, the quality of the online algorithm estimates improves with minibatch size Δ. We observe that for a relatively modest minibatch size of $\Delta = 200$, the online algorithm is more accurate than the EM algorithm while being several orders of magnitude faster.

5 Conclusion

We have implemented an online algorithm for the learning of hidden Markov models on Hadamard spaces, i.e. an algorithm that uses a constant amount of memory instead of one that scales linearly with the amount of available data. We found that our algorithm outperforms a previous algorithm based on expectation-maximization in all measures of accuracy when the right minibatch size is used. Furthermore, this improvement is achieved while running nearly 450 times faster and using 50 times less memory in the example considered.

Future work will consider strategies that would allow the removal of the k-means initialization step and automatically tune the minibatch size without the need for experimentation. Furthermore, significant challenges arise with increasing N, the number of states and hence dimension of the transition matrix. Here difficulties include identifying the correct number of clusters, preventing the algorithm from merging separate clusters (undercounting), and preventing the algorithm from double-counting clusters (overcounting) by assigning two means to the same cluster. These challenges have proven to be the main obstacles in increasing N. By contrast, increasing the dimension of the Hadamard space in which observations take place (e.g. the space of positive definite matrices) does not pose significant challenges and is straightforward using recent work on computing normalization factors of Riemannian Gaussian distributions [4,11].

References

1. Barachant, A., Bonnet, S., Congedo, M., Jutten, C.: Multiclass brain-computer interface classification by Riemannian geometry. IEEE Trans. Biomed. Eng. **59**(4), 920–928 (2012)
2. Bini, D.A., Iannazzo, B.: Computing the Karcher mean of symmetric positive definite matrices. Linear Algebra Appl. **438**(4), 1700–1710 (2013). https://doi.org/10.1016/j.laa.2011.08.052
3. Devijver, P.A.: Baum's forward-backward algorithm revisited. Pattern Recogn. Lett. **3**(6), 369–373 (1985)
4. Heuveline, S., Said, S., Mostajeran, C.: Gaussian distributions on Riemannian symmetric spaces in the large N limit. arXiv preprint arXiv:2102.07556 (2021)
5. Jeuris, B., Vandebril, R.: The Kähler mean of block-Toeplitz matrices with Toeplitz structured blocks. SIAM J. Matrix Anal. Appl. **37**(3), 1151–1175 (2016). https://doi.org/10.1137/15M102112X
6. Krishnamurthy, V., Moore, J.B.: On-line estimation of hidden Markov model parameters based on the Kullback-Leibler information measure. IEEE Trans. Sig. Process. **41**(8), 2557–2573 (1993). https://doi.org/10.1109/78.229888
7. Pennec, X., Fillard, P., Ayache, N.: A Riemannian framework for tensor computing. Int. J. Comput. Vis. **66**(1), 41–66 (2006)
8. Rabiner, L.R.: A tutorial on hidden Markov models and selected applications in speech recognition. Proc. IEEE **77**(2), 257–286 (1989)
9. Said, S., Bombrun, L., Berthoumieu, Y., Manton, J.H.: Riemannian Gaussian distributions on the space of symmetric positive definite matrices. IEEE Trans. Inf. Theory **63**(4), 2153–2170 (2017)
10. Said, S., Bihan, N.L., Manton, J.H.: Hidden Markov chains and fields with observations in Riemannian manifolds. arXiv preprint arXiv:2101.03801 (2021)
11. Santilli, L., Tierz, M.: Riemannian Gaussian distributions, random matrix ensembles and diffusion kernels. arXiv preprint arXiv:2011.13680 (2020)

Sub-Riemannian Geometry and Neuromathematics

Submanifolds of Fixed Degree in Graded Manifolds for Perceptual Completion

Giovanna Citti[1(✉)], Gianmarco Giovannardi[2], Manuel Ritoré[2], and Alessandro Sarti[3]

[1] Department of Mathematics, University of Bologna, Bologna, Italy
giovanna.citti@unibo.it
[2] Department of Mathematics, University of Granada, Granada, Spain
[3] CAMS, EHESS, Paris, France

Abstract. We extend to a Engel type structure a cortically inspired model of perceptual completion initially proposed in the Lie group of positions and orientations with a sub-Riemannian metric. According to this model, a given image is lifted in the group and completed by a minimal surface. The main obstacle in extending the model to a higher dimensional group, which can code also curvatures, is the lack of a good definition of codimension 2 minimal surface. We present here this notion, and describe an application to image completion.

Keywords: Perceptual completion · Graded structures · Fixed degree surfaces · Area formula · Degree preservig variations

1 Introduction

Mathematical models of the visual cortex expressed in terms of differential geometry were proposed for the first time by Hoffman in [17], Mumford in [23], August Zucker in [2] to quote only a few. Petitot and Tondut in 1999 in [26] described the functional architecture of area V1 by a contact structure, and described the propagation in the cortex by a constrained Lagrangian operator. Only in 2003 Sarti and Citti in [6] and J. Petitot in [24] recognized that the geometry of the cortex is indeed sub-Riemannian. In [6], the functional architecture of V1 is described as a Lie group with a sub-Riemannian geometry: if the visual stimulus is corrupted, it is completed via a sub-Riemannian minimal surface. A large literature has been provided on sub-Riemannian models both for image processing or cortical modelling (we refer to the monograph [25] for a list of references). In Sect. 2 we will present the model [6], and its extension in the Engel group provided in [1,25].

The notion of minimal surface in a sub-Riemannian setting as critical points of the first variation of the area functional is well known for co-dimension 1 surfaces (see [11,12] for the area formula, and [7,9,16,18] for the first variation).

Supported by GHAIA, H2020 RICE MCSA project, No 777822.

F. Nielsen and F. Barbaresco (Eds.): GSI 2021, LNCS 12829, pp. 47–55, 2021.
https://doi.org/10.1007/978-3-030-80209-7_6

For higher codimension very few results are available: the notion of area has been introduced in [10,19,20], but the first variation, well-known for curves (see [21]), was studied for surfaces only very recently in [4,5,13]. We will devote Sect. 3 to the description of these results.

We conclude this short presentation with an application of this result to the completion model in the Engel group, contained in Sect. 4.

2 A Subriemannian Model of the Visual Cortex

The primary visual cortex is the first part of the brain processing the visual signal coming from the retina. The receptive profile (RP) $\psi(\xi)$ is the function that models the activation of a cortical neuron when a stimulus is applied to a point $\xi = (\xi_1, \xi_2)$ of the retinal plane. The hypercolumnar structure organizes the cortical cells of V1 in columns corresponding to different features. As a results we will identify cells in the cortex by means of two parameters (x, f), where $x = (x_1, x_2)$ is the position of the point, and f a vector of extracted features. We will denote F the set of features, and consequently the cortical space will be identified with $\mathbb{R}^2 \times F$. In the presence of a visual stimulus $I = I(\xi)$ the whole hypercolumn fires, giving rise to an output

$$O_F(x, f) = \int I(\xi)\psi_{(x,f)}(\xi)d\xi. \tag{1}$$

It is clear that the same image, filtered with a different family of cells, produces a different output.

For every cortical point, the cortical activity, suitably normalized, can be considered a probability density. Hence its maximum over the fibre F can be considered the most probable value of f, and can be considered the feature identified by the system (principle of non maxima suppression):

$$|O_F(x, f_I(x))| = \max_f |O_F(x, f)|. \tag{2}$$

The output of a family of cells is propagated in the cortical space $\mathbb{R}^2 \times F$ via the lateral connectivity.

2.1 Orientation and Curvature Selectivity

In [6] the authors considered only simple cells sensible to a direction $\theta \in \mathbb{S}^1$. Hence the set F becomes in this case $\mathbb{S}^1$ and the underlying manifold reduces to $N := \mathbb{R}^2 \times \mathbb{S}^1$ with a sub-Riemannian metric. The image I is lifted by the procedure (2) to a graph is this structure. If it is corrupted, it is completed via a sub-Riemannian minimal surface.

The geometric description of the cortex was extended by Citti-Petitot-Sarti to a model of orientation and curvature selection, appeared in [25]. In this case the non maxima suppression process (2) selects a function

$$\Psi : \mathbb{R}^2 \to N_2 = \mathbb{R}^2 \times \mathbb{S}^1 \times \mathbb{R}, \quad \Psi(x) = (x, f_I(x)) = (x, \theta(x), k(x)).$$

Level lines of the input I are lifted to integral curves in N of the vector fields

$$X_1 = \cos\theta\frac{\partial}{\partial x_1} + \sin\theta\frac{\partial}{\partial x_2} + k\frac{\partial}{\partial\theta}, \quad X_2 = \frac{\partial}{\partial k} \tag{3}$$

where $x = (x_1, x_2)$. The vector fields define an Hörmander type manifold since the whole space is spanned at every point by the vectors X_1, X_2 and their commutators

$$X_3 := [X_1, X_2] = -\frac{\partial}{\partial\theta}, \quad X_4 := [X_1, X_3] = -\sin\theta\frac{\partial}{\partial x_1} + \cos\theta\frac{\partial}{\partial x_2}. \tag{4}$$

These commutation conditions identify the structure of and Engel-type algebra. Integral curves of this model allow better completion of more complex images. In the image below we represent a grouping performed in [1] with an eigenvalue method in the sub-Riemannian setting. Precisely for every point x_i of the curves, we compute the orientation θ_i and the curvature k_i, and obtain a point $p_i = (x_i, \theta_i, k_i)$ in $\mathbb{R}^2 \times \mathbb{S}^1 \times \mathbb{R}$. Calling d the sub-Riemannian distance induced by the choice of vector fields, we can compute the affinity matrix A, with entries $a_{ij} = d(p_i, p_j)$. The eigenvalues of this matrix, reprojected on $\mathbb{R}^2$ can be identified with the perceptual units present in the image. In Fig. 1 (from [1] the same curve is segmented in the geometry dependent only on orientation, and in the geometry of orientation and curvature: the second method correctly recovers the logarithmic spirals.

Fig. 1. Grouping using orientation (left), or orientation and curvature (right).

It could be nice to see if it is possible to extend in this setting also the minimal surface algorithm for image completion [6]. The main obstacle in doing this was the fact that the notion of area and curvature was not well defined for codimension two surfaces in a sub-Riemannian metric, and that characterizing admissible variations presents intrinsic difficulties.

3 Graded Structures

In the next section we will present the results obtained in [4,5,13] to define the notion of area of high codimension surfaces and its first variation in the setting of a graded structure. A graded structure is defined as following

Definition 31. *Let N be a smooth manifold and let $\mathcal{H}^1 \subset \ldots \subset \mathcal{H}^s$ be an increasing filtration of sub-bundles of the tangent bundle TN s.t.*

$$X \in \mathcal{H}^i, Y \in \mathcal{H}^j \Rightarrow [X, Y] \in \mathcal{H}^{i+j} \text{ and } \mathcal{H}^s_p = T_pN \text{ for all } p$$

We will say that the fibration $\mathcal{H}^1 \subset \ldots \subset \mathcal{H}^s$ is equiregular if $dim(\mathcal{H}^j)$ is constant in N. For such a manifold, we can define an homogeneous dimension $Q = \sum_{i=1}^{s} i\Big(dim(\mathcal{H}^i) - dim(\mathcal{H}^{i-1})\Big)$. We will say that a basis $(X_1, \ldots, X_n)$ of the tangent plane at every point is an adapted basis if $X_1, \ldots, X_{n_1}$ generate $\mathcal{H}^1$, $X_1, \ldots, X_{n_1}, X_{n_1+1}, \ldots, X_{n_2}$ generate $\mathcal{H}^2$, and so on. The presence of a filtration naturally allows to define the degree of a vector field and of an m-vector. In particular, we say that a vector $v \in T_p(M)$ has degree l, and we denote it $deg(v) = l$, if $v \in \mathcal{H}^l \setminus \mathcal{H}^{l-1}$. Given $m < n$, a multi-index $J = (j_1, \cdots, j_m)$, with $1 \leq j_1 < \ldots < j_m \leq n$, and an m-vector field $X_J = X_{j_1} \wedge \ldots \wedge X_{j_m}$ we define $\deg(X_J) = \deg(X_{j_1}) + \ldots + \deg(X_{j_m})$.

If a Riemannian metric g is defined on the graded manifold N we can introduce an orthogonal decomposition of the tangent space, which respects the grading, as follows: $\mathcal{K}^1_p := \mathcal{H}^1_p, \quad \mathcal{K}^{i+1}_p := (\mathcal{H}^i_p)^\perp \cap \mathcal{H}^{i+1}_p, \quad 1 \leq i \leq (s-1)$. So that $T_pN = \mathcal{K}^1_p \oplus \mathcal{K}^2_p \oplus \cdots \oplus \mathcal{K}^s_p$.

Example 1. *A Carnot manifold with a bracket generating distribution is a graded manifold.*

The interest of graded manifolds, is that a submanifold of a sub-Riemannian manifold, is not in general a sub-Riemannian manifold, but submanifolds of graded manifolds are graded.

3.1 Regular Submanifolds

Given an immersion $\Phi : \bar{M} \to N$. $M = \Phi(\bar{M})$. the manifold M inherits the graded structure $\tilde{\mathcal{H}}^i_p = T_pM \cap \mathcal{H}^i_p$. The pointwise degree (introduced in [14]) or local homogenous dimension is defined by

$$\deg_M(p) = \sum_{i=1}^{s} i \dim(\tilde{\mathcal{H}}_i(p) - \tilde{\mathcal{H}}_{i-1}(p)).$$

For submanifolds, it is not possible in general to assume that the local homogeneous dimension is constant so that we will define the degree of M

$$d := \deg(M) = \max_{p \in M} \deg_M(p).$$

Given a graded manifold N with a Riemannian metric g, we are able to introduce a notion of area, as a limit of the corresponding Riemannian areas. To begin with we define a family of Riemannian metrics g_r, adapted to the grading of the manifold as follows:

$$g_r|_{\mathcal{K}^i} = \frac{1}{r^{i-1}} g|_{\mathcal{K}^i}, \qquad i = 1, \ldots, s, \quad \text{for any} \quad r > 0.$$

Now we assume that M is a submanifold of degree $d = deg(M)$, defined by a parametrization $\Phi : (\bar{M}, \mu) \to N$. Also assume that μ is a Riemannian metric on $\bar{M}$. For each $M' \subset \bar{M}$ we consider the Riemannian area, weighted by its degree

$$r^{\frac{d-m}{2}} \int_{M'} |E_1 \wedge \ldots \wedge E_m|_{g_r} d\mu(p),$$

where $E_1, \ldots, E_m$ a μ-orthonormal basis of $T_p M$. If the limit as $r \to 0$ exists, we call it the d-area measure. It can be explicitly expressed as

$$A_d(M') = \int_{M'} |\left(E_1 \wedge \ldots \wedge E_m\right)_d|_g \, d\mu(p).$$

where $(\cdot)_d$ the projection onto the space of m-vectors of degree d:

$$(E_1 \wedge \ldots \wedge E_m)_d = \sum_{X_J, \deg(X_J)=d} \langle E_1 \wedge \ldots \wedge E_m, X_J \rangle X_J.$$

The same formula had been already established by [19].

3.2 Admissible Variations

It would be natural to define the curvature as the first variation of the area. However in this case the area functional depends on the degree of the manifold. Hence we need to ensure that the degree does not change during variation, and we define degree preserving variations.

Definition 32. *A smooth map $\Gamma : \bar{M} \times (-\epsilon, \epsilon) \to N$ is said to be an admissible variation of Φ if $\Gamma_s : \bar{M} \to N$, defined by $\Gamma_s(\bar{p}) := \Gamma(\bar{p}, s)$, satisfies*

(i) $\Gamma_0 = \Phi$,
(ii) $\Gamma_s(\bar{M})$ *is an immersion of the same degree as* $\Phi(\bar{M})$ *for small enough* s,
(iii) $\Gamma_s(\bar{p}) = \Phi(\bar{p})$ *for* $\bar{p}$ *outside a given compact subset of* $\bar{M}$.

We can always choose an adapted frame to a submanifold manifold M . First we choose a tangent basis $(E_1, \cdots, E_m)$, then we complete it to a basis of the space $X_{m+1}, \cdots, X_n$, where $X_{m+1}, \cdots, X_{m+k}$ have degree less or equal to $deg(E_1)$ while $X_{m+k+1}, \cdots, X_n$ have degree bigger than $deg(E_1)$.

Definition 33. *With the previous notation we define the variational vector field W as*

$$W(\bar{p}) = \frac{\partial \Gamma(\bar{p}, 0)}{\partial s}.$$

It is always possible to assume that W has no tangential components, so that it will be represented as

$$W(\bar{p}) = \sum_{i=m+1}^{m+k} h_i X_i + \sum_{r=m+k+1}^{n} v_r X_r = H + V,$$

where $H = \sum_{i=m+1}^{m+k} h_i X_i$ and $V = \sum_{r=m+k+1}^{n} v_r X_r$.

Using the fact that if Γ is an admissible variation, which means that the degree of $\Gamma_s(\bar{M})$ is constant with respect to s, it is possible to prove the following

Proposition 1 *(see [5]). If W is an admissible vector field, then there exist matrices A, B such that*

$$E(V) + BV + AH = 0, \quad \textit{where } E \textit{ is the tangent basis.} \tag{5}$$

This property suggests the following definition

Definition 34. *We say that a compactly supported vector field W is admissible when it satisfies the admissibility system $E(V) + BV + AH = 0$.*

3.3 Variation for Submanifolds

The phenomenon of minima which are isolated and do not satisfy any geodesic equation (abnormal geodesic) was first discovered by Montgomery [22] for geodesic curves. In 1992 Hsu [15] proved a characterization of integrable vector fields along curves. We obtained in [4] a partial analogous of the previous result for manifolds of dimension bigger than one. Precisely we defined

Definition 35. *$\Phi : \bar{M} \to N$ is strongly regular at $\bar{p} \in \bar{M}$ if $A(\bar{p})$ has full rank, where A is defined in Proposition 1.*

Theorem 1 *[4]. $\Phi : \bar{M} \to (N, \mathcal{H}^1 \subset \ldots \subset \mathcal{H}^s)$ with a Riemannian metric g. Assume that Φ of degree d is strongly regular at $\bar{p}$. Then there exists an open neighborhood $U_{\bar{p}}$ of $\bar{p}$ such every admissible vector field W with compact support on $U_{\bar{p}}$ is integrable.*

The following properties are satisfied

Remark 1. The admissibility of a vector field is independent of the Riemannian metric g.

Remark 2. All hypersurfaces in a sub-Riemannian manifold are deformable.

4 Application to Visual Perception

Let us go back to the model of orientation and curvature introduced in Sect. 2.1. The underlying manifold is then $N = \mathbb{R}^2 \times \mathbb{S}^1 \times \mathbb{R}$. Let us call g the metric which makes the vector fields X_1, X_2, X_3, X_4 in (3) and (4) an orthonormal basis. Let us consider a submanifold M defined by the parametrization

$$\Phi : \mathbb{R}^2 \supset \bar{M} \to \mathbb{R}^2 \times \mathbb{S}^1 \times \mathbb{R}, \; M = \Phi(\bar{M}).$$

In particular the surfaces obtained by non maxima suppression are expressed in the form

$$\Phi(x) = (x, \theta(x), \kappa(x)).$$

If we impose the constraint $\kappa = X_1(\theta)$ the tangent vectors to M become $X_1 + X_1(\kappa)X_2$, $X_4 - X_4(\theta)X_3 + X_4(\kappa)X_2$, and the area functional reduces to

$$A_4(M) = \int_{\bar{M}} \sqrt{1 + X_1(\kappa)^2}\, dx. \tag{6}$$

We will denote $\bar{X}_1$ and $\bar{X}_4$ the projection of the vector fields X_1, X_4 onto $\bar{M}$. Then the admissibility system for a variational vector field $W = h_2(X_2 - X_1(\kappa)X_1) + v_3X_3$ is given by

$$\bar{X}_1(v_3) = -\bar{X}_4(\theta)v_3 - Ah_2, \quad \text{where } A = (1 + (\bar{X}_1^2(\theta))^2).$$

Since $rank(A) = 1$ we deduce by Theorem 1 that for this type of surfaces each admissible vector is integrable, which implies that the minimal surfaces can be obtained via variational methods.

This property is quite important, since in [5] the authors proved that the manifold $\{(0, 0, \theta, k)\}$ do not admit degree preserving variation, so that it is isolated. This provides a generalization of the notion of abnormal geodesic.

4.1 Implementation and Results

We directly implement the Euler Lagrangian of the functional (6). Since it is non linear, we compute a step 0 image with Euclidean Laplacian. After that, we compute at each step, orientation and curvature of level lines of the image at the previous step, update curvature and orientation via the linearized Euler Lagrangian equation and complete the 2D image diffusing along the vector field $\bar{X}_1$. We provide here a result in the simplified case with non corrupted points in the occluded region, and a preliminary result for the impainting problem. We plan to study the convergence of the algorithm and compare with existing literature in a forthcoming paper (Fig. 2).

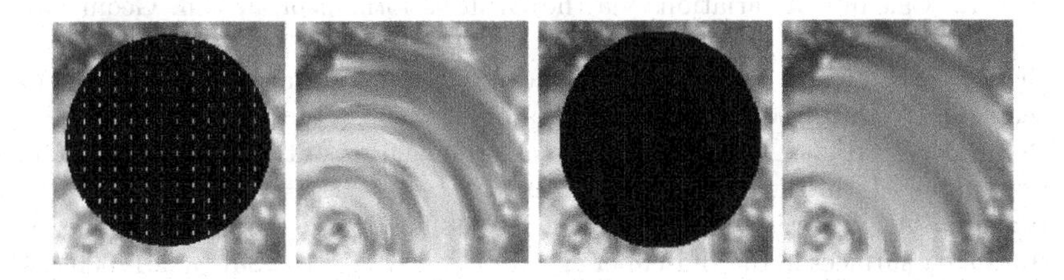

Fig. 2. Two examples of completion using the proposed algorithm.

References

1. Abbasi-Sureshjani, S., Favali, M., Citti, G., Sarti, A., ter Haar Romeny, B.M.:Curvature integration in a 5D Kernel for extracting vessel connections in retinal images. IEEE Trans. Image Process. **27**(2), 606–621 (2017)

2. August, J., Zucker, S.W.: The curve indicator random field: curve organization via edge correlation. In: Boyer, K.L., Sarkar, S. (eds.) Perceptual Organization for Artificial Vision Systems, pp. 265–288. Springer, Boston (2000). https://doi.org/10.1007/978-1-4615-4413-5_15
3. Bianchini, R.M., Stefani, G.: Graded approximations and controllability along a trajectory. SIAM J. Control Optim. **28**(4), 903–924 (1990)
4. Citti, G., Giovannardi, G., Ritoré, M.: Variational formulas for curves of fixed degree, preprint (2019). https://arxiv.org/abs/1902.04015
5. G. Citti, G. Giovannardi M. Ritoré, Variational formulas for submanifolds of fixed degree, preprint (2019). https://arxiv.org/abs/1905.05131
6. Citti, G., Sarti, A.: A cortical based model of perceptual completion in the roto-translation space. J. Math. Imaging Vis. **24**(3), 307–326 (2006). (announced in Proc. Workshop on Second Order Subelliptic Eq. and Appl., Cortona, 2003/6/15)
7. Cheng, J.-H., Hwang, J.-F.: Variations of generalized area functionals and p-area minimizers of bounded variation in the Heisenberg group. Bull. Inst. Math. Acad. Sin. (N.S.) **5**(4), 369–412 (2010)
8. Citti, G., Sarti, A. (eds.): Neuromathematics of Vision. Lecture Notes in Morphogenesis. Springer, Heidelberg (2014). https://doi.org/10.1007/978-3-642-34444-2
9. Danielli, D., Garofalo, N., Nhieu, D.M.: Sub-Riemannian calculus on hypersurfaces in Carnot groups. Adv. Math. **215**(1), 292–378 (2000)
10. Franchi, B., Serapioni, R., Serra Cassano, F.: Regular submanifolds, graphs and area formula in Heisenberg groups. Adv. Math. **211**(1), 152–203 (2007)
11. Franchi, B., Serapioni, R., Serra Cassano, F.: Regular hypersurfaces, intrinsic perimeter and implicit function theorem in Carnot groups. Comm. Anal. Geom. **11**(5), 909–944 (2003)
12. Garofalo, N., Nhieu, D.-M.: Isoperimetric and Sobolev inequalities for Carnot-Carathéodory spaces and the existence of minimal surfaces. Comm. Pure Appl. Math. **49**(10), 1081–1144 (1996)
13. Giovannardi, G.: Higher dimensional holonomy map for ruled submanifolds in graded manifolds. Anal. Geom. Metric Spaces **8**, 68–91 (2020)
14. Gromov, M: Carnot-Carathéodory spaces seen from within. In: Sub-Riemannian geometry, 144 of Progr. Math., Birkhäuser, Basel, pp. 79–323 (1996)
15. Hsu, L.: Calculus of variations via the Griffiths formalism. J. Diff. Geom. **36**(3), 551–589 (1992)
16. Hladky, R.K., Pauls, S.D.: Constant mean curvature surfaces in sub-Riemannian geometry. J. Diff. Geom. **79**(1), 111–139 (2008)
17. Hoffman, W.: Higher visual perception as prolongation of the basic lie transformation group. Math. Biosci. **6**, 437–471 (1970)
18. Hurtado, A., Ritoré, M., Rosales, C.: Classification of complete stable area-stationary surfaces in the Heisenberg group H^1 Adv. Math. **224**(2), 561–600 (2010)
19. Magnani, V., Vittone, D.: An intrinsic measure for submanifolds in stratified groups. Journal für die reine und angewandte Mathematik (Crelles J.) **619**, 203–232 (2008)
20. Magnani, V., Tyson, J.T., Vittone, D.: On transversal submanifolds and their measure. J. Anal. Math. **125**, 319–351 (2015)
21. Montgomery, R.: A tour of subriemannian geometries, their geodesics and applications. Mathematical Surveys and Monographs, AMS, vol. 91 (2002)
22. Montgomery, R.: Abnormal minimizers. SIAM J. Control Optim. **32**(6), 1605–1620 (1994)

23. Mumford, D.: Elastica and computer vision. In: Bajaj, C.L. (eds.) Algebraic Geometry and Its Applications, pp. 491–506, Springer, New York (1994). https://doi.org/10.1007/978-1-4612-2628-4_31
24. Petitot, J.: The neurogeometry of pinwheels as a sub-Riemannian contact structure. J. Physiol. Paris **97**(2–3), 265–309 (2003)
25. Petitot, J.: Neurogeometrie de la Vision Modeles Mathematiques & Physiques des Architectures Fonctionnelles (2009)
26. Petitot, J., Tondut, Y.: Vers une neurogéométrie. Fibrations corticales, structures de contact et contours subjectifs modaux Mathématiques et Sciences humaines, Tome **145**, 5–101 (1999)

An Auditory Cortex Model for Sound Processing

Rand Asswad[1], Ugo Boscain[2], Giuseppina Turco[3], Dario Prandi[1](✉), and Ludovic Sacchelli[4]

[1] Université Paris-Saclay, CNRS, CentraleSupèlec, Laboratoire des Signaux et Systèmes, 91190 Gif-sur-Yvette, France
{rand.asswad,dario.prandi}@centralesupelec.fr
[2] CNRS, LJLL, Sorbonne Université, Université de Paris, Inria, Paris, France
ugo.boscain@upmc.fr
[3] Université de Paris, CNRS, Laboratoire de Linguistique Formelle, UMR 7110, Paris, France
gturco@linguist.univ-paris-diderot.fr
[4] Université Lyon, Université Claude Bernard Lyon 1, CNRS, LAGEPP UMR 5007, 43 Bd du 11 Novembre 1918, 69100 Villeurbanne, France
ludovic.sacchelli@univ-lyon1.fr

Abstract. The reconstruction mechanisms built by the human auditory system during sound reconstruction are still a matter of debate. The purpose of this study is to refine the auditory cortex model introduced in [9], and inspired by the geometrical modelling of vision. The algorithm lifts the time-frequency representation of the degraded sound to the Heisenberg group, where it is reconstructed via a Wilson-Cowan integro-differential equation. Numerical experiments on a library of speech recordings are provided, showing the good reconstruction properties of the algorithm.

Keywords: Auditory cortex · Heisenberg group · Wilson-Cowan equation · Kolmogorov operator

1 Introduction

Human capacity for speech recognition with reduced intelligibility has mostly been studied from a phenomenological and descriptive point of view (see [17] for a review on noise in speech, as well as a wide range of situations in [2]). What is lacking is a proper mathematical model informing us on how the human auditory system is able to reconstruct a degraded speech sound. The aim of this study is to provide a neuro-geometric model for sound reconstruction based on the description of the functional architecture of the auditory cortex.

This study was supported by the IdEx Université de Paris, "ANR-18-IDEX-0001" and by the ANR RUBIN-VASE project, grant ANR-20-CE48-0003 of the French Agence Nationale de la Recherche.

F. Nielsen and F. Barbaresco (Eds.): GSI 2021, LNCS 12829, pp. 56–64, 2021.
https://doi.org/10.1007/978-3-030-80209-7_7

Knowledge on the functional architecture of the auditory cortex and the principles of auditory perception are limited. For that reason, we turn to recent advances in the mathematical modeling of the functional architecture of the primary visual cortex and the processing of visual inputs [8,12,20] (which recently yield very successful applications to image processing [10,14,21]) to extrapolate a model of the auditory cortex. This idea is not new: neuroscientists take models of V1 as a starting point for understanding the auditory system (see, e.g., [18]). Indeed, biological similarities between the structure of the primary visual cortex (V1) and the primary auditory cortex (A1) are well-known to exist. V1 and A1 share a "topographic" organization, a general principle determining how visual and auditory inputs are mapped to those neurons responsible for their processing [23]. Furthermore, the existence of receptive fields of neurons in V1 and A1, allowing for a subdivision of neurons in "simple" and "complex" cells, supports the idea of a "common canonical processing algorithm" [25].

2 The Contact Space Approach in V1

The neuro-geometric model of V1 finds its roots in the experimental results of Hubel and Wiesel [15]. This gave rise to the so-called sub-Riemannian model of V1 in [8,12,20,21]. The main idea behind this model is that an image, seen as a function $f : \mathbb{R}^2 \to \mathbb{R}_+$ representing the grey level, is lifted to a distribution on $\mathbb{R}^2 \times P^1$, the bundle of directions of the plane. Here, P^1 is the projective line, i.e., $P^1 = \mathbb{R}/\pi\mathbb{Z}$. More precisely, the lift is given by $Lf(x, y, \theta) = \delta_{Sf}(x, y, \theta) f(x, y)$ where δ_{S_f} is the Dirac mass supported on the set $S_f \subset \mathbb{R}^2 \times P^1$ of points (x, y, θ) such that θ is the direction of the tangent line to f at (x, y).

When f is corrupted (i.e. when f is not defined in some region of the plane), the reconstruction is obtained by applying a deeply anisotropic diffusion mimicking the flow of information along the horizontal and vertical connections of V1, with initial condition L_f. This diffusion is known as the *sub-Riemannian diffusion* in $\mathbb{R}^2 \times P^1$, cf. [1]. One of the main features of this diffusion is that it is invariant by rototranslation of the plane, a feature that will not be possible to translate to the case of sounds, due to the special role of the time variable.

The V1-inspired pipeline is then the following: first a lift of the input signal to an adequate contact space, then a processing of the lifted signal according to sub-Riemannian diffusions, then projection of the output to the signal space.

3 The Model of A1

The sensory input reaching A1 comes directly from the cochlea. The sensors are tonotopically organized (in a frequency-specific fashion), with cells close to the base of the ganglion being more sensitive to low-frequency sounds and cells near the apex more sensitive to high-frequency sounds. This implies that sound is transmitted to the primary auditory cortex A1 in the form of a 'spectrogram':

when a sound $s : [0,T] \to \mathbb{R}$ is heard, A1 is fed with its time-frequency representation $S : [0,T] \times \mathbb{R} \to \mathbb{C}$. If $s \in L^2(\mathbb{R}^2)$, as given by the short-time Fourier transform of s, that is

$$S(\tau,\omega) := \mathrm{STFT}(s)(\tau,\omega) = \int_{\mathbb{R}} s(t)W(\tau - t)e^{2\pi i t\omega}\, dt. \tag{1}$$

Here, $W : \mathbb{R} \to [0,1]$ is a compactly supported (smooth) window, so that $S \in L^2(\mathbb{R}^2)$. The function S depends on two variables: the first one is time, that here we indicate with the letter τ, and the second one is frequency, denoted by ω. Since S is complex-valued, it can be thought as the collection of two black-and-white images: $|S|$ and $\arg S$. Roughly speaking, $|S(\tau,\omega)|$ represents the strength of the presence of the frequency ω at time τ. In the following, we call S the sound image. The V1-inspired approach is then to apply image perception models on $|S|$. However, in this case time plays a special role: the whole sound image does not reach the auditory cortex simultaneously, but sequentially, and invariance by image rotations is lost.

A more precise modelling of cochlear processing would replace the STFT with a wavelet transform [24,27]. This allows to account for the logarithmic scale for high frequencies and the progressively linear scale for low frequencies.

3.1 The Lift Procedure

We propose an extension of the time-frequency representation of a sound, which is at the core of the proposed algorithm. Sensitivity to variational information, such as the tangent to curves for V1, is now translated in a manner that takes into account the role of time. A regular curve $t \mapsto (t, \omega(t))$ in the time frequency domain (τ, ω) is lifted to a 3-dimensional *augmented space* by adding a new variable $\nu = d\omega/d\tau$. The variation ν in instantaneous frequency is then associated to the chirpiness of the sound. Alternatively, a curve $t \mapsto (\tau(t), \omega(t), \nu(t))$ is a lift of planar curve $t \mapsto (t, \omega(t))$ if $\tau(t) = t$ and if $\nu(t) = d\omega/dt$. Setting $u(t) = d\nu/dt$ we can say that a curve in the contact space $t \mapsto (\tau(t), \omega(t), \nu(t))$ is a lift of a planar curve if there exists a function $u(t)$ such that:

$$\frac{d}{dt}\begin{pmatrix}\tau\\ \omega\\ \nu\end{pmatrix} = \begin{pmatrix}1\\ \nu\\ 0\end{pmatrix} + u\begin{pmatrix}0\\ 0\\ 1\end{pmatrix} = X_0(\tau,\omega,\nu) + uX_1(\tau,\omega,\nu) \tag{2}$$

The vector fields (X_0, X_1) generate the Heisenberg group. However, we are not dealing here with the proper sub-Riemannian distribution, since $\{X_0 + uX_1 \mid u \in \mathbb{R}\}$ is only one-dimensional.

Following [8], when s is a general sound signal, we lift each level line of $|S|$. By the implicit function theorem, this yields the following subset of the augmented space:

$$\Sigma = \left\{(\tau,\omega,\nu) \in \mathbb{R}^3 \mid \nu\partial_\omega|S|(\tau,\omega) + \partial_\tau|S|(\tau,\omega) = 0\right\}. \tag{3}$$

The external input from the cochlea to the augmented space is then given by

$$I(\tau,\omega,\nu) = S(\tau,\omega)\delta_\Sigma(\tau,\omega,\nu) = \begin{cases} S(\tau,\omega) & \text{if } \nu\partial_\omega|S|(\tau,\omega) = -\partial_\tau|S|(\tau,\omega), \\ 0 & \text{otherwise,} \end{cases} \quad (4)$$

with δ_Σ denoting the Dirac delta distribution concentrated on Σ.

3.2 Associated Interaction Kernel

Considering A1 as a slice of the augmented space allows to deduce a natural structure for neuron connections. For a single time-varying frequency $t \mapsto \omega(t)$, its lift is concentrated on the curve $t \mapsto (\omega(t), \nu(t))$, such that

$$\frac{d}{dt}\begin{pmatrix}\omega \\ \nu\end{pmatrix} = Y_0(\omega,\nu) + u(t)Y_1(\omega,\nu), \quad (5)$$

where $Y_0(\omega,\nu) = (\nu, 0)^\top$, $Y_1(\omega,\nu) = (0,1)^\top$, and $u : [0,T] \to \mathbb{R}$.

As in the case of V1 [7], we model neuronal connections via these dynamics. In practice, this amounts to assume that the excitation starting at a neuron $X_0 = (\omega', \nu')$ evolves as the stochastic process $\{A_t\}_{t\geq 0}$ following the SDE $dA_t = Y_0(A_t)dt + Y_1(A_t)dW_T$, $\{W_t\}_{t\geq 0}$ being a Wiener process, with initial condition $A_0 = (\omega', \nu')$. The generator of $\{A_t\}_{t\geq 0}$ is the operator $\mathcal{L} = Y_0 + (Y_1)^2$.

The influence $k_\delta(\omega, \nu \| \omega', \nu')$ of neuron (ω', ν') on neuron (ω, ν) at time $\delta > 0$ is then modeled by the transition density of the process $\{A_t\}_{t\geq 0}$, given by the integral kernel at time δ of the Fokker-Planck equation

$$\partial_t I = \mathcal{L}^* I, \quad \text{where} \quad \mathcal{L}^* = -Y_0 + (Y_1)^2 = -\nu\partial_\omega + b\partial_\nu^2. \quad (6)$$

The vector fields Y_0 and Y_1 are interpreted as first-order differential operators and the scaling parameter $b > 0$ models the relative strength of the two terms. We stress that an explicit expression of k_δ is well-known (see, for instance [3]) and is a consequence of the hypoellipticity of $(\partial_t - \mathcal{L}^*)$.

3.3 Processing of the Lifted Signal

As we already mentioned, the special role played by time in sound signals does not permit to model the flow of information as a pure hypoelliptic diffusion, as was done for static images in V1. We thus turn to a different kind of model: Wilson-Cowan integro-differential equations [26]. This model has been successfully applied to describe the evolution of neural activations, in V1 in particular, where it allowed to predict complex perceptual phenomena such as the emergence of patterns of hallucinatory [11,13] or illusory [4–6] nature. It has also been used in various computational models of the auditory cortex [16,22,28].

Wilson-Cowan equations also present many advantages from the point of view of A1 modelling: i) they can be applied independently of the underlying structure, which is only encoded in the kernel of the integral term; ii) they allow for a natural implementation of delay terms in the interactions; iii) they can be easily tuned via few parameters with a clear effect on the results.

On the basis of these positive results, we emulate this approach in the A1 context. Namely, we consider the lifted sound image $I(\tau,\omega,\nu)$ to yield an A1 activation $a:[0,T]\times\mathbb{R}\times\mathbb{R}\to\mathbb{C}$, solution of

$$\begin{aligned}\partial_t a(t,\omega,\nu) = &-\alpha a(t,\omega,\nu) + \beta I(t,\omega,\nu) \\ &+ \gamma\int_{\mathbb{R}^2} k_\delta(\omega,\nu\|\omega',\nu')\sigma(a(t-\delta,\omega',\nu'))\,d\omega'\,d\nu',\end{aligned} \tag{WC}$$

with initial condition $a(t,\cdot,\cdot)\equiv 0$ for $t\le 0$. Here, $\alpha,\beta,\gamma>0$ are parameters, k_δ is the interaction kernel, and $\sigma:\mathbb{C}\to\mathbb{C}$ is a (non-linear) saturation function, or sigmoid. We pick $\sigma(\rho e^{i\theta})=\tilde\sigma(\rho)e^{i\theta}$ where $\tilde\sigma(x)=\min\{1,\max\{0,\kappa x\}\}$, $x\in\mathbb{R}$, for some fixed $\kappa>0$.

The presence of a delay δ in the activation appearing in the integral interaction term (WC) models the fact that the time-scale of the input signal and of the neuronal activation are comparable. When $\gamma=0$, equation (WC) is a standard low-pass filter $\partial_t a=-\alpha a+I$. Setting $\gamma\neq 0$ adds a non-linear delayed interaction term on top of this exponential smoothing, encoding the inhibitory and excitatory interconnections between neurons.

3.4 Algorithm Pipeline

Both operations in the lift procedure are invertible: the STFT by inverse STFT, and the lift by integration along the ν variable (that is, summation of the discretized solution). The final output signal is thus obtained by applying the inverse of the pre-processing (integration then inverse STFT) to the solution a of (WC). That is, the resulting signal is given by

$$\hat s(t)=\mathrm{STFT}^{-1}\left(\int_{-\infty}^{+\infty} a(t,\omega,\nu)\,\mathrm{d}\,\nu\right). \tag{7}$$

It is guaranteed that $\hat s$ is real-valued and thus correctly represents a sound signal. From the numerical point of view, this implies that we can focus on solutions of (WC) in the half-space $\{\omega\ge 0\}$, which can then be extended to the whole space by mirror symmetry.

The resulting algorithm to process a sound signal $s:[0,T]\to\mathbb{R}$ is then:

A. **Preprocessing:**
   (a) Compute the time-frequency representation $S:[0,T]\times\mathbb{R}\to\mathbb{C}$ of s, via standard short time Fourier transform (STFT);
   (b) Lift this representation to the Heisenberg group, which encodes redundant information about chirpiness, obtaining $I:[0,T]\times\mathbb{R}\times\mathbb{R}\to\mathbb{C}$;

B. **Processing:** Process the lifted representation I via a Wilson-Cowan equation adapted to the Heisenberg structure, obtaining $a:[0,T]\times\mathbb{R}\times\mathbb{R}\to\mathbb{C}$.

C. **Postprocessing:** Project a to the processed time-frequency representation $\hat S:[0,T]\times\mathbb{R}\to\mathbb{C}$ and then apply an inverse STFT to obtain the resulting sound signal $\hat s:[0,T]\to\mathbb{R}$.

4 Numerical Implementation

The discretization of the time and frequency domains is determined by the sampling rate of the original signal and the window size chosen in the STFT procedure. That is, by the Nyquist-Shannon sampling theorem, for a temporal sampling rate δt and a window size of T_w, we consider the frequencies ω such that $|\omega| < 1/(2\delta t)$, with a finest discretization rate of $1/(2T_w)$. Observe, in particular, that the frequency domain is bounded. Nevertheless, the chirpiness ν defined as $\nu\partial_\omega |S|(\tau,\omega) + \partial_\tau |S|(\tau,\omega) = 0$ is unbounded and, since generically there exists points such that $\partial_\omega|S|(\tau_0,\omega_0) = 0$, it stretches over the entire real line.

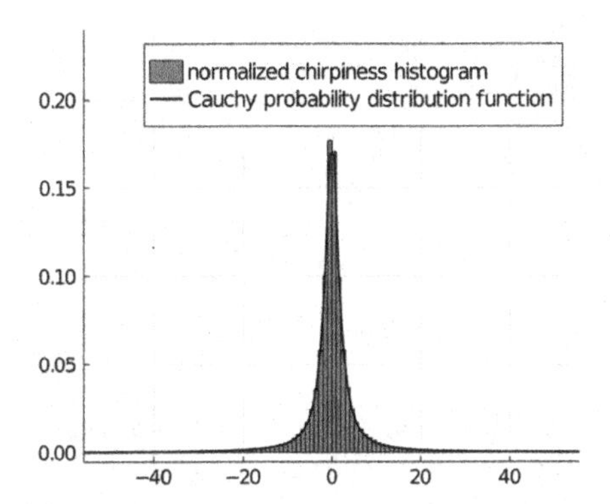

Fig. 1. Chirpiness of a speech signal compared to Cauchy distribution

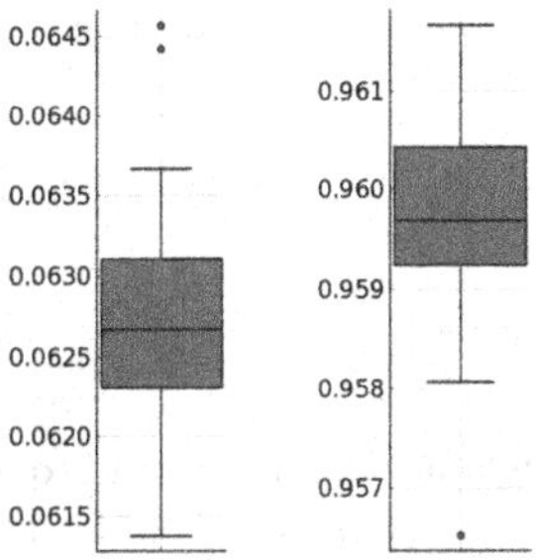

Fig. 2. Box plots for estimated Cauchy distributions of speech signals chirpiness. *Left:* Kolmogorov-Smirnov statistic values. *Right:* percentage of values falling in $I_{0.95}$

To overcome this problem, a natural strategy is to model the chirpiness values as a random variable, and considering only chirpinesses falling inside the confidence interval I_p for some reasonable p-value (e.g., $p = 0.95$). A reasonable assumption for the distribution of X is that it follows a Cauchy distribution $\text{Cauchy}(x_0,\gamma)$. Indeed, this corresponds to assuming that $\partial_\omega|S|$ and $\partial_\tau|S|$ are normal (and independent) distributions [19]. As it is customary, we chose as estimator for the location parameter x_0 the median of X and for the scale parameter γ half the interquartile distance.

Although statistical tests on a library of real-world speech signals[1] rejected the assumption that $X \sim \text{Cauchy}(x_0,\gamma)$, the fit is quite good according to the

[1] The speech material used in the current study is part of an ongoing psycholinguistic project on spoken word recognition. Speech material comprises 49 Italian words and 118 French words. The two sets of words were produced by two (40-year-old) female speakers (a French monolingual speaker and an Italian monolingual speaker) and recorded using a headset microphone AKG C 410 and a Roland Quad Capture audio interface. Recordings took place in the soundproof cabin of the Laboratoire de Phonétique et Phonologie (LPP) of Université de Paris Sorbonne-Nouvelle. Both informants were told to read the set of words as fluently and naturally as possible.

Kolmogorov-Smirnov statistic $D_n = \sup_x |F_n(x) - F_X(x)|$. Here, F_X is the cumulative distribution function of X and F_n is the empirical distribution function evaluated over the chirpiness values (Fig. 1, 2).

5 Denoising Experiments

For simple experiments on synthetic sounds, highligting the characteristics of the proposed algorithm, we refer to [9].

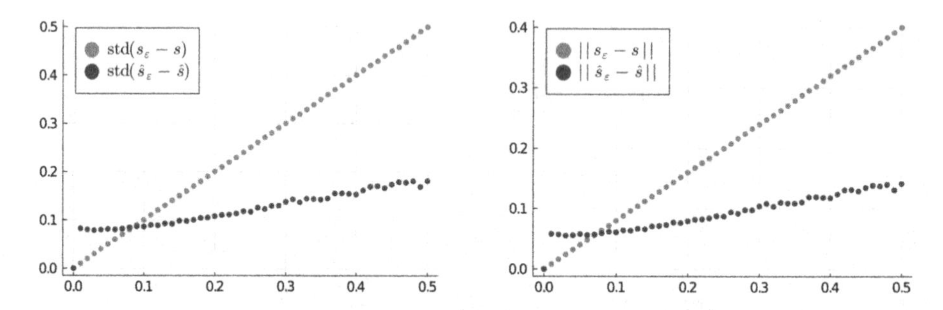

Fig. 3. Distance of noisy sound to original one before (blue) and after (red) the processing, plotted against the standard deviation of the noise (ε). *Left:* standard deviation metric. *Right:* $\|\cdot\|$ norm. (Color figure online)

In Fig. 3 we present the results of the algorithm applied to a denoising task. Namely, given a sound signal s, we let $s_\varepsilon = s + g_\varepsilon$, where $g_\varepsilon \sim \mathcal{N}(0, \varepsilon)$ is a gaussian random variable. We then apply the proposed sound processing algorithm to obtain $\widehat{s_\varepsilon}$. As a reconstruction metric we present both the norm $\|\cdot\|$ where for a real signal s, $\|s\| = \|s\|_1 / \dim(s)$ with $\|\cdot\|_1$ as the L_1 norm and the standard deviation $\mathrm{std}(\widehat{s_\varepsilon} - \widehat{s})$. We observe that according to both metrics the algorithm indeed improves the signal.

References

1. Agrachev, A., Barilari, D., Boscain, U.: A comprehensive introduction to Sub-Riemannian geometry. In: Cambridge Studies in Advanced Mathematics, Cambridge University Press, Cambridge (2019)
2. Assmann, P., Summerfield, Q.: The perception of speech under adverse conditions. In: Assmann, P. (ed.) Speech Processing in the Auditory System. SHAR, vol. 18, pp. 231–308. Springer, New York (2004). https://doi.org/10.1007/0-387-21575-1_5
3. Barilari, D., Boarotto, F.: Kolmogorov-fokker-planck operators in dimension two: heat kernel and curvature (2017)
4. Bertalmío, M., Calatroni, L., Franceschi, V., Franceschiello, B., Gomez Villa, A., Prandi, D.: Visual illusions via neural dynamics: Wilson-cowan-type models and the efficient representation principle. J. Neurophysiol. **123**, 1606 (2020)

5. Bertalmío, M., Calatroni, L., Franceschi, V., Franceschiello, B., Prandi, D.: A cortical-inspired model for orientation-dependent contrast perception: a link with Wilson-Cowan equations. In: Lellmann, J., Burger, M., Modersitzki, J. (eds.) SSVM 2019. LNCS, vol. 11603, pp. 472–484. Springer, Cham (2019). https://doi.org/10.1007/978-3-030-22368-7_37
6. Bertalmío, M., Calatroni, L., Franceschi, V., Franceschiello, B., Prand i, D.: Cortical-inspired Wilson-Cowan-type equations for orientation-dependent contrast perception modelling. J. Math. Imaging Vis. **63**, 263 (2020)
7. Boscain, U., Chertovskih, R., Gauthier, J.P., Remizov, A.: Hypoelliptic diffusion and human vision: a semi-discrete new twist on the petitot theory. SIAM J. Imaging Sci. **7**(2), 669–695 (2014)
8. Boscain, U., Duplaix, J., Gauthier, J.P., Rossi, F.: Anthropomorphic image reconstruction via hypoelliptic diffusion (2010)
9. Boscain, U., Prandi, D., Sacchelli, L., Turco, G.: A bio-inspired geometric model for sound reconstruction. J. Math. Neurosci. **11**(1), 1–18 (2020). https://doi.org/10.1186/s13408-020-00099-4
10. Boscain, U.V., Chertovskih, R., Gauthier, J.P., Prandi, D., Remizov, A.: Highly corrupted image inpainting through hypoelliptic diffusion. J. Math. Imaging Vis. **60**(8), 1231–1245 (2018)
11. Bressloff, P.C., Cowan, J.D., Golubitsky, M., Thomas, P.J., Wiener, M.C.: Geometric visual hallucinations, Euclidean symmetry and the functional architecture of striate cortex. Philos. Trans. R. Soc. Lond. B Biol. Sci. **356**(1407), 299–330 (2001)
12. Citti, G., Sarti, A.: A cortical based model of perceptual completion in the Roto-Translation space. J. Math. Imaging Vis. 24(3), 307–326 (2006)
13. Ermentrout, G.B., Cowan, J.D.: A mathematical theory of visual hallucination patterns. Biol. Cybern. **34**, 137–150 (1979)
14. Franken, E., Duits, R.: Crossing-preserving coherence-enhancing diffusion on invertible orientation scores. Int. J. Comput. Vis. **85**(3), 253–278 (2009)
15. Hubel, D.H., Wiesel, T.N.: Receptive fields of single neurons in the cat's striate cortex. J. Physiol. **148**(3), 574–591 (1959)
16. Loebel, A., Nelken, I., Tsodyks, M.: Processing of sounds by population spikes in a model of primary auditory cortex. Front. Neurosci. **1**(1), 197–209 (2007)
17. Mattys, S., Davis, M., Bradlow, A., Scott, S.: Speech recognition in adverse conditions: a review. Lang. Cogn. Neurosci. **27**(7–8), 953–978 (2012)
18. Nelken, I., Calford, M.B.: Processing strategies in auditory cortex: comparison with other sensory modalities. In: Winer, J., Schreiner, C. (eds.) The Auditory Cortex, pp. 643–656. Springer, Boston (2011). https://doi.org/10.1007/978-1-4419-0074-6_30
19. Papoulis, A.: Probability, Random Variables and Stochastic Processes, 3rd edn., p. 138. McGraw-Hill Companies, New York (1991)
20. Petitot, J., Tondut, Y.: Vers une neurogéométrie. fibrations corticales, structures de contact et contours subjectifs modaux. Mathématiques et Sci. humaines **145**, 5–101 (1999)
21. Prandi, D., Gauthier, J.P.: A Semidiscrete Version of the Citti-petitot-sarti Model as a Plausible Model for Anthropomorphic Image Reconstruction and Pattern Recognition. BRIEFSMATH, Springer, Cham (2017). https://doi.org/10.1007/978-3-319-78482-3
22. Rankin, J., Sussman, E., Rinzel, J.: Neuromechanistic model of auditory bistability. PLoS Comput. Biol. **11**(11), e1004555 (2015)
23. Rauschecker, J.P.: Auditory and visual cortex of primates: a comparison of two sensory systems. Eur. J. Neurosci. **41**(5), 579–585 (2015)

24. Reimann, H.M.: Signal processing in the cochlea: the structure equations. J. Math. Neurosci. **1**(1), 43 (2011)
25. Tian, B., Kuśmierek, P., Rauschecker, J.P.: Analogues of simple and complex cells in rhesus monkey auditory cortex. Proc. Nat. Acad. Sci. **110**(19), 7892–7897 (2013)
26. Wilson, H.R., Cowan, J.D.: Excitatory and inhibitory interactions in localized populations of model neurons. Biophys. J. **12**(1), 1–24 (1972)
27. Yang, X., Wang, K., Shamma, S.: Auditory representations of acoustic signals. IEEE Trans. Inf. Theory **38**, 68 (1992)
28. Zulfiqar, I., Moerel, M., Formisano, E.: Spectro-temporal processing in a two-stream computational model of auditory cortex. Front. Comput. Neurosci. **13**, 95 (2020)

Conformal Model of Hypercolumns in V1 Cortex and the Möbius Group. Application to the Visual Stability Problem

Dmitri Alekseevsky[1,2](✉)

[1] Institute for Information Transmission Problems RAS, Moscow, Russia
[2] University of Gradec Králové, Kralove, Czech Republic
alekseevsky@iitp.ru

Abstract. A conformal spherical model of hypercolumns of primary visual cortex V1 is proposed. It is a modification of the Bressloff-Cowan Riemannian spherical model. The main assumption is that simple neurons of a hypercolumn, considered as Gabor filters, obtained for the mother Gabor filter by transformations from the Möbius group $Sl(2, \mathbb{C})$. It is shown that in a small neighborhood of a pinwheel, which is responsible for detection of high (resp., low) frequency stimuli, it reduces to the Sarti-Citti-Petitot symplectic model of V1 cortex. Application to the visual stability problem is discussed.

Keywords: Conformal model of hypercolumn · Möbius group · Stability problem

1 Introduction

D. Hubel and T. Wiesel [9] described the structure of the primary visual cortex V1 from a neurophysiological point of view and proposed the first model of the V1 cortex ("ice cube model"). Moreover, they justified the existence of hypercolumns, later described as fiber bundles over the retina.

Fiber of the bundle corresponds to different internal parameters (orientation, spatial frequency, ocular dominance, direction of motion, curvature, etc.) that affect the excitation of visual neurons. N.V. Swindale [18] estimated the dimension of the fibers (= the number of internal parameters) as 6–7 or 9–10.

In 1989, W. Hoffman [8] hypothesized that the primary visual cortex is a contact bundle. The realization of this conjecture was started by J. Petitot [16] and continued in [3,4,14,15]. J. Petitot [14] proposed the contact model of V1 cortex as the contact bundle $\pi : P = PT^*R \to R$ of orientations (directions) over the retina $R = \mathbb{R}^2$ (identified with the plane) with circle fibers S^1. The manifold P has coordinates (x, y, θ) where $(x, y) \in \mathbb{R}^2$ and the orientation θ is the angle between the tangent line to a contour in retina and the axis $0x$. The manifold

F. Nielsen and F. Barbaresco (Eds.): GSI 2021, LNCS 12829, pp. 65–72, 2021.
https://doi.org/10.1007/978-3-030-80209-7_8

P is identified with the bundle of (oriented) orthonormal frames in R and with the group $SE(2) = SO(2) \cdot \mathbb{R}^2$ of unimodular Euclidean isometries.

The basic assumption is that simple neurons are parametrized by points of $P = SE(2)$. More precisely, the simple neuron, associated to a frame $f \in P$, is working as the Gabor filter in the Euclidean coordinates defined by the frame f.

Note that in this model, "points" of retina correspond to pinwheels, that is, singular columns of cortex, which contains simple neurons of any orientation. Recall that all simple neurons of a *regular* column act as (almost) identical Gabor filters with (almost) the same receptive field D and receptive profile and they fire only if a contour on the retina, which crosses D, has an appropriate orientation θ.

Recently, this model (equiped with an appropriate sub-Riemannian metric) had been successfully applied by B. Franceschiello, A. Mashtakov, G. Citti and A. Sarti [6] for explanation of optical illusions.

The contact model had been extended by Sarti, Citti and Petitot [17] to a symplectic model, with two-dimensional fibers, associated with the orientation θ and the scaling σ. In this model, simple cells are parametrized by conformal frames or points of the Euclidean conformal group $Sim(E^2) = \mathbb{R}^+ \cdot SE(2)$. More precisely, this means that the simple neuron, associated to a conformal frame f, works as the Gabor filter, obtained form the mother Gabor filter, associated to the standard frame f_0, by the transformation $A_f \in Sim(E^2)$ which maps f_0 onto f. Note that the receptive profiles are transformed as density, see [17].

P. Bressloff and J. Cowan [1,2] proposed a Riemannian spherical model of a hypercolumn H of V1 cortex, associated with two internal parameters: the orientation θ and the spatial frequency p (more precisely, the normalized logarithm σ of p). This means that a simple neuron $n = n(\theta, \sigma)$ fires if the contour on retina which crosses the receptive field of n has (approximately) the orientation θ and the frequency σ. They assume that a hypercolumn H is associated with two pinwheels S, N (singular columns), which has simple neurons of any orientation and the minimum value $\sigma = -\pi/2$ and, respectively, the maximum value $\sigma = \pi/2$ of the normalized frequency. All other columns of the hypercolumn H are regular. Simple neurons of a regular column fire if the contour has fixed value (θ, σ) of the orientation and the normalized frequency. In other words, columns of a hypercolumn are parametrized by coordinates θ, σ and these coordinates may be considered as the spherical coordinates (longitude and latitude) of the sphere. The pinwheels S, N correspond to the south pole and the north pole of the sphere $S^2 = H$. The neighborhood of S detects the low frequency images and the neighborhood of N detects the high frequency images.

We present a modification of this model, based on the assumption that a hypercolumn H is a conformal sphere. In this model, simple neurons of H are parametrized by second order conformal frames, obtained from the standard frame by transformations from the Möbius group $G = SL(2, \mathbb{C})$.

This corresponds to the Cartan approach to conformal geometry, bases on the construction of so-called Cartan connection. For the conformal sphere, the Cartan connection is the principal bundle $G = SL(2, \mathbb{C}) \to S^2 = G/Sim(E^2)$

with the Maurer-Cartan form $\mu : TG \to \mathfrak{sl}(2,\mathbb{C})$ (which identifies tangent spaces T_gG with the Lie algebra $\mathfrak{sl}(2,\mathbb{C})$). Moreover, points of the sphere (which correspond to columns of the hypercolumn H) are parametrized by the stability subgroups. This corresponds to the Tits model of the conformal sphere (where points are defined as stability subgroups).

We show that in a neighborhood of a pinwheel, the conformal model formally reduces to the symplectic model of Sarti, Citti and Petitot.

We discuss the problem of visual stability and relation of the proposed model with stability. The visual stability problem consists in explanation how we perceive stable objects as stable despite the change of their retinal images caused by the rotation of the eyes, see e.g. [19].

2 Riemannian Spinor Model of Conformal Sphere

To describe our conformal model of hypercolumns, which is a conformal modification of the model by Bressloff and Cowan, we recall Riemann spinor model of conformal sphere as Riemann sphere $S^2 = \hat{\mathbb{C}} = \mathbb{C} \cup \{\infty\}$ with two distinguished points $S = 0$ and $N = \infty$ and complex coordinate $z \in S^2 \setminus N$ and $w = \frac{1}{z} \in S^2 \setminus S$. The group $G = SL(2,\mathbb{C})$ acts on S^2 as the conformal group by fractional-linear transformations

$$z \mapsto Az = \frac{az+b}{cz+d},\ a,b,c,d \in \mathbb{C}, \det A = 1.$$

Remark that this group acts non-effectively and the quotient group $PSL(2,\mathbb{C}) = SL(2,\mathbb{C})/\{\pm \mathrm{Id}\}$, which is isomorphic to the connected Lorentz group $SO^0(1,3)$, acts effectively.

Denote by

$$G = G^- \cdot G^0 \cdot G^+ = \begin{pmatrix} 1 & 0 \\ \mathbb{C} & 1 \end{pmatrix} \cdot \left\{ \begin{pmatrix} a & 0 \\ 0 & a^{-1} \end{pmatrix}, a \in \mathbb{C}^* \right\} \cdot \begin{pmatrix} 1 & \mathbb{C} \\ 0 & 1 \end{pmatrix}$$

the Gauss decomposition. Then the stability subgroups G_S, G_N of the points $S = 0$ and $N = \infty$ are $B_{\mp} = G^0 \cdot G^{\mp} \simeq Sim(E^2) = CO_2 \cdot \mathbb{R}^2$. As a homogeneous manifold, the conformal sphere may be writtes as $S^2 = G/B_{\mp} = SL_2(\mathbb{C})/Sim(E^2)$.

3 Conformal Spherical Model of Hypercolumns

We present a conformal modification of the Bressloff-Cowan model. We assume that as in Bressloff-Cowan model the hypercolumn, associated with two pinwheels N, S, is the conformal sphere with the spherical coordinates θ, σ. Simple neurons, considered as Gabor filters, are parametrized by the conformal Möbius group $G \simeq SL(2,\mathbb{C})$ and depend of 6 parameters. More precisely, the Gabor filter $n(g)$ (i.e. the simple neuron), associated with a transformation $g \in G$, is obtained from the mother Gabor filter (associated to the standard coordinate system) by the conformal transformation g.

We show that in a small neighborhood D_S of the south pole (respectively, in a small neighborhood D_N of the north pole N), responsible for perception of low (respectively, high) frequency stimuli, the model reduces to the symplectic Sarti-Citti-Petitot model.

3.1 Relation with Symplectic Sarti-Citti-Petitot Model

Let $\sigma_N : S^2 \to T_S S^2 = E^2$ be the stereographic projection from the north pole N to the tangent space at the south pole. The transitive action of the stability subgroup $G_N = G^0 \cdot G^+ \simeq \mathrm{Sim}(E^2)$ on $S^2 \setminus N$ induces its action on the tangent plane $T_S S^2 = E^2$ as the group $\mathrm{Sim}(E^2)$ of similarity transformations.
More precisely, the subgroup G^+ acts on $T_S S^2 = E^2$ by parallel translations, the group $SO_2 = \{\mathrm{diag}(e^{i\alpha}, e^{-i\alpha})\}$ acts by rotations,
the group $\mathbb{R}^+ = \{\mathrm{diag}(\lambda, \lambda^{-1})\}$ acts by homotheties,
Note also that the subgroup $G^- \subset G_S$ preserves S and acts trivially on $T_S S^2$. Hence in a small neighborhood of S, the action of the conformal group is approximated by the action of the similarity group $G_N \simeq Sim(E^2)$ in $T_S S^2 \simeq E^2$. We conclude:
Simple neurons of a hypercolumn in a small neighbourhood of the south pole S (resp., the north pole N) depends only on 4-parameters and are parametrized by points of the group G_N (resp., G_S) of similarities according to the Sarti-Citti-Petitot model.

3.2 Principle of Invariancy

We will state the following obvious general principle of invariancy:
Let G be a group of transformations of a space V and $\mathcal{O} = Gx$ an orbit.

Principle of Invariancy. The information, which observers, distributed along the orbit $\mathcal{O}$, send to some center is invariant w.r.t. the group G.

Application. The information about the low spatial frequency stimuli, encoded in simple cells of a hypercolumn H near south pinwheel, parametrized by the group G_N, is invariant w.r.t. G_N.

Similarly, the information about the high spatial frequency stimuli is invariant w.r.t. the group G_S.

The information about local structure of a stimulus, encoded in simple cells of a hypercolumn, are parametrized by the conformal group $G = SL_2(\mathbb{C})$ and is invariant with respect to the conformal group G.

In particular, if we assume that the remapping of the retinal image after a microsaccade is carried out by a conformal transformation, then the simple neurons of a hypercolumns contain information, which is sufficient to identify the retina images before and after the saccade.

In the next section, we discuss the conjecture that the remapping after a (micro and macro) saccade is described by a conformal transformation of the retinal image.

4 The Cental Projection and the Shift of Retina Images After Post-saccade Remappling

4.1 The Central Projection onto the Retina

Let $M \subset E^3$ be a surface, whose points are sources of diffuse reflected light. We assume that all the light rays emitted from a point $A \in M$ carry the same energy density $E(A)$. The retina image of the surface is described by the central projection of the surface to the eye sphere S^2 with respect to the nodal point F of the eye.

We will assume that the nodal point F (or optical center) of the eye B (considered as a ball) belongs to the eye sphere $S^2 = \partial B$.

It is not completely true, the nodal point is located inside the eye ball, but very close to the eye sphere.

The **central projection** of a surface $M \subset E^3$ onto a sphere $S^2 \subset E^3$ with center $F \in S^2$ is defined by the map

$$\pi_F : M \ni A \to \bar{A} = \ell_{AF} \cap S^2 \subset S^2$$

where $\ell_{AF} \cap S^2$ is the second point of intersection of the sphere S^2 with the ray ℓ_{AF}, passing through the point F.

We may assume that the central projection is a (local) diffeomorphism and that the energy density $I(\bar{A})$ at a point $\bar{A}$ is proportional to $E(A)$. So the input function $I : S^2 \to \mathbb{R} \subset S^2$ on the retina R contains information about illumination of points of the surface M.

With respect to the retinal coordinates (fixed w.r.t. to the eye ball B^3) the eye rotation R is equivalent to inverse rotation R^{-1} of the external space E^3. It transforms M onto another surface $M' = R^{-1}M$ with the new retina image $\pi_F(M') = \pi_F(R^{-1}(M))$. The transformation $\pi_F(M) \to \pi_F M'$ is called the post-saccade remapping. This means that a visual neuron n with a receptive field $D \subset R$, which detects the retina image $\bar{A} = \pi_F(A)$ of an external point A after a saccade will detect the image $\bar{A}' = \pi_F(A')$ where $A' = R^{-1}(A)$.

J.-R. Duhamel, C.L. Colby and M.E. Goldberg [5], see also [12,13] discovered the fundamental and mysterious fundamental phenomenon of "shift of receptive field" for visual neurons in various systems of visual cortex:

100 ms before the saccade, the neuron n with the receptive field D, which detected the retinal image $\bar{A} \in D$, starts to detect the retina image $\bar{A}'$ of the point $A' = R^{-1}(A)$, which will be in its receptive field after the saccade.

Art historian Ernst Gombrich proposed an important Etcetera Principle, see [5,7]. It states that for saccade remapping, the brain controls information about new retinal image of only 3–4 salient points of the scene. It is sufficient to reconstruct the new retinal image of all scene, using redundancy in the scene and previous experience with the given type of environment.

Gombrich wrote that **"only a few (3–4) salient stimuli are contained in the trans-saccadic visual memory and update."**

We propose an implementation of this idea, based on assumption that remapping of retinal images are described by conformal transformations. First of all, we recall the Möbius projective model of the conformal sphere.

4.2 Möbius Projective Model of Conformal Sphere

Let $(V = \mathbb{R}^{1,3}, g)$ be the Minkowski vector space with the metric of signature $(-,+,+,+)$ and (e_0, e_1, e_2, e_3) an orthonormal basis of V.

Up to scaling, there are three orbits of the connected Lorentz group $G = SO^0(V)$ in V:
$V_T = Ge_0 = G/SO_3$ (the Lobachevski space),
$V_S = Ge_1 = G/SO_{1,2}$ (the De Sitter space),
$V_0 = Gp = G/SE(2)$ - isotropic (light) cone, where $p = e_0 + e_1$ is the null vector, $SE(2) = SO_2 \cdot \mathbb{R}^2$.
Projectivization of these orbits gives three G-orbits in the projective space $P^3 = PV$: the open ball $B^3 = PV_T \simeq V_T$,
its exterior PV_S and
the projective quadric ((or conformal sphere)

$$Q = PV_0 = G/Sim(E^2) = G/(\mathbb{R}^+ \cdot SE(2))$$

with induces conformal metric and conformal action of G.

The metric g defines correspondence $PV \ni [v] \Leftrightarrow P(v^\perp) \subset P^3$ between projective points and projective planes.

The projective plane $P(v^\perp)$ does not intersect (resp., intersect) the quadric Q if $v = n \in V_T$ (resp., $v = m \in V_S$). For $p \in V_0$, the plane $P(p^\perp) = T_{[p]}Q$ is the tangent plane of the quadric.

Euclidean Interpretation
Intersection of these objects with the Euclidean hyperplane $E^3_{e_0} = e_0 + e_0^\perp$ provides an Euclidean interpretation of the quadric Q and projective planes:
$S^2 := Q \cap E^3_{e_0}$ is the unite sphere in the Euclidean space $E^3_{e_0}$,
An Euclidean plane $\Pi_v = v^\perp \cap E^3_{e_0} \subset E^3_{e_0}$ does not intersect (resp, intersect) the sphere S^2 if $v = n \in V_T$ (resp., $v = m \in V_S$) and $\Pi_p = T_F S^2$ is the tangent plane of S^2 at the point $F = \mathbb{R}p \cap E^3_{e_0}$ for $p \in V_0$.

Lemma 1. *The stability group $G_F = Sim(E^2)$ of a point $F = \mathbb{R}p \cap E^3_{e_0}$ acts transitivity on V_T, hence, on the set of planes $\{\Pi_n, n \in V_T\}$ which do not intersect the sphere S^2.*

4.3 Remapping as a Conformal Transformation

Due to Lemma, there is a Lorentz transformation $L \in SO(V)_F$ which fixes the nodal point $F \in S^2$ and transforms the plane $\Pi = \Pi_n$, $n \in V_T$ into any other plane $\Pi' = \Pi_{n'}$ which does not intersect S^2.

Let $\Pi, \subset E^3$ be a plane and $\Pi' = R^{-1}\Pi$ another plane, where $R \in SO(3)$ is the eye rotation, which describes a saccade. Suppose

that the brain identifies the planes by means of the Lorentz transformation L which preserves F and transform Π into Π'. Then the retinal images $\pi_F(\Pi), \pi_F(\Pi') \subset S^2$ before and after a saccade are related by the conformal transformation $L|S^2$.

Such conformal transformation (and the Lorentz transformation L) is determined by the images of three generic points of the eye sphere S^2. Hence, the new position of a three points of retina after a saccade is sufficient for reconstruction of the remapping as Etcetera Principle by E.H. Gombrich states.

Note that is the rotation is small, then the Lorentz transformation L is closed to the rotation R.

Our conjecture reduces the problem of stability for contours to the classical problem of conformal geometry - description of curves on the conformal sphere up to a conformal transformation. (It is the conformal generalisation of the Frenet theory). It was solved by Fialkov, Sulanke, Sharp, Shelechov and others. Recently V. Lychagin and N. Konovenko [11] gave an elegant description of the algebra of conformal differential invariants for the conformal sphere.

References

1. Bressloff, P.C., Cowan, J.D.: A spherical model for orientation as spatial-frequency tuning in a cortical hypercolumn. Philos. Trans. R. Soc. Lond. **358**, 1–22 (2002)
2. Bressloff, P.C., Cowan, J.D.: The visual cortex as a crystal. Physica D **173**, 226–258 (2002)
3. Citti, G., Sarti, A.: A cortical based model of perceptual completion in the roto translation space. In: Birindelli, I., Gutiérrez, C., Lanconelli, E. (eds.) Proceedings workshop on 2D order subelliptic equations and applications (Cortona). Lecture notes of Seminario interdisciplinare di matematica (2003)
4. Citti, G., Sarti, A.: A cortical based model of perceptual completion in the roto-translation space. JMIV **24**(3), 307–326 (2006)
5. Duhamel, J.-R., Colby, C.L., Goldberg, M.E.: The updating of the representation of visual space in parietal cortex by intended eye movements. Science **255**(5040), 90–92 (1992)
6. Franceschiello, B., Mashtakov, A., Citti, G., Sarti, A.: Geometrical optical illusion via Sub-Riemannian geodesics in the roto-translation group. Diff. Geom. Appl. **65**, 55–77 (2019)
7. Gombrich, E.H.: The Sense of Order: A Study in the Psychology of Decorative Art. The Phaidon Press, New York (1979)
8. Hoffman, W.C.: The visual cortex is a contact bundle. Appl. Math. Comput. **32**(2–3), 137–167 (1989)
9. Hubel, D.H., Wiesel, T.N.: Receptive elds, binocular interaction and functional architecture in the cat's visual cortex. J. Physiol. **160**(1), 106–154 (1962)
10. Hubel, D.H.: Eye. Brain and Vision, New York (1988)
11. Lychagin, V., Konovenko, N.: Invariants for primary visual cortex. Diff. Geom. Appl. **60**, 156–173 (2018)
12. Merriam, E.P., Genovese, C.R., Colby, C.L.: Remapping in human visual cortex. J. Neurophys. **97**(2), 1738–1755 (2007)
13. Melcher, D., Colby, C.L.: Trans-saccadic perception. Trends Cogn. Sci. **12**(12), 466–473 (2008)

14. Petitot, J.: The neurogeometry of pinwheels as a sub-Riemannian contact structure. J. Physiol. Paris **97**, 265–309 (2003)
15. Petitot, J.: Elements of neurogeometry (2017)
16. Petitot, J., Tondut, Y.: Vers une neurogéométrie. Fibrations corticales, structures de contact et contours subjectifs modaux Mathématiques et Sciences humaines **145**, 5–101 (1999)
17. Sarti, A., Citti, G., Petitot, J.: The symplectic structure of the primary visual cortex. Biol. Cybern. **98**, 33–48 (2008)
18. Swindale, N.: How many maps are in visual cortex. Cereb. Cortex **10**(7), 633–643 (2000)
19. Wurtz, R.H.: Neuronal mechanisms of visual stability. Vis. Res. **48**, 2070–2089 (2008)

Extremal Controls for the Duits Car

Alexey Mashtakov(✉)

Ailamazyan Program Systems Institute of RAS, Pereslavl-Zalessky, Russia

Abstract. We study a time minimization problem for a model of a car that can move forward on a plane and turn in place. Trajectories of this system are used in image processing for the detection of salient lines. The problem is a modification of a well-known sub-Riemannian problem in the roto-translation group, where one of the controls is restricted to be non-negative. The problem is of interest in geometric control theory as a model example in which the set of admissible controls contains zero on the boundary. We apply a necessary optimality condition—Pontryagin maximum principle to obtain a Hamiltonian system for normal extremals. By analyzing the Hamiltonian system we show a technique to obtain a single explicit formula for extremal controls. We derive the extremal controls and express the extremal trajectories in quadratures.

Keywords: Sub-Riemannian · Geometric control · Reeds-Shepp car

1 Introduction

Consider a model of an idealized car moving on a plane, see Fig. 1. The car has two parallel wheels, equidistant from the axle of the wheelset. Both wheels have independent drives that can rotate forward and backward so that the corresponding rolling of the wheels occurs without slipping. The configuration of the system is described by the triple $q = (x, y, \theta) \in \mathbb{M} = \mathbb{R}^2 \times S^1$, where $(x, y) \in \mathbb{R}^2$ is the central point, and $\theta \in S^1$ is the orientation angle of the car. Note that the configuration space $\mathbb{M}$ forms the Lie group of roto-translations $\mathrm{SE}(2) \simeq \mathbb{M} = \mathbb{R}^2 \times S^1$.

From the driver's point of view, the car has two controls: the accelerator u_1 and the steering wheel u_2. Consider the configuration $e = (0, 0, 0)$, which corresponds to the car located in the origin and oriented along the positive direction of abscissa. An infinitesimal translation is generated by the vector ∂_x and rotation—by ∂_θ. They are possible motions controlled by u_1 and u_2 respectively. The remaining direction ∂_y is forbidden since the immediate motion of the car in direction perpendicular to its wheels is not possible. Thus, the dynamics of the car in the origin is given by $\dot{x} = u_1$, $\dot{y} = 0$, $\dot{\theta} = u_2$.

This work is supported by the Russian Science Foundation under grant 17-11-01387-P and performed in Ailamazyan Program Systems Institute of Russian Academy of Sciences.

F. Nielsen and F. Barbaresco (Eds.): GSI 2021, LNCS 12829, pp. 73–81, 2021.
https://doi.org/10.1007/978-3-030-80209-7_9

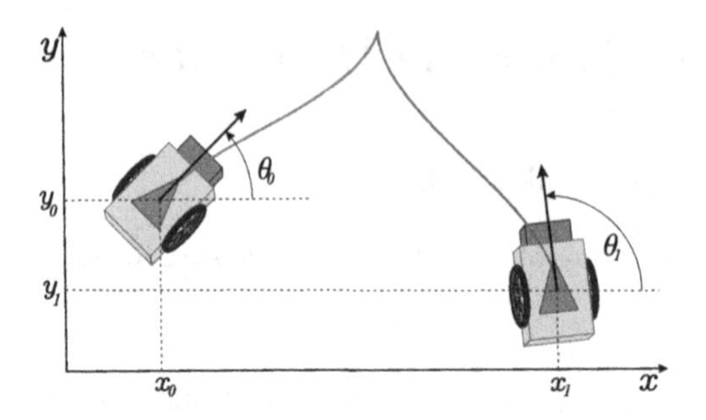

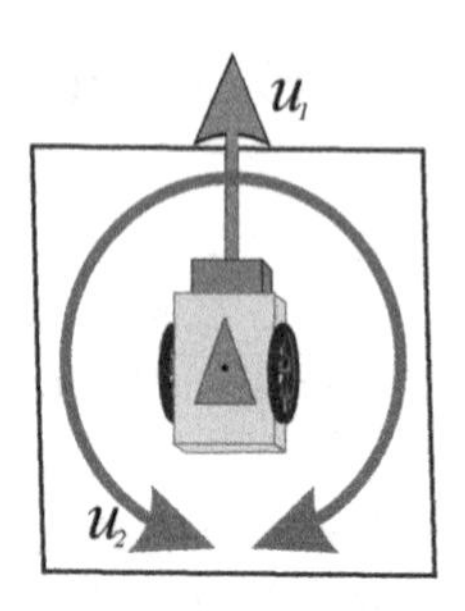

Fig. 1. Left: classical model of a car that can move forward and backward and rotate in place. Its trajectory represents a cusp when the car switches the direction of movement to opposite. Arcs of the trajectory where the car is moving forward/backward are depicted in green/red correspondingly. **Right:** control u_1 is responsible for translations and u_2 for rotations of the car. For the Duits car motion backward is forbidden $u_1 \geq 0$. (Color figure online)

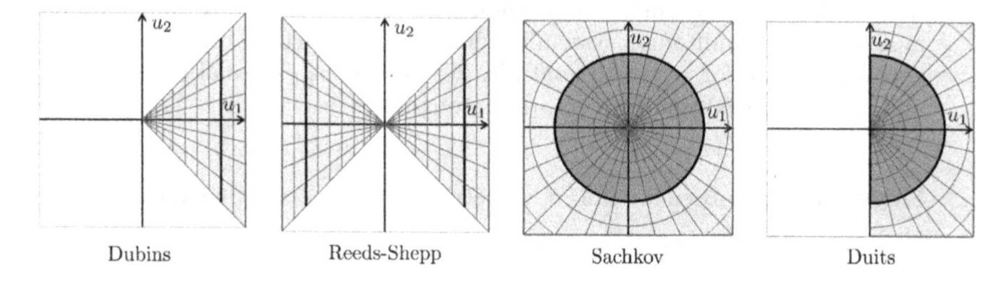

Fig. 2. Set of admissible controls for various models of a car moving on a plane.

The origin e is unit element of the group SE(2). Any other element $q \in \text{SE}(2)$ is generated by left multiplication $L_q e = q \cdot e$. Dynamics in a configuration q is

$$\dot{q} = u_1 X_1(q) + u_2 X_2(q), \tag{1}$$

where the vector fields X_i are obtained via push-forward of left multiplication $X_1(q) = L_{q*}\partial_x$, $X_2 = L_{q*}\partial_\theta$, and the forbidden direction is $X_3(q) = L_{q*}\partial_y$:

$$X_1(q) = \cos\theta\,\partial_x + \sin\theta\,\partial_y, \quad X_2(q) = \partial_\theta, \quad X_3(q) = \sin\theta\,\partial_x - \cos\theta\,\partial_y.$$

Various sets of admissible controls $U \ni (u_1, u_2)$ lead to different models of the car, see [1]. e.g., see Fig. 2, the time minimization problem for

- $u_1 = 1$, $|u_2| \leq \kappa$, $\kappa > 0$ leads to Dubins car [2];
- $|u_1| = 1$, $|u_2| \leq \kappa$, $\kappa > 0$ leads to Reeds-Shepp car [3];
- $u_1^2 + u_2^2 \leq 1$ leads to the model of a car, which trajectories are given by sub-Riemannian length minimizes, studied by Sachkov [4];
- $u_1 \geq 0$, $u_1^2 + u_2^2 \leq 1$ leads to the model of a car moving forward and turning in place, proposed by Duits et al. [5].

System (1) appears in robotics as a model of a car-like robot. The system also arises in the modelling of the human visual system and image processing. A mathematical model of the primary visual cortex of the brain as a sub-Riemannian structure in the space of positions and orientations is developed by Petitot [6], Citti and Sarti [7]. According to this model, contour completion occurs by minimizing the excitation energy of neurons responsible for the area of the visual field, where the contour is hidden from observation.

The principles of biological visual systems are actively used in computer vision. Based on these principles, effective methods of image processing are created, e.g.: image reconstruction [8,9], detection of salient lines in images [10].

In particular, the problem of salient curves detection arises in the analysis of medical images of the human retina when searching for blood vessels. In [10], the set of admissible controls is the disk $u_1^2 + u_2^2 \leq 1$. A disadvantage of this model is the presence of cusps, see Fig. 2. Such curves are not desirable for vessel tracking. To eliminate this drawback, the restriction of the set of admissible control to a half-disc was proposed in [5]. The results of vessel tracking via the minimal paths in this model, which we call the Duits car, can be found in [11].

The problem of optimal trajectories of the Duits car with a given external cost is studied in [5]. In particular, the authors develop a numerical method for finding optimal trajectories using Fast-Marching algorithm [12]. They also study the special case of uniform external cost, which we consider in this paper, and formulate a statement that cusp points are replaced by so-called key points, which are points of in-place rotations, see [5, Theorem 3].

In this paper, we study the time-minimization problem for the Duits car. By direct application of Pontryagin maximum principle (PMP) and analysis of the Hamiltonian system of PMP, we show a formal proof of the statement regarding the replacement of cusps by key points. We also present a technique of obtaining the explicit form of extremal controls by reducing the system to the second-order ODE and solving it in Jacobi elliptic functions.

The problem under consideration is of interest in geometric control theory [13], as a model example of an optimal control problem in which zero control is located on the boundary of the set of admissible controls. A general approach [14] to similar problems is based on convex trigonometry. The approach covers the class of optimal control problems with two-dimensional control belonging to an arbitrary convex compact set containing zero in its interior. However, this approach does not admit immediate generalization to the case when zero lies on the boundary. For systems of this type, the development of new methods is required. This article examines in detail a particular case of such a system.

2 Problem Formulation

We consider the following control system:

$$\begin{cases} \dot{x} = u_1 \cos\theta, \\ \dot{y} = u_1 \sin\theta, \\ \dot{\theta} = u_2, \end{cases} \qquad \begin{array}{l} (x, y, \theta) = q \in \mathrm{SE}(2) = \mathbb{M}, \\ u_1^2 + u_2^2 \leq 1,\ u_1 \geq 0. \end{array} \tag{2}$$

We study a time minimization problem, where for given boundary conditions $q_0, q_1 \in \mathbb{M}$, one aims to find the controls $u_1(t), u_2(t) \in L^\infty([0,T],\mathbb{R})$, such that the corresponding trajectory $\gamma : [0,T] \to \mathbb{M}$ transfers the system from the initial state q_0 to the final state q_1 by the minimal time:

$$\gamma(0) = q_0, \quad \gamma(T) = q_1, \qquad T \to \min. \tag{3}$$

System (2) is invariant under action of SE(2). Thus, w.l.o.g., we set $q_0 = (0,0,0)$.

3 Pontryagin Maximum Principle

It can be shown that, in non-trivial case, problem on the half-disc is equivalent to the problem on the arc $u_1^2 + u_2^2 = 1$. Denote $u = u_2$, then $u_1 = \sqrt{1-u^2}$.

A necessary optimality condition is given by PMP [13]. Denote $h_i = \langle \lambda, X_i \rangle$, $\lambda \in T^*\mathbb{M}$. It can be shown that abnormal extremals are given by in-place rotations $\dot{x} = \dot{y} = 0$. Application of PMP in normal case gives the expression of extremal control $u = h_2$ and leads to the Hamiltonian system

$$\begin{cases} \dot{x} = \sqrt{1-h_2^2}\cos\theta, \\ \dot{y} = \sqrt{1-h_2^2}\sin\theta, \\ \dot{\theta} = h_2, \end{cases} \qquad \begin{cases} \dot{h}_1 = -h_2 h_3, \\ \dot{h}_2 = \sqrt{1-h_2^2}h_3, \\ \dot{h}_3 = h_2 h_1, \end{cases} \tag{4}$$

with the (maximized) Hamiltonian

$$H = 1 = \begin{cases} |h_2|, & \text{for } h_1 \leq 0, \\ \sqrt{h_1^2 + h_2^2}, & \text{for } h_1 > 0, \end{cases} \tag{5}$$

The subsystem for state variables x, y, θ is called the *horizontal* part, and the subsystem for conjugate variables h_1, h_2, h_3 is called the *vertical* part of the Hamiltonian system. An extremal control is determined by a solution of the vertical part, while an extremal trajectory is a solution to the horizontal part.

The vertical part has the first integrals: the Hamiltonian H and the Casimir

$$E = h_1^2 + h_3^2. \tag{6}$$

Remark 1. Casimir functions are universal conservation laws on Lie groups. Connected joint level surfaces of all Casimir functions are coadjoint orbits (see [15]).

In Fig. 3 we show variants of the mutual arrangement of the level surface of the Hamiltonian $H = 1$, which consists of two half-planes glued with half of the cylinder, and the level surface of the Casimir $E \geq 0$, which is a cylinder. In Fig. 4 we show the phase portrait on the surface $H = 1$.

Depending on the sign of h_1, we have two different dynamics. When h_1 switches its sign, the dynamics switches from one to another. We denote by $t_0 \in \{t_0^0 = 0, t_0^1, t_0^2, \ldots\}$ the instance of time when the switching occurs. Note, that at the instances t_0 the extremal trajectory $(x(t), y(t), \theta(t))$ intersects the so-called 'cusp-surface' in SE(2), analytically computed and analysed in [16].

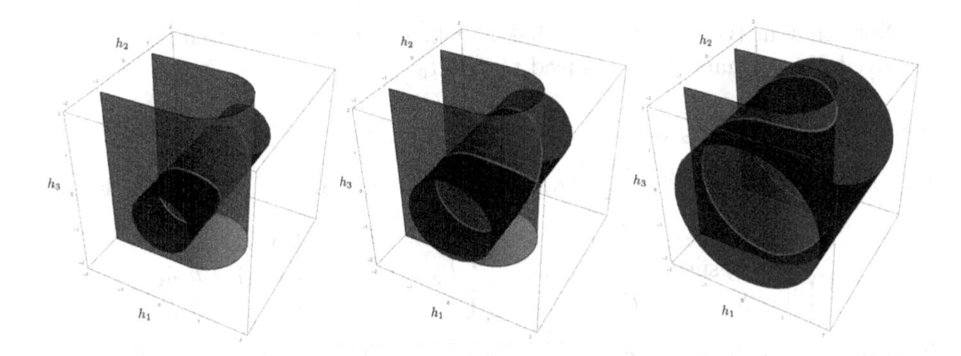

Fig. 3. Level surfaces of the Hamiltonian H (in green) and the Casimir E (in red). The intersection line is highlighted in yellow. **Left:** $E < 1$. **Center:** $E = 1$. **Right:** $E > 1$. (Color figure online)

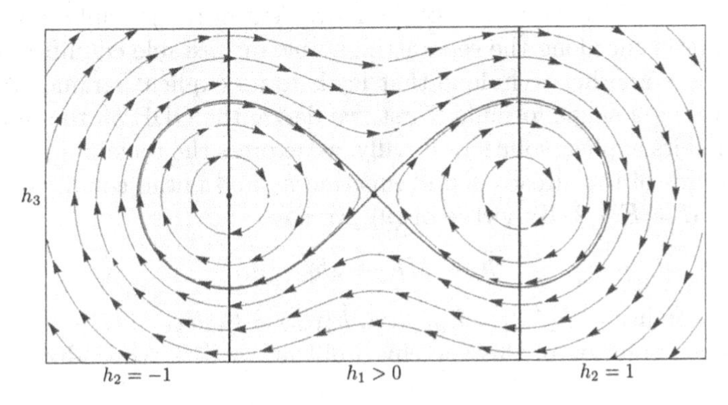

Fig. 4. Phase portrait on the level surface $H = 1$ of the Hamiltonian.

3.1 The Case $h_1 < 0$.

Due to (5), we have $h_2 = h_{20} = \pm 1$. Denote $s_2 = h_2$. We study the system

$$\begin{cases} \dot{x} = 0, & x(t_0) = x_0, \\ \dot{y} = 0, & y(t_0) = y_0, \\ \dot{\theta} = s_2, & \theta(t_0) = \theta_0, \end{cases} \qquad \begin{cases} \dot{h}_1 = -s_2 h_3, & h_1(t_0) = h_{10}, \\ \dot{h}_3 = s_2 h_1, & h_3(t_0) = h_{30}. \end{cases} \tag{7}$$

We immediately see that solutions to the horizontal part (extremal trajectories) are rotations around the fixed point (x_0, y_0) with constant speed $s_2 = \pm 1$.

Solutions to the vertical part are given by arcs of the circles. The motion is clockwise, when $s_2 = -1$, and counterclockwise, when $s_2 = 1$:

$$h_1(t) = h_{10} \cos(t - t_0) - s_2\, h_{30} \sin(t - t_0), \quad h_3(t) = h_{30} \cos(t - t_0) + s_2\, h_{10} \sin(t - t_0).$$

It remains to find the first instance of time $t_1 > t_0$, at which the dynamics switches. That is the moment when the condition $h_1 < 0$ ceases to be met:

$$t_1 - t_0 = \arg\left(-s_2 h_{30} - i h_{10}\right) \in (0, \pi]. \tag{8}$$

Note that in the case $t_0 > 0$, that is, when at least one switch has already occurred, the formula (8) is reduced to $t_1 - t_0 = \pi$.

3.2 The Case $h_1 > 0$

By (5) we have $h_1 = \sqrt{1 - h_2^2}$. The Hamiltonian system of PMP has the form

$$\begin{cases} \dot{x} = h_1 \cos\theta, & x(t_0) = x_0, \\ \dot{y} = h_1 \sin\theta, & y(t_0) = y_0, \\ \dot{\theta} = h_2, & \theta(t_0) = \theta_0, \end{cases} \qquad \begin{cases} \dot{h}_1 = -h_2 h_3, & h_1(t_0) = h_{10}, \\ \dot{h}_2 = h_1 h_3, & h_2(t_0) = h_{20}, \\ \dot{h}_3 = h_2 h_1, & h_3(t_0) = h_{30}. \end{cases} \tag{9}$$

This system is a model example in geometric control theory [13]. An explicit solution in Jacobi elliptic functions was obtained in [4], where the authors reduced the vertical part to the equation of mathematical pendulum. The solution is given by different formulas in different areas of the phase portrait. The specific form is determined by the nature of the movement of the pendulum: oscillation, rotation, movement along the separatrix, stable or unstable equilibrium.

We propose another technique that leads to an explicit parameterization of the solutions by a single formula. First, we derive the ODE on the function h_2. Then we find its explicit solution. Finally, we express the remaining components h_1, h_3 in terms of the already found function h_2 and initial conditions.

Denote $M = E - 2$. By virtue of (9), we have

$$\ddot{h}_2 + M h_2 + 2h_2^3 = 0, \tag{10}$$

with initial conditions $h_2(t_0) = h_{20} =: a$, $\dot{h}_2(t_0) = h_{10} h_{30} =: b$.

An explicit solution of this Cauchy problem, see [18, Appendix A], is given by

$$h_2(t) = s\mathrm{cn}\left(\tfrac{t-t_0}{k} + sF(\alpha, k), k\right), \quad k = \tfrac{1}{\sqrt{E}},$$
$$\alpha = \arg\left(s\,a + \boldsymbol{i}\,s\sqrt{1-a^2}\right) \in (-\pi, \pi], \quad s = -\mathrm{sign}(b),$$

where F denotes the elliptic integral of the first kind in the Legendre form, and cn denotes the elliptic cosine [17].

Next, we express h_1 and h_3 via h_2 and initial conditions.

Since $h_2 \in C(\mathbb{R})$ is bounded, there exists an integral

$$H_2(t) = \int_{t_0}^{t} h_2(\tau) \mathrm{d}\,\tau = s \arccos\left(\mathrm{dn}\left(\frac{t-t_0}{k} + sF(\alpha, k), k\right)\right)\Bigg|_{\tau=t_0}^{\tau=t},$$

where dn denotes the delta amplitude [17].

It can be shown that the Cauchy problem on (h_1, h_3) has a unique solution

$$h_1(t) = h_{10} \cos H_2(t) - h_{30} \sin H_2(t), \quad h_3(t) = h_{30} \cos H_2(t) + h_{10} \sin H_2(t). \tag{11}$$

The extremal trajectories are found by integration of the horizontal part:

$$x(t) = x_0 + \int_{t_0}^{t} h_1(\tau) \cos\theta(\tau)\,\mathrm{d}\tau, \quad y(t) = y_0 + \int_{t_0}^{t} h_1(\tau) \sin\theta(\tau)\,\mathrm{d}\tau, \quad \theta(t) = \theta_0 + H_2(t).$$

4 Extremal Controls and Trajectories

We summarize the previous section by formulating the following theorem

Theorem 1. *The extremal control $u(t)$ in problem (2)–(3) is determined by the parameters* $h_{20} \in [-1, 1]$, $h_{10} \in (-\infty, 0] \cup \left\{\sqrt{1 - h_{20}^2}\right\}$, $h_{30} \in \mathbb{R}$.
Let $s_1 = \mathrm{sign}(h_{10})$ and $\sigma = \frac{s_1+1}{2}$. The function $u(t)$ is defined on time intervals formed by splitting the ray $t \geq 0$ by instances $t_0 \in \{0 = t_0^0, t_0^1, t_0^2, t_0^3, \ldots\}$ as

$$u(t) = \begin{cases} u\left(t_0^{i-\sigma}\right) \in \{-1, 1\}, & \text{for } t \in [t_0^{i-\sigma+s_1}, t_0^{i-\sigma+s_1+1}), \\ \mathrm{scn}\left(\frac{t-t_0}{k} + sF(\alpha, k), k\right), & \text{for } t \in [t_0^{i-\sigma}, t_0^{i-\sigma+1}), \end{cases} \tag{12}$$

where $i \in \{2n - 1 \mid n \in \mathbb{N}\}$,

$$k = \frac{1}{\sqrt{h_{10}^2 + h_{30}^2}}, \quad s = -\mathrm{sign}(h_{30}), \quad \alpha = \arg\left(s\, h_{20} + \boldsymbol{i}\, s\sqrt{1 - h_{20}^2}\right).$$

The corresponding extremal trajectory has the form $\theta(t) = \int_0^t u(\tau)\, \mathrm{d}\tau$,

$$x(t) = \int_0^t \sqrt{1 - u^2(\tau)} \cos\theta(\tau)\, \mathrm{d}\tau, \quad y(t) = \int_0^t \sqrt{1 - u^2(\tau)} \sin\theta(\tau)\, \mathrm{d}\tau.$$

Proof relies on the expression of the extremal controls $u_1 = \sqrt{1 - u^2}$, $u_2 = h_2 = u$, which follows from maximum condition of PMP, see Sect. 3. The index i together with the parameters s_1 and σ specify the dynamics on the corresponding time interval. The dynamics switches, when h_1 changes its sign. Explicit formula (12) for the extremal control is obtained in Sect. 3.1 and Sect. 3.2. The extremal trajectory is obtained by integration of the horizontal part of (4).

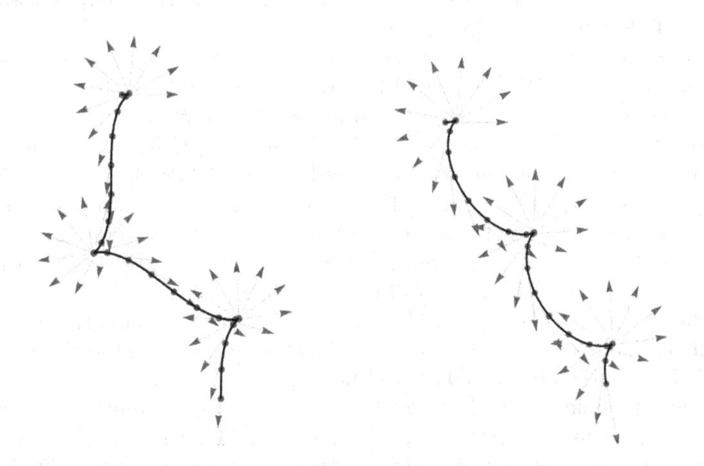

Fig. 5. Projection to the plane (x, y) of two different extremal trajectories. The gray arrow indicates the orientation angle θ at the instances of time $t \in \{0, 0.5, 1, \ldots, 20\}$. **Left:** $h_1^0 = 0.5$, $h_2^0 = \sqrt{3}/2$, $h_3^0 = 1$ **Right:** $h_1^0 = 0.5$, $h_2^0 = \sqrt{3}/2$, $h_3^0 = 0.7$.

By analyzing the solution, we note that there is a relation between the extremal trajectories of the Duits car and the sub-Riemannian geodesics in SE(2). Projection to the plane (x, y) of the trajectories in two models coincide, while the dynamics of the orientation angle θ differs: in Duits model, the angle θ uniformly increases/decreases by π radians at a cusp point. See Fig. 5. Note that an optimal motion of the Duits car can not have internal in-place rotations, see [5]. The in-place rotations may occur at the initial and the final time intervals.

References

1. Laumond, J.-P.: Feasible Trajectories for Mobile Robots with Kinematic and Environment Constraints. IAS (1986)
2. Dubins, L.E.: On curves of minimal length with a constraint on average curvature, and with prescribed initial and terminal positions and tangents. Am. J. Math. **79**(3), 497–516 (1957)
3. Reeds, J.A., Shepp, L.A.: Optimal paths for a car that goes both forwards and backwards. Pacific J. Math. **145**(2), 367–393 (1990)
4. Sachkov, Y.L.: Cut locus and optimal synthesis in the sub-Riemannian problem on the group of motions of a plane. In: ESAIM: COCV, vol. 17, pp. 293–321 (2011)
5. Duits, R., Meesters, S.P.L., Mirebeau, J.-M., Portegies, J.M.: Optimal paths for variants of the 2D and 3D Reeds-Shepp car with applications in image analysis. J. Math. Imaging Vis. **60**, 816–848 (2018)
6. Petitot, J.: The neurogeometry of pinwheels as a sub-Riemannian contact structure. J. Physiol. Paris **97**(2–3), 265–309 (2003)
7. Citti, G., Sarti, A.: A cortical based model of perceptual completion in the roto-translation space. JMIV **24**(3), 307–326 (2006)
8. Mashtakov, A.P., Ardentov, A.A., Sachkov, Y.L.: Parallel algorithm and software for image inpainting via sub-Riemannian minimizers on the group of roto translations. NMTMA **6**(1), 95–115 (2013)
9. Boscain, U., Gauthier, J., Prandi, D., Remizov, A.: Image reconstruction via non-isotropic diffusion in Dubins/Reed-Shepp-like control systems. In: 53rd IEEE Conference on Decision and Control, Los Angeles, CA, USA, pp. 4278–4283 (2014)
10. Bekkers, E.J., Duits, R., Mashtakov, A., Sanguinetti, G.R.: A PDE approach to data-driven sub-Riemannian geodesics in SE(2). SIAM JIS **8**(4), 2740–2770 (2015)
11. Scharpach, W.L.J.: Optimal paths for the Reeds-Shepp car with monotone spatial control and vessel tracking in medical image analysis. Ms. Thesis (2018)
12. Mirebeau, J.-M.: Anisotropic Fast-Marching on cartesian grids using lattice basis reduction. SIAM J. Num. Anal. **52**(4), 1573–1599 (2014)
13. Agrachev, A.A., Sachkov, Y.L.: Control theory from the geometric viewpoint. In: Agrachev, A.A. (ed.) Encyclopaedia of Mathematical Sciences. Springer, Berlin (2004). https://doi.org/10.1007/978-3-662-06404-7
14. Ardentov, A., Lokutsievskiy, L., Sachkov, Yu.L.: Explicit solutions for a series of classical optimization problems with 2-dimensional control via convex trigonometry. Optimization and Control (2020). https://arxiv.org/pdf/2004.10194.pdf
15. Laurent-Gengoux, C., Pichereau, A., Vanhaecke, P.: Poisson structures. In: Pichereau, A. (ed.) Grundlehren der mathematischen Wissenschaften, p. 347. Springer-Verlag, Berlin (2013). https://doi.org/10.1007/978-3-642-31090-4

16. Duits, R., Boscain, U., Rossi, F., Sachkov, Y.: Association fields via cuspless sub-Riemannian geodesics in SE(2). J. Math. Imaging Vis. **49**(2), 384–417 (2013). https://doi.org/10.1007/s10851-013-0475-y
17. Whittaker, E.T., Watson, G.N.: A Course of Modern Analysis. Cambridge University Press, Cambridge (1962)
18. M. Lakshmanan, S. Rajasekar, Nonlinear Dynamics: Integrability Chaos and Patterns. Advanced Texts in Physics. Springer-Verlag, Berlin p. 619 (2003). https://doi.org/10.1007/978-3-642-55688-3

Multi-shape Registration with Constrained Deformations

Rosa Kowalewski[1(✉)] and Barbara Gris[2(✉)]

[1] University of Bath, Bath BA2 7AY, UK
rak53@bath.ac.uk
[2] Laboratoire Jacques-Louis Lions, Sorbonne Université, 75005 Paris, France
barbara.gris@sorbonne-universite.fr

Abstract. Based on a Sub-Riemannian framework, deformation modules provide a way of building large diffeomorphic deformations satisfying a given geometrical structure. This allows to incorporate prior knowledge about object deformations into the model as a means of regularisation [10]. However, diffeomorphic deformations can lead to deceiving results if the deformed object is composed of several shapes which are close to each other but require drastically different deformations. For the related Large Deformation Diffemorphic Metric Mapping, which yields unstructured deformations, this issue was addressed in [2] introducing object boundary constraints. We develop a new registration problem, marrying the two frameworks to allow for different constrained deformations in different coupled shapes.

Keywords: Shape registration · Large deformations · Sub-Riemannian geometry

1 Introduction

The improvement of medical imaging techniques increase the need for automated tools to analyse the generated data. Matching two shapes (curves, images) in order to understand the differences or the evolution between two observations is often a key step. Large deformations, defined as flows of time-dependent vector fields [5,9], allow to perform efficient shape registration [4,6,12]. In some cases, in order to analyse the estimated deformation, it is required to only consider a subset of suitable deformations [10,14–16]. In addition, the regularity of such deformations often prevents the simultaneous registration of several shapes close to each other, but requiring different types of motions, as it appears in medical imaging (with several organs for instance, see [13]). We present in this article a *modular multi-shape registration framework*, combining the deformation module framework [10] with the multi-shape registration developed in [2].

Rosa Kowalewski is supported by a scholarship from the EPSRC Centre for Doctoral Training in Statistical Applied Mathematics at Bath (SAMBa), under the project EP/S022945/1.

F. Nielsen and F. Barbaresco (Eds.): GSI 2021, LNCS 12829, pp. 82–90, 2021.
https://doi.org/10.1007/978-3-030-80209-7_10

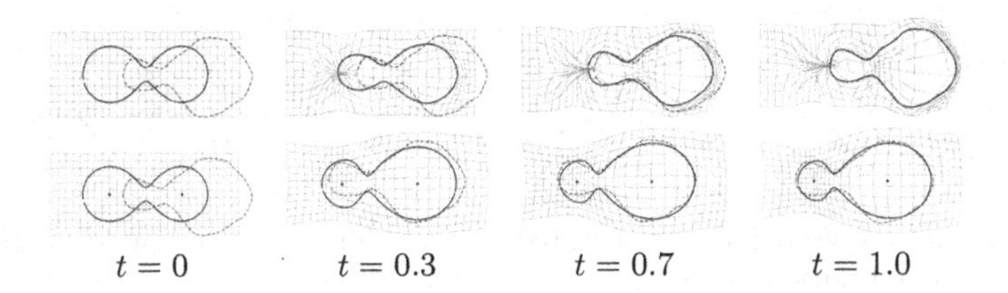

Fig. 1. Registration of unparametrized curves using LDDMM (top) and deformation modules (bottom). The red dotted curve is the target, the black curve the evolving shape. The two points are the centers of the local scaling. (Color figure online)

2 Background

2.1 Structured Large Deformations

A strategy to restrict large deformations to an appropriate subset is to incorporate a structure in the deformation model via chosen field generators [10,14,16,17] or constraints [1]. We will use a simplified version of the deformation module framework [10] which allows to build an appropriate vocabulary of interpretable fields. A deformation module is defined as a quadruple $(\mathcal{O}, H, \zeta, c)$. The field generator $\zeta : \mathcal{O} \times H \mapsto C_0^l(\Omega, \mathbb{R}^d)$, allows to parameterise structured vector fields such as a local scaling, or physically motivated deformation. The *geometrical descriptor* $q \in \mathcal{O}$ contains the 'geometrical information' of the generated vector field, such as its location. In this article we will assume $\mathcal{O}$ is a space of landmarks of $\mathbb{R}^d$, which includes the case of discretised curves. The *control* $h \in H$ selects a particular field amongst the allowed ones. Finally, the *cost* function $c : \mathcal{O} \times H \mapsto \mathbb{R}^+$ associates a cost to each couple in $\mathcal{O} \times H$. Under some conditions (see Sect. 3), the trajectory $\zeta_{q_t}(h_t)$ can be integrated via the flow equation $\dot{\varphi}_t = \zeta_{q_t}(h_t) \circ \varphi_t$, $\varphi_{t=0} = Id$, so that the structure imposed on vector fields can be transferred to a structure imposed on large deformations [10]. The evolution of q is then given by $\dot{q}_t = \zeta_{q_t}(h_t)(q_t)$, where $v(q) = (v(x_1), \dots, v(x_p))$ if q consists of the points $x_i \in \mathbb{R}^d$. The framework comes with the possibility to combine several deformation modules $\mathcal{M}^i = (\mathcal{O}^i, H^i, \zeta^i, c^i)$, $1 \leq i \leq N$, into the *compound module* $C(\mathcal{M}^1, \dots, \mathcal{M}^N) = (\mathcal{O}, H, \zeta, c)$ defined by $\mathcal{O} = \prod_i \mathcal{O}^i$, $H = \prod_i H^i$, $\zeta\colon (q, h) \mapsto \sum_i \zeta^i_{q^i}(h^i)$ and $c\colon (q, h) \mapsto \sum_i c^i_{q^i}(h^i)$ with $q = (q^1, \dots, q^N)$ and $h = (h^1, \dots, h^N)$. This combination enables to define a complex deformation structure as the superimposition of simple ones. Note that in modular large deformations generated by a compound module, each component q^i of the geometrical descriptor is transported by $\sum_i \zeta^i_{q^i}(h^i)$.

Figure 1 shows the result of the registrations of two unparametrized curves[1] (modelled as varifolds [8]) with unstructured large deformations (with the Large Deformation Diffeomorphic Metric Mapping (LDDMM) framework [5]), and

[1] Data courtesy of Alain Trouvé.

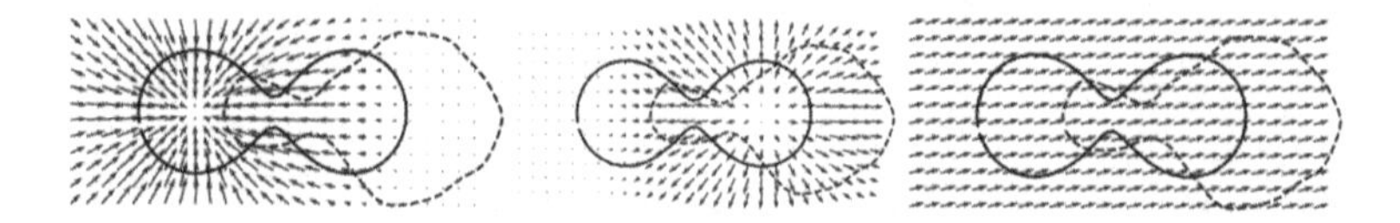

Fig. 2. Examples of vector fields generated by the scaling (left and middle) and translation (right) deformation modules.

modular deformations. The deformation module used here is a compound module generating two scalings (one for each lobe) and a large scale local translation (translating the whole curve)[2]. Figure 2 shows examples of vector fields generated by these modules. Even though both models give good curve registration, the prior in the structure ensures that the modular deformation is composed of uniform scalings. In addition, each component of the global deformation can be studied separately [7,18].

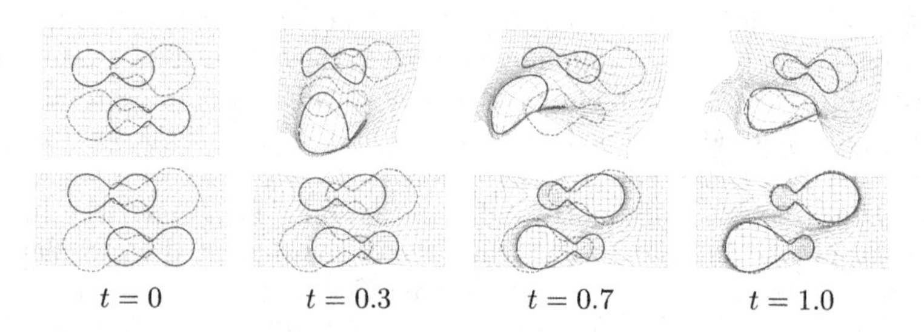

Fig. 3. Simultaneous registration using compound modules with single-shape (top) and multi-shape (bottom) frameworks. The evolving shape is in black, the target in red. (Color figure online)

2.2 Multi-shape Registration

Deformation modules generate smooth deformations, modelling shapes as one homogeneous medium. Studying several shapes using one diffeomorphism (*i.e.* with a single-shape framework) leads to huge interactions between the shapes as they are moving close. Performing a simultaneous registration using a compound module leads to unsatisfying results due to the close location of shapes, as in Fig. 3.

To solve this problem, it would be desirable to have the deformation module for each shape only deform the inside of the curve, without influencing the deformation outside, while maintaining consistency between deformations. The multi-shape registration framework introduced in [2] addresses this problem for LDDMM [5], by considering deformations for each subshape and linking them

[2] For a more detailed description of how these modules are built, we refer to [10].

at the boundaries. In general, N source and corresponding target shapes q_S^i and q_T^i, $i = 1, \dots, N$ are considered. Each shape q^i is associated with an open set U^i of $\mathbb{R}^d$ such that $q^i \subset U^i$, $\cup \bar{U^i} = \mathbb{R}^d$ and $U^i \cap U^j = \emptyset$ if $i \neq j$. For curves, for example, the shape q^i represents the boundary of U^i. The total deformation is defined by $\varphi : x \in \mathbb{R}^d \mapsto \varphi^i(x)$ if $x \in U^i$.

In order to ensure the consistency at boundaries ∂U^i, such as continuity or prevention of overlapping deformations, authors introduce a continuous constraints operator $C\colon \mathcal{O} \to L(V, Y)$ where $V = \prod_i V_i$, V_i is a space of vector fields for each i, Y is a Banach space and $L(V, Y)$ is the set of continuous linear operators between V and Y. Furthermore, consider a continuous function $\mathcal{D}\colon (\prod_i \mathcal{O}^i)^2 \to \mathbb{R}_+$, measuring the similarity of shapes. The multi-shape registration problem [2] reads

$$\min_{v \in L^2([0,1],V)} \frac{1}{2} \sum_{i=1}^{N} \int_0^1 \|v_t^i\|_2^2 \mathrm{d}t + \mathcal{D}(q_{t=1}^i, q_T^i)$$

$$\text{s.t.} \qquad \dot{q}_t^i = v_t^i(q_t^i), \quad q_0 = q_S, \quad \text{and} \quad C_{q_t} v_t = 0.$$

Two different types of constraints C_q have been introduced in [2]: identity constraints (modelling shapes 'stitched' to each other at the boundaries), and sliding constraints (allowing the shapes to slide tangential to the boundary).

3 Modular Multi-shape Registration Problem

In order to address the case where the shapes under study are composed of several subshapes whose deformations should satisfy a certain structure, we design here a modular multi-shape registration framework. Let us consider N shapes and N deformation modules. Following [2], each of these shapes will be displaced by a modular large deformation generated by the associated module. We impose boundary constraints to ensure consistent combination of the deformations. The modular multi-shape registration problem is stated as follows.

Problem 1. Let $\mathcal{M}^i = (\mathcal{O}^i, H^i, \zeta^i, c^i)$, $i = 1, ..., N$, be N deformation modules and $C\colon \prod_i \mathcal{O}^i \mapsto L(V, Y)$ be a continuous constraints operator with $V = \prod_i V_i$, V_i RKHS of vector fields, and Y a finite dimensional vector space. Let q_S and q_T in $\prod_i \mathcal{O}^i$ and let $\mathcal{D}\colon (\prod_i \mathcal{O}^i)^2 \to \mathbb{R}_+$ be continuous. Minimise the functional

$$\mathcal{J}(q, h) = \frac{1}{2} \sum_{i=1}^{N} \int_0^1 c_{q_t^i}^i(h_t^i) \mathrm{d}t + \mathcal{D}(q_{t=1}, q_T) \tag{1}$$

with respect to $q = (q^1, \dots, q^N)$ and $h = (h^1, \dots, h^N)$, where $q_{t=0} = q_S$, and for almost every $t \in [0, 1]$ and for all i, $C_{q_t} \zeta_{q_t}(h_t) = 0$ and $\dot{q}_t^i = \zeta_{q_t^i}^i(h_t^i)(q_t^i)$.

To formalize it, we introduce the *external combination* of deformation modules.

Definition 1 (External Combination of Deformation Modules). *Let* $\mathcal{M}^i = (\mathcal{O}^i, H^i, \zeta^i, c^i)$, $i = 1, ..., N$ *be* C^k*-deformation modules on* $\mathbb{R}^d$ *of order* l. *Then their* external combination $\mathrm{M}(\mathcal{M}^1, \ldots, \mathcal{M}^N)$ *is the quintuple* $(\mathcal{O}, H, \mathrm{F}, \mathrm{X}, c)$ *where* $\mathcal{O} = \prod_i \mathcal{O}^i$, $H = \prod_i H^i$, $\mathrm{F}\colon (q,h) \in \mathcal{O} \times H \mapsto (\zeta^1(q^1,h^1), ..., \zeta^N(q^N,h^N)) \in \prod_i C_0^{l^i}(\mathbb{R}^{d_i}, \mathbb{R}^{d_i})$, $\mathrm{F}\colon (q,h) \in \mathcal{O} \times H \mapsto (\zeta^1(q^1,h^1), ..., \zeta^N(q^N,h^N)) \in \prod_i C_0^{l^i}(\mathbb{R}^{d_i}, \mathbb{R}^{d_i})$, $\mathrm{X}\colon (q,v) \in \mathcal{O} \times \prod_i C_0^{l^i}(\mathbb{R}^{d_i}, \mathbb{R}^{d_i}) \to (v^1(q^1), \ldots, v^N(q^N)) \in \prod_i T\mathcal{O}^i$, *and cost* $c\colon (q,h) \in \mathcal{O} \times H \to c_q(h) = \sum_i c^i_{q^i}(h^i) \in \mathbb{R}_+$, *where* $q = (q^1, \ldots, q^N)$, $h = (h^1, \ldots, h^N)$ *and* $v = (v^1, \ldots, v^N)$.

Remark 1. Unlike the compound module presented in Sect. 2, the external combination does not result in a deformation module.

In the following, we will assume that all the deformation modules $(\mathcal{O}, H, \zeta, c)$ satisfy the *Uniform Embedding Condition* (UEC [10]) for a RKHS V continuously embedded in $C_0^1(\mathbb{R}^d)$, *i.e.* that there exists $\beta > 0$ such that for all $(q,h) \in \mathcal{O} \times H$, $\zeta_q(h)$ is in V and $|\zeta_q(h)|_V^2 \leq \beta c_q(h)$. An easy way to ensure it is to choose $c_q(h) = (1/\beta)|\zeta_q(h)|_V^2 + f(q,h)$ (with $f : \mathcal{O} \times H \longrightarrow \mathbb{R}^+$ smooth), which is the case in the numerical examples in Sect. 4. In order to minimise equation (1), we first need to specify the trajectories (q,h) that will be considered.

Definition 2 (Constrained Controlled Path of Finite Energy). *Let* $\mathrm{M}(\mathcal{M}^1, \ldots, \mathcal{M}^N) = (\mathcal{O}, H, \mathrm{F}, \mathrm{X}, c)$ *be an external combination of the modules* $\mathcal{M}^i$ *and let* $C\colon \prod_i \mathcal{O}^i \mapsto L(V,Y)$ *be a continuous constraints operator with* $V = \prod_i V_i$, V_i *RKHS of vector fields, and* Y *a finite dimensional vector space. We denote by* Ω *the set of measurable curves* $t \mapsto (q_t, h_t) \in \mathcal{O} \times H$ *where* q *is absolutely continuous such that for almost every* $t \in [0,1]$, $\dot{q}_t = F_{q_t}(h_t)(q_t)$, $C_{q_t}\zeta_{q_t}(h_t) = 0$ *and* $E(q,h) := \int_0^1 c_{q_t}(h_t)\mathrm{d}t < \infty$.

Remark 2. If $t \mapsto (q_t, h_t)$ is a constrained controlled path of finite energy of $\mathrm{M}(\mathcal{M}^1, \ldots, \mathcal{M}^n)$, each component q^i is displaced by the action of $\zeta^i_{q^i}(h^i)$ only, instead of the sum of all vector fields like in the compound module (Sect. 2).

If (q,h) is in Ω, for each (q^i, h^i) a modular large deformation φ^i of $\mathbb{R}^d$ as the flow of $\zeta^i_{q^i}(h^i)$ can be built [10,11]. Then (see [3] and Sect. 2.2), we can combine all φ^i into one deformation φ. If the constraints operator C prevents overlapping between sets $\varphi^i(U^i)$ and $\varphi^j(U^j)$, such as the aforementioned identity or sliding constraints, then φ defines a bijective deformation (except potentially on the boundaries) which respects the topology and the desired deformation structure within each set U^i. We show now that Eq. (1) has a minimiser on Ω.

Proposition 1. *Let* $V_1, \ldots, V_N$ *be a* N *RKHS continuously embedded in* $C_0^2(\mathbb{R}^d)$, *and let for* $i = 1, \ldots, N$, $\mathcal{M}^i = (\mathcal{O}^i, H^i, \zeta^i, \xi^i, c^i)$ *be deformation modules satisfying the UEC for* V_i. *Let* $\mathcal{D}\colon \mathcal{O} \times \mathcal{O} \mapsto \mathbb{R}^+$ *be continuous. Let* q_S *and* $q_T \in \mathcal{O} := \prod_i \mathcal{O}^i$ *and* $C\colon \mathcal{O} \times V \mapsto Y$ *be a constraints operator* *with Y a finite dimensional vector space. Then the minimiser of the energy $J(q,h) = \int_0^1 c_{q_t}(h_t)\mathrm{d}t + \mathcal{D}(q_{t=1}, q_T)$ on $\{(q,h) \in \Omega \mid q_{t=0} = q_S\}$ exists.*

We sketch the proof here and refer to [11] for a detailed version.

Proof. Let $(q^j, h^j)_j = ((q^{(i,j)}, h^{(i,j)})_{1\le i\le N})_j$ be a minimising sequence in Ω. Up to successive extractions, we can suppose that for each i, $E^i((q^{(i,j)}, h^{(i,j)})) = \int_0^1 c^i_{q^i}(h^i)$ converges to $E^{i,\infty}$. It can be shown similarly to [10] that $h^{(i,j)} \rightharpoonup h^{(i,\infty)}$ in $L^2([0,1], H^i)$ and $q^{(i,j)} \longrightarrow q^{(i,\infty)}$ uniformly on $[0,1]$ with $(q^{(i,\infty)}, h^{(i,\infty)})$ absolutely continuous such that for almost every t, $\dot{q}^i_t = \zeta^i_{q^i_t}(h_t)(q^i_t)$ and $E^i(q^{(i,\infty)}, h^{(i,\infty)}) \le E^{i,\infty}$. Then with $(q^\infty, h^\infty) = (q^{(i,\infty)}, h^{(i,\infty)})_{1\le j\le N} \in \Omega$, $\inf\{J(q,h) \mid (q,h) \in \Omega\} = \lim J(q^j, h^j) = J(q^\infty, h^\infty)$. In addition, it is shown in [10] that $\mathrm{F}_{q^j}(h^j)$ converges weakly to $\mathrm{F}_{q^\infty}(h^\infty)$, so (see [3]), $C_{q^\infty_t} \circ \mathrm{F}_{q^\infty_t}(h^\infty_t) = 0$ for almost every t in $[0,1]$ which concludes the proof.

Let us define the Hamiltonian function $\mathcal{H}\colon \mathcal{O} \times \mathrm{T}_q\mathcal{O}^* \times H \times Y \to \mathbb{R}$ by

$$\mathcal{H}(q,p,h,\lambda) := (p|\mathrm{F}_q(h)(q))_{\mathrm{T}_q^*\mathcal{O},\mathrm{T}_q\mathcal{O}} - \frac{1}{2}(Z_q h|h)_{H^*,H} - (\lambda|C_q\mathrm{F}_q(h))_{Y^*,Y}\,,$$

where $Z_q\colon H \to H^*$ is the invertible symmetric operator such that $c_q(h) = (Z_q h|h)_{H^*,H}$ ([10]). Similarly to [3], it can be shown that minimising trajectories (q,h) are such that there exists absolutely continuous $p : t \in [0,1] \mapsto T^*_{q_t}\mathcal{O}$ and $\lambda : t \mapsto F$ such that

$$\dot{q} = \frac{\partial}{\partial p}\mathcal{H}(q,p,h,\lambda) \qquad \dot{p} = -\frac{\partial}{\partial q}\mathcal{H}(q,p,h,\lambda)$$
$$\frac{\partial}{\partial h}\mathcal{H}(q,p,h,\lambda) = 0 \qquad \frac{\partial}{\partial \lambda}\mathcal{H}(q,p,h,\lambda) = 0\,.$$

The last two equations can be solved, so that for minimising trajectories $h = Z_q^{-1}(\rho_q^* p - (C_q\mathrm{F}_q)^*\lambda)$ and $\lambda = (C_q\mathrm{F}_q Z_q^{-1}(C_q\mathrm{F}_q)^*)^{-1} C_q\mathrm{F}_q Z_q^{-1}\rho_q^* p$ with $\rho_q : h \in H \mapsto \mathrm{F}_q(h)(q)$ and assuming $C_q\mathrm{F}_q Z_q^{-1}(C_q\mathrm{F}_q)^*$ is invertible. The previous system of equations then has a unique solution associated to each initial condition (q_0, p_0). In order to minimise equation (1), we optimise it with respect to the initial momentum p_0.

Proposition 2 [3,11]. *Let $C_q\zeta_q\colon H \to Y$ be surjective. Then the operator $C_q\mathrm{F}_q Z_q^{-1}(C_q\mathrm{F}_q)^*$ is invertible.*

Remark 3. The surjectivity of $C_q\mathrm{F}_q$ is in general not guaranteed but is easily obtained in practice (see Sect. 4).

Remark 4. Compared to the frameworks of (single-shape) structured deformations [7,10] and multi-shape (unstructured) large deformations [2], it is necessary to invert the operator $C_q\mathrm{F}_q Z_q^{-1}(C_q\mathrm{F}_q)^*$ at each time to compute the final deformation. This increases drastically the time-complexity and the memory print.

4 Numerical Results

In the practical study of multi-shapes, we encounter two main cases where a multi-shape setting is needed. For example, to study the motion of lungs and abdominal organs during the breath cycle, it is natural to model the organs as separate shapes embedded in a background which represents the surrounding tissue. On the other hand, if two organs are in contact, they can be modelled as several subshapes sharing a boundary. Here the boundary constraints are expressed only on the discretisation points (the convergence of this approximation is studied in the case of unstructured deformations in [2]). All the results here were generated using the library IMODAL [7].

Shapes in a Background. We consider again the simultaneous registration of two peanut-shaped curves. They define three open sets (two bounded ones delimited by the curves, and the background) so we need to define three deformation modules. For each curve q^i, $i = 1, 2$, we choose deformation modules modelling scalings of the left and right half, and a translation of the whole shape. We define the background shape as $q^3 = (q^{31}, q^{32})$, where $q^{31} = q^1$ and $q^{32} = q^2$, and the third deformation module is chosen to generate a sum of local translations carried by q^3. We impose identity constraints on corresponding boundary points of each q^i and q^{3i}, with the constraints operator $C_q(v) = (v^1(q^1) - v^3(q^{31}), v^2(q^2) - v^3(q^{32}))$ with $v = (v^1, v^2, v^3)$. The unstructured background deformation module ensures that the operator $C_q\mathrm{F}_q$ is surjective. Figure 3 shows the result of the multi-shape registration. The two curves move closely, without influencing each other. Unlike the compound module, the deformations inside the curves resemble the deformation we obtain from the registration of a single shape, see Fig. 1.

Subshapes with Shared Boundary. We consider two tree-shaped curves, modelled as varifolds [8], where the trunk of the source is thinner than the target's while the crown of the source is larger (see Fig. 4). Such differences require a multi-shape registration where a sliding is allowed at the boundary between the trunk and the crown. We place ourselves in the typical use case of the registration of a synthetic segmented template to a new unsegmented subject. For each subshape (trunk and crown) we define a compound deformation module as the combination of 4 modules generating respectively a uniform horizontal scaling, a uniform vertical scaling, a global translation and a sum of local translations centered at the curve and boundary points (with a high penalty so that the deformation prior is only changed slightly). The last one ensures the surjectivity of the operator $C_q\mathrm{F}_q$. Figure 4 shows the registration using the modular multi-shape framework. The interest of the sliding constraints can be seen in the grid deformation. In addition, similar to [7,18] the trajectory of the controls can then be analysed to understand the difference between the two shapes.

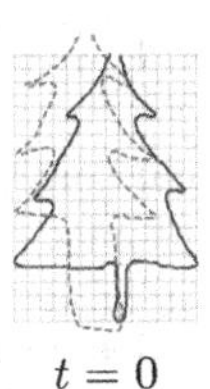

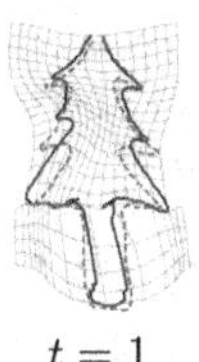

$t = 0$ $t = 0.3$ $t = 0.7$ $t = 1$

Fig. 4. Registration of unparametrized curves with the modular multi-shape framework. The deformed (segmented) source is in black, the (unsegmented) target in red. (Color figure online)

5 Conclusion

We have provided a framework and synthetic examples for using a vocabulary of structured deformations for multiple shape registration. In future work we will optimise our use of the IMODAL tools to improve the computation time and memory print. There are promising perspectives for the application on, e.g., medical images with a known segmentation, in order to incorporate physical properties in different tissues and analyse the resulting registration.

References

1. Arguillere, S., Miller, M.I., Younes, L.: Diffeomorphic surface registration with atrophy constraints. SIAM J. Imag. Sci. **9**(3), 975–1003 (2016)
2. Arguillère, S., Trélat, E., Trouvé, A., Younes, L.: Multiple shape registration using constrained optimal control. arXiv preprint arXiv:1503.00758 (2015)
3. Arguillere, S., Trélat, E., Trouvé, A., Younes, L.: Shape deformation analysis from the optimal control viewpoint. J. de mathématiques pures et appliquées **104**(1), 139–178 (2015)
4. Ashburner, J.: Computational anatomy with the SPM software. Magn. Resonance Imag. **27**(8), 1163–1174 (2009)
5. Beg, M.F., Miller, M.I., Trouvé, A., Younes, L.: Computing large deformation metric mappings via geodesic flows of diffeomorphisms. Int. J. Comput. Vis. **61**(2), 139–157 (2005)
6. Bône, A., Louis, M., Martin, B., Durrleman, S.: Deformetrica 4: an open-source software for statistical shape analysis. In: Reuter, M., Wachinger, C., Lombaert, H., Paniagua, B., Lüthi, M., Egger, B. (eds.) ShapeMI 2018. LNCS, vol. 11167, pp. 3–13. Springer, Cham (2018). https://doi.org/10.1007/978-3-030-04747-4_1
7. Lacroix, L., Charlier, B., Trouvé, A., Gris, B.: Imodal: creating learnable user-defined deformation models. In: CVPR (2021)
8. Charon, N., Trouvé, A.: The varifold representation of nonoriented shapes for diffeomorphic registration. SIAM J. Imag. Sci. **6**(4), 2547–2580 (2013)
9. Christensen, G.E., Rabbitt, R.D., Miller, M.I.: Deformable templates using large deformation kinematics. IEEE Trans. Image Process. **5**(10) (1996)
10. Gris, B., Durrleman, S., Trouvé, A.: A sub-riemannian modular framework for diffeomorphism-based analysis of shape ensembles. SIAM J. Imag. Sci. **11**(1), 802–833 (2018)

11. Kowalewski, R.: Modular Multi-Shape Registration. Master's thesis, Universität zu Lübeck (2019)
12. Miller, M.I., Trouvé, A., Younes, L.: Hamiltonian systems and optimal control in computational anatomy: 100 years since d'arcy thompson. Ann. Rev. Biomed. Eng. **17**, 447–509 (2015)
13. Risser, L., Vialard, F.X., Baluwala, H.Y., Schnabel, J.A.: Piecewise-diffeomorphic image registration: application to the motion estimation between 3D CT lung images with sliding conditions. Med. Image Anal. **17**(2), 182–193 (2013)
14. Seiler, C., Pennec, X., Reyes, M.: Capturing the multiscale anatomical shape variability with polyaffine transformation trees. Med. Image Anal. **16**(7) (2012)
15. Sommer, S., Nielsen, M., Darkner, S., Pennec, X.: Higher-order momentum distributions and locally affine lDDMM registration. SIAM J. Imag. Sci. **6**(1), 341–367 (2013)
16. Srivastava, A., Saini, S., Ding, Z., Grenander, U.: Maximum-likelihood estimation of biological growth variables. In: International Workshop on Energy Minimization Methods in Computer Vision and Pattern Recognition
17. Younes, L.: Constrained diffeomorphic shape evolution. Found. Comput. Math. **12**(3), 295–325 (2012)
18. Younes, L., Gris, B., Trouvé, A.: Sub-Riemannian methods in shape analysis. In: Handbook of Variational Methods for Nonlinear Geometric Data, pp. 463–492 (2020)

Shapes Spaces

Geodesics and Curvature of the Quotient-Affine Metrics on Full-Rank Correlation Matrices

Yann Thanwerdas(✉) and Xavier Pennec

Université Côte d'Azur, and Inria, Epione Project Team, Sophia-Antipolis, France
{yann.thanwerdas,xavier.pennec}@inria.fr

Abstract. Correlation matrices are used in many domains of neurosciences such as fMRI, EEG, MEG. However, statistical analyses often rely on embeddings into a Euclidean space or into Symmetric Positive Definite matrices which do not provide intrinsic tools. The quotient-affine metric was recently introduced as the quotient of the affine-invariant metric on SPD matrices by the action of diagonal matrices. In this work, we provide most of the fundamental Riemannian operations of the quotient-affine metric: the expression of the metric itself, the geodesics with initial tangent vector, the Levi-Civita connection and the curvature.

Keywords: Riemannian geometry · Quotient manifold · Quotient-affine metric · SPD matrices · Correlation matrices

1 Introduction

Correlation matrices are used in many domains with time series data such as functional brain connectivity in functional MRI, electroencephalography (EEG) or magnetoencephalography (MEG) signals. Full-rank correlation matrices form a strict sub-manifold of the cone of Symmetric Positive Definite (SPD) matrices sometimes called the (open) elliptope [1]. However, very few geometric tools were defined for intrinsic computations with correlation matrices. For example, [2] rely on a surjection from a product of n spheres of dimension $n-1$ onto the space of correlation matrices in order to sample valid correlation matrices for financial applications: a point in the former space can be represented by an $n \times n$ matrix A with normed rows and therefore encodes a correlation matrix $AA^\top$. Since low-rank matrices have null measure, one gets a full-rank correlation matrix almost surely. More recently, the open elliptope was endowed with the Hilbert projective geometry [3] which relies on its convexity.

Since there exist efficient tools on SPD matrices (affine-invariant/Fisher-Rao metric, log-Euclidean metric...), correlation matrices are often treated as SPD matrices [4]. Nevertheless, these extrinsic tools do not respect the geometry of correlation matrices. Moreover, most of these tools on SPD matrices are invariant under orthogonal transformations which is not compatible with correlation

F. Nielsen and F. Barbaresco (Eds.): GSI 2021, LNCS 12829, pp. 93–102, 2021.
https://doi.org/10.1007/978-3-030-80209-7_11

matrices. The elliptope is not even stable by the orthogonal action. Hence using the tools designed for SPD matrices may not be relevant for correlation matrices. One could restrict a Riemannian metric from the cone of SPD matrices to the open elliptope to define intrinsic tools by embedding. To the best of our knowledge, this hasn't been studied in depth. However, a key property of correlation matrices in applications with respect to covariance matrices is that the scale is renormalized independently for each axis. This physically corresponds to an invariance under a group action, namely the action of diagonal matrices on SPD matrices. The previously cited structures do not rely on this physical reality. This is why we are more interested in the recently introduced quotient-affine metric which corresponds to the quotient of the affine-invariant metric on SPD matrices by the action of diagonal matrices [5,6].

In this work, we investigate the geometry of this quotient-affine metric and we contribute additional tools to compute on this manifold. Based on the formalization of the quotient manifold with vertical and horizontal distributions, we compute in closed form some fundamental Riemannian operations of the quotient-affine metrics, notably the exponential map and the curvature. In Sect. 2, we recall how quotient-affine metrics are introduced, as well as the basics on quotient manifolds. In Sect. 3, we provide the following fundamental quotient and Riemannian operations of quotient-affine metrics: the vertical and horizontal projections, the metric, the exponential map, the Levi-Civita connection and the sectional curvature. This opens the way to many practical algorithms on the open elliptope for different applications. Considering SPD matrices as the Cartesian product of positive diagonal matrices and full-rank correlation matrices, it also allows to define new Riemannian metrics which preserve the quotient-affine geometry on correlation matrices. Thus in Sect. 4, we illustrate the quotient-affine metric in dimension 2 by coupling it with the diagonal power-Euclidean metrics $g_D^{\mathrm{E}(p)}(\Delta,\Delta) = \mathrm{tr}(D^{2(p-1)}\Delta^2)$ for $p \in \{-1,0,1,2\}$, where D is positive diagonal and Δ diagonal, and then by comparing it with the affine-invariant and the log-Euclidean metrics on SPD matrices.

2 Quotient-Affine Metrics

2.1 The Quotient Manifold of Full-Rank Correlation Matrices

The group of positive diagonal matrices $\mathrm{Diag}^+(n)$ acts on the manifold of SPD matrices $\mathrm{Sym}^+(n)$ via the congruence action $(\Sigma, D) \in \mathrm{Sym}^+(n) \times \mathrm{Diag}^+(n) \longmapsto D\Sigma D \in \mathrm{Sym}^+(n)$. The manifold of full-rank correlation matrices $\mathrm{Cor}^+(n)$ can be seen as the quotient manifold $\mathrm{Sym}^+(n)/\mathrm{Diag}^+(n)$ via the invariant submersion π which computes the correlation matrix from a covariance matrix, $\pi : \Sigma \in \mathrm{Sym}^+(n) \longmapsto \mathrm{Diag}(\Sigma)^{-1/2}\,\Sigma\,\mathrm{Diag}(\Sigma)^{-1/2} \in \mathrm{Cor}^+(n)$, where $\mathrm{Diag}(\Sigma) = \mathrm{diag}(\Sigma_{11}, ..., \Sigma_{nn})$. Hence, any Riemannian metric G on $\mathrm{Sym}^+(n)$ which is invariant under $\mathrm{Diag}^+(n)$ induces a quotient metric g on $\mathrm{Cor}^+(n)$. The steps to define it are the following.

1. Vertical distribution. $\mathcal{V}_\Sigma = \ker d_\Sigma\pi$ for all $\Sigma \in \mathrm{Sym}^+(n)$.
2. Horizontal distribution. $\mathcal{H}_\Sigma := \mathcal{V}_\Sigma^\perp$ for all $\Sigma \in \mathrm{Sym}^+(n)$, where the orthogonality $\perp$ refers to the inner product G_Σ.
3. Horizontal lift. The linear map $d_\Sigma\pi$ restricted to the horizontal space $\mathcal{H}_\Sigma$ is a linear isomorphism onto the tangent space of full-rank correlation matrices $(d_\Sigma\pi)_{|\mathcal{H}_\Sigma} : \mathcal{H}_\Sigma \xrightarrow{\sim} T_{\pi(\Sigma)}\mathrm{Cor}^+(n)$. The horizontal lift $\#$ is its inverse:

$$\# : X \in T_{\pi(\Sigma)}\mathrm{Cor}^+(n) \xrightarrow{\sim} X^\# \in \mathcal{H}_\Sigma. \tag{1}$$

4. Quotient metric. It is defined by pullback through the horizontal lift:

$$\forall C \in \mathrm{Cor}^+(n), \forall X \in T_C\mathrm{Cor}^+(n), g_C(X,X) = G_\Sigma(X^\#, X^\#), \tag{2}$$

where $\Sigma \in \pi^{-1}(C)$ and the definition does not depend on the chosen Σ.

So the only missing ingredient is a Riemannian metric on SPD matrices which is invariant under the congruence action of positive diagonal matrices. In [5,6], the authors chose to use the famous affine-invariant/Fisher-Rao metric.

2.2 The Affine-Invariant Metrics and the Quotient-Affine Metrics

The affine-invariant metric is the Riemannian metric defined on SPD matrices by $G_\Sigma(V,V) = \mathrm{tr}(\Sigma^{-1}V\Sigma^{-1}V)$ for all $\Sigma \in \mathrm{Sym}^+(n)$ and $V \in T_\Sigma\mathrm{Sym}^+(n)$ [7–9]. It is invariant under the congruence action of the whole real general linear group $\mathrm{GL}(n)$ which contains $\mathrm{Diag}^+(n)$ as a subgroup. It provides a Riemannian symmetric structure to the manifold of SPD matrices, hence it is geodesically complete and the geodesics are given by the group action of one-parameter subgroups. We recall the exponential map, the Levi-Civita connection and the sectional curvature below for all $\Sigma \in \mathrm{Sym}^+(n)$ and vector fields $V, W \in \Gamma(T\mathrm{Sym}^+(n))$ where we also denote $V \equiv V_\Sigma$ and $W \equiv W_\Sigma \in T_\Sigma\mathrm{Sym}^+(n)$:

$$\mathrm{Exp}_\Sigma^G(V) = \Sigma^{1/2}\exp(\Sigma^{-1/2}V\Sigma^{-1/2})\Sigma^{1/2}, \tag{3}$$

$$(\nabla_V^G W)_{|\Sigma} = d_\Sigma W(V) - \frac{1}{2}(V\Sigma^{-1}W + W\Sigma^{-1}V), \tag{4}$$

$$\kappa_\Sigma^G(V,W) = \frac{1}{4}\frac{\mathrm{tr}((\Sigma^{-1}V\Sigma^{-1}W - \Sigma^{-1}W\Sigma^{-1}V)^2)}{G(V,V)G(W,W) - G(V,W)^2} \leqslant 0. \tag{5}$$

The metrics that are invariant under the congruence action of the general linear group $\mathrm{GL}(n)$ actually form a two-parameter family of metrics indexed by $\alpha > 0$ and $\beta > -\alpha/n$ [10]: $G_\Sigma^{\alpha,\beta}(V,V) = \alpha\,\mathrm{tr}(\Sigma^{-1}V\Sigma^{-1}V) + \beta\,\mathrm{tr}(\Sigma^{-1}V)^2$. We call them all affine-invariant metrics. In particular, these metrics are invariant under the congruence action of diagonal matrices so they are good candidates to define Riemannian metrics on full-rank correlation matrices by quotient. In [5,6], the authors rely on the "classical" affine-invariant metric ($\alpha = 1$, $\beta = 0$). We generalize their definition below.

Definition 1 (Quotient-affine metrics on full-rank correlation matrices). *The quotient-affine metric of parameters $\alpha > 0$ and $\beta > -\alpha/n$ is the quotient metric $g^{\alpha,\beta}$ on $\mathrm{Cor}^+(n)$ induced by the affine-invariant metric $G^{\alpha,\beta}$ via the submersion $\pi : \Sigma \in \mathrm{Sym}^+(n) \longmapsto \mathrm{Diag}(\Sigma)^{-1/2}\,\Sigma\,\mathrm{Diag}(\Sigma)^{-1/2} \in \mathrm{Cor}^+(n)$.*

3 Fundamental Riemannian Operations

In this section, we detail the quotient geometry of quotient-affine metrics $g^{\alpha,\beta}$. We give the vertical and horizontal distributions and projections in Sect. 3.1. We contribute the formulae of the metric itself in Sect. 3.2, the exponential map in Sect. 3.3, and finally the Levi-Civita connection and the sectional curvature in Sect. 3.4. To the best of our knowledge, all these formulae are new. The proof of the sectional curvature is given in Appendix A while the other proofs can be found on HAL: https://hal.archives-ouvertes.fr/hal-03157992.

3.1 Vertical and Horizontal Distributions and Projections

- Let $\bullet$ be the Hadamard product on matrices defined by $[X \bullet Y]_{ij} = X_{ij}Y_{ij}$.
- Let $A : \Sigma \in \mathrm{Sym}^+(n) \longmapsto A(\Sigma) = \Sigma \bullet \Sigma^{-1} \in \mathrm{Sym}^+(n)$. This smooth map is invariant under the action of positive diagonal matrices. The Schur product theorem ensures that $A(\Sigma) \in \mathrm{Sym}^+(n)$. A fortiori, $I_n + A(\Sigma) \in \mathrm{Sym}^+(n)$.
- Let $\psi : \mu \in \mathbb{R}^n \longmapsto (\mu \mathbb{1}^\top + \mathbb{1}\mu^\top) \in \mathrm{Sym}(n)$. This is an injective linear map.
- Let $\mathcal{S}_\Sigma(V)$ the unique solution of the Sylvester equation $\Sigma\mathcal{S}_\Sigma(V)+\mathcal{S}_\Sigma(V)\Sigma = V$ for $\Sigma \in \mathrm{Sym}^+(n)$ and $V \in \mathrm{Sym}(n)$.
- Let $\mathrm{Hol}^{\mathrm{S}}(n)$ be the vector space of symmetric matrices with vanishing diagonal (symmetric hollow matrices). Each tangent space of the manifold of full-rank correlation matrices can be seen as a copy of this vector space.

Theorem 1 (Vertical and horizontal distributions and projections). *The vertical distribution is given by $\mathcal{V}_\Sigma = \Sigma \bullet \psi(\mathbb{R}^n)$ and the horizontal distribution is given by $\mathcal{H}_\Sigma = \mathcal{S}_{\Sigma^{-1}}(\mathrm{Hol}^{\mathrm{S}}(n))$. The vertical projection is:*

$$\mathrm{ver} : V \in T_\Sigma\mathrm{Sym}^+(n) \longmapsto \Sigma \bullet \psi((I_n + A(\Sigma))^{-1}\mathrm{Diag}(\Sigma^{-1}V)\mathbb{1}) \in \mathcal{V}_\Sigma. \quad (6)$$

Then, the horizontal projection is simply $\mathrm{hor}(V) = V - \mathrm{ver}(V)$.

3.2 Horizontal Lift and Metric

Theorem 2 (Horizontal lift). *Let $\Sigma \in \mathrm{Sym}^+(n)$ and $C = \pi(\Sigma) \in \mathrm{Cor}^+(n)$. The horizontal lift at Σ of $X \in T_C\mathrm{Cor}^+(n)$ is $X^\# = \mathrm{hor}(\Delta_\Sigma X \Delta_\Sigma)$ with $\Delta_\Sigma = \mathrm{Diag}(\Sigma)^{1/2}$. In particular, the horizontal lift at $C \in \mathrm{Sym}^+(n)$ is $X^\# = \mathrm{hor}(X)$.*

Theorem 3 (Expression of quotient-affine metrics). *For all $C \in \mathrm{Cor}^+(n)$ and $X \in T_C\mathrm{Cor}^+(n)$, $g_C^{\alpha,\beta}(X,X) = \alpha\, g_C^{\mathrm{QA}}(X,X)$ (independent from β) where:*

$$g_C^{\mathrm{QA}}(X,X) = \mathrm{tr}((C^{-1}X)^2) - 2\mathbb{1}^\top\mathrm{Diag}(C^{-1}X)(I_n + A(C))^{-1}\mathrm{Diag}(C^{-1}X)\mathbb{1}. \quad (7)$$

3.3 Geodesics

The geodesics of a quotient metric are the projections of the horizontal geodesics of the original metric. This allows us to obtain the exponential map of the quotient-affine metrics.

Theorem 4 (Geodesics of quotient-affine metrics). *The geodesic from $C \in \mathrm{Cor}^+(n)$ with initial tangent vector $X \in T_C\mathrm{Cor}^+(n)$ is:*

$$\forall t \in \mathbb{R}, \gamma^{\mathrm{QA}}_{(C,X)}(t) = \mathrm{Exp}^{\mathrm{QA}}_C(tX) = \pi(C^{1/2}\exp(t\,C^{-1/2}\mathrm{hor}(X)C^{-1/2})C^{1/2}). \quad (8)$$

In particular, the quotient-affine metric is geodesically complete.

The Riemannian logarithm between C_1 and $C_2 \in \mathrm{Cor}^+(n)$ is much more complicated to compute. It amounts to find $\Sigma \in \mathrm{Sym}^+(n)$ in the fiber above C_2 such that $\mathrm{Log}^G_{C_1}(\Sigma)$ is horizontal. Then we have $\mathrm{Log}^{\mathrm{QA}}_{C_1}(C_2) = d_{C_1}\pi(\mathrm{Log}^G_{C_1}(\Sigma))$. This means finding Σ that minimizes the affine-invariant distance in the fiber:

$$D = \arg\min_{D\in\mathrm{Diag}^+(n)} d(C_1, DC_2D),$$

from which we get $\Sigma = DC_2D$. This is the method used in the original paper [5,6]. Note that the uniqueness of the minimizer has not been proved yet.

3.4 Levi-Civita Connection and Sectional Curvature

In this section, we give the Levi-Civita connection and the curvature. The computations are based on the fundamental equations of submersions [11]. We denote $\mathrm{sym}(M) = \frac{1}{2}(M + M^\top)$ the symmetric part of a matrix.

Theorem 5 (Levi-Civita connection and sectional curvature of quotient-affine metrics). *The Levi-Civita connection of quotient-affine metrics is:*

$$\begin{aligned}(\nabla^{\mathrm{QA}}_X Y)_{|C} = d_C Y(X) + \mathrm{sym}[&\mathrm{Diag}(X^\#)Y^\# + \mathrm{Diag}(Y^\#)X^\# + \mathrm{Diag}(X^\# C^{-1}Y^\#)C \\ &- X^\# C^{-1}Y^\# - \tfrac{1}{2}\mathrm{Diag}(X^\#)C\mathrm{Diag}(Y^\#) - \tfrac{3}{2}\mathrm{Diag}(X^\#)\mathrm{Diag}(Y^\#)C]. \quad (9)\end{aligned}$$

The curvature of quotient-affine metrics is:

$$\kappa^{\mathrm{QA}}_C(X,Y) = \kappa^G_C(X^\#,Y^\#) + \frac{3}{4}\frac{G_C(\mathrm{ver}[X^\#,Y^\#],\mathrm{ver}[X^\#,Y^\#])}{g_C(X,X)g_C(Y,Y) - g_C(X,Y)^2}, \quad (10)$$

$$= \frac{2\,\mathrm{tr}((C^{-1}X^\# C^{-1}Y^\# - C^{-1}Y^\# C^{-1}X^\#)^2) + 3\mathbb{1}^\top D(I_n + A(C))^{-1}D\mathbb{1}}{8(g_C(X,X)g_C(Y,Y) - g_C(X,Y)^2)} \quad (11)$$

where $[V,W] = dW(V) - dV(W)$ is the Lie bracket on $\mathrm{Sym}^+(n)$ and $D = D(X,Y) - D(Y,X)$ with $D(X,Y) = \mathrm{Diag}(C^{-1}\mathrm{Diag}(X^\#)Y^\# - C^{-1}Y^\# C^{-1}\mathrm{Diag}(X^\#)C)$. There is a slight abuse of notation because $\mathrm{ver}[X^\#,Y^\#]$ induces that $X^\#$ and $Y^\#$ are vector fields. Indeed here, they are horizontal vector fields extending the horizontal lifts at C.

The first term of the sectional curvature is negative, the second one is positive so we don't know in general the sign of the curvature of quotient-affine metrics.

The quotient-affine metrics not only provide a Riemannian framework on correlation matrices but also provide correlation-compatible statistical tools on SPD matrices if we consider that the space of SPD matrices is the Cartesian product of positive diagonal matrices and full-rank correlation matrices. We give a taste of such construction in the next section.

4 Illustration in Dimension 2

In dimension 2, any correlation matrix $C \in \mathrm{Cor}^+(2)$ writes $C = C(\rho) := \begin{pmatrix} 1 & \rho \\ \rho & 1 \end{pmatrix}$ with $\rho \in (-1, 1)$. In the following theorem, we give explicitly the logarithm, the distance and the interpolating geodesics of the quotient-affine metric.

Theorem 6 (Quotient-affine metrics in dimension 2). *Let* $C_1 = C(\rho_1), C_2 = C(\rho_2) \in \mathrm{Cor}^+(n)$ *with* $\rho_1, \rho_2 \in (-1, 1)$. *We denote* $f : \rho \in (-1, 1) \mapsto \frac{1+\rho}{1-\rho} \in (0, \infty)$ *which is a smooth increasing map. We denote* $\lambda = \lambda(\rho_1, \rho_2) = \frac{1}{2} \log \frac{f(\rho_2)}{f(\rho_1)}$ *which has the same sign as* $\rho_2 - \rho_1$. *Then the quotient-affine operations are:*

1. *(Logarithm)* $\mathrm{Log}_{C_1}^{\mathrm{QA}}(C_2) = \lambda \begin{pmatrix} 0 & 1 - \rho_1^2 \\ 1 - \rho_1^2 & 0 \end{pmatrix}$,
2. *(Distance)* $d^{\mathrm{QA}}(C_1, C_2) = \sqrt{2}|\lambda|$,
3. *(Geodesics)* $\gamma_{(C_1,C_2)}^{\mathrm{QA}}(t) = C(\rho^{\mathrm{QA}}(t))$ *where* $\rho^{\mathrm{QA}}(t) = \frac{\rho_1 \cosh(\lambda t) + \sinh(\lambda t)}{\rho_1 \sinh(\lambda t) + \cosh(\lambda t)} \in (-1, 1)$ *is monotonic (increasing if and only if* $\rho_2 - \rho_1 > 0$*).*

Let $\Sigma_1, \Sigma_2 \in \mathrm{Sym}^+(2)$ and C_1, C_2 their respective correlation matrices. We denote $\gamma^{\mathrm{AI}}, \gamma^{\mathrm{LE}}$ the geodesics between Σ_1 and Σ_2 for the affine-invariant and the log-Euclidean metrics respectively. We define $\rho^{\mathrm{AI}}, \rho^{\mathrm{LE}}$ such that the correlation matrices of $\gamma^{\mathrm{AI}}(t), \gamma^{\mathrm{LE}}(t)$ are $C(\rho^{\mathrm{AI}}(t)), C(\rho^{\mathrm{LE}}(t))$. Figure 1(a) shows $\rho^{\mathrm{AI}}, \rho^{\mathrm{LE}}, \rho^{\mathrm{QA}}$ with $\Sigma_1 = \begin{pmatrix} 4 & 1 \\ 1 & 100 \end{pmatrix}$ and $\Sigma_2 = \begin{pmatrix} 100 & 19 \\ 19 & 4 \end{pmatrix}$. When varying numerically Σ_1 and Σ_2, it seems that ρ^{LE} and ρ^{AI} always have three inflection points. In contrast, ρ^{QA} always has one inflection point since $(\rho^{\mathrm{QA}})'' = -2\lambda^2 \rho^{\mathrm{QA}}(1 - (\rho^{\mathrm{QA}})^2)$. Analogously, we compare the interpolations of the determinant (Fig. 1(b)) and the trace (Fig. 1(c)) using several Riemannian metrics: Euclidean (trace-monotonic); log-Euclidean and affine-invariant (determinant-monotonic); power-Euclidean $\times$ quotient-affine (correlation-monotonic).

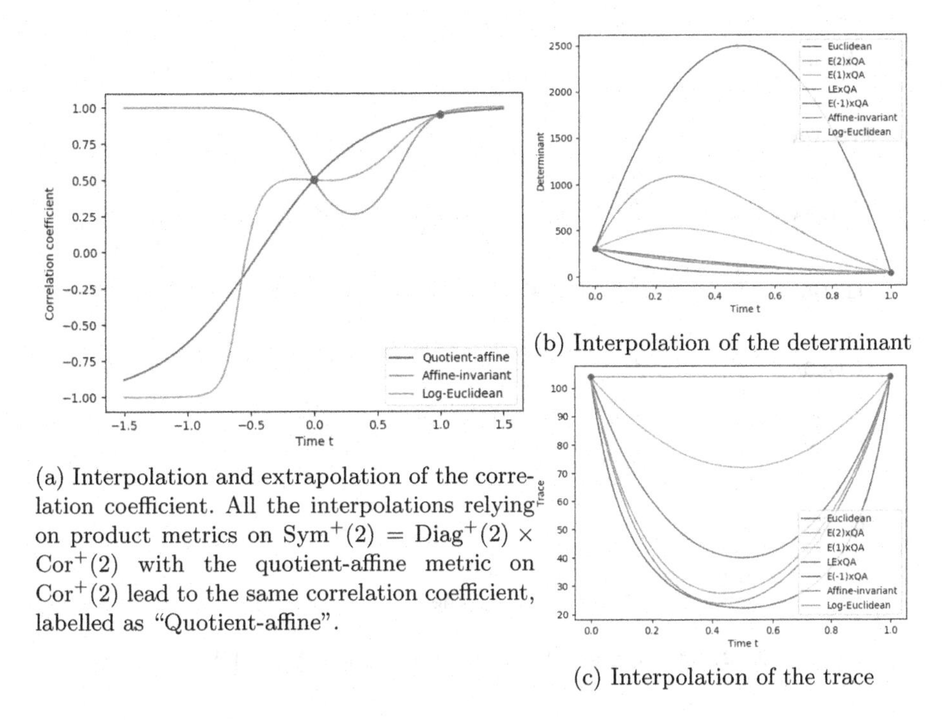

(a) Interpolation and extrapolation of the correlation coefficient. All the interpolations relying on product metrics on $\mathrm{Sym}^+(2) = \mathrm{Diag}^+(2) \times \mathrm{Cor}^+(2)$ with the quotient-affine metric on $\mathrm{Cor}^+(2)$ lead to the same correlation coefficient, labelled as "Quotient-affine".

(b) Interpolation of the determinant

(c) Interpolation of the trace

Fig. 1. Extrapolation and interpolation between the SPD matrices Σ_1 and Σ_2 using various Riemannian metrics. The metrics $\mathrm{E}(p) \times \mathrm{QA}$ refer to the p-power-Euclidean metric on the diagonal part and the quotient-affine metric on the correlation part. When p tends to 0, $\mathrm{E}(p)$ tends to the log-Euclidean metric $\mathrm{LE} \equiv \mathrm{E}(0)$.

On Fig. 2, we compare the geodesics of these metrics. The determinant is the area of the ellipsis and the trace is the sum of the lengths of the axes. Thus the product metrics of the form power-Euclidean on the diagonal part and quotient-affine on the correlation part can be seen as performing a correlation-monotonic trade-off between the trace-monotonicity and the determinant-monotonicity.

5 Conclusion

We investigated in this paper the very nice idea of quotienting the affine-invariant metrics on SPD matrices by the action of the positive diagonal group to obtain the principled quotient-affine metrics on full-rank correlation matrices. The quotient-affine metric with $\alpha = 1$ and $\beta = 0$ was first proposed in Paul David's thesis [6] and in the subsequent journal paper [5]. We contribute here exact formulae for the main Riemannian operations, including the exponential map, the connection and the sectional curvature. The exponential map is particularly interesting to rigorously project tangent computations to the space of full-rank correlation matrices. This opens the way to the implementation of a number of generic algorithms on Riemannian manifolds. However, we could not

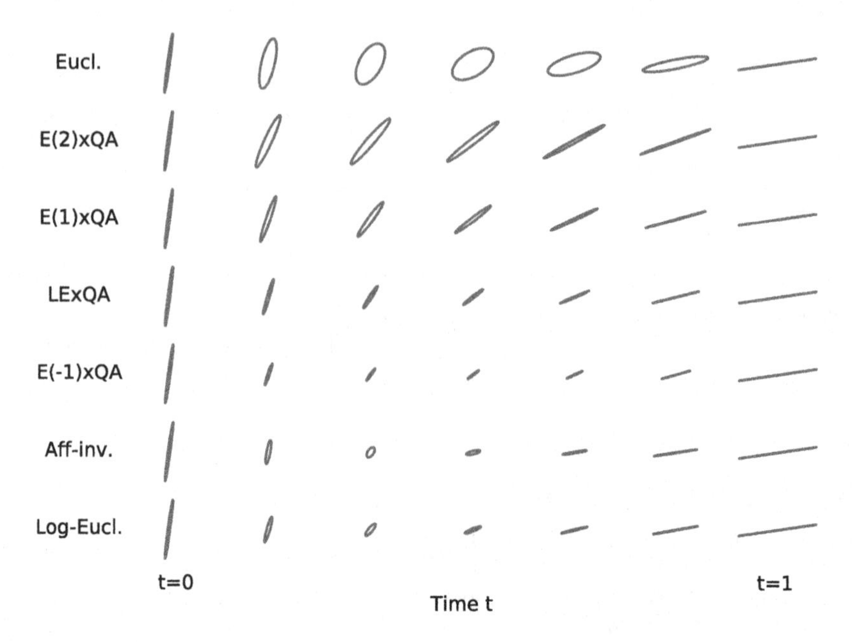

Fig. 2. Interpolations between SPD matrices Σ_1 and Σ_2.

find a closed form expression for the logarithm which remains to be computed through an optimization procedure. Thus, computing distances with these metrics remains computationally expensive. In order to obtain more efficient methods, this leads us to look for other principled Riemannian metrics on correlation matrices for which the logarithm could be expressed in closed form.

Acknowledgements. This project has received funding from the European Research Council (ERC) under the European Union's Horizon 2020 research and innovation program (grant G-Statistics agreement No 786854). This work has been supported by the French government, through the UCAJEDI Investments in the Future project managed by the National Research Agency (ANR) with the reference number ANR-15-IDEX-01 and through the 3IA Côte d'Azur Investments in the Future project managed by the National Research Agency (ANR) with the reference number ANR-19-P3IA-0002. The authors would like to thank Frank Nielsen for insightful discussions on correlation matrices.

A Proof of the Sectional Curvature

The detail of the proofs can be found on the HAL version of the paper: https://hal.archives-ouvertes.fr/hal-03157992. The proofs of Theorem 1 (vertical/horizontal projections), Theorem 2 (horizontal lift) and Theorem 3 (metric) consist in elementary computations, the geodesics of Theorem 4 are simply the projections of horizontal geodesics [12], and the formula of the Levi-Civita connection consists in elementary computations as well. Hence we only give here the proof of the formula of the sectional curvature of quotient-affine metrics.

The curvature in formula (10) directly comes from the fundamental equations of submersions [11]. The curvature of the affine-invariant metric comes from [8]. Hence we only have to compute $G_C(\mathrm{ver}[X^\#, Y^\#], \mathrm{ver}[X^\#, Y^\#])$ where G is the affine-invariant metric, $[\cdot,\cdot]$ is the Lie bracket on the manifold of SPD matrices (it is not the matrix commutator), $C \in \mathrm{Cor}^+(n)$ and $X^\#, Y^\#$ are horizontal vector fields on $\mathrm{SPD}(n)$ extending $X_C^\#$ and $Y_C^\#$ respectively, where $X, Y \in T_C\mathrm{Cor}^+(n)$ are tangent vectors at C. For example, we can consider that X, Y are constant vector fields on $\mathrm{Cor}^+(n)$ and simply define $X^\#, Y^\#$ as their horizontal lifts everywhere:

$$\begin{aligned} Y_\Sigma^\# &= \mathrm{hor}(\Delta_\Sigma Y \Delta_\Sigma), \\ &= \Delta_\Sigma Y \Delta_\Sigma - \mathrm{ver}(\Delta_\Sigma Y \Delta_\Sigma), \\ &= \Delta_\Sigma Y \Delta_\Sigma - 2\Sigma \bullet \psi((I_n + \Sigma \bullet \Sigma^{-1})^{-1}\mathrm{Diag}(\Sigma^{-1}\Delta_\Sigma Y \Delta_\Sigma)\mathbb{1}), \\ &= \Delta_\Sigma Y \Delta_\Sigma - 2\Sigma \bullet \psi((I_n + \Sigma \bullet \Sigma^{-1})^{-1}\mathrm{Diag}(\pi(\Sigma)^{-1} Y)\mathbb{1}), \end{aligned}$$

with $\Delta_\Sigma = \mathrm{Diag}(\Sigma)^{1/2}$ and $\psi(\mu) = \mu\mathbb{1}^\top + \mathbb{1}\mu^\top$ for $\mu \in \mathbb{R}^n$. Now we can compute $[X^\#, Y^\#] = \partial_{X^\#}Y^\# - \partial_{Y^\#}X^\#$ and evaluate it at C. We have:

$$\begin{aligned} \partial_{X^\#}Y^\# &= \frac{1}{2}(\Delta_\Sigma^{-1}\mathrm{Diag}(X^\#)Y\Delta_\Sigma + \Delta_\Sigma Y \mathrm{Diag}(X^\#)\Delta_\Sigma^{-1}) \\ &\quad - X^\# \bullet \psi((I_n + \Sigma \bullet \Sigma^{-1})^{-1}\mathrm{Diag}(\Sigma^{-1}Y)\mathbb{1} + v_\Sigma(X, Y), \end{aligned}$$

where $v_\Sigma(X, Y) = \Sigma \bullet \psi[(I_n + \Sigma \bullet \Sigma^{-1})^{-1}(X^\# \bullet \Sigma^{-1} - \Sigma \bullet \Sigma^{-1}X^\#\Sigma^{-1})(I_n + \Sigma \bullet \Sigma^{-1})^{-1}\mathrm{Diag}(\Sigma^{-1}Y)\mathbb{1} + (I_n + \Sigma \bullet \Sigma^{-1})^{-1}\mathrm{Diag}(\pi(\Sigma)^{-1}X\pi(\Sigma)^{-1}Y)\mathbb{1}] \in \mathcal{V}_\Sigma$.

We evaluate it at $C \in \mathrm{Cor}^+(n)$ so that $\Delta_C = I_n$. We denote $B = (I_n + C \bullet C^{-1})^{-1}$ to simplify the notations. Note that the last term is symmetric in X and Y so we can define $v_C^0(X, Y) = C \bullet \psi[B(X^\# \bullet C^{-1} - C \bullet C^{-1}X^\#C^{-1})B\mathrm{Diag}(C^{-1}Y)\mathbb{1}]$. Then:

$$\begin{aligned} [X^\#, Y^\#] &= \frac{1}{2}(\mathrm{Diag}(X^\#)Y + Y\mathrm{Diag}(X^\#) - \mathrm{Diag}(Y^\#)X - X\mathrm{Diag}(Y^\#)), && (12) \\ &\quad - (X^\# \bullet \psi(B\mathrm{Diag}(C^{-1}Y)\mathbb{1}) - Y^\# \bullet \psi(B\mathrm{Diag}(C^{-1}X)\mathbb{1})), && (13) \\ &\quad + v_C^0(X, Y) - v_C^0(Y, X). && (14) \end{aligned}$$

Fortunately, the computation of $\mathrm{ver}[X^\#, Y^\#] = C \bullet \psi(B\mathrm{Diag}(C^{-1}[X^\#, Y^\#])\mathbb{1})$ brings some simplifications. Indeed, to simplify line (12), we plug $Y = Y^\# - \frac{1}{2}(\mathrm{Diag}(Y^\#)C + C\mathrm{Diag}(Y^\#))$. Firstly, we have $\mathrm{Diag}(C^{-1}\mathrm{Diag}(X^\#)Y) = \mathrm{Diag}(C^{-1}\mathrm{Diag}(X^\#)Y^\#) - \frac{1}{2}\mathrm{Diag}(C^{-1}\mathrm{Diag}(X^\#)\mathrm{Diag}(Y^\#)C) - \frac{1}{2}\mathrm{Diag}(C^{-1}\mathrm{Diag}(X^\#)C\mathrm{Diag}(Y^\#))$ whose second term is symmetric in X and Y. Secondly, the other term is $\mathrm{Diag}(C^{-1}Y\mathrm{Diag}(X^\#)) = \mathrm{Diag}(C^{-1}Y^\#\mathrm{Diag}(X^\#)) - \frac{1}{2}\mathrm{Diag}(C^{-1}\mathrm{Diag}(Y^\#)C\mathrm{Diag}(X^\#)) - \frac{1}{2}\mathrm{Diag}(Y^\#)\mathrm{Diag}(X^\#)$ whose first term is null, whose third term is symmetric in X and Y and whose second term is the symmetric of the third term of the previous expression. Hence, the vertical projection of the line (12) reduces to the vertical projection of $\frac{1}{2}(\mathrm{Diag}(X^\#)Y^\# -$

$\mathrm{Diag}(Y^{\#})X^{\#})$. To simplify line (13), we can show by computing coordinate by coordinate that $\mathrm{Diag}(C^{-1}(X^{\#} \bullet \psi(B\mathrm{Diag}(C^{-1}Y)\mathbb{1})))\mathbb{1} = (C^{-1} \bullet X^{\#})B\mathrm{Diag}(C^{-1}Y)\mathbb{1}$. Hence, line (13) cancels the first term of v^0. Finally, a nicer expression of the second term of v^0 can be obtained by noticing that $\mathrm{Diag}(Y^{\#})\mathbb{1} = -2B\mathrm{Diag}(C^{-1}Y)\mathbb{1}$. After simple calculations, we get $(C \bullet C^{-1}X^{\#}C^{-1})B\mathrm{Diag}(C^{-1}Y)\mathbb{1} = -\frac{1}{2}\mathrm{Diag}(C\mathrm{Diag}(X^{\#})C^{-1}Y^{\#}C^{-1})\mathbb{1}$.

To summarize, we have $\mathrm{ver}[X^{\#}, Y^{\#}] = C \bullet \psi(\frac{1}{2}BD\mathbb{1})$ with $D = D(X,Y) - D(Y,X) \in \mathrm{Diag}(n)$ and $D(X,Y) = \mathrm{Diag}(C^{-1}\mathrm{Diag}(X^{\#})Y^{\#} - C^{-1}Y^{\#}C^{-1}\mathrm{Diag}(X^{\#})C)$. Finally, since $G_C(C \bullet \psi(\mu), C \bullet \psi(\mu)) = 2\mu^\top B\mu$ for any vector $\mu \in \mathbb{R}^n$, we get $G_C(\mathrm{ver}[X^{\#}, Y^{\#}], \mathrm{ver}[X^{\#}, Y^{\#}]) = \frac{1}{2}\mathbb{1}^\top DBD\mathbb{1}$, as expected.

References

1. Tropp, J.A.: Simplicial faces of the set of correlation matrices. Discrete Comput. Geometry **60**(2), 512–529 (2018)
2. Rebonato, R., Jäckel, P.: The most general methodology to create a valid correlation matrix for risk management and option pricing purposes. J. Risk **2**, 17–27 (2000)
3. Nielsen, F., Sun, K.: Clustering in Hilbert's projective geometry: the case studies of the probability simplex and the elliptope of correlation matrices. In: Nielsen, F. (ed.) Geometric Structures of Information. SCT, pp. 297–331. Springer, Cham (2019). https://doi.org/10.1007/978-3-030-02520-5_11
4. Varoquaux, G., Baronnet, F., Kleinschmidt, A., Fillard, P., Thirion, B.: Detection of brain functional-connectivity difference in post-stroke patients using group-level covariance modeling. In: Jiang, T., Navab, N., Pluim, J.P.W., Viergever, M.A. (eds.) MICCAI 2010. LNCS, vol. 6361, pp. 200–208. Springer, Heidelberg (2010). https://doi.org/10.1007/978-3-642-15705-9_25
5. David, P., Gu, W.: A Riemannian structure for correlation matrices. Oper. Matrices **13**(3), 607–627 (2019)
6. David, P.: A Riemannian quotient structure for correlation matrices with applications to data science. Ph.D. thesis, Institute of Mathematical Sciences, Claremont Graduate University (2019)
7. Siegel, C.: Symplectic geometry. Am. J. Math. **65**(1), 1–86 (1943)
8. Skovgaard, L.: A Riemannian geometry of the multivariate normal model. Scand. J. Stat. **11**(4), 211–223 (1984)
9. Amari, S., Nagaoka, H.: Methods of Information Geometry. American Mathematical Society (2000)
10. Pennec, X.: Statistical computing on manifolds: from Riemannian geometry to computational anatomy. Emerg. Trends Visual Comut. **5416**, 347–386 (2008)
11. O'Neill, B.: The fundamental equations of a submersion. Michigan Math. J. **13**(4), 459–469 (1966)
12. Gallot, S., Hulin, D., Lafontaine, J.: Riemannian Geometry. Springer, New York (2004). https://doi.org/10.1007/978-0-387-29403-2

Parallel Transport on Kendall Shape Spaces

Nicolas Guigui[1], Elodie Maignant[1,2](✉), Alain Trouvé[2], and Xavier Pennec[1]

[1] Epione Project Team, Université Côte d'Azur, Inria, Sophia-Antipolis, Nice, France
elodie.maignant@inria.fr
[2] Centre Borelli, Université Paris-saclay, ENS Paris Saclay, Gif-sur-Yvette, France

Abstract. Kendall shape spaces are a widely used framework for the statistical analysis of shape data arising from many domains, often requiring the parallel transport as a tool to normalise time series data or transport gradient in optimisation procedures. We present an implementation of the pole ladder, an algorithm to compute parallel transport based on geodesic parallelograms and compare it to methods by integration of the parallel transport ordinary differential equation.

Keywords: Parallel transport · Shape spaces

1 Introduction

Kendall shape spaces are a ubiquitous framework for the statistical analysis of data arising from medical imaging, computer vision, biology, chemistry and many more domains. The underlying idea is that a shape is what is left after removing the effects of rotation, translation and re-scaling, and to define a metric accounting for those invariances. This involves defining a Riemannian submersion and the associated quotient structure, resulting in non trivial differential geometries with singularities and curvature, requiring specific statistical tools to deal with such data.

In this context parallel transport is a fundamental tool to define statistical models and optimisation procedures, such as the geodesic or spline regression [5,10], intrinsic MANOVA [4], and for the normalisation of time series of shapes [1,7]. However the parallel transport of a tangent vector v along a curve γ is defined by the ordinary differential equation (ODE) $\nabla_{\dot{\gamma}} X = 0, X(0) = v$ (where ∇ is the Levi-Civita connection in our case), and there is usually no closed-form solution. Approximation methods have therefore been derived, either by direct integration [5], or by integration of the geodesic equation to approximate Jacobi fields (the fanning scheme [8]). Another class of approximations, referred to as ladder methods, relies on iterative constructions of geodesic parallelograms that only require approximate geodesics [3].

In this work, we present an implementation of the pole ladder that leverages the quotient structure of Kendall shape spaces and strongly relies on the open-source Python package geomstats. We compare it to the method of Kim et al. [5] by approximate integration.

F. Nielsen and F. Barbaresco (Eds.): GSI 2021, LNCS 12829, pp. 103–110, 2021.
https://doi.org/10.1007/978-3-030-80209-7_12

We first recall the quotient structure of Kendall shape spaces and its use by Kim et al. to compute parallel transport in Sect. 2, then in Sect. 3 we recall the pole ladder scheme and the main result from [3] on its convergence properties. Numerical simulations to compare the two methods are reported in Sect. 4.

2 The Quotient Structure of Kendall Shape Spaces

We first describe the shape space, as quotient of the space of configurations of k points in $\mathbb{R}^m$ (called landmarks) by the groups of translations, re-scaling and rotations of $\mathbb{R}^m$. For a thorough treatment this topic, we refer the reader to [2,6].

2.1 The Pre-shape Space

We define the space of k landmarks of $\mathbb{R}^m$ as the space of $m \times k$ matrices $M(m,k)$. For $x \in M(m,k)$, let x_i denote the columns of x, i.e. points of $\mathbb{R}^m$ and let $\bar{x}$ be their barycentre. We remove the effects of translation by considering the matrix with columns $x_i - \bar{x}$ instead of x and restrict to non-zero centred matrices, i.e. matrices which represent configurations with at least two distinct landmarks. We further remove the effects of scaling by dividing x by its Frobenius norm (written $\|\cdot\|$). This defines the pre-shape space $\mathcal{S}_m^k = \{x \in M(m,k) \mid \sum_{i=1}^k x_i = 0,\ \|x\| = 1\}$, which is identified with the hypersphere of dimension $m(k-1)-1$. The pre-shape space is therefore a differential manifold whose tangent space at any $x \in \mathcal{S}_m^k$ is given by $T_x\mathcal{S}_m^k = \{w \in M(m,k) \mid \sum_{i=1}^k w_i = 0,\ \mathrm{Tr}(w^T x) = 0\}$.

The ambient Frobenius metric $\langle \cdot, \cdot \rangle$ thus defines a Riemannian metric on the pre-shape space, with constant sectional curvature and known geodesics: let $x, y \in \mathcal{S}_m^k$ with $x \neq y$, and $w \in T_x\mathcal{S}_m^k$

$$x_w = \mathrm{Exp}_x(w) = \cos(\|w\|)x + \sin(\|w\|)\frac{w}{\|w\|}, \tag{1}$$

$$\mathrm{Log}_x(y) = \arccos(\langle y, x \rangle)\frac{y - \langle y, x \rangle x}{\|y - \langle y, x \rangle x\|}. \tag{2}$$

Moreover, this metric is invariant to the action of the rotation group $SO(m)$. This allows to define the shape space as the quotient $\Sigma_m^k = \mathcal{S}_m^k / SO(m)$.

2.2 The Shape Space

To remove the effect of rotations, consider the equivalence relation $\sim$ on $\mathcal{S}_m^k$ by $x \sim y \iff \exists R \in SO(m)$ such that $y = Rx$. For $x \in \mathcal{S}_m^k$, let $[x]$ denote its equivalence class for $\sim$. This equivalence relation results from the group action of $SO(m)$ on $\mathbb{R}^m$. This action is smooth, proper but not free everywhere when $m \geq 3$. This makes the orbit space $\Sigma_m^k = \{[x] \mid x \in \mathcal{S}_m^k\}$ a differential manifold with singularities where the action is not free (i.e. at $x \in \mathcal{S}_m^k$ where there exists $R \neq \mathrm{I}_m \in SO(m)$ s.t. $Rx = x$).

One can describe these singularities explicitly: they correspond to the matrices of $\mathcal{S}_m^k$ of rank $m-2$ or less [6]. For $k \geq 3$, the spaces Σ_1^k and Σ_2^k are always smooth. Moreover, as soon as $m \geq k$, the manifold acquires boundaries. As an example, while the space Σ_2^3 of 2D triangles is identified with the sphere $S^2(1/2)$, the space Σ_3^3 of 3D triangles is isometric to a 2-ball [6].

Away from the singularities, the canonical projection map $\pi : x \mapsto [x]$ is a Riemannian submersion, i.e. $d\pi$ is everywhere surjective, and plays a major role in defining the metric on the shape space. Let $d_x\pi$ be its differential map at $x \in \mathcal{S}_m^k$, whose kernel defines the vertical tangent space, which corresponds to the tangent space of the submanifold $\pi^{-1}([x])$, called fiber above $[x]$:

$$\mathrm{Ver}_x = \{Ax \mid A \in \mathrm{Skew}(m)\} = \mathrm{Skew}(m) \mid x$$

where $\mathrm{Skew}(m)$ is the space of skew-symmetric matrices of size m.

2.3 The Quotient Metric

The Frobenius metric on the pre-shape space allows to define the horizontal spaces as the orthogonal complements to the vertical spaces:

$$\begin{aligned}\mathrm{Hor}_x &= \{w \in T_x\mathcal{S}_m^k \mid Tr(Axw^T) = 0 \;\forall A \in \mathrm{Skew}(m)\} \\ &= \{w \in T_x\mathcal{S}_m^k \mid xw^T \in \mathrm{Sym}(m)\}\end{aligned}$$

where $\mathrm{Sym}(m)$ is the space of symmetric matrices of size m. Lemma 1 from [10] allows to compute the vertical component of any tangent vector:

Lemma 1. *For any $x \in \mathcal{S}_m^k$ and $w \in T_x\mathcal{S}_m^k$, the vertical component of w can be computed as $\mathrm{Ver}_x(w) = Ax$ where A solves the Sylvester equation:*

$$Axx^T + xx^TA = wx^T - xw^T \tag{3}$$

If $\mathrm{rank}(x) \geq m-1$, A is the unique skew-symmetric solution of (3).

In practice, the Sylvester equation can be solved by an eigenvalue decomposition of xx^T. This defines ver_x, the orthogonal projection on Ver_x. As $T_x\mathcal{S}_m^k = \mathrm{Ver}_x \oplus \mathrm{Hor}_x$, any tangent vector w at $x \in \mathcal{S}_m^k$ may be decomposed into a horizontal and a vertical component, by solving (3) to compute $\mathrm{ver}_x(w)$, and then $\mathrm{hor}_x(w) = w - \mathrm{ver}_x(w)$.

Furthermore, as $\mathrm{Ver}_x = \ker(d_x\pi)$, $d_x\pi$ is a linear isomorphism from Hor_x to $T_{[x]}\Sigma_m^k$. The metric on Σ_m^k is defined such that this isomorphism is an isometry. Note that the metric does not depend on the choice of the y in the fiber $\pi^{-1}([x])$ since all y in $\pi^{-1}([x])$ may be obtained by a rotation of x, and the Frobenius metric is invariant to the action of rotations. This makes π a Riemannian submersion. Additionally, π is surjective so for every vector field on Σ_m^k there is a unique horizontal lift, i.e. a vector field on $\mathcal{S}_m^k$ whose vertical component is null everywhere. The tangent vectors of Σ_m^k can therefore be identified with horizontal vectors of $\mathcal{S}_m^k$. One of the main characteristics of Riemannian submersions was proved by O'Neill [11, Chapter 7, Lemma 45]:

Theorem 1 (O'Neill). *Let $\pi : M \to B$ be a Riemannian submersion. If γ is a geodesic in M such that $\dot\gamma(0)$ is a horizontal vector, then $\dot\gamma$ is horizontal everywhere and $\pi \circ \gamma$ is a geodesic of B of the same length as γ.*

Remark 1. We emphasise that an equivalent proposition cannot be derived for the parallel transport of a tangent vector. Indeed the parallel transport of a horizontal vector field along a horizontal geodesic may not be horizontal. This will be detailed in the next subsection and constitutes a good example of metric for which computing geodesics is easier than computing parallel transport, although the former is a variational problem and the latter is a linear ODE.

Furthermore, the Riemannian distances d on $\mathcal{S}_m^k$ and d_Σ on Σ_m^k are related by

$$d_\Sigma(\pi(x), \pi(y)) = \inf_{R \in SO(m)} d(x, Ry). \tag{4}$$

The optimal rotation R between any x, y is unique in a subset U of $\mathcal{S}_m^k \times \mathcal{S}_m^k$, which allows to define the *align* map $\omega : U \to \mathcal{S}_m^k$ that maps (x, y) to Ry. In this case, $d_\Sigma(\pi(x), \pi(y)) = d(x, \omega(x, y))$ and $x\omega(x, y)^T \in \mathrm{Sym}(m)$. It is useful to notice that $\omega(x, y)$ can be directly computed by a pseudo-singular value decomposition of xy^T. Finally, x and $\omega(x, y)$ are joined by a horizontal geodesic.

2.4 Implementation in geomstats

The geomstats library [9], available at http://geomstats.ai, implements classes of manifolds equipped with Riemannian metrics. It contains an abstract class for quotient metrics, that allows to compute the Riemannian distance, exponential and logarithm maps in the quotient space from the ones in the top space.

In the case of the Kendall shape spaces, the quotient space cannot be seen explicitly as a submanifold of some $\mathbb{R}^N$ and the projection π and its differential $d\pi$ cannot be expressed in closed form. However, the align map amounts to identifying the shape space with a local horizontal section of the pre-shape space, and thanks to the characteristics of Riemannian submersions mentioned in the previous subsections, all the computations can be done in the pre-shape space.

Recall that Exp, Log, and d denote the operations of the pre-shape space $\mathcal{S}_m^k$ and are given in (1). We obtain from Theorem 1 for any $x, y \in \mathcal{S}_m^k$ and $v \in T_x\mathcal{S}_m^k$

$$\begin{aligned}
\mathrm{Exp}_{\Sigma,[x]}(d_x\pi v) &= \pi(\mathrm{Exp}_x(\mathrm{hor}_x(v))),\\
\mathrm{Log}_{\Sigma,[x]}([y]) &= d_x\pi \mathrm{Log}_x(\omega(x, y)),\\
d_\Sigma([x], [y]) &= d(x, \omega(x, y)).
\end{aligned}$$

2.5 Parallel Transport in the Shape Space

As noticed in Remark 1, one cannot use the projection of the parallel transport in the pre-shape space $\mathcal{S}_m^k$ to compute the parallel transport in the shape space Σ_m^k. Indeed [5] proved the following

Proposition 1 (Kim et al. [5]). *Let γ be a horizontal C^1-curve in $\mathcal{S}_m^k$ and v be a horizontal tangent vector at $\gamma(0)$. Assume that* $\text{rank}(\gamma(s)) \geq m-1$ *except for finitely many s. Then the vector field $s \mapsto v(s)$ along γ is horizontal and the projection of $v(s)$ to $T_{[\gamma(s)]}\Sigma_m^k$ is the parallel transport of $d_{\gamma(0)}\pi v$ along $[\gamma(s)]$ if and only if $s \mapsto v(s)$ is the solution of*

$$\dot{v}(s) = -\text{Tr}(\dot{\gamma}(s)v(s)^T)\gamma(s) + A(s)\gamma(s), \qquad v(0) = v \tag{5}$$

where for every s, $A(s) \in$ Skew(m) *is the unique solution to*

$$A(s)\gamma(s)\gamma(s)^T + \gamma(s)\gamma(s)^T A(s) = \dot{\gamma}(s)v(s)^T - v(s)\dot{\gamma}(s)^T. \tag{6}$$

Equation (5) means that the covariant derivative of $s \mapsto v(s)$ along γ must be a vertical vector at all times, defined by the matrix $A(s) \in \text{Skew}(m)$. These equations can be used to compute parallel transport in the shape space. To compute the parallel transport of $d_{\gamma(0)}\pi w$ along $[\gamma]$, [5] propose the following method: one first chooses a discretization time-step $\delta = \frac{1}{n}$, then repeat for every $s = \frac{i}{n}, i = 0 \dots n$

1. Compute $\gamma(s)$ and $\dot{\gamma}(s)$,
2. Solve the Sylvester Eq. (6) to compute $A(s)$ and the r.h.s. of (5),
3. Take a discrete Euler step to obtain $\tilde{v}(s+\delta)$
4. Project $\tilde{v}(s+\delta)$ to $T_{\gamma(s)}\mathcal{S}_m^k$ to obtain $\hat{v}(s+\delta)$,
5. Project to the horizontal subspace: $v(s+\delta) \leftarrow \text{hor}(\hat{v}(s+\delta))$
6. $s \leftarrow s + \delta$

We notice that this method can be accelerated by a higher-order integration scheme, such as Runge-Kutta (RK) by directly integrating the system $\dot{v} = f(v, s)$ where f is a smooth map given by (5) and (6). In this case, steps 4. and 5. are not necessary. The precision and complexity of this method is then bound to that of the integration scheme used. As ladder methods rely only on geodesics, which can be computed in closed-form and their convergence properties are well understood [3], we compare this method by integration to the pole ladder. We focus on the case where γ is a horizontal geodesic, such that step 1 does not require solving an ODE. Thus, it introduces no additional complexity.

3 The Pole Ladder Algorithm

3.1 Description

The pole ladder is a modification of the Schild's ladder proposed by [7]. The pole ladder is more precise and cheaper to compute as shown by [3]. It is also exact in symmetric spaces [12]. We thus focus on this method. We describe it here in a Riemannian manifold $(M, \langle,\rangle)$.

Consider a geodesic curve $\gamma : t \mapsto \gamma(t) \in M$, with initial conditions $x = \gamma(0) \in M$ and $w = \dot{\gamma}(0) \in T_xM$. In order to compute the parallel transport of $v \in T_xM$ along γ, between times x and $y = \gamma(1)$, the pole ladder consists in first dividing the main geodesic γ in n segments of equal length and computing the geodesic from x with initial velocity $\frac{v}{n^\alpha}$, obtaining $x_v = \text{Exp}_x(\frac{v}{n^\alpha})$. Then for each segment it repeats the following construction (see Fig. 1):

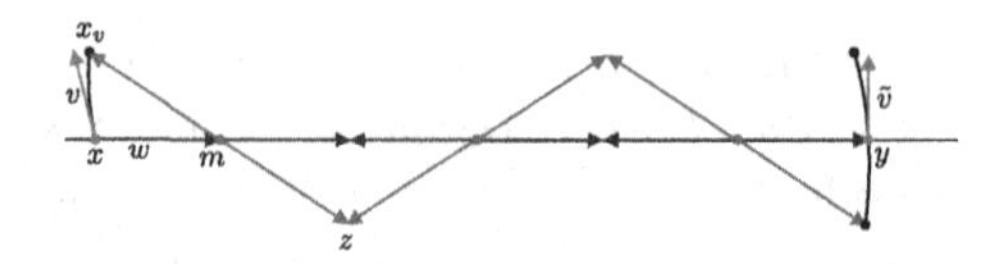

Fig. 1. Schematic representation of the pole ladder

1. Compute the midpoint of the segment $m = \text{Exp}_x(\frac{w}{2n})$ and the initial speed of the geodesic from m to x_v: $a = \text{Log}_m(x_v)$.
2. Extend this diagonal geodesic by the same length to obtain $z = \text{Exp}_m(-a)$.
3. Repeat steps 2 and 3 with $x_v \leftarrow z$ and $m \leftarrow \text{Exp}_m(\frac{w}{n})$.

After n steps, compute $\tilde{v} = n^\alpha(-1)^n\text{Log}_y(z)$. According to [3], $\alpha \geq 1$ can be chosen, and $\alpha = 2$ is optimal. This vector is an approximation of the parallel transport of v along γ, $\Pi_x^{x_w} v$. This is illustrated on the 2-sphere and in the case $k = m = 3$ on Fig. 2.

3.2 Properties

The pole ladder is studied in depth in [3]. We give here the two main properties. Beside its quadratic convergence speed, the main advantage is that this method is available as soon as geodesics are known (even approximately). It is thus applicable very easily in the case of quotient metrics.

Theorem 2 (Guigui and Pennec [3])

- *The pole ladder converges to the exact parallel transport when the number of steps n goes to infinity, and the error decreases in $O(\frac{1}{n^2})$, with rate related to the covariant derivative of the curvature tensor.*
- *If M is a symmetric space, then the pole ladder is exact in just one step.*

For instance, Σ_2^3 is symmetric [6], making pole ladder exact in this case.

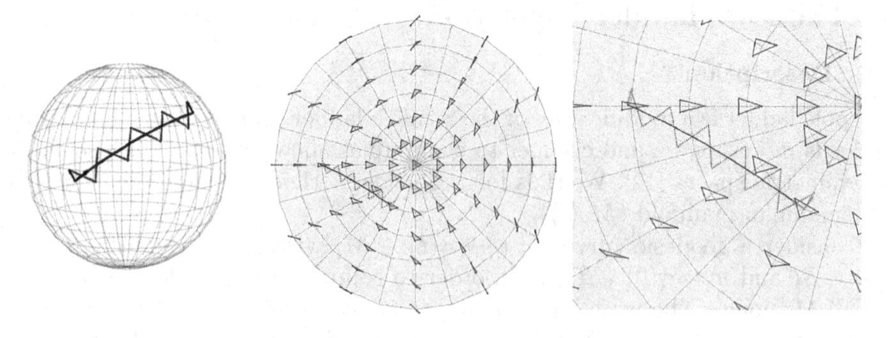

Fig. 2. Visualisation of the pole ladder on S^2 (left) and Σ_3^3 (middle and right)

3.3 Complexity

The main drawback of ladder schemes is that logarithms are required. Indeed the Riemannian logarithm is only locally defined, and often solved by an optimisation problem when geodesics are not known in closed form.

In the case of Kendall shape spaces, it only requires to compute an alignment step, through a singular value decomposition, with usual complexity $O(m^3)$, then the log of the hypersphere, with linear complexity. Moreover, the result of $\log \circ \omega$ is horizontal, so the vertical component needs not be computed for the exponential of step 2, and only the Exp of the hypersphere, also with linear complexity, needs to be computed. The vertical projection needs to be computed for the first step. Solving the Sylvester equation through an eigenvalue decomposition also has complexity m^3. For n rungs of the pole ladder, the overall complexity is thus $O((n+1)(m^3+2mk)) + mk + m^3) = O(nm^3)$.

On the other hand, the method by integration doesn't require logarithms but requires solving a Sylvester equation and a vertical decomposition at every step. The overall complexity is thus $O(2nm^3 + mk)$. Both algorithms are thus comparable in terms of computational cost for a single step.

4 Numerical Simulations and Results

We draw a point x at random in the pre-shape space, along with two orthogonal horizontal unit tangent vectors v, w, and compute the parallel transport of $d_x\pi v$ along the geodesic with initial velocity $d_x\pi w$. We use a number of steps n between 10 and 1000 and the result with $n = 1100$ as the reference value to compute the error made by lower numbers of steps. The results are displayed on Fig. 3 for the cases $k = 4, 6$ and $m = 3$ in log-log plots. As expected, the method proposed by [5] converges linearly, while RK schemes of order two and four show significant acceleration. The pole ladder converges with quadratic speed and thus compares with the RK method of order two, although the complexity of the RK method is multiplied by its order. The implementation of the two methods and

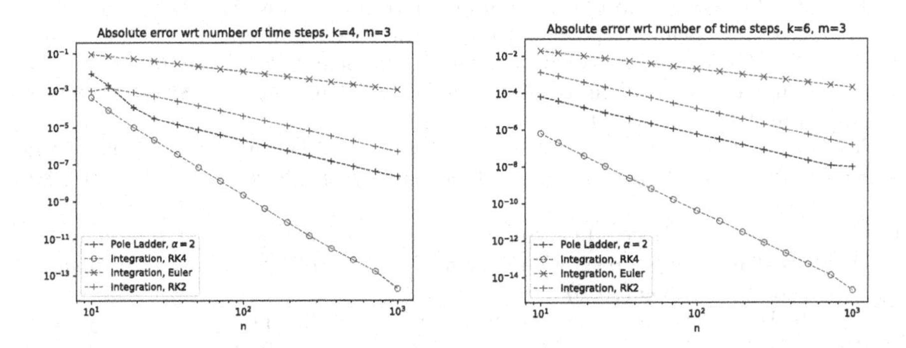

Fig. 3. Error of the parallel transport of v along the geodesic with initial velocity w where v and w are orthonormal.

an illustrative notebook on Kendall shapes of triangles can be found in the geomstats library [9], available at http://geomstats.ai.

5 Conclusion and Future Work

We presented the Kendall shape space and metric, highlighting the properties stemming from its quotient structure. This allows to compute parallel transport with the pole ladder using closed-form solution for the geodesics. This off-the-shelf algorithm can now be used in learning algorithms such as geodesic regression or local non-linear embedding. This will be developed in future works.

Acknowledgments. This work was partially funded by the ERC grant Nr. 786854 G-Statistics from the European Research Council under the European Union's Horizon 2020 research and innovation program. It was also supported by the French government through the 3IA Côte d'Azur Investments ANR-19-P3IA-0002.

References

1. Cury, C., et al.: Spatio-temporal shape analysis of cross-sectional data for detection of early changes in neurodegenerative disease. In: Reuter, M., Wachinger, C., Lombaert, H. (eds.) SeSAMI 2016. LNCS, vol. 10126, pp. 63–75. Springer, Cham (2016). https://doi.org/10.1007/978-3-319-51237-2_6
2. Dryden, I.L., Mardia, K.V.: Statistical Shape Analysis: With Applications in R. Wiley, Hoboken (2016)
3. Guigui, N., Pennec, X.: Numerical accuracy of ladder schemes for parallel transport on manifolds. In: Foundations of Computational Mathematics (Springer, in press)
4. Huckemann, S., Hotz, T., Munk, A.: Intrinsic MANOVA for Riemannian manifolds with an application to Kendall's space of planar shapes. IEEE Trans. Pattern Anal. Mach. Intell. **32**(4), 593–603 (2009)
5. Kim, K.R., Dryden, I.L., Le, H., Severn, K.E.: Smoothing splines on Riemannian manifolds, with applications to 3D shape space. J. Roy. Stat. Soc. Ser. B (Stat. Methodol.) **83**(1), 108–132 (2021). https://doi.org/10.1111/rssb.12402
6. Le, H., Kendall, D.G.: The Riemannian structure of Euclidean shape spaces: a novel environment for statistics. Ann. Stat. **21**(3), 1225–1271 (1993)
7. Lorenzi, M., Pennec, X.: Efficient parallel transport of deformations in time series of images: from Schild to pole ladder. J. Math. Imaging Vision **50**(1), 5–17 (2014)
8. Louis, M., Charlier, B., Jusselin, P., Pal, S., Durrleman, S.: A fanning scheme for the parallel transport along geodesics on Riemannian manifolds. SIAM J. Numer. Anal. **56**(4), 2563–2584 (2018)
9. Miolane, N., Guigui, N., Brigant, A.L., Mathe, J., et al.: Geomstats: a Python package for Riemannian geometry in machine learning. J. Mach. Learn. Res. **21**(223), 1–9 (2020)
10. Nava-Yazdani, E., Hege, H.C., Sullivan, T.J., von Tycowicz, C.: Geodesic analysis in Kendall's shape space with epidemiological applications. J. Math. Imaging Vision **62**(4), 549–559 (2020)
11. O'Neill, B.: Semi-Riemannian Geometry With Applications to Relativity. Academic Press (1983)
12. Pennec, X.: Parallel Transport with Pole Ladder: A Third Order Scheme in Affine Connection Spaces which is Exact in Affine Symmetric Spaces, May 2018

Diffusion Means and Heat Kernel on Manifolds

Pernille Hansen[1], Benjamin Eltzner[2(✉)], and Stefan Sommer[1]

[1] Department of Computer Science, University of Copenhagen, Copenhagen, Denmark
[2] Felix-Bernstein-Institute for Mathematical Statistics in the Biosciences, Georg-August-Universität at Göttingen, Goettingen, Germany
benjamin.eltzner@mpibpc.mpg.de

Abstract. We introduce diffusion means as location statistics on manifold data spaces. A diffusion mean is defined as the starting point of an isotropic diffusion with a given diffusivity. They can therefore be defined on all spaces on which a Brownian motion can be defined and numerical calculation of sample diffusion means is possible on a variety of spaces using the heat kernel expansion. We present several classes of spaces, for which the heat kernel is known and sample diffusion means can therefore be calculated. As an example, we investigate a classic data set from directional statistics, for which the sample Fréchet mean exhibits finite sample smeariness.

1 Introduction

In order to analyze data which are represented not on a vector space but a more general space $\mathcal{M}$, where we focus on manifolds here, it is necessary to generalize concepts from Euclidean spaces to more general spaces. Important examples of data on non-Euclidean spaces include directional data, cf. [14], and landmark shape spaces, see [16].

In the field of general relativity where similar generalizations are required, one usually relies on the *correspondence principle* as a minimal requirement for quantities on curved spaces. It states that an observable in curved space should reduce to the corresponding observable in the flat case. In terms of statistics this means that any generalization $S_{\mathcal{M}}$ of a statistic S on Euclidean space to a manifold $\mathcal{M}$ should reduce to S on Euclidean space.

The mean is the most widely used location statistic and was generalized to metric spaces by [6], who defined it as the minimizer of expected squared distance. This definition of the Fréchet mean satisfies the correspondence principle. However, as is often the case in physics, it is by far not the only parameter that has this property. In order to judge which potential other generalizations of the mean are meaningful, we recall that one of the reasons the mean is so widely used and useful is that it is an estimator of (one of) the model parameter(s) in many parametric families of probability distributions including Normal, Poisson, Exponential, Bernoulli and Binomial distributions. It is therefore useful to

F. Nielsen and F. Barbaresco (Eds.): GSI 2021, LNCS 12829, pp. 111–118, 2021.
https://doi.org/10.1007/978-3-030-80209-7_13

link potential generalizations of the mean to parametric families on general data spaces.

Few parametric families of probability distributions have been widely generalized to non-Euclidean data spaces or even to $\mathbb{R}^m$ with $m > 1$. One class of distributions which can be widely generalized are isotropic normal distributions, as these can be defined as the distribution of a Brownian motion after unit time. Since Brownian motion can be defined on any manifold, this definition is very general. The thus defined probability distributions have one location parameter $\mu^t \in \mathcal{M}$ and a spread parameter $t \in \mathbb{R}^+$. Its clear interpretation in terms of the generalized normal distribution family makes the *diffusion mean* μ^t an interesting contender for a generalization of the Euclidean mean [9].

The location parameter of Brownian motion can be defined using the heat kernels $p(x, y, t)$, which are the transition densities of Brownian motions. For a fixed $t > 0$, we define the *diffusion t-mean set* as the minima

$$E^t(X) = \underset{\mu^t \in P}{\operatorname{argmin}} \, \mathbb{E}[-\ln p(X, \mu^t, t)]. \tag{1}$$

The logarithmic heat kernel is naturally connected to geodesic distance due to the limit $\lim_{t \to 0} -2t \ln p(x, y, t) = \operatorname{dist}(x, y)^2$, cf. [11], which means that the Fréchet mean can be interpreted as the μ^0 diffusion mean, i.e. the limit for $t \to 0$. In Euclidean space, Eq. (1) reduces to the MLE of the normal distribution, which means that μ^t does not depend on t. On other data spaces, the two parameters do not decouple in general and μ^t can be different depending on t. Since all diffusion means satisfy the correspondence principle for the Euclidean mean, it is not immediately clear why the Fréchet mean should be preferred over diffusion means for finite $t > 0$.

Explicit expressions for the heat kernel are known for several classes of manifolds and for a more general class a recursively defined asymptotic series expansion exists. This means that diffusion means can be numerically determined on many data spaces and their dependence on t can be studied. In this article, we focus on *smeariness of means* as described by [4,5,10], more precisely finite sample smeariness as described by [12]. A smeary mean satisfies a modified central limit theorem with a slower asymptotic rate than $n^{-1/2}$. This affects samples drawn from such a population, whose sample means can exhibit *finite sample smeariness.*

After a brief overview of the relevant concepts, we will give a number of examples of data spaces in which diffusion means can be readily computed. Lastly, we investigate the diffusion means for a directional data set on S^1 whose Fréchet mean exhibits finite sample smeariness. We find that the diffusion means exhibit less finite sample smeariness with increasing t.

2 Basic Concepts and Definitions

A Riemannian manifold $(\mathcal{M}, g)$ is a smooth manifold equipped with inner products $\langle .,. \rangle_x$ on the tangent spaces $T_x\mathcal{M}$ for each $x \in \mathcal{M}$ such that $x \mapsto \langle v_x, u_x \rangle_x$ is smooth for all vector fields $u, v \in T\mathcal{M}$. Curves which are locally length minimizing are called *geodesics*. The *Riemannian distance* between two points is defined by the infimum over the lengths of geodesics connecting the points. If a geodesic from x through y is length minimizing exactly up to and including y, but not for any point after that, we say that y is in the cut of x and define the *cut locus* $C(x)$ as the collection of all such points.

For every starting point $x \in \mathcal{M}$ and velocity vector $v \in T_x\mathcal{M}$ there is a unique geodesic $\gamma_{x,v}$ and the exponential map $\exp_x : T_x\mathcal{M} \to \mathcal{M}$ maps v to the point reached in unit time $\gamma_{x,v}(1)$. The exponential map is a diffeomorphism on a subset $D(x) \subset T_x\mathcal{M}$ such that its image coincides with $\mathcal{M}\backslash C(x)$, and the logarithm $\log_x : \mathcal{M}\backslash C(x) \to T_x\mathcal{M}$ is defined as the inverse map.

The heat kernel is the fundamental solution to the heat equation

$$\frac{d}{dt}p(x,y,t) = \frac{1}{2}\Delta_x p(x,y,t)$$

where Δ is the Laplace Beltrami operator on $\mathcal{M}$ and it is also the transition density of Brownian motions on $\mathcal{M}$. A Riemannian manifold is stochastically complete if there exists a minimal solution p satisfying $\int_{\mathcal{M}} p(x,y,t)dy = 1$ for all $x \in \mathcal{M}$ and $t > 0$. The minimal solution is strictly positive, smooth and symmetric in its first two arguments.

Let $(\Omega, \mathcal{F}, \mathbb{P})$ be the underlying probability space of the random variable X on the stochastically complete manifold $\mathcal{M}$ with minimal heat kernel p. We define the negative log-likelihood function $L^t : \mathcal{M} \to \mathbb{R}$ to be

$$L^t(y) = \mathbb{E}[-\ln p(X,y,t)] \tag{2}$$

for $t > 0$. This gives rise to the diffusion means as the global minima of the log-likelihood function, i.e. the points maximizing the log-likelihood of a Brownian motion.

Definition 1. *With the underlying probability space $(\Omega, \mathcal{F}, \mathbb{P})$, let X be a random variable on $\mathcal{M}$ and fix $t > 0$. The* diffusion t-mean set $E^t(X)$ *of X is the set of global minima of the log-likelihood function L^t, i.e.*

$$E^t(X) = \operatorname*{argmin}_{y \in \mathcal{M}} \mathbb{E}[-\ln(p(X,y,t))].$$

If $E^t(X)$ contains a single point μ^t, we say that μ^t is the diffusion t-mean *of X.*

We consider the asymptotic behavior and smeariness of the following estimator.

Definition 2. *For samples $X_1, ..., X_n \overset{i.i.d.}{\sim} X$ on $\mathcal{M}$ we define the* sample log-likelihood function $L_n^t : \mathcal{M} \to \mathbb{R}$,

$$L_n^t(y) = -\frac{1}{n}\ln\Big(\prod_{i=1}^n p(X_i,y,t)\Big) = -\frac{1}{n}\sum_{i=1}^n \ln p(X_i,y,t)$$

for every $n \in \mathbb{N}$ and the sample diffusion t-mean sets $E_{t,n} = \operatorname*{argmin}_{y \in \mathcal{M}} L_n^t(y)$.

Definition 3 (Smeariness of Diffusion Means). *Consider a random variable X on $\mathcal{M}$, an assume that there is $\zeta > 0$ such that for every $x \in B_\zeta(0) \setminus \{0\}$ one has $L^t(\exp_{\mu^t}(x)) > L^t(\mu^t)$. Suppose that the minimal $\kappa \in \mathbb{R}$ such that for fixed constants $C_X > 0$ and for every sufficiently small $\delta > 0$*

$$\sup_{x \in T_{\mu^t}\mathcal{M},\, \|x\| < \delta} \left| L^t(\exp_{\mu^t}(x)) - L^t(\mu^t) \right| \geq C_X \delta^\kappa$$

satisfies $\kappa > 2$. Then we say that the diffusion mean μ^t of X is smeary.

Definition 4 (Finite Sample Smeary Mean). *For the population mean $\mu^t \in \mathcal{M}$ and its corresponding sample estimator $\widehat{\mu}^t_n$ let*

$$\mathfrak{m}^t_n := \frac{n\mathbb{E}[d^2(\widehat{\mu}^t_n, \mu^t)]}{\mathbb{E}[d^2(X, \mu^t)]}$$

be the variance ratio of μ^t. *Then μ^t is called* finite sample smeary, *if*

$$S^t_{\mathrm{FSS}} := \sup_{n \in \mathbb{N}} \mathfrak{m}^t_n > 1 \qquad \text{and} \qquad S^t_{\mathrm{FSS}} < \infty\,.$$

The latter requirement distinguishes finite sample smeariness from smeariness, where $\lim_{n\to\infty} \mathfrak{m}^t_n = \infty$.

A consistent estimator for $\mathfrak{m}^t_n$ can be given using the n-out-of-n bootstrap

$$\widehat{\mathfrak{m}}^t_n := \frac{n\frac{1}{B}\sum_{b=1}^B d^2(\mu^{t,*b}_{n,n}, \widehat{\mu}^t_n)}{\frac{1}{n}\sum_{j=1}^n d^2(\widehat{\mu}^t_n, X_j)}\,.$$

This estimator is used in the application below.

3 Examples of Known Heat Kernels

Example 5. *The heat kernel p on the Euclidean space $\mathbb{R}^m$ is given by the function*

$$p(x, y, t) = \left(\frac{1}{(4\pi t)^{m/2}}\right) e^{\frac{-|x-y|^2}{4t}}\,.$$

for $x, y \in \mathbb{R}^m$ and $t > 0$. The diffusion t-means of a random variable X do not depend on t and coincide with the expected value $\mathbb{E}[X]$ since

$$\operatorname*{argmin}_{y \in \mathbb{R}^m} L^t(y) = \operatorname*{argmin}_{y \in \mathbb{R}^m} \mathbb{E}[(X - y)^2]$$

Thus, $\mu^t = \mathbb{E}[X]$ for all $t > 0$.

Example 6. *The heat kernel on the circle* $\mathcal{S}^1$ *is given by the wrapped Gaussian*

$$p(x,y,t) = \frac{1}{\sqrt{4\pi t}}\Big(\sum_{k\in\mathbb{Z}} \exp\Big(\frac{-(x-y+2\pi k)^2}{4t}\Big)\Big)$$

for $x, y \in \mathbb{R}/\mathbb{Z} \cong \mathcal{S}^1$ *and* $t > 0$, *and the log-likelihood function for the random variable* $X : \Omega \to \mathcal{S}^1$ *becomes*

$$L^t(y) = -\ln\left(\sqrt{4\pi t}\right) + \int_{\mathcal{S}^1} \ln\Big(\sum_{k\in\mathbb{Z}} \exp\Big(\frac{-(x-y+2\pi k)^2}{4t}\Big)\Big) d\mathbb{P}_X(x) \tag{3}$$

Notably, even on this simple space, the t*-dependence in the exponentials is not a simple prefactor and* μ^t *is therefore explicitly dependent on* t.

Example 7. *The heat kernel on the spheres* $\mathcal{S}^m$ *for* $m \geq 2$ *can be expressed as the uniformly and absolutely convergent series, see [19, Theorem 1],*

$$p(x,y,t) = \sum_{l=0}^{\infty} e^{-l(l+m-1)t} \frac{2l+m-1}{m-1} \frac{1}{A_{\mathcal{S}}^m} C_l^{(m-1)/2}(\langle x,y\rangle_{\mathbb{R}^{m+1}})$$

for $x, y \in \mathcal{S}^m$ *and* $t > 0$, *where* C_l^α *are the Gegenbauer polynomials and* $A_{\mathcal{S}}^m = \frac{2\pi^{(m+1)/2}}{\Gamma((m+1)/2)}$ *the surface area of* $\mathcal{S}^m$. *For* $m = 2$, *the Gegenbauer polynomials* $C_l^{1/2}$ *coincide with the Legendre polynomials* P_l^0 *and the heat kernel on* $\mathcal{S}^2$ *is*

$$p(x,y,t) = \sum_{l=0}^{\infty} e^{-l(l+1)t} \frac{2l+1}{4\pi} P_l^0(\langle x,y\rangle_{\mathbb{R}^3}).$$

Again, μ^t *is explicitly dependent on* t *on these spaces.*

Example 8. *The heat kernel on the hyperbolic space* $\mathbb{H}^m$ *can be expressed by the following formulas, see [8], for* $n \geq 1$. *For odd* $m = 2k+1$, *the heat kernel is given by*

$$p(x,y,t) = \frac{(-1)^k}{2^k \pi^k} \frac{1}{\sqrt{4\pi t}} \frac{\rho}{\sinh\rho} \left(\frac{1}{\sinh\rho}\frac{\partial}{\partial\rho}\right)^k e^{-k^2 t - \frac{\rho^2}{4t}}$$

where $\rho = \mathrm{dist}_{\mathbb{H}^m}(x,y)$ *and for even* $m = 2k+2$, *it is given by*

$$p(x,y,t) = \frac{(-1)^k}{2^{k+\frac{5}{2}}\pi^{k+\frac{3}{2}}} t^{-\frac{3}{2}} \frac{\rho}{\sinh\rho} \left(\frac{1}{\sinh\rho}\frac{\partial}{\partial\rho}\right)^k \int_\rho^\infty \frac{s\exp\left(-\frac{s^2}{4t}\right)}{(\cosh s - \cosh\rho)^{\frac{1}{2}}} ds.$$

Again, μ^t *is explicitly dependent on* t *on these spaces.*

Example 9. *The fundamental solution to the heat equation on a Lie group* G *of dimension* m *is*

$$p(x,e,t) = (2\pi t)^{-m/2} \prod_{\alpha\in\Sigma^+} \frac{i\alpha(H)}{2\sin(i\alpha(H)/2)} \exp\left(\frac{\|H\|^2}{2t} + \frac{\|\rho\|^2 t}{2}\right) \cdot E_x(X_\tau > t)$$

where e is the neutral element, $x = \exp(Ad(g)H) \in G \backslash \mathcal{C}(e)$ *for some* $g \in G$, Σ^+ *is the set of positive roots,* $\rho = \sum_{\alpha \in \Sigma^+} \alpha$, *and* τ *is the hitting time of* $\mathcal{C}(e)$ *by* $(x_s)_{0 \leq s \leq t}$. *Lie groups are relevant data spaces for many application, e.g. in the modeling of joint movement for robotics, prosthetic development and medicine.*

The symmetry in the example above is quite noticeable and in fact when $t > 0$ and $x \in \mathcal{M}$ where $\mathcal{M} = \mathbb{R}^m, \mathcal{S}^m$ or $\mathbb{H}^m$, the heat kernel only depends on the geodesic distance, see [1].

Remark 10. *For spaces where a closed form has not been obtained, we can turn to various estimates. We include some of the most well known below.*

1. *For complete Riemannian manifolds of dimension m, we have the asymptotic expansion, see [11],*

$$p(x, y, t) \sim \left(\frac{1}{2\pi t}\right)^{\frac{m}{2}} e^{\frac{-d(x,y)^2}{2t}} \sum_{n=0}^{\infty} H_n(x, y) t^n$$

on compact subset with $x \notin \mathcal{C}(y)$. *Here* H_n *are smooth functions satisfying a recursion formula, see [2], with* $H_0(x, y) = \sqrt{J(\exp_x)(\exp_x^{-1}(y))}$ *and J denoting the Jacobian.*

2. *Assuming also non-negative Ricci curvature, the heat kernel is bounded from both sides, see [7,15],*

$$\frac{c_1}{vol(x, \sqrt{t})} \exp\left(\frac{\text{dist}(x, y)^2}{c_2 t}\right) \leq p(x, y, t) \leq \frac{c_3}{vol(x, \sqrt{t})} \exp\left(\frac{\text{dist}(x, y)^2}{c_4 t}\right)$$

where $vol(x, \sqrt{t})$ *denotes the volume of the ball around x of radius* $\sqrt{t}$ *and positive constants* c_i *for* $i = 1, ..., 4$.

3. *Using bridge sampling, the heat kernel can be estimated by the expectation over guided processes, see [3,13]. An example of this is the estimated heat kernel on landmark manifolds by [17].*

4 Application to Smeariness in Directional Data

In this Section we apply diffusion means to the classic data set denoting the compass directions of sea turtles leaving their nest after egg laying, cf. [18] and Mardia and Jupp [14, p. 9]. The data set is clearly bimodal with two antipodal modes. It was shown in [5] that the Fréchet mean for this data set exhibits pronounced finite sample smeariness. We let the sum in the likelihood (3) run from $k = -3$ to $k = 3$, thus excluded terms are smaller than $e^{-(6\pi)^2/3} < 10^{-50}$. To avoid local minima, we run each optimization with 10 random initial values.

In Fig. 1a, we show that increasing the diffusivity t, the likelihood function at the minimum, starting out unusually flat, approaches a parabolic shape. In Fig. 1b, we show the corresponding curves of estimated variance ratio $\widehat{\mathfrak{m}}_n^t$, whose maxima can be used as estimators for the magnitude S_{FSS}^t of finite sample

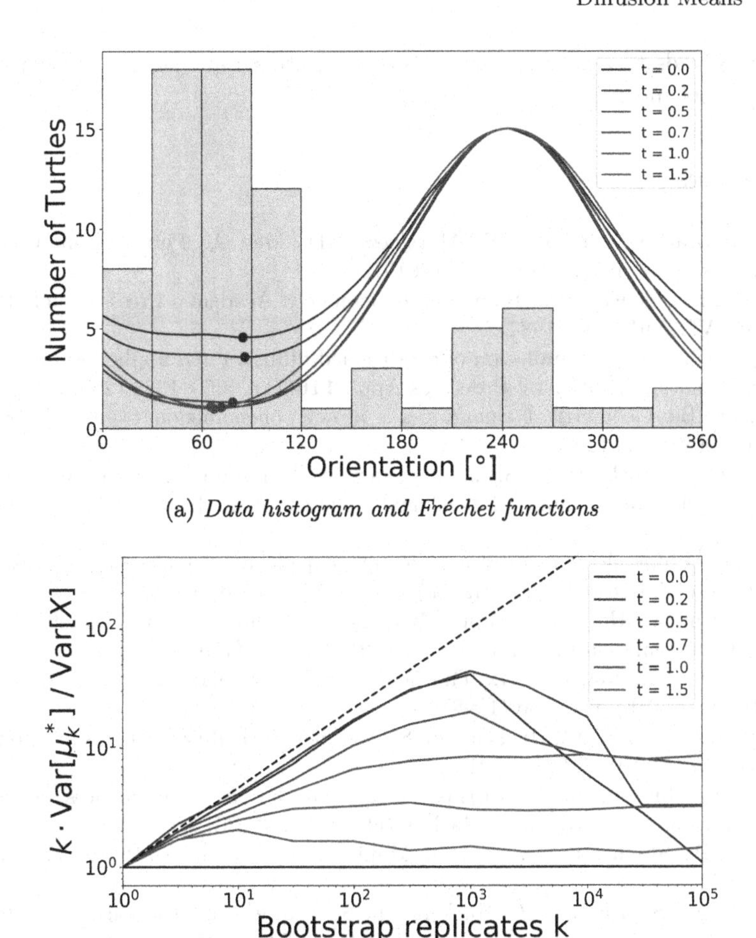

(a) *Data histogram and Fréchet functions*

(b) *Variances of sample means*

Fig. 1. Diffusion means for nesting sea turtles. Panel (a) shows that most turtles head towards the sea shore (east-northeast) while some move in exactly opposite direction. Panel (b) shows that, correspondingly, finite sample smeariness decreases in magnitude with increasing t.

smeariness. The flatter the likelihood is at the minimum, the higher we expect the magnitude of finite sample smeariness to be. From visual inspection of the likelihoods, we thus expect the magnitude of finite sample smeariness to decrease with increasing t, which is confirmed by Fig. 1b.

This behavior of reducing the effects of smeariness has been found analogously in other applications and simulations. These results suggest that diffusion means can provide a more robust location statistic than the Fréchet mean for spread out data on positively curved spaces. This point is reinforced by estimating t and μ^t jointly, which yields $\widehat{t} = 0.963$. As one can see from the case $t = 1$ in

Fig. 1, this leads to a very low magnitude of finite sample smeariness compared to the Fréchet mean.

References

1. Alonso-Orán, D., Chamizo, F., Martínez, A.D., Mas, A.: Pointwise monotonicity of heat kernels. arXiv:1807.11072 (2019)
2. Chavel, I.: Eigenvalues in Riemannian Geometry. Academic Press. Google-Books-ID: 0v1VfTWuKGgC (1984)
3. Delyon, B., Hu, Y.: Simulation of conditioned diffusion and application to parameter estimation. Stochastic Processes Appl. **116**(11), 1660–1675 (2006)
4. Eltzner, B.: Geometrical smeariness - a new phenomenon of Fréchet means. arXiv:1908.04233 (2020)
5. Eltzner, B. Huckemann, S.F.: A smeary central limit theorem for manifolds with application to high dimensional spheres. Ann. Stat. **47**(6), 3360–3381. arXiv: 1801.06581 (2019)
6. Fréchet, M.: Les éléments aléatoires de nature quelconque dans un espace distancié. Annales de l'institut Henri Poincaré **10**(4), 215–310 (1948)
7. Grigor'yan, A.: Heat kernel upper bounds on a complete non-compact manifold. Revista Matemática Iberoamericana **10**(2), 395–452 (1994)
8. Grigor'yan, A., Noguchi, M.: The heat kernel on hyperbolic space. Bull. Lond. Math. Soc. **30**(6), 643–650 (1998)
9. Hansen, P., Eltzner, B., Huckemann, S., Sommer, S.: Diffusion Means in Geometric Spaces. arXiv:2105.12061 (2021)
10. Hotz, T., Huckemann, S.: Intrinsic means on the circle: uniqueness, locus and asymptotics. Ann. Inst. Stat. Math. **67**(1), 177–193 (2015)
11. Hsu, E.P.: Stochastic Analysis on Manifolds. American Mathematical Society (2002)
12. Hundrieser, S., Eltzner, B., Huckemann, S.: Finite sample smeariness of fréchet means and application to climate. arXiv stat.ME 2005.02321 (2020)
13. Jensen, M.H., Sommer, S.: Simulation of conditioned semimartingales on riemannian manifolds. arXiv:2105.13190 (2021)
14. Mardia, K.V., Jupp, P.E.: Directional Statistics. Wiley, New York (2000)
15. Saloff-Coste, L.: A note on Poincaré, Sobolev, and Harnack inequalities. Library Catalog (1992). www.semanticscholar.org
16. Small, C.G.: The Statistical Theory of Shape. SSS, Springer-Verlag, New York (1996). https://doi.org/10.1007/978-1-4612-4032-7
17. Sommer, S., Arnaudon, A., Kuhnel, L., Joshi, S.: Bridge simulation and metric estimation on landmark manifolds. In: Cardoso, M. et al. (Eds.) Graphs in Biomedical Image Analysis, Computational Anatomy and Imaging Genetics. GRAIL 2017, MICGen 2017, MFCA 2017. LNCS, vol. 10551, pp. 79–91. Springer, Cham, September 2017. https://doi.org/10.1007/978-3-319-67675-3_8
18. Stephens, M.: Techniques for directional data. In Technical report 150, Department of Statistics, Stanford University (1969)
19. Zhao, C., Song, J.S.: Exact heat kernel on a hypersphere and its applications in kernel SVM. Front. Appl. Math. Stat. **4**, 1 (2018)

A Reduced Parallel Transport Equation on Lie Groups with a Left-Invariant Metric

Nicolas Guigui(✉) and Xavier Pennec

Université Côte d'Azur and Inria, Epione Project Team, Sophia-Antipolis, France
nicolas.guigui@inria.fr

Abstract. This paper presents a derivation of the parallel transport equation expressed in the Lie algebra of a Lie group endowed with a left-invariant metric. The use of this equation is exemplified on the group of rigid body motions $SE(3)$, using basic numerical integration schemes, and compared to the pole ladder algorithm. This results in a stable and efficient implementation of parallel transport. The implementation leverages the python package geomstats and is available online.

Keywords: Parallel transport · Lie groups

1 Introduction

Lie groups are ubiquitous in geometry, physics and many application domains such as robotics [3], medical imaging [14] or computer vision [9], giving rise to a prolific research avenue. Structure preserving numerical methods have demonstrated significant qualitative and quantitative improvements over extrinsic methods [10]. Moreover, machine learning [2] and optimisation methods [11,15] are being developed to deal with Lie group data.

In this context, parallel transport is a natural tool to define statistical models and optimisation procedures, such as the geodesic or spline regression [12,19], or to normalise data represented by tangent vectors [4,21].

Different geometric structures are compatible with the group structure, such as its canonical Cartan connection, whose geodesics are one-parameter subgroups, or left-invariant Riemannian metrics. In this work we focus on the latter case, that is fundamental in geometric mechanics [13] and has been studied in depth since the foundational papers of Arnold [1] and Milnor [17]. The fundamental idea of Euler-Poincarré reduction is that the geodesic equation can be expressed entirely in the Lie algebra thanks to the symmetry of left-invariance [16], alleviating the burden of coordinate charts.

However, to the best of our knowledge, there is no literature on a similar treatment of the parallel transport equation. We present here a derivation of the parallel transport equation expressed in the Lie algebra of a Lie group endowed with a left-invariant metric. We exemplify the use of this equation on the group

F. Nielsen and F. Barbaresco (Eds.): GSI 2021, LNCS 12829, pp. 119–126, 2021.
https://doi.org/10.1007/978-3-030-80209-7_14

of rigid body motions $SE(3)$, using common numerical integration schemes, and compare it to the pole ladder approximation algorithm. This results in a stable and efficient implementation of parallel transport. The implementation leverages the python package geomstats and is available online at http://geomstats.ai.

In Sect. 2, we give the general notations and recall some basic facts from Lie group theory. Then we derive algebraic expressions of the Levi-Civita connection associated to the left-invariant metric in Sect. 3. The equation of parallel transport is deduced from this expression and its integration is exemplified in Sect. 4.

2 Notations

Let G be a lie group of (finite) dimension n. Let e be its identity element, $\mathfrak{g} = T_eG$ be its tangent space at e, and for any $g \in G$, let $L_g : h \in G \mapsto gh$ denote the left-translation map, and dL_g its differential map. Let $\mathfrak{g}^L$ be the Lie algebra of left-invariant vector fields of G: $X \in \mathfrak{g}^L \iff \forall g \in G, X_g = dL_g X_e$.

$\mathfrak{g}$ and $\mathfrak{g}^L$ are in one-to-one correspondence, and we will write $\tilde{x}$ the left-invariant field generated by $x \in \mathfrak{g}$: $\forall g \in G$, $\tilde{x}_g = dL_g x$. The bracket defined on $\mathfrak{g}$ by $[x, y] = [\tilde{x}, \tilde{y}]_e$ turns $\mathfrak{g}$ into a Lie algebra that is isomorphic to $\mathfrak{g}^L$. One can also check that this bracket coincides with the adjoint map defined by $\mathrm{ad}_x(y) = d_e(g \mapsto \mathrm{Ad}_g y)$, where $Ad_g = d_e(h \mapsto ghg^{-1})$. For a matrix group, it is the commutator.

Let $(e_1, \ldots, e_n)$ be an orthonormal basis of $\mathfrak{g}$, and the associated left-invariant vector fields $\tilde{e_i} = g \mapsto dL_g e_i$. As dL_g is an isomorphism, $(\tilde{e}_{1,g}, \ldots, \tilde{e}_{n,g})$ form a basis of T_gG for any $g \in G$, so one can write $X_g = f^i(g)\tilde{e}_{1,g}$ where for $i = 1, \ldots, n$, $g \mapsto f^i(g)$ is a smooth real-valued function on G. Any vector field on G can thus be expressed as a linear combination of the $\tilde{e}_i$ with function coefficients.

We use Einstein summation convention to express decomposition in a given basis. Summation occurs along the indices that are repeated in sub and superscripts, e.g.

$$\forall x \in \mathfrak{g}, \text{ write } x = \sum_{i=1}^{n} x^i e_i = x^i e_i.$$

Finally, let θ be the Maurer-Cartan form defined on G by:

$$\forall g \in G, \forall v \in T_gG, \; \theta_g(v) = (dL_g)^{-1} v \in \mathfrak{g} \tag{1}$$

It is a $\mathfrak{g}$-valued 1-form and for a vector field X on G we write $\theta(X)_g = \theta_g(X_g)$ to simplify the notations.

3 Left-Invariant Metric and Connection

A Riemannian metric $\langle \cdot, \cdot \rangle$ on G is called left-invariant if the differential map of the left translation is an isometry between tangent spaces, that is

$$\forall g, h \in G, \forall u, v \in T_gG, \; \langle u, v \rangle_g = \langle dL_h u, dL_h v \rangle_{hg}.$$

It is thus uniquely determined by an inner product on the tangent space at the identity $T_eG = \mathfrak{g}$ of G. Furthermore, the metric dual to the adjoint map is defined such that

$$\forall a, b, c \in \mathfrak{g}, \langle \mathrm{ad}_a^*(b), c\rangle = \langle b, \mathrm{ad}_a(c)\rangle = \langle [a, c], b\rangle. \tag{2}$$

As the bracket can be computed explicitly in the Lie algebra, so can ad* thanks to the orthonormal basis of $\mathfrak{g}$. Now let ∇ be the Levi-Civita connection associated to the metric. It is also left-invariant and can be characterised by a bi-linear form on $\mathfrak{g}$ that verifies [6,15,20]:

$$\forall x, y \in \mathfrak{g}, \quad \alpha(x, y) := (\nabla_{\tilde{x}}\tilde{y})_e = \frac{1}{2}\big([x, y] - \mathrm{ad}_x^*(y) - \mathrm{ad}_y^*(x)\big) \tag{3}$$

Indeed by the left-invariance, for two left-invariant vector fields $\tilde{x}, \tilde{y} \in \mathfrak{g}^L$, the map $g \mapsto \langle\tilde{x}, \tilde{y}\rangle_g$ is constant, so for any vector field $\tilde{z}$ we have $\tilde{z}(\langle\tilde{x}, \tilde{y}\rangle) = 0$. Kozsul formula thus becomes

$$\begin{aligned} 2\langle\nabla_{\tilde{x}}\tilde{y}, \tilde{z}\rangle &= \langle[\tilde{x}, \tilde{y}], \tilde{z}\rangle - \langle[\tilde{y}, \tilde{z}], \tilde{x}\rangle - \langle[\tilde{x}, \tilde{z}], \tilde{y}\rangle \\ 2\langle\nabla_{\tilde{x}}\tilde{y}, \tilde{z}\rangle_e &= \langle[x, y], z\rangle_e - \langle \mathrm{ad}_y(z), x\rangle_e - \langle \mathrm{ad}_x(z), y\rangle_e \\ 2\langle\alpha(x, y), z\rangle_e &= \langle[x, y], z\rangle_e - \langle \mathrm{ad}_y^*(x), z\rangle_e - \langle \mathrm{ad}_x^*(y), z\rangle_e. \end{aligned} \tag{4}$$

Note however that this formula is only valid for left-invariant vector fields. We will now generalise to any vector fields defined along a smooth curve on G, using the left-invariant basis $(\tilde{e}_1, \ldots, \tilde{e}_n)$.

Let $\gamma : [0, 1] \to G$ be a smooth curve, and Y a vector field defined along γ. Write $Y = g^i\tilde{e}_i$, $\dot{\gamma} = f^i\tilde{e}_i$. Let's also define the *left-angular velocities* $\omega(t) = \theta_{\gamma(t)}\dot{\gamma}(t) = (f^i \circ \gamma)(t)e_i \in \mathfrak{g}$ and $\zeta(t) = \theta(Y)_{\gamma(t)} = (g^j \circ \gamma)(t)e_j \in \mathfrak{g}$. Then the covariant derivative of Y along γ is

$$\begin{aligned} \nabla_{\dot{\gamma}(t)}Y &= (f^i \circ \gamma)(t)\nabla_{\tilde{e}_i}(g^i\tilde{e}_i) \\ &= (f^i \circ \gamma)(t)\tilde{e}_i(g^j)\tilde{e}_j + (f^i \circ \gamma)(t)(g^j \circ \gamma)(t)(\nabla_{\tilde{e}_i}\tilde{e}_j)_{\gamma(t)} \\ dL_{\gamma(t)}^{-1}\nabla_{\dot{\gamma}(t)}Y &= (f^i \circ \gamma)(t)\tilde{e}_i(g^j)e_j + (f^i \circ \gamma)(t)(g^j \circ \gamma)(t)\nabla_{e_i}e_j \end{aligned}$$

where Leibniz formula and the invariance of the connection is used in $(\nabla_{\tilde{e}_i}\tilde{e}_j) = dL_{\gamma(t)}\nabla_{e_i}e_j$. Therefore for $k = 1..n$

$$\begin{aligned} \langle dL_{\gamma(t)}^{-1}\nabla_{\dot{\gamma}(t)}Y, e_k\rangle &= (f^i \circ \gamma)(t)\tilde{e}_i(g^j)\langle e_j, e_k\rangle \\ &\quad + (f^i \circ \gamma)(t)(g^j \circ \gamma)(t)\langle\nabla_{e_i}e_j, e_k\rangle \end{aligned} \tag{5}$$

but on one hand

$$\zeta(t) = (g^j \circ \gamma)(t)e_j \tag{6}$$

$$\begin{aligned} \dot{\zeta}(t) &= (g^j \circ \gamma)'(t)e_j = d_{\gamma(t)}g^j\dot{\gamma}(t)e_j \\ &= d_{\gamma(t)}g^j\Big((f^i \circ \gamma)(t)\tilde{e}_{i,\gamma(t)}\Big)e_j \\ &= (f^i \circ \gamma)(t)\tilde{e}_i(g^j)e_j \end{aligned} \tag{7}$$

and on the other hand, using (4):

$$\begin{aligned}(f^i \circ \gamma)(g^j \circ \gamma)\langle \nabla_{e_i} e_j, e_k \rangle &= \frac{1}{2}(f^i \circ \gamma)(g^j \circ \gamma)(\langle [e_i, e_j], e_k \rangle \\ &\quad - \langle [e_j, e_k], e_i \rangle - \langle [e_i, e_k], e_j \rangle) \\ &= \frac{1}{2}(\langle [(f^i \circ \gamma)e_i, (g^j \circ \gamma)e_j], e_k \rangle \\ &\quad - \langle [(g^j \circ \gamma)e_j, e_k], (f^i \circ \gamma)e_i \rangle \\ &\quad - \langle [(f^i \circ \gamma)e_i, e_k], (g^j \circ \gamma)e_j \rangle) \\ &= \frac{1}{2}([\omega, \zeta] - \mathrm{ad}^*_\omega \zeta - \mathrm{ad}^*_\zeta \omega) = \alpha(\omega, \zeta) \end{aligned} \tag{8}$$

Thus, we obtain an algebraic expression for the covariant derivative of any vector field Y along a smooth curve γ. It will be the main ingredient of this paper.

$$dL^{-1}_{\gamma(t)} \nabla_{\dot\gamma(t)} Y(t) = \dot\zeta(t) + \alpha(\omega(t), \zeta(t)) \tag{9}$$

A similar expression can be found in [1,7]. As all the variables of the right-hand side are defined in $\mathfrak{g}$, they can be computed with matrix operations and an orthonormal basis.

4 Parallel Transport

We now focus on two particular cases of (9) to derive the equations of geodesics and of parallel transport along a curve.

4.1 Geodesic Equation

The first particular case is for a geodesic curve γ and $Y(t) = \dot\gamma(t)$. It is then straightforward to deduce from (9) the Euler-Poincaré equation for a geodesic curve [5,13]. Indeed in this case, recall that $\omega = \theta|_{\gamma(t)}\dot\gamma(t)$ is the left-angular velocity, $\zeta = \omega$ and $\alpha(\omega, \omega) = -\mathrm{ad}^*_\omega(\omega)$. Hence γ is a geodesic if and only if $dL^{-1}_{\gamma(t)} \nabla_{\dot\gamma(t)}\dot\gamma(t) = 0$ i.e. setting the left-hand side of (9) to 0. We obtain

$$\begin{cases} \dot\gamma(t) &= dL_{\gamma(t)}\omega(t) \\ \dot\omega(t) &= \mathrm{ad}^*_{\omega(t)}\omega(t). \end{cases} \tag{10}$$

Remark 1. One can show that the metric is bi-invariant if and only if the adjoint map is skew-symmetric (see [20] or [6, Prop. 20.7]). In this case $ad^*_\omega(\omega) = 0$ and (10) coincides with the equation of one-parameter subgroups on G.

4.2 Reduced Parallel Transport Equation

The second case is for a vector Y that is parallel along the curve γ, that is, $\forall t, \nabla_{\dot\gamma(t)} Y(t) = 0$. Similarly to the geodesic equation, we deduce from (9) the parallel transport equation expressed in the Lie algebra.

Theorem 1. *Let γ be a smooth curve on G. The vector Y is parallel along γ if and only if it is solution to*

$$\begin{cases} \omega(t) & = dL_{\gamma(t)}^{-1}\dot{\gamma}(t) \\ Y(t) & = dL_{\gamma(t)}\zeta(t) \\ \dot{\zeta}(t) & = -\alpha(\omega(t), \zeta(t)) \end{cases} \tag{11}$$

One would think this result to be well-known, as it is straightforward to deduce from (9), but we are not aware of a reference of its use for a reduced system of equation for the parallel transport of a vector along a curve. Note that in order to parallel transport along a geodesic curve, (10) and (11) are solved jointly.

4.3 Application

We now exemplify Theorem 1 on the group of isometries of $\mathbb{R}^3$, $SE(3)$, endowed with a left-invariant metric g. $SE(3)$, is the semi-direct product of the group of three-dimensional rotations $SO(3)$ with $\mathbb{R}^3$, i.e. the group multiplicative law for $R, R' \in SO(3), t, t' \in \mathbb{R}^3$ is given by

$$(R, t) \cdot (R', t') = (RR', t + Rt').$$

It can be seen as a subgroup of $GL(4)$ and represented by homogeneous coordinates:

$$(R, t) = \begin{pmatrix} R & t \\ 0 & 1 \end{pmatrix},$$

and all group operations then correspond to the matrix operations. Let the metric matrix at the identity be diagonal: $G = \text{diag}(1, 1, 1, \beta, 1, 1)$ for some $\beta > 0$, the anisotropy parameter. An orthonormal basis of the Lie algebra $\mathfrak{se}(3)$ is

$$e_1 = \frac{1}{\sqrt{2}}\begin{pmatrix} 0 & 0 & 0 & 0 \\ 0 & 0 & -1 & 0 \\ 0 & 1 & 0 & 0 \\ 0 & 0 & 0 & 0 \end{pmatrix} \quad e_2 = \frac{1}{\sqrt{2}}\begin{pmatrix} 0 & 0 & 1 & 0 \\ 0 & 0 & 0 & 0 \\ -1 & 0 & 0 & 0 \\ 0 & 0 & 0 & 0 \end{pmatrix} \quad e_3 = \frac{1}{\sqrt{2}}\begin{pmatrix} 0 & -1 & 0 & 0 \\ 1 & 0 & 0 & 0 \\ 0 & 0 & 0 & 0 \\ 0 & 0 & 0 & 0 \end{pmatrix}$$

$$e_4 = \frac{1}{\sqrt{\beta}}\begin{pmatrix} 0 & 0 & 0 & 1 \\ 0 & 0 & 0 & 0 \\ 0 & 0 & 0 & 0 \\ 0 & 0 & 0 & 0 \end{pmatrix} \quad e_5 = \begin{pmatrix} 0 & 0 & 0 & 0 \\ 0 & 0 & 0 & 1 \\ 0 & 0 & 0 & 0 \\ 0 & 0 & 0 & 0 \end{pmatrix} \quad e_6 = \begin{pmatrix} 0 & 0 & 0 & 0 \\ 0 & 0 & 0 & 0 \\ 0 & 0 & 0 & 1 \\ 0 & 0 & 0 & 0 \end{pmatrix}.$$

Define the corresponding structure constants $C_{ij}^k = \langle [e_i, e_j], e_k \rangle$, where the Lie bracket $[\cdot, \cdot]$ is the usual matrix commutator. It is straightforward to compute

$$C_{ij}^k = \frac{1}{\sqrt{2}} \text{ if } ijk \text{ is a direct cycle of } \{1, 2, 3\}; \tag{12}$$

$$C_{15}^6 = -C_{16}^5 = -\sqrt{\beta}C_{24}^6 = \frac{1}{\sqrt{\beta}}C_{26}^4 = \sqrt{\beta}C_{34}^5 = -\frac{1}{\sqrt{\beta}}C_{35}^4 = \frac{1}{\sqrt{2}}. \tag{13}$$

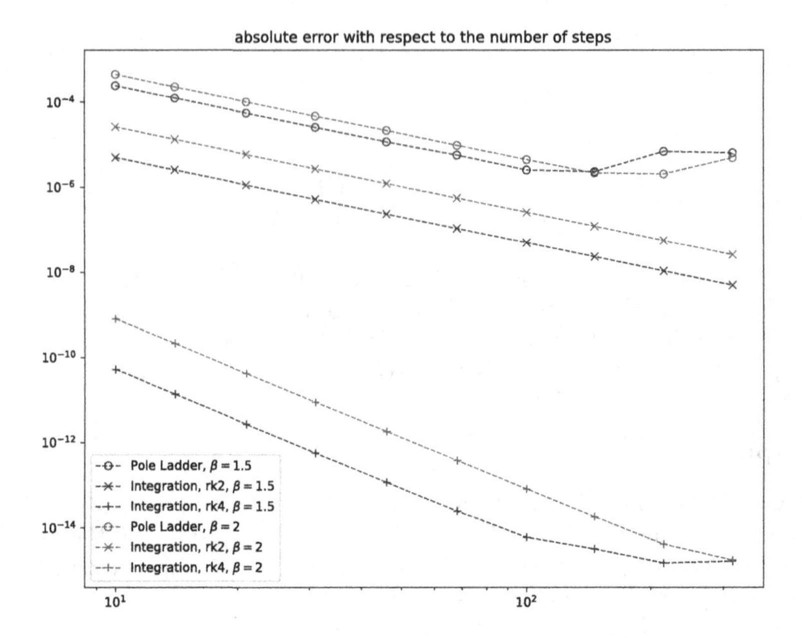

Fig. 1. Comparison of the integration of the reduced equation with the pole ladder. The norm of the absolute error is represented with respect to the number of steps.

and all others that cannot be deduced by skew-symmetry of the bracket are equal to 0. The connection can then easily be computed using

$$\alpha(e_i, e_j) = \nabla_{e_i} e_j = \frac{1}{2} \sum_k (C_{ij}^k - C_{jk}^i + C_{ki}^j) e_k,$$

For $\beta = 1$, $(SE(3), G)$ is a symmetric space and the metric corresponds to the direct product metric of $SO(3) \times \mathbb{R}^3$. However, for $\beta \neq 1$, the geodesics cannot be computed in closed-form and we resort to a numerical scheme to integrate (10). According to [8], the pole ladder can be used with only one step of a fourth-order scheme to compute the exponential and logarithm maps at each rung of the ladder. We use a Runge-Kutta (RK) scheme of order 4. The Riemannian logarithm is computed with a gradient descent on the initial velocity, where the gradient of the exponential is computed by automatic differentiation. All of these are available in the InvariantMetric class of the package geomstats [18].

We now compare the integration of (11) to the pole ladder [8] for $\beta = 1.5, 2$ to parallel transport a tangent vector along a geodesic. We sample a random initial point in $SE(3)$, and two tangent vectors at this point, that are then orthonormalised.

The results are displayed on Fig. 1 in a log-log plot representing the error between a reference value for the parallel transport and the value obtained with n steps in the discretization of the integral scheme, or in the pole ladder. The reference value is the one obtained by the integration scheme with RK steps of order four for $n = 1000$. As expected, we reach convergence speeds of order two for the pole ladder and the RK2 scheme, while the RK4 schemes is of order four. Both integration methods are very stable, while the pole ladder is less stable for $\sim n \geq 200$.

Acknowledgments. This work was partially funded by the ERC grant Nr. 786854 G-Statistics from the European Research Council under the European Union's Horizon 2020 research and innovation program. It was also supported by the French government through the 3IA Côte d'Azur Investments ANR-19-P3IA-0002 managed by the National Research Agency.

References

1. Arnold, V.: Sur la géométrie différentielle des groupes de Lie de dimension infinie et ses applications à l'hydrodynamique des fluides parfaits. Annales de l'institut Fourier **16**(1), 319–361 (1966). https://doi.org/10.5802/aif.233
2. Barbaresco, F., Gay-Balmaz, F.: Lie group cohomology and (Multi) symplectic integrators: new geometric tools for lie group machine learning based on souriau geometric statistical mechanics. Entropy **22**(5), 498 (2020). https://doi.org/10.3390/e22050498
3. Barrau, A., Bonnabel, S.: The invariant extended kalman filter as a stable observer. IEEE Trans. Autom. Control **62**(4), 1797–1812 (2017). https://doi.org/10.1109/TAC.2016.2594085
4. Brooks, D., Schwander, O., Barbaresco, F., Schneider, J.Y., Cord, M.: Riemannian batch normalization for SPD neural networks. In: Wallach, H., Larochelle, H., Beygelzimer, A., d'Alché Buc, F., Fox, E., Garnett, R. (eds.) Advances in Neural Information Processing Systems 32, pp. 15489–15500. Curran Associates, Inc. (2019)
5. Cendra, H., Holm, D.D., Marsden, J.E., Ratiu, T.S.: Lagrangian reduction, the euler-Poincaré equations, and semidirect products. Am. Math. Soc. Translations **186**(1), 1–25 (1998)
6. Gallier, J., Quaintance, J.: Differential Geometry and Lie Groups. GC, vol. 12. Springer, Cham (2020). https://doi.org/10.1007/978-3-030-46040-2
7. Gay-Balmaz, F., Holm, D.D., Meier, D.M., Ratiu, T.S., Vialard, F.X.: Invariant higher-order variational problems II. J Nonlinear Sci. **22**(4), 553–597 (2012). https://doi.org/10.1007/s00332-012-9137-2, http://arxiv.org/abs/1112.6380
8. Guigui, N., Pennec, X.: Numerical Accuracy of Ladder Schemes for Parallel Transport on Manifolds (2021). https://hal.inria.fr/hal-02894783
9. Hauberg, S., Lauze, F., Pedersen, K.S.: Unscented Kalman filtering on Riemannian manifolds. J. Math. Imaging Vis. **46**(1), 103–120 (2013). https://doi.org/10.1007/s10851-012-0372-9
10. Iserles, A., Munthe-Kaas, H., Nørsett, S., Zanna, A.: Lie-group methods. Acta Numerica (2005). https://doi.org/10.1017/S0962492900002154
11. Journée, M., Absil, P.-A., Sepulchre, R.: Optimization on the orthogonal group for independent component analysis. In: Davies, M.E., James, C.J., Abdallah, S.A., Plumbley, M.D. (eds.) ICA 2007. LNCS, vol. 4666, pp. 57–64. Springer, Heidelberg (2007). https://doi.org/10.1007/978-3-540-74494-8_8
12. Kim, K.R., Dryden, I.L., Le, H., Severn, K.E.: Smoothing splines on Riemannian manifolds, with applications to 3D shape space. J. Royal Stat. Soc. Series B (Statistical Methodology) **83**(1), 108–132 (2020)

13. Kolev, B.: Lie Groups and mechanics: an introduction. J. Nonlinear Math. Phys. **11**(4), 480–498 (2004). https://doi.org/10.2991/jnmp.2004.11.4.5. arXiv: math-ph/0402052
14. Lorenzi, M., Pennec, X.: Efficient parallel transport of deformations in time series of images: from schild to pole ladder. J. Math. Imaging Vis. **50**(1), 5–17 (2014). https://doi.org/10.1007/s10851-013-0470-3
15. Mahony, R., Manton, J.H.: The geometry of the Newton method on non-compact lie groups. J. Global Optim. **23**(3), 309–327 (2002). https://doi.org/10.1023/A:1016586831090
16. Marsden, J.E., Ratiu, T.S.: Mechanical systems: symmetries and reduction. In: Meyers, R.A. (ed.) Encyclopedia of Complexity and Systems Science, pp. 5482–5510. Springer, New York (2009). https://doi.org/10.1007/978-0-387-30440-3_326
17. Milnor, J.: Curvatures of left invariant metrics on lie groups. Adv. Math. **21**(3), 293–329 (1976). https://doi.org/10.1016/S0001-8708(76)80002-3
18. Miolane, N., et al.: Geomstats: a python package for riemannian geometry in machine learning. J. Mach. Learn. Res. **21**(223), 1–9 (2020). http://jmlr.org/papers/v21/19-027.html
19. Nava-Yazdani, E., Hege, H.C., Sullivan, T.J., von Tycowicz, C.: Geodesic analysis in Kendall's shape space with epidemiological applications. J. Math. Imaging Vis. **62**(4), 549–559 (2020). https://doi.org/10.1007/s10851-020-00945-w
20. Pennec X., Arsigny, V.: Exponential barycenters of the canonical cartan connection and invariant means on lie groups. In: Nielsen, F., Bhatia, R. (eds.) Matrix Information Geometry, pp. 123–166. Springer, Heidelberg (2013). https://doi.org/10.1007/978-3-642-30232-9_7
21. Yair, O., Ben-Chen, M., Talmon, R.: Parallel transport on the cone manifold of SPD matrices for domain adaptation. IEEE Trans. Sig. Process. **67**, 1797–1811 (2019). https://doi.org/10.1109/TSP.2019.2894801

Currents and K-functions for Fiber Point Processes

Pernille E. H. Hansen[1(✉)], Rasmus Waagepetersen[2], Anne Marie Svane[2], Jon Sporring[1], Hans J. T. Stephensen[1], Stine Hasselholt[3], and Stefan Sommer[1]

[1] Department of Computer Science, University of Copenhagen, Copenhagen, Denmark
{pehh,sporring,hast,sommer}@di.ku.dk

[2] Department of Mathematical Sciences, Aalborg University, Aalborg, Denmark
{rw,annemarie}@math.aau.dk

[3] Stereology and Microscopy, Aarhus University, Aarhus, Denmark
stha@clin.au.dk

Abstract. Analysis of images of sets of fibers such as myelin sheaths or skeletal muscles must account for both the spatial distribution of fibers and differences in fiber shape. This necessitates a combination of point process and shape analysis methodology. In this paper, we develop a K-function for fiber-valued point processes by embedding shapes as currents, thus equipping the point process domain with metric structure inherited from a reproducing kernel Hilbert space. We extend Ripley's K-function which measures deviations from complete spatial randomness of point processes to fiber data. The paper provides a theoretical account of the statistical foundation of the K-function, and we apply the K-function on simulated data and a data set of myelin sheaths. This includes a fiber data set consisting of myelin sheaths configurations at different debts.

Keywords: Point processes · Shape analysis · K-function · Fibers · Myelin sheaths

1 Introduction

We present a generalization of Ripley's K-function for fiber-valued point processes, i.e., processes where each point is a curve in $\mathbb{R}^3$. Fiber structures appear naturally in the human body, for example in tracts in the central nervous system and in skeletal muscles. The introduced K-function captures both spatial and shape clustering or repulsion, thus providing a powerful descriptive statistic for analysis of medical images of sets of fibers or more general shape data. As an example, Fig. 1 displays myelin sheaths in four configurations from different debts in a mouse brain. We develop the methodology to quantify the visually apparent differences in both spatial and shape distribution of the fibers.

Ripley's K-function [6] is a well-known tool for analyzing second order moment structure of point processes [1] providing a measure of deviance from

F. Nielsen and F. Barbaresco (Eds.): GSI 2021, LNCS 12829, pp. 127–134, 2021.
https://doi.org/10.1007/978-3-030-80209-7_15

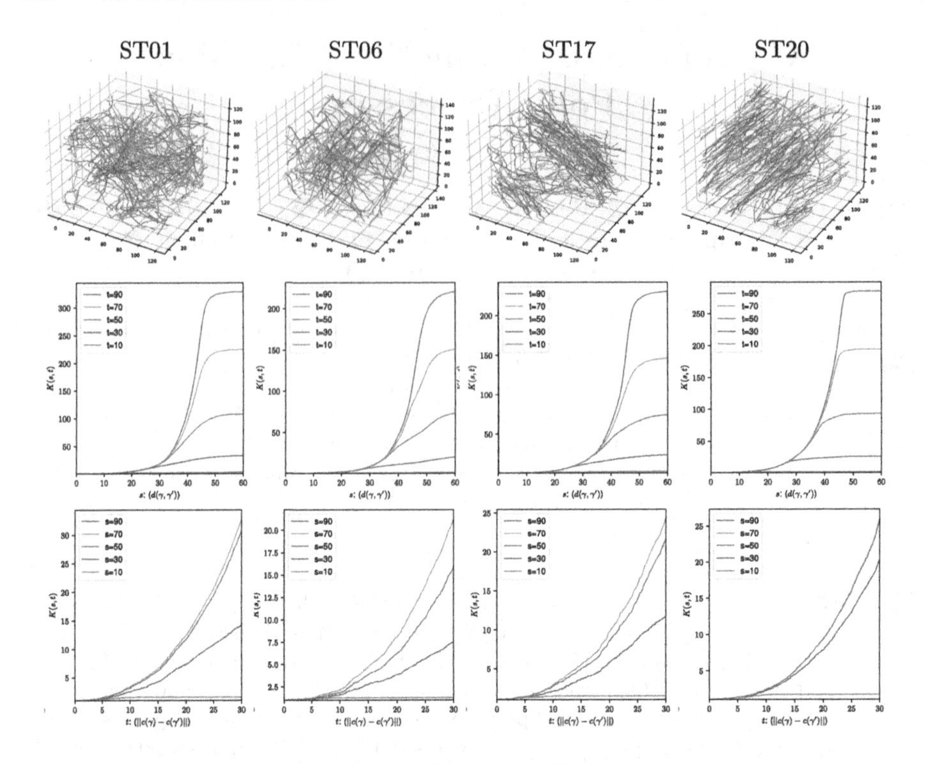

Fig. 1. K-functions for samples of myelin sheaths. Each column corresponds to measured on a data set. (top row): the centerlines of the myelin sheathed axons. (middle row): The K-function for fixed values of t. (bottom row): The K-function for fixed values of s.

complete spatial randomness in point sets. For a stationary point process, $K(t)$ gives the expected number of points within distance t from a typical point. An estimator of Ripley's K-function for a point set $\{p_i\}_{i=1}^n$ inside an observation window W is,

$$\hat{K}(t) = \frac{1}{n\hat{\lambda}} \sum_{i \neq j} 1[\text{dist}(p_i, p_j) < t] \tag{1}$$

where $\hat{\lambda} = \frac{n}{|W|}$ is the sample intensity, $|W|$ is the volume of the observation window, and 1 is the indicator function. By comparing $\hat{K}(t)$ with the K-function corresponding to complete spatial randomness, we can measure the deviation from spatial homogeneity. Smaller values of $\hat{K}(t)$ indicate clustering whereas the points tend to repel each other for greater values.

Generalizations of Ripley's K-function have previously been considered for curve pieces in [4] and several approaches were presented in [7] for space curves. In this paper, we present a K-function inspired by the currents approach from [7] and provide the theoretical account for the statistical foundation. We construct the following K-function for a fiber-valued point process X

$$\hat{K}(t,s) = \frac{1}{|W|\lambda} \sum_{\gamma \in X: c(\gamma) \in W} \sum_{\gamma' \in X \setminus \{\gamma\}} 1[\|c(\gamma) - c(\gamma')\| \leq t, d(\gamma_c, \gamma_c') \leq s] \tag{2}$$

for $t, s > 0$, where $c(\gamma)$ is the center point of the curve γ and γ_c, γ_c' the centered curves, d the currents distance and λ is the spatial intensity of the center points. By introducing a second distance parameter, we separate the spatial distance of the curves from the difference in shape, allowing us to measure both spatial and shape homogeneity. The paper presents the following contributions:

1. A K-function for fiber-valued point processes along with a theoretical account for the statistical foundation.
2. We suggest a certain fiber process which we argue corresponds to complete randomness of points, an analogue of the Poisson process.
3. An application of the K-function to several generated data set and a real data set of myelin sheaths.

2 Fibers as Currents

We model a random collection of fibers as a point process on the space of fibers. Shape spaces are usually defined as the space of embeddings $B_e(\mathcal{M}, \mathbb{R}^d)$ of a manifold $\mathcal{M}$ into $\mathbb{R}^d$ [2], where the space of fibers is $B_e(I, \mathbb{R}^3)$ for some real interval I. A point process X on a metric space S is a measurable map from some probability space $(\Omega, \mathcal{F}, P)$ into the space of locally finite subsets of S. In this paper, we consider how $B_e(I, \mathbb{R}^3)$ is endowed with a metric by considering fibers as currents.

We can characterize a piece-wise smooth curve $\gamma \in B_e(I, \mathbb{R}^d)$ by computing its path-integral of all vector fields w [3,5,8]

$$V_\gamma(w) = \int_\gamma w(x)^t \tau_\gamma(x) d\lambda(x), \tag{3}$$

where $\tau_\gamma(x)$ is the unit tangent of γ at x and λ is the length measure on the curve. This is a way of representing the curve as a current, i.e. as elements in the dual space of the space of vector fields on $\mathbb{R}^d$. Formally, the space of m-currents C_m is the dual space of the space $C^0(\mathbb{R}^d, (\Lambda^m \mathbb{R}^d)^*)$ of differential m-forms.

The space of m-currents C_m is continuously embedded into the dual space of a reproducing kernel Hilbert space (RKHS) H with arbitrary kernel $K_H : \mathbb{R}^d \times \mathbb{R}^d \to \mathbb{R}^{d \times d}$ [5]. It follows from Riesz representation theorem that $v \in H$ can be embedded in the dual space H^* as the functional $\mathcal{L}_H(v) \in H^*$ defined by $\mathcal{L}_H(v)(w) = \langle v, w \rangle_H$ for $w \in H$.

Elements $v(y) = K_H(x,y)\alpha$ form a basis for H where $x, \alpha \in \mathbb{R}^d$, and basis elements in H are lifted to basis elements in H^* as $\delta_x^\alpha := \mathcal{L}_H(v)$ which are called the Dirac delta currents. The element V_γ from (3) can be written in terms of the basis elements $\delta_x^{\tau_\gamma(x)}$ as the 1-current $V_\gamma(w) = \int_\gamma \delta_x^{\tau_\gamma(x)}(w) d\lambda(x)$. It is approximated by the Riemann sum of Dirac delta currents $V_\gamma(w) \approx \tilde{V}_\gamma(w) =$

$\sum_i \delta_{x_i}^{\tau(x_i)\Delta x_i}(w)$ where x_i are sampled points along the curve according to λ. The dual space H^* inherits the inner product from the inner product on the RKHS via the inverse mapping $\mathcal{L}_H^{-1}$, so that the inner product for two curves γ_1 and γ_2 in H^* is

$$\langle V_{\gamma_1}, V_{\gamma_2} \rangle_{H^*} = \int_{\gamma_1} \int_{\gamma_2} \tau_{\gamma_1}^t(x) K_H(x,y) \tau_{\gamma_2}(y) d\lambda_{\gamma_2}(x) d\lambda_{\gamma_1}(y). \tag{4}$$

Writing $||V_\gamma||_{H^*}^2 = \langle V_\gamma, V_\gamma \rangle_{H^*}$, we finally arrive at the currents distance of two curves γ_1 and γ_2

$$d_c(V_{\gamma_1}, V_{\gamma_2}) = ||V_{\gamma_1} - V_{\gamma_2}||_{H^*} = \left(||V_{\gamma_1}||_{H^*}^2 + ||V_{\gamma_2}||_{H^*}^2 - 2\langle V_{\gamma_1}, V_{\gamma_2} \rangle_{H^*} \right)^{1/2}. \tag{5}$$

In practice, we usually don't know the orientation of the curves, thus we choose to consider the minimal distance between them,

$$d(V_{\gamma_1}, V_{\gamma_2}) = \min\{d_c(V_{\gamma_1}, V_{\gamma_2}), d_c(V_{\gamma_1}, V_{-\gamma_2})\} \tag{6}$$

where $-\gamma_2$ denotes the curve with opposite orientation of γ_2.

3 The *K*-function

3.1 Statistical Setup

Let S be the image of the embedding of $B_e(I, \mathbb{R}^d)$ into C_1. For brevity, γ is identified with its representation in C_1. We model a random collection of curves as a point process X in S. Here we endow S with the metric obtained by combining, using the two-norm, Eucledian distance between fiber center points and currents distance between the centered fibers. A subset of S is said to be locally finite if its intersection with any bounded subset of S is finite. We denote by $\mathcal{N}$ the set of locally finite fiber configurations.

Let $c : S \to \mathbb{R}^d$ be a center function on the space of fibers in $\mathbb{R}^d$ that associates a center point to each fiber. A center function should be translation covariant in the sense that $c(\gamma + x) = c(\gamma) + x$ for all $x \in \mathbb{R}^d$. Let S_c denote the space of centered fibers wrt. c, i.e., those $\gamma \in S$ for which $c(\gamma) = 0$. We define $\gamma_c := \gamma - c(\gamma) \in S_c$ to be the centering of γ.

For Borel sets $B_1, A_1 \subset \mathbb{R}^d$ and $B_2, A_2 \subset S_c$, define the first moment measure

$$\mu(B_1 \times B_2) = \mathbb{E} \sum_{\gamma \in X} 1[c(\gamma) \in B_1, \gamma_c \in B_2]$$

and the second moment measure

$$\alpha((A_1 \times A_2) \times (B_1 \times B_2)) = \mathbb{E} \sum_{\gamma, \gamma' \in X}^{\neq} 1[c(\gamma) \in A_1, \gamma_c \in A_2] 1[c(\gamma') \in B_1, \gamma'_c \in B_2].$$

We assume that μ is translation invariant in its first argument. This is for instance the case if the distribution of X is invariant under translations. This implies that $\mu(\cdot \times B_2)$ is proportional to the Lebesgue measure for all B_2. Thus we can write

$$\mu(B_1 \times B_2) = |B_1|\nu(B_2)$$

for some measure $\nu(\cdot)$ on S_c. Note that the total measure $\nu(S_c)$ is the spatial intensity of the center points, i.e. the expected number of center points in a unit volume window. In applications, this will typically be finite. In this case, we may normalize ν to obtain a probability measure which could be interpreted as the distribution of a single centered fiber. For subsets $A_1, A_2 \subset \mathbb{R}^d$ and $F \subseteq \mathcal{N}$, we define the reduced Campbell measure

$$C^!(A_1 \times A_2 \times F) = \mathbb{E} \sum_{\gamma \in X} 1[c(\gamma) \in A_1, \gamma_c \in A_2, X \setminus \{\gamma\} \in F].$$

By disintegration and the standard proof, we get for any measurable function $h : \mathbb{R}^d \times S_c \times \mathcal{N} \to [0, \infty)$

$$\mathbb{E} \sum_{\gamma \in X} h(c(\gamma), \gamma_c, X \setminus \{\gamma\}) = \int_{\mathbb{R}^d \times S_c} \mathbb{E}^!_{c,\gamma_c} h(c, \gamma_c, X) \mu(\mathrm{d}(c, \gamma_c)), \tag{7}$$

where $\mathbb{E}^!_{c,\gamma_c}$ is a point process expectation that can be interpreted as expectation with respect to X conditional on that X contains the fiber γ with $c(\gamma) = c$ and $\gamma - c(\gamma) = \gamma_c$. Assuming that α is invariant under joint translation of the arguments A_1, B_1,

$$\mathcal{K}_{c,\gamma_c}(B_1 \times B_2) := \mathbb{E}^!_{c,\gamma_c} \sum_{\gamma' \in X} 1[c(\gamma') \in B_1, \gamma'_c \in B_2] \tag{8}$$

$$= \mathbb{E}^!_{0,\gamma_0} \sum_{\gamma' \in X} 1[c(\gamma') \in B_1 - c, \gamma'_c \in B_2] \; = \mathcal{K}_{0,\gamma_0}((B_1 - c) \times B_2). \tag{9}$$

Assume that also $\mathbb{E}^!_{c,\gamma_c} h(\gamma_c, X - c) = \mathbb{E}_{0,\gamma_0} h(\gamma_0, X)$ which is true if the distribution of X is invariant over translations. Then, using the above,

$$\mathbb{E} \sum_{\gamma \in X} 1[c(\gamma) \in W] h(\gamma_c, X \setminus \{\gamma\} - c(\gamma)) = \int_{W \times S_c} \mathbb{E}^!_{c,\gamma_c} h(\gamma_c, X - c) \mu(\mathrm{d}(c, \gamma_c))$$

$$= \int_{W \times S_c} \mathbb{E}^!_{0,\gamma_0} h(\gamma_0, X) \mu(\mathrm{d}(c, \gamma_0)) = |W| \int_{S_c} \mathbb{E}^!_{0,\gamma_0} h(\gamma_0, X) \nu(\mathrm{d}\gamma_0). \tag{10}$$

From this it follows that

$$E_h = \frac{1}{|W|} \sum_{\gamma \in X} 1[c(\gamma) \in W] h(\gamma_c, X \setminus \{\gamma\} - c(\gamma))$$

is an unbiased estimator of $\int_{S_c} \mathbb{E}^!_{0,\gamma_0} h(\gamma_0, X) \nu(\mathrm{d}\gamma_0)$. Furthermore, if $\nu(S_c)$ is finite,

$$\int_{S_c} \mathbb{E}^!_{0,\gamma_0} h(\gamma_0, X) \nu(\mathrm{d}\gamma_0) = \nu(S_c) \mathbb{E}_{\tilde{\nu}} \mathbb{E}^!_{0,\Gamma_0} h(\Gamma_0, X)$$

where Γ_0 is a random centered fiber with distribution $\tilde{\nu}(\cdot) = \nu(\cdot)/\nu(S_c)$ and $\mathbb{E}_{\tilde{\nu}}$ is expectation with respect to this distribution of Γ_0.

3.2 K-function for Fibers

In order to define a K-function, we must make an appropriate choice of h. We choose h as

$$h(\gamma_c, X) = \sum_{\gamma' \in X} 1[\|c(\gamma')\| \leq t, d(\gamma_c, \gamma_c') \leq s]$$

for $s, t > 0$, where $\|\cdot\|$ is the usual distance in $\mathbb{R}^d$ between center points. Thus we define the empirical K-function for $t, s > 0$ as

$$\begin{aligned} \hat{K}(t,s) &= \frac{1}{|W|\nu(S_c)} \sum_{\gamma \in X} 1[c(\gamma) \in W] h(\gamma_c, X \setminus \{\gamma\} - c(\gamma)) \\ &= \frac{1}{|W|\nu(S_c)} \sum_{\substack{\gamma \in X : c(\gamma) \in W, \\ \gamma' \in X \setminus \{\gamma\}}} 1[\|c(\gamma) - c(\gamma')\| \leq t, d(\gamma_c, \gamma_c') \leq s]. \end{aligned} \tag{11}$$

Since $\nu(S_c)$ is the intensity of fiber centers, it is estimated by $N/|W|$ where N is the observed number of centers $c(\gamma)$ inside W. The K-function is the expectation of the empirical K-function

$$K(t,s) = \mathbb{E}\hat{K}(t,s) = \mathbb{E}_{\tilde{\nu}}\mathbb{E}^!_{0,\Gamma_0} h(\Gamma_0, X) = \mathbb{E}_{\tilde{\nu}}\mathcal{K}_{0,\Gamma_0}(B(0,t) \times B_c(\Gamma_0, s))$$

where $B(x,r) = \{y : ||x-y|| \leq r\}$ and $B_c(\gamma, s) = \{\gamma' : d(\gamma, \gamma') \leq s\}$.

4 Experiments

To obtain a measure of spatial homogeneity, Ripley's K-function for points is usually compared with the K-function for a Poisson process, $K_P(t) = \text{vol}(B_d(t))$, corresponding to complete spatial randomness. The aim of the experiments on generated data sets is to analyze the behavior of the K-function on different types of distributions and suggest a fiber process that corresponds to complete randomness.

4.1 Generated Data Sets

The four generated data sets X_1, X_2, X_3 and X_4 each contain 500 fibers with curve length $l = 40$ and center points in $[0, 100]^3$ and is visualized in the first row of Fig. 2. For X_1, X_2, X_3 the center points are generated by a Poisson process and the fibers are uniformly rotated lines in X_1, uniformly rotated spirals in X_2 and Brownian motions in X_3. The data set X_4 has clustered center points and within each cluster the fibers are slightly perturbed lines.

To avoid most edge effects, we choose the window $W \approx [13, 87]^3 \subset \mathbb{R}^3$ for the calculation of the K-function. Furthermore, we choose a Gaussian kernel

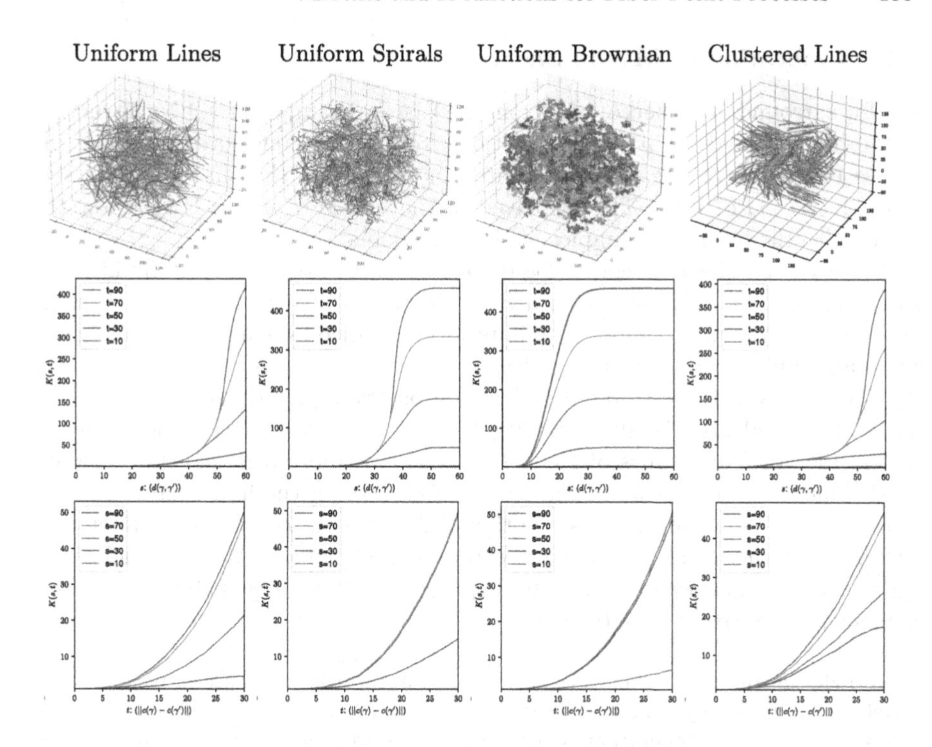

Fig. 2. K-function on the generated data, where each column corresponds to one data set. (top row): The data sets X_1, X_2, X_3 and X_4 described in Sect. 4.1. (middle row): The K-function of the data set above for fixed values of t. (Bottom row): The K-function of the data set above for fixed values of s.

$K_\sigma(x, y) = \exp\left(\frac{-|x-y|^2}{2\sigma}\right)\mathrm{Id}$ with $\sigma = \frac{100}{3}$. Finally, c is defined to be the mass center of the curve.

The first row of Fig. 2 shows the generated data sets and the respective K-functions are visualized in the second and third row. In the second row, $s \mapsto K(t, s)$ is plotted for fixed values of t and in the third row, $t \mapsto K(t, s)$ is plotted for fixed values of s. The graphs in the second row capture the fiber shape difference of each data set whereas the graphs in the third row capture spatial difference.

It is distribution X_3 that we consider to be a natural suggestion for a uniform random distribution of fibers. This is because Brownian motions are well-known for modelling randomness, thus representing shape randomness. And by translating these Brownian motion with a Poisson process, we argue that this distribution is a good choice.

Considering only the data sets with uniformly distributed center points, i.e., the first three columns of Fig. 2, we see a big difference in the second row of plots. This indicates that the K-function is sensitive to the change in shape. The K-function for the Brownian motions captures much more mass for smaller radii compared to the lines, with the spirals being somewhere in-between.

4.2 Application to Myelin Sheaths

Myelin surrounds the nerve cell axons and is an example of a fiber structure in the brain. Based on 3D reconstructions from the region motor cortex of the mouse brain, centre lines were generated in the myelin sheaths. The data sets ST01, ST06, ST17 and ST20 displayed in the first row of Fig. 1 represent the myelin sheaths from four samples at different debts.

Since myelin sheaths tend to be quite long, we chose to divide the fibers of length greater than 40 into several fibers segments of length 40. This has the benefits, that the mass center is an appropriate choice for c and that the results are comparable with the results of Fig. 2, since the curves are of similar length.

The results of the estimated K-function on the four data sets ST01, ST06, ST17 and ST20 are visualized in Fig. 1, where $s \mapsto K(t, s)$ is plotted in the second row for fixed values of t and $t \mapsto K(t, s)$ is plotted in the third row for fixed values of s. The plots in the second row showing the fiber shape are very similar, resembling the fiber distribution of X_2. We notice a slight difference in ST20, where the graphs have a more pronounced cut off. When noticing the scale of the y-axis, we see that the expected number of neighbor fibers vary significantly between the data sets.

Acknowledgements. Zhiheng Xu and Yaqing Wang from Institute of Genetics and Developmental Biology, Chinese Academy of Sciences are thanked for providing the mouse tissue used for generation of the dataset. The work was supported by the Novo Nordisk Foundation grant NNF18OC0052000.

References

1. Baddeley, A., Rubak, E., Turner, R.: Spatial Point Patterns: Methodology and Applications with R. CRC Press, Boca Raton (2015)
2. Bauer, M., Bruveris, M., Michor, P.W.: Overview of the geometries of shape spaces and diffeomorphism groups. J. Math. Imag. Vision **50**(1-2), 60–97 (2014)
3. Charon, N.: Analysis of geometric and functional shapes with extensions of currents : applications to registration and atlas estimation. PhD thesis, Ècole normale supérieure de Cachan - ENS Cachan (2013)
4. Chiu, S.N., Stoyan, D., Kendall, W.S., Mecke, J.: Stochastic Geometry and Its Applications. John Wiley & Sons, Chichester (2013)
5. Durrleman, S., Pennec, X., Trouvé, A., Ayache, N.: Statistical models of sets of curves and surfaces based on currents. Med. Image Anal. **13**(5), 793–808 (2009)
6. Ripley, B.D.: The second-order analysis of stationary point processes. J. Appl. Probal. **13**(2), 255–266 (1976)
7. Sporring, J., Waagepetersen, R., Sommer, S.: Generalizations of Ripley's K-function with application to space curves. In: Chung, A.C.S., Gee, J.C., Yushkevich, P.A., Bao, S. (eds.) IPMI 2019. LNCS, vol. 11492, pp. 731–742. Springer, Cham (2019). https://doi.org/10.1007/978-3-030-20351-1_57
8. Vaillant, M., Glaunès, J.: Surface matching via currents. In: Biennial International Conference on Information Processing in Medical Imaging, pp. 381–392 (2005)

Geometry of Quantum States

Q-Information Geometry of Systems

Christophe Corbier(✉)

Université de Lyon, UJM-Saint-Etienne, LASPI, 42334 IUT de Roanne, France
christophe.corbier@univ-st-etienne.fr

Abstract. This paper proposes a generalization of information geometry named *Q-information geometry* (Q-IG) related to systems described by parametric models, based $\mathbb{R}$-Complex Finsler subspaces and leading to Q-spaces. A Q-space is a pair $\left(\mathrm{M}^n_{S_X}, |F^{Q,n}_{J,x}|\right)$ with $\mathrm{M}^n_{S_X}$ a $\mathbb{C}$-manifold of systems and $|F^{Q,n}_{J,x}|$ a continuous function such that $F^{Q,n}_{3,x} = F^n_{H,x} + \sqrt{-1}F^n_{\overline{H},x}$ with a *signature* $(+,+,-)$ where $F^n_{H,x}$ and $F^n_{\overline{H},x}$ are τ-Hermitian and τ-non-Hermitian metrics, respectively. Experimental results are presented from a semi-finite acoustic waves guide.

Keywords: Information geometry · $\mathbb{R}$-Complex Finsler spaces · Q-space · Parametric model · Maximum phase system

1 Introduction

Among manifolds in information geometry, there exist Kählerian manifolds as an interesting topic in many domains. Kählerian geometry generalizes Hessian geometry since this latter is locally expressed in real coordinates [13]. The tangent bundle of a Hessian manifold naturally admits Kählerian structure, and that a Hessian structure is formally analogous to a Kählerian structure because a Kählerian metric is locally expressed as a complex Hessian. In [9] the aim is to classify compact, simply connected Kählerian manifolds which admit totally geodesic, holomorphic complex homothetic foliations by curves. A relation between sections in globally generated holomorphic vector bundles on Kählerian manifolds, isotropic with respect to a non-degenerate quadratic form and totally geodesic folations on Euclidean open domains is established in [2]. On a Kählerian manifold, the tensor metric and the Levi-Civita connection are determined from a Kählerian potential [14] and the Ricci tensor is yielded from the determinant of the metric tensor. A fundamental work in [5] has been carried out related to the correspondence between the information geometry of a signal filter and a Kähler manifold. This work focused on a minimum phase (mP) signal filter, i.e. when the filter and its inverse are causal and stable or when all zeros (resp poles) are inside the unit disc of the z-plane.

In this paper, the study is extended and focused on the parametric models of linear systems $\mathcal{M}^{\mathcal{L}}_{S}$ both with external input and zeros outside the unit disc of the z-plane in estimated discrete transfer functions. For example acoustic

F. Nielsen and F. Barbaresco (Eds.): GSI 2021, LNCS 12829, pp. 137–144, 2021.
https://doi.org/10.1007/978-3-030-80209-7_16

propagation systems present of such properties (see Carmona and Alvarado in [4]). This movement of zeros outside the unit disc changes Kähler information geometry into a new geometry named *Q-information geometry* (Q-IG) based $\mathbb{R}$-Complex Finsler [1,3,10–12] subspaces denoted $\mathbb{R}^{\mathbb{C}_{\mathrm{w}}(\mathcal{F}^n_{Q,x})}$ and leading to a Q-space. Such a space is a pair $\left(\mathrm{M}^n_{S_X}, |F^{Q,n}_{J,x}|\right)$ with $\mathrm{M}^n_{S_X}$ a $\mathbb{C}$-manifold of systems and $|F^{Q,n}_{J,x}| : \mathrm{T}^Q\mathrm{M}^n_{S_X} \to \mathbb{R}_+$ a Finsler metric with *signature* $(+,+,-)$ where $\mathrm{T}^Q\mathrm{M}^n_{S_X}$ is a holomorphic tangent bundle given by $\mathrm{T}^Q\mathrm{M}^n_{S_X} = \bigoplus_{\mathrm{w}} \mathrm{T}^{\mathrm{w}}\mathrm{M}^n_{S_X}$. For $J = 1$ the corresponding Q-IG is the Kähler information geometry with metric tensor $g^{1,0}_{\mu\overline{\nu},x}$ where $\mathrm{Im}(F^{Q,0}_{1,x}) = 0$ for all minimum phase (mP) systems with $F^{Q,0}_{1,x} = \left(g^{1,0}_{\mu\overline{\nu},x}\Omega^\mu\overline{\Omega}^\nu\right)^{1/2}$. Q-IG for maximum phase (MP) systems corresponds to $J = 3$ with $F^{Q,n}_{3,x} = F^n_{H,x} + \sqrt{-1}F^n_{\overline{H},x}$ where $F^n_{H,x}$ and $F^n_{\overline{H},x}$ are τ-Hermitian and τ-non-Hermitian metrics. Such an information geometry leads to three metric tensors $\left(g^{1,n}_{\mu_o\overline{\nu}_o,x}, g^{2,n}_{\mu\overline{\nu},x}, g^{3,n}_{\mu_o\nu,x}\right)$ where $\left(g^{1,n}_{\mu_o\overline{\nu}_o,x}, g^{2,n}_{\mu\overline{\nu},x}\right)$ are square metric tensors and $g^{3,n}_{\mu_o\nu,x}$ a non-square metric tensor. Table 1 shows the Q-information geometry theory.

Table 1. Q-information geometry theory.

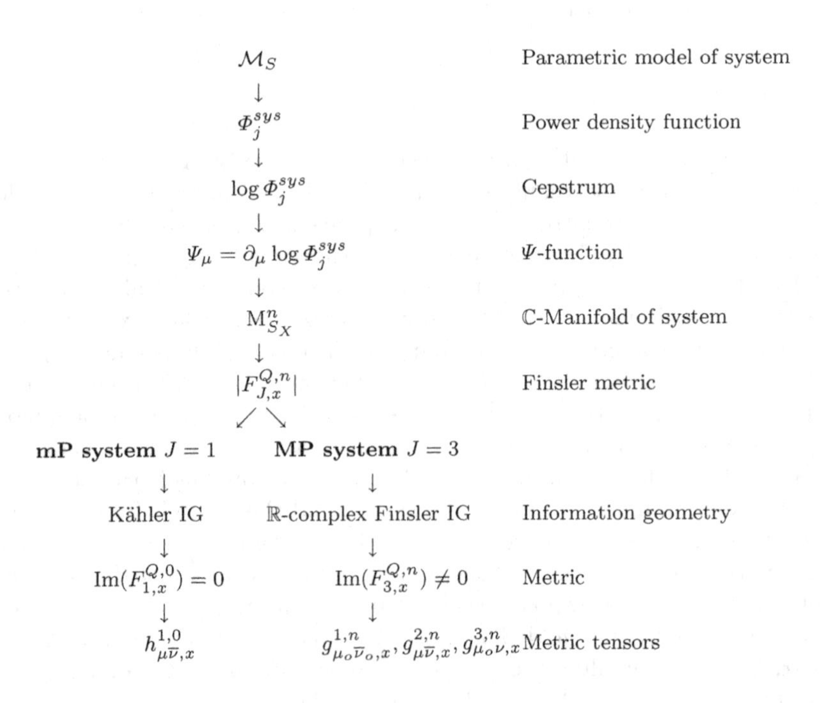

The remainder of this article is structured as follows. Section 2 describes Q-information geometry. Section 3 focuses on experimental results on a real acoustic system. Conclusions and perspectives are drawn in Sect. 4.

2 Q-Information Geometry

Let us consider a stationary and stable discrete-time system S_X with external scalar input $x(t)$, additive scalar disturbance $e(t)$ ($\mathbb{E}e(t) = 0, \mathbb{E}e^2(t) = \lambda$) and the scalar output $s_x(t)$ described as

$$s_x(t) = G(q, \boldsymbol{\theta})x(t) + H(q, \boldsymbol{\theta})e(t) \tag{1}$$

where q is the lag operator such that $q^{-1}x(t) = x(t-1)$ and $\boldsymbol{\theta} = [b_1...b_{n+m+1}a_1...a_s]^T$ the parameter vector. Therefore the power spectral density (psd) of $s_x(t)$ is

$$\Phi^S(e^{i\omega T_s}, \boldsymbol{\theta}) = G(e^{i\omega T_s}, \boldsymbol{\theta})\overline{G}(e^{i\omega T_s}, \boldsymbol{\theta})\Phi_x(e^{i\omega T_s}) + \lambda H(e^{i\omega T_s}, \boldsymbol{\theta})\overline{H}(e^{i\omega T_s}, \boldsymbol{\theta}) \tag{2}$$

Let $\Phi^Y_{j,x}(e^{i\omega T_s}, \boldsymbol{\theta})$ be the *joint-power spectral density* (j-psd) defined as

$$\Phi^Y_{j,x}(e^{i\omega T_s}, \boldsymbol{\theta}) = \sqrt{\Phi^A(e^{i\omega T_s}, \boldsymbol{\theta})\Phi^B(e^{i\omega T_s}, \boldsymbol{\theta})} \tag{3}$$

A correspondence can be used between frequency domain and complex z-domain for information geometry (IG) objects. Therefore the adopted correspondence is $\Phi^Y(e^{i\omega T_s}, \boldsymbol{\theta}) \overset{\text{IG}}{\to} \Phi^Y(z; \boldsymbol{Z}, \overline{\boldsymbol{Z}})$. Some conditions on G and H are necessary for defining the IG of S_X. Stability and minimum phase property remain main conditions. Therefore the $\aleph$-dimentional $\mathbb{C}$-manifold of S_X where S_X is described from parametric model structures can be written as

$$\mathrm{M}^n_{S_X} = \left\{ \Phi^{S,n}_{j,x}(z, \boldsymbol{Z}, \overline{\boldsymbol{Z}}) / \frac{1}{2i\pi} \oint_{|z|=1} \left(\log(\Phi^{S,n}_{j,x}(z, \boldsymbol{Z}, \overline{\boldsymbol{Z}})) \right)^2 \frac{dz}{z} < \infty \right\} \tag{4}$$

where n is the number of zeros outside the unit disc.

Let E^{w} ($\mathrm{w} = \{1, 2, .., J\}$) be a subset related to holomorphic coordinates and $\overline{E}^{\mathrm{w}}$ to antiholomorphic coordinates. Let us denote $\boldsymbol{Z}^{\mathrm{w}} = (Z^\mu)_{\mu \in E^{\mathrm{w}}}$ and $\boldsymbol{\Omega}^{\mathrm{w}} = (\Omega^\mu)_{\mu \in E^{\mathrm{w}}}$ a complex coordinates system on $\mathrm{T}^{\mathrm{w}}\mathrm{M}^n_{S_X}$ a holomorphic tangent subbundle given by $\mathrm{T}^{\mathrm{w}}\mathrm{M}^n_{S_X} = \bigcup_{Z^{\mathrm{w}} \in \mathrm{M}^n_{S_X}} \mathrm{T}^{\mathrm{w}}_Z\mathrm{M}^n_{S_X}$ where $\mathrm{T}^{\mathrm{w}}_Z\mathrm{M}^n_{S_X}$ are holomorphic tangent subspaces in $\boldsymbol{Z}^{\mathrm{w}}$ spanned by $\{\frac{\partial}{\partial Z^\mu}\}_{\mu \in E^{\mathrm{w}}}$. In $\boldsymbol{u}^{\mathrm{w}} = (\boldsymbol{Z}^{\mathrm{w}}, \boldsymbol{\Omega}^{\mathrm{w}})$ the holomorphic tangent space $\mathrm{T}^{\mathrm{w}}_u(\mathrm{T}^{\mathrm{w}}\mathrm{M}^n_{S_X})$ is spanned by $\{\frac{\partial}{\partial Z^\mu}, \frac{\partial}{\partial \Omega^\mu}\}_{\mu \in E^{\mathrm{w}}}$ and each holomorphic tangent subbundle can be decomposed from horizontal and vertical distributions as $\mathrm{T}^{\mathrm{w}}(\mathrm{T}^{\mathrm{w}}\mathrm{M}^n_{S_X}) = \mathrm{H}^{\mathrm{w}}(\mathrm{T}^{\mathrm{w}}\mathrm{M}^n_{S_X}) \oplus \mathrm{V}^{\mathrm{w}}(\mathrm{T}^{\mathrm{w}}\mathrm{M}^n_{S_X})$. A subspace $\mathbb{R}^{\mathbb{C}_{\mathrm{w}}(\mathcal{F}^n_{Q,x})}$ is a pair $\left(\mathrm{M}^n_{S_X}, Q^{\mathrm{w},n}_x\right)$ on $\mathrm{T}^{\mathrm{w}}\mathrm{M}^n_{S_X}$ such that $Q^{\mathrm{w},n}_x : \mathrm{T}^{\mathrm{w}}\mathrm{M}^n_{S_X} \to \mathbb{R}_+$ satisfying

- (i) $L^{\mathrm{w},n}_x(\boldsymbol{Z}^{\mathrm{w}}, \boldsymbol{\Omega}^{\mathrm{w}}, \overline{\boldsymbol{Z}}^{\mathrm{w}}, \overline{\boldsymbol{\Omega}}^{\mathrm{w}}) = (Q^{\mathrm{w},n}_x)^2(\boldsymbol{Z}^{\mathrm{w}}, \boldsymbol{\Omega}^{\mathrm{w}}, \overline{\boldsymbol{Z}}^{\mathrm{w}}, \overline{\boldsymbol{\Omega}}^{\mathrm{w}})$ is smooth on $\tilde{\mathrm{T}}^{\mathrm{w}}\mathrm{M}^n_{S_X} = \mathrm{T}^{\mathrm{w}}\mathrm{M}^n_{S_X} \backslash 0$;
- (ii) $Q^{\mathrm{w},n}_x(\boldsymbol{Z}^{\mathrm{w}}, \boldsymbol{\Omega}^{\mathrm{w}}, \overline{\boldsymbol{Z}}^{\mathrm{w}}, \overline{\boldsymbol{\Omega}}^{\mathrm{w}}) \geq 0$;
- (iii) $Q^{\mathrm{w},n}_x(\boldsymbol{Z}^{\mathrm{w}}, \lambda\boldsymbol{\Omega}^{\mathrm{w}}, \overline{\boldsymbol{Z}}^{\mathrm{w}}, \lambda\overline{\boldsymbol{\Omega}}^{\mathrm{w}}) = \lambda Q^{\mathrm{w},n}_x(\boldsymbol{Z}^{\mathrm{w}}, \boldsymbol{\Omega}^{\mathrm{w}}, \overline{\boldsymbol{Z}}^{\mathrm{w}}, \overline{\boldsymbol{\Omega}}^{\mathrm{w}})$, $\lambda \in \mathbb{R}_+$;

It follows that $L_x^{\mathrm{w},n}(\boldsymbol{Z}^{\mathrm{w}},\boldsymbol{\Omega}^{\mathrm{w}},\overline{\boldsymbol{Z}}^{\mathrm{w}},\overline{\boldsymbol{\Omega}}^{\mathrm{w}})$ is $(2,0)$ homogeneous with respect to λ where identities are fulfilled for $(\mu,\nu,\rho)\in E^{\mathrm{w}}$ (resp. $\overline{E}^{\mathrm{w}}$)

- (a) $\frac{\partial L_x^{\mathrm{w},n}}{\partial\Omega^\mu}\Omega^\mu+\frac{\partial L_x^{\mathrm{w},n}}{\partial\overline{\Omega}^\mu}\overline{\Omega}^\mu=2L_x^{\mathrm{w},n}$; $g_{\nu\mu,x}^{\mathrm{w},n}\Omega^\mu+g_{\nu\overline{\mu},x}^{\mathrm{w},n}\overline{\Omega}^\mu=\frac{\partial L_x^{\mathrm{w},n}}{\partial\Omega^\nu}$
- (b) $2L_x^{\mathrm{w},n}=g_{\mu\nu,x}^{\mathrm{w},n}\Omega^\mu\Omega^\nu+g_{\overline{\mu}\overline{\nu},x}^{\mathrm{w},n}\overline{\Omega}^\mu\overline{\Omega}^\nu+g_{\mu\overline{\nu},x}^{\mathrm{w},n}\Omega^\mu\overline{\Omega}^\nu+g_{\nu\overline{\mu},x}^{\mathrm{w},n}\Omega^\nu\overline{\Omega}^\mu$;
- (c) $\frac{\partial g_{\nu\rho,x}^{\mathrm{w},n}}{\partial\Omega^\mu}\Omega^\mu+\frac{\partial g_{\nu\rho,x}^{\mathrm{w},n}}{\partial\overline{\Omega}^\mu}\overline{\Omega}^\mu=0$; $\frac{\partial g_{\nu\overline{\rho},x}^{\mathrm{w},n}}{\partial\Omega^\mu}\Omega^\mu+\frac{\partial g_{\nu\overline{\rho},x}^{\mathrm{w},n}}{\partial\overline{\Omega}^\mu}\overline{\Omega}^\mu=0$

Let $\mathrm{T}^{\mathrm{w}}\mathrm{M}_{S_X}^n$ be the holomorphic tangent subbundle of $\mathrm{M}_{S_X}^n$ and $\overline{(\mathrm{T}^{\mathrm{w}}\mathrm{M}_{S_X}^n)}$ be its antiholomorphic. Let us define the Ψ-function in the z-domain at $\boldsymbol{Z}$ and $\overline{\boldsymbol{Z}}$ as $\boldsymbol{\Psi}_{\mu,x}^{S,n}(z,\boldsymbol{Z}^{\mathrm{w}},\overline{\boldsymbol{Z}}^{\mathrm{w}})=\frac{\partial\log\Phi_{j,x}^{S,n}(z,\boldsymbol{Z}^{\mathrm{w}},\overline{\boldsymbol{Z}}^{\mathrm{w}})}{\partial Z^\mu}$ where $\mu\in E^{\mathrm{w}}$ (resp. $\overline{E}^{\mathrm{w}}$). For short expressions let us denote $\partial_\mu=\frac{\partial}{\partial Z^\mu}$, $\partial_{\overline{\mu}}=\frac{\partial}{\partial\overline{Z}^\mu}$, $\dot{\partial}_\mu=\frac{\partial}{\partial\Omega^\mu}$ and $\dot{\partial}_{\overline{\mu}}=\frac{\partial}{\partial\overline{\Omega}^\mu}$. The square of the Finsler metric $ds_{\mathrm{w}}(x)$ in $\mathrm{T}^{\mathrm{w}}\mathrm{M}_{S_X}^n$ between two neighboring system model structures $\Phi_{j,x}^{S,n}(z,\boldsymbol{Z}^{\mathrm{w}}(\tau),\overline{\boldsymbol{Z}}^{\mathrm{w}}(\tau))$ and $\Phi_{j,x}^{S,n}(z,\boldsymbol{Z}^{\mathrm{w}}(\tau)+d\boldsymbol{Z}^{\mathrm{w}}(\tau),\overline{\boldsymbol{Z}}^{\mathrm{w}}(\tau)+d\overline{\boldsymbol{Z}}^{\mathrm{w}}(\tau))$ is

$$ds_{\mathrm{w}}^2(x,n)=\left\langle d\log\Phi_{j,x}^{S,n}(z,\boldsymbol{Z}^{\mathrm{w}}(\tau),\overline{\boldsymbol{Z}}^{\mathrm{w}}(\tau)),d\log\Phi_{j,x}^{S,n}(z,\boldsymbol{Z}^{\mathrm{w}}(\tau),\overline{\boldsymbol{Z}}^{\mathrm{w}}(\tau))\right\rangle \quad (5)$$

From $\boldsymbol{X}_{Z,x}^{\mathrm{w},n}(\tau)=\Omega^\mu(\tau)\boldsymbol{\Psi}_{\mu,x}^{S,n}(z,\boldsymbol{Z}^{\mathrm{w}}(\tau),\overline{\boldsymbol{Z}}^{\mathrm{w}}(\tau))$ and the inner product $(P_x^{\mathrm{w},n}(\tau))^2$ $=\left\langle\boldsymbol{X}_{Z,x}^{\mathrm{w},n}(\tau),\boldsymbol{X}_{Z,x}^{\mathrm{w},n}(\tau)\right\rangle$, we get $(P_x^{\mathrm{w},n}(\tau))^2=h_{\mu\overline{\nu},x}^n(\boldsymbol{Z}^{\mathrm{w}}(\tau),\overline{\boldsymbol{Z}}^{\mathrm{w}}(\tau))\Omega^\mu(\tau)\overline{\Omega}^\nu(\tau)$ with $h_{\mu\overline{\nu},x}^n(\boldsymbol{Z}^{\mathrm{w}}(\tau),\overline{\boldsymbol{Z}}^{\mathrm{w}}(\tau))$ the metric tensor given by

$$\frac{1}{2i\pi}\oint_{|z|=1}\boldsymbol{\Psi}_{\mu,x}^{S,n}(z,\boldsymbol{Z}^{\mathrm{w}}(\tau),\overline{\boldsymbol{Z}}^{\mathrm{w}}(\tau))\boldsymbol{\Psi}_{\overline{\nu},x}^{S,n}(z,\boldsymbol{Z}^{\mathrm{w}}(\tau),\overline{\boldsymbol{Z}}^{\mathrm{w}}(\tau))\frac{dz}{z}$$

Since (μ,ν) run over holomorphic and antiholomorphic coordinate systems then $\Omega^\mu(\tau)=\overline{\Omega}^{\mu_Y}(\tau)$ for $\mu=\overline{\mu}_Y$ and $\overline{\overline{\Omega}}^{\nu_Y}(\tau)=\Omega^{\nu_Y}(\tau)$ for $\nu=\overline{\nu}_Y$. Define the Q^{w}-metric by $Q_x^{\mathrm{w},n}(\tau)=|P_x^{\mathrm{w},n}(\tau)|$ with $(P_x^{\mathrm{w},n}(\tau))^2=h_{\mu\nu,x}^n(\boldsymbol{Z}^{\mathrm{w}}(\tau),\overline{\boldsymbol{Z}}^{\mathrm{w}}(\tau))\Omega^\mu(\tau)\Omega^\nu(\tau)$, where $h_{\mu\nu,x}^n(\boldsymbol{Z}^{\mathrm{w}}(\tau),\overline{\boldsymbol{Z}}^{\mathrm{w}}(\tau))$ are components of metric structures given by

$$h_{\mu\nu,x}^{\mathrm{w},n}=\frac{1}{8i\pi}\oint_{|z|=1}(\boldsymbol{\Psi}_{\mu,x}^{\mathrm{w},A,n}(z)\boldsymbol{\Psi}_{\nu,x}^{\mathrm{w},A,n}(z)+\boldsymbol{\Psi}_{\mu,x}^{\mathrm{w},A,n}(z)\boldsymbol{\Psi}_{\nu,x}^{\mathrm{w},B,n}(z)$$
$$+\boldsymbol{\Psi}_{\mu,x}^{\mathrm{w},B,n}(z)\boldsymbol{\Psi}_{\nu,x}^{\mathrm{w},A,n}(z)+\boldsymbol{\Psi}_{\mu,x}^{\mathrm{w},B,n}(z)\boldsymbol{\Psi}_{\nu,x}^{\mathrm{w},B,n}(z))\frac{dz}{z} \quad (6)$$

Let $\mathrm{M}_{S_X}^n$ be a $\aleph$-dimensional $\mathbb{C}$-manifold, $\boldsymbol{Z}^{\mathrm{w}}=(Z^\mu)_{\mu\in E^{\mathrm{w}}}$ be complex coordinates and $\mathrm{T}^{\mathrm{w}}\mathrm{M}_{S_X}^n$ be the holomorphic tangent subbundle which has a natural structure of $\mathbb{C}$-manifold with $\boldsymbol{u}^{\mathrm{w}}=(\boldsymbol{Z}^{\mathrm{w}},\boldsymbol{\Omega}^{\mathrm{w}})$ its local complex coordinates. Let $\mathrm{T}^Q\mathrm{M}_{S_X}^n$ be a $\mathbb{C}$-manifold spanned by $\{\frac{\partial}{\partial Z^\mu},\frac{\partial}{\partial\Omega^\mu}\}_{\mu\in\bigcup_{\mathrm{w}=1}^J E^{\mathrm{w}}}$ given by $\mathrm{T}^Q\mathrm{M}_{S_X}^n=\bigoplus_{\mathrm{w}=1}^J\mathrm{T}^{\mathrm{w}}\mathrm{M}_{S_X}^n$. A Q-space denoted $Q_{J,x}^{S,n}=\bigcup_{\mathrm{w}}^J\mathbb{R}^{\mathrm{C_w}(\mathcal{F}_{Q,x}^n)}$ is a pair $(\mathrm{M}_{S_X}^n,|F_{J,x}^{Q,n}|)$ where $|F_{J,x}^{Q,n}|$ is a Finsler metric $|F_{J,x}^{Q,n}|:\mathrm{T}^Q\mathrm{M}_{S_X}^n\to\mathbb{R}_+$ defined by

$$|F_{J,x}^{Q,n}|^2=\sum_{\mathrm{w}=1}^J\alpha_{\mathrm{w}}h_{\mu\nu,x}^{\mathrm{w},n}\Omega^\mu\Omega^\nu,\ (\mu,\nu)\in E^{\mathrm{w}}(resp.\overline{E}^{\mathrm{w}}) \quad (7)$$

with $F_{J,x}^{Q,n}$ a complex continuous function and $\alpha_{\rm w}$ the *signature* of the Finsler metric with $\alpha_{\rm w} = 1$ for $J = 1$ and $\alpha_{\rm w} = (+,+,-)$ for $J = 3$. Q-spaces generalize spaces related to manifold of systems with minimum and maximum phase properties. For $J = 1$ corresponding to mP systems with a Kähler manifold $\mathrm{M}_{SX}^{\mathrm{K}}$ the Hermitian metric $h_{\mu\overline{\nu},x}^{1,0}$ for an associated 2-form is $F_{1,x}^{Q,0} = \left(g_{\mu\overline{\nu},x}^{1,0}\Omega^{\mu}\overline{\Omega}^{\nu}\right)^{1/2}$, $g_{\mu\overline{\nu},x}^{1,0} = h_{\mu\overline{\nu},x}^{1,0}$. Conversely for $J = 3$ corresponding to MP systems , $F_{3,x}^{Q,n}$ is a complex valued function given by $F_{3,x}^{Q,n} = F_{H,x}^{n} + \sqrt{-1}F_{\overline{H},x}^{n}$ where $F_{H,x}^{n}$ and $F_{\overline{H},x}^{n}$ are $\mathbb{R}$-Complex Hermitian and non-Hermitian Finsler metrics, respectively. For each holomorphic tangent subbundle $\mathrm{T}^{\mathrm{w}}\mathrm{M}_{SX}^{n}$ a Q^{w}-metric is defined and its associated metric tensor $g_{\mu\nu,x}^{\mathrm{w},n}$ is expressed with the fundamental condition $|Z^{\mu}| < 1$ for all $\mu \in E^{\mathrm{w}}$ (resp. $\overline{E}^{\mathrm{w}}$). Therefore $P_x^{1,n}$ is a quadratic form in $\Omega_*^{\mu_o}\overline{\Omega}_*^{\nu_o}$ where $\Omega_*^{\mu_o} = \frac{dZ_*^{\mu_o}}{d\tau}$ with τ named Finsler paramater and $Z_*^{\mu_o} = \frac{1}{Z^{\mu_o}}$ involving $\Omega_*^{\mu_o} = -Z_*^{2\mu_o}\Omega^{\mu_o}$ and $g_{\mu_o\overline{\nu}_o,x}^{1}$ as a function of $Z_*^{\mu_o}$. In the same vein $P_x^{2,n}$ is a quadratic form in $\Omega^{\mu}\overline{\Omega}^{\nu}$ and $P_x^{3,n}$ as a function of $\Omega_*^{\mu_o}\overline{\Omega}^{\nu}$ with $g_{\mu_o\nu,x}^{3,n}$ a mixture of $Z_*^{\mu_o}$ and Z^{ν}. Therefore

$$\left(P_x^{1,n}\right)^2 = 2g_{\mu_o\overline{\nu}_o,x}^{1,n}\Omega_*^{\mu_o}\overline{\Omega}_*^{\nu_o},\ g_{\mu_o\overline{\nu}_o,x}^{1,n} = \frac{h_{\mu_o\overline{\nu}_o,x}^{1}}{(Z_*^{\mu_o}\overline{Z}_*^{\nu_o})^2} \tag{8}$$

$$\left(P_x^{2,n}\right)^2 = 2g_{\mu\overline{\nu},x}^{2,n}\Omega^{\mu}\overline{\Omega}^{\nu},\ g_{\mu\overline{\nu},x}^{2,n} = h_{\mu\overline{\nu},x}^{2} \tag{9}$$

and

$$\left(P_x^{3,n}\right)^2 = i^2\left(g_{\mu_o\nu,x}^{3,n}\Omega_*^{\mu_o}\overline{\Omega}^{\nu} + g_{\overline{\mu}_o\overline{\nu},x}^{3,n}\overline{\Omega}^{\mu_0}\overline{\Omega}^{\nu}\right);\ g_{\mu_o\nu,x}^{3,n} = \frac{h_{\mu_o\nu,x}^{3,n}}{Z_*^{2\mu_o}} \tag{10}$$

From Eq. (8) (9) (10) we have $|F_{3,x}^{Q,n}|^2 = \left(P_x^{1,n}\right)^2 + \left(P_x^{2,n}\right)^2 - \left(P_x^{3,n}\right)^2$ leading to a $(+,+,-)$ signature. Q-information geometry shows a new point of view on the behavior in τ of mP and MP systems. Each system has an evolution in τ due to the motions of its zeros/poles in estimated transfer functions. For $J = 3$

$$F_{H,x}^{Q,n}(\tau) = \sqrt{2g_{\mu_o\overline{\nu}_o,x}^{1,n}(\tau)\Omega_*^{\mu_o}(\tau)\overline{\Omega}_*^{\nu_o}(\tau) + 2g_{\mu\overline{\nu},x}^{2,n}(\tau)\Omega^{\mu}(\tau)\overline{\Omega}^{\nu}(\tau)}$$

$$F_{\overline{H},x}^{Q,n}(\tau) = \sqrt{g_{\mu_o\nu,x}^{3,n}(\tau)\Omega_*^{\mu_o}(\tau)\Omega^{\nu}(\tau) + g_{\overline{\mu}_o\overline{\nu},x}^{3,n}(\tau)\overline{\Omega}_*^{\mu_0}(\tau)\overline{\Omega}^{\nu}(\tau)}$$

Model trajectories in τ of mP and MP systems can be represented in a new graph named Q-graph. Real axis corresponds to Hermitian coordinates, imaginary axis to non-Hermitian coordinates. For mP systems, model trajectories from $|F_{1,x}^{Q,n}(\tau)|$ run over the real plan only. Conversely for MP systems, those trajectories from $|F_{3,x}^{Q,n}(\tau)|$ are in the z-plane.

3 Application

Consider a process model described by an OutPut Error (OE) model with transfer functions $G(z,\boldsymbol{\theta})$ and $H(z,\boldsymbol{\theta})$ given by

$$G(z,\boldsymbol{\theta}) = \frac{\sum_{k=1}^{n_B} b_k z^{-k}}{1+\sum_{k=1}^{n_A} a_k z^{-k}} = b_1 z^{-1} \prod_{k=1}^{n_B-1+n_A} \left(1 - Z^k z^{-1}\right)^{c_k} \quad , \; H(z,\boldsymbol{\theta}) = 1 \tag{11}$$

with $c_k = -1$ for AutoRegressive part and $c_k = +1$ for MovingAverage part, where Z^μ are all zeros-poles of G and H. Let $\mathbb{V}_o$ and $\mathbb{W}_{ip}$ be two sets of holomorphic indices such that $\mathbb{V}_o = \{Z^\mu/|Z^\mu| > 1, \mu = 1_o, 2_o, ..., n_o\}$ and $\mathbb{W}_{ip} = \{Z^\nu/|Z^\nu| < 1, \nu = 1_i, 2_i, ..., m_i, 1_p, 2_p, ..., l_p\}$. Let $\overline{\mathbb{V}}_o$ and $\overline{\mathbb{W}}_{ip}$ be two sets of antiholomorphic indices such that $\overline{\mathbb{V}}_o = \{\overline{Z}^\mu/|\overline{Z}^\mu| > 1, \mu = 1_o, 2_o, ..., n_o\}$ and $\overline{\mathbb{W}}_{ip} = \{\overline{Z}^\nu/|\overline{Z}^\nu| < 1, \nu = 1_i, 2_i, ..., m_i, 1_p, 2_p, ..., l_p\}$. Subscripts o and i, p are related to the zeros outside the unit disc (*exogenous zeros*) and to the (zeros, poles) inside the unit disc (*endogenous zeros, poles*), respectively. Haykin in [8] showed that a discrete transfer functions defined as a backward system-model filters can be changed into discrete transfer functions defined as a forward system-model filters with same frequency responses using $G^n_{\mathbb{V}_o,b}(z,\boldsymbol{Z}) = z^{-n}\overline{G}^n_{\mathbb{V}_o,f}(\frac{1}{\overline{z}},\boldsymbol{Z})$. Therefore

$$G^n(z,\boldsymbol{Z},\overline{\boldsymbol{Z}}) = b_1 z^{1-n} \prod_{\mu=1}^{n}\left(1-\overline{Z}^{\mu_o}z\right) \prod_{\mu=1}^{m}\left(1-Z^{\mu_i}z^{-1}\right) \prod_{\mu=1}^{l}\left(1-Z^{\mu_p}z^{-1}\right)^{-1} \tag{12}$$

From (6) (12) $h^{\mathrm{w},n}_{\mu\nu,X}$ metric tensors are

$$g^{1,n}_{\mu_o\overline{\nu}_o,X} = \frac{1}{8i\pi}\oint_{|z|=1} \partial_{\mu_o}\log\overline{G}^{1,n}(\overline{z})\partial_{\overline{\nu}_o}\log G^{1,n}(z)\frac{dz}{z} = \frac{\alpha^{1,\mathcal{M}_S}_{\mu_o\overline{\nu}_o,X}(Z^{\mu_o}_*\overline{Z}^{\nu_o}_*)^{-2}}{1-(Z^{\mu_o}_*\overline{Z}^{\nu_o}_*)^{-1}} \tag{13}$$

with $\alpha^{1,OE}_{\mu_o\overline{\nu}_o,X} = \frac{-1}{4}$.

$$g^{2,n}_{\mu\overline{\nu},X} = \frac{1}{2i\pi}\oint_{|z|=1} \partial_\mu\log G^{2,n}(z)\partial_{\overline{\nu}}\log\overline{G}^{2,n}(\overline{z})\frac{dz}{z} = \frac{\alpha^{2,\mathcal{M}_S}_{\mu\overline{\nu},X}}{1-Z^\mu\overline{Z}^\nu} \tag{14}$$

with $\alpha^{2,OE}_{\mu\overline{\nu},X} = (1/4; 1/4; -1/4; -1/4)$ for $(\mu\nu) = (\mu_i\nu_i; \mu_p\nu_p; \mu_i\nu_p; \mu_p\nu_i)$, respectively

$$g^{3,n}_{\mu_o\nu,X} = \frac{1}{2i\pi}\oint_{|z|=1} \partial_{\mu_o}\log\overline{G}^{3,n}(z)\partial_\nu\log G^{3,n}(z)\frac{dz}{z} = \frac{\beta^{3,\mathcal{M}_S}_{\mu_o\nu,X}}{1-Z^{\mu_o}_*Z^\nu} \tag{15}$$

with $\beta^{3,OE}_{\mu\nu,X} = (-1/4; -1/4; 1/4; 1/4)$ for $(\mu\nu) = (\mu_o\nu_i; \mu_i\nu_o; \mu_o\nu_p; \mu_p\nu_o)$, respectively.

For application Fig. 1 shows the plant including the secondary loudspeaker, amplifier, acoustic secondary path, measurement microphone and amplifier/anti-aliasing filter [4]. For OE model $n_B = 9$, $n_A = 8$. Finsler parameter τ was equal to

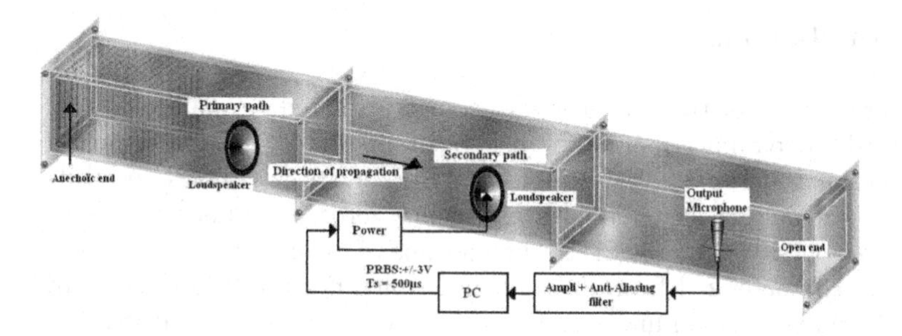

Fig. 1. Experimental setup for identification.

the slope/threshold parameter γ_i in the estimation function [6,7] and was varied from 0.01 to 2 with 50 values such that $\gamma_i = \tau_i$. Finsler complex coordinates system was $\Omega_i^\mu \approx \frac{Z_i^\mu - Z_{i-1}^\mu}{\tau_i - \tau_{i-1}}$ with $\Omega_i^\mu = \Omega^\mu(\tau_i)$ and $Z_i^\mu = Z^\mu(\tau_i)$ for two adjacent system model structures $\mathcal{M}_S^i$ and $\mathcal{M}_S^{i-1}$ $(i = 1..50)$ in the same manifold $\mathrm{M}_{S_X}^n$. Comparisons between two adjacent system models $(\mathrm{oe}_S^i, \mathrm{oe}_S^{i-1})$ in $\mathrm{M}_{S_X}^n$ was led to form two clusters. For OE system models two manifolds was defined: $\mathrm{M}_{S_{\mathrm{oe}}}^3$, $\mathrm{M}_{S_{\mathrm{oe}}}^4$. This means that two clusters have been given: 23 estimated models with 3 exogenous zeros (for example at $\tau_8 = 0.2399$, $Z_8^{1_o} = -2.2296$, $Z_8^{2_o} = -1.0985$, $Z_8^{3_o} = 1.0657$) and 27 with 4 zeros (for example at $\tau_{17} = 0.5378$, $Z_{17}^{1_o} = -1.9844$, $Z_{17}^{2_o} = 1.0579 + i0.0354$, $Z_{17}^{3_o} = 1.0579 - i0.0354$, $Z_{17}^{4_o} = -1.1726$). Figure 2 shows trajectories in the Q-graphs for 3 (left) and 4 (right) exogenous zeros. Remark a different trajectory since for 3 exogenous zeros, the corresponding Finsler trajectory is bounded in $[0, 300] \times [0, 254]$ while for 4 exogenous zeros the Finsler trajectory is in $[0, 800] \times [0, 100]$. In conclusion the Q-graph is a fundamental aspect of this new information geometry and characterizes different propagative systems.

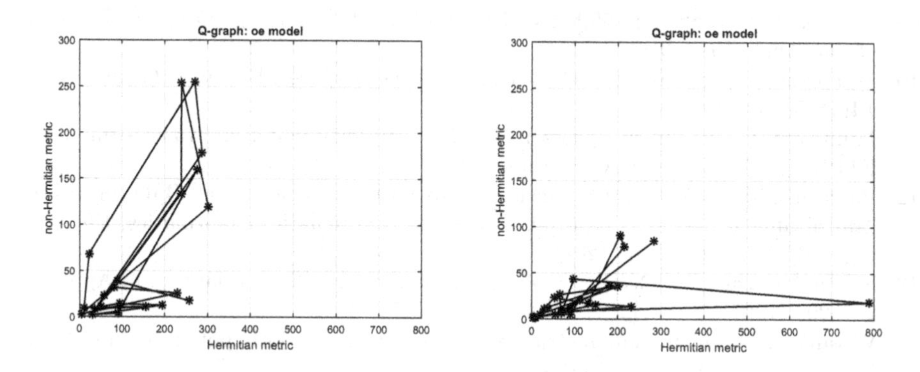

Fig. 2. (left): Q-graph of OE models for 3 exogenous zeros at τ_i $(i = 1..23)$. (right): Q-graph of OE models for 4 exogenous zeros at τ_i $(i = 1..27)$.

4 Conclusions

A generalization of the information geometry related to manifolds of systems described by parametric models has been presented in this paper. The Q-information geometry based $\mathbb{R}$-complex Finsler spaces allows to study information geometry of systems with exogenous zeros in estimated system model structures outside the unit disc. The motions of estimated system model structures are characterized from different Q-graphs. For the future, high investigations should be done on linear and non-linear connections, curvatures, geodesics, Randers metrics, Kropina metrics.

References

1. Aldea, N., Campean, G.: Geodesic curves on $\mathbb{R}$-complex Finsler spaces. Results Math. **70**, 15–29 (2016)
2. Aprodu, M.A., Aprodu, M.: Holomorphic vector bundles on Kahler manifolds and totally geodesic foliations on Euclidean open domains. Differential Geom. Appl. **39**, 10–19 (2015)
3. Campean, G.: Connections on $\mathbb{R}$-complex non-Hermitian Finsler spaces. Differential Geom. Dyn. Syst. **16**, 85–91 (2014)
4. Carmona, J.-C., Alvarado, V.: Active noise control of a duct using Robust control. IEEE Trans. Contr. Syst. Technol. **8**(6), 930–938 (2000)
5. Choi, J., Mullhaupt, A.P.: Kahlerian information geometry for signal processing. Entropy **17**, 1581–1605 (2015)
6. Corbier, C., Carmona, J.-C.: Mixed L_p estimators variety for model order reduction in control oriented system identification. Math. Probl. Eng. (2014). Hindawi Publishing Corporation, Article ID 349070
7. Corbier, C.: Articulated estimator random field and geometrical approach applied in system identification. J. Syst. Sci. Complex. **31**, 1164–1185 (2018)
8. Haykin, S.: Adaptative Filter Theory, 4th edn. Prentice Hall, Upper Saddle River (2002). 07458
9. Jelonek, W.: Kähler manifolds with homothetic foliation by curves. Differential Geom. Appl. **46**, 119–131 (2016)
10. Munteanu, G., Purcaru, M.: On $\mathbb{R}$-complex Finsler spaces. Balkan J. Geom. Appl. **14**(1), 52–59 (2009)
11. Munteanu, G.: Complex Spaces in Finsler, Lagrange and Hamilton Geometries. FTPH, vol. 141. Kluwer Acad. Publ. (2004)
12. Purcaru, M.: On $\mathbb{R}$-complex Finsler spaces with Kropina metric. Bulletin of the Transilvania University of Brasov. Series III: Mathematics, Informatics, Physics, vol. 453, no. 2, pp. 79–88 (2011)
13. Shima, H., Yagi, K.: Geometry of Hessian manifolds. Differential Geom. Appl. **7**, 277–290 (1997)
14. Watanabe, Y.: Kählerian metrics given by certain smooth potential functions. Kodai Math. J. **5**, 329–338 (1982)

Group Actions and Monotone Metric Tensors: The Qubit Case

Florio Maria Ciaglia and Fabio Di Nocera(✉)

Max Planck Institute for Mathematics in the Sciences, Leipzig, Germany
{ciaglia,dinocer}@mis.mpg.de

Abstract. In recent works, a link between group actions and information metrics on the space of faithful quantum states has been highlighted in particular cases. In this contribution, we give a complete discussion of this instance for the particular case of the qubit.

Keywords: Quantum information geometry · Unitary group · Lie algebra · Monotone metrics · Bogoliubov-Kubo-Mori metric · Bures-Helstrom metric · Wigner-Yanase metric · Two-level quantum system

1 Introduction

Because of Wigner's theorem, it is difficult to overestimate the role of the unitary group $\mathcal{U}(\mathcal{H})$ when dealing with symmetries in standard quantum mechanics. Consequently, it is not surprising that, in the context of quantum information geometry, $\mathcal{U}(\mathcal{H})$ again plays a prominent role when dealing with symmetries. More specifically, let $\mathcal{H}$ be the Hilbert space of a finite-level quantum system. The space of quantum states of this system is denoted by $\overline{\mathscr{S}(\mathcal{H})}$ and consists of all density operators on $\mathcal{H}$, that is, $\rho \in \overline{\mathscr{S}(\mathcal{H})}$ is such that $\rho \geq 0$ and $Tr(\rho) = 1$. Clearly, $\overline{\mathscr{S}(\mathcal{H})}$ is a convex set, and its interior is denoted by $\mathscr{S}(\mathcal{H})$ and consists of density operators on $\mathcal{H}$ that are invertible, that is, $\rho \in \mathscr{S}(\mathcal{H})$ is such that $\rho > 0$. It is well-known that $\mathscr{S}(\mathcal{H})$ is a smooth manifold whose topological closure is $\overline{\mathscr{S}(\mathcal{H})}$ [8,11], and the tangent space $T_\rho\mathscr{S}(\mathcal{H})$ may be identified with the space of self-adjoint operators on $\mathcal{H}$ with vanishing trace. In the context of quantum information geometry, if $\mathcal{K}$ is the Hilbert space of another finite-level quantum system, the allowed quantum channels between the two systems are mathematically described by the so-called completely-positive, trace-preserving (CPTP) maps between $\mathcal{B}(\mathcal{H})$ and $\mathcal{B}(\mathcal{K})$ sending $\mathscr{S}(\mathcal{H})$ into $\mathscr{S}(\mathcal{K})$ [12]. These maps are linear on the whole $\mathcal{B}(\mathcal{H})$, and the standard action of $\mathcal{U}(\mathcal{H})$ on $\mathcal{B}(\mathcal{H})$ given by $\mathbf{a} \mapsto \mathbf{U}\mathbf{a}\mathbf{U}^\dagger$ is easily seen to give rise to a CPTP map from $\mathcal{B}(\mathcal{H})$ into itself for every $\mathbf{U} \in \mathcal{U}(\mathcal{H})$. Moreover, it can be proved that, modulo isomorphisms between Hilbert spaces, a CPTP map Φ is invertible if and only if $\Phi(\mathbf{a}) = \mathbf{U}\mathbf{a}\mathbf{U}^\dagger$ for some $\mathbf{U} \in \mathcal{U}(\mathcal{H})$.

F. Nielsen and F. Barbaresco (Eds.): GSI 2021, LNCS 12829, pp. 145–153, 2021.
https://doi.org/10.1007/978-3-030-80209-7_17

Following what Cencov did for the classical case of probability distributions and Markov maps [6], Petz classified the family of all the Riemannian metric tensors $\mathrm{G}^{\mathcal{H}}$ on every $\mathscr{S}(\mathcal{H})$ satisfying the monotonicity property

$$\mathrm{G}^{\mathcal{K}}_{\Phi(\rho)}\left(T_\rho\Phi(\mathbf{a}), T_\rho\Phi(\mathbf{a})\right) \leq \mathrm{G}^{\mathcal{H}}_\rho\left(\mathbf{a},\mathbf{a}\right), \tag{1}$$

where $\Phi\colon \mathcal{B}(\mathcal{H}) \to \mathcal{B}(\mathcal{K})$ is a CPTP map sending $\mathscr{S}(\mathcal{H})$ into $\mathscr{S}(\mathcal{K})$, and $\mathbf{a} \in T_\rho\mathscr{S}(\mathcal{H})$ [15]. Unlike the classical case, it turns out that $\mathrm{G}^{\mathcal{H}}$ is not unique, and there is a one-to-one correspondence between the metric tensors satisfying the monotonicity property in Eq. (1) and the operator monotone functions $f:\mathbb{R}^+ \to \mathbb{R}$ satisfying $f(1)=1$ and $f(t)=tf(t^{-1})$.

Moreover, since $\Phi(\mathbf{a}) = \mathbf{U}\mathbf{a}\mathbf{U}^\dagger$ is an invertible CPTP map for every $\mathbf{U} \in \mathcal{U}(\mathcal{H})$, it follows that every monotone quantum metric tensor must be invariant with respect to the standard action of $\mathcal{U}(\mathcal{H})$, and thus the unitary group may be thought of as a universal symmetry group for quantum information geometry. From the infinitesimal point of view, this means that the fundamental vector fields of the standard action of $\mathcal{U}(\mathcal{H})$ on $\mathscr{S}(\mathcal{H})$ are Killing vector fields for every monotone quantum metric tensor. Since we can write $\mathbf{U} = \mathrm{e}^{\frac{1}{2\imath}\mathbf{a}}$ with $\mathbf{a}$ a self-adjoint operator on $\mathcal{H}$, it follows that the Lie algebra $\mathfrak{u}(\mathcal{H})$ of $\mathcal{U}(\mathcal{H})$ may be identified with the space of self-adjoint operators on $\mathcal{H}$ endowed with the Lie bracket

$$[\mathbf{a},\mathbf{b}] := \frac{1}{2\imath}\left(\mathbf{ab}-\mathbf{ba}\right), \tag{2}$$

and the fundamental vector fields of the action of $\mathcal{U}(\mathcal{H})$ on $\mathscr{S}(\mathcal{H})$ may be labelled by self-adjoint elements in $\mathcal{B}(\mathcal{H})$. We will write these vector fields as $X^{\mathcal{H}}_{\mathbf{b}}$ with $\mathbf{b} \in \mathfrak{u}(\mathcal{H})$.

It is easy to see that the fundamental vector field associated with the identity elements vanishes identically on $\mathscr{S}(\mathcal{H})$. Therefore, in the following, we will focus on the special unitary group $\mathcal{SU}(\mathcal{H})$ rather than on $\mathcal{U}(\mathcal{H})$. This is the Lie subgroup of $\mathcal{U}(\mathcal{H})$ generated by the Lie subalgebra $\mathfrak{su}(\mathcal{H})$ consisting of traceless elements in $\mathfrak{u}(\mathcal{H})$.

Elements in the Lie algebra $\mathfrak{su}(\mathcal{H})$ are not only associated with the fundamental vector fields of $\mathcal{SU}(\mathcal{H})$ on $\mathscr{S}(\mathcal{H})$, they are also associated with the functions $l^{\mathcal{H}}_{\mathbf{a}}$ on $\mathscr{S}(\mathcal{H})$ given by

$$l^{\mathcal{H}}_{\mathbf{a}}(\rho) := Tr_{\mathcal{H}}\left(\rho\,\mathbf{a}\right). \tag{3}$$

According to the standard postulate of quantum mechanics, every such function $l^{\mathcal{H}}_{\mathbf{a}}$ provides the expectation value of the (linear) observable represented $\mathbf{a}$, and thus plays the role of a quantum random variable.

It was observed in [9] that, when the so-called Bures-Helstrom metric tensor $\mathrm{G}^{\mathcal{H}}_{BH}$ is selected among the monotone quantum metric tensors, the gradient vector fields associated with the expectation value functions are complete and, together with the fundamental vector fields of $\mathcal{SU}(\mathcal{H})$, close on an anti-representation of the Lie algebra $\mathfrak{sl}(\mathcal{H})$ integrating to a transitive left action of the special linear group $\mathcal{SL}(\mathcal{H})$ given by

$$\rho \mapsto \frac{\mathrm{g}\rho\mathrm{g}^\dagger}{Tr_{\mathcal{H}}(\mathrm{g}\rho\mathrm{g}^\dagger)}, \tag{4}$$

where $\mathrm{g} = \mathrm{e}^{\frac{1}{2}(\mathbf{a} - \imath \mathbf{b})}$, with $\mathbf{a}, \mathbf{b}$ self-adjoint operators on $\mathcal{H}$.

In [7], it was observed that the Bures-Helstrom metric tensor is not the only monotone quantum metric tensor for which a similar instance is verified. Indeed, if the so-called Wigner-Yanase metric tensor is selected, the gradient vector fields associated with the expectation value functions are complete and, together with the fundamental vector fields of $\mathcal{SU}(\mathcal{H})$, close on an anti-representation of the Lie algebra $\mathfrak{sl}(\mathcal{H})$ integrating to a transitive left action of the special linear group $\mathcal{SL}(\mathcal{H})$ which is different than that in Eq. (4) and is given by

$$\rho \mapsto \frac{\left(\mathrm{g}\sqrt{\rho}\,\mathrm{g}^{\dagger}\right)^{2}}{Tr_{\mathcal{H}}\left(\left(\mathrm{g}\sqrt{\rho}\,\mathrm{g}^{\dagger}\right)^{2}\right)}, \tag{5}$$

where $\mathrm{g} = \mathrm{e}^{\frac{1}{2}(\mathbf{a} - \imath \mathbf{b})}$, with $\mathbf{a}, \mathbf{b}$ self-adjoint operators on $\mathcal{H}$.

Moreover, if the so-called Bogoliubov-Kubo-Mori metric tensor is selected, the gradient vector fields associated with the expectation value functions are complete, commute among themselves, and, together with the fundamental vector fields of $\mathcal{SU}(\mathcal{H})$, close on an anti-representation of the Lie algebra of the cotangent group $T^{*}\mathcal{SU}(\mathcal{H})$ which integrates to a transitive left action given by

$$\rho \mapsto \frac{\mathrm{e}^{\mathbf{U}\ln(\rho)\mathbf{U}^{\dagger} + \mathbf{a}}}{Tr_{\mathcal{H}}\left(\mathrm{e}^{\mathbf{U}\ln(\rho)\mathbf{U}^{\dagger} + \mathbf{a}}\right)}, \tag{6}$$

where $\mathbf{U} = \mathrm{e}^{\frac{1}{2\imath}\mathbf{b}}$, and $\mathbf{a}, \mathbf{b}$ are self-adjoint operators on $\mathcal{H}$.

It is then natural to ask if these three instances are just isolated mathematical coincidences valid only for these very special monotone quantum metric tensors, or if there is an underlying geometric picture waiting to be uncovered and studied. Specifically, we would like to first classify all those monotone quantum metric tensors for which the gradient vector fields associated with the expectation value functions, together with the fundamental vector fields of $\mathcal{SU}(\mathcal{H})$, provide a realization of some Lie algebra integrating to a group action. Then, we would like to understand the geometric significance (if any) of the gradient vector fields thus obtained with respect to the dualistic structures and their geodesics, with respect to quantum estimation theory, and with respect to other relevant structures and tasks in quantum information geometry (a first hint for the Bogoliubov-Kubo-Mori metric tensor may be found in [1,10,13]). Finally, we would like to understand the role (if any) of the "new" Lie groups and Lie algebras appearing in quantum information theory because of these constructions.

At the moment, we do not have a satisfying understanding of the global picture, but we will give a complete classification of the group actions associated with monotone metrics for the case of the simplest quantum system, namely, the qubit.

2 Group Actions and Monotone Metrics for the Qubit

In the case of a qubit, we have $\mathcal{H} \cong \mathbb{C}^2$, and by selecting an orthonormal basis on $\mathcal{H}$, say $\{|1\rangle, |2\rangle\}$, a basis in $\mathfrak{su}(\mathcal{H})$ is given by the Pauli operators

$$\sigma_1 = |1\rangle\langle 2| + |2\rangle\langle 1|, \quad \sigma_2 = \imath|1\rangle\langle 2| - \imath|2\rangle\langle 1|, \quad \sigma_3 = |1\rangle\langle 1| - |2\rangle\langle 2|, \tag{7}$$

so that we have

$$[\sigma_1, \sigma_2] = \sigma_3, \quad [\sigma_3, \sigma_1] = \sigma_2, \quad [\sigma_2, \sigma_3] = \sigma_1. \tag{8}$$

It is well-known that a quantum state ρ may be written as

$$\rho = \frac{1}{2}\left(\sigma_0 + x\sigma_1 + y\sigma_2 + z\sigma_3\right), \tag{9}$$

where σ_0 is the identity operator on $\mathcal{H}$ and x, y, z satisfy $x^2 + y^2 + z^2 \leq 1$. If we want ρ to be invertible, then we must impose $x^2 + y^2 + z^2 < 1$, which means that $\mathscr{S}(\mathcal{H})$ is diffeomorphic to the open interior of a 3-ball of radius 1.

Passing from the Cartesian coordinate system (x, y, z) to the spherical coordinate system (r, θ, ϕ), the monotone quantum metric tensor G_f associated with the operator monotone function f reads

$$\mathrm{G}_f = \frac{1}{1-r^2}\, dr \otimes dr + \frac{r^2}{1+r}\left(f\left(\frac{1-r}{1+r}\right)\right)^{-1} (d\theta \otimes d\theta + \sin^2\theta\, d\phi \otimes d\phi). \tag{10}$$

In the spherical coordinate system, the fundamental vector fields of $\mathcal{SU}(\mathcal{H})$ read

$$\begin{aligned} X_1 &= -\sin\phi\, \partial_\theta - \cot\theta \cos\phi\, \partial_\phi \\ X_2 &= \cos\phi\, \partial_\theta - \cot\theta \sin\phi\, \partial_\phi \\ X_3 &= \partial_\phi \end{aligned} \tag{11}$$

where we have set $X_j \equiv X_{\sigma_j}$ for $j = 1, 2, 3$.

For a generic traceless self-adjoint operator $\mathbf{a} = a_1\sigma_1 + a_2\sigma_2 + a_3\sigma_3$, we have

$$l_a = \frac{1}{2}(a_1 r \sin\theta \cos\phi + a_2 r \sin\theta \sin\phi + a_3 r \cos\theta), \tag{12}$$

and the gradient vector field $Y_{\mathbf{a}}f$ associated with $l_{\mathbf{a}}$ by means of G_f is

$$Y_a = G_f^{-1}(\mathrm{d}l_a, \bullet) = G_f^{-1}(\bullet, \mathrm{d}l_a). \tag{13}$$

Setting $l_j \equiv l_{\sigma_j}$ and $Y_j^f \equiv Y_{\sigma_j}^f$, straightforward computations bring us to

$$\begin{aligned} Y_1^f &= (1-r^2)\sin\theta\cos\phi\, \partial_r + g(r)\left(\cos\theta\cos\phi\, \partial_\theta - \frac{\sin\phi}{\sin\theta}\, \partial_\phi\right) \\ Y_2^f &= (1-r^2)\sin\theta\sin\phi\, \partial_r + g(r)\left(\cos\theta\sin\phi\, \partial_\theta + \frac{\cos\phi}{\sin\theta}\, \partial_\phi\right) \\ Y_3^f &= (1-r^2)\cos\theta\, \partial_r - g(r)\sin\theta\, \partial_\theta, \end{aligned} \tag{14}$$

where

$$g(r) = \frac{1+r}{r} f\left(\frac{1-r}{1+r}\right). \tag{15}$$

Then, the commutators between the gradient vector fields can be computed to be:

$$\begin{aligned} [Y_1^f, Y_2^f] &= F(r)\, X_3 \\ [Y_2^f, Y_3^f] &= F(r)\, X_1 \\ [Y_3^f, Y_1^f] &= F(r)\, X_2 \end{aligned} \tag{16}$$

where we have used the expressions in Eq. (11), and we have set

$$F(r) = (1-r^2)g'(r) + g^2(r) \tag{17}$$

for notational simplicity.

In order to get a representation of a Lie algebra out of the gradient vector fields and the fundamental vector fields of $\mathcal{SU}(\mathcal{H})$ we must set $F(r) = A$ with A a constant, obtaining the following ordinary differential equation

$$(1-r^2)g'(r) + g^2(r) = A \tag{18}$$

for the function $g(r)$. Performing the change of variable $t = \frac{1-r}{1+r}$ and separating variables, we obtain the ordinary differential equation

$$\frac{g'(t)}{g^2(t) - A} = \frac{1}{2t}. \tag{19}$$

This ODE has a different behavior depending on the sign of A, thus we will deal separately with the cases where A is zero, positive and negative.

2.1 First Case ($A = 0$)

This case is peculiar, since it is the only case in which the gradient vector fields close a Lie algebra by themselves, albeit a commutative one. In this case, integration of the ODE in Eq. (19) leads to

$$g_0(t) = \frac{-2}{\log t + c}, \tag{20}$$

with c being an integration constant arising in the solution of the differential equation. Taking into account Eq. (15), we obtain the function

$$f_0(t) = \frac{t-1}{\log t + c}. \tag{21}$$

If we want f_0 to belong to the class of functions appearing in Petz's classification [15], we need to impose $\lim_{t\to 1} f_0(t) = 1$, which is easily seen to impose $c = 0$, and thus we obtain the function

$$f_0(t) = \frac{t-1}{\log t}. \tag{22}$$

This function is operator monotone and gives rise to the *Bogoliubov-Kubo-Mori* (BKM) metric [14]. Following the results presented in [7], we conclude that the fundamental vector fields of $\mathcal{SU}(\mathcal{H})$, together with the gradient vector fields associated with the expectation value functions by means of the BKM metric tensor, provide an anti-representation of the Lie algebra of the cotangent group $T^*\mathcal{SU}(\mathcal{H})$ integrating to the action in Eq. (6).

It is worth noting that the BKM metric tensor also appears as the Hessian metric obtained from the von Neumann entropy following [2–4], and this instance has profound implications from the point of view of Quantum Information Theory.

2.2 Second Case ($A > 0$)

In the case where $A > 0$, integration of the ODE in Eq. (19), together with the expression of $g(r)$ in Eq. (15), leads to

$$f_A(t) = \frac{\sqrt{A}}{2}(1-t)\frac{1 + e^{2c\sqrt{A}}t^{\sqrt{A}}}{1 - e^{2c'\sqrt{A}}t^{\sqrt{A}}}, \tag{23}$$

where c is an integration constant arising in the solution of the differential equation. If we want f_A to belong to the class of functions appearing in Petz's classification [15], we need to impose $\lim_{t\to 1} f_A(t) = 1$, which implies $c = 0$, and thus

$$f_A(t) = \frac{\sqrt{A}}{2}(1-t)\frac{1 + t^{\sqrt{A}}}{1 - t^{\sqrt{A}}}. \tag{24}$$

We now consider the gradient vector fields

$$Y_{\mathbf{a}}^A := G_{f_A}^{-1}\left(\frac{1}{\sqrt{A}}\mathrm{d}l_{\mathbf{a}}\,, \bullet\right) \tag{25}$$

associated with the (renormalized) expectation value functions $\frac{1}{\sqrt{A}}l_{\mathbf{a}}$ by means of the monotone metric tensor G_{f_A} with f_A given in Eq. (24). Recalling Eq. (16) and the fact that $F(r) = A > 0$, we conclude that the fundamental vector fields of the action of $\mathcal{SU}(\mathcal{H})$ given in Eq. (11), together with the gradient vector fields given in Eq. (25), provide an anti-representation of the Lie algebra $\mathfrak{sl}(\mathcal{H})$ of $\mathcal{SL}(\mathcal{H})$. Exploiting theorem 5.3.1 in [5], it is possible to prove that the anti-representation of $\mathfrak{sl}(\mathcal{H})$ integrates to an action of $\mathcal{SL}(\mathcal{H})$ on $\mathscr{S}(\mathcal{H})$ given by

$$\alpha_A(\mathrm{g},\rho) = \frac{\left(\mathrm{g}\rho^{\sqrt{A}}\mathrm{g}^\dagger\right)^{\frac{1}{\sqrt{A}}}}{Tr\left(\mathrm{g}\rho^{\sqrt{A}}\mathrm{g}^\dagger\right)^{\frac{1}{\sqrt{A}}}} \tag{26}$$

Notice that when $A = 1$, this action reduces exactly to the action of $\mathcal{SL}(\mathcal{H})$ in Eq. (4). Accordingly, the associated monotone metric G_{f_1} turns out to be the Bures-Helstrom metric, in agreement with [9]. A similar discussion can be carried out in the case $A = \frac{1}{4}$. In this case, we get the action in Eq. (5) and the monotone metric given by this choice is Wigner-Yanase metric. Again, this is in agreement with results in [7]. The function f_A is continuous on $[0, \infty]$ and differentiable with continuous derivative in $(0, \infty)$ for all $A > 0$. Also, whenever $A > 1$, its derivative tends to the value $-\sqrt{A}/2$ for t approaching 0 from the right. From these two considerations, it follows that f_A can not be operator monotone whenever $A > 1$, since there exists an interval in which its derivative is negative. Let us stress that it is still an open problem to find all values $0 < A < 1$ for which the function f_A is operator monotone.

2.3 Third Case ($A < 0$)

In this case, setting $B = \frac{-A}{4}$, the integration of the ODE in Eq. (19), leads us to

$$f_B(t) = \sqrt{B}\frac{1-t}{2} \tan\left(\sqrt{B}\log t - \sqrt{B}c\right), \tag{27}$$

where c is an integration constant. Since $\mathscr{S}(\mathcal{H})$ is diffeomorphic to the open interior of a 3-ball of radius 1, we have that $0 \leq r < 1$ and thus $0 < t \leq 1$. Consequently, in this range for t, the function $f_B(t)$ presents a countable infinity of points where it is not defined, regardless of the values of B and c. Thus we are led to exclude this case, which would give rise to a metric tensor G_{f_B} that is not defined everywhere on the space of quantum states $\mathscr{S}(\mathcal{H})$.

3 Conclusions

Guided by results presented in [7] and [9], we investigated the relation between the fundamental vector fields of the canonical action of $\mathcal{SU}(\mathcal{H})$ on the space of faithful quantum states and the gradient vector fields associated with the expectation value functions by means of the monotone metric tensors classified by Petz. In particular, we give a complete classification of all those metric tensors for which the above-mentioned gradient vector fields, together with the fundamental vector fields of the canonical action of $\mathcal{SU}(\mathcal{H})$, provide an anti-representation of a Lie algebra having $\mathfrak{su}(\mathcal{H})$ as a Lie subalgebra. We found that there are only two possible such Lie algebras, namely, the Lie algebra of the cotangent Lie group $T^*\mathcal{SU}(\mathcal{H})$, and the Lie algebra of $\mathcal{SL}(\mathcal{H})$ (the complexification of $\mathfrak{su}(\mathcal{H})$).

In the first case, there is only one monotone metric tensor associated with the Lie algebra anti-representation, namely the Bogoliubov-Kubo-Mori metric tensor. Also, this is the only case in which the gradient vector fields close a (commutative) Lie algebra.

In the second case, we found an infinite number of metric tensors associated with an anti-representation of $\mathfrak{sl}(\mathcal{H})$. These metric tensors are labelled by the function f_A, with $A > 0$, given in Eq. 24, and, among them, we recover the Bures-Helstrom metric tensor when $A = 1$, and the Wigner-Yanase metric tensor when $A = \frac{1}{4}$, in accordance with the results in [7]. As already mentioned, we know that f_A is never operator monotone when $A > 1$, however, besides the cases $A = 1$

and $A = \frac{1}{4}$, it is still an open problem to find all values $0 < A < 1$ for which the function f_A is operator monotone. We also found that all these Lie algebra anti-representations integrate to Lie group actions. The explicit forms of these actions are given in Eq. (6) and Eq. (26).

The analysis presented here is clearly preliminar because it addresses only the qubit case. However, building upon the results obtained for the qubit, we have been able to recover the group actions presented here also in the case of a quantum system in arbitrary finite dimension. We are not able to include these cases in this contribution because of space constraints. Also, while for the qubit case we showed that the group actions of $\mathcal{SL}(\mathcal{H})$ and $T^*\mathcal{SU}(\mathcal{H})$ are the only possible, for an arbitrary finite-dimensional quantum system, we are only able to show that these actions are there, but not that they exhaust all the possible group actions compatible with monotone quantum metrics in the sense specified before. This clearly calls for a more deep investigation we plan to pursue in future publications.

References

1. Andai, A., Lovas, A.: Quantum Aitchison geometry. Infin. Dimens. Anal. Quantum Probab. Relat. Top. **24**(01), 2150001 (2021)
2. Balian, R.: Incomplete descriptions and relevant entropies. Am. J. Phys. **67**(12), 1078–1090 (1999)
3. Balian, R.: A metric for quantum states issued from von Neumann's entropy. In: Nielsen, F., Barbaresco, F. (eds.) GSI 2013. LNCS, vol. 8085, pp. 513–518. Springer, Heidelberg (2013). https://doi.org/10.1007/978-3-642-40020-9_56
4. Balian, R.: The entropy-based quantum metric. Entropy **16**(7), 3878–3888 (2014)
5. Bhatia, R.: Positive Definite Matrices. Princeton University Press, Princeton (2007)
6. Cencov, N.N.: Statistical Decision Rules and Optimal Inference. American Mathematical Society, Providence (1982)
7. Ciaglia, F.M.: Quantum states, groups and monotone metric tensors. Eur. Phys. J. Plus **135**(6), 1–16 (2020). https://doi.org/10.1140/epjp/s13360-020-00537-y
8. Ciaglia, F.M., Ibort, A., Jost, J., Marmo, G.: Manifolds of classical probability distributions and quantum density operators in infinite dimensions. Inf. Geom. **2**(2), 231–271 (2019). https://doi.org/10.1007/s41884-019-00022-1
9. Ciaglia, F.M., Jost, J., Schwachhöfer, L.: From the Jordan product to Riemannian geometries on classical and quantum states. Entropy **22**(06), 637–27 (2020)
10. Felice, D., Mancini, S., Ay, N.: Canonical divergence for measuring classical and quantum complexity. Entropy **21**(4), 435 (2019)
11. Grabowski, J., Kuś, M., Marmo, G.: Symmetries, group actions, and entanglement. Open Syst. Inf. Dyn. **13**(04), 343–362 (2006)
12. Holevo, A.S.: Statistical Structure of Quantum Theory. Springer, Heidelberg (2001). https://doi.org/10.1007/3-540-44998-1

13. Naudts, J.: Parameter-free description of the manifold of non-degenerate density matrices. Eur. Phys. J. Plus **136**(1), 1–13 (2021). https://doi.org/10.1140/epjp/s13360-020-01038-8
14. Petz, D.: Geometry of canonical correlation on the state space of a quantum system. J. Math. Phys. **35**, 780–795 (1994)
15. Petz, D.: Monotone metrics on matrix spaces. Linear Algebra Appl. **244**, 81–96 (1996)

Quantum Jensen-Shannon Divergences Between Infinite-Dimensional Positive Definite Operators

Hà Quang Minh(✉)

RIKEN Center for Advanced Intelligence Project, Tokyo, Japan
minh.haquang@riken.jp

Abstract. This work studies a parametrized family of symmetric divergences on the set of Hermitian positive definite matrices which are defined using the α-Tsallis entropy $\forall \alpha \in \mathbb{R}$. This family unifies in particular the Quantum Jensen-Shannon divergence, defined using the von Neumann entropy, and the Jensen-Bregman Log-Det divergence. The divergences, along with their metric properties, are then generalized to the setting of positive definite trace class operators on an infinite-dimensional Hilbert space $\forall \alpha \in \mathbb{R}$. In the setting of reproducing kernel Hilbert space (RKHS) covariance operators, all divergences admit closed form formulas in terms of the corresponding kernel Gram matrices.

Keywords: Quantum Jensen-Shannon divergences · Positive definite operators · Trace class operators

1 Introduction

This work studies the infinite-dimensional generalization of a parametrized family of symmetric divergences on the set of Hermitian positive matrices to the set of Hermitian positive trace class operators on a Hilbert space. These divergence/distance functions have found numerous practical applications, including brain imaging [8], computer vision [4], brain computer interfaces [5], and quantum information theory [2,6,10].

Previous Work. Our starting point is the Quantum Jensen-Shannon divergence (QJSD) [10]. Let $\mathrm{Sym}^+(n)$ and $\mathrm{Sym}^{++}(n)$ denote the sets of $n \times n$ Hermitian positive semi-definite and positive definite matrices, respectively. Then for any pair $A, B \in \mathrm{Sym}^{++}(n)$, $\mathrm{QJSD}(A, B)$ is defined by [10]

$$\mathrm{QJSD}(A, B) = S\left(\frac{A+B}{2}\right) - \frac{1}{2}[S(A) + S(B)], \tag{1}$$

where $S(A) = -\mathrm{tr}[A \log A]$ is the *von Neumann entropy*. This is a special case of the Jensen-Bregman divergences [14], where S is a strictly concave, differentiable function on $\mathrm{Sym}^{++}(n)$. More general Bregman matrix divergences are described

F. Nielsen and F. Barbaresco (Eds.): GSI 2021, LNCS 12829, pp. 154–162, 2021.
https://doi.org/10.1007/978-3-030-80209-7_18

in [15]. QJSD has been applied in quantum information theory [2,6,10], graph theory [1,16], network theory [7], among other fields. In [20], it was proved that $\sqrt{QJSD}$ is a metric on $\mathrm{Sym}^{+}(n)$. In [18], QJSD as defined in (1) was generalized to the following parametrized family of symmetric divergences

$$\mathrm{QJSD}_{\alpha}(A,B) = S_{\alpha}\left(\frac{A+B}{2}\right) - \frac{1}{2}[S_{\alpha}(A) + S_{\alpha}(B)] \quad (2)$$

where S_{α} is the α-*Tsallis entropy*, defined by

$$S_{\alpha}(A) = \frac{\mathrm{tr}A^{\alpha} - \mathrm{tr}(A)}{1-\alpha}, \quad 0 \le \alpha \le 2, \alpha \ne 1, \quad (3)$$

with $\lim_{\alpha\to 1} S_{\alpha}(A) = S(A)$. In particular, $\sqrt{\mathrm{QJSD}_{\alpha}}$ was proved in [18] to be a metric for $0 < \alpha < 2$. For $\alpha = 2$, the metric property is obvious since $\mathrm{QJSD}_2(A,B) = \frac{1}{4}||A-B||_F^2$, with $|| \ ||_F$ being the Frobenius norm. Both proofs in [20] and [17] are based on the fact that the square root of the *symmetric Log-Det divergence* (also called Jensen Bregman Log-Det [4] or S-divergence [17])

$$D_{\mathrm{logdet}}(A,B) = \log\det\left(\frac{A+B}{2}\right) - \frac{1}{2}[\log\det(A) + \log\det(B)] \quad (4)$$

is a metric on $\mathrm{Sym}^{++}(n)$ [17]. For $\alpha = 0$, we have $S_0(A) = n - \mathrm{tr}(A)$ and thus $\mathrm{QJSD}_{\alpha}(A,B) = 0 \ \forall A, B \in \mathrm{Sym}^{+}(n)$, so that QJSD_0 is degenerate.

Contributions of This Work

1. We first give a new definition of QJSD_{α}, instead of Eq. (2), that is a valid symmetric divergence on $\mathrm{Sym}^{++}(n)$ $\forall \alpha \in \mathbb{R}$, unifying in particular the quantum Jensen-Shannon divergence with the symmetric Log-Det divergence.
2. Our main focus is on generalizing this new QJSD_{α} to a family of symmetric divergences, $\forall \alpha \in \mathbb{R}$, on the set of Hermitian positive trace class operators on an *infinite-dimensional Hilbert space*, along with their metric properties.
3. In the setting of reproducing kernel Hilbert space (RKHS) covariance operators, QJSD_{α} is expressed explicitly via the corresponding Gram matrices. Our study of the RKHS setting is motivated in particular by applications in computer vision, see e.g. [9,13].

2 Finite-Dimensional Setting

We start by introducing a new definition of the quantum Jensen-Shannon divergence given in Eq. (2), which is valid $\forall \alpha \in \mathbb{R}$. Throughout the remainder of the paper, QJSD_{α} refers to that given in Definition 1 below.

Definition 1 (Alpha Quantum Jensen-Shannon Divergence - finite-dimensional setting). *Let* $\alpha \in \mathbb{R}$ *be fixed. The* α*-quantum Jensen-Shannon divergence on* $\mathrm{Sym}^{++}(n)$ *is defined to be*

$$\mathrm{QJSD}_{\alpha}(A,B) = \frac{1}{\alpha}\left[S_{\alpha}\left(\frac{A+B}{2}\right) - \frac{1}{2}(S_{\alpha}(A) + S_{\alpha}(B))\right], \quad \alpha \ne 0, 1, \quad (5)$$

$$= \frac{1}{\alpha(1-\alpha)}\left[\mathrm{tr}\left(\frac{A+B}{2}\right)^{\alpha} - \frac{1}{2}\mathrm{tr}(A^{\alpha}) - \frac{1}{2}\mathrm{tr}(B^{\alpha})\right], \tag{6}$$

$$\mathrm{QJSD}_0(A,B) = \lim_{\alpha\to 0}\mathrm{QJSD}_{\alpha}(A,B), \quad \mathrm{QJSD}_1(A,B) = \lim_{\alpha\to 1}\mathrm{QJSD}_{\alpha}(A,B). \tag{7}$$

The validity of QJSD_{α} as a divergence on $\mathrm{Sym}^{++}(n)$ is due to the convexity of the power trace function on $\mathrm{Sym}^{++}(n)$, which follows Theorem 2.10 in [3].

Lemma 1. *The function* $F : \mathrm{Sym}^{++}(n) \to \mathbb{R}$ *defined by* $F(A) = \mathrm{tr}(A^{\alpha})$ *is strictly convex for* $\alpha < 0$ *and* $\alpha > 1$ *and strictly concave for* $0 < \alpha < 1$.

Theorem 1. QJSD_{α} *as defined in Definition 1 is a symmetric divergence on* $\mathrm{Sym}^{++}(n)$ *for all* $\alpha \in \mathbb{R}$. *Specifically,* $\forall A, B \in \mathrm{Sym}^{++}(n)$,

1. $\mathrm{QJSD}_{\alpha}(A,B) = \mathrm{QJSD}_{\alpha}(B,A) \geq 0$, $\mathrm{QJSD}_{\alpha}(A,B) = 0 \Longleftrightarrow A = B$.
2. *At the limiting points* $\alpha = 0$ *and* $\alpha = 1$,

$$\mathrm{QJSD}_0(A,B) = \log\det\left(\frac{A+B}{2}\right) - \frac{1}{2}\log\det(A) - \frac{1}{2}\log\det(B), \tag{8}$$

$$\mathrm{QJSD}_1(A,B) = \mathrm{tr}\left[-\frac{A+B}{2}\log\frac{A+B}{2} + \frac{1}{2}A\log A + \frac{1}{2}B\log B\right]. \tag{9}$$

Special cases. For $\alpha = 2$, $\mathrm{QJSD}_2(A,B) = \frac{1}{8}||A-B||_F^2$. Thus the family QJSD_{α} encompasses (i) squared Euclidean distance ($\alpha = 2$), (ii) quantum Jensen-Shannon divergence ($\alpha = 1$), and (iii) symmetric Log-Det divergence ($\alpha = 0$).

Proposition 1. *For* $-\infty < \alpha < 2$, $\lim_{r\to\infty}\mathrm{QJSD}_{\alpha}[(A+rI),(B+rI)] = 0$ *and*

$$\mathrm{QJSD}_{\alpha}(A,B) = (2-\alpha)\int_0^{\infty}\mathrm{QJSD}_{\alpha-1}[(A+rI),(B+rI)]dr. \tag{10}$$

Theorem 2 (Metric property - finite-dimensional setting). *The distance* $\sqrt{\mathrm{QJSD}_{\alpha}}$ *is a metric on* $\mathrm{Sym}^{++}(n)$ *for* $0 \leq \alpha \leq 2$. *For* $-\infty < \alpha < 2$, $\sqrt{\mathrm{QJSD}_{\alpha}}$ *is a Hilbert space distance on any set* $\mathcal{S}$ *of commuting matrices in* $\mathrm{Sym}^{++}(n)$. *Specifically, there is a Hilbert space* $\mathcal{F}$ *and a map* $\Phi_{\alpha} : \mathrm{Sym}^{++}(n) \to \mathcal{F}$ *such that*

$$\mathrm{QJSD}_{\alpha}(A,B) = ||\Phi_{\alpha}(A) - \Phi_{\alpha}(B)||_{\mathcal{F}}^2 \quad \forall A, B \in \mathcal{S}. \tag{11}$$

3 The Infinite-Dimensional Setting

We now generalize Definition 1 to the setting of positive trace class operators on a Hilbert space. Throughout the following, let $\mathcal{H}$ be a separable Hilbert space with $\dim(\mathcal{H}) = \infty$ unless explicitly stated otherwise. Let $\mathcal{L}(\mathcal{H})$ denote the set of bounded linear operators on $\mathcal{H}$. Let $\mathrm{Sym}(\mathcal{H}) \subset \mathcal{L}(\mathcal{H})$ denote the set of bounded, self-adjoint operators, $\mathrm{Sym}^{+}(\mathcal{H})$ the set of positive operators, i.e.

$\mathrm{Sym}^+(\mathcal{H}) = \{A \in \mathrm{Sym}(\mathcal{H}) : \langle x, Ax\rangle \geq 0 \forall x \in \mathcal{H}\}$. Let $\mathrm{Tr}(\mathcal{H})$ and $\mathrm{HS}(\mathcal{H})$ denote the sets of trace class and Hilbert-Schmidt operators on $\mathcal{H}$, respectively.

For $A \in \mathrm{Sym}^+(\mathcal{H}) \cap \mathrm{Tr}(\mathcal{H})$, let $\{\lambda_k\}_{k\in\mathbb{N}}$ denote its eigenvalues, with corresponding orthonormal eigenvectors $\{\phi_k\}_{k\in\mathbb{N}}$. Using the spectral decomposition $A = \sum_{k=1}^{\infty} \lambda_k \phi_k \otimes \phi_k$, for $\alpha \geq 0$, the power A^α is uniquely defined by $A^\alpha = \sum_{k=1}^{\infty} \lambda_k^\alpha \phi_k \otimes \phi_k \in \mathrm{Sym}^+(\mathcal{H})$. Since $A \in \mathrm{Tr}(\mathcal{H})$, we have $A^\alpha \in \mathrm{Tr}(\mathcal{H})$ for all $\alpha \in \mathbb{R}, \alpha \geq 1$. However, this is no longer true for $0 < \alpha < 1$. For $\alpha = 1/2$, we only have $A^{1/2} \in \mathrm{Sym}^+(\mathcal{H}) \cap \mathrm{HS}(\mathcal{H})$. Furthermore, for $\alpha < 0$, A^α is unbounded.

Extended Trace Class Operators. To obtain the infinite-dimensional generalization of Definition 1 that is valid $\forall \alpha \in \mathbb{R}$, we consider the setting of *positive definite unitized trace class operators* [12]. Let us denote by $\mathbb{P}(\mathcal{H})$ the set of self-adjoint, positive definite operators on $\mathcal{H}$, namely $\mathbb{P}(\mathcal{H}) = \{A \in \mathcal{L}(\mathcal{H}), A^* = A, \exists M_A > 0 \; s.t. \langle x, Ax\rangle \geq M_A||x||^2 \; \forall x \in \mathcal{H}\}$. We write $A > 0 \iff A \in \mathbb{P}(\mathcal{H})$. The set of *extended (unitized) trace class operators* is defined by [12]

$$\mathrm{Tr}_X(\mathcal{H}) = \{A + \gamma I : A \in \mathrm{Tr}(\mathcal{H}), \gamma \in \mathbb{R}\}. \tag{12}$$

For $(A + \gamma I) \in \mathrm{Tr}_X(\mathcal{H})$, the *extended trace* is defined to be [12]

$$\mathrm{tr}_X(A + \gamma I) = \mathrm{tr}(A) + \gamma. \tag{13}$$

By this definition, $\mathrm{tr}_X(I) = 1$, in contrast to the standard trace $\mathrm{tr}(I) = \infty$. Along with the extended trace, [12] introduced the *extended Fredholm determinant*, which, for $\dim(\mathcal{H}) = \infty$ and $\gamma \neq 0$, is given by

$$\det{}_X(A + \gamma I) = \gamma \det\left(I + [A/\gamma]\right), \tag{14}$$

where det is the Fredholm determinant, $\det(I + A) = \prod_{k=1}^{\infty}(1 + \lambda_k)$.

The set of *positive definite unitized trace class operators* is then defined to be

$$\mathscr{PC}_1(\mathcal{H}) = \mathbb{P}(\mathcal{H}) \cap \mathrm{Tr}_X(\mathcal{H}) = \{A + \gamma I > 0 : A \in \mathrm{Tr}(\mathcal{H}), \gamma \in \mathbb{R}\}. \tag{15}$$

In particular, for $A \in \mathrm{Sym}^+(\mathcal{H}) \cap \mathrm{Tr}(\mathcal{H})$, we have $(A + \gamma I) \in \mathscr{PC}_1(\mathcal{H}) \forall \gamma > 0$. With $(A + \gamma I) \in \mathscr{PC}_1(\mathcal{H})$, the following operators are always bounded

$$(A + \gamma I)^\alpha = \sum_{k=1}^{\infty} (\lambda_k + \gamma)^\alpha (\phi_k \otimes \phi_k), \quad \alpha \in \mathbb{R}, \tag{16}$$

$$\log(A + \gamma I) = \sum_{k=1}^{\infty} \log(\lambda_k + \gamma)(\phi_k \otimes \phi_k). \tag{17}$$

Moreover (Lemma 3 in [12], Lemmas 10,11 in [11]), if $(I + A) \in \mathscr{PC}_1(\mathcal{H})$, then $\log(I + A) \in \mathrm{Sym}(\mathcal{H}) \cap \mathrm{Tr}(\mathcal{H})$ and $(I + A)^\alpha - I \in \mathrm{Sym}(\mathcal{H}) \cap \mathrm{Tr}(\mathcal{H}) \; \forall \alpha \in \mathbb{R}$. Thus if $(A + \gamma I) \in \mathscr{PC}_1(\mathcal{H})$, then $\log(A + \gamma I) \in \mathrm{Tr}_X(\mathcal{H})$ and $(A + \gamma I)^\alpha \in \mathrm{Tr}_X(\mathcal{H})$ $\forall \alpha \in \mathbb{R}$. Therefore, the following extended traces are always finite $\forall \gamma > 0$

$$\mathrm{tr}_X[(A + \gamma I)^\alpha] = \gamma^\alpha + \gamma^\alpha \mathrm{tr}[(I + A/\gamma)^\alpha - 1], \tag{18}$$

$$\mathrm{tr_X}[\log(A+\gamma I)] = \log(\gamma) + \mathrm{tr}\left[\log\left(I + \frac{A}{\gamma}\right)\right] = \log\det{}_{\mathrm{X}}(A+\gamma I). \quad (19)$$

Thus the following generalized version of the Tsallis entropy is well-defined.

Definition 2 (Extended α-Tsallis entropy). *Let $\gamma \in \mathbb{R}, \gamma > 0$ be fixed. For $A \in \mathrm{Sym}^+(\mathcal{H}) \cap \mathrm{Tr}(\mathcal{H})$, the extended α-Tsallis entropy is defined to be*

$$S_{\alpha,\gamma}(A) = \frac{\mathrm{tr_X}(A+\gamma I)^\alpha - \mathrm{tr_X}(A+\gamma I)}{1-\alpha}, \quad \alpha \neq 1, \quad (20)$$

$$S_{1,\gamma}(A) = -\mathrm{tr_X}[(A+\gamma I)\log(A+\gamma I)]. \quad (21)$$

The definition of $S_{1,\gamma}$ is motivated by the following limits.

Lemma 2. *Let $A \in \mathrm{Sym}^+(\mathcal{H}) \cap \mathrm{Tr}(\mathcal{H})$, $\gamma \in \mathbb{R}, \gamma > 0$. Then*

$$\lim_{\alpha \to 1} \frac{\mathrm{tr_X}[(A+\gamma I)^\alpha - (A+\gamma I)]}{1-\alpha} = -\mathrm{tr_X}[(A+\gamma I)\log(A+\gamma I)]. \quad (22)$$

In particular, for $\gamma = 1$, $\lim_{\alpha\to 1} \frac{\mathrm{tr}[(I+A)^\alpha - (I+A)]}{1-\alpha} = -\mathrm{tr}[(I+A)\log(I+A)]$.

Lemma 3. *Let $A \in \mathrm{Sym}^+(\mathcal{H}) \cap \mathrm{Tr}(\mathcal{H})$. Let $\gamma \in \mathbb{R}, \gamma > 0$ be fixed. Then*

$$\lim_{\alpha \to 0} \frac{\mathrm{tr_X}[(A+\gamma I)^\alpha - I]}{\alpha} = \log\det{}_{\mathrm{X}}(A+\gamma I). \quad (23)$$

In particular, for $\gamma = 1$, $\lim_{\alpha\to 0} \frac{\mathrm{tr}[(I+A)^\alpha - I]}{\alpha} = \log\det(I+A)$.

Definition 3 (Alpha Quantum Jensen-Shannon Divergence - infinite-dimensional version). *Let $\gamma \in \mathbb{R}, \gamma > 0$ be fixed. The α-Quantum Jensen-Shannon divergence on $\mathrm{Sym}^+(\mathcal{H}) \cap \mathrm{Tr}(\mathcal{H})$ is defined by, for $\alpha \neq 0, \alpha \neq 1$,*

$$\begin{aligned}\mathrm{QJSD}_{\alpha,\gamma}(A,B) &= \frac{1}{\alpha}\left(S_{\alpha,\gamma}\left(\frac{A+B}{2}\right) - \frac{1}{2}[S_{\alpha,\gamma}(A) + S_{\alpha,\gamma}(B)]\right) \quad (24)\\ &= \frac{1}{\alpha(1-\alpha)}\left[\mathrm{tr_X}\left(\frac{A+B}{2} + \gamma I\right)^\alpha - \frac{1}{2}[\mathrm{tr_X}(A+\gamma I)^\alpha + \mathrm{tr_X}(B+\gamma I)^\alpha]\right]\end{aligned}$$

$$\mathrm{QJSD}_{0,\gamma}(A,B) = \lim_{\alpha\to 0} \mathrm{QJSD}_{\alpha,\gamma}(A,B), \quad (25)$$

$$\mathrm{QJSD}_{1,\gamma}(A,B) = \lim_{\alpha\to 1} \mathrm{QJSD}_{1,\gamma}(A,B). \quad (26)$$

The quantities $\mathrm{QJSD}_{\alpha,\gamma}(A,B)$ for $\alpha = 0, 1$ are well-defined, as follows.

Theorem 3 (Limiting behavior). *At the limiting points* $\alpha = 0$ *and* $\alpha = 1$,

$$\mathrm{QJSD}_{0,\gamma}(A, B) = \log \det_X \left(\frac{A+B}{2} + \gamma I\right) - \frac{1}{2}\left[\log \det_X (A + \gamma I) + \log \det_X (B + \gamma I)\right]. \quad (27)$$

$$\mathrm{QJSD}_{1,\gamma}(A, B) = -\mathrm{tr}_X \left[\left(\frac{A+B}{2} + \gamma I\right) \log \left(\frac{A+B}{2} + \gamma I\right)\right] + \frac{1}{2}\mathrm{tr}_X \left[(A + \gamma I) \log(A + \gamma I) + (B + \gamma I) \log(B + \gamma I)\right]. \quad (28)$$

Special Case: Finite-dimensional Setting. For $A, B \in \mathrm{Sym}^{++}(n)$, setting $\gamma = 0$ in Eqs.(24), (27), and (28), we recover Eqs. (6), (8), and (9), respectively.

Special Cases: $\mathrm{QJSD}_{2,\gamma}(A, B) = \frac{1}{8}||A - B||^2_{\mathrm{HS}}\ \forall \gamma \in \mathbb{R}$, and $\mathrm{QJSD}_{0,\gamma}$ is the *infinite-dimensional Symmetric Log-Det divergence* [12].

The validity of QJSD_α as a divergence on $\mathrm{Sym}^+(\mathcal{H}) \cap \mathrm{Tr}(\mathcal{H})$ follows from

Lemma 4. *The function* $F : \mathrm{Sym}^+(\mathcal{H}) \cap \mathrm{Tr}(\mathcal{H}) \to \mathbb{R}$, $F(A) = \mathrm{tr}[(I + A)^\alpha - I]$, *is strictly concave for* $0 < \alpha < 1$ *and strictly convex for* $\alpha > 1$ *and* $\alpha < 0$.

Theorem 4 (Divergence properties). *Let* $\gamma \in \mathbb{R}, \gamma > 0$ *be fixed.* QJSD_α *is a symmetric divergence on* $\mathrm{Sym}^+(\mathcal{H}) \cap \mathrm{Tr}(\mathcal{H})\ \forall \alpha \in \mathbb{R}$. *Specifically,*

1. $\mathrm{QJSD}_{\alpha,\gamma}(A, B) = \mathrm{QJSD}_{\alpha,\gamma}(B, A) \geq 0\ \forall A, B \in \mathrm{Sym}^+(\mathcal{H}) \cap \mathrm{Tr}(\mathcal{H})$.
2. $\mathrm{QJSD}_{\alpha,\gamma}(A, B) = 0 \iff A = B$.
3. **(Unitary invariance)** *Let* U *be any unitary operator on* $\mathcal{H}$. *Then*

$$\mathrm{QJSD}_{\alpha,\gamma}(UAU^*, UBU^*) = \mathrm{QJSD}_{\alpha,\gamma}(A, B). \quad (29)$$

Theorem 5 (Continuity in trace norm). *Let* $\gamma > 0$ *be fixed. Let* $A, B, \{A_n\}_{n\in\mathbb{N}}, \{B_n\}_{n\in\mathbb{N}} \in \mathrm{Sym}(\mathcal{H}) \cap \mathrm{Tr}(\mathcal{H})$, *with* $A + \gamma I > 0, B + \gamma I > 0$, $A_n + \gamma I > 0, B_n + \gamma I > 0\ \forall n \in \mathbb{N}$, *and* $\lim_{n\to\infty} ||A_n - A||_{\mathrm{tr}} = \lim_{n\to\infty} ||B_n - B||_{\mathrm{tr}} = 0$. *Then*

$$\lim_{n\to\infty} \mathrm{QJSD}_{\alpha,\gamma}(A_n, B_n) = \mathrm{QJSD}_{\alpha,\gamma}(A, B). \quad (30)$$

Theorem 6 (Metric property - infinite-dimensional setting). *Let* $\gamma \in \mathbb{R}, \gamma > 0$ *be fixed. For* $0 \leq \alpha \leq 2$, $\sqrt{\mathrm{QJSD}_{\alpha,\gamma}}$ *is a metric on* $\mathrm{Sym}^+(\mathcal{H}) \cap \mathrm{Tr}(\mathcal{H})$. *For* $-\infty < \alpha < 2$, $\sqrt{\mathrm{QJSD}_{\alpha,\gamma}}$ *is a Hilbert space distance on* $\mathcal{S} \cap \mathrm{Tr}(\mathcal{H})$, *where* $\mathcal{S}$ *is any set of commuting operators in* $\mathrm{Sym}^+(\mathcal{H})$.

4 The RKHS Setting

We now present the RKHS setting where $\mathrm{QJSD}_{\alpha,\gamma}$ can be computed in closed form. Let $\mathcal{X}$ be a complete separable metric space and K be a continuous positive definite kernel on $\mathcal{X} \times \mathcal{X}$. Then the reproducing kernel Hilbert space (RKHS) $\mathcal{H}_K$ induced by K is separable ([19], Lemma 4.33). Let $\Phi : \mathcal{X} \to \mathcal{H}_K$ be the corresponding canonical feature map, so that $K(x,y) = \langle \Phi(x), \Phi(y)\rangle_{\mathcal{H}_K}$ $\forall (x,y) \in \mathcal{X} \times \mathcal{X}$. Let ρ be a Borel probability measure on $\mathcal{X}$ such that

$$\int_{\mathcal{X}} ||\Phi(x)||^2_{\mathcal{H}_K} d\rho(x) = \int_{\mathcal{X}} K(x,x) d\rho(x) < \infty. \tag{31}$$

Then the RKHS mean vector $\mu_\Phi \in \mathcal{H}_K$ and covariance operator $C_\Phi : \mathcal{H}_K \to \mathcal{H}_K$ are both well-defined and are given by

$$\mu_\Phi = \int_{\mathcal{X}} \Phi(x) d\rho(x) \in \mathcal{H}_K, \tag{32}$$

$$C_\Phi = \int_{\mathcal{X}} (\Phi(x) - \mu_\Phi) \otimes (\Phi(x) - \mu_\Phi) d\rho(x) : \mathcal{H}_K \to \mathcal{H}_K, \tag{33}$$

where, for $u, v, w \in \mathcal{H}_K, (u \otimes v)w = \langle v, w\rangle_{\mathcal{H}_K} u$. In particular, the covariance operator C_Φ is a positive trace class operator on $\mathcal{H}_K$ (see e.g. [13]).

Let $\mathbf{X} = [x_1, \ldots, x_m], m \in \mathbb{N}$, be a set randomly sampled from $\mathcal{X}$ according to ρ. The feature map Φ on $\mathbf{X}$ defines the bounded linear operator $\Phi(\mathbf{X}) : \mathbb{R}^m \to \mathcal{H}_K, \Phi(\mathbf{X})\mathbf{b} = \sum_{j=1}^m b_j \Phi(x_j), \mathbf{b} \in \mathbb{R}^m$. The corresponding empirical mean vector and empirical covariance operator for $\Phi(\mathbf{X})$ are defined by (see e.g. [12])

$$\mu_{\Phi(\mathbf{X})} = \frac{1}{m}\Phi(\mathbf{X})\mathbf{1}_m = \frac{1}{m}\sum_{j=1}^m \Phi(x_j) \in \mathcal{H}_K, \tag{34}$$

$$C_{\Phi(\mathbf{X})} = \frac{1}{m}\Phi(\mathbf{X}) J_m \Phi(\mathbf{X})^* : \mathcal{H}_K \to \mathcal{H}_K, \tag{35}$$

where $J_m = I_m - \frac{1}{m}\mathbf{1}_m\mathbf{1}_m^T$ is the centering matrix, with $\mathbf{1}_m = (1, \ldots, 1)^T \in \mathbb{R}^m$.

Let $\mathbf{X} = [x_i]_{i=1}^m$, $\mathbf{Y} = [y_i]_{i=1}^m$, be two sets randomly sampled from $\mathcal{X}$ according to two Borel probability distributions and $C_{\Phi(\mathbf{X})}$, $C_{\Phi(\mathbf{Y})}$ be the corresponding empirical covariance operators induced by K. Define the $m \times m$ Gram matrices

$$K[\mathbf{X}] = \Phi(\mathbf{X})^*\Phi(\mathbf{X}),\ K[\mathbf{Y}] = \Phi(\mathbf{Y})^*\Phi(\mathbf{Y}), K[\mathbf{X},\mathbf{Y}] = \Phi(\mathbf{X})^*\Phi(\mathbf{Y}), \tag{36}$$

$$(K[\mathbf{X}])_{ij} = K(x_i, x_j),\ (K[\mathbf{Y}])_{ij} = K(y_i, y_j),\ (K[\mathbf{X},\mathbf{Y}])_{ij} = K(x_i, y_j), \quad i,j = 1, \ldots, m. \tag{37}$$

Define $A = \frac{1}{\sqrt{m}}\Phi(\mathbf{X})J_m : \mathbb{R}^m \to \mathcal{H}_K$, $B = \frac{1}{\sqrt{m}}\Phi(\mathbf{Y})J_m : \mathbb{R}^m \to \mathcal{H}_K$, so that

$$AA^* = \frac{1}{m}\Phi(\mathbf{X})J_m\Phi(\mathbf{X}) = C_{\Phi(\mathbf{X})}, \quad BB^* = \frac{1}{m}\Phi(\mathbf{Y})J_m\Phi(\mathbf{Y}) = C_{\Phi(\mathbf{Y})}, \tag{38}$$

$$A^*A = \frac{1}{m}J_mK[\mathbf{X}]J_m, \; B^*B = \frac{1}{m}J_mK[\mathbf{Y}]J_m, \; A^*B = \frac{1}{m}J_mK[\mathbf{X},\mathbf{Y}]J_m. \tag{39}$$

Theorem 7 (**QJSD$_\alpha$ between RKHS covariance operators**). *Assume that* $\dim(\mathcal{H}_K) = \infty$. *Let* $\mathcal{H}_1 = \mathbb{R}^m$. *Let* $C_{\Phi(\mathbf{X})}, C_{\Phi(\mathbf{Y})}$ *and* A^*A, A^*B, B^*B *be defined as above. Then for* $\alpha \neq 0, 1$,

$$\mathrm{QJSD}_{\alpha,\gamma}(C_{\Phi(\mathbf{X})}, C_{\Phi(\mathbf{Y})}) = \frac{\gamma^\alpha}{\alpha(1-\alpha)}\mathrm{tr}\left(\left[\frac{1}{2\gamma}\begin{pmatrix} A^*A & A^*B \\ B^*A & B^*B\end{pmatrix} + I_{\mathcal{H}_1^2}\right]^\alpha - I_{\mathcal{H}_1^2}\right)$$
$$- \frac{\gamma^\alpha}{2\alpha(1-\alpha)}\mathrm{tr}\left[\left(\frac{A^*A}{\gamma} + I_{\mathcal{H}_1}\right)^\alpha + \left(\frac{B^*B}{\gamma} + I_{\mathcal{H}_1}\right)^\alpha - 2I_{\mathcal{H}_1}\right]. \tag{40}$$

At the limiting points $\alpha = 0$ *and* $\alpha = 1$,

$$\mathrm{QJSD}_{0,\gamma}(C_{\Phi(\mathbf{X})}, C_{\Phi(\mathbf{Y})}) = \log\det\left[\tfrac{1}{2\gamma}\begin{pmatrix} A^*A & A^*B \\ B^*A & B^*B\end{pmatrix} + I_{\mathcal{H}_1^2}\right]$$
$$-\tfrac{1}{2}\left[\log\det\left(\tfrac{A^*A}{\gamma} + I_{\mathcal{H}_1}\right) + \log\det\left(\tfrac{B^*B}{\gamma} + I_{\mathcal{H}_1}\right)\right], \tag{41}$$

$$\mathrm{QJSD}_{1,\gamma}(C_{\Phi(\mathbf{X})}, C_{\Phi(\mathbf{Y})})$$
$$= -\mathrm{tr}\left(\left[\frac{1}{2\gamma}\begin{pmatrix} A^*A & A^*B \\ B^*A & B^*B\end{pmatrix} + I_{\mathcal{H}_1^2}\right]\log\left[\frac{1}{2\gamma}\begin{pmatrix} A^*A & A^*B \\ B^*A & B^*B\end{pmatrix} + I_{\mathcal{H}_1^2}\right]\right) \tag{42}$$
$$+\frac{1}{2}\mathrm{tr}\left[(A^*A + \gamma I_{\mathcal{H}_1})\log\left(\frac{A^*A}{\gamma} + I_{\mathcal{H}_1}\right) + (B^*B + \gamma I_{\mathcal{H}_1})\log\left(\frac{B^*B}{\gamma} + I_{\mathcal{H}_1}\right)\right].$$

Remark 1. A much more detailed mathematical formulation, along with all the proofs, with be presented in the full version of the current paper.

References

1. Bai, L., Rossi, L., Torsello, A., Hancock, E.R.: A quantum Jensen-Shannon graph kernel for unattributed graphs. Pattern Recogn. **48**(2), 344–355 (2015)
2. Briët, J., Harremoës, P.: Properties of classical and quantum Jensen-Shannon divergence. Phys. Rev. A **79**(5), 052311 (2009)
3. Carlen, E.: Trace inequalities and quantum entropy: an introductory course. Entropy Quant. **529**, 73–140 (2010)
4. Cherian, A., Sra, S., Banerjee, A., Papanikolopoulos, N.: Jensen-Bregman logdet divergence with application to efficient similarity search for covariance matrices. TPAMI **35**(9), 2161–2174 (2012)

5. Congedo, M., Barachant, A., Bhatia, R.: Riemannian geometry for EEG-based brain-computer interfaces; a primer and a review. Brain Comput. Interfaces **4**(3), 155–174 (2017)
6. Dajka, J., Luczka, J., Hänggi, P.: Distance between quantum states in the presence of initial qubit-environment correlations: a comparative study. Phys. Rev. A **84**(3), 032120 (2011)
7. De Domenico, M., Nicosia, V., Arenas, A., Latora, V.: Structural reducibility of multilayer networks. Nat. Commun. **6**(1), 1–9 (2015)
8. Dryden, I., Koloydenko, A., Zhou, D.: Non-Euclidean statistics for covariance matrices, with applications to diffusion tensor imaging. Ann. Appl. Stat. **3**, 1102–1123 (2009)
9. Harandi, M., Salzmann, M., Porikli, F.: Bregman divergences for infinite dimensional covariance matrices. In: CVPR (2014)
10. Lamberti, P., Majtey, A., Borras, A., Casas, M., Plastino, A.: Metric character of the quantum Jensen-Shannon divergence. Phys. Rev. A **77**(5), 052311 (2008)
11. Minh, H.Q.: Alpha-Beta Log-Determinant divergences between positive definite trace class operators. Inf. Geom. **2**(2), 101–176 (2019)
12. Minh, H.: Infinite-dimensional Log-Determinant divergences between positive definite trace class operators. Linear Algebra Appl. **528**, 331–383 (2017)
13. Minh, H., Murino, V.: Covariances in computer vision and machine learning. Synth. Lect. Comput. Vis. **7**(4), 1–170 (2017)
14. Nielsen, F., Boltz, S.: The Burbea-Rao and Bhattacharyya centroids. IEEE Trans. Inf. Theor. **57**(8), 5455–5466 (2011)
15. Nock, R., Magdalou, B., Briys, E., Nielsen, F.: Mining matrix data with Bregman matrix divergences for portfolio selection. In: Nielsen, F., Bhatia, R. (eds.) Matrix Information Geometry, pp. 373–402. Springer, Berlin, Heidelberg (2013). https://doi.org/10.1007/978-3-642-30232-9_15
16. Rossi, L., Torsello, A., Hancock, E.R., Wilson, R.C.: Characterizing graph symmetries through quantum Jensen-Shannon divergence. Physi. Rev. E **88**(3), 032806 (2013)
17. Sra, S.: Positive definite matrices and the S-divergence. Proc. Am. Math. Soc. **144**(7), 2787–2797 (2016)
18. Sra, S.: Metrics induced by Jensen-Shannon and related divergences on positive definite matrices. Linear Algebra Appl. **616**, 125–138 (2021)
19. Steinwart, I., Christmann, A.: Support Vector Machines. Springer Science & Business Media, London (2008)
20. Virosztek, D.: The metric property of the quantum Jensen-Shannon divergence. Adv. Math. **380**, 107595 (2021)

Towards a Geometrization of Quantum Complexity and Chaos

Davide Rattacaso[1(✉)], Patrizia Vitale[1,2], and Alioscia Hamma[3]

[1] Dipartimento di Fisica Ettore Pancini, Università degli Studi di Napoli Federico II, Via Cinthia, Fuorigrotta, 80126 Naples, NA, Italy
davide.rattacaso@unina.it
[2] INFN-Sezione di Napoli, Via Cinthia, Fuorigrotta, 80126 Naples, NA, Italy
[3] Physics Department, University of Massachusetts Boston, Boston 02125, USA

Abstract. In this paper, we show how the restriction of the Quantum Geometric Tensor to manifolds of states that can be generated through local interactions provides a new tool to understand the consequences of locality in physics. After a review of a first result in this context, consisting in a geometric out-of-equilibrium extension of the quantum phase transitions, we argue the opportunity and the usefulness to exploit the Quantum Geometric Tensor to geometrize quantum chaos and complexity.

Keywords: Geometry of quantum states · Local interactions · Quantum phase transitions · Quantum complexity · Quantum chaos

1 Introduction

In quantum mechanics, the state of a closed system can be represented by a point in the complex projective space $\mathbb{C}P^n$, also referred to as the *manifold of pure states* $\mathcal{S}$. The tangent space of this manifold can be endowed with an Hermitian structure called *Quantum Geometric Tensor* (QGT) [1]. This geometrical structure arises from the inner product on the Hilbert space and has a meaningful physical interpretation: its real part is the *Fubini-Study metric* and in quantum information theory it encodes the operational distinguishability between two neighboring states of $\mathcal{S}$ [2], while its imaginary part is called *Berry's curvature* and determines the phase acquired by a state during an adiabatic cycle [3].

The Fubini-Study metric on the manifold of pure states is homogeneous and isotropic [4]. This mathematical property reflects the fact that in the space $\mathcal{S}$ there are neither privileged states nor privileged evolutions. On the other side, from the physical point of view, we know that some states are privileged, in the sense that they are more easily experimentally accessible, e.g., ground states of frustration free Hamiltonians, while others can be extremely difficult to both prepare or keep from decohering in microscopic times, e.g., macroscopic superpositions (Schrodinger's cats) [5]. The fundamental constraints on Hamiltonians and the limited resources available in terms of time, energy, and the form of

F. Nielsen and F. Barbaresco (Eds.): GSI 2021, LNCS 12829, pp. 163–171, 2021.
https://doi.org/10.1007/978-3-030-80209-7_19

interactions determine the very small region of the Hilbert space that can actually be explored by a state during its evolution [6]. Since these constraints are at the origin of the difference between a generic quantum state and a physically *accessible quantum state*, we will refer to a manifold $\mathcal{M}$ of accessible quantum states in relation to some reasonably constrained Hamiltonian, e.g., the ground states of a given family of Hamiltonians or the states that can be generated by evolution with these Hamiltonians in a finite amount of time.

Hamiltonians are constrained in the sense that not all the Hermitian operators that could in principle act on the state of a system are actually realized in nature. For example, interactions must respect some symmetry or they must observe some locality constraint. The latter request consists in the fact that the algebraic properties of accessible observables define on the Hilbert space $\mathcal{H}$ a privileged isomorphism $\mathcal{H} \cong \otimes_i^N \mathbb{C}^{d_i}$ [7] such that the interactions only involve some subsets of the local Hilbert spaces $\mathbb{C}^{d_i}$. These subsets can be considered as edges of a graph and therefore such a privileged tensorization of the Hilbert space is endowed with a graph structure, and the number N of these edges represents the size of the physical system in exam. For example, the Ising Hamiltonian only involves two-spin interactions between nearest-neighbors.

One might ask whether the geometry of a manifold of quantum states can be used to investigate some physical phenomena related to the locality of the Hamiltonian exploited to access the manifold. For example, the QGT on the manifold of ground states of an Hamiltonian can reflect the capability of Hamiltonian to generate long range correlations. An insight in this direction comes from the information geometric interpretation of quantum phases by Zanardi et al. [8–10] where it is shown that the scaling of the first energy gap of a local Hamiltonian, and therefore its quantum phase diagram, is reflected in the scaling of the QGT on the manifold of ground states, allowing for a new information-theoretical definition of quantum phases.

In this paper, we demonstrate a formalism to describe the time evolution of quantum geometry [11] defining a protocol to study the evolution of the Fubini-Study geometry on the manifold of ground states of a smooth family of local Hamiltonians, that is, the out-of-equilibrium evolution of the phase diagram. We show that quantum phases are robust after a quantum quench and the geometry of the manifold of evolving states is affected by a process of equilibration.

Our results exemplify how the geometry of a manifold of states evolving with a local Hamiltonian can provide new answers to several problems in local dynamics. This first step serves as motivation to exploit the geometry of these manifolds as a tool to investigate many questions related to locality, like the notion of quantum complexity [12], the emergence of quantum phases [13], equilibration processes [14], operator spreading [15], information scrambling [16–19] and Quantum Darwinism [5]. Motivated by these goals, in the conclusions we outline a path towards a geometric approach to quantum complexity and chaos.

2 Time Evolved QGT and Spreading of Local Correlations

In this section, we demonstrate a protocol to take the manifold of ground states of an Hamiltonian out of equilibrium, obtaining time-evolving manifolds of experimentally accessible quantum states on which out-of-equilibrium quantum phases can be defined through the QGT. We show that, as it happens in the static case [8], the QGT encodes some properties of the system that arise from the locality of the Hamiltonian. In particular, we show that the evolution of this tensor, and therefore the geometry of the evolving manifold, is determined by the spreading of correlations between local operators. This link will be exploited in the next sections to prove some remarkable features of the quantum phases away from equilibrium.

Following the approach of the information-theoretical formulation of quantum phase transitions, we initially consider a manifold $\mathcal{M}_0$ of non-degenerate ground states of a given family of local Hamiltonians $H(\lambda)$, smooth in the control parameters λ^i. The states of this manifold can be experimentally accessed with high accuracy in a tunable synthetic quantum system at low temperature. A coordinate map on $\mathcal{M}_0$ is inherited by the control parameters λ of the corresponding parent Hamiltonian $H(\lambda)$. To take the system out of equilibrium we introduce a family of unitary operators $U_t(\lambda) := e^{-itH^q(\lambda)}$ smooth in the control parameters λ^i. These operators describe the effect of a sudden local perturbation $H^q(\lambda)$, called quantum quench, on the corresponding state $|\psi(\lambda)\rangle$. At each time t we define a manifold of out-of-equilibrium states $\mathcal{M}_t = \{|\psi_{0t}(\lambda)\rangle = U_t(\lambda)|\psi_0(\lambda)\rangle\}$, representing the evolution induced on the initial states of $\mathcal{M}_0$ by the family of quantum quenches $H^q(\lambda)$. As a consequence of the locality of H^q, $\mathcal{M}_t$ is a manifold of dynamically accessible quantum states. The geometrical features of this manifold can be investigated thanks to the restriction of the QGT on $\mathcal{M}_t$, that allows the definition of an out-of-equilibrium phase diagram corresponding to regions in which the QGT diverges most than linearly in the thermodynamic limit.

Given the coordinate map $x^i(\psi)$ for the manifold in exam, the QGT on this manifold can be defined via its coordinate representation

$$q_{ij}(\psi) := \langle\partial_i\psi|(\mathbb{1} - |\psi\rangle\langle\psi|)|\partial_j\psi\rangle. \tag{1}$$

Taking advantage of the fact that for any $t \geq 0$ $\mathcal{M}_t$ is a manifold of non-degenerate ground states for the family of Hamiltonians $H(\lambda)_t := U_t(\lambda)H(\lambda)U_t(\lambda)^\dagger$, one can prove [11] that, in a coordinate map $a^i(\psi)$ which diagonalizes the tensor, the eigenvalues q_a of the QGT are bounded as follows

$$q_0 + q_1(t) - 2\sqrt{q_0 q_1(t)} \leq q(t) \leq q_0 + q_1(t) + 2\sqrt{q_0 q_1(t)} \tag{2}$$

where for the sake of simplicity we have eliminated the index a. In the last expression the term $q_0 := \sum_{n\neq 0} |\langle\psi_0|\partial H|\psi_n\rangle|^2/(E_0 - E_n)^2$ keeps memory of the initial geometry of the manifold, while the term $q_1(t) = \sum_{n\neq 0} |\langle\psi_0|D|\psi_n\rangle|^2 = \langle\psi_0|D|\psi_0\rangle^2 - \langle\psi_0|D^2|\psi_0\rangle$ describes the effect of the quench on the geometry, with $D(t) := \int_0^t dt' U(t')^\dagger \partial H^q U(t')$. We can deduce that the geometry of $\mathcal{M}_t$ is mainly determined by the behaviour of q_1. To represent q_1 as the sum of local correlations, we exploit the locality of the quench Hamiltonian. We can indeed represent the operator D as $D = \sum_i D_i$, where $D_i := \int_0^t dt' U(t')^\dagger \partial H_i^q U(t')$ [11]. As a consequence $q_1 = \sum_{ij} \langle D_i D_j\rangle_C$, where $\langle D_i D_j\rangle_C := \langle\psi_0|D_i D_j|\psi_0\rangle - \langle\psi_0|D_i|\rangle\langle|D_j|\psi_0\rangle$ is the correlation function of a local operator D_i which support spreads in time. Finally, making explicit the operators D_i, we represent the rescaled eigenvalues $q_1(\lambda, t)/N$ as

$$q_1(\lambda, t)/N = \sum_j \int_0^t dt' \int_0^t dt'' \langle \partial H_0^q(\lambda, t') \partial H_j^q(\lambda, t'')\rangle_C, \tag{3}$$

where N is the system size. This equation made explicit the role of the spreading of local correlations in determining the geometry of $\mathcal{M}_t$.

3 The Phase Diagram Away from Equilibrium

3.1 Time Evolution

Here we show that, because of the finite velocity at which local correlations are propagated, the phase diagram, thought of as the analiticity domain of the rescaled QGT Q/N, is robust under a quantum quench. This means that quantum phases are stable when the system is taken out of equilibrium and that a local Hamiltonian is not capable of dynamically generating areas of fast-scaling distinguishability between states.

We suppose that the graph structure defined by the Hamiltonian on the tensorization of the Hilbert space is a lattice and we define the spacing a of this lattice as the length associated with each link of the graph. In this context we say that local correlations between the operators O_i and O_j decay exponentially when $\langle O_i O_j\rangle_C \approx \mathcal{O}\left(e^{-\frac{a|i-j|}{\chi}}\right)$ and we call χ the correlation length of the state. Clearly the scaling in N of Eq. (3) directly depends on whether or not a finite correlation length in the state $|\psi(\lambda, t)\rangle$ exists.

To answer this question we exploit the Lieb-Robinson bound on the spreading of local correlations [15], that states that, if the evolution of the system is generated by a local Hamiltonian H^q acting on a state with exponential decaying correlations, the latter spread out at a finite velocity v_{LR} called *Lieb-Robinson velocity*. From this bound some important results descend, as the existence of a finite correlation length χ for non-degenerate ground states [20]. Generalizing these results it is possible to prove [11] the following:

$$|q_1(t)|/N \leq k \left[4t^2 \exp\left(\frac{2v_{LR}t}{\chi + a}\right) \sum_j \exp\left(-\frac{d_{oj}}{\chi + a}\right) \right] \tag{4}$$

where k does not depend on the time and on the size of the system. The scaling of the above equation for large N depends on the behavior of the correlation length χ for the initial state $|\psi_0(\lambda)\rangle$ and, because of this bound, it is unaffected by the criticalities of the quench Hamiltonian. As a consequence the rescaled QGT, that depends on q_1/N via the triangular inequality Eq. (2), does not diverge in the thermodynamic limit if and only if the correlation length χ is finite. This means that the phase diagram on $\mathcal{M}_t$ is preserved in time and a local quench can not affect the scaling of the QGT inducing new phase transitions for some value of the the control parameters or of the time.

3.2 Equilibration

We are going to show that the correlation function structure of the QGT implies its equilibration in probability in the thermodynamic limit. Given a function $f(t)$ of the state of the system we say that $f(t)$ equilibrates in probability if $\lim_{N\to\infty}\sigma^2[f] \approx \mathcal{O}(e^{-N})$, where $\sigma^2[f] = \int_0^T (f(t) - \overline{f(t)})^2 dt/T$ is the time-variance of the function over the interval of observation T and N is the size of the system. This behavior, that is weaker than the usual equilibration condition in which $\lim_{t\to\infty} f(t) = f_{eq}$, has been proven for the expectation value of local observables evolving with a local Hamiltonian under the non-resonance condition [14]. This consists for the Hamiltonian in having a non-degenerate spectrum with also non-degenerate energy gaps. Since the degeneration is a fine-tuning condition, this request is generally satisfied. Physically this means that in a large-scale quantum system it is extremely unlikely for an experimenter to measure a local expectation value that is different from a fixed equilibration value. To extend this result to the QGT on $\mathcal{M}_t$, we show that equilibration in probability also applies to the correlation functions of local operators. In particular the following holds [11]

Theorem 1. *Consider a Hamiltonian $H^q = \sum E_n^q P_n^q$ satisfying the non-resonance condition and an observable $A(t) = e^{-itH^q} A e^{itH^q}$ evolving under the action of H^q. Then the temporal variance $\sigma^2(C)$ of unequal-time correlation functions $C(t', t'') = \langle A(t')A(t'')\rangle$ is upper-bounded as $\sigma^2(C) \leq \|A\|^4 Tr\overline{\rho}^2$, where $Tr(\overline{\rho}^2)$ is the purity of the completely dephased state $\overline{\rho} = \sum_n P_n^q \rho P_n^q$.*

As a corollary, the same bound holds also for connected correlation functions.

This theorem is a direct extension of a previous result in literature [14], and can be proven by making explicit the role of energy gaps in the time-averages involved in the definition of the time-variance. The equilibration of correlations directly influences the evolution of the QGT. Indeed from Eq. (3) one can easily show that

$$q_1(t) = \alpha t^2 + X(t)t^2 \tag{5}$$

where α is a constant and the error $X(t)$ is a temporal variance for the involved correlation functions, that, as showed above, scales linearly in the dephased state purity. As confirmed for the Cluster-XY model in Sect. 3.3, the dephased purity $Tr(\overline{|\psi_0\rangle\langle\psi_0|}^2)$ generally decays exponentially in the system size N, as a consequence time fluctuations X vanish very fast and q_1 equilibrates to αt^2. An analogous behavior also affects the whole QGT because of the inequality Eq. (2).

3.3 Application to the Cluster-XY Model Phase Diagram

Here we analyze the evolution of the QGT on the manifold of out-of-equilibrium ground states of an exactly solvable model: the Cluster-XY model [21]. In this way we show an example of phase-diagram evolution and how it is affected by our previous statements about scaling and equilibration in the thermodynamic limit. The Cluster-XY model consists in a spin chain evolving with the Hamiltonian

$$H_{CXY}(h,\lambda_x,\lambda_y) = -\sum_{i=1}^{N}\sigma_{i-1}^x\sigma_i^z\sigma_{i+1}^x - h\sum_{i=1}^{N}\sigma_i^z + \lambda_y\sum_{i=1}^{N}\sigma_i^y\sigma_{i+1}^y + \lambda_x\sum_{i=1}^{N}\sigma_i^x\sigma_{i+1}^x$$

This model shows a very rich phase diagram on the equilibrium manifold $\mathcal{M}$.

Here we induce the evolution through a so-called *orthogonal quench*: we prepare the initial manifold $\mathcal{M}_0$ as the ground states manifold of the Hamiltonians $H(\lambda) = H_{CXY}(h = 0, \lambda_x, \lambda_y)$, and consider a sudden change in the control parameter of the transverse field, making the system evolve with the quench Hamiltonian $H^q(\lambda) = H_{CXY}(h = h^q, \lambda_x, \lambda_y)$. $\mathcal{M}_t$ is therefore the manifold of the states $|\Omega(\lambda,t)\rangle := e^{-iH^q(\lambda)t}\,|GS(H(\lambda))\rangle$. As a consequence of the possibility to exactly diagonalize the Cluster-XY model, these states can be represented via the action of suited fermionic creation and annihilation operator on their vacuum state. This allows us to exactly calculate the overlaps between the states of the manifold and, consequently, also the metric g induced from the Fubini-Study metric of the ambient projective space. Indeed, from the squared fidelity $\mathcal{F}^2 \equiv |\langle\Omega(\lambda',t)|\Omega(\lambda,t)\rangle|^2$, one finds that $g_{\mu\nu}d\lambda^\mu d\lambda^\nu := |\langle\Omega(\lambda+d\lambda,t)|\Omega(\lambda,t)\rangle|^2$. The Berry's curvature instead can be easily shown to be zero.

The possible divergences of the metric are related to the energy gaps of the initial Hamiltonian $H(\lambda)$ and of the quench Hamiltonian $H^q(\lambda)$, anyway one can show by standard analytic techniques in [11] that in the thermodynamic limit the only divergences of $g_{\mu\nu}(t)$ are the ones in $g_{\mu\nu}(0)$, determined by the disclosure of the first energy gap of $H(\lambda)$. As a consequence the phase diagram is conserved by temporal evolution, as we have proven in the previous section. In the previous section we have also demonstrated that the equilibration properties for the QGT are determined by the purity of the dephased state $\overline{\rho}$, where $\rho = |\Omega(\lambda,t)\rangle\langle\Omega(\lambda,t)|$. Exploiting the exact diagonalization of the Cluster-XY model one can show that this purity reads $Tr(\overline{\rho}^2) = \prod_k(1 - 1/2\sin^2(2\chi_k))$, where χ_k is different from zero for a non-null quench. As a consequence the purity is exponentially small in N, determining the equilibration of the phase diagram.

4 Conclusions and Outlook

The results that we have shown in this review can be considered as a first step towards the understanding of the role of the natural Hermitian structure of quantum states in determining the properties of dynamically accessible manifold of states. We extended the geometric approach to out-of-equilibrium quantum phases and, exploiting the QGT and its relation with the spreading of local correlations, we have demonstrated the robustness and equilibration of the phase diagram after a quantum quench.

Other aspects of out of equilibrium many-body physics could be investigated exploiting this geometric approach. One of the most important ones is the investigation of quantum chaotic behaviour of the Hamiltonian, which is revealed in butterfly effect and scrambling of local information [16,17]. A unified approach to investigate these behaviors is indeed provided by the study of out-of-time-order correlators (OTOCs) and their generalizations [18,19]. As a direct consequence of Eq. (3), we have proven [11] that the QGT is upper-bounded by a superposition of OTOCs. We can deduce that if a quench Hamiltonian generates higher distinguishability then it also generate a larger spreading of local operators. This is a first evidence of the possibility for a geometric picture of quantum chaos and scrambling. As an example, divergences in a suitable scrambling metric could allow us to identify possible quantum models of black holes thanks to the fast scrambling conjecture [22].

Beyond the geometrization of chaos, the QGT could be related to quantum complexity. A metric representation of complexity has been proposed for the manifold of unitary operators [23,24] and extended to the manifold of states [4]. The latter consist in modified versions of the Fubini-Study metric, in which the matrix elements associated with the infinitesimal evolutions generated only by non-local interactions are enlarged through a penalty factor. As pointed out by Nielsen [23], such a geometric notion of complexity could play a central role in quantum computing, reducing the search for the most efficient algorithm to the search of a geodesic. Here we suggest a different approach, that does not involve an arbitrary penalty. Given a target path of states we consider the best local approximation of its generator. This time-dependent operator, that is an *optimal time-dependent parent Hamiltonian*, applied to the initial point of the target path of states, generates an *accessible approximation* of the latter. The functional distance between the target path and its accessible approximation measures the capability to access the target path exploiting only local interactions, so we expect that it is linked to quantum complexity. Therefore, one of our next goals will be to exploit this distance to find a natural notion of complexity metric, or also *accessibility metric*, on the manifold of quantum states. Our point of view on complexity geometry arises form the search for an optimal parent Hamiltonian [25]. Since this search is directly linked to several practical goals, such as the verification of quantum devices [26] and the quantum control of time-dependent states [27], the accessibility metric could provide a unified framework to understand the limits of our ability to manipulate quantum matter.

References

1. Provost, J.P., Vallee, G.: Riemannian structure on manifolds of quantum states. Commun. Math. Phys. **76**, 289–301 (1980)
2. Wootters, W.K.: Evolution without evolution: dynamics described by stationary observables. Phys. Rev. D **23**, 357 (1981)
3. Berry, M.V.: Quantal phase factors accompanying adiabatic changes. Proc. R. Soc. Lond. A **392**(1802), 47–57 (1984)
4. Brown, A.R., Susskind, L.: Complexity geometry of a single qubit. Phys. Rev. D **100**, 046020 (2019)
5. Zurek, W.H.: Decoherence, einselection, and the quantum origins of the classical. Rev. Mod. Phys. **75**, 715–775 (2003)
6. Poulin, D., Qarry, A., Somma, R., Verstraete, F.: Quantum simulation of time-dependent hamiltonians and the convenient illusion of Hilbert space. Phys. Rev. Lett. **106**, 170501 (2011)
7. Zanardi, P., Lidar, D., Lloyd, S.: Quantum tensor product structures are observable-induced. Phys. Rev. Lett. **92**, 060402 (2004)
8. Zanardi, P., Giorda, P., Cozzini, M.: Information-theoretic differential geometry of quantum phase transitions. Phys. Rev. Lett. **99**, 100603 (2007)
9. Campos Venuti, L., Zanardi, P.: Quantum critical scaling of the geometric tensors. Phys. Rev. Lett. **99**, 095701 (2007)
10. Abasto, D.F., Hamma, A., Zanardi, P.: Fidelity analysis of topological quantum phase transitions. Phys. Rev. A **78**, 010301(R) (2008)
11. Rattacaso, D., Vitale, P., Hamma, A.: Quantum geometric tensor away from equilibrium. J. Phys. Comm. **4**(5), 055017 (2020)
12. Nielsen, M.A., Chuang, I.L.: Quantum Computation and Quantum Information, 10 Anniversary edn. Cambridge University Press, New York (2010)
13. Sachdev, S.: Quantum Phase Transitions, 2nd edn. Cambridge University Press, Cambridge (2011)
14. Reimann, P.: Foundation of statistical mechanics under experimentally realistic conditions. Phys. Rev. Lett. **101**, 190403 (2008)
15. Bravyi, S., Hastings, M.B., Verstraete, F.: Lieb-Robinson bounds and the generation of correlations and topological quantum order. Phys. Rev. Lett. **97**, 050401 (2006)
16. Hosur, P., Qi, X.-L., Roberts, D.A., Yoshida, B.: Chaos in quantum channels. J. High Energy Phys. **2016**(2), 1–49 (2016). https://doi.org/10.1007/JHEP02(2016)004
17. Roberts, D.A., Yoshida, B.: Chaos and complexity by design. J. High Energy Phys. **2017**(4), 1–64 (2017). https://doi.org/10.1007/JHEP04(2017)121
18. Hashimoto, K., Murata, K., Yoshii, R.: Out-of-time-order correlators in quantum mechanics. J. High Energy Phys. **2017**(10), 1–31 (2017). https://doi.org/10.1007/JHEP10(2017)138
19. Gärttner, M., Hauke, P., Rey, A.M.: Relating out-of-time-order correlations to entanglement via multiple-quantum coherences. Phys. Rev. Lett. **120**, 040402 (2018)
20. Hastings, M.B.: Locality in quantum and Markov dynamics on lattices and networks. Phys. Rev. Lett. **93**, 140402 (2004)
21. Montes, S., Hamma, A.: Phase diagram and quench dynamics of the cluster-XY spin chain. Phys. Rev. E **86**, 021101 (2012)

22. Lashkari, N., Stanford, D., Hastings, M., Osborne, T., Hayden, P.: Towards the fast scrambling conjecture. J. High Energy Phys. **2013**, 22 (2013)
23. Nielsen, M.A.: A geometric approach to quantum circuit lower bounds. arXiv:quant-ph/0502070
24. Nielsen, M.A., Dowling, M., Gu, M., Doherty, A.C.: Quantum computation as geometry. Science **311**, 1133 (2006)
25. Chertkov, E., Clark, B.K.: Computational inverse method for constructing spaces of quantum models from wave functions. Phys. Rev. X **8**, 031029 (2018)
26. Carrasco, J., Elben, A., Kokail, C., Kraus, B., Zoller, P.: Theoretical and experimental perspectives of quantum verification. arXiv:2102.05927
27. Werschnik, J., Gross, E.K.U.: Quantum optimal control theory. J. Phys. B Atom. Mol. Opt. Phys. **40**, R175 (2007)

Hunt's Colorimetric Effect from a Quantum Measurement Viewpoint

Michel Berthier[1] and Edoardo Provenzi[2(✉)]

[1] Laboratoire MIA, Pôle Sciences et Technologie, Université de La Rochelle, Avenue Michel Crépeau, 17042 La Rochelle, France
michel.berthier@univ-lr.fr

[2] Université de Bordeaux, CNRS, Bordeaux INP, IMB, UMR 5251, 351 Cours de la Libération, 33400 Talence, France
edoardo.provenzi@math.u-bordeaux.fr

Abstract. The main aim of this paper is to describe how the colorimetric phenomenon known as 'Hunt effect' can be understood in the quantum framework developed by the authors to model color perception of trichromatic observers. In classical colorimetry, the most common definition of the Hunt effect is that the colorfulness of a color increases with its luminance, however several other definitions are available. The need to establish a unique and precise characterization of the features involved in the Hunt effect led us to propose novel mathematical definitions of colorimetric attributes. Within the newly established nomenclature, we show how the Hunt effect can be rigorously explained thanks to the duality between quantum states and effects.

Keywords: Color perception · Quantum states · Quantum effects · Hunt's effect

1 Introduction

The main aim of this paper is to explain the well known Hunt colorimetric effect in the quantum framework first introduced in [4] and then refined and further formalized in [2] and [3]. Due to space limitation, here we will recall only the elements of this framework that are strictly necessary for the following.

The work of the founding fathers of colorimetry, Newton, Maxwell, Grassmann and von Helmholtz, was elegantly resumed by Schrödinger in [13] in a set of axioms which imply that the space of perceptual colors $\mathcal{C}$ is a 3-dimensional convex regular cone. In [12], Resnikoff added an homogeneity axiom which implied that $\mathcal{C}$, as a set, can only take two forms: either $\mathcal{C}_1 = \mathbb{R}^+ \times \mathbb{R}^+ \times \mathbb{R}^+$ or $\mathcal{C}_1 = \mathbb{R}^+ \times \mathbf{H}$, where $\mathbf{H}$ is a 2-dimensional hyperbolic space, see also [11].

The core of our approach is a mathematical axiom called *trichromacy axiom* that summarizes the set of Schrödinger's and Resnokoff's axioms into a single one: *the space of perceptual colors is the positive cone $\mathcal{C}$ of a formally real Jordan algebra of real dimension 3.* Jordan algebras are commutative but non necessarily

F. Nielsen and F. Barbaresco (Eds.): GSI 2021, LNCS 12829, pp. 172–180, 2021.
https://doi.org/10.1007/978-3-030-80209-7_20

associative algebras used as the foundation of the algebraic formulation of quantum theory developed by Jordan, von Neumann and Wigner.

Using the classification theorem of finite dimensional formally real Jordan algebras one can check that $\mathcal{C}_1$ is the positive cone of the commutative and associative Jordan algebra $\mathbb{R} \oplus \mathbb{R} \oplus \mathbb{R}$, and $\mathcal{C}_2$ is $\mathcal{H}^+(2, \mathbb{R})$, the cone of positive-definite 2×2 real matrices, which is the positive cone of the commutative but non-associative Jordan algebra $\mathcal{H}(2, \mathbb{R})$ of 2×2 real symmetric matrices equipped with the Jordan matrix product. The first case corresponds to the *classical* CIE (Commission International de l'Éclairage) colorimetry, whereas the second case corresponds to a *non classical* colorimetric theory. From now on, we focus only on the latter case and we will give several motivations to consider it as a genuine quantum theory of color perception.

Among other reasons, a striking fact leads us to think that the classical CIE approach should be abandoned: all CIE color spaces are founded on either the RGB or the XYZ spaces, which are built with the only aim at codifying the output of the three retinal cone photoreceptors through values that are obtained via color matching experiments. The flaw in this procedure is that the outcome of those experiments is the result of the action of the whole human visual system, not only of the cones. Moreover, the fundamental processing known as Hering's opponency, due to the action of ganglion cells, which detect and magnify the differences between the outputs of the different cones, is not incorporated in the RGB or XYZ spaces. This fact is confirmed by our interpretation of the classical color space as the positive cone of the commutative and associative Jordan algebra $\mathbb{R} \oplus \mathbb{R} \oplus \mathbb{R}$. Instead, as we are going to recall, if we consider as color space the positive cone of the commutative but non-associative Jordan algebra $\mathcal{H}(2, \mathbb{R})$ and we use the self-duality of this cone, then we can intrinsically account for both the trichromatic and Hering's opponency stages.

The key to exhibit Hering's opponency is to exploit the Jordan algebras isomorphism between $\mathcal{H}(2, \mathbb{R})$ and the spin factor $\mathbb{R} \oplus \mathbb{R}^2$, which implies that the trichromacy cone $\mathcal{C}$ can be identified with the 3-dimensional *future lightcone*:

$$\mathcal{L}^+ = \left\{(\alpha + \mathbf{v}) \in \mathbb{R} \oplus \mathbb{R}^2 \ : \ \alpha > 0, \ \|(\alpha + \mathbf{v})\|_{\mathcal{M}} > 0\right\}, \tag{1}$$

where $\|(\alpha + \mathbf{v})\|_{\mathcal{M}}^2 = \alpha^2 - \|\mathbf{v}\|^2$ denotes the squared Minkowski norm, $\| \ \|$ being the Euclidean norm. The state space $\mathcal{S}$ of a *rebit*, the real analog of the usual (complex) qubit, can be identified with $\left\{\mathbf{v} \in \mathbb{R}^2 : \|\mathbf{v}\| \leq 1\right\}$ and, crucially, see e.g. [2–4], it can be naturally embedded in the closure $\overline{\mathcal{L}^+}$ of $\mathcal{L}^+$ via the map

$$\mathcal{S} = \left\{\mathbf{v} \in \mathbb{R}^2 : \|\mathbf{v}\| \leq 1\right\} \ni \mathbf{v} \mapsto \frac{1}{2}(1 + \mathbf{v}) \in \mathcal{D}_{1/2} = \{(\alpha + \mathbf{v}) \in \overline{\mathcal{L}^+} : \alpha = 1/2\}. \tag{2}$$

The properties of this so-called Hering's rebit allow us to give a meaningful mathematical interpretation of Hering's color opponency. Let us consider the embedding

$$\mathcal{S} \ni \mathbf{v} = (r\cos\theta, r\sin\theta) \mapsto \rho(\mathbf{v}) = \rho(r, \theta) = \frac{1}{2}\begin{pmatrix} 1 + r\cos\theta & r\sin\theta \\ r\sin\theta & 1 - r\cos\theta \end{pmatrix} \in \mathcal{DM}, \tag{3}$$

where, $0 \leq \theta < 2\pi$ and $\mathcal{DM}$ denotes the space of density matrices of $\mathcal{H}(2, \mathbb{R})$. One can easily checks that

$$\rho(r,\theta) = \rho_0 + \frac{r\cos\theta}{2}\left[\rho(1,0) - \rho(1,\pi)\right] + \frac{r\sin\theta}{2}\left[\rho\left(1,\pi/2\right) - \rho\left(1,3\pi/2\right)\right]. \quad (4)$$

From a colorimetric viewpoint, this decomposition means that the generic quantum chromatic state represented by a density matrix $\rho(r,\theta)$ can be seen as the superposition of the maximal von Neumann entropy state $\rho_0 = \rho(0,0)$ (see Sect. 3 for more information), which represents the achromatic state, with two diametrical oppositions of pure hues (or pure tints). We refer to [3] or [2] for further information, especially for what concerns the hyperbolic metric aspects of this theory.

The fact that a rebit system emerges naturally from the sole trichromacy axiom is one of the reasons that led us to consider this non-classical colorimetric theory as a *quantum-like* theory in which we assume that:

- a *visual scene* is a setting where we can perform psycho-visual measurements in conditions that are as isolated as possible from external influences;
- a *perceptual chromatic state* is represented by the preparation of a visual scene for psycho-visual experiments;
- a *perceptual color* is the observable identified with a psycho-visual measurement performed on a given perceptual chromatic state.

Of course, the instrument used to measure the observables is the visual system of a human and not a technological device used to perform physical experiments.

So far we have recalled basic information from our quantum-like theory of color perception, in the next sections we are going to introduce in this framework the concept of quantum effect which will allow us to describe the Hunt effect from a mathematical point of view.

2 States and Effects in Colorimetry

Operationally speaking, the concept of a state in quantum mechanics refers to an ensemble of identically prepared systems, while the concept of effect refers to a measurement apparatus that produces an outcome. The duality between states and effects means essentially that when a state and an effect are specified, one can compute a probability distribution which is the only meaningful information that we can obtain about the experiment. In other words, quantum mechanics deals with a collection of possible preparations and measurements, which can be combined together to form an experiment leading to a probability distribution of measurement outcomes [7].

In order to translate in mathematical terms what just stated, let us introduce the *state cone*:

$$\mathcal{C}(\mathcal{S}) = \{\alpha(1,\mathbf{v}), \alpha \geq 0, \mathbf{v} \in \mathcal{S}\}, \quad (5)$$

that, using the embeddings (2) and (3), can be easily checked to satisfy

$$\mathcal{C}(\mathcal{S}) \simeq \left\{2\alpha\left[\frac{1}{2}(1+\mathbf{v})\right], \alpha \geq 0, \mathbf{v} \in \mathcal{S}\right\} = \overline{\mathcal{L}^+}, \quad (6)$$

and

$$\mathcal{C}(\mathcal{S}) \simeq \{2\alpha\rho(\mathbf{v}), \alpha \geq 0, \mathbf{v} \in \mathcal{S}\} = \overline{\mathcal{H}^+(2,\mathbb{R})}, \tag{7}$$

where $\overline{\mathcal{H}^+(2,\mathbb{R})}$ is the set of positive semi-definite 2×2 real matrices.

Consequently, since the positive cone of a formally real Jordan algebra is self dual[1], the state cone $\mathcal{C}(\mathcal{S})$ is itself self dual, *i.e.* $\mathcal{C}^*(\mathcal{S}) = \mathcal{C}(\mathcal{S})$. The effect space of $\mathcal{S}$ is defined as follows:

$$\mathcal{E}(\mathcal{S}) = \{e \in \mathcal{C}^*(\mathcal{S}), e \leq Id\}, \tag{8}$$

where Id is the unit element of either $\mathbb{R}\oplus\mathbb{R}^2$ or $\mathcal{H}(2,\mathbb{R})$, accordingly to the identification (6) or (7), respectively. To obtain a geometric description of $\mathcal{E}(\mathcal{S})$, one can consider an effect $e = \alpha(1,\mathbf{v})$ as an element $2\alpha\rho(\mathbf{v})$ of $\overline{\mathcal{H}^+(2,\mathbb{R})}$ that satisfies the two conditions $\mathrm{Tr}(Id - e) \geq 0$ and $\det(Id - e) \geq 0$. Direct computations show that

$$\mathcal{E}(\mathcal{S}) = \left\{\alpha(1, v_1, v_2),\ \alpha \in [0,1],\ v_1^2 + v_2^2 \leq \min\left\{\frac{(1-\alpha)^2}{\alpha^2}, 1\right\}\right\}. \tag{9}$$

By using self duality, a straightforward computation leads to

$$\mathcal{E}(\mathcal{S}) \simeq \left\{ e = \begin{pmatrix} a_0 \\ a_1 \\ a_2 \end{pmatrix} \ :\ e(1, v_1, v_2) = a_0 + a_1 v_1 + a_2 v_2 \in [0,1],\ \forall (v_1, v_2) \in \mathcal{S} \right\}. \tag{10}$$

Every effect is thus uniquely associated to a $[0,1]$-valued affine function (denoted again with e for simplicity) on the set of states. This is in fact the simplest way to associate a probability to each state. Equation (9) implies that the effect space $\mathcal{E}(\mathcal{S})$ can be represented as a *double cone* in the 3-dimensional Minkowski space, as shown in Fig. 1.

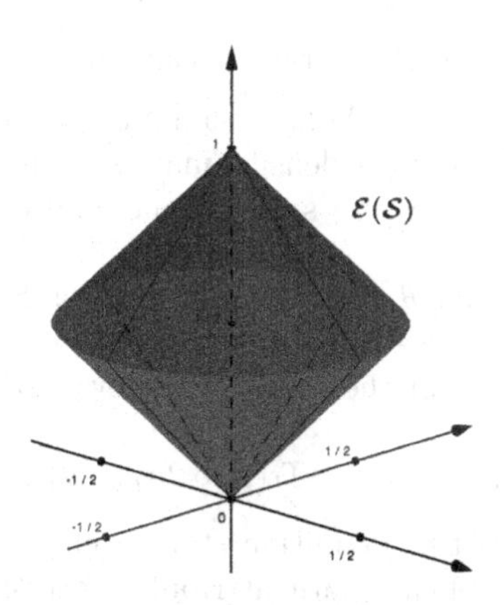

Fig. 1. The convex double cone representing $\mathcal{E}(\mathcal{S})$.

[1] Given a formally real Jordan algebra $\mathcal{A}$, its positive cone agrees with its the dual cone $\mathcal{C}^* := \{a \in \mathcal{A} \ :\ \forall b \in \mathcal{C},\ \langle a, b\rangle > 0\}$, where $\langle a, b\rangle = \mathrm{Tr}(L_{a \circ b})$, $L_{a\circ b}(c) = (a \circ b) \circ c$, $\forall c \in \mathcal{A}$. $\mathcal{C}^*$ can be identified with the set of positive linear functionals on $\mathcal{A}$.

Remarkably, this double cone coincides with that introduced in [5] as a representation of a so-called *perceptual color space* whose shape results from the description of the physiological mechanisms of neural coding of colors. Since this perceptual color space is, up to our knowledge, the only color space that has been proposed to take into account the spectrally opponent and non-opponent interactions of Hering's opponency, we consider this as particularly relevant.

All the previous considerations lead us quite naturally to introduce the following definition: *a color effect* (also called a *perceived color* in [2]), *is an element* c *of the effect space* $\mathcal{E}(\mathcal{S})$.

3 Colorimetric Attributes of Color Effects

In order to give a meaningful mathematical formulation of Hunt's effect it is necessary to introduce supplementary definitions related to the colorimetric attributes of a color effect. These definitions involve the notion of entropy which aims at measuring the mathematical expectation of the possible information gain about a given system. Dealing with quantum-like states, one usually considers the *von Neumann entropy* $S(\rho)$ of a density matrix ρ, which in our context, see e.g. [1], can be taken to be:

$$S(\rho) = -\mathrm{Tr}(\rho \circ \log \rho), \tag{11}$$

where $\circ$ denotes the Jordan product of $\mathcal{H}(2, \mathbb{R})$. By using the parametrization $\rho(r, \theta)$ it can be checked that the von Neumann entropy is given by:

$$S(\rho) = -\log \left(\frac{1-r}{2}\right)^{\frac{1-r}{2}} \left(\frac{1+r}{2}\right)^{\frac{1+r}{2}}, \tag{12}$$

with $S(1) = \lim_{r \to 1^-} S(r) = 0$. Pure state density matrices are characterized by a null entropy: pure states provide the maximum of information. The density matrix $\rho_0 = \rho(0, 0)$ is the unique density matrix with maximal von Neumann entropy: the state $(0, 0) \in \mathcal{S}$ is the state of maximal entropy that provides no chromatic information.

Let $\mathbf{v} = (v_1, v_2) = (r \cos\theta, r \sin\theta)$ be a state of $\mathcal{S}$. Given the color effect $c = \alpha(1, v_1, v_2) \simeq 2\alpha\rho(r, \theta)$, we call the real α, $0 \leq \alpha \leq 1$, the *magnitude* of c. This magnitude is nothing but the result of the evaluation of c on the maximal entropy state ρ_0:

$$\alpha = \langle c \rangle_{\rho_0} = \mathrm{Tr}(\rho_0 \circ 2\alpha\rho(r, \theta)). \tag{13}$$

The extreme simplicity of this definition stands out even more if compared to the plethora of definitions of chromatic attributes corresponding to α in classical colorimetry. Readers can refer for instance to the paper [9] to make up their own mind about this topic.

If we divide the effect c by 2α, we obtain the density matrix $\rho(r, \theta)$ that characterizes the *chromatic state* of c. As its name suggests, this chromatic state retains only the chromatic information of c. In order to recover this latter, one can perform measurements using the two Pauli-like matrices of Hering's rebit:

$$\sigma_1 = \begin{pmatrix} 1 & 0 \\ 0 & -1 \end{pmatrix}, \quad \sigma_2 = \begin{pmatrix} 0 & 1 \\ 1 & 0 \end{pmatrix}. \tag{14}$$

In fact, the expected values

$$\langle \sigma_1 \rangle_{\rho(r,\theta)} = \mathrm{Tr}(\rho(r,\theta) \circ \sigma_1) = r\cos\theta, \quad \langle \sigma_2 \rangle_{\rho(r,\theta)} = \mathrm{Tr}(\rho(r,\theta) \circ \sigma_2) = r\sin\theta, \tag{15}$$

give the degree of opposition between the two pairs of pure chromatic states involved in Hering's opposition, i.e. the two spectrally opponent interactions of the neural coding of colors. Let us insist on the fact that, in practice, performing such measurements is not straightforward: the matrices σ_1 and σ_2 refer to mechanisms implemented by the sensory system of a human being and not by technological devices. The reader can refer for instance to [8] for related psycho-visual experiment descriptions. Due to the lack of space, we do not investigate here the fascinating problem that consists of explaining the relation between the well-known phenomenon of *color indistinguishability*, illustrated by MacAdam ellipses, and the *measurement uncertainty* inherent to the fact that the matrices σ_1 and σ_2 do not commute. This will be the subject of forthcoming studies.

Since the von Neumann entropy $S(\rho)$ of $\rho = \rho(r,\theta)$ allows us to measure the purity degree of the chromatic state ρ of the color effect c, it seems natural to relate it with the classical notion of saturation widely used in colorimetry by defining the *saturation* of the color effect c as follows

$$\sigma(c) = \frac{\log(2) - S(\rho)}{\log(2)}, \tag{16}$$

the graph of $\sigma(c)$ is depicted in Fig. 2.

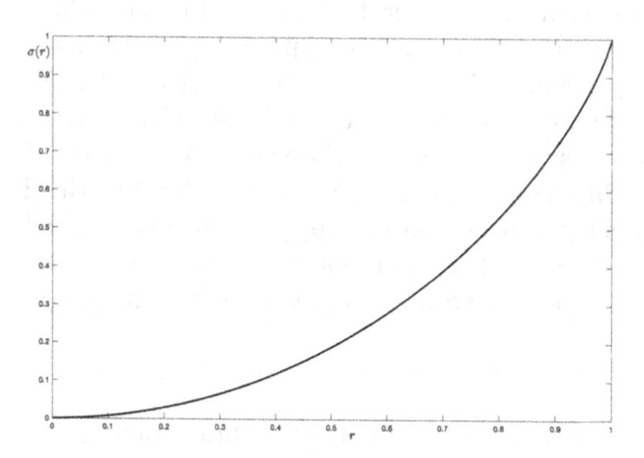

Fig. 2. A proposal for the saturation of a color effect from built from the von Neumann entropy of its chromatic state.

Notice that this proposal is the simplest one which agrees with the classical constraints: 0 for achromatic colors and 1 for pure (spectral) colors.

The final colorimetric attribute that remains to be defined is hue. Actually, this is the only definition that remains identical to that of classical colorimetry, as we will consider the angle θ as the *hue of the color effect* c.

4 Hunt's Colorimetric Effect

In order to introduce this effect (which is *not* a quantum effect, but a psycho-visual effect!) we quote the classical reference on colorimetry [6]: '*The color appearances of objects change significantly when the overall luminance level changes. The Hunt effect can be summarized by the statement that colorfulness of a given stimulus increases with luminance level*'.

To give an idea about the extreme difficulty of discussing color in words, let us quote how the terms involved in this description of Hunt's effect are defined in [6]: '*Colorfulness is the attribute of a visual perception according to which the perceived colour of an area appears to be more or less chromatic. Chroma is the colorfulness of an area judged as a proportion of the brightness of a similarly illuminated area that appears white or highly transmitting. Saturation is the colorfulness of an area judged in proportion to its brightness*'.

In spite of the involved way of dealing with these concepts, it is clear that the Hunt effect reveals that the colorimetric attributes are not independent. This has the immediate consequence that color spaces such as XYZ, RGB, HSV, HCV and so on (see e.g. [6]), in which color is encoded by three uncorrelated values, are not suitable to represent perceived colors. Instead, the category of color spaces as CIELab, CIECAM and so on, issued from the so-called color appearance models, are built in such a way that color coordinates are correlated and, in fact, experimental data about the Hunt effect are available, see [10]. Nonetheless, also these spaces come from XYZ and so their account of Hunt's effect is *a posteriori* and with several ad-hoc parameters difficult to control.

Contrary to this, the quantum-like approach that we have described above allows to derive the Hunt effect intrinsically and rigorously. To show this, consider a visual scene and a chromatic state with density matrix $\rho = \rho(r, \theta)$, i.e. a preparation of the visual scene for psycho-visual experiments. Let $c = 2\alpha\rho(r, \theta)$ be the corresponding color effect, i.e. the observable identified with a psycho-visual measurement performed on ρ. Suppose that c is adapted to the maximal entropy state ρ_0, then, by direct computation, we have that the evaluation of c on ρ_0 and on the chromatic state ρ are, respectively, the magnitude α and

$$\langle c \rangle_\rho = \mathrm{Tr}(\rho \circ c) = \alpha(1 + r^2). \tag{17}$$

Let us discuss particular cases in order to understand this evaluation.

- If $\alpha = 1/2$ then $\langle c \rangle_\rho = (1 + r^2)/2$. This evaluation holds for every ρ and $\langle c \rangle_\rho$ is maximal, and equal to 1, when ρ is a *pure* chromatic state, i.e. when $r = 1$.

- If $\alpha < 1/2$ then $\langle c \rangle_\rho = \alpha(1 + r^2)/2$ holds for every ρ and $\langle c \rangle_\rho < 1$ for every ρ. It is maximal, and equal to $2\alpha < 1$, when $r = 1$, i.e. for pure chromatic states.
- If $\alpha > 1/2$, then the evaluation $\langle c \rangle_\rho$ can only be performed when ρ is such that $r^2 \leq (1 - \alpha)/\alpha$. It is maximal, and equal to 1, when $r^2 = (1 - \alpha)/\alpha$. For instance, if $\alpha = 2/3$, then $\langle c \rangle_\rho = 1$ when $r = \sqrt{2}/2$.

Now, the chromatic state $\rho_1 = \rho(\sqrt{2}/2, \theta)$ is perceived by the effect $c_1 = 4/3\rho_1$ in the same way as the chromatic state $\rho_2 = \rho(1, \theta)$ is perceived by the effect $c_2 = \rho_2$ since $\langle c_1 \rangle_{\rho_1} = \langle c_2 \rangle_{\rho_2} = 1$. This means that the non-pure chromatic state ρ_1 is perceived in the same way as the a pure chromatic state ρ_2, and thus, it is perceived more saturated than it really is. Consequently, increasing the magnitude α from 1/2 to 2/3 increases the perceived saturation of the chromatic state ρ_1. Note that when $\alpha = 1$, the color effect c is the so-called unit effect that gives the same response equal to 1 on all states, this is *the mathematical translation of the glare limit.*

5 Discussion and Future Perspectives

We have shown mathematical objects known as *effects* in quantum theory allow us to mathematically account for the Hunt phenomenon of interdependence between (what are classically called) colorfulness and luminance without the need of introducing any exterior ad-hoc structures or parameters. What we find remarkable is that quantum concepts as rebit and effects, well-known in modern quantum information but less-known in ordinary quantum mechanics, seem to find a natural place in color perception theory.

References

1. Baez, J.C.: Division algebras and quantum theory. Found. Phys. **42**(7), 819–855 (2012)
2. Berthier, M.: Geometry of color perception. Part 2: perceived colors from real quantum states and Hering's rebit. J. Math. Neurosci. **10**(1), 1–25 (2020)
3. Berthier, M., Provenzi, E.: From Riemannian trichromacy to quantum color opponency via hyperbolicity. J. Math. Imag. Vis. **63**, 681–688 (2021)
4. Berthier, M., Provenzi, E.: When geometry meets psycho-physics and quantum mechanics: modern perspectives on the space of perceived colors. In: Nielsen, F., Barbaresco, F. (eds.) GSI 2019. LNCS, vol. 11712, pp. 621–630. Springer, Cham (2019). https://doi.org/10.1007/978-3-030-26980-7_64
5. De Valois, R.L., De Valois, K.D.: Neural coding of color (1997)
6. Fairchild, M.: Color Appearance Models. Wiley, New York (2013)
7. Heinosaari, T., Ziman, M.: The Mathematical Language of Quantum Theory: From Uncertainty to Entanglement. Cambridge University Press, Cambridge (2011)
8. Jameson, D., Hurvich, L.M.: Some quantitative aspects of an opponent-colors theory. i. chromatic responses and spectral saturation. JOSA **45**(7), 546–552 (1955)

9. Kingdom, F.A.: Lightness, brightness and transparency: a quarter century of new ideas, captivating demonstrations and unrelenting controversy. Vis. Res. **51**(7), 652–673 (2011)
10. Luo, M.R., Li, C.: Cie color appearance models and associated color spaces. Colorimetry: Understanding the CIE system, pp. 261–294 (2007)
11. Provenzi, E.: Geometry of color perception. Part 1: structures and metrics of a homogeneous color space. J. Math. Neurosci. **10**(1), 1–19 (2020)
12. Resnikoff, H.: Differential geometry and color perception. J. Math. Biol. **1**, 97–131 (1974)
13. Schrödinger, E.: Grundlinien einer theorie der farbenmetrik im tagessehen (Outline of a theory of colour measurement for daylight vision). Annalen der Physik **63**(4), 397–456 (1920)

Geometric and Structure Preserving Discretizations

The Herglotz Principle and Vakonomic Dynamics

Manuel de León[1,2], Manuel Lainz[1(✉)], and Miguel C. Muñoz-Lecanda[3]

[1] Instituto de Ciencias Matemáticas (CSIC-UAM-UC3M-UCM), Madrid, Spain
{mdeleon,manuel.lainz}@icmat.es
[2] Real Academia de Ciencias Exactas, Físicas y Naturales, Madrid, Spain
[3] Department of Mathematics, Universitat Politècnica de Catalunya, Barcelona, Spain
miguel.carlos.munoz@upc.edu

Abstract. In this paper we study vakonomic dynamics on contact systems with nonlinear constraints. In order to obtain the dynamics, we consider a space of admisible paths, which are the ones tangent to a given submanifold. Then, we find the critical points of the Herglotz action on this space of paths. This dynamics can be also obtained through an extended Lagrangian, including Lagrange multiplier terms.

This theory has important applications in optimal control theory for Herglotz control problems, in which the cost function is given implicitly, through an ODE, instead of by a definite integral. Indeed, these control problems can be considered as particular cases of vakonomic contact systems, and we can use the Lagrangian theory of contact systems in order to understand their symmetries and dynamics.

Keywords: Contact Hamiltonian systems · Constrained systems · Vakonomic dynamics · Optimal control

1 Introduction

Given a Lagrangian $L : TQ \to \mathbb{R}$ and a submanifold $N \subseteq TQ$, one can look for the critical points of the Euler-Lagrange action restricted to the paths which are tangent to N. This critical points are the solutions of the Euler-Lagrange equations for the extended Lagrangian [3] $\mathcal{L}(q^i, \dot{q}^i, \lambda_a) = L(q^i, \dot{q}^i) - \lambda_a \phi^a$, where $\{\phi^a\}$ are a set of independent constraints defining N. We remark that the dynamics obtained from this principle is, in general, different to nonholonomic dynamics [6,13], in which the critical points of the action are computed on the unconstrained space of paths, but the admisible variations are constrained.

While nonholonomic dynamics has applications in engineering problems, vakonomic mechanics can be used to study optimal control problems. This opens up the possibility to apply results and techniques from Lagrangian mechanics to the study of optimal control problems, such as the Noether theorem and its generalizations [15], or variational integrators constructed from the theory of discrete mechanics [4].

F. Nielsen and F. Barbaresco (Eds.): GSI 2021, LNCS 12829, pp. 183–190, 2021.
https://doi.org/10.1007/978-3-030-80209-7_21

On the other hand, in the Herglotz variational principle one considers a Lagrangian $L : TQ \times \mathbb{R} \to \mathbb{R}$, $L(q^i, \dot{q}^i, z)$ that depends not only on the positions and velocities of the system, but also on the action z itself. The action is then defined implicitly, through the ODE $\dot{z} = L(q^i, \dot{q}^i, z)$. The critical points of this action are the solutions of the Herglotz equations [12,14]:

$$\frac{\partial L}{\partial q^i} - \frac{\mathrm{d}}{\mathrm{d}t}\frac{\partial L}{\partial \dot{q}^i} = \frac{\partial L}{\partial \dot{q}^i}\frac{\partial L}{\partial z}. \tag{1}$$

It has recently been acknowledged that the Herglotz principle provides the dynamics for the Lagrangian counterpart of contact Hamiltonian systems [9]. This has allowed the developement of a theory of symmetries [10,11] which in this setting are not related to conserved quantities, but to *dissipated* ones, which decay at the same rate as the energy. Furthermore variational integrators based on the Herglotz principle have been developed [2,19].

Contact dynamics and the Herglotz principle have applications on the description of many physical systems, such as mechanical systems with friction, thermodynamic systems and some cosmological models [5,16–18].

While the dynamics of contact systems with (linear) nonholonomic constraints [7] has been studied, a theory of contact vakonomic dynamics has not still been developed. This theory could be useful for the study of the Herglotz Optimal Control Problem, introduced in [8], in which the cost function is defined by an ODE, instead of an integral. This will allow a new way to obtain the dynamical equations (on [8] they were obtained rather indirectly, throught Pontryaguin maximum principle) and to apply some of the results of contact Lagrangian systems to this situation.

The paper is structured as follows. In Sect. 2 we review the Herglotz principle. Its traditional formulation, in which the action is defined implicitly, makes the implementation of the constraints difficult. We present and alternative in which the action is defined explicitly, but vakonomic constraints are present. This will make the addition of new constraints almost trivial. In Sect. 3 we obtain the vakonomic dynamical equations for a constrained contact system, and see that this dynamics can also be obtained through an extended contact Lagrangian. Finally, in Sect. 4, we sketch the relationship between the vakonomic dynamics of contact Lagrangian systems and the Herglotz optimal control problem.

1.1 The Herglotz Variational Principle, Revisited

Let Q be the configuration manifold and let $L : TQ \times \mathbb{R} \to \mathbb{R}$ be the contact Lagrangian.

Consider the (infinite dimensional) manifold $\Omega(q_0, q_1)$ of curves $c : [0,1] \to Q$ with endpoints $q_0, q_1 \in Q$. That is, $c(0) = q_0$, $c(1) = q_1$. The tangent space of $\Omega(q_0, q_1)$ at the curve c, is the space of vector fields along c vanishing at the endpoints. That is,

$$\begin{aligned} T_c\Omega(q_0, q_1) = \{\delta c : [0,1] \to TQ \;|\delta c(t) \in T_{c(t)}Q \text{ for all t} \in [0,1], \\ \delta c(0) = 0,\, \delta c(1) = 0\}. \end{aligned} \tag{2}$$

We fix a real number $z_0 \in \mathbb{R}$ and consider the following operator:

$$\mathcal{Z} : \Omega(q_0, q_1) \to \mathcal{C}^\infty([0,1] \to \mathbb{R}), \tag{3}$$

which assigns to each curve c the function $\mathcal{Z}(c)$ that solves the following ODE:

$$\begin{cases} \dfrac{\mathrm{d}\mathcal{Z}(c)}{\mathrm{d}t} &= L(c, \dot{c}, \mathcal{Z}(c)), \\ \mathcal{Z}(c)(0) &= z_0, \end{cases} \tag{4}$$

that is, it assigns to each curve on the base space, its action as a function of time.

Now, the *contact action functional* maps each curve $c \in \Omega(q_0, q_1)$ to the increment of the solution of the ODE:

$$\begin{aligned} \mathcal{A} : \Omega(q_0, q_1) &\to \mathbb{R}, \\ c &\mapsto \mathcal{Z}(c)(1) - \mathcal{Z}(c)(0). \end{aligned} \tag{5}$$

Note that, by the fundamental theorem of calculus,

$$\mathcal{A}(c) = \int_0^1 L(c(t), \dot{c}(t), \mathcal{Z}(c)(t))\mathrm{d}t. \tag{6}$$

Thus, in the case that L does not depend on z, this coincides with the classical Euler-Lagrange action.

Remark 1. The Herglotz action is usually defined as $\mathcal{A}_0(c) = \mathcal{Z}(c)(1)$. However, this definition and our definition only differ by a constant. Indeed,

$$\mathcal{A}(c) = \mathcal{A}_0(c) - z_0. \tag{7}$$

In particular they have the same critical points. However the computations in the vakonomic principle are simpler for $\mathcal{A}$.

As it is proved in [9], the critical points of this action functional are precisely the solutions to Herglotz equation:

Theorem 1 (Herglotz variational principle). *Let $L : TQ \times \mathbb{R} \to \mathbb{R}$ be a Lagrangian function and let $c \in \Omega(q_0, q_1)$ and $z_0 \in \mathbb{R}$. Then, $(c, \dot{c}, \mathcal{Z}(c))$ satisfies the Herglotz equations:*

$$\frac{\mathrm{d}}{\mathrm{d}t}\frac{\partial L}{\partial \dot{q}^i} - \frac{\partial L}{\partial q^i} = \frac{\partial L}{\partial \dot{q}^i}\frac{\partial L}{\partial z},$$

if and only if c is a critical point of $\mathcal{A}$.

1.2 An Alternative Formulation of Herglotz Variational Principle

Another way to approach this problem is to consider a constrained variational principle for curves on $Q \times \mathbb{R}$ constrained to a hypersurface N. We see that this is equivalent to the dynamics produced by a Lagrangian L when considering unconstrained curves on Q.

We will work on the manifold $\bar{\Omega}(q_0, q_1, z_0)$ of curves $\bar{c} = (c, c_z) : [0,1] \to Q \times \mathbb{R}$ such that $c(0) = q_0$, $c(1) = q_1$, $c_z(0) = z_0$. We do not constraint $c_z(1)$. The tangent space at the curve c is given by

$$\begin{aligned} T_c\bar{\Omega}(q_0, q_1, z_0) = \{&\delta\bar{c} = (\delta c, \delta c_z) : [0,1] \to T(Q \times \mathbb{R}) \mid \\ &\delta\bar{c}(t) \in T_{c(t)}(Q \times \mathbb{R}) \text{ for all t} \in [0,1], \delta c(0) = 0, \delta c(1) = 0, \delta c_z(0) = 0\}. \end{aligned} \tag{8}$$

In this space, the action functional $\bar{\mathcal{A}}$ can be defined as an integral

$$\begin{aligned} \bar{\mathcal{A}} : \bar{\Omega}(q_0, q_1, z_0) &\to \mathbb{R}, \\ \bar{c} &\mapsto z_1 - z_0 = \int_0^1 \dot{c}_z(t) \mathrm{d}t. \end{aligned} \tag{9}$$

We will restrict this action to the set of paths that satisfy $\dot{c}_z = L$. For this, consider the hypersurface $N \subseteq T(Q \times \mathbb{R})$, which is the zero set of the constraint function ϕ:

$$\phi(q, \dot{q}, z, \dot{z}) = \dot{z} - L(q, \dot{q}, z). \tag{10}$$

Conversely, given any hypersurface N transverse to the $\dot{z}$-parametric curves, by the implicit function theorem there exists locally a function L such that N is given by the equation $\dot{z} = L$. In this sense, we see that an hypersurface $N \subseteq T(Q \times \mathbb{R})$ is roughly equivalent to a Lagrangian $L : TQ \times \mathbb{R} \to \mathbb{R}$.

We consider the submanifold of curves tangent to N

$$\bar{\Omega}_N(q_0, q_1, z_0) = \{\bar{c} \in \bar{\Omega}(q_0, q_1, z_0) \mid \dot{\bar{c}}(t) \in N \text{ for all t}\} \tag{11}$$

Notice that the map $\mathrm{Id} \times \mathcal{Z} : \Omega(q_0, q_1) \to \bar{\Omega}_N(q_0, q_1, z_0)$ given by $(\mathrm{Id} \times \mathcal{Z})(c) = (c, \mathcal{Z}(c))$ is a bijection, with inverse $\mathrm{pr}_Q(c, c_z) = c$. Here, $\mathcal{Z}$, is defined on (3). Moreover, the following diagram commutes

$$\begin{array}{ccc} & \mathbb{R} & \\ {}^{\mathcal{A}}\nearrow & & \nwarrow^{\bar{\mathcal{A}}} \\ \Omega(q_0, q_1) & \xrightarrow{\mathrm{Id} \times \mathcal{Z}} & \bar{\Omega}_N(q_0, q_1, z_0) \end{array} \tag{12}$$

Hence $\bar{c} \in \bar{\Omega}_N(q_0, q_1, z_0)$ is a critical point of $\bar{\mathcal{A}}$ if and only if c is a critical point of $\mathcal{A}$. So the critical points of $\mathcal{A}$ restricted to $\bar{\Omega}(q_0, q_1, z_0)$ are precisely the curves that satisfy the Herglotz equations.

We will also provide an alternate proof. We find directly the critical points of $\bar{\mathcal{A}}$ restricted to $\bar{\Omega}_N(q_0, q_1, z_0) \subseteq \bar{\Omega}(q_0, q_1, z_0)$ using the following infinite-dimensional version of the Lagrange multiplier theorem [1, 3.5.29].

Theorem 2 (Lagrange multiplier Theorem). *Let M be a smooth manifold and let E be a Banach space such that $g : M \to E$ is a smooth submersion, so that $A = g^{-1}(\{0\})$ is a smooth submanifold. Let $f : M \to \mathbb{R}$ be a smooth function. Then $p \in A$ is a critical point of $f|_A$ if and only if there exists $\hat{\lambda} \in E^*$ such that p is a critical point of $f + \hat{\lambda} \circ g$.*

We will apply this result to our situation. In the notation of this last theorem, $M = \bar{\Omega}(q_0, q_1, z_0)$ is the smooth manifold. We pick the Banach space $E = L^2([0,1] \to \mathbb{R})$ of square integrable functions. This space is, indeed, a Hilbert space with inner product

$$\langle \alpha, \beta \rangle = \int_0^1 \alpha(t)\beta(t)\mathrm{d}t. \tag{13}$$

We remind that, by the Riesz representation theorem, there is a bijection between $L^2([0,1] \to \mathbb{R})$ and its dual such that for each $\hat{\alpha} \in L^2([0,1] \to \mathbb{R})^*$ there exists $\alpha \in L^2([0,1] \to \mathbb{R})$ with $\hat{\alpha}(\beta) = \langle \alpha, \beta \rangle$ for all $\beta \in L^2([0,1] \to \mathbb{R})$.

Our constraint function is

$$\begin{aligned} g : \bar{\Omega}(q_0, q_1, z_0) &\to L^2([0,1] \to \mathbb{R}), \\ \bar{c} &\mapsto (\phi) \circ (\bar{c}, \dot{\bar{c}}), \end{aligned} \tag{14}$$

where ϕ is a constraint locally defining N. Note that $A = g^{-1}(0) = \bar{\Omega}_N(q_0, q_1, z_0)$.

By Theorem 2, c is a critical point of $f = \bar{\mathcal{A}}$ restricted to $\bar{\Omega}_N(q_0, q_1, z_0)$ if and only if there exists $\hat{\lambda} \in L^2([0,1] \to \mathbb{R})^*$ (which is represented by $\lambda \in L^2([0,1] \to \mathbb{R})$) such that c is a critical point of $\bar{\mathcal{A}}_\lambda = \bar{\mathcal{A}} + \hat{\lambda} \circ g$.

Indeed,

$$\bar{\mathcal{A}}_\lambda = \int_0^1 \mathcal{L}_\lambda(\bar{c}(t), \dot{\bar{c}}(t))\mathrm{d}t, \tag{15}$$

where

$$\mathcal{L}_\lambda(q, z, \dot{q}, \dot{z}) = \dot{z} - \lambda\phi(q, z, \dot{q}, \dot{z}). \tag{16}$$

Since the endpoint of c_z is not fixed, the critical points of this functional $\bar{\mathcal{A}}_\lambda$ are the solutions of the Euler-Lagrange equations for $\mathcal{L}_\lambda$ that satisfy the natural boundary condition:

$$\frac{\partial \mathcal{L}_\lambda}{\partial \dot{z}}(\bar{c}(1), \dot{\bar{c}}(1)) = 1 - \lambda(1)\frac{\partial \phi}{\partial \dot{z}}(\bar{c}(1), \dot{\bar{c}}(1)) = 0. \tag{17}$$

For $\phi = \dot{z} - L$, this condition reduces to $\lambda(1) = 1$.

The Euler-Lagrange equations of $\mathcal{L}_\lambda$ are given by

$$\frac{\mathrm{d}}{\mathrm{d}t}\left(\lambda(t)\frac{\partial\phi(\bar{c}(t), \dot{\bar{c}}(t))}{\partial \dot{q}^i}\right) - \lambda(t)\frac{\partial\phi(\bar{c}(t), \dot{\bar{c}}(t))}{\partial q^i} = 0 \tag{18a}$$

$$\frac{\mathrm{d}}{\mathrm{d}t}\left(\lambda(t)\frac{\partial\phi(\bar{c}(t), \dot{\bar{c}}(t))}{\partial \dot{z}}\right) - \lambda(t)\frac{\partial\phi(\bar{c}(t), \dot{\bar{c}}(t))}{\partial z} = 0, \tag{18b}$$

since $\phi = \dot{z} - L$, the equation (18b) for z is just

$$\frac{\mathrm{d}\lambda(t)}{\mathrm{d}t} = -\lambda(t)\frac{\partial L}{\partial z}, \tag{19}$$

substituting on (18a) and dividing by λ, we obtain Herglotz equations.

1.3 Vakonomic Constraints

If we have more constraints, we can obtain vakonomic dynamics, just by changing ϕ by ϕ^a and λ by λ_a on (18), where a ranges from 0 to the number of constraints k. Indeed, we restrict our path space to the ones tangent to submanifold $\tilde{N} \subseteq N \subseteq T(Q \times \mathbb{R})$, where N is the zero set of ϕ^0, given by $\phi^0 = \dot{z} - L$. Repeating the similar computations, we would find that the critical points of $\mathcal{A}|_{\Omega(q_0,q_1,\tilde{N})}$ are the solutions of

$$\frac{\mathrm{d}}{\mathrm{d}t}\left(\lambda_a(t)\frac{\partial \phi^a(\bar{c}(t),\dot{\bar{c}}(t))}{\partial \dot{q}^i}\right) - \lambda_a(t)\frac{\partial \phi^a(\bar{c}(t),\dot{\bar{c}}(t))}{\partial q^i} = 0 \tag{20a}$$

$$\frac{\mathrm{d}}{\mathrm{d}t}\left(\lambda_a(t)\frac{\partial \phi^a(\bar{c}(t),\dot{\bar{c}}(t))}{\partial \dot{z}}\right) - \lambda_a(t)\frac{\partial \phi^a(\bar{c}(t),\dot{\bar{c}}(t))}{\partial z} = 0, \tag{20b}$$

$$\phi^a(\bar{c}(t),\dot{\bar{c}}(t)) = 0, \tag{20c}$$

where $(\phi^a)_{a=0}^k$ are constraints defining $\tilde{N}$ as a submanifold of $TQ \times \mathbb{R}$. Since $\frac{\partial \phi^0}{\partial \dot{z}} = 0$, the rest of the constraints can be chosen to be independent of $\dot{z}$. We denote

$$\psi^\alpha(q,\dot{q},z) = \phi^\alpha(q,\dot{q},z,L(q,\dot{q},z)), \tag{21}$$

$$\mu_\alpha = \frac{\lambda_\alpha}{\lambda_0} \tag{22}$$

$$\mathcal{L}_\mu(q,\dot{q},z,t) = L(q,\dot{q},z) - \mu_\alpha(t)\psi^\alpha(q,\dot{q},z) \tag{23}$$

for $\alpha \in \{1,\dots k\}$, provided that $\lambda_0 \neq 0$.

From this, we can write the Eqs. (20) as

$$-\frac{\mathrm{d}}{\mathrm{d}t}\left(\lambda_0(t)\frac{\partial \mathcal{L}_\mu}{\partial \dot{q}^i}\right) + \lambda_0(t)\frac{\partial \mathcal{L}_\mu}{\partial q^i} = 0 \tag{24a}$$

$$\frac{\mathrm{d}\lambda_0(t)}{\mathrm{d}t} = \lambda_0(t)\frac{\partial \mathcal{L}_\mu}{\partial z} = 0, \tag{24b}$$

$$\psi^\alpha(\bar{c}(t),\dot{c}(t)) = 0, \tag{24c}$$

$$\dot{c_z}(t) = \mathcal{L}_\mu(\bar{c}(t),\dot{c}(t),t). \tag{24d}$$

Substituting (24b) onto (24a), dividing by λ_0 and reordering terms, we obtain

$$\frac{\mathrm{d}}{\mathrm{d}t}\left(\frac{\partial \mathcal{L}_\mu}{\partial \dot{q}^i}\right) - \frac{\partial \mathcal{L}_\mu}{\partial q^i} = \frac{\partial \mathcal{L}_\mu}{\partial \dot{q}^i}\frac{\partial \mathcal{L}_\mu}{\partial z} \tag{25a}$$

$$\psi^\alpha(\bar{c}(t),\dot{c}(t)) = 0, \tag{25b}$$

$$\dot{c_z}(t) = \mathcal{L}_\mu(\bar{c}(t),\dot{c}(t),t). \tag{25c}$$

We remark that these equations are just the Herglotz equations for the *extended Lagrangian* $\mathcal{L}$:

$$\begin{aligned} \mathcal{L} : T(Q \times \mathbb{R}^k) \times \mathbb{R} &\to \mathbb{R} \\ (q,\mu,\dot{q},\dot{\mu},z) &\mapsto L(q,\dot{q},z) - \mu_\alpha \psi^\alpha(q,\dot{q},z) \end{aligned} \tag{26}$$

1.4 Applications to Control

The Herglotz optimal control problem [8] can be formulated by working on the control bundle $W \times \mathbb{R} \to Q \times \mathbb{R}$, with local coordinates (x^i, u^a, z): the *variables* x^i, the *controls* u^a and the *action* z.

The problem consists on finding the curves $\gamma : I = [a, b] \to W$, $\gamma = (\gamma_Q, \gamma_U, \gamma_z)$, such that

1) end points conditions: $\gamma_Q(a) = x_a, \gamma_Q(b) = x_b, \gamma_z(a) = z_0$,
2) γ_Q is an integral curve of X: $\dot{\gamma}_Q = X \circ (\gamma_Q, \gamma_U)$,
3) γ_z satifies the differential equation $\dot{z} = F(x, u, z)$, and
4) maximal condition: $\gamma_z(b)$ is maximum over all curves satisfying 1)–3).

We remark that this can be interpreted as a vakonomic Herglotz principle on TW, with constraints given by the control equations $\phi^i = X^i(x, u, z) - \dot{q}^i$ and the Lagrangian being the cost function $L(x^i, u^a, \dot{q}^i, \dot{u}^a, z) = F(x, u, z)$. The equations of motion obtained through the contact vakonomic principle coincides with the ones obtained indirectly through Pontryaguin maximum principle in [8, Eq. 28]

$$\dot{q}^i = X^i, \tag{27a}$$

$$\begin{aligned} \dot{\mu}_i &= \frac{\partial F}{\partial x^i} - \mu_j \frac{\partial X^j}{\partial x^i} - \mu_j \left(\frac{\partial F}{\partial z} - \mu_i \frac{\partial X^j}{\partial z} \right) \\ &= \mu_i \frac{\partial F}{\partial z} - \mu_j \frac{\partial X^j}{\partial x^i} + \frac{\partial F}{\partial x^i} - \frac{\partial X^j}{\partial z} \mu_i \mu_j, \end{aligned} \tag{27b}$$

$$\dot{z} = F \tag{27c}$$

subjected to the constraints

$$\frac{\partial F}{\partial u^a} - \mu_j \frac{\partial X^j}{\partial u^a} = 0. \tag{27d}$$

Acknowledgements. M. de León and M. Lainz acknowledge the partial finantial support from MINECO Grants MTM2016-76-072-P and the ICMAT Severo Ochoa project SEV-2015-0554. M. Lainz wishes to thank MICINN and ICMAT for a FPI-Severo Ochoa predoctoral contract PRE2018-083203. M.C. Muñoz-Lecanda acknowledges the financial support from the Spanish Ministerio de Ciencia, Innovación y Universidades project PGC2018-098265-B-C33 and the Secretary of University and Research of the Ministry of Business and Knowledge of the Catalan Government project 2017-SGR-932.

References

1. Abraham, R., Marsden, J.E., Ratiu, T.: Manifolds, Tensor Analysis, and Applications. Applied Mathematical Sciences, Springer, New York, 2nd edn. (1988). https://doi.org/10.1007/978-1-4612-1029-0
2. Anahory Simoes, A., Martín de Diego, D., Lainz Valcázar, M., de León, M.: On the geometry of discrete contact mechanics. J. Nonlinear Sci. **31**(3), 1–30 (2021). https://doi.org/10.1007/s00332-021-09708-2
3. Arnold, V.I.: Mathematical Methods of Classical Mechanics. Graduate Texts in Mathematics, Springer, New York, 2nd ed edn., vol. 60 (1997). https://doi.org/10.1007/978-1-4757-2063-1
4. Benito, R., Martín de Diego, D.: Discrete vakonomic mechanics. J. Math. Phys. **46(8)**, 083521 (2005)
5. Bravetti, A.: Contact Hamiltonian dynamics: the concept and its use. Entropy **19**(12), 535 (2017). https://doi.org/10.3390/e19100535
6. Cortés, J., de León, M., de Diego, D.M., Martínez, S.: Geometric description of vakonomic and nonholonomic dynamics. Comparison of solutions. SIAM J. Control Optim. **41**(5), 1389–1412 (2002)
7. de León, M., Jiménez, V.M., Valcázar, M.L.: Contact Hamiltonian and Lagrangian systems with nonholonomic constraints. J. Geom. Mech. (2020). https://doi.org/10.3934/jgm.2021001
8. de León, M., Lainz, M., Muñoz-Lecanda, M.C.: Optimal control, contact dynamics and Herglotz variational problem. arXiv:2006.14326. [math-ph], June 2020
9. de León, M., Lainz Valcázar, M.: Singular Lagrangians and precontact Hamiltonian systems. Int. J. Geom. Methods Mod. Phys. **16**(10), 1950158 (2019). https://doi.org/10.1142/S0219887819501585
10. de León, M., Lainz Valcázar, M.: Infinitesimal symmetries in contact Hamiltonian systems. J. Geom. Phys. **153**, 103651 (2020). https://doi.org/10.1016/j.geomphys.2020.103651
11. Gaset, J., Gràcia, X., Muñoz-Lecanda, M.C., Rivas, X., Román-Roy, N.: New contributions to the Hamiltonian and Lagrangian contact formalisms for dissipative mechanical systems and their symmetries. Int. J. Geom. Methods Mod. Phys. (2020). https://doi.org/10.1142/S0219887820500905
12. Georgieva, B., Guenther, R., Bodurov, T.: Generalized variational principle of Herglotz for several independent variables. First Noether-type theorem. J. Math. Phys. **44**(9), 3911–3927 (2003). https://doi.org/10.1063/1.1597419
13. Gracia, X., Marin-Solano, J., Munoz-Lecanda, M.C.: Some geometric aspects of variational calculus in constrained systems. Rep. Math. Phys. **51**(1), 127–148 (2003)
14. Herglotz, G.: Beruhrungstransformationen. In: Lectures at the University of Gottingen, Gottingen (1930)
15. Martínez, S., Cortés, J., De León, M.: Symmetries in vakonomic dynamics: applications to optimal control. J. Geom. Phys. **38**(3–4), 343–365 (2001)
16. Mrugala, R., Nulton, J.D., Christian Schön, J., Salamon, P.: Contact structure in thermodynamic theory. Rep. Math. Phys. **29**(1), 109–121 (1991). https://doi.org/10.1016/0034-4877(91)90017-H
17. Sloan, D.: A new action for cosmology. arXiv preprint arXiv:2010.07329 (2020)
18. Usher, M.: Local rigidity, contact homeomorphisms, and conformal factors. arXiv preprint arXiv:2001.08729 (2020)
19. Vermeeren, M., Bravetti, A., Seri, M.: Contact variational integrators. J. Phys. A Math. Theoret. **52**(44), 445206 (2019). https://doi.org/10.1088/1751-8121/ab4767

Structure-Preserving Discretization of a Coupled Heat-Wave System, as Interconnected Port-Hamiltonian Systems

Ghislain Haine(✉) and Denis Matignon

ISAE-SUPAERO, Université de Toulouse, Toulouse, France
{ghislain.haine,denis.matignon}@isae.fr

Abstract. The heat-wave system is recast as the coupling of port-Hamiltonian subsystems (pHs), and discretized in a structure-preserving way by the Partitioned Finite Element Method (PFEM) [10,11]. Then, depending on the geometric configuration of the two domains, different asymptotic behaviours of the energy of the coupled system can be recovered at the numerical level, assessing the validity of the theoretical results of [22].

Keywords: Port-Hamiltonian Systems · Partitioned finite element method · Long time asymptotics

1 Introduction and Main Results

Multi-physical systems arise in many areas of science, and high-fidelity simulations are often needed to reduce the number of real-life experiments, hence the cost of optimal design in industry. Such problems can be encountered for instance in aeronautics [12], in chemistry [1] and so on [20]. A first difficulty consists in finding an admissible way to model each physical subsystem of the experiment, with a precise identification of negligible phenomena, as well as feasible controls and available measurements. The second difficulty lies in the coupling of the subsystems altogether to build the final system to simulate. And the last difficulties are obviously the construction of appropriate discretization schemes and the simulations themselves.

Port-Hamiltonian systems (pHs) have been extended to infinite-dimensional setting in [21] two decades ago, allowing to tackle Partial Differential Equations (PDE), and especially those appearing in physics. One of the major force of the port-Hamiltonian approach lies in its ease of use for coupling, since the resulting system remains a pHs [13]. Another strength is its versatility with respect to the

This work has been performed in the frame of the Collaborative Research DFG and ANR project INFIDHEM, entitled *Interconnected Infinite-Dimensional systems for Heterogeneous Media*, n° ANR-16-CE92-0028. Further information is available at https://websites.isae-supaero.fr/infidhem/the-project/.

F. Nielsen and F. Barbaresco (Eds.): GSI 2021, LNCS 12829, pp. 191–199, 2021.
https://doi.org/10.1007/978-3-030-80209-7_22

conditions of experiment: on the one hand, axioms of (classical) physics are used to derive an algebraic structure, namely the Stokes-Dirac structure, and on the other hand, physical laws, state equations, and definitions of physical variables close the system (in the algebraic sense).

This formalism seems very appropriate also for structured computer codes dedicated to efficient simulation. Indeed, each subsystem of the general experiment can be separately discretized before the interconnection: provided that the discretization is structured-preserving, meaning that it leads to a finite-dimensional pHs, the final simulation will enjoy the aforementioned advantages. Furthermore, if each subsystem is also well-structured, meaning that it keeps the separation of axioms and laws, the final simulation codes should be easy to enrich with more and more models. This is the purpose of the discretization scheme known as the Partitioned Finite Element Method (PFEM) [11], which can be seen as an adaptation of the well-known, well-proved, and robust Mixed Finite Element Method (MFEM) [5], not to mention that most, if not all, scientific programming languages already propose finite element libraries [9]. PFEM proves to be very well-adapted to discretized pHs and shows an ever-growing range of applications [6,7,17,19]. Nevertheless, only a few of them tackle and test the interconnection problem [8].

The goal of this work is to provide an application of PFEM, together with numerical simulations, for a simplified and linearised system of fluid-structure interaction (FSI), for which long-time behaviour is known [2,22]. Since PFEM aims at mimicking the pHs structure at the discrete level, hence the power balance satisfied by the *energy* of the system, a good approximation of the long-time behaviour provided in [22] is expected.

1.1 A Simplified and Linearised Fluid-Structure Model

Let $\Omega \subset \mathbb{R}^n$ be a bounded domain with a $\mathcal{C}^2$ boundary $\Gamma := \partial\Omega$. Let Ω_1 be a subdomain of Ω and $\Omega_2 := \Omega \setminus \overline{\Omega_1}$. Denote Γ_{int} the interface, $\Gamma_j := \partial\Omega_j \setminus \overline{\Gamma_{\text{int}}}$ $(j = 1, 2)$, and $\boldsymbol{n}_j$ the unit outward normal vector to Ω_j. Note that $\Gamma = \overline{\Gamma_1 \cup \Gamma_2}$.

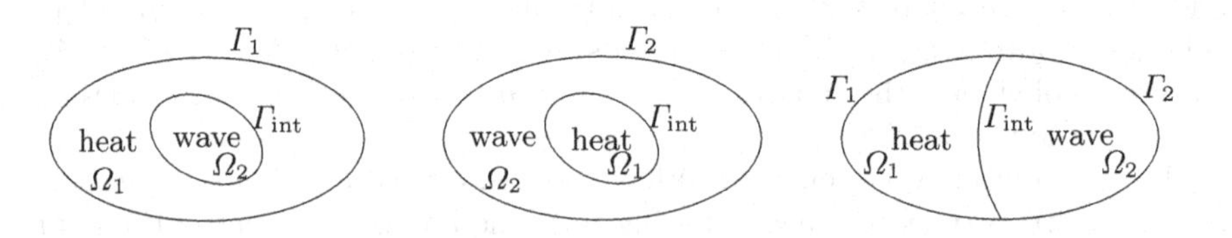

Fig. 1. Different geometrical configurations

We are interested in the structure-preserving discretisation of the following system, $\forall t > 0$:

$$\begin{cases} \partial_t T(t, \boldsymbol{x}) - \Delta T(t, \boldsymbol{x}) = 0, \, \boldsymbol{x} \in \Omega_1, \\ \qquad\qquad\quad T(t, \boldsymbol{x}) = 0, \, \boldsymbol{x} \in \Gamma_1, \end{cases} \quad \begin{cases} \partial_{tt} w(t, \boldsymbol{x}) - \Delta w(t, \boldsymbol{x}) = 0, \, \boldsymbol{x} \in \Omega_2, \\ \qquad\qquad\qquad\quad w(t, \boldsymbol{x}) = 0, \, \boldsymbol{x} \in \Gamma_2, \end{cases} \tag{1}$$

together with transmission conditions across the boundary Γ_{int}:

$$T(t,\boldsymbol{x}) = \partial_t w(t,\boldsymbol{x}), \quad \text{and} \quad \partial_{\boldsymbol{n}_1} T(t,\boldsymbol{x}) = -\partial_{\boldsymbol{n}_2} w(t,\boldsymbol{x}), \quad \forall\, t>0,\, \boldsymbol{x} \in \Gamma_{\text{int}}, \tag{2}$$

and initial data $T(0,\boldsymbol{x}) = T_0(\boldsymbol{x}), \forall \boldsymbol{x} \in \Omega_1$, and $w(0,\boldsymbol{x}) = w_0(\boldsymbol{x})$, $\partial_t w(0,\boldsymbol{x}) = w_1(\boldsymbol{x}), \forall \boldsymbol{x} \in \Omega_2$.

From [22, Thm. 1], it is known that (1)–(2) is well-posed in the *finite energy space* $\mathcal{X} := L^2(\Omega_1) \times H^1_{\Gamma_2}(\Omega_2) \times L^2(\Omega_2)$, endowed with the following norm:

$$\|\boldsymbol{u}\|^2_{\mathcal{X}} := \|u_1\|^2_{L^2(\Omega_1)} + \|u_2\|^2_{L^2(\Omega_2)} + \|\boldsymbol{\nabla}(u_2)\|^2_{(L^2(\Omega_2))^n} + \|u_3\|^2_{L^2(\Omega_1)}. \tag{3}$$

The semi-norm $|\boldsymbol{u}|^2_{\mathcal{X}} := \|u_1\|^2_{L^2(\Omega_1)} + \|\boldsymbol{\nabla}(u_2)\|^2_{(L^2(\Omega_2))^n} + \|u_3\|^2_{L^2(\Omega_1)}$ is a norm on $\mathcal{X}$, equivalent to (3), when Γ_2 has strictly positive measure.

In [22, Thm. 11 & 13], the asymptotic behaviours of these (semi-)norms have been proved, depending on the geometry of Ω (see Fig. 1).

1.2 Main Contributions and Organisation of the Paper

As a major result, we provide a numerical method able to mimick the expected time behaviour in the different geometrical configurations of Fig. 1.

In Sect. 2, the heat and wave PDEs are recast as port-Hamiltonian systems (pHs). In Sect. 3, the Partitioned Finite Element Method (PFEM) is recalled and applied to the coupled system. In Sect. 4, numerical results are provided and compared with the theoretical results of [22].

2 Port-Hamiltonian Formalism

The physical models are recast as port-Hamiltonian systems. However, all physical parameters are taken equal to 1, to stick to system (1)–(2) studied in [22].

2.1 The Fluid Model

The simplified linearised model used for the fluid is the heat equation. The chosen representation corresponds to the *Lyapunov* case already presented in [16,18], with Hamiltonian $\mathcal{H}_1(t) = \frac{1}{2}\int_{\Omega_1} T^2(t,\boldsymbol{x})\, \mathrm{d}\boldsymbol{x}$, where T denotes the temperature. Denoting $\boldsymbol{J}_Q$ the heat flux, the port-Hamiltonian system reads:

$$\begin{pmatrix} \partial_t T \\ -\boldsymbol{\nabla} T \end{pmatrix} = \begin{bmatrix} 0 & -\text{div} \\ -\boldsymbol{\nabla} & 0 \end{bmatrix} \begin{pmatrix} T \\ \boldsymbol{J}_Q \end{pmatrix}, \tag{4}$$

together with boundary ports:

$$\boldsymbol{\nabla} T(t,\boldsymbol{x}) \cdot \boldsymbol{n}_1(\boldsymbol{x}) = u_1(t,\boldsymbol{x}), \quad y_1(t,\boldsymbol{x}) = T(t,\boldsymbol{x}), \quad \forall\, t>0,\, \boldsymbol{x} \in \Gamma_{\text{int}}, \tag{5}$$

$$T(t,\boldsymbol{x}) = 0, \quad y_T(t,\boldsymbol{x}) = \boldsymbol{\nabla} T(t,\boldsymbol{x}) \cdot \boldsymbol{n}_1(\boldsymbol{x}), \quad \forall\, t>0,\, \boldsymbol{x} \in \Gamma_1. \tag{6}$$

To close the system, Fourier's law has been used as constitutive relation: $\boldsymbol{J}_Q = -\boldsymbol{\nabla} T$. The power-balance of the lossy heat subsystem classically reads:

$$\frac{\mathrm{d}}{\mathrm{d}t}\mathcal{H}_1 = -\int_{\Omega_1} |\boldsymbol{\nabla} T|^2 + \langle u_1, y_1 \rangle_{H^{-\frac{1}{2}}(\Gamma_{\text{int}}), H^{\frac{1}{2}}(\Gamma_{\text{int}})}. \tag{7}$$

2.2 The Structure Model

The structure model is the wave equation, a hyperbolic equation widely studied as a pHs (see *e.g.* [19]). The Hamiltonian is the sum of the kinetic and potential energies, $\mathcal{H}_2(t) = \frac{1}{2}\int_{\Omega_2} (\partial_t w(t,\boldsymbol{x}))^2 + |\boldsymbol{\nabla} w(t,\boldsymbol{x})|^2 \, \mathrm{d}\boldsymbol{x}$, w being the deflection and $\partial_t w$ its velocity. The port-Hamiltonian system reads:

$$\begin{pmatrix} \partial_t(\partial_t w) \\ \partial_t(\boldsymbol{\nabla} w) \end{pmatrix} = \begin{bmatrix} 0 & \mathrm{div} \\ \boldsymbol{\nabla} & 0 \end{bmatrix} \begin{pmatrix} \partial_t w \\ \boldsymbol{\nabla} w \end{pmatrix}, \tag{8}$$

together with boundary ports:

$$\partial_t w(t,\boldsymbol{x}) = u_2(t,\boldsymbol{x}), \quad y_2(t,\boldsymbol{x}) = \boldsymbol{\nabla} w(t,\boldsymbol{x}) \cdot \boldsymbol{n}_2(\boldsymbol{x}), \quad \forall\, t>0,\, \boldsymbol{x} \in \Gamma_{\mathrm{int}}, \tag{9}$$

$$\partial_t w(t,\boldsymbol{x}) = 0, \quad y_w(t,\boldsymbol{x}) = \boldsymbol{\nabla} w(t,\boldsymbol{x}) \cdot \boldsymbol{n}_2(\boldsymbol{x}), \quad \forall\, t>0,\, \boldsymbol{x} \in \Gamma_2. \tag{10}$$

The power-balance of the lossless wave subsystem is:

$$\frac{\mathrm{d}}{\mathrm{d}t}\mathcal{H}_2 = \langle y_2, u_2 \rangle_{H^{-\frac{1}{2}}(\Gamma_{\mathrm{int}}), H^{\frac{1}{2}}(\Gamma_{\mathrm{int}})}. \tag{11}$$

2.3 The Coupled System

First note that the homogeneous boundary conditions of (1) have already been taken into account in (6) and (10). The coupling is then obtained by a *gyrator* interconnection of the boundary ports on Γ_{int}, meaning that the input of one system is fully determined by the output of the other one, namely:

$$u_1(t,\boldsymbol{x}) = -y_2(t,\boldsymbol{x}), \qquad u_2(t,\boldsymbol{x}) = y_1(t,\boldsymbol{x}), \quad \forall\, t>0,\, \boldsymbol{x} \in \Gamma_{\mathrm{int}}. \tag{12}$$

As a consequence, the closed coupled system proves dissipative, since the power balance for the global Hamiltonian, $\mathcal{H} = \mathcal{H}_1 + \mathcal{H}_2 := \frac{1}{2}\,|(T, w, \partial_t w)|^2_{\mathcal{X}}$, reads:

$$\frac{\mathrm{d}}{\mathrm{d}t}\mathcal{H} = -\int_{\Omega_1} |\boldsymbol{\nabla} T|^2. \tag{13}$$

3 The Partitioned Finite Element Method

The main idea of PFEM, as in the mixed finite element method, is to integrate by parts only one line of the weak formulation of the system. For our purpose, the choice of which line is to be integrated by parts is dictated by the desired boundary control, see [11] and references therein for more details.

The method leads to a finite-dimensional pHs, which enjoys a discrete power balance, mimicking the continuous one: it is thus structure-preserving. The interconnection of each subsystem is then made using the pHs structure, ensuring an accurate discrete power balance for the coupled system.

3.1 For the Fluid

The heat equation with Hamiltonian $\mathcal{H}_1$ has already been addressed in [17]. The difficulty lies in the mixed Dirichlet–Neumann boundary conditions. Here, following [8], we choose the Lagrange multiplier approach. Let $\boldsymbol{\psi}^1$, φ^1 and ξ^1 be smooth test functions (resp. vectorial, scalar, and scalar at the boundary). Using γ_0 the Dirichlet trace, the weak form of (4)–(5)–(6) reads, after integration by parts of the first line and taking Fourier's law into account in the second:

$$\begin{cases} \int_\Omega \partial_t T \, \varphi^1 = \int_\Omega \boldsymbol{J_Q} \cdot \boldsymbol{\nabla}(\varphi^1) - \int_{\Gamma_1} y_T \, \gamma_0(\varphi^1) - \int_{\Gamma_{\rm int}} u_1 \, \gamma_0(\varphi^1), \\ \int_\Omega \boldsymbol{J_Q} \cdot \boldsymbol{\psi}^1 = - \int_\Omega \boldsymbol{\nabla}(T) \cdot \boldsymbol{\psi}^1, \\ \int_{\Gamma_1} \gamma_0(T) \, \xi^1 = 0, \\ \int_{\Gamma_{\rm int}} y_1 \, \xi^1 = \int_{\Gamma_{\rm int}} \gamma_0(T) \, \xi^1. \end{cases}$$

The output y_T is the Lagrange multiplier of the Dirichlet constraint on 3rd line.

Let $(\boldsymbol{\psi}_i^1)_{1 \le i \le N_Q}$, $(\varphi_j^1)_{1 \le j \le N_T}$, $(\xi_k^1)_{1 \le k \le N_{\Gamma_1}}$ and $(\xi_k^{\rm int})_{1 \le k \le N_{\Gamma_{\rm int}}}$ be finite elements (FE) bases. Note that boundary functions ξ^1 have been divided in two distinct families, on Γ_1 and $\Gamma_{\rm int}$. Approximating each quantity in the appropriate basis leads to a port-Hamiltonian Differential Algebraic Equation (pHDAE) [4,15]:

$$\begin{bmatrix} M_T & 0 & 0 & 0 \\ 0 & M_Q & 0 & 0 \\ 0 & 0 & M_1 & 0 \\ 0 & 0 & 0 & M_{\rm int} \end{bmatrix} \begin{pmatrix} \underline{\dot{T}} \\ \underline{J_Q} \\ \underline{0} \\ -\underline{y_1} \end{pmatrix} = \begin{bmatrix} 0 & D_1 & B_1 & B_{\rm int} \\ -D_1^\top & 0 & 0 & 0 \\ -B_1^\top & 0 & 0 & 0 \\ -B_{\rm int}^\top & 0 & 0 & 0 \end{bmatrix} \begin{pmatrix} \underline{T} \\ \underline{J_Q} \\ \underline{y_T} \\ \underline{u_1} \end{pmatrix}, \tag{14}$$

where $M_\star$ is the mass matrix of the FE basis corresponding to $\star$, $\underline{\star}$ is the column vector collecting the coefficients of the approximation of $\star$ in its FE basis, and:

$$(D_1)_{j,i} = \int_{\Omega_1} \boldsymbol{\psi}_i^1 \cdot \boldsymbol{\nabla}(\varphi_j^1) \in \mathbb{R}^{N_T \times N_Q}, \quad (B_\star)_{j,k} = - \int_{\Gamma_\star} \xi_k^\star \, \gamma_0(\varphi_j^1) \in \mathbb{R}^{N_T \times N_{\Gamma_\star}}.$$

Defining the discrete Hamiltonian $\mathcal{H}_1^d$ as the evaluation of $\mathcal{H}_1$ on $T^d := \sum_{j=1}^{N_T} T_j \varphi_j^1$, one gets $\mathcal{H}_1^d(t) = \frac{1}{2} \underline{T}^\top M_T \underline{T}$. Thanks to the structure of (14), it satisfies a perfectly mimicking discrete counterpart of (7):

$$\frac{\rm d}{\rm dt} \mathcal{H}_1^d = - \underline{J_Q}^\top M_Q \underline{J_Q} + \underline{u_1}^\top M_{\rm int} \underline{y_1}. \tag{15}$$

3.2 For the Structure

The wave equation, studied in *e.g.* [19], does not present difficulty here, since only Dirichlet boundary conditions are considered. One can integrate by parts the second line of the weak form of (8)–(9)–(10) and project on FE bases $(\boldsymbol{\psi}_i^2)_{1 \le i \le N_q}$, $(\varphi_j^2)_{1 \le j \le N_p}$, $(\xi_k^2)_{1 \le k \le N_{\Gamma_2}}$ and $(\xi_k^{\rm int})_{1 \le k \le N_{\Gamma_{\rm int}}}$, where q-type quantities are related to stress and strain ($\boldsymbol{\alpha}_q := \boldsymbol{\nabla}(w)$) and p-type quantities to linear momentum and velocity ($\alpha_p := \partial_t w$). The finite-dimensional pHs then reads:

$$\begin{bmatrix} M_p & 0 & 0 & 0 \\ 0 & M_q & 0 & 0 \\ 0 & 0 & M_2 & 0 \\ 0 & 0 & 0 & M_{\text{int}} \end{bmatrix} \begin{pmatrix} \dot{\underline{\alpha_p}} \\ \dot{\underline{\boldsymbol{\alpha}_q}} \\ -\underline{y_q} \\ -\underline{y_2} \end{pmatrix} = \begin{bmatrix} 0 & D_2 & 0 & 0 \\ -D_2^\top & 0 & B_2 & B_{\text{int}} \\ 0 & -B_2^\top & 0 & 0 \\ 0 & -B_{\text{int}}^\top & 0 & 0 \end{bmatrix} \begin{pmatrix} \underline{\alpha_p} \\ \underline{\boldsymbol{\alpha}_q} \\ \underline{0} \\ \underline{u_2} \end{pmatrix}, \tag{16}$$

with $(D_2)_{j,i} = \int_{\Omega_2} \operatorname{div}(\boldsymbol{\psi}_j^2)\, \varphi_i^2 \in \mathbb{R}^{N_p \times N_q}$, $\quad (B_\star)_{j,k} = -\int_{\Gamma_\star} \xi_k^\star\, \boldsymbol{\phi}_j^2 \cdot \boldsymbol{n}_2 \in \mathbb{R}^{N_q \times N_{\Gamma_\star}}$.

Defining the discrete Hamiltonian $\mathcal{H}_2^d(t) = \frac{1}{2}\left(\underline{\alpha_p}^\top M_p \underline{\alpha_p} + \underline{\boldsymbol{\alpha}_q}^\top M_q \underline{\boldsymbol{\alpha}_q}\right)$, thanks to the structure of (16), one easily gets the discrete counterpart of (11):

$$\frac{\mathrm{d}}{\mathrm{d}t}\mathcal{H}_2^d = \underline{u_2}^\top M_{\text{int}} \underline{y_2}. \tag{17}$$

3.3 Coupling by Gyrator Interconnection

The gyrator interconnection (12) is now imposed weakly, at the discrete level: for all ξ^{int}, smooth boundary test functions on Γ_{int}:

$$\int_{\Gamma_{\text{int}}} \xi^{\text{int}} u_1 = -\int_{\Gamma_{\text{int}}} \xi^{\text{int}} y_2, \quad \int_{\Gamma_{\text{int}}} \xi^{\text{int}} u_2 = \int_{\Gamma_{\text{int}}} \xi^{\text{int}} y_1.$$

After projection on the FE basis $(\xi_k^{\text{int}})_{1 \le k \le N_{\Gamma_{\text{int}}}}$, this becomes:

$$M_{\text{int}} \underline{u_1} = -M_{\text{int}} \underline{y_2}, \quad M_{\text{int}} \underline{u_2} = M_{\text{int}} \underline{y_1}. \tag{18}$$

In Sect. 4, the interconnection matrix $C := B_1 M_{\text{int}}^{-1} B_2^\top$ (and its transpose) is used. However, the inverse M_{int}^{-1} appearing is not a drawback, since only the interface mass matrix is involved, which is of very small size.

The total discrete Hamiltonian $\mathcal{H}^d = \mathcal{H}_1^d + \mathcal{H}_2^d$ then satisfies thanks to (18):

$$\frac{\mathrm{d}}{\mathrm{d}t}\mathcal{H}^d = -\underline{\boldsymbol{J}_Q}^\top M_Q \underline{\boldsymbol{J}_Q} - \underline{y_2}^\top M_{\text{int}} \underline{y_1} + \underline{y_1}^\top M_{\text{int}} \underline{y_2} = -\underline{\boldsymbol{J}_Q}^\top M_Q \underline{\boldsymbol{J}_Q},$$

which perfectly mimics (13) with no approximation (compare with [3,14]): hence, PFEM also proves to be a reliable structure-preserving method for coupled pHs.

4 Numerical Simulations

In order to test numerically the behaviours proved in [22], two geometries are chosen. Switching the domains Ω_1 and Ω_2, four cases are covered. These geometries are given in Fig. 2. The time integration is performed by IDA SUNDIALS (BDF adaptative scheme) *via* `assimulo`. Moreover, $\mathbb{P}^1$ Lagrange elements of order 1 (both distributed and at the boundary) have been used for scalar fields, while RT1 Raviart-Thomas elements of order 1 have been used for vector fields.

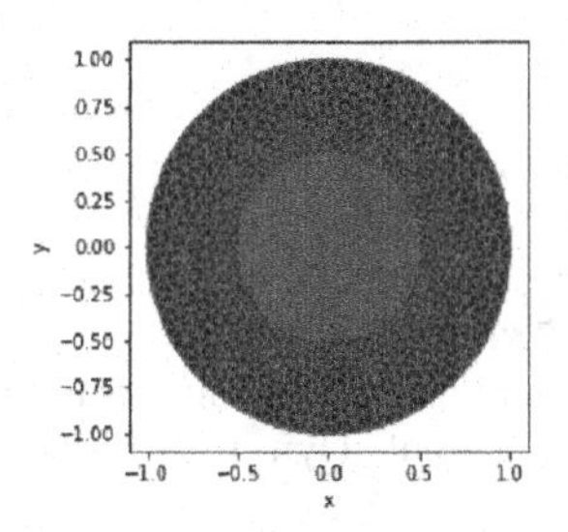

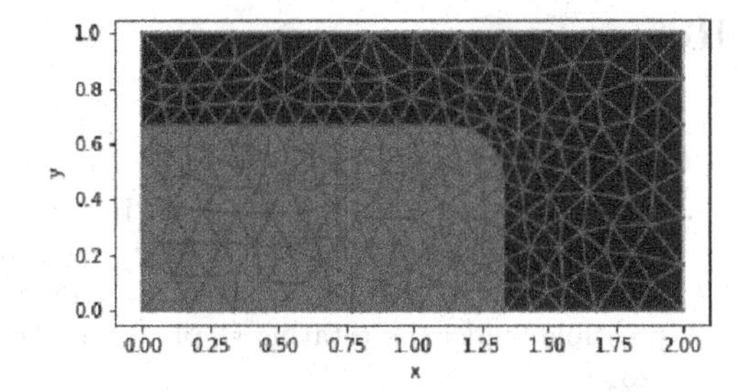

Fig. 2. The colors define the subdomains Ω_1 and Ω_2. On the left, either Γ_1 or Γ_2 is empty. On the right, either Ω_1 is the L-shape subdomain and the geometric optic condition (GCC) is satisfied in Ω, or Ω_1 is the rectangle and the GCC fails.

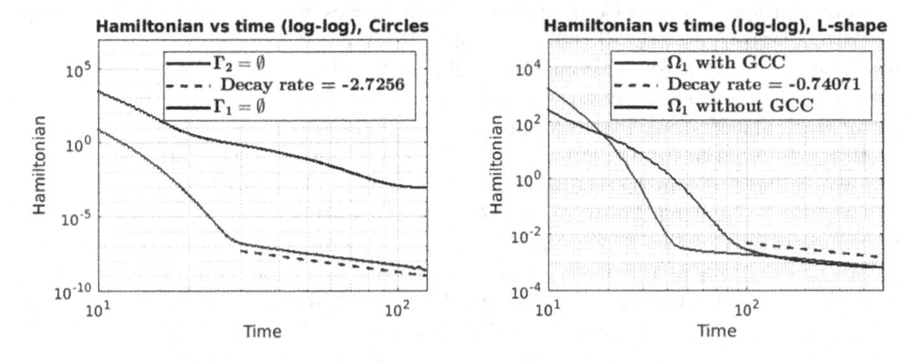

Fig. 3. On the left, the "circles" cases. On the right, the "L-shape" cases.

On the one hand, when $\Gamma_2 = \emptyset$ or when the GCC is satisfied by Ω_1, [22, Thm. 11] asserts a polynomial decay, of rate at least $-1/3$. One can appreciate the blue curves on both plots of Fig. 3, showing this polynomial decay (dashed blue lines) in the long time range. On the other hand, when $\Gamma_1 = \emptyset$ or when the GCC fails, [22, Thm. 13, Rem. 20 & 22] assert that the best decay that could be expected is logarithmic. This is indeed the asymptotic behaviour of the red curves on both plots of Fig. 3.

Note that even in the case of a very coarse mesh (see the "L-shape" case on Fig. 2) and the apparent loss of precision on the right plot of Fig. 3, PFEM still captures the underlying structure, hence the expected decays.

5 Conclusion

PFEM also proves efficient for the structure-preserving simulation of coupled pHs, and long time (polynomial) behaviour can be recovered in many cases.

Further works concern more realistic fluid-structure interactions, *i.e.* non-linear ones, also moving body, hence moving interface, applied *e.g.* to piston dynamics.

References

1. Altmann, R., Schulze, P.: A port-Hamiltonian formulation of the Navier-Stokes equations for reactive flows. Syst. Control Lett. **100**, 51–55 (2017)
2. Avalos, G., Lasiecka, I., Triggiani, R.: Heat-wave interaction in 2–3 dimensions: optimal rational decay rate. J. Math. Anal. Appl. **437**, 782–815 (2016)
3. Bauer, W., Gay-Balmaz, F.: Towards a geometric variational discretization of compressible fluids: the rotating shallow water equations. J. Comput. Dyn. **6**(1), 1–37 (2019)
4. Beattie, C., Mehrmann, V., Xu, H., Zwart, H.: Linear port-Hamiltonian descriptor systems. Math. Control Signals Syst. **30**(4), 1–27 (2018). https://doi.org/10.1007/s00498-018-0223-3
5. Boffi, D., Brezzi, F., Fortin, M.: Mixed Finite Element Methods and Applications. Springer Series in Computational Mathematics, vol. 44. Springer, Heidelberg (2013). https://doi.org/10.1007/978-3-642-36519-5
6. Brugnoli, A., Alazard, D., Pommier-Budinger, V., Matignon, D.: Port-Hamiltonian formulation and symplectic discretization of plate models Part I: Mindlin model for thick plates. Appl. Math. Model. **75**, 940–960 (2019)
7. Brugnoli, A., Alazard, D., Pommier-Budinger, V., Matignon, D.: Port-Hamiltonian formulation and symplectic discretization of plate models Part II: Kirchhoff model for thin plates. Appl. Math. Model. **75**, 961–981 (2019)
8. Brugnoli, A., Cardoso-Ribeiro, F.L., Haine, G., Kotyczka, P.: Partitioned finite element method for power-preserving structured discretization with mixed boundary conditions. IFAC-PapersOnLine **53**(2), 7647–7652 (2020)
9. Brugnoli, A., Haine, G., Serhani, A., and Vasseur, X.: Numerical approximation of port-Hamiltonian systems for hyperbolic or parabolic PDEs with boundary control. J. Appl. Math. Phys. **9**(6), 1278–1321 (2021). Supplementary material https://doi.org/10.5281/zenodo.3938600
10. Cardoso-Ribeiro, F.L., Matignon, D., Lefèvre, L.: A structure-preserving Partitioned Finite Element Method for the 2D wave equation. IFAC-PapersOnLine **51**(3), 119–124 (2018)
11. Cardoso-Ribeiro, F.L., Matignon, D., Lefèvre, L.: A partitioned finite element method for power-preserving discretization of open systems of conservation laws. IMA J. Math. Control Inf. (2020). https://doi.org/10.1093/imamci/dnaa038
12. Cardoso-Ribeiro, F.L., Matignon, D., Pommier-Budinger, V.: Port-Hamiltonian model of two-dimensional shallow water equations in moving containers. IMA J. Math. Control Inf. **37**, 1348–1366 (2020)
13. Cervera, J., van der Schaft, A.J., Baños, A.: Interconnection of port-Hamiltonian systems and composition of Dirac structures. Automatica **43**(2), 212–225 (2007)
14. Egger, H.: Structure-preserving approximation of dissipative evolution problems. Numerische Mathematik **143**(1), 85–106 (2019)
15. Mehrmann, V., Morandin, R.: Structure-preserving discretization for port-Hamiltonian descriptor systems. In: IEEE 58th Conference on Decision and Control (CDC), Nice, France, pp. 6863–6868. IEEE (2019)
16. Serhani, A., Haine, G., Matignon, D.: Anisotropic heterogeneous n-D heat equation with boundary control and observation: I. Modeling as port-Hamiltonian system. FAC-PapersOnLine **52**(7), 51–56 (2019)
17. Serhani, A., Haine, G., Matignon, D.: Anisotropic heterogeneous n-D heat equation with boundary control and observation: II. Structure-preserving discretization. IFAC-PapersOnLine **52**(7), 57–62 (2019)

18. Serhani, A., Matignon, D., Haine, G.: A partitioned finite element method for the structure-preserving discretization of damped infinite-dimensional port-Hamiltonian systems with boundary control. In: Nielsen, F., Barbaresco, F. (eds.) GSI 2019. LNCS, vol. 11712, pp. 549–558. Springer, Cham (2019). https://doi.org/10.1007/978-3-030-26980-7_57
19. Serhani, A., Matignon, D., Haine, G.: Partitioned Finite Element Method for port-Hamiltonian systems with boundary damping: anisotropic heterogeneous 2-D wave equations. IFAC-PapersOnLine **52**(2), 96–101 (2019)
20. van der Schaft, A., Jeltsema, D.: Port-Hamiltonian systems theory: an introductory overview. Found. Trends Syst. Control **1**(2–3), 173–378 (2014)
21. van der Schaft, A.J., Maschke, B.: Hamiltonian formulation of distributed-parameter systems with boundary energy flow. J. Geom. Phys. **42**(1–2), 166–194 (2002)
22. Zhang, X., Enrique, Z.: Long-time behavior of a coupled heat-wave system arising in fluid-structure interaction. Arch. Rat. Mech. Anal. **184**, 49–120 (2007)

Examples of Symbolic and Numerical Computation in Poisson Geometry

Miguel Evangelista-Alvarado, José Crispín Ruíz-Pantaleón(✉), and Pablo Suárez-Serrato

Instituto de Matemáticas, Universidad Nacional Autónoma de México, Circuito Exterior, Ciudad Universitaria, Coyoacán, 04510 Mexico City, Mexico
miguel.eva.alv@matem.unam.mx, {jcpanta,pablo}@im.unam.mx

Abstract. We developed two Python modules for symbolic and numerical computation in Poisson Geometry. We explain how to use them through illustrative examples, which include the computation of Hamiltonian and modular vector fields.

Keywords: Poisson geometry · Computer algebra · Python

1 Introduction

This paper is invitation to use our two Python modules for symbolic and numerical computation in Poisson Geometry. Leisurely introductions to the history, relevance and multiple applications of this branch of Differential Geometry can be found in our two papers that detail these algorithms, and the references therein [2,3]. Comprehensive treatments of Poisson Geometry are available [1,5,9].

Our Python module PoissonGeometry[1] may be used for investigating theoretical properties that occur symbolically (and depends almost only on SymPy). Our module NumPoissonGeometry[2] integrates with machine learning frameworks (and might be prone to follow the future changes in these too). The theoretical motivation behind these implementations, a list of current and potential applications, and an overview of the code structure is available [3].

We will first explain the syntax required to code objects such as functions and bivector fields, so that they can be used as inputs. This is covered in Sect. 2. Then we start with an overview of examples in Sect. 3, serving as simple illustrations of how our two modules can be used. We hope to encourage users to perform their own experiments and computations. The results can be included in research papers with LaTeX code as output, or be used in numerical experiments with NumPy, TensorFlow or PyTorch outputs, as we explain below.

This research was partially supported by CONACyT and UNAM-DGAPA-PAPIIT-IN104819. JCRP thanks CONACyT for a postdoctoral fellowship held during the production of this work.

[1] https://github.com/appliedgeometry/PoissonGeometry.

[2] https://github.com/appliedgeometry/NumericalPoissonGeometry.

F. Nielsen and F. Barbaresco (Eds.): GSI 2021, LNCS 12829, pp. 200–208, 2021.
https://doi.org/10.1007/978-3-030-80209-7_23

2 Installation and Syntax

Our modules work with Python 3. To download the latest stable versions via pip[3,4], run: `>>> pip install poissongeometry` for PoissonGeometry, and `>>> pip install numericalpoissongeometry` for NumPoissonGeometry.

The syntax of PoissonGeometry and NumPoissonGeometry are the same.

Scalar Functions. A scalar function is written using string type expressions.

Example 1. The function $f = ax_1 + bx_2^2 + cx_3^3$ on $\mathbf{R}^3$, with $a, b, c \in \mathbf{R}$, should be written exactly as follows: `>>> "a * x1 + b * x2**2 + c * x3**3"` .

Here, `x1`, `x2`, `x3` are symbolic variables that our modules define by default and that here represent the coordinates (x_1, x_2, x_3) on $\mathbf{R}^3$. A character that is not a coordinate is treated as a symbolic parameter, like `a`, `b`, `c` above.

Multivector Fields. A multivector field is written using python dictionaries where all the keys are tuples of integers and all the values are string type expressions, i.e., when we have a (non–trivial) multivector A of degree a on $\mathbf{R}^m$

$$A = \sum_{1 \leq i_1 < i_2 < \cdots < i_a \leq m} A^{i_1 i_2 \cdots i_a} \frac{\partial}{\partial x_{i_1}} \wedge \frac{\partial}{\partial x_{i_2}} \wedge \cdots \wedge \frac{\partial}{\partial x_{i_a}},$$

the keys of the dictionary are tuples $(i_1, i_2, \ldots, i_a)$ corresponding to the ordered indices $i_1 i_2 \cdots i_a$ of $A^{i_1 i_2 \cdots i_a}$ and the values are the corresponding string expression of the coefficient (scalar function) $A^{i_1 i_2 \cdots i_a}$.

Example 2. The vector field $x_1 \frac{\partial}{\partial x_1} + x_2 \frac{\partial}{\partial x_2} + x_3 \frac{\partial}{\partial x_3}$ on $\mathbf{R}^3$ it should be written as: `>>> {(1,): "x1", (2,): "x2", (3,): "x3"}` .

Writing keys with null values is not needed.

Example 3. The bivector field $x_3 \frac{\partial}{\partial x_1} \wedge \frac{\partial}{\partial x_2} - x_2 \frac{\partial}{\partial x_1} \wedge \frac{\partial}{\partial x_3} + x_1 \frac{\partial}{\partial x_2} \wedge \frac{\partial}{\partial x_3}$ on $\mathbf{R}^m$, $m \geq 3$, may be written as: `>>> {(1,2): "x3", (1,3): "-x2", (2,3): "x1"}`
.

The order of the indices in each tuple can be changed, according to the skew–symmetry of the exterior algebra operations.

Example 4. The bivector field $x_1 x_2 \frac{\partial}{\partial x_1} \wedge \frac{\partial}{\partial x_2}$ on $\mathbf{R}^m$, $m \geq 2$, may be written as `>>> {(1,2): "x1 * x2"}` or as `>>> {(2,1): "-x1 * x2"}` , where this last dictionary corresponds to the bivector field $-x_1 x_2 \frac{\partial}{\partial x_2} \wedge \frac{\partial}{\partial x_1}$.

Differential Forms. The syntax for differential forms is the same as for multivectors fields.

Example 5. The differential form $-\mathrm{d}x_1 \wedge \mathrm{d}x_2 - (x_3 + x_4)\, \mathrm{d}x_5 \wedge \mathrm{d}x_6$ on $\mathbf{R}^m$, $m \geq 6$, is written as `>>> {(1,2): "-1", (5,6): "-(x3 + x4)"}` .

[3] https://pypi.org/project/poissongeometry/.
[4] https://pypi.org/project/numericalpoissongeometry/.

3 Examples of Our Symbolic and Numerical Methods

First instantiate PoissonGeometry or NumPoissonGeometry by entering the dimension and the letter/word for the symbolic variable to be used.

Example 6. To work with PoissonGeometry in dim = 6 and coordinates $(x_1, \ldots, x_6)$:

```
# Import the module PoissonGeometry with the alias pg for simplicity
>>> from poisson.poisson import PoissonGeometry as pg
# Instantiate the module to work in dimension 6 and with the symbolic variable x (by default)
>>> pg6 = pg(6)
```

Example 7. To work with NumPoissonGeometry in dim = 6 and coordinates $(z_1, \ldots, z_6)$:

```
# Import the module PoissonGeometry with the alias npg for simplicity
>>> from numpoisson.numpoisson import NumPoissonGeometry as npg
# Instantiate the module to work in dimension 6 and with the symbolic variable z
>>> npg6 = npg(6, variable="z")
```

In `NumPoissonGeometry` it is necessary to define *meshes* to evaluate the objects described in Sect. 2. The inputs have to be written as lists of lists. For example, as a NumPy array.

Example 8. To create a $(10^6, 6)$ NumPy array with random samples from a uniform distribution over $[0, 1)$ execute `>>> numpy.random.rand(10**6, 6)` .

3.1 Poisson Brackets

Consider the Poisson bivector field on $\mathbf{R}^6$, used in the context of the infinitesimal geometry of Poisson submanifolds [7]:

$$\begin{aligned}\Pi = x_3\frac{\partial}{\partial x_1}\wedge\frac{\partial}{\partial x_2} - x_2\frac{\partial}{\partial x_1}\wedge\frac{\partial}{\partial x_3} + x_6\frac{\partial}{\partial x_1}\wedge\frac{\partial}{\partial x_5} - x_5\frac{\partial}{\partial x_1}\wedge\frac{\partial}{\partial x_6} + x_1\frac{\partial}{\partial x_2}\wedge\frac{\partial}{\partial x_3}\\ - x_6\frac{\partial}{\partial x_2}\wedge\frac{\partial}{\partial x_4} + x_4\frac{\partial}{\partial x_2}\wedge\frac{\partial}{\partial x_6} + x_5\frac{\partial}{\partial x_3}\wedge\frac{\partial}{\partial x_4} - x_4\frac{\partial}{\partial x_3}\wedge\frac{\partial}{\partial x_5}\end{aligned}$$

Example 9. We can verify that Π is a Poisson bivector field with the function `is_poisson_bivector` as follows:

```
>>> P = {(1,2): "x3", (1,3): "-x2", (1,5): "x6", (1,6): "-x5",
         (2,3): "x1", (2,4): "-x6", (2,6): "x4", (3,4):  "x5",
         (3,5): "-x4"}                              # dictionary encoding Π
>>> pg6.is_poisson_bivector(P)                      # run is_poisson_bivector function

True
```

Example 10. The Poisson bracket of $f = x_1^2 + x_2^2 + x_3^2$ and $g = x_4 + x_5 + x_6$, induced by Π, can be computed using the function poisson_bracket:

```
>>> f, g = "x1**2 + x2**2 + x3**2", "x4 + x5 + x6"
                                        # string variables encoding f and g
>>> pg6.poisson_bracket(P, f, g)         # run poisson_bracket function

-2*x1*x5 + 2*x1*x6 + 2*x2*x4 - 2*x2*x6 - 2*x3*x4 + 2*x3*x5
```

Hence, $\{f, g\}_\Pi = -2x_1x_5 + 2x_1x_6 + 2x_2x_4 - 2x_2x_6 - 2x_3x_4 + 2x_3x_5$. We can easily conclude that $\{f, g\}_\Pi = 0$ at points $x \in \mathbf{R}^6$ such that $x_4 = x_5 = x_6$.

Example 11. We can check this last fact with num_poisson_bracket, using a random mesh of shape $\{a_1, b_1\} \times \{a_2, b_2\} \times \{a_3, b_3\} \times \{1\} \times \{1\} \times \{1\}$. Here, the samples a_i, b_i are uniformly randomly selected from $[0, 1)$. In this example, the output is a PyTorch tensor (using the flag `>>> torch_output=True`):

```
>>> npg6.num_poisson_bracket(P, f, g, mesh, torch_output=True)
                    # run num_poisson_bracket function with torch_output flag

tensor(-0.0, -0.0, -0.0, -0.0, -0.0, -0.0, -0.0, -0.0, -0.0,
       -0.0, -0.0, -0.0, -0.0, -0.0, -0.0, -0.0, -0.0, -0.0,
       -0.0, -0.0 dtype=torch.float64)
```

Example 12. The output of Example 10 can be obtained in LaTeX code, adding the flag `>>> latex_format=True` :

```
>>> pg6.poisson_bracket(P, f, g, latex_format=True)
                        # run poisson_bracket function with latex_format flag

"- 2x_{1}x_{5} + 2x_{1}x_{6} + 2x_{2}x_{4} - 2x_{2}x_{6}
 - 2x_{3}x_{4} + 2x_{3}x_{5}"
```

3.2 Hamiltonian Vector Fields

Consider the Poisson bivector field $\Pi = \frac{\partial}{\partial x_1} \wedge \frac{\partial}{\partial x_4} + \frac{\partial}{\partial x_2} \wedge \frac{\partial}{\partial x_5} + \frac{\partial}{\partial x_3} \wedge \frac{\partial}{\partial x_6}$ on $\mathbf{R}^6$, and $h = \frac{1}{2}(x_4^2 + x_5^2 + x_6^2) - \frac{1}{|x_2 - x_1|} - \frac{1}{|x_3 - x_1|} - \frac{1}{|x_3 - x_2|}$ on $\mathbf{R}^6 \setminus_{\{x_1 = x_2 = x_3\}}$.

Example 13. Computes the Hamiltonian vector field X_h with respect to Π:

```
>>> P = {(1,4): "1", (2,5): "1", (3,6): "1"}    # dictionary encoding Π
```

```
>>> h = "
        1/2 * (x4**2 + x5**2 + x6**2)                         # string variable encoding h
        - 1 / sqrt((x2 - x1)**2) - 1 / sqrt((x3 - x1)**2)
        - 1 / sqrt((x3 - x2)**2)"
>>> pg6.hamiltonian_vf(P, h)                                  # run hamiltonian_vf method

{(1,): "-x4", (2,): "-x5", (3,): -x6,
 (4,): "  1 / ((x1 - x3) * sqrt((x1 - x3)**2))
        + 1 / ((x1 - x2) * sqrt((x1 - x2)**2))",
 (5,): "  1 / ((x2 - x3) * sqrt((x2 - x3)**2))
        + 1 / ((x1 - x2) * sqrt((x1 - x2)**2))",
 (6,): "  1 / ((x2 - x3) * sqrt((x2 - x3)**2))
        + 1 / ((x1 - x3) * sqrt((x1 - x3)**2))"}
```

Therefore, the Hamiltonian vector field X_h can be written as:

$$X_h = -x_4\frac{\partial}{\partial x_1} - x_5\frac{\partial}{\partial x_2} - x_6\frac{\partial}{\partial x_3} + \left[\tfrac{1}{(x_1-x_3)|x_1-x_3|} + \tfrac{1}{(x_1-x_2)|x_1-x_2|}\right]\frac{\partial}{\partial x_4}$$
$$+ \left[\tfrac{1}{(x_2-x_3)|x_2-x_3|} + \tfrac{1}{(x_1-x_2)|x_1-x_2|}\right]\frac{\partial}{\partial x_5} - \left[\tfrac{1}{(x_2-x_3)|x_2-x_3|} + \tfrac{1}{(x_1-x_3)|x_1-x_3|}\right]\frac{\partial}{\partial x_6}$$

Example 14. To evaluate X_h on a mesh in $\mathbf{R}^6$ we can use the method `num_hamiltonian_vf`. First we generate a $(10^6, 6)$ NumPy array following Example 8:

```
>>> mesh = numpy.random.rand(10**6, 6)
              # numpy array with random samples from a uniform distribution over [0,1)
>>> npg6.num_hamiltonian_vf(P, h, mesh, torch_output=True)
                              # run num_hamiltonian_vf method with torch_output flag

tensor([ [[-5.07], [-3.30], [-9.95], [-1.34], [ 1.28], [5.43]],
        ...
        [[-9.42], [-3.60], [-2.44], [-1.08], [-1.14], [2.23]]],
        dtype=torch.float64)
```

Observe that we only need the algebraic expression for Π and h to carry out the numerical evaluation (and not X_h explicitly).

3.3 Modular Vector Fields

Consider the Poisson bivector field $\Pi = 2x_4\frac{\partial}{\partial x_1}\wedge\frac{\partial}{\partial x_3} + 2x_3\frac{\partial}{\partial x_1}\wedge\frac{\partial}{\partial x_4} - 2x_4\frac{\partial}{\partial x_2}\wedge\frac{\partial}{\partial x_3} + 2x_3\frac{\partial}{\partial x_2}\wedge\frac{\partial}{\partial x_4} + (x_1-x_2)\frac{\partial}{\partial x_3}\wedge\frac{\partial}{\partial x_4}$ on $\mathbf{R}^4$.

Our method `modular_vf` computes the modular vector field of a Poisson bivector field on $\mathbf{R}^m$ with respect to a volume form $f\Omega_0$, where f is a non-vanishing function and Ω_0 the Euclidean volume form on $\mathbf{R}^m$.

Example 15. Computing the modular vector field Z of Π on $(\mathbf{R}^4, \Omega_0)$:

```
>>> P = {(1,3): "2 * x4", (1,4): "2 * x3", (2,3): "-2 * x4",
         (2,4): "2 * x3", (3,4):"x1 - x2"}          # dictionary encoding Π
>>> pg4.modular_vf(P, 1)                             # run modular_vf method

{}
```

We thus conclude that $Z = 0$, and hence Π is unimodular on all of $\mathbf{R}^4$.

Example 16. We can use random meshes to numerically test the triviality of Z:

```
>>> mesh = numpy.random.rand(10**6, 4)
        # (10^6, 4) numpy array with random samples from a uniform distribution over [0,1)
>>> npg4.num_modular_vf(P, 1, mesh, torch_output=True)
                                  # run modular_vf method with torch_output flag

tensor([ [[-0.0],[-0.0],[-0.0],[-0.0]], [[-0.0],[-0.0],[-0.0],[-0.0]],
        ...,
        [[-0.0],[-0.0],[-0.0],[-0.0]], [[-0.0],[-0.0],[-0.0],[-0.0]],
        dtype=torch.float64)
```

3.4 Poisson Bivector Fields with Prescribed Casimirs

Consider the functions $K_1 = \frac{1}{2}x_4$ and $K_2 = -x_1^2 + x_2^2 + x_3^2$ on $\mathbf{R}^4$.

Example 17. We can construct a Poisson bivector field Π on $\mathbf{R}^4$ that admits K_1 and K_2 as Casimir functions using the `flaschka_ratiu_bivector` function (with respect to the canonical volume form):

```
>>> casimirs = ["1/2 * x4", "-x1**2 + x2**2 + x3**2"]
                              # list containing string expressions for K1 and K2
>>> pg4.flaschka_ratiu_bivector(casimirs)
                                        # run flaschka_ratiu_bivector function

{(1,2): "x3", (1,3): "-x2", (2,3): "-x1"}
```

This yields $\Pi = x_3 \frac{\partial}{\partial x_1} \wedge \frac{\partial}{\partial x_2} - x_2 \frac{\partial}{\partial x_1} \wedge \frac{\partial}{\partial x_3} - x_1 \frac{\partial}{\partial x_2} \wedge \frac{\partial}{\partial x_3}$.

Example 18. The `is_casimir` method verifies that K_1, K_2 are Casimirs of Π:

```
>>> P = {(1,2): "x3", (1,3): "-x2", (2,3): "-x1"}
                                               # dictionary encoding Π
>>> pg4.is_casimir(P, "1/2 * x4"),              # run is_casimir function
    pg4.is_casimir(P, "-x1**2 + x2**2 + x3**2")
```

```
True, True
```

Example 19. To evaluate Π of Example 17 at the 'corners' $Q^4 := \{0,1\}^4$ of the unit cube in $\mathbf{R}^4$ we can use the function `num_flaschka_ratiu_bivector`. In this example the output is—by default—a list of dictionaries:

```
>>> npg4.num_flaschka_ratiu_bivector(functions, Qmesh_4)
        # run num_flaschka_ratiu_bivector function with Qmesh_4 a NumPy array encoding Q^4

[{(1, 2): 0.0, (1, 3): 0.0, (2, 3): 0.0}, {(1, 2): 0.0, (1, 3): 0.0, (2, 3): 0.0},
 ...
 {(1, 2): 1.0, (1, 3): -1.0, (2, 3): -1.0}, {(1, 2): 1.0, (1, 3): -1.0, (2, 3): -1.0}]
```

Example 20. The bivector Π from Example 17 can be evaluated on Q^4 using `num_bivector`:

```
>>> npg4.num_bivector(P, Qmesh_4, tf_output=True)
    # run num_bivector method with Qmesh_4 a NumPy array encoding Q^4 with tf_output flag

tensor([ [[-0.0, 1.0,-1.0,-0.0], [-1.0,-0.0,-0.0,-0.0],
          [ 1.0,-0.0,-0.0,-0.0], [-0.0,-0.0,-0.0,-0.0]],
         ...
         [[-0.0, 1.0,-1.0,-0.0], [-1.0,-0.0,-1.0,-0.0],
          [ 1.0, 1.0,-0.0,-0.0], [-0.0,-0.0,-0.0,-0.0]]],
         dtype=float64)
```

The output here is a TensorFlow tensor, invoked via `>>> tf_output=True` .

3.5 Classification of Lie–Poisson Bivector Fields on $\mathbf{R}^3$

A Lie–Poisson structure is a Poisson structure induced by the Lie bracket on a given Lie algebra (see [1,5]).

Consider the Lie–Poisson bivector $\Pi = -10x_3\frac{\partial}{\partial x_1} \wedge \frac{\partial}{\partial x_2} + 10x_2\frac{\partial}{\partial x_1} \wedge \frac{\partial}{\partial x_3} - 10x_1\frac{\partial}{\partial x_2} \wedge \frac{\partial}{\partial x_3}$ on $\mathbf{R}^3$.

Example 21. To compute a normal form Π_0 of Π, run:

```
>>> P = {(1,2): "-10*x3", (1,3): "10*x2", (2,3): "-10*x1"}
                                                  # dictionary encoding Π
>>> pg3.linear_normal_form_R3(P),          # run linear_normal_form_R3 function

{(1,2): "x3", (1,3): "-x2", (2,3): "x1"}
```

Yielding $\Pi_0 = x_3\frac{\partial}{\partial x_1} \wedge \frac{\partial}{\partial x_2} - x_2\frac{\partial}{\partial x_1} \wedge \frac{\partial}{\partial x_3} + x_1\frac{\partial}{\partial x_2} \wedge \frac{\partial}{\partial x_3}$, which simplifies Π.

Example 22. Π is isomorphic to Π_0, verified by `isomorphic_lie_poisson_R3`:

```
>>> P0 = {(1,2): "x3", (1,3): "-x2", (2,3): "x1"}
                                          # dictionary encoding Π_so(3)
>>> pg3.isomorphic_lie_poisson_R3(P, P_so3)
                                      # run isomorphic_lie_poisson function
True
```

Example 23. To evaluate Π_0 at points of $Q^3 := \{0, 1\}^3$ we can use the method `num_linear_normal_form_R3`:

```
>>> npg4.num_linear_normal_form_R3(P, Qmesh_3)
          # run num_linear_normal_form_R3 method with Qmesh_3 a NumPy array encoding Q^3
[{(1, 2): 0.0, (1, 3): -1.0, (2, 3): 0.0}, {(1, 2): 0.0, (1, 3): 0.0, (2, 3): 1.0},
 ...
 {(1, 2): 1.0, (1, 3): -1.0, (2, 3): -1.0}, {(1, 2): 1.0, (1, 3): -1.0, (2, 3): -1.0}]
```

Related Work and Conclusion

As far as we know this is the first comprehensive implementation of these methods. The only other related works are, a Sage implementation of the Schouten bracket, a test of the Jacobi identity for a given generalized Poisson bracket [4], and symbolical and numerical analysis of Hamiltonian systems on the symplectic vector space $\mathbf{R}^{2n}$ [8]. However, they only address specific problems of Poisson Geometry. We systematically generalised and unified these efforts within the same scripting language and modules. In terms of development of computational tools from Differential Geometry, the package GeomStats [6] is a great resource for computations and statistics on manifolds, implementing concepts such as Riemannian metrics, estimation, and clustering.

Our work differs in that we compute in (local) coordinates, which is sufficient for analysing some physical systems or with global operations, like the trace or coboundary operators. Furthermore, the objects can be defined symbolically for both modules, allowing exterior calculus operations to be worked out using a simple syntax. Our modules contribute to the development of computational methods. They can assist the progress of theoretical aspects of Poisson Geometry and abet the creation of novel applications. We hope to stimulate the use of these methods in Machine Learning, Information Geometry, and allied fields.

References

1. Dufour, J.P., Zung, N.T.: Poisson Structures and their Normal Forms. Birkhäuser Basel (2005)
2. Evangelista-Alvarado, M., Ruíz-Pantaleón, J.C., Suárez-Serrato, P.: On computational Poisson geometry I: symbolic foundations. J. Geom. Mech. (to appear)
3. Evangelista-Alvarado, M., Ruíz-Pantaleón, J.C., Suárez-Serrato, P.: On computational Poisson geometry II: numerical methods. J. Comput. Dyn. (in press). https://doi.org/10.3934/jcd.2021012
4. Kröger, M., Hütter, M., Öttinger, H.C.: Symbolic test of the Jacobi identity for given generalized 'Poisson' bracket. Comput. Phys. Commun. **137**, 325–340 (2001)
5. Laurent-Gengoux, C., Pichereau, A., Vanhaecke, P.: Poisson Structures. Springer, Heidelberg (2013). https://doi.org/10.1007/978-3-642-31090-4
6. Miolane, N., et al.: Geomstats: a Python package for Riemannian geometry in machine learning. J. Mach. Learn. Res. **21**(223), 1–9 (2020)
7. Ruíz-Pantaleón, J.C., García-Beltrán, D., Vorobiev, Yu.: Infinitesimal Poisson algebras and linearization of Hamiltonian systems. Ann. Glob. Anal. Geom. **58**(4), 415–431 (2020). https://doi.org/10.1007/s10455-020-09733-6
8. Takato, S., Vallejo, J.A.: Hamiltonian dynamical systems: symbolical, numerical and graphical study. Math. Comput. Sci. **13**, 281–295 (2019). https://doi.org/10.1007/s11786-019-00396-6
9. Weinstein, A.: Poisson geometry. Differ. Geom. Appl. **9**, 213–238 (1998)

New Directions for Contact Integrators

Alessandro Bravetti[1], Marcello Seri[2(✉)], and Federico Zadra[2]

[1] Instituto de Investigaciones en Matemáticas Aplicadas y en Sistemas (IIMAS–UNAM), Mexico City, Mexico
alessandro.bravetti@iimas.unam.mx

[2] Bernoulli Institute for Mathematics, Computer Science and Artificial Intelligence, University of Groningen, Groningen, The Netherlands
{m.seri,f.zadra}@rug.nl

Abstract. Contact integrators are a family of geometric numerical schemes which guarantee the conservation of the contact structure. In this work we review the construction of both the variational and Hamiltonian versions of these methods. We illustrate some of the advantages of geometric integration in the dissipative setting by focusing on models inspired by recent studies in celestial mechanics and cosmology.

Keywords: Contact geometry · Contact mechanics · Geometric integrators · Numerical methods · Celestial mechanics

1 Introduction

With the range of applications of contact geometry growing rapidly, geometric numerical integrators that preserve the contact structure have gained increasing attention [5,6,13,14,17,19]. Deferring to the above literature for detailed presentations of contact systems, their properties and many of their uses, in this work we will present new applications of the contact geometric integrators introduced by the authors in [6,17,19] to two particular classes of examples inspired by celestial mechanics and cosmology.

A contact manifold is a pair (M, ξ) where M is a $(2n+1)$-dimensional manifold and $\xi \subset TM$ is a contact structure, that is, a maximally non-integrable distribution of hyperplanes. Locally, such distribution is given by the kernel of a one form η satisfying $\eta \wedge (\mathrm{d}\eta)^n \neq 0$ (see e.g. [8] for more details). The 1-form η is called the contact form. Darboux's theorem for contact manifolds states that for any point on M there exist local coordinates $(q_1, \dots, q_n, p_1, \dots, p_n, s)$ such that the contact 1-form can be written as $\eta = \mathrm{d}s - \sum_i p_i \,\mathrm{d}q_i$. Moreover, given η, we can associate to any smooth function $\mathcal{H} : M \to \mathbb{R}$ a contact Hamiltonian vector field $X_{\mathcal{H}}$, defined by

$$\mathcal{L}_{X_{\mathcal{H}}}\eta = -R_\eta(\mathcal{H})\eta \qquad \text{and} \qquad \eta(X_{\mathcal{H}}) = -\mathcal{H},$$

where $\mathcal{L}$ is the Lie derivative and R_η is the Reeb vector field corresponding to η [8]. In canonical coordinates the flow of $X_{\mathcal{H}}$ is given by

$$\dot{q} = \frac{\partial \mathcal{H}}{\partial p}, \qquad \dot{p} = -\frac{\partial \mathcal{H}}{\partial q} - p\frac{\partial \mathcal{H}}{\partial s}, \qquad \dot{s} = p\frac{\partial \mathcal{H}}{\partial p} - \mathcal{H}.$$

F. Nielsen and F. Barbaresco (Eds.): GSI 2021, LNCS 12829, pp. 209–216, 2021.
https://doi.org/10.1007/978-3-030-80209-7_24

The flow of a contact Hamiltonian system preserves the contact structure, but it does not preserve the Hamiltonian:

$$\frac{\mathrm{d}}{\mathrm{d}t}\mathcal{H} = -\mathcal{H}\frac{\partial \mathcal{H}}{\partial s}. \tag{1}$$

Using the this differential equation we can split the contact manifold in invariant parts for the Hamiltonian dynamics separated by the sub-manifold $\mathcal{H} = 0$, unique situation in which the Hamiltonian is conserved.

Contact Hamiltonian systems, like symplectic Hamiltonian systems, benefit from an associated variational principle, which is due to Herglotz: let Q be an n-dimensional manifold with local coordinates q^i and let $\mathcal{L} : \mathbb{R} \times TQ \times \mathbb{R} \to \mathbb{R}$. For any given smooth curve $q : [0, T] \to Q$ we consider the initial value problem

$$\dot{s} = \mathcal{L}(t, q(t), \dot{q}(t), s), \qquad s(0) = s_{\text{init}}.$$

Then the value $s(T)$ is a functional of the curve q. We say that q is *a critical curve* if $s(T)$ is invariant under infinitesimal variations of q that vanish at the boundary of $[0, T]$. It can be shown that critical curves for the Herglotz' variational principle are characterised by the following generalised Euler-Lagrange equations:

$$\frac{\partial \mathcal{L}}{\partial q^a} - \frac{\mathrm{d}}{\mathrm{d}t}\frac{\partial \mathcal{L}}{\partial \dot{q}^a} + \frac{\partial \mathcal{L}}{\partial s}\frac{\partial \mathcal{L}}{\partial \dot{q}^a} = 0.$$

Furthermore, the corresponding flow consists of contact transformations with respect to the 1-form $\eta = \mathrm{d}s - p_a \mathrm{d}q^a$.

This work is structured as follows: after a brief recap of the construction of contact integrators in Sect. 2, we showcase in Sect. 3 some interesting properties of the numerical integrators on some explicit examples. Finally, in Sect. 4 we present some considerations and ideas for future explorations.

2 Contact Integrators: Theory

In this section we summarize the main results of [6,17] on the development of variational and Hamiltonian integrators for contact systems.

2.1 Contact Variational Integrators (CVI)

There is a natural discretisation of Herglotz' variational principle [14,17].

Definition 1 (Discrete Herglotz' variational principle). *Let Q be an n-dimensional manifold with local coordinates q^i and let $L : \mathbb{R} \times Q^2 \times \mathbb{R}^2 \to \mathbb{R}$. For any given discrete curve $q : \{0, \ldots, N\} \to Q$ we consider the initial value problem $s_{k+1} = s_k + \tau L(k\tau, q_k, q_{k+1}, s_k, s_{k+1})$, $s_0 = s_{\text{init}}$. Then the value s_N is a functional of the discrete curve q. We say that q is* a critical curve *if*

$$\frac{\partial s_N}{\partial q_k} = 0 \qquad \forall k \in \{1, \ldots, N-1\}.$$

From this, one can derive the discrete generalised Euler-Lagrange equations. As in the conventional discrete calculus of variations, they can be formulated as the equality of two formulas for the momentum [12].

Proposition 1. *Let*

$$p_k^- = \frac{\partial_{q_k} L((k-1)\tau, q_{k-1}, q_k, s_{k-1}, s_k)}{1 - \tau \partial_{s_k} L((k-1)\tau, q_{k-1}, q_k, s_{k-1}, s_k)}, \quad p_k^+ = -\frac{\partial_{q_k} L(k\tau, q_k, q_{k+1}, s_k, s_{k+1})}{1 + \tau \partial_{s_k} L(k\tau, q_k, q_{k+1}, s_k, s_{k+1})}.$$

Then solutions to the discrete Herglotz variational principle are characterised by $p_k^- = p_k^+$. *Furthermore, the map* $(q_k, p_k, s_k) \mapsto (q_{k+1}, p_{k+1}, s_{k+1})$ *induced by a critical discrete curve preserves the contact structure* $\ker(\mathrm{d}s - p_a \mathrm{d}q^a)$.

Without loss of generality it is possible to take the discrete Lagrange function depending on only one instance of s: $L(k\tau, q_k, q_{k+1}, s_k)$. Then the discrete generalised Euler-Lagrange equations read

$$\partial_{q_k} L(k\tau, q_k, q_{k+1}, s_k) + \partial_{q_k} L((k-1)\tau, q_{k-1}, q_k, s_{k-1}) \left(1 + \tau \partial_{s_k} L(k\tau, q_k, q_{k+1}, s_k)\right) = 0.$$

For a discrete Lagrangian of the form

$$L(k\tau, q_k, q_{k+1}, s_k) = \frac{1}{2}\left(\frac{q_{k+1} - q_k}{\tau}\right)^2 - \frac{V(q_k, k\tau) + V(q_{k+1}, (k+1)\tau)}{2} - F(s_k, k\tau),$$

which includes all the examples treated below, the CVI is explicit and takes the remarkably simple form

$$\begin{aligned}
q_{k+1} &= q_k - \frac{\tau^2}{2}\partial_{q_k} V(q_k, k\tau) + p_k\left(\tau - \frac{\tau^2}{2}\partial_{s_k} F(s_k, k\tau)\right), \\
s_{k+1} &= s_k + \tau L(k\tau, q_k, q_{k+1}, s_k), \\
p_{k+1} &= \frac{\left(1 - \frac{\tau}{2}\partial_{s_k} F(s_k, k\tau)\right) p_k - \frac{\tau}{2}\left(\partial_{q_k} V(q_k, k\tau) + \partial_{q_{k+1}} V(q_{k+1}, (k+1)\tau)\right)}{1 + \frac{\tau}{2}\partial_{s_{k+1}} F(s_{k+1}, (k+1)\tau)}.
\end{aligned}$$

Higher order CVIs can be constructed with a Galerkin discretisation [6,14].

2.2 Contact Hamiltonian Integrators (CHI)

In [6], the authors derive a contact analogue of the symplectic integrators introduced by Yoshida [18] for separable Hamiltonians.

Proposition 2 [6]. *Let the contact Hamiltonian* $\mathcal{H}$ *be separable into the sum of* n *functions* $\phi_j(q, p, s)$, $j = 1, \ldots, n$. *Assume that each of the vector fields* X_{ϕ_j} *is exactly integrable and denote its flow by* $\exp(tX_{\phi_j})$. *Then the integrator*

$$S_2(\tau) = e^{\frac{\tau}{2}X_{\phi_1}} e^{\frac{\tau}{2}X_{\phi_2}} \cdots e^{\tau X_{\phi_n}} \cdots e^{\frac{\tau}{2}X_{\phi_2}} e^{\frac{\tau}{2}X_{\phi_1}},$$

is a second-order contact Hamiltonian integrator (CHI) for the flow of $\mathcal{H}$.

Note that exact integrability is not a necessary condition in general. The same construction gives in fact rise to *contact compositional integrators* in a straightforward manner, but we will not digress further on this.

Starting from Proposition 2, it is possible to construct CHIs of any even order: the construction is presented in detail in [6], where it was also shown how to use the *modified Hamiltonian* in order to obtain error estimates for the integrator.

For a contact Hamiltonian of the form

$$\mathcal{H} = \underbrace{\frac{p^2}{2}}_{C} + \underbrace{V(q,t)}_{B} + \underbrace{f(s,t)}_{A}, \tag{2}$$

one obtains for a time step τ the following discrete maps

$$A: \begin{cases} q_i = q_{i-1} \\ p_i = p_{i-1}\frac{f(s_i,t_{i-1})}{f(s_{i-1},t_{i-1})} \\ \int_{s_{i-1}}^{s_i} \frac{d\xi}{f(\xi,t_{i-1})} = -\tau \end{cases} \quad B: \begin{cases} q_i = q_{i-1} \\ p_i = -V'(q_{i-1},t_{i-1})\tau + p_{i-1} \\ s_i = s_{i-1} - \frac{V(q_{i-1},t_{i-1})}{2}\tau \end{cases} \quad C: \begin{cases} q_i = p_{i-1}\tau + q_{i-1} \\ p_i = p_{i-1} \\ s_i = s_{i-1} + \frac{p_{i-1}^2}{2}\tau \end{cases}$$

The time is advanced by an extra map, $D: t_i = t_{i-1} + \tau$. Altogether, this can be seen in fluid dynamical terms as a streamline approximation of each piece of the Hamiltonian. All the examples in this paper are of the form (2) with an explicit map A.

3 Contact Integrators: Applications

Until recently, the main focus in the literature on contact mechanical systems has been on models with linear dependence on the action, see e.g. [4,17]. In general, however, for a contact system to have non-trivial periodic trajectories in the contact space, $\frac{\partial \mathcal{H}}{\partial s}$ has to change sign. This can be achieved either with a time-varying damping coefficient, as is the case in [11] and in our first example below, or by including a non-linear dependence on the action, as is done in [10] and in our second example.

Even though the contact oscillator that we use in the simulation is of purely theoretical interest and not associated to physical systems (that we know of), the reduction presented in [15,16] shows that contact Hamiltonians with quadratic dependence on the action appear naturally in the study of the intrinsic dynamics of Friedman-Lemaitre-Robertson-Walker and Bianchi universes in cosmology.

The source code for all the simulations is provided in [7].

3.1 Perturbed Kepler Problem

Many relevant systems in celestial mechanics fall in the realm of Newtonian mechanics of systems with time-varying non-conservative forces [6]. Their equations of motion are the solution of the Newton equations

$$\ddot{q} + \frac{\partial V(q,t)}{\partial q} + f(t)\dot{q} = 0\,. \tag{3}$$

A direct computation, shows that (3) coincide with the equations of motion of the contact Hamiltonian

$$\mathcal{H}(p,q,s,t) = \sum_{a=1}^{n} \frac{p_a^2}{2} + V(q,t) + f(t)\,s. \tag{4}$$

Since (4) is separable in the sense of Proposition 2, one can directly apply the CVI and CHI to such Hamiltonian systems. Even in presence of singularities, like in the perturbed Kepler problem re-discussed here, contact integrators show a remarkable stability also for large time steps.

Here we consider a Kepler potential $V(q^a,t) = -\alpha/|q|$, $\alpha \in \mathbb{R}$, with an external periodic forcing $f(t)\dot{q} = \alpha \sin(\Omega t)\dot{q}$. This choice of the perturbation is selected in order to have some sort of energy conservation on average, to emphasise the stability of the methods and their applicability to time-dependent problems.

It is well-known that, in presence of Keplerian potentials, Euclidean integrators become unstable for long integration times or large time steps: a Runge-Kutta integrator drifts toward the singularity and explodes in a rather short amount of iterations.

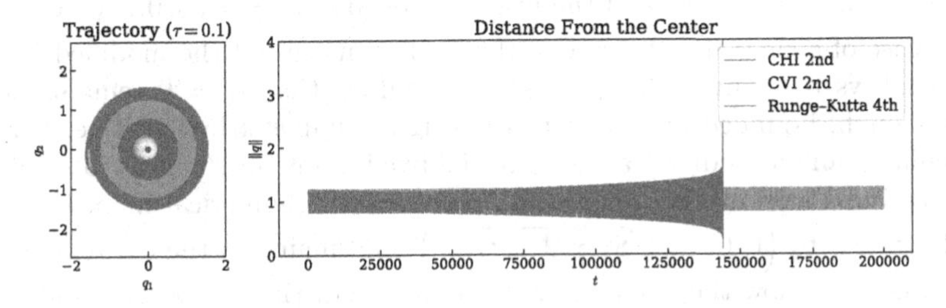

Fig. 1. Long-time integration of a perturbed Kepler problem with time step $\tau \in \{0.1\}$.

In Fig. 1, the same perturbed Kepler problem with $\Omega = \pi$ and $\alpha = 0.01$ is integrated over a very long time interval $[0, 200.000]$ with the second-order contact integrators and a fixed-step fourth-order Runge-Kutta (RK4). One can clearly observe the long-time stability of the contact method. The price to pay for this is the introduction of an artificial precession of the trajectory. This is more evident in Fig. 2: here the problem with $\Omega = \pi$ and $\alpha = 0.05$ is integrated with time step $\tau = 0.3$ with the aforementioned integrators and with a 6th order CHI from [6].

3.2 Contact Oscillator with Quadratic Action

Motivated by the analysis in [10] and [15,16], in this section we study contact Hamiltonians of the form $\mathcal{H}(p,q,s) = \sum_{a=1}^{n} \frac{p_a^2}{2} + \gamma\frac{s^2}{2} + V(q^a)$. In particular, we consider the 1-dimensional quadratic contact harmonic oscillator, $V(q) = +\frac{q^2}{2} - C$, $\gamma, C > 0$. As shown by Eq. (1), the value of the contact Hamiltonian is

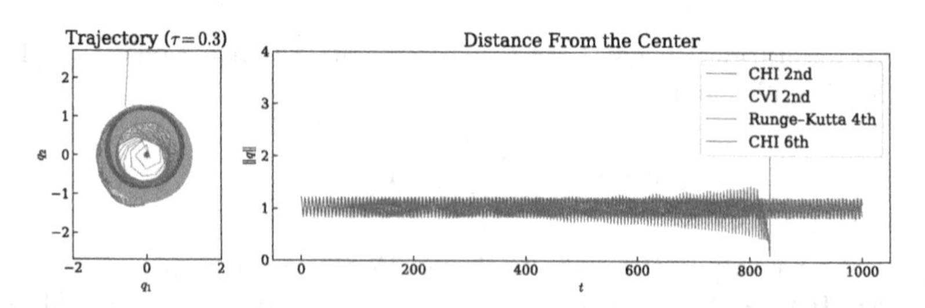

Fig. 2. Integration of a perturbed Kepler problem with time step $\tau = 0.3$. The plot is truncated after the first 1.000 units of time to emphasize the blow-up of the Runge-Kutta 4 method. The contact integrators present no visible difference for the whole integration time.

not preserved unless its initial value is equal to zero [3]. This generally defines an (hyper)surface in the contact manifold that separates two invariant basins for the evolution. In the case at hand, the surface $\mathcal{H} = 0$ is an ellipsoid, or a sphere with radius $\sqrt{2C}$ if $\gamma = 1$. Furthermore, the quadratic contact oscillator presents two equilibrium points of different nature on $\mathcal{H} = 0$: the stable north pole $N = \left(0, 0, \sqrt{2C\gamma^{-1}}\right)$ and the unstable south pole $S = \left(0, 0, -\sqrt{2C\gamma^{-1}}\right)$. In the case of geometric integrators, the explicit nature of the modified Hamiltonian allows to analyze this process and confirm that $\mathcal{H} = 0$ remains, for τ small enough, bounded and close to the original unperturbed surface. For our particular example, a direct analysis of the fixed points of the integrator allows to control analytically the equilibrium points of the numerical map: these are simply shifted to $\left(0, 0, \pm\frac{1}{2}\sqrt{8C\gamma^{-1} + \tau^2 C^2}\right)$, and maintain their stability.

In Fig. 3 we show the dynamics of trajectories starting on, outside and inside $\mathcal{H} = 0$. The deformed invariant surface is so close to the sphere that they are

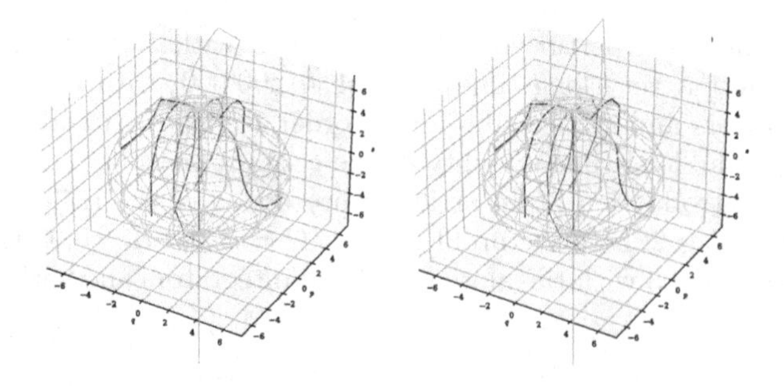

Fig. 3. The yellow sphere of radius 6 is the invariant surface $\mathcal{H} = 0$, for $\gamma = 1$ and $C = 18$. The trajectories are coloured according to their initial conditions: blue starts on the surface, purple outside and green inside. The half-line below S is an unstable invariant submanifold of the system. Left is the CHI, right the CVI, both of 2nd order. (Color figure online)

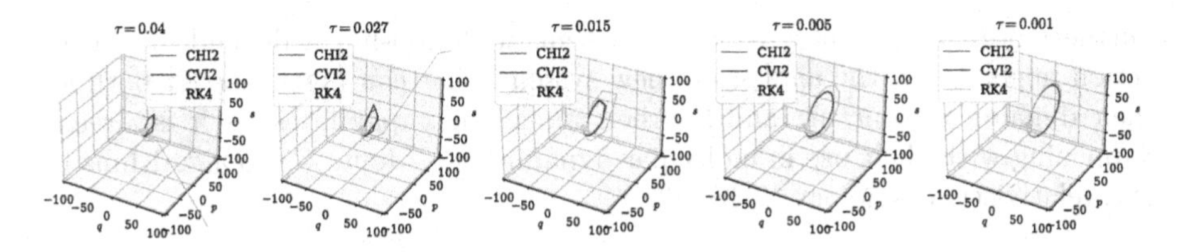

Fig. 4. Trajectory with initial conditions $(q_0, p_0, s_0) = (0, -1, -7)$ integrated with different time steps. Close to the unstable fixed point $(0, 0, -6)$, the system becomes stiff: here the Runge-Kutta 4 integrator shows a higher degree of instability compared to the contact methods.

Table 1. Integration of a contact oscillator with $\tau = 0.1$ and $t \in [0, 500]$.

Integrator type (order)	Mean time (from 10 runs)	Standard deviation
CHI (2nd)	0.0986	±0.0083
CVI (2nd)	0.0724	±0.0052
Runge-Kutta (4th)	0.1375	±0.0075
Midpoint (2nd)	0.0363	±0.0031

practically indistinguishable even from a close analysis of the three dimensional dynamical plot.

The unstable south pole provides an excellent opportunity to compare the performance of our low-order integrators against the common Runge-Kutta 4 method: initial conditions close to the unstable point are subject to fast accelerations away from the sphere, making the problem stiff. One can clearly see this in Fig. 4: for very small time steps we see that the trajectories are all converging towards the north pole, but as the time step gets larger we see the instability overcoming the Runge-Kutta integrator first and for larger τ also the CVI. For this problem, in fact, the CHI displays a remarkable stability for larger times steps. Moreover, in terms of vector field evaluations, the low-order contact integrators are comparable to an explicit midpoint integrator [19] and this reflects also in the comparable running time (see Table 1).

4 Conclusions

In this manuscript we discussed some new directions for contact integrators. Even though from a physical perspective we presented basic examples, they show some of the generic advantages provided by contact integrators. In particular, they show the remarkable stability of our low-order integrators in comparison to standard higher-order methods.

The study of contact integrators started in [6,17,19] is still at its early stages. The recent work [14] has shown that they are underpinned by a beautiful geometric construction which appears to be closely related to non-holonomic and sub-Riemannian systems and which will require further investigation on its own.

Furthermore, contact integrators for systems with non-linear dependence in the action might boost the analysis of new systems that can be of interest both for their dynamical structure and for their modelling capabilities, as it is already happening with their use in molecular dynamics [2], Monte Carlo algorithms [1] and relativistic cosmology [9,15,16].

References

1. Betancourt, M.: Adiabatic Monte Carlo. arXiv:1405.3489 (2014)
2. Bravetti, A., Tapias, D.: Thermostat algorithm for generating target ensembles. Phys. Rev. E **93**(2), 022139 (2016)
3. Bravetti, A.: Contact Hamiltonian dynamics: the concept and its use. Entropy **19**(10), 678 (2017)
4. Bravetti, A., Cruz, H., Tapias, D.: Contact Hamiltonian mechanics. Ann. Phys. NY **376**, 17–39 (2017)
5. Bravetti, A., Daza-Torres, M.L., Flores-Arguedas, H., Betancourt, M.: Optimization algorithms inspired by the geometry of dissipative systems. arXiv:1912.02928 (2019)
6. Bravetti, A., Seri, M., Vermeeren, M., Zadra, F.: Numerical integration in Celestial mechanics: a case for contact geometry. Celest. Mech. Dyn. Astr. **132**, 34 (2020)
7. Bravetti, A., Seri, M., Zadra, F.: New directions for contact integrators: support code (2021). https://doi.org/10.5281/zenodo.4751141
8. Geiges, H.: An Introduction to Contact Topology, vol. 109. Cambridge University Press, Cambridge (2008)
9. Gryb, S., Sloan, D.: When scale is surplus. arXiv:2005.03951 (2021)
10. Huang, Y., Jia, L., Sun, X., Li, Z.: Stable and unstable periodic solutions for quadratic contact Hamiltonians with a small parameter. J. Phys. Conf. Ser. **1324**(1), 012009 (2019)
11. Liu, Q., Torres, P.J., Wang, C.: Contact Hamiltonian dynamics: variational principles, invariants, completeness and periodic behavior. Ann. Phys. NY **395**, 26–44 (2018)
12. Marsden, J.E., West, M.: Discrete mechanics and variational integrators. Acta Numer. **10**, 357–514 (2001)
13. Pozsár, Á., Szücs, M., Kovács, R., Fülöp, T.: Four spacetime dimensional simulation of rheological waves in solids and the merits of thermodynamics. Entropy **22**(12), 1376 (2020)
14. Simoes, A.A., de Diego, D.M., de León, M., Valcázar, M.L.: On the geometry of discrete contact mechanics. arXiv:2003.11892 (2020)
15. Sloan, D.: Dynamical similarity. Phys. Rev. D **97**, 123541 (2018)
16. Sloan, D.: New action for cosmology. Phys. Rev. D **103**(4), 043524 (2021)
17. Vermeeren, M., Bravetti, A., Seri, M.: Contact variational integrators. J. Phys. A Math. Theor. **52**, 445206 (2019)
18. Yoshida, H.: Construction of higher order symplectic integrators. Phys. Lett. A **150**(5–7), 262–268 (1990)
19. Zadra, F., Bravetti, A., Seri, M.: Geometric numerical integration of Liénard systems via a contact Hamiltonian approach. arXiv:2005.03951 (2020)

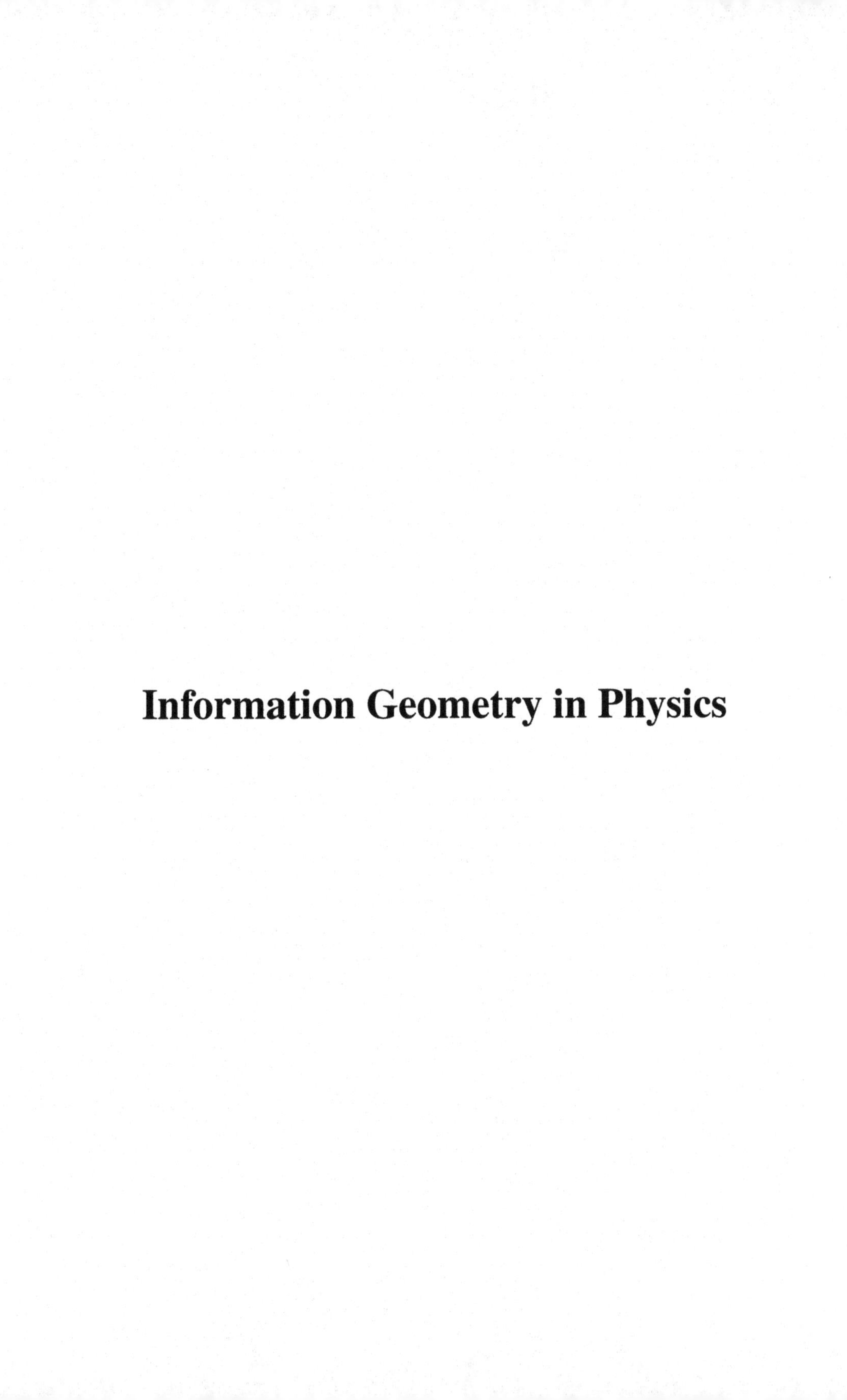

Information Geometry in Physics

Space-Time Thermo-Mechanics for a Material Continuum

Emmanuelle Rouhaud[1(✉)], Richard Kerner[2], Israa Choucair[1], Roula El Nahas[1], Alexandre Charles[1], and Benoit Panicaud[1]

[1] Université de Technologie de Troyes, Troyes, France
rouhaud@utt.fr
[2] Sorbonne Université, Paris, France

Abstract. The object of this article is to introduce a *covariant* description for a thermo-mechanical continuum. The conservation equations are written in this context and a constitutive model is derived for a reversible transformation. It is then possible to formulate a weak form of the problem to be solved for the finite transformations of a solid. A finite-element computation is next proposed using this approach for thermo-mechanical problems. The advantages of such a formulation are highlighted throughout the article.

Keywords: Space-time formalism · Covariance · Thermo-mechanics of continua

1 Introduction

The notion of frame-indifference, classically used in continuum mechanics, can be advantageously replaced with the principle of covariance introduced by Einstein in his theory of relativity. Since the transformations corresponding to changes of observer involve space and time, we explore the possibilities offered by a space-time description of thermo-mechanical problems. This approach, developed in previous articles [6,7,9] is first briefly introduced. Then, we express the first and second principles of thermodynamics in this framework and derive a covariant form for an equivalent to Clausius-Duhem Inequality. A constitutive model is deduced and a weak form of the problem is proposed. Then, space-time numerical simulations are performed.

The advantages of the approach are: i) the space-time approach enables to write covariant (= frame indifferent) conservation and constitutive equations, as opposed to Galilean relativity, which is limited to inertial frame ii) the energy conservation and the momentum equation correspond to the conservation of energy-momentum: the numerical solution verifies thus automatically the conservation of energy; iii) the numerical resolution is performed in one step, time being considered as another mesh coordinate.

Supported by Région Grand Est, FEDER, Safran.

F. Nielsen and F. Barbaresco (Eds.): GSI 2021, LNCS 12829, pp. 219–226, 2021.
https://doi.org/10.1007/978-3-030-80209-7_25

2 Framework for a Space-Time Formulation of Continuum Mechanics

Consider a space-time domain Ω of a four-dimensional differentiable manifold $\mathcal{M}$ with an ambient metric tensor $\boldsymbol{g}$ of signature $(1,-1,-1,-1)$. Let M be a point in Ω, called an event and corresponding to a representative elementary hyper-volume of the system. Let $\{x^\mu\}, \mu = 1,2,3,4$, be a local coordinate system in an open neighborhood containing M. Continuously distributed state variables and functions are introduced to represent the thermodynamic state of the system under consideration.

A motion is defined by world-lines in Ω, the flow associated to the four-velocity field $\boldsymbol{u}$, the vector tangent to the world-line at each event:

$$u^\mu = \frac{dx^\mu}{ds} \quad \text{with} \quad ds^2 = g_{\mu\nu}\, dx^\mu dx^\nu. \tag{1}$$

The principle of covariance states that the formulation of the laws of physics should be independent of observers. The most appropriate geometrical framework to describe this situation is the so-called fiber bundle structure.

A natural basis in the tangent space at M, $T_M(\mathcal{M})$, is given by the set of 4 derivations $(\partial/\partial x^\mu)$. Under a local change of the coordinate system, $x^\mu \to y^\nu(x^\mu)$, a 4D basis, defining an observer, transforms as $\frac{\partial}{\partial x^\mu} \to \left(\frac{\partial x^\kappa}{\partial y^\nu}\right)\frac{\partial}{\partial x^\kappa} = \left(\frac{\partial}{\partial y^\nu}\right)$. Consider now *all* possible local bases defined as $e_\mu = X^\nu_\mu(x)\frac{\partial}{\partial x^\nu}$. The matrix X^ν_μ has to be non-singular and therefore belongs to the group $GL(4,\mathbb{R})$. This group acts naturally on the right by the group multiplication of the matrices. Therefore, the set $\mathcal{M} \times GL(4,\mathbb{R})$ has the natural structure of a principal fibre bundle over $\mathcal{M}$, named the principle bundle of frames. Its dimension is equal to $\dim \mathcal{M} + \dim(GL(4,\mathbb{R})) = 20$.

We then define two specific observers: the *inertial* observer, an orthonormal basis for which the components of the metric tensor $\boldsymbol{g}$ are equal to the components of Minkowski's metric and a *proper* observer related to the evolving matter for which the components of the four-velocity are $(1,0,0,0)$ *for all events.*

By construction, the velocity vector, is a unit vector in the direction of time. The projection on time of a vector $\boldsymbol{\alpha}$ is thus $\alpha^\mu u_\mu$; the projection on space of $\boldsymbol{\alpha}$ is then $\overline{\alpha}^\mu = \alpha_\nu(g^{\mu\nu} - u^\mu u^\nu)$. These projection operations can be generalized to be applied to tensors of higher ranks; in this document, an over-line indicates that the quantity is a projection on space.

We will use the Lie derivative with respect to the velocity field noted $\mathscr{L}_u(.)$ and the covariant derivative with a metric connection noted $\nabla_\mu(.)$.

We define a body $\mathcal{B}$ as a 3D domain of Ω. Consider the congruence of the world-lines that cross the domain $\mathcal{B}$. This set of world-lines is called a *world-tube* and defines the motion of $\mathcal{B}$. We define the *inertial motion* of a body $\mathcal{B}$ corresponding to a motion for which the proper observer is inertial. The events of this motion are noted X^μ. Define a bijective transformation ϕ such that $x^\mu = \phi(X^\nu)$ with existing and continuous derivatives. The tangent application of ϕ is given by $dx^\mu = \dfrac{\partial x^\mu}{\partial X^\nu} dX^\nu$. The transformation ϕ is a diffeomorphism that defines another motion of $\mathcal{B}$. Define the four-dimensional Cauchy deformation tensor $\boldsymbol{b}$ as:

$$b_{\mu\nu} = g_{\alpha\beta} \frac{\partial X^\alpha}{\partial x^\mu} \frac{\partial X^\beta}{\partial x^\nu}. \tag{2}$$

Define the 4D Eulerian strain tensor $\boldsymbol{e}$ as $e_{\mu\nu} = \dfrac{1}{2}(g_{\mu\nu} - b_{\mu\nu})$ and the rate of deformation $\boldsymbol{d}$ as $d_{\mu\nu} = \dfrac{1}{2}\mathscr{L}_u(g_{\mu\nu})$; it can be verified that $\boldsymbol{d} = \overline{\boldsymbol{d}}$ and $\mathscr{L}_u(\overline{e}_{\mu\nu}) = \frac{1}{2}\mathscr{L}_u(\overline{g}_{\mu\nu})$.

3 Energy-Momentum Tensor

The energy-momentum tensor $\boldsymbol{T}$, a symmetric covariant second-rank tensor, is introduced, defined for each event [5]. The projection on space and time of this tensor leads to the decomposition:

$$T^{\mu\nu} = \mathcal{U} u^\mu u^\nu + (q^\mu u^\nu + q^\nu u^\mu) + T_\sigma^{\mu\nu}. \tag{3}$$

The scalar density $\mathcal{U}$ is the energy density, the result of the double projection of $\boldsymbol{T}$ on the time axis: $\mathcal{U} = T^{\mu\nu} u_\mu u_\nu$. The vector $\boldsymbol{q}$ is the result of the projection of $\boldsymbol{T}$ on space and time: $q^\mu = (\delta^\mu{}_\alpha - u^\mu u_\alpha) T^{\alpha\beta} u_\beta$; its interpretation is discussed in the next Section. The tensor $\boldsymbol{T_\sigma}$ is the result of the double projection of $\boldsymbol{T}$ on space: $T_\sigma^{\mu\nu} = (\delta^\mu{}_\alpha - u^\mu u_\alpha)(\delta^\nu{}_\beta - u^\nu u_\beta) T^{\alpha\beta}$, interpreted as a stress tensor. Because mass is equivalent to energy in relativistic theories [8], the tensor $\mathcal{U} u^\mu u^\nu$ can be interpreted as the mass energy-momentum tensor and we have $\mathcal{U} = \widetilde{\rho}_c \left(c^2 + e_{int}\right)$ where $\widetilde{\rho}_c$ is the rest mass density, a scalar density of weight one and e_{int} is a scalar quantity, the specific internal energy. The principle of conservation of the energy-momentum tensor states that:

$$\nabla_\nu T^{\mu\nu} = 0. \tag{4}$$

The projection on time of the equation above corresponds to the conservation of energy while the projection on space corresponds to the momentum principle [4]. Also, we consider that there is no coupling between the energy-momentum tensor and the metric tensor. The space is thus flat: without space-time curvature.

4 Space-Time Thermodynamics

We propose to construct a space-time thermodynamics in the context of the theory of irreversible processes. We therefore introduce the entropy current as a field on space-time, represented by the vector $\boldsymbol{S}$. The decomposition of this vector on time and space, $S^\nu = \widetilde{\rho}_c \eta u^\mu + \overline{S}^\mu$, leads to the definition of the quantity $\widetilde{\rho}_c \eta$, the entropy per unit volume, a scalar density where η is the specific entropy;

$\overline{S}^{\mu}$ is the projection on space of the entropy vector. The generalization of the second principle of thermodynamics is:

$$\nabla_{\mu} S^{\mu} \geq 0. \tag{5}$$

Following the 3D classical approach, it is supposed that the transfer of entropy to a 3D volume is only due to heat. We further make the hypothesis that $\overline{S}^{\mu} = (\delta^{\mu}{}_{\alpha} - u^{\mu} u_{\alpha}) T^{\alpha\beta} u_{\beta} = \frac{q^{\mu}}{\theta}$ where θ is the absolute temperature, a strictly positive scalar; q^{μ}, the projection of the energy-momentum tensor (see Eq. 3) is thus the heat flux. These definitions are equivalent to the one proposed in [3] in the case of Galilean transformations. After some derivation, the generalized second principle of thermodynamics may be written, with the use of the conservation of the rest mass ($\nabla_{\mu}(\widetilde{\rho}_c u^{\mu}) = 0$):

$$\theta \widetilde{\rho}_c \mathscr{L}_u(\eta) + \nabla_{\mu} q^{\mu} - \frac{q^{\mu}}{\theta} \nabla_{\mu} \theta \geq 0. \tag{6}$$

To construct a covariant Clausius-Duhem inequality we substract Eq. 6 from the equation of conservation of energy-momentum projected on time ($u_{\mu} \nabla_{\nu} T^{\mu\nu} = 0$) to obtain:

$$\widetilde{\rho}_c \big(\theta \mathscr{L}_u(\eta) - \mathscr{L}_u(e_{int})\big) + T^{\mu\nu}_{\sigma} \overline{d}_{\mu\nu} + q^{\mu} \big(a_{\mu} - \frac{\nabla_{\mu} \theta}{\theta}\big) \geq 0 \tag{7}$$

where $a^{\mu} = u^{\lambda} \nabla_{\lambda}(u^{\mu})$ is the acceleration. We next introduce the Helmholtz specific free energy $\Psi = e_{int} - \theta \eta$ and Eq. 7 becomes:

$$T^{\mu\nu}_{\sigma} \overline{d}_{\mu\nu} - \widetilde{\rho}_c \big(\mathscr{L}_u(\Psi) + \eta \mathscr{L}_u(\theta)\big) + q^{\mu} \big(a_{\mu} - \frac{\nabla_{\mu} \theta}{\theta}\big) \geq 0. \tag{8}$$

5 A Space-Time Constitutive Model for Thermo-Elastic Solids

A constitutive model has next to be written, describing the behavior of a material to close the thermo-mechanical problem. In this work, we limit the scope to isotropic thermo-elastic transformations and consider a specific free energy of the form [2]:

$$\Psi(\theta, \overline{I_I}, \overline{I_{II}}) = -\frac{\mathcal{C}}{2\theta_0}(\theta - \theta_0)^2 - 3\frac{\kappa\alpha}{\widetilde{\rho}_c}(\theta - \theta_0)\overline{I_I} + \frac{\lambda}{2\widetilde{\rho}_c}(\overline{I_I})^2 + \frac{\mu}{\widetilde{\rho}_c}\overline{I_{II}} \tag{9}$$

where θ_0 is a reference temperature, $\mathcal{C}$ is a constant scalar equivalent to the specific heat and where the coefficients $(\kappa\alpha)$, λ and μ are scalar densities; the quantities $\overline{I_I} = \overline{e}_{\mu\nu} \overline{g}^{\mu\nu}$ and $\overline{I_{II}} = \overline{e}^{\mu\nu} \overline{e}_{\mu\nu}$ are two invariants of the projected strain tensor $\overline{e}$. With the fact that $\mathscr{L}_u(\overline{I_I}) = (\overline{g}^{\mu\nu} - 2\overline{e}^{\mu\nu})\overline{d}_{\mu\nu}$, $\mathscr{L}_u(\overline{I_{II}}) = 2(\overline{e}^{\mu\nu} - \overline{e}^{\mu}_{\beta}\, \overline{e}^{\beta\nu} - \overline{e}^{\nu}_{\beta}\, \overline{e}^{\mu\beta})\overline{d}_{\mu\nu}$ and $\mathscr{L}_u(\mathcal{C}) = 0$, Clausius-Duhem Inequality (Eq. 8) is verified for any path thus:

$$q^{\mu} \ (\nabla_{\mu}\theta - \theta a_{\mu}) \leq 0 \tag{10}$$

$$\eta = 3\frac{\kappa\alpha}{\widetilde{\rho}_c}\overline{e}_{\mu\nu}\overline{g}^{\mu\nu} + \frac{\mathcal{C}}{\theta_0}(\theta - \theta_0) \tag{11}$$

$$T_{\sigma}^{\mu\nu} = -3\kappa\alpha(\theta-\theta_0)\overline{g}^{\mu\nu} + \lambda\overline{e}_{\gamma\beta}\overline{g}^{\gamma\beta}\overline{g}^{\mu\nu} + 2\mu\overline{e}^{\mu\nu} - 3\kappa\alpha(\theta-\theta_0)\left(\overline{e}_{\gamma\beta}\overline{g}^{\gamma\beta}\overline{g}^{\mu\nu} - 2\overline{e}^{\mu\nu}\right)$$
$$+ \lambda\left(-2\overline{e}_{\gamma\beta}\overline{g}^{\gamma\beta}\overline{e}^{\mu\nu} + \frac{\left(\overline{e}_{\gamma\beta}\overline{g}^{\gamma\beta}\right)^2}{2}\overline{g}^{\mu\nu}\right) + 2\mu\left(\overline{e}_{\gamma\beta}\overline{e}^{\gamma\beta}\overline{g}^{\mu\nu} - 2\left(\overline{e}^{\mu}_{\beta}\overline{e}^{\beta\nu}\right)^{sym}\right) \tag{12}$$

because a *reversible* case is considered. Equation 12 corresponds to a thermodynamically compatible constitutive model for isotropic thermo-elastic transformations.

6 Formulation of the Space-Time Problem for Isotropic Thermo-Elastic Transformations

We now propose a four-dimensional weak formulation for the case of the isotropic reversible material presented in Sect. 5. Define a virtual vector r^*. The conservation of the energy momentum tensor is then written:

$$\int_{\Omega} r^*_{\nu}\nabla_{\mu}T^{\mu\nu}d\Omega = 0 \quad \forall r^*. \tag{13}$$

To solve the problem, we need to find $\theta(x^{\mu})$ and $\boldsymbol{u}(x^{\mu})$ such that:

$$\int_{\partial\Omega} [T^{\mu\nu}n_{\mu}r^*_{\nu}]\, dV - \int_{\Omega} [T^{\mu\nu}\nabla_{\mu}r^*_{\nu}]\, d\Omega = 0 \quad \forall r^* \tag{14}$$

where $\boldsymbol{T}$ is given by Eq. 3, $\boldsymbol{T_{\sigma}}$ is given by Eq. 12, the internal energy is (with the definition of the free energy in Sect. 4 and Eq. 11):

$$e_{int} = \frac{\mathcal{C}}{2\theta_0}(\theta^2 - \theta_0^2) + 3\theta_0\frac{\kappa\alpha}{\widetilde{\rho}_c}(\overline{e}_{\mu\nu}\overline{g}^{\mu\nu}) + \frac{\lambda}{2\widetilde{\rho}_c}(\overline{e}_{\mu\nu}\overline{g}^{\mu\nu})^2 + \frac{\mu}{\widetilde{\rho}_c}(\overline{e}^{\mu\nu}\overline{e}_{\mu\nu}) \tag{15}$$

and we choose for the heat flux $q^{\mu} = K\overline{g}^{\mu\nu}(\nabla_{\nu}\theta - \theta a_{\nu})$ where K a constant scalar, the thermal conductivity [4]. The Dirichlet boundary conditions are given by $\theta(x^{\mu}_D)$ and $\boldsymbol{u}(x^{\mu}_D)$ on $\partial\Omega_D$; the Neumann boundary conditions are given by $T^{\mu\nu}(x^{\mu}_N)n_{\nu}(x^{\mu}_N) = T^{\mu}_N$ on $\partial\Omega_N$.

The equations above correspond to a non-linear covariant weak formulation of a problem for the finite transformations of an isotropic thermo-elastic solid. When solved, we obtain the field of temperature and velocity that verify both the conservation of energy and the principle of momentum.

7 Space-Time Finite Element Computation

To illustrate the approach, we propose here a space-time finite-element resolution of a thermo-mechanical problem. The system under consideration is 2D plate made of Aluminum with a hole in its center; the material is thermo-elastic. A displacement in the y direction is imposed on two opposite edges of the plate and an increase in temperature is imposed on the edge of the hole (see Fig. 1). Both the displacement amplitude and the temperature increase linearly with time. The corresponding finite-element discretization is presented in Fig. 1. The space-time volume is constructed with the 2D plate and the third direction represents time. Note that the mesh is not regular in time: it is refined where the gradients are expected to be important (both in time and space) and coarse where the solution is expected to be smooth. The resolution is performed in one step, for space and time: there is no need for a finite discretization in time. We use Fenics [1] for the resolution.

The results of the finite-element computation are presented in Fig. 2 and 3. The solution gives the evolution of the temperature and the displacements in space and time. We present here the internal energy and the yy component of the stress, two components of the energy-momentum tensor; a comparison is proposed with a solution obtained without thermal loading.

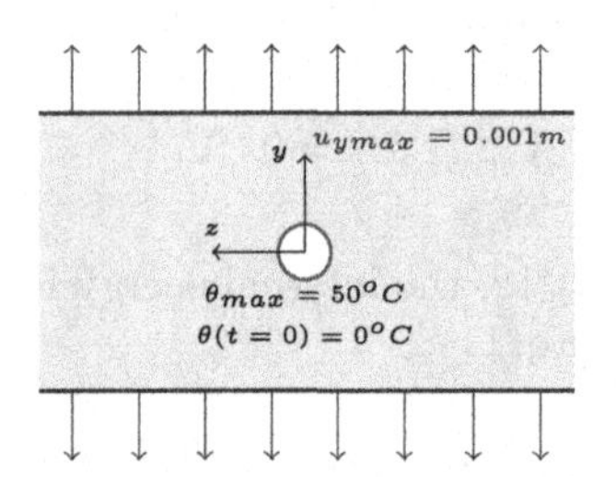

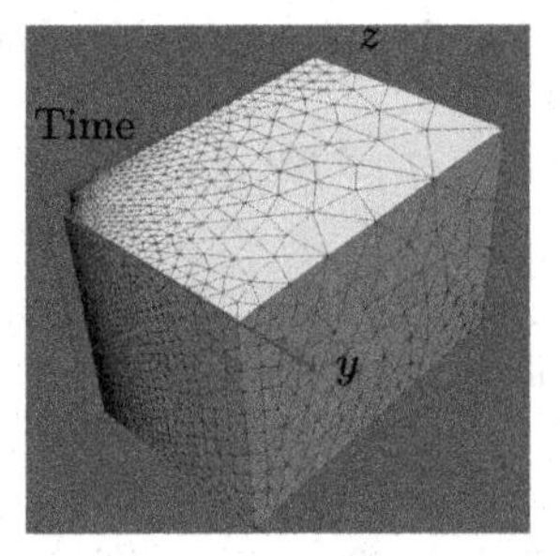

Fig. 1. Left: geometry of the plate (2 m high and 4 m wide), the hole has a radius of 10 cm; a displacement is imposed in the y direction and an increase in temperature is imposed on the edge of the hole. Right: mesh of the system; only one quarter of the plate is discretized using the symmetries of the problem; the 2D plate is extruded in the time direction to construct a space-time volume; the final time is 10 000 s.

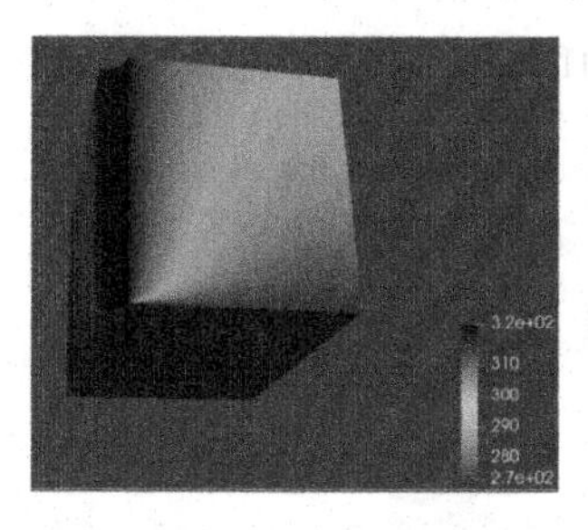

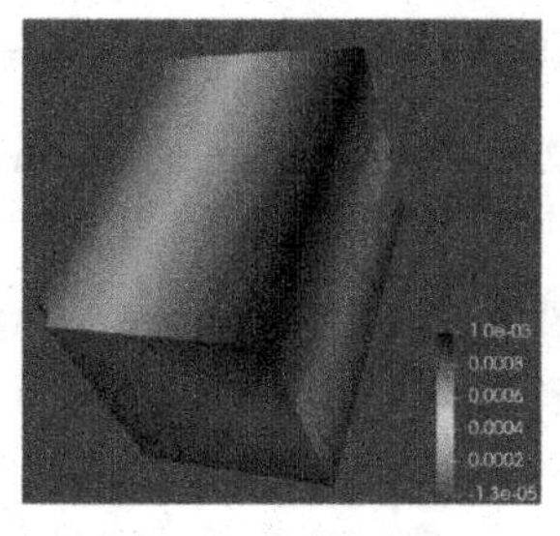

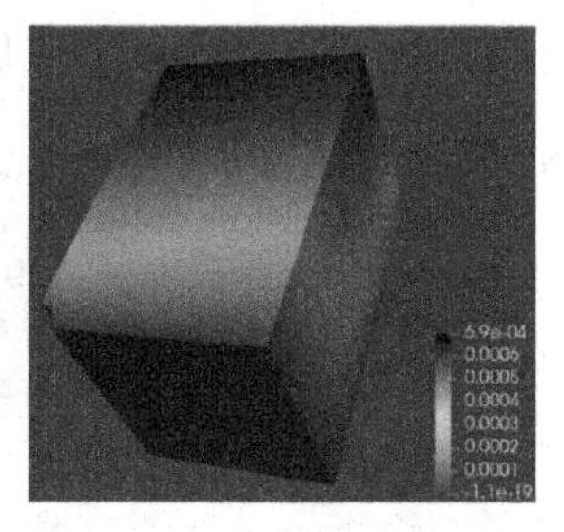

Fig. 2. Space-time solution fields for, from left to right, the temperature and displacements (y and z direction) in the plate.

Fig. 3. Space-time solution fields for components of the energy-momentum tensor in the plate; from left to right: the internal energy and the yy component of the stress. The figure on the right presents the yy component of the stress for a computation for which there is no temperature variation imposed on the edge of the hole.

References

1. Alnæs, M., Blechta, J., et al.: The FEniCS project version 1.5. Arch. Numer. Softw. **3**(100), 9–23 (2015)
2. Chrysochoos, A.: Effets de couplage et effets dissipatifs accompagnant la déformation des matériaux solides, September 2018. https://hal.archives-ouvertes.fr/cel-02057720, lecture - Cours donné dans le cadre d'une école d'été de mécanique théorique
3. De Saxcé, G., Vallée, C.: Bargmann group, momentum tensor and Galilean invariance of Clausius-Duhem inequality. Int. J. Eng. Sci. **50**(1), 216–232 (2011)
4. Grot, R.A., Eringen, A.: Relativistic continuum mechanics part I-mechanics and thermodynamics. Int. J. Eng. Sci. **4**(6), 611–638 (1966)
5. Misner, C.W., Thorne, K.S., Wheeler, J.A., et al.: Gravitation. Macmillan (1973)
6. Panicaud, B., Rouhaud, E.: A frame-indifferent model for a thermo-elastic material beyond the three-dimensional Eulerian and Lagrangian descriptions. Continuum Mech. Thermodyn. **26**(1), 79–93 (2014)
7. Rouhaud, E., Panicaud, B., Kerner, R.: Canonical frame-indifferent transport operators with the four-dimensional formalism of differential geometry. Comput. Mater. Sci. **77**, 120–130 (2013)

8. Synge, J.L.: Relativity: The General Theory. North-Holland Publishing Company Amsterdam (1960)
9. Wang, M., Panicaud, B., et al.: Incremental constitutive models for elastoplastic materials undergoing finite deformations by using a four-dimensional formalism. Int. J. Eng. Sci. **106**, 199–219 (2016)

Entropic Dynamics Yields Reciprocal Relations

Pedro Pessoa(✉)

Department of Physics, University at Albany (SUNY), Albany, NY 12222, USA
ppessoa@albany.edu

Abstract. Entropic dynamics is a framework for defining dynamical systems that is aligned with the principles of information theory. In an entropic dynamics model for motion on a statistical manifold, we find that the rate of changes for expected values is linear with respect to the gradient of entropy with reciprocal (symmetric) coefficients. Reciprocity principles have been useful in physics since Onsager. Here we show how the entropic dynamics reciprocity is a consequence of the information geometric structure of the exponential family, hence it is a general property that can be extended to a broader class of dynamical models.

Keywords: Exponential family · Information theory · Information geometry · Entropic dynamics

1 Introduction

Entropic dynamics (EntDyn) is a method to characterize dynamical processes that is aligned to the principles of information theory taking into account the differential geometric structure of probability distributions given by information geometry (IG) [2,3,6,16,17]. EntDyn has been successful in deriving dynamical processes in fundamental physics [5,7,11], but has also been expanded to other fields such as renormalization groups [20], finance [1], machine learning [8] and network science [10] some advances have also been made towards EntDyn applications in ecology [21]. Particularly in [22] EntDyn is presented as a general method for dynamical systems of probability distributions in the exponential family – also referred to as Gibbs or canonical distributions. Under a particular choice of sufficient statistics, the space of exponential families is parametrized by the expected values of these sufficient statistics and the Riemmanian metric obtained from IG is the covariance matrix between the sufficient statistics [6,17]. The EntDyn obtained in [22] is a diffusion process in the exponential family statistical manifold.

Here we focus on a particular consequence of the EntDyn presented in [22]: the geometrical reciprocity in the linear regime. Meaning, as the dynamics evolves, the time derivative of the expected value of parameters is proportional

Author was financed in part by CNPq (scholarship GDE 249934/2013-2).

F. Nielsen and F. Barbaresco (Eds.): GSI 2021, LNCS 12829, pp. 227–234, 2021.
https://doi.org/10.1007/978-3-030-80209-7_26

to the gradient of entropy with reciprocal (symmetric) values. In the present article, we show how this reciprocity is a consequence of IG. This relates to what is known, in physics, as Onsager reciprocity principle (ORP) [18] – also referred to as Onsager reciprocal relations – for non-equilibrium statistical mechanics. A brief review of ORP is presented in Sect. 4 below. Moreover, a diffusion process that reduces to a reciprocal relation in the linear regime is known in physics as the Onsager-Machlup process [19]. Unlike in EntDyn, the reciprocity found by Onsager is a consequence of the time reversal symmetry in an underlying dynamics, hence ORP is not trivially generalizable to dynamical processes beyond physics.

Gibbs distributions arise in equilibrium statistical mechanics as the maximum entropy distributions obtained when one chooses the conserved quantities in the Hamiltonian dynamics as sufficient statistics [12]. The generality of maximum entropy and the concept of sufficiency ensures that the exponential family is not limited to thermodynamics. Since both the information theory and the IG approach guiding EntDyn are consequences of fundamental concepts in probability theory, the reciprocity presented here is applicable for dynamical systems beyond physics. Just like ORP found applicability in non-equilibrium statistical mechanics[1], awarding Onsager with a Nobel prize in 1968, it is remarkable that a similar principle arises from a geometric framework.

The layout of the present article is as follows: In the following section we will establish notation and give a brief review on exponential family/Gibbs distributions and their geometric structure endowed by IG. In Sect. 3 we will review the assumptions and results in [22], obtaining the transition probabilities and expressing EntDyn as a diffusion process on the statistical manifold. In Sect. 4 we comment on the similarities and differences in the reciprocity emerging from EntDyn and the one found by Onsager.

2 Background - Exponential Family

We start by defining the exponential family of distributions ρ for a given set $\mathcal{X}$ with elements $x \in \mathcal{X}$:

$$\rho(x|\lambda) \doteq \frac{q(x)}{Z(\lambda)} \exp\left[-\sum_{i=1}^{n} \lambda_i a^i(x)\right], \tag{1}$$

which can be obtained either as a result of maximization of entropy (as in [22]) or as a consequence of sufficiency (as in [17]). In (1) the functions a^i – indexed up to a finite number n – are the sufficient statistics, the set of real values $\lambda = \{\lambda_i\}$ are the parameters for the exponential family (or the Lagrange multipliers arising

[1] It is also relevant to say that information theory ideas have been applied to nonequilibrum statistical mechanics (see [13] and posterior literature on maximum caliber e.g. [23]) which, unlike EntDyn, are approaches based on Hamiltonian dynamics.

from maximization of entropy), $q(x)$ is an underlying measure (or the maximum entropy prior) and $Z(\lambda)$ is a normalization factor computed as

$$Z(\lambda) = \int \mathrm{d}x \; q(x) \exp\left[-\lambda_i a^i(x)\right]. \tag{2}$$

Here and for the remainder of this article we use the Einstein's summation notation, $A^i B_i = \sum_{i=1}^{n} A^i B_i$. Details on how many well-known distributions can be put in the exponential form (1) are found in [17].

The expected values of sufficient statistics $a^i(x)$ in the exponential family can be found using

$$A^i \doteq \langle a^i(x) \rangle = -\frac{\partial}{\partial \lambda_i} \log Z(\lambda). \tag{3}$$

Moreover, the entropy for ρ in (1) with a prior given by $q(x)$ is

$$\begin{aligned} S(A) \doteq S[\rho|q] &= -\int dx \; \rho(x|\lambda(A)) \log \frac{\rho(x|\lambda(A))}{q(x)} \\ &= \lambda_i(A) A^i + \log Z(\lambda(A)). \end{aligned} \tag{4}$$

Thus, the entropy of the exponential family, as a function of the expected values $A = \{A^i\}$, is the Legendre transform of $-\log Z$. It also follows that $\lambda_i = \frac{\partial S}{\partial A^i}$. Hence, each set of expected values A corresponds to a single λ thus we will use the expected values A as coordinates for the space of probability distributions $\rho(x|\lambda(A))$ written, for simplicity, as just $\rho(x|A)$. IG consists of assigning a Riemmanian geometry structure to the space of probability distributions. Meaning, if we have a family of probability distributions parametrized by a set of coordinates, $P(x|\theta)$ where, $\theta = \{\theta^i\}$, the distance $\mathrm{d}\ell$ between the neighbouring distributions $P(x|\theta + \mathrm{d}\theta)$ and $P(x|\theta)$ is given by $\mathrm{d}\ell^2 = g_{ij} \mathrm{d}\theta^i \mathrm{d}\theta^j$ where g_{ij} is the Riemmanian metric. For probability distributions the only metric that is appropriate[2] is the Fisher-Rao information metric (FRIM)

$$g_{ij} = \int \mathrm{d}x \; P(x|\theta) \frac{\partial \log P(x|\theta)}{\partial \theta^i} \frac{\partial \log P(x|\theta)}{\partial \theta^j}. \tag{5}$$

For the exponential family – $\theta = A$ and $P(x|\theta) = \rho(x|A)$ in (5) – FRIM can be expressed in the useful form

$$g_{ij} = -\frac{\partial^2 S}{\partial A^i \partial A^j}. \tag{6}$$

Having the exponential family and its IG established, we will explain the EntDyn obtained from it.

[2] The formal argument is that, as proven by Čencov (Ченцов, also written in Latin alphabet as Chentsov or Cencov) [9], FRIM is the only metric that is invariant under Markov embeddings, see also [3]. Therefore, it is the only metric leading to a geometry consistent with the grouping property of probability distributions.

3 Entropic Dynamics on the Exponential Family Manifold

The dynamical process in EntDyn consists of evolving the probability distribution for a system in an instant $\rho(x|A)$ to the distribution describing a later instant $\rho(x'|A')$. This is done by assigning a probability $P(A)$ from which the system's state probabilities can be obtained as

$$P(x) = \int \mathrm{d}A \; P(A)\rho(x|A), \tag{7}$$

which means that the conditional probability of x given A has to be of the exponential form defined in (1). Under EntDyn, we find the transition probabilities $P(A'|A)$ through the methods of information theory – maximizing an entropy. As we will see below, the constraint presented establishes that the motion does not leave the manifold of exponential distributions.

3.1 Obtaining the Transition Probability

The entropy we opt to maximize must be so that the dynamical process accounts for both the change in state of the system – evolving from x to x' – and the change in the underlying distribution – from $\rho(x|A)$ to $\rho(x'|A')$. Hence, it is the entropy for the joint transition, $P(x', A'|x, A)$, given by

$$\mathsf{S}[P|Q] = -\int \mathrm{d}x'\mathrm{d}A' \; P(x', A'|x, A) \log\left(\frac{P(x', A'|x, A)}{Q(x', A'|x, A)}\right), \tag{8}$$

where Q is the prior which we will determine below. Since the maximization of S will define the dynamics, we will call it the dynamical entropy. One should not confuse the entropy for the exponential distribution, S in (4), with the dynamical entropy, S in (8).

The prior for the maximization of (8) must implement a continuous motion – meaning implementing short steps so that A' converges in probability to A. As explained in [22], the least informative prior that does so is

$$Q(x', A'|x, A) \propto g^{1/2}(A') \exp\left(-\frac{1}{2\tau} g_{ij} \Delta A^i \Delta A^j\right), \tag{9}$$

where $\Delta A^i = A'^i - A^i$ and $g = \det g_{ij}$. The parameter τ establishes continuity – ΔA converges to 0 in probability when $\tau \to 0$. Later we will see how τ takes the role of time.

The constraint for the maximization of (8) that implements a motion that does not leave the exponential manifold is

$$P(x', A'|x, A) = P(x'|x, A, A')P(A'|x, A) = \rho(x'|A')P(A'|x, A). \tag{10}$$

which means that the posterior distribution for x' has to be conditioned only on A' and to be in the exponential form (1).

The transition that maximizes (8) with the prior (9) and under (10) is

$$P(x',A'|x,A) \propto \rho(x|A')e^{-S(A)}g^{1/2}(A')\exp\left(-\frac{1}{2\tau}g_{ij}\Delta A^i\Delta A^j\right). \tag{11}$$

In order to obtain the transition probability from A to A', we must identify that from conditional probabilities in (10) we have $P(A'|x,A) = \frac{P(x,A'|x,A)}{\rho(x'|A')}$ then apply a linear approximation of $S(A')$ near $S(A)$. As explained in [22] this results in

$$P(A'|A) = \frac{g^{1/2}(A')}{\mathcal{Z}(A)}\exp\left(\frac{\partial S}{\partial A^i}\Delta A^i - \frac{1}{2\tau}g_{ij}\Delta A^i\Delta A^j\right), \tag{12}$$

where $\mathcal{Z}(A)$ is a normalization factor. In the following subsection, we will explain how this transition probability results in a diffusion process on the statistical manifold.

3.2 Entropic Dynamics as a Diffusion Process

Since according to (9) $\tau \to 0$ for short steps, we will compute the moments for ΔA in (12) up to order τ obtaining, as in [22],

$$\langle \Delta A^i \rangle = \tau g^{ij}\frac{\partial S}{\partial A^j} - \frac{\tau}{2}\Gamma^i + o(\tau), \tag{13}$$

where the upper indexes in g to represent the inverse matrix, $g_{ij}g^{jk} = \delta^k_i$, $\Gamma^i = \Gamma^i_{jk}g^{jk}$ and Γ^i_{jk} are the Christoffel symbols. In [22] we also show that the second and third moments of ΔA are

$$\langle \Delta A^i\Delta A^j \rangle = \tau g^{ij} + o(\tau) \quad \text{and} \quad \langle \Delta A^i\Delta A^j\Delta A^k \rangle = o(\tau). \tag{14}$$

The rules for a change from A to A' must be the same for a posterior change from A' to A''. To keep track of the accumulation of changes, we will design the transitions as a Markovian process. Time is introduced so that it parametrizes different instants, meaning $P(A) = P_t(A)$ and $P(A') = P_{t'}(A)$ for $t' > t$. Two important consequences of this dynamical design are: (i) time is defined as an emerging parameter for the motion – the system is its own clock – and (ii) the dynamic is Markovian by design, rather than proven to be a Markov chain in an existing time. The way to assign the time duration, $\Delta t = t' - t$, so that motion looks the simplest is in terms of its fluctuations, meaning $\Delta t \doteq \tau \propto g_{ij}\Delta A^i\Delta A^j$.

Using τ as the time duration, the moments calculated in (13) and (14) are what is known as smooth diffusion (see e.g. [15]). Therefore the dynamics can be written as a Fokker-Planck (diffusion) equation which written in invariant form is (see [22])

$$\frac{\partial p}{\partial t} = -\frac{1}{g^{1/2}}\frac{\partial}{\partial A^i}\left(g^{1/2}pv^i\right), \quad \text{where} \quad v^i = g^{ij}\frac{\partial S}{\partial A^j} - \frac{g^{ij}}{2p}\frac{\partial p}{\partial A^j}, \tag{15}$$

and p is the invariant probability density $p(A) \doteq \frac{P(A)}{\sqrt{g(A)}}$. The first term for v^i in (15) is the diffusion drift velocity – related to the average motion in the diffusion process – while the second term is the osmotic velocity – guided by the differences in probability density. The drift velocity is linear with respect to the gradient of entropy, $\frac{\partial S}{\partial A^i}$, with symmetric proportionality factors – since the metric tensor is symmetric – hence, we say that the drift velocity is reciprocal to the gradient of entropy. In the following section we will make further comments on reciprocity in both EntDyn and ORP.

4 Reciprocity

In Onsager's approach to nonequilibrim statistical mechanics [18], we suppose that a thermodynamical system is fully described by a set of parameters ξ evolving in time $\xi(t) = \{\xi^i(t)\}$. ORP states that, near an equilibrium point ξ_0, the rate of change for ξ is reciprocal to the entropy gradient, meaning

$$\frac{\mathrm{d}\xi^i}{\mathrm{d}t} = \gamma^{ij}\frac{\partial S}{\partial \xi^j} \tag{16}$$

where $\gamma^{ij} = \gamma^{ji}$ and S is the thermodynamical entropy – in mathematical terms it is the Gibbs distribution entropy (4) written in terms of ξ.

In ORP, the equilibrium value, ξ_0, has to both be a local maxima of entropy and a fixed point for the dynamics of ξ. That means

$$\left.\frac{\mathrm{d}\xi^i}{\mathrm{d}t}\right|_{\xi=\xi_0} = 0 \quad \text{and} \quad \left.\frac{\partial S}{\partial \xi^i}\right|_{\xi=\xi_0} = 0. \tag{17}$$

If we make a linear approximation for both S and $\frac{\mathrm{d}\xi^i}{\mathrm{d}t}$ around ξ_0 we obtain

$$\frac{\mathrm{d}\xi^i}{\mathrm{d}t} = -L^i_k\beta^{kj}\frac{\partial S}{\partial \xi^j}, \tag{18}$$

where

$$L^i_j \doteq \left.\frac{\partial}{\partial \xi^j}\frac{\mathrm{d}\xi^i}{\mathrm{d}t}\right|_{\xi=\xi_0} \quad \text{and} \quad \beta_{ij} \doteq -\left.\frac{\partial^2 S}{\partial \xi^i \partial \xi^j}\right|_{\xi=\xi_0}. \tag{19}$$

which, when compared to (16), yields $\gamma^{ij} = -L^i_k\beta^{kj}$. Hence proving ORP equates to proving $L^i_j = L^j_i$, since β is already symmetric[3] for being defined as a second derivative (19). The usual proof of the symmetry of L (see e.g. [14]) consists of assuming that the fluctuations of ξ are symmetric through time reversal, meaning $\langle \xi^i(0)\xi^j(t)\rangle = \langle \xi^i(t)\xi^j(0)\rangle$. Unlike the formalism derived by EntDyn, this cannot be expected to be general for dynamical systems. For further discussion on how the macrostates dynamics arise from Hamiltonian dynamics see e.g. [4].

[3] Note that although in (6) β matches FRIM for $\xi = A$, one cannot say that β is FRIM in general since (6) does not transform covariantly for an arbitrary change of coordinates $A \to \xi(A)$.

A direct comparison with (13) means that when the system is described by the expected value parameters, $\xi = A$, EntDyn reduces, in the first moment, to a motion that is reciprocal to the gradient of entropy. The second term for $\langle \Delta A^i \rangle$ in (13) is a correction for motion that probes the manifold's curvature. Also, as discussed, in (15) we see that the diffusion process resulting from EntDyn has drift velocity reciprocal to the gradient of entropy, analogous to the Onsager-Machlup process [19].

Although it would make a trivial connection between EntDyn and nonequilibrium statistical mechanics, one might find the rates of change described by it to be a physically unrealistic model since substituting γ^{ij} by g^{ij} for $\xi = A$ in (16) and comparing to (18) implies:

$$L^i_j = -\delta^i_j \quad \rightarrow \quad \frac{d\xi^i}{dt} \propto (\xi^i_o - \xi^i). \tag{20}$$

The right hand side above is obtained by substituting the left hand side into the definition of L in (19) and integration. In thermodynamics, one can not expect macroscopic parameters, such as internal energy and number of particles, to evolve independently.

5 Conclusion and Perspectives

An entropic framework for dynamics on the exponential family manifold was presented here. The change in the expected values in (13) is reciprocal to the gradient of entropy in the manifold, while (15) gives a drift velocity that is also reciprocal to the gradient of entropy. The proportionality factors in this reciprocal motion is given by the inverse FRIM, therefore reciprocity is a consequence of IG.

From (20), one should not consider the dynamics developed here as a method for nonequilibrum statistical mechanics. The reciprocity found here does not necessarily substitute ORP. That said, EntDyn offers a systematic method to find dynamics aligned with information theory, while Onsager's approach is based solely on calculus considerations around a supposed fixed point. Another way to present it is to say that ORP is based on an understanding of thermodynamics guided by Hamiltonian dynamics, while the EntDyn developed here is inspired by the generality of exponential families, not relying on an underlying Hamiltonian dynamics. Both arrive at reciprocal relations.

Acknowledgments. I thank N. Caticha, A. Caticha, F. Xavier Costa, B. Arderucio Costa, and M. Rikard for insightful discussions in the development of this article. I was financed in part by CNPq – Conselho Nacional de Desenvolvimento Científico e Tecnológico– (scholarship GDE 249934/2013-2).

References

1. Abedi, M., Bartolomeo, D.: Entropic dynamics of exchange rates and options. Entropy **21**(6), 586 (2019). https://doi.org/10.3390/e21060586

2. Amari, S.: Information Geometry and Its Applications. AMS, vol. 194. Springer, Tokyo (2016). https://doi.org/10.1007/978-4-431-55978-8
3. Ay, N., Jost, J., Lê, H.V., Schwachhöfer, L.: Information Geometry. EMG-FASMSM, vol. 64. Springer, Cham (2017). https://doi.org/10.1007/978-3-319-56478-4
4. Balian, R.: Introduction a la thermodynamique hors équilibre. In: Ecole e2phy de Bordeaux (2003). http://e2phy.in2p3.fr/2003/actesBalian.pdf
5. Caticha, A.: Entropic dynamics, time and quantum theory. J. Phys. A Math. Theoret. **44**(22), 225303 (2011). https://doi.org/10.1088/1751-8113/44/22/225303
6. Caticha, A.: The basics of information geometry. In: AIP Conference Proceedings. vol. 1641, pp. 15–26. American Institute of Physics (2015). https://doi.org/10.1063/1.4905960
7. Caticha, A.: Entropic dynamics: quantum mechanics from entropy and information geometry. Annalen der Physik **531**(3), 1700408 (2018). https://doi.org/10.1002/andp.201700408
8. Caticha, N.: Entropic dynamics in neural networks, the renormalization group and the Hamilton-Jacobi-Bellman equation. Entropy **22**(5), 587 (2020). https://doi.org/10.3390/e22050587
9. Cencov, N.N.: Statistical decision rules and optimal inference. Transl. Math. Monographs, vol. 53, Amer. Math. Soc., Providence-RI (1981)
10. Costa, F.X., Pessoa, P.: Entropic dynamics of networks. Northeast J. Complex Syst. **3**(1), 5 (2021). https://doi.org/10.22191/nejcs/vol3/iss1/5
11. Ipek, S., Abedi, M., Caticha, A.: Entropic dynamics: reconstructing quantum field theory in curved space-time. Class. Quantum Gravity **36**(20), 205013 (2019). https://doi.org/10.1088/1361-6382/ab436c
12. Jaynes, E.T.: Information theory and statistical mechanics: I. Phys. Rev. **106**(4), 620–630 (1957). https://doi.org/10.1103/PhysRev.106.620
13. Jaynes, E.T.: Where do we stand on maximum entropy? In: Levine, R.D., Tribus, M. (eds.) The Maximum Entropy Principle. MIT Press (1979). https://doi.org/10.1007/978-94-009-6581-2_10
14. Landau, L., , Lifshitz, E.: Statistical Physics – Course of Theoretical Physics, vol. 5. Butterworth-Heinemann (1980)
15. Nelson, E.: Quantum Fluctuations. Princeton University Press (1985)
16. Nielsen, F.: An elementary introduction to information geometry. Entropy **22**(10), 1100 (2020). https://doi.org/10.3390/e22101100
17. Nielsen, F., Garcia, V.: Statistical exponential families: a digest with flash cards (2011). https://arxiv.org/abs/0911.4863
18. Onsager, L.: Reciprocal relations in irreversible processes. I. Phys. Rev. **37**(4), 405–426 (1931). https://doi.org/10.1103/PhysRev.37.405
19. Onsager, L., Machlup, S.: Fluctuations and irreversible processes. Phys. Rev. **91**(6), 1505–1512 (1953). https://doi.org/10.1103/physrev.91.1505
20. Pessoa, P., Caticha, A.: Exact renormalization groups as a form of entropic dynamics. Entropy **20**(1), 25 (2018). https://doi.org/10.3390/e20010025
21. Pessoa, P.: Legendre transformation and information geometry for the maximum entropy theory of ecology (2021). Under review. https://arxiv.org/abs/2103.11230
22. Pessoa, P., Costa, F.X., Caticha, A.: Entropic dynamics on Gibbs statistical manifolds. Entropy **23**(5), 494 (2021). https://doi.org/10.3390/e23050494
23. Pressé, S., Ghosh, K., Lee, J., Dill, K.A.: Principles of maximum entropy and maximum caliber in statistical physics. Rev. Mod. Phys. **85**(3), 1115–1141 (2013). https://doi.org/10.1103/revmodphys.85.1115

Lie Group Machine Learning

Gibbs States on Symplectic Manifolds with Symmetries

Charles-Michel Marle[1,2(✉)]

[1] Université Pierre et Marie Curie, Paris, France
charles-michel.marle@math.cnrs.fr
[2] Sorbonne Université, Paris, France

Abstract. In this paper, (M, Ω) is a connected symplectic manifold on which a Lie group G acts by a Hamitonian action, with a moment map $J : M \to \mathfrak{g}^*$. A short reminder of the definitions of statistical states, Gibbs states, entropy, generalized temperatures and associated thermodynamic functions is first given. Then several examples of such Gibbs states are presented, together with the associated thermodynamic functions. Examples are given too of symplectic manifolds on which a Lie group acts by a Hamiltonian action on which no Gibbs state built with a moment map of this action can exist. Most of these examples are obtained with the use of a remarkable isomorphism of a fully oriented three-dimensional Euclidean or pseudo-Euclidean vector space onto the Lie algebra of its Lie group of symmetries.

Keywords: Gibbs states · Symplectic manifolds · Moment map · Coadjoint orbits · Poincaré disk · Poincaré half-plane

1 Introduction

The French mathematician and physicist Jean-Marie Souriau (1922–2012) considered, first in [14], then in his book [15], Gibbs states on a symplectic manifold built with the moment map of the Hamiltonian action of a Lie group, and the associated thermodynamic functions. In several later papers [16–18], he developed these concepts and considered their possible applications in Physics and in Cosmology. More recently, under the name *Souriau's Lie groups thermodynamics*, these Gibbs states and the associated thermodynamic functions were considered by several scientists, notably by Frédéric Barbaresco, for their possible applications in Geometric Information Theory, Deep Learning and Machine Learning [1–5,12,13].

Long before the works of Souriau, in his book [6] published in 1902 (formula (98), chapter IV, page 43), the American scientist Josiah Willard Gibbs (1839–1903) clearly described Gibbs states in which the components of the total angular

Not supported by any organization.
C.-M. Marle—Retired from Université Pierre et Marie Curie, today Sorbonne Université.

F. Nielsen and F. Barbaresco (Eds.): GSI 2021, LNCS 12829, pp. 237–244, 2021.
https://doi.org/10.1007/978-3-030-80209-7_27

momentum (which are the components of the moment map of the action of the group of rotations on the phase space of the considered system) appear, on the same footing as the Hamiltonian. He did not discuss the properties of these Gibbs states as thoroughly as Souriau did much later, but presented an example of mechanical system, made of point particles contained in a spherical shell, for which such a Gibbs state appears as a state of thermodynamic equilibrium. He even considered Gibbs states involving conserved quantities more general than those associated with the Hamiltonian action of a Lie group. Souriau's main merits do not lie, in my opinion, in the consideration of these Gibbs states, but rather in the use of the *manifold of motions* of a Hamiltonian system instead of the use of its *phase space*, and his introduction, under the name of *Maxwell's principle*, of the idea that a symplectic structure should exist on the manifold of motions of systems encountered as well in classical Mechanics as in relativistic Physics.

For physicists, Gibbs states are states of thermodynamic equilibrium of a physical system, mathematically described by the flow of a Hamiltonian vector field on a symplectic manifold. Such Gibbs states are built with the Hamiltonian of the considered system, infinitesimal generator of the action, on the symplectic manifold, of the one dimensional group of translations in time. They are indexed by a single real parameter, interpreted in Physics as the inverse of the absolute temperature, which must lie in some open interval, corresponding to the range of temperatures for which the considered physical system exists. Gibbs states on a symplectic manifold built with the moment map of the Hamiltonian action of a Lie group are indexed by an element of the Lie algebra of this group, which must lie in some open subset of this Lie algebra, for which integrals which appear in the definition of thermodynamic functions are normally convergent. They were already presented in several papers [3,4,7], sometimes with full proofs of all the stated results [9]. Their definition and main properties are briefly presented in Sect. 2. Then in Sect. 3 examples of such Gibbs states are presented, together with examples of symplectic manifolds on which a Lie group acts by a Hamiltonian action, on which no Gibbs state can exist. More details about some of these examples can be found in [10].

2 Gibbs States Built with the Moment Map of the Hamiltonian Action of a Lie Group

In this section (M, ω) is a connected, $2n$-dimensional symplectic manifold and G is a Lie group, whose Lie algebra is denoted by $\mathfrak{g}$, which acts on the left on this manifold by a Hamiltonian action Φ. We denote by λ_ω the Liouville measure on the Borel σ-algebra of M. Let $J : M \to \mathfrak{g}^*$ be a moment map of this action (see for example [8,9,11] for the definitions of Hamiltonian actions, moment maps, Liouville measure, ...).

Definitions. 1. A *statistical state* on M is a probability measure μ on the Borel σ-algebra of M. The statistical state μ is said to be *continuous* (respectively, *smooth*) when it can be written as $\mu = \rho\lambda_\omega$, where ρ is a continuous function (respectively, a smooth function) defined on M, called the *probability density* (or simply the *density*) of μ with respect to λ_ω.

2. The *entropy* of a continuous statistical state of density ρ with respect to λ_ω is the quantity

$$s(\rho) = \int_M \log\left(\frac{1}{\rho(x)}\right)\rho(x)\lambda_\omega(\mathrm{d}x) = -\int_M \log\bigl(\rho(x)\bigr)\rho(x)\lambda_\omega(\mathrm{d}x). \tag{1}$$

with the conventions $0\log 0 = 0$ and $s(\rho) = -\infty$ when the integral in the right hand side of the above equality is divergent. The map $\rho \mapsto s(\rho)$ on M is called the *entropy functional.*

3. For each $\boldsymbol{\beta} \in \mathfrak{g}$, the integral in the right hand side of the equality

$$P(\boldsymbol{\beta}) = \int_M \exp\bigl(-\langle J(x), \boldsymbol{\beta}\rangle\bigr)\lambda_\omega(\mathrm{d}x) \tag{2}$$

is said to be *normally convergent* when there exists an open neighbourhood U of $\boldsymbol{\beta}$ in $\mathfrak{g}$ and a function $f : M \to \mathbb{R}^+$, integrable on M with respect to the Liouville measure λ_ω, such that for any $\boldsymbol{\beta}' \in U$, the inequality $\exp\bigl(-\langle J(x), \boldsymbol{\beta}'\rangle\bigr) \leq f(x)$ is satisfied for all $x \in M$. In this case $\boldsymbol{\beta}$ is called a *generalized temperature.* When the set Ω of generalized temperatures is not empty, the function $P : \Omega \to \mathbb{R}$ defined by the above equality is called the *partition function.*

4. Let $\boldsymbol{\beta} \in \Omega$ be a generalized temperature. The statistical state on M whose probability density, with respect to the Liouville measure λ_ω, is expressed as

$$\rho_{\boldsymbol{\beta}}(x) = \frac{1}{P(\boldsymbol{\beta})}\exp\bigl(-\langle J(x), \boldsymbol{\beta}\rangle\bigr), \quad x \in M, \tag{3}$$

is called the *Gibbs state* associated to (or indexed by) $\boldsymbol{\beta}$.

The reader is referred to [10, 11] for comments about the links between the definition of entropy of a statistical state given here and Claude Shannon's entropy.

Proposition. *The assumptions and notations are those of the above Definitions.*
1. *The set Ω of generalized temperatures does not depend on the choice of the moment map J. When it is not empty, this set is an open convex subset of the Lie algebra $\mathfrak{g}$, invariant by the adjoint action of G. The partition function P is of class C^∞ and its differentials of all orders can be calculated by differentiation under the integration sign $\int$. The mean value of J, denoted by E_J, and the entropy S (for the Gibbs state indexed by $\boldsymbol{\beta}$), are other thermodynamic functions, whose expressions are*

$$E_J(\boldsymbol{\beta}) = \frac{1}{P(\boldsymbol{\beta})}\int_M J(x)\exp\bigl(-\langle J(x), \boldsymbol{\beta}\rangle\bigr)\lambda_\omega(\mathrm{d}x), \quad S(\boldsymbol{\beta}) = s(\rho_{\boldsymbol{\beta}}), \tag{4}$$

which are of class C^∞ on Ω and can be expressed in terms of the partition function P and its differentials.

2. *For any other continuous statistical state with a probability density ρ_1 with respect to the Liouville measure λ_ω, such that the mean value of J in this statistical state is equal to $E_J(\boldsymbol{\beta})$, the entropy functional s satisfies the inequality $s(\rho_1) \leq s(\rho_{\boldsymbol{\beta}}) = S(\boldsymbol{\beta})$, and the equality $s(\rho_1) = s(\rho_{\boldsymbol{\beta}})$ occurs if and only if $\rho_1 = \rho_{\boldsymbol{\beta}}$.*

3 Examples of Gibbs States on Symplectic Manifolds with Symmetries

3.1 Three-Dimensional Real Oriented Vector Spaces with a Scalar Product

In what follows, ζ is a real integer whose value is either $+1$ or -1, and $\mathbf{F}$ is a three-dimensional real vector space endowed with a scalar product denoted by $(\mathbf{v}, \mathbf{w}) \mapsto \mathbf{v} \cdot \mathbf{w}$, with $\mathbf{v}$ and $\mathbf{w} \in \mathbf{F}$, whose signature is $(+,+,+)$ when $\zeta = 1$ and $(+,+,-)$ when $\zeta = -1$. A basis $(\mathbf{e}_x, \mathbf{e}_y, \mathbf{e}_z)$ of $\mathbf{F}$ is said to be *orthonormal* when $\mathbf{e}_x \cdot \mathbf{e}_x = \mathbf{e}_y \cdot \mathbf{e}_y = 1$, $\mathbf{e}_z \cdot \mathbf{e}_z = \zeta$, $\mathbf{e}_x \cdot \mathbf{e}_y = \mathbf{e}_y \cdot \mathbf{e}_z = \mathbf{e}_z \cdot \mathbf{e}_x = 0$. When $\zeta = -1$, the vector space $\mathbf{F}$ is called a *three-dimensional Minkowski vector space.* A non-zero element $\mathbf{v} \in \mathbf{F}$ is said to be *space-like* when $\mathbf{v} \cdot \mathbf{v} > 0$, *time-like* when $\mathbf{v} \cdot \mathbf{v} < 0$ and *light-like* when $\mathbf{v} \cdot \mathbf{v} = 0$. The subset of $\mathbf{F}$ made of non-zero time-like or light-like elements has two connected components. A *temporal orientation* of $\mathbf{F}$ is the choice of one of these connected components whose elements are said to be *directed towards the future*, while those of the other component are said to be *directed towards the past.* The words *future* and *past*, purely conventional, do not refer to the flow of physical time. Both when $\zeta = 1$ and when $\zeta = -1$, we will assume in what follows that an orientation of $\mathbf{F}$ in the usual sense is chosen, and when $\zeta = -1$, that a temporal orientation of $\mathbf{F}$ is chosen too. The orthonormal bases of $\mathbf{F}$ used will always be chosen positively oriented and, when $\zeta = -1$, their third element $\mathbf{e}_z$ will be chosen time-like and directed towards the future. Such bases of $\mathbf{F}$ will be called *admissible bases.* We denote by G the subgroup of $\mathrm{GL}(\mathbf{F})$ made of linear automorphisms g of $\mathbf{F}$ which transform any admissible basis into an admissible basis. It is a connected Lie group isomorphic to the rotation group $\mathrm{SO}(3)$ when $\zeta = 1$, and to the restricted three-dimensional Lorentz group $\mathrm{SO}(2,1)$ when $\zeta = -1$. Its Lie algebra, denoted by $\mathfrak{g}$, is therefore isomorphic to $\mathfrak{so}(3)$ when $\zeta = 1$, and to $\mathfrak{so}(2,1)$ when $\zeta = -1$. The scalar product determines an isomorphism, denoted by scal, of $\mathbf{F}$ onto its dual vector space $\mathbf{F}^*$, such that, for all $\mathbf{u}$ and $\mathbf{v} \in \mathbf{F}$, $\langle \mathrm{scal}(\mathbf{u}), \mathbf{v} \rangle = \mathbf{u} \cdot \mathbf{v}$. Moreover, there exists a smooth vector spaces isomorphism j of $\mathbf{F}$ onto the Lie algebra $\mathfrak{g}$, considered as a vector subspace of the space $\mathcal{L}(\mathbf{F}, \mathbf{F})$ of linear endomorphisms of $\mathbf{F}$. Given an admissible basis $(\mathbf{e}_x, \mathbf{e}_y, \mathbf{e}_z)$ of $\mathbf{F}$, for any $(a, b, c) \in \mathbb{R}^3$, the isomorphism j is such that

$$\text{matrix of } j(a\mathbf{e}_x + b\mathbf{e}_y + c\mathbf{e}_z) = \begin{pmatrix} 0 & -c & b \\ c & 0 & -a \\ -\zeta b & \zeta a & 0 \end{pmatrix}. \tag{5}$$

The isomorphism $j : \mathbf{F} \to \mathfrak{g}$ does not depend on the admissible basis of $\mathbf{F}$ used for stating the above equality and can be expressed in terms of the Hodge star operator [10,11]. There exists a unique bilinear and skew-symmetric map $\mathbf{F} \times \mathbf{F} \to \mathbf{F}$, $(\mathbf{v}, \mathbf{w}) \mapsto \mathbf{v}\dot{\times}\mathbf{w}$ called the *cross product* on $\mathbf{F}$, which turns $\mathbf{F}$ into a Lie algebra, such that j becomes a Lie algebras isomorphism. We have indeed, for all $\mathbf{v}$ and $\mathbf{w} \in \mathbf{F}$, $j(\mathbf{v}\dot{\times}\mathbf{w}) = [j(\mathbf{v}), j(\mathbf{w})] = j(\mathbf{v}) \circ j(\mathbf{w}) - j(\mathbf{w}) \circ j(\mathbf{v})$. These properties, well known when $\zeta = 1$, still hold when $\zeta = -1$ and allow the definition on $\mathbf{F}$ of a rich structure. Of course $\mathbf{F}$ has a Riemannian or pseudo-Riemannian metric determined by its scalar product. Moreover, when it is identified with $\mathfrak{g}$ by means of j, $\mathbf{F}$ has a Lie algebra structure and the natural action of G becomes the adjoint action of this group on its Lie algebra. When $\mathbf{F}$ is identified with the dual vector space $\mathfrak{g}^*$ of this Lie algebra by means of $(j^{-1})^T \circ \text{scal}$ (composed of scal : $\mathbf{F} \to \mathbf{F}^*$ with the transpose $(j^{-1})^T : \mathbf{F}^* \to \mathfrak{g}^*$ of j), is becomes endowed with a Lie-Poisson structure, the natural action of G on F becomes the coadjoint action on the left of G on the dual $\mathfrak{g}^*$ of its Lie algebra, so the coadjoint orbits of G can be considered as submanifolds of F. With the coordinate functions x, y and z in an admissible basis $(\mathbf{e}_x, \mathbf{e}_y, \mathbf{e}_z)$ of $\mathbf{F}$, coadjoint orbits are connected submanifolds of $\mathbf{F}$ determined by an equation

$$x^2 + y^2 + \zeta z^2 = \text{Constant}, \tag{6}$$

for some Constant $\in \mathbb{R}$. The singleton $\{0\}$, whose unique element is the origin of $\mathbf{F}$, is a zero-dimensional coadjoint orbit. All other coadjoint orbits are two-dimensional.

When $\zeta = 1$, all two-dimensional coadjoint orbits are spheres centered on the origin 0 of $\mathbf{F}$. Their radius can be any real $R > 0$. We will denote by S_R the sphere of radius R centered on 0.

When $\zeta = -1$, there are three kinds of two-dimensional coadjoint orbits, described below.

- The orbits, denoted by P_R^+ and P_R^-, whose respective equations are

$$z = \sqrt{R^2 + x^2 + y^2} \text{ for } P_R^+ \text{ and } z = -\sqrt{R^2 + x^2 + y^2} \text{ for } P_R^-, \tag{7}$$

  with $R > 0$. They are called *pseudo-spheres* of radius R. Each one is a sheet of a two-sheeted two-dimensional hyperboloid with the z axis as revolution axis. They are said to be *space-like* submanifolds of $\mathbf{F}$, since all their tangent vectors are space-like vectors.
- The orbits, denoted by H_R, defined by the equation

$$x^2 + y^2 = z^2 + R^2, \quad \text{with } R > 0. \tag{8}$$

  Each of these orbits is a single-sheeted hyperboloid with the z axis as revolution axis. The tangent space at any point to such an orbit is a two-dimensional Minkowski vector space.
- The two orbits, denoted by C^+ and C^-, defined respectively by

$$z^2 = x^2 + y^2 \text{ and } z > 0, \qquad z^2 = x^2 + y^2 \text{ and } z < 0. \tag{9}$$

  They are the cones in $\mathbf{F}$ (without their apex, the origin 0 of $\mathbf{F}$), made of light-like vectors directed, respectively, towards the future and towards the past.

3.2 Gibbs States on Two-Dimensional Spheres

We assume here that $\zeta = 1$. The Lie group G is therefore isomorphic to SO(3) and its Lie algebra $\mathfrak{g}$ is isomorphic to $\mathfrak{so}(3)$. Let us consider the sphere S_R of radius $R > 0$ centered on the origin 0 of the vector space $\mathbf{F} \equiv \mathfrak{g}^*$. This sphere is a coadjoint orbit, and the moment map of the Hamiltonian action of G on it is its canonical injection into $\mathbf{F} \equiv \mathfrak{g}^*$.

The open subset Ω of generalized temperatures, for the Hamiltonian action of $G \equiv \mathrm{SO}(3)$ on its coadjoint orbit S_R, is the whole Lie algebra $\mathfrak{g}$. For each $\boldsymbol{\beta} \in \mathbf{F} \equiv \mathfrak{g}$, we can choose an admissible basis $(\mathbf{e}_x, \mathbf{e}_y, \mathbf{e}_z)$ of $\mathbf{F}$ such that $\mathbf{e}_z$ and $\boldsymbol{\beta}$ are parallel and directed in the same direction. Then $\boldsymbol{\beta} = \beta\mathbf{e}_z$, with $\beta \geq 0$. The partition function P and the probability density $\rho_{\boldsymbol{\beta}}$ of the Gibbs state indexed by $\boldsymbol{\beta}$ are expressed as

$$P(\boldsymbol{\beta}) = \begin{cases} \dfrac{4\pi\sinh(R\beta)}{\beta} & \text{if } \beta > 0, \\ 4\pi R & \text{if } \beta = 0, \end{cases} \tag{10}$$

$$\rho_{\boldsymbol{\beta}}(\mathbf{r}) = \begin{cases} \dfrac{\beta\exp(-\beta z)}{4\pi\sinh(R\beta)} & \text{if } \beta > 0, \\ \dfrac{1}{4\pi R} & \text{if } \beta = 0, \end{cases} \quad \text{with } \mathbf{r} = x\mathbf{e}_x + y\mathbf{e}_y + z\mathbf{e}_z \in S_R. \tag{11}$$

When $\beta > 0$, the thermodynamic functions mean value of J and entropy are

$$E_J(\boldsymbol{\beta}) = \frac{1 - R\beta\coth(R\beta)}{\beta^2}\boldsymbol{\beta}, \tag{12}$$

$$S(\boldsymbol{\beta}) = 1 + \log\left(\frac{4\pi\sinh(R\beta)}{\beta}\right) - R\beta\coth(R\beta). \tag{13}$$

3.3 Gibbs States on Pseudo-spheres and Other Coadjoint Orbits

We assume here that $\zeta = -1$. The Lie group G is therefore isomorphic to SO(2, 1) and its Lie algebra $\mathfrak{g}$ is isomorphic to $\mathfrak{so}(2,1)$.

We first consider the pseudo-sphere P_R^+ defined, in the coordinates x, y and z associated to any admissible basis, by the equation $z = \sqrt{R^2 + x^2 + y^2}$, for some $R > 0$. The set Ω of generalized temperatures is the subset of $\mathbf{F} \equiv \mathfrak{g}$ made of time-like vectors directed towards the past. For each $\boldsymbol{\beta} \in \Omega$, let $(\mathbf{e}_x, \mathbf{e}_y, \mathbf{e}_z)$ be an admissible basis of $\mathbf{F}$ such that $\boldsymbol{\beta} = \beta\mathbf{e}_z$, with $\beta < 0$. The partition function P and the probability density $\rho_{\boldsymbol{\beta}}$ of the Gibbs state indexed by $\boldsymbol{\beta}$, with respect to the Liouville measure $\lambda_{\omega_{P_{R+}}}$, are given by the formulae, where we have set $\|\boldsymbol{\beta}\| = \sqrt{-\boldsymbol{\beta}\cdot\boldsymbol{\beta}}$, since $\boldsymbol{\beta}\cdot\boldsymbol{\beta} < 0$,

$$P(\boldsymbol{\beta}) = \frac{2\pi}{\|\boldsymbol{\beta}\|}\exp\bigl(-\|\boldsymbol{\beta}\|R\bigr), \ \boldsymbol{\beta} \in \mathbf{F} \text{ timelike directed towards the past,} \tag{14}$$

$$\rho_{\boldsymbol{\beta}}(\mathbf{r}) = \frac{\|\boldsymbol{\beta}\|\exp\Bigl(-\|\boldsymbol{\beta}\|\bigl(z(\mathbf{r}) - R\bigr)\Bigr)}{2\pi}, \quad \mathbf{r} \in P_R^+. \tag{15}$$

The thermodynamic functions mean value of J and entropy are

$$E_J(\boldsymbol{\beta}) = -\frac{1 + R\|\boldsymbol{\beta}\|}{\|\boldsymbol{\beta}\|^2}\,\boldsymbol{\beta}, \quad S(\boldsymbol{\beta}) = 1 + \log\frac{2\pi}{\|\boldsymbol{\beta}\|}. \tag{16}$$

When the considered coadjoint orbit is the pseudo-sphere P_R^-, with $R > 0$, the open subset Ω of generalized temperatures is the subset of $\mathbf{F}$ made of time-like vectors directed towards the future. The probability density of Gibbs states and the corresponding thermodynamic functions are given by the same formulae as those indicated above, of course with the appropriate sign changes.

When the considered coadjoint orbit is either a one-sheeted hyperboloid H_R, for some $R > 0$, or one of the light cones C^+ and C^-, the set Ω of generalized temperatures is empty. Therefore no Gibbs state, for the coadjoint action of the Lie group $G \equiv \mathrm{SO}(2,1)$, can exist on these coadjoint orbits.

4 Final Comments

For spheres and pseudo-spheres, and more genearlly when the Hamiltonian action of the Lie group G is effective, the map E_J is a diffeomorphism of the set Ω of generalized temperatures onto an open subset Ω^* of $\mathfrak{g}^*$. It is therefore possible to express the Gibbs probability density $\rho_{\boldsymbol{\beta}}$ as a funcion of $E_J(\boldsymbol{\beta})$

The choice of any admissible basis $(\mathbf{e}_x, \mathbf{e}_y, \mathbf{e}_z)$ of $\mathbf{F}$ determines, for each $R > 0$, a diffeomorphism ψ_R of the pseudo-sphere P_R^+ onto the Poincaré disk D_P, subset of the complex plane $\mathbb{C}$ whose elements w satisfy $|w| < 1$. Its expression is

$$\psi_R(\mathbf{r}) = \frac{x + iy}{R + \sqrt{R^2 + x^2 + y^2}}, \quad \mathbf{r} = x\mathbf{e}_x + y\mathbf{e}_y + \sqrt{R^2 + x^2 + y^2}\,\mathbf{e}_z \in P_R^+. \tag{17}$$

Using this diffeomorphism and an appropriate Möbius transformation, one can easily deduce the Gibbs states and the thermodynamic functions on the Poincaré disk and on the Poincaré half-plane from those on the pseudo-sphere P_R^+. From the non-existence of Gibbs states on the light cones C^+ an C^-, one can deduce that no Gibbs state can exist on a two-dimensional symplectic vector space, for the Hamiltonian action of $\mathrm{SL}(2, \mathbb{R})$. Similar calculations allow the determination of Gibbs states and thermodynamic functions on an Euclidean and symplectic plane, for the Hamiltonian action of the group of its displacements (translations and rotations) [10,11]. Extensions of these results for the Siegel upper half-spaces seem to be possible.

References

1. Barbaresco, F.: Koszul information geometry and Souriau geometric temperature/capacity of Lie group thermodynamics. Entropy **16**, 4521–4565 (2014)
2. Barbaresco, F.: Symplectic structure of information geometry: fisher metric and Euler-Poincaré equation of Souriau Lie group thermodynamics. In: Nielsen, F., Barbaresco, F. (eds.) GSI 2015. LNCS, vol. 9389, pp. 529–540. Springer, Cham (2015). https://doi.org/10.1007/978-3-319-25040-3_57

3. Barbaresco, F.: Geometric Theory of Heat from Souriau Lie Groups thermodynamics and Koszul Hessian geometry: applications in information geometry for exponential families. Entropy **18**(11), 386 (2016). https://doi.org/10.3390/e18110386. In the Special Issue "Differential Geometrical Theory of Statistics", MDPI
4. Barbaresco, F.: Lie group statistics and Lie group machine learning based on Souriau lie groups thermodynamics and Koszul-Souriau-fisher metric: new entropy definition as Casimir invariant function in coadjoint representation. Entropy **22**, 642 (2020). https://doi.org/10.3390/e22060642
5. Barbaresco, F., Gay-Balmaz, F.: Lie group cohomology and (multi)symplectic integrators: new geometric tools for Lie group machine learning based on Souriau geometric statistical mechanics. Entropy **22**, 498 (2020). https://doi.org/10.3390/e22050498
6. Gibbs, W.: Elementary Principles in Statistical Mechanics, Developed with Especial Reference to the Rational Foundation of Thermodynamics. Charles Scribner's Sons, New York, Edward Arnold, London (1902). This public-domain ebook may be downloaded gratis at www.gutenberg.org/ebooks/50992
7. Marle, C.-M.: From tools in symplectic and Poisson geometry to J.-M. Souriau's theories of statistical mechanics and thermodynamics. Entropy **18**, 370 (2016). https://doi.org/10.3390/e18100370
8. Marle, C.-M.: Géométrie Symplectique et Géométrie de Poisson. Calvage & Mounet, Paris (2018)
9. Marle, C.-M.: On Gibbs states of mechanical systems with symmetries. J. Geom. Symm. Phys. **57**, 45–85 (2020). https://doi.org/10.7546/jgsp-57-2020-45-85
10. Marle, C.-M.: Examples of Gibbs states of mechanical systems with symmetries. J. Geom. Symm. Phys. **58**, 55–79 (2020). https://doi.org/10.7546/jgsp-58-2020-55-79
11. Marle, C.-M.: On Gibbs states of mechanical systems with symmetries, 13 January 2021. arXiv:2012.00582v2
12. Saxcé, G.: Entropy and structure of the thermodynamical systems. In: Nielsen, F., Barbaresco, F. (eds.) GSI 2015. LNCS, vol. 9389, pp. 519–528. Springer, Cham (2015). https://doi.org/10.1007/978-3-319-25040-3_56
13. de Saxcé, G.: Link between Lie group statistical mechanics and thermodynamics of continua. Entropy **18**, 254 (2016). https://doi.org/10.3390/e18070254. In the special Issue "Differential Geometrical Theory of Statistics, MDPI
14. Souriau, J.-M.: Définition Covariante des Équilibres Thermodynamiques, Supplemento al Nuovo cimento, vol. IV n.1, pp. 203–216 (1966)
15. Souriau, J.-M.: Structure des Systèmes Dynamiques, Dunod, Paris, 1969. English translation: Structure of Dynamical Systems, a Symplectic View of Physics, translated by C. H. Cushman-de Vries, (Transl. Cushman, R.H., Tuynman) G.M. Progress in Mathematics volume 149, Birkhäuser Boston (1997)
16. Souriau, J.-M.: Mécanique Statistique, Groupes de Lie et Cosmologie, Colloques internationaux du CNRS numéro 237 Géométrie Symplectique et Physique Mathématique, pp. 59–113 (1974)
17. Souriau, J.-M.: Géométrie Symplectique et Physique Mathématique, deux conférences de Jean-Marie Souriau, Colloquium de la Société Mathématique de France, 19 février et 12 novembre 1975
18. Souriau, J.-M.: Mécanique Classique et Géométrie Symplectique, preprint, Université de Provence et Centre de Physique Théorique (1984)

Gaussian Distributions on the Space of Symmetric Positive Definite Matrices from Souriau's Gibbs State for Siegel Domains by Coadjoint Orbit and Moment Map

Frédéric Barbaresco(✉)

THALES Land and Air Systems, Limours, France
frederic.barbaresco@thalesgroup.com

Abstract. We will introduce Gaussian distribution on the space of Symmetric Positive Definite (SPD) matrices, through Souriau's covariant Gibbs density by considering this space as the pure imaginary axis of the homogeneous Siegel upper half space where Sp (2n,R)/U(n) acts transitively. Gauss density of SPD matrices is computed through Souriau's moment map and coadjoint orbits. We will illustrate the model first for Poincaré unit disk, then Siegel unit disk and finally upper half space. For this example, we deduce Gauss density for SPD matrices.

Keywords: Symmetric positive definite matrices · Lie groups thermodynamics · Symplectic geometry · Maximum entropy · Exponential density family

1 Statistics on Lie Groups Based on Souriau Model of Gibbs Density

"There is nothing more in physical theories than symmetry groups, except the mathematical construction which allows precisely to show that there is nothing more"

Jean-Marie Souriau

In this chapter, we will explain how to define statistics on Lie Groups and its Symplectic manifold associated to the coadjoint orbits, and more especially how to define extension of Gauss density as Gibbs density in the Framework of Geometric Statistical Mechanics. Amari has proved that the Riemannian metric in an exponential family is the Fisher information matrix defined by:

$$g_{ij} = -\left[\frac{\partial^2 \Phi}{\partial \theta_i \partial \theta_j}\right]_{ij} \text{ with } \Phi(\theta) = -\log \int_{\mathbb{R}} e^{-\langle \theta, y \rangle} dy \tag{1}$$

and the dual potential, the Shannon entropy, is given by the Legendre transform:

F. Nielsen and F. Barbaresco (Eds.): GSI 2021, LNCS 12829, pp. 245–255, 2021.
https://doi.org/10.1007/978-3-030-80209-7_28

$$S(\eta) = \langle \theta, \eta \rangle - \Phi(\theta) \text{ with } \eta_i = \frac{\partial \Phi(\theta)}{\partial \theta_i} \quad \text{and} \quad \theta_i = \frac{\partial S(\eta)}{\partial \eta_i} \tag{2}$$

We can observe that $\Phi(\theta) = -\log \int\limits_R e^{-\langle \theta, y \rangle} dy = -\log \psi(\theta)$ is related to the classical cumulant generating function. J.L. Koszul and E. Vinberg have introduced a generalization through an affinely invariant Hessian metric on a sharp convex cone by its characteristic function [17]:

$$\begin{aligned} \Phi_\Omega(\theta) &= -\log \int\limits_{\Omega^*} e^{-\langle \theta, y \rangle} dy = -\log \psi_\Omega(\theta) \text{ with } \theta \in \Omega \text{ sharp convex cone} \\ \psi_\Omega(\theta) &= \int\limits_{\Omega^*} e^{-\langle \theta, y \rangle} dy \text{ with Koszul-Vinberg Characteristic function} \end{aligned} \tag{3}$$

The name "characteristic function" come from the link underlined by Ernest Vinberg:

Let Ω be a cone in U and Ω^* its dual, for any $\lambda > 0, H_\lambda(x) = \{ y \in U / \langle x, y \rangle = \lambda \}$ and let $d^{(\lambda)}y$ denote the Lebesgue measure on $H_\lambda(x)$:

$$\psi_\Omega(x) = \int\limits_{\Omega^*} e^{-\langle x, y \rangle} dy = \frac{(m-1)!}{\lambda^{m-1}} \int_{\Omega^* \cap H_\lambda(x)} d^{(\lambda)}y \tag{4}$$

There exist a bijection $x \in \Omega \mapsto x^* \in \Omega^*$, satisfying the relation $(gx)^* = {}^t g^{-1} x^*$ for all $g \in G(\Omega) = \{ g \in GL(U) / g\Omega = \Omega \}$ the linear automorphism group of Ω and x^* is:

$$x^* = \int_{\Omega^* \cap H_\lambda(x)} y d^{(\lambda)}y \Big/ \int_{\Omega^* \cap H_\lambda(x)} d^{(\lambda)}y \tag{5}$$

We can observe that x^* is the center of gravity of $\Omega^* \cap H_\lambda(x)$. We have the property that $\psi_\Omega(gx) = |\det(g)|^{-1} \psi_\Omega(x)$ for all $x \in \Omega, g \in G(\Omega)$ and then that $\psi_\Omega(x)dx$ is an invariant measure on Ω. Writing $\partial_a = \sum\limits_{i=1}^{m} a^i \frac{\partial}{\partial x^i}$, one can write:

$$\partial_a \Phi_\Omega(x) = \partial_a(-\log \psi_\Omega(x)) = \psi_\Omega(x)^{-1} \int\limits_{\Omega^*} \langle a, y \rangle e^{-\langle x, y \rangle} dy = \langle a, x^* \rangle \ , \ a \in U, x \in \Omega \tag{6}$$

Then, the tangent space to the hypersurface $\{y \in U / \psi_\Omega(y) = \psi_\Omega(x)\}$ at $x \in \Omega$ is given by $\{y \in U / \langle x^*, y \rangle = m\}$. For $x \in \Omega, a, b \in U$, the bilinear form $\partial_a \partial_b \log \psi_\Omega(x)$ is symmetric and positive definite, so that it defines an invariant Riemannian metric on Ω.

These relations have been extended by Jean-Marie Souriau in geometric statistical mechanics, where he developed a "Lie groups thermodynamics" of dynamical systems where the (maximum entropy) Gibbs density is covariant with respect to the action of the Lie group. In the Souriau model, previous structures are preserved:

$$I(\beta) = -\frac{\partial^2 \Phi}{\partial \beta^2} \text{ with } \Phi(\beta) = -\log \int_M e^{-\langle U(\xi), \beta \rangle} d\lambda_\omega \text{ and } U : M \to \mathfrak{g}^* \tag{7}$$

We preserve the Legendre transform:

$$S(Q) = \langle Q, \beta \rangle - \Phi(\beta) \text{ with } Q = \frac{\partial \Phi(\beta)}{\partial \beta} \in \mathfrak{g}^* \text{ and } \beta = \frac{\partial S(Q)}{\partial Q} \in \mathfrak{g} \tag{8}$$

In the Souriau Lie groups thermodynamics model [1, 2, 9, 10], β is a "geometric" (Planck) temperature, element of Lie algebra $\mathfrak{g}$ of the group, and Q is a "geometric" heat, element of the dual space of the Lie algebra $\mathfrak{g}^*$ of the group. Souriau has proposed a Riemannian metric that we have identified as a generalization of the Fisher metric:

$$I(\beta) = [g_\beta] \text{ with } g_\beta([\beta, Z_1], [\beta, Z_2]) = \tilde{\Theta}_\beta(Z_1, [\beta, Z_2]) \tag{9}$$

$$\text{with } \tilde{\Theta}_\beta(Z_1, Z_2) = \tilde{\Theta}(Z_1, Z_2) + \langle Q, ad_{Z_1}(Z_2) \rangle \text{ where } ad_{Z_1}(Z_2) = [Z_1, Z_2] \tag{10}$$

Souriau has proved that all co-adjoint orbit of a Lie Group given by $O_F = \left\{ Ad_g^* F, g \in G \right\}$ subset of $\mathfrak{g}^*, F \in \mathfrak{g}^*$ carries a natural homogeneous symplectic structure by a closed G-invariant 2-form. If we define $K = Ad_g^* = \left(Ad_{g^{-1}}\right)^*$ and $K_*(X) = -(ad_X)^*$ with:

$$\left\langle Ad_g^* F, Y \right\rangle = \langle F, Ad_{g^{-1}} Y \rangle, \forall g \in G, Y \in \mathfrak{g}, \; F \in \mathfrak{g}^* \tag{11}$$

where if $X \in \mathfrak{g}$, $Ad_g(X) = gXg^{-1} \in \mathfrak{g}$, the G-invariant 2-form is given by the following expression:

$$\sigma_\Omega(ad_X F, ad_Y F) = B_F(X, Y) = \langle F, [X, Y] \rangle, X, Y \in \mathfrak{g} \tag{12}$$

Souriau Foundamental Theorem is that «Every symplectic manifold on which a Lie group acts transitively by a Hamiltonian action is a covering space of a coadjoint orbit». We can observe that for Souriau model, Fisher metric is an extension of this 2-form in non-equivariant case:

$$g_\beta([\beta,Z_1],[\beta,Z_2]) = \tilde{\Theta}(Z_1,[\beta,Z_2]) + \langle Q,[Z_1,[\beta,Z_2]]\rangle \tag{13}$$

The Souriau additional term $\tilde{\Theta}(Z_1,[\beta,Z_2])$ is generated by non-equivariance through Symplectic cocycle. The tensor $\tilde{\Theta}$ used to define this extended Fisher metric is defined by the moment map $J(x)$, application from M(homogeneous symplectic manifold) to the dual space of the Lie algebra $\mathfrak{g}^*$, given by:

$$\tilde{\Theta}(X,Y) = J_{[X,Y]} - \{J_X, J_Y\} \tag{14}$$

$$\text{with } J(x): M \to \mathfrak{g}^* \text{ such that } J_X(x) = \langle J(x), X\rangle,\ X \in \mathfrak{g}$$

This tensor $\tilde{\Theta}$ is also defined in tangent space of the cocycle $\theta(g) \in \mathfrak{g}^*$ (this cocycle appears due to the non-equivariance of the coadjoint operator Ad^*_g, action of the group on the dual space of the lie algebra; the action of the group on the dual space of the Lie algebra is modified with a cocycle so that the momentu map becomes equivariant relative to this new affine action):

$$Q\big(Ad_g(\beta)\big) = Ad^*_{\mathfrak{g}}(Q) + \theta(g) \tag{15}$$

$\theta(g) \in \mathfrak{g}^*$ is called nonequivariance one-cocycle, and it is a measure of the lack of equivariance of the moment map.

$$\begin{aligned} \tilde{\Theta}(X,Y): \mathfrak{g} \times \mathfrak{g} \to \mathbb{R} \text{ with } \Theta(X) = T_e\theta(X(e)) \\ \text{X,Y} \mapsto \langle \Theta(X), Y\rangle \end{aligned} \tag{16}$$

Souriau has then defined a Gibbs density that is covariant under the action of the group:

$$p_{Gibbs}(\xi) = e^{\Phi(\beta)-\langle U(\xi),\beta\rangle} = \frac{e^{-\langle U(\xi),\beta\rangle}}{\int_M e^{-\langle U(\xi),\beta\rangle} d\lambda_\omega} \text{ with } \Phi(\beta) = -\log \int_M e^{-\langle U(\xi),\beta\rangle} d\lambda_\omega \tag{17}$$

$$Q = \frac{\partial \Phi(\beta)}{\partial \beta} = \frac{\int_M U(\xi) e^{-\langle U(\xi),\beta\rangle} d\lambda_\omega}{\int_M e^{-\langle U(\xi),\beta\rangle} d\lambda_\omega} = \int_M U(\xi) p(\xi) d\lambda_\omega \tag{18}$$

We can express the Gibbs density with respect to Q by inverting the relation $Q = \frac{\partial\Phi(\beta)}{\partial\beta} = \Theta(\beta)$. Then $p_{Gibbs,Q}(\xi) = e^{\Phi(\beta)-\langle U(\xi),\Theta^{-1}(Q)\rangle}$ with $\beta = \Theta^{-1}(Q)$.

2 Gauss Density on Poincaré Unit Disk

We will introduce Souriau moment map for SU (1,1)/U(1) group that acts transitively on Poincaré Unit Disk, based on moment map. Considering the Lie group:

$$SU(1,1) = \left\{ \begin{pmatrix} a & b \\ b^* & a^* \end{pmatrix} = \begin{pmatrix} 1 & ba^{*-1} \\ 0 & 1 \end{pmatrix} \begin{pmatrix} a^{*-1} & 0 \\ 0 & a^* \end{pmatrix} \begin{pmatrix} 1 & 0 \\ a^{*-1}b^* & 1 \end{pmatrix} / a, b \in \mathbb{C},\ |a|^2 - |b|^2 = 1 \right\} \tag{19}$$

and its Lie algebra given by elements:

$$su(1,1) = \left\{ \begin{pmatrix} ir & \eta \\ \eta^* & -ir \end{pmatrix} / r \in \mathbb{R}, \eta \in \mathbb{C} \right\} \tag{20}$$

A basis for this Lie algebra $su(1,1)$ is $(u_1, u_2, u_3) \in \mathfrak{g}$ with

$$u_1 = \frac{i}{2}\begin{pmatrix} 1 & 0 \\ 0 & -1 \end{pmatrix},\ u_2 = -\frac{1}{2}\begin{pmatrix} 0 & 1 \\ 1 & 0 \end{pmatrix} \text{ and } u_3 = \frac{1}{2}\begin{pmatrix} 0 & -i \\ i & 0 \end{pmatrix} \tag{21}$$

with $[u_1, u_3] = -u_2, [u_1, u_2] = u_3, [u_2, u_3] = -u_1$. The Harish-Chandra embedding is given by $\varphi(gx_0) = \zeta = ba^{*-1}$. From $|a|^2 - |b|^2 = 1$, one has $|\zeta| < 1$. Conversely, for any $|\zeta| < 1$, taking any $a \in \mathbb{C}$ such that $|a| = \left(1 - |a|^2\right)^{-1/2}$ and putting $b = \zeta a^*$, one obtains $g \in G$ for which $\varphi(gx_0) = \zeta$. The domain $D = \varphi(M)$ is the unit disc $D = \{\zeta \in \mathbb{C} / |\zeta| < 1\}$.

The compact subgroup is generated by u_1, while u_2 and u_3 generate a hyperbolic subgroup. The dual space of the Lie algebra is given by:

$$su(1,1)^* = \left\{ \begin{pmatrix} z & x + iy \\ -x + iy & -z \end{pmatrix} / x, y, z \in \mathbb{R} \right\} \tag{22}$$

with the basis $(u_1^*, u_2^*, u_3^*) \in \mathfrak{g}^* : u_1^* = \begin{pmatrix} 1 & 0 \\ 0 & -1 \end{pmatrix},\ u_2^* = \begin{pmatrix} 0 & i \\ i & 0 \end{pmatrix}$ and

$$u_3^* = \begin{pmatrix} 0 & 1 \\ -1 & 0 \end{pmatrix} \tag{23}$$

Let us consider $D = \{z \in \mathbb{C} / |z| < 1\}$ be the open unit disk of Poincaré. For each $\rho > 0$, the pair (D, ω_ρ) is a symplectic homogeneous manifold with $\omega_\rho = 2i\rho \frac{dz \wedge dz^*}{\left(1 - |z|^2\right)^2}$, where ω_ρ is invariant under the action:

$$\begin{gathered} SU(1,1) \times D \to D \\ (g, z) \mapsto g.z = \frac{az + b}{b^* z + a^*} \end{gathered} \tag{24}$$

This action is transitive and is globally and strongly Hamiltonian. Its generators are the hamiltonian vector fields associated to the functions:

$$J_1(z,z^*) = \rho \frac{1+|z|^2}{1-|z|^2}, \; J_2(z,z^*) = \frac{\rho}{i} \frac{z-z^*}{1-|z|^2}, \; J_3(z,z^*) = -\rho \frac{z+z^*}{1-|z|^2} \quad (25)$$

The associated moment map $J : D \to su^*(1,1)$ defined by $J(z).u_i = J_i(z,z^*)$, maps D into a coadjoint orbit in $su^*(1,1)$. Then, we can write the moment map as a matrix element of $su^*(1,1)$ [4–7]:

$$J(z) = J_1(z,z^*)u_1^* + J_2(z,z^*)u_2^* + J_3(z,z^*)u_3^* = \rho \begin{pmatrix} \frac{1+|z|^2}{1-|z|^2} & -2\frac{z^*}{1-|z|^2} \\ 2\frac{z}{1-|z|^2} & -\frac{1+|z|^2}{1-|z|^2} \end{pmatrix} \in \mathfrak{g}^* \quad (26)$$

The moment map J is a diffeomorphism of D onto one sheet of the two-sheeted hyperboloid in $su^*(1,1)$, determined by the following equation $J_1^2 - J_2^2 - J_3^2 = \rho^2$, $J_1 \geq \rho$ with $J_1 u_1^* + J_2 u_2^* + J_3 u_3^* \in su^*(1,1)$. We note O_ρ^+ the coadjoint orbit $Ad^*_{SU(1,1)}$ of $SU(1,1)$, given by the upper sheet of the two-sheeted hyperboloid given by previous equation. The orbit method of Kostant-Kirillov-Souriau associates to each of these coadjoint orbits a representation of the discrete series of $SU(1,1)$, provided that ρ is a half integer greater or equal than 1 ($\rho = \frac{k}{2}, k \in N$ and $\rho \geq 1$). When explicitly executing the Kostant-Kirillov construction, the representation Hilbert spaces H_ρ are realized as closed reproducing kernel subspaces of $L^2(D, \omega_\rho)$. The Kostant-Kirillov-Souriau orbit method shows that to each coadjoint orbit of a connected Lie group is associated a unitary irreducible representation of G acting in a Hilbert space H.

Souriau has oberved that action of the full Galilean group on the space of motions of an isolated mechanical system is not related to any equilibrium Gibbs state (the open subset of the Lie algebra, associated to this Gibbs state is empty). The main Souriau idea was to define the Gibbs states for one-parameter subgroups of the Galilean group. We will use the same approach, in this case We will consider action of the Lie group $SU(1,1)$ on the symplectic manifold (M,ω) (Poincaré unit disk) and its momentum map J are such that the open subset $\Lambda_\beta = \left\{ \beta \in \mathfrak{g} / \int_D e^{-\langle J(z),\beta \rangle} d\lambda(z) < +\infty \right\}$ is not empty. This condition is not always satisfied when *(M, ω)* is a cotangent bundle, but of course it is satisfied when it is a compact manifold. The idea of Souriau is to consider a one parameter subgroup of $SU(1,1)$. To parametrize elements of $SU(1,1)$ is through its Lie algebra. In the neighborhood of the identity element, the elements of $g \in SU(1,1)$ can be written as the exponential of an element β of its Lie algebra:

$$g = \exp(\varepsilon\beta) \text{ with } \beta \in \mathfrak{g} \quad (27)$$

The condition $g^+ Mg = M$ for $M = \begin{pmatrix} 1 & 0 \\ 0 & -1 \end{pmatrix}$ can be expanded for $\varepsilon << 1$ and is equivalent to $\beta^+ M + M\beta = 0$ which then implies $\beta = \begin{pmatrix} ir & \eta \\ \eta^* & -ir \end{pmatrix}$, $r \in \mathbb{R}$, $\eta \in \mathbb{C}$. We can observe that r and $\eta = \eta_R + i\eta_I$ contain 3 degrees of freedom, as required. Also because $\det g = 1$, we get $Tr(\beta) = 0$. We exponentiate β with exponential map to get:

$$g = \exp(\varepsilon\beta) = \sum_{k=0}^{\infty} \frac{(\varepsilon\beta)^k}{k!} = \begin{pmatrix} a_\varepsilon(\beta) & b_\varepsilon(\beta) \\ b_\varepsilon^*(\beta) & a_\varepsilon^*(\beta) \end{pmatrix} \tag{28}$$

If we make the remark that we have the following relation $\beta^2 = \begin{pmatrix} ir & \eta \\ \eta^* & -ir \end{pmatrix}\begin{pmatrix} ir & \eta \\ \eta^* & -ir \end{pmatrix} = \left(|\eta|^2 - r^2\right)I$, we develop the exponential map:

$$g = \exp(\varepsilon\beta) = \begin{pmatrix} \cosh(\varepsilon R) + ir\frac{\sinh(\varepsilon R)}{R} & \eta\frac{\sinh(\varepsilon R)}{R} \\ \eta^*\frac{\sinh(\varepsilon R)}{R} & \cosh(\varepsilon R) - ir\frac{\sinh(\varepsilon R)}{R} \end{pmatrix} \tag{29}$$

with $R^2 = |\eta|^2 - r^2$

We can observe that one condition is that $|\eta|^2 - r^2 > 0$ then the subset to consider is given by the subset $\Lambda_\beta = \left\{\beta = \begin{pmatrix} ir & \eta \\ \eta^* & -ir \end{pmatrix}, r \in \mathbb{R}, \eta \in \mathbb{C} / |\eta|^2 - r^2 > 0\right\}$ such that $\int_D e^{-\langle J(z),\beta\rangle} d\lambda(z) < +\infty$. The generalized Gibbs states of the full $SU(1,1)$ group do not exist. However, generalized Gibbs states for the one-parameter subgroups $\exp(\alpha\beta)$, $\beta \in \Lambda_\beta$, of the $SU(1,1)$ group do exist. The generalized Gibbs state associated to β remains invariant under the restriction of the action to the one-parameter subgroup of $SU(1,1)$ generated by $\exp(\varepsilon\beta)$.

To go futher, we will develop the Souriau Gibbs density from the Souriau moment map $J(z)$ and the Souriau temperature $\beta \in \Lambda_\beta$. If we note $b = \frac{1}{1-|z|^2}\begin{bmatrix} 1 \\ -z \end{bmatrix}$, we can write the moment map:

$$J(z) = \rho(2Mbb^+ - Tr(Mbb^+)I) \text{ with } M = \begin{bmatrix} 1 & 0 \\ 0 & -1 \end{bmatrix} \tag{30}$$

We can the write the covariant Gibbs density in the unit disk given by moment map of the Lie group $SU(1,1)$ and geometric temperature in its Lie algebra $\beta \in \Lambda_\beta$:

$$p_{Gibbs}(z) = \frac{e^{-\langle J(z),\beta\rangle}}{\int_D e^{-\langle J(z),\beta\rangle} d\lambda(z)} \text{ with } d\lambda(z) = 2i\rho\frac{dz \wedge dz^*}{\left(1-|z|^2\right)^2} \tag{31}$$

$$p_{Gibbs}(z) = \frac{e^{-\langle \rho(2\Im bb^+ - Tr(\Im bb^+)I),\beta\rangle}}{\int_D e^{-\langle J(z),\beta\rangle} d\lambda(z)} = \frac{e^{-\left\langle \rho\begin{pmatrix} \frac{1+|z|^2}{(1-|z|^2)} & \frac{-2z^*}{(1-|z|^2)} \\ \frac{2z}{(1-|z|^2)} & -\frac{1+|z|^2}{(1-|z|^2)} \end{pmatrix}, \begin{pmatrix} ir & \eta \\ \eta^* & -ir \end{pmatrix}\right\rangle}}{\int_D e^{-\langle J(z),\beta\rangle} d\lambda(z)} \tag{32}$$

To write the Gibbs density with respect to its statistical moments, we have to express the density with respect to $Q = E[J(z)]$. Then, we have to invert the relation between Q and β, to replace this last variable $\beta = \begin{pmatrix} ir & \eta \\ \eta^* & -ir \end{pmatrix} \in \Lambda_\beta$ by $\beta = \Theta^{-1}(Q) \in \mathfrak{g}$ where $Q = \frac{\partial \Phi(\beta)}{\partial \beta} = \Theta(\beta) \in \mathfrak{g}^*$ with $\Phi(\beta) = -\log \int\limits_D e^{-\langle J(z), \beta \rangle} d\lambda(z)$, deduce from Legendre tranform. The mean moment map is given by:

$$Q = E[J(z)] = E\left[\rho \begin{pmatrix} \frac{1+|w|^2}{\left(1-|w|^2\right)} & \frac{-2w^*}{\left(1-|w|^2\right)} \\ \frac{2w}{\left(1-|w|^2\right)} & -\frac{1+|w|^2}{\left(1-|w|^2\right)} \end{pmatrix}\right] \text{ where } w \in D \tag{33}$$

3 Gauss Density on Siegel Unit Disk

To address computation of covariant Gibbs density for Siegel Unit Disk $SD_n = \{Z \in Mat(n, \mathbb{C}) / I_n - ZZ^+ > 0\}$, we will consider in this section $SU(p, q)$ Unitary Group [3, 8, 12, 13, 15, 16]:

$$G = SU(n,n) \text{ and } K = S(U(n) \times U(n)) = \left\{ \begin{pmatrix} A & 0 \\ 0 & D \end{pmatrix} / A \in U(n), D \in U(n), \det(A)\det(D) = 1 \right\}$$

We can use the following decomposition for $g \in G^{\mathbb{C}}$(complexification of g), and consider its action on Siegel Unit Disk given by:

$$g = \begin{pmatrix} A & B \\ C & D \end{pmatrix} \in G^{\mathbb{C}}, g = \begin{pmatrix} I_n & BD^{-1} \\ 0 & I_n \end{pmatrix} \begin{pmatrix} A - BD^{-1}C & 0 \\ 0 & D \end{pmatrix} \begin{pmatrix} I_n & 0 \\ D^{-1}C & I_n \end{pmatrix} \tag{34}$$

Benjamin Cahen has studied this case and introduced the moment map by identifing G-equivariantly $\mathfrak{g}^*$ with $\mathfrak{g}$ by means of the Killing form β on $\mathfrak{g}^{\mathbb{C}}$:

$$\mathfrak{g}^* \text{ G-equivariant with by } \mathfrak{g} \text{ Killing form } \beta(X, Y) = 2(p+q)Tr(XY)$$

The set of all elements of $\mathfrak{g}$ fixed by K is $\mathfrak{h}$:

$$\begin{aligned} \mathfrak{h} &= \{\text{element of } G \text{ fixed by } K\}, \ \xi_0 \in \mathfrak{h}, \xi_0 = i\lambda \begin{pmatrix} -nI_p & 0 \\ 0 & nI_q \end{pmatrix} \\ &\Rightarrow \langle \xi_0, [Z, Z^+] \rangle = -2i\lambda (2n)^2 Tr(ZZ^+), \forall Z \in D \end{aligned} \tag{35}$$

Then, we the equivariant moment map is given by:

$$\begin{array}{l}\forall X \in g^C \ , \ Z \in D, \ \psi(Z) = Ad^*(\exp(-Z^+)\zeta(\exp Z^+ \exp Z))\xi_0 \\ \forall g \in G, Z \in D \text{ then } \psi(g.Z) = Ad_g^*\psi(Z) \\ \psi \text{ is a diffeomorphism from } SD \text{ onto orbit } O(\xi_0)\end{array} \tag{36}$$

$$\psi(Z) = i\lambda\begin{pmatrix} (I_n - ZZ^+)^{-1}(-nZZ^+ - nI_n) & (2n)Z(I_n - Z^+Z)^{-1} \\ -(2n)(I_n - Z^+Z)^{-1}Z^+ & (nI_q + nZ^+Z)(I_n - Z^+Z)^{-1} \end{pmatrix} \tag{37}$$

and

$$\zeta(\exp Z^+ \exp Z) = \begin{pmatrix} I_p & Z(I_q - Z^+Z)^{-1} \\ 0 & I_q \end{pmatrix} \tag{38}$$

The moment map is then given by:

$$J(Z) = \rho n\begin{pmatrix} (I_n - ZZ^+)^{-1}(I_n + ZZ^+) & -2Z^+(I_n - ZZ^+)^{-1} \\ 2(I_n - ZZ^+)^{-1}Z & (I_n + ZZ^+)(I_n - ZZ^+)^{-1} \end{pmatrix} \in \mathfrak{g}^* \tag{39}$$

Souriau Gibbs density is then given with $\beta, M \in \mathfrak{g}$ and $Z \in SD_n$ by:

$$p_{Gibbs}(Z) = \frac{e^{-\left\langle \rho n\begin{pmatrix} (I_n - ZZ^+)^{-1}(I_n + ZZ^+) & -2Z^+(I_n - ZZ^+)^{-1} \\ 2(I_n - ZZ^+)^{-1}Z & (I_n + ZZ^+)(I_n - ZZ^+)^{-1} \end{pmatrix}, \beta \right\rangle}}{\int\limits_{SD_n} e^{-\langle J(Z), \beta \rangle} d\lambda(Z)} \text{ with } \begin{cases} \beta = \Theta^{-1}(Q) \in \mathfrak{g} \\ Q = E[J(Z)] \\ Q = \frac{\partial \Phi(\beta)}{\partial \beta} = \Theta(\beta) \in \mathfrak{g}^* \end{cases} \tag{40}$$

Gauss density of SPD matrix is then given by $Z = (Y - I)(Y + I)^{-1}, Y \in Sym(n)^+$

4 Gauss Density on Siegel Upper Half Plane

We conclude by considering $H_n = \{W = U + iV / U, V \in Sym(n), V > 0\}$ Siegel Upper Half Space, related to Siegel Unit Disk by Cayley transform $Z = (W + iI_n)^{-1}(W - iI_n)$, that we will analyze as an homogeneous space with respect to $Sp(2n, \mathbb{R})/U(n)$ group action where the Sympletic group is given by the definition $Sp(2n, \mathbb{R}) = \left\{ \begin{pmatrix} A & B \\ C & D \end{pmatrix} \in Mat(2n, \mathbb{R}) / \begin{array}{c} A^TC = C^TA, B^TD = D^TB \\ A^TD - C^TB = I_n \end{array} \right\}$ and the left action $\Phi :$ $Sp(2n, \mathbb{R}) \times H_n \to H_n$ with $\Phi\left(\begin{pmatrix} A & B \\ C & D \end{pmatrix}, W\right) = (C + DW)(A + BW)^{-1}$ that is transitive $\Phi : Sp(2n, \mathbb{R}) \times H_n \to H_n$ with $\Phi\left(\begin{pmatrix} V^{-1/2} & 0 \\ UV^{-1/2} & V^{1/2} \end{pmatrix}, iI_n\right) = U + iV$. The isotropy subgroup of the element $iI_n \in H_n$ is $Sp(2n,\mathbb{R}) \cap O(2n) = \left\{ \begin{bmatrix} X & Y \\ -Y & X \end{bmatrix} \in Mat(2n,\mathbb{R}) / X^T X + Y^T Y = I_n, X^T Y = Y^T X \right\}$ that is identified with $U(n)$ by $Sp(2n,R) \cap O(2n) \to U(n); \begin{bmatrix} X & Y \\ -Y & X \end{bmatrix} \mapsto X + iY$ and then $H_n \cong Sp(2n,\mathbb{R})/U(n)$ that is also a symplectic manifold with symplectic form $\Omega_{H_n} = -d\Theta_{H_n}$ with one-form $\Theta_{H_n} = -tr(UdV^{-1})$.

By identifying the symplectic algebra $sp(2n,\mathbb{R})$ with $sym\,(2n,\mathbb{R})$ [14] via $\varepsilon = \begin{bmatrix} \varepsilon_{11} & \varepsilon_{12} \\ \varepsilon_{12}^T & \varepsilon_{22} \end{bmatrix} \in sym(2n,\mathbb{R}) \mapsto \tilde{\varepsilon} = J_n^T \varepsilon = \begin{bmatrix} -\varepsilon_{12}^T & -\varepsilon_{22} \\ \varepsilon_{11} & \varepsilon_{12} \end{bmatrix} \in sp(2n,\mathbb{R})$ with $[\varepsilon,\delta]_{sym} = \varepsilon J_n^T \delta - \delta J_n^T \varepsilon$ and $\left[\tilde{\varepsilon},\tilde{\delta}\right]_{sp} = \tilde{\varepsilon}\tilde{\delta} - \tilde{\delta}\tilde{\varepsilon}$, and the associated inner products $\langle \varepsilon,\delta \rangle_{sym} = tr(\varepsilon\delta)$ and $\left\langle \tilde{\varepsilon},\tilde{\delta} \right\rangle_{sp} = tr\left(\tilde{\varepsilon}^T \tilde{\delta}\right)$. With these inner products, we identify their dual spaces with themselves $sym(2n,\mathbb{R}) = sym(2n,\mathbb{R})^*$, $sp(2n,\mathbb{R}) = sp(2n,\mathbb{R})^*$.

We define adjoint operator $Ad_G\varepsilon = (G^{-1})^T \varepsilon G^{-1}, Ad_G\tilde{\varepsilon} = G\tilde{\varepsilon}G^{-1}$ with $G \in Sp(2n,\mathbb{R})$ and $ad_\varepsilon \delta = \varepsilon J_n^T \delta - \delta J_n^T \varepsilon = [\varepsilon,\delta]_{sym}$, and co-adjoint operators $Ad_{G^{-1}}^* \eta = G\eta G^T$ and $ad_\varepsilon^* \eta = J_n \varepsilon \eta - \eta \varepsilon J_n$. To coadjoint orbit $O = \left\{ Ad_G^* \eta \in sym(2n,\mathbb{R})^* / G \in Sp(2n,\mathbb{R}) \right\}$ for each $\eta \in sym(2n,\mathbb{R})^*$, we can associate a symplectic manifold with the KKS (Kirillov-Kostant-Souriau) 2-form $\Omega_O(\eta)\left(ad_\varepsilon^*\eta, ad_\delta^*\eta\right) = \left\langle \eta, [\varepsilon,\delta]_{sym} \right\rangle = tr(\eta[\varepsilon,\delta]_{sym})$. We then compute the moment map $J : H_n \to sp(2n,\mathbb{R})^*$ such that $i_\varepsilon \Omega_{H_n} = d\langle J(.),\varepsilon \rangle$, given by $J(W) = \begin{bmatrix} V^{-1} & V^{-1}U \\ UV^{-1} & UV^{-1}U + V \end{bmatrix}$ for $W = U + iV \in H_n$, we deduce:

$$p_{Gibbs}(W) = \frac{e^{-\langle J(W),\varepsilon \rangle}}{\int\limits_{H_n} e^{-\langle J(W),\varepsilon \rangle} d\lambda(W)} \text{with } d\lambda(W) = 2i\rho\left(V^{-1}dW \wedge V^{-1}dW^+\right) \quad (41)$$

With

$$\langle J(W),\varepsilon \rangle = Tr(J(W)\varepsilon) = Tr\left(\begin{bmatrix} V^{-1} & V^{-1}U \\ UV^{-1} & UV^{-1}U + V \end{bmatrix} \begin{bmatrix} \varepsilon_{11} & \varepsilon_{12} \\ \varepsilon_{12}^T & \varepsilon_{22} \end{bmatrix} \right) \quad (42)$$

And then

$$\langle J(W),\varepsilon \rangle = Tr\left[V^{-1}\varepsilon_{11} + 2UV^{-1}\varepsilon_{12} + \left(UV^{-1}U + V\right)\varepsilon_{22}\right] \quad (43)$$

To consider Gibbs density and Gauss density for Symmetric Positive Definite Matrix, we have to consider the case $W = iV$ with $U = 0$ and $\langle J(W),\varepsilon \rangle = Tr\left[V^{-1}\varepsilon_{11} + V\varepsilon_{22}\right]$.

References

1. Barbaresco, F.: Lie group statistics and lie group machine learning based on Souriau lie groups Thermodynamics & Koszul-Souriau-Fisher metric: new entropy definition as generalized Casimir invariant function in Coadjoint representation. Entropy **22**, 642 (2020)
2. Barbaresco, F., Gay-Balmaz, F.: Lie group cohomology and (multi)symplectic integrators: new geometric tools for lie group machine learning based on Souriau geometric statistical mechanics. Entropy **22**, 498 (2020)
3. Chevallier, E., Forget, T., Barbaresco, F., Angulo, J.: Kernel density estimation on the Siegel space with an application to radar processing. Entropy **18**, 396 (2016)
4. Cishahayo, C., de Bièvre, S.: On the contraction of the discrete series of SU(1;1). Annales de l'institut Fourier, tome **43**(2), 551–567 (1993)
5. Renaud, J.: The contraction of the SU(1, 1) discrete series of representations by means of coherent states. J. Math. Phys. **37**(7), 3168–3179 (1996)
6. Cahen, B.: Contraction de SU(1,1) vers le groupe de Heisenberg, Travaux de Mathématiques. Fascicule **XV**, 19-43 (2004)
7. Cahen, B.: Global Parametrization of Scalar Holomorphic Coadjoint Orbits of a Quasi-Hermitian Lie Group, Acta. Univ. Palacki. Olomuc. Fac. rer. Nat., Mat. 52, n°&, pp.35–48, 2013
8. Berezin, F.A.: Quantization in complex symmetric space. Math, USSR Izv **9**, 341–379 (1975)
9. Marle, C.-M.: On gibbs states of mechanical systems with symmetries. JGSP **57**, 45–85 (2020)
10. Marle, C.-M.: On Generalized Gibbs states of mechanical systems with symmetries, arXiv: 2012.00582v2 [math.DG] (2021)
11. Marle, C.-M. Projection Stereographique et moments, Hal-02157930, Version 1; June 2019. https://hal.archives-ouvertes.fr/hal-02157930/. Accessed 31 May 2020
12. Hua, L.K.: Harmonic Analysis of Functions of Several Complex Variables in the Classical Domains, Translations of Mathematical Monographs, vol. 6, American Mathematical Society, Providence (1963)
13. Nielsen, F.: The Siegel-Klein disk. Entropy **22**, 1019 (2020)
14. Ohsawa, T., Tronci, C.: Geometry and dynamics of Gaussian wave packets and their Wigner transforms. J. Math. Phys. **58**, 092105 (2017)
15. Siegel, C.L.: Symplectic geometry. Am. J. Math. **65**(1), 1–86 (1943)
16. Leverrier, A.: SU(p, q) coherent states and a Gaussian de Finetti theorem. J. Math. Phys. **59**, 042202 (2018)
17. Satake I.: Algebraic Structures of Symmetric Domains, Princeton University Press, Princeton (1980)

On Gaussian Group Convex Models

Hideyuki Ishi(✉)

Osaka City University, OCAMI, 3-3-138 Sugimoto, Sumiyoshi-ku,
558-8585 Osaka, Japan
hideyuki@sci.osaka-cu.ac.jp

Abstract. The Gaussian group model is a statistical model consisting of central normal distributions whose precision matrices are of the form $gg^\top$, where g is an element of a matrix group G. When the set of $gg^\top$ is convex in the vector space of real symmetric matrices, the set forms an affine homogeneous convex domain studied by Vinberg. In this case, we give the smallest number of samples such that the maximum likelihood estimator (MLE) of the parameter exists with probability one. Moreover, if the MLE exists, it is explicitly expressed as a rational function of the sample data.

Keywords: Gaussian group model · Affine homogeneous convex domain · Maximum likelihood estimator · Riesz distribution

1 Introduction

Let G be a subgroup of $GL(N, \mathbb{R})$, and $\mathcal{M}_G$ the set $\{ gg^\top ; g \in G \} \subset \mathrm{Sym}(N, \mathbb{R})$. The Gaussian group model associated to G is a statistical model consisting of central multivariate normal laws whose precision (concentration) matrices belong to $\mathcal{M}_G$ ([1]). It is an example of a transformation family, that is, an exponential family whose parameter space forms a group (see [2]). A fundamental problem is to estimate an unknown precision matrix $\theta = gg^\top \in \mathcal{M}_G$ from samples $X_1, X_2, \ldots, X_n \in \mathbb{R}^N$. In [1], the existence and uniqueness of the maximum likelihood estimator (MLE) $\hat{\theta}$ of θ are discussed in connection with Geometric Invariant Theory. In the present paper, under the assumption that $\mathcal{M}_G \neq \{I_N\}$ is a convex set, we compute the number n_0 for which MLE exists uniquely with probability one if and only if $n \geq n_0$ (Theorem 4). In this case, an expression of the MLE as a rational function of the samples $X_1, \ldots, X_n$ is given (Theorem 3).

Since $\mathcal{M}_G$ is convex, it is regarded as a convex domain in an affine space $I_N + V$, where V is a linear subspace of the vector space $\mathrm{Sym}(N, \mathbb{R})$ of real symmetric matrices of size N. Moreover $\mathcal{M}_G$ is contained in the cone $\mathrm{Sym}^+(N, \mathbb{R})$ of positive definite symmetric matrices, so that $\mathcal{M}_G$ does not contain any straight line (actually, we shall see that $\mathcal{M}_G$ is exactly the intersection

Partially supported by KAKENHI 20K03657 and Osaka City University Advanced Mathematical Institute (MEXT Joint Usage/Research Center on Mathematics and Theoretical Physics JPMXP0619217849).

F. Nielsen and F. Barbaresco (Eds.): GSI 2021, LNCS 12829, pp. 256–264, 2021.
https://doi.org/10.1007/978-3-030-80209-7_29

$(I_N+V)\cap \mathrm{Sym}^+(N,\mathbb{R})$, see Proposition 2 and Theorem 2). Thus $\mathcal{M}_G$ is an affine homogeneous convex domain on which G acts transitively as affine transforms. Then we can apply Vinberg's theory [9] to $\mathcal{M}_G$. In particular, using the left-symmetric algebra structure on V explored in [6], we give a specific description of $\mathcal{M}_G$ as in Theorem 2, where the so-called real Siegel domain appears naturally. On the other hand, every homogeneous cone is obtained as $\mathcal{M}_G$ by [6]. In particular, all the symmetric cones discussed in [3] as well as the homogeneous graphical models in [8, Section 3.3] appear in our setting.

The rational expression of MLE is obtained by using the algebraic structure on V, whereas the existence condition is deduced from the previous works [4] and [5] about the Wishart and Riesz distributions on homogeneous cones. It seems feasible to generalize the results of this paper to a wider class of Gaussian models containing decomposable graphical models based on [7] in future.

The author should like to express his gratitude to Professor Fumihiro Sato for the information of the paper [1] as well as the interest to the present work. He is also grateful to Professors Piotr Graczyk, Bartosz Kołodziejek, Yoshihiko Konno, and Satoshi Kuriki for the collaboration about applications of homogeneous cones to mathematical statistics. The discussions with them yield most of the featured ideas and techniques in this paper. Finally, the author sincerely thanks the referees for their valuable comments and suggestions.

2 Structure of the Convex Parameter Set

In what follows, we assume that $\mathcal{M}_G$ is a convex set in the vector space $\mathrm{Sym}(N,\mathbb{R})$ and that $\mathcal{M}_G \neq \{I_N\}$. The affine subspace spanned by elements of $\mathcal{M}_G$ is of the form I_N+V, where V is a linear subspace of $\mathrm{Sym}(N,\mathbb{R})$, and $\mathcal{M}_G$ is an open connected set in I_N+V. Let $\overline{G}$ be the closure of G in $GL(N,\mathbb{R})$. Then we have $\mathcal{M}_{\overline{G}} = \mathcal{M}_G$. Indeed, for $\tilde{g} \in \overline{G}$, the set $\tilde{g}\mathcal{M}_G\tilde{g}^\top = \{\, \tilde{g}\theta\tilde{g}^\top \,;\, \theta \in \mathcal{M}_G \,\}$ is contained in the closure $\overline{\mathcal{M}_G}$ of $\mathcal{M}_G$. On the other hand, since $\tilde{g}$ is invertible, the set $\tilde{g}\mathcal{M}_G\tilde{g}^\top$ is open in the affine space $I_N + V$. Thus the point $\tilde{g}\tilde{g}^\top \in \tilde{g}\mathcal{M}_G\tilde{g}^\top \subset \overline{\mathcal{M}_G}$ is an relatively interior point of the convex set $\overline{\mathcal{M}_G}$, so that $\tilde{g}\tilde{g}^\top \in \mathcal{M}_G$, which means that $\mathcal{M}_{\overline{G}} \subset \mathcal{M}_G$. Therefore we can assume that G is a closed linear Lie group without loss of generality. Furthermore, we can assume that G is connected. Following Vinberg's argument [9, Chapter 1, Sect. 6], we shall show that $\mathcal{M}_{G^0_{\mathrm{alg}}} = \mathcal{M}_G$, where G^0_{alg} is the identity component (in the classical topology) of the real algebraic hull of G in $GL(N,\mathbb{R})$. Let $\mathfrak{g}$ be the Lie algebra of G. For $h \in G^0_{\mathrm{alg}}$, we have $hh^\top \in I_N + V$ and $h\,\mathfrak{g}\,h^{-1} = \mathfrak{g}$ because these are algebraic conditions that are satisfied by all the elements of G. The second condition together with the connectedness of G tells us that $hGh^{-1} = G$. Let $\mathcal{U}$ be a neighborhood of I_N in G^0_{alg} such that $hh^\top \in \mathcal{M}_G$ for all $h \in \mathcal{U}$. Then for any $\theta = gg^\top \in \mathcal{M}_G$ $(g \in G)$, we have $h\theta h^\top = (hgh^{-1})hh^\top(hgh^{-1})^\top \in g_0\mathcal{M}_G g_0^\top = \mathcal{M}_G$, where $g_0 := hgh^{-1} \in G$. Thus, if h is the product $h_1h_2\cdots h_m \in G^0_{\mathrm{alg}}$ with $h_1,\ldots,h_m \in \mathcal{U}$, we see that $hh^\top \in \mathcal{M}_G$ inductively. Therefore we conclude that $\mathcal{M}_{G^0_{\mathrm{alg}}} = \mathcal{M}_G$.

By Vinberg [9], we have the generalized Iwasawa decomposition

$$G^0_{\mathrm{alg}} = \mathcal{T} \cdot (G^0_{\mathrm{alg}} \cap O(N)),$$

where $\mathcal{T}$ is a maximal connected split solvable Lie subgroup of G^0_{alg}. Moreover, $\mathcal{T}$ is triangularized simultaneously by an orthogonal matrix $U \in O(N)$, which means that a group $\mathcal{T}^U = \{ U^{-1}hU \,;\, h \in \mathcal{T} \}$ is contained in the group $\mathcal{T}_N$ of lower triangular matrices of size N with positive diagonal entries. Let $\mathcal{M}_G^U$ be the set $\{ U^\top \theta U \,;\, \theta \in \mathcal{M}_G \}$, which is equal to $\mathcal{M}_{\mathcal{T}^U}$. By the uniqueness of the Cholesky decomposition, we have a bijection $\mathcal{T}^U \ni h \mapsto hh^\top \in \mathcal{M}_G^U$. The tangent space of $\mathcal{M}_G^U$ at I_N is naturally identified with $V^U := \{ U^\top y U \,;\, y \in V \}$.

In general, for $x \in \mathrm{Sym}(N, \mathbb{R})$, we denote by $\underset{\vee}{x}$ the lower triangular matrix for which $x = \underset{\vee}{x} + (\underset{\vee}{x})^\top$. Actually, we have $(\underset{\vee}{x})_{ij} = \begin{cases} x_{ij} & (i > j) \\ x_{ii}/2 & (i = j) \\ 0 & (i < j). \end{cases}$ Then we define a bilinear product $\triangle$ on $\mathrm{Sym}(N, \mathbb{R})$ by

$$x \triangle y := \underset{\vee}{x} y + y (\underset{\vee}{x})^\top \qquad (x, y \in \mathrm{Sym}(N, \mathbb{R})).$$

The algebra $(\mathrm{Sym}(N, \mathbb{R}), \triangle)$ forms a compact normal left-symmetric algebra (CLAN), see [9, Chapter 2] and [6].

Lemma 1. *The space V^U is a subalgebral of $(\mathrm{Sym}(N, \mathbb{R}), \triangle)$. Namely, for any $x, y \in V^U$, one has $x \triangle y \in V^U$.*

Proof. Let $\mathfrak{t}^U$ be the Lie algebra of $\mathcal{T}^U$. In view of the Cholesky decomposition mentioned above, we have a linear isomorphism $\mathfrak{t}^U \ni T \mapsto T + T^\top \in V^U$. Thus we obtain

$$\mathfrak{t}^U = \left\{ \underset{\vee}{x} \,;\, x \in V^U \right\}. \tag{1}$$

Let us consider the action of $h \in \mathcal{T}^U$ on the set $\mathcal{M}_G^U$ given by $\mathcal{M}_G^U \ni \theta \mapsto h\theta h^\top \in \mathcal{M}_G^U$. This action is naturally extended to the affine space $I_N + V^U$ as affine transformations. The infinitesimal action of $T = \underset{\vee}{x} \in \mathfrak{t}^U$ on $I_N + y \in I_N + V^U$ $(y \in V^U)$ is equal to $T(I_N + y) + (I_N + y)T^\top = x + x \triangle y$ which must be an element of V^U. Therefore $x \triangle y \in V^U$. □

Proposition 1 ([6, Theorem 2 and Proposition 2]). *If V^U contains the identity matrix I_N, after an appropriate permutation of the rows and columns, V^U becomes the set of symmetric matrices of the form*

$$y = \begin{pmatrix} Y_{11} & Y_{21}^\top & \dots & Y_{r1}^\top \\ Y_{21} & Y_{22} & & Y_{r2}^\top \\ \vdots & & \ddots & \vdots \\ Y_{r1} & Y_{r2} & \dots & Y_{rr} \end{pmatrix} \quad \begin{pmatrix} Y_{kk} = y_{kk} I_{\nu_k},\ y_{kk} \in \mathbb{R}, \quad (k = 1, \dots, r) \\ Y_{lk} \in V_{lk} \quad (1 \le k < l \le r) \end{pmatrix},$$

where $N = \nu_1 + \dots + \nu_r$, and V_{lk} are subspaces of $\mathrm{Mat}(\nu_l, \nu_k; \mathbb{R})$ satisfying
(V1) *$A \in V_{lk} \Rightarrow AA^\top \in \mathbb{R} I_{\nu_l}$ for $1 \le k < l \le r$,*

(V2) $A \in V_{lj}, B \in V_{kj} \Rightarrow AB^\top \in V_{lk}$ *for* $1 \le j < k < l \le r$,
(V3) $A \in V_{lk}, B \in V_{kj} \Rightarrow AB \in V_{lj}$ *for* $1 \le j < k < l \le r$.

Clearly $I_N \in V^U$ if and only if $I_N \in V$. In this case, Proposition 1 together with [6, Theorem 3] tells us that $\mathcal{T}^U = \mathfrak{t}^U \cap \mathcal{T}_N$ and $\mathcal{M}_G^U = \mathcal{M}_{\mathcal{T}^U} = V^U \cap \mathrm{Sym}^+(N, \mathbb{R})$. It follows that we obtain:

Proposition 2. *If* $I_N \in V$, *one has* $\mathcal{M}_G = V \cap \mathrm{Sym}^+(N, \mathbb{R})$.

Here we remark that every homogeneous cone is realized as $\mathcal{M}_G$ this way by [6]. For example, if $U = I_N$, $\nu_1 = \cdots = \nu_r = 2$ and $V_{lk} = \left\{ \begin{pmatrix} a & -b \\ b & a \end{pmatrix} ; a, b \in \mathbb{R} \right\}$ $(1 \le k < l \le r)$, then $\mathcal{M}_G = V \cap \mathrm{Sym}^+(N, \mathbb{R})$ is linearly isomorphic to the cone $\mathrm{Herm}^+(r, \mathbb{C})$ of positive definite $r \times r$ Hermitian matrices, which is a realization of the symmetric space $GL(r, \mathbb{C})/U(r)$. See [4] and [6] for other examples.

If V^U does not contain I_N, we consider the direct sum $\tilde{V}^U := \mathbb{R}I_N \oplus V^U$, which is also a subalgebra of $(\mathrm{Sym}(N, \mathbb{R}), \triangle)$. Then we apply Proposition 1 to $\tilde{V}^U$. Since V^U is a two-sided ideal of $\tilde{V}^U$ of codimension one, after some renumbering of indices k and l, we see that V^U equals the subspace of $\tilde{V}^U$ costing of elements y with $y_{11} = 0$. Namely, we have the following.

Theorem 1. *If* V^U *does not contain the identity matrix* I_N, *after an appropriate permutation of the rows and columns,* V^U *becomes the set of symmetric matrices of the form*

$$y = \begin{pmatrix} 0 & Y_{21}^\top & \dots & Y_{r1}^\top \\ Y_{21} & Y_{22} & & Y_{r2}^\top \\ \vdots & & \ddots & \vdots \\ Y_{r1} & Y_{r2} & \dots & Y_{rr} \end{pmatrix} \quad \begin{pmatrix} Y_{kk} = y_{kk} I_{\nu_k}, \; y_{kk} \in \mathbb{R}, \quad k = 2, \dots, r \\ Y_{lk} \in V_{lk}, \quad 1 \le k < l \le r \end{pmatrix},$$

where V_{lk} *are the same as in Proposition 1.*

In what follows, we shall consider the case where $I_N \notin \mathcal{M}_G^U$ because our results below for the case $I_N \in \mathcal{M}_G^U$ will be obtained formally by putting $\nu_1 = 0$ and $V_{k1} = \{0\}$, as is understood by comparing Proposition 1 and Theorem 1. Put $N' := \nu_2 + \cdots + \nu_r = N - \nu_1$. Let W and V' be the vector spaces of matrices w and y' respectively of the forms

$$w = \begin{pmatrix} Y_{21} \\ \vdots \\ Y_{r1} \end{pmatrix} \in \mathrm{Mat}(N', \nu_1; \mathbb{R}), \quad y' = \begin{pmatrix} Y_{22} & & Y_{r2}^\top \\ \vdots & \ddots & \\ Y_{r2} & \dots & Y_{rr} \end{pmatrix} \in \mathrm{Sym}(N', \mathbb{R}).$$

Let $\mathcal{P}'$ be the set $V' \cap \mathrm{Sym}^+(N', \mathbb{R})$. Then $\mathcal{P}'$ forms a pointed open convex cone in the vector space V'.

Theorem 2. *Under the assumptions above, one has*

$$\mathcal{M}_G = \left\{ U \begin{pmatrix} I_{\nu_1} & w^\top \\ w & y' \end{pmatrix} U^\top ; w \in W, \; y' \in V', \; y' - ww^\top \in \mathcal{P}' \right\}.$$

Moreover $\mathcal{M}_G$ *equals the intersection of* $\mathrm{Sym}^+(N, \mathbb{R})$ *and the affine space* $I_N + V$.

Proof. Let $\mathfrak{t}'$ be the set of the lower triangular matrices $\underset{\vee}{y'}$ with $y' \in V'$. Then we see from [6] that $\mathfrak{t}'$ is a Lie algebra, and that the corresponding Lie group $\mathcal{T}' := \exp \mathfrak{t}' \subset GL(N', \mathbb{R})$ equals $\mathfrak{t}' \cap \mathcal{T}_{N'}$. Namely, $\mathcal{T}'$ is the set of h' of the form

$$h' = \begin{pmatrix} T_{22} & & \\ \vdots & \ddots & \\ T_{r2} & \dots & T_{rr} \end{pmatrix} \in \mathcal{T}_{N'} \quad \begin{pmatrix} T_{kk} = t_{kk} I_{\nu_k},\ t_{kk} > 0, \quad k = 2, \dots, r \\ T_{lk} \in V_{lk}, \quad 2 \le k < l \le r \end{pmatrix}.$$

Since we have $\mathfrak{t}^U = \left\{ \begin{pmatrix} 0 & \\ L & T' \end{pmatrix} ;\ L \in W,\ T' \in \mathfrak{t}' \right\}$ by (1), the corresponding Lie group $\mathcal{T}^U$ is the set of $\begin{pmatrix} I_{\nu_1} & 0 \\ L & h' \end{pmatrix}$ with $L \in W$ and $h' \in \mathcal{T}'$. Therefore, $\mathcal{M}_G^U = \mathcal{M}_{\mathcal{T}^U}$ is the set of matrices $\begin{pmatrix} I_{\nu_1} & 0 \\ L & h' \end{pmatrix} \begin{pmatrix} I_{\nu_1} & L^\top \\ 0 & (h')^\top \end{pmatrix} = \begin{pmatrix} I_{\nu_1} & L^\top \\ L & LL^\top + a' \end{pmatrix}$ with $a' = h'(h')^\top \in \mathcal{P}'$. On the other hand, as is discussed in [6, Theorem 3], the map $\mathcal{T}' \ni h' \mapsto h'(h')^\top \in \mathcal{P}'$ is bijective, which completes the proof. □

3 Existence Condition and an Explicit Expression of MLE

Let $X_1, \dots, X_n$ be independent random vectors obeying the central multivariate normal law $N(0, \Sigma)$ with $\theta := \Sigma^{-1} \in \mathcal{M}_G$. The density function f of $(X_1, \dots, X_n)$ is given by

$$f(x_1, \dots, x_n; \theta) := (2\pi)^{-nN/2} (\det \theta)^{n/2} \prod_{j=1}^{N} e^{-x_j^\top \theta x_j / 2} \quad (x_j \in \mathbb{R}^N,\ j = 1, \dots, n).$$

Let $\pi : \mathrm{Sym}(N, \mathbb{R}) \to V$ be the orthogonal projection with respect to the trace inner product. Putting $y := \pi(\sum_{j=1}^N x_j x_j^\top / 2) \in V$, we have $f(x_1, \dots, x_n; \theta) = c(x)(\det \theta)^{n/2} e^{-\operatorname{tr} \theta y}$, where $c(x) := (2\pi)^{-nN/2} \exp(\operatorname{tr} \{(I_N - \pi(I_N)) \sum_{j=1}^N x_j x_j^\top / 2\})$, which is independent of θ. Thus, given $Y := \pi(\sum_{k=1}^n X_j X_j^\top / 2) \in V$, the maximum likelihood estimator $\hat{\theta}$ is an element of $\mathcal{M}_G$ at which the log likelihood function $F(\theta; Y) := (n/2) \log \det \theta - \operatorname{tr}(Y\theta)$ attains the maximum value.

In what follows, we shall assume that $U = I_N$. Indeed, a general case is easily reduced to this case. For $k = 2, \dots, r$, let $V_{[k]}$ and W_{k-1} be the vector spaces of matrices $y_{[k]}$ and Z_{k-1} respectively of the forms

$$y_{[k]} = \begin{pmatrix} Y_{kk} & & Y_{rk}^\top \\ \vdots & \ddots & \\ Y_{rk} & \dots & Y_{rr} \end{pmatrix} \in \mathrm{Sym}(N_k, \mathbb{R}), \quad Z_{k-1} = \begin{pmatrix} Y_{k,k-1} \\ \vdots \\ Y_{r,k-1} \end{pmatrix} \in \mathrm{Mat}(N_k, \nu_{k-1}; \mathbb{R}),$$

where $N_k := \nu_k + \cdots + \nu_r$. Note that $V' = V_{[2]}$ and $W = W_1$ in the previous notation. We have an inductive expression of $y \in V$ as

$$y = \begin{pmatrix} 0 & Z_1^\top \\ Z_1 & y_{[2]} \end{pmatrix}, \quad y_{[k]} = \begin{pmatrix} y_{kk} I_{\nu_k} & Z_k^\top \\ Z_k & y_{[k+1]} \end{pmatrix} \quad (k = 2, \ldots r-1), \quad y_{[r]} = y_{rr} I_{\nu_r}. \tag{2}$$

Let $\mathcal{T}_{[k]} \subset \mathcal{T}_{N_k}$ be the group of lower triangular matrices $\underset{\vee}{y_{[k]}}$ $(y_{[k]} \in V_{[k]})$ with positive diagonal entries. Then $\mathcal{T}' = \mathcal{T}_{[2]}$. Any element $h \in \mathcal{T}$ is expressed as

$$h = \begin{pmatrix} I_{\nu_1} & 0 \\ L_1 & h_{[2]} \end{pmatrix}, \quad h_{[k]} = \begin{pmatrix} t_{kk} I_{\nu_k} & 0 \\ L_k & h_{[k+1]} \end{pmatrix} \in \mathcal{T}_{[k]} \quad (k = 2, \ldots, r-1)$$

with $L_k \in W_k$ and $h_{[r]} = t_{rr} I_{\nu_r} \in \mathcal{T}_{[r]}$. We observe that

$$h_{[k]} h_{[k]}^\top = \begin{pmatrix} t_{kk}^2 I_{\nu_k} & t_{kk} L_k^\top \\ t_{kk} L_k & L_k L_k^\top + h_{[k+1]} h_{[k+1]}^\top \end{pmatrix} \in V_{[k]}.$$

We shall regard $V_{[k]}$ as a subspace of $V' = V_{[2]}$ by zero-extension. Define a map $q_k : \mathbb{R}_{>0} \times W_k \to V'$ for $k = 2, \ldots, r-1$ by $q_k(t_{kk}, L_k) := \begin{pmatrix} t_{kk}^2 & t_{kk} L_k^\top \\ t_{kk} L_k & L_k L_k^\top \end{pmatrix} \in V_{[k]} \subset V'$, and define also $q_r(t_{rr}) := t_{rr}^2 I_{n_r} \in V_{[r]} \subset V'$. If $\theta = hh^\top$, we have

$$\theta = \begin{pmatrix} I_{\nu_1} & L_1 \\ L_1 & L_1 L_1^\top + \theta' \end{pmatrix}, \quad \theta' = \sum_{k=2}^{r-1} q_k(t_{kk}, L_k) + q_r(t_{rr}). \tag{3}$$

For $y' \in V' = V_{[2]}$ and $k = 2, \ldots, r-1$, we have

$$\operatorname{tr}(y' q_k(t_{kk}, L_k)) = \nu_k y_{kk} t_{kk}^2 + 2 t_{kk} \operatorname{tr}(Z_k^\top L_k) + \operatorname{tr}(y_{[k+1]} L_k L_k^\top). \tag{4}$$

Let m_k $(k = 1, \ldots, r-1)$ be the dimension of the vector space W_k, and take an orthonormal basis $\{e_{k\alpha}\}_{\alpha=1}^{m_k}$ of W_k with respect to the trace inner product. For $L_k \in W_k$, let $\lambda_k := (\lambda_{k1}, \ldots, \lambda_{km_k})^\top \in \mathbb{R}^{m_k}$ be the column vector for which $L_k = \sum_{\alpha=1}^{m_k} \lambda_{k\alpha} e_{k\alpha}$. Defining $\zeta_k \in \mathbb{R}^{m_k}$ for $Z_k \in W_k$ similarly, we have $\zeta_k^\top \lambda_k = \operatorname{tr}(Z_k^\top L_K)$. Let $\psi_k : V' \to \mathrm{Sym}(m_k, \mathbb{R})$ be a linear map defined in such a way that $\operatorname{tr}(y' L_k L_k^\top) = \lambda_k^\top \psi_k(y') \lambda_k$, and define $\phi_k : V' \to \mathrm{Sym}(1 + m_k, \mathbb{R})$ by $\phi_k(y') := \begin{pmatrix} \nu_k y_{kk} & \zeta_k^\top \\ \zeta_k & \psi_k(y) \end{pmatrix}$. In view of (4), we have

$$\operatorname{tr}(y' q_k(t_{kk}, L_k)) = \operatorname{tr}\left(\phi_k(y') \begin{pmatrix} t_{kk}^2 & t_{kk} \lambda_k^\top \\ t_{kk} \lambda_k & \lambda_k \lambda_k^\top \end{pmatrix} \right). \tag{5}$$

Let $\phi_k^* : \mathrm{Sym}(1 + m_k, \mathbb{R}) \to V'$ be the adjoint map of ϕ_k, which means that $\operatorname{tr} \phi_k^*(S) y' = \operatorname{tr} S \phi(y')$ for $S \in \mathrm{Sym}(1 + m_k, \mathbb{R})$. Define also $\phi_r(y') := \nu_r y_{rr} \in \mathbb{R} \equiv \mathrm{Sym}(1, \mathbb{R})$ and $\phi_r^*(c) = c I_{\nu_r} \in V_{[r]} \subset V'$ for $c \in \mathbb{R}$.

Let $\mathcal{Q}' \subset V'$ be the dual cone of $\mathcal{P}' \subset V'$, that is, the set of $y' \in V'$ such that $\operatorname{tr}(y'a) > 0$ for all $a \in \overline{\mathrm{P}'} \setminus \{0\}$. If $y' \in \mathcal{Q}'$, then $\phi_k(y')$ and $\psi_{k-1}(y')$ are positive definite for $k = 2, \ldots, r$. Moreover, it is known (see [4, Proposition 3.4 (iii)]) that

$$\mathcal{Q}' = \{ y' \in V' \,;\, \det \phi_k(y') > 0 \text{ for all } k = 2, \ldots, r \}. \tag{6}$$

If $Y := \pi(\sum_{j=1}^n X_j X_j^\top / 2)$ is expressed as in (2), then we can show that $Y' := Y_{[2]}$ belongs to the closure of $\mathcal{Q}'$, so that $\phi_k(Y')$ and $\psi_{k-1}(Y')$ are positive semidefinite for $k = 2, \ldots r$.

Theorem 3. (i) *If $Y' \in \mathcal{Q}'$, then $\hat{\theta} = \arg \max_{\theta \in \mathcal{M}_G} F(\theta; Y)$ exists uniquely, and it is expressed as $\hat{\theta} = \begin{pmatrix} I_{\nu_1} & \hat{L}_1^\top \\ \hat{L}_1 & \hat{L}_1 \hat{L}_1^\top + \hat{\theta}' \end{pmatrix}$ with $\hat{\lambda}_1 = -\psi_1(Y')^{-1}\zeta_1 \in \mathbb{R}^{m_1}$, which is the column vector corresponding to $\hat{L}_1$, and*

$$\hat{\theta}' = \frac{n\nu_r}{2} \phi_r^*(\phi_r(Y')^{-1}) + \sum_{k=2}^{r-1} \frac{n\nu_k}{2} \phi_k^* \left(\phi_k(Y')^{-1} - \begin{pmatrix} 0 & 0 \\ 0 & \psi_k(Y')^{-1} \end{pmatrix} \right). \tag{7}$$

(ii) *If $Y' \notin \mathcal{Q}'$, then $F(\theta; Y)$ is unbounded, so that $\hat{\theta}$ does not exist.*

Proof. (i) Keeping (2), (3) and $(\det \theta)^{n/2} = \prod_{k=2}^r (t_{kk})^{n\nu_k}$ in mind, we define

$$\begin{aligned} F_1(L_1; Y) &:= -2\mathrm{tr}(Z_1^\top L_1) - \mathrm{tr}(Y' L_1 L_1^\top), \\ F_k(t_{kk}, L_k; Y) &:= n\nu_k \log t_{kk} - \mathrm{tr}\,(q_k(t_{kk}, L_k) Y') \quad (k = 2, \ldots, r-1), \\ F_r(t_{rr}; Y) &:= n\nu_r \log t_{rr} - \nu_r y_{rr} t_{rr}^2, \end{aligned}$$

so that $F(\theta; Y) = F_1(L_1; Y) + F_r(t_{rr}; Y) + \sum_{k=2}^{r-1} F_k(t_{kk}, L_k; Y)$. It is easy to see that $F_r(t_{rr}; Y)$ takes a maximum value at $\hat{t}_{rr} = \sqrt{\frac{n}{2y_{rr}}}$. Then $q_r(\hat{t}_{rr}) = \frac{n}{2y_{rr}} I_{\nu_r} = \frac{n\nu_r}{2} \phi_r^*(\phi_r(Y')^{-1})$. On the other hand, $F_1(L_1; Y) = -2\zeta_1^\top \lambda_1 - \lambda_1^\top \psi_1(Y')\lambda_1$ equals

$$\zeta_1^\top \psi_1(Y')^{-1}\zeta_1 - (\lambda_1 + \psi_1(Y')^{-1}\zeta_1)^\top \psi_1(Y')(\lambda_1 + \psi_1(Y')^{-1}\zeta_1),$$

which attains a maximum value when $\lambda_1 = -\psi_1(Y')^{-1}\zeta_1$ because $\psi_1(Y')$ is positive definite. Similarly, we see from (5) that

$$\begin{aligned} F_k(t_{kk}, L_k) = &\, n\nu_k \log t_{kk} - (\nu_k y_{kk} - \zeta_k^\top \psi_k(Y')^{-1}\zeta_k) t_{kk}^2 \\ &- (\lambda_k + t_{kk}\psi_k(Y')^{-1}\zeta_k)^\top \psi_k(Y')(\lambda_k + t_{kk}\psi_k(Y')^{-1}\zeta_k). \end{aligned} \tag{8}$$

Since $\nu_k y_{kk} - \zeta_k^\top \psi_k(Y')^{-1}\zeta_k = \det \phi_k(Y') / \det \psi_k(Y') > 0$, we see that $F_k(t_{kk}, L_k; Y')$ attains at $(\hat{t}_{kk}, \hat{L}_k)$ with $\hat{t}_{kk} = \sqrt{\frac{n\nu_k}{2(\nu_k y_{kk} - \zeta_k^\top \psi_k(Y')^{-1}\zeta_k)}}$ and $\hat{\lambda}_k = -\hat{t}_{kk}\psi_k(Y')^{-1}\zeta_k \in \mathbb{R}^{m_k}$, which is the column vector corresponding to $\hat{L}_k$. By a straightforward calculation, we have

$$\begin{pmatrix} \hat{t}_{kk}^2 & \hat{t}_{kk}\hat{\lambda}_k^\top \\ \hat{t}_{kk}\hat{\lambda}_k & \hat{\lambda}_k \hat{\lambda}_k^\top \end{pmatrix} = \frac{n\nu_k}{2} \left(\phi_k(Y')^{-1} - \begin{pmatrix} 0 & 0 \\ 0 & \psi_k(Y')^{-1} \end{pmatrix} \right) \in \mathrm{Sym}(1 + m_k, \mathbb{R}),$$

which maps to $q_k(\hat{t}_{kk}, \hat{L}_k)$ by $\phi_k^* : \mathrm{Sym}(1+m_k, \mathbb{R}) \to V'$ thanks to (5). Therefore the assertion (i) is verified.
(ii) By (6), there exists k for which $\det \phi_k(Y') = 0$. If $\det \phi_r(Y') = y_{rr}^{\nu_r} = 0$, then $F_r(t_{rr}; Y') = n\nu_r \log t_{rr} \to +\infty$ in $t_{rr} \to +\infty$. Let us consider the case where $\det \phi_l(Y') > 0$ for $l = k+1, \ldots, r$ and $\det \phi_k(Y') = 0$. Then one can show that $\psi_k(Y')$ is positive definite because in this case $Y_{[k+1]} \in V_{[k+1]}$ belongs to the dual cone of $\mathcal{P}_{[k+1]} := V_{[k+1]} \cap \mathrm{Sym}^+(N_k, \mathbb{R})$ by the same reason as (6). Since $\nu_k y_{kk} - \zeta_k^\top \psi_k(Y')^{-1} \zeta_k = \det \phi_k(Y') / \det \psi_k(Y') = 0$, we see from (8) that

$$\max_{L_k \in W_k} F_k(t_{kk}, L_k; Y') = n\nu_k \log t_{kk} \to +\infty \quad (t_{kk} \to +\infty),$$

which completes the proof. □

We remark that a generalization of the formula (7) is found in [5, Theorem 5.1]. By Theorem 3, the existence of the MLE $\hat{\theta}$ is equivalent to that the random matrix Y' belongs to the cone $\mathcal{Q}'$ with probability one. On the other hand, the distribution of Y' is nothing else but the Wishart distribution on $\mathcal{Q}'$ studied in [4] and [5]. The Wishart distribution is obtained as a natural exponential family generated by the Riesz distribution μ_n on V' characterized by its Laplace transform: $\int_{V'} e^{-\mathrm{tr}(y'a)} \mu_n(dy') = (\det a)^{-n/2}$ for all $a \in \mathcal{P}'$. Note that, if $a \in \mathcal{P}'$ is a diagonal matrix, we have $(\det a)^{-n/2} = \prod_{k=2}^{r} a_{kk}^{-n\nu_k/2}$. As is seen in [5, Theorem 4.1 (ii)], the support of the Riesz distribution μ_n is determined from the parameter $(n\nu_2/2, \ldots, n\nu_r/2) \in \mathbb{R}^{r-1}$. In fact, $\mathrm{supp}\, \mu_n = \overline{\mathcal{Q}'}$ if and only if $n\nu_k/2 > m_k/2$ for $k = 2, \ldots, r$, and $Y' \in \mathcal{Q}'$ almost surely in this case. Otherwise, $\mathrm{supp}\, \mu_n$ is contained in the boundary of $\mathcal{Q}'$, so that Y' never belongs to $\mathcal{Q}'$. Therefore, if n_0 is the smallest integer that is greater than $\max\left\{ \frac{m_k}{\nu_k} \,;\, k = 2, \ldots, r \right\}$, we have the following final result.

Theorem 4. *The maximum likelihood estimator* $\hat{\theta} = \arg\max_{\theta \in \mathcal{M}_G} F(\theta; Y)$ *exists with probability one if and only if* $n \geq n_0$. *If* $n < n_0$, *then the log likelihood function* $F(\theta; Y)$ *of* $\theta \in \mathcal{M}_G$ *is unbounded.*

References

1. Améndola, C., Kohn, K., Reichenbach, P., Seigal, A.: Invariant theory and scaling algorithms for maximum likelihood estimation. arXiv:2003.13662 (2020)
2. Barndorff-Nielsen, O., Blaesild, P., Ledet Jensen, J., Jørgensen, B.: Exponential transformation models. Proc. Royal Soc. London Ser. A **379**(1776), 41–65 (1982)
3. Faraut, J., Korányi, A.: Analysis on Symmetric Cones. Clarendon Press, New York (1994)
4. Graczyk, P., Ishi, H.: Riesz measures and Wishart laws associated to quadratic maps. J. Math. Soc. Japan **66**, 317–348 (2014)
5. Graczyk, P., Ishi, H., Kołodziejek, B.: Wishart laws and variance function on homogeneous cones. Probab. Math. Statist. **39**, 337–360 (2019)
6. Ishi, H.: Matrix realization of a homogeneous cone. In: Nielsen, F., Barbaresco, F. (eds.) GSI 2015. LNCS, vol. 9389, pp. 248–256. Springer, Cham (2015). https://doi.org/10.1007/978-3-319-25040-3_28

7. Ishi, H.: Explicit formula of Koszul-Vinberg characteristic functions for a wide class of regular convex cones. Entropy **18**(11), 383 (2016). https://doi.org/10.3390/e18110383
8. Letac, G., Massam, H.: Wishart distributions for decomposable graphs. Ann. Statist. **35**, 1278–1323 (2007)
9. Vinberg, E.B.: The theory of convex homogeneous cones. Trans. Mosc. Math. Soc. **12**, 340–403 (1963)

Exponential-Wrapped Distributions on $SL(2, \mathbb{C})$ and the Möbius Group

Emmanuel Chevallier(✉)

Aix Marseille Univ, CNRS, Centrale Marseille, Institut Fresnel, Marseille, France
emmanuel.chevallier@univ-amu.fr

Abstract. In this paper we discuss the construction of probability distributions on the group $SL(2, \mathbb{C})$ and the Möbius group using the exponential map. In particular, we describe the injectivity and surjectivity domains of the exponential map and provide its Jacobian determinant. We also show that on $SL(2, \mathbb{C})$ and the Möbius group, there are no isotropic distributions in the group sense.

Keywords: Statistics on lie groups · Exponential map · Wrapped distributions · Killing form

1 Introduction

Modelling and estimating probability densities on manifolds raises several difficulties and is still an active research topic, see for instance [1,2]. In most cases the manifolds studied are endowed with a Riemannian metric. It is for instance the case for the manifold of positive definite matrices, or the rotations group of an Euclidean space. In these contexts all the structures used in the statistical analysis are related to the distance. The other important structure addressed in the literature is the Lie group structure. Despite being a large class of manifolds, a reason why it is less often studied than the Riemannian setting, is because compact Lie groups admit a bi-invariant metric. In that case, the statistical analysis based on the Riemannian distance satisfies all the group requirements. Most examples of statistics problems on non-compact groups studied in the literature arise from rigid and affine deformations of the physical space $\mathbb{R}^3$. Due to their role in polarization optics, see [7,8], we study here the group $SL(2, \mathbb{C})$ and its quotient, the Möbius group.

In Sect. 2, we review the main important facts about $SL(2, \mathbb{C})$ and the Möbius group. In Sect. 3, we describe the construction of exponential-wrapped distributions. In particular, we show that the non-surjectivity of the exponential map on $SL(2, \mathbb{C})$ is not a major obstacle and give an expression of its Jacobian. In Sect. 4, we make a parallel between the notion of isotropy on a Riemannian manifold and a notion of isotropy on a group. We show that unfortunately, except the Dirac on the identity, there are no isotropic probability distributions on $SL(2, \mathbb{C})$ and the Möbius group in that sense.

F. Nielsen and F. Barbaresco (Eds.): GSI 2021, LNCS 12829, pp. 265–272, 2021.
https://doi.org/10.1007/978-3-030-80209-7_30

2 The Group SL(2, $\mathbb{C}$) and the Möbius Group

SL(2, $\mathbb{C}$) is the group of 2 by 2 complex matrices of determinant 1. Since it is defined be the polynomial equation

$$\det(M) = 1, \quad M \in M_2(\mathbb{C}), \tag{1}$$

it is a complex Lie group. Recall that a complex Lie group is a complex manifold such that the group operations are holomorphic. Recall also that a complex manifold is a manifold whose charts are open sets of $\mathbb{C}^d$ and whose transition are holomorphic.

In this paper, we define the Möbius group, noted *Möb*, as the quotient of SL(2, $\mathbb{C}$) by the group $+I, -I$, where I is the identity matrix. Since

$$M\ddot{o}b = \mathrm{SL}(2,\mathbb{C})/\{+I,-I\} \sim \mathrm{PGL}(2,\mathbb{C}),$$

the Möbius group can also be seen as a projective linear group. Recall that PGL(2, $\mathbb{C}$) is defined as a quotient of $GL(2,\mathbb{C})$ by the multiples of the identity matrix.

Since *Möb* is a quotient of SL(2, $\mathbb{C}$) by a discrete subgroup, they have the same Lie algebra, noted $\mathfrak{sl}(2,\mathbb{C})$. The complex structure of SL(2, $\mathbb{C}$) makes $\mathfrak{sl}(2,\mathbb{C})$ a complex vector space. By differentiating Eq. 1 around the identity matrix, we can check that $\mathfrak{sl}(2,\mathbb{C})$ is the vector space of complex matrices with zero trace. It is easy to see that $\mathfrak{sl}(2,\mathbb{C})$ is the complexification of real traceless matrices, hence

$$e = \begin{pmatrix} 0 & 1 \\ 0 & 0 \end{pmatrix}, \quad h = \begin{pmatrix} 1 & 0 \\ 0 & -1 \end{pmatrix}, \quad f = \begin{pmatrix} 0 & 0 \\ 1 & 0 \end{pmatrix}$$

generates $\mathfrak{sl}(2,\mathbb{C})$ and it is of complex dimension 3. The Lie brackets are given by

$$[h,e] = 2e, \quad [h,f] = -2f, \quad [e,f] = h,$$

hence their adjoints in the basis (e,h,f) are

$$ad_e = \begin{pmatrix} 0 & -2 & 0 \\ 0 & 0 & 1 \\ 0 & 0 & 0 \end{pmatrix} \quad ad_h = \begin{pmatrix} 2 & 0 & 0 \\ 0 & 0 & 0 \\ 0 & 0 & -2 \end{pmatrix}, \quad ad_f = \begin{pmatrix} 0 & 0 & 0 \\ -1 & 0 & 0 \\ 0 & 2 & 0 \end{pmatrix}.$$

The Killing form on a Lie algebra is given by

$$\kappa(X,Y) = \mathrm{Tr}(ad_X ad_Y).$$

It can be checked that on $\mathfrak{sl}(2,\mathbb{C})$,

$$\kappa(X,Y) = 4\,\mathrm{Tr}(XY),$$

and that for a matrix $X = a.e + b.h + c.f$,

$$\kappa(X,X) = 8(b^2 + ac) = -8\det(X) = 8\lambda^2,$$

where λ is an eigenvalue of X. Recall that the eigenvalues of $X \in \mathrm{SL}(2,\mathbb{C})$ are either distinct and opposite or null.

3 Exponential-Wrapped Distributions

In this section, we determine domains on the Lie groups and Lie algebra involved in the construction of exponential-wrapped distributions on SL(2, ℂ) and the Möbius group, and provide the expression of their densities.

Let us recall how the definition of the exponential map is defined on the Lie algebra, and how it is extended to every tangent spaces. The exponential map exp on a Lie group G maps a vector in the Lie algebra $X \in T_eG$ to $\gamma(1)$, where γ is the one parameter subgroup with $\gamma'(0) = X$. This map is defined on T_eG but can be extended to every tangent spaces T_gG using pushforwards of the group multiplications. The following identity

$$g \exp_e(X) g^{-1} = \exp_e(dL_g dR_g^{-1}(X)) = \exp_e(dR_g^{-1} dL_g(X)),$$

ensures that the definition of the exponential at g is independent of the choice of left or right multiplication:

$$\begin{aligned} \exp_g :& T_gG \to G \\ & u \mapsto \exp_g(X) = g. \exp\left(dL_{g^{-1}} X\right) = \exp\left(dR_{g^{-1}} X\right) g. \end{aligned}$$

In the rest of the paper, exponentials without subscript refer to exponentials at identity. The definition of exponential maps on arbitrary tangent space has a deep geometric interpretation, as exponentials maps of an affine connection. See [5] for a detailed description of this point of view.

Exponential-wrapped distributions refers to distributions pushed forward from a tangent space to the group, by an exponential map. Given a probability distribution $\tilde{\mu}$ on T_gG,

$$\mu = \exp_{g*}(\tilde{\mu})$$

is a probability distribution on G. Consider a left or right invariant field of basis on G and the associated Haar measure. Recall that since SL(2, ℂ) is unimodular, left invariant and right invariant measure are bi-invariant. When $\tilde{\mu}$ has a density $\tilde{f}$, the density f of μ with respect to the Haar measure is the density $\tilde{f}$ divided by the absolute value of the Jacobian determinant of the differential of the exponential map. If $\tilde{f}$ is vanishing outside an injectivity domain,

$$f(\exp_g(X)) = \frac{\tilde{f}(X)}{J(X)}, \quad \text{with } J(X) = |\det(d\exp_{g,X})|. \tag{2}$$

The interest of this construction is dependent on the injectivity and surjectivity properties of the exponential map. The situation is ideal when the exponential map is bijective: exponential-wrapped distributions can model every distribution on the group. Non-surjectivity is a potentially bigger issue that non-injectivity. As we will see, the injectivity can be forced by restricting the tangent space to an injectivity domain. However when the ranges of the exponential maps are too small, arbitrary distributions on the group might only be modeled by a mixture of exponential-wrapped distributions involving a large number of tangent spaces. On SL(2, ℂ), the exponential map is unfortunately neither injective nor surjective. However, we have the following.

Fact 1. *The range of the exponential map of the group* $SL(2,\mathbb{C})$ *is*

$$U = \{M \in SL(2,\mathbb{C}) | \operatorname{Tr}(M) \neq -2\} \cup \{-I\},$$

where I *is the identity matrix. As a result,* $SL(2,\mathbb{C}) \setminus \exp(\mathfrak{sl}(2,\mathbb{C}))$ *has zero measure.*

Fact 1 can be checked using the Jordan decomposition of $\mathrm{SL}(2,\mathbb{C})$ matrices. A detailed study of the surjectivity of the exponential for $\mathrm{SL}(2,\mathbb{R})$ and $\mathrm{SL}(2,\mathbb{C})$ can be found in [6]. This fact shows that the non surjectivity does not significantly affect the modeling capacities of exponential-wrapped distributions. As a direct consequence of Fact 1, we have

Corollary 1.

- $SL(2,\mathbb{C})$ *is covered by* $\exp_I$ *and* $\exp_{-I}$
- *The exponential map of the Möbius group is surjective.*

Proof. For $M \in \mathrm{SL}(2,\mathbb{C})$, M or $-IM$ has a positive trace and is in the range of the $\mathrm{SL}(2,\mathbb{C})$ exponential. Since the Möbius group is $\mathrm{SL}(2,\mathbb{C})$ quotiented by the multiplication by $-I$, at least one element of the equivalent classes is reached by the exponential.

We now provide injectivity domains of the exponentials, which enable the definition of the inverses.

Fact 2. *The exponential of* $SL(2,\mathbb{C})$ *is a bijection between* $\mathfrak{U}$ *and* U *with,*

$$\mathfrak{U} = \{X \in \mathfrak{sl}(2,\mathbb{C}) | \quad \operatorname{Im}(\lambda) \in]-\pi,\pi] \, \textit{for any eigenvalue}\, \lambda \, \textit{of}\, X\}.$$

Recall that eigenvalues of X *are always opposite. The exponential of the Möbius group is bijective on*

$$\mathfrak{U}_{M\ddot{o}b} = \left\{X \in \mathfrak{sl}(2,\mathbb{C}) | \quad \operatorname{Im}(\lambda) \in \left]-\frac{\pi}{2},\frac{\pi}{2}\right] \textit{for any eigenvalue}\, \lambda \, \textit{of}\, X\right\}.$$

This is a direct consequence of Theorem 1.31 of [4], which defines the principal matrix logarithm for matrices with no eigenvalues in $\mathbb{R}_-$, by setting the eigenvalues of the logarithm to the interval $]-\pi,\pi[$. In order to have a bijective maps, the intervals should contains exactly one of their extremities. $\mathfrak{U}_{M\ddot{o}b}$ is obtained by noting that the quotient by $\{I,-I\}$ in $\mathrm{SL}(2,\mathbb{C})$ translates to an quotient by $i\pi\mathbb{Z}$ on eigenvalues in the Lie algebra. Note that since $\sqrt{\frac{\kappa(X,X)}{2}} = \pm\lambda$, the injectivity domains can be expressed as the inverse image of subset of $\mathbb{C}$ by the quadratic form $X \mapsto \kappa(X,X)$.

Hence, when the support of $\tilde{f}$ is included in $\mathfrak{U}$ or $\mathfrak{U}_{M\ddot{o}b}$, Eq. 2 holds. We address now the computation of the volume change term. Note that in Eq. 2, the determinant is seen as a volume change between real vector spaces. Recall also that for a complex linear map A on $\mathbb{C}^n$, the real determinant over $\mathbb{R}^{2n}$ is given by $\det_{\mathbb{R}}(A) = |\det_{\mathbb{C}}(A)|^2$.

The differential in a left invariant field of basis of the exponential map at identity e evaluated on the vector $X \in T_eG$ is given by the following formula, see [3],

$$d\exp_X = dL_{\exp(X)} \circ \left(\sum_{k \geq 0} \frac{(-1)^k}{(k+1)!} ad_X^k \right),$$

where subscript e is dropped. Using the Jordan decomposition of ad_X, authors of [2] pointed out the fact that this Jacobian can be computed for every Lie group, even when the adjoint endmorphisms are not diagonalizable. We have,

$$\det_{\mathbb{C}} (d\exp_X) = \prod_{\lambda \in \Lambda_X} \left(\frac{1 - e^{\lambda}}{\lambda} \right)^{d_\lambda},$$

where Λ_X is the set of nonzero eigenvalues of ad_X and d_λ the algebraic multiplicity of the eigenvalue λ. For the group SL(2, ℂ), a calculation shows at the matrix X of coordinates (a, b, c) in the basis (e, h, f), the eigenvalues of

$$ad_X = a. \begin{pmatrix} 0 & -2 & 0 \\ 0 & 0 & 1 \\ 0 & 0 & 0 \end{pmatrix} + b. \begin{pmatrix} 2 & 0 & 0 \\ 0 & 0 & 0 \\ 0 & 0 & -2 \end{pmatrix} + c. \begin{pmatrix} 0 & 0 & 0 \\ -1 & 0 & 0 \\ 0 & 2 & 0 \end{pmatrix},$$

are

$$\lambda_1 = 0, \quad \lambda_2 = 2\sqrt{ac + b^2} = \sqrt{\frac{\kappa(X, X)}{2}}, \quad \lambda_3 = -\lambda_2.$$

It is interesting to note that eigenvalues of X are also eigenvalues of ad_X. The absolute value of the Jacobian becomes

$$J(a, b, c) = |\det_{\mathbb{C}} (d\exp_X)|^2 = \left| \frac{(1 - e^{-\lambda_2})(1 - e^{\lambda_2})}{\lambda_2^2} \right|^2 = \left| 2\frac{1 - \cosh(\lambda_2)}{\lambda_2^2} \right|^2.$$

J is extended by continuity by $J(0, 0, 0) = 1$: it is not surprising since the differential of the exponential at zero is the identity. Equation 2 can be rewritten as,

$$f(\exp(X)) = \left| \frac{\kappa(X, X)}{4\left(1 - \cosh\left(\sqrt{\frac{1}{2}\kappa(X, X)}\right)\right)} \right| \tilde{f}(X). \tag{3}$$

Recall that $\lambda_{2,3}$ are complex numbers, and that $\cosh(ix) = \cos(x)$. It is interesting to note that on the one parameter subgroup generated by $(0, 1, 0)$ the Jacobian is increasing, which is a sign of geodesic spreading. On the other hand on the one parameter subgroup generated by $(0, i, 0)$ the Jacobian is decreasing over $[0, \pi]$, which is a sign of geodesic focusing.

4 There Are No Group-Isotropic Probability Distributions

When the underlying manifold is equipped with a Riemannian metric, it is possible to define the notion of isotropy of a measure. A measure μ is isotropic if there is a point on the manifold such that μ is invariant by all the isometries which preserve the point. They form an important class of probability, due to their physical interpretation, and to the fact that their high degree of symmetries enable to parametrize them with a small number of parameters.

On compact Lie groups, there exists Riemannian metrics such that left and right translations are isometries, and the notion of isotropy can hence be defined in term of the distance. Unfortunately there are no Riemannian metric on $\mathrm{SL}(2,\mathbb{C})$ compatible with the group multiplications. This comes from the fact that the scalar product at identity of such a metric should be invariant by the adjoint action of the group. Since the adjoint representation of $\mathrm{SL}(2,\mathbb{C})$ is faithful, this scalar product should be invariant by a non compact group, which is not possible. Hence, isotropy cannot be defined by Riemannian distance.

However the role of the distance in the definition of isotropy is not crucial: isotropy is defined by the invariance with respect to a set of transformations. When the manifold is Riemannian, this set is the set of isometries that fix a given point, when the manifold is a Lie group, the relevant set becomes a set of group operations.

Definition 1 (group-isotropy). *Let G be a Lie group and $\mathcal{T}$ be the set of all maps $T : G \to G$ obtained by arbitrary compositions of left multiplications and right multiplication. A measure μ on a Lie group G is group-isotropic with respect to an element g if*

$$T_*\mu = \mu, \quad \forall T \in \mathcal{T}, T(g) = g,$$

where $T_*\mu$ is the pushforward of μ by T.

It is easy to checked that the elements of $\mathcal{T}$ which preserve the identity are the conjugations. Recall that the Killing form is invariant under the differential of conjugations, and that the Jacobian of the exponantial is a function of the Killing form, see Eq. 3. It can be checked that if a density $\tilde{f}$ on the Lie algebra is a function of the Killing form, the push forward on the group is group-isotropic. Hence wrapping measures to the group with the exponential map is a natural way to construct group-isotropic measures on $\mathrm{SL}(2,\mathbb{C})$. However, the following results shows that unfortunately, the group-isotropy notion is not relevant on $\mathrm{SL}(2,\mathbb{C})$ for probability distributions: it contains only Dirac distributions.

Theorem 1. *If μ is a finite positive measure on $SL(2,\mathbb{C})$ isotropic with respect to the identity, then μ is a Dirac at the identity.*

Proof. Assume that μ is a measure on $\mathrm{SL}(2,\mathbb{C})$ whose support contains a matrix $M = \begin{pmatrix} a & b \\ c & d \end{pmatrix}$ different from the identity matrix. Since

$$\begin{pmatrix} 1 & t \\ 0 & 1 \end{pmatrix} \cdot \begin{pmatrix} a & b \\ c & d \end{pmatrix} \cdot \begin{pmatrix} 1 & t \\ 0 & 1 \end{pmatrix}^{-1} = \begin{pmatrix} a+tc & b+t(d-a)-ct^2 \\ c & -ct+d \end{pmatrix},$$

M is always conjugated to a matrix whose upper right coefficient is not zero. By the isotropic assumption, this matrix is still in the support of μ. Hence we can suppose that $b \neq 0$. Let $B(M)$ be the open ball centered on M:

$$B(M) = \left\{ \begin{pmatrix} a+\epsilon_1 & b+\epsilon_2 \\ c+\epsilon_3 & d+\epsilon_4 \end{pmatrix}, |\epsilon_i| < \frac{|b|}{2} \right\}.$$

Since M is in the support of μ, $\mu(B(M)) > 0$. Let $g = \begin{pmatrix} 2 & 0 \\ 0 & \frac{1}{2} \end{pmatrix} \in \mathrm{SL}(2,\mathbb{C})$. If we can show that $g^n B(M) g^{-n}$ and $g^{n+k} B(M) g^{-(n+k)}$ for all $n \in \mathbb{N}$ and $k \in \mathbb{N}_*$ are disjoint, μ has to be infinite since there are countable disjoint sets of identical nonzero mass. We have

$$\begin{aligned} g^n M g^{-n} &= \begin{pmatrix} 2 & 0 \\ 0 & \frac{1}{2} \end{pmatrix}^n \begin{pmatrix} a & b \\ c & d \end{pmatrix} \cdot \begin{pmatrix} \frac{1}{2} & 0 \\ 0 & 2 \end{pmatrix}^{-n} \\ &= \begin{pmatrix} a & 2^{2n} b \\ \frac{c}{2^{2n}} & d \end{pmatrix}. \end{aligned}$$

Hence,

$$\begin{gathered} g^n B(M) g^{-n} \cap g^{n+k} B(M) g^{-(n+k)} \neq \emptyset \\ \Rightarrow \exists \epsilon, \epsilon', \text{ with } |\epsilon| \text{ and } |\epsilon'| < \tfrac{|b|}{2}, \text{ such that, } 2^{2n}(b+\epsilon) = 2^{2n} 2^{2k} (b+\epsilon') \end{gathered}$$

Using the triangular inequalities $|b+\epsilon| < |b| + |\epsilon|$ and $|b| - |\epsilon'| < |b+\epsilon'|$, we see that such ϵ and ϵ' do not exist when k is a positive integer. Hence the images of $B(M)$ by the conjugations by $g^{n\in\mathbb{N}}$ are disjoints and the measure is infinite.

5 Conclusion

In this paper, we laid the foundations for density modeling using exponential-wrapped distributions on $\mathrm{SL}(2,\mathbb{C})$ and the Möbius group. The Möbius group plays an important role in polarization optics due to its action on wave polarization states. Future works will focus on applications of density modeling on the Möbius group to the propagation of light through random media.

References

1. Brofos, J.A., Brubaker, M.A., Lederman, R.R.: Manifold density estimation via generalized dequantization. arXiv:2102.07143 [cs, stat], February 2021
2. Falorsi, L., Haan, P.d., Davidson, T.R., Forré, P.: Reparameterizing distributions on lie groups. In: The 22nd International Conference on Artificial Intelligence and Statistics, pp. 3244–3253. PMLR, April 2019. http://proceedings.mlr.press/v89/falorsi19a.html. ISSN 2640–3498
3. Helgason, S.: Differential Geometry, Lie Groups, and Symmetric Spaces. Academic Press, New York (1978). http://site.ebrary.com/id/10259443. OCLC 316566751
4. Higham, N.J.: Functions of Matrices. Other Titles in Applied Mathematics, Society for Industrial and Applied Mathematics (2008). https://doi.org/10.1137/1.9780898717778. https://epubs.siam.org/doi/book/10.1137/1.9780898717778
5. Pennec, X., Lorenzi, M.: 5 - Beyond Riemannian geometry: the affine connection setting for transformation groups. In: Pennec, X., Sommer, S., Fletcher, T. (eds.) Riemannian Geometric Statistics in Medical Image Analysis, pp. 169–229. Academic Press, January 2020. https://doi.org/10.1016/B978-0-12-814725-2.00012-1. https://www.sciencedirect.com/science/article/pii/B9780128147252000121
6. Qiao, Z., Dick, R.: Matrix logarithms and range of the exponential maps for the symmetry groups SL(2, R), SL(2, C), and the Lorentz group. J. Phys. Commun. **3**(7), 075008 (2019)
7. Takenaka, H.: A unified formalism for polarization optics by using group theory I (theory). Japanese J. Appl. Phys. **12**(2), 226 (1973)
8. Takenaka, H.: A unified formalism for polarization optics by using group theory II (generator representation). Japanese J. Appl. Phys. **12**(11), 1729 (1973)

Information Geometry and Hamiltonian Systems on Lie Groups

Daisuke Tarama[1] and Jean-Pierre Françoise[2(✉)]

[1] Department of Mathematical Sciences, Ritsumeikan University, 1-1-1 Nojihigashi, Kusatsu, Shiga 525-8577, Japan
dtarama@fc.ritsumei.ac.jp

[2] Sorbonne Université, Laboratoire Jacques-Louis Lions, UMR 7598 CNRS, 4 Place Jussieu, 75252 Paris, France
jean-pierre.francoise@sorbonne-universite.fr

Abstract. The present paper deals with a class of left-invariant semi-definite metrics, called Fisher-Rao semi-definite metrics, on Lie groups appearing in transformation models. It is assumed that a family of invariant probability density functions on the sample manifold is given and that these probability density functions are invariant under a smooth Lie group action. As have been studied by Barndorff-Nielsen and his coauthors, as well as Amari and his collaborators, the Fisher-Rao semi-definite metric is naturally induced as a left-invariant semi-definite metric on the Lie group, which is regarded as the parameter space of the family of probability density functions. For a specific choice of family of probability density functions on compact semi-simple Lie group, the equation for the geodesic flow is derived through the Euler-Poincaré reduction. Certain perspectives are mentioned about the geodesic equation on the basis of its similarity with the Brockett double bracket equation and with the Euler-Arnol'd equation for a generalized free rigid body dynamics.

Keywords: Lie group · Information geometry · Hamiltonian systems

1 Introduction

For a statistical model, the Fisher-Rao metric is naturally defined on the parameter space for a family of probability density functions on the sample space. The Fisher-Rao metric, as well as Amari-Chentsov cubic tensor, has given rise to many important studies in information geometry from the viewpoint of differential geometry, which can now be formulated by the theory of dual connections, α-connections, α-geodesics, and statistical manifolds. See e.g. [1,2].[1] Particularly many researches have been done for dually flat manifolds, formulated also in the framework of the Hessian geometry. See e.g. [13,14,29].[2]

[1] Note that the Fisher-Rao metric itself was first introduced by Fréchet in [11].

[2] The Hessian geometry originates in the work by Koszul.

Partially supported by JSPS KAKENHI Grant Number JP19K14540.

F. Nielsen and F. Barbaresco (Eds.): GSI 2021, LNCS 12829, pp. 273–280, 2021.
https://doi.org/10.1007/978-3-030-80209-7_31

In the theory of dynamical systems, the complete integrability of the geodesic flows is one of the fundamental problems. In information geometry, the e- and m-geodesics for finite sample spaces are particularly well known. The relations between information geometry and completely integrable systems have been studied by Nakamura [20,22,23] and can now be found also in the textbooks [3, §8.3] and [1, §13]. However, further studies are still possible in these directions and may results in discoveries of interesting dynamical phenomena.

The present paper concerns a sample manifold admitting a smooth action of a Lie group. Given a family of invariant probability density functions on the sample manifold, the Lie group can be regarded as the parameter space of the statistical model. Such a statistical model has already been studied by Barndorff-Nielsen and his coauthors as the *transformation models.* In [3, §8.3], such statistical models are also mentioned and analogues of the Fisher-Rao metric and the Amari-Chentsov cubic tensor are considered with further perspectives of studies. It should also be pointed out that the exponential families on vector spaces invariant under group actions were studied by Casalis e.g. in [9,10].

Section 2 deals with the Fisher-Rao semi-definite metric on Lie groups appearing in transformation models, following [3, §8.3]. By left-invariance, the metric is a positive-semi-definite bilinear form on the Lie algebra. Although it is not necessarily positive-definite, one can define a Riemannian metric, called *Fisher-Rao Riemannian metric*, on a homogeneous space, assuming that a Lie subgroup preserves the Fisher-Rao semi-definite metric and that the restriction of the metric to the complement to the Lie algebra of the Lie subgroup is positive-definite.

Section 3 deals with the case where the sample manifold coincides with the parameter Lie group and a specific family of probability density functions is introduced under the influence of the objective functions which appear in the formulation of the generalized Toda lattice equations as the Brockett double bracket equations for optimization problems. See e.g. [8] for the details of the generalized Toda lattice equations. As a main result of the present paper, it is shown that the Fisher-Rao Riemannian metric is induced on an adjoint orbit and the geodesic equation is derived through the Euler-Poincaré reduction [17,26]. In view of the similarities with the Brockett double bracket equations (cf. [8]) and the Euler-Arnol'd equations for the generalized free rigid bodies (cf. [18,27,28]), certain future perspectives of this geodesic equation are mentioned.

2 Fisher-Rao Semi-definite Metric and Amari-Chentsov Cubic Tensor for the Transformation Models

In this section, we consider a sample manifold admitting a Lie group action. Given an invariant family of probability density functions on the sample manifold, we introduce a Fisher-Rao (semi-definite) metric on the Lie group, regarding the Lie group as the parameter space of a family of certain probability density functions on the sample manifold. These statistical models have been considered by Barndorff-Nielsen and his collaborators as the *transformation models.* See e.g. [5–7]. In this section, we discuss the Fisher-Rao (semi-definite) metric and the Amari-Chentsov cubic tensor in the framework proposed in [3, §8.3].

Let M be a compact smooth manifold and G a Lie group acting smoothly on M from the left: $G \times M \ni (g, x) \mapsto g \cdot x \in M$. We assume that M is endowed with an invariant volume element dvol_M with respect to the G-action. Consider a probability density function $\rho : G \times M \to \mathbb{R}$ on M parameterized by elements in G, such that $\rho(g, x) > 0$ for all $(g, x) \in G \times M$. Note that

$$\int_M \rho(g, x)\mathsf{dvol}_M(x) = 1 \tag{1}$$

for any $g \in G$. Now, we assume the probability density function ρ is G-invariant:

$$\rho(g \cdot h, x) = \rho(g, h \cdot x), \qquad \forall g, h \in G, \quad \forall x \in M. \tag{2}$$

Remark 1. The condition (2) may seem to be different from the one in [3, §7.3]. However, the Lie group action in [3] is from the right as $G \times M \ni (g, x) \mapsto g^{-1} \cdot x \in M$. As this is a conventional difference, the formulation in the present article is essentially the same as in [3, §8.3]. ♦

We denote the Lie algebra of G by $\mathfrak{g}$. For any $X \in \mathfrak{g}$, we have the induced vector fields X^M on M defined through $X^M_x[f] = \left.\frac{\mathsf{d}}{\mathsf{d}t}\right|_{t=0} f\left(\exp(-tX) \cdot x\right), \forall f \in C^\infty(M)$. For a smooth function f on M, the expectation with respect to ρ given as $E[f] := \int_M f(x)\rho(g, x)\mathsf{dvol}_M(x)$ is a smooth function in $g \in G$ and in particular we have $E[1] = 1$ by (1). For $X \in \mathfrak{g}$, we have

$$X^M_x[\rho(g, x)] = \left.\frac{\mathsf{d}}{\mathsf{d}t}\right|_{t=0} \rho\left(g, e^{-tX} \cdot x\right) = \left.\frac{\mathsf{d}}{\mathsf{d}t}\right|_{t=0} \rho\left(g \cdot e^{-tX}, x\right) = -X^{(L)}_g[\rho(g, x)],$$

at $g \in G$, $x \in M$, where $X^{(L)}_g$ stands for the left-invariant vector field on G induced by $X \in \mathfrak{g}$ evaluated at $g \in G$. This yields the formula

$$\begin{aligned}
E\left[X^M[\log \rho]\right] &= \int_M X^M_x[\log \rho(g, x)]\rho(g, x)\mathsf{dvol}_M(x) = \int_M X^M_x[\rho(g, x)]\mathsf{dvol}_M(x) \\
&= -\int_M X^{(L)}_g[\rho(g, x)]\mathsf{dvol}_M(x) = -X^{(L)}_g\left[\int_M \rho(g, x)\mathsf{dvol}_M(x)\right] \\
&= -X^{(L)}_g\left[E[1]\right] = -X^{(L)}_g[1] = 0.
\end{aligned}$$

For $X, Y \in \mathfrak{g}$, we similarly have

$$\begin{aligned}
0 &= X^{(L)}_g\left[Y^{(L)}_g\left[E[1]\right]\right] = -X^{(L)}_g\left[\int_M Y^M_x[\log \rho(g, x)]\rho(g, x)\mathsf{dvol}_M(x)\right] \\
&= \int_M X^M_x\left[Y^M_x[\log \rho(g, x)]\rho(g, x)\right]\mathsf{dvol}_M(x) \\
&= \int_M X^M_x[\log \rho(g, x)] \cdot Y^M_x[\log \rho(g, x)]\rho(g, x)\mathsf{dvol}_M(x) \\
&+ \int_M X^M_x\left[Y^M_x[\log \rho(g, x)]\right]\rho(g, x)\mathsf{dvol}_M(x) \\
&= E\left[X^M[\log \rho] \cdot Y^M[\log \rho]\right] + E\left[X^M\left[Y^M[\log \rho]\right]\right].
\end{aligned}$$

Taking these variation formulas into account, we introduce the Fisher-Rao bilinear form on the Lie algebra $\mathfrak{g}$ as well as Fisher-Rao (semi-definite) metric on G. The following proposition is pointed out in [3, §8.3].

Proposition 1. *The integral*

$$E\left[X^M\left[\log\rho\right]Y^M\left[\log\rho\right]\right] = \int_M X^M_x\left[\log\rho(g,x)\right]Y^M_x\left[\log\rho(g,x)\right]\rho(g,x)\mathrm{dvol}_M(x)$$

is constant for all $g \in G$. □

Proof. This can be proved by a straightforward computation:

$$\begin{aligned}
&E\left[X^M\left[\log\rho\right]Y^M\left[\log\rho\right]\right](g)\\
&= \int_M X^M_x\left[\log\rho(g,x)\right]Y^M_x\left[\log\rho(g,x)\right]\rho(g,x)\mathrm{dvol}_M(x)\\
&= \int_M X^M_{h\cdot x}\left[\log\rho(g,h\cdot x)\right]Y^M_{h\cdot x}\left[\log\rho(g,h\cdot x)\right]\rho(g,x)\mathrm{dvol}_M(h\cdot x)\\
&= \int_M X^M_x\left[\log\rho(g\cdot h,x)\right]Y^M_x\left[\log\rho(g\cdot h,x)\right]\rho(g\cdot h,x)\mathrm{dvol}_M(x)\\
&= E\left[X^M\left[\log\rho\right]Y^M\left[\log\rho\right]\right](g\cdot h),
\end{aligned}$$

where $g, h \in G$ are arbitrary elements. ■

Definition 1. *For $X, Y \in \mathfrak{g}$, the Fisher-Rao bilinear form on $\mathfrak{g}$ for the family of probability density functions ρ is defined through*

$$\langle X, Y\rangle = E\left[X^M\left[\log\rho\right]Y^M\left[\log\rho\right]\right] = -E\left[X^M\left[Y^M[\log\rho]\right]\right]. \tag{3}$$

□

Note that the bilinear form $\langle\cdot,\cdot\rangle$ is positive-semi-definite, since $\langle X, X\rangle = E\left[\left(X^M\left[\log\rho\right]\right)^2\right] \geq 0$. However, it is not guaranteed that the Fisher-Rao bilinear form be positive-definite as is concretely discussed in the next section. By the right-trivialization of the tangent bundle TG to the Lie group G given as

$$TG \supset T_gG \ni X_g \leftrightarrow \left(g, \mathrm{d}L_{g^{-1}}X_g\right) \in G \times \mathfrak{g},$$

the scalar product (3) can naturally be extended to a left-invariant $(0,2)$-tensor on G, which we denote by the same symbol. Here, $L_g : G \ni h \mapsto gh \in G$ is the left-translation by $g \in G$.

Definition 2. *The left-invariant $(0,2)$-tensor $\langle\cdot,\cdot\rangle$ on G is called Fisher-Rao semi-definite metric with respect to the invariant family of the probability density functions ρ.* □

Similarly to the Fisher-type bilinear form on $\mathfrak{g}$, we introduce the following $(0,3)$-tensor on G.

Definition 3. *The left-invariant* $(0,3)$*-tensor*

$$C(X,Y,Z) = E\left[X_x^M[\log\rho]\cdot Y_x^M[\log\rho]\cdot Z_x^M[\log\rho]\right]$$
$$= \int_M X_x^M[\log\rho(g,x)]\cdot Y_x^M[\log\rho(g,x)]\cdot Z_x^M[\log\rho(g,x)]\rho(g,x)\mathsf{dvol}_M(x) \quad (4)$$

is called Amari-Chentsov cubic tensor. □

By a similar computation to the proof for Proposition 1, we can easily prove that the Amari-Chentsov cubic tensor is left-invariant.

Now, we assume that there is a Lie subgroup $H \subset G$ whose Lie algebra $\mathfrak{h}$ gives the direct sum decomposition $\mathfrak{g} = \mathfrak{m}\dot{+}\mathfrak{h}$ and that the Fisher-Rao bilinear form $\langle\cdot,\cdot\rangle$ is positive-definite on $\mathfrak{m}$ and invariant under the adjoint action by H: $\langle \mathrm{Ad}_h(X), \mathrm{Ad}_h(Y)\rangle = \langle X, Y\rangle$ for all $h \in H$, $X, Y \in \mathfrak{g}$. As G is compact, so is H and hence the homogeneous space G/H is reductive in the sense of [25, §7] and the Fisher-Rao bilinear form $\langle\cdot,\cdot\rangle$ induces a G-invariant Riemannian metric on G/H, which we call Fisher-Rao Riemannian metric and denote by the same symbol $\langle\cdot,\cdot\rangle$. By [25, Theorem 13.1], the Levi-Civita connection ∇ on $(G/H, \langle\cdot,\cdot\rangle)$ is given as

$$\nabla_X Y = \frac{1}{2}[X,Y]_{\mathfrak{m}} + U(X,Y),$$

where $U(X,Y)$ is defined through $\langle U(X,Y), Z\rangle = \frac{1}{2}\langle [X,Z], Y\rangle + \langle X, [Y,Z]_{\mathfrak{m}}\rangle$ for $X, Y, Z \in \mathfrak{m}$, with $[Y,Z]_{\mathfrak{m}}$ being the $\mathfrak{m}$-component of $[Y,Z]$.

The α-connection $\nabla^{(\alpha)}$ is defined through

$$\langle \nabla_X^{(\alpha)} Y, Z\rangle = \langle \nabla_X Y, Z\rangle - \frac{\alpha}{2}C(X,Y,Z),$$

where $X, Y, Z \in \mathfrak{m}$. See e.g. [12,15].

The α-geodesic corresponding to $\nabla^{(\alpha)}$ is of much interest, which will be studied in future studies. Here, in the present paper, we rather consider the geodesic flow with respect to the Fisher-Rao Riemannian metric in a specific situation as is discussed in the next section.

3 Fisher-Rao Geodesic Flow and Euler-Poincaré Equation on Compact Semi-simple Lie Algebra

In this section, we consider the case where the parameter Lie group and the sample manifold M is a compact semi-simple Lie group G. We denote the Killing form of $\mathfrak{g}$ by κ, which is a negative-definite symmetric bilinear form by the semi-simplicity and the compactness of the Lie group G. We focus on the specific family of probability density functions $\rho : G \times G \ni (g,h) \mapsto c\cdot\exp(F(gh)) \in \mathbb{R}$, where $F(\theta) = \kappa(Q, \mathrm{Ad}_\theta N)$, $\theta \in G$, with Q, N being fixed elements in $\mathfrak{g}$. Here, the constant $c > 0$ is chosen such that $\int_G \rho(g,h)\mathsf{dvol}_G(h) = 1$. The volume form dvol_G is chosen to be the Haar measure on G.

The function F is earlier studied in [8] to investigate the generalized Toda lattice equation and its expression in the Brockett double bracket equation. It is shown for example that the generalized Toda lattice equations can be regarded as gradient flows on a special adjoint orbits. In the present paper, we rather consider the information geometry in relation to the probability density functions arising from the same function.

For the above specific family of probability density functions, the corresponding Fisher-Rao bilinear form on $\mathfrak{g}$ is calculated by (3) as

$$\begin{aligned}\langle X,Y\rangle &= c\cdot\int_G X_h^G\left[Y_h^G[\kappa\left(Q,\mathrm{Ad}_{gh}N\right)]\right]\rho(g,h)\mathrm{dvol}_G(h)\\ &= c\cdot\kappa\left(Q,\left[X,\left[Y,\int_G \mathrm{Ad}_{gh}N\rho(g,h)\mathrm{dvol}_M(h)\right]\right]\right)\\ &= -c\cdot\kappa\left(\left[X,Q\right],\left[Y,N'\right]\right),\end{aligned} \tag{5}$$

where

$$\begin{aligned}N' &:= \int_G \mathrm{Ad}_{gh}N\rho(g,h)\mathrm{dvol}_M(h) = \int_G \mathrm{Ad}_{gh}N\rho(e,gh)\mathrm{dvol}_M(gh)\\ &= c\cdot\int_G \mathrm{Ad}_h N\exp(\kappa(Q,\mathrm{Ad}_h N))\mathrm{dvol}_M(h).\end{aligned}$$

Note that the vector fields X^G, Y^G are nothing but the left-invariant vector fields on G associated to $X,Y\in\mathfrak{g}$.

In order for the scalar product (5) to be symmetric, we assume that the elements N',Q both belong to a Cartan subalgebra $\mathfrak{h}\subset\mathfrak{g}$ and regular.

Remark 2. By the decomposition of (quasi-)invariant measures discussed in [6, §5], we can rewrite the integral formula for the element N' as follows:

$$N' = c\cdot\int_{\mathcal{O}_N} x\exp\left(\kappa\left(Q,x\right)\right)\mathrm{dvol}_{\mathcal{O}_N}(x))$$

where $\mathcal{O}_N$ denotes the adjoint orbit through N. This integral can be regarded as orbital integral for a vector-valued function. ♦

We consider the Cartan subgroup $H\subset G$ whose Lie algebra is $\mathfrak{h}$. The Fisher-Rao bilinear form $\langle\cdot,\cdot\rangle$ is invariant with respect to Ad_H and hence induces Fisher-Rao Riemannian metric on the adjoint orbit $\mathcal{O}_N = G/H$.

The geodesic equation can be formulated through Euler-Poincaré reduction, cf. [17,26], with respect to the Hamiltonian function $\mathcal{H}(X) := \frac{1}{2}\langle X,X\rangle = -c\kappa\left([X,Q],[X,N']\right)$, $X\in\mathfrak{g}$.

Theorem 1. *The Euler-Poincaré equation for the geodesic flow with respect to the above Fisher-Rao Riemannian metric is given as*

$$\dot{X} = c\cdot\left[X,\left[\left[X,Q\right],N'\right]\right]. \tag{6}$$

□

Note that the right hand side of the equation is nothing but $\nabla_X X$ with respect to the Levi-Civita connection appearing in the previous section.

We finally put several perspectives around the above Euler-Poincaré equation, which will be studied in the future works.

- The Eq. (6) has similar expression to the Brockett double bracket equation discussed e.g. in [8,19,21,23,24], as well as Euler-Arnol'd equation for generalized free rigid body dynamics studied e.g. in [18]. See also [3, §8.3] and [1, §13]. It might be of interest to investigate the complete integrability of (6) when restricted to an adjoint orbit in $\mathfrak{g}$.
- Besides the integrability, the stability of equilibria for (6) should be investigated. If the system is integrable as in the above perspectives, some methods in [27,28] might also be applicable.
- It is of much interests to compare the geodesic flows in the present paper with the Souriau's thermodynamical formulation of information geometry and the associated Euler-Poincaré equations in [4,16].

References

1. Amari, S.-I.: Information Geometry and its Applications. Applied Mathematical Sciences, vol. 194, Springer, Tokyo (2016)
2. Ay, N., Jost, J., Schwachhöfer, Hông Vân Lê, L.: Information Geometry. Ergebnisse der Mathematik und ihrer Grenzgebiete. 3. Folge. A Series of Modern Surveys in Mathematics, vol. 64, Springer, Cham (2017)
3. Amari, S.-I., Nagaoka, H.: Methods of Information Geometry. Translated from the 1993 Japanese original by D. Harada. Translations of Mathematical Monographs, vol. 191. American Mathematical Society, Providence, RI, Oxford University Press, Oxford (2000)
4. Barbaresco, F., Gay-Balmaz, F.: Lie group cohomology and (multi)symplectic integrators: new geometric tools for Lie group machine learning based on Souriau geometric statistical mechanics. Entropy **22**, 498 (2020)
5. Barndorff-Nielsen, O.E.: Recent developments in statistical inference using differential and integral geometry. In: Amari, S.-I., Barndorff-Nielsen, O.E., Kass, R.E., Lauritzen, S.L., Rao, C.R. (eds.) Differential Geometry in Statistical Inference, pp. 95–161. Institute of Mathematical Statistics, Hayward, CA (1987)
6. Barndorff-Nielsen, O.E., Blaesild, P., Eriksen, P.S.: Decomposition and Invariance of Measures, and Statistical Transformation Models. Lecture Notes in Statistics, vol. 58. Springer-Verlag, New York (1989)
7. Barndorff-Nielsen, O.E., Blaesild, P., Jensen, J.L., Jorgensen, B.: Exponential transformation models. Proc. Roy. Soc. London Ser. A **379**(1776), 41–65 (1982)
8. Bloch, A., Brockett, R.W., Ratiu, T.S.: Completely integrable gradient flows. Comm. Math. Phys. **147**(1), 57–74 (1992)
9. Casalis, M.: Familles exponentielles naturelles sur $\mathbb{R}^d$ invariantes par un groupe. Int. Stat. Rev. **59**(2), 241–262 (1991)
10. Casalis, M.: Les familles exponentielles à variance quadratique homogène sont des lois de Wishart sur un cône symétrique, C. R. Acad. Sci. Paris 312, Série I, 537–540 (1991)

11. Fréchet, M.: Sur l'existence de certaines évaluations statistiques au cas de petits échantillons. Rev. Int. Statist. **11**, 182–205 (1943)
12. Furuhata, H., Inoguchi, J.-I., Kobayashi, S.: A characterization of the alpha-connections on the statistical manifold of normal distributions. Inf. Geo. (2020). https://doi.org/10.1007/s41884-020-00037-z
13. Ishi, H.: Special Issue Affine differential geometry and Hesse geometry: a tribute and memorial to Jean-Louis Koszul. Inf. Geo. (2021). https://doi.org/10.1007/s41884-020-00042-2
14. Koszul, J.-L.: Domaines bornés homogènes et orbites de groupes de transformations affines. Bull. Soc. Math. France **89**, 515–533 (1961)
15. Kobayashi, S., Ohno, Y.: On a constant curvature statistical manifold. preprint, arXiv:2008.13394v2 (2020)
16. Marle, Ch.-M.: On Gibbs states of mechanical systems with symmetries, preprint, arXiv: 2012.00582 (2021)
17. Marsden, J.E., Ratiu, T.S.: Introduction to Mechanics and Symmetry. 2nd edn., Texts in Applied Mathematics vol. 17. Springer, New York (2003)
18. Mishchenko, A.S., Fomenko, A.T.: Euler equations on finite-dimensional Lie groups. Math. USSR-Izv. **12**(2), 371–389 (1978). Translated from Russian, Izvest. Akad. Nauk SSSR, Ser. Matem. 42, 1978, 396–415
19. Moser, J.: Finitely many mass points on the line under the influence of an exponential potential-an integrable system. In: Moser J. (ed.) Dynamical Systems, Theory and Applications. Lecture Notes in Physics, vol. 38. Springer, Heidelberg (1975). https://doi.org/10.1007/3-540-07171-7_12
20. Nakamura, Y.: Completely integrable gradient systems on the manifolds of Gaussian and multinomial distributions. Jpn. J. Ind. Appl. Math. **10**(2), 179–189 (1993)
21. Nakamura, Y.: Neurodynamics and nonlinear integrable systems of Lax type. Jpn J. Ind. Appl. Math. **11**(1), 11–20 (1994)
22. Nakamura, Y.: Gradient systems associated with probability distributions. Jpn. J. Ind. Appl. Math. **11**(1), 21–30 (1994)
23. Nakamura, Y.: A tau-function for the finite Toda molecule, and information spaces. In: Maeda, Y., Omori, H., Weinstein, A. (eds.), Symplectic Geometry and Quantization (Sanda and Yokohama, 1993), pp. 205–211, Contemp. Math. vol. 179, Amer. Math. Soc., Providence, RI (1994)
24. Nakamura, Y., Kodama, Y.: Moment problem of Hamburger, hierarchies of integrable systems, and the positivity of tau-functions. KdV '95 (Amsterdam, 1995). Acta Appl. Math. **39**(1–3), 435–443 (1995)
25. Nomizu, K.: Affine connections on homogeneous spaces. Amer. J. Math. **76**(1), 33–65 (1954)
26. Ratiu, T.S., et al.: A Crash Course in Geometric Mechanics. In: Montaldi, J., Ratiu, T. (eds.) Geometric Mechanics and Symmetry: the Peyresq Lectures. Cambridge University Press, Cambridge (2005)
27. Ratiu, T.S., Tarama, D.: The $U(n)$ free rigid body: integrability and stability analysis of the equilibria. J. Diff. Equ. **259**, 7284–7331 (2015)
28. Ratiu, T.S., Tarama, D.: Geodesic flows on real forms of complex semi-simple Lie groups of rigid body type. Research in the Mathematical Sciences **7**(4), 1–37 (2020). https://doi.org/10.1007/s40687-020-00227-2
29. Shima, H.: The Geometry of Hessian Structures. World Scientific Publishing, Singapore (2007)

Geometric and Symplectic Methods for Hydrodynamical Models

Multisymplectic Variational Integrators for Fluid Models with Constraints

François Demoures[1] and François Gay-Balmaz[2](✉)

[1] EPFL, Doc & Postdoc Alumni, Av Druey 1, Lausanne 1018, Switzerland
francois.demoures@alumni.epfl.ch

[2] CNRS & Laboratoire de Météorologie Dynamique, École Normale Supérieure, Paris, France
francois.gay-balmaz@lmd.ens.fr

Abstract. We present a structure preserving discretization of the fundamental spacetime geometric structures of fluid mechanics in the Lagrangian description in 2D and 3D. Based on this, multisymplectic variational integrators are developed for barotropic and incompressible fluid models, which satisfy a discrete version of Noether theorem. We show how the geometric integrator can handle regular fluid motion in vacuum with free boundaries and constraints such as the impact against an obstacle of a fluid flowing on a surface. Our approach is applicable to a wide range of models including the Boussinesq and shallow water models, by appropriate choice of the Lagrangian.

Keywords: Structure preserving discretization · Barotropic fluid · Incompressible ideal fluid · Multisymplectic integrator · Noether theorem

1 Introduction

This abstract presents a multisymplectic variational integrator for barotropic fluids and incompressible fluids with free boundaries in the Lagrangian description. The integrator is derived from a spacetime discretization of the Hamilton principle of fluid dynamics and is based on a discrete version of the multisymplectic geometric formulation of continuum mechanics. As a consequence of its variational nature, the resulting scheme preserves exactly the momenta associated to symmetries, it is symplectic in time, and energy is well conserved. In addition to its conservative properties, the variational scheme can be naturally extended to handle constraints, by augmenting the discrete Lagrangian with penalty terms.

Multisymplectic variational integrators were developed in [10, 11] via a spacetime discretization of the Hamilton principle of field theories, which results in numerical schemes that satisfy a discrete version of the multisymplectic form formula and a discrete covariant Noether theorem. We refer to [2–5, 10] for the development of multisymplectic variational integrators for several mechanical systems of interest in engineering.

F. Nielsen and F. Barbaresco (Eds.): GSI 2021, LNCS 12829, pp. 283–291, 2021.
https://doi.org/10.1007/978-3-030-80209-7_32

2 Barotropic and Incompressible Fluids

Assume that the reference configuration of the fluid is a compact domain $\mathcal{B} \subset \mathbb{R}^n$ with piecewise smooth boundary, and the fluid moves in the ambient space $\mathcal{M} = \mathbb{R}^n$. We denote by $\varphi : (t, X) \in \mathbb{R} \times \mathcal{B} \mapsto \varphi(t, X) \in \mathcal{M}$ the fluid configuration map. The deformation gradient is denoted $\mathbf{F}(t, X)$, given in coordinates by $\mathbf{F}^a{}_i = \varphi^a{}_{,i}$, with $X^i, i = 1, ..., n$ the Cartesian coordinates on $\mathcal{B}$ and $m^a, a = 1, ..., n$ the Cartesian coordinates on $\mathcal{M}$.

2.1 Barotropic Fluids

A fluid is *barotropic* if it is compressible and the surfaces of constant pressure p and constant density ρ coincide, i.e., we have a relation

$$p = p(\rho). \tag{1}$$

The internal energy W of barotropic fluids in the material description depends on the deformation gradient $\mathbf{F}$ only through the Jacobian J of φ, given in Cartesian coordinates by $J(t, m) = \det(\mathbf{F}(t, X))$, hence in the material description we have $W = W(\rho_0, J)$, with $\rho_0(X)$ the mass density of the fluid in the reference configuration. The pressure in the material description is

$$P_W(\rho_0, J) = -\rho_0 \frac{\partial W}{\partial J}(\rho_0, J). \tag{2}$$

The *continuity equation for mass* is written as $\rho_0(X) = \rho(t, \varphi(t, X))J(t, X)$, with $\rho(t, m)$ the Eulerian mass density.

The Lagrangian of the barotropic fluid evaluated on a fluid configuration map $\varphi(t, X)$ has the standard form

$$L(\varphi, \dot{\varphi}, \nabla\varphi) = \frac{1}{2}\rho_0|\dot{\varphi}|^2 - \rho_0 W(\rho_0, J) - \rho_0 \Pi(\varphi), \tag{3}$$

with Π a potential energy, such as the gravitational potential $\Pi(\varphi) = \mathbf{g}\cdot\varphi$. From the Hamilton principle we get the barotropic fluid equations in the Lagrangian description as

$$\rho_0 \ddot{\varphi} + \frac{\partial}{\partial x^i}\left(P_W J \mathbf{F}^{-1}\right)^i = -\rho_0 \frac{\partial \Pi}{\partial \varphi}, \tag{4}$$

together with the natural boundary conditions

$$P_W J\, n_i (\mathbf{F}^{-1})^i_a \delta\varphi^a = 0 \;\text{ on }\; \partial\mathcal{B}, \quad \forall\, \delta\varphi. \tag{5}$$

Example 1. Isentropic Perfect Gas. For this case, we can use the general expression

$$W(\rho_0, J) = \frac{A}{\gamma - 1}\left(\frac{J}{\rho_0}\right)^{1-\gamma} + B\left(\frac{J}{\rho_0}\right), \tag{6}$$

for constants A, B, and adiabatic coefficient γ, see [1]. The boundary condition (5) yields $P|_{\partial\mathcal{B}} = B$, with $P = A\left(\frac{\rho_0}{J}\right)^\gamma$ the pressure of the isentropic perfect gas.

2.2 Incompressible Fluid Models

Incompressible models are obtained by inserting the constraint $J = 1$ with the Lagrange multiplier $\lambda(t, X)$ in the Hamilton principle, as explained in [12],

$$\delta \int_0^T \int_{\mathcal{B}} \big(L(\varphi, \dot{\varphi}, \nabla\varphi) + \lambda(J-1)\big) \mathrm{d}t\, \mathrm{d}X = 0. \tag{7}$$

3 2D Discrete Barotropic and Incompressible Fluid Models

3.1 Multisymplectic Discretizations

The geometric setting of continuum mechanics underlying the formulation recalled in Sect. 2 is based on the configuration bundle $\mathcal{Y} = \mathcal{X} \times \mathcal{M} \to \mathcal{X} = \mathbb{R} \times \mathcal{B}$, with the fluid configuration map $\varphi(t, X) \in \mathbb{R} \times \mathcal{B} \mapsto \varphi(t, X) \in \mathcal{M}$ naturally identified with a section of the configuration bundle. The discrete geometric setting is described as follows

$$\begin{array}{ccc} & & \mathcal{Y}_d = \mathcal{X}_d \times \mathcal{M} \\ & \overset{\phi_d}{\nearrow} & \varphi_d \uparrow\downarrow \pi_d \\ \mathcal{U}_d & \xrightarrow[\phi_{\mathcal{X}_d}]{} & \phi_{\mathcal{X}_d}(\mathcal{U}_d) = \mathcal{X}_d \subset \mathcal{X} \end{array}$$

In this figure $\mathcal{U}_d := \{0, ..., j, ..., N\} \times \mathbb{B}_d$ is the *discrete parameter space* where $\{0, ..., j, ..., N\}$ encodes an increasing sequence of time and $\mathbb{B}_d$ parameterizes the nodes and simplexes of the discretization of $\mathcal{B}$. We focus on the 2D case and assume $\mathcal{B}$ rectangular so that $\mathbb{B}_d = \{0, \ldots, A\} \times \{0, \ldots, B\}$. From the discretization we get a set of parallelepipeds, denoted $\boxdot^j_{a,b}$, and defined by

$$\begin{aligned} \boxdot^j_{a,b} = \big\{&(j, a, b), (j+1, a, b), (j, a+1, b), (j, a, b+1), (j, a+1, b+1), \\ &(j+1, a+1, b), (j+1, a, b+1), (j+1, a+1, b+1)\big\}, \end{aligned} \tag{8}$$

$j = 0, ..., N-1$, $a = 0, ..., A-1$, $b = 0, ..., B-1$. The *discrete base space configuration* $\phi_{\mathcal{X}_d}$ is of the form $\phi_{\mathcal{X}_d}(j, a, b) = s^j_{a,b} = (t^j, z^j_a, z^j_b) \in \mathbb{R} \times \mathcal{B}$, see bellow:

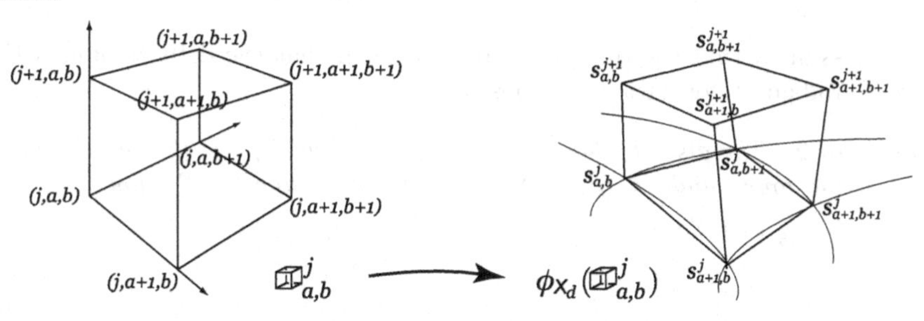

Given a discrete base space configuration $\phi_{\mathcal{X}_d}$ and a discrete field φ_d, we define the following four vectors $\mathbf{F}^j_{\ell;a,b} \in \mathbb{R}^2$, $\ell = 1,2,3,4$ at each node $(j,a,b) \in \mathcal{U}_d$.

$$\mathbf{F}^j_{1;a,b} = \frac{\varphi^j_{a+1,b} - \varphi^j_{a,b}}{|s^j_{a+1,b} - s^j_{a,b}|} \quad \text{and} \quad \mathbf{F}^j_{2;a,b} = \frac{\varphi^j_{a,b+1} - \varphi^j_{a,b}}{|s^j_{a,b+1} - s^j_{a,b}|} :$$

$$\mathbf{F}^j_{3;a,b} = \frac{\varphi^j_{a-1,b} - \varphi^j_{a,b}}{|s^j_{a,b} - s^j_{a-1,b}|} = -\mathbf{F}^j_{1;a-1,b} \quad \text{and} \quad \mathbf{F}^j_{4;a,b} = \frac{\varphi^j_{a,b-1} - \varphi^j_{a,b}}{|s^j_{a,b} - s^j_{a,b-1}|} = -\mathbf{F}^j_{2;a,b-1}.$$

We refer to Fig. 1 on the right for an illustration of these definitions.

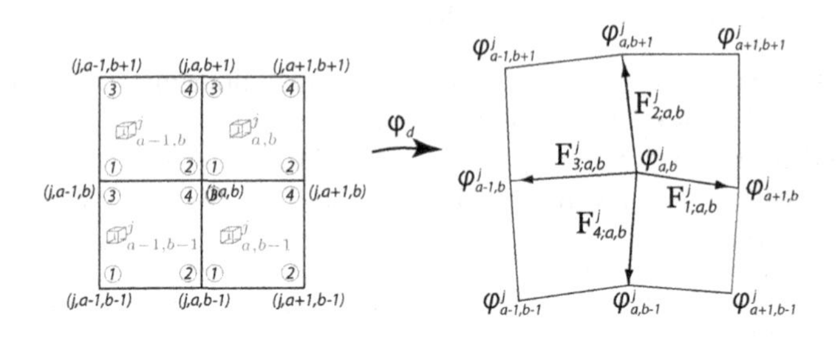

Fig. 1. Discrete field $\phi_d = \varphi_d \circ \phi_{\mathcal{X}_d}$ evaluated on $\square^j_{a,b}$, $\square^j_{a,b}$, $\square^j_{a,b}$, $\square^j_{a,b}$ at time t^j.

Based on these definitions, the discrete deformation gradient is constructed as follows.

Definition 1. *The discrete deformation gradients of a discrete field φ_d at the parallelepiped $\square^j_{a,b}$ are the four 2×2 matrices $\mathbf{F}^\ell(\square^j_{a,b})$, $\ell = 1,2,3,4$, defined at the four nodes at time t^j of $\square^j_{a,b}$, as follows:*

$$\mathbf{F}_1(\square^j_{a,b}) = \left[\mathbf{F}^j_{1;a,b}\ \mathbf{F}^j_{2;a,b}\right], \qquad \mathbf{F}_2(\square^j_{a,b}) = \left[\mathbf{F}^j_{2;a+1,b}\ \mathbf{F}^j_{3;a+1,b}\right],$$

$$\mathbf{F}_3(\square^j_{a,b}) = \left[\mathbf{F}^j_{4;a,b+1}\ \mathbf{F}^j_{1;a,b+1}\right], \qquad \mathbf{F}_4(\square^j_{a,b}) = \left[\mathbf{F}^j_{3;a+1,b+1}\ \mathbf{F}^j_{4;a+1,b+1}\right].$$

The ordering $\ell = 1$ to $\ell = 4$ is respectively associated to the nodes (j,a,b), $(j,a,b+1)$, $(j,a+1,b)$, $(j,a+1,b+1)$, see Fig. 1 on the left.

It is assumed that the discrete field φ_d is such that the determinant of the discrete gradient deformations are positive.

Definition 2. *The discrete Jacobians of a discrete field φ_d at the parallelepiped $\square^j_{a,b}$ are the four numbers $J_\ell(\square^j_{a,b})$, $\ell = 1,2,3,4$, defined at the four nodes at time t^j of $\square^j_{a,b}$ as follows:*

$$
\begin{aligned}
J_1(\boxplus^j_{a,b}) &= |\mathbf{F}^j_{1;a,b} \times \mathbf{F}^j_{2;a,b}| = \det\big(\mathbf{F}_1(\boxplus^j_{a,b})\big),\\
J_2(\boxplus^j_{a,b}) &= |\mathbf{F}^j_{2;a+1,b} \times \mathbf{F}^j_{3;a+1,b}| = \det\big(\mathbf{F}_2(\boxplus^j_{a,b})\big)\\
J_3(\boxplus^j_{a,b}) &= |\mathbf{F}^j_{4;a,b+1} \times \mathbf{F}^j_{1;a,b+1}| = \det\big(\mathbf{F}_3(\boxplus^j_{a,b})\big)\\
J_4(\boxplus^j_{a,b}) &= |\mathbf{F}^j_{3;a+1,b+1} \times \mathbf{F}^j_{4;a+1,b+1}| = \det\big(\mathbf{F}_4(\boxplus^j_{a,b})\big).
\end{aligned}
\tag{9}
$$

As a consequence, from relations (9), the variation of the discrete Jacobian is given at each $\boxplus^j_{a,b}$ by

$$
\delta J_\ell = \frac{\partial \det(\mathbf{F}_\ell)}{\partial \mathbf{F}_\ell} : \delta \mathbf{F}_\ell = J_\ell(\mathbf{F}_\ell)^{-\mathsf{T}} : \delta \mathbf{F}_\ell,
$$

which is used in the derivation of the discrete Euler-Lagrange equations.

3.2 Discrete Lagrangian and Discrete Hamilton Principle

Let $\mathcal{U}_d^{\boxplus}$ be the set of all parallelepipeds $\boxplus^j_{a,b}$ in $\mathcal{U}_d$. Its image in $\mathcal{X}_d$ by $\phi_{\mathcal{X}_d}$ is $\mathcal{X}_d^{\boxplus} := \phi_{\mathcal{X}_d}(\mathcal{U}_d^{\boxplus})$. The *discrete version of the first jet bundle* is given by

$$
J^1\mathcal{Y}_d := \mathcal{X}_d^{\boxplus} \times \underbrace{\mathcal{M} \times \ldots \times \mathcal{M}}_{8\ \text{times}} \to \mathcal{X}_d^{\boxplus}. \tag{10}
$$

Given a discrete field φ_d, its *first jet extension* is the section of (10) defined by

$$
j^1\varphi_d(\boxplus^j_{a,b}) = \big(\varphi^j_{a,b}, \varphi^{j+1}_{a,b}, \varphi^j_{a+1,b}, \varphi^{j+1}_{a+1,b}, \varphi^j_{a,b+1}, \varphi^{j+1}_{a,b+1}, \varphi^j_{a+1,b+1}, \varphi^{j+1}_{a+1,b+1}\big), \tag{11}
$$

which associates to each parallelepiped $\boxplus^j_{a,b}$, the values of the field at its nodes.

We consider a class of discrete Lagrangians associated to (3) of the form

$$
\mathcal{L}_d(j^1\varphi_d(\boxplus)) = \mathrm{vol}(\boxplus)\Big(\rho_0 K_d(j^1\varphi_d(\boxplus)) - \rho_0 W_d(\rho_0, j^1\varphi_d(\boxplus)) - \rho_0 \Pi_d(j^1\varphi_d(\boxplus))\Big),
$$

where $\mathrm{vol}(\boxplus)$ is the volume of the parallelepiped $\boxplus \in \mathcal{X}_d^{\boxplus}$. The discrete internal energy $W_d : J^1\mathcal{Y}_d \to \mathbb{R}$ is defined as

$$
W_d(\rho_0, j^1\varphi_d(\boxplus^j_{a,b})) := \frac{1}{4}\sum_{\ell=1}^{4} W\big(\rho_0, J_\ell(\boxplus^j_{a,b})\big), \tag{12}
$$

where W is the material internal energy of the continuous fluid model. To simplify the exposition, we assume that the discrete base space configuration is fixed and given by $\phi_{\mathcal{X}_d}(j,a,b) = (j\Delta t, a\Delta s_1, b\Delta s_2)$, for given Δt, Δs_1, Δs_2. In this case, we have $\mathrm{vol}(\boxplus) = \Delta t \Delta s_1 \Delta s_2$ in the discrete Lagrangian and the mass of each $2D$ cell in $\phi_{\mathcal{X}_d}(\mathbb{B}_d)$ is $M = \rho_0 \Delta s_1 \Delta s_2$.

The discrete action functional associated to $\mathcal{L}_d$ is obtained as

$$S_d(\varphi_d) = \sum_{\square \in \mathcal{X}_d^{\square}} \mathcal{L}_d\big(j^1\varphi_d(\square)\big) = \sum_{j=0}^{N-1}\sum_{a=0}^{A-1}\sum_{b=0}^{B-1} \mathcal{L}_d\big(j^1\varphi_d(\square^j_{a,b})\big). \tag{13}$$

We compute the variation $\delta S_d(\varphi_d)$ of the action sum and we get

$$\begin{aligned}
&\sum_{j=0}^{N-1}\sum_{a=0}^{A-1}\sum_{b=0}^{B-1}\Big[\frac{M}{4}\left(v^j_{a,b}\cdot\delta\varphi^{j+1}_{a,b} + v^j_{a+1,b}\cdot\delta\varphi^{j+1}_{a+1,b} + v^j_{a,b+1}\cdot\delta\varphi^{j+1}_{a,b+1} + v^j_{a+1,b+1}\cdot\delta\varphi^{j+1}_{a+1,b+1}\right)\\
&-\frac{M}{4}\left(v^j_{a,b}\cdot\delta\varphi^j_{a,b} + v^j_{a+1,b}\cdot\delta\varphi^j_{a+1,b} + v^j_{a,b+1}\cdot\delta\varphi^j_{a,b+1} + v^j_{a+1,b+1}\cdot\delta\varphi^j_{a+1,b+1}\right)\\
&+A(\square^j_{a,b})\cdot\delta\varphi^j_{a,b} + B(\square^j_{a,b})\cdot\delta\varphi^j_{a+1,b} + C(\square^j_{a,b})\cdot\delta\varphi^j_{a,b+1} + D(\square^j_{a,b})\cdot\delta\varphi^j_{a+1,b+1}\\
&+E(\partial\square^j_{a,b})\cdot\delta\varphi^j_{a,b} + F(\partial\square^j_{a,b})\cdot\delta\varphi^j_{a+1,b} + G(\partial\square^j_{a,b})\cdot\delta\varphi^j_{a,b+1} + H(\partial\square^j_{a,b})\cdot\delta\varphi^j_{a+1,b+1}\Big],
\end{aligned}$$

where we have used the following expressions of the partial derivative of $\mathcal{L}_d$:

$$D_2\mathcal{L}^j_{a,b} = \frac{M}{4}v^j_{a,b} \qquad D_6\mathcal{L}^j_{a,b} = \frac{M}{4}v^j_{a,b+1}$$
$$D_4\mathcal{L}^j_{a,b} = \frac{M}{4}v^j_{a+1,b} \qquad D_8\mathcal{L}^j_{a,b} = \frac{M}{4}v^j_{a+1,b+1}$$

and the following notations for the other partial derivatives

$$D_1\mathcal{L}^j_{a,b} = -\frac{M}{4}v^j_{a,b} + A(\square^j_{a,b}) + E(\partial\square^j_{a,b}) \qquad D_5\mathcal{L}^j_{a,b} = -\frac{M}{4}v^j_{a,b+1} + C(\square^j_{a,b}) + G(\partial\square^j_{a,b})$$
$$D_3\mathcal{L}^j_{a,b} = -\frac{M}{4}v^j_{a+1,b} + B(\square^j_{a,b}) + F(\partial\square^j_{a,b}) \qquad D_7\mathcal{L}^j_{a,b} = -\frac{M}{4}v^j_{a+1,b+1} + D(\square^j_{a,b}) + H(\partial\square^j_{a,b}).$$

The quantities $E(\partial\square^j_{a,b})$, $F(\partial\square^j_{a,b})$, $G(\partial\square^j_{a,b})$, $H(\partial\square^j_{a,b})$ denote the boundary terms associated to the outward unit normal to $\square^j_{a,b}$ in (t^j, s_a, s_b), (t^j, s_{a+1}, s_b), (t^j, s_a, s_{b+1}), (t^j, s_{a+1}, s_{b+1}), respectively.

From the discrete Hamilton principle $\delta S_d(\varphi_d) = 0$, we get the discrete Euler-Lagrange equations by collecting the terms associated to variations $\delta\varphi^j_{a,b}$ at the interior of the domain:

$$\begin{aligned}
Mv^j_{a,b} - Mv^{j-1}_{a,b} - A(\square^j_{a,b}) - B(\square^j_{a-1,b}) - C(\square^j_{a,b-1}) - D(\square^j_{a-1,b-1})&\\
- E(\partial\square^j_{a,b}) - F(\partial\square^j_{a-1,b}) - G(\partial\square^j_{a,b-1}) - H(\partial\square^j_{a-1,b-1}) = 0.&
\end{aligned} \tag{14}$$

3.3 Discrete Multisymplectic Form Formula and Discrete Noether Theorem

The solution of the discrete Euler-Lagrange equations (14) exactly satisfies a notion of discrete multisymplecticity as well as a discrete Noether theorem, see [10]. See also [3,4,6] for more explanations concerning discrete conservation laws for multisymplectic variational discretizations.

3.4 Numerical Simulations

We consider a fluid subject to gravity and flowing without friction on a surface until it comes into contact with an obstacle. See [6] for barotropic and incompressible fluid numerical simulations in 2D.

Example 2. Impact against an obstacle of a barotropic fluid flowing on a surface. (Fig. 2)

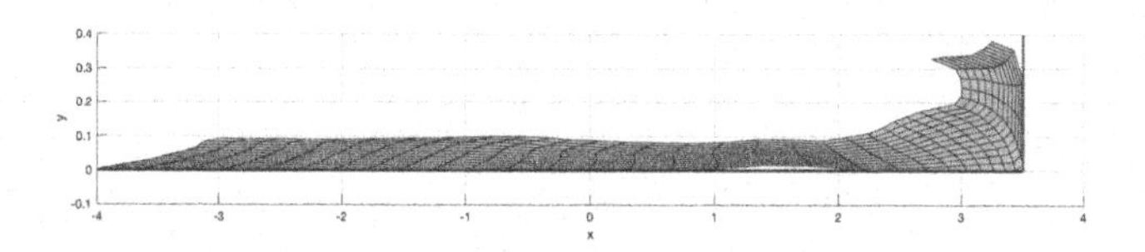

Fig. 2. Barotropic fluid, after 1.6 s, with $\rho_0 = 997\,\mathrm{kg/m}^2$, $\gamma = 6$, $A = \tilde{A}\rho_0^{-\gamma}$ with $\tilde{A} = 3.041 \times 10^4$ Pa, and $B = 3.0397 \times 10^4$ Pa. Time-step $\Delta t = 10^{-4}$.

Example 3. Barotropic and incompressible fluid motion in vacuum with free boundaries. (Fig. 3)

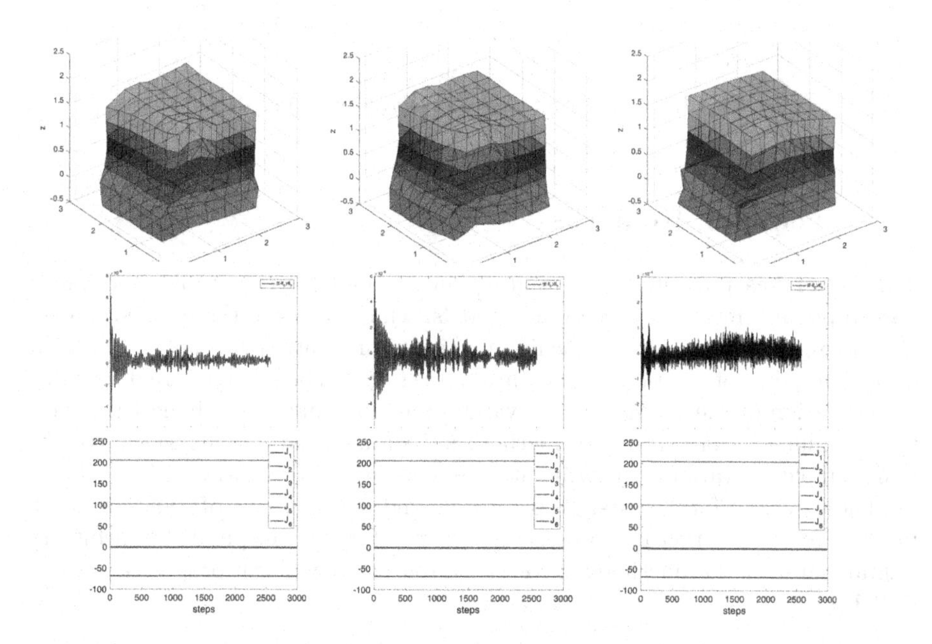

Fig. 3. *Left to right*: Discrete barotropic and incompressible ideal fluid model (Lagrangian multipliers $r = 10^5$ and $r = 10^7$). *Top to bottom*: Configuration after 2 s, relative energy, and momentum map evolution during 3 s.

4 3D Discrete Barotropic and Incompressible Fluid Models

The general discrete multisymplectic framework explained earlier can be adapted to 3D fluids. The main step is the definition of the discrete Jacobian of a discrete field φ_d at the eight nodes at time t^j of $\boxdot^j_{a,b,c}$, based on the same principles as those used previously for the 2D case. The discrete Euler-Lagrange equations in 3D are obtained in a similar way as in (10), (11), (12), (13), (14), see [6].

Example 4. Impact against an obstacle of an incompressible fluid flowing on a surface. (Fig. 4)

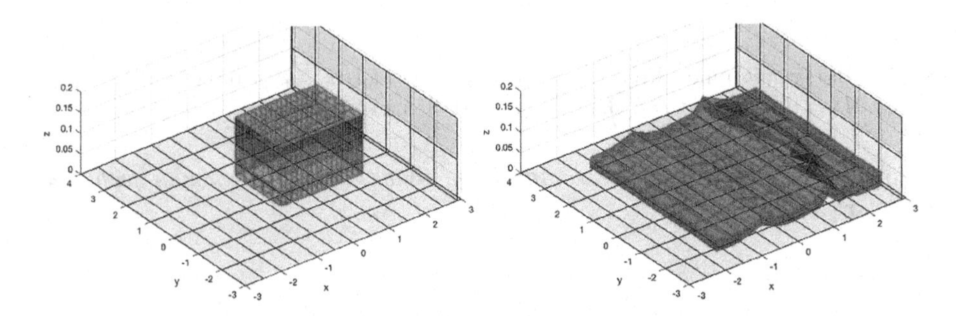

Fig. 4. Fluid impact. *Left to right*: after 0.4 s, 1.1 s.

5 Concluding Remarks

This paper has presented new Lagrangian schemes for the regular motion of barotropic and incompressible fluid models, which preserve the momenta associated to symmetries, up to machine precision, and satisfy the nearly constant energy property of symplectic integrators, see Fig. 3. The schemes are derived by discretization of the geometric and variational structures underlying the space-time formulation of continuum mechanics, seen as a particular instance of the multisymplectic variational formulation of classical field theories.

These results form a first step towards the development of dynamic mesh update from a structure preserving point of view, inspired by arbitrary Lagrangian-Eulerian methods. Several approaches have been proposed, such as [7–9,13].

References

1. Courant, R., Friedrichs, K.O.: Supersonic Flow and Shock Waves. Institute for Mathematics and Mechanics, New York University, New York (1948)

2. Demoures, F., Gay-Balmaz, F., Desbrun, M., Ratiu, T.S., Alejandro, A.: A multisymplectic integrator for elastodynamic frictionless impact problems. Comput. Methods Appl. Mech. Eng. **315**, 1025–1052 (2017)
3. Demoures, F., Gay-Balmaz, F., Kobilarov, M., Ratiu, T.S.: Multisymplectic Lie group variational integrators for a geometrically exact beam in $\mathbb{R}^3$. Commun. Nonlinear Sci. Numer. Simulat. **19**(10), 3492–3512 (2014)
4. Demoures, F., Gay-Balmaz, F., Ratiu, T.S.: Multisymplectic variational integrator and space/time symplecticity. Anal. Appl. **14**(3), 341–391 (2014)
5. Demoures, F., Gay-Balmaz, F., Ratiu, T.S.: Multisymplectic variational integrators for nonsmooth Lagrangian continuum mechanics. Forum Math. Sigma **4**, e19, 54 p. (2016)
6. Demoures, F., Gay-Balmaz, F.: Multisymplectic variational integrators for barotropic and incompressible fluid models with constraints (2020, submitted)
7. Donea, J., Giuliani, S., Halleux, J.P.: An arbitrary Lagrangian-Eulerian finite element method for transient dynamic fluid-structure interactions. Comput. Meth. Appl. Mech. Eng. **33**, 689–723 (1982)
8. Farhat, C., Rallu, A., Wang, K., Belytschko, T.: Robust and provably second-order explicit-explicit and implicit-explicit staggered time-integrators for highly non-linear compressible fluid-structure interaction problems. Int. J. Numer. Meth. Eng. **84**, 73–107 (2010)
9. Hughes, T.J.R., Liu, W.K., Zimmerman, T.: Lagrangian-Eulerian finite element formulation for incompressible viscous flow. Comput. Meth. Appl. Mech. Eng. **29**, 329–349 (1981)
10. Lew, A., Marsden, J.E., Ortiz, M., West, M.: Asynchronous variational integrators. Arch. Ration. Mech. Anal. **167**(2), 85–146 (2003)
11. Marsden, J.E., Patrick, G.W., Shkoller, S.: Multisymplectic geometry, variational integrators and nonlinear PDEs. Comm. Math. Phys. **199**, 351–395 (1998)
12. Marsden, J.E., Pekarsky, S., Shkoller, S., West, M.: Variational methods, multisymplectic geometry and continuum mechanics. J. Geom. Phys. **38**, 253–284 (2001)
13. Soulaimani, A., Fortin, M., Dhatt, G., Ouetlet, Y.: Finite element simulation of two- and three-dimensional free surface flows. Comput. Meth. Appl. Mech. Eng. **86**, 265–296 (1991)

Metriplectic Integrators for Dissipative Fluids

Michael Kraus[1,2(✉)]

[1] Max-Planck-Institut für Plasmaphysik, Boltzmannstraße 2, 85748 Garching, Germany
[2] Technische Universität München, Zentrum Mathematik, Boltzmannstraße 3, 85748 Garching, Germany
michael.kraus@ipp.mpg.de

Abstract. Many systems from fluid dynamics and plasma physics possess a so-called metriplectic structure, that is the equations are comprised of a conservative, Hamiltonian part, and a dissipative, metric part. Consequences of this structure are conservation of important quantities, such as mass, momentum and energy, and compatibility with the laws of thermodynamics, e.g., monotonic dissipation of entropy and existence of a unique equilibrium state.

For simulations of such systems to deliver accurate and physically correct results, it is important to preserve these relations and conservation laws in the course of discretisation. This can be achieved most easily not by enforcing these properties directly, but by preserving the underlying abstract mathematical structure of the equations, namely their metriplectic structure. From that, the conservation of the aforementioned desirable properties follows automatically.

This paper describes a general and flexible framework for the construction of such metriplectic structure-preserving integrators, that facilitates the design of novel numerical methods for systems from fluid dynamics and plasma physics.

Keywords: Fluid dynamics · Geometric numerical integration · Metriplectic dynamics

1 Metriplectic Dynamics

Metriplectic dynamics [6–8,11,12,14–16] provides a convenient framework for the description of systems that encompass both Hamiltonian and dissipative parts. The Hamiltonian evolution of such a system is determined by a Poisson bracket $\{\cdot,\cdot\}$ and the Hamiltonian functional $\mathcal{H}$, usually the total energy of the system. The dissipative evolution is determined by a metric bracket $(\cdot,\cdot)$ and some functional $\mathcal{S}$ that evolves monotonically in time, usually some kind of entropy.

F. Nielsen and F. Barbaresco (Eds.): GSI 2021, LNCS 12829, pp. 292–301, 2021.
https://doi.org/10.1007/978-3-030-80209-7_33

Let us denote by $u(t,x) = (u^1, u^2, ..., u^m)^T$ the dynamical variables, defined over the domain $\mathcal{D}$ with coordinates x. The evolution of any functional $\mathcal{F}$ of the dynamical variables u is given by

$$\frac{\mathrm{d}\mathcal{F}}{\mathrm{d}t} = \{\mathcal{F}, \mathcal{G}\} + (\mathcal{F}, \mathcal{G}), \tag{1}$$

with $\mathcal{G} = \mathcal{H} - \mathcal{S}$ a generalised free energy functional, analogous with the Gibb's free energy from thermodynamics, $\{\cdot,\cdot\}$ a Poisson bracket, and $(\cdot,\cdot)$ a metric bracket. The Poisson bracket, describing the ideal, conservative evolution of the system, is a bilinear, anti-symmetric bracket of the form

$$\{\mathcal{A}, \mathcal{B}\} = \int_{\mathcal{D}} \frac{\delta \mathcal{A}}{\delta u^i} \, \mathcal{J}^{ij}(u) \, \frac{\delta \mathcal{B}}{\delta u^j} \, \mathrm{d}x, \tag{2}$$

where $\mathcal{A}$ and $\mathcal{B}$ are functionals of u and $\delta \mathcal{A}/\delta u^i$ is the functional derivative, defined by

$$\frac{d}{d\epsilon} \mathcal{A}\big[u^1, \, ..., \, u^i + \epsilon v^i, \, ..., u^m\big]\bigg|_{\epsilon=0} = \int_{\mathcal{D}} \frac{\delta \mathcal{A}}{\delta u^i} \, v^i \, \mathrm{d}x. \tag{3}$$

The kernel of the bracket, $\mathcal{J}(u)$, is an anti-self-adjoint operator, which has the property that

$$\sum_{l=1}^{m} \left(\frac{\partial \mathcal{J}^{ij}(u)}{\partial u^l} \, \mathcal{J}^{lk}(u) + \frac{\partial \mathcal{J}^{jk}(u)}{\partial u^l} \, \mathcal{J}^{li}(u) + \frac{\partial \mathcal{J}^{ki}(u)}{\partial u^l} \, \mathcal{J}^{lj}(u) \right) = 0 \quad \text{for} \quad 1 \leq i,j,k \leq m, \tag{4}$$

ensuring that the bracket $\{\cdot,\cdot\}$ satisfies the Jacobi identity,

$$\{\{\mathcal{A}, \mathcal{B}\}, \mathcal{C}\} + \{\{\mathcal{B}, \mathcal{C}\}, \mathcal{A}\} + \{\{\mathcal{C}, \mathcal{A}\}, \mathcal{B}\} = 0, \tag{5}$$

for arbitrary functionals $\mathcal{A}, \mathcal{B}, \mathcal{C}$ of u. Apart from that, $\mathcal{J}(u)$ is not required to be of any particular form, and most importantly it is allowed to depend on the fields u. If $\mathcal{J}(u)$ has a non-empty nullspace, there exist so-called Casimir invariants, that is functionals $\mathcal{C}$ for which $\{\mathcal{A}, \mathcal{C}\} = 0$ for all functionals $\mathcal{A}$. The monotonic entropy functional $\mathcal{S}$ is usually one of these Casimir invariants.

The metric bracket $(\cdot,\cdot)$, describing non-ideal, dissipative effects, is a symmetric bracket, defined in a similar way as the Poisson bracket by

$$(\mathcal{A}, \mathcal{B}) = \int_{\mathcal{D}} \frac{\delta \mathcal{A}}{\delta u^i} \, \mathcal{G}^{ij}(u) \, \frac{\delta \mathcal{B}}{\delta u^j} \, \mathrm{d}x, \tag{6}$$

where $\mathcal{G}(u)$ is now a self-adjoint operator with an appropriate nullspace such that $(\mathcal{H}, \mathcal{G}) = 0$. All Casimirs $\mathcal{C}$ of the Poisson bracket should be Casimirs also of the metric bracket, except the entropy functional $\mathcal{S}$ which is explicitly required not to be a Casimir of the metric bracket.

In the following, we choose a convention that dissipates entropy and conserves the Hamiltonian, so that the equilibrium state is reached when entropy is at minimum. For this framework and our conventions to be consistent, it is essential that (i) $\mathcal{H}$ is a Casimir of the metric bracket, (ii) $\mathcal{S}$ is a Casimir of the Poisson bracket, and that (iii) the metric bracket is positive semi-definite. With respect to these choices, we then have

$$\frac{\mathrm{d}\mathcal{H}}{\mathrm{d}t} = \{\mathcal{H}, \mathcal{G}\} + (\mathcal{H}, \mathcal{G}) = \{\mathcal{H}, -\mathcal{S}\} = 0, \quad \frac{\mathrm{d}\mathcal{S}}{\mathrm{d}t} = \{\mathcal{S}, \mathcal{G}\} + (\mathcal{S}, \mathcal{G}) = -(\mathcal{S}, \mathcal{S}) \leq 0, \tag{7}$$

reproducing the First and Second Law of Thermodynamics.

For an equilibrium state u_{eq}, the time evolution of any functional $\mathcal{F}$ stalls and the entropy functional reaches its minimum. If the metriplectic system has no Casimirs, the equilibrium state satisfies an energy principle, according to which the first variation of the free energy vanishes, $\delta\mathcal{G}[u_{eq}] = 0$, and its second variation is strictly positive, $\delta^2\mathcal{G}[u_{eq}] > 0$ (for details see e.g. [10]). If Casimirs $\mathcal{C}_i$ exist, the equilibrium state becomes degenerate, and the energy principle must be modified to account for the existing Casimirs. This leads to the so-called energy–Casimir principle [17]. In this case the the equilibrium state satisfies

$$\delta\mathcal{G}[u_{eq}] + \sum_i \lambda_i \delta\mathcal{C}_i[u_{eq}] = 0, \tag{8}$$

where λ_i act as Lagrange multipliers and are determined uniquely from the values of the Casimirs at the initial conditions for u. Lastly, for each $x \in \mathcal{D}$ the equilibrium state of the free-energy functional $\mathcal{G}$ is unique. This can be verified by employing a convexity argument, namely if $\mathcal{D}$ is a convex domain and $\mathcal{G}$ is strictly convex, then $\mathcal{G}$ has at most one critical point (see e.g. [5] for details). This is the case if

$$\delta^2(\mathcal{G} + \sum_i \lambda_i \mathcal{C}_i) > 0, \tag{9}$$

for the non-vanishing field u_{eq}.

2 Dissipative Hydrodynamics

Many dissipative systems from fluid dynamics and plasma physics feature a metriplectic structure. Examples include Korteweg-type fluids [9,19–22], geophysical fluid flows [3,4] and extended magnetohydronamics [2]. In the following we consider the example of viscous heat-conductive flows, whose state is described in terms of the variables (ρ, m, σ), where ρ is the mass density, m is the momentum density, and σ is the entropy density. The ideal dynamics of this system are described by the Poisson bracket [1,9,17–19,21]

$$\begin{aligned}\{\mathcal{A}, \mathcal{B}\} = \int_{\mathcal{D}} m(x,t) \cdot \left[\frac{\delta\mathcal{B}}{\delta m} \cdot \nabla \frac{\delta\mathcal{A}}{\delta m} - \frac{\delta\mathcal{A}}{\delta m} \cdot \nabla \frac{\delta\mathcal{B}}{\delta m}\right] + \int_{\mathcal{D}} \rho(x,t) \left[\frac{\delta\mathcal{B}}{\delta m} \cdot \nabla \frac{\delta\mathcal{A}}{\delta \rho} - \frac{\delta\mathcal{A}}{\delta m} \cdot \nabla \frac{\delta\mathcal{B}}{\delta \rho}\right] \\ + \int_{\mathcal{D}} \sigma(x,t) \left[\frac{\delta\mathcal{B}}{\delta m} \cdot \nabla \frac{\delta\mathcal{A}}{\delta \sigma} - \frac{\delta\mathcal{A}}{\delta m} \cdot \nabla \frac{\delta\mathcal{B}}{\delta \sigma}\right],\end{aligned} \tag{10}$$

and the Hamiltonian functional

$$\mathcal{H} = \int_{\mathcal{D}} \left[\frac{1}{\rho} \frac{|m|^2}{2} + \rho U(\rho, \sigma/\rho) \right] \mathrm{d}x. \tag{11}$$

Here, U denotes the internal energy, from which the pressure and temperature follow as $p = \rho^2 U_\rho$ and $T = \rho U_\sigma$, respectively. The ideal dynamics preserves the total mass, momentum and entropy,

$$\frac{d\mathcal{M}}{dt} = \{\mathcal{M}, \mathcal{H}\} = 0, \qquad \mathcal{M} = \int_{\mathcal{D}} \rho(x,t)\,\mathrm{d}x, \tag{12}$$

$$\frac{d\mathcal{P}}{dt} = \{\mathcal{P}, \mathcal{H}\} = 0, \qquad \mathcal{P} = \int_{\mathcal{D}} m(x,t)\,\mathrm{d}x, \tag{13}$$

$$\frac{d\mathcal{S}}{dt} = \{\mathcal{S}, \mathcal{H}\} = 0, \qquad \mathcal{S} = \int_{\mathcal{D}} \sigma(x,t)\,\mathrm{d}x, \tag{14}$$

where mass and entropy are Casimir invariants of the Poisson bracket (10), i.e., $\{\mathcal{M}, \mathcal{B}\} = 0$ and $\{\mathcal{S}, \mathcal{B}\} = 0$ for any $\mathcal{B}$, and conservation of momentum follows from a symmetry of the Hamiltonian (11), i.e., $\{\mathcal{M}, \mathcal{H}\} = 0$ only for $\mathcal{H}$ (for details see e.g. Reference [17]). Dissipation due to viscous friction and heat conduction is modelled by the metric bracket [19–21]

$$\begin{aligned}
(\mathcal{A}, \mathcal{B}) = & \int_{\mathcal{D}} 2\mu \left[T \left\langle \nabla \frac{\delta \mathcal{A}}{\delta m} \right\rangle - \frac{\delta \mathcal{A}}{\delta \sigma} \langle \nabla v \rangle \right] : \left[T \left\langle \nabla \frac{\delta \mathcal{B}}{\delta m} \right\rangle - \frac{\delta \mathcal{B}}{\delta \sigma} \langle \nabla v \rangle \right] \mathrm{d}x \\
& + \int_{\mathcal{D}} \lambda \left[T \nabla \cdot \frac{\delta \mathcal{A}}{\delta m} - \frac{\delta \mathcal{A}}{\delta \sigma} \nabla \cdot v \right] \cdot \left[T \nabla \cdot \frac{\delta \mathcal{B}}{\delta m} - \frac{\delta \mathcal{B}}{\delta \sigma} \nabla \cdot v \right] \mathrm{d}x \\
& + \int_{\mathcal{D}} \kappa T^2 \, \nabla \left[\frac{1}{T} \frac{\delta \mathcal{A}}{\delta \sigma} \right] \cdot \nabla \left[\frac{1}{T} \frac{\delta \mathcal{B}}{\delta \sigma} \right] \mathrm{d}x,
\end{aligned} \tag{15}$$

and the entropy functional from (14). Here, $v = m/\rho$ denotes the velocity, μ the coefficient of shear viscosity, λ the coefficient of bulk viscosity, κ the thermal conductivity, and $\langle \cdot \rangle$ denotes the projection

$$\langle \nabla v \rangle = \frac{1}{2} (\nabla v + \nabla v^T) - \frac{1}{3} (\operatorname{trace} \nabla v) \mathbb{1}, \tag{16}$$

with $\mathbb{1}$ the 3×3 identity matrix. Note that all of the coefficients μ, λ and κ need to be non-negative for the metric bracket (15) to be positive semi-definite. It is straightforward to verify that the metric bracket (15) preserves mass (12), momentum (13) and energy (11). The equations of motion are computed using (1) for $\mathcal{F} \in \{\rho, m, \sigma\}$ and $\mathcal{G} = \mathcal{H} - \mathcal{S}$ with $\mathcal{H}$ from (11) and $\mathcal{S}$ from (14) as

$$\frac{\partial \rho}{\partial t} = -\nabla \cdot m, \tag{17}$$

$$\frac{\partial m}{\partial t} = -\nabla \cdot (m \otimes v) + \nabla \cdot \big[(\rho \nabla \cdot \xi - p)1 - \nabla \rho \otimes \xi + 2\mu \langle \nabla v \rangle + \lambda (\nabla \cdot v) 1\big], \tag{18}$$

$$\frac{\partial \sigma}{\partial t} = -\nabla \cdot \left[\sigma v - \frac{\kappa \nabla T}{T}\right] + \frac{1}{T}\left[2\mu \langle \nabla v \rangle \cdot \langle \nabla v \rangle + \lambda (\nabla \cdot v)^2 + \frac{\kappa}{T} |\nabla T|^2\right]. \tag{19}$$

With standard approaches such as finite element or discontinuous Galerkin methods, numerical algorithms are obtained by direct discretisation of these equations. The idea of metriplectic integrators [13] instead is to construct discrete expressions of the brackets (10) and (15) and to obtain semi-discrete equations of motion from the discrete brackets in analogy to (1). The advantage of the latter approach is that important properties like energy conservation and monotonic dissipation of entropy will automatically be preserved.

3 Metriplectic Integrators

In the following, we describe a general framework for the construction of structure-preserving integrators for dissipative fluids and similar systems that have a metriplectic structure. This framework facilitates the construction of novel numerical methods that automatically preserve important physical quantities, such as mass, energy and the laws of thermodynamics, independently of the particular discretisation framework. Due to limited space, we do not go into the details of a specific discretisation technique and we only consider the spatial discretisation and its properties independently of a particular temporal discretisation. The fully discrete setting together with numerical examples will be described elsewhere.

The spatial discretisation consists of the following components: (a) choosing approximations of the function spaces of the dynamical variables, (b) choosing approximations of the inner products on these spaces, (c) choosing an approximation of functionals, (d) choosing a finite-dimensional representation of the functional derivative. The first step implies the choice of an appropriate finite element or similar space. The second as well as the third step tend to boil down to choosing a quadrature formula. And the last step usually amounts to the most simple choice of a plain partial derivative. Apart from these degrees of freedom, the whole construction is systematic and automatic. Everything follows from these choices and the metriplectic structure of the system of equations.

In the following, we explain this methodology using a rather abstract continuous Galerkin approach, where we do not go into the details of an actual discretisation of the domain or a specific choice of basis (that is we do not concretise component (a) from above). This avoids certain technicalities that arise in practical applications and specific frameworks such as discontinuous Galerkin methods, and thus allows us to focus on conveying the general idea of metriplectic integrators. Nonetheless, the following treatment should provide enough insight for applying and adapting the framework to other discretisation techniques. Concrete examples of discretisations using discontinuous Galerkin methods will be discussed in consecutive publications.

Let v and w be elements of $\mathbb{V} = L^2(\mathcal{D})$, that is the space of square integrable functions over $\mathcal{D}$ with scalar product

$$\langle v \, , \, w \rangle_{\mathcal{D}} = \int_{\mathcal{D}} v(x) \, w(x) \, \mathrm{d}x. \tag{20}$$

Let $\mathbb{V}_h \subset \mathbb{V}$ denote some finite dimensional subspace and $\{\varphi_i\}_{i=1}^N$ a basis in $\mathbb{V}_h$. Then $v_h \in \mathbb{V}_h$ can be written as

$$v_h(t,x) = \sum_{i=1}^{N} v_i(t) \, \varphi_i(x). \tag{21}$$

The discretisation of the density ρ, the momentum m and the entropy density σ follow in full analogy with the function spaces $\mathbb{V}$ chosen appropriately. Note that while ρ and σ are scalar quantities, the momentum m is a three-vector.

In simple situations, we can retain the inner product $\langle \cdot \, , \, \cdot \rangle$ on $\mathbb{V}$ and work with the continuous functionals $\mathcal{F}$ evaluated on (ρ_h, m_h, σ_h), however, in many situations the resulting integrals cannot be computed easily. Therefore we introduce an approximate inner product $\langle \cdot \, , \, \cdot \rangle_h$ as well as approximate functionals $\mathcal{F}_h$ by utilising some quadrature rule $\{(b_n, c_n)\}_{n=1}^R$, where c_n denotes the nodes of the quadrature and b_n the weights, such that

$$\langle v_h \, , \, w_h \rangle_h = \sum_{n=1}^{R} b_n v_h(c_n) \, w_h(c_n) = V^T \mathbb{M} W, \tag{22}$$

with $V = (v_1, v_2, ..., v_N)^T$ the coefficient vector of v_h when expressed in the basis $\{\varphi_i\}_{i=1}^N$ according to (21), analogously W for w_h, and $\mathbb{M}$ the mass matrix

$$\mathbb{M}_{ij} = \sum_{n=1}^{R} b_n \, \varphi_i(c_n) \, \varphi_j(c_n). \tag{23}$$

In the same fashion, functionals of (ρ, m, σ) are approximated using the quadrature rule $\{(b_n, c_n)\}_{n=1}^R$, e.g., for the Hamiltonian (11) we obtain

$$\mathcal{H}_h[\rho_h, m_h, \sigma_h] = \sum_{n=1}^{R} b_n \left[\frac{|m_h(c_n)|^2}{2\rho_h(c_n)} + \rho_h(c_n) \, U\big(\rho_h(c_n), \, \sigma_h(c_n)/\rho_h(c_n)\big) \right] \equiv \mathsf{H}(\hat{\rho}, \hat{m}, \hat{\sigma}), \tag{24}$$

for the mass (12), momentum (13), and entropy (14), respectively, we get

$$\mathcal{M}_h[\rho_h] = \sum_{n=1}^{R} b_n \, \rho_h(c_n) = 1_N^T \mathbb{M}^\rho \hat{\rho} \equiv \mathsf{M}(\hat{\rho}), \tag{25}$$

$$\mathcal{P}_h[\rho_h] = \sum_{n=1}^{R} b_n \, m_h(c_n) = 1_{3N}^T \mathbb{M}^m \hat{m} \equiv \mathsf{P}(\hat{m}), \tag{26}$$

$$\mathcal{S}_h[\sigma_h] = \sum_{n=1}^{R} b_n \, \sigma_h(c_n) = 1_N^T \mathbb{M}^\sigma \hat{\sigma} \equiv \mathsf{S}(\hat{\sigma}), \tag{27}$$

where $\hat{\rho} = (\rho_1, \rho_2, ..., \rho_N)$ are the coefficients of ρ_h, cf. Equation (21), and analogously $\hat{m} = (m_1, m_2, ..., m_{3N})$ and $\hat{\sigma} = (\sigma_1, \sigma_2, ..., \sigma_N)$ are the coefficients of the discrete momentum m_h and entropy density σ_h. By $1_N \in \mathbb{R}^N$ and $1_{3N} \in \mathbb{R}^{3N}$ we denote the respective vectors with all components being equal to 1, and $\mathbb{M}^a$ denotes the mass matrices for $a \in \{\rho, m, \sigma\}$ corresponding to the respective basis functions φ^a following (23). Finally, we need to construct a discrete equivalent to the functional derivative. To that end, we set $\mathsf{A}(\hat{v}) = \mathcal{A}[v]$ and require that

$$\left\langle \frac{\delta \mathcal{A}}{\delta v}[v_h] \, , \, w_h \right\rangle_h = \left\langle \frac{\partial \mathsf{A}}{\partial \hat{v}} \, , \, \hat{w} \right\rangle_N , \tag{28}$$

where $\langle \cdot \, , \, \cdot \rangle_N$ denotes the inner product in $\mathbb{R}^N$, i.e., the scalar product. The functional derivative $\delta \mathcal{A}/\delta v$ is an element of the dual space $\mathbb{V}^*$ of $\mathbb{V}$. Restricting $\mathcal{A}$ to elements v_h of $\mathbb{V}_h$, we can express $\delta \mathcal{A}/\delta v[v_h]$ in the basis $\{\psi_i\}_{i=1}^N$ of $\mathbb{V}_h^*$ as

$$\frac{\delta \mathcal{A}}{\delta v}[v_h] = \sum_{i=1}^{N} a_i \, \psi_i(x). \tag{29}$$

By the Riesz representation theorem we can express the basis functions ψ_i in terms of the basis $\{\varphi_i\}_{i=1}^N$, and using the duality between the bases, i.e., $\langle \psi_i \, , \, \varphi_j \rangle = \delta_{ij}$, as well as (28), we find that

$$\frac{\delta \mathcal{A}}{\delta v}[v_h] = \sum_{i,j=1}^{N} \frac{\partial \mathsf{A}}{\partial v_i} \, \mathbb{M}_{ij}^{-1} \, \varphi_j(x). \tag{30}$$

In order to obtain a semi-discrete metriplectic system, discretised in space but not yet time, all that needs to be done is replace the functional derivatives in (10) and (15) and replace the integrals with the quadrature rule $\{(b_n, c_n)\}_{n=1}^R$. Setting $\mathsf{A}(\hat{\rho}, \hat{m}, \hat{\sigma}) = \mathcal{A}[\rho_h, m_h, \sigma_h]$ and $\mathsf{B}(\hat{\rho}, \hat{m}, \hat{\sigma}) = \mathcal{B}[\rho_h, m_h, \sigma_h]$, for the Poisson bracket (10) this yields

$$\begin{aligned} \{\mathsf{A}, \mathsf{B}\}_h = \sum_{i,j,k}^{3N} \mathbb{P}_{ijk}^{mm} \, m_k \left[\frac{\partial \mathsf{B}}{\partial m_i} \frac{\partial \mathsf{A}}{\partial m_j} - \frac{\partial \mathsf{A}}{\partial m_i} \frac{\partial \mathsf{B}}{\partial m_j} \right] &+ \sum_{i}^{3N} \sum_{j,k=1}^{N} \mathbb{P}_{ijk}^{m\rho} \, \rho_k \left[\frac{\partial \mathsf{B}}{\partial m_i} \frac{\partial \mathsf{A}}{\partial \rho_j} - \frac{\partial \mathsf{A}}{\partial m_i} \frac{\partial \mathsf{B}}{\partial \rho_j} \right] \\ &+ \sum_{i}^{3N} \sum_{j,k=1}^{N} \mathbb{P}_{ijk}^{m\sigma} \, \sigma_k \left[\frac{\partial \mathsf{B}}{\partial m_i} \frac{\partial \mathsf{A}}{\partial \sigma_j} - \frac{\partial \mathsf{A}}{\partial m_i} \frac{\partial \mathsf{B}}{\partial \sigma_j} \right], \end{aligned} \tag{31}$$

with

$$\mathbb{P}_{ijk}^{mm} = \sum_{p,q=1}^{3N} \mathbb{L}_{pqk}^{mm} \, (\mathbb{M}^m)_{ip}^{-1} \, (\mathbb{M}^m)_{qj}^{-1} , \qquad \mathbb{L}_{ijk}^{mm} = \sum_{n=1}^{R} b_n \, \varphi_k^m(c_n) \, \varphi_i^m(c_n) \cdot \nabla \varphi_j^m(c_n), \tag{32}$$

$$\mathbb{P}^{m\rho}_{ijk} = \sum_{p=1}^{3N} \sum_{q=1}^{N} \mathbb{L}^{m\rho}_{pqk} \, (\mathbb{M}^{m})^{-1}_{ip} \, (\mathbb{M}^{\rho})^{-1}_{qj}, \quad \mathbb{L}^{m\rho}_{ijk} = \sum_{n=1}^{R} b_n \, \varphi^{\rho}_{k}(c_n) \, \varphi^{m}_{i}(c_n) \cdot \nabla \varphi^{\rho}_{j}(c_n), \tag{33}$$

$$\mathbb{P}^{m\sigma}_{ijk} = \sum_{p=1}^{3N} \sum_{q=1}^{N} \mathbb{L}^{m\sigma}_{pqk} \, (\mathbb{M}^{m})^{-1}_{ip} \, (\mathbb{M}^{\sigma})^{-1}_{qj}, \quad \mathbb{L}^{m\sigma}_{ijk} = \sum_{n=1}^{R} b_n \, \varphi^{\sigma}_{k}(c_n) \, \varphi^{m}_{i}(c_n) \cdot \nabla \varphi^{\sigma}_{j}(c_n). \tag{34}$$

The discrete bracket (31) is anti-symmetric, but it does not satisfy the Jacobi identity. In the context of metriplectic systems, this is not critical as the full metriplectic bracket does not satisfy the Jacobi identity either. More importantly, the discrete bracket (31) preserves the discrete energy (24) as well as mass (25) and entropy (27), and depending on the actual discretisation possibly also momentum (26). Energy conservation follows immediately from the anti-symmetry of (31), so that $\{\mathsf{H}, \mathsf{H}\}_h = 0$. Mass and entropy conservation can be seen by inserting (25) or (27) into (31), e.g.,

$$\{\mathsf{M}, \mathsf{B}\}_h = \sum_{i,l}^{3N} \sum_{j,k=1}^{N} \mathbb{L}^{m\rho}_{ijk} \, \rho_k \, 1_j \, (\mathbb{M}^{m})^{-1}_{il} \, \frac{\partial \mathsf{B}}{\partial m_l} = 0 \qquad \text{as} \qquad \sum_{j=1}^{N} \mathbb{L}^{m\rho}_{ijk} \, 1_j = 0. \tag{35}$$

Note that this holds for any B. Momentum conservation, on the other hand, relies on the specific form of H and is not warranted for any discrete Hamiltonian of the form (24), but only for specific choices of basis functions and quadrature rules.

The discretisation of the metric bracket (15) and the proof of its conservation properties follow exactly along the same lines. The semi-discrete equations of motion are obtained by utilising Eq. (1) and thus computing

$$\frac{du}{dt} = \{u, \mathsf{G}\}_h + (u, \mathsf{G})_h \quad \text{for} \quad u \in \{\rho_1, ..., \rho_N, m_1, ..., m_{3N}, \sigma_1, ..., \sigma_N\}, \tag{36}$$

where $\mathsf{G} = \mathsf{H} - \mathsf{S}$ approximates the free energy $\mathcal{G}$.

4 Summary

The preceding paper outlines a flexible framework for the construction of structure-preserving algorithms for the numerical integration of dissipative systems from fluid dynamics and plasma physics. This framework is very general with respect to both, the system of equations and the particular numerical method applied. Although here it was applied to the specific example of Korteweg-type fluids, it is equally applicable e.g. to geophysical fluid flows, magnetohydrodynamics or kinetic systems. While in the above construction a continuous Galerkin discretisation was assumed, this approach is equally well applicable to discontinuous Galerkin approximations, isogeometric analysis, spectral methods or even finite differences.

Note that in contrast to typical finite volume or discontinuous Galerkin methods, the metriplectic framework and the conservation properties of the resulting schemes do not depend on the coordinate representation of the system. Although conservative coordinates have been used here, this is not required in order to obtain conservation of the corresponding invariants. The same can have been achieved using, for example, the density ρ, velocity $v = m/\rho$ and entropy $s = \sigma/\rho$ instead.

An important ingredient for a fully discrete algorithm is the temporal discretisation. It has not been discussed here due to the constrained space and will be explained elsewhere, together with a thorough discussion of the discrete conservation laws and the discrete H-theorem.

References

1. Beris, A.N., Edwards, B.J.: Thermodynamics of Flowing Systems. Oxford University Press, Oxford (1994)
2. Coquinot, B., Morrison, P.J.: A general metriplectic framework with application to dissipative extended magnetohydrodynamics. J. Plasma Phys. **86**, 835860302 (2020)
3. Eldred, C., Gay-Balmaz, F.: Single and double generator bracket formulations of multicomponent fluids with irreversible processes. J. Phys. A Math. Theor. **53**, 395701 (2020)
4. Gay-Balmaz, F., Yoshimura, H.: From variational to bracket formulations in nonequilibrium thermodynamics of simple systems. J. Geom. Phys. **158**, 103812 (2020)
5. Giaquinta, M., Hildebrandt, S.: Calculus of Variations I. Springer, Dordrecht (2004). https://doi.org/10.1007/978-3-662-03278-7
6. Grmela, M.: Bracket formulation of dissipative fluid mechanics equations. Phys. Lett. A **102**(8), 355–358 (1984)
7. Grmela, M.: Particle and bracket formulations of kinetic equations. In: Fluids and Plasmas: Geometry and Dynamics, pp. 125–132. American Mathematical Society, Providence (1984)
8. Grmela, M.: Bracket formulation of dissipative time evolution equations. Phys. Lett. A **111**(1–2), 36–40 (1985)
9. Grmela, M., Öttinger, H.C.: Dynamics and thermodynamics of complex fluids. I. Development of a general formalism. Phys. Rev. E **56**, 6620–6632 (1997)
10. Holm, D.D., Marsden, J.E., Ratiu, T., Weinstein, A.: Nonlinear stability of fluid and plasma equilibria. Phys. Rep. **123**(1), 1–116 (1985)
11. Kaufman, A.N.: Dissipative Hamiltonian systems: a unifying principle. Phys. Lett. A **100**(8), 419–422 (1984)
12. Kaufman, A.N., Morrison, P.J.: Algebraic structure of the plasma quasilinear equations. Phys. Lett. A **88**(8), 405–406 (1982)
13. Kraus, M., Hirvijoki, E.: Metriplectic integrators for the landau collision operator. Phys. Plasmas **24**, 102311 (2017)
14. Morrison, P.J.: Bracket formulation for irreversible classical fields. Phys. Lett. A **100**(8), 423–427 (1984)
15. Morrison, P.J.: Some observations regarding brackets and dissipation. Tech. rep., Center for Pure and Applied Mathematics Report PAM-228, University of California, Berkeley (1984)

16. Morrison, P.J.: A paradigm for joined Hamiltonian and dissipative systems. Phys. D Nonlinear Phenom. **18**, 410–419 (1986)
17. Morrison, P.J.: Hamiltonian description of the ideal fluid. Rev. Mod. Phys. **70**(2), 467–521 (1998)
18. Morrison, P.J., Greene, J.M.: Noncanonical Hamiltonian density formulation of hydrodynamics and ideal magnetohydrodynamics. Phys. Rev. Lett. **45**, 790–794 (1980)
19. Öttinger, H.C.: Beyond Equilibrium Thermodynamics. John Wiley & Sons, New York(2005)
20. Öttinger, H.C., Grmela, M.: Dynamics and thermodynamics of complex fluids. II. Illustrations of a general formalism. Phys. Rev. E **56**, 6633–6655 (1997)
21. Suzuki, Y.: A GENERIC formalism for Korteweg-type fluids: I. A comparison with classical theory. Fluid Dyn. Res. **52**, 015516 (2020)
22. Suzuki, Y.: A GENERIC formalism for Korteweg-type fluids: II. Higher-order models and relation to microforces. Fluid Dyn. Res. **52**, 025510 (2020)

From Quantum Hydrodynamics to Koopman Wavefunctions I

François Gay-Balmaz[1](✉) and Cesare Tronci[2,3]

[1] CNRS & École Normale Supérieure, Paris, France
francois.gay-balmaz@lmd.ens.fr

[2] Department of Mathematics, University of Surrey, Guildford, UK
c.tronci@surrey.ac.uk

[3] Department of Physics and Engineering Physics, Tulane University, New Orleans, USA

Abstract. Based on Koopman's theory of classical wavefunctions in phase space, we present the Koopman-van Hove (KvH) formulation of classical mechanics as well as some of its properties. In particular, we show how the associated classical Liouville density arises as a momentum map associated to the unitary action of strict contact transformations on classical wavefunctions. Upon applying the Madelung transform from quantum hydrodynamics in the new context, we show how the Koopman wavefunction picture is insufficient to reproduce arbitrary classical distributions. However, this problem is entirely overcome by resorting to von Neumann operators. Indeed, we show that the latter also allow for singular δ-like profiles of the Liouville density, thereby reproducing point particles in phase space.

Keywords: Hamiltonian dynamics · Koopman wavefunctions · Momentum map · Prequantization

1 Introduction

Koopman wavefunctions on phase-space have a long history going back to Koopman's early work [6] and their unitary evolution was revisited by van Hove [10], who unfolded their role within a niche area of symplectic geometry now known as *prequantization* [5,7,9]. In this context, classical mechanics possesses a Hilbert-space formulation similar to quantum mechanics and our recent work [2] provided a new geometric insight on the correspondence between Koopman wavefunctions and classical distributions on phase-space. Then, classical wavefunctions were renamed *Koopman-van Hove* wavefunctions to stress van Hove's contribution.

Here, we show how the evolution of Koopman wavefunctions can be described in terms of the same geometric quantities already emerging in Madelung's hydrodynamic formulation of standard quantum mechanics [8]. In particular, after

CT acknowledges partial support by the Royal Society and the Institute of Mathematics and its Applications, UK.

F. Nielsen and F. Barbaresco (Eds.): GSI 2021, LNCS 12829, pp. 302–310, 2021.
https://doi.org/10.1007/978-3-030-80209-7_34

reviewing the geometry of quantum hydrodynamics, we shall show how the introduction of density and momentum variables on phase-space leads to an alternative geometric description of Koopman classical mechanics. While most of our discussion is based on [4] (where further details can be found), here we shall also show that this picture naturally allows for δ-like particle solutions, whose occurrence is instead questionable in the Hilbert-space wavefunction picture.

2 Geometry of Quantum Hydrodynamics

Madelung's equation of quantum hydrodynamics are obtained by replacing the polar form $\psi(t,x) = \sqrt{D(t,x)}e^{-iS(t,x)/\hbar}$ of the wavefunction into Schrödinger's equation $i\hbar\partial_t\psi = -m^{-1}\hbar^2\Delta\psi/2 + V\psi$. Then, defining the velocity vector field $v = \nabla S/m$ leads to the well-known set of hydrodynamic equations

$$\partial_t v + v\cdot\nabla v = -\frac{1}{m}\nabla\left(V + \frac{\hbar^2}{2m}\frac{\Delta\sqrt{D}}{\sqrt{D}}\right) \qquad \partial_t D + \operatorname{div}(Dv) = 0\,. \tag{1}$$

Madelung's equations were the point of departure for Bohm's interpretation of quantum dynamics [1], in which the integral curves of $v(x,t)$ are viewed as the genuine trajectories in space of the physical quantum particles. The fact that a hydrodynamic system arises from the Schrödinger equation is actually a consequence of an intrinsic geometric structure underlying the hydrodynamic density and momentum variables D and $\mu = D\nabla S$.

Madelung's Momentum Map. Madelung's equations (1) can be geometrically explained by noticing that the map

$$\begin{aligned}&\mathcal{J} : L^2(M,\mathbb{C}) \to \mathfrak{X}(M)^* \times \operatorname{Den}(M)\,,\\ &\mathcal{J}(\psi) = \left(\hbar\operatorname{Im}(\psi^*\nabla\psi), |\psi|^2\right) = (mDv, D)\end{aligned} \tag{2}$$

is an equivariant momentum map. Here, $L^2(M,\mathbb{C})$ is the Hilbert space of quantum wavefunctions, while $\mathfrak{X}(M)^* \times \operatorname{Den}(M)$ is the dual space to the semidirect-product Lie algebra $\mathfrak{X}(M) \circledS \mathcal{F}(M)$. The momentum map structure of (2) is associated to the following unitary representation of the semidirect-product group $\operatorname{Diff}(M) \circledS \mathcal{F}(M, S^1)$ on $L^2(M,\mathbb{C})$:

$$\psi \mapsto \sqrt{J_\chi^{-1}}\,(e^{-i\varphi/\hbar}\psi)\circ\chi^{-1}, \qquad (\chi, e^{i\varphi}) \in \operatorname{Diff}(M) \circledS \mathcal{F}(M,S^1)\,, \tag{3}$$

where $J_\chi = \det\nabla\chi$ is the Jacobian determinant of χ. The notation is as follows: $\operatorname{Diff}(M)$ is the group of diffeomorphisms of M, $\mathfrak{X}(M)$ is its Lie algebra of vector fields, and $\mathcal{F}(M,S^1)$ is the space of S^1-valued functions on M. Being an equivariant momentum map, $\mathcal{J}$ is a Poisson map hence mapping Schrödinger's Hamiltonian dynamics on $L^2(M,\mathbb{C})$ to Lie-Poisson fluid dynamics on $\mathfrak{X}(M)^* \times \operatorname{Den}(M)$.

Definition of Momentum Maps. Without entering the technicalities involved in the construction of semidirect-product groups, here we simply recall the definition of momentum maps. Given a Poisson manifold $(P, \{,\})$ and a Hamiltonian action of a Lie group G on P, we call $\mathcal{J} : P \to \mathfrak{g}^*$ a *momentum map* if it satisfies $\{f, \langle \mathcal{J}, \xi \rangle\} = \xi_P[f]$. Here, $\mathfrak{g}$ denotes the Lie algebra of G and $\mathfrak{g}^*$ its dual space, ξ_P is the infinitesimal generator of the G–action $\Phi_g : P \to P$ and $\langle \cdot, \cdot \rangle$ is the natural duality pairing on $\mathfrak{g}^* \times \mathfrak{g}$. A momentum map is equivariant if $\mathrm{Ad}^*_g \mathcal{J}(p) = \mathcal{J}(\Phi_g(p))$ for all $g \in G$. When P is a symplectic vector space carrying a (symplectic) G-representation with respect to the symplectic form Ω, the momentum map $\mathcal{J}(p)$ is given by $2\langle \mathcal{J}(p), \xi \rangle := \Omega(\xi_P(p), p)$, for all $p \in P$ and all $\xi \in \mathfrak{g}$. Here we specialize this definition to the symplectic Hilbert space $P = L^2(M, \mathbb{C})$ of wavefunctions, endowed with the standard symplectic form $\Omega(\psi_1, \psi_2) = 2\hbar \,\mathrm{Im} \int_M \bar{\psi}_1 \psi_2 \,\mathrm{d}x$. Then, a momentum map associated to a unitary G–representation on $L^2(M, \mathbb{C})$ is the map $\mathcal{J}(\psi) \in \mathfrak{g}^*$ given by

$$\langle \mathcal{J}(\psi), \xi \rangle = -\hbar \,\mathrm{Im} \int_M \bar{\psi}\, \xi_P(\psi) \,\mathrm{d}x \,. \tag{4}$$

It is easily checked that, if $G = \mathrm{Diff}(M) \circledS \mathcal{F}(M, S^1)$ acts symplectically on $P = L^2(M, \mathbb{C})$ via (3), expression (4) yields the Madelung momentum map.

3 Koopman-van Hove Formulation of Classical Mechanics

As anticipated in the introduction, we want to show how the geometric setting of quantum hydrodynamics transfers to the case of classical wavefunctions in their Koopman-van Hove (KvH) formulation. The latter possesses a deep geometric setting which is reviewed in the present section.

3.1 Evolution of Koopman Wavefunctions

Let Q be the configuration manifold of the classical mechanical system and T^*Q its phase space, given by the cotangent bundle of Q. We assume that the manifold Q is connected. The phase space is canonically endowed with the one-form $\mathcal{A} = p_i \mathrm{d}q^i$ and the symplectic form $\omega = -\mathrm{d}\mathcal{A} = \mathrm{d}q^i \wedge \mathrm{d}p_i$. For later purpose, it is also convenient to consider the trivial circle bundle

$$T^*Q \times S^1 \to T^*Q \,, \tag{5}$$

known as *prequantum bundle*. Then, $\mathcal{A}$ identifies a principal connection $\mathcal{A} + \mathrm{d}s$ with curvature given by (minus) the symplectic form ω. A classical wavefunction Ψ is an element of the classical Hilbert space $\mathscr{H}_C = L^2(T^*Q, \mathbb{C})$ with standard Hermitian inner product $\langle \Psi_1 | \Psi_2 \rangle := \int_{T^*Q} \bar{\Psi}_1 \Psi_2 \,\mathrm{d}z$, where $\mathrm{d}z$ denotes the Liouville volume form on T^*Q. The corresponding real-valued pairing and symplectic form on $\mathscr{H}_C$ are defined by

$$\langle \Psi_1, \Psi_2 \rangle = \mathrm{Re} \langle \Psi_1 | \Psi_2 \rangle \quad \text{and} \quad \Omega(\Psi_1, \Psi_2) = 2\hbar \,\mathrm{Im} \langle \Psi_1 | \Psi_2 \rangle \,. \tag{6}$$

The KvH Equation. Given a classical Hamiltonian function $H \in C^\infty(T^*Q)$, the *KvH equation for classical wavefunctions* is (see [2] and references therein)

$$i\hbar \partial_t \Psi = i\hbar \{H, \Psi\} - L_H \Psi \qquad \text{with} \qquad L_H := \mathcal{A} \cdot X_H - H = p_i \partial_{p_i} H - H \,. \tag{7}$$

Here, X_H is the Hamiltonian vector field associated to H so that $\mathbf{i}_{X_H}\omega = \mathrm{d}H$, and $\{H, K\} = \omega(X_H, X_K)$ is the canonical Poisson bracket, extended to $\mathbb{C}$-valued functions by $\mathbb{C}$-linearity. In addition, we recognize that L_H identifies the Lagrangian function (on phase-space) associated to the Hamiltonian H. The right hand side of (7) defines the *covariant Liouvillian operator*

$$\widehat{\mathcal{L}}_H = \mathrm{i}\hbar\{H, \;\} - L_H \tag{8}$$

also known as *prequantum operator*, which is easily seen to be an unbounded Hermitian operator on $\mathscr{H}_C$. As a consequence, the KvH equation (7) is a Hamiltonian system with respect to the symplectic form (6) and the Hamiltonian

$$h(\Psi) = \int_{T^*Q} \bar{\Psi}\widehat{\mathcal{L}}_H\Psi \,\mathrm{d}z\,. \tag{9}$$

Lie Algebraic Structure. We remark that the correspondence $H \mapsto \widehat{\mathcal{L}}_H$ satisfies

$$[\widehat{\mathcal{L}}_H, \widehat{\mathcal{L}}_F] = \mathrm{i}\hbar\widehat{\mathcal{L}}_{\{H,F\}}\,, \tag{10}$$

for all $H, F \in C^\infty(T^*Q)$. Hence, it follows that on its domain, the operator

$$\Psi \mapsto -\mathrm{i}\hbar^{-1}\widehat{\mathcal{L}}_H\Psi \tag{11}$$

defines a skew-Hermitian (or, equivalently, symplectic) left representation of the Lie algebra $(C^\infty(T^*Q), \{\;,\;\})$ of Hamiltonian functions on $\mathscr{H}_C$. We shall show that the Lie algebra action (11) integrates into a unitary action of the group of strict contact transformations of (5) on $\mathscr{H}_C$, when the first cohomology group $\mathsf{H}^1(T^*Q, \mathbb{R}) = 0$ (or, equivalently, $\mathsf{H}^1(Q, \mathbb{R}) = 0$).

Madelung Transform. As discussed in the introduction, the KvH equation possesses an alternative formulation arising from the Madelung transform. By mimicking the quantum case, we write $\Psi = \sqrt{D}e^{\mathrm{i}S/\hbar}$ so that (7) yields

$$\partial_t S + \{S, H\} = L_H\,, \qquad \partial_t D + \{D, H\} = 0\,. \tag{12}$$

Then, the first Madelung equation is revealing of the dynamics of the classical phase, which reads

$$\frac{\mathrm{d}}{\mathrm{d}t}S(t, \eta(t, z)) = L_H(\eta(t, z))\,, \tag{13}$$

where $\eta(t)$ is the flow of X_H. We remark that taking the differential of the first equation in (12) leads to

$$(\partial_t + \pounds_{X_H})(\mathrm{d}S - \mathcal{A}) = 0\,, \tag{14}$$

which is written in terms of the Lie derivative $\pounds_{X_H} = \mathrm{d}\mathbf{i}_{X_H} + \mathbf{i}_{X_H}\mathrm{d}$. At this point, one would be tempted to set $\mathrm{d}S = \mathcal{A}$ so that the second equation in (12) simply acquires the same meaning as the classical Liouville equation. However, things are not that easy: unless one allows for topological singularities in the phase variable S, the relation $\mathrm{d}S = \mathcal{A}$ cannot hold and therefore the link between the KvH equation (7) and the classical Liouville equation needs extra care. However, we shall come back to the Madelung picture later on to show how this may be extended beyond the current context.

3.2 Strict Contact Transformations

Having characterized the KvH equation and the Lie algebraic structure of prequantum operators, we now move on to characterizing their underlying group structure in terms of (strict) contact transformations and their unitary representation on Koopman wavefunctions.

Connection Preserving Automorphisms. Given the trivial prequantum bundle (5), we consider its automorphism group given by $\operatorname{Diff}(T^*Q)$ Ⓢ $\mathcal{F}(T^*Q, S^1)$. There is a unitary representation of this group on $\mathscr{H}_C$ given by

$$\Psi \mapsto U_{(\eta, e^{i\varphi})}\Psi = \sqrt{J_\eta^{-1}}(e^{-i\varphi/\hbar}\Psi) \circ \eta^{-1}, \tag{15}$$

where $(\eta, e^{i\varphi}) \in \operatorname{Diff}(T^*Q)$ Ⓢ $\mathcal{F}(T^*Q, S^1)$. This is essentially the same representation as in (3), upon replacing the quantum configuration space M with the classical phase space T^*Q. A relevant subgroup of the automorphism group $\operatorname{Diff}(T^*Q)$ Ⓢ $\mathcal{F}(T^*Q, S^1)$ is given by the group of connection-preserving automorphisms of the principal bundle (5): this group is given by

$$\operatorname{Aut}_{\mathcal{A}}(T^*Q \times S^1) = \left\{(\eta, e^{i\varphi}) \in \operatorname{Diff}(T^*Q) \circledS \mathcal{F}(T^*Q, S^1) \;\middle|\; \eta^*\mathcal{A} + d\varphi = \mathcal{A}\right\}, \tag{16}$$

where η^* denotes pullback. The above transformations were studied extensively in van Hove's thesis [10] and are known as forming the group of *strict contact diffeomorphisms.* Note that the relation $\eta^*\mathcal{A} + d\varphi = \mathcal{A}$ implies

$$\eta^*(d\mathcal{A}) = 0\,, \qquad\qquad \varphi(z) = \theta + \int_{z_0}^{z} (\mathcal{A} - \eta^*\mathcal{A})\,, \tag{17}$$

so that η is a symplectic diffeomorphism, i.e. $\eta \in \operatorname{Diff}_\omega(T^*Q)$. Also, φ is determined up to a constant phase $\theta = \varphi(z_0)$ since $\mathsf{H}^1(T^*Q, \mathbb{R}) = 0$ and thus the line integral above does not depend on the curve connecting z_0 to z. The Lie algebra of (16) is

$$\mathfrak{aut}_{\mathcal{A}}(T^*Q \times S^1) = \left\{(X, \nu) \in \mathfrak{X}(T^*Q) \circledS \mathcal{F}(T^*Q) \;\middle|\; \pounds_X\mathcal{A} + d\nu = 0\right\}$$

and we notice that this is isomorphic to the Lie algebra $\mathcal{F}(T^*Q)$ endowed with the canonical Poisson structure. The Lie algebra isomorphism is $H \in \mathcal{F}(T^*Q) \longmapsto (X_H, -L_H) \in \mathfrak{aut}_{\mathcal{A}}(T^*Q \times S^1)$.

The van Hove Representation. As a subgroup of the semidirect product $\operatorname{Diff}(T^*Q)$ Ⓢ $\mathcal{F}(T^*Q, S^1)$, the group $\operatorname{Aut}_{\mathcal{A}}(T^*Q \times S^1)$ inherits from (15) a unitary representation, which is obtained by replacing (17) in (15). First appeared in van Hove's thesis [10], we shall call this the *van Hove representation.* Then, a direct computation shows that $-i\hbar^{-1}\widehat{\mathcal{L}}_H$ emerges as the infinitesimal generator of this representation, i.e., we have

$$\left.\frac{d}{d\epsilon}\right|_{\epsilon=0} U_{(\eta_\epsilon, e^{i\varphi_\epsilon})}\Psi = -i\hbar^{-1}\widehat{\mathcal{L}}_H\Psi, \tag{18}$$

for a path $(\eta_\epsilon, e^{i\varphi_\epsilon}) \in \mathrm{Aut}_{\mathcal{A}}(T^*Q\times S^1)$ tangent to $(X_H, -L_H)$ at $(id, 1)$. Being an infinitesimal generator of the representation (15) restricted to $\mathrm{Aut}_{\mathcal{A}}(T^*Q \times S^1)$, $\widehat{\mathcal{L}}_H$ is equivariant, namely

$$U^\dagger_{(\eta,e^{i\varphi})}\widehat{\mathcal{L}}_H U_{(\eta,e^{i\varphi})} = \widehat{\mathcal{L}}_{H\circ\eta}\,, \quad \forall\,(\eta, e^{i\varphi}) \in \mathrm{Aut}_{\mathcal{A}}(T^*Q \times S^1)\,. \tag{19}$$

3.3 Momentum Maps and the Classical Liouville Equation

Having discussed the geometry underlying the KvH equation (7), we are now in the position of presenting its relation to classical mechanics in terms of a momentum map taking Koopman wavefunctions to distributions on phase-space [2]. Since the representation (15) is unitary, it is symplectic with respect to the symplectic form (6) and admits a momentum map $\mathcal{J} : \mathscr{H}_C \to \mathrm{Den}(T^*Q)$, where $\mathrm{Den}(T^*Q)$ denotes the space of density distributions on T^*Q. From the general formula (4) for momentum maps for unitary representations, we have $\langle \mathcal{J}(\Psi), F\rangle = \int_{T^*Q} \bar\Psi\, \widehat{\mathcal{L}}_F\Psi\,\mathrm{d}z$, where $\langle\ ,\ \rangle$ denotes the L^2-pairing between $\mathcal{F}(T^*Q)$ and its dual $\mathrm{Den}(T^*Q)$. This yields the expression

$$\begin{aligned}\mathcal{J}(\Psi) &= |\Psi|^2 - \mathrm{div}\big(\mathbb{J}\mathcal{A}|\Psi|^2\big) + i\hbar\{\Psi, \bar\Psi\}\\ &= |\Psi|^2 - \mathrm{div}\big(\bar\Psi\mathbb{J}(\mathcal{A}\Psi + i\hbar\nabla\Psi)\big)\,.\end{aligned} \tag{20}$$

Here, the divergence is associated to the Liouville form and $\mathbb{J} : T^*(T^*Q) \to T(T^*Q)$ is defined by $\{F, H\} = \langle \mathrm{d}F, \mathbb{J}(\mathrm{d}H)\rangle$. This equivariant momentum map is a Poisson map with respect to the symplectic Poisson structure $\{\!\{f, h\}\!\}(\Psi)$ on $\mathscr{H}_C$ and the Lie-Poisson structure $\{\!\{f, h\}\!\}(\rho)$ on $\mathrm{Den}(T^*Q)$. With an abuse of notation, we write

$$\{\!\{f,h\}\!\}(\Psi) = \frac{1}{2\hbar}\,\mathrm{Im}\int_{T^*Q}\overline{\frac{\delta f}{\delta\psi}}\,\frac{\delta h}{\delta\psi}\,\mathrm{d}z \quad\longrightarrow\quad \{\!\{f,h\}\!\}(\rho) = \int_{T^*Q}\rho\left\{\frac{\delta f}{\delta\rho}, \frac{\delta h}{\delta\psi}\right\}\mathrm{d}z\,.$$

Hence, if $\Psi(t)$ is a solution of the KvH equation, the density $\rho(t) = \mathcal{J}(\Psi(t))$ in (20) solves the Liouville equation $\partial_t\rho = \{H, \rho\}$. A density of the form (20) is not necessarily positive definite. However, the Liouville equation generates the sign-preserving evolution $\rho(t) = \eta(t)_*\rho_0$, where $\eta(t)$ is the flow of X_H, thereby recovering the usual probabilistic interpretation. In summary, we have the following scheme:

Koopman-van Hove equation (7) for $\Psi \in \mathscr{H}_C$	—— Momentum map $\mathcal{J}$ for $\mathrm{Aut}_{\mathcal{A}}(T^*Q \times S^1)$ ——→	Classical Liouville equation for $\rho \in \mathrm{Den}(T^*Q)$.

Having characterized the classical phase-space density in terms of Koopman wavefunctions (20), we ask if this representation is sufficient to reproduce all the possible classical densities. For example, Gaussian densities allow for this representation [2]. However, this may not be true in more general cases for which KvH theory needs to be extended appropriately.

4 Hydrodynamic Quantities and von Neumann Operators

Let us start our discussion by introducing Madelung's hydrodynamic quantities

$$\big(\hbar\,\mathrm{Im}(\Psi^*\nabla\Psi), |\Psi|^2\big) = (D\nabla S, D) =: (\sigma, D)$$

as they arise from applying the momentum map (2) to the Koopman wavefunction. We notice that the KvH Hamiltonian (9) is rewritten in terms of these variables as

$$h(\sigma, D) = \int_{T^*Q} (X_H \cdot \sigma - D L_H)\,\mathrm{d}z\,. \tag{21}$$

Then, upon using the results in Sect. 3.1, we write the evolution equations as

$$(\partial_t + \pounds_{X_H})(\sigma - D\mathcal{A}) = 0\,, \qquad \partial_t D + \mathrm{div}(D X_H) = 0\,. \tag{22}$$

These variables provide an alternative representation of the classical density via the momentum map (20), that is

$$\rho = D + \mathrm{div}(\mathbb{J}\sigma - \mathbb{J}\mathcal{A}D)\,. \tag{23}$$

Once again, one is tempted to set $\sigma = D\mathcal{A}$ so that $\rho = D$, thereby eliminating all possible restrictions arising from specific representations of the classical density. However, we have $\mathrm{d}(\sigma/D) = 0$ and even if one allows for topological singularities in the phase so that $\nabla S = \mathcal{A}$ and $\rho = |\Psi|^2$, the last relation prevents the existence of δ-like particle solutions. These observations lead us to the necessity of extending the present construction in order to include more general classical distributions.

Von Neumann Operator. Here, we exploit the analogy between classical and quantum wavefunctions to introduce a self-adjoint von Neumann operator $\widehat{\Theta}$. We write the von Neumann evolution equation [2].

$$i\hbar\frac{\partial\widehat{\Theta}}{\partial t} = \big[\widehat{\mathcal{L}}_H, \widehat{\Theta}\,\big], \tag{24}$$

so that $\widehat{\Theta} = \Psi\Psi^\dagger$ recovers that KvH equation (7). Here, the adjoint is defined so that $\Psi_1^\dagger\Psi_2 = \langle\Psi_1|\Psi_2\rangle$. However, in the present discussion, we allow for a general von Neumann operator $\widehat{\Theta}$ with unit trace, while we do not necessarily restrict $\widehat{\Theta}$ to be positive-definite, although this can be set as a convenient initial condition. As is well known, equation (24) is a Hamiltonian system with the following Lie-Poisson structure:

$$\{\!\{f, h\}\!\}(\widehat{\Theta}) = -i\hbar^{-1}\,\mathrm{Tr}\left(\widehat{\Theta}\left[\frac{\delta f}{\delta\widehat{\Theta}}, \frac{\delta h}{\delta\widehat{\Theta}}\right]\right), \qquad h(\widehat{\Theta}) = \mathrm{Tr}\big(\widehat{\mathcal{L}}_H\widehat{\Theta}\big)\,. \tag{25}$$

Momentum Map. As shown in [3], the representation (3) on the wavefunction transfers naturally to a unitary action on the von Neumann operator that is conveniently written in terms of its integral kernel $\mathcal{K}_{\widehat{\Theta}}(z, z')$. Indeed, one writes

$$\mathcal{K}_{\widehat{\Theta}}(z, z') \mapsto \sqrt{J_\chi^{-1}(z) J_\chi^{-1}(z')}\, e^{-i\hbar^{-1}(\varphi(z) - \varphi(z'))}\mathcal{K}_{\widehat{\Theta}}(\chi^{-1}(z), \chi^{-1}(z'))\,,$$

where $(\chi, e^{i\varphi}) \in \text{Diff}(T^*Q) \circledS \mathcal{F}(T^*Q, S^1)$. In turn, this action produces the equivariant momentum map

$$\mathcal{J}(\widehat{\Theta})(z) = \left(\frac{i\hbar}{2}\frac{\partial}{\partial z}\mathcal{K}_{\widehat{\Theta}}(z', z) - \frac{i\hbar}{2}\frac{\partial}{\partial z}\mathcal{K}_{\widehat{\Theta}}(z, z'),\, \mathcal{K}_{\widehat{\Theta}}(z, z')\right)\Big|_{z'=z} = (\sigma(z), D(z))$$

and we verify that the Hamiltonian $h(\widehat{\Theta})$ in (25) may be rewritten exactly as in (21). Then, as in the case of quantum hydrodynamics, $\mathcal{J}(\widehat{\Theta})$ is a Poisson map hence mapping the von Neumann equation (24) to the Lie-Poisson dynamics (22) on $\mathfrak{X}(T^*Q)^* \times \text{Den}(T^*Q)$.

Point Particles in Phase-Space. At this point, we have $\mathrm{d}(\sigma/D) \neq 0$ and the relation $\sigma = D\mathcal{A}$ can be adequately set as an initial condition that is preserved in time. In this case, the momentum map (23) representing the classical density reduces to $\rho = D$, which now is allowed to have the δ-like expression $D(z) = \delta(z-\zeta)$ reproducing a point particle located at the phase-space coordinates ζ. Then, one may wish to identify a possible solution of the von Neumann equation (24) ensuring that the relation $\sigma = D\mathcal{A}$ is indeed satisfied. Following the discussion in [3], one can see that this is given by the integral kernel

$$\mathcal{K}_{\widehat{\Theta}}(z, z') = D\Big(\frac{z+z'}{2}\Big) \exp\Big[\frac{i}{\hbar}(z-z')\cdot\mathcal{A}\Big(\frac{z+z'}{2}\Big)\Big] = D\Big(\frac{z+z'}{2}\Big)\, e^{\frac{i}{2\hbar}(p+p')\cdot(q-q')},$$

where the second step uses $\mathcal{A} = p_k \mathrm{d}q^k$. The origin of this (unsigned) form of the von Neumann operator may be unfolded by resorting to the Wigner function formalism, although this discussion would take us beyond the purpose of the present work.

References

1. Bohm, D.A.: Suggested interpretation of the quantum theory in terms of "hidden" variables. I. Phys. Rev. **85**, 166–179 (1952)
2. Bondar, D.I., Gay-Balmaz, F., Tronci, C.: Koopman wavefunctions and classical-quantum correlation dynamics. Proc. R. Soc. A **475**, 20180879 (2019)
3. Foskett, M.S., Holm, D.D., Tronci, C.: Geometry of nonadiabatic quantum hydrodynamics. Acta Appl. Math. **162**, 63–103 (2019)
4. Gay-Balmaz, F., Tronci, C.: Madelung transform and probability densities in hybrid classical-quantum dynamics. Nonlinearity **33**, 5383–5424 (2020)
5. Kirillov, A.A.: Geometric quantization. In: Arnold, V.I., Novikov S.P. (eds.) Dynamical Systems IV. Encyclopaedia of Mathematical Sciences, vol. 4, pp. 139–176. Springer, Heidelberg (2001). https://doi.org/10.1007/978-3-662-06791-8_2
6. Koopman, B.O.: Hamiltonian systems and transformations in Hilbert space. Proc. Nat. Acad. Sci. **17**, 315–318 (1931)
7. Kostant, B.: Quantization and unitary representations. In: Taam, C.T. (ed.) Lectures in Modern Analysis and Applications III. LNM, vol. 170, pp. 87–208. Springer, Heidelberg (1970). https://doi.org/10.1007/BFb0079068
8. Madelung, E.: Quantentheorie in hydrodynamischer Form. Z. Phys. **40**, 322–326 (1927)

9. Souriau, J.M.: Quantification géométrique. Commun. Math. Phys. **1**, 374–398 (1966)
10. van Hove, L.: On certain unitary representations of an infinite group of transformations. Ph.D. thesis. Word Scientific 2001 (1951)

From Quantum Hydrodynamics to Koopman Wavefunctions II

Cesare Tronci[1,2(✉)] and François Gay-Balmaz[3]

[1] Department of Mathematics, University of Surrey, Guildford, UK
c.tronci@surrey.ac.uk
[2] Department of Physics and Engineering Physics, Tulane University, New Orleans, USA
[3] CNRS & École Normale Supérieure, Paris, France
francois.gay-balmaz@lmd.ens.fr

Abstract. Based on the Koopman-van Hove (KvH) formulation of classical mechanics introduced in Part I, we formulate a Hamiltonian model for hybrid quantum-classical systems. This is obtained by writing the KvH wave equation for two classical particles and applying canonical quantization to one of them. We illustrate several geometric properties of the model regarding the associated quantum, classical, and hybrid densities. After presenting the quantum-classical Madelung transform, the joint quantum-classical distribution is shown to arise as a momentum map for a unitary action naturally induced from the van Hove representation on the hybrid Hilbert space. While the quantum density matrix is positive by construction, no such result is currently available for the classical density. However, here we present a class of hybrid Hamiltonians whose flow preserves the sign of the classical density. Finally, we provide a simple closure model based on momentum map structures.

Keywords: Mixed quantum-classical dynamics · Koopman wavefunctions · Hamiltonian systems · Momentum maps

1 Introduction

Following the first part of this work, here we deal with the dynamics of hybrid systems where one subsystem is classical while the other is quantum. While concrete occurrences of quantum-classical physical systems still need to be identified, quantum-classical models have been sought for decades in the field of quantum chemistry. In this context, nuclei are treated as classical while electrons retain their fully quantum nature. A celebrated example is given by Born-Oppenheimer molecular dynamics. The formulation of hybrid classical-quantum dynamics is usually based on fully quantum treatments, in which some kind of factorization

CT acknowledges partial support by the Royal Society and the Institute of Mathematics and its Applications, UK.

F. Nielsen and F. Barbaresco (Eds.): GSI 2021, LNCS 12829, pp. 311–319, 2021.
https://doi.org/10.1007/978-3-030-80209-7_35

ansatz is invoked on the wavefunction. This ansatz is then followed by a classical limit on the factor that is meant to model the classical particle.

In our recent work [2,5], we followed a reverse route consisting in starting with a fully classical formulation and then quantizing one of the subsystems. The starting point is the Koopman-van Hove (KvH) equation considered in Part I. After writing the KvH equation for e.g. a two-particle wavefunction $\Psi(z_1, z_2)$, we enforce $\partial\Psi/\partial p_2 = 0$ and then apply Dirac's canonical quantization rule $p_2 \to -\mathrm{i}\hbar\partial/\partial q_2$. Here, the notation is such that $z_i = (q^i, p_i)$. In the case of only one particle, one verifies that this process leads to the standard Schrödinger equation $\mathrm{i}\hbar\partial_t\Psi = -(m^{-1}\hbar^2/2)\Delta\Psi + V\Psi$ [6]. On the other hand, in the two-particle case one obtains the *quantum-classical wave equation*

$$\mathrm{i}\hbar\partial_t\Upsilon = \{\mathrm{i}\hbar\widehat{H}, \Upsilon\} - L_{\widehat{H}}\Upsilon, \qquad \text{with} \qquad L_{\widehat{H}} := \mathcal{A}\cdot X_{\widehat{H}} - \widehat{H} = p\cdot\partial_p\widehat{H} - \widehat{H}\,. \quad (1)$$

Here, $\Upsilon \in \mathscr{H}_{CQ} = L^2(T^*Q \times M, \mathbb{C})$ is a wavefunction on the hybrid coordinate space $T^*Q \times M$ comprising both the classical and the quantum coordinates z_1 and q_2, now relabelled simply by $z = (q, p) \in T^*Q$ and $x \in M$, respectively. The function $\widehat{H}(z)$ is defined on T^*Q and takes values in the space of unbounded Hermitian operators on the quantum Hilbert space $\mathscr{H}_Q = L^2(M, \mathbb{C})$. Also, we recall that $\mathcal{A} = p_i\mathrm{d}q^i \in \Lambda^1(T^*Q)$ is the canonical one form on T^*Q.

After reviewing the main aspects of the quantum-classical wave equation, here we shall extend the treatment in Part I by extending Madelung's hydrodynamic description to hybrid quantum-classical systems. Importantly, this unlocks the door to the hybrid continuity equation for mixed quantum-classical densities. Eventually, we shall extend the treatment to von Neumann operators and show how the latter may be used to treat more general classes on mixed quantum-classical systems.

2 The Quantum-Classical Wave Equation

This section reviews the algebraic and geometric structure of the quantum-classical wave equation (1). First, we introduce the *hybrid Liouvillian operator*

$$\widehat{\mathcal{L}}_{\widehat{H}} = \{\mathrm{i}\hbar\widehat{H},\ \} - L_{\widehat{H}}, \quad (2)$$

which is an unbounded self-adjoint operator on $\mathscr{H}_{CQ}$ thereby making the hybrid wavefunction $\Upsilon(z,x)$ undergo unitary dynamics. The name is due to the fact that in the absence of the hybrid Lagrangian $L_{\widehat{H}}$, this reduces to the classical Liouvillian operator appearing in Koopman's early work. Also, we notice that in the absence of quantum coordinates the hybrid Liouvillian reduces to the prequantum operator from the classical KvH theory.

At this point, it is evident that Eq. (1) is Hamiltonian as it carries the same canonical Poisson structure as the quantum Schrödinger equation. In addition, its Hamiltonian functional (total energy) is conveniently written as

$$h(\Upsilon) = \int_{T^*Q} \langle\Upsilon|\widehat{\mathcal{L}}_{\widehat{H}}\Upsilon\rangle\,\mathrm{d}z = \int_{T^*Q}\int_M (\bar{\Upsilon}\,\widehat{\mathcal{L}}_{\widehat{H}}\,\Upsilon)\,\mathrm{d}z\wedge\mathrm{d}x\,. \quad (3)$$

Here, the L^2-inner product is the immediate extension of the classical and quantum cases. We proceed by emphasizing certain specific aspects of the quantum-classical wave equation.

Properties of the Hybrid Liouvillian. As opposed to the prequantum operator from the KvH theory the hybrid Liouvillian (2) does not comprise a Lie algebra structure. Indeed, while prequantum operators satisfy $[\widehat{\mathcal{L}}_H, \widehat{\mathcal{L}}_F] = i\hbar\widehat{\mathcal{L}}_{\{H,F\}}$, hybrid Liouvillians possess the following noncommutative variant:

$$[\widehat{\mathcal{L}}_{\widehat{H}}, \widehat{\mathcal{L}}_{\widehat{F}}] + [\widehat{\mathcal{L}}_{\bar{H}}, \widehat{\mathcal{L}}_{\bar{F}}]^{\mathsf{T}} = i\hbar\widehat{\mathcal{L}}_{\{\widehat{H},\widehat{F}\}-\{\widehat{F},\widehat{H}\}}. \tag{4}$$

Here, the bar denotes conjugate operators while T stands for transposition [5]. In addition, the hybrid Liouvillian enjoys relevant equivariance properties with respect to classical and quantum transformations. In particular, if $(\eta, e^{i\varphi}) \in \mathrm{Aut}_{\mathcal{A}}(T^*Q \times S^1)$, we have:

$$U^\dagger_{(\eta,e^{i\varphi})}\widehat{\mathcal{L}}_{\widehat{A}}U_{(\eta,e^{i\varphi})} = \widehat{\mathcal{L}}_{\eta^*\widehat{A}}, \quad \forall\, (\eta, e^{i\theta}) \in \widehat{\mathrm{Diff}}_\omega(T^*Q). \tag{5}$$

Here, $\mathrm{Aut}_{\mathcal{A}}(T^*Q \times S^1)$ denotes the group of connection-preserving automorphisms of the prequantum bundle, as explained in Part I. Alternatively, one also has equivariance under the group $\mathcal{U}(\mathscr{H}_Q)$ unitary transformations of the quantum Hilbert space space $\mathscr{H}_Q$, namely

$$U^\dagger\widehat{\mathcal{L}}_{\widehat{A}}U = \widehat{\mathcal{L}}_{U^\dagger\widehat{A}U}, \quad \forall\, U \in \mathcal{U}(\mathscr{H}_Q). \tag{6}$$

These equivariance relations also apply to a hybrid density operator extending the classical Liouville density as well as the quantum density matrix.

The Hybrid Density Operator. We define the *hybrid quantum-classical density operator* $\widehat{\mathcal{D}}$ associated to Υ in such a way that the identity

$$\int_{T^*Q}\langle\Upsilon|\widehat{\mathcal{L}}_{\widehat{A}}\Upsilon\rangle\,\mathrm{d}z = \mathrm{Tr}\int_{T^*Q}\widehat{A}\widehat{\mathcal{D}}\,\mathrm{d}z,$$

holds for any hybrid quantum-classical observable $\widehat{A}(z)$. One gets the expression

$$\widehat{\mathcal{D}}(z) = \Upsilon(z)\Upsilon^\dagger(z) - \mathrm{div}\left(\mathbb{J}\mathcal{A}(z)\Upsilon(z)\Upsilon^\dagger(z)\right) + i\hbar\{\Upsilon(z), \Upsilon^\dagger(z)\}, \tag{7}$$

so that $\mathrm{Tr}\int_{T^*Q}\widehat{\mathcal{D}}(z)\,\mathrm{d}z = 1$. Here, the superscript $\dagger$ denotes the quantum adjoint so that $\Upsilon_1^\dagger(z)\Upsilon_2(z) = \int_M \bar{\Upsilon}_1(z,x)\Upsilon_2(z,x)\,\mathrm{d}x$. Even if unsigned, the operator $\widehat{\mathcal{D}}(z)$ represents the hybrid counterpart of the Liouville density in classical mechanics and of the density matrix in quantum mechanics. This hybrid operator enjoys the following classical and quantum equivariance properties

$$\widehat{\mathcal{D}}(U_{(\eta,e^{i\varphi})}\Upsilon) = \widehat{\mathcal{D}}(\Upsilon)\circ\eta^{-1}, \quad \forall\, (\eta, e^{i\varphi}) \in \mathrm{Aut}_{\mathcal{A}}(T^*Q \times S^1). \tag{8}$$

$$\widehat{\mathcal{D}}(\widehat{U}\Upsilon) = \widehat{U}\widehat{\mathcal{D}}(\Upsilon)\widehat{U}^\dagger, \quad \forall\, \widehat{U} \in \mathcal{U}(\mathscr{H}_Q). \tag{9}$$

The equivariance properties (8)–(9) of the hybrid density operator under both classical and quantum transformations have long been sought in the theory of hybrid classical-quantum systems [3] and stand as one of the key geometric properties of the present construction.

Quantum and Classical Densities. Given the hybrid density operator, the classical density and the quantum density matrix are obtained as

$$\rho_c(z) = \operatorname{Tr}\widehat{\mathcal{D}}(z)\,, \qquad\qquad \hat{\rho} := \int_{T^*Q} \widehat{\mathcal{D}}(z)\,\mathrm{d}z\,, \tag{10}$$

respectively. While $\hat{\rho} = \int_{T^*Q} \Upsilon\Upsilon^\dagger\,\mathrm{d}z$ is positive by construction, at present there is no criterion available to establish whether the dynamics of ρ_c preserves its positivity, unless one considers the trivial case of absence of coupling, that is $\widehat{H}(z) = H(z) + \widehat{H}$. Possible arguments in favor of Wigner-like negative distributions are found in Feynman's work [4]. So far, all we know is that the hybrid classical-quantum theory presented here is the only available Hamiltonian theory beyond the mean-field approximation that (1) retains the quantum uncertainty principle and (2) allows the mean-field factorization $\Upsilon(z,x) = \Psi(z)\psi(x)$ as an exact solution in the absence of quantum-classical coupling. However, we recently identified infinite families of hybrid Hamiltonians for which both $\hat{\rho}$ and ρ_c are positive in time. We will get back to this point later on.

3 Madelung Equations and Quantum-Classical Trajectories

In this section we extend the usual Madelung transformation from quantum mechanics to the more general setting of coupled quantum–classical systems. Following Madelung's quantum treatment, we shall restrict to consider hybrid Hamiltonians of the type

$$\widehat{H}(q,p,x) = -\frac{\hbar^2}{2m}\Delta_x + \frac{1}{2M}|p|^2 + V(q,x)\,, \tag{11}$$

thereby ignoring the possible presence of magnetic fields. Here Δ_x and the norm $|p|$ are given with respect to Riemannian metrics on M and Q. In this case, the hybrid quantum–classical wave equation (1) reads

$$\mathrm{i}\hbar\partial_t\Upsilon = -\left(L_I + \frac{\hbar^2}{2m}\Delta_x\right)\Upsilon + \mathrm{i}\hbar\left\{H_I, \Upsilon\right\}\,, \tag{12}$$

where we have defined the following scalar functions L_I, H_I on the hybrid space $T^*Q \times M$:

$$H_I(q,p,x) := \frac{1}{2M}|p|^2 + V(q,x)\,, \qquad L_I(q,p,x) := \frac{1}{2M}|p|^2 - V(q,x)\,. \tag{13}$$

These are respectively the classical Hamiltonian and Lagrangian both augmented by the presence of the interaction potential.

Madelung Transform. We apply the Madelung transform by writing the hybrid wavefunction in polar form, that is $\Upsilon = \sqrt{D}e^{\mathrm{i}S/\hbar}$. Then, the quantum–classical wave equation (12) produces the following hybrid dynamics

$$\frac{\partial S}{\partial t} + \frac{|\nabla_x S|^2}{2m} - \frac{\hbar^2}{2m}\frac{\Delta_x\sqrt{D}}{\sqrt{D}} = L_I + \{H_I, S\}\,, \tag{14}$$

$$\frac{\partial D}{\partial t} + \frac{1}{m}\operatorname{div}_x(D\nabla_x S) = \{H_I, D\}\,, \tag{15}$$

where the operators ∇_x, div_x, and $\Delta_x = \operatorname{div}_x \nabla_x$ are defined in terms of the Riemannian metric on M. Each equation carries the usual quantum terms on the left-hand side, while the terms arising from KvH classical dynamics appear on the right-hand side (see the corresponding equations in Part I). We observe that (14) can be written in Lie derivative form as follows:

$$(\partial_t + \pounds_{\mathsf{X}})\, S = \mathscr{L}\,, \qquad \text{with} \qquad \mathsf{X} = (X_{H_I}, \nabla_x S/m)\,. \tag{16}$$

Here, X_{H_I} is the x-dependent Hamiltonian vector field on (T^*Q, ω) associated to H_I. Moreover, we have defined the (time-dependent) hybrid Lagrangian

$$\mathscr{L} := L_I + \frac{|\nabla_x S|^2}{2m} + \frac{\hbar^2}{2m}\frac{\Delta_x\sqrt{D}}{\sqrt{D}}\,,$$

in analogy to the so-called *quantum Lagrangian* given by the last two terms above. Then, upon taking the total differential d (on $T^*Q \times M$) of (16) and rewriting (15) in terms of the density D, we may rewrite (14)–(15) as follows:

$$(\partial_t + \pounds_{\mathsf{X}})\, \mathrm{d}S = \mathrm{d}\mathscr{L}\,, \qquad \partial_t D + \operatorname{div}(D\mathsf{X}) = 0\,. \tag{17}$$

Here, the operator div denotes the divergence operator induced on $T^*Q \times M$ by the Liouville form on T^*Q and the Riemannian metric on M. This form of the hybrid Madelung equations has important geometric consequences which we will present below.

Hybrid Quantum-Classical Trajectories. Although the hybrid Madelung equations (14)–(15) are distinctively different from usual equations of hydrodynamic type, the equations (17) still lead to a similar continuum description to that obtained in the quantum case. For example, the second equation in (17) still yields hybrid trajectories, which may be defined by considering the density evolution $D(t) = (D_0 \circ \Phi(t)^{-1})\sqrt{J_{\Phi(t)^{-1}}}$, where $\Phi(t)$ is the flow of the vector field X and J_Φ the Jacobian determinant. Then, this flow is regarded as a Lagrangian trajectory obeying the equation

$$\dot{\Phi}(t, z, x) = \mathsf{X}(\Phi(t, z, x))\,, \tag{18}$$

which is the hybrid quantum–classical extension of quantum trajectories [1]. In turn, the hybrid trajectories (18) are also useful to express (14) in the form $\frac{\mathrm{d}}{\mathrm{d}t}(S(t, \Phi(t, z, x))) = \mathscr{L}(t, \Phi(t, z, x))$, which is the hybrid analogue of the KvH phase evolution; see Part I. Additionally, in the absence of classical degrees of freedom, this picture recovers the quantum Bohmian trajectories since in that case the coordinate z plays no role. Similar arguments apply in the absence of quantum degrees of freedom.

The Symplectic Form. The first equation in (17) can be rewritten in such a way to unfold the properties of the flow $\Phi \in \mathrm{Diff}(T^*Q \times M)$ on the quantum-classical coordinate space. Since $\mathcal{A} \in \Lambda^1(T^*Q)$ and $\Lambda^1(T^*Q) \subset \Lambda^1(T^*Q \times M)$, we denote $\mathsf{A} = p \cdot dq \in \Lambda^1(T^*Q \times M)$. Likewise, the differential $\mathrm{d}_x : \Lambda^n(M) \to \Lambda^{n+1}(M)$ on M induces a one-form $\mathrm{d}_x V = \nabla_x V \cdot \mathrm{d}x \in \Lambda^1(T^*Q \times M)$ on the hybrid coordinate space. Then, with a slight abuse of notation, one can rewrite the first in (17) as

$$(\partial_t + \pounds_{\mathsf{X}})(\mathrm{d}S - \mathsf{A}) = \mathrm{d}(\mathscr{L} - L_I) - \mathrm{d}_x V\,.$$

This equation is crucially important for deducing the quantum-classical dynamics of the Poincaré invariant. Indeed, integrating over a loop γ_0 in $T^*Q \times M$ leads to

$$\frac{\mathrm{d}}{\mathrm{d}t}\oint_{\gamma(t)} p \cdot \mathrm{d}q = \oint_{\gamma(t)} \nabla_x V \cdot \mathrm{d}x\,, \tag{19}$$

where $\gamma(t) = \Phi(t) \circ \gamma_0$. Then, standard application of Stoke's theorem yields the following relation involving the classical symplectic form $\Omega = -\mathrm{d}\mathsf{A}$:

$$\frac{\mathrm{d}}{\mathrm{d}t}\Phi(t)^*\Omega = -\Phi(t)^*\mathrm{d}\big(\mathrm{d}_x V\big)\,, \qquad \text{with} \qquad \mathrm{d}(\mathrm{d}_x V) = \frac{\partial^2 V}{\partial q^j \partial x^k}\,\mathrm{d}q^j \wedge \mathrm{d}x^k\,, \tag{20}$$

which reduces to the usual conservation $\Phi(t)^*\Omega = \Omega$ in the absence of coupling.

4 Joint Quantum-Classical Distributions

In standard quantum mechanics, the emergence of a probability current gives to the evolution of the quantum density distribution $\rho_q(x) = |\psi(x)|^2$ the structure of a continuity equation. This structure transfers to the hybrid case for which $\partial_t \rho_q = -\,\mathrm{div}_x \int_{T^*Q} (D\nabla_x S/m)\,\mathrm{d}z$, which follows from (15). Here, we will present the analogue of the continuity equation for joint quantum-classical distributions.

Hybrid Joint Distribution. Denoting by $\mathcal{K}_{\widehat{\mathcal{D}}}(z; x, y)$ the kernel of the hybrid density operator (7), we first introduce the *joint quantum-classical distribution*:

$$\mathcal{D}(z,x) := \mathcal{K}_{\widehat{\mathcal{D}}}(z; x, x) \tag{21}$$

$$= |\Upsilon|^2 - \mathrm{div}\big(\mathbb{J}\mathcal{A}|\Upsilon|^2\big) + i\hbar\{\Upsilon, \bar{\Upsilon}\}\,. \tag{22}$$

This represents the joint density for the position of the system in the hybrid space $T^*Q \times M$. Then, the quantum probability density is given by $\rho_q = \int_{T^*Q} \mathcal{D}(z,x)\,\mathrm{d}z$. As shown in [5], Eq. (22) identifies a momentum map $\Upsilon \mapsto \mathcal{D}(\Upsilon)$ for the natural action on $\mathscr{H}_{CQ}$ of a slight generalization of the group

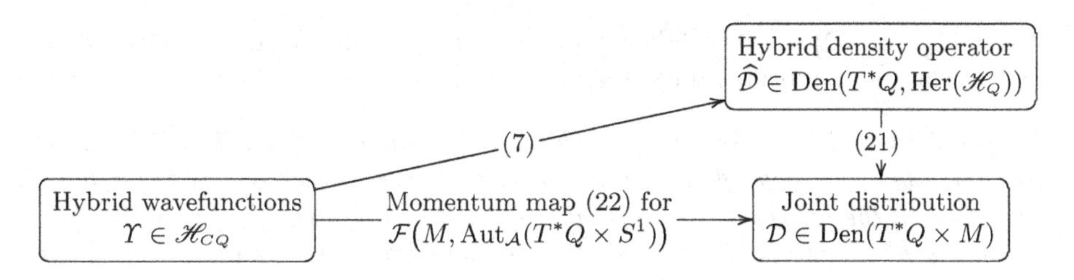

Fig. 1. Relations between hybrid wavefunctions Υ, the hybrid operator $\widehat{\mathcal{D}}$, and the joint density $\mathcal{D}$.

$\mathrm{Aut}_{\mathcal{A}}(T^*Q \times S^1)$, which we are now going to introduce. Denote by $\mathcal{F}(M, G)$ the space of smooth mappings from M to a group G; this space is naturally endowed with a group structure inherited by G. Then, we let $G = \mathrm{Aut}_{\mathcal{A}}(T^*Q \times S^1)$ so that $\mathcal{F}(M, \mathrm{Aut}_{\mathcal{A}}(T^*Q \times S^1))$ possesses a unitary representation on $\mathscr{H}_{CQ}$ that is inherited from the van Hove representation of $\mathrm{Aut}_{\mathcal{A}}(T^*Q \times S^1)$ discussed in Part I. We refer the reader to [5] for further details. In particular, the Lie algebra of $\mathcal{F}(M, \mathrm{Aut}_{\mathcal{A}}(T^*Q \times S^1))$ coincides with the Poisson algebra of smooth phase-space functions $A(z; x)$ that are parameterized by the quantum coordinates $x \in M$, that is $\mathcal{F}(M, \mathfrak{aut}_{\mathcal{A}}(T^*Q \times S^1)) \simeq \mathcal{F}(M, C^\infty(T^*Q))$. Given $A \in \mathcal{F}(M, C^\infty(T^*Q))$, the corresponding infinitesimal action on $\mathscr{H}_{CQ}$ is then $\Upsilon \mapsto -\mathrm{i}\hbar^{-1}\widehat{\mathcal{L}}_A \Upsilon$, so that the associated momentum map $\Upsilon \mapsto \mathcal{D}(\Upsilon)$ is given by (22) (Fig. 1).

Quantum-Classical Continuity Equation. In analogy to the evolution of the quantum mechanical probability density, the joint distribution satisfies a continuity equation generally involving the hybrid wavefunction. Upon using the polar form $\Upsilon = \sqrt{D}e^{\mathrm{i}S/\hbar}$, the continuity equation for the joint distribution reads

$$\partial_t \mathcal{D} = -\operatorname{div} \mathbf{J} = -\operatorname{div}_z J_C - \operatorname{div}_x J_Q\,, \tag{23}$$

where the classical and quantum components of the hybrid current $\mathbf{J} = (J_C, J_Q)$ are expressed in terms of the vector field X in (16) as follows:

$$J_C := \mathcal{D} X_{H_I}\,, \tag{24}$$

$$J_Q := \frac{1}{m}\Big(D\nabla_x S + \partial_{p_i}(p_i D \nabla_x S) + \{D\nabla_x S, S\} - \frac{\hbar^2}{4mD}\{D, \nabla_x D\}\Big). \tag{25}$$

Notice the emergence of the last term in J_Q, which is reminiscent of the quantum potential from standard quantum hydrodynamics. Now, does $\mathcal{D}$ conserve its initial sign during its evolution? Obviously, this is what happens in the absence of coupling, when $\partial^2_{q^j x^k} V = 0$ and the symplectic form Ω is preserved in (20). In addition, if the quantum kinetic energy is absent in (11), then $J_Q = 0$ and the hybrid continuity equation assumes the characteristic form $\partial_t \mathcal{D} = \{H_I, \mathcal{D}\}$ thereby preserving the sign of $\mathcal{D}$. Whether this sign preservation extends to more general situations is a subject of current studies. Alternatively, one may ask if the classical density $\rho_c = \int_M \mathcal{D}\,\mathrm{d}x$ is left positive in time by its equation of motion

$\partial_t\rho_c = \int_M\{H_I,\mathcal{D}\}\,\mathrm{d}x$. The following statement provides a positive answer for certain classes of hybrid Hamiltonians.

Proposition 1 ([5]). *Assume a hybrid Hamiltonian of the form* $\widehat{H}(z) = H(z,\widehat{\alpha})$*, where the dependence on the purely quantum observable* $\widehat{\alpha}$ *is analytic. Assume that the hybrid density operator* $\widehat{\mathcal{D}}(z)$ *is initially positive, then the density* ρ_c *is also initially positive and its sign is preserved by the hybrid wave equation* (1).

5 A Simple Closure Model

The high dimensionality involved in mixed quantum-classical dynamics poses formidable computational challenges that are typically addressed by devising appropriate closure schemes. The simplest closure is derived by resorting to the mean-field factorization $\Upsilon(z,x) = \Psi(z)\psi(x)$, as described in [2]. Alternatively, one can resort to the von Neumann operator description: after defining a quantum-classical von Neumann operator $\widehat{\Xi}$ satisfying $i\hbar\partial_t\widehat{\Xi} = [\widehat{\mathcal{L}}_{\widehat{H}},\widehat{\Xi}]$, a mean-field factorization is obtained by writing $\widehat{\Xi} = \widehat{\Theta}\hat{\rho}$. Here, $\widehat{\Theta}$ and $\hat{\rho}$ are von Neumann operators on the classical and the quantum Hilbert spaces $\mathscr{H}_C$ and $\mathscr{H}_Q$, respectively. Then, as discussed in Part I, one may set the integral kernel of $\widehat{\Theta}$ to be $\mathcal{K}_{\widehat{\Theta}}(z,z') = D(z/2+z'/2)\,e^{\frac{i}{2\hbar}(p+p')\cdot(q-q')}$.

Here, we will illustrate a slight extension of the mean-field factorization that may again be obtained by following the final discussion in Part I. First, we perform a convenient abuse of notation by writing $\widehat{\Xi}(z,x,z',x') := \mathcal{K}_{\widehat{\Xi}}(z,x,z',x')$. Then, the quantity $i\hbar\tilde{\rho} := i\hbar\widehat{\Xi}(z,x,z',x')|_{z=z'}$ emerges as a momentum map for the natural action of $\mathcal{F}(T^*Q,\mathcal{U}(\mathscr{H}_Q))$ on the Hilbert space $\mathscr{H}_{CQ}$, while the quantum density matrix is $\int\widehat{\Xi}(z,x,z',x')|_{z=z'}\,\mathrm{d}z$ (again, notice the slight abuse of notation on the integral symbol). Likewise, the hydrodynamic variables

$$(\sigma(z),D(z)) = \left(\frac{i\hbar}{2}\frac{\partial}{\partial z}\int\widehat{\Xi}(z',x,z,x)\,\mathrm{d}x - \frac{i\hbar}{2}\frac{\partial}{\partial z}\int\widehat{\Xi}(z,x,z',x)\,\mathrm{d}x,\ \widehat{\Xi}(z,z')\right)\Big|_{z'=z}$$

comprise a momentum map structure associated to the action of the semidirect-product group $\mathrm{Diff}(T^*Q)\circledS\mathcal{F}(T^*Q,S^1)$ discussed in Part I. Overall, the map $\widehat{\Xi}\mapsto(\sigma,D,i\hbar\tilde{\rho})$ identifies an equivariant momentum map for the action of the semidirect-product $\mathrm{Diff}(T^*Q)\circledS\big(\mathcal{F}(T^*Q,S^1)\times\mathcal{F}(T^*Q,\mathcal{U}(\mathscr{H}_Q))\big)$. At this point, we express the total energy $h(\widehat{\Xi}) = \mathrm{Tr}(\widehat{\Xi}\widehat{\mathcal{L}}_{\widehat{H}})$ in terms of the latter momentum map by using the following ansatz (denote $u=\sigma/D$)

$$\widehat{\Xi}(z,x,z',x') = \tilde{\rho}(z/2+z'/2,x,x')\,e^{\frac{i}{\hbar}(z-z')\cdot u(z/2+z'/2)}\,. \tag{26}$$

Then, the Hamiltonian functional reads $h(\sigma,D,\tilde{\rho}) = \mathrm{Tr}\int\tilde{\rho}(D^{-1}X_{\widehat{H}}\cdot\sigma - L_{\widehat{H}})\,\mathrm{d}z$ and the resulting system for the dynamics of $(\sigma,D,i\hbar\tilde{\rho})$ is Lie-Poisson on the dual of the semidirect-product Lie algebra $\mathfrak{X}(T^*Q)\circledS\big(\mathcal{F}(T^*Q)\times\mathcal{F}(T^*Q,\mathfrak{u}(\mathscr{H}_Q))\big)$.

Upon introducing $\hat{\rho} = \tilde{\rho}/D$ and $\langle \widehat{A} \rangle = \operatorname{Tr}(\widehat{A}\hat{\rho})$, the equations of motion read

$$
\begin{aligned}
&\left(\partial_t + \pounds_{\langle X_{\widehat{H}} \rangle}\right)(u - \mathcal{A}) = (u - \mathcal{A}) \cdot \operatorname{Tr}\big(X_{\widehat{H}} \nabla_z \hat{\rho}\big)\,, \\
&\partial_t D + \operatorname{div}_z\big(\langle X_{\widehat{H}} \rangle D\big) = 0\,, \qquad \partial_t \hat{\rho} + \langle X_{\widehat{H}} \rangle \cdot \nabla_z \hat{\rho} = \big[u \cdot X_{\widehat{H}} - L_{\widehat{H}}, \hat{\rho}\big]\,,
\end{aligned}
$$

where we used $\pounds_{\langle X_{\widehat{H}} \rangle} \mathcal{A} = \nabla_z \langle L_{\widehat{H}} \rangle + \operatorname{Tr}(\widehat{H} \nabla_z \hat{\rho})$. Then, setting $u = \mathcal{A}$ yields

$$
\partial_t D + \operatorname{div}_z(D \langle X_{\widehat{H}} \rangle) = 0\,, \qquad \partial_t \hat{\rho} + \langle X_{\widehat{H}} \rangle \cdot \nabla_z \hat{\rho} = \big[\widehat{H}, \hat{\rho}\big]\,.
$$

Notice that, while $\hat{\rho}$ must be positive, the operator $\widehat{\Xi}$ in (26) is unsigned. Also, the classical density $\rho_c = D$ does not follow a Hamiltonian flow, while the quantum density operator $\hat{\rho}$ evolves unitarily in the frame moving with velocity $\langle X_{\widehat{H}} \rangle$.

References

1. Bohm, D.: A suggested interpretation of the quantum theory in terms of "hidden" variables. I. Phys. Rev. **85**, 166–179 (1952)
2. Bondar, D.I., Gay-Balmaz, F., Tronci, C.: Koopman wavefunctions and classical-quantum correlation dynamics. Proc. R. Soc. A **475**, 20180879 (2019)
3. Boucher, W., Traschen, J.: Semiclassical physics and quantum fluctuations. Phys. Rev. D **37**, 3522–3532 (1988)
4. Feynman, R.: Negative probability. In: Hiley, B., Peat, F.D. (eds.) Quantum Implications: Essays in Honour of David Bohm, pp. 235–248. Routledge & Kegan Paul Ltd. (1987)
5. Gay-Balmaz, F., Tronci, C.: Madelung transform and probability densities in hybrid classical-quantum dynamics. Nonlinearity **33**, 5383–5424 (2020)
6. Klein, U.: From Koopman-von Neumann theory to quantum theory. Quantum Stud. Math. Found. **5**, 219–227 (2018)

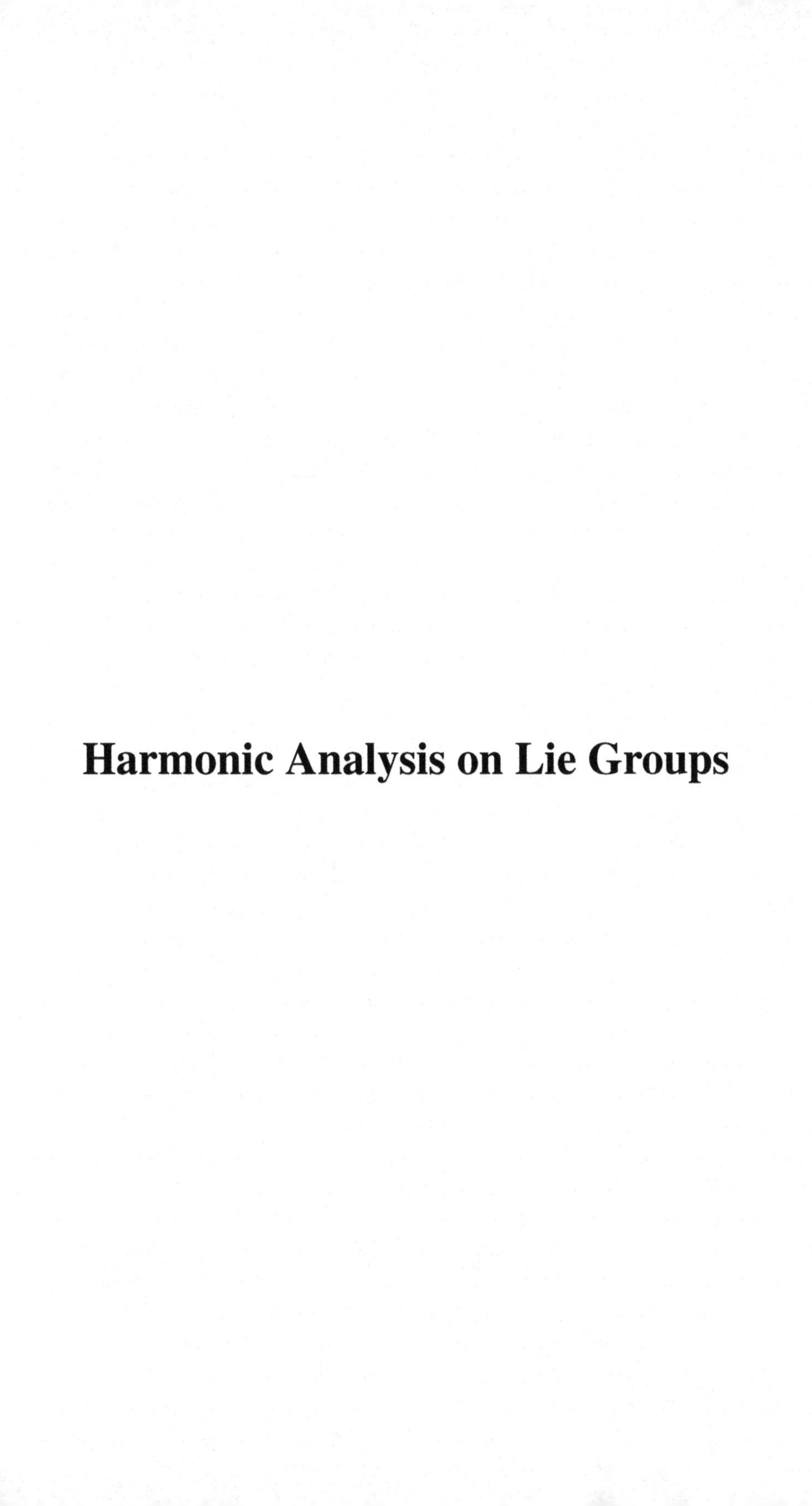

Harmonic Analysis on Lie Groups

The Fisher Information of Curved Exponential Families and the Elegant Kagan Inequality

Gérard Letac(✉)

Université de Toulouse, Institut de Mathématiques, 118 route de Narbonne, 31062 Toulouse, France
gerard.letac@math.univ-toulouse.fr

Abstract. Curved exponential families are so general objects that they seem to have no interesting universal properties. However Abram Kagan [1] discovered in 1985 a remarkable inequality on their Fisher information. This note gives a modern presentation of this result and examples, comparing in particular noncentral and central Wishart distributions.

1 Introduction: Fisher Information

FISHER MODELS. A Fisher model

$$P = \{P_\theta(dw) = e^{\ell_w(\theta)}\nu(dw) \ ; \ \theta \in \Theta\} \tag{1}$$

on the measured space (Ω, ν) is defined with the help of a function $(w, \theta) \mapsto \ell_w(\theta)$ on $\Omega \times \Theta$ where

1. Θ is a connected open subset of $\mathbb{R}^d$.
2. For ν almost w then ℓ_w is twice continously differentiable on Θ.
3. $P_\theta(dw) = e^{\ell_w(\theta)}\nu(dw)$ is a probability.

Actually a more general definition could be that a Fisher model $(P_\theta(dw))_{\theta\in\Theta}$ is a model such that for all θ and $\theta' \in \Theta$, we have that P_θ is absolutely continuous with respect to $P_{\theta'}$. The existence of ν and of $\ell_w(\theta)$ are consequences of this more general choice, but we would have anyway to add condition (2) to this second definition for being more useful and for being equivalent to the first.

Comments

1. A traditional presentation of a Fisher model is

$$(P_\theta(dx))_{\theta\in\Theta} = (f(x, \theta)\nu(dx))_{\theta\in\Theta}$$

where in most of the cases $\nu(dx)$ is the Lebesgue measure on $\mathbb{R}^n$. It could also be for instance the counting measure on the set $\mathbb{N}$ of non negative integers. We part here from the tradition for several reasons: it is interesting to

F. Nielsen and F. Barbaresco (Eds.): GSI 2021, LNCS 12829, pp. 323–330, 2021.
https://doi.org/10.1007/978-3-030-80209-7_36

replace $(\mathbb{R}^n, dx)$ by an abstract measured space (Ω, ν) which will simplify the notations. It will avoid artificial distractions like the consideration of the set $S_\theta = \{x \in \mathbb{R}^n; f(x,\theta) > 0\}$. Actually, since we have to impose to the set $S = S_\theta$ to be constant with respect to $\theta \in \Theta$, the formalism allows to replace $(\mathbb{R}^n, dx)$ by $(S, \nu(dx))$ where ν is simply the restriction to S of the Lebesgue measure. Dealing with $e^{\ell_x(\theta)}$ instead of $f(x,\theta)$ will also avoid the plague of the partial derivatives $\frac{\partial}{\partial\theta_i} f(x,\theta)$ or $\frac{\partial^2}{\partial\theta_i\partial\theta_j} f(x,\theta)$, large Hessian matrices and their cumbersome consequences. With the present formalism the partial derivatives will never be used, since only differentials with respect to θ will occur.

2. The representation (1) of a Fisher model is not unique, since $\nu(dw)$ can be replaced by some $f(w)\nu_1(dw)$ and thus the $\ell_w(\theta)$ is replaced by $\ell_w(\theta) + \log f(w)$. However, when $\theta \mapsto \ell_w(\theta)$ is differentiable, the ℓ'_w are invariant when replacing the reference measure ν by ν_1.
3. Not all models are Fisher models: if $\Omega = \Theta = (0, \infty)$ and P_θ is the uniform distribution on $(0, \theta)$ it is false that P_θ is absolutely continuous with respect to $P_{\theta'}$ when $0 < \theta' < \theta$.

Fisher Information. Of course the pair $(\nu(dw), \ell_w(\theta))$ is not completely arbitrary since it satisfies

$$\int_\Omega P_\theta(dw) = 1 = \int_\Omega e^{\ell_w(\theta)} \nu(dw) \tag{2}$$

Suppose now that $\theta \mapsto \ell_w(\theta)$ is differentiable and that there exists a positive and ν integrable function f such that $\|\ell'_w(\theta)\| e^{\ell_w(\theta)} \leq f(w)$ for all $\theta \in \Theta$. In these circumstances we can differentiate (2) and we get the important vector equality

$$\int_\Omega \ell'_w(\theta) P_\theta(dw) = 0 \tag{3}$$

Most of the authors call $(w, \theta) \mapsto \ell'_w(\theta)$ *the score function.* Therefore $\ell'_w(\theta)$ is a centered random variable valued in $\mathbb{R}^d$ with respect to the probability $P_\theta(dw)$. The Fisher information $I_P(\theta)$ is nothing but the covariance matrix of $\ell'_w(\theta)$ with respect to the probability $P_\theta(dw)$. It has also an important representation by the Hessian matrix $\ell''_w(\theta)$ of ℓ_w, namely

$$I_P(\theta) = \int_\Omega \ell'_w(\theta) \otimes \ell'_w(\theta) P_\theta(dw) \overset{(*)}{=} -\int_\Omega \ell''_w(\theta) P_\theta(dw) \tag{4}$$

As usual, for a and b in $\mathbb{R}^d$ we write $a \otimes b$ for the matrix $d \times d$ representing the linear endomorphism of $\mathbb{R}^d$ defined by $x \mapsto (a \otimes b)(x) = a\langle b, x\rangle$. Equality (*) in (4) is obtained by differentiating (3) with respect to θ.

Finally one needs quite often a formula describing what happens to the Fisher information while replacing the initial parameter $\theta \in \Theta \subset \mathbb{R}^d$ by a new parameter $\lambda \in \Lambda \subset \mathbb{R}^k$ with $k \leq d$. One assumes that Λ is open and that the map $\lambda \mapsto \theta(\lambda)$ is differentiable: thus one can imagine $\theta'(\lambda)$ as a matrix with d rows and

k columns. Denote $Q_\lambda(dw) = P_{\theta(\lambda)}(dw)$ and consider the Fisher model $Q = \{Q_\lambda; \lambda \in \Lambda\}$. The formula of change of variable for the Fisher models is easy to prove from the definitions and is

$$I_Q(\lambda) = \theta'(\lambda)^T I_P(\theta(\lambda))\theta'(\lambda) \tag{5}$$

Next proposition gives an example of calculation of Fisher information. Here we have denoted the parameter of the model by t instead of θ for clarity while using this example later in Sect. 3.

Proposition 1.1. Let (a,b) be an interval and $t \mapsto (m(t), A(t))$ a C^2 application of (a,b) to $\mathbb{R}^n \times \mathbb{P}(n)$ where $\mathbb{P}(n)$ is the convex cone of positive definite of matrices of order n. Consider the following Fisher model of the Gaussian distributions $F = \{N(m(t), A^{-1}(t))\ ;\ a < t < b\}$. Then the Fisher information of F is

$$I_F = m'^T Am' + \frac{1}{2}\operatorname{tr}((A^{-1}A')^2).$$

Examples. For $n = 1$, classical examples for F are $F_1 = \{N(t,t);\ t > 0\}$ and $F_2 = \{N(t,t^2);\ t > 0\}$. The proposition implies that $I_{F_1}(t) = (2t+1)/2t^2$, $I_{F_2}(t) = 3/t^2$.

Proof. With the above notations, the Fisher model is described by

$$(\Omega, \nu) = (\mathbb{R}^n, dx/(\sqrt{2\pi})^n),\ \ell_x = -\frac{1}{2}(x-m)^T A(x-m) + \frac{1}{2}\log\det A.$$

Denote the trace of a square matrix X by $\operatorname{tr}(X)$. Recall that the first and second differentials of the function $f(X) = \log\det X$ defined on the space $S(n)$ of real symmetric matrices are the linear and bilinear forms on $S(n)$ given by $f'(X)(h) = \operatorname{tr}(X^{-1}h)$ and $f''(X)(h,k) = -\operatorname{tr}(X^{-1}hX^{-1}k)$ for h and k in $S(n)$. This implies that

$$\frac{d}{dt}\log\det A = \operatorname{tr}(A^{-1}A'),\ \frac{d^2}{dt^2}\log\det A = \operatorname{tr}(A^{-1}A'') - \operatorname{tr}((A^{-1}A')^2).$$

This allows to compute ℓ'_x and ℓ''_x

$$\begin{aligned}
\ell'_x &= -\frac{1}{2}(x-m)^T A'(x-m) + (x-m)^T Am' + \frac{1}{2}\operatorname{tr}(A^{-1}A') \\
\ell''_x &= -\frac{1}{2}(x-m)^T A''(x-m) + 2(x-m)^T A'm' - m'Am' \\
&\quad -\frac{1}{2}\operatorname{tr}((A^{-1}A')^2) + \frac{1}{2}\operatorname{tr}(A^{-1}A'').
\end{aligned}$$

Now, using (4) we compute $\mathbb{E}(\ell''_X(t)) = -I_F(t)$ if $X \sim N(m, A^{-1})$. A quick way to proceed is to use $X - m = A^{-1/2}Z$ where $Z \sim N(0, I_n)$ and, by diagonalization of $P \in \mathbb{P}(n)$, to observe that $\mathbb{E}(Z^T PZ) = \operatorname{tr}(P)$. Applying this to $P = A^{-1/2}A''A^{-1/2}$ and $\operatorname{tr}(P) = \operatorname{tr}(A^{-1}A'')$ one gets the given value of $I_F(t)$.

2 Exponential Families

NATURAL EXPONENTIAL FAMILIES. A positive measure μ on the Euclidean space $\mathbb{R}^d$ (not necessarily bounded) has the Laplace transform $L_\mu(\theta) = \int_{\mathbb{R}^d} e^{\langle\theta,x\rangle}\mu(dx) \leq \infty$. If $D(\mu) = \{\theta \in \mathbb{R}^d; L_\mu(\theta) < \infty\}$ the Hölder inequality shows that $D(\mu)$ is convex and that $k_\mu = \log L_\mu$ is a convex function. Denote by $\Theta(\mu)$ the interior of $D(\mu)$ and assume now two things: $\Theta(\mu)$ is not empty and μ is not concentrated on an affine hyperplane of $\mathbb{R}^d$ (let us call $\mathcal{M}(\mathbb{R}^d)$ the set of measures μ fulfilling these two conditions: the second one implies that k_μ is now strictly convex). With such a μ and with θ in $\Theta(\mu)$ we create the probability $P(\theta,\mu)(dx) = e^{\langle\theta,x\rangle - k_\mu(\theta)}\mu(dx)$. The model $F(\mu) = \{P(\theta,\mu)\ ;\ \theta \in \Theta(\mu)\}$ is called the *natural exponential family* generated by μ.

FISHER INFORMATION OF A NATURAL EXPONENTIAL FAMILY. Here it is necessary to point out that the Fisher information of the natural exponential family $F(\mu)$ considered as a Fisher model parameterized by θ (called the natural parameter) is the Hessian matrix $k''_\mu(\theta)$, which is the covariance of $X \sim P(\theta,\mu)$. However, among other possible parameterizations of $F(\mu)$ a frequent one is the parameterization by the mean $m = k'_\mu(\theta) = \int_{\mathbb{R}^d} xP(\theta,\mu)(dx)$. Its set $M(F(\mu))$ of possible values is an open subset of $\mathbb{R}^d$ called the domain of the means: since k_μ is strictly convex, the inverse function $m \mapsto \theta = \psi_\mu(m)$ of $\theta \mapsto m$ is well defined from $M(F(\mu))$ to $\Theta(\mu)$. Denote for simplicity $Q_m(dx) = P(\psi_\mu(m),\mu)(dx)$ and $V(m)$ the covariance matrix of Q_m. Observe that $V(m) = k''_\mu(\psi_\mu(m))$ and that $\psi'_\mu(m) = V(m)^{-1}$ (this inverse exists since Q_m is not concentrated on a strict affine subspace of $\mathbb{R}^d$)

Now, using the change of variable formula (5), the Fisher information of the Fisher model $Q = (Q_m \quad m \in M_{F(\mu)})$ is given by the somewhat paradoxical formula:

$$I_Q(m) = V(m)^{-1} \tag{6}$$

which is exactly the inverse of the Fisher information of $F(\mu)$ when parameterized by its natural parameter.

GENERAL EXPONENTIAL FAMILIES. More generally consider an abstract measured space (Ω,ν) and a map $w \mapsto x = t(w)$ from Ω to $\mathbb{R}^d$ such that the image $\mu(dx)$ of ν by t belongs to $\mathcal{M}(\mathbb{R}^d)$. Therefore for $\theta \in \Theta(\mu)$ one considers the probability $P(t,\nu,\theta)(dw)$ on Ω defined by $P(\theta,t,\nu)(dw) = e^{\langle\theta,t(w)\rangle - k_\mu(\theta)}\nu(dw)$. The model $F(t,\nu) = \{P(\theta,t,\nu)\ ;\theta \in \Theta(\mu)\}$ is called the *general exponential family* generated by (t,ν). The model $F(\mu)$ is called the associated natural exponential family of $F(t,\nu)$. It is easy to check that the Fisher information $I_{F(t,\nu)}(\theta)$ coincides with the Fisher information $I_{F(\mu)}(\theta)$ of the associated natural exponential family.

CURVED EXPONENTIAL FAMILIES. If Λ is an open subset of $\mathbb{R}^k$ with $k \leq d$ and if $\lambda \mapsto \theta(\lambda)$ is a map from Λ to $\Theta(\mu)$ some authors consider the set of probabilities

$$P(\theta(\lambda),t,\nu)(dw) = e^{\langle\theta(\lambda),t(w)\rangle - c(\lambda)}\nu(dw)$$

where $c(\lambda) = k_\mu(\theta(\lambda))$ and when λ describes Λ. If $k = d$ it is a reparameterisation of $F(t, \nu)$. If $k < d$ this submodel has few general properties. When $k = 1$ we will called it a *curved exponential family*, providing a model still valued on $\mathbb{R}^d$ but with a one dimensional parameter. In general, replacing the general exponential family- from which the curved exponential family is issued- by its natural exponential family is harmless and clearer.

3 The Kagan Inequality

In this section we state in Proposition 3.1 and we illustrate by examples the original Kagan inequality [1]. It considers actually rather a curved exponential family. We eliminate the artificial role of general exponential families by going to the core of the proof and by dealing only with a natural exponential family and the curved one issued from it.

Proposition 3.1. Consider the natural exponential family $(G_\theta)_{\theta\in\Theta}$ in $\mathbb{R}^d$ defined by $G_\theta(dx) = e^{\langle x,\theta\rangle - k(\theta)}\mu(dx)$, an open interval $I \subset \mathbb{R}$, a smooth function $t \mapsto \theta(t)$ from I to $\Theta \subset \mathbb{R}^d$ and the curved exponential family $P = (P_t)_{t\in I}$ defined by $P_t(dx) = G_{\theta(t)}(dx)$. Consider also a Fisher model $F = (F_t)_{t\in I}$ on $\mathbb{R}^d$ defined $F_t(dy) = e^{\ell_y(t)}\nu(dy)$ where ν and μ are absolutely continuous with respect to each other.

For all $t \in I$ we consider the random variables $X \sim P_t$ and $Y \sim F_t$ and we assume that

$$\mathbb{E}(X) = \mathbb{E}(Y), \ \ \mathbb{E}(X \otimes X) = \mathbb{E}(Y \otimes Y), \tag{7}$$

or equivalently, X and Y have same means and covariances. Then the respective Fisher informations satisfy $I_P(t) \leq I_F(t)$ for all $t \in I$. Furthermore $P = F$ if and only if $I_P = I_F$.

Proof. Since μ and ν are absolutely continuous with respect to each other without loss of generality we assume that $\mu = \nu$, by Comment 2 of Sect. 1. We fix $t \in I$ for a while. We consider the two Hilbert spaces $L^2(P_t)$ and $L^2(F_t)$ of real functions defined on $\mathbb{R}^d$. We consider also their subspaces H_x and H_y respectively generated by $1, x_1, \ldots, x_d$ and $1, y_1, \ldots, y_d$. Hypothese (7) implies that they are isometric.

Consider now the element of $L^2(F_t)$ defined by $\ell'_y(t)$. Since $\int_{\mathbb{R}^d} \ell'_y(t)F_t(dy) = 0$, we can claim that there exist $a = (a_1, \ldots, a_d)^T \in \mathbb{R}^d$ and $c \in L^2(F_t)$ such that c is orthogonal to H_y and that

$$\ell'_y(t) = a_1(y_1 - \mathbb{E}(Y_1)) + \cdots + a_d(y_d - \mathbb{E}(Y_d)) + c(y) = (y - \mathbb{E}(Y))^T a + c(y).$$

We are going to prove that $a = \theta'(t)$. For this we observe that $k'(\theta) = \int_{\mathbb{R}^d} xG_\theta(dx)$ gives

$$k'(\theta(t)) = \int_{\mathbb{R}^d} xP_t(dx) = \mathbb{E}(X) = \mathbb{E}(Y) = \int_{\mathbb{R}^d} ye^{\ell_y(t)}\mu(dy).$$

Now we stop to have t fixed and we take derivative of the above line with respect to t. We get the equality in $\mathbb{R}^d$:

$$\begin{aligned} k''(\theta(t))(\theta'(t)) &= \int_{\mathbb{R}^d} y\ell'_y(t)F_t(dy) \\ &= \int_{\mathbb{R}^d} (a_1(y_1 - \mathbb{E}(Y_1) + \cdots + a_d(y_d - \mathbb{E}(Y_d))yF_t(dy) \\ &= \mathrm{Cov}(Y)(a) = \mathrm{Cov}(X)(a) \end{aligned}$$

Since $k''(\theta)$ is the covariance of $X \sim G_\theta$, hence $k''(\theta(t))$ is the covariance of $X \sim P_t$ and therefore $k''(\theta(t))(\theta'(t)) = k''(\theta(t))(a)$. Since by hypothesis G is a natural exponential family on $\mathbb{R}^d$ the matrix $k''(\theta(t))$ is invertible and $a = \theta'(t)$. Finally we write

$$\begin{aligned} I_F(t) &= \int_{\mathbb{R}^d} (\ell'_y(t))^2 F_t(dy) = \int_{\mathbb{R}^d} \langle y - \mathbb{E}(Y), \theta'(t)\rangle^2 F_t(dy) + \int_{\mathbb{R}^d} c^2(y)F_t(dy) \\ &= \langle \theta'(t), \mathrm{Cov}(Y)(\theta'(t)\rangle + \int_{\mathbb{R}^d} c^2(y)F_t(dy) \\ &= \langle \theta'(t), \mathrm{Cov}(X)(\theta'(t))\rangle + \int_{\mathbb{R}^d} c^2(y)F_t(dy) \geq \langle \theta'(t), \mathrm{Cov}(X)(\theta'(t))\rangle = I_P(t) \end{aligned}$$

If equality occurs for one t then $c = 0$. If equality occurs for all t then

$$\ell'_y(t) = \langle y - \mathbb{E}(Y), \theta'(t)\rangle = \langle y - k'(\theta(t), \theta'(t)\rangle.$$

Since a primitive of the right hand side is $\langle y, \theta(t)\rangle - k(\theta(t))$, we get $P = F$.

Example 1. Consider the exponential family $P = (P_t; t \in (-\frac{1}{2}\pi, \frac{1}{2}\pi))$ defined by

$$P_t(dx) = e^{tx + \log\cos(t)} \frac{dx}{2\cosh(\frac{\pi x}{2})}.$$

Here the curved exponential family is the exponential family itself, with $\theta = t$.

If $X \sim P_t$ the expectation and the variance of X are $m(t) = -\tan(t)$ and $\sigma^2(t) = 1/\cos^2(t)$. As recalled in Sect. 2, since P is an exponential family parameterized by its natural parameter t its Fisher information $I_P(t)$ is the variance of P_t.

Now, consider another Fisher model $F = (F_t; t \in (-\frac{1}{2}\pi, \frac{1}{2}\pi))$ such that the reference measures ν of F and μ of P are absolutely continuous with respect to each other. Consider $Y \sim F_t$. Assume that the expectation and the variance of Y are also $m(t) = -\tan(t)$ and $\sigma^2(t) = 1/\cos^2(t)$. Then $I_F \geq I_P$. As an example let us choose $F_t = N(m(t), A^{-1}(t))$ as considered in Proposition 1.1, with $m(t) = \tan(t)$ and $A(t) = \cos^2(t)$. From Proposition 1.1 we can check Proposition 3.1 as follows:

$$I_F(t) = m'(t)^2 A(t) + \frac{1}{2}(A^{-1}(t)A'(t))^2 = \frac{1}{\cos^2(t)} + 2\tan^2(t) \geq \frac{1}{\cos^2(t)}.$$

One can also consider the same pair exponential family- Fisher model but using the parameterization by the mean. With the help of (6) the Kagan inequality can again be easily checked.

Example 2. Consider the exponential family

$$G = \{G_\theta = \gamma(p, a, \sigma); \sigma \in \mathbb{P}_n\}$$

of noncentral Wishart distributions with shape parameter $p > \frac{n-1}{2}$, non centrality parameter $a \in \mathbb{P}_n$ and $\sigma = (-\theta)^{-1} \in \mathbb{P}_n$ defined by its Laplace transform

$$\int_{\mathbb{P}_n} e^{-\operatorname{tr}(sx)}\gamma(p, a, \sigma)(dx) = \frac{1}{\det(I_n + \sigma s)^p} e^{-\operatorname{tr}(s(I_n+\sigma s)^{-1}w)}$$

where $w = \sigma a \sigma$. Note that the ordinary Wishart correspond to the case $a = 0$. If $X \sim \gamma(p, a, \sigma)$ it is proved for instance in Letac and Massam [2] page 1407 that mean and covariance of X are given by

$$\mathbb{E}(X) = w + p\sigma, \ \operatorname{cov}(X) : h \mapsto wh\sigma + \sigma hw + p\sigma h\sigma$$

Of course, since the space on which the exponential family G is defined is the space S_n of symmetric matrices of order n, it is simpler to present the covariance as a linear operator $h \mapsto \operatorname{cov}(X)(h)$ on S_n instead of representing it by a matrix of matrices. Our restriction to $p > (n-1)/2$ and $a \in \mathbb{P}(n)$ insures the existence of G: see Letac and Massam [3] for details and references on the delicate problem of the existence of G for other values of (p, a). Since G is an exponential family parameterized by its natural parameter θ, its Fisher information I_{G_θ} is the covariance of G_θ, as explained in Sect. 2 above.

Next, from this natural exponential family, we extract the curved exponential family $P = (P_t;\ t_0 < t < t_1)$ defined by the curve $t \mapsto \theta(t) \in -\mathbb{P}(n)$ and $P_t = G_{\theta(t)}$, we use the formula giving the Fisher information for the change of parameter $t \mapsto \theta(t)$. As a consequence, the Fisher information $I_P(t)$ is obtained from I_{G_θ} by the formula (5), leading to

$$\begin{aligned} I_P(t) &= \operatorname{tr}(\theta'(t) I_G(\theta(t))(\theta'(t)) \\ &= 2\operatorname{tr}(\theta'(t)w(t)\theta'(t)\sigma(t) + p\operatorname{tr}(\theta'(t)\sigma(t)\theta'(t)\sigma(t)), \end{aligned} \tag{8}$$

where we have denoted with some abuse $\sigma(t) = (-\theta(t))^{-1}$ and $w(t) = \sigma(t)a\sigma(t)$.

Now we introduce the Fisher model $F = (F_t;\ 0 < t < 1)$ such that F_t is the Wishart distribution $\gamma(p^*(t), 0, \sigma^*(t))(dx)$ where the functions $p^*(t)$ and $\sigma^*(t)$ are chosen such that $Y \sim F_t$ has the mean and covariance of $X \sim P_t$. In other terms:

$$w(t) + p\sigma(t) = p^*(t)\sigma^*(t), \tag{9}$$

$$w(t)h\sigma(t) + \sigma(t)hw(t) + p\sigma(t)h\sigma(t) = p^*(t)\sigma^*(t)h\sigma^*(t) \tag{10}$$

Clearly (10) cannot be realized for all $h \in S_n$ without extra conditions on $t \mapsto \theta(t)$ and on $a \in \mathbb{P}_n$. For this reason we take $a = I_n$, $\theta(t) = -I_n/t$ and $0 < t < 1$. With such a choice we get from (8), (9) and (10)

$$I_P(t) = n\frac{2t+p}{t^2}, \quad p^*(t) = \frac{(t+p)^2}{t(2t+p)}, \quad \sigma^*(t) = \frac{t(2t+p)}{t+p}I_n.$$

The restriction $0 < t < 1$ insures that $p^*(t) > (n-1)/2$. To see this observe that $t \mapsto p^*(t)$ is decreasing and that $p^*(1) = \frac{(p+1)^2}{p+2} > p > \frac{n-1}{2}$.

On the other hand, the expression of $I_F(t)$ is too complicated to be displayed here, since it starts from

$$\begin{aligned} \ell_x(t) = & -\frac{t+p}{t(2t+p)}\,\mathrm{tr}\,(x) - np^*(t)\log\frac{t(2t+p)}{(t+p)} \\ & + \left(p^*(t) - \frac{n+1}{2}\right)\log\det x - \log\Gamma_n(p^*(t)) \end{aligned}$$

while the Kagan inequality says $I_F(t) \geq I_P(t)$.

References

1. Kagan, A.: An information property of exponential families. Teor. Veroyatnost. i Primenen. **30**(4), 783–786 (1985), Theory Probab. Appl. 30(4), 831–835 (1986)
2. Letac, G., Massam, H.: The noncentral Wishart as an exponential family and its moments. J. Multivariate Anal. **99**, 1393–1417 (2008)
3. Letac, G., Massam, H.: The Laplace transform $(\det s)^{-p} \exp \mathrm{trace}\,(\mathrm{s}^{-1})$ and the existence of the non-central Wishart distributions. J. Multivariate Anal. **163**, 96–110 (2018)

Continuous Wavelet Transforms for Vector-Valued Functions

Hideyuki Ishi(✉) and Kazuhide Oshiro

OCAMI, Osaka City University, 3-3-138 Sugimoto, Sumiyoshi-ku, Osaka 558-8585, Japan
hideyuki@sci.osaka-cu.ac.jp

Abstract. We consider continuous wavelet transforms associated to unitary representations of the semi-direct product of a vector group with a linear Lie group realized on the Hilbert spaces of square-integrable vector-valued functions. In particular, we give a concrete example of an admissible vector-valued function (vector field) for the 3-dimensional similitude group.

Keywords: Continuous wavelet transforms · Unitary representations · Similitude group · Divergence-free vector fields

1 Introduction

Let ϕ be a square-integrable function on $\mathbb{R}$ such that $2\pi \int_{\mathbb{R}} |\mathcal{F}\phi(\xi)|^2 \frac{d\xi}{|\xi|} = 1$ (see below for the definition of the Fourier transform $\mathcal{F}\phi$). Then for any $f \in L^2(\mathbb{R})$, we have the reproducing formula $f = \int_{\mathbb{R}\times(\mathbb{R}\setminus\{0\})} (f|\phi_{b,a})_{L^2} \phi_{b,a} \frac{dbda}{|a|^2}$, where $\phi_{b,a}(x) := |a|^{-1/2}\phi(\frac{x-b}{a})$ $(a \neq 0,\, b \in \mathbb{R})$, and the integral is interpreted in the weak sense. Let $\mathrm{Aff}(\mathbb{R})$ be the group of invertible affine transforms $g_{b,a}(x) := b + ax$ $(a \neq 0,\, b \in \mathbb{R})$. We have a natural unitary representation π of $\mathrm{Aff}(\mathbb{R})$ on $L^2(\mathbb{R})$ given by $\pi(g_{b,a})f := f_{b,a}$. Then the reproducing formula above means that the continuous wavelet transform $W_\phi : L^2(\mathbb{R}) \ni f \mapsto (f|\pi(\cdot)\phi) \in L^2(G)$ is an isometry, and as was pointed out in [6], the isometry property follows from the orthogonal relation for square-integrable irreducible unitary representations. After [6], theory of continuous wavelet transforms is generalized to a wide class of groups and representations (cf. [1,3]). In particular, the continuous wavelet transform associated to the quasi-regular representation of the semi-direct product $G = \mathbb{R}^n \rtimes H$, where H is a Lie subgroup of $GL(n, \mathbb{R})$, defined on the Hilbert space $L^2(\mathbb{R}^n)$ is studied thoroughly, whereas the assumption of the irreducibility of the representation is relaxed by [5].

In the present paper, we consider continuous wavelet transforms for vector-valued functions in a reasonably general setting. We show in Theorem 3 that,

Partially supported by KAKENHI 20K03657 and Osaka City University Advanced Mathematical Institute (MEXT Joint Usage/Research Center on Mathematics and Theoretical Physics JPMXP0619217849).

F. Nielsen and F. Barbaresco (Eds.): GSI 2021, LNCS 12829, pp. 331–339, 2021.
https://doi.org/10.1007/978-3-030-80209-7_37

under certain assumptions (A1), (A2) and (A3), one can construct a continuous wavelet transform with an appropriate admissible vector. The conditions (A1) and (A2) seem common in the wavelet theory, whereas the condition (A3) is specific in our setting. The results can be applied to the natural action of the similitude group $\mathbb{R}^n \rtimes (\mathbb{R}_{>0} \times SO(n))$ on vector fields and tensor fields on $\mathbb{R}^n$ (see Sect. 3). In particular, we present a concrete example of admissible vector for the 3-dimensional vector field (see (11)). Our example can be used for the decomposition of vector fields into a sum of divergence-free and curl-free vector fields. This result will shed some insight to the wavelet theory for such vector fields [7,9]. A part of the contents of this paper is contained in the second author's doctoral thesis [8], where one can find other concrete examples of admissible vectors.

Let us fix some notations used in this paper. The transposition of a matrix A is denoted by $A^\top$. For $w, w' \in \mathbb{C}^m$, the standard inner product is defined by $(w|w') := w^\top \bar{w}' = \sum_{i=1}^m w_i \bar{w}'_i$, where w and w' are regarded as column vectors. Write $\|w\| := \sqrt{(w|w)}$. A $\mathbb{C}^m$-valued function f on $\mathbb{R}^n$ is said to be square-integrable if $\|f\|_{L^2}^2 := \int_{\mathbb{R}^n} \|f(x)\|^2 \, dx < \infty$, and the condition is equivalent to that each of its components is square-integrable. Similarly, we say that f is rapidly decreasing if so is each of its components. The space $L^2(\mathbb{R}^n, \mathbb{C}^m)$ of square-integrable $\mathbb{C}^m$-valued functions forms a Hilbert space, where the inner product is given by $(f_1|f_2)_{L^2} = \int_{\mathbb{R}^n} (f_1(x)|f_2(x)) \, dx \quad (f_1, f_2 \in L^2(\mathbb{R}^n, \mathbb{C}^m))$. The Fourier transform $\mathcal{F} : L^2(\mathbb{R}^n, \mathbb{C}^m) \to L^2(\mathbb{R}^n, \mathbb{C}^m)$ is defined as a unitary isomorphism such that $\mathcal{F}f(\xi) = (2\pi)^{-n/2} \int_{\mathbb{R}^n} e^{i(\xi|x)} f(x) \, dx \quad (\xi \in \mathbb{R}^n)$ if f is rapidly decreasing. For two unitary representations π_1 and π_2 of a group, we write $\pi_1 \simeq \pi_2$ if π_1 and π_2 are equivalent.

The authors would like to thank the referees' comments and suggestions, which are quite helpful for the improvement of the paper.

2 General Results

Let $H \subset GL(n, \mathbb{R})$ be a linear Lie group. Then the semi-direct product $G := \mathbb{R}^n \rtimes H$ acts on $\mathbb{R}^n$ as affine transforms:

$$g \cdot x := v + hx \qquad (g = (v, h) \in G, \ x, v \in \mathbb{R}^n, \ h \in H).$$

Let $\sigma : H \to U(m)$ be a unitary representation of H on $\mathbb{C}^m$. We define a unitary representation π of G on the Hilbert space $L^2(\mathbb{R}^n, \mathbb{C}^m)$ by

$$\pi(g)f(x) := |\det h|^{-1/2} \sigma(h) f(h^{-1}(x - v)) \qquad (g = (v, h) \in G, \ f \in L^2(\mathbb{R}^n, \mathbb{C}^m), \ x \in \mathbb{R}^n).$$

We introduce another unitary representation $\hat{\pi}$ of G on $L^2(\mathbb{R}^n, \mathbb{C}^m)$ defined by $\hat{\pi}(g) := \mathcal{F} \circ \pi(g) \circ \mathcal{F}^{-1}$ for $g \in G$. By a straightforward calculation, we have

$$\hat{\pi}(g)\varphi(\xi) = e^{i(\xi|v)} |\det h|^{1/2} \sigma(h) \varphi(h^\top \xi) \qquad (g = (v, h) \in G, \ \varphi \in L^2(\mathbb{R}^n, \mathbb{C}^m), \ \xi \in \mathbb{R}^n). \tag{1}$$

This formula implies that the contragredient action ρ of H on $\mathbb{R}^n$ given by $\rho(h)\xi := (h^\top)^{-1}\xi$ $(h \in H, \xi \in \mathbb{R}^n)$ plays crucial roles for the study of the representation $\pi \simeq \hat{\pi}$ (cf. [1,5]). For $\xi \in \mathbb{R}^n$, we write $\mathcal{O}_\xi$ for the orbit $\rho(H)\xi = \{\, \rho(h)\xi\,;\, h \in H \,\} \subset \mathbb{R}^n$. Now we pose the first assumption:

(A1): There exists an element $\xi \in \mathbb{R}^n$ for which $\mathcal{O}_\xi$ is an open set in $\mathbb{R}^n$.

Let $\mathfrak{h} \subset \mathrm{Mat}(n, \mathbb{R})$ be the Lie algebra of H, and $\{X_1, \ldots, X_N\}$ $(N := \dim \mathfrak{h})$ be a basis of $\mathfrak{h}$. For $\xi \in \mathbb{R}^n$, we consider the matrix $R(\xi) \in \mathrm{Mat}(n, N, \mathbb{R})$ whose i-th column is $X_i^\top \xi \in \mathbb{R}^n$ for $i = 1, \ldots, N$. Then the tangent space of $\mathcal{O}_\xi$ at ξ equals $\mathrm{Image}\, R(\xi) \subset \mathbb{R}^n$. In particular, $\mathcal{O}_\xi$ is open if and only if the matrix $R(\xi)$ is of rank n, and the condition is equivalent to $\det(R(\xi)R(\xi)^\top) \neq 0$. Since $p(\xi) := \det(R(\xi)R(\xi)^\top)$ is a polynomial function of ξ, the set $\{\, \xi \in \mathbb{R}^n\,;\, p(\xi) \neq 0 \,\}$ has a finite number of connected components with respect to the classical topology by [10]. On the other hand, we see from a connectedness argument that each of the connected components is an open $\rho(H)$-orbit. Summarizing these observations, we have the following.

Lemma 1. *There exist elements $\xi^{[1]}, \ldots \xi^{[K]} \in \mathbb{R}^n$ such that*

(a) The orbit $\mathcal{O}_k := \rho(H)\xi^{[k]}$ is open for $k = 1, \ldots, K$.
(b) If $k \neq l$, then $\mathcal{O}_k \cap \mathcal{O}_l = \emptyset$.
(c) The disjoint union $\bigsqcup_{k=1}^K \mathcal{O}_k$ is dense in $\mathbb{R}^n$.

For $k = 1, \ldots, K$, put $L_k := \left\{\, f \in L^2(\mathbb{R}^n, \mathbb{C}^m)\,;\, \mathcal{F}f(\xi) = 0 \text{ (a.a. } \xi \in \mathbb{R}^n \backslash \mathcal{O}_k) \,\right\}$. By (1) and Lemma 1, we have an orthogonal decomposition $L^2(\mathbb{R}^n, \mathbb{C}^m) = \sum_{1 \le k \le K}^{\oplus} L_k$, which gives also a decomposition of the unitary representation π. We shall decompose each (π, L_k) further into a direct sum of irreducible subrepresentations. Let H_k be the isotropy subgroup $\left\{\, h \in H\,;\, \rho(h)\xi^{[k]} = \xi^{[k]} \,\right\}$ of H at $\xi^{[k]}$, and decompose the restriction $\sigma|_{H_k}$ of the representation σ orthogonally as

$$\mathbb{C}^m = \sum_{\alpha \in A_k}^{\oplus} W_{k,\alpha}, \tag{2}$$

where A_k is a finite index set, and $W_{k,\alpha}$ is an irreducible subspace of $\mathbb{C}^m$. We write $\sigma_{k,\alpha}$ for the irreducible representation $H_k \ni h \mapsto \sigma(h)|_{W_{k,\alpha}} \in U(W_{k,\alpha})$. For $\xi \in \mathcal{O}_k$, take $h \in H$ for which $\xi = \rho(h)\xi^{[k]}$ and put $W_{\xi,\alpha} := \sigma(h)W_{k,\alpha}$. Note that the space $W_{\xi,\alpha}$ is independent of the choice of $h \in H$. For $\alpha \in A_k$, define

$$L_{k,\alpha} = \{f \in L_k\,;\, \mathcal{F}f(\xi) \in W_{\xi,\alpha} \text{ (a.a. } \xi \in \mathcal{O}_k)\}\,. \tag{3}$$

Then we have an orthogonal decomposition $L_k = \sum_{\alpha \in A_k}^{\oplus} L_{k,\alpha}$, and each $L_{k,\alpha}$ is an invariant subspace thanks to (1). Moreover, we can show that the subrepresentation $(\pi, L_{k,\alpha})$ is equivalent to the induced representation $\mathrm{Ind}_{\mathbb{R}^n \rtimes H_k}^G \tilde{\sigma}_{k,\alpha}$, where $(\tilde{\sigma}_{k,\alpha}, W_{k,\alpha})$ is a unitary representation of $\mathbb{R}^n \rtimes H_k$ defined by $\tilde{\sigma}_{k,\alpha}(v, h) := e^{i(\xi^{[k]}|v)}\sigma_{k,\alpha}(h)$ $(v \in \mathbb{R}^n, h \in H_k)$. It follows from the Mackey theory (see [4, Theorem 6.42]) that $(\pi, L_{k,\alpha})$ is irreducible. Therefore we obtain:

Theorem 1. *Under the assumption* (A1), *an irreducible decomposition of the representation* $(\pi, L^2(\mathbb{R}^n, \mathbb{C}^m))$ *is given by*

$$L^2(\mathbb{R}^n, \mathbb{C}^m) = \sum_{1\le k\le K}^{\oplus} \sum_{\alpha\in A_k}^{\oplus} L_{k,\alpha}. \tag{4}$$

Let d_H be a left Haar measure on the Lie group H. Then a measure d_G on G defined by $d_G(g) := |\det h|^{-1} dv\, d_H(h)$ $(g = (v, h) \in G)$ is a left Haar measure.

Theorem 2. *If the group* H_k *is compact, then the representation* $(\pi, L_{k,\alpha})$ *of* G *is square-integrable for all* $\alpha \in A_k$. *In this case, one has*

$$\int_G |(f|\pi(g)\phi)_{L^2}|^2\, d_G(g) = (2\pi)^n (\dim W_{k,\alpha})^{-1} \|f\|_{L^2}^2 \int_H \|\mathcal{F}\phi(h^\top \xi^{[k]})\|^2\, d_H(h) \tag{5}$$

for $f, \phi \in L_{k,\alpha}$.

Proof. In order to show (5), it is sufficient to consider the case where f and ϕ are rapidly decreasing. For $g = (v, h) \in G$, the isometry property of $\mathcal{F}$ together with (1) implies that

$$\begin{aligned}(f|\pi(v,h)\phi)_{L^2} &= \int_{\mathbb{R}^n} (\mathcal{F}f(\xi) \,|\, \mathcal{F}[\pi(v,h)\phi](\xi))\, d\xi \\ &= \int_{\mathbb{R}^n} e^{-i(\xi|v)} |\det h|^{1/2} \left(\mathcal{F}f(\xi) \,|\, \sigma(h)\mathcal{F}\phi(h^\top\xi)\right) d\xi.\end{aligned}$$

Note that the right-hand is the inverse Fourier transform of $(2\pi)^{n/2} |\det h|^{1/2}$ $(\mathcal{F}f(\xi)|\sigma(h)\mathcal{F}\phi(h^\top\xi))$, which is a rapidly decreasing function of $\xi \in \mathbb{R}^n$, so that the Plancherel formula tells us that

$$\int_{\mathbb{R}^n} |(f|\pi(v,h)\phi)_{L^2}|^2 dv = (2\pi)^n |\det h| \int_{\mathbb{R}^n} |(\mathcal{F}f(\xi) \,|\, \sigma(h)\mathcal{F}\phi(h^\top\xi))|^2 d\xi. \tag{6}$$

Thus, to obtain the formula (5), it is enough to show that

$$\begin{aligned}&\int_H |(\mathcal{F}f(\xi) \,|\, \sigma(h)\mathcal{F}\phi(h^\top\xi))|^2 d_H(h) \\ &\quad = (\dim W_{k,\alpha})^{-1} \|\mathcal{F}f(\xi)\|^2 \int_H \|\mathcal{F}\phi(h^\top \xi^{[k]})\|^2 d_H(h)\end{aligned} \tag{7}$$

for each $\xi \in \mathcal{O}_k$. Let us take $h_1 \in H$ for which $\xi = \rho(h_1)\xi^{[k]}$. Then $h^\top\xi = \rho(h^{-1}h_1)\xi^{[k]} = \rho(h_1^{-1}h)^{-1}\xi^{[k]}$. By the left-invariance of the Haar measure d_H, the left-hand side of (7) equals

$$\begin{aligned}&\int_H |(\mathcal{F}f(\xi) \,|\, \sigma(h_1 h)\mathcal{F}\phi(\rho(h)^{-1}\xi^{[k]}))|^2\, d_H(h) \\ &= \int_H |(\sigma(h_1)^{-1}\mathcal{F}f(\xi) \,|\, \sigma(h)\mathcal{F}\phi(\rho(h)^{-1}\xi^{[k]}))|^2\, d_H(h).\end{aligned}$$

Writing d_{H_k} for the normalized Haar measure on the compact group H_k, we rewrite the right-hand side as

$$\int_H \int_{H_k} |(\sigma(h_1)^{-1}\mathcal{F}f(\xi) \,|\, \sigma(h'h)\mathcal{F}\phi(\rho(h'h)^{-1}\xi^{[k]}))|^2 \, d_{H_k}(h')d_H(h)$$
$$= \int_H \int_{H_k} |(\sigma(h_1)^{-1}\mathcal{F}f(\rho(h_1)\xi^{[k]}) \,|\, \sigma(h')\sigma(h)\mathcal{F}\phi(\rho(h)^{-1}\xi^{[k]}))|^2 \, d_{H_k}(h')d_H(h).$$

Now put $w_1 := \sigma(h_1)^{-1}\mathcal{F}f(\rho(h_1)\xi^{[k]})$ and $w_2 := \sigma(h)\mathcal{F}\phi(\rho(h)^{-1}\xi^{[k]})$. Since $f, \phi \in L_{k,\alpha}$, we see that w_1 and w_2 belong to $W_{k,\alpha}$ for almost all $h, h_1 \in H$, so that the Schur orthogonality relation for the representation $\sigma_{k,\alpha}$ yields

$$\int_{H_k} |(\sigma(h_1)^{-1}\mathcal{F}f(\rho(h_1)\xi^{[k]}) \,|\, \sigma(h')\sigma(h)\mathcal{F}\phi(\rho(h)^{-1}\xi^{[k]}))|^2 \, d_{H_k}(h')$$
$$= \int_{H_k} |(w_1|\sigma_{k,\alpha}(h')w_2)|^2 \, d_{H_k}(h') = (\dim W_{k,\alpha})^{-1}\|w_1\|^2\|w_2\|^2$$
$$= (\dim W_{k,\alpha})^{-1}\|\sigma(h_1)^{-1}\mathcal{F}f(\rho(h_1)\xi^{[k]})\|^2\|\sigma(h)\mathcal{F}\phi(\rho(h)^{-1}\xi^{[k]})\|^2$$
$$= (\dim W_{k,\alpha})^{-1}\|\mathcal{F}f(\xi)\|^2\|\mathcal{F}\phi(h^\top\xi^{[k]})\|^2,$$

which leads us to (7).

Note that the formula (5) is valid whether both sides converge or not. On the other hand, we can define a function $\phi \in L_{k,\alpha}$ for which (5) converges for all $f \in L_{k,\alpha}$ as follows. In view of the homeomorphism $H/H_k \ni hH_k \mapsto \rho(h)\xi^{[k]} \in \mathcal{O}_k$, we take a Borel map $h_k : \mathcal{O}_k \to H$ such that $\xi = \rho(h_k(\xi))\xi^{[k]}$ for all $\xi \in \mathcal{O}_k$. Let ψ be a non-negative continuous function on H with compact support such that $\int_H \psi(h^{-1})^2 \, d_H(h) = 1$. Take a unit vector $e_{k,\alpha} \in W_{k,\alpha}$ and define $\phi_{k,\alpha} \in L^2(\mathbb{R}^n, \mathbb{C}^m)$ by

$$\phi_{k,\alpha}(x) := (\dim W_{k,\alpha})^{1/2}(2\pi)^{-n}\int_{\mathcal{O}_k} e^{-i(\xi|x)}\psi(h_k(\xi))\sigma(h_k(\xi))e_{k,\alpha}\,d\xi \quad (x \in \mathbb{R}^n). \tag{8}$$

Note that the integral above converges because the support of the integrand is compact. We can check that $\phi_{k,\alpha} \in L_{k,\alpha}$ and

$$(2\pi)^n(\dim W_{k,\alpha})^{-1}\int_H \|\mathcal{F}\phi_{k,\alpha}(h^\top\xi^{[k]})\|^2 \, d_H(h) = 1. \tag{9}$$

The formula (5) together with (9) tells us that

$$\int_G |(f|\pi(g)\phi_{k,\alpha})_{L^2}|^2 \, d_G(g) = \|f\|_{L^2}^2 \qquad (f \in L_{k,\alpha}). \tag{10}$$

In particular, the left-hand side converges for all $f \in L_{k,\alpha}$, which means that the unitary representation $(\pi, L_{k,\alpha})$ is square-integrable. □

Let us recall the decomposition (2). The multiplicity of the representation $\sigma_{k,\alpha}$ of H_k in $\sigma|_{H_k}$, denoted by $[\sigma|_{H_k} : \sigma_{k,\alpha}]$, is defined as the number of $\beta \in A_k$

for which $\sigma_{k,\beta}$ is equivalent to $\sigma_{k,\alpha}$. If $\sigma_{k,\beta}$ is equivalent to $\sigma_{k,\alpha}$, there exists a unitary intertwining operator $U_{k,\alpha,\beta} : W_{k,\beta} \to W_{k,\alpha}$ between $\sigma_{k,\beta}$ and $\sigma_{k,\alpha}$, and $U_{k,\alpha,\beta}$ is unique up to scalar multiples by the Schur lemma. Moreover, by the Schur orthogonality relation, we have

$$\int_{H_k} (w_\alpha|\sigma(h)w'_\alpha)\,(\sigma(h)w'_\beta|w_\beta)d_{H_k}(h) = (\dim W_{k,\alpha})^{-1}(w_\alpha|U_{k,\alpha,\beta}w_\beta)\,(U_{k,\alpha,\beta}w'_\beta|w'_\alpha).$$

Now we assume:

(A2) The group H_k is compact for all $k = 1, \ldots, K$.
(A3) $[\sigma|_{H_k} : \sigma_{k,\alpha}] \leq \dim W_{k,\alpha}$ for all $k = 1, \ldots, K$ and $\alpha \in A_k$.

Because of (A3), we can take a unit vector $e_{k,\alpha} \in W_{k,\alpha}$ for each $k = 1, \ldots, K$ and $\alpha \in A_k$ in such a way that if $\sigma_{k,\alpha}$ and $\sigma_{k,\beta}$ are equivalent with $\alpha \neq \beta$, then $(e_{k,\alpha}|U_{k,\alpha,\beta}e_{k,\beta}) = 0$. As in (8), we construct $\phi_{k,\alpha} \in L_{k,\alpha}$ from the vector $e_{k,\alpha}$. Put $\tilde{\phi} := \sum_{k=1}^{K} \sum_{\alpha\in A_k} \phi_{k,\alpha}$.

Theorem 3. *The transform* $W_{\tilde{\phi}} : L^2(\mathbb{R}^n, \mathbb{C}^m) \ni f \mapsto (f|\pi(\cdot)\tilde{\phi})_{L^2} \in L^2(G)$ *is an isometry. In other words, one has*

$$f = \int_G (f|\pi(g)\tilde{\phi})_{L^2}\, \pi(g)\tilde{\phi}\, d_G(g) \qquad (f \in L^2(\mathbb{R}^n, \mathbb{C}^m)),$$

where the integral of the right-hand side is defined in the weak sense.

Proof. Along (4), we write $f = \sum_{k=1}^{K} \sum_{\alpha\in A_k} f_{k,\alpha}$ with $f_{k,\alpha} \in L_{k,\alpha}$. Then

$$\begin{aligned}\int_G |(f|\pi(g)\tilde{\phi})_{L^2}|^2\, d_G(g) &= \int_G \left|\sum_{k=1}^{K} \sum_{\alpha\in A_k} (f_{k,\alpha}|\pi(g)\phi_{k,\alpha})_{L^2}\right|^2 d_G(g)\\ = \sum_{(k,\alpha)} \int_G |(f_{k,\alpha}|\pi(g)\phi_{k,\alpha})_{L^2}|^2\, d_G(g)&\\ + \sum_{(k,\alpha)\neq(l,\beta)} \int_G (f_{k,\alpha}|\pi(g)\phi_{k,\alpha})_{L^2}\overline{(f_{l,\beta}|\pi(g)\phi_{l,\beta})_{L^2}}\, d_G(g).&\end{aligned}$$

By (10), we have $\int_G |(f_{k,\alpha}|\pi(g)\phi_{k,\alpha})_{L^2}|^2\, d_G(g) = \|f_{k,\alpha}\|^2$. On the other hand, if $(k,\alpha) \neq (l,\beta)$, we have

$$\int_G (f_{k,\alpha}|\pi(g)\phi_{k,\alpha})_{L^2}\overline{(f_{l,\beta}|\pi(g)\phi_{l,\beta})_{L^2}}\, d\mu_G(g) = 0.$$

In fact, similarly to (6) we have for all $h \in H$

$$\begin{aligned}&\int_{\mathbb{R}^n} (f_{k,\alpha}|\pi(v,h)\phi_{k,\alpha})_{L^2}\overline{(f_{l,\beta}|\pi(v,h)\phi_{l,\beta})_{L^2}}\, dv\\ &= (2\pi)^n|\det h| \int_{\mathbb{R}^n} (\mathcal{F}f_{k,\alpha}(\xi)|\sigma(h)\mathcal{F}\phi_{k,\alpha}(h^\top\xi))\overline{(\mathcal{F}f_{l,\beta}(\xi)|\sigma(h)\mathcal{F}\phi_{l,\beta}(h^\top\xi))}\, d\xi.\end{aligned}$$

If $k \neq l$, this quantity vanishes because the integrand in the right-hand side equals 0 by $\mathcal{O}_k \cap \mathcal{O}_l = \emptyset$. If $k = l$ and $\alpha \neq \beta$, by the same argument as the proof of Theorem 2, we have for all $\xi \in \mathcal{O}_k$

$$\int_H (\mathcal{F}f_{k,\alpha}(\xi)|\sigma(h)\mathcal{F}\phi_{k,\alpha}(h^\top\xi))\overline{(\mathcal{F}f_{k,\beta}(\xi)|\sigma(h)\mathcal{F}\phi_{k,\beta}(h^\top\xi))}\, d_H(h) = 0$$

whether $\sigma_{k,\alpha} \simeq \sigma_{k,\beta}$ or not, by the Schur orthogonality relation. Thus we get

$$\int_G |(f|\pi(g)\tilde{\phi})_{L^2}|^2\, d_G(g) = \sum_{(k,\alpha)} \|f_{k,\alpha}\|^2 = \|f\|^2,$$

which completes the proof. □

3 The 3-Dimensional Similitude Group Case

We consider the case where H is the group $\{\, cA\,;\, c > 0, A \in SO(n)\,\} \simeq \mathbb{R}_{>0} \times SO(n)$ on $\mathbb{R}^n$. Let σ be an irreducible unitary representation of $SO(n)$. We extend σ to a representation of H by $\sigma(cA) := \sigma(A)$. The contragredient action ρ of H on $\mathbb{R}^n$ has one open orbit $\mathcal{O} := \mathbb{R}^n \backslash \{0\}$, and the isotropy subgroup H_1 at any $\xi^{[1]} \in \mathcal{O}$ is isomorphic to $SO(n-1)$. Moreover, since $SO(n-1)$ is a multiplicity free subgroup of $SO(n)$ by [2], we have $[\sigma|_{H_1} : \sigma_{1,\alpha}] = 1$ for all $\alpha \in A_1$. Thus the assumptions (A1), (A2) and (A3) are satisfied, and we can find an admissible vector $\tilde{\phi}$ in Theorem 3. A natural unitary representation of G on the space of square-integrable (r, s)-tensor fields on $\mathbb{R}^n$ corresponds to the case where σ is the tensor product $\sigma_0 \otimes \cdots \otimes \sigma_0$ ($r + s$ times) of the natural representation σ_0 of $SO(n)$. Then the assumption (A3) is not satisfied in general. In this case, we decompose the value of $f \in \mathrm{L}^2(\mathbb{R}^n, (\mathbb{C}^n)^{\otimes(r+s)})$ into the sum of irreducible summands of $(\sigma, (\mathbb{C}^n)^{\otimes(r+s)})$ as a pretreatment, and we can consider a continuous wavelet transform for each component.

Now let us present a concrete example of $\tilde{\phi}$ for the case where $n = 3$ and σ is the natural representation of $SO(3)$. For $1 \leq i < j \leq 3$, put $X_{ij} := -E_{ij} + E_{ji} \in \mathrm{Mat}(3, \mathbb{R})$. The Lie algebra of $SO(3)$ is spanned by X_{12}, X_{23}, and X_{13}. Put $\xi^{[1]} := (0, 0, 1)^\top$. Then $H_1 = \exp \mathbb{R}X_{12} \simeq SO(2)$. Define $e_{1,\pm1} := \frac{1}{\sqrt{2}}(1, \mp i, 0)^\top$ and $e_{1,0} := (0, 0, 1)^\top$, so that we have $\sigma(\exp tX_{12})e_{1,\alpha} = e^{i\alpha t}e_{1,\alpha}$ for $\alpha = 0, \pm1$ and $t \in \mathbb{R}$, which means that we have the decomposition (2) with $A_1 = \{0, \pm1\}$ and $W_{1,\alpha} := \mathbb{C}e_{1,\alpha}$. Note that $\dim W_{1,\alpha} = 1$ in this case.

Let us consider the spherical coordinate $\xi = (r \sin\theta \cos\varphi, r \sin\theta \sin\varphi, r \cos\theta)$ with $r \geq 0$, $\theta \in [0, \pi]$, $\varphi \in [0, 2\pi)$. If $\xi \in \mathbb{R}^3 \setminus \mathbb{R}\xi^{[1]}$, then θ and φ are uniquely determined. In this case, we put $h_1(\xi) := r^{-1} \exp(\varphi X_{12}) \exp(-\theta X_{13}) \in H$ so that $\rho(h_1(\xi))\xi^{[1]} = \xi$. Then $e_{\xi,\alpha} := \sigma(h_1(\xi))e_{1,\alpha}$ for $\alpha \in A_1$ is computed as

$$e_{\xi,0} = \frac{1}{\|\xi\|}\begin{pmatrix} \xi_1 \\ \xi_2 \\ \xi_3 \end{pmatrix}, \quad e_{\xi,\pm1} = \frac{1}{\|\xi\|\sqrt{2(\xi_1^2 + \xi_2^2)}}\begin{pmatrix} \xi_1\xi_3 \pm i\xi_2\|\xi\| \\ \xi_2\xi_3 \mp i\xi_1\|\xi\| \\ -(\xi_1^2 + \xi_2^2) \end{pmatrix}.$$

As in (3), the irreducible summand $L_{1,\alpha}$ is the space of $f \in L^2(\mathbb{R}^3, \mathbb{C}^3)$ such that $\mathcal{F}f(\xi) \in \mathbb{C}e_{\xi,\alpha}$ for almost all $\xi \in \mathbb{R}^3 \setminus \mathbb{R}\xi^{[1]}$. We see easily that a vector field belonging to $L_{1,0}$ is curl-free, while one belonging to $L_{1,1} \oplus L_{1,-1}$ is divergence-free. Putting $\psi_0(\xi) := C_0 e^{-\|\xi\|^2/2}\|\xi\|^3$, $\psi_{\pm1}(\xi) := \pm iC_1 e^{-\|\xi\|^2/2}\|\xi\|^2\sqrt{\xi_1^2+\xi_2^2}$ with $C_0 := (4\pi)^{-1}$ and $C_1 := (32\pi^2/3)^{-1/2}$, we define $\phi_{1,\alpha} \in L_{1,\alpha}$ for $\alpha \in A_1$ by

$$\phi_{1,\alpha}(x) := (2\pi)^{-3} \int_{\mathbb{R}^3\setminus\mathbb{R}\xi^{[1]}} e^{-i(x|\xi)}\psi_\alpha(\xi)e_{\xi,\alpha}\,d\xi \quad (x \in \mathbb{R}^3).$$

We write $d_{SO(3)}$ for the normalized Haar measure on $SO(3)$, and define $d_H(h) := \frac{dc}{c}d_{SO(3)}(A)$ for $h = cA \in H$ with $c > 0$ and $A \in SO(3)$. Then we have $\int_H \varphi(h^\top \xi^{[1]})\,d_H(h) = 4\pi \int_{\mathbb{R}^3} \varphi(\xi)\frac{d\xi}{\|\xi\|^3}$ for a non-negative measurable function φ on $\mathbb{R}^3$. Thus the left-hand side of (9) equals $4\pi\int_{\mathbb{R}^3}|\psi_\alpha(\xi)|^2\frac{d\xi}{\|\xi\|^3}$. If $\alpha = 0$, this becomes $4\pi C_0^2\int_{\mathbb{R}^3} e^{-\|\xi\|^2}\|\xi\|^3\,d\xi = 1$, and if $\alpha = \pm1$, it is $4\pi C_1^2\int_{\mathbb{R}^3} e^{-\|\xi\|^2}\|\xi\|(\xi_1^2+\xi_2^2)\,d\xi = 1$. Namely (9) holds for $\alpha \in A_1$. We compute

$$\phi_{1,0}(x) = (2\pi)^{-3}C_0\int_{\mathbb{R}^3} e^{-i(\xi|x)}e^{-\|\xi\|^2/2}\|\xi\|^2\begin{pmatrix}\xi_1\\ \xi_2\\ \xi_3\end{pmatrix}d\xi = C(\|x\|^2-5)e^{-\|x\|^2/2}\begin{pmatrix}x_1\\ x_2\\ x_3\end{pmatrix}$$

with $C := 2^{-7/2}\pi^{-5/2}$. On the other hand, although we don't have explicit expression of $\phi_{1,\pm1}$, we see that $\phi_{1,\sharp}(x) := \phi_{1,1}(x) + \phi_{1,-1}(x)$ equals

$$(2\pi)^{-3}\frac{C_1}{\sqrt{2}}\int_{\mathbb{R}^3} e^{-i(\xi|x)}e^{-\|\xi\|^2/2}\|\xi\|^2\begin{pmatrix}-2\xi_2\\ 2\xi_1\\ 0\end{pmatrix}d\xi = C(\|x\|^2-5)e^{-\|x\|^2/2}\begin{pmatrix}-\sqrt{3}x_2\\ \sqrt{3}x_1\\ 0\end{pmatrix}.$$

Therefore we obtain a concrete admissible vector field $\tilde{\phi} := \phi_{1,0} + \phi_{1,\sharp}$ given by

$$\tilde{\phi}(x) = 2^{-7/2}\pi^{-5/2}(\|x\|^2-5)e^{-\|x\|^2}\begin{pmatrix}x_1-\sqrt{3}x_2\\ x_2+\sqrt{3}x_1\\ x_3\end{pmatrix} \quad (x \in \mathbb{R}^3). \tag{11}$$

If a vector field $f \in L^2(\mathbb{R}^3, \mathbb{C}^3)$ is decomposed as $\sum_{\alpha\in A_1} f_{1,\alpha}$, then $f_{1,1} + f_{1,-1}$ is the divergence-free part of f, which equals $\int_G (f|\pi(g)\phi_{1,\sharp})_{L_2}\pi(g)\phi_{1,\sharp}\,d_G(g)$. A similar formula holds for the curl-free part $f_{1,0}$ and $\phi_{1,0}$.

References

1. Ali, S.T., Antoine, J.-P., Gazeau, J.-P.: Coherent States, Wavelets and Their Generalizations. Graduate Texts in Contemporary Physics, Springer, New York (2000). https://doi.org/10.1007/978-1-4612-1258-4
2. Diximier, J.: Sur les représentations de certains groupes orthogonaux. C.R. Acad. Sci. Paris **250**, 3263–3265 (1960)
3. Esmaeelzadeh, F., Kamyabi Gol, R.A.: On the C.W.T on homogeneous spaces associated to quasi invariant measure. J. Pseudo-Differ. Oper. Appl. **9**, 487–494 (2018). https://doi.org/10.1007/s11868-018-0255-y

4. Folland, G.B.: A Course in Abstract Harmonic Analysis. CRC Press, Boca Raton (1995)
5. Führ, H.: Abstract Harmonic Analysis of Continuous Wavelet Transform. Lecture Notes in Mathematics, vol. 1863. Springer, Heidelberg (2005). https://doi.org/10.1007/b104912
6. Grossmann, A., Morlet, J., Paul, T.: Transforms associated to square integrable group representations I. General results. J. Math. Phys. **26**, 2473–2479 (1985)
7. Lessig, C.: Divergence free polar wavelets for the analysis and representation of fluid flows. J. Math. Fluid Mech. **21**(1), 20 (2019). https://doi.org/10.1007/s00021-019-0408-7. Article number: 18
8. Ohshiro, K.: Construction of continuous wavelet transforms associated to unitary representations of semidirect product groups. Doctoral thesis, Nagoya University (2017)
9. Urban, K.: Wavelet bases in H(div) and H(curl). Math. Comput. **70**, 739–766 (2001)
10. Whitney, H.: Elementary structure of real algebraic varieties. Ann. Math. **66**, 545–556 (1957)

Entropy Under Disintegrations

Juan Pablo Vigneaux[1,2(✉)]

[1] Institut de Mathématiques de Jussieu–Paris Rive Gauche (IMJ-PRG), Université de Paris, 8 place Aurélie Némours, 75013 Paris, France
[2] Max-Planck-Institut für Mathematik in den Naturwissenschaften, Inselstraße 22, 04103 Leipzig, Germany

Abstract. We consider the differential entropy of probability measures absolutely continuous with respect to a given σ-finite "reference" measure on an arbitrary measure space. We state the asymptotic equipartition property in this general case; the result is part of the folklore but our presentation is to some extent novel. Then we study a general framework under which such entropies satisfy a chain rule: disintegrations of measures. We give an asymptotic interpretation for conditional entropies in this case. Finally, we apply our result to Haar measures in canonical relation.

Keywords: Generalized entropy · Differential entropy · AEP · Chain rule · Disintegration · Topological group · Haar measure · Concentration of measure

1 Introduction

It is part of the "folklore" of information theory that given any measurable space $(E, \mathfrak{B})$ with reference measure μ, and a probability measure ρ on E that is absolutely continuous with respect to μ (i.e. $\rho \ll \mu$), one can define a *differential entropy* $S_\mu(\rho) = -\int_E \log(\frac{\mathrm{d}\rho}{\mathrm{d}\mu})\,\mathrm{d}\rho$ that gives the exponential growth rate of the $\mu^{\otimes n}$-volume of a typical set of realizations of $\rho^{\otimes n}$. Things are rarely treated at this level of generality in the literature, so the first purpose of this article is to state the *asymptotic equipartition property (AEP)* for $S_\mu(\rho)$. This constitutes a unified treatment of the discrete and euclidean cases, which shows (again) that the differential entropy introduced by Shannon is not an unjustified *ad hoc* device as some still claim.

Then we concentrate on a question that has been largely neglected: what is the most general framework in which one can make sense of the chain rule? This is at least possible for any disintegration of a measure.

Definition 1 (Disintegration). *Let $T : (E, \mathfrak{B}) \to (E_T, \mathfrak{B}_T)$ be a measurable map, ν a σ-finite measure on $(E, \mathfrak{B})$, and ξ a σ-finite measure on $(E_T, \mathfrak{B}_T)$. The measure ν has a disintegration $\{\nu_t\}_{t \in E_T}$ with respect to T and ξ, or a (T, ξ)-disintegration, if*

F. Nielsen and F. Barbaresco (Eds.): GSI 2021, LNCS 12829, pp. 340–349, 2021.
https://doi.org/10.1007/978-3-030-80209-7_38

1. *ν_t is a σ-finite measure on $\mathfrak{B}$ concentrated on $\{T = t\}$, which means that $\nu_t(T \neq t) = 0$ for ξ-almost every t;*
2. *for each measurable nonnegative function $f : E \to \mathbb{R}$,*
   (a) $t \mapsto \int_E f \, \mathrm{d}\nu_t$ is measurable,
   (b) $\int_E f \, \mathrm{d}\nu = \int_{E_T} \left(\int_E f(x) \, \mathrm{d}\nu_t(x) \right) \mathrm{d}\xi(t)$.

We shall see that if the reference measure μ has a (T, ξ)-disintegration $\{\mu_t\}_{t \in E_T}$, then any probability ρ absolutely continuous with respect to it has a $(T, T_*\rho)$-disintegration; each ρ_t is absolutely continuous with respect to μ_t, and its density can be obtained normalizing the restriction of $\frac{\mathrm{d}\rho}{\mathrm{d}\mu}$ to $\{T = t\}$. Moreover, the following chain rule holds:

$$S_\mu(\rho) = S_\xi(T_*\rho) + \int_{E_T} S_{\mu_t}(\rho_t) \, \mathrm{d}T_*\rho(t). \tag{1}$$

We study the meaning of $\int_{E_T} S_{\mu_t}(\rho_t) \, \mathrm{d}T_*\rho(t)$ in terms of asymptotic volumes. Finally, we show that our generalized chain rule can be applied to Haar measures in canonical relation.

2 Generalized Differential Entropy

2.1 Definition and AEP

Let $(E_X, \mathfrak{B})$ be a measurable space, supposed to be the range of some random variable X, and let μ be a σ-finite measure μ on it. In applications, several examples appear:

1. E_X a countable set, $\mathfrak{B}$ the corresponding atomic σ-algebra, and μ the counting measure;
2. E_X euclidean space, $\mathfrak{B}$ its Borel σ-algebra, and μ the Lebesgue measure;
3. More generally: E_X a locally compact topological group, $\mathfrak{B}$ its Borel σ-algebra, and μ some Haar measure;
4. $(E_X, \mathfrak{B})$ arbitrary and μ a probability measure on it, that might be a prior in a Bayesian setting or an initial state in a physical/PDE setting.

The reference measure μ gives the relevant notion of volume.

Let ρ is a probability measure on $(E_X, \mathfrak{B})$ absolutely continuous with respect to μ, and f a representative of the Radon-Nikodym derivative $\frac{\mathrm{d}\rho}{\mathrm{d}\mu} \in L^1(E_X, \mu_X)$. The *generalized differential entropy* of ρ with respect to (w.r.t.) μ is defined as

$$S_\mu(\rho) := \mathbb{E}_\rho \left(-\ln \frac{\mathrm{d}\rho}{\mathrm{d}\mu} \right) = -\int_{E_X} f(x) \log f(x) \, \mathrm{d}\mu(x). \tag{2}$$

This was introduced by Csiszár in [5], see also Eq. (8) in [7]. Remark that the set where $f = 0$, hence $\log(f) = -\infty$, is ρ-negligible.

Let $\{X_i : (\Omega, \mathfrak{F}, \mathbb{P}) \to (E_X, \mathfrak{B}, \mu)\}_{i\in\mathbb{N}}$ be a collection of i.i.d random variables with law ρ. The density of the joint variable $(X_1, ..., X_n)$ w.r.t. $\mu^{\otimes n}$ is given by $f_{X_1,...,X_n}(x_1, ..., x_n) = \prod_{i=1}^n f(x_i)$. If the Lebesgue integral in (2) is finite, then

$$-\frac{1}{n}\log f_{X_1,...,X_n}(X_1, ..., X_n) \to S_\mu(\rho) \tag{3}$$

$\mathbb{P}$-almost surely (resp. in probability) as a consequence of the strong (resp. weak) law of large numbers. The convergence in probability is enough to establish the following result.

Proposition 1 (Asymptotic Equipartition Property). *Let $(E_X, \mathfrak{B}, \mu)$ be a σ-finite measure space, and ρ a probability measure on $(E_X, \mathfrak{B})$ such that $\rho \ll \mu$ and $S_\mu(\rho)$ is finite. For every $\delta > 0$, set*

$$A_\delta^{(n)}(\rho;\mu) := \left\{ (x_1, ..., x_n) \in E_X^n \;\middle|\; \left| -\frac{1}{n}\log f_{X_1,...,X_n}(X_1, ..., X_n) - S_\mu(\rho)\right| \le \delta \right\}.$$

Then,

1. *for every $\varepsilon > 0$, there exists $n_0 \in \mathbb{N}$ such that, for all $n \ge n_0$,*
$$\mathbb{P}\left(A_\delta^{(n)}(\rho;\mu)\right) > 1 - \varepsilon;$$
2. *for every $n \in \mathbb{N}$,*
$$\mu^{\otimes n}(A_\delta^{(n)}(\rho;\mu)) \le \exp\{n(S_\mu(\rho) + \delta)\};$$
3. *for every $\varepsilon > 0$, there exists $n_0 \in \mathbb{N}$ such that, for all $n \ge n_0$,*
$$\mu^{\otimes n}(A_\delta^{(n)}(\rho;\mu)) \ge (1-\varepsilon)\exp\{n(S_\mu(\rho) - \delta)\}.$$

We proved these claims in [11, Ch. 12]; our proofs are very similar to the standard ones for (euclidean) differential entropy, see [4, Ch. 8].

Below, we write $A_\delta^{(n)}$ if ρ and μ are clear from context.

When E_X is a countable set and μ the counting measure, every probability law ρ on E_X is absolutely continuous with respect to μ; if $p : E_X \to \mathbb{R}$ is its density, $S_\mu(\rho)$ corresponds to the familiar expression $-\sum_{x\in E_X} p(x)\log p(x)$.

If $E_X = \mathbb{R}^n$, μ is the corresponding Lebesgue measure, and ρ a probability law such that $\rho \ll \mu$, then the derivative $\mathrm{d}\rho/\mathrm{d}\mu \in L^1(\mathbb{R}^n)$ corresponds to the elementary notion of density, and the quantity $S_\mu(\rho)$ is the *differential entropy* that was also introduced by Shannon in [10]. He remarked that the covariance of the differential entropy under diffeomorphisms is consistent with the measurement of randomness "*relative to an assumed standard.*" For example, consider a linear automorphism of $\mathbb{R}^n$, $\varphi(x_1, ..., x_n) = (y_1, ..., y_n)$, represented by a matrix A. Set $\mu = \mathrm{d}x_1 \cdots \mathrm{d}x_n$ and $\nu = \mathrm{d}y_1 \cdots \mathrm{d}y_n$. It can be easily deduced from the change-of-variables formula that $\nu(\varphi(V)) = |\det A|\mu(V)$. Similarly, $\varphi_*\rho$ has density

$f(\varphi^{-1}(y))|\det A|^{-1}$ w.r.t. ν, and this implies that $S_\nu(\varphi_*\rho) = S_\mu(\rho) + \log|\det A|$, cf. [4, Eq. 8.71]. Hence

$$\left| -\frac{1}{n}\log\prod_{i=1}^{n}\frac{\mathrm{d}\varphi_*\rho}{\mathrm{d}\nu}(y_i) - S_\nu(\varphi_*\rho)\right| = \left| -\frac{1}{n}\log\prod_{i=1}^{n}\frac{\mathrm{d}\rho}{\mathrm{d}\mu}(\varphi^{-1}(y_i)) - S_\mu(\rho)\right|, \quad (4)$$

from which we deduce that $A_\delta^{(n)}(\varphi_*\rho, \nu) = \varphi^{\times n}(A_\delta^{(n)}(\rho;\mu))$ and consequently

$$\nu^{\otimes n}(A_\delta^{(n)}(\varphi_*\rho;\nu)) = |\det A|^n \mu^{\otimes n}(A_\delta^{(n)}(\rho;\mu)), \quad (5)$$

which is consistent with the corresponding estimates given by Proposition 1.

In the discrete case one could also work with any multiple of the counting measure, $\nu = \alpha\mu$, for $\alpha > 0$. In this case, the chain rule for Radon-Nikodym derivatives (see [6, Sec. 19.40]) gives

$$\frac{\mathrm{d}\rho}{\mathrm{d}\mu} = \frac{\mathrm{d}\rho}{\mathrm{d}\nu}\frac{\mathrm{d}\nu}{\mathrm{d}\mu} = \alpha\frac{\mathrm{d}\rho}{\mathrm{d}\nu}, \quad (6)$$

and therefore $S_\mu(\rho) = S_\nu(\rho) - \log\alpha$. Hence the discrete entropy depends on the choice of reference measure, contrary to what is usually stated. This function is invariant under a bijection of finite sets, but taking on both sides the counting measure as reference measure. The proper analogue of this in the euclidean case is a measure-preserving transformation (e.g. $|\det A| = 1$ above), under which the differential entropy *is* invariant.

For any E_X, if μ is a probability law, the expression $S_\mu(\rho)$ is the opposite of the *Kullback-Leibler divergence* $D_{KL}(\rho\|\mu) := -S_\mu(\rho)$. The positivity of the divergence follows from a customary application of Jensen's inequality or from the asymptotic argument given in the next subsection.

The asymptotic relationship between volume and entropy given by the AEP can be summarized as follows:

Corollary 1.

$$\lim_{\delta\to 0}\lim_{n\to\infty}\frac{1}{n}\log\mu^{\otimes n}(A_\delta^{(n)}(\rho;\mu)) = S_\mu(\rho).$$

2.2 Certainty, Positivity and Divergence

Proposition 1 gives a meaning to the divergence and the positivity/negativity of $S_\mu(\rho)$.

1. Discrete case: let E_X be a countable set and μ be the counting measure. Irrespective of ρ, the cardinality of $\mu^{\otimes n}(A_\delta^{(n)}(\rho;\mu))$ is at least 1, hence the limit in Corollary 1 is always positive, which establishes $S_\mu(\rho) \geq 0$. The case $S_\mu(\rho) = 0$ corresponds to certainty: if $\rho = \delta_{x_0}$, for certain $x_0 \in E_X$, then $A_\delta^{(n)} = \{(x_0, ..., x_0)\}$.

2. Euclidean case: E_X Euclidean space, μ Lebesgue measure. The differential entropy is negative if the volume of the typical set is (asymptotically) smaller than 1. Moreover, the divergence of the differential entropy to $-\infty$ correspond to asymptotic concentration on a μ-negligible set. For instance, if ρ has $\mu(B(x_0,\varepsilon))^{-1}\chi_{B(x_0,\varepsilon)}$, then $S_{\lambda_d}(\rho) = \log(|B(x_0,\varepsilon)|) = \log(c_d\varepsilon^d)$, where c_d is a constant characteristic of each dimension d. By part (2) of Proposition 1, $|A_\delta^{(n)}| \leq \exp(nd\log\varepsilon + Cn)$, which means that, for fixed n, the volume goes to zero as $\varepsilon \to 0$, as intuition would suggest. Therefore, *the divergent entropy is necessary to obtain the good volume estimates.*
3. Whereas the positivity of the (discrete) entropy arises from a *lower bound* to the volume of typical sets, the positivity of the Kullback-Leibler is of a different nature: it comes from an *upper bound.* In fact, when μ and ρ are probability measures such that $\rho \ll \mu$, the inequality $\mu^{\otimes n}(A_\delta^{(n)}(\rho;\mu)) \leq 1$ holds for any δ, which translates into $S_\mu(\rho) \leq 0$ and therefore $D_{KL}(\rho||\mu) \geq 0$. In general, there is no upper bound for the divergence.

Remark that entropy maximization problems are well defined when the reference measure is a finite measure.

3 Chain Rule

3.1 Disintegration of Measures

We summarize in this section some fundamental results on disintegrations as presented in [2]. Throughout it, $(E,\mathfrak{B})$ and $(E_T,\mathfrak{B}_T)$ are measurable spaces equipped with σ-finite measures ν and ξ, respectively, and $T : (E,\mathfrak{B}) \to (E_T,\mathfrak{B}_T)$ is a measurable map.

Definition 1 is partly motivated by the following observation: when E_T is finite and $\mathfrak{B}_T$ is its algebra of subsets 2^{E_T}, we can associate to any probability P on $(E,\mathfrak{B})$ a (T,T_*P)-disintegration given by the conditional measures $P_t : \mathfrak{B} \to \mathbb{R}, B \mapsto P(B\cap\{T=t\})/P(T=t)$, indexed by $t\in E_T$. In particular,

$$P(B) = \sum_{t\in E_T} P(T=t)P_t(B). \tag{7}$$

Remark that P_t is only well defined on the maximal set of $t\in E_T$ such that $T_*P(t) > 0$, but only these t play a role in the disintegration (7).

General disintegrations give *regular versions* of conditional expectations. Let ν be a probability measure, $\xi = T_*\nu$, and $\{\nu_t\}$ the corresponding (T,ξ)-disintegration. Then the function $x\in E\mapsto \int_E \chi_B(x)\,\mathrm{d}\nu_{T(x)}$—where χ_B denotes the characteristic function—is σ_T measurable and a regular version of the conditional probability $\nu_{\sigma(T)}(B)$ as defined by Kolmogorov.

Disintegrations exist under very general hypotheses. For instance, if ν is Radon, $T_*\nu \ll \xi$, and $\mathfrak{B}_T$ is countably generated and contains all the singletons $\{t\}$, then ν has a (T,ξ)-disintegration. The resulting measures ν_t measures are uniquely determined up to an almost sure equivalence. See [2, Thm. 1].

As we explained in the introduction, a disintegration of a reference measure induces disintegrations of all measures absolutely continuous with respect to it.

Proposition 2. *Let ν have a (T,ξ)-disintegration $\{\nu_t\}$ and let ρ be absolutely continuous with respect to ν with finite density $r(x)$, with each ν, ξ and ρ σ-finite.*

1. *The measure ρ has a (T,ξ)-disintegration $\{\tilde{\rho}_t\}$ where each $\tilde{\rho}_t$ is dominated by the corresponding ν_t, with density $r(x)$.*
2. *The image measure $T_*\rho$ is absolutely continuous with respect to ξ, with density $\int_E r\,\mathrm{d}\nu_t$.*
3. *The measures $\{\tilde{\rho}_t\}$ are finite for ξ-almost all t if and only if $T_*\rho$ is σ-finite.*
4. *The measures $\{\tilde{\rho}_t\}$ are probabilities for ξ-almost all t if and only if $\xi = T_*\rho$.*
5. *If $T_*\rho$ is σ-finite then $0 < \nu_t r < \infty$ $T_*\nu$-almost surely, and the measures $\{\rho_t\}$ given by*
$$\int_E f\,\mathrm{d}\rho_t = \frac{\int_E fr\,\mathrm{d}\nu_t}{\int_E r\,\mathrm{d}\nu_t}$$
are probabilities that give a $(T,T_*\rho)$-disintegration of ρ.

Example 1 (Product spaces). We suppose that $(E,\mathfrak{B},\nu)$ is the product of two measured spaces spaces $(E_T,\mathfrak{B}_T,\xi)$ and $(E_S,\mathfrak{B}_S,\nu)$, with ξ and ν both σ-finite. Let ν_t be the image of ν under the inclusion $s \mapsto (t,s)$. Then Fubini's theorem implies that ν_t is a (T,ξ)-disintegration of ν. (Remark that $\xi \neq T_*\nu$. In general, the measure $T_*\nu$ is not even σ-finite.) If $r(t,s)$ is the density of a probability ρ on $(E,\mathfrak{B})$, then $\rho_t \ll \nu_t$ with density $r(t,s)$—the value of t being fixed—and $\tilde{\rho}_t$ is a probability supported on $\{T=t\}$ with density $r(t,s)/\int_{E_S} r(t,s)\,\mathrm{d}\nu(s)$.

3.2 Chain Rule Under Disintegrations

Any disintegration gives a chain rule for entropy.

Proposition 3 (Chain rule for general disintegrations). *Let $T : (E_X,\mathfrak{B}_X) \to (E_Y,\mathfrak{B}_Y)$ be a measurable map between arbitrary measurable spaces, μ (respectively ν) a σ-finite measure on $(E_X,\mathfrak{B}_X)$ (resp. $(E_Y,\mathfrak{B}_Y)$), and $\{\mu_y\}$ a (T,ν)-disintegration of μ. Then any probability measure ρ absolutely continuous w.r.t. μ, with density r, has a (T,ν)-disintegration $\{\tilde{\rho}_y\}_{y\in Y}$ such that for each y, $\tilde{\rho}_y = r\cdot\mu_y$. Additionally, ρ has a $(T,T_*\rho)$-disintegration $\{\rho_y\}_{y\in Y}$ such that each ρ_y is a probability measure with density $r/\int_{E_X} r\,\mathrm{d}\mu_y$ w.r.t. μ_y, and the following chain rule holds:*

$$S_\mu(\rho) = S_\nu(T_*\rho) + \int_{E_Y} S_{\mu_y}(\rho_y)\,\mathrm{d}T_*\rho(y). \tag{8}$$

Proof. For convenience, we use here linear-functional notation: $\int_X f(x)\,\mathrm{d}\mu(x)$ is denoted $\mu(f)$ or $\mu^x(f(x))$ if we want to emphasize the variable integrated.

Almost everything is a restatement of Proposition 2. Remark that $y \mapsto \mu_y(r)$ is the density of $T_*\rho$ with respect to ν.

Equation (8) is established as follows:

$$S_\mu(\rho) \overset{(def)}{=} \rho\left(-\log\frac{\mathrm{d}\rho}{\mathrm{d}\mu}\right) = T_*\rho^y\left(\rho_y\left(-\log\frac{\mathrm{d}\rho}{\mathrm{d}\mu}\right)\right) \tag{9}$$

$$= T_*\rho^y\left(\rho_y\left(-\log\frac{\mathrm{d}\rho_y}{\mathrm{d}\mu_y} - \log\mu_y(r)\right)\right) \tag{10}$$

$$= T_*\rho^y\left(\rho_y\left(-\log\frac{\mathrm{d}\rho_y}{\mathrm{d}\mu_y}\right)\right) + T_*\rho^y\left(-\log\frac{\mathrm{d}T_*\rho}{\mathrm{d}\nu}\right) \tag{11}$$

$$= T_*\rho^y\left(S_{\mu_y}(\rho_y)\right) + S_\nu(T_*\rho), \tag{12}$$

where (9) is the fundamental property of the T-disintegration $\{\rho_y\}_y$ and (10) is justified by the equalities

$$\frac{\mathrm{d}\rho}{\mathrm{d}\mu} = \frac{\mathrm{d}\tilde{\rho}_y}{\mathrm{d}\mu_y} = m_y(r)\frac{\mathrm{d}\rho_y}{\mathrm{d}\mu_y}.$$

Example 2. From the computations of Example 1, it is easy to see that if $E_X = \mathbb{R}^n \times \mathbb{R}^m$, μ is the Lebesgue measure, and T is the projection on the $\mathbb{R}^n$ factor, then (8) corresponds to the familiar chain rule for Shannon's differential entropy.

Example 3 (Chain rule in polar coordinates). Let $E_X = \mathbb{R}^2\setminus\{0\}$, μ be the Lebesgue measure $\mathrm{d}x\,\mathrm{d}y$ on $\mathbb{R}^2$, and $\rho = f\,\mathrm{d}x\,\mathrm{d}y$ a probability measure. Every point $\boldsymbol{v} \in E_X$ can be parametrized by cartesian coordinates (x, y) or polar coordinates (r, θ), i.e. $\boldsymbol{v} = \boldsymbol{v}(x, y) = \boldsymbol{v}(r, \theta)$. The parameter r takes values from the set $E_R =]0, \infty[$, and θ from $E_\Theta = [0, 2\pi[$; the functions $R : E_X \to E_R$, $v \mapsto r(\boldsymbol{v})$ and $\Theta : E_X \to E_\Theta$, $\boldsymbol{v} \mapsto \theta(\boldsymbol{v})$ can be seen as random variables with laws $R_*\rho$ and $\Theta_*\rho$, respectively. We equip E_R (resp. E_Θ) with the Lebesgue measure $\mu_R = \mathrm{d}r$ (resp. $\mu_H = \mathrm{d}\theta$).

The measure μ has a (R, μ_R)-disintegration $\{r\,\mathrm{d}\theta\}_{r\in E_R}$; here $r\,\mathrm{d}\theta$ is the uniform measure on $R^{-1}(r)$ of total mass $2\pi r$. This is a consequence of the change-of-variables formula:

$$\int_{\mathbb{R}^2} \varphi(x, y)\,\mathrm{d}x\,\mathrm{d}y = \int_{[0,\infty[}\left(\int_0^{2\pi} \varphi(r, \theta) r\,\mathrm{d}\theta\right)\mathrm{d}r, \tag{13}$$

which is precisely the disintegration property. Hence, according to Proposition 2, ρ disintegrates into probability measures $\{\rho_r\}_{r\in E_R}$, with each ρ_r concentrated on $\{R = r\}$, absolutely continuous w.r.t. $\mu_r = r\,\mathrm{d}\theta$ and with density $f / \int_0^{2\pi} f(r, \theta) r\,\mathrm{d}\theta$. The exact chain rule (8) holds in this case.

This should be compared with Lemma 6.16 in [8]. They consider the random vector (R, Θ) as an $\mathbb{R}^2$ valued random variable, and the reference measure to be $\nu = \mathrm{d}r\,\mathrm{d}\theta$. The change-of-variables formula implies that (R, Θ) has density $rf(r, \theta)$ with respect to ν, so $S_\nu(\rho) = S_\mu(\rho) - \mathbb{E}_\rho(\log R)$. Then they apply the standard chain rule to $S_\nu(\rho)$, i.e. as in Examples 1 and 2, to obtain a deformed chain rule for $S_\mu(\rho)$:

$$S_\mu(\rho) = S_{\mu_R}(R_*\rho) + \int_0^\infty\left(-\int_0^{2\pi}\log\left(\frac{f}{\int_0^{2\pi} f\,\mathrm{d}\theta}\right)\frac{f\,\mathrm{d}\theta}{\int_0^{2\pi} f\,\mathrm{d}\theta}\right) + \mathbb{E}_\rho(\log R). \tag{14}$$

Our term $\int_{E_R} S_{\mu_r}(\rho_r)\,\mathrm{d}R_*\rho(r)$ comprises the last two terms in the previous equation.

Remark 1. Formula (13) is a particular case of the *coarea formula* [1, Thm. 2.93], which gives a disintegration of the Hausdorff measure $\mathcal{H}^N$ restricted to a countably $\mathcal{H}^N$-rectifiable subset E of $\mathbb{R}^M$ with respect to a Lipschitz map $f:\mathbb{R}^M \to \mathbb{R}^k$ (with $k \leq N$) and the Lebesgue measure on $\mathbb{R}^k$. So the argument of the previous example also applies to the extra term $-\mathbb{E}_{(\mathbf{x},\mathbf{y})}[\log J^{\mathcal{E}}_{\mathbf{p_y}}(\mathbf{x},\mathbf{y})]$ in the chain rule of [7, Thm. 41], which could be avoided by an adequate choice of reference measures.

Combining Corollary 1 and the preceding proposition, we get a precise interpretation of the conditional term in terms of asymptotic growth of the volume of *slices* of the typical set.

Proposition 4. *Keeping the setting of the previous proposition,*

$$\lim_{\delta\to 0}\lim_{n\to\infty}\frac{1}{n}\log\left(\frac{\int_{E_Y}\mu_y^{\otimes n}(A_\delta^{(n)}(\rho;\mu))\,\mathrm{d}\nu^{\otimes n}(y)}{\nu^{\otimes n}(A_\delta^{(n)}(T_*\rho;\nu))}\right)=\int_{E_Y}S_{\mu_y}(\rho_y)\,\mathrm{d}T_*\rho(y).$$

Proof. It is easy to prove that if $\{\mu_y\}_y$ is a (T,ν)-disintegration of μ, then $\{\mu_y^{\otimes n}\}_y$ is a $(T^{\times n},\nu^{\otimes n})$-disintegration of $\mu^{\otimes n}$. The disintegration property reads

$$\mu^{\otimes n}(A)=\int_{E_Y}\mu_y^{\otimes n}(A)\,\mathrm{d}\nu^{\otimes n}(y), \tag{15}$$

for any measurable set A. Hence

$$\log\mu^{\otimes n}(A_\delta^{(n)}(\rho;\mu))=\log\nu^{\otimes n}(A_\delta^{(n)}(T_*\rho;\nu))+\log\frac{\int_{E_Y}\mu_y^{\otimes n}(A_\delta^{(n)}(\rho;\mu))\,\mathrm{d}\nu^{\otimes n}(y)}{\nu^{\otimes n}(A_\delta^{(n)}(T_*\rho;\nu))}. \tag{16}$$

The results follows from the application of $\lim_{\delta\to 0}\lim_n\frac{1}{n}$ to this equality and comparison of the result with the chain rule.

In connection to this result, remark that $(T_*\rho)^{\otimes n}$ concentrates on $A_\delta^{(n)}(T_*\rho;\nu)$ and has approximately density $1/\nu^{\otimes n}(A_\delta^{(n)}(T_*\rho;\nu))$, so

$$\int_{E_Y}\mu_y^{\otimes n}(A_\delta^{(n)}(\rho;\mu))\frac{1}{\nu^{\otimes n}(A_\delta^{(n)}(T_*\rho;\nu))}\,\mathrm{d}\nu^{\otimes n}(y)$$

is close to an average of $\mu_y^{\otimes n}(A_\delta^{(n)}(\rho;\mu)\cap T^{-1}(y))$, the "typical part" of each fiber $T^{-1}(y)$, according to the "true" law $(T_*\rho)^{\otimes n}$.

3.3 Locally Compact Topological Groups

Given a locally compact topological group G, there is a unique left-invariant positive measure (left Haar measure) up to a multiplicative constant [3, Thms. 9.2.2 & 9.2.6]. A particular choice of left Haar measure will be denoted by λ with superscript G e.g. λ^G. The disintegration of Haar measures is given by Weil's formula.

Proposition 5 (Weil's formula). *Let G be a locally compact group and H a closed normal subgroup of G. Given Haar measures on two groups among G, H and G/H, there is a Haar measure on the third one such that, for any integrable function $f : G \to \mathbb{R}$,*

$$\int_G f(x) d\lambda^G(x) = \int_{G/H} \left(\int_H f(xy)\, \mathrm{d}\lambda^H(y) \right) \mathrm{d}\lambda^{G/H}(xH). \tag{17}$$

The three measures are said to be in *canonical relation*, which is written $\lambda^G = \lambda^{G/H}\lambda^H$. For a proof of Proposition 5, see pp. 87–88 and Theorem 3.4.6 of [9].

For any element $[g]$ of G/H, representing a left coset gH, let us denote by $\lambda^H_{[g]}$ the image of λ^H under the inclusion $\iota_g : H \to G$, $h \mapsto gH$. This is well defined i.e. does not depend on the chosen representative g: the image of ι_g depends only on the coset gH, and if g_1, g_2 are two elements of G such that $g_1H = g_2H$, and A is subset of G, the translation $h \mapsto g_2^{-1}g_1h$ establishes a bijection $\iota_{g_1}^{-1}(A) \xrightarrow{\sim} \iota_{g_2}^{-1}(A)$; the left invariance of the Haar measure implies that $\lambda^H(\iota_{g_1}^{-1}(A)) = \lambda^H(\iota_{g_2}^{-1}(A))$ i.e. $(\iota_{g_1})_*\lambda^H = (\iota_{g_2})_*\lambda^H$ as claimed. Proposition 5 shows then that $\{\lambda^H_{[g]}\}_{[g]\in G/H}$ is a $(T, \lambda^{G/H})$-disintegration of λ^G. In view of this and Proposition 3, the following result follows.

Proposition 6 (Chain rule, Haar case). *Let G be a locally compact group, H a closed normal subgroup of G, and λ^G, λ^H, and $\lambda^{G/H}$ Haar measures in canonical relation. Let ρ be a probability measure on G. Denote by $T : G \to G/H$ the canonical projection. Then, there is T-disintegration $\{\rho_{[g]}\}_{[g]\in G/H}$ of ρ such that each $\rho_{[g]}$ is a probability measure, and*

$$S_{\lambda^G}(\rho) = S_{\lambda^{G/H}}(\pi_*\rho) + \int_{G/H} S_{\lambda^H_{[g]}}(\rho_{[g]})\, \mathrm{d}\pi_*\rho([g]). \tag{18}$$

References

1. Ambrosio, L., Fusco, N., Pallara, D.: Functions of Bounded Variation and Free Discontinuity Problems. Oxford Science Publications/Clarendon Press (2000)
2. Chang, J.T., Pollard, D.: Conditioning as disintegration. Stat. Neerl. **51**(3), 287–317 (1997)
3. Cohn, D.: Measure Theory. Birkhäuser Advanced Texts Basler Lehrbücher, 2nd edn. Springer, New York (2013). https://doi.org/10.1007/978-1-4614-6956-8
4. Cover, T., Thomas, J.: Elements of Information Theory. A Wiley-Interscience Publication, Hoboken (2006)
5. Csiszár, I.: Generalized entropy and quantization problems. In: Transactions of the 6th Praque Conference on Information Theory, Statistical Decision Functions, Random Processes. Academia, Praque (1973)
6. Hewitt, E., Stromberg, K.: Real and Abstract Analysis: A Modern Treatment of the Theory of Functions of a Real Variable. Springer, Heidelberg (1965). https://doi.org/10.1007/978-3-642-88044-5
7. Koliander, G., Pichler, G., Riegler, E., Hlawatsch, F.: Entropy and source coding for integer-dimensional singular random variables. IEEE Trans. Inf. Theory **62**(11), 6124–6154 (2016)

8. Lapidoth, A., Moser, S.M.: Capacity bounds via duality with applications to multiple-antenna systems on flat-fading channels. IEEE Trans. Inf. Theory **49**(10), 2426–2467 (2003)
9. Reiter, H., Stegeman, J.D.: Classical Harmonic Analysis and Locally Compact Groups. No. 22 in London Mathematical Society. Monographs New Series, Clarendon Press (2000)
10. Shannon, C.: A mathematical theory of communication. Bell Syst. Tech. J. **27**, 379–423, 623–656 (1948)
11. Vigneaux, J.P.: Topology of statistical systems: a cohomological approach to information theory. Ph.D. thesis, Université de Paris (2019). https://hal.archives-ouvertes.fr/tel-02951504v1

Koszul Information Geometry, Liouville-Mineur Integrable Systems and Moser Isospectral Deformation Method for Hermitian Positive-Definite Matrices

Frédéric Barbaresco(✉)

THALES Land and Air Systems, Limours, France
frederic.barbaresco@thalesgroup.com

Abstract. As soon as 1993, A. Fujiwara and Y. Nakamura have developed close links between Information Geometry and integrable system by studying dynamical systems on statistical models and completely integrable gradient systems on the manifolds of Gaussian and multinomial distributions, Recently, Jean-Pierre Françoise has revisited this model and extended it to the Peakon systems. In parallel, in the framework of integrable dynamical systems and Moser isospectral deformation method, originated in the work of P. Lax, A.M. Perelomov has applied models for space of Positive Definite Hermitian matrices of order 2, considering this space as an homogeneous space. We conclude with links of Lax pairs with Souriau equation to compute coefficients of characteristic polynomial of a matrix. Main objective of this paper is to compose a synthesis of these researches on integrability and Lax pairs and their links with Information Geometry, to stimulate new developments at the interface of these disciplines.

Keywords: Information geometry · Complete integrable systems · Isospectral deformation · Lax pairs

1 Liouville Complete Integrability and Information Geometry

Integrability refers to the existence of invariant, regular foliations, related to the degree of integrability, depending on the dimension of the leaves of the invariant foliation.

In the case of Hamiltonian systems, this integrability is called "complete integrability" in the sense of Liouville (Liouville-Mineur Theorem) [1–9].

In case of Liouville integrability, a regular foliation of the phase space by invariant manifolds such that the Hamiltonian vector fields associated to the invariants of the foliation span the tangent distribution. In this case, a maximal set of Poisson commuting invariants exist (functions on the phase space whose Poisson brackets with the Hamiltonian of the system, and with each other, vanish).

If the phase space is symplectic, the leaves of the foliation are totally isotropic with respect to the symplectic form and such a maximal isotropic foliation is called Lagrangian.

F. Nielsen and F. Barbaresco (Eds.): GSI 2021, LNCS 12829, pp. 350–359, 2021.
https://doi.org/10.1007/978-3-030-80209-7_39

In 1993, in the framework of Liouville Completely Integrable Systems on Statistical Manifolds, Y. Nakamura developed gradient systems on manifolds of various probability distributions. He proved the following results [10–17]:

- the gradient systems on the even-dimensional manifolds of Gaussian and multinomial distributions are completely integrable Hamiltonian systems.
- the gradient systems can always be linearized by using Information Geometry dual coordinate system of the original coordinates.
- the gradient flows on the statistical manifolds converge to equilibrium points exponentially.
- the gradient systems are completely integrable Hamiltonian systems if the manifolds have even dimensions. There is a 2 m-dimensional Hamiltonian System which is integrable so that the Dynamics restricted to the common level sets of first integrals is a gradient. This type of dynamics (sometimes called Brockett flows) has been studied in itself [47].
- the gradient system associated with the Gaussian distribution can be related to an Orstein-Ulhenbeck process

We consider the statistical Manifolds with Fisher Metric

$$\begin{gathered} p(x,\theta) \text{ with } \theta = (\theta_1, \ldots, \theta_n)^T \quad \text{and} \quad l(x,\theta) = \log p(x,\theta) \\ G = [g_{i,j}] \text{ with } g_{i,j} = E\left[\frac{\partial l(x,\theta)}{\partial \theta_i}\frac{\partial l(x,\theta)}{\partial \theta_j}\right] = \frac{\partial^2 \Psi(\theta)}{\partial \theta_i \partial \theta_j} \end{gathered} \tag{1}$$

Y. Nakamura has considered the following gradient flow on Statistical Manifolds

$$\frac{d\theta}{dt} = -G^{-1}\frac{\partial \Psi}{\partial \theta} \text{ with } \frac{\partial \Psi}{\partial \theta} = \left(\frac{\partial \Psi}{\partial \theta_1}, \ldots, \frac{\partial \Psi}{\partial \theta_n}\right)^T \tag{2}$$

First, Y. Nakamura proved the first theorem:

- The gradient system (2) is always linearizable. The induced flow on an open subset of the Riemannian Statistical Manifold such that there exists a potential function satisfying (2) converges to equilibrium points exponentially.
  Second Y. Nakamura Theorem states:
- If n is even (n = 2 m), then the gradient system (2) is a completely integrable Hamiltonian system:

$$\begin{aligned} &\tfrac{d\eta}{dt} = -\eta \Rightarrow \tfrac{dH_j}{dt} = \tfrac{dP_j}{dt}Q_j + P_j\tfrac{dQ_j}{dt} \\ &\tfrac{dH_j}{dt} = \left(-\eta_j^{-2}\tfrac{d\eta_j}{dt}\right)\eta_{m+j} + \eta_j^{-1}\tfrac{d\eta_{m+j}}{dt} = \eta_j^{-1}\eta_{m+j} - \eta_j^{-1}\eta_{m+j} = 0 \\ &\Rightarrow \tfrac{dH_j}{dt} = 0 \end{aligned} \tag{3}$$

Define the Hamiltonian $H = \sum_{j=1}^{m} H_j$, then the corresponding equations of motion,

$$\frac{d\Xi_k}{dt} = \{\Xi_k, H\} \text{ with } \begin{cases} \Xi_k = P_k \ , \ k = 1, ..., m \\ \Xi_k = Q_{k-m} \ , \ k = m+1, ..., n \end{cases} \tag{4}$$

coincides with $\frac{d\eta}{dt} = -\eta$ and $\begin{cases} \{H_j, H\} = 0 \\ \{H_i, H_j\} = 0 \end{cases}$.

Nakamura model is illustrated for monovariate Gaussian law $p(x,\theta) = \frac{1}{\sqrt{2\pi}\sigma} e^{-\cdot\frac{(x-m)^2}{2\sigma^2}}$:

$$\theta = (\theta_1, \theta_2) = \left(\frac{m}{\sigma^2}, \frac{1}{2\sigma^2}\right) \quad \text{and} \quad E[x] = \frac{1}{2}\theta_1\theta_2^{-1} \ , \ E[x^2] = \frac{1}{4}\theta_1^2\theta_2^{-2} + \frac{1}{2}\theta_2^{-1} \tag{5}$$

$$G = \frac{1}{2}\begin{pmatrix} \theta_2^{-1} & \theta_1\theta_2^{-2} \\ \theta_1\theta_2^{-2} & \theta_2^{-2} + \theta_1^2\theta_2^{-3} \end{pmatrix} = \left[\frac{\partial^2\Psi}{\partial\theta_i\partial\theta_j}\right] , \ \Psi(\theta) = \frac{\theta_1^2}{4\theta_2} - \log\left(\sqrt{\theta_2}\right) + \log\sqrt{\pi} \tag{6}$$

$$\frac{d\theta}{dt} = -G^{-1}\frac{\partial\Psi}{\partial\theta} \Rightarrow \begin{pmatrix} \frac{d\theta_1}{dt} \\ \frac{d\theta_2}{dt} \end{pmatrix} = -\frac{1}{2}\begin{pmatrix} \theta_1^3\theta_2^{-1} \\ \theta_1^2 - 2\theta_2 \end{pmatrix} \Rightarrow \begin{cases} \theta_1(t) = \dfrac{\eta_1(0)}{\eta_2(0) - \eta_1(0)^2 e^{-t}} \\ \theta_2(t) = \dfrac{1}{2\left(\eta_2(0)e^{-t} - \eta_1(0)^2 e^{-2t}\right)} \end{cases} \tag{7}$$

$$\frac{dH}{dt} = 0 \text{ where } H = PQ = \theta_1^{-1} + \frac{1}{2}\theta_1\theta_2^{-1} \tag{8}$$

with $P = 2\theta_2\theta_1^{-1} = E[x]^{-1}, Q = \frac{1}{4}\theta_1^2\theta_2^{-2} + \frac{1}{2}\theta_2^{-1} = E[x^2]$

$$\frac{dQ}{dt} = -\frac{\partial H}{\partial P} \ , \ \frac{dP}{dt} = \frac{\partial H}{\partial Q} \tag{9}$$

Nakamura also made the link with the Lax pairs:

$$\frac{dL}{dt} = [[L, D], L] \tag{10}$$

$$L = \frac{1}{2\theta_1\theta_2}\begin{pmatrix} 2\theta_2 & \sqrt{2\theta_2}\theta_1 \\ \sqrt{2\theta_2}\theta_1 & \theta_1^2 \end{pmatrix} , \ D = \frac{1}{2}\begin{pmatrix} 0 & 0 \\ 0 & \theta_1 \end{pmatrix} \tag{11}$$

$$\begin{aligned} Tr(L) &= \theta_1^{-1} + \tfrac{1}{2}\theta_1\theta_2^{-1} = -\tfrac{\partial\Psi(\theta)}{\partial\theta_2}\left(\tfrac{\partial\Psi(\theta)}{\partial\theta_1}\right)^{-1} = \eta_2\eta_1^{-1} \\ \tfrac{dTr(L)}{dt} &= Tr[[L,D],L] = 0 \end{aligned} \tag{12}$$

Recently, Jean-Pierre Françoise has revisited this model and extended it to the Peakon systems [18].

We make also reference to works of W.W. Symes [25], A. Thimm [30–32] and others references on integrable systems [33, 35–37, 42, 43, 46, 48]

2 The Isospectral Deformation Method for Hermitian Positive Definite Matrices

Integrable dynamical systems can be studied with the efficient isospectral deformation method, originated in the work of P. Lax. These tools have been developed by Russian A.M. Perelomov [26–29, 44, 45] and applied for space of Positive Definite Hermitian matrices of order 2, considering this space as an homogeneous space with the group $G = SL(2,C)$ of complex 2×2 matrices that acts $gxg^+, g \in G, x \in \{x \in GL(2)/x^+ = x, x > 0\}$ transitively on it. For this space, the G-invariant metric and geodesics are given by:

$ds^2 = Trace(x^{-1}dxx^{-1}dx^+)$ with geodesics equation

$$\frac{d}{dt}\left(x^{-1}\frac{dx}{dt} + \frac{dx}{dt}x^{-1}\right) \tag{13}$$

One solution of this geodesics equation is:

$$x(t) = \beta e^{2\alpha t}\beta^+, \beta \in SL(2,C), \alpha^+ = \alpha, Trace(\alpha) = 0 \tag{14}$$

But Perelomov has observed that Hermitian Positive Definite Matrices could be expressed with respect to a unitary matrix $u(t)$:

$$x(t) = u(t)e^{2aQ(t)}u^{-1}(t) \tag{15}$$

Given by the following relations:

$$x^{-1}\frac{dx}{dt} + \frac{dx}{dt}x^{-1} = 2au(t)L(t)u^{-1}(t)$$
$$\text{with} \begin{cases} L(t) = P + \dfrac{i}{4a}\left(e^{-2aQ(t)}Me^{2aQ(t)} - e^{2aQ(t)}Me^{-2aQ(t)}\right) \\ M(t) = -u^{-1}(t)\dfrac{du(t)}{dt} \end{cases} \tag{16}$$

By differentiating the previous equation with respect to time t, a Lax equation can be deduced from an Hamiltonian $H = \frac{1}{2}(m^2 + k^2.sh^{-2}(n))$:

$$\frac{dL}{dt} = [M, L] \text{ with } \begin{cases} L = \begin{pmatrix} m & k.cth(n) \\ k.cth(n) & -m \end{pmatrix} \\ M = \dfrac{k}{2.sh^2(n)} \begin{pmatrix} 0 & 1 \\ -1 & 0 \end{pmatrix} \end{cases} \tag{17}$$

Similar construction have been studied by J. Moser in 1975 [34] for a system of n particles with pair interaction and a potential given by $v(n) = k^2.sh^{-2}n$ [19–24].

Another system of coordinate could be also considered with an upper triangular matrix $w(t)$ with unit diagonal:

$$x(t) = w(t)e^{2Q(t)}w^+(t) \text{ with } Q = \begin{pmatrix} n & 0 \\ 0 & -n \end{pmatrix} \tag{18}$$

We can then compute:

$$\frac{dx}{dt}x^{-1} = 2w(t)Lw^{-1}(t) \text{ with } \begin{cases} L = P + \dfrac{M}{2} + \dfrac{1}{2}e^{2Q}M^+e^{-2Q} \\ M = w^{-1}(t)\dfrac{dz}{dt} \end{cases} \tag{19}$$

By differentiation (19) with respect to time t and using the property that $x(t)$ is a geodesic, a new Lax equation appear:

$$\frac{dL}{dt} = [M, L] \text{ with } \begin{cases} L = \begin{pmatrix} m & e^{2n} \\ 1 & -m \end{pmatrix} \\ M = 2\begin{pmatrix} 0 & e^{2n} \\ 0 & 0 \end{pmatrix} \end{cases} \tag{20}$$

This Lax equation is associated to the equations of motion induced by the following Hamiltonian

$$H = \frac{1}{2}\left(m^2 + e^{2n}\right) \tag{21}$$

3 First Integrals Associated to Lax Pairs, Characteristic Polynomial and Souriau Algorithm

In previous chapters, we have considered Lax Pairs equation $\frac{dL}{dt} = [M, L]$. This equation is a powerful method to build first integrals of a Hamiltonian system in the framework of integrable systems. Two matrices $L = \left(L_{ij}(x)\right)$ and $M = \left(M_{ij}(x)\right)$ are Lax pairs for a dynamical system $\frac{dx}{dt} = f(x)$ if for every solutions, the matrices the equation $\frac{dL}{dt} = [M, L] = LM - ML$ and the eigenvalues $\lambda_1(x), \cdots, \lambda_m(x)$ of $L(x)$ are integrals of motion for the dynamical system. Coefficients of the characteristic polynomial of the matrix $L(x)$, $\det(L-\lambda I)=(-1)^m(\lambda^m-\alpha_1(x)\lambda^{m-1}+\alpha_2(x)\lambda^{m-2}+\cdots+(-1)^m\alpha_m(x))$, are polynomials in $tr(L^i), i=1,...,m$:

$$\alpha_1 = tr(L), \alpha_2 = \frac{1}{2}\left[(tr(L))^2 - tr(L^2)\right], \alpha_3 = \cdots \tag{22}$$

We can observe that $tr(L^i), i=1,...,m$ are first integrals of the dynamical system, due to the consequence of:

$$\frac{d}{dt}tr(L) = tr\left(\frac{dL}{dt}\right) = tr(ML-LM) = tr(ML) - tr(LM) = 0 \tag{23}$$

$$\frac{d}{dt}tr(L^i) = itr\left(L^{i-1}\frac{dL}{dt}\right) = itr(L^{i-1}[L,M]) = i(tr(L^{i-1}ML - L^iM)) = 0 \tag{24}$$

As the coefficients of the characteristic polynomials $L(x)$ are constants of motion, it induces that its eigenvalues $\lambda_1(x),\cdots,\lambda_m(x)$ are also constants of motion.

At this step, we make the link with a work of Souriau to compute characteristic polynomial of a matrix, initially discovered by Urbain Jean Joseph Le Verrier in 1840, and extended in 1948 by Jean-Marie Souriau [38–41]. By using the following relation:

$$vol(adj(A)Av_1)(v_2)...(v_n) = vol(Av_1)(Av_2)..(Av_n) = \det(A)vol(v_1)(v_2)...(v_n) \tag{25}$$

If A is invertible, we recover classical equations:

$$adj(A)A = \det(A)I \text{ and } A^{-1} = [\det(A)]^{-1}adj(A) \tag{26}$$

Using these formulas, we can try to invert $[\lambda I - A]$ assuming that $\det(\lambda I - A) \neq 0$. If we use previous determinant definition, we have:

$$\begin{aligned}\det(\lambda I - A)vol(v_1)(v_2)...(v_n) &= vol(\lambda v_1 - Av_1)(\lambda v_2 - Av_2)...(\lambda v_n - Av_n)\\ &= \lambda^n vol(v_1)(v_2)...(v_n) + ...\end{aligned} \tag{27}$$

where $\det(\lambda I - A)$ is the characteristic polynomial of A, a polynomial in λ of degree n, with:

$$adj(\lambda I - A)[\lambda I - A] = \det(\lambda I - A)I \Leftrightarrow A.Q(\lambda) = \lambda Q(\lambda) - P(\lambda)I \tag{28}$$

(if λ is an eigenvalue of A, the nonzero columns of $Q(\lambda)$ are corresponding eigenvectors). We can then observe that $adj(\lambda I - A)$ is a polynomial of degree *n−1*. We can then define both $P(\lambda)$ and $Q(\lambda)$ by polynomials:

$$P(\lambda) = \det(\lambda I - A) = \sum_{i=0}^{n} k_i \lambda^{n-i} \text{ and } Q(\lambda) = adj(\lambda I - A) = \sum_{i=0}^{n-1} \lambda^{n-i-1} B_i \quad (29)$$

$$\text{With } k_0 = 1,\ k_n = (-1)^n \det(A),\ B_0 = I, \text{ and } B_{n-1} = (-1)^{n-1} adj(A) \quad (30)$$

By developing equation $adj(\lambda I - A)[\lambda I - A] = \det(\lambda I - A)I$, we can write:

$$\sum_{i=0}^{n} k_i \lambda^{n-i} I = \sum_{i=0}^{n-1} \lambda^{n-i-1} B_i [\lambda I - A] = \lambda B_{n-1} + \sum_{i=1}^{n-1} \lambda^{n-i} [B_i - B_{i-1}A] - B_{n-1}A \quad (31)$$

By identification term by term, we find the expression of matrices B_i:

$$\begin{cases} B_0 = I \\ B_i = B_{i-1}A + k_i I\ ,\ i = 1, \ldots, n-1 \\ B_{n-1}A + k_n I = 0 \end{cases} \quad (32)$$

We can observe that $A^{-1} = -\frac{B_{n-1}}{k_n}$ and also the Cayley-Hamilton theorem:

$$k_0 A^n + k_1 A^{n-1} + \ldots + k_{n-1} A + k_n I = 0 \quad (33)$$

To go further, we have to use this classical result from analysis on differentiation given by $\delta[\det(G)] = tr(adj(G)\delta G)$. If we set $G = (\lambda I - A)$ and $\delta = \frac{d}{d\lambda}$, we then obtain $tr(adj(\lambda I - A)) = \frac{d}{d\lambda} \det(\lambda I - A)$ providing:

$$\sum_{i=0}^{n-1} \lambda^{n-i-1} tr(B_i) = \frac{d}{d\lambda} \left(\sum_{i=0}^{n} k_i \lambda^{n-i} \right) = \sum_{i=0}^{n-1} (n-i) k_i \lambda^{n-i-1} \quad (34)$$

We can then deduce that $tr(B_i) = (n-i)k_i\ ,\ i = 0, \ldots, n-1$.

$$\text{As } B_i = B_{i-1}A + k_i I,\ tr(B_i) = tr(B_{i-1}A) + n.k_i \text{ and then } k_i = -\frac{tr(B_{i-1}A)}{i} \quad (35)$$

We finally obtain the Souriau Algorithm:

$$\begin{array}{l} k_0 = 1 \text{ and } B_0 = I \\ \begin{cases} A_i = B_{i-1}A, k_i = -\frac{1}{i} tr(A_i),\ i = 1, \ldots, n-1 \\ B_i = A_i + k_i I \text{ or } B_i = B_{i-1}A - \frac{1}{i} tr(B_{i-1}A) I \end{cases} \\ A_n = B_{n-1}A \text{ and } k_n = -\frac{1}{n} tr(A_n) \end{array} \quad (36)$$

This Souriau algorithm for characteristic polynomial computation of a matrix is a new way to address complete integral systems by mean of Lax pairs and to build first integrals of a Hamiltonian system and especially in the case of Symplectic Geometry.

References

1. Liouville, J.: Note de l'Editeur. J. Math. Pures Appl. **5**, 351–355 (1840)
2. Liouville, J.: Sur quelques cas particuliers où les equations du mouvement d'un point matériel peuvent s'intégrer. J. Math. Pure Appl. **11**, 345–378 (1846)
3. Liouville, J.: L'intégration des équations différentielles du mouvement d'un nombre quelconque de points matérielles. J. Math. Pures Appl. **14**, 257–299 (1849)
4. Liouville, J. : Note sur l'intégration des équations différentielles de la Dynamique. J. Math. Pures Appl. **20**, 137–138(1855). Présentée en 1853
5. Liouville, J.: Note sur les équations de la dynamique. J. Math. Pures Appl. **20**, 137–138 (1855)
6. Tannery, J.: Sur une surface de révolution du quatrième degré dont les lignes géodesiques sont algébriques. Bull. Sci. Math. **16**, 190–192 (1892)
7. Mineur, H. : Sur les systèmes mécaniques admettant n intégrales premières uniformes et l'extension à ces systèmes de la méthode de quantification de Sommerfeld. C. R. Acad. Sci., Paris **200**, 1571–1573(1935)
8. Mineur, H. : Réduction des systèmes mécaniques à n degrés de liberté admettant n intégrales premières uniformes en involution aux systèmes à variables séparées. J. Math Pure Appl. **IX** (15), 385–389 (1936)
9. Mineur, H. : Sur les systèmes mécaniques dans lesquels figurent des paramètres fonctions du temps. Etude des systèmes admettant n intégrales premières uniformes en involution. Extension à ces systèmes des conditions de quantification de Bohr-Sommerfeld. J. de l'Ecole Polytech. **III**(143), 173–191 and 237–270 (1937)
10. Obata, T., Hara, H.: Differential geometry of nonequilibrium processes. Phys. Rev. A **45** (10), 6997–7001 (1992)
11. Fujiwara, A.: Dynamical Systems on Statistical Models. Recent Developments and Perspectives in Nonlinear Dynamical Systems, RIMS Kokyuroku No. 822, pp. 32–42. Kyoto Univ., Kyoto (1993)
12. Nakamura, Y.: Completely integrable gradient systems on the manifolds of Gaussian and multinomial distributions. Jpn. J. Ind. Appl. Math. **10**, 79–189 (1993)
13. Nakamura, Y.: Lax pair and fixed point analysis of Karmarkar's projective scaling trajectory for linear programming. Jpn. J. Ind. Appl. Math. **11**, 1–9 (1994)
14. Nakamura, Y.: Neurodynamics and nonlinear integrable systems of Lax type. Jpn. J. Ind. Appl. Math. **11**, 11–20 (1994)
15. Nakamura, Y.: Gradient systems associated with probability distributions. Jpn. J. Ind. Appl. Math. **11**, 21–30 (1994)
16. Nakamura, Y., Faybusovich, L.: On explicitly solable gradient systems of Moser-Karmarkar type. J. Math. Anal. Appl. **205**, 88–108 (1997)
17. Fujiwara, A., Shigeru, S.: Hereditary structure in Hamiltonians: information geometry of Ising spin chains. Phys. Lett. A **374**, 911–916 (2010)
18. Françoise, J.P.: Information Geometry and Integrable Systems, Les Houches Summer Week, SPIGL 2020, Les Houches, July 2020
19. Siegel, C.L.: On the integrals of canonical systems. Ann. Math. **2**(2), 806–822, MR 3-214 (1942), Matematika 5, 2 (1961), 103–117, Matematika 5, 2, 103–117 (1961)

20. Siegel, C.L., Moser, J.: Lectures on Celestial Mechanics. Springer, Berlin (1971). https://doi.org/10.1007/978-3-642-87284-6
21. Moser, J.: Stable and random motions in dynamical systems. Ann. Math. Stud. **77** (1973). https://press.princeton.edu/books/paperback/9780691089102/stable-and-random-motions-in-dynamical-systems
22. Moser, J.: Finitely many mass points on the line under the influence of an exponential potential - an integrable system. In: Moser, J. (eds.) Dynamical Systems, Theory and Applications. Lecture Notes in Physics, vol. 38. Springer, Heidelberg (1975). https://doi.org/10.1007/3-540-07171-7_12
23. Moser, J.: Various aspects of integrable Hamiltonian systems. In: Dynamical Systems. CIME Summer school, Bressanone 1978, pp. 233–289. Birkhauser, Boston, MA (1980)
24. Moser, J.: Geometry of quadrics and spectral theory. In: Proceedings, Chern Symposium, Berkeley 1979, pp. 147–188. Springer, Heidelberg (1980). https://doi.org/10.1007/978-1-4613-8109-9_7
25. Symes, W.W.: Hamiltonian group actions and integrable systems. Phys. Dl **1**, 339–374 (1980)
26. Kozlov, V.V.: Integrability and non-integrability in Hamiltonian mechanics. Uspekhi Mat. Nauk **38**(1), 3–67 (1983). [Russian]; English translation: Russian Math. Surveys 38, No.1, 1–76, 1983
27. Mishchenko, A.S., Fomenko, A.T.: A generalized Liouville method for the integration of Hamiltonian systems. Funktsional Anal. i Prilozhen. **12**(2), 46–56 (1978). MR 58 # 24357. Funct. Anal. Appl. **12**, 113–121 (1978)
28. Fomenko, A.T.: Topological classification of integrable systems, In: Fomenko, A.T. (ed.) Advances in Soviet Mathematics, vol. 6. AMS, Providence (1991)
29. Perelomov, A.M.: Integrable Systems of Classical Mechanics and Lie Algebras, vol. I. Birkhäuser, Boston (1990)
30. Thimm, A.: Integrable geodesic flows on homogeneous spaces. Ergodic Theor. Dyn. Syst. **1** (4), 495–517 (1981)
31. Thimm, A.: Integrabilitat beim geodatischen FlujJ. Bonner Math. Schriften **103**, 32 (1978)
32. Thimm, A.: Uber die Integrabilitat geodatischer Fliisse auf homogenen Raumen. Dissertation. Bonn (1980)
33. Guillemin, V., Sternberg, S.: On collective complete integrability according to the method of Thimm. Ergodic Theor. Dyn. Syst. **3**(2), 219–230 (1983)
34. Moser, J.: Three integrable Hamiltonian systems connected with isospectral deformations. Adv. Math. **16**, 197–220 (1975)
35. Calogero, F.: Exactly solvable one-dimensional many-body problems. Lettere al Nuovo Cimento (1971-1985) **13**(11), 411–416 (1975). https://doi.org/10.1007/BF02790495
36. Grava, T.: Notes on Integrable Systems and Toda Lattice. Lecture ISRM (2018)
37. Correa, E.M., Grama, L.: Lax formalism for Gelfand-Tsetlin integrable systems. arXiv:1810.08848v2 [math.SG] (2020)
38. Souriau, J.-M.: Une méthode pour la décomposition spectrale et l'inversion des matrices. CRAS **227**(2), 1010–1011. Gauthier-Villars, Paris (1948)
39. Souriau, J.-M.: Calcul Linéaire, vol. 1, EUCLIDE, Introduction aux études Scientifiques, Presses Universitaires de France, Paris (1959)
40. Souriau, J.-M., Vallée, C., Réaud, K., Fortuné, D. : Méthode de Le Verrier–Souriau et équations différentielles linéaires. CRASs IIB Mechan. **328**(10), 773–778 (2000)
41. Barbaresco, F.: Souriau exponential map algorithm for machine learning on matrix lie groups. In: Nielsen, F., Barbaresco, F. (eds.) GSI 2019. LNCS, vol. 11712, pp. 85–95. Springer, Cham (2019). https://doi.org/10.1007/978-3-030-26980-7_10

42. Kosmann-Schwarzbach, Y., Magri, F.: Lax-Nijenhuis operators for integrable systems. J. Math. Phys. **37**, 6173 (1996)
43. Magri, F.: A simple model of the integrable Hamiltonian equation. J. Math. Phys. **19**, 1156–1162 (1978)
44. Olshanetsky, M.A., Perelomov, A.M.: Completely integrable Hamiltonian systems connected with semisimple Lie algebras. Invent. Math. **37**, 93–108 (1976)
45. Olshanetsky, M., Perelomov, A.: Classical integrable finite-dimensional systems related to Lie algebras. Phys. Rep. C **71**, 313–400 (1981)
46. Zotov, A.V.: Classical integrable systems and their field-theoretical generalizations. Phys. Part. Nucl. **37**, 400–443 (2006)
47. Zotov, A.V.: Introduction to Integrable Systems, Steklov Mathematical Institute of Russian Academy of Sciences, Moscow (2021)
48. Bloch, A., Brockett, R.W., Ratiu, T.: Completely integrable gradient flows. Comm. Math. Phys. **147**(1), 57–74 (1992)

Flapping Wing Coupled Dynamics in Lie Group Setting

Zdravko Terze(✉), Viktor Pandža, Marijan Andrić, and Dario Zlatar

Faculty of Mechanical Engineering and Naval Architecture, University of Zagreb, Ivana Lučića 5, Zagreb, Croatia
zdravko.terze@fsb.hr

Abstract. In order to study dynamics of flapping wing moving in ambient fluid, the geometric modeling approach of fully coupled fluid-solid system is adopted, incorporating boundary integral method and time integrator in Lie group setting. If the fluid is assumed to be inviscid and incompressible, the configuration space of the fluid-solid system is reduced by eliminating fluid variables via symplectic reduction. Consequently, the equations of motion for the flapping wing are formulated without explicitly incorporating fluid variables, while effect of the fluid flow to the flapping wing overall dynamics is accounted for by the added mass effect only (computed by the boundary integral functions of the fluid density and the flow velocity potential). In order to describe additional viscous effects and include fluid vorticity and circulation in the system dynamics, vortex shedding mechanism is incorporated by enforcing Kutta conditions on the flapping wing sharp edges. In summary, presented approach exhibits significant computational advantages in comparison to the standard numerical procedures that - most commonly - comprise inefficient discretization of the whole fluid domain. Most importantly, due to its 'mid-fidelity' computational efficiency, presented approach allows to be embedded in the 'automated' optimization procedure for the multi-criterial flapping wing flight design.

Keywords: Fluid-structure interaction · Flapping wing · Geometric mechanics modeling

1 Introduction

Although designing of flapping wing vehicles (FWV) has reached certain level of maturity, current design of FWV is still very much dependent on the experimental activities, not relying extensively on the computational models. Certainly, one of the reasons for this is challenging modeling of the FWV flight physics (that includes highly unsteady aerodynamics, wing-fluid interactions, high flapping

This work has been fully supported by Croatian Science Foundation under the project IP-2016-06-6696.

F. Nielsen and F. Barbaresco (Eds.): GSI 2021, LNCS 12829, pp. 360–367, 2021.
https://doi.org/10.1007/978-3-030-80209-7_40

frequency effects etc.), but also numerical inefficiency of the standard computational procedures, mostly based on the continuum discretization of the ambient fluid as well as on the loosely coupled co-simulation routines between solid and fluid part of the system. Indeed, numerical inefficiency and time-consuming characteristics of such 'high-fidelity' numerical methods restrict their utilization mostly to case-by-case simulation campaigns, preventing their usage within the effective optimization procedures that would allow for multicriterial design optimization of FWV.

To this end, specially tailored reduced-order numerical models of the flapping wing dynamics, that allow for numerically efficient simulation of the coupled fluid-structure-interaction (FSI) phenomena pertinent for the FWV dynamics, will be presented in this paper. The method is based on geometric modeling approach of the fully coupled multibody-dynamics-fluid system (MBS-fluid system), incorporating boundary integral method and time integrator in Lie group setting.

By firstly assuming inviscid and incompressible fluid, the configuration space of the MBS-fluid system is reduced by eliminating fluid variables via symplectic reduction without compromising any accuracy. The reduction exploits 'particle relabeling' symmetry, associated with the conservation of circulation: fluid kinetic energy, fluid Lagrangian and associated momentum map are invariant with respect to this symmetry [1]. Additionally, in order to allow for modeling of flapping wing vorticity effects, the point vortex shedding and evolution mechanism is incorporated in the fluid-structure dynamical model. The vortices are assumed to be irrotational, and are being modeled by unsteady potential flow method enforcing Kutta condition at sharp edges. By using the proposed framework it is possible to include an arbitrary combination of smooth and sharp-edged bodies in a coupled MBS-fluid system, as long as the major viscosity effects of the fluid on the body can be described by shedding and evolution of the irrotational point vortices. However, in this paper we will focus on the coupled dynamics of a single flapping wing, moving in the ambient fluid at the prescribed trajectory.

2 Reduced Fluid-Solid Interaction Model

We consider rigid flapping wing submerged in ideal fluid, which is at rest at the infinity. In other words, at any time t, the system consisting of the rigid body and the fluid occupies an open connected region $\mathcal{M}$ of the Euclidean space, which we identify with $\mathbb{R}^3$. More specifically, the rigid body occupies region $\mathcal{B}$ and the fluid occupies a connected region $\mathcal{F} \subset \mathcal{M}$ such that $\mathcal{M}$ can be written as a disjoint union of open sets as $\mathcal{M} = \mathcal{B} \cup \mathcal{F}$. Configuration space $\mathcal{Q}$ of such a system is the set of all appropriately smooth maps from $\mathcal{M}$ to $\mathcal{M}$, where fluid part $\mathcal{Q}_{\mathrm{f}} \in \mathrm{Diff}_{\mathrm{vol}}(\mathcal{F})$ represents position field of the fluid particles. $\mathrm{Diff}_{\mathrm{vol}}(\mathcal{F})$ is the set of volume preserving diffeomorphisms, possessing properties of the Lie group. Body part $\mathcal{Q}_{\mathrm{B}}$, represents motion of the rigid body $\mathcal{B} \subset \mathcal{M}$ with boundary $\partial\mathcal{B}$, meaning that the configuration space of a rigid body is modeled as Lie group $G = \mathbb{R}^3 \times SO(3)$ with the elements of the form $p = (\boldsymbol{x}, \boldsymbol{R})$. G is a Lie group of dimension $n = 6$. The left multiplication in the group is given as $L_p : G \to G$, $p \to p \cdot p$, where origin of the group G is given as $e = (\boldsymbol{0}, \boldsymbol{I})$. With G so defined, its Lie algebra is given as $g = \mathbb{R}^3 \times so(3)$ with the elements of the form $v = (\boldsymbol{v}, \tilde{\boldsymbol{\omega}})$, $\boldsymbol{v}$ being velocity of body mass center and $\tilde{\boldsymbol{\omega}}$ body angular velocity in skew symmetric matrix form [10].

Moreover, the state space of the body is then $S = G \times g$, i.e. $S = \mathbb{R}^3 \times SO(3) \times \mathbb{R}^3 \times so(3) \cong TG$ with the elements $x = (\boldsymbol{x}, \boldsymbol{R}, \boldsymbol{v}, \tilde{\boldsymbol{\omega}})$ [10]. S is the left-trivialization of the tangent bundle TG. This is a Lie group itself that possesses the Lie algebra $S = \mathbb{R}^3 \times so(3) \times \mathbb{R}^3 \times \mathbb{R}^3$ with the element $z = (\boldsymbol{v}, \tilde{\boldsymbol{\omega}}, \dot{\boldsymbol{v}}, \dot{\boldsymbol{\omega}})$.

If kinematical constraints are imposed on the system - meaning that flapping wing is not unconstrained any more, but rheonomically driven along the specified trajectory - it can be shown that wing coupled dynamics in fluid flow can be expressed as differential-algebraic system of equations (DAE) formulated as [10]

$$\begin{bmatrix} \boldsymbol{M} & \boldsymbol{C}^{\mathrm{T}} \\ \boldsymbol{C} & \boldsymbol{0} \end{bmatrix} \begin{bmatrix} \dot{\boldsymbol{z}} \\ \boldsymbol{\lambda} \end{bmatrix} = \begin{bmatrix} \boldsymbol{Q} \\ \boldsymbol{\xi} \end{bmatrix}, \tag{1}$$

where $\boldsymbol{M}$ represents inertia matrix and $\boldsymbol{Q}$ is the force vector. Inertia matrix $\boldsymbol{M}$ contains standard inertial properties of the rigid wing [10], supplemented by added mass that the body perceives due to interaction with the fluid [8]. Due to the coupling, added mass effect needs to be calculated via boundary element method (BEM), taking into account changes in the body velocity and potential flow velocity field, after symplectic reduction on $\mathcal{Q}$ is performed [9].

The force vector $\boldsymbol{Q}$ can be written as

$$\boldsymbol{Q} = \boldsymbol{Q}_{\mathrm{Kirch_{ext}}} + \boldsymbol{Q}_{\mathrm{vort}}, \tag{2}$$

where $\boldsymbol{Q}_{\mathrm{Kirch_{ext}}}$ represents (possible) general external forces and torques acting on the wing according to the extended Kirchhoff equations [5], while $\boldsymbol{Q}_{\mathrm{vort}}$ represents additional forces and torques imposed on the system due to the flow circulatory part, which are consistent with the system vortex shedding mechanism. Modeling of the vortex shedding can be performed - for the example of the flapping wing - by the unsteady vortex lattice method, which utilizes shedding of an irrotational point vortex in each time step, such that the Kutta conditions are satisfied at sharp edges (here, shedding only on the wing trailing edge is considered; for two edges vortex shedding, see [9]). Note that added mass momentum rate of change should also be computed and introduced in (1) through the force vector $\boldsymbol{Q}_{\mathrm{Kirch_{ext}}}$, as well as standard Kirchhoff equations right hand side expressions (external product of angular momentum and linear/angular velocity of the body).

Other parts of the equation stem from the kinematical generalized position constraint equation

$$\boldsymbol{\Phi}(q, t) = \boldsymbol{0}, \tag{3}$$

which is differentiated twice to obtain a constraint equation at the acceleration level

$$\boldsymbol{C}(q)\,\dot{\boldsymbol{z}} = \boldsymbol{\xi}(q, \boldsymbol{z}),$$

where $\boldsymbol{C}$ represents constraint Jacobian, while $\boldsymbol{\lambda}$ is associated Lagrange multiplier vector, see [10]. The flapping wing trajectory can be represented by three flapping angles, as illustrated in Fig. 1 and Fig. 2, where θ represents deviation angle, η represents pitching angle, while ϕ represents stroking angle. It should be noted that there is one to one correspodence between three flapping angles θ, η, ϕ and wing orientation matrix $R \in SO(3)$ in the flapping motion studied here.

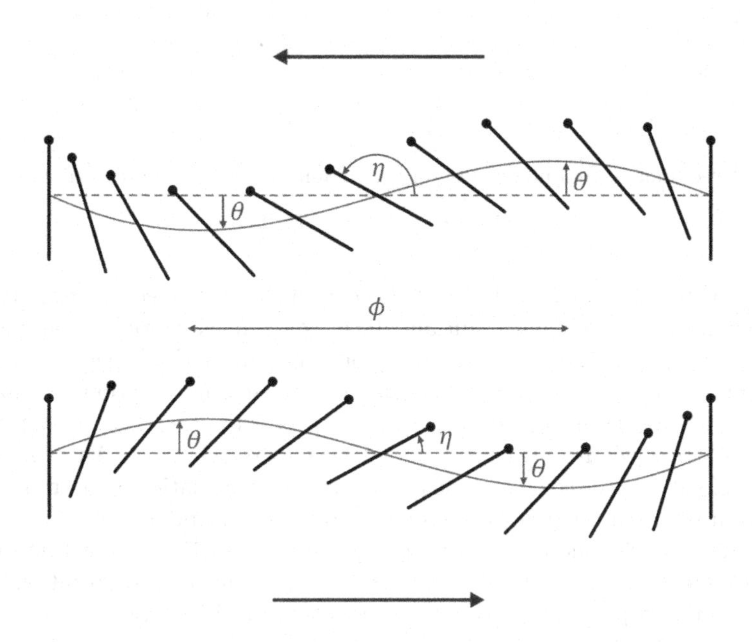

Fig. 1. Illustration of the flapping wing motion, described by three flapping angles.

Mathematical model of the coupled wing-fluid dynamics formulated as (1) is basically governed by the rigid wing variables only i.e. fluid terms vanish from the model, except for the vector $\boldsymbol{Q}_{\text{vort}}$.

Indeed, if vorticity effects are excluded from the analysis, it can be shown that symplectic reduction at zero vorticity yields

$$J_{\mathcal{F}}^{-1}(0)/\text{Diff}_{\text{vol}}(\mathcal{F}) = T^*\left(\mathcal{Q}/\text{Diff}_{\text{vol}}(\mathcal{F})\right) = T^*G, \tag{4}$$

indicating that the whole dynamics of the system evolves in T^*G (fluid variables are reduced out) and fluid influence on the wing dynamics is governed by added mass effect only [3]. In (4), $J_{\mathcal{F}}$ is momentum map associated with the action of $\text{Diff}_{\text{vol}}(\mathcal{F})$ on $\mathcal{Q}$ representing vorticity advection [11] that is conserved since Lagrangian of the fluid [2] is invariant on $\text{Diff}_{\text{vol}}(\mathcal{F})$ ('particle relabeling' symmetry). However, if vorticity effects are to be introduced, it can be shown that symplectic reduction, similar to the one introduced in (4), can also be performed when vorticity level is not set to zero, but possesses certain value that equals to

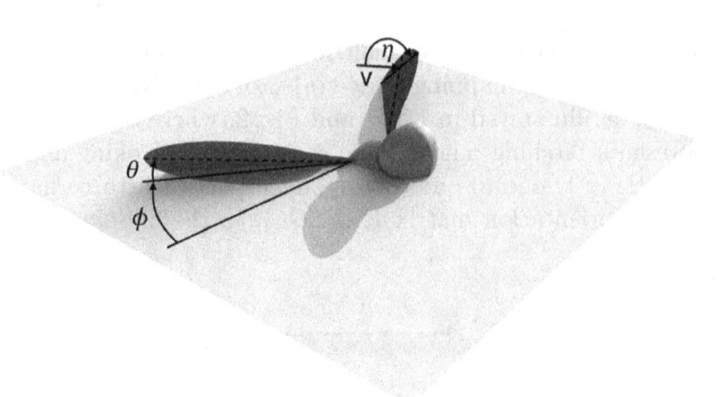

Fig. 2. Three dimensional illustration of the flapping wing motion, described by three flapping angles.

the system circulation that is in consistence with the introduced vortex shedding mechanism [6,12] (producing additional term $\boldsymbol{Q}_{\text{vort}}$ in the motion equations (1)).

Similarly as it is shown in [10], Lie group integrator that operates in S can be utilized to obtain system velocities $\boldsymbol{z}$, while constraint violation stabilization algorithm for constraint equation (3) needs to be simultaneously performed as well as BEM procedure [9] for the added mass determination. After obtaining system velocities, a reconstruction of the system translation is a trivial task, while rotational part requires a non-linear (Lie group) update.

To this end, Munthe-Kaas based Lie group method [7,10] is utilized, where kinematic reconstruction is performed by seeking an incremental rotation vector $\boldsymbol{\theta}$ in each time step by integrating ODE equation in Lie algebra

$$\dot{\boldsymbol{\theta}} = \operatorname{dexp}_{-\boldsymbol{\theta}}^{-1}(\boldsymbol{\omega}), \quad \boldsymbol{\theta}(0) = \mathbf{0}.$$

The wing rotation matrix in time step $n+1$ is then reconstructed as

$$\boldsymbol{R}_{n+1} = \boldsymbol{R}_n \exp(\boldsymbol{\theta}_{n+1}),$$

where exp represents exponential mapping on $SO(3)$ and $\operatorname{dexp}_{-\boldsymbol{\theta}}^{-1}$ represents inverse differential exponential operator (more details can be found in [10]).

As already mentioned, the added inertia occurring due to the interaction of rigid wing with the fluid can be computed via BEM. The assumptions of irrotational and incompressible flow lead to

$$u = \nabla \psi \quad , \quad \Delta \psi = 0, \tag{5}$$

where u represents fluid velocity and ψ represents velocity potential.

On the other hand, kinetic energy of the surrounding fluid is equal to

$$T_{\mathcal{F}} = \frac{1}{2} \int_{\mathcal{F}} \rho_{\mathcal{F}} |u|^2 \, \mathrm{d}V, \tag{6}$$

where $\rho_{\mathcal{F}}$ represents fluid density. After applying Green's theorem and fluid assumptions (5), the kinetic energy expression can be rewritten as

$$T_{\mathcal{F}} = \frac{1}{2}\rho_{\mathcal{F}} \int_{\mathcal{F}} \nabla\psi \cdot \nabla\psi \, \mathrm{d}V = \frac{1}{2}\rho_{\mathcal{F}} \int_{\mathcal{B}} \psi \nabla\psi \cdot n \, \mathrm{d}s, \tag{7}$$

where n represents the unit normal pointing to the inside of the body. This leads to integration over the surface of body, instead of the conventional integration over the fluid domain. The values of $\nabla\psi \cdot n$ at the body surface are defined by the impenetrability boundary conditions, requiring that the fluid and body velocities in normal direction at the body surface have to be equal, leading to the following Neumann boundary conditions

$$\nabla\psi \cdot n = v \cdot n. \tag{8}$$

The values of velocity potential ψ at the surface of the body can then be computed by any boundary element method.

3 Viscosity Effects

Modeling of viscosity effects via irrotational point vortices (generating thus circulatory flow part of the external force vector $\boldsymbol{Q}_{\text{vort}}$, as introduced in (2)) is achieved by applying Kutta condition at the sharp trailing edge of thin airfoil, enforcing physical meaningfulness of the velocities in the vicinity of the sharp edge.

The circulatory flow field around moving body in an incompressible and irrotational fluid flow can be obtained by using unsteady potential flow methods. Since these methods are based on the continuity equation, which does not directly include unsteady terms, time dependency can be introduced through changing boundary conditions: zero normal fluid velocity on the surface of the body and Kelvin condition (which is direct consequence of the vorticity advection condition, i.e. $\mathrm{Diff}_{\mathrm{vol}}(\mathcal{F})$ invariance of the momentum map $J_{\mathcal{F}}$ introduced in (4)).

The wake behind the sharp edge of the body can be modeled by vortex distribution, but vortex shedding occurs only if circulation of the body varies in time. By applying Kutta condition for specifying shedding position of each wake vortex, a solution in the form of the velocity field can be obtained, which can be used to calculate pressure field around body by using the modified unsteady Bernoulli equation

$$\frac{p_{\mathrm{inf}} - p}{\rho} = \frac{1}{2}\left[\left(\frac{\partial \Psi}{\partial x}\right)^2 + \left(\frac{\partial \Psi}{\partial y}\right)^2 + \left(\frac{\partial \Psi}{\partial z}\right)^2\right] - v \cdot \nabla\Psi + \frac{\partial \Psi}{\partial t}, \tag{9}$$

where Ψ represents a fluid velocity potential, while v represents the body velocity.

The numerical solution used in this paper is based on 2D Unsteady Lumped-Vortex Element Method described in [4]. In the chosen numerical example, geometry of the thin airfoil is discretized with a set of vortex singularities placed along the airfoil chord. Vortices are also shed from the trailing edge (sharp edge) of the airfoil when the circulation changes with time, which allows the wake to be modeled by the same vortex model.

Each discretized airfoil subpanel has a vortex point placed at its quarter chord and collocation point at its three-quarter chord, where the zero normal flow boundary condition must be fulfilled. In order to find the solution of the problem, in addition to the prescribed boundary conditions, the kinematic conditions need to be known. In the tested numerical example, the kinematics is reconstructed from the prescribed movement of the thin airfoil. At each time-step new lumped-vortex element is shed from the trailing edge satisfying Kutta condition by being placed along the chord behind the trailing edge at the 0.3 of the distance covered by the trailing edge during previous time-step.

By solving a system consisting of vorticity influence coefficients and induced normal velocity components, intensity of each panel vortex and last shed vortex can be obtained. At the end of each time-step force free vortex wake rollup, induced by the current flow field, needs to be updated. A snapshot of wake vortices moving in fluid is shown in Fig. 3.

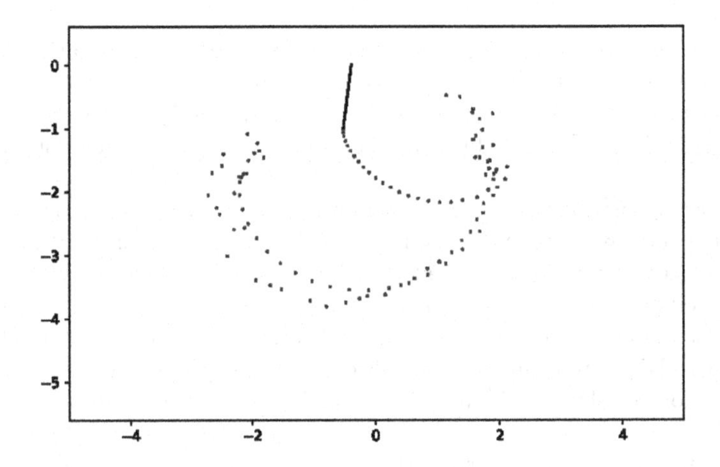

Fig. 3. Snapshot of wake vortices shed from the sharp trailing edge.

It is important to note that vorticity effects modeled as described in this chapter are used only for formulating circulatory flow part of the vector $\boldsymbol{Q}$ in (1), while the added mass effect introduced in $\boldsymbol{M}$ stays unchanged.

4 Concluding Remarks

The approach presented above enables modeling of the coupled wing-fluid system by discretizing wing boundary, instead of a conventional discretization of the whole fluid domain. When vorticity effects are not considered in the analysis (i.e. the wing moves in potential flow without circulation and no vorticies are being modeled), the formulation results in a coupled wing-fluid system without explicit fluid variables. In the case of flapping wing with circulation, which is usually characterized by having both leading and trailing sharp edge, the fluid variables are needed to describe vorticity effects (in the numerical example presented in this paper, however, only vortices shed from the trailing edge were presented). However, these variables include only the position and strength of each irrotational point vortex, excluding discretisation necessity of the whole fluid domain. This leads to the significantly improved computational efficiency when compared to conventional models and therefore this approach is well suited for design optimization and optimal control problems, involving flapping wing in the fluid flow.

References

1. Arnold, V.I., Khesin, B.A.: Topological Methods in Hydrodynamics. Springer, Heidelberg (1998). https://doi.org/10.1007/b97593
2. García-Naranjo, L.C., Vankerschaver, J.: Nonholonomic ll systems on central extensions and the hydrodynamic Chaplygin sleigh with circulation. J. Geom. Phys. **73**, 56–69 (2013). https://doi.org/10.1016/j.geomphys.2013.05.002
3. Kanso, E., Marsden, J.E., Rowley, C.W., Melli-Huber, J.B.: Locomotion of articulated bodies in a perfect fluid. J. Nonlinear Sci. **15**(4), 255–289 (2005). https://doi.org/10.1007/s00332-004-0650-9
4. Katz, J., Plotkin, A.: Low-Speed Aerodynamics. Cambridge Aerospace Series. Cambridge University Press, 2nd edn. (2001). https://doi.org/10.1017/CBO9780511810329
5. Leonard, N.E.: Stability of a bottom-heavy underwater vehicle. Automatica **33**(3), 331–346 (1997). https://doi.org/10.1016/S0005-1098(96)00176-8
6. Marsden, J., Misiolek, G., Ortega, J.P., Perlmutter, M., Ratiu, T.: Hamiltonian Reduction by Stages. Springer, Heidelberg (2007)
7. Munthe-Kaas, H.: Runge-Kutta methods on Lie groups. BIT Numer. Math. **38**(1), 92–111 (1998). https://doi.org/10.1007/BF02510919
8. Shashikanth, B.N., Marsden, J.E., Burdick, J.W., Kelly, S.D.: The Hamiltonian structure of a two-dimensional rigid circular cylinder interacting dynamically with n point vortices. Phys, Fluids **14**(3), 1214–1227 (2002). https://doi.org/10.1063/1.1445183
9. Terze, Z., Pandža, V., Andrić, Zlatar, D.: Computational dynamics of reduced coupled multibody-fluid system in lie group setting. In: Geometric Structures of Statistical Physics, Information Geometry, and Learning. Springer Proceedings in Mathematics & Statistics (2020)
10. Terze, Z., Müller, A., Zlatar, D.: Lie-group integration method for constrained multibody systems in state space. Multibody Syst. Dyn. **34**(3), 275–305 (2014). https://doi.org/10.1007/s11044-014-9439-2
11. Vankerschaver, J., Kanso, E., Marsden, J.E.: The geometry and dynamics of interacting rigid bodies and point vortices. J. Geom. Mech. **1** (2009). https://doi.org/10.3934/jgm.2009.1.223
12. Vankerschaver, J., Kanso, E., Marsden, J.E.: The dynamics of a rigid body in potential flow with circulation. Regular Chaotic Dyn. **15**, 606–629 (2010). https://doi.org/10.1134/S1560354710040143

Statistical Manifold and Hessian Information Geometry

Canonical Foliations of Statistical Manifolds with Hyperbolic Compact Leaves

Michel Boyom[1], Emmanuel Gnandi[2], and Stéphane Puechmorel[2](✉)

[1] Institut Alexandre Grothendieck, IMAG-UMR CNRS 5149-c.c.051, Université de Montpellier, Montpellier, France
boyom@math.univ-montp2.fr
[2] Ecole Nationale de l'Aviation Civile, Université de Toulouse, Toulouse, France
stephane.puechmorel@enac.fr

Abstract. The sheaf of solutions $\mathcal{J}_\nabla$ of the Hessian equation on a gauge structure (M, ∇) is a key ingredient for understanding important properties from the cohomological point of view. In this work, a canonical representation of the group associated by Lie third's theorem to the Lie algebra formed by the sections of $\mathcal{J}_\nabla$ is introduced. On the foliation it defines, a characterization of compact hyperbolic leaves is then obtained.

1 Introduction

Hyperbolicity is quite an important notion in geometry and can be tackled using different approaches. In its original work [6], Koszul introduces the development map $\mathcal{Q}$ to define hyperbolic manifold as those for which the image by $\mathcal{Q}$ of their universal coverings is an open convex domain without straight line. In [7], a cohomological characterization is given, stating that a necessary condition for a gauge structure (M, ∇) to be hyperbolic is the existence of a closed 1-form admitting a positive definite covariant derivative. Finally, if (M, g, ∇) is a gauge structure on an Hessian manifold, then (M, ∇) is hyperbolic iff $[g] = 0$ in Koszul-Vinberg cohomology [1]. The aim of this paper is to study hyperbolicity of the leaves of canonical foliations in statistical manifolds using representations in the affine group of a finite dimensional vector space. The paper is organized as follows: in Sect. 2, the Hessian equation on a gauge structure is introduced along with some basic facts. In Sect. 3, the canonical group representation is defined, from which a vanishing condition in cohomology is given in Theorem 32. Finally, the case of statistical manifolds is treated in Sect. 4 and a characterization of hyperbolic compact leaves is given.

2 Hessian Equation on Gauge Structures

Most of the material presented here can be found in greater detail in [1]. Only the required notions will be introduce in this section. In the sequel, $\chi(M)$ stands for

F. Nielsen and F. Barbaresco (Eds.): GSI 2021, LNCS 12829, pp. 371–379, 2021.
https://doi.org/10.1007/978-3-030-80209-7_41

the real Lie algebra of smooth vector fields on M. The convention of summation on repeated indices will be used thorough the document.

Definition 21. *A gauge structure is a couple (M, ∇) where M is a smooth manifold and ∇ a Koszul connection on TM.*

For a given gauge structure, the Koszul-Vinberg complex is defined as follows:

Definition 22. *Let $C_q(\nabla), q \geq 0$ be the vector spaces:*

$$\begin{cases} C^0(\nabla) = \{f \in C^\infty(M), \, \nabla^2 f = 0\} \\ C^q(\nabla) = Hom_{\mathbb{R}}(\otimes^q \chi(M), C^\infty(M)) \end{cases}$$

and let $\delta \colon C^q(M) \to C^{q+1}(M)$ be the coboundary operator:

$$\begin{cases} \delta f = df \, \forall f \in C^0(\nabla) \\ \delta f(X_1 \otimes ... \otimes X_{q+1}) = \Sigma_1^q (-1)^i [d(f(.. \otimes \hat{X}_i \otimes .. \otimes X_{q+1}))(X_i) \\ \qquad\qquad\qquad\qquad - \Sigma_{j \neq i} f(.. \otimes \hat{X}_i \otimes .. \otimes \nabla_{X_i} X_j \otimes ..] \end{cases}$$

The complex $(C(\nabla), \delta)$ is the Koszul-Vinberg complex of (M, ∇), with cohomology groups denoted by $H^q_{KV}(\nabla)$.

Definition 23. *An element g in H^2_{KV} is called a Hessian (resp. Hessian non degenerate, Hessian positive definite) class if it contains a symmetric (resp symmetric non degenerate, symmetric positive definite) cocycle.*

A pair (M, ∇) is locally flat if its curvature tensor vanishes. The next proposition relates cohomology to a metric property. used thorough the document.

Proposition 21 ([8])**.** *A locally flat manifold (M, g, ∇) is Hessian in the sense of [9] if δg vanishes identically.*

The next theorem relates hyperbolicity to KV cohomology. used thorough the document.

Theorem 24. *Let (M, g, ∇) be a compact Hessian manifold then the following assertions are equivalent:*

- $[g] = 0 \in H^2_{KV}(\nabla)$.
- *(M, ∇) is hyperbolic.*

Definition 25. *The Hessian operator $\nabla^2 \colon \chi(M) \to T^1_2(M)$ is, for a fixed $Z \in \chi(M)$, the covariant derivative of the $T^1_1(M)$-tensor ∇Z. For any triple (X, Y, Z) of vector fields, its expression is given by:*

$$\nabla^2_{X,Y} Z = \nabla_X(\nabla_Y Z) - \nabla_{\nabla_X Y} Z.$$

Proposition 22. *The product* $(X, Y) \in \chi(M) \mapsto \nabla_X Y$ *has associator* ∇^2.

Proof. This a direct consequence of Definition 25, since:

$$\nabla^2_{X,Y} Z = X.(Y.Z) - (X.Y).Z$$

Let $(x_1, .., x_m)$ be a system of local coordinate functions of M and let $X \in \mathcal{X}(M)$. We set

$$\partial i = \frac{\partial}{\partial x_i},$$

$$X = \sum X^k \partial k$$

$$\nabla_{\partial i} \partial j = \Sigma_k \Gamma^k_{i.j}$$

The principal of symbol of the Hessian differential operator can be expressed as

$$(\nabla^2 X)(\partial i, \partial j) = \Sigma_k \Omega^k_{i,j} \partial k \tag{2.1}$$

where

$$\begin{aligned} \Omega^l_{ij} &= \frac{\partial^2 X^l}{\partial x_i \partial x_j} + \Gamma^l_{ik} \frac{\partial X^k}{\partial x_j} + \Gamma^l_{jk} \frac{\partial X^k}{\partial x_i} - \Gamma^k_{ij} \frac{\partial X^l}{\partial x_k} \\ &+ \frac{\partial \Gamma^l_{jk}}{\partial x_i} + \Gamma^m_{jk} \Gamma^k_{im} - \Gamma^m_{ij} \Gamma^l_{mk} \end{aligned} \tag{2.2}$$

Definition 26 ([2]). *Let* (M, ∇) *be a gauge structure. The sheaf of solutions of its Hessian equation, denoted by* $\mathcal{J}_\nabla(M)$, *is the sheaf of associative algebras:*

$$U \mapsto \{X \in \chi(U), \nabla^2 X = 0\}$$

with product defined in Proposition 22.

The space of sections of $\mathcal{J}_\nabla(M)$ will be denoted by J_∇.

Proposition 23. *The pair* (J_∇, ∇) *is an associative algebra with commutator Lie algebra* $(J_\nabla, [-,-]_\nabla)$ *where the bracket* $[X, Y]_\nabla$ *is:*

$$[X, Y]_\nabla = \nabla_X Y - \nabla_Y X.$$

When ∇ has vanishing torsion, (M, ∇) is said to be symmetric and the Lie algebra $(J_\nabla, [-,-]_\nabla)$ is obviously Lie subalgebra of the Lie algebra of vector fields

$$(\mathcal{X}(M), [-,-]);$$

Proposition 24 ([2]). *If (M, ∇) is symmetric then the Lie subalgebra $J_\nabla \subset \mathcal{X}(M)$ is finite-dimensional over the field of real numbers.*

The next proposition shows that J_∇ may be trivial.

Proposition 25. *Let (M, g) be a Riemannian manifold and let Ric be its Ricci curvature tensor. Then $J_\nabla \subset Ker(Ric)$*

Proof. Let (X, Y) be a couple of vector fields and let $\xi \in J_\nabla$. It comes:

$$R^\nabla(Y, X)\xi = (\nabla^2\xi)(X, Y) - (\nabla^2\xi)(Y, X). \tag{2.3}$$

Taking the trace we deduce that:

$$Ric(X, \xi) = 0 \tag{2.4}$$

Recalling that a Riemannian manifold (M, g) is called Einstein if:

$$\mathrm{Ric} = \lambda g$$

for some constant λ, the next proposition is a consequence of the previous result.

Corollary 21. *If ∇^{LC} is the Levi-Civita connection of an Einstein Riemannian manifold, then $J_{\nabla^{LC}} = \{0\}$.*

As an easy consequence of Lie's third theorem, it comes:

Theorem 27. *Up to a Lie group isomorphism, it exists a unique simply connected Lie group G_∇ whose Lie algebra is isomorphic to the Lie algebra J_∇.*

3 A Canonical Representation of G_∇

Definition 31. *W be a finite dimensional real vector space. Its group of affine isomorphisms, denoted by Aff(W), is defined as the semi-direct product:*

$$Aff(W) = GL(W) \ltimes W$$

where W is by abuse of notation the group of translations of W.

There is a natural affine representation of the Lie algebra J_∇ in itself as a vector space:

$$J_\nabla \ni X \to \rho(X) = (\nabla_X, X) \in gl(J_\nabla) \times J_\nabla = aff(J_\nabla).$$

with affine action given by:

$$\rho(X).Y = \nabla_X Y + X, \quad \forall Y \in J_\nabla.$$

By virtue of the universal property of simply connected finite dimensional Lie groups, there exist a unique continuous affine representation

$$\gamma \in G_\nabla \to (f_\nabla(\gamma), q_\nabla(\gamma)) \in Aff(J_\nabla).$$

Proposition 31. *With the above notations, f_∇ is a linear representation of G_∇ in J_∇ and q_∇ is a J_∇ valued 1-cocycle of f.*

Proof. The couple (f, q) is a continuous homomorphism of the Lie group G_∇ on the Lie group $Aff(J_\nabla)$, thus, for any $\gamma_1, \gamma_2 \in G_\nabla$:

$$(f(\gamma_1).f(\gamma_2), f(\gamma_1)q(\gamma_2) + q(\gamma_1)) = (f(\gamma_1.\gamma_2), q(\gamma_1.\gamma_2)).$$

The cohomology of the Lie group G_∇ value in his Lie algebra J_∇ is defined by the complex [3–5]:

$$... \to C^q(G_\nabla, J_\nabla) \to C^{q+1}(G_\nabla, J_\nabla) \to C^{q+2}(G_\nabla, J_\nabla) \to$$

with differential operator D:

$$\begin{aligned} D\theta(\gamma_1, ..., \gamma_{q+1}) &= f_\nabla(\gamma_1).\theta(\gamma_2, ..., \gamma_{q+1}) \\ &\quad + \sum_{i \leqslant q} (-1)^i \theta(.., \gamma_i\gamma_{i+1}, ...) + (-1)^q \theta(\gamma_1, ..., \gamma_q) \end{aligned}$$

The condition:

$$q_\nabla(\gamma_1.\gamma_2) = f_\nabla(\gamma_1)q_\nabla(\gamma_2) + q_\nabla(\gamma_1)$$
$$\forall \gamma_1, \gamma_2 \in G_\nabla$$

is equivalent to $q_\nabla \in Z^1(G_\nabla, J_\nabla)$. The next definition thus makes sens: The cohomology class $[q_\nabla] \in H^1(G_\nabla, J_\nabla)$ is called the radiant class of the affine representation (f, q). We are now in position to state one of the main results of the article:

Theorem 32. *The following statements are equivalent:*

1. *The affine action*

$$G_\nabla \ltimes J_\nabla \ni (\gamma, X) \to f_\nabla(\gamma).X + q_\nabla(\gamma) \in J_\nabla$$

   has a fixed point;

2. *The cohomology class $[q_\nabla]$ vanishes;*
3. *The affine representation*

$$G_\nabla \ni \gamma \to (f_\nabla(\gamma), q_\nabla(\gamma) \in Aff(J_\nabla)$$

   is conjugated to the linear representation

$$G_\nabla \ni \gamma \to f_\nabla(\gamma) \in GL(J_\nabla).$$

Proof. Let us first show that (1) implies (2). Let $-Y_0$ be a fixed point of the affine action (f_∇, q_∇), then

$$f_\nabla(\gamma)(-Y_0) + q_\nabla(\gamma) = -Y_0, \quad \forall \gamma \in G_\nabla.$$

Therefore one has:

$$q_\nabla(\gamma) = f_\nabla(\gamma)(Y_0) - Y_0, \quad \forall \gamma \in G_\nabla.$$

So the cocycle q is exact. To prove that (2) implies (3), consider the affine isomorphism (e, Y_0). It is nothing but the translation by Y_0

$$X \to X + Y_0;$$

We calculate

$$(e, Y_0)\dot{(}f_\nabla(\gamma), q_\nabla(\gamma))\dot{(}e, Y_0)^{-1} = (f_\nabla(\gamma), 0_\nabla)$$

where O_∇ stands for the zero element of the vector space J_∇. Finally, (3) implies (1). This assertion means that there exists an affine isomorphism

$$J_\nabla \ni Y \to L(Y) + X_0 \in J_\nabla$$

such that

$$(L, X_0)\dot{(}f_\nabla(\gamma), q_\nabla(\gamma))\dot{(}L, X_0)^{-1} = (f_\nabla(\gamma), 0_\nabla), \quad \forall \gamma \in G_\nabla$$

The calculation of the left member yields the following identities:

$$(a): \quad L\dot{f}_\nabla(\gamma)\dot{L} = f_\nabla(\gamma) \tag{3.1}$$

$$(b): \quad L(q_\nabla(\gamma)) + X_0 - [L\dot{(}f_\nabla(\gamma)\dot{L}^{-1}](X_0) = 0_\nabla. \tag{3.2}$$

The identity (b) yields

$$q_\nabla(\gamma) = f_\nabla(\gamma)(L^{-1}(X_0)) - L^{-1}(X_0), \quad \forall \gamma \in G_\nabla.$$

Taking into account identity (a), we obtain:

$$q_\nabla(\gamma) = f_\nabla(\gamma)(X_0) - X_0, \quad \forall \gamma \in G_\nabla.$$

So the vector $-X_0$ is a fixed point of the affine representation (f_∇, q_∇).

Definition 33. *The affine representation (f_∇, q_∇) is called the canonical affine representation of the gauge structure (M, ∇).*

Remark 1. When the infinitesimal action J_∇ is integrable, the proposition above is a key tool to relate the canonical affine representation of (M, ∇) and the hyperbolicity problem for the orbits of G_∇.

4 Application to Statistical Manifolds

Let (M, g, ∇, ∇^*) be a statistical manifold and ∇^{LC} be its Levi-Civita connection. From now, we assume that either both (M, ∇), (M, ∇^*) and (M, ∇^{LC}) are geodesically complete or M is compact. Therefore J_∇, J_{∇^*} and $J_{\nabla^{LC}}$ are the infinitesimal counterpart of the following locally effective differentiable dynamical systems:

$$G_\nabla \times M \to M \tag{4.1}$$

$$G_{\nabla^*} \times M \to M \tag{4.2}$$

$$G_{\nabla^{LC}} \times M \to M \tag{4.3}$$

Notation 41. *$\mathcal{F}_\nabla$ is the foliation whose leaves are orbits of the Lie group G_∇ and g_∇ the restriction of g to $\mathcal{F}_\nabla$.*

Theorem 42. *Let (M, g, ∇, ∇^*) be a compact statistical manifold. Then the foliation $\mathcal{F}_\nabla$ (resp. $\mathcal{F}_{\nabla^*}$) is a Hessian foliation in (M, g, ∇) (resp. (M, g, ∇^*)).*

Theorem 43. *In a statistical manifold (M, g, ∇, ∇^*) the following assertions are equivalent:*

1. $[g_\nabla] = 0 \in H^2_{KV}(\mathcal{F}_\nabla)$.
2. $[q_{\nabla^*}] = 0 \in H^1(G_{\nabla^*}, J_{\nabla^*})$.
3. *$(f_{\nabla^*}, q_{\nabla^*})$ has a fixed point.*
4. *$(f_{\nabla^*}, q_{\nabla^*})$ is affinely conjugated to its linear component f_{∇^*}.*

Proof. According to Theorem 32, assertions (2), (3) and (4) are equivalent. Therefore, it is sufficient to prove that assertions (1) and (2) are equivalent.

Let us demonstrate first that (1) implies (2).

Since the class $[g_\nabla]$ vanishes, it exist a de Rham closed differential 1-form θ such that:

$$g_\nabla(X, Y) = -X\theta(Y) + \theta(\nabla_X Y), \quad \forall (X, Y).$$

By the defining property of statistical manifolds, it comes:

$$Xg_\nabla(Y, Z) = g_\nabla(\nabla^*_X Y, Z) + g_\nabla(Y, \nabla_X Z), \quad \forall (X, Y, Z).$$

Let H be the unique vector field satisfying:

$$\theta(X) = g_\nabla(H, X), \quad \forall X.$$

Using once again the defining property of statistical manifolds, we get:

$$Xg_\nabla(H, Y) - g_\nabla(\nabla^*_X H, Y) - g_\nabla(H, \nabla_X Y) = 0.$$

Since left hand member is $C^\infty(M)$-multilinear we can assume that $H \in J_\nabla$ and we get the following identity

$$g_\nabla(\nabla^\star_X H, Y) = Xg_\nabla(H,Y) - g_\nabla(H, \nabla_X Y) = g_\nabla(X,Y), \quad \forall (X,Y) \subset \mathcal{X}(M).$$

Thus one has:

$$\nabla^\star_X(-H) - X = 0_\nabla, \quad \forall X \in \mathcal{X}(M).$$

So the $-H$ is a fixed point of $(f_{\nabla^\star}, q_{\nabla^\star})$.

Let us demonstrate now that (2) implies (1).

Let us assume that $(f_{\nabla^\star}, q_{\nabla^\star})$ has a fixed point $Y_0 \in J_{\nabla^\star}$. Then:

$$f_{\nabla^\star}(\gamma)(Y_0) + q_{\nabla^\star}(\gamma) = Y_0, \quad \forall \gamma \in G_{\nabla^\star}.$$

To every $X \in J_{\nabla^\star}$, we assign the one parameter subgroup

$$\{Exp(tX), t \in \mathbb{R}\} \subset G_{\nabla^\star}.$$

We have:

$$f_{\nabla^\star}(Exp(tX))(Y_0) + q_{\nabla^\star}(Exp(tX)) = Y_0, \quad \forall t \in \mathbb{R}.$$

Calculating the derivative at $t = 0$, one obtains:

$$\nabla^\star_X Y_0 + X = 0.$$

Finally:

$$Xg_\nabla(Y_0, Y) = g_\nabla(\nabla^\star_X Y_0, Y) + g_\nabla(Y_0, \nabla_X Y)$$

Using $\nabla^\star_X Y_0 = -X, \forall X$, one obtains the following identity:

$$Xg_\nabla(Y_0, Y) = -g_\nabla(X,Y) + g_\nabla(Y_0, \nabla_X Y).$$

By putting $\theta(Y) = -g_\nabla(Y_0, Y)$, it comes:

$$g_\nabla(X,Y) = X\theta(Y) - \theta(\nabla_X Y), \quad \forall (X,Y)$$

Then one has $[g_\nabla] = 0 \in H^2_{KV}(\nabla)$, concluding the proof.

Corollary 41. *Let $(M, g, \nabla, \nabla^\star)$ such that both J_∇ and $J_{\nabla^\star}$ be integrable. We assume that leaves of both $\mathcal{F}_\nabla$ and $\mathcal{F}_{\nabla^\star}$ satisfy one among the assertions of Theorem 43. Then every compact leaf of $\mathcal{F}_\nabla$ (resp. $\mathcal{F}_{\nabla^\star}$) is hyperbolic.*

5 Conclusion and Future Works

The characterization of hyperbolicity given in Theorem 43 has an important application in elucidating the structure of some statistical manifolds as statistical models. In a future work, an explicit construction of an exponential family on leaves of a statistical manifold will be given along with an application to the classification of compact exponential models in dimension 4.

References

1. Boyom, M.N.: Foliations-webs-Hessian geometry-information geometry-entropy and cohomology. Entropy **18**(12), 433 (2016)
2. Boyom, M.N.: Numerical properties of Koszul connections. arXiv preprint arXiv:1708.01106 (2017)
3. Boyom, N.B., et al.: Sur les structures affines homotopes à zéro des groupes de lie. J. Differ. Geom. **31**(3), 859–911 (1990)
4. Cartan, H., Eilenberg, S.: Homological Algebra, vol. 41. Princeton University Press, Princeton (1999)
5. Chevalley, C., Eilenberg, S.: Cohomology theory of lie groups and lie algebras. Trans. Am. Math. Soc. **63**(1), 85–124 (1948)
6. Koszul, J.L.: Domaines bornés homogenes et orbites de groupes de transformations affines. Bull. Soc. Math. France **89**, 515–533 (1961)
7. Koszul, J.L.: Déformations de connexions localement plates. Annales de l'institut Fourier **18**(1), 103–114 (1968)
8. Nguiffo Boyom, M.: The cohomology of Koszul-Vinberg algebras. Pac. J. Math. **225**(1), 119–153 (2006)
9. Shima, H., Yagi, K.: Geometry of Hessian manifolds. Differential Geom. Appl. **7**(3), 277–290 (1997)

Open Problems in Global Analysis. Structured Foliations and the Information Geometry

Michel Boyom(✉)

Institut Alexandre Grothendieck, IMAG-UMR CNRS 5149-c.c.051, Université de Montpellier, Montpellier, France
boyom@math.univ-montp2.fr

Abstract. The global analysis deals with the algebraic topology (e.g. the formalism of Spencer) and the Sternberg geometry (e.g. the formalism of Guillemin-Sternberg) of differential equations and their formal solutions. An immediate constant is the impacts of the global analysis on the quantitative global differential geometry. In this area the question of the existence of certain geometric structures is still open for lack of creteria of the integrability of the equations which define them. The purpose of this talk is to survey recent advances on some of these open questions. These advances basically concern the geometry of Koszul and the symplectic geometry. The methods of the information geometry have inspired the introduction of new numerical invariants and new homological invariants. These new invariants have informed innovative approaches to those old open problems.

1 Notation and Basic Notions

Throughout this talk the following notation is used.
$\mathbb{Z}$ is the ring of integers.
$\mathbb{R}$ is the field of real numbers.
$\mathbb{C}$ is the field of complex numbers.
Topological spaces are assumed to be connected, Hausdorff and paracompact.
A pair (x, X) is formed of a differentiable manifold X and $x \in X$.

1.1 Differential Structure, Versus Parameterization

Definition 11. *Given a pair (x, X) and a non negative integer n, an n-dimensional parameterization at x is a local homeomorphism of $(O, \mathbb{R}^n)$ in (x, X)*♣.

F. Nielsen and F. Barbaresco (Eds.): GSI 2021, LNCS 12829, pp. 380–388, 2021.
https://doi.org/10.1007/978-3-030-80209-7_42

1.2 C^∞ Relation

Definition 12. *Two n-dimensional parameterizations at x, namely ϕ and $\phi^\star$ are C^∞-related if $\phi^{-1} \circ \phi^\star$ is a local C^∞ diffeomorphism of $(O, \mathbb{R}^n)$♣*

Definition 13. *An n-dimensional parameterization of X is a datum*

$$\mathcal{P}(X) = \cup_{x \in X} \mathcal{P}_x(X)$$

where $\mathcal{P}_x(M)$ is a family of n-dimensional parameterizations at x which are pairwise C^∞ related ♣.

Given a n-dimensional parameterization $\mathcal{P}(X)$ and $x \in X$, an n-dimensional parameterization at x ϕ is C^∞ related with $\mathcal{P}(X)$ if

$$\mathcal{P}(X) \cup \{\phi\}$$

is an C^∞ parameterization of X.

This yields the notion of maximal C^∞ n-dimensional parameterization of X. More precisely, a C^∞ n-dimensional parameterization of X, $\mathcal{P}(X)$ is called maximal (or complete) if for all $x \in X$ every n-dimensional parameterization at x which is C^∞ related with $\mathcal{P}(X)$ belongs to $\mathcal{P}(X)$.

Definition 14. *A set E is called a torsor for a group G if G acts simply transitively on E♣.*

Definition 15. *A maximal n-dimensional C^∞ parameterization*

$$\mathcal{P}(X) = \cup_x \mathcal{P}_x(X)$$

is geometrically closed if it satisfies the following requirements:

(1) $\mathcal{P}_x(X)$ is a torsor of a subgroup $\mathcal{G}(x)$ of local C^∞ diffeomorphisms of $(0, \mathcal{R}^n)$.
(2) $\mathcal{G}(x) = \mathcal{G}(x^\star)$ $\forall (x, x^\star) \subset X$.

Definition 16. *Put $\mathcal{G} = \mathcal{G}(x)$, the couple $(X, \mathcal{P}(M))$ is called a C^∞ $\mathcal{G}$-structure in X♣.*

Henceforth differentiable always means of class C^∞.

1.3 Geometric Structure, Versus Topology of Fibrations

The Notion of k-jet, Functional Version. The set of all differentiable mappings of $\mathbb{R}^n$ in itself is denoted by $C^\infty(\mathbb{R}^n, \mathbb{R}^n)$.

Given $(f, F) \in C^\infty(\mathbb{R}^n) \times C^\infty(\mathbb{R}^n, \mathbb{R}^n)$ the mapping $f.F$ is defined as

$$f.F(y) = f(f(y))F(y)).$$

Thus the vector space $C^\infty(\mathbb{R}^n, \mathbb{R}^n)$ is a left algebra of $C^\infty(\mathbb{R}^n)$ whose addition is defined as

$$(F + F*)(y) = F(y) + F*(y);$$

and the product is defined as

$$(F \circ F*)(y) = F(F*(y)).$$

Let $I_0(\mathbb{R}^n) \subset C^\infty(\mathbb{R}^n)$ be the ideal formed of functions which vanish at $0 \in \mathbb{R}^n$. For every non negative integer k, one consider the quotient vector space

$$(NJ): \quad J_0^k(\mathbb{R}^n) = \frac{C^\infty(\mathbb{R}^n, \mathbb{R}^n)}{I_0^{k+1}(\mathbb{R}^n).C^\infty(\mathbb{R}^n, \mathbb{R}^n)}.$$

This quotient vector space is called the space of k-jets of differentiable mappings of $\mathbb{R}^n$ in itself. The canonical projection of $C^\infty(\mathbb{R}^n, \mathbb{R}^n)$ on $J_0^k(\mathbb{R}^n)$ is denoted by j_0^k.

Given a subgroup $\mathcal{G} \subset Diff(O, \mathbb{R}^n)$ and a differentiable geometrically complete $\mathcal{G}$-structure $(X, \mathcal{P}(X))$, one defines

$$J_0^1\mathcal{P}_x(X) = \{d\phi(0) : \mathbb{R}^n \to T_xX, \quad \forall \phi \in \mathcal{P}_x(X)\},$$
$$J_0^1\mathcal{P}(X) = \cup_{x \in X} J_0^1\mathcal{P}_x(X)$$

and

$$G = \{d\gamma(0), \quad \forall \gamma \in \mathcal{G}\}.$$

Of course one has

$$G = j_0^1(\mathcal{G}).$$

It is clear that $J_0^1\mathcal{P}_x(X)$ is a **torsor for the linear group** G.

The Notion of Jet, Calculus Version. By the abuse of notation the k-th Taylor expansion at 0 of $\phi \in \mathcal{P}_x(X)$ is denoted by $j_0^k\phi$. Thus we set

$$J_0^k\mathcal{P}_x(X) = \left\{j_0^k\phi : \mathbb{R}^n \to T_xX, \quad \phi \in \mathcal{P}_x(X)\right\};$$
$$G^k = \left\{j_0^k\gamma, \quad \gamma \in \mathcal{G}\right\}.$$

As said above $j_0^k\phi$ stands for the Taylor expansion of ϕ up to order k; it is a polynomial isomorphism of $\mathbb{R}^n$ in T_xX.

At the level k, for all $x \in X$ $J_0^k\mathcal{P}(x)$ **is a torsor for the group** G^k.

Put

$$J^k\mathcal{P}(X) = \cup_{x \in X} \left\{J_0^k\mathcal{P}_x(X)\right\}.$$

Definition 17. *Elements of $J^k\mathcal{P}(X)$ are called $k-th$ order frames of X.*

It is easy to check that $J^k\mathcal{P}(X)$ is a principal G^k-bundle over X.

1.4 Geometric Structure Versus Global Analysis: Equations of $(X, \mathcal{P}(X))$

It is easy to see that elements of $\mathcal{P}(X)$ are solutions of the following differential equation

$$\mathcal{E}: \quad j_0^1(\phi^{-1} \circ \phi^\star) \in G.$$

Roughly speaking $\mathcal{E}$ is the equation of the $\mathcal{G}$-structure $(X, \mathcal{P}(X))$.

A Few Examples. (1) If G is the orthogonal group $O(n)$ then the structure $(X, \mathcal{P}(X))$ is a flat Riemannian structure in X.
(2) If G is the linear symplectic group then $(X, \mathcal{P}(X))$ is a symplectic structure in X.
(3) If G is the real representation in $GL(2m, \mathbb{R})$ of the complex linear group $GL(m, \mathbb{C})$ then $(X, \mathcal{P}(X))$ is a complex analytic structure in X. According to Jean-Pierre Serre that is the same as a structure of complex algebraic variety, see GAGA.

Regarding the notions just mentioned, the aim is to introduce non specialists to two of the classical approaches to the quantitative global differential geometry, viz the theory of **Geometric structures.**

Let us focus on the viewpoint of the **global analysis which** is regarded as **the study of Geometry and Topology of Differential Equations and of their Formal Solutions.**

1.5 Structured Foliations

Definition 18. *A structured foliation is a foliation whose leaves carry (uniformly) a prescribed* ***mathematical structure*** *S, i.e. S-foliation* ♣.

Examples. (1) The Euclidean space $\mathbb{R}^3$ is regarded as the Cartesian as

$$\mathbb{R}^3 = \cup_{y \in \mathbb{R}} \{y\} \times \mathbb{C}.$$

Thus it carries a foliation whose leaves are Kaehlerian manifolds.
(2) Let ω be a nongenerate differential 2-form in a manifold M. Then the couple

$$(TM, \omega) = \cup_{x \in M} (T_x M, \omega(x)).$$

Thus TM carries a foliation whose leaves are symplectic vector spaces.
Generically a differentiable manifolds is (rich in) foliations. The reasons is the richness in statistical structures. For instance, in a differential manifold M every couple (E, g) formed of a Riemannian metric tensor g whose Levi Civita connection is ∇^{LC}, and a vector field E defines the statistical structure (M, g, ∇^E) where the connection ∇^E is defined as

$$\nabla^E_X Y = \nabla^{LC}_X Y + g(X, E)g(Y, E)E.$$

The question of whether a manifold M admits a $\mathcal{S}$ foliation is denoted by $EXF(\mathcal{S} > M)$, see [Nguiffo Boyom (1)], and [Nguiffo Boyom (2)]. The problem $EX(\mathcal{S} > M)$ is among the difficult problems in the differential topology, see Ghys.

1.6 Open Problems

A fundamental qualitative-quantitative question in the global analysis is the question of whether a given manifold M admits a prescribed Geometric structure $\mathcal{S}$. This problem is denoted by EX($\mathcal{S}$>M).
From the viewpoint of the applied information geometry, the question is whether a given complex system (or given Big Data) admits a prescribed geometry structure or a prescribed topological structure.

(1) **A fundamental problem in the structured differential topology is the question of whether a manifold M admits a structured $\mathcal{S}$-foliation.**

(2) *WARNING. The existence of structured foliations is highly relevant in the Applied Information Geometry. In an abstract situation, (viz in discrete topology,) Pool Samples are nothing but Leaves of Coarse Foliations.*
Therefore from the machine learning viewpoint the question is whether given poll samples can be regarded a Discretization of a Prescribed structured foliation.

2 Relevant Geometric Structures Which Impact the Information Geometry

In the precedent section two fundamental problems are raised.

The first problem is a question in the quantitative global analysis; the question of existence.

The second problem is a question in the qualitative differential topology.

Up to nowadays these problems are still open if $\mathcal{S}$ is one of the structures which are listed below.

Locally flat structure in a manifold M
[Benzecri, Goldman, Koszul, Vinberg, Auslander, Sullivan-Thurston, Milnor, Margulis and others].

Smplectic structure in a manifold M
[Weinstein, Libermann-Marle, Lichnerowicz, Gompf, Hajduk, Banyaga and others].

Almost complex structure in a manifold M
[Bryant, Gromov and others].

Hessian structure in a Riemannian manifold (M, g)
[Koszul, Vinberg, Matsushima, Guts, Amari-Armonstrong and others].

Hessian structure in a locally flat manifold (M, ∇)
[Kaup, Vey, Vinberg, Shima and others].

Left invariant locally flat structure in a finite dimensional Lie G
[Goldman, Milnor, Matsushima, Katsumi, Fired-Goldman-Hirsh and others].

Left invariant symplectic structure in a finite dimensional Lie group G
[Lichnerowicz, Arnold, Marsden-Weinstein and others].

Left invariant Hessian structure in a finite dimensional Lie group G
[Matsushima, Katsumi and others].

Two-sided invariant Riemannian structure in a finite dimensional Lie Group G
[Ghanam-Hondeleh-Thompson and others].

Structured foliations in a finite dimensional manifolds
[Molino, Reihnardt, Ghys and others].

3 The Aim of the Talk

The present talk is devoted to survey recent advances on these issues.

3.1 Long-Awaited Characteristic Obstructions

Concerning the conceptual tools which are mentioned in the list above, it is useful to underline that their use guarantees the relevance of choice of models and ad hoc algorithms.

Definition 31. *A characteristic obstruction is an invariant which gives a necessary and sufficient condition.*

Regarding the open problems listed in the latter subsection above, **two families of long time expected geometric invariants are introduced.**

(1) The first family is formed of **new numerical geometric invariants.**

(2) The second family is formed of **new homological invariants**.

Regarding the notion of **structured foliation** the new numerical invariants are interpreted as codimensions of **optimal structured foliations**.

By the way these new invariants yield the complete solutions of the fundamental question raised in the structured differential topology:

How to Construct all Riemannian foliations, (Etienne Ghys).

The new homological invariants and new numerical invariants have the mathematical meaning of being characteristic obstructions to solve these quantitative open problems

Ideas of Constructions of New Invariants. *To introduce new powerful invariants our arsenal is the category of triples (M, g, ∇) formed of a (pseudo) Riemannian structure (M, g) and a gauge structure (M, ∇). Below is the sketch of our strategy.*

With a triple (M, g, ∇) is associated the following data:
(D1) the Koszul connection $\nabla^\star$ defined by

$$g(\nabla^\star_X Y, Z) = Xg(Y, Z) - g(Y, \nabla_X Z);$$

(D2) the $T_1^2(M)$-valued first order differential operator defined in $T^\star M \otimes TM$ as it follows,

$$D^{\nabla,g}\phi = \nabla^\star \otimes \phi - \phi \otimes \nabla.$$

More explicitly given two vector fields X and Y one has

$$D^{\nabla,g}\phi(X,Y) = \nabla^\star_X\phi(Y) - \phi(\nabla_X Y).$$

The sheaf of ∇-invariant differential 2-forms is denoted by $\Omega_2^\nabla(M)$; the sheaf of ∇-invariant symmetric 2-forms is denoted by $S_2^\nabla(M)$. What we call gauge equation is the differential equation

$$D^{\nabla,g}\phi = 0.$$

The sheaf of solutions of the gauge equation is denoted by $\mathcal{J}_{\nabla,g}(M)$. The constructions of new numerical invariants and new homological invariants involve the following (splitting) short exact sequences of sheaves,

$$0 \to \Omega_2^\nabla(M) \to \mathcal{J}_{\nabla,g}(M) \to S_2^\nabla(M) \to 0.$$

Every section of $\mathcal{J}_{\nabla,g}(M)$ gives rise to two other sections of $\mathcal{J}_{\nabla,g}(M)$, Φ and $\Phi^\star$. At one side interests are carried by the following family of numerical invariants

$$s^B(\nabla,g) = rank(Ker(\Phi)),$$
$$s^{\star B}(\nabla,g) = rank(Ker(\Phi^\star)).$$

At another side $A^\star(\nabla,g)$ is the Lie subalgebra of vector fields which is spanned by the sections of the distribution $Ker(\Phi^\star)$.
Then other interests are carried by the family of canonical Koszul classes

$$[k_\infty(\nabla,g)] \in H^1_{CE}(A^\star(\nabla,g), T_1^2(M))$$

which represented by any Chevalley-Eilenberg cocycle defined as

$$A^\star(\nabla,g) \ni \xi \to L_\xi D \in T_1^2(M)$$

where D is an arbitrary Koszul connection in TM and $L_\xi D$ is the Lie derivative of D is the direction ξ.

Details of these constructions will appear elsewhere, [Nguiffo Boyom (2)].

References

1. Koszul, J.-L.: Variétés localement plates et hyperbolicité. Osaka. Math. **2**, 285–290 (1968)
2. Koszul, J.-L.: Deformations des connexions localement plates. Ann. Institut. Fourier **18**(1), 103–114 (1968)

3. Milnor, J.: On the fundamental group of complete affinely flat manifolds. Adv. Math. **25**, 178–187 (1977)
4. Malgrange, B.: Equations de Lie, I. J. Diff. Geom. **6**, 503–522 (1972) . II **7**, 117–141 (1972)
5. Guillemin, V., Sternberg, S.: An algebraic model for transitive differential geometry. Bull. Am. Math. Soc. **70**, 16–47 (1964)
6. Gromov, M.: Solf and hard symplectic geometry. In: Proceedings of International Congress of Mathematicians, vol. 1 (1986)
7. Katsumi, Y.: On Hessian structure on an affine manifolds. In: Hano, J., Morimoto, A., Murakami, S., Okamoto, K., Ozeki, H. (eds.) Manifolds and Lie Groups. Progress in Mathematics, vol. 14, pp. 449–459 (1981). https://doi.org/10.1007/978-1-4612-5987-9_24
8. McCullagh, P.: What is a statistical model. Ann. Stat. **30**(5), 1225–1310 (2002)
9. Vinberg, E.B.: The theory of homogeneous cones. Trans. Moscow Math. Soc. **12**, 340–403 (1963)
10. Reinhardt, B.L.: Foliated manifolds with bundle-like metrics. Ann. Math. **69**(2), 119–132 (1959)
11. Ghys, E.: How to construct all Riemannian foliations (see appendix of) Molino: Riemannian foliations, Birkhauser, Boston Massachusetts
12. Goldman, W.M.: Complete affine manifolds: a survey (researchGate)
13. Kumperan, A., Spencer, D.: Lie Equations, the General Theory. Annals of Mathematics Studies. Princeton University Press, Princeton (1972)
14. Nirenberg, L.: Lectures on linear partial differential equations. In: Conference Board of the Mathematical Sciences, vol. 17. AMS (1973)
15. Nguiffo Boyom, M.: Numerical properties of Koszul connections. 3188. https://www.researchgate.ne. arXiv.1708.01106 (Math)
16. Nguiffo Boyom, M.: New geometric invariants and old open problems in the global analysis (in preparation)
17. Margulis, G.: Complete affine locally flat manifolds with a free fundamental group. J. Sovet Math. **134**, 129–134 (1987)
18. Singier, I., Sternberg, S.: The infinite groups of Lie and Cartan. Journal d'Analyse Mathematique (15), 1–114 (1965)
19. Serre, J.-P.: GAGA: Geometrie algebrique et Geometrie analystique. Annales de l'Institut Fourier **6**, 1–42 (1956)
20. Spencer, D.: Overdeterminated system s of partial differential equations. Bull. Am. Math. Soc. **75**, 179–223 (1964)

Curvature Inequalities and Simons' Type Formulas in Statistical Geometry

Barbara Opozda(✉)

Faculty of Mathematics and Computer Science, Jagiellonian University, ul. Łojasiewicza 6, 30-348 Cracow, Poland
barbara.opozda@im.uj.edu.pl

Abstract. We present some inequalities for curvatures and some applications of Simons' formulas to statistical structures.

Keywords: Statistical structure · Simons' formula · Maximum principle

1 Introduction

The paper deals with statistical structures treated from a purely geometrical viewpoint. By a statistical manifold we mean a differentiable manifold M equipped with a positive-definite metric tensor g and a torsion-free connection ∇ for which $(\nabla_X g)(Y, Z)$ is symmetric for X and Y. For such a structure many geometrical objects can be studied. Of course, the geometry of the metric tensor field g must be in use. But the objects relating the metric g and the connection ∇ are the most important. First of all we have the cubic form $A = -1/2\nabla g$ and the Czebyshev form τ defined as $\tau(X) = \mathrm{tr}\,_g A(X, \cdot, \cdot)$. In this paper we put the emphasis on the two objects. Even if the statistical connection is very special, for instance flat – as in the case of Hessian structures, the cubic form and the Czebyshev form may have various properties which influence even the geometry of the metric g. One can study such properties and quantities like: $\|\tau\|$, $\hat{\nabla}\tau = 0$, $\hat{\nabla}^2\tau = 0$, τ is harmonic, $\|A\|$, $\hat{\nabla}A = 0$, $\|\hat{\nabla}A\|$, $\|\hat{\nabla}^2 A\|$, where $\hat{\nabla}$ is the Levi-Civita connection for g.

We shall first briefly collect known facts concerning curvature tensors appearing in statistical geometry and needed in this paper. Then we present some new inequalities and their interpretation in the context of the geometry of curvatures.

We shall consider the following Simons' formula

$$\frac{1}{2}\Delta(\|s\|^2) = g(\Delta s, s) + \|\hat{\nabla}s\|^2. \tag{1}$$

This formula can be applied, for instance, to the cubic form or to the Czebyshev form of a statistical structure. This gives, in particular, some theorems concerning curvatures for various types of conjugate symmetric statistical structures. Using Yau's maximum principle and inequalities derived from the Simons

F. Nielsen and F. Barbaresco (Eds.): GSI 2021, LNCS 12829, pp. 389–396, 2021.
https://doi.org/10.1007/978-3-030-80209-7_43

formula one can prove global theorems for trace-free statistical structures of constant curvature and with complete metric.

The notion of a statistical structure of constant curvature was introduced by Kurose in [2]. Such structures can be locally realized on equiaffine spheres, so one can say that they belong to affine geometry of hypersurfaces (even for global results one can usually go to the universal cover). But it is worth to understand their properties in the framework of abstract statistical manifolds. Firstly, because the specific apparatus of the geometry of hypersurfaces is often too much for the considerations, secondly, the affine geometry of hypersurfaces is very well developed for Blaschke (i.e. trace-free) affine hypersurfaces and not so much for equiaffine hypersurfaces, including equiaffine spheres, and, finally, the equiaffine hypersurfaces form a very thin subset in the category of statistical manifold and treating specific statistical structures from the viewpoint of general statistical structures gives a better point of reference for possible generalizations within the category of all statistical manifolds.

The paper does not contain proofs of the presented results. All details concerning this paper can be found in [10].

2 Preliminaries

A statistical structure on a differentiable manifold M is a pair (g, ∇), where g is a positive-definite metric tensor and ∇ is a torsion-free connection on M such that the cubic form $\nabla g(X, Y, Z) = (\nabla_X g)(Y, Z)$ is symmetric. The difference tensor defined by

$$K(X, Y) = K_X Y = \nabla_X Y - \hat{\nabla}_X Y \tag{2}$$

is symmetric and symmetric relative to g. It is clear that a statistical structure can be also defined as a pair (g, K), where K is a symmetric and symmetric relative to g $(1, 2)$-tensor field. Namely, the connection ∇ is defined by (2). Alternatively, instead of K one can use the symmetric cubic form $A(X, Y, Z) = g(K(X, Y)Z)$. The relation between the cubic forms ∇g and A is the following $\nabla g = -2A$.

A statistical structure is called trace-free if $\mathrm{tr}\,_g K = 0$. Having a metric tensor g and a connection ∇ on a manifold M one defines a conjugate connection $\overline{\nabla}$ by the formula

$$g(\nabla_X Y, Z) + g(Y, \overline{\nabla}_X Z) = Xg(Y, Z). \tag{3}$$

A pair (g, ∇) is a statistical structure if and only if $(g, \overline{\nabla})$ is. The pairs are also simultaneously trace-free because if K is the difference tensor for (g, ∇), then $-K$ is the difference tensor for $\overline{\nabla}$. A statistical structure is called trivial, if $\nabla = \hat{\nabla}$, that is, $K = 0$.

The curvature tensors for ∇, $\overline{\nabla}$, $\hat{\nabla}$ will be denoted by R, $\overline{R}$ and $\hat{R}$ respectively. The corresponding Ricci tensors will be denoted by Ric, $\overline{\mathrm{Ric}}$ and $\widehat{\mathrm{Ric}}$. In general, the curvature tensor R does not satisfy the equality $g(R(X, Y)Z, W) = -g(R(X, Y)W, Z)$. It it does, we say that a statistical structure is conjugate symmetric. For a statistical structure we always have

$$g(R(X, Y)Z, W) = -g(\overline{R}(X, Y)W, Z). \tag{4}$$

It is known that the following conditions are equivalent:

1) $R = \overline{R}$,
2) $\hat{\nabla} K$ is symmetric (equiv. $\hat{\nabla} A$ is symmetric),
3) $g(R(X,Y)Z,W)$ is skew-symmetric relative to Z, W.

Each of the above three conditions characterizes a conjugate symmetric statistical structure. The following formula holds

$$R + \overline{R} = 2\hat{R} + 2[K,K], \tag{5}$$

where $[K,K]$ is a $(1,3)$-tensor field defined as $[K,K](X,Y)Z = [K_X, K_Y]Z$. In particular, if $R = \overline{R}$, then

$$R = \hat{R} + [K,K]. \tag{6}$$

The $(1,3)$-tensor field $[K,K]$ is a curvature tensor, that is, it satisfies the first Bianchi identity and has the same symmetries as the Riemannian curvature tensor. In [7] the following equality was proved

$$\operatorname{tr}\{X \to [K,K](X,Y)Z\} = \tau(K(Y,Z)) - g(K_Y, K_Z), \tag{7}$$

where τ is the Czebyshev 1-form defined as $\tau(X) = \operatorname{tr} K_X = g(E,X)$.

The 1-form τ is closed (i.e. $\hat{\nabla}\tau$ is symmetric) if and only if the Ricci tensor Ric is symmetric. For a conjugate symmetric statistical structure the Ricci tensor is symmetric. More generally, if for a statistical structure $\mathrm{Ric} = \overline{\mathrm{Ric}}$, then $d\tau = 0$. We have

$$\mathrm{Ric} + \overline{\mathrm{Ric}} = 2\widehat{\mathrm{Ric}} + 2\tau \circ K - 2g(K_\cdot, K_\cdot). \tag{8}$$

Therefore, if (g, ∇) is trace-free, then

$$2\widehat{\mathrm{Ric}} \geq \mathrm{Ric} + \overline{\mathrm{Ric}}. \tag{9}$$

For any statistical structure we have the Riemannian scalar curvature $\hat{\rho}$ for g and the scalar curvature ρ for ∇: $\rho = \operatorname{tr}_g \mathrm{Ric}$. The following important relation holds

$$\hat{\rho} = \rho + \|K\|^2 - \|E\|^2. \tag{10}$$

Recall now some sectional curvatures of statistical structures. Of course, we have the usual Riemannian sectional curvature for the metric tensor g. We shall denote it by $\hat{k}(\pi)$ for a vector plane π in a tangent space. In [7] another sectional curvature was introduced. It was called the sectional ∇-curvature. Namely, the tensor field $R + \overline{R}$ has all algebraic properties needed to produce the sectional curvature. The sectional ∇-curvature is defined as follows

$$k(\pi) = \frac{1}{2} g((R + \overline{R})(e_1, e_2)e_2, e_1), \tag{11}$$

where e_1, e_2 is an orthonormal basis of π. But, in general, this sectional curvature does not have the same properties as the Riemannian sectional curvature.

For instance, Schur's lemma does not hold, in general. But it holds for conjugate symmetric statistical structures. Another good curvature tensor, which we have in statistical geometry is $[K, K]$. It also allows to produce a sectional curvature, see [8].

A statistical structure is called Hessian if ∇ is flat, that is, if $R = 0$. Since $\mathrm{Ric} = 0$ for a Hessian structure and ∇ is torsion-free, we have that the Koszul form $\nabla\tau$ is symmetric. A Hessian structure is called Einstein-Hessian if $\nabla\tau = \lambda g$, see [11]. For a Hessian structure we have

$$\beta = \hat{\nabla}\tau - \tau \circ K, \qquad \mathrm{tr}\,_g\beta = \delta\tau - \|\tau\|^2, \tag{12}$$

where δ denotes the codifferential relative to g, i.e. $\delta = \mathrm{tr}\,_g\hat{\nabla}\tau$. By (8) we also have

$$\widehat{\mathrm{Ric}} = g(K.,K.) - \tau \circ K \tag{13}$$

A statistical structure is said to have constant curvature k if

$$R(X,Y)Z = k\{g(Y,Z)X - g(X,Z)Y\} \tag{14}$$

for some constant number k, see [2]. From a local viewpoint statistical manifolds of constant curvature are nothing but equiaffine spheres in the affine space $\mathbf{R}^{n+1}$. Statistical structures of constant curvature are clearly conjugate symmetric and have constant sectional ∇-curvature. Conversely, if a statistical structure is conjugate symmetric and has constant sectional ∇-curvature, then it has constant curvature in the sense of (14). Let $R_0(X,Y)Z = g(Y,Z)X - g(X,Z)Y$. If a statistical structure has constant curvature k, then $R = kR_0$. If a statistical structure on a 2-dimensional manifold is conjugate symmetric, then $R = kR_0$ and k may be a function.

3 Algebraic and Curvature Inequalities

In this section two algebraic inequalities for statistical structures are presented. Some applications are mentioned. All details can be found in [10].

Theorem 1. *For every statistical structure we have*

$$(\tau \circ K)(U,U) - g(K_U, K_U) \leq \frac{1}{4}\|\tau\|^2 g(U,U) \tag{15}$$

for every vector U. If $U \neq 0$, the equality holds if and only if $\tau = 0$ and $K_U = 0$.

As a consequence of the above inequality and (8) we get the following generalization of (9)

Theorem 2. *For any statistical structure we have*

$$2\widehat{\mathrm{Ric}} \geq \mathrm{Ric} + \overline{\mathrm{Ric}} - \frac{1}{2}\|\tau\|^2 g. \tag{16}$$

For a statistical structure with constant curvature k we have

$$\widehat{\mathrm{Ric}} \geq \left(k(n-1) - \frac{1}{4}\|\tau\|^2 \right) g. \tag{17}$$

For a Hessian structure the following inequality holds

$$\widehat{\mathrm{Ric}} \geq -\frac{1}{4}\|\tau\|^2 g. \tag{18}$$

Theorem 3. *For any statistical structure we have*

$$\frac{n+2}{3}\|A\|^2 - \|E\|^2 \geq 0. \tag{19}$$

Using this theorem, (10) and other formulas from Sect. 2 we get

Theorem 4. *For any statistical structure we have*

$$\hat{\rho} - \rho = \|A\|^2 - \|E\|^2 \geq -\frac{n-1}{3}\|A\|^2, \tag{20}$$

$$\hat{\rho} - \rho = \|A\|^2 - \|E\|^2 \geq -\frac{n-1}{n+2}\|E\|^2. \tag{21}$$

For a statistical structure with constant sectional curvature k we have

$$\hat{\rho} \geq \frac{n-1}{n+2}\left(kn(n+2) - \|E\|^2\right), \tag{22}$$

$$\hat{\rho} \geq \frac{n-1}{3}\left(3kn - \|A\|^2\right). \tag{23}$$

In particular, for a Hessian structure we have

$$\hat{\rho} \geq -\frac{n-1}{n+2}\|E\|^2, \tag{24}$$

$$\hat{\rho} \geq -\frac{n-1}{3}\|A\|^2. \tag{25}$$

4 Simons' Formula

For any tensor field s of type $(0, r)$ on a manifold M the following Simons' formula holds

$$\frac{1}{2}\Delta(\|s\|^2) = g(\Delta s, s) + \|\hat{\nabla} s\|^2, \tag{26}$$

where the Laplacian on the left hand side is the Hodge one and the Laplacian on the right hand side is defined as follows

$$(\Delta s)(X_1, ..., X_r) = \sum_{i=1}^{n} (\hat{\nabla}^2 s)(e_i, e_i, X_1, ..., X_r), \tag{27}$$

where $e_1, ..., e_n$ is an orthonormal basis of a tangent space T_xM for $x \in M$.

If τ is a 1-form on M, then by the Ricci identity we have

$$(\hat{\nabla}^2\tau)(X,Y,Z) - (\hat{\nabla}^2\tau)(Y,X,Z) = -\tau(\hat{R}(X,Y)Z). \tag{28}$$

Hence we have

Proposition 1. *If $\hat{\nabla}^2\tau$ is a symmetric cubic form, then $\operatorname{im}\hat{R} \subset \ker\tau$. In particular $\widehat{\mathrm{Ric}}\,(X,E) = 0$ for every X. If $\widehat{\mathrm{Ric}}$ is non-degenerate, then $\tau = 0$.*

By (28) we obtain

$$\sum_i(\hat{\nabla}^2\tau)(e_i,e_i,X) = (d\delta\tau + \delta d\tau)(X) + \widehat{\mathrm{Ric}}\,(X,E), \tag{29}$$

Simons' formula (26) now yields the following classical Bochner's formula

$$\frac{1}{2}\Delta(\|\tau\|^2) = g((d\delta + \delta d)\tau,\tau) + \widehat{\mathrm{Ric}}\,(E,E) + \|\hat{\nabla}\tau\|^2. \tag{30}$$

Using the above facts one can prove

Theorem 5. *Let τ be a 1-form such that $\hat{\nabla}^2\tau = 0$. Then τ is harmonic and $\widehat{\mathrm{Ric}}\,(X,E) = 0$ for every X.*
If, additionally, M is compact, or $\|\tau\|$ is constant, then $\hat{\nabla}\tau = 0$. If $\widehat{\mathrm{Ric}}$ is non-degenerate at a point of M, then $\tau = 0$.

Applying this theorem to the Czebyshev form we get

Corollary 1. *Let (g,∇) be a statistical structure on a manifold M and τ its Czebyshev form. Assume that $\hat{\nabla}^2\tau = 0$, $\|\tau\|$ is constant or M is compact and, moreover, the Ricci tensor of the metric g is non-degenerate at a point of M. Then the structure is trace-free.*

We now interpret the Simons formula for the statistical cubic form A. By computing the right hand side of (26) in various ways we obtain, see [10],

Theorem 6. *For any conjugate symmetric statistical structure we have*

$$\frac{1}{2}\Delta(\|A\|^2) = \|\hat{\nabla}A\|^2 + g(\hat{\nabla}^2\tau, A) - g([K,K],\hat{R}) + g(\widehat{\mathrm{Ric}}\,, g(K.,K.)), \tag{31}$$

$$\frac{1}{2}\Delta(\|A\|^2) = \|\hat{\nabla}A\|^2 + g(\hat{\nabla}^2\tau, A) + g(\hat{R} - R,\hat{R}) + g(\widehat{\mathrm{Ric}}\,, g(K.,K.)), \tag{32}$$

$$\begin{aligned}\tfrac{1}{2}\Delta(\|A\|^2) = \|\hat{\nabla}A\|^2 &+ g(\hat{\nabla}^2\tau, A)\\ &+ \hat{R}^2 + \widehat{\mathrm{Ric}}^{\,2} - g(R,\hat{R}) - g(\mathrm{Ric}\,,\widehat{\mathrm{Ric}}\,)\\ &\qquad + g(\widehat{\mathrm{Ric}}\,,\tau\circ K).\end{aligned} \tag{33}$$

If we assume that the curvature tensors have special properties, some terms in the above formulas can be simplified. As a result, one gets

Theorem 7. *Let (g, ∇) be a statistical structure of constant curvature k and $k \leq 0$. If $\hat{\nabla}\tau = 0$, $\widehat{\mathrm{Ric}} \geq 0$ and $\hat{\rho}$ is constant, then $\hat{R} = 0$ and $\hat{\nabla}A = 0$.*

Theorem 8. *Let (g, ∇) be a conjugate symmetric statistical structure on a compact manifold M. If the sectional ∇-curvature is constant non-positive, $\widehat{\mathrm{Ric}} \geq 0$ and $\hat{\nabla}^2\tau = 0$, then $\hat{R} = 0$ and $\hat{\nabla}A = 0$.*

Proposition 2. *Let (M, g, ∇) be a conjugate symmetric statistical manifold such that $\hat{\nabla}^2\tau = 0$, $\hat{R} = 0$ and $\|A\|$ is constant (or M is compact). Then $\hat{\nabla}A = 0$.*

Proposition 3. *For a Hessian structure (g, A) if $\hat{\nabla}A = 0$ and $\widehat{\mathrm{Ric}} \geq 0$, then $\hat{R} = 0$.*

We shall now consider the function $u = \|A\|^2$. The following inequalities were proved by Blaschke, Calabi and An-Min Li for Blaschke affine hypersurfaces. We formulate them in the language of statistical structures.

Theorem 9. *Let (g, ∇) be a trace-free statistical structure on an n-dimensional manifold M and $R = kR_0$. We have*

$$(n+1)ku + \frac{n+1}{n(n-1)}u^2 + \|\hat{\nabla}A\|^2 \leq \frac{1}{2}\Delta u \leq (n+1)ku + \frac{3}{2}u^2 + \|\hat{\nabla}A\|^2. \quad (34)$$

If $n = 2$, the equalities hold.

As an immediate consequence of this theorem we have

Corollary 2. *Let (g, ∇) be a trace-free statistical structure on an n-dimensional manifold M and $R = kR_0$. Assume that $\hat{\nabla}A = 0$ on M.*
If $k \geq 0$, then $A = 0$ on M.
If $k < 0$, then either $A = 0$ on M or

$$\frac{2}{3}(n+1)(-k) \leq u \leq n(n-1)(-k). \quad (35)$$

If $n = 2$ and $\sup k = 0$, then $A = 0$.

For treating the case, where g is complete one can use the following maximum principle

Lemma 1 [12]. *Let (M, g) be a complete Riemannian manifold, whose Ricci tensor is bounded from below and let f be a C^2-function bounded from above on M. For every $\varepsilon > 0$ there exists a point $x \in M$ at which*

i) $f(x) > \sup f - \varepsilon$
ii) $\Delta f(x) < \varepsilon$.

The following theorem and its generalizations can be found in [1,5,9].

Theorem 10. *Let* (g, ∇) *be a trace-free statistical structure with* $R = kR_0$ *and complete* g *on a manifold* M. *If* $k \geq 0$, *then the structure is trivial. If* k *is constant and negative, then the Ricci tensor of* g *is non-positive and consequently*

$$u \leq n(n-1)(-k). \tag{36}$$

More delicate estimations can be established. Namely, the term $\|\hat{\nabla} A\|^2$ in (34) is usually seen as just nonnegative. However, we can also take into considerations $\inf \|\hat{\nabla} A\|^2$ or $\sup \|\hat{\nabla} A\|^2$. Using (34) and the maximum principle one can prove, for example, see [10],

Theorem 11. *Let* (g, ∇) *be a trace-free statistical structure on an* n*-dimensional manifold* M *and* $R = kR_0$, *where* k *is a negative constant. If* g *is complete and*

$$\sup \|\hat{\nabla} A\|^2 < \frac{k^2(n+1)^2}{6}, \tag{37}$$

then

$$\inf u \geq \frac{(n+1)(-k) + \sqrt{(n+1)^2k^2 - 6N_2}}{3}, \tag{38}$$

or

$$\inf u \leq \frac{(n+1)(-k) - \sqrt{(n+1)^2k^2 - 6N_2}}{3}, \tag{39}$$

where $u = \|A\|^2$ *and* $N_2 = \sup \|\hat{\nabla} A\|^2$.

References

1. Calabi, E.: Complete affine hypersurfaces I. Symposia Math. **10**, 19–38 (1972)
2. Kurose, T.: Dual connections and affine geometry. Math. Z. **203**, 115–121 (1990). https://doi.org/10.1007/BF02570725
3. Li, A.-M.: Some theorems in affine differential geometry. Acta Math. Sin. **5**, 345–354 (1989). https://doi.org/10.1007/BF02107712
4. Li, A.-M.: An intrinsic rigidity theorem for minimal submanifolds in a sphere. Arch. Math. **58**, 582–594 (1992). https://doi.org/10.1007/BF01193528
5. Li, A.-M., Simon, U., Zhao, G.: Global Affine Differential Geometry of Hypersurfaces. Walter de Gruyter, Berlin (1993)
6. Nomizu, K., Sasaki, T.: Affine Differential Geometry. Cambridge University Press, Cambridge (1994)
7. Opozda, B.: Bochner's technique for statistical structures. Ann. Glob. Anal. Geom. **48**, 357–395 (2015). https://doi.org/10.1007/s10455-015-9475-z
8. Opozda, B.: A sectional curvature for statistical structures. Linear Algebra Appl. **497**, 134–161 (2016)
9. Opozda, B.: Curvature bounded conjugate symmetric statistical structures with complete metric. Ann. Glob. Anal. Geom. **55**(4), 687–702 (2019). https://doi.org/10.1007/s10455-019-09647-y
10. Opozda, B.: Some inequalities and Simons' type formulas in Riemannian, affine and statistical geometry, arXiv:2105.04676
11. Shima, H.: The Geometry Hessian Structure. World Scientific Publishing, London (2007)
12. Yau, S.T.: Harmonic functions on complete Riemannian manifolds. Commun. Pure Appl. Math. **28**, 201–228 (1975)

Harmonicity of Conformally-Projectively Equivalent Statistical Manifolds and Conformal Statistical Submersions

T. V. Mahesh and K. S. Subrahamanian Moosath(✉)

Department of Mathematics, Indian Institute of Space Science and Technology, Thiruvanathapuram 695547, Kerala, India
maheshtv.16@res.iist.ac.in, smoosath@iist.ac.in

Abstract. In this paper, a condition for harmonicity of conformally-projectively equivalent statistical manifolds is given. Then, the conformal statistical submersion is introduced which is a generalization of the statistical submersion and prove that harmonicity and conformality cannot coexist.

Keywords: Harmonic maps · Conformally-projectively equivalent · Conformal statistical submersion

1 Introduction

The motivation to study harmonic maps comes from the applications of Riemannian submersion in theoretical physics [7]. Harmonic maps formulation of field theories lead to a geometrical description in the unified field theory program. Presently, we see an increasing interest in harmonic maps between statistical manifolds [8,9]. In [9], Keiko Uohashi studied harmonic maps related to α-connections on statistical manifolds. He obtained a condition for the harmonicity on α-conformally equivalent statistical manifolds. In this paper, we study the harmonicity of conformally-projectively equivalent statistical manifolds. Also, we introduce the conformal statistical submersion which is a generalization of the statistical submersion and studies the harmonicity of a conformal statistical submersion.

In Sect. 2, we recall the necessary basic concepts. Section 3, deals with the harmonicity of conformally-projectively equivalent statistical manifolds. A necessary and sufficient condition is obtained for the harmonicity of the identity map between conformally-projectively equivalent statistical manifolds. In Sect. 4, we introduce the conformal statistical submersion and prove that harmonicity and conformality cannot coexist. Throughout this paper, all the objects are assumed to be smooth.

F. Nielsen and F. Barbaresco (Eds.): GSI 2021, LNCS 12829, pp. 397–404, 2021.
https://doi.org/10.1007/978-3-030-80209-7_44

2 Preliminaries

In this section, we recall the necessary basic concepts [1,2,5,6,9].

A pseudo-Riemannian manifold $(\mathbf{M}, g)$ with a torsion-free affine connection ∇ is called a statistical manifold if ∇g is symmetric. For a statistical manifold $(\mathbf{M}, \nabla, g)$ the dual connection ∇^* is defined by

$$Xg(Y,Z) = g(\nabla_X Y, Z) + g(Y, \nabla^*_X Z) \tag{1}$$

for X, Y and Z in $\mathcal{X}(\mathbf{M})$, where $\mathcal{X}(\mathbf{M})$ denotes the set of all vector fields on $\mathbf{M}$. If (∇, g) is a statistical structure on $\mathbf{M}$, so is (∇^*, g). Then $(\mathbf{M}, \nabla^*, g)$ becomes a statistical manifold called the dual statistical manifold of $(\mathbf{M}, \nabla, g)$.

Definition 1. *Two statistical manifolds $(\mathbf{M}, \nabla, g)$ and $(\mathbf{M}, \tilde{\nabla}, \tilde{g})$ are said to be α-conformally equivalent if there exists a real valued function ϕ on $\mathbf{M}$ such that*

$$\tilde{g}(X,Y) = e^{\phi} g(X,Y) \tag{2}$$

$$\begin{aligned} g(\tilde{\nabla}_X Y, Z) &= g(\nabla_X Y, Z) - \frac{1+\alpha}{2} d\phi(Z) g(X,Y) \\ &+ \frac{1-\alpha}{2}\{d\phi(X) g(Y,Z) + d\phi(Y) g(X,Z)\}, \end{aligned} \tag{3}$$

where X, Y and Z in $\mathcal{X}(\mathbf{M})$ and α is a real number.

Note 1. Two statistical manifolds $(\mathbf{M}, \nabla, g)$ and $(\mathbf{M}, \tilde{\nabla}, \tilde{g})$ are α-conformally equivalent if and only if the dual statistical manifolds $(\mathbf{M}, \nabla^*, g)$ and $(\mathbf{M}, \tilde{\nabla}^*, \tilde{g})$ are $(-\alpha)$-conformally equivalent.

Definition 2. *Two statistical manifolds $(\mathbf{M}, \nabla, g)$ and $(\mathbf{M}, \tilde{\nabla}, \tilde{g})$ are said to be conformally-projectively equivalent if there exist two real valued functions ϕ and ψ on $\mathbf{M}$ such that*

$$\tilde{g}(X,Y) = e^{\phi+\psi} g(X,Y) \tag{4}$$

$$\tilde{\nabla}_X Y = \nabla_X Y - g(X,Y) grad_g \psi + d(\phi)(X)Y + d(\phi)(Y)X, \tag{5}$$

where X, Y and Z in $\mathcal{X}(\mathbf{M})$.

Let $(\mathbf{M}, \nabla, g)$ and $(\mathbf{B}, \nabla', g')$ be two statistical manifolds of dimensions n and m, respectively. Let $\{x^1, x^2, ...x^n\}$ be a local co-ordinate system on $\mathbf{M}$. We set $g_{ij} = g(\frac{\partial}{\partial x^i}, \frac{\partial}{\partial x^j})$ and $[g^{ij}] = [g_{ij}]^{-1}$.

Definition 3. *A smooth map $\pi : (\mathbf{M}, \nabla, g) \longrightarrow (\mathbf{B}, \nabla', g')$ is said to be a harmonic map relative to (g, ∇, ∇') if the tension field $\tau_{(g,\nabla,\nabla')}(\pi)$ of π vanishes at each point $p \in \mathbf{M}$, where $\tau_{(g,\nabla,\nabla')}(\pi)$ is defined as*

$$\tau_{(g,\nabla,\nabla')}(\pi) = \sum_{i,j=1}^{n} g^{ij} \left\{ \overline{\nabla}_{\frac{\partial}{\partial x^i}} \left(\pi_*(\frac{\partial}{\partial x^j}) \right) - \pi_* \left(\nabla_{\frac{\partial}{\partial x^i}} \frac{\partial}{\partial x^j} \right) \right\}, \tag{6}$$

where $\overline{\nabla}$ is the pullback of the connection ∇' of $\mathbf{B}$ to the induced vector bundle $\pi^{-1}(T\mathbf{B}) \subset T(\mathbf{M})$ and $\overline{\nabla}_{\frac{\partial}{\partial x^i}} \left(\pi_*(\frac{\partial}{\partial x^j}) \right) = \nabla'_{\pi_*(\frac{\partial}{\partial x^i})} \pi_*(\frac{\partial}{\partial x^j})$.

Let $(\mathbf{M}, \nabla, g)$ and $(\mathbf{M}, \tilde{\nabla}, \tilde{g})$ be statistical manifolds. Then the identity map $id : (\mathbf{M}, \nabla, g) \longrightarrow (\mathbf{M}, \tilde{\nabla}, \tilde{g})$ is a harmonic map relative to $(g, \nabla, \tilde{\nabla})$ if

$$\tau_{(g,\nabla,\tilde{\nabla})}(id) = \sum_{i,j=1}^{n} g^{ij}(\tilde{\nabla}_{\frac{\partial}{\partial x^i}} \frac{\partial}{\partial x^j} - \nabla_{\frac{\partial}{\partial x^i}} \frac{\partial}{\partial x^j}) \tag{7}$$

vanishes identically on $\mathbf{M}$.

Let $\mathbf{M}$ and $\mathbf{B}$ be Riemannian manifolds with dimensions n and m respectively with $n > m$. An onto map $\pi : \mathbf{M} \longrightarrow \mathbf{B}$ is called a submersion if $\pi_{*p} : T_p\mathbf{M} \longrightarrow T_{\pi(p)}\mathbf{B}$ is onto for all $p \in \mathbf{M}$. For a submersion $\pi : \mathbf{M} \longrightarrow \mathbf{B}$, $\pi^{-1}(b)$ is a submanifold of $\mathbf{M}$ of dimension $(n-m)$ for each $b \in \mathbf{B}$. These submanifolds $\pi^{-1}(b)$ are called the fibers. Set $\mathcal{V}(\mathbf{M})_p = Ker(\pi_{*p})$ for each $p \in \mathbf{M}$.

Definition 4. *A submersion $\pi : \mathbf{M} \longrightarrow \mathbf{B}$ is called a submersion with horizontal distribution if there is a smooth distribution $p \longrightarrow \mathcal{H}(M)_p$ such that*

$$T_p\mathbf{M} = \mathcal{V}(\mathbf{M})_p \bigoplus \mathcal{H}(M)_p. \tag{8}$$

We call $\mathcal{V}(\mathbf{M})_p$ ($\mathcal{H}(M)_p$) the vertical (horizontal) subspace of $T_p\mathbf{M}$. $\mathcal{H}$ and $\mathcal{V}$ denote the projections of the tangent space of $\mathbf{M}$ onto the horizontal and vertical subspaces, respectively.

Note 2. Let $\pi : \mathbf{M} \longrightarrow \mathbf{B}$ be a submersion with the horizontal distribution $\mathcal{H}(M)$. Then $\pi_* |_{\mathcal{H}(M)_p} : \mathcal{H}(M)_p \longrightarrow T_{\pi(p)}\mathbf{B}$ is an isomorphism for each $p \in \mathbf{M}$.

Definition 5. *A vector field Y on $\mathbf{M}$ is said to be projectable if there exists a vector field Y_* on $\mathbf{B}$ such that $\pi_*(Y_p) = Y_{*\pi(p)}$ for each $p \in \mathbf{M}$, that is Y and Y_* are π-related. A vector field X on $\mathbf{M}$ is said to be basic if it is projectable and horizontal. Every vector field X on $\mathbf{B}$ has a unique smooth horizontal lift, denoted by $\tilde{X}$, to $\mathbf{M}$.*

Definition 6. *Let ∇ and ∇^* be affine connections on $\mathbf{M}$ and $\mathbf{B}$, respectively. $\pi : (\mathbf{M}, \nabla) \longrightarrow (\mathbf{B}, \nabla^*)$ is said to be an affine submersion with horizontal distribution if $\pi : \mathbf{M} \longrightarrow \mathbf{B}$ is a submersion with horizontal distribution and satisfies $\mathcal{H}(\nabla_{\tilde{X}}\tilde{Y}) = (\nabla^*_X Y)\tilde{}$, for X, Y in $\mathcal{X}(\mathbf{B})$.*

Note 3. Abe and Hasegawa [1] introduced an affine submersion with horizontal distribution and they considered general affine connection instead of Levi-Civita connection on Riemannian manifolds. On the other hand, Fuglede [3] and Ishihara [4] in their independent work introduced horizontally conformal submersion as a generalization of Riemannian submersion. Their study focuses on the conformality relation between the metrics on Riemannian manifolds and the Levi-Civita connections. In this paper, we study conformal submersion between Riemannian manifolds $\mathbf{M}$ and $\mathbf{B}$ and the conformality relation between any two affine connections ∇ and ∇' (not necessarily Levi-Civita connections) on $\mathbf{M}$ and $\mathbf{B}$, respectively.

3 Harmonic Maps of Conformally-Projectively Equivalent Statistical Manifolds

In this section, we describe harmonic maps with respect to the conformally-projectively equivalence of statistical manifolds. In [9], Keiko Uohashi studied harmonic maps for α-conformal equivalence. He proved that, if $\alpha = -\frac{(n-2)}{(n+2)}$ or ϕ is a constant function on $\mathbf{M}$, then the identity map $id : (\mathbf{M}, \nabla, g) \longrightarrow (\mathbf{M}, \tilde{\nabla}, \tilde{g})$ is a harmonic map relative to $(g, \nabla, \tilde{\nabla})$ for α-conformally equivalent statistical manifolds $(\mathbf{M}, \nabla, g)$ and $(\mathbf{M}, \tilde{\nabla}, \tilde{g})$ of dimension $n \geq 2$.

Now, we have the following theorem.

Theorem 1. *Let* $(\mathbf{M}, \nabla, g)$ *and* $(\mathbf{M}, \tilde{\nabla}, \tilde{g})$ *be conformally-projectively equivalent statistical manifolds of dimension n. Then, the identity map* $id : (\mathbf{M}, \nabla, g) \longrightarrow (\mathbf{M}, \tilde{\nabla}, \tilde{g})$ *is a harmonic map if and only if* $\phi = \frac{n}{2}\psi + c$*, where c is some constant.*

Proof. By Eqs. (5) and (7), for k = 1, 2, .. n

$$\begin{aligned}
\tau_{(g,\nabla,\tilde{\nabla})}(id) &= \sum_{i,j=1}^{n} g^{ij}\left(\tilde{\nabla}_{\frac{\partial}{\partial x^i}}\frac{\partial}{\partial x^j} - \nabla_{\frac{\partial}{\partial x^i}}\frac{\partial}{\partial x^j}\right) \\
&= \sum_{i,j=1}^{n} g^{ij}\left(-g(\frac{\partial}{\partial x_i}, \frac{\partial}{\partial x_j})grad_g\psi + d\phi\left(\frac{\partial}{\partial x_j}\right)\frac{\partial}{\partial x_i} + d\phi\left(\frac{\partial}{\partial x_i}\right)\frac{\partial}{\partial x_j}\right) \\
&= \sum_{i,j=1}^{n} g^{ij}\left((-g_{ij})grad_g\psi + \frac{\partial\phi}{\partial x_j}\frac{\partial}{\partial x_i} + \frac{\partial\phi}{\partial x_i}\frac{\partial}{\partial x_j}\right).
\end{aligned}$$

Now, we have

$$\begin{aligned}
g\left(\tau_{(g,\nabla,\tilde{\nabla})}(id), \frac{\partial}{\partial x_k}\right) &= -ng\left(grad_g\psi, \frac{\partial}{\partial x_k}\right) + \sum_{i=1}^{k}\frac{\partial\phi}{\partial x_i}\delta_{ik} + \sum_{j=1}^{k}\frac{\partial\phi}{\partial x_j}\delta_{jk} \\
&= -n\frac{\partial\psi}{\partial x_k} + \sum_{i=1}^{k}\frac{\partial\phi}{\partial x_i}\delta_{ik} + \sum_{j=1}^{k}\frac{\partial\phi}{\partial x_j}\delta_{jk} \\
&= -n\frac{\partial\psi}{\partial x_k} + 2\frac{\partial\phi}{\partial x_k}. \qquad (9)
\end{aligned}$$

Now from (9), id is harmonic if and only if $\frac{\partial\phi}{\partial x_k} = \frac{n}{2}\frac{\partial\psi}{\partial x_k}$ for all $k \in \{1, 2, ...n\}$. Hence, id is harmonic if and only if $\phi = \frac{n}{2}\psi + c$, where c is some constant.

4 Conformal Statistical Submersion and Harmonic Map

In this section, we generalize the concept of an affine submersion with horizontal distribution to the conformal submersion with horizontal distribution and prove a necessary condition for the existence of such maps. Also, we introduce the conformal statistical submersion, which is a generalization of the statistical submersion and prove that harmonicity and conformality cannot coexist.

Definition 7. *Let* $(\mathbf{M}, g_M)$ *and* $(\mathbf{B}, g_B)$ *be Riemannian manifolds. A submersion* $\pi : (\mathbf{M}, g_M) \longrightarrow (\mathbf{B}, g_B)$ *is called a conformal submersion if there exists a* $\phi \in C^\infty(\mathbf{M})$ *such that*

$$g_M(X, Y) = e^{2\phi} g_B(\pi_* X, \pi_* Y), \tag{10}$$

for all horizontal vector fields X, Y *on* $\mathbf{M}$.

For $\pi : (\mathbf{M}, \nabla) \longrightarrow (\mathbf{B}, \nabla^*)$ an affine submersion with horizontal distribution, $\mathcal{H}(\nabla_{\tilde{X}}\tilde{Y}) = \tilde{(\nabla^*_X Y)}$, for $X, Y \in \mathcal{X}(\mathbf{B})$. In the case of a conformal submersion, we prove the following theorem which is the motivation for us to generalize the concept of an affine submersion with horizontal distribution.

Theorem 2. *Let* $\pi : (\mathbf{M}, g_M) \longrightarrow (\mathbf{B}, g_B)$ *be a conformal submersion. If* ∇ *on* $\mathbf{M}$ *and* ∇^* *on* B *are the Levi-Civita connections then*

$$\mathcal{H}(\nabla_{\tilde{X}}\tilde{Y}) = (\tilde{\nabla^*_X Y}) + \tilde{X}(\phi)\tilde{Y} + \tilde{Y}(\phi)\tilde{X} - \mathcal{H}(grad_\pi \phi) g_m(\tilde{X}, \tilde{Y}),$$

where $X, Y \in \mathcal{X}(\mathbf{B})$ *and* $\tilde{X}, \tilde{Y}$ *denote its unique horizontal lifts on* $\mathbf{M}$ *and* $grad_\pi \phi$ *denotes the gradient of* ϕ *with respect to* g_M.

Proof. We have the Koszul formula for the Levi-Civita connection,

$$\begin{aligned} 2g_M(\nabla_{\tilde{X}}\tilde{Y}, \tilde{Z}) = {} & \tilde{X} g_M(\tilde{Y}, \tilde{Z}) + \tilde{Y} g_M(\tilde{Z}, \tilde{X}) \\ & - \tilde{Z} g_M(\tilde{X}, \tilde{Y}) - g_M(\tilde{X}, [\tilde{Y}, \tilde{Z}]) \\ & + g_M(\tilde{Y}, [\tilde{Z}, \tilde{X}]) + g_M(\tilde{Z}, [\tilde{X}, \tilde{Y}]). \end{aligned} \tag{11}$$

Now, consider

$$\begin{aligned} \tilde{X} g_M(\tilde{Y}, \tilde{Z}) &= \tilde{X}(e^{2\phi} g_B(Y, Z)) \\ &= \tilde{X}(e^{2\phi}) g_B(Y, Z) + e^{2\phi} \tilde{X}(g_B(Y, Z)) \\ &= 2e^{2\phi} \tilde{X}(\phi) g_B(Y, Z) + e^{2\phi} X g_B(Y, Z). \end{aligned}$$

Similarly,

$$\begin{aligned} \tilde{Y} g_M(\tilde{X}, \tilde{Z}) &= 2e^{2\phi} \tilde{Y}(\phi) g_B(X, Z) + e^{2\phi} Y g_B(X, Z) \\ \tilde{Z} g_M(\tilde{X}, \tilde{Y}) &= 2e^{2\phi} \tilde{Z}(\phi) g_B(X, Y) + e^{2\phi} Z g_B(X, Y). \end{aligned}$$

Also, we have

$$\begin{aligned} g_M(\tilde{X}, [\tilde{Y}, \tilde{Z}]) &= e^{2\phi} g_B(X, [Y, Z]) \\ g_M(\tilde{Y}, [\tilde{Z}, \tilde{X}]) &= e^{2\phi} g_B(Y, [Z, X]) \\ g_M(\tilde{Z}, [\tilde{X}, \tilde{Y}]) &= e^{2\phi} g_B(Z, [X, Y]). \end{aligned}$$

Then, from Eq. (11) and the above equations, we get

$$\begin{aligned} 2g_M(\nabla_{\tilde{X}}\tilde{Y}, \tilde{Z}) = {} & 2\tilde{X}(\phi) e^{2\phi} g_B(Y, Z) + 2\tilde{Y}(\phi) e^{2\phi} g_B(X, Z) \\ & - 2\tilde{Z}(\phi) e^{2\phi} g_B(X, Y) + 2e^{2\phi} g_B(\nabla^*_X Y, Z). \end{aligned}$$

Since, $\tilde{Z}(\phi) = e^{2\phi} g_B(\pi_*(grad_\pi \phi), Z)$ we get

$$\pi_*(\nabla_{\tilde{X}} \tilde{Y}) = \nabla^*_X Y + \tilde{X}(\phi)Y + \tilde{Y}(\phi)X - e^{2\phi}\pi_*(grad_\pi \phi) g_B(X, Y).$$

Hence,

$$\mathcal{H}(\nabla_{\tilde{X}} \tilde{Y}) = (\tilde{\nabla^*_X Y}) + \tilde{X}(\phi)\tilde{Y} + \tilde{Y}(\phi)\tilde{X} - \mathcal{H}(grad_\pi \phi) g_m(\tilde{X}, \tilde{Y}).$$

Now, we generalize the concept of an affine submersion with horizontal distribution.

Definition 8. *Let* $\pi : (\mathbf{M}, g_M) \longrightarrow (\mathbf{B}, g_B)$ *be a conformal submersion and let* ∇ *and* ∇^* *be affine connections on* **M** *and* **B**, *respectively. Then,* $\pi : (\mathbf{M}, \nabla) \longrightarrow (\mathbf{B}, \nabla^*)$ *is said to be a conformal submersion with the horizontal distribution* $\mathcal{H}(\mathbf{M}) = \mathcal{V}(\mathbf{M})^\perp$ *if*

$$\mathcal{H}(\nabla_{\tilde{X}} \tilde{Y}) = (\tilde{\nabla^*_X Y}) + \tilde{X}(\phi)\tilde{Y} + \tilde{Y}(\phi)\tilde{X} - \mathcal{H}(grad_\pi \phi) g_m(\tilde{X}, \tilde{Y}),$$

for some $\phi \in C^\infty(\mathbf{M})$ *and for* $X, Y \in \mathcal{X}(\mathbf{B})$.

Note 4. If ϕ is a constant, it turns out to be an affine submersion with horizontal distribution.

Now, a necessary condition is obtained for the existence of a conformal submersion with horizontal distribution.

Theorem 3. *Let* $\pi : (\mathbf{M}, g_M) \longrightarrow (\mathbf{B}, g_B)$ *be a conformal submersion and* ∇ *be an affine connection on* **M**. *Assume that* $\pi : \mathbf{M} \longrightarrow \mathbf{B}$ *is a submersion with horizontal distribution. If* $\mathcal{H}(\nabla_{\tilde{X}} \tilde{Y})$ *is projectable for vector fields* X *and* Y *on* **B**, *then there exists a unique connection* ∇^* *on* **B** *such that* $\pi : (\mathbf{M}, \nabla) \longrightarrow (\mathbf{B}, \nabla^*)$ *is a conformal submersion with horizontal distribution.*

Proof. Setting $\nabla^*_X Y = \pi_*(\nabla_{\tilde{X}} \tilde{Y}) - \tilde{X}(\phi)Y - \tilde{Y}(\phi)X + e^{2\phi}\pi_*(grad_\pi \phi) g_B(X, Y)$, we show that ∇^* is an affine connection on **B**. Since $(\tilde{fX}) = (f \circ \pi)\tilde{X}$,

$$\nabla^*_X(fY) = \pi_*(\nabla_{\tilde{X}}(f \circ \pi)\tilde{Y}) - \tilde{X}(\phi) fY - (\tilde{fY})(\phi)X \\ + e^{2\phi}\pi_*(grad_\pi \phi) g_B(X, fY). \tag{12}$$

Now, consider

$$\pi_*(\nabla_{\tilde{X}}(f \circ \pi)\tilde{Y}) = \pi_*((\tilde{X}(f \circ \pi))\tilde{Y} + (f \circ \pi)\nabla_{\tilde{X}} \tilde{Y}) \\ = X(f)Y + f\nabla^*_X Y + \tilde{X}(\phi) fY + f\tilde{Y}(\phi)X \\ - e^{2\phi}\pi_*(grad_\pi \phi) f g_B(X, Y). \tag{13}$$

From (12) and (13), we get

$$\nabla^*_X(fY) = X(f)Y + f\nabla^*_X Y.$$

The other conditions for affine connection can be proved similarly. The uniqueness is clear from the definition.

4.1 Harmonicity of Conformal Statistical Submersion

Definition 9. *Let* $(\mathbf{M}, \nabla, g)$ *and* $(\mathbf{B}, \nabla', \tilde{g})$ *be two statistical manifolds. Then, a semi-Riemannian submersion* $\pi : \mathbf{M} \to \mathbf{B}$ *is said to be statistical submersion if*

$$\pi_*(\nabla_X Y)_p = (\nabla'_{X'} Y')_{\pi(p)},$$

for basic vector fields X, Y *on* $\mathbf{M}$ *which are* π*-related to* X' *and* Y' *on* $\mathbf{B}$ *and* $p \in \mathbf{M}$.

Note 5. Note that every statistical submersion is an affine submersion with horizontal distribution and also every statistical submersion is a harmonic map.

Now, we define the conformal statistical submersion which is a generalization of the statistical submersion.

Definition 10. *Let* $(\mathbf{M}, \nabla, g_m)$ *and* $(\mathbf{B}, \nabla', g_b)$ *be two statistical manifolds of dimensions* n *and* m *respectively* $(n \geq m)$*. A submersion* $\pi : (\mathbf{M}, \nabla, g_m) \longrightarrow (\mathbf{B}, \nabla', g_b)$ *is called a conformal statistical submersion if there exists a smooth function* ϕ *on* $\mathbf{M}$ *such that*

$$g_m(X, Y) = e^{2\phi} g_b(\pi_* X, \pi_* Y) \tag{14}$$

$$\pi_*(\nabla_X Y) = \nabla'_{\pi_* X} \pi_* Y + X(\phi)\pi_* X + Y(\phi)\pi_* X - \pi_*(grad_\pi \phi) g_m(X, Y), \tag{15}$$

for basic vector fields X *and* Y *on* $\mathbf{M}$.

Note 6. If ϕ is a constant, then π is a statistical submersion. Also, note that conformal statistical submersions are conformal submersions with horizontal distribution.

Next, we prove that harmonicity and conformality cannot coexist.

Theorem 4. *Let* $\pi : (\mathbf{M}, \nabla, g_m) \longrightarrow (\mathbf{B}, \nabla', g_b)$ *be a conformal statistical submersion. Then,* π *is a harmonic map if and only if* ϕ *is constant.*

Proof. Assume that ϕ is a constant, then by Eqs. (15) and (6) we get π is a harmonic map. Conversely, assume π is harmonic. Now, consider the equations

$$\begin{aligned}
\tau(\pi) &= \sum_{i,j=1}^{n} g_m^{ij} \left\{ \tilde{\nabla}_{\frac{\partial}{\partial x^i}} \left(\pi_*(\frac{\partial}{\partial x^j}) \right) - \pi_* \left(\nabla_{\frac{\partial}{\partial x^i}} \frac{\partial}{\partial x^j} \right) \right\} \\
&= n\pi_*(grad_\pi \phi) - 2 \sum_{i,j=1}^{n} \frac{\partial \phi}{\partial x_i} g_m^{ij} \pi_*(\frac{\partial}{\partial x_j}).
\end{aligned} \tag{16}$$

and

$$\begin{aligned}
g_b(\tau(\pi), \pi_* \frac{\partial}{\partial x_k}) &= ne^{-2\phi} g_m(grad_\pi \phi, \frac{\partial}{\partial x_k}) - 2e^{-2\phi} \sum_{i,j=1}^{m} \frac{\partial \phi}{\partial x_i} g_m^{ij} g_{mjk} \\
&= ne^{-2\phi} g_m(grad_\pi \phi, \frac{\partial}{\partial x_k}) - 2e^{-2\phi} \left(\sum_{i=1}^{n} \frac{\partial \phi}{\partial x_i} \right) \\
&= ne^{-2\phi} \frac{\partial \phi}{\partial x_k} - 2e^{-2\phi} \left(\sum_{i=1}^{n} \frac{\partial \phi}{\partial x_i} \right).
\end{aligned} \tag{17}$$

Since π is harmonic, from (17) we get

$$n\frac{\partial \phi}{\partial x_k} = 2\sum_{i=1}^{n} \frac{\partial \phi}{\partial x_i}, \tag{18}$$

for each $k \in \{1, 2, ..n\}$. That is, we have the system of equations

$$\begin{bmatrix} 2-n & 2 & 2 \dots & 2 \\ 2 & 2-n & 2 \dots & 2 \\ \dots & \dots & \dots\dots & \dots \\ 2 & 2 & 2 \dots & 2-n \end{bmatrix} \begin{bmatrix} \frac{\partial \phi}{\partial x_1} \\ \frac{\partial \phi}{\partial x_2} \\ \dots \\ \frac{\partial \phi}{\partial x_n} \end{bmatrix} = \begin{bmatrix} 0 \\ 0 \\ \dots \\ 0 \end{bmatrix}.$$

Since for each fixed n the above nxn matrix is invertible, we get $\frac{\partial \phi}{\partial x_k} = 0$ for all k. Hence, ϕ is constant. Thus, π is a statistical submersion.

Acknowledgements. The first named author was supported by the Doctoral Research Fellowship from the Indian Institute of Space Science and Technology (IIST), Department of Space, Government of India.

References

1. Abe, N., Hasegawa, K.: An affine submersion with horizontal distribution and its applications. Differ. Geom. Appl. **14**(3), 235–250 (2001)
2. Amari, S.: Information Geometry and Its Applications, vol. 194. Springer, Heidelberg (2016). https://doi.org/10.1007/978-4-431-55978-8
3. Fuglede, B.: Harmonic morphisms between Riemannian manifolds. In: Annales de l'institut Fourier, vol. 28, pp. 107–144 (1978)
4. Ishihara, T.: A mapping of Riemannian manifolds which preserves harmonic functions. J. Math. Kyoto Univ. **19**, 215–229 (1979)
5. Kurose, T.: Dual connections and affine geometry. Math. Z. **203**(1), 115–121 (1990). https://doi.org/10.1007/BF02570725
6. Matsuzoe, H., et al.: On realization of conformally-projectively flat statistical manifolds and the divergences. Hokkaido Math. J. **27**(2), 409–421 (1998)
7. Pastore, A.M., Falcitelli, M., Ianus, S.: Riemannian Submersions and Related Topics. World Scientific, London (2004)
8. Şimşir, F.M.: A note on harmonic maps of statistical manifolds. Commun. Fac. Sci. Univ. Ankara Ser. A1 Math. Stat. **68**(2), 1370–1376 (2019)
9. Uohashi, K.: Harmonic maps relative to α-connections. In: Nielsen, F. (ed.) Geometric Theory of Information. SCT, pp. 81–96. Springer, Cham (2014). https://doi.org/10.1007/978-3-319-05317-2_4

Algorithms for Approximating Means of Semi-infinite Quasi-Toeplitz Matrices

Dario A. Bini[1], Bruno Iannazzo[2], and Jie Meng[1(✉)]

[1] Dipartimento di Matematica, Università di Pisa, Pisa, Italy
dario.bini@unipi.it, jie.meng@dm.unipi.it
[2] Dipartimento di Matematica e Informatica, Università di Perugia, Perugia, Italy
bruno.iannazzo@unipg.it

Abstract. We provide algorithms for computing the Karcher mean of positive definite semi-infinite quasi-Toeplitz matrices. After showing that the power mean of quasi-Toeplitz matrices is a quasi-Toeplitz matrix, we obtain a first algorithm based on the fact that the Karcher mean is the limit of a family of power means. A second algorithm, that is shown to be more effective, is based on a generalization to the infinite-dimensional case of a reliable algorithm for computing the Karcher mean in the finite-dimensional case. Numerical tests show that the Karcher mean of infinite-dimensional quasi-Toeplitz matrices can be effectively approximated with a finite number of parameters.

Keywords: Karcher mean · Power mean · Geometric mean · Quasi-Toeplitz matrices · Toeplitz algebra

1 Introduction

Let i be the complex unit, $z = e^{\mathrm{i}\vartheta}$ and $a(\vartheta) = \sum_{i\in\mathbb{Z}} a_i z^i$ be a complex valued function defined for $0 \le \vartheta \le 2\pi$. A semi-infinite Toeplitz matrix $T(a)$ associated with the function a is defined by $(T(a))_{i,j} = a_{j-i}$, where a is said to be the symbol of $T(a)$. Toeplitz matrices have been widely analyzed in the literature due to their elegant theoretical properties and the important role played in the applications (see, for instance, [2,13–16]).

Let $\mathcal{B}(\ell^2)$ be the C^*-algebra of linear bounded operators in $\ell^2 = \{(v_i)_i : \sum_{i=1}^{\infty} |v_i|^2 < \infty\}$. A Toeplitz matrix $T(a)$ can be associated with a bounded operator in $\mathcal{B}(\ell^2)$ if and only if a is essentially bounded, but the set of Toeplitz matrices is not closed by multiplication. Here, we are interested in the set $\mathcal{QT} = \{T(a) + K,\ a \in C_{2\pi},\ K \in \mathcal{K}\}$, where $C_{2\pi}$ are the continuous functions $f : [0, 2\pi] \to \mathbb{C}$ such that $f(0) = f(2\pi)$ and $\mathcal{K}$ is the set of compact operators in ℓ^2. The set $\mathcal{QT}$ is the smallest C^*-subalgebra of $\mathcal{B}(\ell^2)$ containing all Toeplitz operators $T(a)$ with $a \in C_{2\pi}$ [15]. In what follows, we call quasi-Toeplitz matrix

The work of the first and second authors is partly supported by the INdAM through a GNCS project.

F. Nielsen and F. Barbaresco (Eds.): GSI 2021, LNCS 12829, pp. 405–414, 2021.
https://doi.org/10.1007/978-3-030-80209-7_45

any operator in $\mathcal{QT}$, since it can be represented by a semi-infinite matrix and when we write $A = T(a) + K \in \mathcal{QT}$, we will assume implicitly that $a \in C_{2\pi}$ and $K \in \mathcal{K}$. An operator A is said to be positive definite if there exists $\gamma > 0$ such that $x^*Ax \geq \gamma \sum_{i=1}^{\infty} |x_i|^2$ for $x \in \ell^2$, where x^* is the conjugate transpose of x.

Quasi-Toeplitz matrices, have received considerable attention due to their applications in the analysis of stochastic process and in the numerical solutions of certain PDE problems and have been investigated in [6–9,11,12] in terms of their theoretical and computational properties.

Geometric means of positive definite quasi-Toeplitz matrices have been investigated recently in [5], where it has been shown that the techniques developed for the infinite-dimensional case are effective for large-scale finite size Toeplitz matrices as well. This enlarges the range of applications including all cases where large-scale Toeplitz matrices should be averaged, such as a radar detection model discussed in [22,27] (see also [18], and the references therein, for applications where means of large-scale matrices are required).

One of the results of [5] is that the quasi-Toeplitz structure is preserved by the classical geometric means of operators, such as the ALM, the NBMP, and the Karcher mean[1]: if G is the ALM, NBMP, or the Karcher mean of positive definite matrices $A_i = T(a_i) + K_i \in \mathcal{QT}$ for $i = 1, \dots, p$, then $G = T(g) + K_G \in \mathcal{QT}$. Moreover, it turns out that $g(z) = (a_1(z) \dots a_p(z))^{\frac{1}{p}}$.

The geometric mean of two positive definite matrices A and B is established as $A \#_{1/2} B$, where $A \#_t B := A^{1/2}(A^{-1/2}BA^{-1/2})^t A^{1/2}$, for $t \in [0, 1]$. For more than two matrices, there are several different definitions. For positive matrices A, B and C of the same finite size, the ALM mean, proposed by Ando, Li and Mathias [1], is the common limit of the sequences

$$A_{k+1} = B_k \#_{\frac{1}{2}} C_k, \quad B_{k+1} = C_k \#_{\frac{1}{2}} A_k, \quad C_{k+1} = A_k \#_{\frac{1}{2}} B_k, \tag{1}$$

with $A_0 = A, B_0 = B, C_0 = C$. The NBMP mean introduced independently by Bini, Meini, Poloni [10] and Nakamura [26], is the common limit of the sequences

$$A_{k+1} = A_k \#_{\frac{2}{3}} (B_k \#_{\frac{1}{2}} C_k),\ B_{k+1} = B_k \#_{\frac{2}{3}} (C_k \#_{\frac{1}{2}} A_k),\ C_{k+1} = C_k \#_{\frac{2}{3}} (A_k \#_{\frac{1}{2}} B_k), \tag{2}$$

with $A_0 = A$, $B_0 = B$, $C_0 = C$. The sequences (1) and (2) are well-defined and converge in norm also when $A, B, C \in \mathcal{B}(\ell^2)$, and can be extended to more than three matrices and operators and, specifically, to quasi-Toeplitz matrices. We refer the reader to [5] for the details.

Here we focus, in particular, on the Karcher mean $\Lambda := \Lambda(A_1, \dots, A_p)$, of the positive definite operators $A_1, \dots, A_p$, that is the unique positive definite solution of the operator equation

$$\sum_{i=1}^{p} \log(X^{-1/2} A_i X^{-1/2}) = 0. \tag{3}$$

[1] Recently, the term "Karcher mean" for matrices is falling out of use, in favor of "Riemannian center of mass" or "matrix geometric mean". We kept the former here, because we deal with operators, with no underlying Riemannian structure.

Notice that the operator $X^{-1/2}A_iX^{-1/2}$ is positive definite and the logarithm and square root can be defined using continuous functional calculus. The theoretical background on the Karcher mean can be found in [24] and [23].

This paper may act as a complementary to [5]. Indeed, we start by showing that not only the Karcher mean, but also more general two variable means and the power mean of positive definite quasi-Toeplitz matrices belong to $\mathcal{QT}$. The power mean $P_s := P_s(A_1, \dots, A_p)$, with $s \in (0, 1]$, is the unique solution of the equation [24]

$$\sum_{i=1}^{p} \frac{1}{p}(X \#_s A_i) = X. \tag{4}$$

Showing that P_s with $s \in (0, 1]$ belongs to $\mathcal{QT}$, besides being interesting per se, allows us to get a constructive proof that the Karcher mean $\Lambda(A_1, \dots, A_p)$ belongs to $\mathcal{QT}$, based on the convergence of P_s to the Karcher mean in norm as $s \to 0$ [25]. This may be the basis of an algorithm for approximating the Karcher mean of quasi-Toeplitz matrices, by computing a power mean with a small s.

However, the convergence of P_s to Λ is very slow as $s \to 0$, and while the iteration $X_{k+1} = \sum_{i=1}^{p} \frac{1}{p}(X_k \#_s A_i)$ is globally convergent to the power mean, with at least linear rate $1 - s$, the rate is near to 1 as $s \to 0$. These facts make it difficult in practice to approximate the Karcher mean of quasi-Toeplitz matrices using power means.

A remedy that we propose is to use an algorithm that works well in the finite-dimensional case, that is the Richardson-like iteration of [4]. In our numerical tests the Richardson-like iteration is able to approximate the Karcher mean with a reasonable accuracy and CPU time.

2 Means of Two Quasi-Toeplitz Matrices

Using continuous function calculus, a scalar function $f : \mathcal{I} \to \mathbb{R}$ yields an operator function $f(M)$, when $M \in \mathcal{B}(\ell^2)$ is self-adjoint and has spectrum in $\mathcal{I}$.

In this way one can give a meaning to A^t, for $t \in \mathbb{R}$, when A is positive definite, and to the weighted geometric mean $A \#_t B$ of two positive definite operators A and B. In [5] we have proved that if $A, B \in \mathcal{QT}$ and positive, then $A \#_t B \in \mathcal{QT}$. We will prove that several means or objects with this form are indeed in $\mathcal{QT}$. The following result is an immediate consequence of [5, Theorem 7].

Theorem 1. *Let $A, B \in \mathcal{QT}$, A positive definite and B self-adjoint, $f : \mathcal{U} \subset \mathbb{R} \to \mathbb{R}$. If $\mathcal{U} = \mathbb{R}$ then $f(A; B) := A^{1/2} f(A^{-1/2} B A^{-1/2}) A^{1/2} \in \mathcal{QT}$. If B is positive definite and $\mathcal{U} = \mathbb{R}_0^+$, then $f(A; B) \in \mathcal{QT}$, moreover, if $f(\mathbb{R}_0^+) \subset \mathbb{R}_0^+$ then $f(A; B)$ is positive definite. In both cases, $f(A; B) = T(g) + K$, with $g = af(b/a)$, where a and b are symbols associated with A and B, respectively.*

In this way we have $A \#_t B = f(A; B)$ for $f(x) = x^t$, and we can define $\exp(A; B) := A^{1/2} \exp(A^{-1/2} B A^{-1/2}) A^{1/2} \in \mathcal{QT}$, that is positive definite if A is positive definite and B is self-adjoint; $\log(A; B) := A^{1/2} \log(A^{-1/2} B A^{-1/2}) A^{1/2} \in \mathcal{QT}$, that is self-adjoint for A, B positive definite.

Theorem 1 allows one to define a mean in the sense of Kubo and Ando [21] of quasi-Toeplitz matrices as $f(A; B)$, where $f : \mathbb{R}^+ \to \mathbb{R}^+$ is a normalized operator monotone function, that is $f(1) = 1$ and $f(M_1) - f(M_2)$ is positive definite when $M_1 - M_2$ is positive definite. The symbol associated with the mean is $g := af(b/a)$. This class of means includes the most common two-variable operator means.

3 Power Mean

For $s \in (0, 1]$, the power mean $P_s := P_s(A_1, \dots, A_p)$ of the positive definite operators $A_1, \dots, A_p \in \mathcal{B}(\ell^2)$, is the unique positive definite solution to (4). The uniqueness has been proved in [24], where the authors show that the iteration

$$X_{k+1} = \sum_{i=1}^{p} \frac{1}{p}(X_k \#_s A_i), \qquad X_0 \text{ positive definite,} \tag{5}$$

converges in the Thompson metric on the cone of positive operators of $\mathcal{B}(\ell^2)$ to a limit X and the limit is the unique solution to (4). Note that for matrices there is an alternative definition of power mean [17] but it does not readily extend to operators, thus it will not be discussed here.

The equivalence between convergence in the Thompson metric and norm convergence in $\mathcal{B}(\ell^2)$, and the fact that $\mathcal{QT}$ is a C^*-algebra, gives us the proof that the power mean belongs to $\mathcal{QT}$, but first we need a technical result (see [5, Lemma 9]). For $a \in C_{2\pi}$ we denote $\|a\|_\infty = \max_{x \in [0,2\pi]} |a(x)|$.

Lemma 1. *Let $\{A_k\}_k$ be a sequence of quasi-Toeplitz matrices such that $A_k = T(a_k) + K_k \in \mathcal{QT}$ for $k \in \mathbb{Z}^+$. Let $A \in \mathcal{B}(\ell^2)$ be such that $\lim_k \|A - A_k\| = 0$. Then $A = T(a) + K \in \mathcal{QT}$, moreover, $\lim_k \|a - a_k\|_\infty = 0$ and $\lim_k \|K - K_k\| = 0$.*

We are ready to prove that the power mean of more than two quasi-Toeplitz matrices is in $\mathcal{QT}$. For two matrices the result follows from Sect. 2, since the power mean is $f(A; B)$, with $f(x) = (\frac{1+x^s}{2})^{1/s}$.

Theorem 2. *Let $A_i = T(a_i) + E_i \in \mathcal{QT}$, for $i = 1, \dots, p$, $p \geq 3$, be positive definite. Then the matrices $\{X_k\}$ generated by (5) with $X_0 = \frac{1}{p}\sum_{i=1}^{p} A_i$, and the power mean P_s, for $s \in (0, 1]$, satisfy the following properties*

1. *$X_k = T(x_k) + K_k \in \mathcal{QT}$, for any $k \geq 1$;*
2. *$P_s = T(x^{(s)}) + K^{(s)} \in \mathcal{QT}$, where $x^{(s)} = \left(\frac{\sum_{i=1}^{p} a_i^s}{p}\right)^{1/s}$;*
3. *the symbols x_k associated with the Toeplitz part of X_k satisfy the equation $x_{k+1} = (x^{(s)})^s x_k^{1-s}$, $x_0 = \frac{1}{p}\sum_{i=1}^{p} a_i$;*
4. *$\lim_k \|x_k - x^{(s)}\|_\infty = 0$, $\lim_k \|K_k - K^{(s)}\| = 0$.*

Proof. Concerning part 1, since $X_0 = \frac{1}{p}\sum_{i=1}^{p} A_i \in \mathcal{QT}$, we have that $X_1 \in \mathcal{QT}$ because of [5, Corollary 8] and the fact that $\mathcal{QT}$ is an algebra. An induction argument shows that $X_k = T(x_k)+K_k \in \mathcal{QT}$ for $k \geq 1$. Using the same corollary and the arithmetic of quasi-Toeplitz matrices, one proves that for the symbol x_k we have $x_{k+1} = \frac{1}{p}\sum_{i=1}^p a_i^s x_k^{1-s}$, with $x_0 = \frac{1}{p}\sum_{i=1}^p a_i$, that is part 3. Since X_k converges to P_s in the Thompson metric [24] then it converges in norm that is $\lim_k \|X_k - P_s\| = 0$, according to Lemma 1, we have that $P_s = T(\varphi)+K^{(s)} \in \mathcal{QT}$, moreover, $\lim_k \|x_k - \varphi\|_\infty = 0$ and $\lim_k \|K_k - K^{(s)}\| = 0$. It remains to prove that $\varphi = x^{(s)}$, but it is easily seen that the sequence $\{x_k\}_k$ pointwise converges to $x^{(s)}$ and thus $\varphi = x^{(s)}$.

For computational purposes, it might be convenient to get a bound for the convergence rate of $\{x_k\}$ to $x^{(s)}$. Setting $y_k := x_k/x^{(s)}$, we obtain $y_{k+1} = y_k^{1-s}$, that gives $y_k = y_0^{(1-s)^k}$. Since $y_0 \geq 1$ by the classical inequalities, we have that y_k is monotonically decreasing to 1 and, by the mean value theorem, there exists $\xi_k \in (0, (1-s)^k)$ such that $y_k - 1 = y_0^{\xi_k} \log(y_0)(1-s)^k \leq y_0 \log(y_0)(1-s)^k$, thus

$$\|x_k - x^{(s)}\|_\infty \leq \|x^{(s)}\|_\infty \|y_0 \log(y_0)\|_\infty (1-s)^k. \tag{6}$$

This proves that the convergence is at least linear with rate $(1-s)$.

The approximation through power means yields a constructive proof that the Karcher mean of quasi-Toeplitz matrices belongs to $\mathcal{QT}$ (a non-constructive proof can be found in [5]). It has been proved in [25, Theorem 7.7] that, in norm, $\lim_{s\to 0^+} P_s(A_1,\dots,A_p) = \Lambda(A_1,\dots,A_p)$. If $A_1,\dots,A_p \in \mathcal{QT}$, then, since $P_{1/n} = T(x^{(1/n)}) + K_n \in \mathcal{QT}$ for any positive integer n, according to Lemma 1, the limit of $P_{1/n}$ belongs to $\mathcal{QT}$, that is $\Lambda(A_1,\dots,A_p) = T(g) + K_g \in \mathcal{QT}$. Lemma 1 guarantees that $\lim_k \|x^{(1/n)} - g\|_\infty = 0$, and since the scalar power mean converges pointwise to the geometric mean, we have that $g = (a_1 \cdots a_p)^{1/p}$.

In order to use these results to approximate the Karcher mean, it is useful to get a bound for the uniform convergence of the sequence $\{x^{(1/n)}\}_n$ to g.

For any $s \in (0,1]$ we find that $(x^{(s)})^s = \frac{1}{p}\sum_{i=1}^p e^{s\log a_i} \leq \frac{1}{p}\sum_{i=1}^p (1 + s\log a_i + \frac{s^2}{2}\log^2 a_i a_i^s) = 1 + s\log g + s^2\sigma(s)$, where $\sigma(s) := \frac{1}{2}\sum_{i=1}^p a_i^s \log^2 a_i$, and thus $\log x^{(s)} - \log g = \frac{1}{s}\log(x^{(s)})^s - \log g \leq s\sigma(s)$. Finally, since $x^{(s)} \geq g > 0$, we have $\frac{x^{(s)}}{g} - 1 = \exp(\log x^{(s)} - \log g) - 1 \leq (\log x^{(s)} - \log g)\frac{x^{(s)}}{g} \leq s\sigma(s)\frac{x^{(s)}}{g}$, from which, since $\sigma(s)$ and $x^{(s)}$ are non decreasing with respect to s, we have

$$\|x^{(s)} - g\|_\infty \leq s\|\sigma(1)\|_\infty \|x^{(1)}\|_\infty. \tag{7}$$

The latter bound suggests a sublinear rate of convergence, for $s = 1/n$.

4 Computing the Karcher Mean in $\mathcal{QT}$

Let $A_1,\dots,A_p \in \mathcal{QT}$ be positive definite, with $A_i = T(a_i) + E_i$, $a_i \in C_{2\pi}$, for $i = 1,\dots,p$. The convergence of the power mean $P_{1/n} := P_{1/n}(A_1,\dots,A_p)$ to the Karcher mean $\Lambda := \Lambda(A_1,\dots,A_p)$ in the $\mathcal{QT}$ algebra, allows one to approximate Λ with $P_{1/n}$ for a sufficiently large n. In turn, the power mean $P_{1/n}$ can be computed using iteration (5), for an initial value X_0.

The computation can be made practical using the Matlab CQT-Toolbox [11], that implements the main operations between quasi-Toeplitz matrices and can be easily extended to other operations such as the sharp operator in (5).

Unfortunately the convergence of $P_{1/n}$ to Λ and the convergence of the iteration (5) can be very slow. Indeed, considering just the convergence of the symbols, the bound (7) shows that we need to choose a large value of n to get a good approximation of Λ. On the other hand, the bound (6) suggests linear convergence, with rate $(1 - 1/n)$, of the iteration for computing $P_{1/n}$.

As a remedy, we propose to approximate the Karcher mean by using the Richardson-like iteration

$$X_{k+1} = X_k + \vartheta_k X_k^{1/2}\Big(\sum_{i=1}^{p} \log(X_k^{-1/2} A_i X_k^{-1/2})\Big)X_k^{1/2}, \tag{8}$$

where $\{\vartheta_k\}_k$ is a given sequence. The method has been proposed in [4] for the finite-dimensional case, but using the arithmetic of quasi-Toeplitz matrices and continuous functional calculus on quasi-Toeplitz matrices (see [5, Theorem 7]), the same iteration can be applied to positive definite quasi-Toeplitz matrices.

Two important issues in the use of iteration (8) are the choice of the starting point X_0 and the choice of ϑ_k. The latter task has been worked out in [4] by an asymptotic analysis. Concerning the choice of X_0, a value closer to the limit is recommended to provide a faster convergence. A possible choice is the Cheap mean [3] that, for positive definite matrices, is the common limit of the sequences

$$X_{k+1}^{(\ell)} = (X_k^{(\ell)})^{1/2}\Big(\sum_{i=1}^{p} \log((X_k^{(\ell)})^{-1/2} X_k^{(i)} (X_k^{(\ell)})^{-1/2})\Big)(X_k^{(\ell)})^{1/2}, \quad \ell = 1, \dots, p,$$

for $X_0^{(\ell)} = A_\ell \in \mathcal{QT}$. The iteration yields a sequence in $\mathcal{QT}$.

We have implemented the Richardson-like iteration with the Cheap mean as initial value and the optimal choice of ϑ_k from [4] in Test 1. The implementations, performed in Matlab v.9.5, have been run on an Intel i3 CPU with 4 Gb memory.

To measure errors, we have used the norm $\|A\|_\infty = \sup_i \sum_j |a_{ij}|$, that is well defined for $A = T(a) + K \in \mathcal{QT}$ with absolutely convergent Fourier series.

Test 1. We compute the Karcher mean of the six tridiagonal Toeplitz matrices $A_i = T(a_i)$, for $i = 1, \dots, 6$, where a_i is a trigonometric polynomial of the type $f_0 + 2f_1 \cos(t) + 2f_2 \cos(2t)$, with the following choices for f_0, f_1, f_2:

$$\{3, 1, 0\}, \quad \{10, 4, 4\}, \quad \{4, 2, 1\}, \quad \{2.5, 1, 0\}, \quad \{9.5, 4, 4\}, \quad \{3.5, 2, 1\}.$$

The computation is made by using the Richardson-like iteration, where X_0 is the Cheap mean. The computation of the Cheap mean required only three steps to arrive at numerical convergence while the Richardson-like iteration required 22 steps for an overall time of 131 seconds of CPU time. The Karcher mean $\Lambda = T(g) + E_\Lambda$ is such that the symbol $g(\vartheta) = \sum_{i\in\mathbb{Z}} g_i z^i$, has coefficients $|g_i| < \varepsilon$ for $i > 150$, where $\varepsilon = 10^{-15}$. While the compact correction E_Λ has entries $e_{i,j}$ such that $|e_{i,j}| < \varepsilon$ for $i > 151$ or $j > 151$.

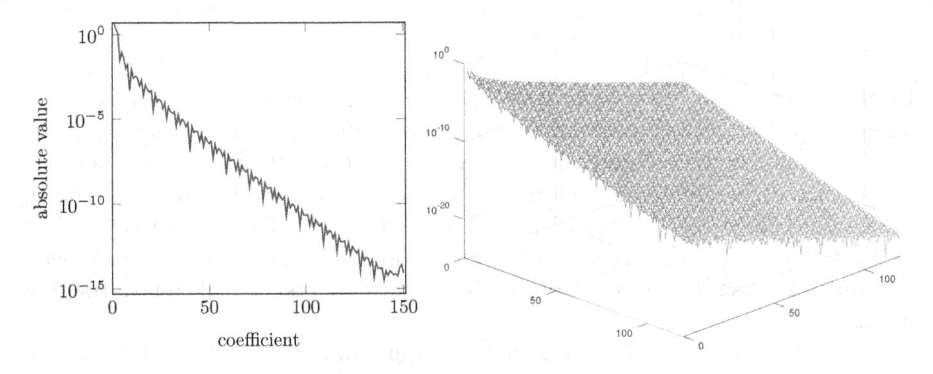

Fig. 1. Log-plot of the symbol g (left) and of the correction E_Λ (right) of the Karcher mean $\Lambda = T(g) + E_\Lambda$ of the dataset of Test 1.

Figure 1 displays a log plot of the modulus of the symbol and of the correction.

Test 2. We compute the power mean P_s of the dataset of Test 1, for $s = 1/2^\ell$, for different values of ℓ, using iteration (5) and compare P_s with the Karcher mean.

Figure 2 shows the convergence of the value of $\|X_{k+1}-X_k\|_\infty$ in iteration (5), starting with the arithmetic mean. Theoretically, this quantity tends to 0, and the obtained bound (6) is linear with parameter $1-s$. Indeed, in our experiment the convergence follows quite exactly the predicted rate.

Moreover, we measure the norm $\|P_s - \Lambda\|_\infty/\|\Lambda\|_\infty$, where Λ is the one computed in Test 1. We observe in Fig. 2 that also the bound (7) is respected, and the errors decreases as s.

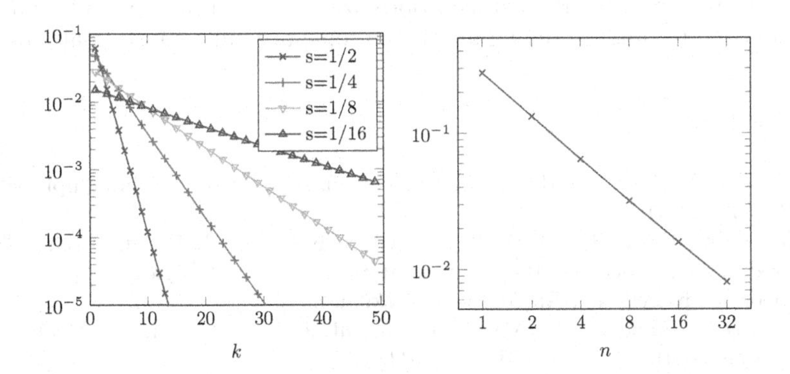

Fig. 2. Left: convergence of the sequence $\{X_k\}_k$ from (5) with the arithmetic mean as initial value, in terms of $\|X_{k+1} - X_k\|_\infty$. Right: error in the approximation of the Karcher mean with the Power mean $P_{1/n}$, in logarithmic scale, measured as $\|P_{1/n} - \Lambda\|_\infty/\|\Lambda\|_\infty$.

5 Conclusions

In previous work [5] we have proved that most common geometric means, such as the ALM, the NBMP and the Karcher mean can be extended to semi-infinite quasi-Toeplitz matrices, and that the ALM and the NBMP can be efficiently approximated by a finite number of parameters. The same techniques were shown to be useful also when dealing with large-scale finite size Toeplitz matrices. Here we have shown that any two variable mean of operators and also the power mean by Lim and Palfia if applied to quasi-Toeplitz arguments, provide quasi-Toeplitz means.

We have considered also the numerical approximation of the Karcher mean of positive definite quasi-Toeplitz matrices, considering two approaches: one is based on approximating the mean as the limit of power means and the other is based on extending to operators the Richardson-like iteration, a reliable algorithm for finite-dimensional matrices.

Some numerical results show that the power mean with parameter s of quasi-Toeplitz matrices can be effectively computed by an iteration but it is not recommended as an approximation of the Karcher mean, since in order to get a good approximation, a small s should be chosen and this leads to a large computational cost of the iteration, for a given tolerance.

On the other hand, in our numerical tests, the Richardson-like algorithm shows a linear convergence, as in the finite dimensional case. While in the finite dimensional case the algorithm has been understood as an approximate Riemannian gradient descent with respect to a Riemannian structure in the cone of positive definite matrices, in the infinite dimensional case this interpretation fails because of the lack of Riemannian structure, but the algorithm has a similar behavior. This fact, besides being interesting per se, opens in principle the possibility to use faster Riemannian optimization algorithms [19,20], and thus deserves a better understanding and is the topic of a future investigation.

References

1. Ando, T., Li, C.-K., Mathias, R.: Geometric means. Linear Algebra Appl. **385**(1), 305–334 (2004)
2. Bini, D.A., Ehrhardt, T., Karlovich, A.Y., Spitkovsky, I.M. (eds.): Large Truncated Toeplitz Matrices, Toeplitz Operators, and Related Topics, Series: Operator Theory: Advances and Applications. Birkhäuser (2017)
3. Bini, D.A., Iannazzo, B.: A note on computing Matrix Geometric Means. Adv. Comput. Math. **35–2**(4), 175–192 (2011)
4. Bini, D.A., Iannazzo, B.: Computing the Karcher mean of symmetric positive definite matrices. Linear Algebra Appl. **438**(4), 1700–1710 (2013)
5. Bini, D.A., Iannazzo, B., Meng, J.: Geometric mean of quasi-Toeplitz matrices: arXiv preprint. https://arxiv.org/abs/2102.04302 (2021)
6. Bini, D.A., Massei, S., Meini, B.: Semi-infinite quasi-Toeplitz matrices with applications to QBD stochastic processes. Math. Comput. **87**(314), 2811–2830 (2018)

7. Bini, D.A., Massei, S., Meini, B.: On functions of quasi-Toeplitz matrices. Sb. Math. **208**(11), 56–74 (2017). 1628
8. Bini, D.A., Massei, S., Meini, B., Robol, L.: On quadratic matrix equations with infinite size coefficients encountered in QBD stochastic processes. Numer. Linear Algebra Appl. **25**(6), e2128 (2018)
9. Bini, D.A., Massei, S., Meini, B., Robol, L.: A computational framework for two-dimensional random walks with restarts. SIAM J. Sci. Comput. **42**(4), A2108–A2133 (2020)
10. Bini, D.A., Meini, B., Poloni, F.: An effective matrix geometric mean satisfying the Ando-Li-Mathias properties. Math. Comput. **79**(269), 437–452 (2010)
11. Bini, D.A., Massei, S., Robol, L.: Quasi-Toeplitz matrix arithmetic: a MATLAB toolbox. Numer. Algorithms **81**(2), 741–769 (2019). https://doi.org/10.1007/s11075-018-0571-6
12. Bini, D.A., Meini, B., Meng, J.: Solving quadratic matrix equations arising in random walks in the quarter plane. SIAM J. Matrix Anal. Appl. **41**(2), 691–714 (2020)
13. Böttcher, A., Grudsky, S.M.: Toeplitz Matrices, Asymptotic Linear Algebra, and Functional Analysis. Birkhäuser Verlag, Basel (2000)
14. Böttcher, A., Grudsky, S.M.: Spectral Properties of Banded Toeplitz Matrices. SIAM (2005)
15. Böttcher, A., Silbermannn, B.: Introduction to Large Truncated Toeplitz Matrices. Springer, New York (1999). https://doi.org/10.1007/978-1-4612-1426-7
16. Chan, R.H.-F., Jin, X.-Q.: An Introduction to Iterative Toeplitz Solvers, SIAM (2007)
17. Chouaieb, N., Iannazzo, B., Moakher M.: Geometries on the cone of positive-definite matrices derived from the power potential and their relation to the power means: arXiv preprint https://arxiv.org/abs/2102.10279 (2021)
18. Fasi, M., Iannazzo, B.: Computing the weighted geometric mean of two large-scale matrices and its inverse times a vector. SIAM J. Matrix Anal. Appl. **39**(1), 178–203 (2018)
19. Yuan, X., Huang, W., Absil, P.-A., Gallivan, K.A.: Computing the matrix geometric mean: Riemannian versus Euclidean conditioning, implementation techniques, and a Riemannian BFGS method. Numer. Linear Algebra Appl. **27**(5), e2321 (2020)
20. Iannazzo, B., Porcelli, M.: The Riemannian Barzilai–Borwein method with nonmonotone line search and the matrix geometric mean computation. IMA J. Numer. Anal. **38**(1), 495–517 (2018)
21. Kubo, F., Ando, T.: Means of positive Linear Operators. Math. Ann. **246**(3), 205–224 (1980)
22. Lapuyade-Lahorgue, J., Barbaresco, F.: Radar detection using Siegel distance between autoregressive processes, application to HF and X-band radar. In: 2008 IEEE Radar Conference, pp. 1–6 (2008)
23. Lawson, J., Lim, Y.: Karcher means and Karcher equations of positive definite operators. Trans. Amer. Math. Soc. Ser. B **1**, 1–22 (2014)
24. Lim, Y., Pálfia, M.: Matrix power means and the Karcher mean. J. Funct. Anal. **262**(4), 1498–1514 (2012)
25. Lim, Y., Pálfia, M.: Existence, uniqueness and an ODE approach to the L^1 Karcher mean. Adv. Math. **376**, 107435 (2021)

26. Nakamura, N.: Geometric means of positive operators. Kyungpook Math. J. **49**(1), 167–181 (2009)
27. Yang, L., Arnaudon, M., Barbaresco, F.: Geometry of covariance matrices and computation of median. In: Bayesian Inference and Maximum Entropy Methods in Science and Engineering, volume 1305 of AIP Conference Proceedings, pp. 479–486. American Institute of Physics, Melville (2010)

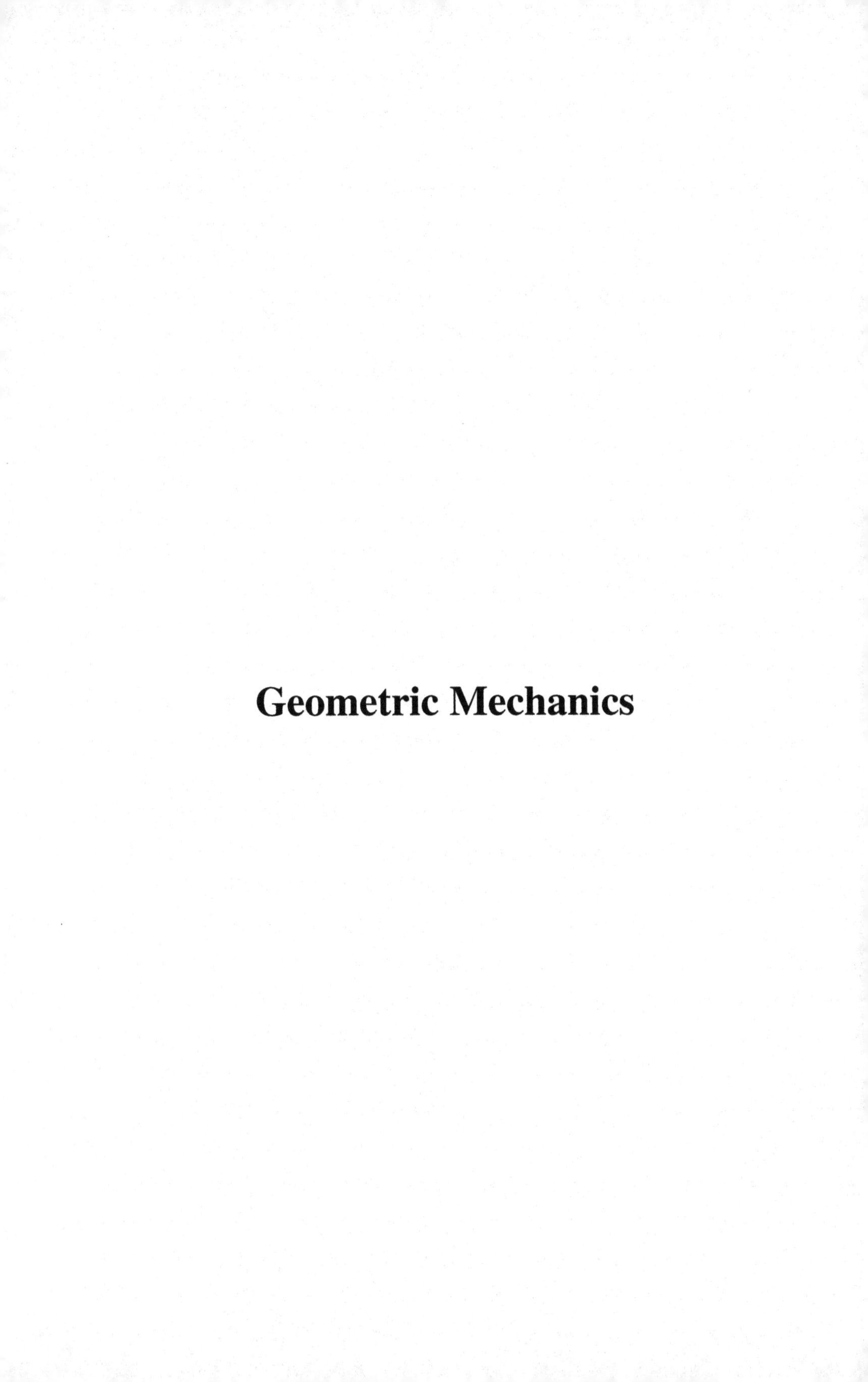

Geometric Mechanics

Archetypal Model of Entropy by Poisson Cohomology as Invariant Casimir Function in Coadjoint Representation and Geometric Fourier Heat Equation

Frédéric Barbaresco(✉)

THALES Land and Air Systems, Meudon, France
frederic.barbaresco@thalesgroup.com

Abstract. In 1969, Jean-Marie Souriau introduced a "Lie Groups Thermodynamics" in the framework of Symplectic model of Statistical Mechanics. Based on this model, we will introduce a geometric characterization of Entropy as a generalized Casimir invariant function in coadjoint representation, where Souriau cocycle is a measure of the lack of equivariance of the moment mapping. The dual space of the Lie algebra foliates into coadjoint orbits that are also the level sets on the entropy that could be interpreted in the framework of Thermodynamics by the fact that motion remaining on these surfaces is non-dissipative, whereas motion transversal to these surfaces is dissipative. We will also explain the 2nd Principle in thermodynamics by definite positiveness of Souriau tensor extending the Koszul-Fisher metric from Information Geometry, and introduce a new geometric Fourier heat equation with Souriau-Koszul-Fisher tensor. In conclusion, Entropy as Casimir function is characterized by Koszul Poisson Cohomology.

Keywords: Symplectic geometry · Casimir function · Entropy · Lie groups thermodynamics · Koszul-Fisher metric · Heat equation · Poisson cohomology

1 Introduction

"Ayant eu l'occasion de m'occuper du mouvement de rotation d'un corps solide creux, dont la cavité est remplie de liquide, j'ai été conduit à mettre les équations générales de la Mécanique sous une forme que je crois nouvelle et qu'il peut être intéressant de faire connaître" – Henri Poincaré [23]

We will observe that Souriau Entropy $S(Q)$ defined on affine coadjoint orbit of the group (where Q is a "geometric" heat, element of the dual space of the Lie algebra $\mathfrak{g}^*$of the group) has a property of invariance $S\left(Ad_g^{\#}(Q)\right) = S(Q)$ with respect to Souriau affine definition of coadjoint action $Ad_g^{\#}(Q) = Ad_g^*(Q) + \theta(g)$ where $\theta(g)$ is called the Souriau cocyle, and is associated to the default of equivariance of the moment map [2, 18, 20, 21]. In the framework of Souriau Lie groups Thermodynamics [2–8], we will then characterize the Entropy as a generalized Casimir invariant function [1] in coadjoint representation. When M is a Poisson manifold, a function on M is a Casimir

F. Nielsen and F. Barbaresco (Eds.): GSI 2021, LNCS 12829, pp. 417–429, 2021.
https://doi.org/10.1007/978-3-030-80209-7_46

function if and only if this function is constant on each symplectic leaf (the non-empty open subsets of the symplectic leaves are the smallest embedded manifolds of M which are Poisson submanifolds). Classically, the Entropy is defined axiomatically as Shannon or von Neumann Entropies without any geometric structures constraints. In this paper, the Entropy will be characterized as solution of the Casimir equation given for affine equivariance by

$$\left(ad^*_{\frac{\partial S}{\partial Q}}Q\right)_j + \Theta\left(\frac{\partial S}{\partial Q}\right)_j = C^k_{ij} ad^*_{\left(\frac{\partial S}{\partial Q}\right)^i} Q_k + \Theta_j = 0$$

where $\Theta(X) = T_e\theta(X(e))$ with $\tilde{\Theta}(X,Y) = \langle\Theta(X), Y\rangle = J_{[X,Y]} - \{J_X, J_Y\}$ in non-null cohomology case (non-equivariance of coadjoint operator on the moment map), with $\theta(g)$ the Souriau Symplectic cocycle. The KKS (Kostant-Kirillov Souriau) 2-form that associates a structure of homogeneous symplectic manifold to coadjoint orbits, will be linked with extension of Koszul-Fisher metric. The information manifold foliates into level sets of the entropy that could be interpreted in the framework of Thermodynamics by the fact that motion remaining on these surfaces is non-dissipative, whereas motion transversal to these surfaces is dissipative, where **the dynamics is given by** $\frac{dQ}{dt} = \{Q,H\}_{\tilde{\Theta}} = ad^*_{\frac{\partial H}{\partial Q}}Q + \Theta\left(\frac{\partial H}{\partial Q}\right)$ with the existence of a stable equilibrium when $H = S \Rightarrow \frac{dQ}{dt} = \{Q,S\}_{\tilde{\Theta}} = ad^*_{\frac{\partial S}{\partial Q}}Q + \Theta\left(\frac{\partial S}{\partial Q}\right) = 0$ (algorithm described in [30, 31] preserves coadjoint orbits and Casimirs of the Lie–Poisson equation by construction).

We will also observe that $dS = \tilde{\Theta}_\beta\left(\frac{\partial H}{\partial Q}, \beta\right)dt$ where $\tilde{\Theta}_\beta\left(\frac{\partial H}{\partial Q}, \beta\right) = \tilde{\Theta}\left(\frac{\partial H}{\partial Q}, \beta\right) + \left\langle Q, \left[\frac{\partial H}{\partial Q}, \beta\right]\right\rangle$, showing that 2nd principle is linked to positive definiteness of Souriau tensor related to Fisher Information or Koszul 2-form extension. We can extend **Affine Lie-Poisson equation** $\frac{dQ}{dt} = ad^*_{\frac{\partial H}{\partial Q}}Q + \Theta\left(\frac{\partial H}{\partial Q}\right)$ to a new Stratonovich differential equation for the stochastic process given by the following relation by mean of Souriau's symplectic cocycle:

$$dQ + \left[ad^*_{\frac{\partial H}{\partial Q}}Q + \Theta\left(\frac{\partial H}{\partial Q}\right)\right]dt + \sum_{i=1}^{N}\left[ad^*_{\frac{\partial H_i}{\partial Q}}Q + \Theta\left(\frac{\partial H_i}{\partial Q}\right)\right] \circ dW_i(t) = 0.$$

More details on Souriau Lie Groups Thermodynamics are available in author paper [9–13] or Charles-Michel Marle papers [14–16].

2 Lie Groups Thermodynamics and Souriau-Fisher Metric

Amari has proved that the Riemannian metric in an exponential family is the Fisher information matrix defined by:

$$g_{ij} = -\left[\frac{\partial^2 \Phi}{\partial \theta_i \partial \theta_j}\right]_{ij} \quad \text{with } \Phi(\theta) = -\log \int_R e^{-\langle \theta, y \rangle} dy \tag{1}$$

and the dual potential, the Shannon entropy, is given by the Legendre transform:

$$S(\eta) = \langle \theta, \eta \rangle - \Phi(\theta) \quad \text{with} \quad \eta_i = \frac{\partial \Phi(\theta)}{\partial \theta_i} \quad \text{and} \quad \theta_i = \frac{\partial S(\eta)}{\partial \eta_i} \tag{2}$$

We can observe that $\Phi(\theta) = -\log \int_R e^{-\langle \theta, y \rangle} dy = -\log \psi(\theta) = -\log \psi(\theta)$ is linked with the cumulant generating function. In the Souriau model, the structure of information geometry is preserved and extended on the Symplectic manifolds associated to coadjoint orbits:

$$I(\beta) = -\frac{\partial^2 \Phi}{\partial \beta^2} \text{ with } \Phi(\beta) = -\log \int_M e^{-\langle U(\xi), \beta \rangle} d\lambda_\omega \text{ and } U : M \to \mathfrak{g}^* \tag{3}$$

$$S(Q) = \langle Q, \beta \rangle - \Phi(\beta) \text{ with } Q = \frac{\partial \Phi(\beta)}{\partial \beta} \in \mathfrak{g}^* \text{ and } \beta = \frac{\partial S(Q)}{\partial Q} \in \mathfrak{g} \tag{4}$$

In the Souriau Lie groups thermodynamics model, β is a "geometric" (Planck) temperature, element of Lie algebra $\mathfrak{g}$of the group, and Q is a "geometric" heat, element of the dual space of the Lie algebra $\mathfrak{g}^*$of the group. Souriau has proposed a Riemannian metric that we have identified as a generalization of the Fisher metric:

$$I(\beta) = [g_\beta] \quad \text{with} \quad g_\beta([\beta, Z_1], [\beta, Z_2]) = \tilde{\Theta}_\beta(Z_1, [\beta, Z_2]) \tag{5}$$

$$\text{with } \tilde{\Theta}_\beta(Z_1, Z_2) = \tilde{\Theta}(Z_1, Z_2) + \langle Q, ad_{Z_1}(Z_2) \rangle \text{ where } ad_{Z_1}(Z_2) = [Z_1, Z_2] \tag{6}$$

Souriau Foundamental Theorem is that « Every symplectic manifold on which a Lie group acts transitively by a Hamiltonian action is a covering space of a coadjoint orbit». We can observe that for Souriau model, Fisher metric is an extension of this 2-form in non-equivariant case $g_\beta([\beta, Z_1], [\beta, Z_2]) = \tilde{\Theta}(Z_1, [\beta, Z_2]) + \langle Q, [Z_1, [\beta, Z_2]] \rangle$.

The Souriau additional term $\tilde{\Theta}(Z_1, [\beta, Z_2])$ is generated by non-equivariance through Symplectic cocycle. The tensor $\tilde{\Theta}$ used to define this extended Fisher metric is defined by the moment map $J(x)$, application from M (homogeneous symplectic manifold) to the dual space of the Lie algebra $\mathfrak{g}^*$, given by:

$$\tilde{\Theta}(X, Y) = J_{[X,Y]} - \{J_X, J_Y\} \tag{7}$$

$$\text{with } J(x): M \to \mathfrak{g}^* \text{ such that } J_X(x) = \langle J(x), X \rangle, X \in \mathfrak{g} \,. \tag{8}$$

This tensor $\tilde{\Theta}$ is also defined in tangent space of the cocycle $\theta(g) \in \mathfrak{g}^*$ (this cocycle appears due to the non-equivariance of the coadjoint operator , action of the group on the dual space of the lie algebra; the action of the group on the dual space of the Lie algebra is modified with a cocycle so that the momentum map becomes equivariant relative to this new affine action):

$$Q(Ad_g(\beta)) = Ad_g^*(Q) + \theta(g) \tag{9}$$

I use notation $Ad_g^* = (Ad_{g^{-1}})^*$ with: $\langle Ad_g^* F, Y \rangle = \langle F, Ad_{g^{-1}} Y \rangle, \forall g \in G, Y \in \mathfrak{g}, F \in \mathfrak{g}^*$, as used by Koszul and Souriau. $\theta(g) \in \mathfrak{g}^*$ is called nonequivariance one-cocycle, and it is a measure of the lack of equivariance of the moment map.

$$\begin{array}{ll} \tilde{\Theta}(X,Y): \mathfrak{g} \times \mathfrak{g} \to \Re & \text{with } \Theta(X) = T_e\theta(X(e)) \\ X,Y \mapsto \langle \Theta(X), Y \rangle & \end{array} \tag{10}$$

It can be then deduced that the tensor could be also written by (with cocyle relation):

$$\tilde{\Theta}(X,Y) = J_{[X,Y]} - \{J_X, J_Y\} = -\langle d\theta(X), Y \rangle \ , \ X,Y \in \mathfrak{g} \tag{11}$$

$$\tilde{\Theta}([X,Y],Z) + \tilde{\Theta}([Y,Z],X) + \tilde{\Theta}([Z,X],Y) = 0 \ , \ X,Y,Z \in \mathfrak{g} \tag{12}$$

This study of the moment map J equivariance, and the existence of an affine action of G on $\mathfrak{g}^*$, whose linear part is the coadjoint action, for which the moment J is equivariant, is at the cornerstone of Souriau theory of geometric mechanics and Lie groups thermodynamics. When an element of the group g acts on the element $\beta \in \mathfrak{g}$ of the Lie algebra, given by adjoint operator Ad_g. With respect to the action of the group $Ad_g(\beta)$, the entropy $S(Q)$ and the Fisher metric $I(\beta)$ are invariant:

$$\beta \in \mathfrak{g} \to Ad_g(\beta) \Rightarrow \begin{cases} S\left[Q(Ad_g(\beta))\right] = S(Q) \\ I\left[Ad_g(\beta)\right] = I(\beta) \end{cases} \tag{13}$$

Souriau completed his "geometric heat theory" by introducing a 2-form in the Lie algebra, that is a Riemannian metric tensor in the values of adjoint orbit of β, $[\beta, Z]$ with Z an element of the Lie algebra. This metric is given for (β, Q):

$$g_\beta = ([\beta, Z_1], [\beta, Z_2]) = \langle \Theta(Z_1), [\beta, Z_2] \rangle + \langle Q, [Z_1, [\beta, Z_2]] \rangle \tag{14}$$

where Θ is a cocycle of the Lie algebra, defined by $\Theta = T_e\theta$ with θ a cocycle of the Lie group defined by $\theta(M) = Q(Ad_M(\beta)) - Ad_M^* Q$. We observe that Souriau Riemannian metric, introduced with symplectic cocycle, is a generalization of the Fisher metric, that we call the Souriau-Fisher metric, that preserves the property to be defined as a Hessian of the partition function logarithm $g_\beta = -\frac{\partial^2 \Phi}{\partial \beta^2} = \frac{\partial^2 \log \psi_\Omega}{\partial \beta^2}$ as in classical information geometry. We will establish the equality of two terms, between Souriau definition based on Lie group cocycle Θ and parameterized by "geometric heat" Q (element of the dual space of the Lie algebra) and "geometric temperature" β (element of Lie algebra) and hessian of characteristic function $\Phi(\beta) = -\log \psi_\Omega(\beta)$ with respect to the variable β:

$$g_\beta([\beta, Z_1], [\beta, Z_2]) = \langle \Theta(Z_1), [\beta, Z_2] \rangle + \langle Q, [Z_1[\beta, Z_2]] \rangle = \frac{\partial^2 \log \psi_\Omega}{\partial \beta^2} \quad (15)$$

If one assumes that $U(g\xi) = Ad_g^* U(\xi) + \theta(g), g \in G, \xi \in M$ which means that the energy $U : M \to \mathfrak{g}^*$ satisfies the same equivariance condition as the moment map $\mu : M \to \mathfrak{g}^*$, then one has for $g \in G$ and $\beta \in \Omega$:

$$\begin{aligned}
&\psi_\Omega(Ad_g \beta) = \int_M e^{-\langle U(\xi), Ad_g \beta \rangle} d\lambda(\xi) = \int_M e^{-\left\langle Ad^*_{g^{-1}} U(\xi), \beta \right\rangle} d\lambda(\xi) \\
&\psi_\Omega(Ad_g \beta) = \int_M e^{-\langle U(g^{-1}\xi) - \theta(g^{-1}), \beta \rangle} d\lambda(\xi) = e^{\langle \theta(g^{-1}), \beta \rangle} \int_M e^{-\langle U(g^{-1}\xi), \beta \rangle} d\lambda(\xi) \\
&\psi_\Omega(Ad_g \beta) = e^{\langle \theta(g^{-1}), \beta \rangle} \psi_\Omega(\beta) \\
&\Phi(Ad_g \beta) = -\log \psi_\Omega(Ad_g \beta) = \Phi(\beta) - \langle \theta(g^{-1}), \beta \rangle
\end{aligned} \quad (16)$$

To consider the invariance of Entropy, we have to use the property that

$$Q(Ad_g \beta) = Ad_g^* Q(\beta) + \theta(g) = g.Q(\beta), \beta \in \Omega, g \in G \quad (17)$$

We can prove the invariance under the action of the group:

$$\begin{aligned}
&s(Q(Ad_g \beta)) = \langle Q(Ad_g \beta), Ad_g \beta \rangle - \Phi(Ad_g \beta) \\
&s(Q(Ad_g \beta)) = \left\langle Ad_g^* Q(\beta) + \theta(g), Ad_g \beta \right\rangle - \Phi(\beta) + \langle \theta(g^{-1}), \beta \rangle \\
&s(Q(Ad_g \beta)) = \left\langle Ad_g^* Q(\beta), Ad_g \beta \right\rangle - \Phi(\beta) + \left\langle Ad^*_{g^{-1}} \theta(g) + \theta(g^{-1}), \beta \right\rangle \\
&s(Q(Ad_g \beta)) = s(Q(\beta)) \text{ using } Ad^*_{g^{-1}} \theta(g) + \theta(g^{-1}) = \theta(g^{-1} g) = 0
\end{aligned} \quad (18)$$

For $\beta \in \Omega$, let g_β be the Hessian form on $T_\beta \Omega \equiv \mathfrak{g}$ with the potential $\Phi(\beta) = -\log \psi_\Omega(\beta)$ For $X, Y \in \mathfrak{g}$, we define:

$$g_\beta(X, Y) = \frac{\partial^2 \Phi}{\partial \beta^2}(X, Y) = \left(\frac{\partial^2}{\partial s \partial t} \right)_{s=t=0} \log \psi_\Omega(\beta + sX + tY) \quad (19)$$

The positive definitiveness is given by Cauchy-Schwarz inequality:

$$g_\beta(X,Y) = \frac{1}{\psi_\Omega(\beta)^2}\left\{\begin{array}{l}\int\limits_M e^{-\langle U(\xi),\beta\rangle}d\lambda(\xi)\cdot\int\limits_M \langle U(\xi),X\rangle^2 e^{-\langle U(\xi),\beta\rangle}d\lambda(\xi)\\ -\left(\int\limits_M \langle U(\xi),X\rangle e^{-\langle U(\xi),\beta\rangle}d\lambda(\xi)\right)^2\end{array}\right\}$$
$$= \frac{1}{\psi_\Omega(\beta)^2}\left\{\begin{array}{l}\int\limits_M \left(e^{-\langle U(\xi),\beta\rangle/2}\right)^2 d\lambda(\xi)\cdot\int\limits_M \left(\langle U(\xi),X\rangle e^{-\langle U(\xi),\beta\rangle/2}\right)^2 d\lambda(\xi)\\ -\left(\int\limits_M e^{-\langle U(\xi),\beta\rangle/2}\cdot\langle U(\xi),X\rangle e^{-\langle U(\xi),\beta\rangle/2}d\lambda(\xi)\right)^2\end{array}\right\}\geq 0 \tag{20}$$

We observe that $g_\beta(X,X)=0$ if and only if $\langle U(\xi),X\rangle$ is independent of $\xi\in M$, which means that the set $\{U(\xi);\ \xi\in M\}$ is contained in an affine hyperplane in $\mathfrak{g}^*$ perpendicular to the vector $X\in\mathfrak{g}$. We have seen that $g_\beta=\frac{\partial^2\Phi}{\partial\beta^2}$, that is a **generalization of classical Fisher metric from Information geometry**, and will give the relation the Riemannian metric introduced by Souriau.

$$g_\beta(X,Y)=\left\langle -\frac{\partial Q}{\partial\beta}(X),Y\right\rangle \text{ for } X,Y\in\mathfrak{g} \tag{21}$$

we have for any $\beta\in\Omega, g\in G$ and $Y\in\mathfrak{g}$:

$$\langle Q(Ad_g\beta),Y\rangle = \langle Q(\beta),Ad_{g^{-1}}Y\rangle + \langle\theta(g),Y\rangle \tag{22}$$

Let us differentiate the above expression with respect to g. Namely, we substitute $g=\exp(tZ_1), t\in R$ and differentiate at t = 0. Then the left-hand side of (22) becomes

$$\left(\frac{d}{dt}\right)_{t=0}\langle Q(\beta+t[Z_1,\beta]+o(t^2)),Y\rangle = \left\langle\frac{\partial Q}{\partial\beta}([Z_1,\beta]),Y\right\rangle \tag{23}$$

and the right-hand side of (22) is calculated as:

$$\begin{array}{l}\left(\frac{d}{dt}\right)_{t=0}\langle Q(\beta),Y-t[Z_1,Y]+o(t^2)\rangle + \langle\theta(I+tZ_1+o(t^2)),Y\rangle\\ = -\langle Q(\beta),[Z_1,Y]\rangle + \langle d\theta(Z_1),Y\rangle\end{array} \tag{24}$$

Therefore,

$$\left\langle\frac{dQ}{d\beta}([Z_1,\beta]),Y\right\rangle = \langle d\theta(Z_1),Y\rangle - \langle Q(\beta),[Z_1,Y]\rangle \tag{25}$$

Substituting $Y=-[\beta,Z_2]$to the above expression:

$$\begin{array}{l}g_\beta([\beta,Z_1],[\beta,Z_2]) = \left\langle -\frac{dQ}{d\beta}([Z_1,\beta]),[\beta,Z_2]\right\rangle\\ g_\beta([\beta,Z_1],[\beta,Z_2]) = \langle d\theta(Z_1),[\beta,Z_2]\rangle + \langle Q(\beta),[Z_1,[\beta,Z_2]]\rangle\end{array} \tag{26}$$

We define then symplectic 2-cocycle and the tensor:

$$\begin{aligned}\Theta(Z_1) &= d\theta(Z_1)\\ \widetilde{\Theta}(Z_1, Z_2) &= \langle \Theta(Z_1), Z_2 \rangle = J_{[Z_1, Z_2]} - \{J_{Z_1}, J_{Z_2}\}\end{aligned} \tag{27}$$

Considering $\widetilde{\Theta}_\beta(Z_1, Z_2) = \langle Q(\beta_1), [Z_1, Z_2] \rangle + \widetilde{\Theta}(Z_1, Z_2)$, that is an extension of KKS (Kirillov-Kostant-Souriau) 2 form in the case of non-null cohomology. Introduced by Souriau, we can define this extension Fisher metric with 2-form of Souriau:

$$g_\beta([\beta, Z_1], [\beta, Z_2]) = \widetilde{\Theta}_\beta(Z_1, [\beta, Z_2]) \tag{28}$$

As the entropy is defined by the Legendre transform of the characteristic function, a dual metric of the Fisher metric is also given by the hessian of "geometric entropy" $S(Q)$ with respect to the dual variable given by Q: $\frac{\partial^2 S(Q)}{\partial Q^2}$. The Fisher metric has been considered by Souriau as a ***generalization of* "*heat capacity*"**. Souriau called it K the **"*geometric capacity*"**:$I(\beta) = -\frac{\partial^2 \Phi(\beta)}{\partial \beta^2} = -\frac{\partial Q}{\partial \beta}$.

3 Entropy Characterization as Generalized Casimir Invariant Function in Coadjoint Representation

In his 1974 paper, Jean-Marie Souriau has written $\langle Q, [\beta, Z] \rangle + \widetilde{\Theta}(\beta, Z) = 0$. To prove this equation, we have to consider the parametrized curve $t \mapsto Ad_{\exp(tZ)}\beta$ with $Z \in \mathfrak{g}$ and $t \in R$. The parameterized curve $Ad_{\exp(tZ)}\beta$ passes, for $t = 0$, through the point β, since $Ad_{\exp(0)}$ is the identical map of the Lie Algebra $\mathfrak{g}$. This curve is in the adjoint orbit of β. So by taking its derivative with respect to t, then for $t = 0$, we obtain a tangent vector in β at the adjoint orbit of this point. When Z takes all possible values in $\mathfrak{g}$, the vectors thus obtained generate all the vector space tangent in β to the orbit of this point:

$$\left.\frac{d\Phi\left(Ad_{\exp(tZ)}\beta\right)}{dt}\right|_{t=0} = \left\langle \frac{d\Phi}{d\beta}, \left(\left.\frac{d\left(Ad_{\exp(tZ)}\beta\right)}{dt}\right|_{t=0} \right) \right\rangle = \langle Q, ad_Z \beta \rangle = \langle Q, [Z, \beta] \rangle \tag{29}$$

As we have seen before $\Phi\left(Ad_g \beta\right) = \Phi(\beta) - \left\langle \theta(g^{-1}), \beta \right\rangle$. If we set $g = \exp(tZ)$, we obtain $\Phi\left(Ad_{\exp(tZ)}\beta\right) = \Phi(\beta) - \langle \theta(\exp(-tZ)), \beta \rangle$ and by derivation with respect to t at $t = 0$, we finally recover the equation given by Souriau:

$$\left.\frac{d\Phi\left(Ad_{\exp(tZ)}\beta\right)}{dt}\right|_{t=0} = \langle Q, [Z, \beta] \rangle = -\langle d\theta(-Z), \beta \rangle \text{ with } \tilde{\Theta}(X, Y) = -\langle d\theta(X), Y \rangle \tag{30}$$

Souriau has stopped by this last equation, the characterization of Group action on $Q = \frac{d\Phi}{d\beta}$. Souriau has observed that $S\left[Q\left(Ad_g(\beta)\right)\right] = S\left[Ad_g^*(Q) + \theta(g)\right] = S(Q)$. We propose to characterize more explicitly this invariance, by **characterizing Entropy as an invariant Casimir function in coadjoint representation**. From last Souriau equation, if we use the identities $\beta = \frac{\partial S}{\partial Q}$, $ad_\beta Z = [\beta, Z]$ and $\widetilde{\Theta}(\beta, Z) = \langle \Theta(\beta), Z \rangle$, then we can deduce that $\left\langle ad^*_{\frac{\partial S}{\partial Q}} Q + \Theta\left(\frac{\partial S}{\partial Q}\right), Z \right\rangle = 0, \quad \forall Z$. So, Entropy $S(Q)$ should verify $ad^*_{\frac{\partial S}{\partial Q}} Q + \Theta\left(\frac{\partial S}{\partial Q}\right) = 0$, characterizes **an invariant Casimir function in case of non-null cohomology**, that we propose to write with Poisson brackets, where:

$$\{S, H\}_{\tilde{\Theta}}(Q) = \left\langle Q, \left[\frac{\partial S}{\partial Q}, \frac{\partial H}{\partial Q}\right]\right\rangle + \tilde{\Theta}\left(\frac{\partial S}{\partial Q}, \frac{\partial H}{\partial Q}\right) = 0, \ \forall H : \mathfrak{g}^* \to R, \ Q \in \mathfrak{g}^* \quad (31)$$

We have observed that Souriau Entropy is a Casimir function in case with non-null cohomology when an additional cocycle should be taken into account. Indeed, infinitesimal variation is characterized by the following differentiation:

$$\begin{aligned}\frac{d}{dt} S\left(Q\left(Ad_{\exp(tx)}\beta\right)\right)\Big|_{t=0} &= \frac{d}{dt} S\left(Ad^*_{\exp(tx)} Q + \theta(\exp(tx))\right)\Big|_{t=0} \\ &= -\left\langle ad^*_{\frac{\partial S}{\partial Q}} Q + \Theta\left(\frac{\partial S}{\partial Q}\right), x \right\rangle.\end{aligned}$$

We recover extended Casimir equation in case of non-null cohomology verified by Entropy, $ad^*_{\frac{\partial S}{\partial Q}} Q + \Theta\left(\frac{\partial S}{\partial Q}\right) = 0$, and then the generalized Casimir condition $\{S, H\}_{\widetilde{\Theta}}(Q) = 0$. Hamiltonian motion on these affine coadjoint orbits is given by the solutions of the Lie-Poisson equations with cocycle.

The identification of Entropy as an Invariant Casimir Function in Coadjoint representation is also important in Information Theory, because classically Entropy is introduced axiomatically. With this new approach, we can build Entropy by constructing the Casimir Function associated to the Lie group and also in case of non-null cohomology.

This new equation is important because it introduces new structure of differential equation in case of non-null cohomology. For an arbitrary Hamiltonian $H : \mathfrak{g}^* \to R : \frac{dQ}{dt} = ad^*_{\frac{\partial H}{\partial Q}} Q + \Theta\left(\frac{\partial H}{\partial Q}\right)$ because it allows extending stochastic perturbation of the Lie-Poisson equation with cocycle within the setting of stochastic Hamiltonian dynamics, which preserves the affine coadjoint orbits [22]. This previous Lie-Poisson equation is equivalent to the **modified Lie-Poisson variational principle**:

$$\delta\int_0^\tau\left(\left\langle Q(t),\frac{\partial H}{\partial Q}(t)\right\rangle - H(Q(t))\right)dt = 0 \text{ where } \begin{cases} \frac{\partial H}{\partial Q} = g^{-1}\dot{g}\in\mathfrak{g}, g\in G \\ \frac{\partial^2 H}{\partial Q^2}\delta Q = \delta\left(\frac{\partial H}{\partial Q}\right), \eta = g^{-1}\delta g \\ \left\langle Q,\delta\left(\frac{\partial H}{\partial Q}\right)\right\rangle = \left\langle Q,\dot{\eta}+\left[\frac{\partial H}{\partial Q},\eta\right]\right\rangle + \tilde{\Theta}\left(\frac{\partial H}{\partial Q},\eta\right) \end{cases}$$

$$\delta\int_0^\tau\left(\left\langle Q(t),\frac{\partial H}{\partial Q}(t)\right\rangle - H(Q(t))\right)dt = \int_0^\tau\left(\left\langle \delta Q,\frac{\partial H}{\partial Q}\right\rangle + \left\langle Q,\delta\left(\frac{\partial H}{\partial Q}\right)\right\rangle - \left\langle \delta Q,\frac{\partial H}{\partial Q}\right\rangle\right)dt = 0$$

$$= \int_0^\tau\left(\left\langle Q,\frac{d\eta}{dt}+\left[\frac{\partial H}{\partial Q},\eta\right]\right\rangle + \tilde{\Theta}\left(\frac{\partial H}{\partial Q},\eta\right)\right)dt = \int_0^\tau\left(\left\langle Q,\frac{d\eta}{dt}\right\rangle + \left\langle Q, ad_{\frac{dH}{dQ}}\eta\right\rangle + \left\langle \Theta\left(\frac{\partial H}{\partial Q}\right),\eta\right\rangle\right)dt$$

$$\underset{\substack{Int.\\ by\\ parts}}{=} \int_0^\tau\left\langle -\frac{dQ}{dt} + ad^*_{\frac{dH}{dQ}}Q + \Theta\left(\frac{\partial H}{\partial Q}\right),\eta\right\rangle dt + \left.\langle Q,\eta\rangle\right|_0^\tau = 0$$

From this Lie-Poisson equation, we can introduce a ***Geometric Heat Fourier Equation***:

$$\frac{\partial Q}{\partial t} = \frac{\partial Q}{\partial \beta}\cdot\frac{\partial \beta}{\partial t} = ad^*_{\frac{\partial H}{\partial Q}}Q + \Theta\left(\frac{\partial H}{\partial Q}\right) = \{Q,H\}_{\tilde{\Theta}} \tag{32}$$

where $\frac{\partial Q}{\partial \beta}$ geometric heat capacity is given by $g_\beta(X,Y) = \left\langle -\frac{\partial Q}{\partial \beta}(X), Y\right\rangle$ for $X,Y\in\mathfrak{g}$ with $g_\beta(X,Y) = \tilde{\Theta}_\beta(X,Y) = \langle Q(\beta),[X,Y]\rangle + \tilde{\Theta}(X,Y)$ related to Souriau-Fisher tensor Heat Eq. (32) is the PDE for (calorific) Energy density where the nature of material is characterized by the geometric heat capacity. In the Euclidean case with homogeneous material, we recover classical equation [24]:

$$\frac{\partial \rho_E}{\partial t} = div\left(\frac{\lambda}{C}\nabla\rho_E\right) \text{with } \frac{\partial \rho_E}{\partial t} = C\frac{\partial T}{\partial t}.$$

The link with **2nd principle of Thermodynamics** will be deduced from positivity of Souriau-Fisher metric:

$$
\begin{aligned}
& S(Q) = \langle Q, \beta \rangle - \Phi(\beta) \text{ with } \tfrac{dQ}{dt} = ad^*_{\frac{\partial H}{\partial Q}} Q + \Theta\left(\tfrac{\partial H}{\partial Q}\right) \\
& \tfrac{dS}{dt} = \left\langle Q, \tfrac{d\beta}{dt} \right\rangle + \left\langle ad^*_{\frac{\partial H}{\partial Q}} Q + \Theta\left(\tfrac{\partial H}{\partial Q}\right), \beta \right\rangle - \tfrac{d\Phi}{dt} = \left\langle Q, \tfrac{d\beta}{dt} \right\rangle + \left\langle ad^*_{\frac{\partial H}{\partial Q}} Q, \beta \right\rangle + \left\langle \Theta\left(\tfrac{\partial H}{\partial Q}\right), \beta \right\rangle - \tfrac{d\Phi}{dt} \\
& \tfrac{dS}{dt} = \left\langle Q, \tfrac{d\beta}{dt} \right\rangle + \left\langle Q, \left[\tfrac{\partial H}{\partial Q}, \beta\right] \right\rangle + \tilde{\Theta}\left(\tfrac{\partial H}{\partial Q}, \beta\right) - \tfrac{d\Phi}{dt} = \left\langle Q, \tfrac{d\beta}{dt} \right\rangle + \tilde{\Theta}_\beta\left(\tfrac{\partial H}{\partial Q}, \beta\right) - \left\langle \tfrac{\partial \Phi}{\partial \beta}, \tfrac{d\beta}{dt} \right\rangle \\
& \tfrac{dS}{dt} = \left\langle Q, \tfrac{d\beta}{dt} \right\rangle + \tilde{\Theta}_\beta\left(\tfrac{\partial H}{\partial Q}, \beta\right) - \left\langle \tfrac{\partial \Phi}{\partial \beta}, \tfrac{d\beta}{dt} \right\rangle \text{ with } \tfrac{\partial \Phi}{\partial \beta} = Q \\
& \tfrac{dS}{dt} = \tilde{\Theta}_\beta\left(\tfrac{\partial H}{\partial Q}, \beta\right) \geq 0, \forall H \text{ (link to positivity of Fisher metric)} \\
& \text{if } H = S \underset{\frac{\partial S}{\partial Q} = \beta}{\Rightarrow} \tfrac{dS}{dt} = \tilde{\Theta}_\beta(\beta, \beta) = 0 \text{ because } \beta \in Ker \tilde{\Theta}_\beta
\end{aligned}
\tag{33}
$$

Entropy production is then linked with Souriau-Fisher structure, $dS = \widetilde{\Theta}_\beta\left(\frac{\partial H}{\partial Q}, \beta\right) dt$ with $\widetilde{\Theta}_\beta\left(\frac{\partial H}{\partial Q}, \beta\right) = \widetilde{\Theta}\left(\frac{\partial H}{\partial Q}, \beta\right) + \left\langle Q, \left[\frac{\partial H}{\partial Q}, \beta\right] \right\rangle$ Souriau-Fisher tensor.

The 2 equations characterizing Entropy as invariant Casimir function are related by:

$$
\begin{aligned}
& \{S, H\}_{\widetilde{\Theta}}(Q) = \left\langle Q, \left[\tfrac{\partial S}{\partial Q}, \tfrac{\partial H}{\partial Q}\right] \right\rangle + \left\langle \Theta\left(\tfrac{\partial S}{\partial Q}\right), \tfrac{\partial H}{\partial Q} \right\rangle = 0 \\
& \{S, H\}_{\widetilde{\Theta}}(Q) = \left\langle Q, ad_{\frac{\partial S}{\partial Q}} \tfrac{\partial H}{\partial Q} \right\rangle + \left\langle \Theta\left(\tfrac{\partial S}{\partial Q}\right), \tfrac{\partial H}{\partial Q} \right\rangle = 0 \\
& \{S, H\}_{\widetilde{\Theta}}(Q) = \left\langle ad^*_{\frac{\partial S}{\partial Q}} Q, \tfrac{\partial H}{\partial Q} \right\rangle + \left\langle \Theta\left(\tfrac{\partial S}{\partial Q}\right), \tfrac{\partial H}{\partial Q} \right\rangle = 0 \\
& \forall H, \{S, H\}_{\widetilde{\Theta}}(Q) = \left\langle ad^*_{\frac{\partial S}{\partial Q}} Q + \Theta\left(\tfrac{\partial S}{\partial Q}\right) +, \tfrac{\partial H}{\partial Q} \right\rangle = 0 \Rightarrow ad^*_{\frac{\partial S}{\partial Q}} Q + \Theta\left(\tfrac{\partial S}{\partial Q}\right) = 0
\end{aligned}
\tag{34}
$$

This equation was observed by Souriau in his paper of 1974, where he has written that geometric temperature β is a kernel of $\widetilde{\Theta}_\beta$, that is written:

$$
\beta \in Ker \widetilde{\Theta}_\beta \Rightarrow \langle Q, [\beta, Z] \rangle + \widetilde{\Theta}(\beta, Z) = 0 \tag{35}
$$

That we can develop to recover the Casimir equation:

$$
\begin{aligned}
& \Rightarrow \langle Q, ad_\beta Z \rangle + \widetilde{\Theta}(\beta, Z) = 0 \Rightarrow \left\langle ad^*_\beta Q, Z \right\rangle + \widetilde{\Theta}(\beta, Z) = 0 \\
& \beta = \tfrac{\partial S}{\partial Q} \Rightarrow \left\langle ad^*_{\frac{\partial S}{\partial Q}} Q, Z \right\rangle + \widetilde{\Theta}\left(\tfrac{\partial S}{\partial Q}, Z\right) = \left\langle ad^*_{\frac{\partial S}{\partial Q}} Q + \Theta \tfrac{\partial S}{\partial Q}, Z \right\rangle = 0 \, \forall Z \\
& \Rightarrow ad^*_{\frac{\partial S}{\partial Q}} Q + \Theta\left(\tfrac{\partial S}{\partial Q}\right) = 0
\end{aligned}
\tag{36}
$$

4 Koszul Poisson Cohomology and Entropy Characterization

Poisson Cohomology was introduced by A. Lichnerowicz [26] and J.L. Koszul [27]. Koszul made reference to seminal E. Cartan paper [28, 31] "*Elie Cartan does not explicitly mention $\Lambda(g')$ [the complex of alternate forms on a Lie algebra], because he treats groups as symmetrical spaces and is therefore interested in differential forms which are invariant to both by the translations to the left and the translations to the right, which corresponds to the elements of $\Lambda(g')$ invariant by the prolongation of the coadjoint representation. Nevertheless, it can be said that by 1929 an essential piece of the cohomological theory of Lie algebras was in place*". A. Lichnerowics defined Poisson structure by considering generalization of Symplectic structures which involve contravariant tensor fields rather than differential forms and by observing that the Schouten-Nijenhuis bracket allows to write an intrinsic, coordinate free form, Poisson structure. Let's consider $ad_P Q = [P, Q]$, where ad_P is a graded linear endomorphism of degree $p - 1$ of $A(M)$. From graded Jacobi identity we can write [29]:

$$\begin{aligned} ad_P[Q,R] &= [ad_P Q, R] + (-1)^{(p-1)(q-1)}[Q, ad_p R] \\ ad_{[P,Q]} &= ad_P \circ ad_Q - (-1)^{(p-1)(q-1)} ad_Q \circ ad_P \end{aligned} \quad (37)$$

First equation of (37) means that the graded endormorphims $ad_P Q = [P, Q]$, of degree $p - 1$, is a derivation of the graded Lie algebra $A(M)$ with the **Schouten-Nijenhuis bracket** as composition law. The second equation of (37) means that the endomorphism $ad_{[P,Q]}$ is the graded commutator of endomorphisms ad_P and ad_Q. Y. Vorob'ev and M.V. Karasev [30] have suggested cohomology classification in terms of closed forms and de Rham Cohomology of coadjoint orbits Ω (called Euler orbits by authors), symplectic leaves of a Poisson manifold N. Let $Z^k(\Omega)$ and $H^k(\Omega)$ be the space of closed k-forms on Ω and their de Rham cohomology classes. Considering the base of the fibration of N by these orbits as N/Ω, they have introduced the smooth mapping $Z^k[\Omega] = C^\infty(N/\Omega \to Z^k(\Omega))$ and $H^k[\Omega] = C^\infty(N/\Omega \to H^k(\Omega))$. The elements of $Z^k[\Omega]$ are closed forms on Ω, depending on coordinates on N/Ω. **Then** $H^0[\Omega] = Casim(N)$ **is the set of Casimir functions on N, of functions which are constant on all Euler orbits. In the framework of Lie Groups Thermodynamics, Entropy is then characterized by zero-dimensional de Rham Cohomology.** For arbitrary $v \in V(N)$, with the set $V(N)$ of all vector field on N, the tensor Dv defines a closed 2-form $\alpha(Dv) = Z^2[\Omega]$, and if $v \in V(N)$ annihilates $H^0[\Omega] = Casim(N)$, then this form is exact. The center of Poisson algebra [25] induced from the symplectic structure is the zero-dimensional de Rham cohomology group, the Casimir functions.

References

1. Casimir, H.G.B.: Uber die konstruktion einer zu den irreduziblen darstellungen halbeinfacher kontinuierlicher gruppen gehörigen differentialgleichung. Proc. R. Soc. Amsterdam **30**, 4 (1931)
2. Souriau, J.-M.: Structure des systèmes dynamiques. Dunod (1969)

3. Souriau, J.-M.: Structure of Dynamical Systems: A Symplectic View of Physics. The Progress in Mathematics Book Series, vol. 149. Springer, Boston (1997). https://doi.org/10.1007/978-1-4612-0281-3
4. Souriau, J.-M.: Mécanique statistique, groupes de Lie et cosmologie. Colloque International du CNRS "Géométrie symplectique et physique Mathématique", Aix-en-Provence (1974)
5. Souriau, J.-M.: Géométrie Symplectique et Physique Mathématique. In: Deux Conférences de Jean-Marie Souriau, Colloquium de la SMF, 19 Février 1975 - 12 Novembre 1975
6. Souriau, J.-M.: Mécanique Classique et Géométrie Symplectique, CNRS-CPT-84/PE-1695, November 1984
7. Souriau, J.-M.: Définition covariante des équilibres thermodynamiques. Supplemento al Nuovo cimento **IV**(1), 203–216 (1966)
8. Souriau, J.-M.: Thermodynamique et Geometrie. In: Bleuler, K., Reetz, A., Petry, H.R. (eds.) Differential Geometrical Methods in Mathematical Physics II, pp. 369–397. Springer, Heidelberg (1978). https://doi.org/10.1007/BFb0063682
9. Barbaresco, F.: Lie group statistics and Lie group machine learning based on Souriau Lie groups thermodynamics & Koszul-Souriau-fisher metric: new entropy definition as generalized casimir invariant function in coadjoint representation. Entropy **22**, 642 (2020)
10. Barbaresco, F.: Lie groups thermodynamics & Souriau-Fisher Metric. In: SOURIAU 2019 Conference. Institut Henri Poincaré, 31st May 2019
11. Barbaresco, F.: Souriau-Casimir Lie groups thermodynamics & machine learning. In: Les Houches SPIGL 2020 Proceedings. Springer Proceedings in Mathematics & Statistics (2021)
12. Barbaresco, F.: Jean-Marie Souriau's symplectic model of statistical physics: seminal papers on Lie groups thermodynamics - Quod Erat demonstrandum. In: Les Houches SPIGL 2020 Proceedings. Springer Proceedings in Mathematics & Statistics (2021)
13. Barbaresco, F.: Koszul lecture related to geometric and analytic mechanics, Souriau's Lie group thermodynamics and information geometry. Inf. Geom. (2021). https://doi.org/10.1007/s41884-020-00039-x
14. Marle, C.-M.: From tools in symplectic and poisson geometry to J.-M. Souriau's theories of statistical mechanics and thermodynamics. Entropy **18**, 370 (2016)
15. Marle, C.-M.: Projection Stéréographique et Moments, Hal-02157930, Version 1, June 2019
16. Marle, C.-M.: On Gibbs states of mechanical systems with symmetries. JGSP **57**, 45–85 (2020)
17. Koszul, J.-L.: Introduction to Symplectic Geometry. Springer, Heidelberg (2019). https://doi.org/10.1007/978-981-13-3987-5
18. Cartier, P.: Some fundamental techniques in the theory of integrable systems, IHES/M/94/23, SW9421 (1994)
19. Stratonovich, R.L.: On Distributions in representation space. Soviet Phys. JETP **4**(6), 891–898 (1957)
20. De Saxcé, G., Vallée, C.: Galilean Mechanics and Thermodynamics of Continua (2016)
21. Mikami, K.: Local Lie algebra structure and momentum mapping. J. Math. Soc. Jpn. **39**(2), 233–246 (1987)
22. Engo, K., Faltinsen, S.: Numerical integration of Lie-Poisson systems while preserving coadjoint orbits and energy. SIAM J. Numer. Anal. **39**(1), 128–145 (2002)
23. Poincaré, H.: Sur une forme nouvelle des équations de la Mécanique. Compte-rendus des séances de l'Académie des Sciences, pp. 48–51, lundi 18 Février 1901
24. Balian, R.: Introduction à la thermodynamique hors-équilibre. CEA report (2003)
25. Berezin, F.A.: Some remarks about the associated envelope of a Lie algebra. Funct. Anal. Appl. **1**(2), 91–102 (1968). https://doi.org/10.1007/BF01076082
26. Lichnerowicz, A.: Les variétés de Poisson et leurs algèbres de Lie associées. J. Differ. Geom. **12**, 253–300 (1977)

27. Koszul, J.L.: Crochet de Schouten-Nijenhuis et cohomologie. Astérisque, numéro hors-série Élie Cartan et les mathématiques d'aujourd'hui, Lyon, 25–29 juin 1984, pp. 257–271 (1985)
28. Cartan, E.: Sur les invariants intégraux de certains espaces homogènes clos et les propriétés topologiques de ces espaces. Ann. Soc. Pol. Math. **8**, 181–225 (1929)
29. Marle, C.-M.: The Schouten-Nijenhuis bracket and interior products. J. Geom. Phys. **23**(3–4), 350–359 (1997)
30. Vorob'ev, Y.M., Karasev, M.V.: Poisson manifolds and the Schouten bracket. Funktsional. Anal. i Prilozhen. **22**(1), 1–11, 96 (1988)
31. Vialatte, A.: Les gloires silencieuses: Elie Cartan, Journalismes, Le Petit Dauphinois 1932–1944, Cahiers Alexandre Vialatte n°36, pp. 150–160 (2011)

Bridge Simulation and Metric Estimation on Lie Groups

Mathias Højgaard Jensen[1(✉)], Sarang Joshi[2], and Stefan Sommer[1]

[1] Department of Computer Science, University of Copenhagen, Universitetsparken 1, 2100 Copenhagen, Denmark
{matje,sommer}@di.ku.dk

[2] Department of Biomedical Engineering, University of Utah, 72 S Central Campus Drive, Salt Lake City, UT 84112, USA
sjoshi@sci.utah.edu

Abstract. We present a simulation scheme for simulating Brownian bridges on complete and connected Lie groups. We show how this simulation scheme leads to absolute continuity of the Brownian bridge measure with respect to the guided process measure. This result generalizes the Euclidean result of Delyon and Hu to Lie groups. We present numerical results of the guided process in the Lie group SO(3). In particular, we apply importance sampling to estimate the metric on SO(3) using an iterative maximum likelihood method.

Keywords: Brownian motion · Brownian bridge simulation · Importance sampling · Lie groups · Metric estimation

1 Introduction

Bridge simulation techniques are known to play a fundamental role in statistical inference for diffusion processes. Diffusion bridges in manifolds have mainly been used to provide gradient and hessian estimates. To the best of our knowledge, this paper is the first to describe a simulation technique for diffusion bridges in the context of Lie groups.

The paper is organized as follows. In Sect. 2, we describe some background theory of Lie groups, Brownian motions, and Brownian bridges in Riemannian manifolds. Section 3 presents the theory and results. Section 4 shows in practice the simulation scheme in the Lie group SO(3). Using importance sampling, we obtain an estimate of the underlying unknown metric.

2 Notation and Background

Lie Groups. Throughout, we let G denote a connected Lie Group of dimension d, i.e., a smooth manifold with a group structure such that the group operations $G \times G \ni (x, y) \overset{\mu}{\mapsto} xy \in G$ and $G \ni x \overset{\iota}{\mapsto} x^{-1} \in G$ are smooth maps. If $x \in G$, the

F. Nielsen and F. Barbaresco (Eds.): GSI 2021, LNCS 12829, pp. 430–438, 2021.
https://doi.org/10.1007/978-3-030-80209-7_47

left-multiplication map, $L_x y$, defined by $y \mapsto \mu(x, y)$, is a diffeomorphism from G to itself. Similarly, the right-multiplication map $R_x y$ defines a diffeomorphism from G to itself by $y \mapsto \mu(y, x)$. We assume throughout that G acts on itself by left-multiplication. Let $dL_x \colon TG \to TG$ denote the pushforward map given by $(dL_x)_y \colon T_yG \to T_{xy}G$. A vector field V on G is said to be left-invariant if $(dL_x)_y V(y) = V(xy)$. The space of left-invariant vector fields is linearly isomorphic to T_eG, the tangent space at the identity element $e \in G$. By equipping the tangent space T_eG with the Lie bracket we can identify the Lie algebra $\mathcal{G}$ with T_eG. The group structure of G makes it possible to define an action of G on its Lie algebra $\mathcal{G}$. The conjugation map $C_x := L_x \circ R_x^{-1} \colon y \mapsto xyx^{-1}$, for $x \in G$, fixes the identity e. Its pushforward map at e, $(dC_x)_e$, is then a linear automorphism of $\mathcal{G}$. Define $\mathrm{Ad}(x) := (dC_x)_e$, then $\mathrm{Ad} \colon x \mapsto \mathrm{Ad}(x)$ is the adjoint representation of G in $\mathcal{G}$. The map $G \times \mathcal{G} \ni (x, v) \mapsto \mathrm{Ad}(x)v \in \mathcal{G}$ is the adjoint action of G on $\mathcal{G}$. We denote by $\langle \cdot, \cdot \rangle$ a Riemannian metric on G. The metric is said to be left-invariant if $\langle u, v \rangle_y = \langle (dL_x)_y u, (dL_x)_y v \rangle_{L_x(y)}$, for every $u, v \in T_yG$, i.e., the left-multiplication maps are isometries, for every $x \in G$. In particular, we say that the metric is $\mathrm{Ad}(G)$-invariant if $\langle u, v \rangle_e = \langle \mathrm{Ad}(x)u, \mathrm{Ad}(x)v \rangle_e$, for every $u, v \in \mathcal{G}$. Note that an $\mathrm{Ad}(G)$-invariant inner on $\mathcal{G}$ induces a bi-invariant (left- and right-invariant) metric on G.

Brownian Motion. Endowing a smooth manifold M with a Riemannian metric, g, allows us to define the Laplace-Beltrami operator, $\Delta_M f = \operatorname{div} \operatorname{grad} f$. This operator is the generalization of the Euclidean Laplacian operator to manifolds. In terms of local coordinates $(x_1, \ldots, x_d)$ the expression for the Laplace-Beltrami operator becomes $\Delta_M f = \det(g)^{-1/2} \left(\frac{\partial}{\partial x_j} g^{ji} \det(g)^{1/2} \frac{\partial}{\partial x_i} \right) f$, where $\det(g)$ denotes the determinant of the Riemannian metric g and g^{ij} are the coefficients of the inverse of g. An application of the product rule implies that Δ_M can be rewritten as $\Delta_{M f} = a^{ij} \frac{\partial}{\partial x_i} \frac{\partial}{\partial x_j} f + b^j \frac{\partial}{\partial x_j} f$, where $a^{ij} = g^{ij}$, $b^k = -g^{ij} \Gamma^k_{ij}$, and Γ denote the Christoffel symbols related to the Riemannian metric. This diffusion operator defines a Brownian motion on the M, valid up to its first exit time of the local coordinate chart.

In the case of the Lie group G, the identification of the space of left-invariant vector fields with the Lie algebra $\mathcal{G}$ allows for a global description of Δ_G. Indeed, let $\{v_1, \ldots v_d\}$ be an orthonormal basis of T_eG. Then $V_i(x) = (dL_x)_e v_i$ defines left-invariant vector fields on G and the Laplace-Beltrami operator can be written as (cf. [6, Proposition 2.5]) $\Delta_G f(x) = \sum_{i=1}^d V_i^2 f(x) - V_0 f(x)$, where $V_0 = \sum_{i,j=1}^d C^j_{ij} V_j$ and C^k_{ij} denote the structure coefficients given by $[V_i, V_j] = C^k_{ij} V_k$. The corresponding stochastic differential equation (SDE) for the Brownian motion on G, in terms of left-invariant vector fields, then becomes

$$dX_t = -\frac{1}{2} V_0(X_t) dt + V_i(X_t) \circ dB^i_t, \qquad X_0 = e, \tag{1}$$

where $\circ$ denotes integration in the Stratonovich sense. By [6, Proposition 2.6], if the inner product is $\mathrm{Ad}(G)$ invariant, then $V_0 = 0$. The solution of (1) is conservative or non-explosive and is called the left-Brownian motion on G (see [8] and references therein).

Riemannian Brownian Bridges. In this section, we briefly review some classical facts on Brownian bridges on Riemannian manifolds. As Lie groups themselves are manifolds, the theory carries over *mutatis mutandis.* However, Lie groups' group structure allows the notion of left-invariant vector fields. The identification of the Lie algebra with the vector space of left-invariant vector fields makes Lie groups parallelizable. Thus, the frame bundle construction for developing stochastic processes on manifolds becomes superfluous since left-invariant vector fields ensure stochastic parallel displacement.

Let $\mathbb{P}_x^t$ be the measure of a Riemannian Brownian motion, X_t, at some time t started at point x. Suppose p denotes the transition density of the Riemannian Brownian motion. In that case, $d\mathbb{P}_x^t = p(t,x,y)d\,\mathrm{Vol}(y)$ describes the measure of the Riemannian Brownian motion, where $d\,\mathrm{Vol}(y)$ is the Riemannian volume measure. Conditioning the Riemannian Brownian motion to hit some point v at time $T > 0$ results in a Riemannian Brownian bridge. Here, $\mathbb{P}_{x,v}^T$ denotes the corresponding probability measure. The two measures are equivalent over the time interval $[0,T)$, however mutually singular at time $t = T$. The initial enlargement of the filtration remedies the singularity. The corresponding Radon-Nikodym derivative is given by

$$\frac{d\mathbb{P}_{x,v}^T}{d\mathbb{P}_x^T}\Big|_{\mathcal{F}_s} = \frac{p(T-s, X_s, v)}{p(T,x,v)} \qquad \text{for } 0 \leq s < T,$$

which is a martingale for $s < T$. The Radon-Nikodym derivative defines the density for the change of measure and provides the basis for the description of Brownian bridges. In particular, it provides the conditional expectation defined by

$$\mathbb{E}[F(X_t)|X_T = v] = \frac{\mathbb{E}[p(T-t, X_t, v)F(X_t)]}{p(T,x,v)},$$

for any bounded and $\mathcal{F}_s$-measurable random variable $F(X_s)$. As described in [3], the Brownian bridge yields an SDE in the frame bundle, $\mathcal{FM}$, given by

$$dU_t = H_i(U_t) \circ \left(dB_t^i + H_i \log \tilde{p}(T-t, U_t, v)dt\right), \qquad U_0 = u_0, \tag{2}$$

in terms of the horizontal vector fields (H_i), which is the lifted M-valued Brownian bridge, $X_t := \pi(U_t)$, where $\pi\colon \mathcal{FM} \to M$.

3 Simulation of Bridges on Lie Groups

In this section, we consider the task of simulating (1) conditioned to hit $v \in G$, at time $T > 0$. The potentially intractable transition density for the solution of (1) inhibits simulation directly from (2). Instead, we propose to add a guiding term mimicking that of Delyon and Hu [2], i.e., the guiding term becomes the gradient of the distance to v divided by the time to arrival. The SDE for the guided diffusion becomes

$$dY_t = -\frac{1}{2}V_0(Y_t)dt + V_i(Y_t) \circ \left(dB_t^i - \frac{\left(\nabla_y d(Y_t, v)^2\right)^i}{2(T-t)} dt \right), \qquad Y_0 = e, \tag{3}$$

where d denotes the Riemannian distance function. Note that we can always, for convenience, take the initial value to be the identity e.

Radial Process. We denote by $r_v(\cdot) := d(\cdot, v)$ the radial process. Due to the radial process's singularities on $\mathrm{Cut}(v) \cup \{v\}$, the usual Itô's formula only applies on subsets away from the cut-locus. The extension beyond the cut-locus of a Brownian motion's radial process was due to Kendall [4]. Barden and Le [1,5] generalized the result to M-semimartingales. The radial process of the Brownian motion (1) is given by

$$r(X_t) = r(X_0)^2 + \int_0^t \langle \nabla r(X_s), V(X_s)dB_s \rangle + \frac{1}{2}\int_0^t \Delta_G r(X_s)ds - L_s(X), \tag{4}$$

where L is the geometric local time of the cut-locus $\mathrm{Cut}(v)$, which is non-decreasing continuous random functional increasing only when X is in $\mathrm{Cut}(v)$ (see [1,4,5]). Let $W_t := \int_0^t \left\langle \frac{\partial}{\partial r}, V_i(X_s) \right\rangle dB_s^i$, which is the local-martingale part in the above equation. The quadratic variation of W_t satisfies $d[W, W]_t = dt$, by the orthonormality of $\{V_1, \dots, V_d\}$, thus W_t is a Brownian motion by Levy's characterization theorem. From the stochastic integration by parts formula and (4) the squared radial process of X satisfies

$$r(X_t)^2 = r(X_0)^2 + 2\int_0^t r(X_s)dW_s + \int_0^t r(X_s)\Delta_G r(X_s)ds - 2\int_0^t r(X_s)dL_s, \tag{5}$$

where dL_s is the random measure associated to $L_s(X)$.

Similarly, we obtain an expression for the squared radial process of Y. Using the shorthand notation $r_t := r_v(Y_t)$ the radial process then becomes

$$r_t^2 = r_0^2 + 2\int_0^t r_s dW_s + \int_0^t \frac{1}{2}\Delta_G r_s^2 ds - \int_0^t \frac{r_s^2}{T-s} ds - 2\int_0^t r_s dL_s. \tag{6}$$

Imposing a growth condition on the radial process yields an L^2-bound on the radial process of the guided diffusion, [10]. So assume there exist constants $\nu \geq 1$ and $\lambda \in \mathbb{R}$ such that $\frac{1}{2}\Delta_G r_v^2 \leq \nu + \lambda r_v^2$ on $D \setminus \mathrm{Cut}(v)$, for every regular domain $D \subseteq G$. Then (6) satisfies

$$\mathbb{E}[1_{t<\tau_D} r_v(Y_t)^2] \leq \left(r_v^2(e) + \nu t \left(\frac{t}{T-t} \right) \right) \left(\frac{T-t}{t} \right)^2 e^{\lambda t}, \tag{7}$$

where τ_D is the first exit time of Y from the domain D.

Girsanov Change of Measure. Let B be the Brownian motion in $\mathbb{R}^d$ defined on the filtered probability space $(\Omega, \mathcal{F}, (\mathcal{F}_s), \mathbb{P})$ and X the solution of (1). The process $\frac{\nabla r_v(X_t)^2}{2(T-t)}$ is an adapted process. As X is non-explosive, we see that

$$\int_0^t \left\| \frac{\nabla r(X_s)^2}{2(T-s)} \right\|^2 ds = \int_0^t \frac{r(X_s)^2}{(T-s)^2} ds \leq C, \tag{8}$$

for every $0 \leq t < T$, almost surely, and for some fixed constant $C > 0$. Define a new measure $\mathbb{Q}$ by

$$Z_t := \frac{d\mathbb{Q}}{d\mathbb{P}}\bigg|_{\mathcal{F}_t}(X) = \exp\left\{ -\int_0^t \left\langle \frac{\nabla r(X_s)^2}{2(T-s)}, V(X_t)dB_s \right\rangle - \frac{1}{2}\int_0^t \frac{r(X_s)^2}{(T-s)^2} ds \right\}. \tag{9}$$

From (8), the process Z_t is a martingale, for $t \in [0, T)$, and $\mathbb{Q}_t$ defines a probability measure on each $\mathcal{F}_t$ absolutely continuous with respect to $\mathbb{P}$. By Girsanov's theorem (see e.g. [3, Theorem 8.1.2]) we get a new process b_s which is a Brownian motion under the probability measure $\mathbb{Q}$. Moreover, under the probability $\mathbb{Q}$, Eq. (1) becomes

$$dY_t = -\frac{1}{2}V_0(Y_t)dt + V_i(Y_t) \circ \left(db_t^i - \frac{r(Y_t)}{T-t} \left(\frac{\partial}{\partial r} \right)^i dt \right), \tag{10}$$

where $\left(\frac{\partial}{\partial r}\right)^i$ is the i'th component of the unit radial vector field in the direction of v. The squared radial vector field is smooth away from $\mathrm{Cut}(v)$ and thus we set it to zero on $\mathrm{Cut}(v)$. Away from $\mathrm{Cut}(v)$, the squared radial vector field is $2\,\mathrm{Log}_v$, which is the inverse exponential at v. The added drift term acts as a guiding term, which pulls the process towards v at time $T > 0$.

From (9), we see that $\mathbb{E}[f(Y_t)] = \mathbb{E}[f(X_t)Z_t]$. Using (5) and the identity $\Delta_G r_v = \frac{d-1}{r_v} + \frac{\partial}{\partial r_v} \log \Theta_v$ (see [9]), we equivalently write $\mathbb{E}[f(Y_t)\varphi_t] = \mathbb{E}[f(X_t)\psi_t]$, with

$$\psi_t := \exp\left\{ \frac{-r(X_t)^2}{2(T-t)} \right\} \qquad \varphi_t := \exp\left\{ \int_0^t \frac{r_v(Y_s)^2}{T-s} (dA_s + dL_s) \right\}, \tag{11}$$

where $dA_s = \frac{\partial}{\partial r_v} \log \Theta_v$ is a random measure supported on $G \setminus \mathrm{Cut}(v)$ and Θ_v is the Jacobian determinant of Exp_v.

Delyon and Hu in Lie Groups. This section generalizes the result of Delyon and Hu [2, Theorem 5] to the Lie group setting. The result can be modified to incorporate a generalization of [2, Theorem 6].

Theorem 1. *Let X be the solution of* (1)*. The SDE* (3) *yields a strong solution on* $[0, T)$ *and satisfies* $\lim_{t \uparrow T} Y_t = v$ *almost surely. Moreover, the conditional expectation of X given $X_T = v$ is*

$$\mathbb{E}[f(X)|X_T = v] = C\mathbb{E}\left[f(Y)\varphi_T\right], \tag{12}$$

for every $\mathcal{F}_t$-measurable non-negative function f on G, $t < T$, where φ_t is given in (11).

Proof. The result is a consequence of the change of measure together with Lemma 1, Lemma 2, and Lemma 3.

Lemma 1. *The solution of SDE* (3) *satisfies* $\lim_{t\to T} Y_t = v$ *almost surely.*

Proof. Let $\{D_n\}_{n=1}^{\infty}$ be an exhaustion of G, that is, the sequence consists of open, relatively compact subsets of M such that $\bar{D}_n \subseteq D_{n+1}$ and $G = \bigcup_{n=1}^{\infty} D_n$. Furthermore, let τ_{D_n} denote the first exit time of Y from D_n, then from (7) we have that the sequence $\left(\mathbb{E}[1_{\{t<\tau_{D_n}\}} r_v^2(Y_t)]\right)_{n=1}^{\infty}$ is non-decreasing and bounded, hence from the monotone convergence theorem, it has a limit which is bounded by the right-hand side of (7). Applying Jensen's inequality to the left-hand side of (7)

$$\mathbb{E}[r_v(Y_t)] \leq \left(r_v^2(e) + \nu t\left(\frac{t}{T-t}\right)\right)^{\frac{1}{2}} \left(\frac{T-t}{t}\right) e^{\frac{\lambda t}{2}}.$$

Since obviously $\mathbb{E}[r_v(Y_T)] = r_v(Y_T)\mathbb{Q}(r_v(Y_T) \neq 0)$, by Fatou's lemma $\mathbb{E}[r_v(Y_T)] \leq \liminf_{t\to T} \mathbb{E}[r(Y_t)] = 0$, we conclude that $r(Y_t) \to 0$, $\mathbb{Q}$-almost surely.

Lemma 2. *Let* $0 < t_1 < t_2 < \cdots < t_N < T$ *and* h *be a continuous bounded function on* G^N. *With* ψ_t *as in* (11), *then*

$$\lim_{t\to T} \frac{\mathbb{E}\left[h\left(X_{t_1}, X_{t_2}, \ldots, X_{t_N}\right)\psi_t\right]}{\mathbb{E}[\psi_t]} = \mathbb{E}\left[h\left(X_{t_1}, X_{t_2}, \ldots, X_{t_N}\right) | X_T = v\right]. \quad (13)$$

Proof. The proof is similar to that of [2, Lemma 7]. Let (U, ϕ) be a normal chart centered at $v \in G$. First, since the cut locus of any complete connected manifold has (volume) measure zero, we can integrate indifferently in any normal chart. For any $t \in (t_N, T)$ we have

$$\mathbb{E}[h(x_{t_1}, ..., x_{t_N})\psi_t] = \int_G \Phi_h(t,z) e^{-\frac{r_v(z)^2}{2(T-t)}} d\,\mathrm{Vol}(z) \quad (14)$$

where $d\,\mathrm{Vol}(z) = \sqrt{\det(A(z))}dz$ denotes the volume measure on G, dz the Lebesgue measure, and A the metric tensor. Moreover,

$$\Phi_h(t,z) = \int_{G^N} h(z_1, ..., z_N) p(t_1, u, z_1) \cdots p(t - t_N, z_N, z) d\,\mathrm{Vol}(z_1) \cdots d\,\mathrm{Vol}(z_N),$$

and of course $\Phi_1(t,z) = p(t,e,z)$. Using the normal chart and applying the change of variable $x = (T-t)^{1/2}y$ we get

$$(T-t)^{-\frac{d}{2}} \mathbb{E}[h(x_{t_1}, ..., x_{t_N})\psi_t] \overset{t\to T}{\to} \Phi_h(T,v) \det(A(v))^{\frac{d}{2}} \int_{\phi(G)} e^{-\frac{r_v(\phi^{-1}(y))^2}{2}} dy.$$

The conclusion follows from Bayes' formula.

Lemma 3. *With* φ_t *as defined above then* $\varphi_t \overset{L_1}{\to} \varphi_T$.

Proof. Note that for each $t \in [0,T)$ we have $\mathbb{E}^{\mathbb{Q}}[\varphi_t] < \infty$ as well as $\varphi_t \to \varphi_T$ almost surely by Lemma 1. The result then follows from the uniform integrability of $\{\varphi_t\colon t \in [0,T)\}$, which can be found in Appendix C.2 in [9].

4 Importance Sampling and Metric Estimation on SO(3)

This section takes G to be the special orthogonal group of rotation matrices, SO(3), a compact connected matrix Lie group. In the context of matrix Lie groups, computing left-invariant vector fields is straightforward.

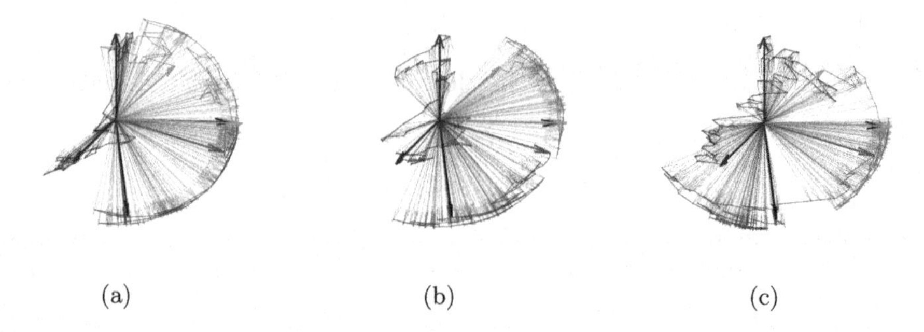

Fig. 1. Three sample paths (a)–(c) of the guided diffusion process on SO(3) visualized by its action on the basis vectors $\{e_1, e_2, e_3\}$ (red, green, blue) of $\mathbb{R}^3$. The sample paths are conditioned to hit the rotation represented by the black vectors. (Color figure online)

Numerical Simulations. The Euler-Heun scheme leads to approximation of the Stratonovich integral. With a time discretization $t_1, \ldots, t_k$, $t_k - t_{k-1} = \Delta t$ and corresponding noise $\Delta B_{t_i} \sim N(0, \Delta t)$, the numerical approximation of the Brownian motion (1) takes the form

$$x_{t_{k+1}} = x_{t_k} - \frac{1}{2}\sum_{j,i} C^j{}_{ij} V_i(x_{t_k})\Delta t + \frac{v_{t_{k+1}} + V_i(v_{t_{k+1}} + x_{t_{k+1}})\Delta B^i_{t_k}}{2} \tag{15}$$

where $v_{t_{k+1}} = V_i(x_{t_k})\Delta B^i_{t_k}$ is only used as an intermediate value in integration. Adding the logarithmic term in (10) to (15) we obtain a numerical approximation of a guided diffusion (3). Figure 1 shows three different sample paths from the guided diffusion conditioned to hit the rotation represented by the black vectors.

Metric Estimation on SO(3). In the d-dimensional Euclidean case, importance sampling yields the estimate [7]

$$p(T, u, v) = \left(\frac{\det\left(A(T, v)\right)}{2\pi T}\right)^{\frac{d}{2}} e^{-\frac{\|u-v\|^2_A}{2T}}\mathbb{E}[\varphi_T],$$

where $\|x\|_A = x^T A(0, u)x$. Thus, from the output of the importance sampling we get an estimate of the transition density. Similar to the Euclidean case, we obtain an expression for the heat kernel $p(T, e, v)$ as $p(T, e, v) = q(T, e)\mathbb{E}\left[\varphi_T\right]$, where

$$q(T,e) = \left(\frac{\det A(v)}{2\pi T}\right)^{\frac{3}{2}} \exp\left(-\frac{d(e,v)^2}{2T}\right) = \left(\frac{\det A(v)}{2\pi T}\right)^{\frac{3}{2}} \exp\left(-\frac{\|\mathrm{Log}_v(e)\|_A^2}{2T}\right), \tag{16}$$

where the equality holds almost everywhere and $A \in \mathrm{Sym}^+(\mathcal{G})$ denotes the metric $A(e)$. The Log_v map in (16) is the Riemannian inverse exponential map.

Figure 2 illustrate how importance sampling on SO(3) leads to metric estimation of the underlying true metric, from which the Brownian motion was generated. We sampled 128 points as endpoints of a Brownian motion from the metric diag(0.2, 0.2, 0.8). We used 20 time steps and sampled 4 bridges per observation. An iterative maximum likelihood method using gradient descent with a learning rate of 0.2, and initial guess of the metric being diag(1, 1, 1) yielded a convergence to the true metric. Note that in iteration the logarithmic map changes.

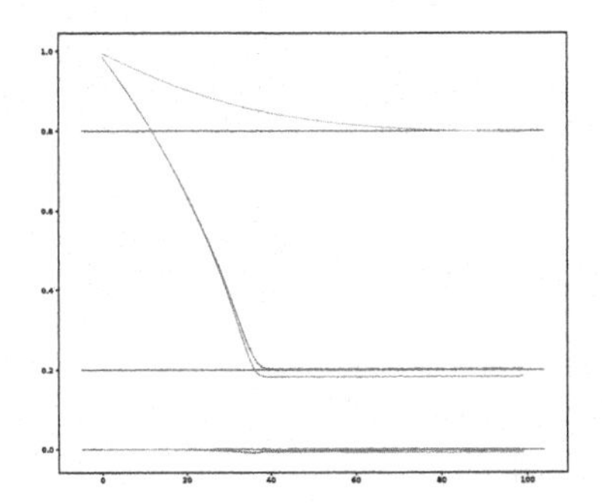

(a) Estimation of the unknown underlying metric using bridge sampling. Here the true metric is the diagonal matrix diag(0.2, 0.2, 0.8).

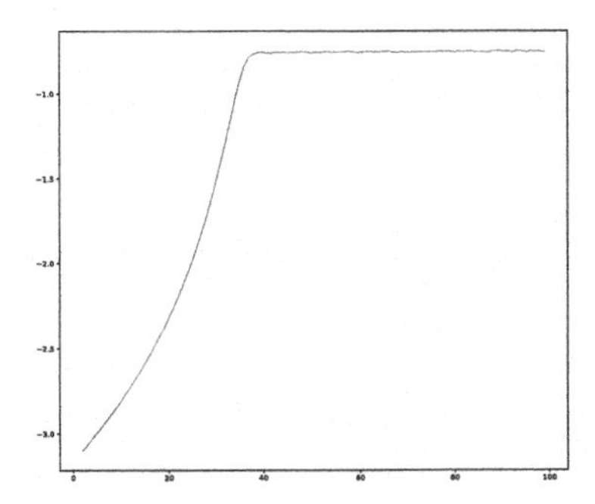

(b) The iterative log-likelihood.

Fig. 2. The importance sampling technique applies to metric estimation on the Lie group SO(3). Sampling a Brownian motion from an unknown underlying metric we obtain a convergence to the true underlying metric using an iterative maximum-likelihood method. Here we sampled 4 bridge processes per observation, starting from the metric diag(1, 1, 1), providing a relatively smooth iterative likelihood in 2b.

References

1. Barden, D., Le, H.: Some consequences of the nature of the distance function on the cut locus in a Riemannian manifold. J. LMS **56**(2), 369–383 (1997)
2. Delyon, B., Hu, Y.: Simulation of conditioned diffusion and application to parameter estimation. Stochast. Processes Appl. **116**(11), 1660–1675 (2006)
3. Hsu, E.P.: Stochastic Analysis on Manifolds, vol. 38. AMS, Providence (2002)
4. Kendall, W.S.: The radial part of Brownian motion on a manifold: a semimartingale property. Ann. Probab. **15**(4), 1491–1500 (1987)

5. Le, H., Barden, D.: Itô correction terms for the radial parts of semimartingales on manifolds. Probab. Theory Relat. Fields **101**(1), 133–146 (1995). https://doi.org/10.1007/BF01192198
6. Liao, M.: Lévy Processes in Lie Groups, vol. 162. Cambridge University Press, Cambridge (2004)
7. Papaspiliopoulos, O., Roberts, G.: Importance sampling techniques for estimation of diffusion models. In: Statistical Methods for Stochastic Differential Equations (2012)
8. Shigekawa, I.: Transformations of the Brownian motion on a riemannian symmetric space. Zeitschrift für Wahrscheinlichkeitstheorie und Verwandte Gebiete **65**, 493–522 (1984). https://doi.org/10.1007/BF00531836
9. Thompson, J.: Submanifold bridge processes. Ph.D. thesis, University of Warwick (2015)
10. Thompson, J.: Brownian bridges to submanifolds. Potential Anal. **49**(4), 555–581 (2018). https://doi.org/10.1007/s11118-017-9667-1

Constructing the Hamiltonian from the Behaviour of a Dynamical System by Proper Symplectic Decomposition

Nima Shirafkan[1], Pierre Gosselet[2], Franz Bamer[1], Abdelbacet Oueslati[2], Bernd Markert[1], and Géry de Saxcé[2](✉)

[1] Institute of General Mechanics, RWTH, Aachen, Germany
[2] Univ. Lille, CNRS, Centrale Lille, UMR 9013 LaMcube Laboratoire de mécanique multiphysique multiéchelle, 59000 Lille, France
gery.de-saxce@univ-lille.fr
https://www.iam.rwth-aachen.de
http://lamcube.univ-lille.fr

Abstract. The modal analysis is revisited through the symplectic formalism, what leads to two intertwined eigenproblems. Studying the properties of the solutions, we prove that they form a canonical basis. The method is general and works even if the Hamiltonian is not the sum of the potential and kinetic energies. On this ground, we want to address the following problem: data being given in the form of one or more structural evolutions, we want to construct an approximation of the Hamiltonian from a covariant snapshot matrix and to perform a symplectic decomposition. We prove the convergence properties of the method when the time discretization is refined. If the data cloud is not enough rich, we extract the principal component of the Hamiltonian corresponding to the leading modes, allowing to perform a model order reduction for very high dimension models. The method is illustrated by a numerical example.

Keywords: Symplectic mechanics · Modal analysis · Model order reduction · Principal component analysis

1 Introduction

In structural mechanics, the modal analysis coupled with the finite element method is widely used by engineers to determine the eigenmodes and eigenfrequencies of linear dynamical systems [2].

Very large numerical models are ubiquitous in structural dynamics. Working in high dimension spaces is time-consuming, often intractable and requires storing big pieces of data, hence the need to simplify the models to make them easier and faster to interpret by users. The Proper Orthogonal Decomposition (POD) is one of the most successfully used model reduction technique for nonlinear systems [1]. Nevertheless, it is based on the metric structure of the configuration

F. Nielsen and F. Barbaresco (Eds.): GSI 2021, LNCS 12829, pp. 439–447, 2021.
https://doi.org/10.1007/978-3-030-80209-7_48

space, while for dynamical systems the phase space is naturally equipped with a symplectic structure [3,5].

The aim of the present work is, for large scale conservative systems, to develop a new method of Proper Symplectic Decomposition (PSD) able to extract the leading eigenmodes of the modal analysis and the principal component of the Hamiltonian of the system from a data cloud comprised of one or many evolutions of the system subjected to external excitations.

In modern literature, a PSD-based model reduction technique has been proposed by Peng and. Mohseni in [4] where the symplectic projection is determined from a snapshot matrix solving a nonlinear optimization problem for linear systems. Our strategy is to develop an alternative PSD method leading to a linear eigenproblem for linear structures, the nonlinear problem being set aside for the modeling of dissipative systems.

2 The Modal Analysis as an Equivalence Problem

The phase space $V = \mathbb{R}^{2N}$ being endowed with the symplectic form:

$$\omega(\boldsymbol{x}, \boldsymbol{x}') = \boldsymbol{x}^T \boldsymbol{J}\, \boldsymbol{x}' = \boldsymbol{q}^T \boldsymbol{p}' - \boldsymbol{p}^T \boldsymbol{q}'$$

where $\boldsymbol{q}$ are the degrees of freedom, $\boldsymbol{p}$ are the moments, $\boldsymbol{x}^T = (\boldsymbol{q}^T, \boldsymbol{p}^T)$ and $\boldsymbol{J}$ is a skew-symmetric matrix. The motion of the system is governed by the canonical equations

$$\dot{\boldsymbol{x}} = \nabla^{\omega} h = \boldsymbol{J} \nabla h$$

where $\nabla^{\omega} h$ is the symplectic gradient of the Hamiltonian h (or Hamiltonian vector field). The symplectic matrices $\boldsymbol{S}$ that leave invariant ω (*i.e.* $\boldsymbol{S}^T \boldsymbol{J}\, \boldsymbol{S} = \boldsymbol{J}$) form the symplectic group $Sp(2\,N, \mathbb{R})$. The dual space V^* is equipped with a Poisson bracket $\{\cdot, \cdot\}$ such that $\{f, g\} = \omega(\nabla^{\omega} g, \nabla^{\omega} f)$. The modal analysis can be rewritten saying there is a symplectic matrix $\boldsymbol{S}$ mapping $h(\boldsymbol{x})$ and its Hessian matrix $\boldsymbol{H} \in V^* \otimes V^*$ onto the Hamiltonian $h'(\boldsymbol{x}')$ of a reduced system of independent harmonic oscillators and its diagonal Hessian matrix $\boldsymbol{H}'$. The equivalence problem consists in finding a symplectic matrix $\boldsymbol{S}$ such that:

$$h' = h \circ \boldsymbol{S}, \quad \text{then} \quad \boldsymbol{S}^T \boldsymbol{H}\, \boldsymbol{S} = \boldsymbol{H}'$$

3 Intertwined Eigenproblems and Spectral Decomposition

As the symplectic form is nondegenerate, there is a natural isomorphism $V \rightarrow V^* : \boldsymbol{x} \mapsto \boldsymbol{x}^* = \boldsymbol{J}\, \boldsymbol{x}$ that gives rise, in indicial notations, to the operations of lowering and raising the indices:

$$x^*_{\alpha} = J_{\alpha\beta} x^{\beta}, \qquad x^{\beta} = (J^{-1})^{\beta\alpha} x^*_{\alpha}$$

These operations can be extended to any type of tensors. For instance, as the Hamiltonian is a quadratic form on the symplectic vector space V, the Hessian matrix $\boldsymbol{H}$, of components $H_{\alpha\beta}$, represents a 2-covariant tensor on V, element of $V^* \otimes V^*$. By raising the first index:

$$(H_\omega)^\alpha_\beta = (J^{-1})^{\alpha\mu} H_{\mu\beta}$$

we obtain the components of the matrix $\boldsymbol{H}_\omega$ representing a linear map from V into V, then a mixed 1-covariant 1-contravariant tensor, element of $V \otimes V^*$.

$$\boldsymbol{H}_\omega = \boldsymbol{J}^{-1}\boldsymbol{H} = -\boldsymbol{J}\,\boldsymbol{H}$$

called an Hamiltonian matrix. Decomposing the symplectic matrix into columns ($\boldsymbol{S} = [\boldsymbol{u}_1, \cdots, \boldsymbol{u}_N, \boldsymbol{v}_1, \cdots, \boldsymbol{v}_N]$) leads to two **intertwined eigenproblems**:

$$\boldsymbol{H}_\omega \boldsymbol{u}_i = k_i \boldsymbol{v}_i, \quad \boldsymbol{H}_\omega \boldsymbol{v}_i = -g_i \boldsymbol{u}_i \tag{1}$$

that can be transformed into a classical eigenproblem:

$$\boldsymbol{H}^2_\omega \boldsymbol{u}_i = \lambda_i \boldsymbol{u}_i \tag{2}$$

where $\lambda_i = -k_i g_i$. The properties of the eigenmodes are given by the following result:

Theorem 1. *If the Hessian matrix* $\boldsymbol{H}$ *is positive definite:*

- *the eigenvalues* λ *of* $\boldsymbol{H}^2_\omega$ *are negative.*
- *the twin eigenvectors* $\boldsymbol{u}_i$ *and* $\boldsymbol{v}_i$ *of the eigenproblem (1) are not orthogonal:* $\omega(\boldsymbol{u}_i, \boldsymbol{v}_i) \neq 0$.

Scaling the eigenvectors by

$$\omega(\boldsymbol{u}_i, \boldsymbol{v}_i) = 1 \tag{3}$$

they form a canonical basis of V.

Remark 1: in practice, if the structure has suffisant supports to avert rigid displacements, the Hessian matrix is positive definite.

Remark 2: for the particular case of the standard modal analysis where the Hamiltonian is decoupled:

$$h(\boldsymbol{x}) = \frac{1}{2}\boldsymbol{p}^T \boldsymbol{M}^{-1}\boldsymbol{p} + \frac{1}{2}\boldsymbol{q}^T \boldsymbol{K}\boldsymbol{q}, \quad \boldsymbol{u}_i^T = \left[\boldsymbol{a}_i^T, \boldsymbol{0}^T\right], \quad \boldsymbol{v}_i^T = \left[\boldsymbol{0}^T, (\boldsymbol{M}\boldsymbol{a}_i)^T\right] \tag{4}$$

we recover the eigenproblem $\boldsymbol{K}\,\boldsymbol{a}_i = \omega_i^2 \boldsymbol{M}\,\boldsymbol{a}_i$ with $\lambda_i = -\omega_i^2$ and the normalization condition $\omega(\boldsymbol{u}_i, \boldsymbol{v}_i) = \boldsymbol{a}_i^T \boldsymbol{M}\,\boldsymbol{a}_i = 1$.

Remark 3: Our method is more general and allows to treat also cases where there are terms coupling $\boldsymbol{q}$ and $\boldsymbol{p}$, for instance in problems with Coriolis' force or electromagnetic fields.

Another result of interest is:

Theorem 2. *If* $(\boldsymbol{u}_1, \cdots, \boldsymbol{u}_N, \boldsymbol{v}_1, \cdots, \boldsymbol{v}_N)$ *is a canonical basis,* $\boldsymbol{u}_i^* = -\boldsymbol{J}\,\boldsymbol{u}_i$ *and* $\boldsymbol{v}_i^* = \boldsymbol{J}\,\boldsymbol{v}_i$ *, then* $(\boldsymbol{v}_1^*, \cdots, \boldsymbol{v}_N^*, \boldsymbol{u}_1^*, \cdots, \boldsymbol{u}_N^*)$ *is its dual basis.*

Indeed, it leads to the **spectral decomposition** of the identity of $\mathbb{R}^{2N}$, next of the Hamiltonian matrix $\boldsymbol{H}_\omega$ and of the Hessian matrix $\boldsymbol{H}$:

$$\boldsymbol{I}_{2\,N} = \boldsymbol{v}_j \otimes \boldsymbol{u}_j^* + \boldsymbol{u}_j \otimes \boldsymbol{v}_j^*, \qquad \boldsymbol{H}_\omega = g_j \boldsymbol{v}_j \otimes \boldsymbol{u}_j^* - k_j \boldsymbol{u}_j \otimes \boldsymbol{v}_j^*$$

$$\boldsymbol{H} = g_j \boldsymbol{u}_j^* \otimes \boldsymbol{u}_j^* + k_j \boldsymbol{v}_j^* \otimes \boldsymbol{v}_j^* \tag{5}$$

4 Proper Symplectic Decomposition (PSD)

We hope to address the following problem: data being given in the form of a structural evolution $[0, T] \to V : t \mapsto \boldsymbol{x}(t)$ (or a concatenation of structural evolutions), we want to construct an approximation of the Hamiltonian of the system. The functional space of the components of these evolutions is endowed with the metrics:

$$(f_1 \mid f_2) = \frac{1}{T} \int_0^T f_1(t)\, f_2(t)\, dt \tag{6}$$

Our method consists in decomposing the interval $[0, T]$ into m subintervals I_k of timestep Δt_k and reference point $t_k \in I_k$. Starting from a structural evolution $[0, T] \to V : t \mapsto \boldsymbol{x}(t)$ as data, we construct, through the isomorphism $\boldsymbol{J}$ from V into its dual V^*, the **covariant snapshot matrix**:

$$\boldsymbol{X}^* = [\boldsymbol{J}\boldsymbol{x}(t_1), \cdots, \boldsymbol{J}\boldsymbol{x}(t_m)] \tag{7}$$

representing a map of codomain the dual space V^* and of domain the Euclidean approximation space W of dimension m, equipped with the scalar product between snapshot vectors $\boldsymbol{f}_j = [f_j(t_1), \cdots, f_j(t_m)]^T$:

$$(\boldsymbol{f}_1, \boldsymbol{f}_2) = \frac{1}{T} \sum_{k=1}^{m} f_1(t_k)\, f_2(t_k) \Delta t_k$$

a discretized version of (6). The corresponding Gram's matrix of the metrics being:

$$\boldsymbol{G}_t = \frac{1}{T}\, diag(\Delta t_1, \cdots, \Delta t_m)$$

Inspiring from the Singular Value Decomposition (SVD), we introduce the symmetric matrix:

$$\boldsymbol{H} = 2\, \boldsymbol{X}^*\, \boldsymbol{G}_t\, (\boldsymbol{X}^*)^T \tag{8}$$

which represents a 2-covariant tensor on V and is positive semi-definite because the metrics of W is positive:

$$\boldsymbol{x}^T\, \boldsymbol{X}^* \boldsymbol{G}_t\, (\boldsymbol{X}^*)^T \boldsymbol{x} = ((\boldsymbol{X}^*)^T \boldsymbol{x}, (\boldsymbol{X}^*)^T \boldsymbol{x}) \geq 0$$

Next an approximation of the eigenmodes $\boldsymbol{u}_i$ and $\boldsymbol{v}_i$ of the system can be obtained by solving the eigenproblem (2).

Remark: in the SVD, the domain and codomain of the snapshot matrix are Euclidean spaces, the codomain is not necessarily of even dimension and Gram's matrix is the identity. The factor 2 in (8) is required because of the factor 1/2 in the second order term of the Taylor expansion of the Hamiltonian around $\boldsymbol{x} = \mathbf{0}$.

5 Convergence Properties of the Method

We would like to show that $\boldsymbol{H}$ given by (8) converges to a matrix which allows to find the exact eigenvectors and eigenvalues when the time interval of the data increases and the time discretization is refined. To set these ideas down on a simple problem of standard modal analysis, we consider the free vibrations of a discrete system with non null initial velocity. According to the modal decomposition, we have:

$$\boldsymbol{q}(t) = \sum_{i=1}^{N} \frac{\dot{q}_i'(0)}{\omega_i} \sin(\omega_i t)\, \boldsymbol{a}_i = \sum_{i=1}^{N} \alpha_i\, f_i(t)\, \boldsymbol{a}_i,$$

$$\boldsymbol{p}(t) = \boldsymbol{M}\, \dot{\boldsymbol{q}}(t) = \sum_{i=1}^{N} \alpha_i\, \dot{f}_i(t)\, \boldsymbol{M}\, \boldsymbol{a}_i$$

with $f_i(t) = \sin(\omega_i t)$ and $\dot{f}_i(t) = \omega_i \cos(\omega_i t)$. The time evolution of the structure in terms of covariant vector is:

$$\boldsymbol{x}^*(t) = \boldsymbol{J}\, \boldsymbol{x}(t) = \begin{bmatrix} \boldsymbol{p}(t) \\ -\boldsymbol{q}(t) \end{bmatrix} = \sum_{i=1}^{N} \alpha_i \left[\dot{f}_i(t)\, \boldsymbol{v}_i^* - f_i(t)\, \boldsymbol{u}_i^* \right]$$

where $(\boldsymbol{u}_i^*)^T = \left[\mathbf{0}^T, \boldsymbol{a}_i^T\right]$, $(\boldsymbol{v}_i^*)^T = \left[(\boldsymbol{M}\boldsymbol{a}_i)^T, \mathbf{0}^T\right]$ are the elements of the dual basis. Owing to (7) and (8) and refining the time discretization, one has:

$$\lim_{m\to\infty} \boldsymbol{H} = \lim_{m\to\infty} \frac{2}{T} \sum_{k=1}^{m} \Delta t_k \boldsymbol{x}^*(t_k) \otimes \boldsymbol{x}^*(t_k) = \frac{2}{T} \int_0^T \boldsymbol{x}^*(t) \otimes \boldsymbol{x}^*(t)\, dt$$

$$= 2 \sum_{i,j=1}^{N} \alpha_i\, \alpha_j \Big[(\dot{f}_i \mid \dot{f}_j)\, \boldsymbol{v}_i^* \otimes \boldsymbol{v}_j^* - (\dot{f}_i \mid f_j)\, \boldsymbol{v}_i^* \otimes \boldsymbol{u}_j^*$$

$$-(f_i \mid \dot{f}_j)\, \boldsymbol{u}_i^* \otimes \boldsymbol{v}_j^* + (f_i \mid f_j)\, \boldsymbol{u}_i^* \otimes \boldsymbol{u}_j^* \Big]$$

where:

$$(f_i \mid f_i) = \frac{1}{2}\left[1 - \frac{\sin(2\,\omega_i T)}{2\,\omega_i T}\right], \qquad (\dot{f}_i \mid \dot{f}_i) = \frac{\omega_i^2}{2}\left[1 + \frac{\sin(2\,\omega_i T)}{2\,\omega_i T}\right]$$

When T approaches $+\infty$, one has:

$$\lim_{T\to\infty} (f_i \mid f_i) = \frac{1}{2}, \qquad \lim_{T\to\infty} (\dot{f}_i \mid \dot{f}_i) = \frac{\omega_i^2}{2}$$

The other scalar product above approaching zero, then it remains:

$$\lim_{T\to\infty}\lim_{m\to\infty} \boldsymbol{H} = 2\sum_{i=1}^{N} \alpha_i^2 \left[(f_i \mid f_i)\, \boldsymbol{u}_i^* \otimes \boldsymbol{u}_i^* + (\dot{f}_i \mid \dot{f}_i)\, \boldsymbol{v}_i^* \otimes \boldsymbol{v}_i^*\right]$$

$$\lim_{T\to\infty}\lim_{m\to\infty} \boldsymbol{H} = \sum_{i=1}^{N} \alpha_i^2 \left[\boldsymbol{u}_i^* \otimes \boldsymbol{u}_i^* + \omega_i^2 \boldsymbol{v}_i^* \otimes \boldsymbol{v}_i^*\right]$$

Comparing to the spectral decomposition (5) we obtain by identification:

$$g_i = \alpha_i^2, \qquad k_i = \alpha_i^2\omega_i^2, \qquad \lambda_i = -k_i g_i = -\alpha_i^4\omega_i^2 \tag{9}$$

Solving the eigenproblem (2), we would find the exact orthogonal modes. With sufficiently large values of T and m, good approximations of these modes are expected.

6 Numerical Application

To illustrate the method, let us consider an elastic bar in traction-compression of length L, cross-section area S, made of a material of elasticity modulus E and mass μ per length unit. The truss is clamped at the extremity $x = 0$ and free at the extremity $x = L$. We approximate the displacement field by a polynomial function of degree two. Taking into account the support condition, it reads:

$$u(x) = \frac{x}{L}\, q_1 + \left(\frac{x}{L}\right)^2 q_2$$

where q_1 and q_2 are the components of the vector $\boldsymbol{q}$. The stiffness and mass matrix are:

$$\boldsymbol{K} = \frac{E\,S}{L}\begin{bmatrix} 1 & 1 \\ 1 & \frac{4}{3} \end{bmatrix}, \qquad \boldsymbol{M} = \mu\,L\begin{bmatrix} \frac{1}{3} & \frac{1}{4} \\ \frac{1}{4} & \frac{1}{5} \end{bmatrix}$$

For sake of easiness, the units are choosen in such way that $E\,S/L = \mu\,L = 1$. Solving the eigenvalue problem (2), we obtain two negative eigenvalues with multiplicity 2:

$$\lambda_1 = -32.18070, \quad \lambda_2 = -2.48596 \tag{10}$$

In terms of of circular frequency $\omega_i = \sqrt{-\lambda_i}$ and period T_i, we have $\omega_1 = 5.672$, $\omega_2 = 1.556$, $T_1 = 1.107$, $T_2 = 3.985$. The corresponding twin eigenvectors are for λ_1:

$$\boldsymbol{u}_1 = \begin{bmatrix} -6.4220 \\ 8.8665 \\ 0.0 \\ 0.0 \end{bmatrix}, \qquad \boldsymbol{v}_1 = \begin{bmatrix} 0.0 \\ 0.0 \\ 0.075962 \\ 0.16780 \end{bmatrix} \tag{11}$$

and for λ_2:

$$\boldsymbol{u}_2 = \begin{bmatrix} -19.586 \\ 8.8665 \\ 0.0 \\ 0.0 \end{bmatrix}, \qquad \boldsymbol{v}_2 = \begin{bmatrix} 0.0 \\ 0.0 \\ -0.57233 \\ -0.41453 \end{bmatrix} \tag{12}$$

Table 1. Sample $\alpha_1 = \alpha_2 = 1$, eigenvalues of $\boldsymbol{H}_\omega^2$

T	m	Value of λ'_1	Value of λ'_2	Error on λ'_1	Error on λ'_2
10	10	−31.83290	−2.21129	$1.08\,10^{-2}$	0.11
10	40	−32.23037	−2.46422	$1.54\,10^{-3}$	$8.74\,10^{-3}$
20	100	−32.18026	−2.48581	$1.35\,10^{-5}$	$5.90\,10^{-5}$
$+\infty$	$+\infty$	−32.18070	−2.48596	–	–

They are of the form (4) for a standard Hamiltonian. They form a canonical basis of V.

That being said, we examine three data samples:

– **Sample 1: equilibrated combination of the two modes** ($\alpha_1 = \alpha_2 = 1$) as data:

$$\boldsymbol{q}(t) = \begin{bmatrix} -6.4220 \\ 8.8665 \end{bmatrix} \sin(5.672\,t) + \begin{bmatrix} -19.586 \\ 8.8665 \end{bmatrix} \sin(1.556\,t)$$

We divide the time interval form 0 to T into m subintervals of same timestep. The snapshots are provided at the middle point of each subinterval. Computing the matrices (7) and (8), next solving the eigenvalue problem (2), we obtain two eigenvalues λ_i of multiplicity 2. Their numerical values are given in Table 1 for some values of T and m. The last row gives the reference values (10). The two latter columns provide the relative error $\mid \lambda'_i - \lambda_i \mid / \mid \lambda_i \mid$ with respect to these reference values λ_i. We observe the convergence when increasing T and m. The corresponding eigenvectors are given within a factor. After normalization according to the condition (3), we obtain the approximations $\boldsymbol{u}'_i, \boldsymbol{v}'_i$ of the twin vectors. The relative errors:

$$\| \boldsymbol{u}'_i - \boldsymbol{u}_i \| / \| \boldsymbol{u}_i \|, \qquad \| \boldsymbol{v}'_i - \boldsymbol{v}_i \| / \| \boldsymbol{v}_i \|$$

with respect to the expected values (11) and (12) are given in Table 2 for $T = 20$ and $m = 100$. The corresponding approximation of the Hessian matrix of the Hamiltonian is:

$$\boldsymbol{H}' = \begin{bmatrix} 0.99872 & 0.99634 & -3.7182\,10^{-3} & 4.6590\,10^{-3} \\ 0.99634 & 1.3301 & 3.6907\,10^{-3} & -4.7010\,10^{-3} \\ -3.7182\,10^{-3} & 3.6907\,10^{-3} & 47.809 & -59.767 \\ 4.659\,10^{-3} & -4.7010\,10^{-3} & -59.767 & 79.699 \end{bmatrix} \tag{13}$$

Table 2. Sample $\alpha_1 = \alpha_2 = 1$, error on the eigenvectors of $\boldsymbol{H}_\omega^2$

Eigenvectors	$\boldsymbol{u}_1$	$\boldsymbol{v}_1$	$\boldsymbol{u}_2$	$\boldsymbol{v}_2$
Relative error	3.18%	0.64%	0.88%	0.02%

By comparison with the exact matrix, the maximum error on the entries is 0.397%. In a nutshell, when the data is the sum of the eigenmodes, we are able to deduce from the snapshot matrix a very accurate expression of the Hamiltonian.

- **Sample 2: non equilibrated combination of the two modes** ($\alpha_1 = 1/2$, $\alpha_2 = 1$) as data:

$$\boldsymbol{q}(t) = 0.5 \begin{bmatrix} -6.4220 \\ 8.8665 \end{bmatrix} \sin(5.672\, t) + \begin{bmatrix} -19.586 \\ 8.8665 \end{bmatrix} \sin(1.556\, t)$$

In Table 3, we compare the numerical values of the eigenvalues λ'_i to the reference values λ_i given by (9). The twin eigenvectors corresponding to λ'_1 (resp. λ'_2) are very close to the reference values (11), (12).

Table 3. Sample $\alpha_1 = 1/2$, $\alpha_2 = 1$, eigenvalues of $\boldsymbol{H}_\omega^2$

T	m	Value of λ'_1	Value of λ'_2	Error on λ'_1	Error on λ'_2
20	100	−2.01087	−2.48629	$2.07\, 10^{-4}$	$1.34\, 10^{-4}$
$+\infty$	$+\infty$	−2.01129	−2.48596	–	–

- **Sample 3: only the second eigenmode** ($\alpha_1 = 0, \alpha_2 = 1$) as data:

$$\boldsymbol{q}(t) = \begin{bmatrix} -19.586 \\ 8.8665 \end{bmatrix} \sin(1.556\, t)$$

With $T = 20$ and $m = 100$, we obtain two eigenvalues of multiplicity 2:

$$\lambda'_1 = -5.5511 \cdot 10^{-17}, \qquad \lambda'_2 = -2.48592$$

In terms of absolute values, the latter eigenvalue overwhelms the former one, that is expected because the data are provided only by the second eigenmode.

7 Conclusions and Perspectives

The main interest of the method is, for large scale systems, to extract from experimental data or numerical simulations the principal component of the Hamiltonian, operation that can be done offline from a big data cloud. Next, the reduced system can be used online to predict the response to given excitations by solving a canonical equation system of small size. Besides, whenever the data cloud is enriched, the Hamiltonian can be updated, according to the machine learning process.

The application realm of the proposed method is not limited to the structural mechanics but can be extended to the homogenization of materials to find the effective properties by considering a reference elementary volume [6]. In the future, we hope to extend the approach also to dissipative dynamical systems, first in the linear case of damping, next to the nonlinear case of elastoplasticity and viscoelastoplasticity.

References

1. Chinesta, F., Ladevèze, P.: Separated Representations and PGD-Based Model Reduction, Fundamentals and Applications. Springer, Vienna (2014). https://doi.org/10.1007/978-3-7091-1794-1
2. Géradin, M., Rixen, D.: Mechanical Vibrations: Theory and Application to Structural Dynamics, 3rd edn. Wiley, Hoboken (2015)
3. Libermann, P., Marle, C.-M.: Symplectic Geometry and Analytical Mechanics. D. Reidel Publishing Company, Dordrecht (1987)
4. Peng, L., Mohseni, K.: Symplectic model reduction of Hamiltonian systems. SIAM J. Sci. Comput. **38**(1), A1–A27 (2016)
5. Souriau, J.-M.: Structure of Dynamical Systems, a Symplectic View of Physics. Birkhäuser Verlag, New York (1997)
6. Willis, J.: The construction of effective relations for waves in a composite. C.R. Mecanique **340**, 181–192 (2012)

Non-relativistic Limits of General Relativity

Eric Bergshoeff[1(✉)], Johannes Lahnsteiner[1], Luca Romano[1], Jan Rosseel[2], and Ceyda Şimşek[1]

[1] Van Swinderen Institute for Particle Physics and Gravity, Nijenborgh 4, 9747 AG Groningen, The Netherlands
{E.A.Bergshoeff,c.simsek}@rug.nl

[2] Faculty of Physics, University of Vienna, Boltzmanngasse 5, 1090 Vienna, Austria

Abstract. We discuss non-relativistic limits of general relativity. In particular, we define a special fine-tuned non-relativistic limit, inspired by string theory, where the Einstein-Hilbert action has been supplemented by the kinetic term of a one-form gauge field. Taking the limit, a crucial cancellation takes place, in an expansion of the action in terms of powers of the velocity of light, between a leading divergence coming from the spin-connection squared term and another infinity that originates from the kinetic term of the one-form gauge field such that the finite invariant non-relativistic gravity action is given by the next subleading term. This non-relativistic action allows an underlying torsional Newton-Cartan geometry as opposed to the zero torsion Newton-Cartan geometry that follows from a more standard limit of General Relativity but it lacks the Poisson equation for the Newton potential. We will mention extensions of the model to include this Poisson equation.

Keywords: Newton-Cartan geometry · Torsion

1 Introduction

Eight years after the invention of General Relativity by Albert Einstein giving a frame-independent formulation of relativistic gravity [1] it was Élie Cartan who applied the same geometric formulation used by Einstein to give a frame-independent reformulation of Newtonian gravity called Newton-Cartan (NC) gravity [2]. The geometry underlying NC gravity that plays the same role as the Riemannian geometry in General Relativity is called NC geometry. The distinguishing feature of NC geometry is that it allows an absolute time direction, upon which all observers agree.

To obtain NC gravity as a non-relativistic (NR) limit of General Relativity, one usually introduces, beyond the relativistic Vierbein field, an additional one-form gauge field M_μ which, before taking the limit, plays the role of the

Supported by the Dutch organization FOM/NWO.

F. Nielsen and F. Barbaresco (Eds.): GSI 2021, LNCS 12829, pp. 448–455, 2021.
https://doi.org/10.1007/978-3-030-80209-7_49

U(1) gauge field corresponding to particle-antiparticle conservation. In the usual approach, see, e.g., [4], one does not associate a physical degree to M_μ and, correspondingly, assumes that M_μ is closed, i.e. $\partial_{[\mu}M_{\nu]} = 0$. Since this constraint cannot be derived from an action one can only take this limit at the level of the equations of motion. Using a second-order formulation of general relativity, the NR limit is then defined by redefining the Vierbein components plus the additional U(1) gauge field M_μ as follows:

$$E_\mu{}^0 = c\tau_\mu + \frac{1}{c}\, m_\mu; \quad E_\mu{}^{A'} = e_\mu{}^{A'}; \quad M_\mu = c\tau_\mu \tag{1}$$

and taking the limit that c goes to infinity. After taking this limit, one can identify τ_μ and $e_\mu{}^{A'}$ with the NC Vierbein fields. The NR gauge field m_μ plays the role of the NR central charge gauge field corresponding to particle conservation thereby extending the underlying Galilei algebra to a centrally extended Bargmann algebra.

Substituting the last expression of (1) into the condition $\partial_{[\mu}M_{\nu]} = 0$ and taking the limit that c goes to infinity, one obtains the zero torsion constraint $\partial_{[\mu}\tau_{\nu]} = 0$. This is a geometric constraint that defines the standard NC geometry with zero torsion. It has the effect that the time difference ΔT measured by an observer following a path between two time slices $t = t_1$ and $t = t_2$

$$\Delta T = \int_C \tau_\mu dx^\mu \tag{2}$$

is independent of the path C between $t = t_1$ and $t = t_2$ taken by the observer, i.e. all observers agree about the time difference ΔT. This is the defining property of a Newtonian spacetime with a one-dimensional foliation. Taking the NR limit of the Einstein equations using the zero torsion constraint leads to the usual NC gravity equations of motion whose number is the same as the Einstein equations of motion. These equations involve the Poisson equation of the Newton potential which is identified with the time component $\tau^\mu m_\mu$ of the central charge gauge field m_μ.

The procedure described above is the standard way to derive the NC gravity equations of motion. A characteristic feature of this limit is that it is only defined at the level of the equations of motion and that one automatically ends up with the zero torsion constraint. It is the purpose of this work to show that, in a sense to be discussed below, a limit at the level of the action can be defined without imposing any torsion constraint provided that one adds a separate kinetic term for the one-form gauge field M_μ to the Einstein-Hilbert action. This limit is inspired by the recent result [5] in string theory[1] that a NR limit of the so-called NS-NS gravity action

$$S_{\rm NSNS} = \frac{1}{2\kappa^2}\int \mathrm{d}^{10}x\, E\, \mathrm{e}^{-2\Phi}\left(\mathcal{R} + 4\,\partial_\mu\Phi\,\partial^\mu\Phi - \frac{1}{12}\mathcal{H}^{\mu\nu\rho}\mathcal{H}_{\mu\nu\rho}\right) \tag{3}$$

[1] String theory aims to unify gravity with quantum mechanics. At low energies, it gives rise to Einstein's gravity coupled to certain matter fields as given in Eq. (3).

allows for a NR limit at the level of the action without imposing any geometric constraints. Here, $\kappa^2 = 8\pi G_N$ is the gravitational coupling constant, E is the determinant of the Vierbein field, Φ is the dilaton field and $\mathcal{R}$ is the Ricci scalar. Furthermore, $\mathcal{H}_{\mu\nu\rho}$ is the curvature of the Kalb-Ramond (KR) 2-form gauge field. The existence of this NR limit is based upon a crucial cancellation of divergences originating form the Einstein-Hilbert term and the Kalb-Ramond field.

Before proceeding, it is instructive to first discus the NR limit at the level of an action describing a particle moving in a background of relativistic Vierbein fields together with the one-form gauge field M_μ.

2 The Non-relativistic Particle Action

One way to obtain a NR action for a particle in a NC background is by taking the NR limit of a relativistic particle moving in a Lorentzian background together with the U(1) gauge field M_μ. The gauge field M_μ couples to the particle via a Wess-Zumino (WZ) term.[2] The so-called Nambu-Goto (NG) formulation of the action is given by

$$S_{\rm NG} = -M \int d\sigma \sqrt{-(\dot{X}^\mu E_\mu{}^{\hat{A}})(\dot{X}^\nu E_\nu{}^{\hat{B}})\eta_{\hat{A}\hat{B}}} + Q \int d\sigma \dot{X}^\mu M_\mu . \tag{4}$$

Here σ parameterizes the worldline of the particle and $\dot{X}^\mu = \frac{dX^\mu}{d\sigma}$ where $X^\mu(\sigma)$ are the embedding coordinates. This action describes a particle of mass M and charge Q. The important thing is that the limit defined by (1) only works if one fine-tunes the M and Q parameters by redefining $M = Q = cm$. Upon expanding the action in powers of c, this fine-tuning implies a cancellation of a leading divergence with contributions coming from the time component of the Vierbein and the one form gauge field M_μ. The NR action is then given by the next subleading term in the expansion of the action of order c^0 which is given by

$$S_{\rm NR} = \frac{m}{2} \int d\sigma \, \frac{\dot{X}^\mu \dot{X}^\nu}{\tau_\rho \dot{X}^\rho} h_{\mu\nu} . \tag{5}$$

3 A Non-relativistic Target Space Action

In the previous two sections we considered

1. the standard NR limit of the Einstein equations leading to the NC gravity equations of motion and
2. the NR limit of the NG particle action leading to the NR particle action given in Eq. (5).

[2] A Wess-Zumino term is characterized by the fact that it is invariant under a gauge transformation only up to a total derivative.

In the first case we obtained the NC gravity equations of motion for zero torsion only. This was due to the fact that we assumed that the relativistic gauge field M_μ related to particle-antiparticle conservation was closed. This condition cannot be derived by the variation of an action and that is why we took the limit at the level of the equations of motion. In the second case we did not need to impose any geometric constraint. Here, the limit was made possible by a crucial cancellation of the leading divergences coming from the kinetic and WZ terms.

The above suggests that, in order to define a NR limit of the Einstein-Hilbert term, we need to give up the condition that M_μ is closed. Since this field played an important role in taking the NR limit of the particle action, where it occurred via a WZ term pairing up with the Vierbein field in the kinetic term, it is natural to similarly combine the Einstein-Hilbert action, which contains derivatives of the Vierbein squared terms, with a kinetic term for M_μ containing derivatives of M_μ squared terms. This leads us to consider the following combination of the Einstein-Hilbert action and a kinetic term for M_μ:

$$S = \frac{1}{2\kappa^2}\int d^4x\, E\left(\mathcal{R} - \frac{1}{4}F_{\mu\nu}F^{\mu\nu}\right), \tag{6}$$

where $F_{\mu\nu} = \partial_\mu M_\nu - \partial_\nu M_\mu$. We now expand this action in powers of c using the following slightly different version of the rescalings (1):

$$E_\mu{}^0 = c\,\tau_\mu; \quad E_\mu{}^{A'} = e_\mu{}^{A'}; \quad M_\mu = c\,\tau_\mu - \frac{1}{c}\,m_\mu. \tag{7}$$

Moreover, we define an effective Newton constant $\kappa_{NR}^2 \equiv c^{-1}\kappa^2$. The rescalings (1) and (7) are equivalent because in both cases the solution of the NR central charge gauge field m_μ in terms of the relativistic fields is given by

$$m_\mu = c\left(E_\mu{}^0 - M_\mu\right). \tag{8}$$

The reason that we wish to use the expressions (7) is that the m_μ field now occurs in the expansion of the kinetic term of the M_μ field which is easier to expand than the spin-connection squared terms in the Einstein-Hilbert term due to the presence of inverse Vierbein fields there. Of course, both expressions lead to the same answer, the reason is just to make the calculation easier.

One may verify that the Einstein-Hilbert term gives rise to the following leading quadratic divergence

$$\frac{c^2}{2\kappa_{NR}^2}\int d^4x\, e\,\tau_{A'B'}\tau^{A'B'}, \qquad \tau_{A'B'} \equiv e_{A'}{}^\mu e_{B'}{}^\nu \tau_{\mu\nu}, \tag{9}$$

where $\tau_{\mu\nu} \equiv \partial_{[\mu}\tau_{\nu]}$, $e = \det(\tau_\mu, e_\mu{}^{A'})$, and $e_{A'}{}^\mu$ is the projective inverse of the spatial Vierbein $e_\mu{}^{A'}$. The remarkable thing now is that the kinetic term for M_μ gives rise to the same quadratic divergence but with an opposite sign. Therefore, the two contributions to the leading divergence of the action precisely cancel.

Expanding the action we therefore find

$$S = c^2\, S^{(2)} + c^0\, S^{(0)} + c^{-2} S^{(-2)}, \tag{10}$$

where $S^{(0)} \neq 0$ and $S^{(-2)} \neq 0$ but $S^{(2)} = 0$ due to the cancellation of quadratic divergences just mentioned. We thus find that the NR limit of the action (6) is given by the terms of degree zero in c, i.e. $S^{(0)}$. After a straightforward calculation we find

$$S_{\rm NR} = \frac{1}{2\kappa_{NR}^2} \int d^4x\, e \left(R(J) + 4\nabla_{A'}\tau_0{}^{A'} - 6\,\tau_{A'0}\tau^{A'0} \right), \tag{11}$$

where $\tau_{A'0} = e_{A'}{}^\mu \tau^\nu \tau_{\mu\nu}$ and the derivative $\nabla_{A'}$ is covariant with respect to spatial rotations and Galilean boosts. The Ricci scalar $R(J)$ corresponding to the curvature tensor for spatial rotations:[3]

$$R(J) = -e_{A'}{}^\mu e_{B'}{}^\nu R_{\mu\nu}{}^{A'B'}(J) \quad \text{with} \tag{12}$$

$$R_{A'B'}{}^{A'B'}(J) = 2\, e_{A'}{}^\mu e_{B'}{}^\nu \left(\partial_{[\mu}\omega_{\nu]}{}^{A'B'} + \omega_{[\mu}{}^{A'}{}_{C'}\omega_{\nu]}{}^{B'C'}\right) + 4\,\omega_{A'B'}\tau^{A'B'} \tag{13}$$

with $\omega_{A'B'} = e_{A'}{}^\mu \omega_{\mu B'}$ and the dependent spatial rotation spin-connection field $\omega_\mu{}^{A'B'}$ and Galilean boost connection field $\omega_\mu{}^{A'}$ given by

$$\omega_\mu{}^{A'B'} = -2\, e_\mu{}^{[A'B']} + e_{\mu C'} e^{A'B'C'} - \frac{1}{2}\tau_\mu f^{A'B'}, \tag{14}$$

$$\omega_\mu{}^{A'} = \tau^\nu e_{\mu\nu}{}^{A'} - e_{\mu B'}\tau^\nu e_\nu{}^{A'B'} + \frac{1}{2} f_\mu{}^{A'} + \frac{1}{2}\tau_\mu \tau^\nu f_\nu{}^{A'} \qquad \text{with} \tag{15}$$

$$e_{\mu\nu}{}^{A'} = \frac{1}{2}\left(\partial_\mu e_\nu{}^{A'} - \partial_\nu e_\mu{}^{A'}\right), \quad f_{\mu\nu} = \partial_\mu m_\nu - \partial_\nu m_\mu. \tag{16}$$

Summarizing, by making use of a crucial cancellation of divergences, we are able to construct a non-relativistic Galilei-invariant action as a limit of General Relativity without imposing any geometric constraint.

4 Equations of Motion

The NR action (11) leads to equations of motion for $\tau_\mu, e_\mu{}^{A'}$ and m_μ which we denote by $\langle\tau\rangle^\mu$, $\langle e\rangle^\mu{}_{A'}$ and $\langle m\rangle^\mu$, respectively. They are defined by varying $S_{\rm NR}$ as follows:

$$\delta S_{\rm NR} = \frac{1}{2\kappa_{NR}^2} \int d^4x\, e \left(\langle\tau\rangle^\mu \delta\tau_\mu + \langle e\rangle_{A'}{}^\mu \delta e_\mu{}^{A'} + \langle m\rangle^\mu \delta m_\mu \right). \tag{17}$$

In total, the equations of motion $\langle\tau\rangle^\mu$, $\langle e\rangle_{A'}{}^\mu$ and $\langle m\rangle^\mu$ consist of $4+12+4=20$ components. Not all of these components are independent however. Indeed, the invariance of the action (11) under spatial rotations and Galilean boosts implies 6 Noether identities. We are therefore left with 14 independent equations.

It is instructive to see how these 14 equations of motion are related to the 14 relativistic equations of motion that follow from the relativistic action (6). Upon varying the relativistic action (6) we denote the resulting equations by

$$\delta\, S = \frac{1}{2\,\kappa^2} \int d^4x\, E([G]_{\mu\nu}\delta\, G^{\mu\nu} + [M]^\mu \delta\, M_\mu). \tag{18}$$

[3] The letter J refers to the spatial rotation generators $J_{A'B'}$ in the Bargmann algebra.

They are given by

$$[G]_{\mu\nu} \equiv \mathcal{R}_{\mu\nu} - \frac{1}{2} F_{\mu\rho} F_{\nu}{}^{\rho} - \frac{1}{2} G_{\mu\nu} (\mathcal{R} - \frac{1}{4} F_{\rho\sigma} F^{\rho\sigma}) = 0, \tag{19}$$

$$[M]^{\mu} \equiv \nabla_{\nu} F^{\nu\mu} = 0. \tag{20}$$

In order to make contact with the non-relativistic structure, one can furthermore organize the relativistic equations as representations of local $\mathrm{SO}(3) \subset \mathrm{SO}(1,3)$ rotations

$$[G]_{00}, \qquad [G]_{0A'}, \qquad [G]_{A'B'}, \qquad [M]_{0}, \qquad [M]_{A'}, \tag{21}$$

by contracting with $(E_0{}^{\mu}, E_{A'}{}^{\mu})$. Given the rescalings (7) one can write the equations of motion (19) and (20) as a power series in c where the terms at every order are expressions in terms of $(\tau_{\mu}, e_{\mu}{}^{A'}, m_{\mu})$. From general arguments it is clear that the set of leading order terms in these expressions form a closed set under Galilei transformations. We will use the following notation

$$[X] = c^{n} \langle X \rangle + \mathcal{O}(c^{n-2}), \tag{22}$$

where n is the highest order in the expansion of a generic equation of motion $[X]$. The leading orders $\langle X \rangle$ provide a self-consistent set of equations for NC geometry. However, there is a redundancy in this prescription since the leading orders of the singlets and vector $\mathrm{SO}(1,3)$ representations in (21) are identical, i.e.,

$$\langle G \rangle_{00} = \frac{1}{2} \langle M \rangle_{0} \qquad \text{and} \qquad \langle G \rangle_{0A'} = \frac{1}{2} \langle M \rangle_{A'}. \tag{23}$$

In other words, the above prescription fails to give the same number of independent equations as in the relativistic parent theory. Even though this shorter set of equations is consistent with Galilean symmetries, it does not extract the full information about the non-relativistic theory. This shortcoming can be circumvented by reformulating the relativistic theory (21) in terms of the following linear combinations

$$[S_{\pm}] \equiv [G]_{00} \pm \frac{1}{2} [M]_{0}, \qquad [V_{\pm}]_{A'} \equiv [G]_{0A'} \pm \frac{1}{2} [M]_{A'}, \qquad [G]_{A'B'}, \tag{24}$$

where $[S_{\pm}]$ and $[V_{\pm}]_{A'}$ have different leading orders. We make here use of the fact that the leading and sub-leading order terms in $[G]_{00}$, $[M]_0$ and $[G]_{A'0}$, $[M]_{A'}$ are not both the same. Clearly, the set of equations given in (24) is equivalent to the original one (21) relativistically. When it comes to restricting to the leading order, however, the resulting set of equations is different. More specifically, $\langle S_+ \rangle$ and $\langle S_- \rangle$ occur at different leading orders ($n = 2$ and $n = 0$, respectively). Similarly for $\langle V_{\pm} \rangle_{A'}$ (occurring at $n = \pm 1$). This, in turn, means that the set of 14 equations

$$\langle S_{\pm} \rangle = 0, \qquad \langle V_{\pm} \rangle_{A'} = 0, \qquad \langle G \rangle_{A'B'} = 0, \tag{25}$$

is equal in number to the relativistic ones (19), (20) while still being closed under Galilei boosts. The set (25) is furthermore related to the equations of motion following from the non-relativistic action (11) (as defined in Eq. (17)) by

$$\langle\tau\rangle^{\mu} = 2\,\langle S_{-}\rangle\,\tau^{\mu} - 2\,\langle V_{-}\rangle_{A'}\,e^{A'\mu}, \tag{26}$$

$$\langle e\rangle_{A'}{}^{\mu} = \langle V_{+}\rangle_{A'}\,\tau^{\mu} - 2\,\langle G\rangle_{A'B'}\,e^{B'\mu}, \tag{27}$$

$$\langle m\rangle^{\mu} = \langle S_{+}\rangle\,\tau^{\mu} - \langle V_{+}\rangle_{A'}e^{A'\mu}. \tag{28}$$

These equations of motion form a reducible but indecomposable representation of the Galilei algebra. Under Galilean boosts they transform to each other, as indicated by the arrows, as follows:

$$\begin{array}{ccccc} & \langle G\rangle_{A'B'} & & & \\ & \nearrow \quad \searrow & & & \\ \langle V_{-}\rangle_{A'} & & \langle V_{+}\rangle_{A'} & \longrightarrow & \langle S_{+}\rangle\,. \\ & \searrow \quad \nearrow & & & \\ & \langle S_{-}\rangle & & & \end{array} \tag{29}$$

5 Discussion

We have presented a new NR limit of General Relativity that avoids the need to impose any geometric constraint. This was achieved by adding to the Einstein-Hilbert term a kinetic term for the one-form gauge field M_μ as discussed in the main text. This additional term led, when taking the NR limit, to a cancellation of leading divergences such that the NR gravity action was given by the next sub-leading term in the expansion. The resulting action is given in Eq. (11). This invariant is based upon a generalized NC geometry with non-zero torsion that could have interesting phenomenological applications for strong gravity effects.[4]

The limit we discussed in this work was inspired by a similar NR limit in string theory where the limit was taken of the NS-NS gravity action given in Eq. (3) with the role of M_μ being played by the KR 2-form gauge field [5]. Although a similar cancellation of divergences takes place in both models, there is an important difference. It turns out that both in the particle and the string case, the NR action one obtains does not give rise to the Poisson equation for the Newton potential. In the string case, this is due to the fact that the NR action exhibits an emerging local dilatation symmetry. However, one can show that the Poisson equation can be obtained as a NR limit of the NS-NS gravity equations of motion. In fact, the Poisson equation forms, together with the equations of motion that follow from the NR NS-NS gravity action a reducible indecomposable representation of the Galilei symmetries. It means that the Poisson equation transforms under Galilei boost transformations to the equations of motion corresponding the NR NS-NS gravity action similar to how $\langle V_{-}\rangle_{A'}$ transforms to the other equations in (29).

[4] For a related discussion where this was emphasized, see [6].

The situation for the particle case is different. Identifying the Newton potential with the time component of m_μ, it is easy to see that the action (11) does not give rise to the Poisson equation for the Newton potential, i.e., the linearized equations of motion do not contain a term of the form $\partial^{A'}\partial_{A'}m_0$. However, in this case, there is no emerging dilatation symmetry and, therefore, the Poisson equation neither follows by taking the NR limit of the equations of motion. Therefore, as it stands, the action (11) does not give a full description of NC gravity.

It is suggestive that, in order to obtain a local emerging dilatation symmetry, we should introduce the analog of the dilaton and Kalb-Ramond 2-form field. In four spacetime dimensions this KR 2-form field is dual to an axionic scalar. This suggests that a proper description of torsional NC gravity can be obtained by adding to the Einstein-Hilbert term the vector field M_μ together with a complex axion-dilaton scalar. At the time of writing these proceedings, such an extended particle model is under investigation.

Acknowledgments. The work of CŞ is part of the research programme of the Foundation for Fundamental Research on Matter (FOM), which is financially supported by the Netherlands Organisation for Science Research (NWO). The work of LR is supported by the FOM/NWO free program *Scanning New Horizons*.

References

1. Einstein, A.: Die Feldgleichungen der Gravitation, Sitzungsberichte der Königlich Preußischen Akademie der Wissenschaften, Berlin (1915)
2. Cartan, E.: Sur les variétés à connexion affine et la théorie de la relativité généralisée (première partie). In: Annales scientifiques de l'École Normale Supérieure, Série 3, Tome, vol. 40, pp. 325–412 (1923)
3. Bergshoeff, E., Gomis, J., Rollier, B., Rosseel, J., ter Veldhuis, T.: Carroll versus Galilei Gravity. JHEP **03**, 165 (2017). https://doi.org/10.1007/JHEP03(2017)165. arXiv:1701.06156 [hep-th]
4. Bergshoeff, E., Rosseel, J., Zojer, T.: Non-relativistic fields from arbitrary contracting backgrounds. Class. Quant. Grav. **33**(17), 175010 (2016). https://doi.org/10.1088/0264-9381/33/17/175010. arXiv:1512.06064 [hep-th]. [5]
5. Bergshoeff, E., Lahnsteiner, J., Romano, L., Rosseel, J., Şimşek, C.: A non-relativistic limit of NS-NS gravity. J. High Energ. Phys. **2021** (2021). https://doi.org/10.1007/JHEP06(2021)021
6. Van den Bleeken, D.: Torsional Newton-Cartan gravity from the large c expansion of general relativity. Class. Quant. Grav. **34**(18), 185004 (2017). https://doi.org/10.1088/1361-6382/aa83d4. arXiv:1703.03459 [gr-qc]

Deformed Entropy, Cross-Entropy, and Relative Entropy

A Primer on Alpha-Information Theory with Application to Leakage in Secrecy Systems

Olivier Rioul(✉)

LTCI, Télécom Paris, Institut Polytechnique de Paris, 91120 Palaiseau, France
olivier.rioul@telecom-paris.fr

Abstract. We give an informative review of the notions of Rényi's α-entropy and α-divergence, Arimoto's conditional α-entropy, and Sibson's α-information, with emphasis on the various relations between them. All these generalize Shannon's classical information measures corresponding to $\alpha = 1$. We present results on data processing inequalities and provide some new generalizations of the classical Fano's inequality for any $\alpha > 0$.

This enables one to α-information as a information theoretic metric of leakage in secrecy systems. Such metric can bound the gain of an adversary in guessing some secret (any potentially random function of some sensitive dataset) from disclosed measurements, compared with the adversary's prior belief (without access to measurements).

Keywords: Rényi entropy and divergence · Arimoto conditional entropy · Sibson's information · Data processing inequalities · Fano's inequality · Information leakage · Side-Channel analysis

1 Introduction

Shannon's information theory is based on the classical notions of entropy $H(X)$, relative entropy $D(P\|Q)$ a.k.a. divergence, conditional entropy $H(X|Y)$ and mutual information $I(X;Y)$. The fundamental property that makes the theory so powerful is that all these informational quantities satisfy *data processing inequalities*[1].

An increasingly popular generalization of entropy is *Rényi entropy* of order α, or α-entropy $H_\alpha(X)$. Many compatible generalizations of relative and conditional entropies and information have been proposed, yet the only suitable quantities that do satisfy the correct data processing inequalities are Arimoto's conditional entropy $H_\alpha(X|Y)$ and Sibson's α-information $I_\alpha(X;Y)$. In this paper, we first review the corresponding α-information theory that beautifully generalizes Shannon's classical information theory (which is recovered as the limiting case $\alpha \to 1$). For $\alpha \neq 1$, however, α-information is no longer *mutual*: $I_\alpha(X;Y) \neq I_\alpha(Y;X)$ in general.

[1] See [11] for a review of several data processing results.

F. Nielsen and F. Barbaresco (Eds.): GSI 2021, LNCS 12829, pp. 459–467, 2021.
https://doi.org/10.1007/978-3-030-80209-7_50

The classical *Fano inequality* $H(X|Y) \leq h(\mathbb{P}_e) + \mathbb{P}_e \log(M-1)$ relating conditional entropy and probability of error $\mathbb{P}_e$ for a M-ary variable X can then be appropriately generalized using the data processing inequality for α-divergence. We present an appealing application to side-channel analysis.

2 A Primer on α-Information Theory

2.1 Notations

Probability Distributions. In this paper, probability distributions such as P, Q are such that $P \ll \mu$ and $Q \ll \mu$ where μ is a given reference (σ-finite) measure. The corresponding lower-case letters denote the Radon-Nykodym derivatives $p = \frac{\mathrm{d}P}{\mathrm{d}\mu}$, $q = \frac{\mathrm{d}Q}{\mathrm{d}\mu}$. The probability distribution P of random variable X is sometimes noted P_X. The reference measure is then noted μ_X and the Radon-Nykodym derivative is $p_X = \frac{\mathrm{d}P_X}{\mathrm{d}\mu_X}$.

Discrete and Continuous Random Variables. If X is a discrete random variable (taking values in a discrete set $\mathcal{X}$), then μ_X can be taken as the counting measure on $\mathcal{X}$; any integral over μ_X then reduces to a discrete sum over $x \in \mathcal{X}$, and we have $p_X(x) = \mathbb{P}(X=x)$. When X is a continuous random variable taking values in $\mathbb{R}^n$, μ_X is the Lebesgue measure on $\mathbb{R}^n$ and $p_X(x)$ is the corresponding probability density function. The notation $\mathbb{E}_x$ denotes expectation with respect to μ_X: $\mathbb{E}_x f(x) = \mathbb{E}\, f(X) = \int_{\mathcal{X}} f(x) p_X(x) \mathrm{d}\mu_X(x)$.

Uniform Distributions. As an example, we write $X \sim \mathcal{U}(M)$ if X is uniformly distributed over a set $\mathcal{X}$ of finite measure $M = \int_{\mathcal{X}} \mathrm{d}\mu$. The corresponding density is $p_X(x) = \frac{1_{\mathcal{X}}(x)}{M}$. In the discrete case, this means that X takes M equiprobable values in the set $\mathcal{X}$ of cardinality M.

Escort Distributions. For any distribution $p = \frac{\mathrm{d}P}{\mathrm{d}\mu}$, its *escort* distribution P_α of exponent α is given by the normalized α-power:

$$p_\alpha(x) = \frac{p^\alpha(x)}{\int_{\mathcal{X}} p^\alpha(x) \mathrm{d}\mu(x)}. \tag{1}$$

Joint Distributions. A joint distribution $P_{X,Y}$ of two random variables X, Y is such that $P_{X,Y} \ll \mu_X \otimes \mu_Y$ where μ_X is the reference measure for X and μ_Y is the reference measure for Y. In this paper, all functions of (x, y) integrated over $\mu_X \otimes \mu_Y$ will always be measurable and nonnegative so that all the integrals considered in this paper exist (with values in $[0, +\infty]$) and Fubini's theorem always applies. The conditional distributions $P_{Y|X}$ and $P_{X|Y}$ are such that $p_{X,Y}(x, y) = p_X(x) p_{Y|X}(y|x) = p_Y(y) p_{X|Y}(x|y)$ $(\mu_X \otimes \mu_Y)$-a.e.. We write $P_{X,Y} = P_X P_{Y|X} = P_{X|Y} P_Y$. In particular, $P_X \otimes P_Y$ is simply noted $P_X P_Y$.

Random Transformations A random transformation $P_{Y|X}$ applies to any input distribution P_X and provides an output distribution P_Y, which satisfies $p_Y(y) = \int p_{Y|X}(y|x)p_X(x)\mathrm{d}\mu_X(x)$. We write $P_X \rightarrow \boxed{P_{Y|X}} \rightarrow P_Y$. The same random transformation can be applied to another distribution Q_X. We then write $Q_X \rightarrow \boxed{P_{Y|X}} \rightarrow Q_Y$ where the corresponding output distribution Q_Y is such that $q_Y(y) = \int p_{Y|X}(y|x)q_X(x)\mathrm{d}\mu_X(x)$.

Deterministic Transformations. Any deterministic function $Y = f(X)$ taking discrete values can be seen as a particular case of a random transformation $P_{Y|X}$ where $p_{Y|X}(y|x) = \delta_{f(x)}(y)$.

2.2 Definitions

Throughout this paper we consider a Rényi order $\alpha \in (0,1) \cup (1,+\infty)$. In the following definitions, the corresponding quantities for $\alpha = 0, 1$ and $+\infty$ will be obtained by taking limits.

Definition 1 (Rényi Entropy [10]). *The α-entropy of $X \sim p$ is*

$$H_\alpha(X) = \frac{1}{1-\alpha} \log \int_{\mathcal{X}} p^\alpha(x)\,\mathrm{d}\mu(x). \tag{2}$$

It can also be noted $H_\alpha(P)$ or $H_\alpha(p)$. When X is binary with distribution $(p, 1-p)$, the α-entropy reduces to

$$h_\alpha(p) = \frac{1}{1-\alpha} \log\bigl(p^\alpha + (1-p)^\alpha\bigr). \tag{3}$$

Definition 2 (Rényi Divergence [5,10]). *The α-divergence or relative α-entropy of P and Q is*

$$D_\alpha(P\|Q) = \frac{1}{\alpha-1} \log \int_{\mathcal{X}} p^\alpha(x) q^{1-\alpha}(x)\,\mathrm{d}\mu(x). \tag{4}$$

For binary distributions $(p, 1-p)$ and $(q, 1-q)$, it reduces to

$$d_\alpha(p\|q) = \frac{1}{\alpha-1} \log\bigl(p^\alpha q^{1-\alpha} + (1-p)^\alpha (1-q)^{1-\alpha}\bigr). \tag{5}$$

Definition 3 (Arimoto-Rényi Entropy [1,7]). *The conditional α-entropy of X given Y is*

$$H_\alpha(X|Y) = \frac{\alpha}{1-\alpha} \log \int_{\mathcal{Y}} p_Y(y) \Bigl(\int_{\mathcal{X}} p^\alpha_{X|Y}(x|y)\,\mathrm{d}\mu_X(x) \Bigr)^{1/\alpha} \mathrm{d}\mu_Y(y). \tag{6}$$

Definition 4 (Sibson's mutual information [4,14,15]). *The α-mutual information (or simply α-information) of X and Y is*

$$I_\alpha(X;Y) = \frac{\alpha}{\alpha-1} \log \int_{\mathcal{Y}} \Bigl(\int_{\mathcal{X}} p_X(x) p^\alpha_{Y|X}(y|x)\,\mathrm{d}\mu_X(x) \Bigr)^{1/\alpha} \mathrm{d}\mu_Y(y) \tag{7}$$

$$= \frac{\alpha}{\alpha-1} \log \int_{\mathcal{Y}} p_Y(y) \Big(\int_{\mathcal{X}} p^{\alpha}_{X|Y}(x|y) p_X^{1-\alpha}(x) \, \mathrm{d}\mu_X(x) \Big)^{1/\alpha} \mathrm{d}\mu_Y(y). \tag{8}$$

Notice that $I_\alpha(X;Y) \neq I_\alpha(Y;X)$ in general.

By continuous extension of the above quantities, we recover the classical entropy $H_1(X) = H(X)$, divergence $D_1(P\|Q) = D(P\|Q)$, conditional entropy $H_1(X|Y) = H(X|Y)$, and mutual information $I_1(X;Y) = I(X;Y)$.

2.3 Basic Properties

Several known properties of α-divergence, α-entropies, and α-information are needed in the sequel. For completeness we list them as lemmas along with proof sketches.

Lemma 1 (α-information inequality [5, Thm. 8][2]**).**

$$D_\alpha(P\|Q) \geq 0 \tag{9}$$

with equality if and only if $P = Q$.

Proof. Apply Hölder's inequality $\int p^\alpha q^{1-\alpha} \leq (\int p)^\alpha (\int q)^{1-\alpha}$ for $\alpha < 1$, or the reverse Hölder inequality for $\alpha > 1$. Equality holds when $p = q$ μ-a.e., that is, $P = Q$. An alternative proof uses Jensen's inequality applied to $x \mapsto x^\alpha$, which is concave for $\alpha < 1$ and convex for $\alpha > 1$.

Lemma 2 (link between α-divergence and α-entropy [5, Eq. (2)]**).** *Letting $U = \mathcal{U}(M)$ be the uniform distribution,*

$$D_\alpha(P\|U) = \log M - H_\alpha(P) \tag{10}$$

Proof. Set $Q = U$ and $q = 1/M$ in (4).

Lemma 2 holds for discrete or continuous distributions. In particular from (9), (10) implies that if $M = \int_{\mathcal{X}} \mathrm{d}\mu < \infty$ then

$$H_\alpha(X) \leq \log M \tag{11}$$

with equality if and only if $X \sim \mathcal{U}(M)$.

The following is the natural generalization for $\alpha \neq 1$ of the well-known Gibbs inequality $H(X) \leq -\mathbb{E} \log q(X)$.

Lemma 3 (α-Gibbs inequality [12, Thm 1]**).** *Let $X \sim p$. For any probability distribution $\phi(x)$,*

$$H_\alpha(X) \leq \frac{\alpha}{1-\alpha} \log \mathbb{E}\big[\phi^{\frac{\alpha-1}{\alpha}}(X)\big] \tag{12}$$

[2] We name this "information inequality" after Cover and Thomas which used this terminology for the usual divergence $D_1(P\|Q) = D(P\|Q)$ [3, Theorem 2.6.3].

with equality if and only if $\phi = p_\alpha$*, the escort distribution* (1) *of* p*. For any family of conditional densities* $\phi(x|y)$*,*

$$H_\alpha(X|Y) \leq \frac{\alpha}{1-\alpha} \log \mathbb{E}\big[\phi^{\frac{\alpha-1}{\alpha}}(X|Y)\big] \tag{13}$$

with equality if and only if $\phi(x|Y) = p_\alpha(x|Y)$ μ_Y*-a.e. where* $p_\alpha(x|y)$ *is the escort distribution of* $p(x|y) = p_{X|Y}(x|y)$*.*

For uniform $q(x) \sim \mathcal{U}(M)$ in (12) we recover (11).

Proof. Let $q = \phi_{1/\alpha}$ so that $\phi = q_\alpha$. An easy calculation gives $\frac{\alpha}{1-\alpha} \log \mathbb{E}\big[\phi^{\frac{\alpha-1}{\alpha}}(X)\big] = D_{1/\alpha}(P_\alpha \| Q_\alpha) + H_\alpha(X)$. The first assertion then follows from Lemma 1.

Now for fixed $y \in \mathcal{Y}$, one has $H_\alpha(X|Y=y) \leq \frac{\alpha}{1-\alpha} \log \mathbb{E}\big[\phi^{\frac{\alpha-1}{\alpha}}(X|Y=y)\big]$ with equality if and only if $\phi(x|y) = p_\alpha(x|y)$. In both cases $\alpha < 1$ and $\alpha > 1$, it follows that $H_\alpha(X|Y) = \frac{\alpha}{1-\alpha} \log \mathbb{E}_y \exp \frac{1-\alpha}{\alpha} H_\alpha(X|Y=y) \leq \frac{\alpha}{1-\alpha} \log \mathbb{E}\big[\phi^{\frac{\alpha-1}{\alpha}}(X|Y)\big]$.

Lemma 4 (conditioning reduces α-entropy [1,7]).

$$H_\alpha(X|Y) \leq H_\alpha(X) \tag{14}$$

with equality if and only if X *and* Y *are independent.*

Proof. Set $\phi(x|y) = p_\alpha(x)$ in (13), with equality iff $p_\alpha(x|Y) = p_\alpha(x)$ a.e., i.e., X and Y are independent. (An alternate proof [7] uses Minkowski's inequality.)

Lemma 5 (α-information and conditional α-entropy [13, Eq. (47)]). *If* $X \sim \mathcal{U}(M)$*,*

$$I_\alpha(X;Y) = \log M - H_\alpha(X|Y). \tag{15}$$

Proof. Set $p_X(x) = 1/M$ in (8).

It is not true in general that $I_\alpha(X;Y) = H_\alpha(X) - H_\alpha(X|Y)$ for nonuniform X [15].

Lemma 6 (α-information and α-divergence). *Sibson's identity [14, Thm. 2.2], [4, Eq. (20)], [15, Thm. 1]:*

$$D_\alpha(P_{X,Y} \| P_X Q_Y) = I_\alpha(X;Y) + D_\alpha(Q_Y^* \| Q_Y) \tag{16}$$

for any distribution Q_Y*, where* Q_Y^* *is given by*

$$q_Y^*(y) = \frac{\big(\int_\mathcal{X} p_X(x) p_{Y|X}^\alpha(y|x)\,\mathrm{d}\mu(x)\big)^{1/\alpha}}{\int_\mathcal{Y} \big(\int_\mathcal{X} p_X(x) p_{Y|X}^\alpha(y|x)\,\mathrm{d}\mu(x)\big)^{1/\alpha} \mathrm{d}\mu_Y(y)} \tag{17}$$

In particular from (9)*,*

$$I_\alpha(X;Y) = \min_{Q_Y} D_\alpha(P_{X,Y} \| P_X Q_Y) = D_\alpha(P_{X,Y} \| P_X Q_Y^*) \tag{18}$$

hence $I_\alpha(X;Y) \geq 0$ *with equality if and only if* X *and* Y *are independent.*

Proof. $D_\alpha(P_{X,Y}\|P_X Q_Y) = \frac{1}{\alpha-1}\log\int p_X(x)\Big(\int p_{Y|X}^{\alpha}(y|x)q_Y^{1-\alpha}(y)\mathrm{d}\mu_Y(y)\Big)\mathrm{d}\mu_X(x) = \frac{1}{\alpha-1}\log\int q_Y^{1-\alpha}(y)\Big(\int p_X(x)p_{Y|X}^{\alpha}(y|x)\mathrm{d}\mu_X(x)\Big)\mathrm{d}\mu_Y(y)$ where we applied Fubini's theorem in the second equality. Substituting the expression inside parentheses using (17) gives (16). Using Lemma 1 it follows as expected that $I_\alpha(X;Y)\geq 0$ with equality if and only if X and Y are independent (in which case $Q_Y^* = P_Y$).

Lemma 7 (Convexity of α-divergence [4, Appendix], [5, Thm. 12]**).** *$Q \mapsto D_\alpha(P\|Q)$ is convex: For any $\lambda\in(0,1)$,*

$$D_\alpha(P\|\lambda Q_1 + (1-\lambda)Q_2) \leq \lambda D_\alpha(P\|Q_1) + (1-\lambda)D_\alpha(P\|Q_2). \tag{19}$$

Notice that $D_\alpha(P\|Q)$ is not convex in P in general (when $\alpha > 1$) [4,5].

2.4 Data Processing Inequalities

Lemma 8 (Data processing reduces α-divergence [9], [5, Thm. 1]**).** *For random transformations $P_X \to \boxed{P_{Y|X}} \to P_Y$ and $Q_X \to \boxed{P_{Y|X}} \to Q_Y$, we have the data processing inequality:*

$$D_\alpha(P_X\|Q_X) \geq D_\alpha(P_Y\|Q_Y) \tag{20}$$

with equality if and only if $P_{X|Y} = Q_{X|Y}$, where $P_{X|Y}P_Y = P_{Y|X}P_X$ and $Q_{X|Y}Q_Y = P_{Y|X}Q_X$.

In particular, for any μ_X-measurable set $A\subset\mathcal{X}$,

$$D_\alpha(P_X\|Q_X) \geq d_\alpha(p_A\|q_A) \tag{21}$$

where $p_A = \mathbb{P}(X\in A)$, $q_A = \mathbb{Q}(X\in A)$, and d_α is the binary α-divergence (5)*. Equality in* (21) *holds if and only if $P_{X|X\in A} = Q_{X|X\in A}$ and $P_{X|X\notin A} = Q_{X|X\notin A}$.*

Proof. Write $D_\alpha(P_X\|Q_X) = D_\alpha(P_X P_{Y|X}\|Q_X P_{Y|X}) = D_\alpha(P_Y P_{X|Y}\|Q_Y Q_{X|Y}) = \frac{1}{\alpha-1}\log\int p_Y^\alpha q_Y^{1-\alpha}\Big(\int p_{X|Y}^{\alpha} q_{X|Y}^{1-\alpha}\mathrm{d}\mu_X\Big)\mathrm{d}\mu_Y \geq D_\alpha(P_Y\|Q_Y)$ where we used (9) in the form $\int p^\alpha q^{1-\alpha} \leq (\int p)^\alpha(\int q)^{1-\alpha}$ for $\alpha<1$ and the opposite inequality for $\alpha>1$, applied to $p = p_{X|Y}$ and $q = q_{X|Y}$. The equality condition in (9) implies $P_{X|Y} = Q_{X|Y}$, which in turns implies equality in (20). Applying the statement to the deterministic transformation $Y = 1_{X\in A}$ gives (21).

Lemma 9 (Data processing reduces α-information [9, Thm. 5 (2)]**).** *For any Markov chain*[3] *$W - X - Y - Z$,*

$$I_\alpha(X;Y) \geq I_\alpha(W;Z) \tag{22}$$

[3] We do not specify a direction since reversal preserves the Markov chain property: $X_1 - X_2 - \cdots - X_n$ is Markov if and only if $X_n - X_{n-1} - \cdots - X_1$ is Markov.

Proof. Let $P_{X,Y} \to \boxed{P_{X,Z|X,Y}} \to P_{X,Z} \to \boxed{P_{W,Z|X,Z}} \to P_{W,Z}$. By the Markov condition, $P_{X,Z|X,Y} = P_{X|X}P_{Z|X,Y} = P_{X|X}P_{Z|Y}$ where $P_{X|X}$ is the identity operator; similarly $P_{W,Z|X,Z} = P_{W|X,Z}P_{Z|Z} = P_{W|X}P_{Z|Z}$. Thus if $Q_Y \to \boxed{P_{Z|Y}} \to Q_Z$, we find $P_X Q_Y \to \boxed{P_{X,Z|X,Y}} \to P_X Q_Z \to \boxed{P_{W,Z|X,Z}} \to P_W Q_Z$. By the data processing inequality for α-divergence (20), $D_\alpha(P_{X,Y}\|P_X Q_Y) \geq D_\alpha(P_{W,Z}\|P_W Q_Z) \geq I_\alpha(W;Z)$. Minimizing over Q_Y gives (22).

Lemma 10 (Data processing increases conditional α-entropy [7, Cor. 1]**).** *For any Markov chain $X - Y - Z$,*

$$H_\alpha(X|Y) \leq H_\alpha(X|Z) \tag{23}$$

with equality if and only if $X - Z - Y$ forms a Markov chain.

When $X \sim \mathcal{U}(M)$, inequality (23) also follows from (15) and the data processing inequality for α-information (22) where $W = X$.

Proof. Since $X - Y - Z$ forms a Markov chain, $H_\alpha(X|Y) = H_\alpha(X|Y,Z)$. Now set $\phi(x|y,z) = p_\alpha(x|z)$ in (13) to obtain $H_\alpha(X|Y,Z) \leq H_\alpha(X|Z)$, with equality iff $p_\alpha(x|Y,Z) = p_\alpha(x|Z)$ a.e., i.e., $X - Z - Y$ is Markov. (Alternate proof: see [7].)

Lemma 11 (Data processing inequalities for binary α-divergence). *Suppose $0 \leq p \leq q \leq r \leq 1$ (or that $1 \geq p \geq q \geq r \geq 0$). Then*

$$d_\alpha(p\|r) \geq d_\alpha(p\|q) \quad \text{and} \quad d_\alpha(p\|r) \geq d_\alpha(q\|r). \tag{24}$$

Proof. Consider an arbitrary binary channel with with parameters $\delta, \epsilon \in [0,1]$ and transition matrix $P_{Y|X} = \left(\begin{smallmatrix} 1-\delta & \delta \\ \epsilon & 1-\epsilon \end{smallmatrix}\right)$. By Lemma 8 applied to $P_X = (1-p, p)$ and $Q_X = (1-q, q)$, one obtains $d_\alpha(p\|q) \geq d_\alpha(\delta(1-p)+(1-\epsilon)p \,\|\, \delta(1-q)+(1-\epsilon)q)$ for any $p, q \in [0,1]$. Specializing to values of δ, ϵ such that $\epsilon p = \delta(1-p)$ or $\epsilon q = \delta(1-q)$ gives (24).

3 Fano's Inequality Applied to a Side-Channel Attack

We now apply α-information theory to any Markov chain X–Y–$\hat{X}$, modeling a side-channel attack using leaked information in a secrecy system. Here X is a sensitive data that depends on some secret (cryptographic key or password), input to a side channel $P_X \to \boxed{P_{Y|X}} \to P_Y$ through which information leaks, and Y is disclosed to the attacker (e.g., using some sniffer or probe measurements), and the attack provides $\hat{X}$ as a function of Y estimating X using the maximum a posteriori probability (MAP) rule so as to maximize the probability of success $\mathbb{P}_s = \mathbb{P}_s(X|Y) = \mathbb{P}(\hat{X} = X \mid Y)$, or equivalently, minimize the probability of error $\mathbb{P}_e = \mathbb{P}_e(X|Y) = \mathbb{P}(\hat{X} \neq X \mid Y)$.

The classical Fano inequality [6] then writes $H(X|Y) \leq h(\mathbb{P}_e) + \mathbb{P}_e \log(M-1)$ when $X \sim \mathcal{U}(M)$. It was generalized by Han and Verdú [8] as a lower bound on the mutual information $I(X;Y) \geq d(\mathbb{P}_s(X|Y)\|\mathbb{P}_s(X)) = d(\mathbb{P}_e(X|Y)\|\mathbb{P}_e(X))$

where $d(p\|q)$ denotes discrete divergence, and where $\mathbb{P}_s(X) = 1 - \mathbb{P}_e(X)$ corresponds to the case where X (possibly nonuniform) is guessed without even knowing Y. Using the MAP rule it can be easily seen that

$$\begin{cases} \mathbb{P}_s(X|Y) = \mathbb{E}\big[\max_{x\in\mathcal{X}} p_X(x|Y)\big] = \exp\big(-H_\infty(X|Y)\big) \\ \mathbb{P}_s(X) = \sup_{x\in\mathcal{X}}\ p_X(x) = \exp\big(-H_\infty(X)\big). \end{cases} \tag{25}$$

We now generalize the (generalized) Fano inequality to any value of $\alpha > 0$.

Theorem 1 (Generalized Fano's Inequality for α-Information).

$$I_\alpha(X;Y) \geq d_\alpha(\mathbb{P}_s(X|Y)\|\mathbb{P}_s(X)) = d_\alpha(\mathbb{P}_e(X|Y)\|\mathbb{P}_e(X)) \tag{26}$$

Proof. By the data processing inequality for α-information (Lemma 9), $I_\alpha(X;Y) \geq I_\alpha(X;\hat{X})$. Then by (18), $I_\alpha(X;\hat{X}) = D_\alpha(P_{X,\hat{X}}\|P_X Q^*_{\hat{X}}) \geq d_\alpha(\mathbb{P}_s(X|Y)\|\mathbb{P}'_s)$ where we have used the data processing inequality for α-divergence (Lemma 8, inequality (21)) to the event $A = \{\hat{X} = X\}$. Here $\mathbb{P}(\hat{X} = X) = \mathbb{P}_s(X|Y)$ by definition and $\mathbb{P}'_s = \sum_x p_X(x) q^*_{\hat{X}}(x) \leq \max_x p_X(x) = \mathbb{P}_s(X)$. Now by the binary data processing inequality (Lemma 11), $d_\alpha(\mathbb{P}_s(X|Y)\|\mathbb{P}'_s) \geq d_\alpha(\mathbb{P}_s(X|Y)\|\mathbb{P}_s(X))$.

Our main Theorem 1 states that α-information $I_\alpha(X;Y)$ bounds the gain $d_\alpha(\mathbb{P}_s(X|Y)\|\mathbb{P}_s(X))$ of any adversary in guessing secret X from disclosed measurements Y with success $\mathbb{P}_s(X|Y)$, compared with the adversary's prior belief without access to measurements, with lower success $\mathbb{P}_s(X)$. It additionally provides an implicit upper bound on $\mathbb{P}_s(X|Y)$ (or lower bound on $\mathbb{P}_e(X|Y)$) as a function of α-information—which can be loosened to obtain explicit bounds on success or error by further lower bounding the binary α-divergence.

Also, bounding α-information (by some "α-capacity" of the side channel) one can obtain bounds on the success of any possible attack for a given secrecy system based on some leakage model, in a similar fashion as what was made in the classical case $\alpha = 1$ in [2]. This is particularly interesting for the designer who needs to evaluate the robustness of a given implementation to any type of side-channel analysis, regardless of the type of attacker.

References

1. Arimoto, S.: Information measures and capacity of order α for discrete memoryless channels. In: Topics in Information Theory, Proceedings of the 2nd Colloqium Mathematical Societatis János Bolyai. North Holland, Keszthely, Hungary, 1975, vol. 2, No. 16, pp. 41–52 (1977)
2. de Chérisey, E., Guilley, S., Piantanida, P., Rioul, O.: Best information is most successful: Mutual information and success rate in side-channel analysis. In: IACR Transactions on Cryptographic Hardware and Embedded Systems (TCHES 2019) vol. 2, pp. 49–79 (2019)
3. Cover, T., Thomas, J.: Elements of Information Theory. J. Wiley & Sons, Hoboken (2006)

4. Csiszár, I.: Generalized cutoff rates and Rényi's information measures. IEEE Trans. Inf. Theor. **41**(1), 26–34 (1995)
5. van Erven, T., Harremoës, P.: Rényi divergence and Kullback-Leibler divergence. IEEE Trans. Inf. Theor **60**(7), 3797–3820 (2014)
6. Fano, R.M.: Class Notes for Course 6.574: Transmission of Information. MIT, Cambridge, MA (1952)
7. Fehr, S., Berens, S.: On the conditional Rényi entropy. IEEE Trans. Inf. Theor. **60**(11), 6801–6810 (2014)
8. Han, T.S., Verdú, S.: Generalizing the Fano inequality. IEEE Trans. Inf. Theor **40**(4), 1247–1251 (1994)
9. Polyanskiy, Y., Verdú, S.: Arimoto channel coding converse and Rényi divergence. In: Proceedings of Forty-Eighth Annual Allerton Conference. Allerton House, UIUC, IL, pp. 1327–1333, October 2010
10. Rényi, A.: On measures of entropy and information. In: Proceedings of the Fourth Berkeley Symposium on Mathematical Statistics and Probability, University of California Press, Berkeley, CA, vol. 1, pp. 547–561 (1961)
11. Rioul, O.: Information theoretic proofs of entropy power inequalities. IEEE Trans. Inf. Theor **57**(1), 33–55 (2011)
12. Rioul, O.: Rényi entropy power inequalities via normal transport and rotation. Entropy **20**(9, 641), 1–17 (2018)
13. Sason, I., Verdú, S.: Arimoto-Rényi conditional entropy and Bayesian M-ary hypothesis testing. IEEE Trans. Inf. Theor. **64**(1), 4–25 (2018)
14. Sibson, R.: Information radius. Zeitschrift Wahrscheinlichkeitstheorie Verwandte Gebiete **14**, 149–160 (1969)
15. Verdú, S.: α-mutual information. In: Proceedings of the Information Theory and Applications Workshop. La Jolla, CA, February 2015

Schrödinger Encounters Fisher and Rao: A Survey

Léonard Monsaingeon[1,2(✉)] and Dmitry Vorotnikov[3]

[1] IECL Nancy, Université de Lorraine, Nancy, France
leonard.monsaingeon@univ-lorraine.fr
[2] GFM, Universidade de Lisboa, Lisbon, Portugal
[3] CMUC, Universidade de Coimbra, Coimbra, Portugal
mitvorot@mat.uc.pt

Abstract. In this short note we review the dynamical Schrödinger problem on the non-commutative Fisher-Rao space of positive semi-definite matrix-valued measures. The presentation is meant to be self-contained, and we discuss in particular connections with Gaussian optimal transport, entropy, and quantum Fisher information.

Keywords: Schrödinger problem · Fisher-Rao space · Optimal transport · Fisher information · Entropy

The abstract Schrödinger problem on a generic Riemannian manifold (M, g) can be formulated as the dynamical minimization problem

$$\min_q \left\{ \int_0^1 |\dot{q}_t|^2 \mathrm{d}t + \varepsilon^2 \int_0^1 |\nabla V(q_t)|^2 \mathrm{d}t \qquad \text{s.t.} \qquad q|_{t=0,1} = q_0, q_1 \right\}. \tag{1}$$

The unknown curves $q = (q_t)_{t \in [0,1]}$ take values in M and interpolate between the prescribed endpoints $q_0, q_1 \in M$, the potential $V : M \to \mathbb{R}$ is given, and $\varepsilon > 0$ is a regularization parameter. Clearly when $\varepsilon \to 0$ this ε-problem is expected to converge in some sense to the geodesic problem on M.

The original Schrödinger problem [17], or rather, its dynamical equivalent (sometimes referred to as the Yasue problem [5,9]), can be rewritten as a particular instanciation of (1) when (M, g) is the Wasserstein space of probability measures, the potential V is given by the Boltzmann entropy $H(\rho) = \int \rho \log \rho$, and its Wasserstein gradient $I(\rho) = \|\nabla_{\mathcal{W}} H(\rho)\|^2 = \int |\nabla \log \rho|^2 \rho$ is the Fisher information functional. This was only recently recognized as an *entropic regularization* of the Monge-Kantorovich problem [10] and led to spectacular developments in computational optimal transport [16]. Recently, attempts have been made to develop an optimal transport theory for *quantum* objects, namely (symmetric positive-definite) matrix-valued measures, see [6] and references therein. In [13] we studied (1) precisely in the corresponding noncommutative Fisher-Rao space of matrix-measures. In this note we aim at providing a comprehensive

F. Nielsen and F. Barbaresco (Eds.): GSI 2021, LNCS 12829, pp. 468–476, 2021.
https://doi.org/10.1007/978-3-030-80209-7_51

introduction to this specific setting, in particular we wish to emphasize the construction based on Gaussian optimal transport that will lead to some specific entropy and quantum Fisher information functionals later on.

Section 1 briefly reviews classical optimal transport and the original Schrödinger problem. In Sect. 2 we discuss Gaussian optimal transport and the corresponding Bures-Wasserstein geometry. This then suggests the natural construction of the Fisher-Rao space detailed in Sect. 3. We introduce in Sect. 4 the resulting entropy and Fisher information functionals dictated by the previous construction. We also compute explicitly the induced heat flow, and finally discuss the Fisher-Rao-Schrödinger problem.

1 Optimal Transport and the Schrödinger Problem

In its fluid-mechanical *Benamou-Brenier* formulation [3], the quadratic *Wasserstein distance* between probability measures $\rho_0, \rho_1 \in \mathcal{P}(\Omega)$ over a smooth domain $\Omega \subset \mathbb{R}^d$ reads

$$\left.\begin{aligned} \mathcal{W}^2(\rho_0,\rho_1) = \min_{\rho,v} \Bigg\{ & \int_0^1 \int_\Omega |v_t(x)|^2 \rho_t(x) \mathrm{d}x\, \mathrm{d}t \\ & \text{s.t.} \quad \partial_t \rho_t + \operatorname{div}(\rho_t v_t) = 0 \text{ in } (0,1)\times\Omega \quad \text{and} \quad \rho|_{t=0,1} = \rho_{0,1} \end{aligned}\right\} \quad (2)$$

(supplemented with homogeneous no-flux boundary conditions on $\partial\Omega$ if needed in order to ensure mass conservation). This can be seen as an optimal control problem, where the control $v = v_t(x)$ is used to drive the system from ρ_0 to ρ_1 while minimizing the overall kinetic energy. We refer to [19] for a gentle yet comprehensive introduction to optimal transport.

As originally formulated by E. Schrödinger himself in [17], the Schrödinger problem roughly consists in determining the most likely evolution of a system for $t \in [0,1]$ in an ambient noisy environment at temperature $\varepsilon > 0$, given the observation of its statistical distribution at times $t = 0$ and $t = 1$. Here we shall rather focus on the equivalent dynamical problem

$$\left.\begin{aligned} \min_{\rho,v} \Bigg\{ & \int_0^1 \int_\Omega |v_t(x)|^2 \rho_t(x) \mathrm{d}x\, \mathrm{d}t + \varepsilon^2 \int_0^1 \int_\Omega |\nabla \log \rho_t(x)|^2 \rho_t(x) \mathrm{d}x\, \mathrm{d}t \\ & \text{s.t.} \quad \partial_t \rho_t + \operatorname{div}(\rho_t v_t) = 0 \text{ in } (0,1)\times\Omega \quad \text{and} \quad \rho|_{t=0,1} = \rho_{0,1} \end{aligned}\right\} \quad (3)$$

(again supplemented with Neumann boundary conditions on $\partial\Omega$ if needed). This is sometimes called the Yasue problem, cf. [5,9]. It is known [2,10,11] that the noisy problem Gamma-converges towards deterministic optimal transport in the small-temperature limit $\varepsilon \to 0$, and this actually holds in a much more general metric setting than just optimal transport [14]. We also refer to [1,20] for connections with Euclidean quantum mechanics, and to the survey [5] for an optimal-control perspective.

Following F.Otto [15], an important feature of optimal transport is that one can view $\mathcal{P}(\Omega)$ as a (formal) Riemannian manifold, whose Riemannian distance coincides with the Wasserstein distance (2). This relies upon the identification of infinitesimal variations ξ with (gradients of) Kantorovich potentials $\phi : \Omega \to \mathbb{R}$ by uniquely selecting the velocity field $v = \nabla\phi$ with minimal kinetic energy $\int_\Omega \rho|v|^2$, given $-\operatorname{div}(\rho v) = \xi$. Practically speaking, this means that tangent vectors $\xi \in T_\rho\mathcal{P}$ at a point ρ are identified with scalar potentials ϕ_ξ via the Onsager operator $\Delta_\rho = \operatorname{div}(\rho\nabla\cdot)$ and the elliptic equation

$$-\operatorname{div}(\rho\nabla\phi_\xi) = \xi, \tag{4}$$

see [19] for details. Accordingly, the linear heat flow $\partial_t\rho = \Delta\rho$ can be identified [8] as the Wasserstein gradient flow $\frac{d\rho}{dt} = -\operatorname{grad}_{\mathcal{W}} H(\rho)$ of the *Boltzmann entropy*

$$H(\rho) = \int_\Omega \rho(x)\log(\rho(x))\mathrm{d}x.$$

Here $\operatorname{grad}_{\mathcal{W}}$ denotes the gradient computed with respect to Otto's Riemannian structure. Moreover, the *Fisher information* functional

$$F(\rho) = \int_\Omega |\nabla \log\rho(x)|^2\rho(x)\,\mathrm{d}x = \|\operatorname{grad}_{\mathcal{W}} H(\rho)\|^2 \tag{5}$$

appearing in (3) coincides with the squared gradient of the entropy, or equivalently with the dissipation rate $-\frac{dH}{dt} = F$ of H along its own gradient flow. This identification (5) shows that (3) can indeed be written as a particular case of (1) with respect to the Wasserstein geometry.

2 The Bures-Wasserstein Distance and Gaussian Optimal Transport

When $\rho_0 = \mathcal{N}(m_0, A_0), \rho_1 = \mathcal{N}(m_1, A_1)$ are Gaussian measures with means $m_i \in \mathbb{R}^d$ and covariance matrices $A_i \in \mathcal{S}_d^{++}(\mathbb{R})$ (the space of symmetric positive definite matrices), the optimal transport problem (2) is explicitly solvable [18] and the Wasserstein distance can be computed as

$$\mathcal{W}^2(\rho_0, \rho_1) = |m_1 - m_0|^2 + \mathfrak{B}^2(A_0, A_1). \tag{6}$$

Here $\mathfrak{B}$ is the *Bures distance*

$$\mathfrak{B}^2(A_0, A_1) = \min_{RR^t=\mathrm{Id}} |A_1^{\frac12} - RA_0^{\frac12}|^2 = \operatorname{tr} A_0 + \operatorname{tr} A_1 - 2\operatorname{tr}\left(\left(A_0^{\frac12}A_1A_0^{\frac12}\right)^{\frac12}\right), \tag{7}$$

and $|M|^2 = \operatorname{tr}(MM^t)$ is the Euclidean norm of a matrix $M \in \mathbb{R}^{d\times d}$ corresponding to the Frobenius scalar product $\langle M, N\rangle = \operatorname{tr}(MN^t)$. The Bures distance, sometimes also called the *Helstrom metric*, can be considered as a quantum generalization of the usual Fisher information metric [4]. Since the Euclidean geometry of the translational part $|m_1 - m_0|^2$ in (6) is trivially flat, we restrict for simplicity to centered Gaussians $m_i = 0$. We denote accordingly

$$\mathfrak{G}_0 = \left\{ \rho \in \mathcal{P}(\mathbb{R}^d) : \qquad \rho = \mathcal{N}(0, A) \text{ for some } A \in \mathcal{S}_d^{++}(\mathbb{R}) \right\}$$

the *statistical manifold* of centered Gaussians and simply write $\rho = \mathcal{N}(0, A) = \mathcal{N}(A)$. Note that, for *positive*-definite matrices $A \in \mathcal{S}_d^{++}(\mathbb{R})$, the space of infinitesimal perturbations (the tangent plane $T_A\mathcal{S}_d^{++}(\mathbb{R})$) is the whole space of symmetric matrices $U \in \mathcal{S}_d(\mathbb{R})$. *Semi*-definite matrices are somehow degenerate (extremal) points of $\mathcal{S}_d^{++}(\mathbb{R})$, but it is worth stressing that (7) makes sense even for *semi*-definite $A_0, A_1 \in \mathcal{S}_d^{+}(\mathbb{R})$.

Going back to optimal transport, it is well-known [12,18] that the statistical manifold $\mathfrak{G}_0$ is a totally geodesic submanifold in the Wasserstein space. In other words, if $\rho_0 = \mathcal{N}(A_0)$ and $\rho_1 = \mathcal{N}(A_1)$ are Gaussians, then minimizers of the dynamical problem (2) remain Gaussian, i-e $\rho_t = \mathcal{N}(0, A_t)$ for some suitable $A_t \in \mathcal{S}_d^{++}$. Thus one expects that $\mathfrak{G}_0$, or rather $\mathcal{S}_d^{++}(\mathbb{R})$, can be equipped with a well-chosen, finite-dimensional Riemannian metric such that the induced Riemannian distance coincides with the Wasserstein distance between corresponding Gaussians. This is indeed the case [18], and the Riemannian metric is explicitly given at a point $\rho = \mathcal{N}(A)$ by

$$g_A(U, V) = \operatorname{tr}(UAV) = \langle AU, V\rangle = \langle AV, U\rangle, \qquad U, V \in T_A\mathcal{S}_d^{++}(\mathbb{R}) \cong \mathcal{S}_d(\mathbb{R}).$$

Note that, if $g_{Eucl}(\xi, \zeta) = \langle \xi, \zeta\rangle = \operatorname{tr}(\xi\zeta^t)$ is the Euclidean scalar product, the corresponding Riesz isomorphism $U \mapsto \xi_U$ (at a point $A \in \mathcal{S}_d^{++}$) identifying $g_{Eucl}(\xi_U, \xi_V) = g_A(U, V)$ is given by $\xi_U = (AU)^{sym} = \frac{AU+UA}{2}$. (This is the exact equivalent of the elliptic correspondence (4) for optimal transport.) As a consequence the Bures-Wasserstein distance can be computed [13,18] as

$$4\mathfrak{B}^2(A_0, A_1) = \min_{A,U} \left\{ \int_0^1 \langle A_tU_t, U_t\rangle \, dt \quad \text{s.t.} \quad \frac{dA_t}{dt} = (A_tU_t)^{sym} \right\}, \qquad A_0, A_1 \in \mathcal{S}_d^{++} \tag{8}$$

Up to the scaling factor 4 this an exact counterpart of (2), where $\frac{dA_t}{dt} = (A_tU_t)^{sym}$ plays the role of the continuity equation $\partial_t\rho_t + \operatorname{div}(\rho_t v_t) = 0$ and $\langle A_tU_t, U_t\rangle$ substitutes for the kinetic energy density $\int_\Omega \rho_t|v_t|^2$.

3 The Non-commutative Fisher-Rao Space

Given two probability densities $\rho_0, \rho_1 \in \mathcal{P}(\mathcal{D})$ on a domain $\mathcal{D} \subset \mathbb{R}^N$ (not to be confused with the previous $\Omega = \mathbb{R}^d$ for Gaussian optimal transport, codomain over which our matrices $A \in \mathcal{S}_d^{++}$ were built), the *scalar* Fisher-Rao distance is classically defined as the Riemannian distance induced by the Fisher information metric

$$\mathsf{FR}^2(\rho_0,\rho_1) = \min_{\rho}\left\{\int_0^1\int_{\mathcal{D}}\left|\frac{\partial \log \rho_t(x)}{\partial t}\right|^2 \rho_t(x)\,\mathrm{d}x\,\mathrm{d}t, \quad \text{s.t. } \rho_t\in\mathcal{P}(\mathcal{D})\right\}$$
$$= \min_{\rho,u}\left\{\int_0^1\int_{\mathcal{D}}|u_t(x)|^2\rho_t(x)\,\mathrm{d}x\,\mathrm{d}t \quad \text{s.t. } \partial_t\rho_t=\rho_t u_t \text{ and } \rho_t\in\mathcal{P}(\mathcal{D})\right\}. \quad (9)$$

Note that the unit-mass condition $\rho_t \in \mathcal{P}(\mathcal{D})$ is enforced here as a hard constraint. (Without this constraint, (9) actually defines the *Hellinger distance* H between arbitrary nonnegative measures [13].)

The strong structural similarity between (8) (9) suggest a natural extension of the latter scalar setting to *matrix-valued* measures. More precisely, the *Fisher-Rao space* is the space of (d-dimensional) positive *semi*-definite valued measures over $\mathcal{D}$ with unit mass

$$\mathbb{P}(\mathcal{D}) = \left\{A\in\mathcal{M}(\mathcal{D};\mathcal{S}_d^+(\mathbb{R})) \quad \text{s.t.} \quad \int_{\mathcal{D}}\operatorname{tr}A(x)\,\mathrm{d}x = 1\right\}.$$

This can be thought of as the space of noncommutative probability measures (see [13] for a short discussion on the connections with *free probability* theory and C^*-algebras). The *matricial* Fisher-Rao distance is then defined as

$$\mathsf{FR}^2(A_0,A_1) = \min_{A,U}\left\{\int_0^1\int_{\mathcal{D}}\langle A_t(x)U_t(x),U_t(x)\rangle\,\mathrm{d}x\,\mathrm{d}t \right.$$
$$\left. \text{s.t.} \quad \partial_t A_t = (A_tU_t)^{sym} \text{ in } (0,1)\times\mathcal{D} \quad \text{and} \quad A_t\in\mathbb{P}(\mathcal{D})\right\}, \quad (10)$$

which is indeed a higher-dimensional extension of (9) and should be compared again with (2). This clearly suggests viewing $\mathbb{P}(\mathcal{D})$ as a (formal, infinite-dimensional) Riemannian manifold with tangent space and norm

$$T_A\mathbb{P} = \left\{\xi_U=(AU)^{sym}: \quad U\in L^2\left(A(x)\mathrm{d}x;\ \mathcal{S}_d^+\right) \text{ and } \int_{\mathcal{D}}\langle A(x),U(x)\rangle\,\mathrm{d}x = 0\right\}$$

$$\|\xi_U\|_A^2 = \|U\|_{L_A^2}^2 = \int_{\mathcal{D}}\langle A(x)U(x),U(x)\rangle\,\mathrm{d}x. \quad (11)$$

Note once again that we imposed the unit-mass condition $A_t\in\mathbb{P}$ as a hard constraint in (10), which results in the zero-average condition $\int\langle A,U\rangle=0$ for tangent vectors U. Removing the mass constraint leads to (an extension of) the Hellinger distance H between arbitrary PSD matrix-valued measures. (The Hellinger space $(\mathbb{H}^+,\mathsf{H})$ is a metric cone over the "unit-sphere" $(\mathbb{P},\mathsf{FR})$ but we shall not discuss this rich geometry any further, see again [13].)

The corresponding Riemannian gradients of internal-energy functionals $\mathcal{F}(A) = \int_{\mathcal{D}} F(A(x))\,dx$ can be explicited [13] as

$$\operatorname{grad}_{\mathsf{FR}} \mathcal{F}(A) = (A(x)F'(A(x)))^{sym} - \operatorname{tr}\left(\int_{\mathcal{D}} A(y)F'(A(y))dy\right) A(x). \quad (12)$$

Here $F'(A)$ stands for the usual first variation of the matricial function $F(A)$, computed with respect to the standard Euclidean (Frobenius) scalar product.

4 Extended Entropy, the Heat Flow, and the Schrödinger Problem

From now on we always endow $\mathbb{P}$ with the Riemannian metric given by (11). Once this metric is fixed, the geometric Schrödinger problem (1) will be fully determined as soon as we choose a driving potential V on $\mathbb{P}$. One should ask now:

What is a good choice of "the" canonical entropy on the Fisher-Rao space?

A first natural guess would be based on the classical (negative) *von Neumann entropy* $S(A) = \operatorname{tr}(A \log A)$ from quantum statistical mechanics. However, the functional $\mathcal{S}(A) = \int_{\mathcal{D}} S(A(x))\,dx$ lacks geodesic convexity with respect to our ambient Fisher-Rao metric [13], and this makes it unfit for studying a suitable Schrödinger problem (see [14] for the connection between the validity of the Gamma-convergence in the limit $\varepsilon \to 0$ in (1) and the necessity of geodesic convexity of V in a fairly general metric context).

It turns out that there is a second natural choice, dictated by our previous construction based on Gaussian optimal transport and the canonical Boltzmann entropy $H(\rho) = \int \rho \log \rho$ on $\mathcal{P}(\mathbb{R}^d)$. Indeed, consider $A \in \mathcal{S}_d^{++}$ and write $\rho_A = \mathcal{N}(A)$ for the corresponding Gaussian. Assuming for simplicity that $\mathcal{D}$ has measure $|\mathcal{D}| = \frac{1}{d}$ so that $\operatorname{tr} \int_{\mathcal{D}} \operatorname{Id} dx = 1$, we choose as a reference measure the generalized uniform Lebesgue measure $\operatorname{Id} \in \mathbb{P}$ (more general reference measures can also be covered, see [13]). An explicit computation then gives the Boltzmann entropy of ρ_A relatively to $\rho_I = \mathcal{N}(\operatorname{Id})$ as

$$E(A) = H(\rho_A|\rho_I) = \int_{\mathbb{R}^d} \frac{\rho_A(y)}{\rho_I(y)} \log\left(\frac{\rho_A(y)}{\rho_I(y)}\right) \rho_I(y)\,dy = \frac{1}{2}\operatorname{tr}[A - \log A - \operatorname{Id}]. \quad (13)$$

Note carefully that $E(A) = +\infty$ as soon as A is only positive *semi*-definite (due to $-\operatorname{tr}\log A = -\sum \log \lambda_i = +\infty$ if any of the eigenvalues $\lambda_i = 0$), and that by convexity $E(A) \geq E(\operatorname{Id}) = 0$ is minimal only when $A = \operatorname{Id}$. Our canonical definition of *the* entropy on the Fisher-Rao space is then simply

$$\mathcal{E}(A) = \int_{\mathcal{D}} E(A(x))\,dx = -\frac{1}{2}\operatorname{tr}\int_{\mathcal{D}} \log A(x)\,dx, \qquad \text{for } A = A(x) \in \mathbb{P}. \quad (14)$$

(The terms $\int \operatorname{tr} A = 1 = \int \operatorname{tr} \mathrm{Id}$ cancel out in (13) due to our mass normalization.) Since the (Euclidean) first variation of (13) is $E'(A) = \frac{1}{2}(\mathrm{Id} - A^{-1})$, (12) shows that the Fisher-Rao gradient flow $\frac{dA_t}{dt} = -\operatorname{grad}_{\mathsf{FR}} \mathcal{E}(A_t)$ reads

$$\begin{aligned} \partial_t A_t &= -\left(A_t \frac{1}{2}[\mathrm{Id} - A_t^{-1}]\right)^{sym} + \operatorname{tr}\left(\int_{\mathcal{D}} A_t \frac{1}{2}[\mathrm{Id} - A_t^{-1}] \mathrm{d}y\right) A_t \\ &= \frac{1}{2}(\mathrm{Id} - A_t)^{sym} + \frac{1}{2}\operatorname{tr}\left(\int_{\mathcal{D}} (A_t - \mathrm{Id}) \mathrm{d}y\right) A_t = \frac{1}{2}(\mathrm{Id} - A_t). \end{aligned} \tag{15}$$

This generalized "heat flow" is consistent with the observation that, whenever $\rho_0 = \mathcal{N}(A_0)$ is Gaussian, the solution $\rho_t = \mathcal{N}(A_t)$ of the Fokker-Planck equation (the Wasserstein gradient flow of the relative entropy $\rho \mapsto H(\rho|\rho_I)$)

$$\partial_t \rho = \Delta \rho - \operatorname{div}(\rho \nabla \log \rho_I) \tag{16}$$

remains Gaussian with precisely $\frac{dA_t}{dt} = \frac{1}{2}(\mathrm{Id} - A_t)$. Our extended Fisher information functional can then be defined as the dissipation rate of the entropy $\mathcal{E}$ along the "heat flow" $\frac{dA_t}{dt} = -\operatorname{grad}_{\mathsf{FR}} \mathcal{E}(A_t)$ given by (15), namely

$$\begin{aligned} \mathcal{F}(A_t) &= -\frac{d}{dt}\mathcal{E}(A_t) = \frac{1}{2}\frac{d}{dt}\operatorname{tr}\int_{\mathcal{D}} \log A_t(x)\, \mathrm{d}x = \frac{1}{2}\operatorname{tr}\int_{\mathcal{D}} A_t^{-1}(x)\partial_t A_t(x)\, \mathrm{d}x \\ &= \frac{1}{2}\operatorname{tr}\int_{\mathcal{D}} A_t^{-1}(x)\frac{1}{2}[\mathrm{Id} - A_t(x)]\, \mathrm{d}x = \frac{1}{4}\left(\operatorname{tr}\int_{\mathcal{D}} A_t^{-1}(x)\, \mathrm{d}x - 1\right). \end{aligned} \tag{17}$$

Equivalently and consistently, $\mathcal{F}(A) = \|\operatorname{grad}_{\mathsf{FR}} \mathcal{E}(A)\|^2$. Note that $\mathcal{F}(A_t) > F(\mathrm{Id}) = 0$ and $\mathcal{E}(A_t) > \mathcal{E}(\mathrm{Id}) = 0$ unless $A_t(x) \equiv \mathrm{Id}$, which is of course consistent with the expected long-time behavior $A_t \to \mathrm{Id}$ for (15) as $t \to \infty$ (or equivalently $\rho_t \to \rho_I = \mathcal{N}(\mathrm{Id})$ for Gaussian solutions of the Fokker-Planck equation (16)).

With these explicit representations (11)(17) of the quantum (i-e matricial) Fisher-Rao metric and Fisher information, we can now make sense of the geometric Schrödinger problem (1) in the noncommutative Fisher-Rao space as

$$\min_A \left\{ \int_0^1 \int_{\mathcal{D}} \langle A_t(x) U_t(x), U_t(x) \rangle \, \mathrm{d}x\mathrm{d}t + \frac{\varepsilon^2}{4} \int_0^1 \left(\operatorname{tr} \int_{\mathcal{D}} A_t^{-1}(x)\, \mathrm{d}x - 1 \right) \right.$$
$$\left. \text{s.t.} \qquad \partial_t A_t = (A_t U_t)^{sym} \quad \text{and} \quad A_t \in \mathbb{P} \text{ for all } t \in [0,1] \right\}, \tag{18}$$

with fixed endpoints $A_0, A_t \in \mathbb{P}(\mathcal{D})$. Let us mention at this stage that a different entropic regularization of Gaussian optimal transport was investigated in [7] in a static framework and for a much simpler setting, namely when $\mathcal{D} = \{x\}$ is a single point

As could be expected from the above geometric machinery, we have now

Theorem *([13]). In the limit* $\varepsilon \to 0$ *and for fixed endpoints* $A_0, A_1 \in \mathbb{P}$, *the* ε*-functional in* (18) *Gamma-converges towards the kinetic functional in* (10) *for the uniform convergence on* $C([0,1];\mathbb{P})$.

We omit the details for the sake of brevity, but let us point out that our proof leverages fully explicit properties of the geometric heat flow (15). A particular byproduct of our analysis is the $\frac{1}{2}$-geodesic convexity of the entropy (14) in the Fisher-Rao geometry, which was already established in [12] by formal Riemannian computations. The key argument builds up on a Lagrangian construction originally due to A. Baradat in [2]. Finally, we recently extended the result to arbitrary metric spaces and general entropy functionals [14] (provided a suitable heat flow is available), and we showed moreover that geodesic convexity is actually necessary (and almost sufficient) for the Γ-convergence.

References

1. Albeverio, S., Yasue, K., Zambrini, J.C.: Euclidean quantum mechanics: analytical approach. Annales de l'IHP Physique théorique **50**(3), 259–308 (1989)
2. Aymeric, B., Léonard, M.: Small noise limit and convexity for generalized incompressible flows, Schrödinger problems, and optimal transport. Arch. Ration. Mech. Anal. **235**(2), 1357–1403 (2019). https://doi.org/10.1007/s00205-019-01446-w
3. Benamou, J.D., Brenier, Y.: A computational fluid mechanics solution to the Monge-Kantorovich mass transfer problem. Num. Math. **84**(3), 375–393 (2000)
4. Bures, D.: An extension of Kakutani's theorem on infinite product measures to the tensor product of semifinite ω^*-algebras. Trans. AMS **135**, 199–212 (1969)
5. Chen, Y., Georgiou, T.T., Pavon, M.: Stochastic control liaisons: Richard Sinkhorn meets Gaspard Monge on a Schroedinger bridge. arXiv preprint arXiv:2005.10963 (2020)
6. Chen, Y., Gangbo, W., Georgiou, T., Tannenbaum, A.: On the matrix Monge-Kantorovich problem. Eur. J. Appl. Math. **31**(4), 574–600 (2020)
7. Janati, H., Muzellec, B., Peyré, G., Cuturi, M.: Entropic optimal transport between (unbalanced) Gaussian measures has a closed form. arXiv preprint arXiv:2006.02572 (2020)
8. Jordan, R., Kinderlehrer, D., Otto, F.: The variational formulation of the Fokker-Planck equation. SIAM J. Math. Anal. **29**(1), 1–17 (1998)
9. Léger, F.: A geometric perspective on regularized optimal transport. J. Dyn. Diff. Equ. **31**(4), 1777–1791 (2018). https://doi.org/10.1007/s10884-018-9684-9
10. Léonard, C.: A survey of the Schrödinger problem and some of its connections with optimal transport. Discrete Contin. Dyn. Syst. A **34**, 1533–1574 (2014)
11. Mikami, T.: Monge's problem with a quadratic cost by the zero-noise limit of h-path processes. Probabl. Theor. Relat. Fields **129**(2), 245–260 (2004)
12. Modin, K.: Geometry of matrix decompositions seen through optimal transport and information geometry. J. Geom. Mech. **9**(3), 335–390 (2017)
13. Monsaingeon, L., Vorotnikov. D.: The Schrödinger problem on the non-commutative Fisher-Rao space. Calculus Variat. Part. Diff. Equ. **60(1)** (2021). art. no. 14
14. Monsaingeon, L., Vorotnikov, D., Tamanini, L.: The dynamical Schrödinger problem in abstract metric spaces. Preprint arXiv:2012.12005 (2020)
15. Otto, F.: The geometry of dissipative evolution equations: the porous medium equation. Commun. Part. Diff. Equ. **23**(1–2), 101–174 (2001)
16. Peyré, G., Cuturi. M.: Computational optimal transport: with applications to data science. Found. Trends® in Mach. Learn. **11**(5–6), 355–607 (2019)

17. Schrödinger. E.: Über die umkehrung der naturgesetze. Sitzungsberichte Preuss. Akad. Wiss. Berlin. Phys. Math. Klasse, **144**, 144–153 (1931)
18. Takatsu, A.: Wasserstein geometry of Gaussian measures. Osaka J. Math. **48**(4), 1005–1026 (2011)
19. Villani, C.: Topics in optimal transportation. AMS Graduate Studies in Mathematics, vol. 58. American Mathematical Society, Providence (2003)
20. Zambrini, J.C.: Variational processes and stochastic versions of mechanics. J. Math. Phys. **27**(9), 2307–2330 (1986)

Projections with Logarithmic Divergences

Zhixu Tao and Ting-Kam Leonard Wong(✉)

University of Toronto, Toronto, Canada
zhixu.tao@mail.utoronto.ca, tkl.wong@utoronto.ca

Abstract. In information geometry, generalized exponential families and statistical manifolds with curvature are under active investigation in recent years. In this paper we consider the statistical manifold induced by a logarithmic $L^{(\alpha)}$-divergence which generalizes the Bregman divergence. It is known that such a manifold is dually projectively flat with constant negative sectional curvature, and is closely related to the $\mathcal{F}^{(\alpha)}$-family, a generalized exponential family introduced by the second author [16]. Our main result constructs a dual foliation of the statistical manifold, i.e., an orthogonal decomposition consisting of primal and dual autoparallel submanifolds. This decomposition, which can be naturally interpreted in terms of primal and dual projections with respect to the logarithmic divergence, extends the dual foliation of a dually flat manifold studied by Amari [1]. As an application, we formulate a new $L^{(\alpha)}$-PCA problem which generalizes the exponential family PCA [5].

Keywords: Logarithmic divergence · Generalized exponential family · Dual foliation · Projection · Principal component analysis

1 Introduction

A cornerstone of information geometry [2,4] is the dually flat geometry induced by a Bregman divergence [10]. This geometry underlies the exponential and mixture families and explains their efficacy in statistical applications. A natural direction is to study extensions of the dually flat geometry and exponential/mixture families as well as their applications. Motivated by optimal transport, in a series of papers [12,13,16,17] Pal and the second author developed the $L^{(\alpha)}$-divergence which is a logarithmic extension of the Bregman divergence. Let $\alpha > 0$ be a constant. Given a convex domain $\Theta \subset \mathbb{R}^d$ and a differentiable α-exponentially concave function $\varphi : \Theta \to \mathbb{R}$ (i.e., $\Phi = e^{\alpha\varphi}$ is concave), we define the $L^{(\alpha)}$-divergence $\mathbf{L}_{\varphi}^{(\alpha)} : \Theta \times \Theta \to [0, \infty)$ by

$$\mathbf{L}_{\varphi}^{(\alpha)}[\theta : \theta'] = \frac{1}{\alpha} \log(1 + \alpha \mathrm{D}\varphi(\theta') \cdot (\theta - \theta')) - (\varphi(\theta) - \varphi(\theta')), \tag{1}$$

This research is supported by NSERC Discovery Grant RGPIN-2019-04419 and a Connaught New Researcher Award.

F. Nielsen and F. Barbaresco (Eds.): GSI 2021, LNCS 12829, pp. 477–486, 2021.
https://doi.org/10.1007/978-3-030-80209-7_52

where $\mathsf{D}\varphi$ is the Euclidean gradient and $\cdot$ is the Euclidean dot product. We recover the Bregman divergence (of a concave function) by letting $\alpha \to 0^+$. Throughout this paper we work under the regularity conditions stated in [16, Condition 7]; in particular, φ is smooth and the Hessian $\mathsf{D}^2 \Phi$ is strictly negative definite (so Φ is strictly concave). By the general theory of Eguchi [7], the divergence $\mathbf{L}_{\varphi}^{(\alpha)}$ induces a dualistic structure (g, ∇, ∇^*) consisting of a Riemannian metric g and a pair (∇, ∇^*) of torsion-free affine connections that are dual with respect to g. It was shown in [13,16] that the induced geometry is dually projectively flat with constant negative sectional curvature $-\alpha$ (see Sect. 2 for a brief review). Moreover, the $L^{(\alpha)}$-divergence is a canonical divergence (in the sense of [3]) for such a geometry. Thus, this geometry can be regarded as the constant-curvature analogue of the dually flat manifold. The geometric meaning of the curvature $-\alpha$ was investigated further in [18,19]. Also see [13,14,19] for connections with the theory of optimal transport. In [16] we also introduced the $\mathcal{F}^{(\alpha)}$-family, a parameterized density of the form

$$p(x;\theta) = (1 + \alpha\theta \cdot F(x))^{-1/\alpha} e^{\varphi(\theta)}, \tag{2}$$

and showed that it is naturally compatible with the $L^{(\alpha)}$-divergence. To wit, the potential function φ in (2) can be shown to be α-exponentially concave, and its $L^{(\alpha)}$-divergence is the Rényi divergence of order $1+\alpha$. Note that letting $\alpha \to 0^+$ in (2) recovers the usual exponential family. In a forthcoming paper we will study in detail the relationship between the $\mathcal{F}^{(\alpha)}$-family and the q-exponential family [11], where $q = 1 + \alpha$.

To prepare for statistical applications of the logarithmic divergence and the $\mathcal{F}^{(\alpha)}$-family, in this paper we study primal and dual projections with respect to the $L^{(\alpha)}$-divergence. These are divergence minimization problems of the form

$$\inf_{Q \in \mathcal{A}} \mathbf{L}_{\varphi}^{(\alpha)}[P : Q] \quad \text{and} \quad \inf_{Q \in \mathcal{A}} \mathbf{L}_{\varphi}^{(\alpha)}[Q : P], \tag{3}$$

where P is a given point and $\mathcal{A}$ is a submanifold of the underlying manifold. The optimal solutions in (3) are respectively the primal and dual projections of P onto $\mathcal{A}$. Our main result, presented in Sect. 3, provides an orthogonal foliation of the manifold in terms of ∇ and ∇^*-autoparallel submanifolds. This extends the orthogonal foliation of a dually flat manifold constructed by Amari [1], and shows that projections with respect to an $L^{(\alpha)}$-divergence are well-behaved. As an application, we formulate in Sect. 4 a nonlinear dimension reduction problem that we call the $L^{(\alpha)}$-PCA. In a nutshell, the $L^{(\alpha)}$-PCA problem extends the exponential family PCA [5] to the $\mathcal{F}^{(\alpha)}$-family (2). In future research we will study further properties of the $L^{(\alpha)}$-PCA problem, including optimization algorithms and statistical applications.

2 Preliminaries

Consider an $L^{(\alpha)}$-divergence $\mathbf{L}_{\varphi}^{(\alpha)}$ as in (1). We regard the convex set Θ as the domain of the (global) primal coordinate system θ. We denote the underlying manifold by $\mathcal{S}$. For $P \in \mathcal{S}$, we let $\theta_P \in \Theta$ be its primal coordinates. We define the dual coordinate system by

$$\eta_P = \mathsf{T}(\theta_P) := \frac{\mathsf{D}\varphi(\theta_P)}{1 - \alpha \mathsf{D}\varphi(\theta_P) \cdot \theta_P}. \tag{4}$$

We call the mapping $\theta \mapsto \eta = \mathsf{T}(\theta)$ the α-Legendre transformation and let Ω be the range of η. This transformation corresponds to the α-conjugate defined by

$$\psi(y) = \inf_{\theta \in \Theta} \left(\frac{1}{\alpha} \log(1 + \alpha \theta \cdot y) - \varphi(\theta) \right).$$

It can be shown that ψ is also α-exponentially concave. The inverse of T is the α-Legendre transform of ψ, and we have the self-dual expression

$$\mathbf{D}[P : Q] := \mathbf{L}_{\varphi}^{(\alpha)}[\theta_P : \theta_Q] = \mathbf{L}_{\psi}^{(\alpha)}[\eta_Q : \eta_P]. \tag{5}$$

We refer the reader to [16, Section 3] for more details about this generalized duality. Note that (5) defines a divergence $\mathbf{D}$ on $\mathcal{S}$.

Let (g, ∇, ∇^*) be the duallistic structure, in the sense of [7], induced by the divergence $\mathbf{D}$ on $\mathcal{S}$. The following lemma expresses the Riemannian metric g in terms of the coordinate frames $(\frac{\partial}{\partial \theta_i})$ and $(\frac{\partial}{\partial \eta_j})$ (also see [16, Remark 6]).

Lemma 1. *[16, Proposition 8] Let $\langle \cdot, \cdot \rangle$ be the Riemannian inner product. Then*

$$\left\langle \frac{\partial}{\partial \theta_i}, \frac{\partial}{\partial \eta_j} \right\rangle = \frac{-1}{\Pi} \delta_{ij} + \frac{\alpha}{\Pi^2} \theta_j \eta_i, \quad \Pi = 1 + \alpha \theta \cdot \eta. \tag{6}$$

We call ∇ the primal connection and ∇^* the dual connection. A primal geodesic is a curve γ such that $\nabla_{\dot{\gamma}} \dot{\gamma} = 0$, and a dual geodesic is defined analogously. As shown in [16, Section 6], the geometry is dually projectively flat in the following sense: If $\gamma : [0, 1] \to \mathcal{S}$ is a primal geodesic, then under the primal coordinate system, $\theta_{\gamma(t)}$ is a straight line up to a time reparameterization. Similarly, a dual geodesic is a time-changed straight line under the dual coordinate system. We also have the following generalized Pythagorean theorem.

Theorem 1 (Generalized Pythagorean theorem). *[16, Theorem 16] For $P, Q, R \in \mathcal{S}$, the generalized Pythagorean relation*

$$\mathbf{D}[Q : P] + \mathbf{D}[R : Q] = \mathbf{D}[R : P]$$

holds if and only if the primal geodesic from Q to R meets g-orthogonally to the dual geodesic from Q to P.

3 Dual Foliation and Projection

In this section we construct an orthogonal decomposition of $\mathcal{S}$. We begin with some notations. Let $\mathcal{A} \subset \mathcal{S}$ be a submanifold of $\mathcal{S}$. Thanks to the dual projective flatness, we say that $\mathcal{A}$ is ∇-autoparallel (resp. ∇^*-autoparallel) if it is a convex set in the θ coordinates (resp. η coordinates). Consider a maximal ∇-autoparallel submanifold $\mathcal{E}_k$ with dimension $k \leq d = \dim \mathcal{S}$. Given $P_0 \in \mathcal{E}_k$, we may write

$$\mathcal{E}_k = \{P \in \mathcal{S} : \theta_P - \theta_{P_0} \in A_0\}, \tag{7}$$

where $A_0 \subset \mathbb{R}^d$ is a vector subspace with dimension k. Dually, we may consider maximal ∇^*-autoparallel submanifolds $\mathcal{M}_k$ with dimension $d - k$ (so k is the codimension).

We are now ready to state our first result. Here we focus on projections onto a ∇-autoparallel submanifold. The other case is similar and is left to the reader.

Theorem 2. *Fix $1 \leq k < d$. Consider a ∇-autoparallel submanifold $\mathcal{E}_k$ and $P_0 \in \mathcal{S}$. Then there exists a unique $\mathcal{M}_k$ such that $\mathcal{E}_k \cap \mathcal{M}_k = \{P_0\}$ and the two submanifolds meet orthogonally, i.e., if $u \in T_{P_0}\mathcal{E}_k$ and $v \in T_{P_0}\mathcal{M}_k$, then $\langle u, v\rangle = 0$. We call $\mathcal{M}_k = \mathcal{M}_k(P_0)$ the dual complement of $\mathcal{E}_k$ at P_0.*

Proof. Let $\mathcal{E}_k$ be given by (7). We wish to find a submanifold $\mathcal{M}_k$ of the form

$$\mathcal{M}_k = \{P \in \mathcal{S} : \eta_P - \eta_{P_0} \in B_0\}, \tag{8}$$

where $B_0 \subset \mathbb{R}^d$ is a vector subspace of dimension $d-k$, such that $\mathcal{E}_k \cap \mathcal{M}_k = \{P_0\}$ and the two submanifolds meet orthogonally.

Consider the tangent spaces of $\mathcal{E}_k$ and $\mathcal{M}_k$ at P_0. Since $\mathcal{E}_k$ is given by an affine constraint in the θ coordinates, we have

$$T_{P_0}\mathcal{E}_k = \left\{ u = \sum_i a_i \frac{\partial}{\partial \theta_i} \in T_{P_0}\mathcal{S} : (a_1, \ldots, a_d) \in A_0 \right\}.$$

Similarly, we have

$$T_p\mathcal{M}_k = \left\{ v = \sum_j b_j \frac{\partial}{\partial \eta_j} \in T_{P_0}\mathcal{S} : (b_1, \ldots, b_d) \in B_0 \right\}.$$

Given the subspace A_0, our first task is to choose a vector subspace B_0 such that $T_{P_0}\mathcal{S} = T_{P_0}\mathcal{E}_k \oplus T_{P_0}\mathcal{M}_k$ and $T_{P_0}\mathcal{E}_k \perp T_{P_0}\mathcal{M}_k$.

Let $u = \sum_i a_i \frac{\partial}{\partial \theta_i}, v = \sum_j b_j \frac{\partial}{\partial \eta_j} \in T_{P_0}\mathcal{S}$. Regard $a = (a_1, \ldots, a_d)^\top$ and $b = (b_1, \ldots, b_d)^\top$ as column vectors. Writing (6) in matrix form, we have

$$\langle u, v\rangle = \sum_{i,j} a_i b_j \left\langle \frac{\partial}{\partial \theta_i}, \frac{\partial}{\partial \eta_j} \right\rangle = a^\top G b, \tag{9}$$

where the matrix $G = G(P_0)$ is invertible and is given by

$$G(P_0) = \frac{-1}{\Pi} I + \frac{\alpha}{\Pi^2} \theta_{P_0} \eta_{P_0}^\top. \tag{10}$$

With this notation, we see that by letting B_0 be the subspace

$$B_0 = \{b \in \mathbb{R}^d : a^\top G b = 0 \ \forall a \in A_0\}, \tag{11}$$

and defining $\mathcal{M}_k$ by (8), we have $T_{P_0}\mathcal{S} = T_{P_0}\mathcal{E}_k \oplus T_{P_0}\mathcal{M}_k$ and $T_{P_0}\mathcal{E}_k \perp T_{P_0}\mathcal{M}_k$. Regarding $(a, b) \mapsto a^\top G b$ as a nondegenerate bilinear form, we may regard $B_0 = A_0^\perp$ as the (right) orthogonal complement of A_0 with respect to G.

It remains to check that $\mathcal{E}_k \cap \mathcal{M}_k = \{P_0\}$. Suppose on the contrary that $\mathcal{E}_k \cap \mathcal{M}_k$ contains a point P which is different from P_0. From the definition of $\mathcal{E}_k$ and $\mathcal{M}_k$, we have

$$\theta_P - \theta_{P_0} \in A_0 \setminus \{0\}, \quad \eta_P - \eta_{P_0} \in B_0 \setminus \{0\}.$$

Let γ be the primal geodesic from P_0 to P which is a straight line from θ_{P_0} to θ_P after a time change. Writing $\dot{\gamma}(0) = \sum_i a_i \frac{\partial}{\partial \theta_i} \in T_{P_0}\mathcal{E}_k$, we have $a = \lambda(\theta_P - \theta_{P_0})$ for some $\lambda > 0$. Dually, if γ^* is the dual geodesic from P_0 to P and $\dot{\gamma}^*(0) = \sum_j b_j \frac{\partial}{\partial \eta_j} \in T_{P_0}M_k$, then $b = \lambda'(\eta_P - \eta_{P_0})$ for some $\lambda' > 0$. It follows that

$$\langle \dot{\gamma}(0), \dot{\gamma}^*(0) \rangle = \lambda\lambda'(\theta_P - \theta_{P_0})^\top G(\eta_P - \eta_{P_0}) = 0.$$

Thus the two geodesics meet orthogonally at P_0. By the generalized Pythagorean theorem, we have

$$\mathbf{D}[P_0 : P] + \mathbf{D}[P : P_0] = \mathbf{D}[P_0 : P_0] = 0,$$

which is a contradiction since $P \neq P_0$ implies $\mathbf{D}[P_0 : P], \mathbf{D}[P : P_0] > 0$. □

Remark 1 (Comparison with the dually flat case). Theorem 2 is different from the corresponding result in [1, Section 3B]. In [1], given a dually flat manifold, Amari considered a k-cut coordinate system and showed that the e-flat $\mathcal{E}_k(c_k+)$ is orthogonal to the m-flat $\mathcal{M}_k(d_k-)$ for any values of c_k and d_k. Since the Riemannian metric (6) is different, the construction using k-cut coordinates no longer works in our setting.

As in the dually flat case, the orthogonal complement can be interpreted in terms of primal and dual projections.

Definition 1 (Primal and dual projections). *Let $\mathcal{A} \subset \mathcal{S}$ and $P \in \mathcal{S}$. Consider the problem*

$$\mathbf{D}[\mathcal{A} : P] := \inf_{Q \in \mathcal{A}} \mathbf{D}[Q : P]. \tag{12}$$

An optimal solution to (12) *is called a dual projection of P onto $\mathcal{A}$ and is denoted by* $\mathrm{proj}^*_{\mathcal{A}}(P)$. *Similarly, the primal projection of P onto $\mathcal{A}$ is defined by* $\arg\min_{Q \in \mathcal{A}} \mathbf{D}[P : Q]$ *and is denoted by* $\mathrm{proj}_{\mathcal{A}}(P)$.

The following theorem gives a geometric interpretation of the dual complement $\mathcal{M}_k$.

Theorem 3. *Consider a submanifold $\mathcal{E}_k$ and let $P_0 \in \mathcal{E}_k$. Let $\mathcal{M}_k = \mathcal{M}_k(P_0)$ be the dual complement of $\mathcal{E}_k$ at P_0 given by Theorem 2. Then, for any $P \in \mathcal{M}_k$ we have $P_0 = \mathrm{proj}^*_{\mathcal{E}_k}(P)$. In fact, we have*

$$\mathcal{M}_k(P_0) = (\mathrm{proj}^*_{\mathcal{E}_k})^{-1}(P_0) = \{P \in \mathcal{S} : \mathrm{proj}^*_{\mathcal{E}_k}(P) = P_0\}. \tag{13}$$

Consequently, for $P_0, P_1 \in E_k$ with $P_0 \neq P_1$, we have $M_k(P_0) \cap M_k(P_1) = \emptyset$.

Proof. Let $P \in \mathcal{M}_k$. By definition of $\mathcal{M}_k$, the dual geodesic γ^* from P to P_0 is orthogonal to $\mathcal{E}_k$. Since $\mathcal{E}_k$ is ∇-autoparallel, for any $Q \in \mathcal{E}_k$, $Q \neq P$, the primal geodesic γ from Q to P_0 lies in $\mathcal{E}_k$, and its orthogonal to γ^*. By the generalized Pythagorean theorem, we have

$$\mathbf{D}[Q : P] = \mathbf{D}[P_0 : P] + \mathbf{D}[Q : P_0] > \mathbf{D}[P_0 : P].$$

It follows that $P_0 = \mathrm{proj}^*_{\mathcal{E}_k}(P)$. This argument also shows that the dual projection onto $\mathcal{E}_k$, if exists, is unique.

Conversely, suppose $P \in \mathcal{S} \backslash \mathcal{M}_k$. Then, by definition of $\mathcal{M}_k$, the dual geodesic from P to P_0 is not orthogonal to $\mathcal{E}_k$. This violates the first order condition of the optimization problem (12). Hence P_0 is not the dual projection of P onto $\mathcal{E}_k$ and we have $\mathcal{M}_k = (\mathrm{proj}^*_{\mathcal{E}_k})^{-1}(P_0)$. □

We have shown that the dual complements are disjoint and correspond to preimages of the dual projections. We complete the circle of ideas by stating the dual foliation. See Fig. 2 for a graphical illustration in the context of principal component analysis with respect to an $L^{(\alpha)}$-divergence.

Corollary 1 (Dual foliation). *Let $\mathcal{E}_k$ be given. Suppose that for each $P \in \mathcal{S}$ the infimum in* (12) *is attained. Then*

$$\mathcal{S} = \bigcup_{P_0 \in \mathcal{E}_k} \mathcal{M}_k(P_0), \tag{14}$$

where the union is disjoint. We call (14) *a dual foliation of $\mathcal{S}$.*

Remark 2. In [9] the dual foliation derived from a Bregman divergence was used to study a $\mathcal{U}$-boost algorithm. It is interesting to see if the logarithmic divergence – which satisfies a generalized Pythagorean theorem and induces a dual foliation – leads to a class of new algorithms.

4 PCA with Logarithmic Divergences

Motivated by the success of Bregman divergence, it is natural to consider statistical applications of the $L^{(\alpha)}$-divergence. In this section we consider dimension reduction problem with the logarithmic divergences. Principal component analysis (PCA) is a fundamental technique in dimension reduction [8]. The most basic and well-known version of PCA operates by projecting orthogonally the data onto a lower dimensional affine subspace in order to minimize the sum of

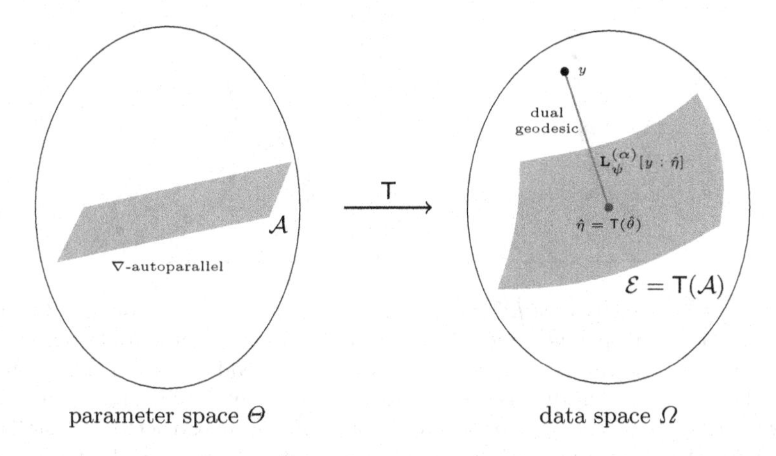

Fig. 1. Geometry of $L^{(\alpha)}$-PCA. Here $\hat{\eta}$ is the dual projection of y onto $\mathcal{E}$.

squared errors. Equivalently, this can be phrased as a maximum likelihood estimation problem where each data point is normally distributed and the mean vectors are constrained to lie on a lower dimensional affine subspace. Using the duality between exponential family and Bregman divergence (which can be regarded as a generalization of the quadratic loss), Collins et al. [5] formulated a nonlinear extension of PCA to exponential family. Here, we propose to replace the Bregman divergence by an $L^{(\alpha)}$-divergence, and the exponential family by the $\mathcal{F}^{(\alpha)}$-family (2). We call the resulting problem $L^{(\alpha)}$-PCA.

We assume that an α-exponentially concave function ψ is given on the dual domain Ω which we also call the data space (state space of data). We define the primal parameter by $\theta = \mathsf{T}^{-1}(\eta) = \frac{\mathsf{D}\psi(\eta)}{1-\alpha \mathsf{D}\psi(\eta)\cdot\eta}$ which takes values in the primal domain Θ (parameter space). For $k \leq d = \dim \Theta$ fixed, let $\mathsf{A}_k(\Theta)$ be the set of all $A \cap \Theta$ where $A \subset \mathbb{R}^d$ is a k-dimensional affine subspace.

Definition 2. ($L^{(\alpha)}$-PCA). *Let data points $y(1), \ldots, y(N) \in \Omega$ be given, and let $k \leq d$. The $L^{(\alpha)}$-PCA problem is*

$$\min_{\mathcal{A} \in \mathsf{A}_k(\Theta)} \min_{\theta(i) \in \mathcal{A}} \sum_{i=1}^{N} \mathbf{L}_{\psi}^{(\alpha)}[y(i) : \eta(i)], \quad \eta(i) = \mathsf{T}(\theta(i)). \tag{15}$$

The geometry of $L^{(\alpha)}$-PCA is illustrated in Fig. 1. A k-dimensional affine subspace $\mathcal{A} \subset \Theta$ is given in the parameter space Θ and provides dimension reduction of the data. The α-Legendre transform T can be regarded as a "link function" that connects the parameter and data spaces. Through the mapping T we obtain a submanifold $\mathcal{E} = \mathsf{T}(\mathcal{A})$ which is typically curved in the data space but is ∇-autoparallel under the induced geometry. From (5), the inner minimization in (15) is solved by letting $\eta(i) = \mathsf{T}(\theta(i))$ be the dual projection of $y(i)$ onto $\mathcal{E}$. Consequently, the straight line (dual geodesic) between $y(i)$ and $\eta(i)$ in the data space is g-orthogonal to $\mathcal{E}$. Finally, we vary $\mathcal{A}$ over $\mathsf{A}_k(\Theta)$ to minimize the total divergence.

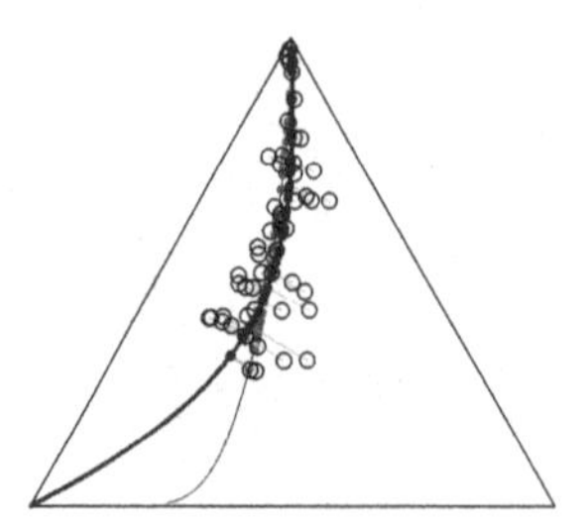
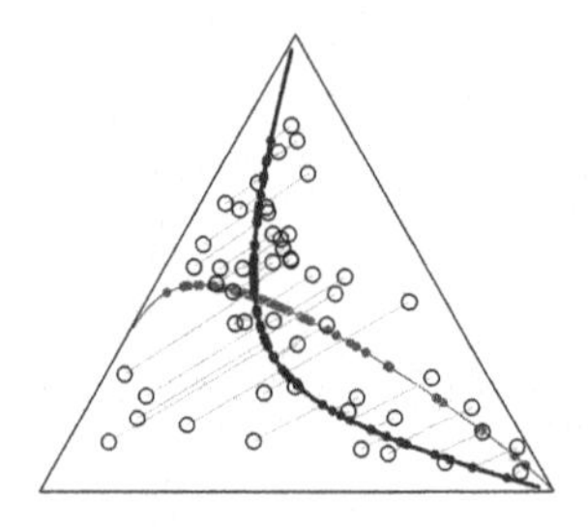

Fig. 2. Two sample outputs of $L^{(\alpha)}$-PCA and Aitchison-PCA in the context of Example 1. Black circles: Data points. Blue (thick) curve: Optimal ∇-autoparallel submanifold $\mathcal{E} = \mathsf{T}(\mathcal{A})$ shown in the state space Δ_n. Grey: Dual geodesics from data points to projected points on $\mathcal{E}$. The dual geodesics, which are straight lines in Δ_n, are orthogonal to $\mathcal{E}$ by construction and form part of the dual foliation constructed in Corollary 1. Red (thin) curve: First principal component from Aitchison-PCA. (Color figure online)

Remark 3 (Probabilistic interpretation). Consider an $\mathcal{F}^{(\alpha)}$-family (2). Under suitable conditions, the density can be expressed in the form

$$p(x;\theta) = e^{-\mathbf{L}_{\psi}^{(\alpha)}[F(x):\eta]-\psi(y)}, \tag{16}$$

where ψ is the α-conjugate of φ. Taking logarithm, we see that $L^{(\alpha)}$-PCA can be interpreted probabilistically in terms of maximum likelihood estimation, where each $\theta(i)$ is constrained to lie in an affine subspace of Θ. In this sense the $L^{(\alpha)}$-PCA extends the exponential PCA to the $\mathcal{F}^{(\alpha)}$-family.

To illustrate the methodology we give a concrete example.

Example 1 (Dirichlet perturbation). Let $\Delta_n = \{p \in (0,1)^n : \sum_i p_i = 1\}$ be the open unit simplex in $\mathbb{R}^n$. Consider the divergence

$$c(p,q) = \log\left(\frac{1}{n}\sum_{i=1}^{n}\frac{q_i}{p_i}\right) - \frac{1}{n}\sum_{i=1}^{n}\log\frac{q_i}{p_i}, \tag{17}$$

which is the cost function of the Dirichlet transport problem [14] and corresponds to the negative log-likelihood of the Dirichlet perturbation model $Q = p \oplus D$, where $p \in \Delta_n$, D is a Dirichlet random vector and $\oplus$ is the Aitchison perturbation (see [14, Section 3]). It is a multiplicative analogue of the additive Gaussian model $Y = x + \epsilon$. To use the framework of (15), write $\eta_i = \frac{p_i}{p_n}$ and $y_i = \frac{q_i}{q_n}$ for $i = 1, \ldots, d := n-1$, so that the (transformed) data space is $\Omega = (0,\infty)^d$, the positive quadrant. Then (17) can be expressed as an $L^{(1)}$-divergence by $c(p,q) = \mathbf{L}_{\psi}^{(1)}[y:\eta]$, where $\psi(y) = \frac{1}{n}\sum_{i=1}^{n}\log y_i$. The 1-Legendre transform is given by $\theta_i = \frac{1}{\eta_i} = \frac{p_n}{p_i}$, so the (transformed) parameter space is $\Theta = (0,\infty)^d$. In Fig. 2 we perform $L^{(\alpha)}$-PCA for two sets of simulated data, where $d = 2$ (or $n = 3$) and $k = 1$. Note that the traces of dual geodesics are straight lines in the (data) simplex Δ_n [13]. In Fig. 2 we also show the outputs of the popular Aitchison-PCA based on the ilr-transformation [6,15]. It is clear that the geometry of Dirichlet-PCA, which is non-Euclidean, can be quite different.

In future research, we plan to develop the $L^{(\alpha)}$-PCA carefully, including a rigorous treatment of (16), optimization algorithms and statistical applications.

References

1. Amari, S.I.: Information geometry on hierarchy of probability distributions. IEEE Trans. Inf. Theor. **47**(5), 1701–1711 (2001)
2. Amari, S.: Information Geometry and Its Applications. AMS, vol. 194. Springer, Tokyo (2016). https://doi.org/10.1007/978-4-431-55978-8
3. Ay, N., Amari, S.I.: A novel approach to canonical divergences within information geometry. Entropy **17**(12), 8111–8129 (2015)
4. Ay, N., Jost, J., Lê, H.V., Schwachhöfer, L.: Information Geometry. EMG-FASMSM, vol. 64. Springer, Cham (2017). https://doi.org/10.1007/978-3-319-56478-4
5. Collins, M., Dasgupta, S., Schapire, R.E.: A generalization of principal components analysis to the exponential family. In: Advances in Neural Information Processing Systems, pp. 617–624 (2002)
6. Egozcue, J.J., Pawlowsky-Glahn, V., Mateu-Figueras, G., Barcelo-Vidal, C.: Isometric logratio transformations for compositional data analysis. Math. Geol. **35**(3), 279–300 (2003)
7. Eguchi, S.: Second order efficiency of minimum contrast estimators in a curved exponential family. Ann. Statist. **11** 793–803 (1983)
8. Jolliffe, I.T., Cadima, J.: Principal component analysis: a review and recent developments. Philos. Trans. R. Soc. A Math. Phys. Eng. Sci. **374**(2065), 20150202 (2016)
9. Murata, N., Takenouchi, T., Kanamori, T., Eguchi, S.: Information geometry of u-boost and bregman divergence. Neural Comput. **16**(7), 1437–1481 (2004)
10. Nagaoka, H., Amari, S.I.: Differential geometry of smooth families of probability distributions. Tech. Rep. pp. 82–87. METR, University of Tokyo (1982)
11. Naudts, J.: Generalised Thermostatistics. Springer, London (2011). https://doi.org/10.1007/978-0-85729-355-8
12. Pal, S., Wong, T.-K.L.: The geometry of relative arbitrage. Math. Fin. Econ. **10**(3), 263–293 (2015). https://doi.org/10.1007/s11579-015-0159-z
13. Pal, S., Wong, T.K.L.: Exponentially concave functions and a new information geometry. Ann. Probabl. **46**(2), 1070–1113 (2018)
14. Pal, S., Wong, T.K.L.: Multiplicative Schröodinger problem and the Dirichlet transport. Probabl. Theor. Relat.Fields **178**(1), 613–654 (2020)
15. Pawlowsky-Glahn, V., Buccianti, A.: Compositional Data Analysis: Theory and Applications. John Wiley & Sons, Chichester (2011)
16. Wong, T.-K.L.: Logarithmic divergences from optimal transport and Rényi geometry. Inf. Geom. **1**(1), 39–78 (2018). https://doi.org/10.1007/s41884-018-0012-6
17. Wong, T.-K.L.: Information geometry in portfolio theory. In: Nielsen, F. (ed.) Geometric Structures of Information. SCT, pp. 105–136. Springer, Cham (2019). https://doi.org/10.1007/978-3-030-02520-5_6

18. Wong, T.-K.L., Yang, J.: Logarithmic divergences: geometry and interpretation of curvature. In: Nielsen, F., Barbaresco, F. (eds.) GSI 2019. LNCS, vol. 11712, pp. 413–422. Springer, Cham (2019). https://doi.org/10.1007/978-3-030-26980-7_43
19. Wong, T.K.L., Yang, J.: Pseudo-Riemannian geometry embeds information geometry in optimal transport. arXiv preprint. arXiv:1906.00030 (2019)

Chernoff, Bhattacharyya, Rényi and Sharma-Mittal Divergence Analysis for Gaussian Stationary ARMA Processes

Eric Grivel(✉)

Bordeaux University - INP Bordeaux ENSEIRB-MATMECA - IMS - UMR CNRS 5218, Talence, France
eric.grivel@ims-bordeaux.fr

Abstract. The purpose of this paper is first to derive the expressions of the Chernoff, Bhattacharyya, Rényi and Sharma-Mittal divergences when comparing two probability density functions of vectors storing k consecutive samples of Gaussian ARMA processes. This can be useful for process comparison or statistical change detection issue. Then, we analyze the behaviors of the divergences when k increases and tends to infinity by using some results related to ARMA processes such as the Yule-Walker equations. Comments and illustrations are given.

Keywords: Divergence rate · Sharma-Mittal divergence · Chernoff divergence · Bhattacharyya divergence · Rényi divergence · ARMA processes

1 Introduction

The properties of the entropies and the related divergences can be of interest in different fields, from information theory to change detection, passing by the comparison between model parameters and their estimates. Thus the entropy rate, defined as the entropy per unit time, has been studied for instance in [7]. The divergence rates for stationary Gaussian processes have been addressed in [1,4], where the authors respectively provide the expression of the Kullback-Leibler (KL) divergence rate as well as the Rényi divergence rate for zero-mean Gaussian processes by using the theory of Toeplitz matrices and the properties of the asymptotic distribution of the eigenvalues of Toeplitz forms. The expressions they gave were integrals of functions depending on the power spectral densities of the processes to be compared. More recently, combining the definitions of the divergences in the Gaussian case and the statistical properties of the processes to be compared, we derived [2,5,6,8] the expressions of the divergence rates of the Jeffreys divergence between the probability density functions (pdfs) of vectors storing consecutive samples of Gaussian autoregressive moving average (ARMA) or AR fractionally integrated MA (ARFIMA) processes -noisy or not-. In every case, the divergence rates depend on the process parameters and can be interpreted in terms of powers of the processes that have been filtered.

F. Nielsen and F. Barbaresco (Eds.): GSI 2021, LNCS 12829, pp. 487–495, 2021.
https://doi.org/10.1007/978-3-030-80209-7_53

In this paper, we propose to analyze how the Chernoff, Bhattacharyya, Rényi and Sharma-Mittal divergences between the pdfs of vectors storing k consecutive samples of Gaussian ARMA processes evolve when k increases. We derive different properties that can be useful to compare ARMA processes for instance.

The rest of the paper is organized as follows: after recalling the definition of the divergences in the Gaussian case and the properties of Gaussian ARMA processes, we analyze how the increments of the divergences evolve when k increases. Some illustrations as well as comments are then given.

2 Chernoff, Bhattacharyya, Rényi and Sharma-Mittal Divergences. Application to the Gaussian Case

2.1 Definitions

Let us introduce the pdfs related to two real Gaussian random vectors defined from k consecutive samples, *i.e.* $X_{k,i} = [x_{t,i}\ x_{t-1,i} \cdots x_{t-k+1,i}]^T$ for $i = 1, 2$:

$$p_i(X_{k,i}) = \frac{1}{(\sqrt{2\pi})^k |Q_{k,i}|^{1/2}} exp\Big(- \frac{1}{2}[X_{k,i} - \mu_{k,i}]^T Q_{k,i}^{-1} [X_{k,i} - \mu_{k,i}]\Big) \quad (1)$$

with $\mu_{k,i} = E[X_{k,i}]$ the mean, $|Q_{k,i}|$ the determinant of the covariance matrix $Q_{k,i} = E[(X_{k,i} - \mu_{k,i})(X_{k,i} - \mu_{k,i})^T]$ and $E[\cdot]$ the expectation.

The Chernoff coefficient of order α is then defined as follows:

$$C_\alpha(p_1(X_k), p_2(X_k)) = \int_{X_k} p_1^\alpha(X_k) p_2^{1-\alpha}(X_k) dX_k \quad (2)$$

To study the dissimilarities between two real Gaussian random processes, denoted as $x_{t,1}$ and $x_{t,2}$, various divergences can be considered as an alternative to the KL divergence. More particularly, various divergences can be deduced and defined as a function f of the Chernoff coefficient. Thus, when $f(t) = -\ln(t)$, the Chernoff divergence between $p_1(X_k)$ and $p_2(X_k)$ is given by:

$$CD_k^{(1,2)}(\alpha) = -\ln\Big(C_\alpha(p_1(X_k), p_2(X_k))\Big) \quad (3)$$

When $\alpha = \frac{1}{2}$, this leads to the Bhattacharyya distance, which is symmetric:

$$\begin{aligned} BD_k^{(1,2)} &= CD_k^{(1,2)}(\frac{1}{2}) = -\ln\Big(C_{\frac{1}{2}}(p_1(X_k), p_2(X_k))\Big) \\ &= -\ln\Big(\int_{X_k} \sqrt{p_1(X_k) p_2(X_k)} dX_k\Big) = BD_k^{(2,1)} \end{aligned} \quad (4)$$

When $f(t) = -\frac{1}{1-\alpha} ln(t)$, the Rényi divergence of order α is deduced:

$$RD_k^{(1,2)}(\alpha) = -\frac{1}{1-\alpha} \ln \int_{X_k} p_1^\alpha(X_k)\, p_2^{1-\alpha}(X_k)\, \mathrm{d}X_k = \frac{1}{1-\alpha} CD_k^{(1,2)}(\alpha) \quad (5)$$

Note that when α tends to 1, using L'Hospital rules, one can show that the Rényi divergence tends to the KL divergence. Finally, the Sharma-Mittal divergence of parameters β and α between $p_1(X_k)$ and $p_2(X_k)$ is given by:

$$SM_k^{(1,2)}(\beta,\alpha) = \frac{1}{\beta-1}\Big(C_\alpha^{\frac{1-\beta}{1-\alpha}}(p_1(X_k),p_2(X_k)) - 1\Big) \tag{6}$$

Using (5) and (6), one can see that:

$$SM_k^{(1,2)}(\beta,\alpha) = \frac{1}{\beta-1}\Big(exp\Big[(\beta-1)RD_k^{(1,2)}\Big] - 1\Big) \tag{7}$$

When β tends to 1, $SM_k^{(1,2)}(\beta,\alpha)$ tends to $RD_k^{(1,2)}(\alpha)$. So, when β tends to 1 and α tends to 1, $SM_k^{(1,2)}(\beta,\alpha)$ tends to the KL divergence. It should be noted that symmetric versions of these divergences can be also deduced either by taking the minimum value, the sum or the mean of the divergence computed between $p_1(X_k)$ and $p_2(X_k)$ and the one between $p_2(X_k)$ and $p_1(X_k)$.

2.2 Expressions of the Divergences in the Gaussian Case

Given[1] $Q_{k,\alpha} = \alpha Q_{k,2} + (1-\alpha)Q_{k,1}$ and $\Delta\mu_k = \mu_{k,2} - \mu_{k,1}$ we can show, by substituting the pdfs by their expressions in the definition (3) and after some mathematical developments, that the Chernoff divergence is the following:

$$CD_k^{(1,2)}(\alpha) = \frac{1}{2}\ln\Big(\frac{|Q_{k,\alpha}|}{|Q_{k,1}|^{1-\alpha}|Q_{k,2}|^{\alpha}}\Big) + \frac{\alpha(1-\alpha)}{2}\Delta\mu_k^T Q_{k,\alpha}^{-1}\Delta\mu_k \tag{8}$$

Therefore, the Bhattacharyya distance can be deduced as follows:

$$BD_k^{(1,2)} = CD_k^{(1,2)}(\frac{1}{2}) = \frac{1}{2}\ln\Big(\frac{|Q_{k,\frac{1}{2}}|}{|Q_{k,1}|^{\frac{1}{2}}|Q_{k,2}|^{\frac{1}{2}}}\Big) + \frac{1}{8}\Delta\mu_k^T Q_{k,\frac{1}{2}}^{-1}\Delta\mu_k \tag{9}$$

Concerning the expression of the Rényi divergence, given (5), this leads to:

$$RD_k^{(1,2)}(\alpha) = \frac{1}{2(1-\alpha)}\ln\Big(\frac{|Q_{k,\alpha}|}{|Q_{k,1}|^{1-\alpha}|Q_{k,2}|^{\alpha}}\Big) + \frac{\alpha}{2}\Delta\mu_k^T Q_{k,\alpha}^{-1}\Delta\mu_k \tag{10}$$

Given (7), this means that the expression of the SM divergence becomes:

$$SM_k^{(1,2)}(\beta,\alpha) = \frac{1}{\beta-1}\times \tag{11}$$
$$\Big(exp\Big[(\beta-1)\Big(\frac{1}{2(1-\alpha)}\ln\Big(\frac{|Q_{k,\alpha}|}{|Q_{k,1}|^{1-\alpha}|Q_{k,2}|^{\alpha}}\Big) + \frac{\alpha}{2}\Delta\mu_k^T Q_{k,\alpha}^{-1}\Delta\mu_k\Big)\Big] - 1\Big)$$

In the following, let us analyze how the divergences evolve when k increases when dealing two wide-sense stationary (w.s.s.) Gaussian ARMA processes. To this end, we need to recall some properties of ARMA processes.

[1] The definition of $Q_{k,\alpha}$ amounts to saying that a third process is introduced and corresponds to a linear combination of the two processes to be compared.

3 About w.s.s. Gaussian ARMA Processes

The tth sample $x_{t,i}$ of the ith ARMA process of order (p_i, q_i) is defined by:

$$x_{t,i} = -\sum_{j=1}^{p_i} a_{j,i} x_{t-j,i} + \sum_{j=0}^{q_i} b_{j,i} u_{t-j,i} \text{ with } i = 1, 2 \tag{12}$$

where $\{a_{j,i}\}_{j=1,\dots,p_i}$ and $\{b_{j,i}\}_{j=0,\dots,q_i}$ are the ARMA parameters with $b_{0,i} = 1$ and the driving process $u_{t,i}$ is a zero-mean w.s.s. Gaussian white sequence with variance $\sigma^2_{u,i}$. In the following, a non zero-mean w.s.s. ARMA process is obtained by adding $\mu_{x,i}$ to $x_{t,i}$, once generated.

The process $x_{t,i}$ is interpreted as the output of an infinite-impulse-response stable linear filter whose input is $u_{t,i}$. Using the z transform, the transfer function $H_i(z) = \frac{\prod_{l=1}^{q_i}(1-z_{l,i}z^{-1})}{\prod_{l=1}^{p_i}(1-p_{l,i}z^{-1})}$ is defined by its zeros $\{z_{l,i}\}_{l=1,\dots,q_i}$ and poles $\{p_{l,i}\}_{l=1,\dots,p_i}$.

When the moduli of the zeros are smaller than 1, one speaks of minimum-phase ARMA processes. For every non-minimum phase ARMA process, the equivalent minimum-phase process is obtained by replacing the zeros $\{z_{l,i}\}_{l=1,\dots,m_i \leq q_i}$ whose modulus is larger than 1 by $1/z^*_{l,i}$ for $l = 1, \dots, m_i$ to get the transfer function $H_{min,i}(z)$. Then, by introducing $K_{l,i} = |z_{l,i}|^2$, one substitutes the variance $\sigma^2_{u,i}$ of the initial ARMA process with the product:

$$\sigma^2_{u,min,i} = \sigma^2_{u,i} \prod_{l=1}^{q_i} K_{l,i} \tag{13}$$

All the above ARMA processes lead to the same power spectral density (PSD) equal to $S_{x,i}(\theta) = \sigma^2_{u,i} |H_i(e^{j\theta})|^2$ with θ the normalized angular frequency and consequently to the same correlation and covariance functions.

As this paper aims at comparing the pdfs of ARMA processes characterized by their means and their covariance matrices, one can always consider the ARMA processes associated to the minimum-phase filter $H_i(z)$ [9].

Comparing the minimum-phase ARMA processes is of interest because they can be represented by an infinite-order AR process. The latter can be approximated by a AR model of finite-order $\tau > \max(p_i, q_i)$:

$$x_{t,i} \approx -\sum_{j=1}^{\tau} \alpha_{j,\tau,i} x_{t-j,i} + u_{t,\tau,i} \tag{14}$$

where $u_{t,\tau,i}$ is the driving process of the ith process associated with the order τ and $\{\alpha_{j,\tau,i}\}_{j=1,\dots,\tau}$ are the AR parameters. They can be stored in the column vector $\Theta_{\tau,i}$ that can be estimated using the Yule-Walker equations, given by:

$$\begin{cases} Q_{\tau,i}\Theta_{\tau,i} = -\underline{r}_{\tau,i} \\ \sigma^2_{u,\tau,i} = r_{0,i} - \underline{r}^T_{\tau,i}\Theta_{\tau,i} \end{cases} \tag{15}$$

with $\underline{r}_{\tau,i}$ is the column vector storing the values of the covariance function $r_{k,i} = E[(x_{t,i} - E(x_{t,i}))\,(x_{t-k,i} - E(x_{t-k,i}))]$ for lags equal to $k = 1, ..., \tau + 1$.

The Toeplitz covariance matrix $Q_{k,i}$ of $X_{k,i}$, whose elements are defined by the covariance function, is non singular even if the PSD of the process is equal to zero at some frequencies. It is hence invertible. Only the infinite-size Toeplitz covariance matrix is not invertible when the corresponding transfer function of the ARMA process has unit roots.

There are various ways to compute the inverse of the covariance matrix of the AR process. In this paper, we suggest a LDL factorization of $Q_{k,i}$. To this end, each element of the column vector $X_{k,i}$ can be expressed from $U_{k,i} = [u_{t,k-1,i} \quad u_{t-1,k-2,i} \quad \dots \quad u_{t-k+1,0,i}]^T$ by using (14). One has:

$$L_{k,i}^{-1} X_{k,i} = U_{k,i} \tag{16}$$

where:

$$L_{k,i}^{-1} = \begin{bmatrix} 1 & \alpha_{1,k-1,i} & \alpha_{2,k-1,i} & \dots & \alpha_{k-1,k-1,i} \\ 0 & 1 & \alpha_{1,k-2,i} & \dots & \alpha_{k-2,k-2,i} \\ \vdots & 0 & 1 & \ddots & \vdots \\ \vdots & \vdots & \ddots & \ddots & \alpha_{1,1} \\ 0 & 0 & \dots & 0 & 1 \end{bmatrix} \tag{17}$$

Thus, if $diag(x)$ is the diagonal matrix whose main diagonal is x, post multiplying each part of the equality in (16) by their transpose conjugate and taking the expectation leads to:

$$L_{k,i}^{-1} Q_{k,i} (L_{k,i}^H)^{-1} = D_{k,i} = diag(\sigma_{u,k-1,i}^2 \cdots \sigma_{u,0,i}^2) \tag{18}$$

The above equality has two consequences. On the one hand, the inverse of the covariance matrix is given by:

$$Q_{k,i}^{-1} = (L_{k,i}^H)^{-1} D_{k,i}^{-1} L_{k,i}^{-1} \tag{19}$$

On the other hand, taking the determinant of (18), one has:

$$|L_{k,i}^{-1}||Q_{k,i}||(L_{k,i}^H)^{-1}| = |D_{k,i}| = \prod_{n=0}^{k-1} \sigma_{u,n,i}^2 \tag{20}$$

As $|L_{k,i}^{-1}|$ and $|(L_{k,i}^H)^{-1}|$ are equal to 1, one obtains $\frac{|Q_{k,i}|}{|Q_{k-1,i}|} = \sigma_{u,k-1,i}^2$. Therefore, one has for the ith minimum-phase ARMA process:

$$\lim_{k \to +\infty} \frac{|Q_{k,i}|}{|Q_{k-1,i}|} = \sigma_{u,min,i}^2 \tag{21}$$

Finally, the linear combination $x_{t,\alpha}$ of stationary independent ARMA processes $x_{t,1}$ and $x_{t,2}$, respectively of orders (p_1, q_1) and (p_2, q_2), is an ARMA(p, q) process with $p \le p_1 + p_2$, $q \le max(p_1 + q_2, p_2 + q_1)$.

In the next section, let us study how the divergences evolve when k increases and tends to infinity, when comparing Gaussian w.s.s. ARMA processes.

4 Analysis of the Increments of the Divergences

Let us first express $\Delta CD_k^{(1,2)} = CD_{k+1}^{(1,2)} - CD_k^{(1,2)}$, $\Delta BD_k^{(1,2)}$ and $\Delta RD_k^{(1,2)}$ which are similarly defined. Given (8), one has:

$$\Delta CD_k^{(1,2)} = \frac{\alpha(1-\alpha)}{2}\Big(\Delta\mu_{k+1}^T Q_{k+1,\alpha}^{-1}\Delta\mu_{k+1} - \Delta\mu_k^T Q_{k,\alpha}^{-1}\Delta\mu_k\Big) + \frac{1}{2}\ln\Big(\frac{|Q_{k,2}|^\alpha}{|Q_{k+1,2}|^\alpha}\frac{|Q_{k+1,\alpha}|}{|Q_{k,\alpha}|}\frac{|Q_{k,1}|^{1-\alpha}}{|Q_{k+1,1}|^{1-\alpha}}\Big) \tag{22}$$

On the one hand, taking into account (21), one has:

$$\lim_{k\to+\infty}\Big(\frac{|Q_{k,2}|^\alpha}{|Q_{k+1,2}|^\alpha}\frac{|Q_{k+1,\alpha}|}{|Q_{k,\alpha}|}\frac{|Q_{k,1}|^{1-\alpha}}{|Q_{k+1,1}|^{1-\alpha}}\Big) = \frac{\sigma_{u,min,\alpha}^2}{(\sigma_{u,min,2}^2)^\alpha(\sigma_{u,min,1}^2)^{1-\alpha}} \tag{23}$$

On the other hand, given $A_{k+1} = \Delta\mu_{k+1}^T Q_{k+1,\alpha}^{-1}\Delta\mu_{k+1}$ and $\Delta\mu_x$ the difference between the process means, one can show by using the LDL decomposition:

$$A_{k+1} = \sum_{\tau=0}^{k}\frac{|\sum_{i=0}^{\tau}\alpha_{i,\tau,\alpha}|^2}{\sigma_{u,\tau,1}^2}\Delta\mu_x^2 = A_k + \frac{|\sum_{i=0}^{k}\alpha_{i,k,\alpha}|^2}{\sigma_{u,k,1}^2}\Delta\mu_x^2 \tag{24}$$

Let us take the limit of the difference $\Delta A_k = A_{k+1} - A_k$ when k tends to infinity. As $\lim_{k\to+\infty}\frac{|\sum_{i=0}^{k}\alpha_{i,k,i}|^2}{\sigma_{u,k,1}^2}\Delta\mu_x^2$ corresponds to the power $P^{\Delta\mu_x,i}$ of a constant signal equal to $\Delta\mu_x$ that has been filtered by the inverse filter associated with the ith minimum-phase ARMA process, one obtains:

$$\lim_{k\to+\infty}\Delta A_k = P^{\Delta\mu_x,\alpha} \tag{25}$$

Therefore, the asymptotic increment for the Chernoff divergence is given by:

$$\Delta CD^{(1,2)} = \lim_{k\to+\infty}\Delta CD_k^{(1,2)} = \frac{\alpha(1-\alpha)}{2}P^{\Delta\mu_x,\alpha} + \frac{1}{2}\ln\Big(\frac{\sigma_{u,\alpha,min}^2}{(\sigma_{u,2,min}^2)^\alpha(\sigma_{u,1,min}^2)^{1-\alpha}}\Big) \tag{26}$$

Consequently, one has:

$$\Delta BD^{(1,2)} = \lim_{k\to+\infty}\Delta BD_k^{(1,2)} = \frac{1}{8}P^{\Delta\mu_x,\frac{1}{2}} + \frac{1}{2}\ln\Big(\frac{\sigma_{u,\frac{1}{2},min}^2}{\sigma_{u,2,min}\sigma_{u,1,min}}\Big) \tag{27}$$

and

$$\Delta RD^{(1,2)} = \lim_{k\to+\infty}\Delta RD_k^{(1,2)} = \frac{\alpha}{2}P^{\Delta\mu_x,\alpha} - \frac{1}{2(\alpha-1)}\ln\Big(\frac{\sigma_{u,\alpha,min}^2}{(\sigma_{u,2,min}^2)^\alpha(\sigma_{u,1,min}^2)^{1-\alpha}}\Big) \tag{28}$$

$\Delta CD_k^{(1,2)}$, $\Delta BD_k^{(1,2)}$ and $\Delta RD_k^{(1,2)}$ depend on the ARMA parameters because $\sigma^2_{u,\alpha,min}$ depend on the transfer functions and the variances of the driving processes of the two processes to be compared. The above quantities correspond to the divergence rates. Thus, $RD_k^{(1,2)} \approx k\Delta RD^{(1,2)} + \gamma$ when k tends to infinity, with γ a constant. So, $\lim_{k\to+\infty} RD_k^{(1,2)} = +\infty$ and $\lim_{k\to+\infty} \frac{\Delta RD_k^{(1,2)}}{RD_k^{(1,2)}} = 0$. Note that a detailed analysis of the RD divergence rate has been recently presented in [3].

Let us now express $\Delta SM_k^{(1,2)} = SM_{k+1}^{(1,2)} - SM_k^{(1,2)}$. After simplifications, one has:

$$\Delta SM_k^{(1,2)} = \frac{exp\Big[(\beta-1)RD_k^{(1,2)}\Big]}{(\beta-1)}\Big(exp\Big[(\beta-1)\Delta RD_k^{(1,2)}\Big]-1\Big) \tag{29}$$

Therefore, the normalized increment is equal to:

$$\frac{\Delta SM_k^{(1,2)}}{SM_k^{(1,2)}} = \frac{exp\Big[(\beta-1)\Delta RD_k^{(1,2)}\Big]-1}{1-exp\Big[-(\beta-1)RD_k^{(1,2)}\Big]} \tag{30}$$

As said in Sect. 2, $\lim_{\beta\to1} \Delta SM_k^{(1,2)} = RD_{k+1}^{(1,2)} - RD_k^{(1,2)} = \Delta RD_k^{(1,2)}$ and $\lim_{\beta\to1} \frac{\Delta SM_k^{(1,2)}}{SM_k^{(1,2)}} = \frac{\Delta RD_k^{(1,2)}}{RN_k^{(1,2)}}$. In addition, remembering (28), one has:

$$\lim_{k\to+\infty} \frac{\Delta SM_k^{(1,2)}}{SM_k^{(1,2)}} = \frac{exp\Big[(\beta-1)\Delta RD^{(1,2)}\Big]-1}{1-\lim_{k\to+\infty} exp\Big[-(\beta-1)RD_k^{(1,2)}\Big]} \tag{31}$$

As $RD_k^{(1,2)} \geq 0$ necessarily increases when k increases, $\lim_{k\to+\infty} exp\Big[(1-\beta)RD_k^{(1,2)}\Big]$ depends on the values of β. This leads to three cases:

$$\begin{cases} \lim_{k\to+\infty} \frac{\Delta SM_k^{(1,2)}}{SM_k^{(1,2)}} = 0 \text{ if } 0<\beta<1 \\ \lim_{k\to+\infty} \frac{\Delta SM_k^{(1,2)}}{SM_k^{(1,2)}} = \lim_{k\to+\infty} \frac{\Delta RD_k^{(1,2)}}{RD_k^{(1,2)}} = 0 \text{ when } \beta \text{ tends to } 1 \\ \lim_{k\to+\infty} \frac{\Delta SM_k^{(1,2)}}{SM_k^{(1,2)}} = exp\Big[(\beta-1)\Delta RD^{(1,2)}\Big]-1 \text{ if } \beta>1 \end{cases} \tag{32}$$

Two remarks can be made: $\lim_{\beta\to1} exp\Big[(\beta-1)\Delta RD^{(1,2)}\Big]-1 = 0$. Moreover, for a given value of α, as $exp\Big[(\beta-1)\Delta RD^{(1,2)}\Big]$ is an increasing function of β when $\beta>1$, $\lim_{k\to+\infty} \frac{\Delta SM_k^{(1,2)}}{SM_k^{(1,2)}}$ is also an increasing function of β for $\beta>1$.

5 Illustrations

Let us give one example and study some divergences analyzed in this paper (for the lack of space). The first process is defined by its zeros equal to $0.8e^{j\frac{\pi}{4}}$, $0.9e^{j\frac{3\pi}{4}}$ and their conjugates and its poles equal to $0.9e^{j\frac{\pi}{6}}$, $0.7e^{j\frac{\pi}{5}}$, $0.9e^{\frac{j\pi}{4}}$ and their conjugates. In addition, $\sigma_{u,1}^2 = 1$ and $\mu_1 = 5$. The second process is defined by its zeros equal to $0.8e^{j\frac{2\pi}{3}}$, $0.6e^{j\frac{3\pi}{7}}$ and their conjugates and its poles equal to $0.95e^{j\frac{5\pi}{6}}$, $0.7e^{j\frac{3\pi}{5}}$, $0.9e^{j\frac{\pi}{4}}$ and their conjugates. In addition, $\sigma_{u,2}^2 = 4$ and $\mu_2 = 1$. Based on the data generated, the means and the covariance functions are

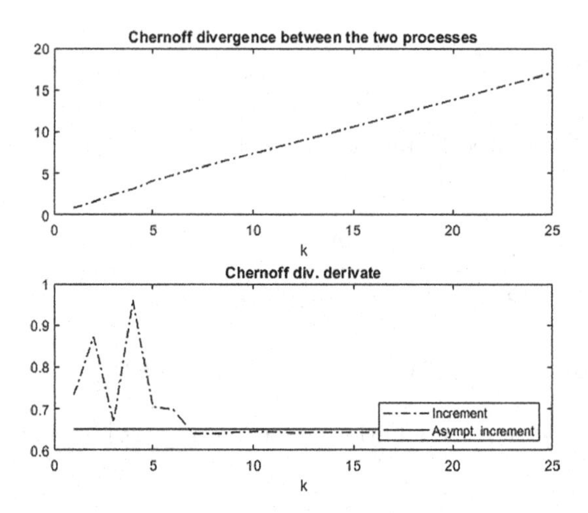

Fig. 1. Chernoff divergence evolution and divergence rate for the given example

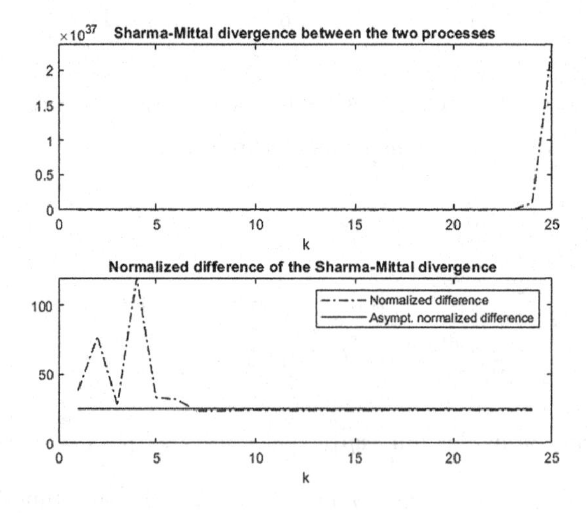

Fig. 2. Sharma-Mittal divergence and normalized difference of the increment with $\alpha = 0.9$ and $\beta = 1.5$

then estimated. The divergences for different values of k and the increments are computed. Results in Fig. 1 and 2 confirm our analysis and the convergence of the increment or the normalized increment to the proposed asymptotic expressions (26) and (32).

6 Conclusions and Perspectives

We have shown that expressions of various divergences between Gaussian ARMA processes can be obtained with the Yule-Walker equations. When dealing with Sharma-Mittal divergence, it is of interest to use the normalized increment either to compare the ARMA processes or to detect a statistical change in one of them over time while the divergence rates can be considered for the other divergences. We plan to analyze the case of long-memory processes.

References

1. Gil, M.: On Rényi divergence measures for continuous alphabet sources. Ph.D. thesis, Department of Mathematics and Statistics, Queen's University, Kingston, Ontario, Canada (2011)
2. Grivel, E., Saleh, M., Omar, S.-M.: Interpreting the asymptotic increment of Jeffrey's divergence between some random processes. Digital Signal Process. **75**, 120–133 (2018)
3. Grivel, E., Diversi, R., Merchan, F.: Kullback-Leibler and Rényi divergence rate for Gaussian stationary ARMA processes comparison. Digital Signal Process. (2021, in Press)
4. Ihara, S.: Information Theory for Continuous Systems. World Scientific (1993)
5. Legrand, L., Grivel, E.: Jeffrey's divergence between autoregressive processes disturbed by additive white noises. Signal Process. **149**, 162–178 (2018)
6. Magnant, C., Giremus, A., Grivel, E.: Jeffrey's divergence for state-space model comparison. Signal Process. **114**(9), 61–74 (2015)
7. Regnault, P., Girardin, V., Lhote, L.: Weighted Closed Form Expressions Based on Escort Distributions for Rényi Entropy Rates of Markov Chains GSI 2017
8. Saleh, M., Grivel, E., Omar, S.-M.: Jeffrey's divergence between ARFIMA processes. Digital Signal Process. **82**, 175–186 (2018)
9. Soderstrom, T., Stoica, P.: System Identification. Prentice Hall (1989)

Transport Information Geometry

Wasserstein Statistics in One-Dimensional Location-Scale Models

Shun-ichi Amari and Takeru Matsuda(✉)

RIKEN Center for Brain Science, 2-1 Hirosawa, Wako, Saitama 351-0198, Japan
amari@brain.riken.jp, takeru.matsuda@riken.jp

Abstract. In this study, we analyze statistical inference based on the Wasserstein geometry in the case that the base space is one-dimensional. By using the location-scale model, we derive the W-estimator that explicitly minimizes the transportation cost from the empirical distribution to a statistical model and study its asymptotic behaviors. We show that the W-estimator is consistent and explicitly give its asymptotic distribution by using the functional delta method. The W-estimator is Fisher efficient in the Gaussian case.

Keywords: Information geometry · Location-scale model · Optimal transport · Wasserstein distance

1 Introduction

Wasserstein geometry and information geometry are two important structures to be introduced in a manifold of probability distributions. Wasserstein geometry is defined by using the transportation cost between two distributions, so it reflects the metric of the base manifold on which the distributions are defined. Information geometry is defined to be invariant under reversible transformations of the base space. Both have their own merits for applications. In particular, statistical inference is based upon information geometry, where the Fisher metric plays a fundamental role, whereas Wasserstein geometry is useful in computer vision and AI applications.

Both geometries have their own histories (see e.g., Villani 2003, 2009; Amari 2016). Information geometry has been successful in elucidating statistical inference, where the Fisher information metric plays a fundamental role. It has successfully been applied to, not only statistics, but also machine learning, signal processing, systems theory, physics, and many other fields (Amari 2016). Wasserstein geometry has been a useful tool in geometry, where the Ricci flow has played an important role (Villani 2009; Li et al. 2020). Recently, it has found a widened scope of applications in computer vision, deep learning, etc. (e.g., Fronger et al. 2015; Arjovsky et al. 2017; Montavon et al. 2015; Peyré and Cuturi 2019). There have been attempts to connect the two geometries (see Amari et al. (2018, 2019) and Wang and Li (2020) for examples), and Li et al. (2019) has proposed a unified theory connecting them.

F. Nielsen and F. Barbaresco (Eds.): GSI 2021, LNCS 12829, pp. 499–506, 2021.
https://doi.org/10.1007/978-3-030-80209-7_54

It is natural to consider statistical inference from the Wasserstein geometry point of view (Li et al. 2019) and compare its results with information-geometrical inference based on the likelihood. The present article studies the statistical inference based on the Wasserstein geometry from a point of view different from that of Li et al. (2019). Given a number of independent observations from a probability distribution belonging to a statistical model with a finite number of parameters, we define the W-estimator that minimizes the transportation cost from the empirical distribution $\hat{p}(x)$ derived from observed data to the statistical model. This is the approach taken in many studies (see e.g., Bernton et al. 2019; Bassetti et al. 2006). In contrast, the information geometry estimator is the one that minimizes the Kullback–Leibler divergence from the empirical distribution to the model, and it is the maximum likelihood estimator. Note that Matsuda and Strawderman (2021) investigated predictive density estimation under the Wasserstein loss.

The present W-estimator is different from the estimator of Li et al. (2019), which is based on the Wasserstein score function. While their fundamental theory is a new paradigm connecting information geometry and Wasserstein geometry, their estimator does not minimize the W-divergence from the empirical one to the model. It is an interesting problem to compare these two frameworks of Wasserstein statistics.

This paper is based on Amari and Matsuda (2021).

2 W-Estimator

The optimal transport from $p(x)$ to $q(x)$ for $x \in \boldsymbol{R}^1$ is explicitly obtained when the transportation cost from x to y is $(x - y)^2$. Let $P(x)$ and $Q(x)$ be the cumulative distribution functions of p and q, respectively, defined by

$$P(x) = \int_{-\infty}^{x} p(y)dy, \quad Q(x) = \int_{-\infty}^{x} q(y)dy.$$

Then, it is known (Santambrogio 2015; Peyré and Cuturi 2019) that the optimal transportation plan is to send mass of $p(x)$ at x to x' in a way that satisfies $P(x) = Q(x')$. See Fig. 1. Thus, the total cost sending p to q is

$$C(p, q) = \int_{0}^{1} \left| P^{-1}(u) - Q^{-1}(u) \right|^2 du,$$

where P^{-1} and Q^{-1} are the inverse functions of P and Q, respectively.

We consider a regular statistical model

$$S = \{p(x, \boldsymbol{\theta})\},$$

parametrized by a vector parameter $\boldsymbol{\theta}$, where $p(x, \boldsymbol{\theta})$ is a probability density function of a random variable $x \in \boldsymbol{R}^1$ with respect to the Lebesgue measure of

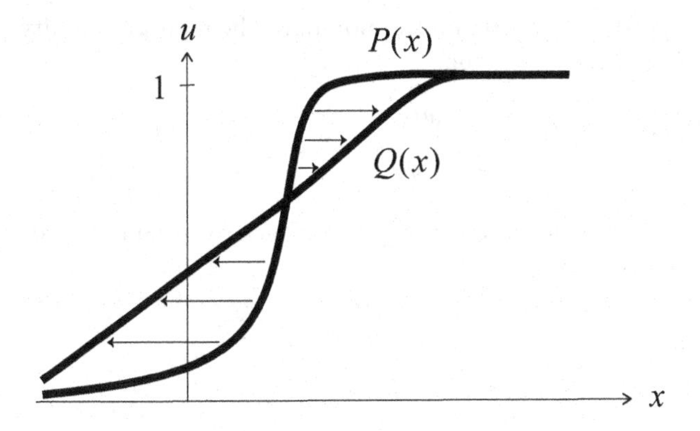

Fig. 1. Optimal transportation plan from p to q

$\boldsymbol{R}^1$. Let $D = \{x_1, \cdots, x_n\}$ be n independent samples from $p(x, \boldsymbol{\theta})$. We denote the empirical distribution by

$$\hat{p}(x) = \frac{1}{n} \sum_i \delta\left(x - x_i\right),$$

where δ is the Dirac delta function. We rearrange $x_1, \cdots, x_n$ in the increasing order,

$$x_{(1)} \leq x_{(2)} \leq \cdots \leq x_{(n)},$$

which are order statistics.

The optimal transportation plan from $\hat{p}(x)$ to $p(x, \boldsymbol{\theta})$ is explicitly solved when x is one-dimensional, $x \in \boldsymbol{R}^1$. The optimal plan is to transport mass at x to those points x' satisfying

$$\hat{P}(x_-) \leq P(x', \boldsymbol{\theta}) \leq \hat{P}(x),$$

where $\hat{P}(x)$ and $P(x, \boldsymbol{\theta})$ are the (right-continuous) cumulative distribution functions of $\hat{p}(x)$ and $p(x, \boldsymbol{\theta})$, respectively:

$$\hat{P}(x) = \int_{-\infty}^{x} \hat{p}(y) dy, \quad P(x, \boldsymbol{\theta}) = \int_{-\infty}^{x} p(y, \boldsymbol{\theta}) dy,$$

and $\hat{P}(x_-) = \lim_{y \to x-0} \hat{P}(y)$. The total cost C of optimally transporting $\hat{p}(x)$ to $p(x, \boldsymbol{\theta})$ is given by

$$C(\boldsymbol{\theta}) = C(\hat{p}, p_{\boldsymbol{\theta}}) = \int_0^1 \left| \hat{P}^{-1}(u) - P^{-1}(u, \boldsymbol{\theta}) \right|^2 du,$$

where $\hat{P}^{-1}$ and P^{-1} are inverse functions of $\hat{P}$ and P, respectively. Note that

$$\hat{P}^{-1}(u) = \inf\{y \mid P(y) \geq u\}.$$

Let $z_0(\boldsymbol{\theta}), z_1(\boldsymbol{\theta}), \cdots, z_n(\boldsymbol{\theta})$ be the points of the equi-probability partition of the distribution $p(x, \boldsymbol{\theta})$ such that

$$\int_{z_{i-1}(\boldsymbol{\theta})}^{z_i(\boldsymbol{\theta})} p(x, \boldsymbol{\theta}) dx = \frac{1}{n}, \tag{1}$$

where $z_0(\boldsymbol{\theta}) = -\infty$ and $z_n(\boldsymbol{\theta}) = \infty$. In terms of the cumulative distribution, $z_i(\boldsymbol{\theta})$ can be written as

$$P\left(z_i(\boldsymbol{\theta}), \boldsymbol{\theta}\right) = \frac{i}{n}$$

and

$$z_i(\boldsymbol{\theta}) = P^{-1}\left(\frac{i}{n}, \boldsymbol{\theta}\right).$$

See Fig. 2.

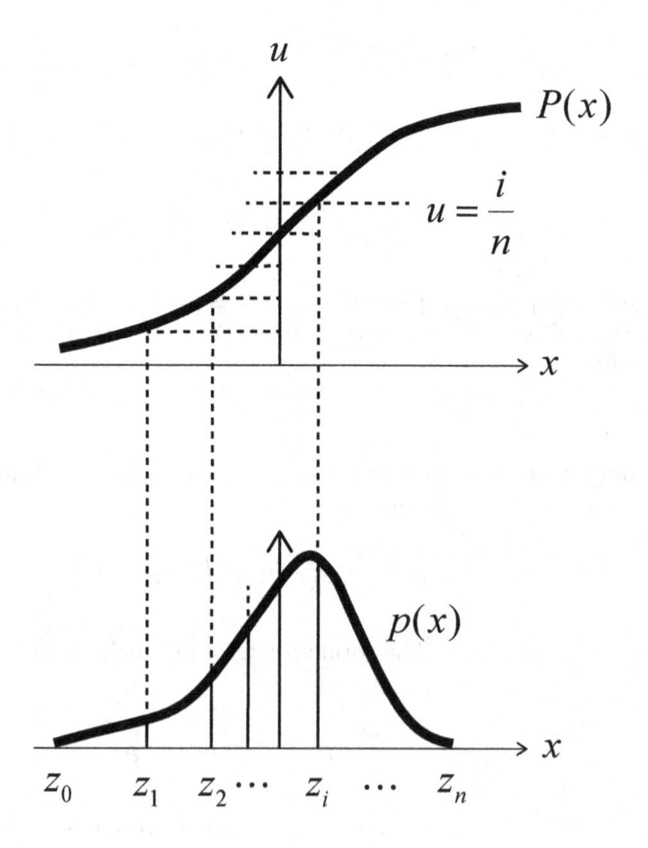

Fig. 2. Equi-partition points $z_0, z_1, \ldots, z_n$ of $p(x)$

The optimal transportation cost $C(\boldsymbol{\theta}) = C(\hat{p}, p_{\boldsymbol{\theta}})$ is rewritten as

$$C(\boldsymbol{\theta}) = \sum_i \int_{z_{i-1}(\boldsymbol{\theta})}^{z_i(\boldsymbol{\theta})} (x_{(i)} - y)^2 p(y, \boldsymbol{\theta}) dy = \frac{1}{n} \sum_i x_{(i)}^2 - 2 \sum_i k_i(\boldsymbol{\theta}) x_{(i)} + S(\boldsymbol{\theta}),$$

where we have used (1) and put

$$k_i(\boldsymbol{\theta}) = \int_{z_{i-1}(\boldsymbol{\theta})}^{z_i(\boldsymbol{\theta})} y p(y, \boldsymbol{\theta}) dy,$$

$$S(\boldsymbol{\theta}) = \sum_i \int_{z_{i-1}(\boldsymbol{\theta})}^{z_i(\boldsymbol{\theta})} y^2 p(y, \boldsymbol{\theta}) dy = \int_{-\infty}^{\infty} y^2 p(y, \boldsymbol{\theta}) dy.$$

By using the mean and variance of $p(x, \boldsymbol{\theta})$,

$$\mu(\boldsymbol{\theta}) = \int_{-\infty}^{\infty} y p(y, \boldsymbol{\theta}) dy,$$

$$\sigma^2(\boldsymbol{\theta}) = \int_{-\infty}^{\infty} y^2 p(y, \boldsymbol{\theta}) dy - \mu(\boldsymbol{\theta})^2,$$

we have

$$S(\boldsymbol{\theta}) = \mu(\boldsymbol{\theta})^2 + \sigma^2(\boldsymbol{\theta}).$$

The W-estimator $\hat{\boldsymbol{\theta}}$ is the minimizer of $C(\boldsymbol{\theta})$. Differentiating $C(\boldsymbol{\theta})$ with respect to $\boldsymbol{\theta}$ and putting it equal to 0, we obtain the estimating equation as follows.

Theorem 1. The W-estimator $\hat{\boldsymbol{\theta}}$ satisfies

$$\frac{\partial}{\partial \boldsymbol{\theta}} \sum_i k_i(\boldsymbol{\theta}) x_{(i)} = \frac{1}{2} \frac{\partial}{\partial \boldsymbol{\theta}} S(\boldsymbol{\theta}).$$

It is interesting to see that the estimating equation is linear in n observations $x_{(1)}, \cdots, x_{(n)}$ for any statistical model. This is quite different from the maximum likelihood estimator or Bayes estimator.

3 W-Estimator in Location-Scale Model

Now, we focus on location-scale models. Let $f(z)$ be a standard probability density function, satisfying

$$\int_{-\infty}^{\infty} f(z) dz = 1, \quad \int_{-\infty}^{\infty} z f(z) dz = 0, \quad \int_{-\infty}^{\infty} z^2 f(z) dz = 1,$$

that is, its mean is 0 and the variance is 1. The location-scale model $p(x, \boldsymbol{\theta})$ is written as

$$p(x, \boldsymbol{\theta}) = \frac{1}{\sigma} f\left(\frac{x - \mu}{\sigma}\right),$$

where $\boldsymbol{\theta} = (\mu, \sigma)$ is a parameter for specifying the distribution.

We define the equi-probability partition points z_i for the standard $f(z)$ as

$$z_i = F^{-1}\left(\frac{i}{n}\right),$$

where F is the cumulative distribution function

$$F(z) = \int_{-\infty}^{z} f(x)dx.$$

Then, the equi-probability partition points $y_i = y_i(\boldsymbol{\theta})$ of $p(x, \boldsymbol{\theta})$ are given by $y_i = \sigma z_i + \mu$.

The cost of the optimal transport from the empirical distribution $\hat{p}(x)$ to $p(x, \boldsymbol{\theta})$ is then written as

$$\begin{aligned} C(\mu, \sigma) &= \sum_i \int_{y_{i-1}}^{y_i} \left(x_{(i)} - x\right)^2 p(x, \mu, \sigma)dx \\ &= \mu^2 + \sigma^2 + \frac{1}{n}\sum_i x_{(i)}^2 - 2\sum_i x_{(i)} \int_{z_{i-1}}^{z_i} (\sigma z + \mu) f(z)dz. \end{aligned} \quad (2)$$

By differentiating (2), we obtain

$$\frac{1}{2}\frac{\partial}{\partial \mu}C = \mu - \frac{1}{n}\sum_i x_{(i)}, \quad \frac{1}{2}\frac{\partial}{\partial \sigma}C = \sigma - \sum_i k_i x_{(i)},$$

where

$$k_i = \int_{z_{i-1}}^{z_i} z f(z)dz,$$

which does not depend on μ or σ and depends only on the shape of f. By putting the derivatives equal to 0, we obtain the following theorem.

Theorem 2. The W-estimator of a location-scale model is given by

$$\hat{\mu} = \frac{1}{n}\sum_i x_{(i)}, \quad \hat{\sigma} = \sum_i k_i x_{(i)}.$$

The W-estimator of the location parameter μ is the arithmetic mean of the observed data irrespective of the form of f. The W-estimator of the scale parameter σ is also a linear function of the observed data $x_{(1)}, \cdots, x_{(n)}$, but it depends on f through k_i.

4 Asymptotic Distribution of W-Estimator

Here, we derive the asymptotic distribution of the W-estimator in location-scale models. Our derivation is based on the fact that the W-estimator has the form of L-statistics (van der Vaart 1998), which is a linear combination of order statistics.

Theorem 3. *The asymptotic distribution of the W-estimator $(\hat{\mu}, \hat{\sigma})$ is*

$$\sqrt{n}\begin{pmatrix}\hat{\mu}-\mu\\ \hat{\sigma}-\sigma\end{pmatrix} \Rightarrow \mathrm{N}\left(\begin{pmatrix}0\\0\end{pmatrix}, \begin{pmatrix}\sigma^2 & \frac{1}{2}m_3\sigma^2\\ \frac{1}{2}m_3\sigma^2 & \frac{1}{4}(m_4-1)\sigma^2\end{pmatrix}\right),$$

where

$$m_4 = \int_{-\infty}^{\infty} z^4 f(z)dz, \quad m_3 = \int_{-\infty}^{\infty} z^3 f(z)dz,$$

are the fourth and third moments of $f(z)$, respectively.

Proof. See Amari and Matsuda (2021).

In particular, the W-estimator is Fisher efficient for the Gaussian model, but it is not efficient for other models.

Corollary 1. *For the Gaussian model, the asymptotic distribution of the W-estimator $(\hat{\mu}, \hat{\sigma})$ is*

$$\sqrt{n}\begin{pmatrix}\hat{\mu}-\mu\\ \hat{\sigma}-\sigma\end{pmatrix} \Rightarrow \mathrm{N}\left(\begin{pmatrix}0\\0\end{pmatrix}, \begin{pmatrix}\sigma^2 & 0\\ 0 & \frac{1}{2}\sigma^2\end{pmatrix}\right),$$

which attains the Cramer–Rao bound.

Figure 3 plots the ratio of the mean square error $\mathrm{E}[(\hat{\mu}-\mu)^2+(\hat{\sigma}-\sigma)^2]$ of the W-estimator to that of the MLE for the Gaussian model with respect to n. The ratio converges to one as n goes to infinity, which shows that the W-estimator has statistical efficiency.

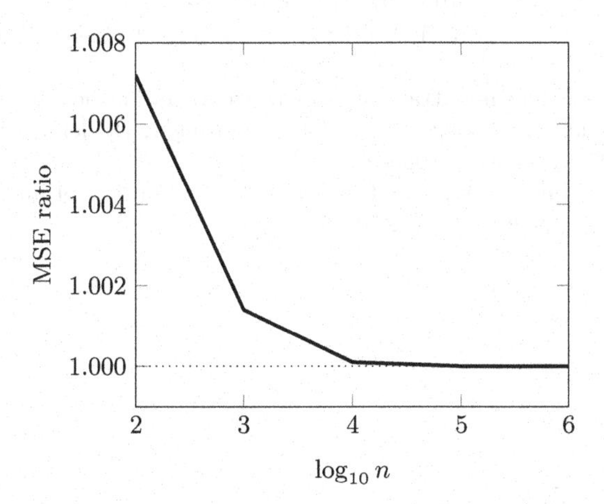

Fig. 3. Ratio of mean square error of W-estimator to that of MLE for the Gaussian model.

References

Amari, S.: Information Geometry and Its Applications. Springer, Tokyo (2016). https://doi.org/10.1007/978-4-431-55978-8
Amari, S., Karakida, R., Oizumi, M.: Information geometry connecting Wasserstein distance and Kullback-Leibler divergence via the entropy-relaxed transportation problem. Inf. Geom. **1**, 13–37 (2018)
Amari, S., Karakida, R., Oizumi, M., Cuturi, M.: Information geometry for regularized optimal transport and barycenters of patterns. Neural Comput. **31**, 827–848 (2019)
Amari, S., Matsuda, T.: Wasserstein statistics in one-dimensional location-scale models. Ann. Inst. Stat. Math. (2021, to appear)
Arjovsky, M., Chintala, S., Bottou, L.: Wasserstein GAN (2017). arXiv:1701.07875
Bernton, E., Jacob, P.E., Gerber, M., Robert, C.P.: On parameter estimation with the Wasserstein distance. Inf. Infer. J. IMA **8**, 657–676 (2019)
Bassetti, F., Bodini, A., Regazzini, E.: On minimum Kantorovich distance estimators. Stat. Prob. Lett. **76**, 1298–1302 (2006)
Fronger, C., Zhang, C., Mobahi, H., Araya-Polo, M., Poggio, T.: Learning with a Wasserstein loss. In: Advances in Neural Information Processing Systems 28 (NIPS 2015) (2015)
Li, W., Montúfar, G.: Ricci curvature for parametric statistics via optimal transport. Inf. Geom. **3**(1), 89–117 (2020). https://doi.org/10.1007/s41884-020-00026-2
Li, W., Zhao, J.: Wasserstein information matrix (2019). arXiv:1910.11248
Matsuda, T., Strawderman, W.E.: Predictive density estimation under the Wasserstein loss. J. Stat. Plann. Infer. **210**, 53–63 (2021)
Montavon, G., Müller, K.R., Cuturi, M.: Wasserstein training for Boltzmann machine. In: Advances in Neural Information Processing Systems 29 (NIPS 2016) (2015)
Peyré, G., Cuturi, M.: Computational optimal transport: with applications to data science. Found. Trends® Mach. Learn. **11**, 355–607 (2019)
Santambrogio, F.: Optimal Transport for Applied Mathematicians. Springer, Cham (2015). https://doi.org/10.1007/978-3-319-20828-2
van der Vaart, A.W.: Asymptotic Statistics. Cambridge University Press, Cambridge (1998)
Villani, C.: Topics in Optimal Transportation. American Mathematical Society (2003)
Villani, C.: Optimal Transport: Old and New. Springer, Heidelberg (2009). https://doi.org/10.1007/978-3-540-71050-9
Wang, Y., Li, W.: Information Newton's flow: Second-order optimization method in probability space (2020). arXiv:2001.04341

Traditional and Accelerated Gradient Descent for Neural Architecture Search

Nicolás García Trillos[1(✉)], Félix Morales[2], and Javier Morales[3]

[1] University of Wisconsin Madison, Madison, WI 53711, USA
garciatrillo@wisc.edu
[2] Departamento de Ingenería Informática, Universidad Católica Andrés Bello, Capital District, Caracas 1000, Venezuela
famorales.14@est.ucab.edu.ve
[3] Center for Scientific Computation and Mathematical Modeling (CSCAMM), University of Maryland, College Park, MD 20742, USA
javier.moralesdelgado16@gmail.com

Abstract. In this paper we introduce two algorithms for neural architecture search (NASGD and NASAGD) following the theoretical work by two of the authors [4] which used the geometric structure of optimal transport to introduce the conceptual basis for new notions of traditional and accelerated gradient descent algorithms for the optimization of a function on a semi-discrete space. Our algorithms, which use the network morphism framework introduced in [1] as a baseline, can analyze forty times as many architectures as the hill climbing methods [1,10] while using the same computational resources and time and achieving comparable levels of accuracy. For example, using NASGD on CIFAR-10, our method designs and trains networks with an error rate of 4.06 in only 12 h on a single GPU.

Keywords: Neural networks · Neural architecture search · Gradient flows · Optimal transport · Second order dynamics · Semi-discrete optimization

1 Introduction

Motivated by the success of neural networks in applications such as image recognition and language processing, in recent years practitioners and researchers have devoted great efforts in developing computational methodologies for the

NGT was supported by NSF-DMS 2005797. The work of JM was supported by NSF grants DMS16-13911, RNMS11-07444 (KI-Net) and ONR grant N00014-1812465. The authors are grateful to the CSCAMM, to Prof. P.-E. Jabin and Prof. E. Tadmor for sharing with them the computational resources used to complete this work. Support for this research was provided by the Office of the Vice Chancellor for Research and Graduate Education at the University of Wisconsin-Madison with funding from the Wisconsin Alumni Research Foundation.

F. Nielsen and F. Barbaresco (Eds.): GSI 2021, LNCS 12829, pp. 507–514, 2021.
https://doi.org/10.1007/978-3-030-80209-7_55

automatic design of neural architectures in order to use deep learning methods in further applications. There is an enormous literature on neural architecture search methodologies and some of its applications (see [11] for an overview on the subject), but roughly speaking, most approaches for *neural architecture search* (NAS) found in the literature build on ideas from reinforcement learning [12], evolutionary algorithms [8,9], and hill-climbing strategies based on network morphisms [1,10]. All NAS approaches attempt to address a central difficulty: the high computational burden of training multiple architecture models. Several developments in the design of algorithms, implementation, and computational power have resulted in methodologies that are able to produce neural networks that outperform the best networks designed by humans. Despite all the recent exciting computational developments in NAS, we believe that it is largely of interest to propose sound mathematical frameworks for the design of new computational strategies that can better explore the architecture space and ultimately achieve higher accuracy rates in learning while reducing computational costs.

In this paper we propose two new algorithms for NAS: *neural architecture search gradient descent* **(NASGD)** and *neural architecture search accelerated gradient descent* **(NASAGD)**. These algorithms are based on: 1) the mathematical framework for semi-discrete optimization (deeply rooted in the geometric structure of optimal transport) that two of the authors have introduced and motivated in their theoretical work [4], and 2) the neural architecture search methods originally proposed in [1,10]. We have chosen the *network morphism* framework from [1] because it allows us to illustrate the impact that our mathematical ideas can have on existing NAS algorithms without having to introduce the amount of background that other frameworks like those based on reinforcement learning would require. We emphasize that our high level ideas are not circumscribed to the network morphism framework.

In the morphism framework from [1] an iterative process for NAS is considered. In a first step, the parameters/weights of a collection of architectures are optimized for a *fixed time*, and in a second step the set of architectures are updated by applying network morphisms to the best performing networks in the previous stage; these two steps are repeated until some stopping criterion is reached. In both NASGD and NASAGD we also use the concept of network morphism, but now the time spent in training a given set of networks is *dynamically chosen* by an evolving particle system. In our numerical experiments we observe that our algorithms change architectures much earlier than the fixed amount of time proposed in [1], while achieving error rates of 4.06% for the CIFAR-10 data set trained over 12 h with a single GPU for NASGD, and of 3.96% on the same training data set trained over 1 day with one GPU for NASAGD.

2 Our Algorithms

In this section we introduce our algorithms NASGD and NASAGD. In order to motivate them, for pedagogical purposes we first consider an idealized setting where we imagine that NAS can be seen as a tensorized semi-discrete optimization problem of the form:

$$\min_{(x,g)\in\mathbb{R}^d\times\mathcal{G}} V(x,g). \tag{1}$$

In the above, it will be useful to think of the g coordinate as an *architecture* and the x coordinate as the *parameters* of that architecture. It will also be useful to think of $\mathcal{G}=(\mathcal{G},K)$ as a finite similarity graph of architectures with K a matrix of positive weights characterizing a *small* neighborhood of a given architecture (later on $\mathcal{G}$ is defined in terms of network morphisms around a given architecture –see Sect. 2.3), and V as a loss function (for concreteness cross-entropy loss) which quantifies how well an architecture with given parameters performs in classifying a given training data set. Working in this ideal setting, in the next two subsections we introduce particle systems that aim at solving (1). These particle systems are inspired by the gradient flow equations derived in [4] that we now discuss.

2.1 First Order Algorithm

The starting point of our discussion is a modification of equation (2.13) in [4] now reading:

$$\begin{aligned}\partial_t f_t(x,g) = \sum_{g'\in\mathcal{G}} &\left[\log f_t(g) + V(x,g) - (\log f_t(g') + V(x,g'))\right]\\ &\cdot K(g,g')\theta_{x,g,g'}(f_t(x,g), f_t(x,g')) + \mathrm{div}_x(f_t(x,g)\nabla_x V(x,g)),\end{aligned} \tag{2}$$

for all $t>0$. In the above, $f_t(x,g)$ must be interpreted as a probability distribution on $\mathbb{R}^d\times\mathcal{G}$ and $f_t(g)$ as the corresponding marginal distribution on g. ∇_x denotes the gradient in $\mathbb{R}^d$ and div_x the divergence operator acting on vector fields on $\mathbb{R}^d$. The first term on the right hand side of (2) is a divergence term on the graph acting on graph vector fields (i.e. real valued functions defined on the set of edges of the graph). The term $\theta_{x,g,g'}(f_t(x,g), f_t(x,g'))$ plays the role of interpolation between the masses located at the points (x,g) and (x,g'), and it provides a simple way to define induced masses on the edges of the graph. With induced masses on the set of edges one can in turn define fluxes along the graph that are in close correspondence with the ones found in the dynamic formulation of optimal transport in the Euclidean space setting (see [6]).

The relevance of the evolution of distributions (2) is that it can be interpreted as a continuous time *steepest descent equation* for the minimization of the energy:

$$\widetilde{\mathcal{E}}(f) := \sum_{g\in\mathcal{G}} \log f(g) f(g) + \sum_g \int_{\mathbb{R}^d} V(x,g) f(x,g) \tag{3}$$

with respect to the geometric structure on the space of probability measures on $\mathbb{R}^d\times\mathcal{G}$ that was discussed in section 2.3 in [4]. Naturally, the choice of different interpolators θ endow the space of measures with a different geometry. In [4] the emphasis was given to choices of θ that give rise to a Riemannian structure on the space of measures, but alternative choices of θ, like the one made in [2], induce a general *Finslerian* structure instead. In this paper we work with an interpolator inducing a Finslerian structure, and in particular define

$$\theta_{x,g,g'}(s,s') := s\mathbb{1}_{U(x,g,g')>0} + s'\mathbb{1}_{U(x,g,g')<0}, \quad s,s'>0, \tag{4}$$

where $U(x,g,g') := \log f_t(g) + V(x,g) - (\log f_t(g') + V(x,g'))$. We notice that with the entropic term used in (3) we only allow "wandering" in the g coordinate. This term encourages exploration of the architecture space.

We now consider a collection of moving particles on $\mathbb{R}^d \times \mathcal{G}$ whose evolving empirical distribution aims at mimicking the evolution described in (2). Initially the particles have locations (x_i, g_i) $i = 1, \ldots, N$ where we assume that if $g_i = g_j$ then $x_i = x_j$ (see Remark 1 below). For fixed time step $\tau > 0$, particle locations are updated by repeatedly applying the following steps:

- **Step 1: Updating parameters (Training)**: For each particle i with position (x_i, g_i) we update its parameters by setting:

$$x_i^\tau = x_i - \tau \nabla_x V(x_i, g_i).$$

- **Step 2: Moving in the architecture space (Mutation)**: First, for each of the particles i with position (x_i, g_i) we decide to change its g coordinate with probability:

$$\tau \sum_j (\log f(g_j) + V(x_j, g_j)) - (\log f(g_i) + V(x_i, g_i))^- K(g_i, g_j),$$

or 1 if the above number is greater than 1. If we decide to move particle i, we move it to the position of particle j, i.e. (x_j, g_j), with probability p_j:

$$p_j \propto \left[\log f(g_j) + V(x_j, g_j) - (\log f(g_i) + V(x_i, g_i))\right]^- K(g_i, g_j).$$

In the above, $f(g)$ denotes the ratio of particles that are located at g. Additionally, $a^- = \max\{0, -a\}$ denotes the negative part of the quantity a.

Remark 1. Given the assumptions on the initial locations of the particles, throughout all the iterations of Step 1 and Step 2 it is true that if $g_i = g_j$ then $x_i = x_j$. This is convenient from a computational perspective because in this way the number of architectures that need to get trained is equal to the number of nodes in the graph (which we recall should be interpreted as a small local graph) and not to the number of particles in our scheme.

Remark 2. By modifying the energy $\widetilde{\mathcal{E}}(f)$ replacing the entropic term with an energy of the form $\frac{1}{\beta+1}\sum_g (f(g))^{\beta+1}$ for some parameter $\beta > 0$, one can motivate a new particle system where in Step 2 every appearance of $\log f$ is replaced with f^β. The effect of this change is that the resulting particle system moves at a slower rate than the version of the particle system as described in Step 2.

2.2 Second Order Algorithm

Our *second order algorithm* is inspired by the system of equations (2.17) in [4] which now reads:

$$\begin{cases} \partial_t f_t(x,g) + \sum_{g'}(\varphi_t(x,g') - \varphi_t(x,g))K(g,g')\theta_{x,g,g'}(f_t(x,g), f_t(x,g)) \\ \qquad\qquad\qquad\qquad + \mathrm{div}_x(f_t(x,g)\nabla_x \varphi_t) = 0 \\ \partial_t \varphi_t + \frac{1}{2}|\nabla_x \varphi_t|^2 + \sum_{g'}\left(\varphi_t(x,g) - \varphi_t(x,g')\right)^2 K(g,g')\partial_s \theta_{x,g,g'}(f_t(x,g), f_t(x,g')) \\ \qquad\qquad\qquad\qquad = -[\gamma\varphi_t(x,g) + \log f_t(g) + V(x,g)], \end{cases} \tag{5}$$

for $t > 0$. We use θ as in (4) except that now we set $U(x,g,g') := \varphi_t(x,g') - \varphi_t(x,g)$. System (5) describes a second order algorithm for the optimization of $\widetilde{\mathcal{E}}$ – see sections 2.4 and 3.3 in [4] for a detailed discussion. Here, the function φ_t is a real valued function over $\mathbb{R}^d \times \mathcal{G}$ that can be interpreted as *momentum* variable. $\gamma \geq 0$ is a friction parameter.

System (5) motivates the following particle system, where now we think that the position of a particle is characterized by the tuple (x_i, g_i, v_i) where $x_i, v_i \in \mathbb{R}^d$, $g_i \in \mathcal{G}$, and in addition we have a potential function $\varphi : \mathcal{G} \to \mathbb{R}$ that also gets updated. Initially, we assume that if $g_i = g_j$ then $x_i = x_j$ and $v_i = v_j$. We also assume that initially φ is identically equal to zero.

We summarize the *second order gradient flow dynamics* as the iterative application of three steps:

- **Step 1: Updating parameters (Training):** For each particle i located at (x_i, g_i, v_i) we update its parameters x_i, v_i by setting

$$x_i^\tau = x_i + \tau v_i, \quad v_i^\tau = v_i - \tau(\gamma v_i + \nabla_x V(x_i, g_i)).$$

- **Step 2: Moving in the architecture space (Mutation):** First, for each of the particles i with position (x_i, g_i, v_i) we decide to move it with probability

$$\tau \sum_j (\varphi(g_i) - \varphi(g_j))^- K(g_i, g_j),$$

or 1 if the above quantity is greater than 1. Then, if we decided to move the particle i we move it to location of particle j, (x_j, g_j, v_j) with probability p_j

$$p_j \propto (\varphi(g_i) - \varphi(g_j))^- K(g_i, g_j).$$

- **Step 3: Updating momentum on the g coordinate**: We update φ according to:

$$\varphi^\tau(g_i) = \varphi(g_i) - \frac{\tau}{2}|v_i|^2 - \tau\left(\sum_j \left([\varphi(g_i) - \varphi(g_j)]^-\right)^2 K(g_i, g_j)\right)$$
$$- \tau(\gamma\varphi(g_i) + \log f(g_i) + V(x_i, g_i)),$$

for every particle i. Here, $f(g)$ represents the ratio of particles located at g.

Remark 3. Notice that given the assumption on the initial locations of the particles, throughout all the iterations of Step 1 and Step 2 and Step 3 we make sure that if $g_i = g_j$ then $x_i = x_j$ and $v_i = v_j$.

2.3 NASGD and NASAGD

We are now ready to describe our algorithm NASGD:

1. Load an initial architecture g_0 with initial parameters x_0 and set $r = 0$.
2. Construct a graph $\mathcal{G}_r$ around g_r using the notion of network morphism introduced in [1]. More precisely, we produce n_{neigh} new architectures with associated parameters, each new architecture is constructed by modifying g_r using a *single* network morphism from [1]. Then define $\mathcal{G}_r$ as the set consisting of the loaded n_{neigh} architectures and the architecture g_r. Set the graph weights $K(g, g')$ (for example, setting all weights to one).
3. Put N particles on (x_r, g_r) and put 1 "ghost" particle on each of the remaining architectures in $\mathcal{G}_r$. The architectures for these ghost particles are never updated (to make sure we always have at least one particle in each of the architectures in $\mathcal{G}_r$), but certainly their parameters will.
   Then, run the dynamics discussed in Sect. 2.1 on the graph $\mathcal{G}_r$ (or the modified dynamics see Remark 2, and Appendix of the ArXiv version of this paper [3]) until the node in $\mathcal{G}_r \setminus \{g_r\}$ with the most particles g^{max} has twice as many particles as g_r.
   Set $r = r + 1$. Set $g_r = g^{max}$ and $x_r = x^{max}$, where x^{max} are the parameters of architecture g^{max} at the moment of stopping the particle dynamics.
4. If size of g_r exceeds a prespecified threshold (in terms of number of convolutional layers for example) go to 5. If not go back to 2.
5. Train g_r until convergence.

The algorithm NASAGD is defined similarly with the natural adjustments to account for the momentum variables. Details can be found in sections 2 and 4 from the ArXiv version of this paper [3].

3 Experiments

We used NASGD and NASAGD on the CIFAR-10 data set to obtain two architecture models NASGD1 and NASAGD1 respectively (see Appendix in the ArXiv version of this paper [3]). In the next table we compare the performance of NASGD1 and NASAGD1 against our benchmark architectures NASH2 and NasGraph produced by the methodologies proposed in [1,10].

Numerical experiments				
CIFAR 10	Model	Resources	# params $\times 10^6$	Error
	NASH2	1 GPU, 1 day	19.7	5.2
	NASGraph	1 GPU, 20 h	?	4.96
	NASGD1	1 GPU, 12 h	25.4	4.06
	NASAGD1	1 GPU, 1 day	22.9	3.96

Besides producing better accuracy rates, it is worth highlighting that our algorithms can explore many more architectures (about 40 times more) than in [1] with the same computational resources. We took advantage of this faster exploration and considered positive as well as negative architecture mutations, i.e., mutations that can increase or decrease the number of filters, layers, skip, and dimension of convolutional kernels.

Here are some extra details on the implementation of our algorithms. For further details we refer the reader to the ArXiv version of this work [3]. In a similar way to [1] and [10], we pre-train an initial network g_0 with the structure Conv-MaxPool-Conv-MaxPool-Conv-Softmax for 20 epochs using cosine aliasing that interpolates between 0.5 and 10^{-7}; here Conv is interpreted as Conv+batchnorm+Relu. We use g_0 with parameters x_0 as the initial data for our gradient flow dynamics introduced in Sect. 2.1 for the first-order algorithm NASGD and Sect. 5 for the second-order algorithm NASAGD. During the NASGD and NASAGD algorithms, we use cosine aliasing interpolating the learning rate from λ_{start} to λ_{final} with a restart period of $epochs_{neigh}$. In contrast to the NASH approach from [1], since we initialize new architectures, we do not reset the time step along with the interpolation for the $epochs_{neighs}$ epochs. Our particle system *dynamically* determines the number of initialization. We continue this overall dynamics, resetting the learning rate from λ_{start} to λ_{final} every $epochs_{neigh}$ at most n_{steps} times. We perform several experiments letting the first and second-order gradient flow dynamics run for different lengths of time. Finally, we train the found architectures until convergence.

In the table below, we display the rest of the parameters used to find these models.

Variable	NASH2	NASGraph	NASGD1	NASAGD1
n_{steps}	8	10	0.89	2.54
n_{NM}	5	5	*dynamic*	*dynamic*
n_{neigh}	8	8	8	8
$epoch_{neigh}$	17	16	18	18
λ_{start}	0.05	0.1	0.05	0.05
λ_{final}	0	0	10^{-7}	10^{-7}
Gradient stopping	No	Yes	No	No

Here, n_{steps} denotes the number of restart cycles for the cosine aliasing; n_{NM} is the number of morphism operations applied on a given restart cycle; n_{neigh} is the number of children architectures generated every time the current best model changes; $epoch_{neigh}$ is the number of epochs that go by before the cosine aliasing is restarted; λ_{final} and λ_{start} are the parameters required for SGDR.

4 Conclusions and Discussion

In this work we have proposed novel first and second order gradient descent algorithms for neural architecture search: NASGD and NASAGD. The theoretical gradient flow structures in the space of probability measures over a semi-discrete space introduced in [4] serve as the primary motivation for our algorithms. Our numerical experiments illustrate the positive effect that our mathematical perspective has on the performance of NAS algorithms.

The methodologies introduced in this paper are part of a first step in a broader program where we envision the use of well defined mathematical structures to motivate new learning algorithms that use neural networks. Although here we have achieved competitive results, we believe that there are still several possible directions for improvement that are worth exploring in the future. Some of these directions include: a further analysis of the choice of hyperparameters for NASGD and NASAGD, the investigation of the synergy that our dynamic perspective may have with reinforcement learning approaches (given that more architectures can be explored with our dynamic approach), the adaptation of our geometric and analytic insights to other NAS paradigms such as parameter sharing [7] and differential architecture search [5].

References

1. Elsken, T., Metzen, J.-H., Hutter, F.: Simple and efficient architecture search for convolutional neural networks. arXiv:1711.04528 (2017)
2. Esposito, A., Patacchini, F.S., Schlichting, A., Slepčev, D.: Nonlocal-interaction equation on graphs: gradient flow structure and continuum limit. arXiv:1912.09834 (2019)
3. García-Trillos, N., Morales, F., Morales, J.: Traditional and accelerated gradient descent for neural architecture search. arXiv:2006.15218 (2020)
4. García-Trillos, N., Morales, J.: Semi-discrete optimization through semi-discrete optimal transport: a framework for neural architecture search. arXiv:2006.15221 (2020)
5. Liu, H., Simonyan, K., Yang, Y.: Darts: differentiable architecture search. arXiv:1806.09055 (2018)
6. Maas, J.: Gradient flows of the entropy for finite Markov chains. J. Funct. Anal. **261**(8), 2250–2292 (2011)
7. Pham, H., Guan, M., Zoph, B., Le, Q., Dean, J.: Efficient neural architecture search via parameters sharing. In: Dy, J., Krause, A. (eds.) Proceedings of the 35th International Conference on Machine Learning, volume 80 of Proceedings of Machine Learning Research, pp. 4095–4104. Stockholmsmässan, Stockholm Sweden, PMLR, 10–15 July 2018
8. Real, E., Aggarwal, A., Huang, Y., Le, Q.V.: Regularized evolution for image classifier architecture search. In: AAAI (2018)
9. Stanley, K.O., Miikkulainen, R.: Evolving neural networks through augmenting topologies. Evol. Comput. **10**(2), 99–127 (2002)
10. Verma, M., Sinha, K., Goyal, A., Verma, S., Susan, P.: A novel framework for neural architecture search in the hill climbing domain. In: 2019 IEEE Second International Conference on Artificial Intelligence and Knowledge Engineering (AIKE), pp 1–8 (2019)
11. Yu, T., Zhu, H.: Hyper-parameter optimization: a review of algorithms and applications. arXiv:2003.05689 (2020)
12. Zoph, B., Le, Q.V.: Neural architecture search with reinforcement learning. arXiv:1611.01578 (2016)

Recent Developments on the MTW Tensor

Gabriel Khan[1,2(✉)] and Jun Zhang[1,2(✉)]

[1] Iowa State University, Ames, USA
gkhan@iastate.edu
[2] University of Michigan, Ann Arbor, USA
junz@umich.edu

Abstract. We survey some recent research related to the regularity theory of optimal transport and its associated geometry. We discuss recent progress and pose some open questions and a conjecture related to the MTW tensor, which provides a local obstruction to the smoothness of the Monge transport maps.

In this paper we survey some recent progress on the Monge problem of optimal transport and its associated geometry. The main goal is to discuss some connections between the MTW tensor and the curvature of pseudo-Riemannian and complex manifolds.

1 A Short Background on Optimal Transport

Optimal transport is a classic field of mathematics which studies the most economical way to allocate resources. More precisely, the Kantorovich problem of optimal transport seeks to find a coupling between two measures which minimizes the total cost. In other words, we consider two probability spaces (X, μ) and (Y, ν) and a lower semi-continuous cost function c and find the coupling γ which minimizes the integral

$$\min_{\gamma \in \Gamma(\mu,\nu)} \int_{X \times Y} c(x, y) d\gamma(x, y). \tag{1}$$

Here, $\Gamma(\mu, \nu)$ denotes the set of all couplings between μ and ν and, heuristically, the transport plan at a point $x \in X$ is given by the disintegration of γ.

The Monge problem of optimal transport imposes the additional assumption that γ is concentrated on the graph of a measurable map $T : X \to Y$. In other words, each point in X is sent to a unique point in Y. A solution to the Monge problem need not exist in general. For instance, if μ is atomic but ν is continuous, then there are no transport maps at all, let alone an optimal one. However, one great success of the modern theory of optimal transport is to develop a satisfactory existence theory for the Monge problem. In particular, Brenier [1], Gangbo-McCann [8], and Knott-Smith [19] established that for fairly general cost functions and measures, the solution to the Kantorovich problem is actually a solution to the Monge problem. Furthermore, the transport map T

F. Nielsen and F. Barbaresco (Eds.): GSI 2021, LNCS 12829, pp. 515–523, 2021.
https://doi.org/10.1007/978-3-030-80209-7_56

is induced by the c-subdifferential of a potential function u, which is defined as follows

$$\partial_c u := \{(x,y) \in X \times Y; \quad u^c(y) - u(x) = c(x,y)\}, \tag{2}$$

where $u^c(y)$ is the c-transform of u, defined as $u^c(y) = \inf_{x \in X}(u(x) + c(x,y))$. For non-degenerate costs and measures, u will be an Alexandrov solution to a Monge-Ampère type equation

$$\det(\nabla^2 u + A(x, \nabla u)) = B(x, \nabla u). \tag{3}$$

In this formula, A is a matrix-valued function which depends on the cost function and B is a scalar valued function which depends on the cost function and the two probability measures. For a more complete reference, see Chapters 10–12 of Villani's text [32].

A priori, the function u will be c-convex[1] and uniformly Lipschitz, which implies that its c-subdifferential (and thus the transport map) is defined Lebesgue almost-everywhere. However, the transport map need not be continuous, even for smooth measures. For example, for the cost function $c(x,y) = \|x-y\|^2$ (i.e., the squared-distance cost), Caffarelli showed that for the transport between arbitrary smooth measures to be continuous, it is necessary to assume that the target domain Y is convex [2]. In other words, non-convexity of Y is a *global obstruction* to regularity.

From an analytic perspective, the reason for this somewhat pathological behavior is that Eq. 3 is *degenerate elliptic*. To explain this, let us specialize to the squared-distance cost in Euclidean space, in which case the matrix $A \equiv 0$ and c-convexity of u is convexity in the usual sense. For a convex function u, the linearized Monge-Ampére operator

$$Lf = \sum_{i,j} \left(\left(\det D^2 u\right) \left(D^2 u\right)^{-1} \right)^{ij} \frac{\partial^2}{\partial x^i \partial x^j} f$$

is non-negative definite. However, without an a priori bound on $D^2 u$, L can have arbitrarily small eigenvalues, which prevents the use of "standard elliptic theory" to prove regularity of u. On the other hand, if one can establish an a priori C^2 estimate on u, then the linearized operator is uniformly elliptic, and we can use Schauder estimates to show that the transport is smooth.

This phenomena gives rise to a striking dichotomy. For smooth (i.e., C^∞) and non-degenerate measures and cost functions, the optimal transport map is either discontinuous, or C^∞-smooth. In the former case, work of Figalli and De Phillippis [5] shows that the discontinuity occurs on a set of measure zero and that the transport is smooth elsewhere. Kim and Kitagawa [15] showed that there cannot be isolated singularities, at least for costs which satisfy the MTW(0) condition (Definition 1). However, there are many open questions about the structure of the singular set. For example, it is not known whether the Hausdorff dimension of this set is at most $n-1$ or, if so, whether the set is necessarily rectifiable.

[1] A function $u : X \to \mathbb{R}$ is c-convex if there is some function $v : Y \to \mathbb{R} \cup \{\pm\infty\}$ so that $u(x) = \sup_{y \in Y}(v(y) - c(x,y))$.

2 The Ma-Trudinger-Wang Theory

For a fairly wide class of optimal transport problems, by the 1990s it was well-understood that the solution to the Kantorovich problem is in fact given by a transport map. However, as we previously discussed, this map need not be continuous. One question of considerable interest was to find conditions on the cost function and the measures so that the transport has better regularity. For the squared-distance cost in Euclidean space, Caffarelli and others developed such a theory when the support of the target measure is convex (see, e.g., [2,4,31]). However, for more general cost functions, the problem of regularity remained open.

In 2005, a breakthrough paper by Ma, Trudinger, and Wang established smoothness for the transport under two additional assumptions [22]. The first was a global condition, that the supports of the initial and target measures are *relatively c-convex*. The second was a local condition, that a certain fourth-order quantity, known as the *MTW tensor*, is positive. In 2009, Trudinger and Wang [30] extended their previous work to the case when this tensor is non-negative (i.e., when $\mathfrak{S}(\xi, \eta) \geq 0$).

Definition 1 (MTW(κ) condition). *A cost function c is said to satisfy the MTW(κ) condition if for all vector-covector pairs (ξ, η) with $\eta(\xi) = 0$, the following inequality holds.*

$$\mathfrak{S}(\xi, \eta) := \sum_{i,j,k,l,p,q,r,s} (c_{ij,p}c^{p,q}c_{q,rs} - c_{ij,rs})c^{r,k}c^{s,l}\xi^i\xi^j\eta^k\eta^l \geq \kappa\|\eta\|^2\|\xi\|^2 \quad (4)$$

Here, the notation $c_{I,J}$ denotes $\partial_{x^I}\partial_{y^J}c$ for multi-indices I and J. Furthermore, $c^{i,j}$ denotes the matrix inverse of the mixed derivative $c_{i,j}$.

2.1 Insights from Convex Analysis and Pseudo-Riemannian Geometry

Although the Ma-Trudinger-Wang theory was a significant breakthrough, it also raised many questions. For instance, the geometric significance of $\mathfrak{S}$ was not well understood and it was unclear whether its non-negativity plays an essential role in optimal transport or if it was merely a technical assumption in Ma-Trudinger-Wang's work.

These questions were studied by Loeper [20], who showed that for costs which are C^4, the MTW(0) condition is equivalent to requiring that the c-subdifferential of an arbitrary c-convex function be connected. This provides a geometric interpretation of the MTW tensor in terms of convex analysis, and shows that the failure of MTW condition is a *local obstruction* to regularity. More precisely, given a cost which fails to satisfy the MTW(0) condition, it is possible to find smooth measures μ and ν whose optimal transport is discontinuous[2].

[2] For non-degenerate cost functions, the c-subdifferential of a smooth potential consists of a single point, and thus is connected. As such, Loeper's result shows that, for cost functions which do not satisfy the MTW condition, the space of smooth optimal transports is not dense within the space of all optimal transports.

Furthermore, Loeper developed a geometric maximum principle, known as "double above sliding mountains" (DASM) which is equivalent to the MTW(0) condition for costs which are sufficiently smooth. Very recently, Loeper and Trudinger found a weakened version of this maximum principle which is also equivalent to the MTW(0) condition for C^4 costs [21]. Heuristically, one can interpret these conditions as "synthetic curvature bounds," akin to how Alexandrov spaces generalize the notion of sectional curvature bounds (see Section 6 of [13] for more details). In fact, there are multiple ways to formulate the MTW(0) condition synthetically (see Guillen-Kitagawa [9] for another formulation and Jeong [11] for its relationship to DASM).

Somewhat unexpectedly, another insight into the MTW tensor came from pseudo-Riemannian geometry. In particular, Kim and McCann [16] developed a formulation of optimal transport in which the cost function defines a pseudo-metric h of signature (n, n) on the space $X \times Y$. In this geometry, the MTW tensor is the curvature of certain light-like planes. This interpretation gives intrinsic differential geometric structure to the regularity problem and immediately explains many of the properties of $\mathfrak{S}$, such as why it transforms tensorially under change of coordinates. Furthermore, the optimal map T induces a maximal space-like hypersurface for a conformal pseudo-metric ρh where ρ is a function induced by the probability measures μ and ν [17]. Recently, Wong and Yang discovered an interesting link between this geometry and information geometry. In particular, they found a relationship between the Levi-Civita connection on $X \times Y$ and the dual connections on the space X, when interpreted as a submanifold of $X \times Y$ via the embedding $x \mapsto (x, T(x))$ [34].

2.2 The Squared-Distance on a Riemannian Manifold

The squared-distance cost plays a special and important role within optimal transport, and any discussion of the regularity theory would be incomplete without mentioning it. Unfortunately, there are several factors which make analysis of optimal transport on Riemannian manifolds very difficult, so we will not be able to discuss this line of work in depth.

One basic challenge is that there are very few Riemannian manifolds whose distance function can be computed explicitly. Even for these manifolds, the MTW tensor can be very complicated, since it involves calculating four derivatives of the distance (see Equation 5.7 of [7] for the MTW tensor of a round sphere). However, Loeper observed that for the squared distance, the MTW tensor restricted to the diagonal (i.e., $x = y$) is two-thirds the sectional curvature, so the MTW(0) condition is a non-local strengthening of non-negative sectional curvature [20].

Despite this difficulty, there are several Riemannian manifolds which are known to satisfy the MTW condition. For instance, spheres with the round metric do, as do complex projective space and quaternionic projective space. Furthermore, C^4-small perturbations of a round sphere satisfy MTW [7] (see also [33] for more details on the stability theory). The proof of this fact is extremely technical and relies on the fact that in the MTW tensor there are *fifteen terms* which combine to form a perfect square. One question of considerable interest is to find a conceptual argument to explain this seemingly miraculous cancellation.

Another central difficulty in studying optimal transport on Riemannian manifolds is the presence of the cut locus (and conjugate points in particular). On a compact Riemannian manifold, the distance function fails to be smooth along the cut locus, and the shape of the injectivity domains[3] can present a global obstruction to smoothness of the transport. However, the MTW condition exerts strong control over geometry of the injectivity domains, which lead Villani to pose the following conjecture.

Conjecture 1 (Villani's Conjecture). If (M, g) satisfies the MTW condition, then all of its injectivity domains are convex.

3 Kähler Geometry

For cost functions which are induced by a convex potential (such as $c(x, y) = \Psi(x - y)$ or a $\mathcal{D}_{\Psi}^{(\alpha)}$-divergence [36]), the authors found a separate geometric formulation in terms of Kähler/information geometry [12]. In particular, given such a cost function, one can use the convex potential Ψ to define a Kähler metric whose curvature encodes the MTW tensor.

Theorem 1. *For a cost which is induced by a convex potential $\Psi : \Omega \to \mathbb{R}$, the MTW tensor $\mathfrak{S}(\xi, \eta)$ is proportional[4] to the anti-bisectional curvature of an associated Kähler-Sasaki metric on the tangent bundle $T\Omega$.*

$$\mathfrak{S}(\xi, \eta) \propto R\left(\xi^{pol.}, \overline{(\eta^{\sharp})^{pol.}}, \xi^{pol.}, \overline{(\eta^{\sharp})^{pol.}}\right) \tag{5}$$

In this theorem, the superscript *pol.* denotes the polarized lift, which takes $\eta^{\sharp}$ and ξ to polarized $(1, 0)$ vectors in $T\Omega$. For a more precise definition, see page 7 of [14]. Furthermore, the conjugate connection (of the associated Hessian manifold) encodes the notion of relative c-convexity (for more details, see Proposition 8 of [12]).

At present, this result only holds for cost functions which are induced by a convex potential. However, it seems likely there is a non-Kähler version for costs on a Lie group.

Question 1. Is there a non-Kähler version of Theorem 1 for costs of the form $c(x, y) = \Psi(xy^{-1})$ where x and y are elements of a Lie group? What are some examples of such costs satisfying MTW(0)? In particular, which left-invariant metrics on $SO(3)$ have non-negative MTW tensor?

---

[3] For a point x, the injectivity domain $I(x) \subset T_x M$ is defined as the subset of the tangent bundle whose exponentials are distance-minimizing.

[4] The proportionality constant depends on how the convex potential is used to induce the cost function.

Another question is how to interpret the solution of an optimal transport problem in the complex setting. One can encode displacement interpolation in terms of sections of the tangent space TX, where (X, μ) is the initial probability measure in the optimal transport problem. However, it is not clear geometrically when a section corresponds to an optimal map.

Question 2. When does a section of TX induce an optimal flow (in the sense of displacement interpolation)? What is the relationship between these sections and Theorem 1?

As a related problem, the connection between the complex geometry and the Kim-McCann geometry is not well understood.

Question 3. Is it possible to transform the pseudo-Riemannian geometry into complex geometry via a transformation akin to the way that the Wick rotation transforms Minkowski $(3,1)$ geometry into Euclidean 4-space?

Kähler-Ricci Flow. The main advantage of the complex perspective is that we can use techniques from complex geometry, such as the Kahler-Ricci flow, to study optimal transport. The Ricci flow [10] is a geometric flow which deforms a Riemannian manifold by its Ricci curvature:

$$\frac{\partial g}{\partial t} = -2\,\mathrm{Ric}(g). \tag{6}$$

When the underlying manifold is Kähler, this flow preserves the complex structure and the Kählerity of the metric [3], so is known as the Kähler-Ricci flow. The connection between Kähler-Ricci flow and the MTW tensor was explored by the first named author and Fangyang Zheng [14].

Theorem 2. *Suppose that $\Omega \subset \mathbb{R}^n$ is a convex domain and $\Psi : \Omega \to \mathbb{R}$ is a strongly convex function so that the associated Kähler manifold $(T\Omega, \omega_0)$ is complete and has bounded curvature.*

1. *For $n = 2$, non-negative orthogonal anti-bisectional curvature is preserved by Kähler-Ricci flow.*
2. *For all n, non-positive anti-bisectional curvature is preserved under Kähler-Ricci flow.*
3. *When Ω is bounded and $T\Omega$ has negative holomorphic sectional curvature (or when Ω is a compact Hessian manifold whose first affine Chern class is negative), a stronger version of non-positive anti-bisectional curvature known as negative cost-curvature is preserved.*

The proofs of the first two claims use Shi's generalization [27] of the tensor maximum principle developed by Hamilton. The third uses some careful tensor analysis and recent work of Tong [28] on the convergence of Kähler-Ricci flow. We refer to the paper for the full argument.

Smoothing Flows in Optimal Transport. Using Theorem 2, we also proved a Hölder estimate for the transport map when the cost functions is $W^{2,p}$ (with $p > 2$) and satisfies the MTW(0) condition in some weak sense (see Corollary 6 [14] for details). This result suggests that smoothing flows can play a role in optimal transport, especially in cases with low regularity. Parabolic flows for the transport potential u had previously been considered [18] but in this context Kähler-Ricci flow is used to deform the *cost function.*

For general cost functions, it is not possible to define a Ricci flow naively (for pseudo-Riemannian metrics, Ricci flow is no longer weakly parabolic). As such, Theorem 2 is limited to costs which are induced by a convex potential. However, when the cost function is the squared distance, one natural idea is to evolve the underlying space using Ricci flow (see, e.g., [23,29]). This raises the following questions.

Question 4. For the squared distance cost on a Riemannian manifold, does the Ricci flow preserve non-negativity of the MTW tensor? For more general cost functions, is it possible to define a smoothing flow?

We suspect that the answer to the first question is positive, at least in dimensions two and three. In higher dimensions, Ricci flow need not preserve positivity of the sectional curvature, so perhaps the additional assumption of positive curvature operator is needed.

4 Applications of the MTW Tensor to Hessian Geometry

In this final section, we discuss some developments which come from studying the MTW tensor as a complex curvature. Independent of optimal transport, we can use anti-bisectional curvature to understand the geometry of Hessian manifolds and their tangent bundles. In particular, we pose the following conjecture, which can be roughly considered as a Hessian analogue to the Frankel conjecture.

Conjecture 1. The affine universal cover $\widetilde{M}$ of a compact negatively cost-curved Hessian manifold (M, g, D) factors into the product of domains admitting metrics of constant Hessian sectional curvature. In other words, the Hesse-Einstein metric on $\widetilde{M}$ is the product of Hessian space forms.

In order to prove this conjecture, one approach would be to use the Hesse-Koszul flow [24], which deforms the Hessian potential according to the parabolic Monge-Ampère equation

$$\frac{\partial}{\partial t}\Psi = \log(\det(D^2\Psi)). \tag{7}$$

The Hesse-Koszul flow on a Hessian manifold is equivalent to Kähler-Ricci flow on its tangent bundle, so the distinction between these two flows is largely one of terminology. Puechmorel and Tô studied this flow and showed that for Hessian manifolds with negative first affine Chern class, the normalized flow converges to a Hesse-Einstein metric (i.e., a Hessian metric whose tangent bundle is Kähler-Einstein). When combined with Theorem 2, this greatly restrict the geometry of compact negatively cost-curved Hessian manifolds, which provides strong evidence for Conjecture 1. However, the conjecture fails if one considers non-compact complete Hessian manifolds (see Example 6.2.1 of [14] for a counter-example). As such, any proof must use compactness in an essential way.

4.1 Complex Surfaces Satisfying MTW(0)

Finally, it would be of interest to better understand complete complex surfaces which admit metrics whose orthogonal anti-bisectional curvature is non-negative (i.e., which satisfy the MTW(0) condition). We are aware of three such examples; $\mathbb{C}^2$ with its flat metric, the Siegel-Jacobi half-plane with its Bergman metric and the Siegel half-space with an $SL(2,\mathbb{R}) \ltimes \mathbb{R}^2$-invariant metric [35]. It would be of interest to find others, since each such surface induces cost functions which have a "good" regularity theory.

References

1. Brenier, Y.: Décomposition polaire et réarrangement monotone des champs de vecteurs. C.R. Acad. Sci. Paris Sér. I Math. **305**, 805–808 (1987)
2. Caffarelli, L.A.: The regularity of mappings with a convex potential. J. Amer. Math. Soc. **5**(1), 99–104 (1992)
3. Cao, H.D.: Deformation of Kahler metrics to Kahler-Einstein metrics on compact Kahler manifolds. Invent. Math. **81**, 359–372 (1986)
4. Delanoë, P.: Classical solvability in dimension two of the second boundary-value problem associated with the Monge-Ampere operator. Ann I H Poincarè-AN, vol. 8, No. 5, pp. 443–457. Elsevier Masson (1991)
5. De Philippis, G., Figalli, A.: Partial regularity for optimal transport maps. Publications mathématiques de l'IHÈS **121**(1), 81–112 (2015)
6. De Philippis, G., Figalli, A.: The Monge-Ampère equation and its link to optimal transportation. Bull. Am. Math. Soc. **51**(4), 527–580 (2014)
7. Figalli, A., Rifford, L., Villani, C.: Nearly round spheres look convex. Am. J. Math. **134**(1), 109–139 (2012)
8. Gangbo, W., McCann, R.J.: The geometry of optimal transportation. Acta Mathematica **177**(2), 113–161 (1996)
9. Guillen, N., Kitagawa, J.: On the local geometry of maps with c-convex potentials. Calc. Var. Partial Differ. Equ. **52**(1–2), 345–387 (2015)
10. Hamilton, R.S.: Three-manifolds with positive Ricci curvature. J. Differ. Geom. **17**(2), 255–306 (1982)
11. Jeong, S.: Synthetic MTW conditions and their equivalence under mild regularity assumption on the cost function (2020). arXiv preprint arXiv:2010.14471
12. Khan, G., Zhang, J.: The Kähler geometry of certain optimal transport problems. Pure Appl. Anal. **2**(2), 397–426 (2020)
13. Khan, G., Zhang, J., Zheng, F.: The geometry of positively curved Kähler metrics on tube domains (2020). arXiv preprint arXiv:2001.06155
14. Khan, G., Zheng, F.: Kähler-Ricci Flow preserves negative anti-bisectional curvature (2020). arXiv preprint arXiv:2011.07181

15. Kim, Y.H., Kitagawa, J.: Prohibiting isolated singularities in optimal transport. Ann. Scoula Norm-Sci. **16**(1), 277–290 (2016)
16. Kim, Y.H., McCann, R.: Continuity, curvature, and the general covariance of optimal transportation. J. Eur. Math. Soc. **12**(4), 1009–1040 (2010)
17. Kim, Y.H., McCann, R.J., Warren, M.: Pseudo-Riemannian geometry calibrates optimal transportation. Math. Res. Lett. **17**(6), 1183–1197 (2010)
18. Kitagawa, J.: A parabolic flow toward solutions of the optimal transportation problem on domains with boundary. J. Reine Angew. Math. **2012**(672), 127–160 (2012)
19. Knott, M., Smith, C.S.: On the optimal mapping of distributions. J. Optim. Theory App. **43**(1), 39–49 (1984)
20. Loeper, G.: On the regularity of solutions of optimal transportation problems. Acta Mathematica **202**(2), 241–283 (2009)
21. Loeper, G., Trudinger, N.S.: Weak formulation of the MTW condition and convexity properties of potentials (2020). arXiv preprint arXiv:2007.02665
22. Ma, X.N., Trudinger, N.S., Wang, X.J.: Regularity of potential functions of the optimal transportation problem. Arch. Ration. Mech. Anal. **177**(2), 151–183 (2005)
23. McCann, R.J., Topping, P.M.: Ricci flow, entropy and optimal transportation. Am. J. Math. **132**(3), 711–730 (2010)
24. Mirghafouri, M., Malek, F.: Long-time existence of a geometric flow on closed Hessian manifolds. J. Geom. Phys. **119**, 54–65 (2017)
25. Monge, G.: Mémoire sur la théorie des déblais et des remblais. Histoire de l'Académie Royale des Sciences de Paris (1781)
26. Puechmorel, S., Tô, T.D.: Convergence of the Hesse-Koszul flow on compact Hessian manifolds (2020). arXiv preprint arXiv:2001.02940
27. Shi, W.X.: Ricci flow and the uniformization on complete noncompact Kähler manifolds. J. Differ. Geom. **45**(1), 94–220 (1997)
28. Tong, F.: The Kähler-Ricci flow on manifolds with negative holomorphic curvature (2018). arXiv preprint arXiv:1805.03562
29. Topping, P.: $\mathcal{L}$-optimal transportation for Ricci flow. J. Reine Angew. Math. **2009**(636), 93–122 (2009)
30. Trudinger, N.S., Wang, X.J.: On the second boundary value problem for Monge-Ampere type equations and optimal transportation. Ann. Scoula Norm-Sci.-Serie IV **8**(1), 143 (2009)
31. Urbas, J.: On the second boundary value problem for equations of Monge-Ampere type. J. Reine Angew. Math. **487**, 115–124 (1997)
32. Villani, C.: Optimal Transport: Old and New, vol. 338. Springer, Heidelberg (2008). https://doi.org/10.1007/978-3-540-71050-9
33. Villani, C.: Stability of a 4th-order curvature condition arising in optimal transport theory. J. Funct. Anal. **255**(9), 2683–2708 (2008)
34. Wong, T.K.L., Yang, J.: Optimal transport and information geometry (2019). arXiv preprint arXiv:1906.00030
35. Yang, J.H.: Invariant metrics and Laplacians on Siegel-Jacobi space. J. Number Theory **127**(1), 83–102 (2007)
36. Zhang, J.: Divergence function, duality, and convex analysis. Neural Comput. **16**(1), 159–195 (2004)

Wasserstein Proximal of GANs

Alex Tong Lin[1], Wuchen Li[2(✉)], Stanley Osher[1], and Guido Montúfar[1,3]

[1] University of California, Los Angeles, Los Angeles, CA 90095, USA
[2] University of South Carolina, Columbia, SC 29208, USA
wuchen@mailbox.sc.edu
[3] Max Planck Institute for Mathematics in Sciences, 04103 Leipzig, Germany

Abstract. We introduce a new method for training generative adversarial networks by applying the Wasserstein-2 metric proximal on the generators. The approach is based on Wasserstein information geometry. It defines a parametrization invariant natural gradient by pulling back optimal transport structures from probability space to parameter space. We obtain easy-to-implement iterative regularizers for the parameter updates of implicit deep generative models. Our experiments demonstrate that this method improves the speed and stability of training in terms of wall-clock time and Fréchet Inception Distance.

Keywords: Generative-Adversarial Networks · Wasserstein metric · Natural gradient

1 Introduction

Generative Adversarial Networks (GANs) [6] are a powerful approach to learning generative models. Here, a discriminator tries to tell apart the data generated by a real source and the data generated by a generator, whereas the generator tries to fool the discriminator. This adversarial game is formulated as an optimization problem over the discriminator and an implicit generative model for the generator. An implicit generative model is a parametrized family of functions mapping a noise source to sample space. In trying to fool the discriminator, the generator should try to recreate the real source.

The problem of recreating a target density can be formulated as the minimization of a discrepancy measure. The Kullback–Leibler (KL) divergence is known to be difficult to work with when the densities have a low dimensional support set, as is commonly the case in applications with structured data and high dimensional sample spaces. An alternative is to use the Wasserstein distance or Earth Mover's distance, which is based on optimal transport theory.

A. Lin, W. Li and S. Osher were supported by AFOSR MURI FA 9550-18-1-0502, AFOSR FA 9550-18-0167, ONR N00014-18-2527 and NSF DMS 1554564 (STROBE). G. Montúfar has received funding from the European Research Council (ERC) under the European Union's Horizon 2020 research and innovation programme (grant agreement n° 757983).

F. Nielsen and F. Barbaresco (Eds.): GSI 2021, LNCS 12829, pp. 524–533, 2021.
https://doi.org/10.1007/978-3-030-80209-7_57

This has been used recently to define the loss function for learning generative models [5,11]. In particular, the Wasserstein GAN [3] has attracted much interest in recent years.

Besides defining the loss function, optimal transport can also be used to introduce structures serving the *optimization* itself, in terms of the gradient operator. In full probability space, this method is known as the Wasserstein steepest descent flow [7,13]. In this paper we derive the Wasserstein steepest descent flow for deep generative models in GANs. We use the Wasserstein-2 metric function, which allows us to obtain a Riemannian structure and a corresponding natural (i.e., Riemannian) gradient. A well known example of a natural gradient is the Fisher-Rao natural gradient, which is induced by the KL-divergence. In learning problems, one often finds that the natural gradients offer advantages compared to the Euclidean gradient [1,2,12]. In GANs, the densities under consideration typically have a small support set, which prevents implementations of the Fisher-Rao natural gradient. Therefore, we propose to use the gradient operator induced by the Wasserstein-2 metric on probability models [8–10].

We propose to compute the parameter updates of the generators in GANs by means of a proximal operator where the proximal penalty is a squared constrained Wasserstein-2 distance. In practice, the constrained distance can be approximated by a neural network. In implicit generative models, the constrained Wasserstein-2 metric exhibits a simple structure. We generalize the Riemannian metric and introduce two methods: the relaxed proximal operator for generators and the semi-backward Euler method. Both approaches lead to practical numerical implementations of the Wasserstein proximal operator for GANs. The method can be easily implemented as a drop-in regularizer for the generator updates. Experiments demonstrate that this method improves the stability of training and reduces the training time. We remark that this paper is the first paper to apply Wasserstein gradient operator in GANs.

This paper is organized as follows. In Sect. 2 we introduce the Wasserstein natural gradient and proximal optimization methods. In Sect. 3 we review basics of implicit generative models. In Sect. 4 we derive practical computational methods.

2 Wasserstein Natural Proximal Optimization

In this section, we present the Wasserstein natural gradient and the corresponding proximal method.

2.1 Motivation and Illustration

The natural gradient method is an approach to parameter optimization in probability models, which has been promoted especially within information geometry [2,4]. This method chooses the steepest descent direction when the size of the step is measured by means of a metric on probability space.

If $F(\theta)$ is the loss function, the steepest descent direction is the vector $d\theta$ that solves

$$\min_{d\theta} F(\theta + d\theta) \quad \text{subject to} \quad D(\rho_\theta, \rho_{\theta+d\theta}) = \epsilon, \tag{1}$$

for a small enough ϵ. Here D is a divergence function on probability space. Expanding the divergence to second order and solving leads to an update of the form

$$d\theta \propto G(\theta)^{-1}\nabla_\theta F(\theta),$$

where G is the Hessian of D. Usually the Fisher-Rao metric is considered for G, which corresponds to having D as the KL-divergence.

In this work, we use structures derived from optimal transport. Concretely, we replace D in Eq. (1) with the Wasserstein-p distance. This is defined as

$$W_p(\rho_\theta, \rho_{\theta^k})^p = \inf \int_{\mathbb{R}^n \times \mathbb{R}^n} \|x - y\|^p \pi(x, y) dx dy, \tag{2}$$

where the infimum is over all joint probability densities $\pi(x, y)$ with marginals ρ_θ, ρ_{θ^k}. We will focus on $p = 2$. The Wasserstein-2 distance introduces a metric tensor in probability space, making it an infinite dimensional Riemannian manifold. We will introduce a finite dimensional metric tensor G on the parameter space of a generative model.

The Wasserstein metric allows us to define a natural gradient even when the support of the distributions is low dimensional and the Fisher-Rao natural gradient is not well defined. We will use the proximal operator, which computes the parameter update by minimizing the loss function plus a penalty on the step size. This saves us the need to compute the matrix G and its inverse explicitly. As we will show, the Wasserstein metric can be translated to practical proximal methods for implicit generative models. We first present a toy example, with explicit calculations, to illustrate the effectiveness of Wasserstein proximal operator.

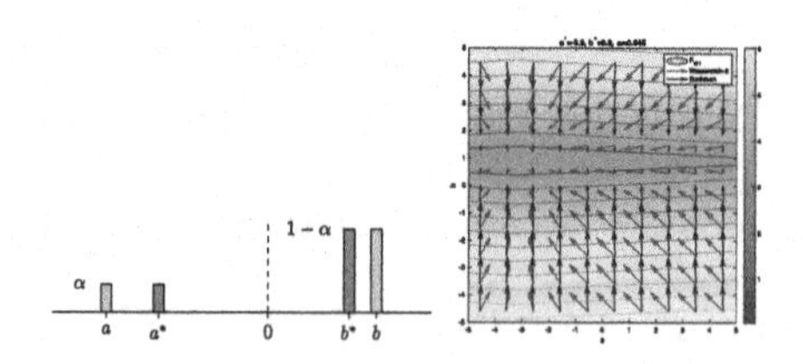

Fig. 1. Illustration of the Wasserstein proximal operator. Here the Wasserstein proximal penalizes parameter steps in proportion to the mass being transported, which results in updates pointing towards the minimum of the loss function. The Euclidean proximal penalizes all parameters equally, which results in updates naively orthogonal to the level sets of the loss function.

Example 1. Consider a probability model consisting of mixtures of pairs of delta measures. Let $\Theta = \{\theta = (a, b) \in \mathbb{R}^2 \colon a < 0 < b\}$, and define

$$\rho(\theta, x) = \alpha\delta_a(x) + (1 - \alpha)\delta_b(x),$$

where $\alpha \in [0, 1]$ is a given ratio and $\delta_a(x)$ is the delta measure supported at point a. See Fig. 1. For a loss function F, the proximal update is

$$\theta^{k+1} = \arg\min_{\theta \in \Theta} F(\theta) + \frac{1}{2h} D(\rho_\theta, \rho_{\theta^k}).$$

We check the following common choices for the function D to measure the distance between θ and θ^k, $\theta \neq \theta^k$.

1. Wasserstein-2 distance:

$$W_2(\rho_\theta, \rho_{\theta^k})^2 = \alpha(a - a^k)^2 + (1 - \alpha)(b - b^k)^2;$$

2. Euclidean distance:

$$\|\theta - \theta^k\|^2 = (a - a^k)^2 + (b - b^k)^2;$$

3. Kullback–Leibler divergence:

$$D_{\mathrm{KL}}(\rho_\theta \| \rho_{\theta^k}) = \int_{\mathbb{R}^n} \rho(\theta, x) \log \frac{\rho(\theta, x)}{\rho(\theta^k, x)} dx = \infty;$$

4. L^2-distance:

$$L^2(\rho_\theta, \rho_{\theta^k}) = \int_{\mathbb{R}^n} |\rho(\theta, x) - \rho(\theta^k, x)|^2 dx = \infty.$$

As we see, the KL-divergence and L^2-distance take value infinity, which tells the two parameters apart, but does not quantify the difference in a useful way. The Wasserstein-2 and Euclidean distances still work in this case. The Euclidean distance captures the difference in the locations of the delta measures, but not their relative weights. On the other hand, the Wasserstein-2 takes these into account. The right panel of Fig. 1 illustrates the loss function $F(\theta) = W_1(\rho_\theta, \rho_{\theta^*})$ for a random choice of θ^*, alongside with the Euclidean and Wasserstein-2 proximal parameter updates. The Wasserstein proximal update points more consistently in the direction of the global minimum.

2.2 Wasserstein Natural Gradient

We next present the Wasserstein natural gradient operator for parametrized probability models.

Definition 1 (Wasserstein natural gradient operator). *Given a model of probability densities $\rho(\theta, x)$ over $x \in \mathbb{R}^n$, with locally injective parametrization by $\theta \in \Theta \subseteq \mathbb{R}^d$, and a loss function $F\colon \Theta \to \mathbb{R}$, the Wasserstein natural gradient operator is given by*

$$\operatorname{grad} F(\theta) = G(\theta)^{-1}\nabla_\theta F(\theta).$$

Here $G(\theta) = (G(\theta)_{ij})_{1\le i,j\le d} \in \mathbb{R}^{d\times d}$ is the matrix with entries

$$G(\theta)_{ij} = \int_{\mathbb{R}^n} \Big(\nabla_{\theta_i}\rho(\theta, x), \mathcal{G}(\rho_\theta)\nabla_{\theta_j}\rho(\theta, x)\Big)dx,$$

where $\mathcal{G}(\rho_\theta)$ is the Wasserstein-2 metric tensor in probability space. More precisely, $\mathcal{G}(\rho) = (-\Delta_\rho)^{-1}$ is the inverse of the elliptic operator $\Delta_\rho := \nabla\cdot(\rho\nabla)$.

For completeness, we briefly explain the definition of the Wasserstein natural gradient. The gradient operator on a Riemannian manifold (Θ, g) is defined as follows. For any $\sigma \in T_\theta\Theta$, the Riemannian gradient $\nabla^g_\theta F(\theta) \in T_\theta\Theta$ satisfies $g_\theta(\sigma, \operatorname{grad}F(\theta)) = (\nabla_\theta F(\theta), \sigma)$. In other words, $\sigma^\top G(\theta)\operatorname{grad}F(\theta) = \nabla_\theta F(\theta)^\top\sigma$. Since $\theta \in \mathbb{R}^d$ and $G(\theta)$ is positive definite, $\operatorname{grad}F(\theta) = G(\theta)^{-1}\nabla_\theta F(\theta)$.

Our main focus will be in deriving practical computational methods that allow us to apply these structures to optimization in GANs. Consider the gradient flow of the loss function:

$$\frac{d\theta}{dt} = -\operatorname{grad}F(\theta) = -G(\theta)^{-1}\nabla_\theta F(\theta). \tag{3}$$

There are several discretization schemes for a gradient flow of this type. One of them is the forward Euler method, known as the steepest descent method:

$$\theta^{k+1} = \theta^k - hG(\theta^k)^{-1}\nabla_\theta F(\theta^k), \tag{4}$$

where $h > 0$ is the learning rate (step size). In practice we usually do not have a closed formula for the metric tensor $G(\theta)$. In (4), we need to solve for the inverse Laplacian operator, the Jacobian of the probability model, and compute the inverse of $G(\theta)$. When the parameter $\theta \in \Theta$ is high dimensional, these computations are impractical. Therefore, we will consider a different approach based on the proximal method.

2.3 Wasserstein Natural Proximal

To practically apply the Wasserstein natural gradient, we present an alternative way to discretize the gradient flow, known as the proximal method or backward Euler method. The proximal operator computes updates of the form

$$\theta^{k+1} = \arg\min_\theta\ F(\theta) + \frac{\operatorname{Dist}(\theta, \theta^k)^2}{2h}, \tag{5}$$

where Dist is an iterative regularization term, given by the Riemannian distance

$$\operatorname{Dist}(\theta, \theta^k)^2 = \inf\Big\{\int_0^1 \dot\theta_t^\top G(\theta_t)\dot\theta_t dt\colon \theta_0 = \theta,\ \theta_1 = \theta^k\Big\}.$$

Here the infimum is taken among all continuously differentiable parameter paths $\theta_t = \theta(t) \in \Theta$, $t \in [0, 1]$. The proximal operator is defined implicitly, in terms of a minimization problem, but in some cases it can be written explicitly. Interestingly, it allows us to consider an iterative regularization term in the parameter update.

We observe that there are two time variables in the proximal update (5). One is the time discretization of gradient flow, known as the learning rate $h > 0$; the other is the time variable in the definition of the Riemannian distance $\mathrm{Dist}(\theta, \theta^k)$. The variation in the time variable of the Riemannian distance can be further simplified as follows.

Proposition 1 (Semi-backward Euler method). *The iteration*

$$\theta^{k+1} = \arg\min_{\theta} F(\theta) + \frac{\tilde{D}(\theta, \theta^k)^2}{2h}, \tag{6}$$

with

$$\tilde{D}(\theta, \theta^k)^2 = \int_{\mathbb{R}^n} \Big(\rho_\theta - \rho_{\theta^k}, \mathcal{G}(\rho_{\tilde{\theta}})(\rho_\theta - \rho_{\theta^k})\Big) dx,$$

and $\tilde{\theta} = \frac{\theta+\theta^k}{2}$, *is a consistent time discretization of the Wassserstein natural gradient flow (3).*

Here the distance term in (5) is replaced by $\tilde{D}$, which is obtained by a mid-point approximation in time. The mid-point $\tilde{\theta}$ can be chosen in many ways between θ and θ^k. For simplicity and symmetry, we let $\tilde{\theta} = \frac{\theta+\theta^k}{2}$. In practice, we also use $\tilde{\theta} = \theta^k$, since in this case $\mathcal{G}(\rho_{\tilde{\theta}})$ can be held fixed when iterating over θ to obtain (6). Formula (6) is called the semi-backward Euler method (SBE), because it can also be expressed as

$$\theta^{k+1} = \theta^k - hG(\tilde{\theta})^{-1}\nabla_\theta F(\theta^{k+1}) + o(h).$$

The proof is contained in the appendix.

We point out that all methods described above, i.e., the forward Euler method (4), the backward Euler method (5), and the semi-backward Euler method (6), are time consistent discretizations of the Wasserstein natural gradient flow (3) with first order accuracy in time. We shall focus on the semi-backward Euler method and derive practical formulas for the iterative regularization term.

3 Implicit Generative Models

Before proceeding, we briefly recall the setting of Generative Adversarial Networks (GANs). GANs consist of two parts: the *generator* and the *discriminator*. The generator is a function $g_\theta \colon \mathbb{R}^\ell \to \mathbb{R}^n$ that takes inputs z in latent space $\mathbb{R}^\ell$ with distribution $p(z)$ (a common choice is a Gaussian) to outputs $x = g_\theta(z)$ in sample space $\mathbb{R}^n$ with distribution $\rho(\theta, x)$. The objective of training is to find a value of the parameter θ so that $\rho(\theta, x)$ matches a given target distribution, say $\rho_{\text{target}}(x)$. The discriminator is merely an assistance during optimization of a GAN in order to obtain the right parameter value for the generator. It is a function $f_\omega : \mathbb{R}^n \to \mathbb{R}$, whose role is to discriminate real images (sampled from the target distribution) from fake images (produced by the generator).

To train a GAN, one works on min-maxing a function such as

$$\inf_\theta \sup_\omega \mathbb{E}_{x \sim \rho_{\text{target}}(x)} \left[\log f_\omega(x)\right] + \mathbb{E}_{z \sim p(z)} \left[\log(1 - f_\omega(g_\theta(z))\right].$$

The specific loss function can be chosen in many different ways (including the Wasserstein-1 loss [3]), but the above is the one that was first considered for GANs, and is a common choice in applications. Practically, to perform the optimization, we adopt an alternating gradient optimization scheme for the generator parameter θ and the discriminator parameter ω.

4 Computational Methods

In this section, we present two methods for implementing the Wasserstein natural proximal for GANs. The first method is based on solving the variational formulation of the proximal penalty over an affine space of functions. This leads to a low-order version of the Wasserstein metric tensor $\mathcal{G}(\rho_\theta)$. The second method is based on a formula for the Wasserstein metric tensor for 1-dimensional sample spaces, which we relax to sample spaces of arbitrary dimension.

4.1 Affine Space Variational Approximation

The mid point approximation $\tilde{D}$ from Proposition 1 can be written using dual coordinates (cotangent space) of probability space in the variational form

$$\tilde{D}(\theta, \theta^k)^2 = \sup_{\Phi \in C^\infty(\mathbb{R}^n)} \Big\{ \int_{\mathbb{R}^n} \Phi(x)(\rho(\theta, x) - \rho(\theta^k, x)) - \frac{1}{2} \|\nabla \Phi(x)\|^2 \rho(\tilde{\theta}, x)\, dx \Big\}.$$

In order to obtain an explicit formula, we consider a function approximator of the form

$$\Phi_\xi(x) := \sum_j \xi_j \psi_j(x) = \xi^\top \Psi(x),$$

where $\Psi(x) = (\psi_j(x))_{j=1}^K$ are given basis functions on sample space $\mathbb{R}^n$, and $\xi = (\xi_j)_{j=1}^K \in \mathbb{R}^K$ is the parameter. In other words, we consider

$$\tilde{D}(\theta, \theta^k)^2 = \sup_{\xi \in \mathbb{R}^K} \Big\{ \int_{\mathbb{R}^n} \Phi_\xi(x)(\rho(\theta, x) - \rho(\theta^k, x)) - \frac{1}{2} \|\nabla \Phi_\xi(x)\|^2 \rho(\tilde{\theta}, x)\, dx \Big\}. \tag{7}$$

Theorem 1 (Affine metric function $\tilde{D}$). *Consider some $\Psi = (\psi_1, \ldots, \psi_K)^\top$ and assume that $M(\theta) = (M_{ij}(\theta))_{1\leq i,j\leq K} \in \mathbb{R}^{K\times K}$ is a regular matrix with entries*

$$M_{ij}(\theta) = \mathbb{E}_{Z\sim p}\Big(\sum_{l=1}^{n} \partial_{x_l}\psi_i(g(\tilde{\theta}, Z))\partial_{x_l}\psi_j(g(\tilde{\theta}, Z))\Big),$$

where $\tilde{\theta} = \frac{\theta+\theta^k}{2}$. Then,

$$\tilde{D}(\theta, \theta^k)^2 = \Big(\mathbb{E}_{Z\sim p}[\Psi(g(\theta, Z)) - \Psi(g(\theta^k, Z))]\Big)^\top M(\tilde{\theta})^{-1}\Big(\mathbb{E}_{Z\sim p}[\Psi(g(\theta, Z)) - \Psi(g(\theta^k, Z))]\Big).$$

The proof is contained in the appendix. There are many possible choices for the basis Ψ. For example, if $K = n$ and $\psi_k(x) = x_k$, $k = 1, \ldots, n$, then $M(\theta)$ is the identity matrix. In this case,

$$\tilde{D}(\theta, \tilde{\theta})^2 = \|\mathbb{E}_{Z\sim p}(g(\theta, Z) - g(\theta^k, Z))\|^2.$$

We will focus on degree one and degree two polynomials. The algorithms are presented in Sect. 4.3.

4.2 Relaxation from 1-D

Now we present a second method for approximating $\tilde{D}$. In the case of implicit generative models with 1-dimensional sample space, the constrained Wasserstein-2 metric tensor has an explicit formula. This allows us to define a relaxed Wasserstein metric for implicit generative models with sample spaces of arbitrary dimension.

Theorem 2 (1-D sample space). *If $n = 1$, then*

$$\mathrm{Dist}(\theta_0, \theta_1)^2 = \inf\Big\{\int_0^1 \mathbb{E}_{Z\sim p}\|\frac{d}{dt}g(\theta(t), Z)\|^2\, dt\colon \theta(0) = \theta_0, \theta(1) = \theta_1\Big\},$$

where the infimum is taken over all continuously differentiable parameter paths. Therefore, we have

$$\tilde{D}(\theta, \theta^k)^2 = \mathbb{E}_{Z\sim p}\|g(\theta, Z) - g(\theta^k, Z)\|^2.$$

In sample spaces of dimension higher than one, we no longer have the explicit formula for $\tilde{D}$. The relaxed metric consists of using the same formulas from the theorem. Later on, we show that this formulation of $\tilde{D}$ still provides a metric with parameterization invariant properties in the proximal update.

Algorithm 1. Wasserstein Natural Proximal

Require: F_ω, a parameterized function to minimize (e.g., the Wasserstein-1 with a parameterized discriminator); g_θ, the generator.
Require: Optimizer$_{F_\omega}$; Optimizer$_{g_\theta}$.
Require: h proximal step-size; B mini-batch size; max iterations; generator iterations.

1: **for** $k = 0$ **to** max iterations **do**
2: Sample real data $\{x_i\}_{i=1}^B$ and latent data $\{z_i\}_{i=1}^B$
3: $\omega^k \leftarrow \text{Optimizer}_{F_\omega}\left(\frac{1}{B}\sum_{i=1}^B F_\omega(g_\theta(z_i))\right)$
4: **for** $\ell = 0$ **to** generator iterations **do**
5: Sample latent data $\{z_i\}_{i=1}^B$
6: $\tilde{D} =$ RWP, or O1-SBE, or O2Diag-SBE (Sec. 4.3)
7: $\theta^k \leftarrow \text{Optimizer}_{g_\theta}\left(\frac{1}{B}\sum_{i=1}^B F_\omega(g_\theta(z_i)) + \frac{1}{2h}\tilde{D}(\theta,\theta^k)^2\right)$.
8: **end for**
9: **end for**

4.3 Algorithms

The Wasserstein natural proximal method for GANs optimizes the parameter θ of the generator by the proximal iteration (6). We implement this in the following ways:

RWP Method. The first and simplest method follows Sect. 4.2, and updates the generator by:

$$\theta^{k+1} = \arg\min_{\theta\in\Theta} F(\theta) + \frac{1}{2h}\mathbb{E}_{Z\sim p}\|g(\theta,Z) - g(\theta^k,Z)\|^2.$$

We call this the Relaxed Wasserstein Proximal (RWP) method.

SBE Order 1 Method. The second method is based on the discussion from Sect. 4.1, approximating Φ by linear functions. We update the generator by:

$$\theta^{k+1} = \arg\min_{\theta\in\Theta} F(\theta) + \frac{1}{2h}\|\mathbb{E}_{Z\sim p}\Big(g(\theta,Z) - g(\theta^k,Z)\Big)\|^2,$$

We call this the Order-1 SBE (O1-SBE) method.

SBE Order 2 Method. In an analogous way to the SBE order 1 method, we can approximate Φ by quadratic functions, to obtain the Order-2 SBE (O2Diag-SBE) method:

$$\begin{aligned}\theta^{k+1} =& \arg\min_{\theta\in\Theta} F(\theta) + \frac{1}{h}\bigg(\frac{1}{2}\|\mathbb{E}_{Z\sim p}[g(\theta,Z) - g(\theta^k,Z)] - \mathbb{E}_{Z\sim p}[Qg(\theta^k,Z)]\|^2 \\ &+ \frac{1}{2}\mathbb{E}_{Z\sim p}[\langle g(\theta,Z), Qg(\theta,Z)\rangle] - \frac{1}{2}\mathbb{E}_{Z\sim p}[\langle g(\theta^k,Z), Qg(\theta^k,Z)\rangle] \\ &- \frac{1}{2}\mathbb{E}_{Z\sim p}[\|Qg(\theta^k,Z)\|^2]\bigg),\end{aligned}$$

where $Q = \mathrm{diag}(q_i)_{i=1}^n$ is the diagonal matrix with diagonal entries

$$q_i = \frac{1}{2}\frac{\mathbb{E}_{Z\sim p}[(g(\theta, Z)_i - g(\theta^k, Z)_i)^2]}{\mathrm{Var}(g(\theta^k, Z)_i)} + \frac{\mathrm{Cov}_{Z\sim p}(g(\theta, Z)_i, g(\theta^k, Z)_i)}{\mathrm{Var}_{Z\sim p}(g(\theta^k, Z)_i)} - 1,$$

where $g(\theta, Z)_i$ is the ith coordinate of the samples.

The methods described above can be regarded as iterative regularizers. RWP penalizes the expected squared norm of the differences between samples (second moment differences). O1-SBE penalizes the squared norm of the expected differences between samples. O2Diag-SBE penalizes a combination of squared norm of the expected differences plus variances. They all encode statistical information of the generators. All these approaches *regularize the generator* by the expectation and variance of the samples. The implementation is shown in Algorithm 1.

Because of the page limit, we leave all detailed proofs and numerical experiments in [14].

References

1. Amari, S.: Natural gradient works efficiently in learning. Neural Comput. **10**(2), 251–276 (1998)
2. Amari, S.: Information Geometry and Its Applications. AMS, vol. 194. Springer, Tokyo (2016). https://doi.org/10.1007/978-4-431-55978-8
3. Arjovsky, M., Chintala, S., Bottou, L.: Wasserstein GAN. arXiv:1701.07875 [cs, stat] (2017)
4. Ay, N., Jost, J., Lê, H.V., Schwachhöfer, L.: Information Geometry. EMG-FASMSM, vol. 64. Springer, Cham (2017). https://doi.org/10.1007/978-3-319-56478-4
5. Frogner, C., Zhang, C., Mobahi, H., Araya-Polo, M., Poggio, T.: Learning with a Wasserstein Loss. arXiv:1506.05439 [cs, stat] (2015)
6. Goodfellow, I., et al.: Generative adversarial nets. In: Advances in Neural Information Processing Systems, vol. 27, pp. 2672–2680. Curran Associates Inc. (2014)
7. Jordan, R., Kinderlehrer, D., Otto, F.: The variational formulation of the Fokker-Planck equation. SIAM J. Math. Anal. **29**(1), 1–17 (1998)
8. Li, W.: Transport information geometry: Riemannian calculus on probability simplex. arXiv:1803.06360 [math] (2018)
9. Li, W., Montúfar, G.: Natural gradient via optimal transport. Inf. Geom. **1**(2), 181–214 (2018)
10. Li, W., Montúfar, G.: Ricci curvature for parametric statistics via optimal transport. arXiv:1807.07095 [cs, math, stat] (2018)
11. Montavon, G., Müller, K.-R., Cuturi, M.: Wasserstein training of restricted Boltzmann machines. In: Lee, D.D., Sugiyama, M., Luxburg, U.V., Guyon, I., Garnett, R. (eds.) Advances in Neural Information Processing Systems, vol. 29, pp. 3718–3726. Curran Associates Inc. (2016)
12. Ollivier, Y.: Riemannian metrics for neural networks I: feedforward networks. Inf. Infer. J. IMA **4**(2), 108–153 (2015)
13. Otto, F.: The geometry of dissipative evolution equations the porous medium equation. Commun. Partial Differ. Equ. **26**(1–2), 101–174 (2001)
14. Tong Lin, A., Li, W., Osher, S., Montúfar, G.: Wasserstein proximal of GANs. arXiv:2102.06862 (2018)

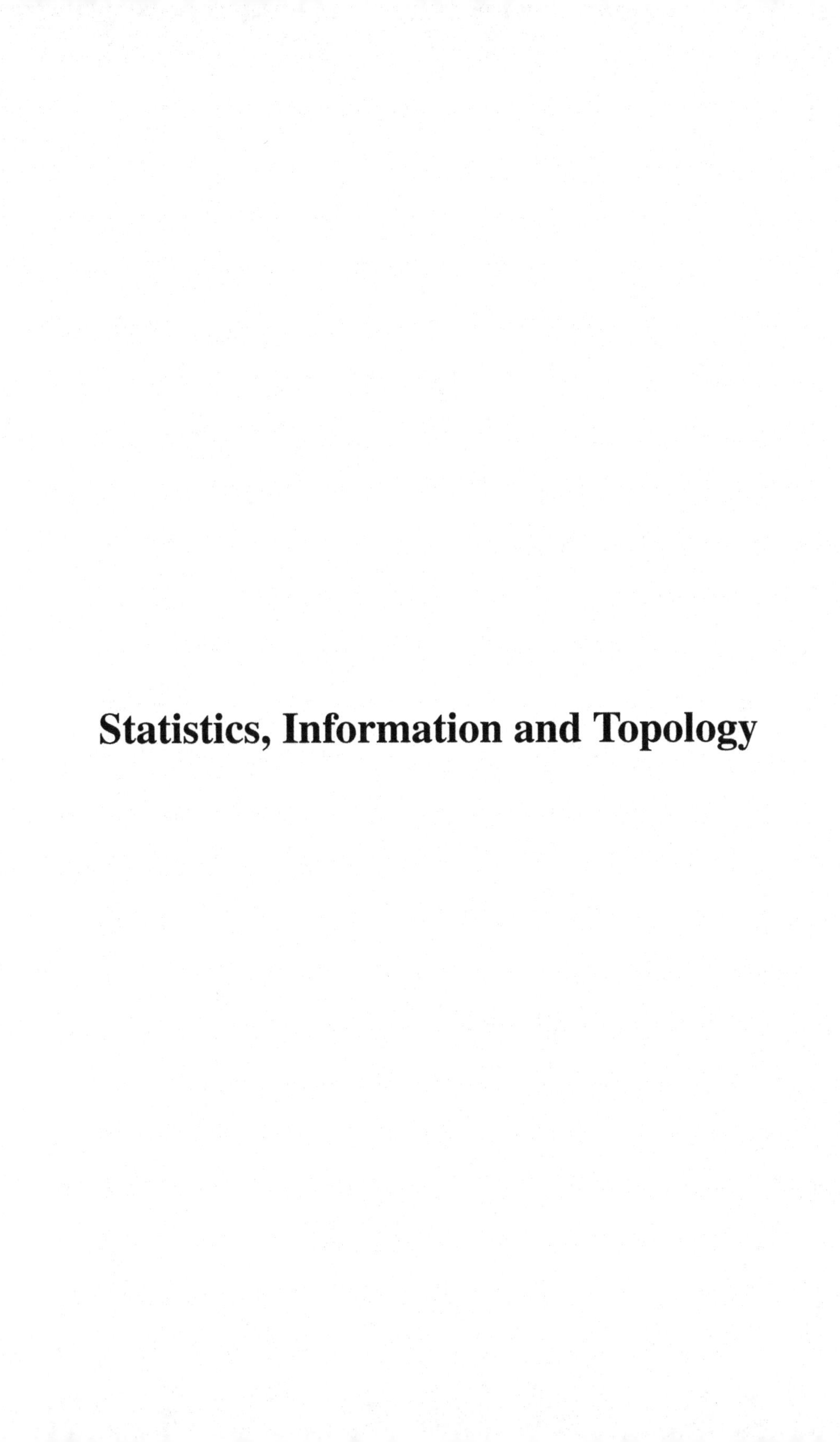

Statistics, Information and Topology

Information Cohomology of Classical Vector-Valued Observables

Juan Pablo Vigneaux[1,2(✉)]

[1] Institut de Mathématiques de Jussieu–Paris Rive Gauche (IMJ-PRG), Université de Paris, 8 place Aurélie Némours, 75013 Paris, France
[2] Max-Planck-Institut für Mathematik in den Naturwissenschaften, Inselstraße 22, 04103 Leipzig, Germany

Abstract. We provide here a novel algebraic characterization of two *information measures* associated with a vector-valued random variable, its differential entropy and the dimension of the underlying space, purely based on their recursive properties (the chain rule and the nullity-rank theorem, respectively). More precisely, we compute the information cohomology of Baudot and Bennequin with coefficients in a module of continuous probabilistic functionals over a category that mixes discrete observables and continuous vector-valued observables, characterizing completely the 1-cocycles; evaluated on continuous laws, these cocycles are linear combinations of the differential entropy and the dimension.

Keywords: Information cohomology · Entropy · Dimension · Information measures · Topos theory

1 Introduction

Baudot and Bennequin [2] introduced *information cohomology*, and identified Shannon entropy as a nontrivial cohomology class in degree 1. This cohomology has an explicit description in terms of *cocycles* and *coboundaries*; the *cocycle equations* are a rule that relate different values of the cocycle. When its coefficients are a module of measurable *probabilistic functionals* on a category of *discrete* observables, Shannon's entropy defines a 1-cocycle and the aforementioned rule is simply the chain rule; moreover, the cocycle equations in every degree are systems of functional equations and one can use the techniques developed by Tverberg, Lee, Knappan, Aczél, Daróczy, etc. [1] to show that, in degree one, the entropy is the *unique* measurable, nontrivial solution. The theory thus gave a new algebraic characterization of this information measure based on topos theory *à la* Grothendieck, and showed that *the chain rule is its defining algebraic property.*

It is natural to wonder if a similar result holds for the *differential entropy.* We consider here information cohomology with coefficients in a presheaf of continuous probabilistic functionals on a category that mixes discrete and continuous (vector-valued) observables, and establish that every 1-cocycle, when evaluated

F. Nielsen and F. Barbaresco (Eds.): GSI 2021, LNCS 12829, pp. 537–546, 2021.
https://doi.org/10.1007/978-3-030-80209-7_58

on probability measures absolutely continuous with respect to the Lebesgue measure, is a linear combination of the differential entropy and the dimension of the underlying space (the term continuous has in this sentence three different meanings). We already showed that this was true for gaussian measures [9]; in that case, there is a finite dimensional parametrization of the laws, and we were able to use Fourier analysis to solve the 1-cocycle equations. Here we exploit that result, expressing any density as a limit of gaussian mixtures (i.e. convex combinations of gaussian densities), and then using the 1-cocycle condition to compute the value of a 1-cocycle on gaussian mixtures in terms of its value on discrete laws and gaussian laws. The result depends on the conjectural existence of a "well-behaved" class of probabilities and probabilistic functionals, see Sect. 3.2.

The dimension appears here as an information quantity in its own right: its "chain rule" is the nullity-rank theorem. In retrospective, its role as an information measure is already suggested by old results in information theory. For instance, the expansion of Kolmogorov's ε-entropy $H_\varepsilon(\xi)$ of a continuous, $\mathbb{R}^n$-valued random variable ξ "is determined first of all by the dimension of the space, and the differential entropy $h(\xi)$ appears only in the form of the second term of the expression for $H_\varepsilon(\xi)$." [7, Paper 3, p. 22]

2 Some Known Results About Information Cohomology

Given the severe length constraints, it is impossible to report here the motivations behind information cohomology, its relationship with traditional algebraic characterizations of entropy, and its topos-theoretic foundations. For that, the reader is referred to the introductions of [9] and [10]. We simply remind here a minimum of definitions in order to make sense of the characterization of 1-cocycles that is used later in the article.

Let $\mathbf{S}$ be a partially ordered set (poset); we see it as a category, denoting the order relation by an arrow. It is supposed to have a terminal object $\top$ and to satisfy the following property: whenever $X, Y, Z \in \operatorname{Ob}\mathbf{S}$ are such that $X \to Y$ and $X \to Z$, the categorical product $Y \wedge Z$ exists in $\mathbf{S}$. An object of X of $\mathbf{S}$ (i.e. $X \in \operatorname{Ob}\mathbf{S}$) is interpreted as an *observable*, an arrow $X \to Y$ as Y being coarser than X, and $Y \wedge Z$ as the joint measurement of Y and Z.

The category $\mathbf{S}$ is just an algebraic way of encoding the relationships between observables. The measure-theoretic "implementation" of them comes in the form of a functor $\mathcal{E} : \mathbf{S} \to \mathbf{Meas}$ that associates to each $X \in \operatorname{Ob}\mathbf{S}$ a measurable set $\mathcal{E}(X) = (E_X, \mathfrak{B}_X)$, and to each arrow $\pi : X \to Y$ in $\mathbf{S}$ a measurable *surjection* $\mathcal{E}(\pi) : \mathcal{E}(X) \to \mathcal{E}(Y)$. To be consistent with the interpretations given above, one must suppose that $E_\top \cong \{*\}$ and that $\mathcal{E}(Y \wedge Z)$ is mapped *injectively* into $\mathcal{E}(Y) \times \mathcal{E}(Z)$ by $\mathcal{E}(Y \wedge Z \to Y) \times \mathcal{E}(Y \wedge Z \to Z)$. We consider mainly two examples: the discrete case, in which E_X finite and $\mathfrak{B}_X$ the collection of its subsets, and the Euclidean case, in which E_X is a Euclidean space and $\mathfrak{B}_X$ is its Borel σ-algebra. The pair $(\mathbf{S}, \mathcal{E})$ is an *information structure*.

Throughout this article, conditional probabilities are understood as disintegrations. Let ν a σ-finite measure on a measurable space $(E, \mathfrak{B})$, and ξ a σ-finite measure on $(E_T, \mathfrak{B}_T)$. The measure ν has a *disintegration* $\{\nu_t\}_{t\in E_T}$ with respect to a measurable map $T : E \to E_T$ and ξ, or a (T, ξ)-disintegration, if each ν_t is a σ-finite measure on $\mathfrak{B}$ concentrated on $\{T = t\}$—i.e. $\nu_t(T \neq t) = 0$ for ξ-almost every t—and for each measurable nonnegative function $f : E \to \mathbb{R}$, the mapping $t \mapsto \int_E f \,\mathrm{d}\nu_t$ is measurable and $\int_E f \,\mathrm{d}\nu = \int_{E_T} \left(\int_E f(x) \,\mathrm{d}\nu_t(x)\right) \mathrm{d}\xi(t)$ [3].

We associate to each $X \in \operatorname{Ob}\mathbf{S}$ the set $\Pi(X)$ of probability measures on $\mathcal{E}(X)$ i.e. of possible *laws* of X, and to each arrow $\pi : X \to Y$ the *marginalization map* $\pi_* := \Pi(\pi) : \Pi(X) \to \Pi(Y)$ that maps ρ to the image measure $\mathcal{E}(\pi)_*(\rho)$. More generally, we consider any subfunctor $\mathcal{Q}$ of Π that is stable under conditioning: for all $X \in \operatorname{Ob}\mathbf{S}$, $\rho \in \mathcal{Q}(X)$, and $\pi : X \to Y$, $\rho|_{Y=y}$ belongs to $\mathcal{Q}(X)$ for $\pi_*\rho$-almost every $y \in E_Y$, where $\{\rho|_{Y=y}\}_{y\in E_Y}$ is the $(\mathcal{E}\pi, \pi_*\rho)$-disintegration of ρ.

We associate to each $X \in \operatorname{Ob}\mathbf{S}$ the set $\mathcal{S}_X = \{Y \,|\, X \to Y\}$, which is a monoid under the product $\wedge$ introduced above. The assignment $X \mapsto \mathcal{S}_X$ defines a contravariant functor (presheaf). The induced algebras $\mathcal{A}_X = \mathbb{R}[\mathcal{S}_X]$ give a presheaf $\mathcal{A}$. An $\mathcal{A}$-module is a collection of modules $\mathcal{M}_X$ over $\mathcal{A}_X$, for each $X \in \operatorname{Ob}\mathbf{S}$, with an action that is "natural" in X. The main example is the following: for any adapted probability functor $\mathcal{Q} : \mathbf{S} \to \mathbf{Meas}$, one introduces a *contravariant* functor $\mathcal{F} = \mathcal{F}(\mathcal{Q})$ declaring that $\mathcal{F}(X)$ are the measurable functions on $\mathcal{Q}(X)$, and $\mathcal{F}(\pi)$ is precomposition with $\mathcal{Q}(\pi)$ for each morphism π in $\mathbf{S}$. The monoid $\mathcal{S}_X$ acts on $\mathcal{F}(X)$ by the rule:

$$\forall Y \in \mathcal{S}_X, \forall \rho \in \mathcal{Q}(X), \quad Y.\phi(\rho) = \int_{E_Y} \phi(\rho|_{Y=y}) \,\mathrm{d}\pi_*^{YX}\rho(y) \tag{1}$$

where π_*^{YX} stands for the marginalization $\mathcal{Q}(\pi^{YX})$ induced by $\pi^{YX} : X \to Y$ in $\mathbf{S}$. This action can be extended by linearity to $\mathcal{A}_X$ and is natural in X.

In [10], the information cohomology $H^\bullet(\mathbf{S}, \mathcal{F})$ is defined using *derived functors* in the category of $\mathcal{A}$-modules, and then described explicitly, for each degree $n \geq 0$, as a quotient of n-cocycles by n-coboundaries. For $n = 1$, the coboundaries vanish, so we simply have to describe the cocycles. Let $\mathcal{B}_1(X)$ be the $\mathcal{A}_X$-module freely generated by a collection of bracketed symbols $\{[Y]\}_{Y\in\mathcal{S}_X}$; an arrow $\pi : X \to Y$ induces an inclusion $\mathcal{B}_1(Y) \hookrightarrow \mathcal{B}_1(X)$, so $\mathcal{B}_1$ is a presheaf. A 1-*cochain* is a natural transformations $\varphi : \mathcal{B}_1 \Rightarrow \mathcal{F}$, with components $\varphi_X : \mathcal{B}_1(X) \to \mathcal{F}(X)$; we use $\varphi_X[Y]$ as a shorthand for $\varphi_X([Y])$. The naturality implies that $\varphi_X[Z](\rho)$ equals $\varphi_Z[Z](\pi_*^{ZX}\rho)$, a property that [2] called *locality*; sometimes we write Φ_Z instead of $\varphi_Z[Z]$. A 1-cochain φ is a 1-cocycle iff

$$\forall X \in \operatorname{Ob}\mathbf{S}, \forall X_1, X_2 \in \mathcal{S}_X, \quad \varphi_X[X_1 \wedge X_2] = X_1.\varphi_X[X_2] + \varphi_X[X_1]. \tag{2}$$

Remark that this is an equality *of functions* in $\mathcal{Q}(X)$.

An information structure is *finite* if for all $X \in \operatorname{Ob} S$, E_X is finite. In this case, [10, Prop. 4.5.7] shows that, whenever an object X can be written as a product $Y \wedge Z$ and E_X is "close" to $E_Y \times E_Z$, as formalized by the definition of *nondegenerate product* [10, Def. 4.5.6], then there exists $K \in \mathbb{R}$ such that for all $W \in \mathcal{S}_X$ and ρ in $\mathcal{Q}(Z)$

$$\Phi_W(\rho) = -K \sum_{w \in E_W} \rho(w) \log \rho(w). \tag{3}$$

The continuous case is of course more delicate. In the case of $\mathcal{E}$ taking values in vector spaces, and $\mathcal{Q}$ made of gaussian laws, [9] treated it as follows. We start with a vector space E with Euclidean metric M, and a poset $\mathbf{S}$ of vector subspaces of E, ordered by inclusion, satisfying the hypotheses stated above; remark that $\wedge$ corresponds to intersection. Then we introduce $\mathcal{E}$ by $V \in \operatorname{Ob}\mathbf{S} \mapsto E_V := E/V$, and further identify E_V is $V^\perp$ using the metric (so that we only deal with vector subspaces of E). We also introduce a sheaf $\mathcal{N}$, such that $\mathcal{N}(X)$ consists of affine subspaces of E_X and is closed under intersections; the sheaf is supposed to be closed under the projections induced by $\mathcal{E}$ and to contain the fibers of all these projections. On each affine subspace $N \in \mathcal{N}(X)$ there is a unique Lebesgue measure $\mu_{X,N}$ induced by the metric M. We consider a sheaf $\mathcal{Q}$ such that $\mathcal{Q}(X)$ are probabilities measures ρ that are absolutely continuous with respect to $\mu_{X,N}$ for some $N \in \mathcal{N}(X)$ and have a gaussian density with respect to it. We also introduce a subfunctor $\mathcal{F}'$ of $\mathcal{F}$ made of functions that *grow moderately* (i.e. at most polynomially) with respect to the mean, in such a way that the integral (1) is always convergent. Reference [9] called a triple $(\mathbf{S}, \mathcal{E}, \mathcal{N})$ *sufficiently rich* when there are "enough supports", in the sense that one can perform marginalization and conditioning with respect to projections on subspaces generated by elements of at least two different bases of E. In this case, we showed that every 1-cocycle φ, with coefficients in $\mathcal{F}'(\mathcal{Q})$, there are real constants a and c such that, for every $X \in \operatorname{Ob}\mathbf{S}$ and every gaussian law ρ with support E_X and variance Σ_ρ (a nondegenerate, symmetric, positive bilinear form on E_X^*),

$$\Phi_X(\rho) = a \det(\Sigma_\rho) + c. \dim(E_X). \tag{4}$$

Moreover, φ its completely determined by its behavior on nondegenerate laws. (The measure $\mu_X = \mu_{X,E_X}$ is enough to define the determinant $\det(\Sigma_\rho)$ [9, Sec. 11.2.1], but the latter can also be computed w.r.t. a basis of E_X such that $M|_{E_X}$ is represented by the identity matrix.)

3 An Extended Model

3.1 Information Structure, Supports, and Reference Measures

In this section, we introduce a more general model, that allows us to "mix" discrete and continuous variables. It is simply the product of a structure of discrete observables and a structure of continuous ones.

Let $(\mathbf{S}_d, \mathcal{E}_d)$ be a finite information structure, such that for every $n \in \mathbb{N}$, there exist $Y_n \in \operatorname{Ob}\mathbf{S}_d$ with $|E_{Y_n}| \cong \{1, 2, ..., n\} =: [n]$, and for every $X \in \operatorname{Ob}\mathbf{S}_d$, there is a $Z \in \operatorname{Ob}\mathbf{S}_d$ that can be written as non-degenerate product and such that

$Z \to X$; this implies that (3) holds for every $W \in \operatorname{Ob} \mathbf{S}_d$. Let $(\mathbf{S}_c, \mathcal{E}_c, \mathcal{N}_c)$ be a *sufficiently rich triple* in the sense of the previous section, associated to a vector space E with metric M, so that (4) holds for every $X \in \operatorname{Ob} \mathbf{S}_c$.

Let $(\mathbf{S}, \mathcal{E} : \mathbf{S} \to \mathbf{Meas})$ be the product $(\mathbf{S}_c, \mathcal{E}_c) \times (\mathbf{S}_d, \mathcal{E}_d)$ in the category of information structures, see [10, Prop. 2.2.2]. By definition, every object $X \in \mathbf{S}$ has the form $\langle X_c, X_d \rangle$ for $X_c \in \operatorname{Ob} \mathbf{S}_c$ and $X_d \in \operatorname{Ob} \mathbf{S}_d$, and $\mathcal{E}(X) = \mathcal{E}(X_1) \times \mathcal{E}(X_2)$, and there is an arrow $\pi : \langle X_1, X_2 \rangle \to \langle Y_1, Y_2 \rangle$ in $\mathbf{S}$ if and only if there exist arrows $\pi_1 : X_1 \to Y_1$ in $\mathbf{S}_c$ and $\pi_2 : X_2 \to Y_2$ in $\mathbf{S}_d$; under the functor $\mathcal{E}$, such π is mapped to $\mathcal{E}_c(\pi_1) \times \mathcal{E}_d(\pi_2)$. There is an embedding in the sense of information structures, see [9, Sec. 1.4], $\mathbf{S}_c \hookrightarrow \mathbf{S}$, $X \to \langle X, \mathbf{1} \rangle$; we call its image the "continuous sector" of $\mathbf{S}$; we write X instead of $\langle X, \mathbf{1} \rangle$ and $\mathcal{E}(X)$ instead of $\mathcal{E}(\langle X, \mathbf{1} \rangle) = E(X) \times \{*\}$. The "discrete sector" is defined analogously.

We extend the sheaf of supports $\mathcal{N}_c$ to the whole $\mathbf{S}$ setting $\mathcal{N}_d(Y) = 2^Y \setminus \{\emptyset\}$ when $Y \in \operatorname{Ob} \mathbf{S}_d$, and then $\mathcal{N}(Z) = \{ A \times B \mid A \in N_c(X), B \in N_d(Y) \}$ for any $Z = \langle X, Y \rangle \in \operatorname{Ob} \mathbf{S}$. The resulting $\mathcal{N}$ is a functor on $\mathbf{S}$ closed under projections and intersections, that contains the fibers of every projection.

For every $X \in \operatorname{Ob} \mathbf{S}$ and $N \in \mathcal{N}(X)$, there is a unique reference measure $\mu_{X,N}$ compatible with M: on the continuous sector it is the Lebesgue measure given by the metric M on the affine subspaces of E, on the discrete sector it is the counting measure restricted to N, and for any product $A \times B \subset E_X \times E_Y$, with $X \in \operatorname{Ob} \mathbf{S}_c$ and $Y \in \operatorname{Ob} \mathbf{S}_d$, it is just $\sum_{y \in B} \mu_A^y$, where μ_A^y is the image of μ_A under the inclusion $A \hookrightarrow A \times B$, $a \mapsto (a, y)$. We write μ_X instead of μ_{X, E_X}.

The disintegration of the counting measure into counting measures is trivial. The disintegration of the Lebesgue measure $\mu_{V,N}$ on a support $N \subset E_V$ under the projection $\pi^{WV} : E_V \to E_W$ of vector spaces is given by the Lebesgue measures on the fibers $(\pi^{W,V})^{-1}(w)$, for $w \in E_W$. We recall that we are in a framework where E_V and E_W are identified with subspaces of E, which has a metric M; the disintegration formula is just Fubini's theorem.

To see that disintegrations exist under any arrow of the category, consider first an object $Z = \langle X, Y \rangle$ and arrows $\tau : Z \to X$ and $\tau' : Z \to Y$, when $X \in \operatorname{Ob} \mathbf{S}_c$ and $Y \in \operatorname{Ob} \mathbf{S}_d$. By definition $E_Z = E_Y \times E_X = \cup_{y \in E_Y} E_X \times \{y\}$, and the canonical projections $\pi^{YZ} : E_Z \to E_Y$ and $\pi^{XZ} : E_Z \to E_X$ are the images under $\mathcal{E}$ of τ and τ', respectively. Set $E_{Z,y} := (\pi^{YZ})^{-1}(y) = E_X \times \{y\}$ and $E_{Z,x} := (\pi^{XZ})^{-1}(x) = \{x\} \times E_Y$. According to the previous definitions, E_Z has reference measure $\mu_Z = \sum_{y \in E_Y} \mu_X^y$, where μ_X^y is the image of μ_X under the inclusion $E_X \to E_X \times \{y\}$. Hence by definition, $\{\mu_X^y\}_{y \in E_Y}$ is (π^{YZ}, μ_Y)-disintegration of μ_Z. Similarly, μ_Z has as (π^{XZ}, μ_X)-disintegration the family of measures $\{\mu_Y^x\}$, where each μ_Y^x is the counting measure restricted on the fiber $E_{Z,x} \cong E_Y$.

More generally, the disintegration of reference measure $\mu_{Z, A \times B} = \sum_{y \in B} \mu_A^y$ on a support $A \times B$ of $Z = \langle X, Y \rangle$ under the arrow $\langle \pi_1, \pi_2 \rangle : Z \to Z' = \langle X', Y' \rangle$ is the collection of measures $(\mu_{Z, A \times B, x', y'})_{(x', y') \in \pi_1(A) \times \pi_2(B)}$ such that

$$\mu_{Z, A \times B, x', y'} = \sum_{y \in \pi_2^{-1}(y')} \mu_{X, A, x'}^y \tag{5}$$

where $\mu^y_{X,A,x'}$ is the image measure, under the inclusion $E_X \to E_X \times \{y\}$, of the measure $\mu_{X,A,x'}$ that comes from the $(\pi_1, \mu_{X',\pi_1(A)})$-disintegration of $\mu_{X,A}$.

3.2 Probability Laws and Probabilistic Functionals

Consider the subfunctor $\Pi(\mathcal{N})$ of Π that associates to each $X \in \operatorname{Ob}\mathbf{S}$ the set $\Pi(X;\mathcal{N})$ of probability measures on $\mathcal{E}(X)$ that are absolutely continuous with respect to the reference measure $\mu_{X,N}$ on some $N \in \mathcal{N}(X)$. We define the *(affine) support* or *carrier* of ρ, denoted $A(\rho)$, as the unique $A \in \mathcal{N}(X)$ such that $\rho \ll \mu_{X,A}$.

In this work, we want to restrict our attention to subfunctors $\mathcal{Q} \subset \Pi(\mathcal{N})$ of probability laws such that:

1. $\mathcal{Q}$ is adapted;
2. for each $\rho \in \mathcal{Q}(X)$, the differential entropy $S_{\mu_{A(\rho)}}(\rho) := -\int \log \frac{\mathrm{d}\rho}{\mathrm{d}\mu_{A(\rho)}}\,\mathrm{d}\rho$ exists i.e. it is finite;
3. when restricted to probabilities in $\mathcal{Q}(X)$ with the same carrier A, the differential entropy is a continuous functional in the total variation norm;
4. for each $X \in \operatorname{Ob}\mathbf{S}_c$ and each $N \in \mathcal{N}(X)$, the *gaussian mixtures* carried by N are contained in $\mathcal{Q}(X)$—cf. next section.

Problem 1. The characterization of functors $\mathcal{Q}$ that satisfy properties 1–4.

Below we use kernel estimates, which interact nicely with the total variation norm. This norm is defined for every measure ρ on $(E_X, \mathfrak{B}_X)$ by $\|\rho\|_{TV} = \sup_{A\in\mathfrak{B}_X} |\rho(A)|$. Let φ be a real-valued functional defined on $\Pi(E_X, \mu_A)$, and $L^1_1(A, \mu_A)$ the space of functions $f \in L^1(A, \mu_A)$ with total mass 1 i.e. $\int_A f\,\mathrm{d}\mu_A = 1$; the continuity of φ in the total variation distance is equivalent to the continuity of $\tilde{\varphi} : L^1_1(A, \mu_A) \to \mathbb{R}, f \mapsto \varphi(f \cdot \mu_A)$ in the L^1-norm, because of Scheffe's identity [5, Thm. 45.4].

The characterization referred to in Problem 1 might involve the densities or the moments of the laws. It is the case with the main result that we found concerning convergence of the differential entropy and its continuity in total variation [6]. Or it might resemble Otáhal's result [8]: if densities $\{f_n\}$ tend to f in $L^\alpha(\mathbb{R}, \mathrm{d}x)$ and $L^\beta(\mathbb{R}, \mathrm{d}x)$, for some $0 < \alpha < 1 < \beta$, then $-\int f_n(x) \log f_n(x)\,\mathrm{d}x \to -\int f(x)\log f(x)\,\mathrm{d}x$.

For each $X \in \operatorname{Ob}\mathbf{S}$, let $\mathcal{F}(X)$ be the vector space of measurable functions of (ρ, μ_M), equivalently $(\mathrm{d}\rho/d\mu_{A(\rho)}, \mu_M)$, where ρ is an element of $\mathcal{Q}(X)$, μ_M is a global determination of reference measure on any affine subspace given by the metric M, and $\mu_{A(\rho)}$ is the corresponding reference measure on the carrier of ρ under this determination.

We want to restrict our attention to functionals for which the action (1) is integrable. Of course, these depends on the answer to Problem 1.

Problem 2. What are the appropriate restrictions on the functionals $\mathcal{F}(X)$ to guarantee the convergence of (1)?

4 Computation of 1-cocycles

4.1 A Formula for Gaussian Mixtures

Let $\mathcal{Q}$ be a probability functor satisfying conditions (1)–(4) in Subsect. 3.2, and $\mathcal{G}$ a linear subfunctor of $\mathcal{F}$ such that (1) converges for laws in $\mathcal{Q}$. In this section, we compute $H^1(\mathbf{S}, \mathcal{G})$.

Consider a generic object $Z = \langle X, Y\rangle = \langle X, 1\rangle \wedge \langle 1, Y\rangle$ of $\mathbf{S}$; we write everywhere X and Y instead of $\langle X, 1\rangle$ and $\langle 1, Y\rangle$. We suppose that E_X is an Euclidean space of dimension d. Remind that E_Y is a finite set and $E_Z = E_X \times E_Y$. Let $\{G_{M_y,\Sigma_y}\}_{y\in E_Y}$ be gaussian densities on E_X (with mean M_y and covariance Σ_y), and $p : E_Y \to [0, 1]$ a density on E_Y; then $\rho = \sum_{y\in E_Y} p(y)G_{M_y,\Sigma_y}\mu_X^y$ is a probability measure on E_Z, absolutely continuous with respect to μ_Z, with density $r(x, y) := p(y)G_{M_y,\Sigma_y}(x)$. We have that $\pi_*^{YZ}\rho$ has density p with respect to the counting measure μ_Y, whereas $\pi_*^{XZ}\rho$ is absolutely continuous with respect to μ_X (see [3, Thm. 3]) with density

$$\mu_Z^x(r) = \int_{E_X} \frac{\mathrm{d}\rho}{\mathrm{d}\mu_Z}\,\mathrm{d}\mu_Z^x = \sum_{y\in E_Y} p(y)G_{M_y,\Sigma_y}(x). \tag{6}$$

i.e. it is a *gaussian mixture*. For conciseness, we utilize here linear-functional notation for some integrals, e.g. $\mu_Z^x(r)$. The measure ρ has a π^{XZ}-disintegration into probability laws $\{\rho_x\}_{x\in E_X}$, such that each ρ_x is concentrated on $E_{Z,x} \cong E_Y$ and

$$\rho_x(x, y) = \frac{p(y)G_{M_y,\Sigma_y}(x)}{\sum_{y\in E_Y} p(y)G_{M_y,\Sigma_y}(x)}. \tag{7}$$

In virtue of the cocycle condition (remind that $\Phi_Z = \varphi_Z[Z]$),

$$\Phi_Z(\rho) = \varphi_Z[Y](\rho) + \sum_{y\in E_Y} p(y)\varphi_Z[X](G_{M_y,\Sigma_y}\mu_X^y) \tag{8}$$

Locality implies that $\varphi_Z[Y](\rho) = \Phi_Y(\pi_*^{YZ}\rho)$, and the later equals $-b\sum_{y\in E_Y} p(y)\log p(y)$, for some $b \in \mathbb{R}$, in virtue of the characterization of cocycles restricted to the discrete sector. Similarly, $\varphi_Z[X](G_{M_y,\Sigma_y}\mu_X^y) = \Phi_X(G_{M_y,\Sigma_y}) = a\log\det(\Sigma_y) + c\dim E_X$ for some $a, c \in \mathbb{R}$, since our hypotheses are enough to characterize the value of any cocycle restricted to the gaussian laws on the continuous sector. Hence

$$\Phi_Z(\rho) = -b\sum_{y\in E_Y} p(y)\log p(y) + \sum_{y\in E_Y} p(y)(a\log\det(\Sigma_y) + c\dim E_X). \tag{9}$$

Remark that the same argument can be applied to any densities $\{f_y\}_{y\in Y}$ instead $\{G_{M_y,\Sigma_y}\}$. Thus, it is enough to determine Φ_X on general densities, for each $X \in \mathbf{S}_c$, to characterize completely the cocycle φ.

The cocycle condition also implies that

$$\Phi_Z(\rho) = \Phi_X(\pi^{XZ}\rho) + \int_{E_X} \varphi_Z[Y](\rho_x)\,\mathrm{d}\pi_*^{XZ}\rho(x). \tag{10}$$

The law $\pi_*^{XZ}\rho$ is a composite gaussian, and the law ρ_x is supported on the discrete space $E_{Z,x} \cong E_Y$, with density

$$(x,y) \mapsto \rho_x(y) = \frac{p(y)G_{M_y,\Sigma_y}(x)}{\sum_{y' \in E_Y} p(y')G_{M_y,\Sigma_y}(x)} = \frac{r(x,y)}{\mu_Z^x(r)}. \tag{11}$$

Using again locality and the characterization of cocycles on the discrete sector, we deduce that

$$\varphi_Z[Y](\rho_x) = \Phi_Y(\pi_*^{YZ}\rho_x) = -b \sum_{y \in E_Y} \rho_x(y) \log \rho_x(y). \tag{12}$$

A direct computation shows that $\sum_{y \in E_Y} \rho_x(y) \log \rho_x(y)$ equals:

$$\sum_{y \in E_Y} p(y) \log p(y) - \sum_{y \in E_Y} p(y) S_{\mu_X}(G_{M_y,\Sigma_y}) + S_{\mu_X}(\mu_Z^x(r)), \tag{13}$$

where S_{μ_X} is the differential entropy, defined for any $f \in L^1(E_X, \mu_X)$ by $S_{\mu_X}(f) = -\int_{E_X} f(x) \log f(x)\, \mathrm{d}\mu_X(x)$.

It is well known that $S_{\mu_X}(G_{M,\Sigma}) = \frac{1}{2}\log\det\Sigma + \frac{\dim E_X}{2}\log(2\pi e)$.

Equating the right hand sides of (9) and (10), we conclude that

$$\Phi_X(\pi^{XZ}\rho) = \sum_{y \in E_Y} p(y)\left((a - \frac{b}{2})\log\det\Sigma_y + cd - \frac{bd}{2}\log(2\pi e)\right) + bS_{\mu_X}(\mu_Z^x(r)). \tag{14}$$

4.2 Kernel Estimates and Main Result

Equation 17 gives an explicit value to the functional $\Phi_X \in \mathcal{F}(X)$ evaluated on a gaussian mixture. Any density in $L^1(E_X, \mu_X)$ can be approximated by a random mixture of gaussians. This approximation is known as a kernel estimate.

Let $X_1, X_2, ...$ be a sequence of independently distributed random elements of $\mathbb{R}^d$, all having a common density f with respect to the Lebesgue measure λ_d. Let K be a nonnegative Borel measurable function, called *kernel*, such that $\int K \,\mathrm{d}\lambda_d = 1$, and $(h_n)_n$ a sequence of positive real numbers. The *kernel estimate* of f is given by

$$f_n(x) = \frac{1}{nh_n^d} \sum_{i=1}^n K\left(\frac{x - X_i}{h_n}\right). \tag{15}$$

The distance $J_n = \int_E |f_n - f| \,\mathrm{d}\lambda_d$ is a random variable, invariant under arbitrary automorphisms of E. The key result concerning these estimates [4, Ch. 3, Thm. 1] says, among other things, that $J_n \to 0$ in probability as $n \to \infty$ for *some* f if and only if $J_n \to 0$ almost surely as $n \to \infty$ for *all* f, which holds if and only if $\lim_n h_n = 0$ and $\lim_n nh_n^d = \infty$.

Theorem 1. *Let φ be a 1-cocycle on* $\mathbf{S}$ *with coefficients in* $\mathcal{G}$*,* X *an object in* $\mathbf{S}_c$*, and* ρ *a probability law in* $\mathcal{Q}(X)$ *absolutely continuous with respect to* μ_X*. Then, there exist real constants* c_1, c_2 *such that*

$$\Phi_X(\rho) := \varphi_X[X](\rho) = c_1 S_{\mu_A}(\rho) + c_2 \dim E_X. \tag{16}$$

Proof. By hypothesis. $X = \langle X, \mathbf{1} \rangle$ is an object in the continuous sector of $\mathbf{S}$. Let f be any density of ρ with respect to μ_X, $(X_n)_{n\in\mathbb{N}}$ an i.i.d sequence of points of E_X with law ρ, and (h_n) any sequence such that $h_n \to 0$ and $nh_n^d \to \infty$. Let $(X_n(\omega))_n$ be any realization of the process such that f_n tend to f in L^1. We introduce, for each $n \in \mathbb{N}$, the kernel estimate (15) evaluated at $(X_n(\omega))_n$, taking K equal to the density of a standard gaussian. Each f_n is the density of a composite gaussian law ρ_n equal to $n^{-1}\sum_{i=1}^n G_{X_i(\omega),h_n^2 I}$. This can be "lifted" to $Z = \langle X, Y_n \rangle$, this is, there exists a law $\tilde{\rho}$ on $E_Z := E_X \times [n]$ with density $r(x,i) = p(i) G_{X_i(\omega),h_n^2 I}(x)$, where $p : [n] \to \mathbb{R}$ is taken to be the uniform law. The arguments of Sect. 4 then imply that

$$\Phi_X(\rho_n) = 2d\left(a - \frac{b}{2}\right)\log h_n + cd - \frac{bd}{2}\log(2\pi e) + bS_{\mu_A}(\rho_n). \tag{17}$$

In virtue of the hypotheses on $\mathcal{Q}$, $S_{\mu_A}(\rho)$ is finite and $S_{\mu_A}(\rho_n) \to S_{\mu_A}(\rho)$. Since Φ_X is continuous when restricted to $\Pi(A, \mu_A)$ and $\Phi_X(f)$ is a real number, we conclude that necessarily $a = b/2$. The statement is then just a rewriting of (17).

This is the best possible result: the dimension is an invariant associated to the reference measure, and the entropy depends on the density.

References

1. Aczél, J., Daróczy, Z.: On Measures of Information and Their Characterizations. Mathematics in Science and Engineering. Academic Press, Cambridge (1975)
2. Baudot, P., Bennequin, D.: The homological nature of entropy. Entropy **17**(5), 3253–3318 (2015)
3. Chang, J.T., Pollard, D.: Conditioning as disintegration. Statistica Neerlandica **51**(3), 287–317 (1997)
4. Devroye, L., Györfi, L.: Nonparametric Density Estimation: The L_1 View. Wiley Series in Probability and Mathematical Statistics. Wiley, Hoboken (1985)
5. Devroye, L., Györfi, L.: Distribution and density estimation. In: Györfi, L. (ed.) Principles of Nonparametric Learning. ICMS, vol. 434, pp. 211–270. Springer, Vienna (2002). https://doi.org/10.1007/978-3-7091-2568-7_5
6. Ghourchian, H., Gohari, A., Amini, A.: Existence and continuity of differential entropy for a class of distributions. IEEE Commun. Lett. **21**(7), 1469–1472 (2017)
7. Kolmogorov, A., Shiryayev, A.: Selected Works of A. N. Kolmogorov. Volume III: Information Theory and the Theory of Algorithms. Mathematics and Its Applications, Kluwer Academic Publishers (1993)
8. Otáhal, A.: Finiteness and continuity of differential entropy. In: Mandl P., Huskovsa, M. (eds.) Asymptotic Statistics, pp. 415–419. Springer, Heidelberg (1994). https://doi.org/10.1007/978-3-642-57984-4_36

9. Vigneaux, J.P.: Topology of statistical systems: a cohomological approach to information theory. Ph.D. thesis, Université de Paris (2019). https://hal.archives-ouvertes.fr/tel-02951504v1
10. Vigneaux, J.P.: Information structures and their cohomology. Theory Appl. Categ. **35**(38), 1476–1529 (2020)

Belief Propagation as Diffusion

Olivier Peltre(✉)

Faculté Jean Perrin (LML), Université d'Artois, Rue Jean Souvraz,
62307 Lens Cedex, France

Abstract. We introduce novel belief propagation algorithms to estimate the marginals of a high dimensional probability distribution. They involve natural (co)homological constructions relevant for a localised description of statistical systems.

Introduction

Message-passing algorithms such as belief propagation (BP) are parallel computing schemes that try to estimate the marginals of a high dimensional probability distribution. They are used in various areas involving the statistics of a large number of interacting random variables, such as computational thermodynamics [5,10], artificial intelligence [11,15,21], computer vision [18] and communications processing [3,4].

We have shown the existence of a non-linear correspondence between BP algorithms and discrete integrators of a new form of continuous-time diffusion equations on belief networks [13,14]. Practical contributions include (a) regularised BP algorithms for any time step or *diffusivity*[1] coefficient $0 < \varepsilon < 1$, and (b) a canonical *Bethe diffusion flux* that regularises GBP messages by new Möbius inversion formulas in degree 1[2].

The purpose of this text is to describe the structure of belief networks as concisely as possible, with the geometric operations that appear in our rewriting of BP equations. An open-sourced python implementation, hosted on github at opeltre/topos was also used to conduct benchmarks showing the importance of chosing $\varepsilon < 1$.

In the following, we denote by:

- $\Omega = \{i, j, k, \dots\}$ a finite set of indices (e.g. atoms, neurons, pixels, bits ...)
- x_i the microstate of atom i, valued in a finite set E_i
- x_Ω the microstate of the global system, valued in $E_\Omega = \prod_{i \in \Omega} E_i$

Part of this work was supported by a PhD funding from Université de Paris.

[1] This coefficient ε would appear as an exponent of messages in the usual multiplicative writing of BP equations. Diffusivity relates energy density gradients to heat fluxes in physics, as in $\varphi = -\varepsilon \cdot \nabla(u)$.

[2] Generalised belief propagation = BP on hypergraphs, see [21] for the algorithm. Our algorithm 2 exponentiates their messages $m_{\alpha\beta}$ by the coefficients $c_\alpha \in \mathbb{Z}$ appearing in the Bethe-Kikuchi local approximation of free energy.

F. Nielsen and F. Barbaresco (Eds.): GSI 2021, LNCS 12829, pp. 547–556, 2021.
https://doi.org/10.1007/978-3-030-80209-7_59

The statistical state of the system is described by a probability distribution p_Ω on E_Ω. We write $\Delta_\Omega = \mathrm{Prob}(E_\Omega)$ for the convex space of statistical states.

1 Graphical Models

Definition 1. *A* hypergraph (Ω, K) *is a set of* vertices Ω *and a set of* faces[3] $K \subseteq \mathcal{P}(\Omega)$.

Let us denote by x_α the microstate of a face $\alpha \subseteq \Omega$, valued in $E_\alpha = \prod_{i\in\alpha} E_\alpha$. For every $\beta \subseteq \alpha$ in $\mathcal{P}(\Omega)$, we have a canonical projection or *restriction*[4] map:

$$\pi^{\beta\alpha} : E_\alpha \to E_\beta$$

We simply write x_β for the restriction of x_α to a subface β of α.

Definition 2. *A* graphical model $p_\Omega \in \Delta_\Omega$ *on the hypergraph* (Ω, K) *is a positive probability distribution on* E_Ω *that factorises as a product of positive local factors over faces:*

$$p_\Omega(x_\Omega) = \frac{1}{Z_\Omega} \prod_{\alpha\in K} f_\alpha(x_\alpha) = \frac{1}{Z_\Omega} \mathrm{e}^{-\sum_\alpha h_\alpha(x_\alpha)}$$

We denote by $\Delta_K \subseteq \Delta_\Omega$ *the subspace of graphical models on* (Ω, K).

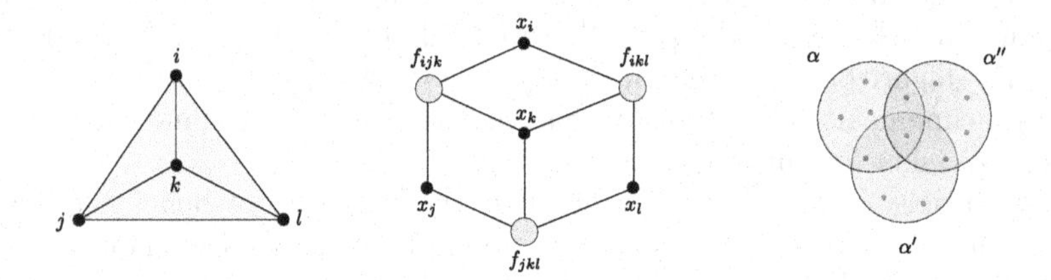

Fig. 1. Graphical model $p_{ijkl}(x_{ijkl}) = f_{ijk}(x_{ijk}) \cdot f_{ikl}(x_{ikl}) \cdot f_{jkl}(x_{jkl})$ with its factor graph representation (middle) on a simplicial complex K formed by joining 3 triangles at a common vertex and called 2-*horn* Λ^2 of the 3-simplex (left). The situation is equivalent when K is a three-fold covering of Ω by intersecting regions $\alpha, \alpha', \alpha''$ (right).

[3] Also called hyperedges, or regions. A *graph* is a hypergraph with only hyperedges of cardinality 2. A *simplicial complex* is a hypergraph such that any subset of a face is also a face. A *lattice* is a hypergraph closed under $\cap$ and $\cup$. We shall mostly be interested in *semi-lattices*, closed only under intersection, of which simplicial complexes are a special case.

[4] The contravariant functor $E : \mathcal{P}(\Omega)^{op} \to \mathbf{Set}$ of microstates defines a sheaf of sets over Ω.

A graphical model p_Ω for (Ω, K) is also called *Gibbs state* of the associated energy function or hamiltonian $H_\Omega : E_\Omega \to \mathbb{R}$:

$$H_\Omega(x_\Omega) = \sum_{\alpha \in K} h_\alpha(x_\alpha)$$

The normalisation factor of the Gibbs density e^{-H_Ω} is computed by the partition function $Z_\Omega = \sum_{x_\Omega} \mathrm{e}^{-H_\Omega(x_\Omega)}$. The free energy $F_\Omega = -\ln Z_\Omega$ and partition function generate most relevant statistical quantities in their derivatives[5]. They are however not computable in practice, the sum over microstates scaling exponentially in the number of atoms.

Message-passing algorithms rely on local structures induced by K to estimate marginals, providing with an efficient alternative [5,10] to Markov Chain Monte Carlo methods such as Hinton's contrastive divergence algorithm commonly used for training restricted Boltzmann machines [15,17]. They are also related to local variational principles involved in the estimation of F_Ω [13,14,21] by Bethe approximation [2,5].

We showed in [14] that message-passing explores a subspace of potentials (u_α) related to equivalent factorisations of p_Ω, until an associated collection of local probabilities (q_α) is consistent. Two fundamental operations constraining this non-linear correspondence are introduced below. They consist of a differential d associated to a *consistency* constraint, and its adjoint boundary $\delta = d^*$ enforcing a dual *energy conservation* constraint. These operators relate graphical models to a statistical (co)homology theory, in addition to generating the BP equations.

2 Marginal Consistency

In the following, we suppose given a hypergraph (Ω, K) closed under intersection:

$$\alpha \cap \beta \in K \qquad \text{for all} \qquad \alpha, \beta \in K$$

We denote by Δ_α the space of probability distributions on E_α for all $\alpha \in K$. Given a graphical model $p_\Omega \in \Delta_\Omega$, the purpose of belief propagation algorithms is to efficiently approximate the collection of true marginals $p_\alpha \in \Delta_\alpha$ for $\alpha \in K$ by local beliefs $q_\alpha \in \Delta_\alpha$, in a space Δ_0 of dimension typically much smaller than Δ_Ω.

Definition 3. *We call* belief *over* (Ω, K) *a collection* $q \in \Delta_0$ *of local probabilities over faces, where:*

$$\Delta_0 = \prod_{\alpha \in K} \Delta_\alpha$$

[5] Letting μ_H denote the image by H of the counting measure on microstates, $Z_\Omega^\theta = \int_{\lambda \in \mathbb{R}} \mathrm{e}^{-\theta\lambda} \mu_H(d\lambda)$ is the Laplace transform of μ_H with respect to inverse temperature $\theta = 1/k_B T$. In [14] we more generally consider free energy as a functional $A_\Omega \to \mathbb{R}$ whose differential at $H_\Omega \in A_\Omega$ is the Gibbs state $p_\Omega \in A_\Omega^*$.

Definition 4. *For every $\beta \subseteq \alpha$ the* marginal *or partial integration map* $\Sigma^{\beta\alpha}$: $\Delta_\alpha \to \Delta_\beta$ *is defined by:*

$$\Sigma^{\beta\alpha} q_\alpha(x_\beta) = \sum_{y \in E_{\alpha \setminus \beta}} q_\alpha(x_\beta, y)$$

Definition 5. Consistent *beliefs span the convex subset* $\Gamma \subseteq \Delta_0$ *defined by marginal consistency constraints*[6]*:*

$$q_\beta = \Sigma^{\beta\alpha}(q_\alpha) \qquad \text{for all} \quad \beta \subseteq \alpha$$

The true marginals $(p_\alpha) \in \Delta_0$ of a global density $p_\Omega \in \Delta_\Omega$ are *always consistent.* However their symbolic definition $p_\alpha = \Sigma^{\alpha\Omega} p_\Omega$ involves a sum over fibers of $E_{\Omega\setminus\alpha}$, not tractable in practice. Message-passing algorithms instead explore a parameterised family of beliefs $q \in \Delta_0$ until meeting the consistency constraint surface $\Gamma \subseteq \Delta_0$.

Let us denote by A^*_α the space of linear measures on E_α for all $\alpha \subseteq \Omega$, and by:

$$\Sigma^{\beta\alpha} : A^*_\alpha \to A^*_\beta$$

the partial integration map.

Definition 6. *We call n*-density *over* (Ω, K) *an element* $\lambda \in A^*_n$ *of local measures indexed by ordered chains of faces, where:*

$$A^*_n = \prod_{\alpha_0 \supset \cdots \supset \alpha_n} A^*_{\alpha_n}$$

The marginal consistency constraints are expressed by a differential operator[7] d on the graded vector space $A^*_\bullet = \prod_n A^*_n$ of densities over (Ω, K):

$$A^*_0 \xrightarrow{\ d\ } A^*_1 \xrightarrow{\ d\ } \dots \xrightarrow{\ d\ } A^*_n$$

Definition 7. *The* differential $d : A^*_0 \to A^*_1$ *acts on a density* $(\lambda_\alpha) \in A^*_0$ *by:*

$$d(\lambda)_{\alpha\beta} = \lambda_\beta - \Sigma^{\beta\alpha}\lambda_\alpha$$

Consistent *densities* $\lambda \in [A^*_0]$ *satisfy* $d\lambda = 0$, *and called* 0*-cocycles.*

The space of consistent beliefs $\Gamma \subseteq [A^*_0]$ is the intersection of Ker(d) with $\Delta_0 \subseteq A^*_0$. True marginals define a convex map $\Delta_\Omega \to \Gamma$, restriction[8] of a linear surjection $A^*_\Omega \to [A^*_0]$. Consistent beliefs $q \in \Gamma$ acting as for global distributions $p_\Omega \in \Delta_\Omega$, marginal diffusion iterates over a smooth subspace of Δ_0, diffeomorphic to equivalent parameterisations of a graphical model p_Ω, until eventually reaching Γ.

[6] Equivalently, Γ is the projective limit of the functor $\Delta : K^{op} \to$ **Top** defined by local probabilities and marginal projections, or space of global sections of the sheaf of topological spaces Δ over (Ω, K).

[7] Cohomology sequences of this kind were considered by Grothendieck and Verdier [19], see also [8].

[8] Note the image of Δ_Ω inside Γ can be a strict convex polytope of Γ, and consistent $q \in \Gamma$ do not always admit a positive preimage $q_\Omega \in \Delta_\Omega$ [1,20].

3 Energy Conservation

Graphical models parameterise a low dimensional subspace of Δ_Ω, but Definition 2 is not injective in the local factors f_α or local potentials $u_\alpha = -\ln f_\alpha$. The fibers of this parameterisation can be described linearly at the level of potentials, and correspond to homology classes of the codifferential operator $\delta = d^*$.

We denote by A_α the algebra of real functions on E_α for all $\alpha \subseteq \Omega$, and by:

$$j_{\alpha\beta} : A_\alpha \to A_\beta$$

the natural extension[9] of functions pulled from E_β to E_α by the restriction $x_\alpha \mapsto x_\beta$.

Definition 8. *We let $\delta = d^*$ denote the adjoint of d, defined by duality:*

$$A_0 \xleftarrow{\delta} A_1 \xleftarrow{\delta} \dots \xleftarrow{\delta} A_n$$

Proposition 1. *The* divergence $\delta : A_1 \to A_0$ *dual of $d : A_0^* \to A_1^*$, acts on $\varphi \in A_1$ by:*

$$\delta(\varphi)_\beta = \sum_{\alpha \supseteq \beta} \varphi_{\alpha\beta} - \sum_{\gamma \subseteq \beta} j_{\beta\gamma} \varphi_{\beta\gamma}$$

Proof. Let $\lambda \in A_0^*$ and $\varphi \in A_1$. The duality bracket $A_0^* \otimes A_0 \to \mathbb{R}$ is naturally defined by sum of local duality brackets $A_\beta^* \otimes A_\beta \to \mathbb{R}$, which correspond to integration of local measures against observables:

$$\langle \lambda \,|\, \delta\varphi \rangle = \sum_{\beta \in K} \langle \lambda_\beta \,|\, \delta\varphi_\beta \rangle = \sum_{\beta \in K} \sum_{x_\beta \in E_\beta} \lambda_\beta(x_\beta) \delta\varphi_\beta(x_\beta)$$

Substituting with the expression of $\delta\varphi$ we get[10]:

$$\begin{aligned} \langle \lambda \,|\, \delta\varphi \rangle &= \sum_{\beta \in K} \sum_{x_\beta \in E_\beta} \lambda_\beta(x_\beta) \Big(\sum_{\alpha \supseteq \beta} \varphi_{\alpha\beta}(x_\beta) - \sum_{\gamma \subseteq \beta} \varphi_{\beta\gamma}(x_\gamma) \Big) \\ &= \sum_{\alpha \supseteq \beta} \sum_{x_\beta \in E_\beta} \varphi_{\alpha\beta}(x_\beta) \lambda_\beta(x_\beta) - \sum_{\beta \supseteq \gamma} \sum_{x_\gamma \in E_\gamma} \varphi_{\beta\gamma}(x_\gamma) \sum_{y \in E_{\beta \setminus \gamma}} \lambda_\beta(x_\gamma, y) \end{aligned}$$

The factorisation of the rightmost sum by $\varphi_{\beta\gamma}(x_\gamma)$ reflects the duality of $\Sigma^{\beta\alpha}$ with $j_{\beta\gamma}$. Relabeling summation indices $\beta \supseteq \gamma$ as $\alpha \supseteq \beta$, we finally get:

$$\sum_{\alpha \supseteq \beta} \langle \lambda_\beta \,|\, \varphi_{\alpha\beta} \rangle - \sum_{\beta \supseteq \gamma} \langle \Sigma^{\gamma\beta} \lambda_\beta \,|\, \varphi_{\beta\gamma} \rangle = \sum_{\alpha \supseteq \beta} \langle \lambda_\beta - \Sigma^{\beta\alpha} \lambda_\alpha \,|\, \varphi_{\alpha\beta} \rangle$$

So that $\langle \lambda \,|\, \delta\varphi \rangle = \langle d\lambda \,|\, \varphi \rangle$ for all $\lambda \in A_0^*$ and all $\varphi \in A_1$.

[9] Functions on $E_\beta = \prod_{j \in \beta} E_j$ can be viewed as functions on $E_\alpha = \prod_{i \in \alpha}$ that do not depend on the state of x_i for $i \in \alpha \setminus \beta$. Therefore A_β is essentially a subspace of A_α and $j_{\alpha\beta}$ an inclusion.

[10] In this substitution, we simply wrote $\varphi_{\beta\gamma}(x_\gamma)$ for $j_{\beta\gamma}(\varphi_{\beta\gamma})(x_\beta)$, as $j_{\beta\gamma} : A_\gamma \to A_\beta$ is an inclusion.

Consider the total energy map $\zeta_\Omega : A_0 \to A_\Omega$ defined by:

$$\zeta_\Omega(u) = \sum_{\alpha \in K} u_\alpha$$

We have left injections $j_{\Omega\alpha}$ implicit, viewing each $A_\alpha \subseteq A_\Omega$ as a subalgebra of A_Ω. Denoting by $A_K \subseteq A_\Omega$ the image of ζ_Ω, a graphical model $p_\Omega \in \Delta_K$ is then associated to $u \in A_0$ by normalising the Gibbs density $\mathrm{e}^{-\zeta_\Omega(u)}$, as in 2.

Theorem 1. *For all $u, u' \in A_0$ the following are equivalent [14, Chapter 5]:*

- *conservation of total energy $\sum_\alpha u'_\alpha = \sum_\alpha u_\alpha$ in A_Ω,*
- *there exists $\varphi \in A_1$ such that $u' = u + \delta\varphi$ in A_0.*

Theorem 1 states that $\mathrm{Ker}(\zeta_\Omega)$ coincides with the image of the divergence $\delta A_1 \subseteq A_0$. The subspace of total energies $\mathrm{Im}(\zeta_\Omega) \simeq A_0/\mathrm{Ker}(\zeta_\Omega)$ is therefore isomorphic to the quotient $[A_0] = A_0/\delta A_1$, formed by homology classes of potentials $[u] = u + \delta A_1 \subseteq A_0$. Global observables of $A_K \subseteq A_\Omega$ can thus be represented by equivalence classes of local potentials in $[A_0]$, homology under δ giving a local characterisation for the fibers of ζ_Ω.

4 Diffusions

The local approach to the marginal estimation problem, given $p_\Omega = \frac{1}{Z_\Omega}\mathrm{e}^{-H_\Omega}$, consists of using a low dimensional map $A_0 \to \Delta_0$ as substitute for the high dimensional parameterisation $A_\Omega \to \Delta_\Omega$, until parameters $u \in A_0$ define a consistent belief $q \in \Gamma$ whose components $q_\alpha \in \Delta_\alpha$ estimate the true marginals p_α of p_Ω.

$$\begin{array}{ccccc}
\Delta_\Omega & \longrightarrow & \Gamma & \hookrightarrow & \Delta_0 \\
\uparrow & & \uparrow & & \uparrow \\
A_\Omega & \hookleftarrow & [A_0] & \twoheadleftarrow & A_0
\end{array}$$

Assume the hamiltonian is defined by $H_\Omega = \sum_\alpha h_\alpha$ for given $h \in A_0$. According to Theorem 1, parameters $u \in A_0$ will define the same total energy if and only if:

$$u = h + \delta\varphi$$

for some heat flux $\varphi \in \delta A_1$. The energy conservation constraint $[u] = [h]$ therefore restricts parameters to fibers of the bottom-right arrow in the above diagram. The rightmost arrow $A_0 \to \Delta_0$ is given by the equations:

$$q_\alpha = \frac{1}{Z_\alpha}\mathrm{e}^{-U_\alpha} \qquad \text{where} \qquad U_\alpha = \sum_{\beta \subseteq \alpha} u_\beta \tag{1}$$

The image of $[h]$ in Δ_0 is a smooth non-linear manifold of $\Delta_0 \subseteq A_0^*$, which may intersect the convex polytope $\Gamma = \mathrm{Ker}(d) \cap \Delta_0$ of consistent beliefs an unknown number of times. Such consistent beliefs in $\Gamma \subseteq \Delta_0$ are the fixed points of belief propagation algorithms. The central dashed vertical arrow therefore represents what they try to compute, although no privileged $q \in \Gamma$ may be defined from $[h] \in A_0$ in general.

Definition 9. *Given a* flux functional $\Phi : A_0 \to A_1$, *we call* diffusion *associated to* Φ *the vector field* $\delta\Phi$ *on* A_0 *defined by:*

$$\frac{du}{dt} = \delta\Phi(u) \tag{2}$$

Letting $q \in \Delta_0$ *be defined by (1), we say that* Φ *is* consistent *if* $q \in \Gamma \Rightarrow \Phi(u) = 0$, *and that* Φ *is* faithful *if it is consistent and* $\Phi(u) = 0 \Rightarrow q \in \Gamma$.

Consistent flux functionals Φ are constructed by composition with two remarkable operators $\zeta : A_0 \to A_0$, mapping potentials to local hamiltonians $u \mapsto U$, and $\mathcal{D} : A_0 \to A_1$, a non-linear analog of the differential $d : A_0^* \to A_1^*$, measuring inconsistency of the local beliefs defined by $U \mapsto q$ in (1). The definition of $\mathcal{D}$ involves a conditional form of free energy $\mathbb{F}^{\beta\alpha} : A_\alpha \to A_\beta$, which generates conditional expectation maps with respect to local beliefs by differentiation[11].

Definition 10. *We call* effective energy *the smooth map* $\mathbb{F}^{\beta\alpha} : A_\alpha \to A_\beta$ *defined by:*

$$\mathbb{F}^{\beta\alpha}(U_\alpha \mid x_\beta) = -\ln \sum_{y \in E_{\alpha\setminus\beta}} \mathrm{e}^{-U_\alpha(x_\beta, y)}$$

and effective energy gradient *the smooth map* $\mathcal{D} : A_0 \to A_1$ *defined by:*

$$\mathcal{D}(U)_{\alpha\beta} = U_\beta - \mathbb{F}^{\beta\alpha}(U_\alpha)$$

Letting $q = \mathrm{e}^{-U}$ denote local Gibbs densities, note that $q \in \Gamma \Leftrightarrow \mathcal{D}(U) = 0$ by:

$$\mathcal{D}(U)_{\alpha\beta} = \ln \left[\frac{\Sigma^{\beta\alpha} q_\alpha}{q_\beta} \right]$$

The map $u \mapsto U$ is a fundamental automorphism ζ of A_0, inherited from the partial order structure of K. Möbius inversion formulas define its inverse $\mu = \zeta^{-1}$ [7,14,16]. We have extended ζ and μ to automorphisms on the full complex $A_\bullet$ in [14, Chapter 3], in particular, ζ and μ also act naturally on A_1.

Definition 11. *The* zeta *transform* $\zeta : A_0 \to A_0$ *is defined by:*

$$\zeta(u)_\alpha = \sum_{\beta \subseteq \alpha} u_\beta$$

[11] The tangent map of $\mathcal{D}$ in turn yields differential operators $\nabla_q : A_0 \to A_1 \to \dots$ for all $q \in \Gamma$, whose kernels characterise tangent fibers $\mathrm{T}_q\Gamma$ pulled by the non-linear parameterisation (1), see [14, Chapter 6].

Algorithms. GBP and Bethe Diffusions[12].

Input: potential $u \in A_0$, diffusivity $\varepsilon > 0$, number of iterations n_{it}

Output: belief $q \in \Delta_0$

A. GBP ε-diffusion

```
1: for i = 0 ... n_it do
2:     U_α ← ζ(u)_α
3:     U_α ← U_α + ln Σe^{-U_α}
4:     Φ_αβ ← −𝒟(U)_αβ
5:
6:     u_α ← u_α + ε · δ(Φ)_α
7: end for
8: q_α ← e^{-U_α}
9: return q
```

B. Bethe ε-diffusion

```
1: for i = 0 ... n_it do
2:     U_α ← ζ(u)_α
3:
4:     Φ_αβ ← −𝒟(U)_αβ
5:     φ_αβ ← c_α · Φ_αβ
6:     u_α ← u_α + ε · δ(φ)_α
7: end for
8: q_α ← e^{-U_α}
9: return q
```

The flux functional $\Phi = -\mathcal{D} \circ \zeta$ is consistent and faithful [14], meaning that $\delta\Phi$ is stationary on $u \in A_0$ if and only if associated beliefs $q \in \Delta_0$ are consistent. This flux functional yields the GBP equations of algorithm A (up to the normalisation step of line 3, ensuring normalisation of beliefs). It may however not be optimal.

We propose another flux functional $\phi = -\mu \circ \mathcal{D} \circ \zeta$ by degree-1 Möbius inversion on heat fluxes in algorithm B. It is remarkable that the associated diffusion $\delta\phi$ involves only the coefficients $c_\alpha \in \mathbb{Z}$ originally used by Bethe [2] to estimate the free energy of statistical systems close to their critical temperature. These coefficients also appear in the cluster variational problem [5,9,12] on free energy, solved by fixed points of belief propagation and diffusion algorithms [14,21].

It remains open whether fixed points of Bethe diffusion are always consistent. We were only able to prove this in a neighbourhood of the consistent manifold, a property we called local faithfulness of the Bethe diffusion flux ϕ, see [14, Chapter 5]. Faithfulness proofs are non-trivial and we conjecture the global faithfulness of ϕ.

Definition 12. *The* Bethe numbers $(c_\alpha) \in \mathbb{Z}^K$ *are uniquely defined by the equations:*

$$\sum_{\alpha \supseteq \beta} c_\alpha = 1 \quad \text{for all} \quad \beta \in K$$

Both algorithms consist of time-step ε discrete Euler integrators of diffusion equations of the form (2), for two different flux functionals. Generalised belief propagation (GBP) is usually expressed multiplicatively for $\varepsilon = 1$ in terms of beliefs $q_\alpha = \frac{1}{Z_\alpha} \mathrm{e}^{-U_\alpha}$ and messages $m_{\alpha\beta} = \mathrm{e}^{-\varphi_{\alpha\beta}}$. A choice of $\varepsilon < 1$ would

[12] Note the normalisation operation $U_\alpha \leftarrow U_\alpha + \ln Z_\alpha$ line 3 in A. It is replaced by line 4 in B, which takes care of harmonising normalisation factors by eliminating redundancies in Φ. The arrows $U_\alpha \leftarrow \dots$ suggest `map` operations that may be efficiently parallelised through asynchronous streams, by locality of the associated operators $\zeta, \mathcal{D}, \delta \dots$. Each stream performs local operations over tensors in A_α, whose dimensions depend on the cardinality of local configuration spaces $E_\alpha = \prod_{i \in \alpha} E_i$.

appear as an exponent in the product of messages by this substitution. This is different from *damping* techniques [6] and has not been previously considered to our knowledge.

Bethe numbers c_α would also appear as exponents of messages in the multiplicative formulation of algorithm B. The combinatorial regularisation offered by Bethe numbers stabilises divergent oscillations in non-constant directions on hypergraphs, improving convergence of GBP diffusion at higher diffusivities. When K is a graph, the two algorithms are actually equivalent, so that Bethe numbers only regularise normalisation factors in the degree ≥ 2 case.

DECAY OF INCONSISTENCY (after 10 iterations - averaged over 20 runs)

GBP diffusion

diffusivity / temperature	0.1	0.2	0.3	0.4	0.5	0.6	0.7	0.8
1E-06	0.3	0.2	0.4	0.7	1.8	4.7	23.0	58.2
0.001	0.3	0.2	0.4	0.7	1.8	5.2	21.9	60.3
0.01	0.3	0.2	0.4	0.7	1.8	5.4	21.4	60.6
0.1	0.3	0.2	0.4	0.7	1.8	5.3	20.7	58.1
1	0.4	0.2	0.4	0.8	2.1	5.7	24.6	60.8
10	0.4	0.3	0.5	0.9	2.3	6.0	26.3	64.5
100	0.4	0.3	0.5	0.9	2.3	6.0	26.4	64.5
1,000	0.4	0.3	0.5	0.9	2.3	6.0	26.4	64.5
1E+06	0.4	0.3	0.5	0.9	2.3	6.0	26.4	64.5

Bethe diffusion

diffusivity / temperature	0.1	0.2	0.3	0.4	0.5	0.6	0.7	0.8
1E-06	0.3	0.1	0.0	0.0	0.0	0.1	1.3	11.3
0.001	0.3	0.1	0.0	0.0	0.0	0.1	1.2	10.6
0.01	0.3	0.1	0.0	0.0	0.0	0.1	1.4	10.5
0.1	0.3	0.1	0.0	0.0	0.0	0.1	1.4	10.7
1	0.2	0.2	0.3	0.3	0.3	0.3	1.1	9.5
10	0.2	0.2	0.3	0.4	0.4	0.4	1.2	8.7
100	0.2	0.2	0.3	0.4	0.4	0.4	1.2	8.7
1,000	0.2	0.2	0.3	0.4	0.4	0.4	1.2	8.7
1E+06	0.2	0.2	0.3	0.4	0.4	0.4	1.2	8.7

Fig. 2. Convergence of GBP and Bethe diffusions for different values of diffusivity $0 < \varepsilon < 1$ and energy scales on the 2-horn, depicted in Fig. 1. Both diffusions almost surely diverge for diffusivities $\varepsilon \geq 1$, so that the usual GBP algorithm is not represented in this table.

Figure 2 shows the results of experiments conducted on the simplest hypergraph K for which GBP does not surely converge to the unique solution $q \in [u] \cap A_0^\Gamma$, the horn Λ^2 depicted in Fig. 1. Initial potentials $u \in A_0$ were normally sampled according to $h_\alpha(x_\alpha) \sim \frac{1}{T}\mathcal{N}(0,1)$ at different temperatures or energy scales $T > 0$. For each value of T and for each fixed diffusivity $\varepsilon > 0$, GBP and Bethe diffusion algorithms were run on random initial conditions for $\mathtt{n_{it}} = 10$ iterations. Consistency of the returned beliefs, if any, was assessed in the effective gradient Φ to produce the represented decay ratios. Diffusivity was then increased until the drop in Bethe diffusion convergence, occuring significantly later than GBP diffusion but *before* $\varepsilon < 1$, reflecting the importance of using finer integrators than usual $\varepsilon = 1$ belief propagation algorithms.

The discretised diffusion $(1 + \varepsilon\delta\Phi)^n$ may be compared to the approximate integration of $\exp(-n\varepsilon x)$ as $(1 - \varepsilon x)^n$, which should only be done under the constraint $\varepsilon|x| < 1$. Assuming all eigenvalues of the linearised diffusion flow $\delta\Phi_*$ are negative (as is the case in the neighbourhood of a stable potential), one should still ensure $\varepsilon|\delta\Phi_*| < 1$ to confidently estimate the large time asymptotics of diffusion as $\exp(n\varepsilon\delta\Phi) \simeq (1 + \varepsilon\delta\Phi)^n$ and reach Γ.

An open-sourced python implementation of the above algorithms, with implementations of the (co)-chain complex $A_\bullet(K)$ for arbitrary hypergraphs K, Bethe numbers, Bethe entropy and free energy functionals, and other operations for designing marginal estimation algorithms is on github at opeltre/topos.

References

1. Abramsky, S., Brandenburger, A.: The Sheaf-theoretic structure of non-locality and contextuality. New J. Phys. 13 (2011)
2. Bethe, H.A., Bragg, W.L.: Statistical theory of superlattices. Proc. Roy. Soc. Lond. Ser. A Math. Phys. Sci. **150**, 552–575 (1935)
3. Gallager, R.G.: Low-Density Parity-Check Codes. MIT Press, Cambridge (1963)
4. Jego, C., Gross, W.J.: Turbo decoding of product codes using adaptive belief propagation. IEEE Trans. Commun. **57**, 2864–2867 (2009)
5. Kikuchi, R.: A theory of cooperative phenomena. Phys. Rev. **81**, 988–1003 (1951)
6. Knoll, C., Pernkopf, F.: On Loopy Belief Propagation - Local Stability Analysis for Non-Vanishing Fields, in Uncertainty in Artificial Intelligence (2017)
7. Leinster, T.: The Euler characteristic of a category. Documenta Mathematica **13**, 21–49 (2008)
8. Moerdijk, I.: Classifying Spaces and Classifying Topoi. LNM, vol. 1616. Springer, Heidelberg (1995). https://doi.org/10.1007/BFb0094441
9. Morita, T.: Cluster variation method of cooperative phenomena and its generalization I. J. Phys. Soc. Japan **12**, 753–755 (1957)
10. Mézard, M., Montanari, A.: Information, Physics and Computation. Oxford University Press, Oxford (2009)
11. Pearl, J.: Reverend Bayes on inference engines: a distributed hierachical approach. In: AAAI 1982 Proceedings (1982)
12. Pelizzola, A.: Cluster variation method in statisical physics and probabilistic graphical models. J. Phys. A Math. General 38 (2005)
13. Peltre, O.: A homological approach to belief propagation and Bethe approximations. In: Nielsen, F., Barbaresco, F. (eds.) GSI 2019. LNCS, vol. 11712, pp. 218–227. Springer, Cham (2019). https://doi.org/10.1007/978-3-030-26980-7_23
14. Peltre, O.: Message-passing algorithms and homology. Ph.D. preprint arXiv:2009.11631 (2020)
15. Ping, W., Ihler, A.: Belief propagation in conditional RBMs for structured prediction. In: Proceedings of the 20th International Conference on Artificial Intelligence and Statistics, vol. 54, pp. 1141–1149 (2017)
16. Rota, G.-C.: On the foundations of combinatorial theory - I. Theory of Möbius functions. Z. Warscheinlichkeitstheorie **2**, 340–368 (1964)
17. Salakhutdinov, R., Hinton, G.: Deep Boltzmann machines. In: van Dyk, D., Welling, M. (eds.) Proceedings of the Twelth International Conference on Artificial Intelligence and Statistics, Proceedings of Machine Learning Research, vol. 5, pp. 448–455 (2009)
18. Sun, J., Zheng, N.-N., Shum, H.-Y.: Stereo matching using belief propagation. IEEE Trans. Pattern Anal. Mach. Intellig. **25**, 787–800 (2003)
19. Verdier, J.-L., Grothendieck, A.V: Cohomologie dans les Topos, SGA-4, 2 (1972)
20. Vorob'ev, N.: Consistent families of measures and their extensions. Theory Probab. Appl. **7**, 147–164 (1962)
21. Yedidia, J., Freeman, W., Weiss, Y.: Constructing free energy approximations and generalized belief propagation algorithms. IEEE Trans. Inf. Theory **51**, 2282–2312 (2005)

Towards a Functorial Description of Quantum Relative Entropy

Arthur J. Parzygnat(✉)

Institut des Hautes Études Scientifiques, 35 Route de Chartres 91440, Bures-sur-Yvette, France
parzygnat@ihes.fr

Abstract. A Bayesian functorial characterization of the classical relative entropy (KL divergence) of finite probabilities was recently obtained by Baez and Fritz. This was then generalized to standard Borel spaces by Gagné and Panangaden. Here, we provide preliminary calculations suggesting that the finite-dimensional quantum (Umegaki) relative entropy might be characterized in a similar way. Namely, we explicitly prove that it defines an affine functor in the special case where the relative entropy is finite. A recent non-commutative disintegration theorem provides a key ingredient in this proof.

Keywords: Bayesian inversion · Disintegration · Optimal hypothesis

1 Introduction and Outline

In 2014, Baez and Fritz provided a categorical Bayesian characterization of the relative entropy of finite probability measures using a category of hypotheses [1]. This was then extended to standard Borel spaces by Gagné and Panangaden in 2018 [5]. An immediate question remains as to whether or not the quantum (Umegaki) relative entropy [12] has a similar characterization.[1] The purpose of the present work is to begin filling this gap by using the recently proved non-commutative disintegration theorem [9].

The original motivation of Baez and Fritz came from Petz' characterization of the quantum relative entropy [11], which used a quantum analogue of hypotheses known as conditional expectations. Although Petz' characterization had some minor flaws, which were noticed in [1], we believe Petz' overall idea is correct when formulated on an appropriate category of non-commutative probability spaces and non-commutative hypotheses. In this article, we show how the Umegaki relative entropy defines an affine functor that vanishes on the subcategory of non-commutative optimal hypotheses for faithful states. The chain rule for quantum conditional entropy is a consequence of functoriality. The non-faithful case will be addressed in future work, where we hope to provide a characterization of the quantum relative entropy as an affine functor.

[1] The ordinary Shannon and von Neumann entropies were characterized in [2] and [7], respectively, in a similar categorical setting.

F. Nielsen and F. Barbaresco (Eds.): GSI 2021, LNCS 12829, pp. 557–564, 2021.
https://doi.org/10.1007/978-3-030-80209-7_60

2 The Categories of Hypotheses and Optimal Hypotheses

In this section, we introduce non-commutative analogues of the categories from [1]. All C^*-algebras here are finite-dimensional and unital. All $*$-homomorphisms are unital unless stated otherwise. $\mathcal{M}_n$ denotes the algebra of $n \times n$ matrices. If $V : \mathbb{C}^m \to \mathbb{C}^n$ is a linear map, $\mathrm{Ad}_V : \mathcal{M}_n \rightsquigarrow \mathcal{M}_m$ denotes the linear map sending A to $VAV^\dagger$, where $V^\dagger$ is the adjoint (conjugate transpose) of V. Linear maps between algebras are written with squiggly arrows $\rightsquigarrow$, while $*$-homomorphisms are written as straight arrows $\to$. The acronym CPU stands for "completely positive unital." If $\mathcal{A}$ and $\mathcal{B}$ are matrix algebras, then $\mathrm{tr}_{\mathcal{A}} : \mathcal{A} \otimes \mathcal{B} \rightsquigarrow \mathcal{B}$ denotes the ***partial trace over*** $\mathcal{A}$ and is the unique linear map determined by $\mathrm{tr}_{\mathcal{A}}(A \otimes B) = \mathrm{tr}(A)B$ for $A \in \mathcal{A}$ and $B \in \mathcal{B}$. If ω is a state on $\mathcal{A}$, its quantum entropy is denoted by $S(\omega)$ (cf. [7, Definition 2.20]).

Definition 1. *Let* **NCFS** *be the category of* **non-commutative probability spaces**, *whose objects are pairs $(\mathcal{A}, \omega)$, with $\mathcal{A}$ a C^*-algebra and ω a state on $\mathcal{A}$. A morphism $(\mathcal{B}, \xi) \to (\mathcal{A}, \omega)$ is a pair (F, Q) with $F : \mathcal{B} \to \mathcal{A}$ a $*$-homomorphism and $Q : \mathcal{A} \rightsquigarrow \mathcal{B}$ a CPU map (called a* **hypothesis**), *such that*

$$\omega \circ F = \xi \qquad \text{and} \qquad Q \circ F = \mathrm{id}_{\mathcal{B}}.$$

The composition rule in **NCFS** *is given by*

$$(\mathcal{C}, \zeta) \xrightarrow{(G,R)} (\mathcal{B}, \xi) \xrightarrow{(F,Q)} (\mathcal{A}, \omega) \quad \mapsto \quad (\mathcal{C}, \zeta) \xrightarrow{(F \circ G, R \circ Q)} (\mathcal{A}, \omega).$$

Let **NCFP** *be the subcategory of* **NCFS** *with the same objects but whose morphisms are pairs (F, Q) as above and Q is an* **optimal hypothesis**, *i.e.* $\xi \circ Q = \omega$.

The subcategories of $\mathbf{NCFS}^{\mathrm{op}}$ and $\mathbf{NCFP}^{\mathrm{op}}$ consisting of *commutative* C^*-algebras are equivalent to the categories **FinStat** and **FP** from [1] by stochastic Gelfand duality (cf. [6, Sections 2.5 and 2.6], [4], and [9, Corollary 3.23]).

Notation 1. *On occasion, the notation $\mathcal{A}$, $\mathcal{B}$, and $\mathcal{C}$ will be used to mean*

$$\mathcal{A} := \bigoplus_{x \in X} \mathcal{M}_{m_x}, \qquad \mathcal{B} := \bigoplus_{y \in Y} \mathcal{M}_{n_y}, \qquad \text{and} \qquad \mathcal{C} := \bigoplus_{z \in Z} \mathcal{M}_{o_z},$$

where X, Y, Z are finite sets, often taken to be ordered sets $X = \{1, \ldots, s\}$, $Y = \{1, \ldots, t\}$, $Z = \{1, \ldots, u\}$ for convenience (cf. [9, Section 5] and/or [7, Example 2.2]). Note that every element of $\mathcal{A}$ (and analogously for $\mathcal{B}$ and $\mathcal{C}$) is of the form $A = \bigoplus_{x \in X} A_x$, with $A_x \in \mathcal{M}_{m_x}$. Furthermore, ω, ξ, and ζ will refer to states on $\mathcal{A}$, $\mathcal{B}$, and $\mathcal{C}$, respectively, with decompositions of the form

$$\omega = \sum_{x \in X} p_x \mathrm{tr}(\rho_x \,\cdot\,), \qquad \xi = \sum_{y \in Y} q_y \mathrm{tr}(\sigma_y \,\cdot\,), \qquad \text{and} \qquad \zeta = \sum_{z \in Z} r_z \mathrm{tr}(\tau_z \,\cdot\,).$$

If $Q : \mathcal{A} \rightsquigarrow \mathcal{B}$ is a linear map, its yx component Q_{yx} is the linear map obtained from the composite $\mathcal{M}_{m_x} \hookrightarrow \mathcal{A} \overset{Q}{\rightsquigarrow} \mathcal{B} \twoheadrightarrow \mathcal{M}_{n_y}$, where the first and last maps are the (non-unital) inclusion and projection, respectively.

Definition 2. *Let $\mathcal{A}$ and $\mathcal{B}$ be as in Notation 1. A morphism $(\mathcal{B},\xi) \xrightarrow{(F,Q)} (\mathcal{A},\omega)$ in* **NCFS** *is said to be in* **standard form** *iff there exist non-negative integers c^F_{yx} such that (cf. [7, Lemma 2.11] and [3, Theorem 5.6])*

$$F(B) = \bigoplus_{x\in X}\boxplus_{y\in Y}\left(\mathbb{1}_{c^F_{yx}} \otimes B_y\right) \equiv \bigoplus_{x\in X}\operatorname{diag}\left(\mathbb{1}_{c^F_{1x}} \otimes B_1,\ldots,\mathbb{1}_{c^F_{tx}} \otimes B_t\right) \quad \forall\, B\in\mathcal{B},$$

which is a direct sum of block diagonal matrices. The number c^F_{yx} is called the **multiplicity** *of $\mathcal{M}_{n_y}$ in $\mathcal{M}_{m_x}$ associated to F. In this case, each $A_x \in \mathcal{M}_{m_x}$ will occasionally be decomposed as $A_x = \sum_{y,y'\in Y} E^{(t)}_{yy'} \otimes A_{x;yy'}$, where $\{E^{(t)}_{yy'}\}$ denote the matrix units of $\mathcal{M}_t$ and $A_{x;yy'}$ is a $(c^F_{yx}n_y)\times(c^F_{y'x}n_{y'})$ matrix.*

If F is in standard form and if ω and ξ are states on $\mathcal{A}$ and $\mathcal{B}$ (as in Notation 1) such that $\xi = \omega\circ F$, then (cf. [7, Lemma 2.11] and [9, Proposition 5.67])

$$q_y\sigma_y = \sum_{x\in X} p_x \operatorname{tr}_{\mathcal{M}_{c^F_{yx}}}(\rho_{x;yy}). \tag{2.1}$$

The standard form of a morphism will be useful later for proving functoriality of relative entropy, and it will allow us to formulate expressions more explicitly in terms of matrices.

Lemma 1. *Given a morphism $(\mathcal{B},\xi) \xrightarrow{(F,Q)} (\mathcal{A},\omega)$ in* **NCFS**, *with $\mathcal{A}$ and $\mathcal{B}$ be as in Notation 1, there exists a unitary $U\in\mathcal{A}$ such that $(\mathcal{B},\xi) \xrightarrow{(\mathrm{Ad}_{U^\dagger}\circ F, Q\circ\mathrm{Ad}_U)} (\mathcal{A},\omega\circ\mathrm{Ad}_U)$ is a morphism in* **NCFS** *that is in standard form. Furthermore, if (F,Q) is in* **NCFP**, *then $(\mathrm{Ad}_{U^\dagger}\circ F, Q\circ\mathrm{Ad}_U)$ is also in* **NCFP**.

Proof. First, $(\mathrm{Ad}_{U^\dagger}\circ F, Q\circ\mathrm{Ad}_U)$ is in **NCFS** for *any* unitary U because

$$(\omega\circ\mathrm{Ad}_U)\circ(\mathrm{Ad}_{U^\dagger}\circ F) = \xi \quad \text{and} \quad (Q\circ\mathrm{Ad}_U)\circ(\mathrm{Ad}_{U^\dagger}\circ F) = \mathrm{id}_{\mathcal{B}},$$

so that the two required conditions hold. Second, the fact that a unitary U exists such that F is in the form in Definition 2 is a standard fact regarding (unital) $*$-homomorphisms between direct sums of matrix algebras [3, Theorem 5.6]. Finally, if (F,Q) is in **NCFP**, which means $\xi\circ Q = \omega$, then $(\mathrm{Ad}_{U^\dagger}\circ F, Q\circ\mathrm{Ad}_U)$ is also in **NCFP** because $\xi\circ(Q\circ\mathrm{Ad}_U) = (\xi\circ Q)\circ\mathrm{Ad}_U = \omega\circ\mathrm{Ad}_U$.

Although the composite of two morphisms in standard form is *not* necessarily in standard form, a permutation can always be applied to obtain one. Furthermore, a pair of composable morphisms in **NCFS** can also be simultaneously rectified. This is the content of the following lemmas.

Lemma 2. *Given a composable pair $(\mathcal{C},\zeta) \xrightarrow{(G,R)} (\mathcal{B},\xi) \xrightarrow{(F,Q)} (\mathcal{A},\omega)$ in* **NCFS**, *there exist unitaries $U\in\mathcal{A}$ and $V\in\mathcal{B}$ such that*

$$(\mathcal{C},\zeta) \xrightarrow{(\mathrm{Ad}_{V^\dagger}\circ G, R\circ\mathrm{Ad}_V)} (\mathcal{B},\xi\circ\mathrm{Ad}_V) \xrightarrow{(\mathrm{Ad}_{U^\dagger}\circ F\circ\mathrm{Ad}_V, \mathrm{Ad}_{V^\dagger}\circ Q\circ\mathrm{Ad}_U)} (\mathcal{A},\omega\circ\mathrm{Ad}_U)$$

is a pair of composable morphisms in **NCFS** *that are* both *in standard form.*

Proof. By Lemma 1, there exists a unitary $V \in \mathcal{B}$ such that

$$(\mathcal{C},\zeta) \xrightarrow{(\mathrm{Ad}_{V^\dagger}\circ G, R\circ \mathrm{Ad}_V)} (\mathcal{B},\xi\circ\mathrm{Ad}_V) \xrightarrow{(F\circ\mathrm{Ad}_V, \mathrm{Ad}_{V^\dagger}\circ Q)} (\mathcal{A},\omega)$$

is a composable pair of morphisms in **NCFS** with the left morphism in standard form. The right morphism is indeed in **NCFS** because

$$\omega\circ(F\circ\mathrm{Ad}_V) = (\omega\circ F)\circ\mathrm{Ad}_V = \xi\circ\mathrm{Ad}_V \qquad \text{and}$$

$$(\mathrm{Ad}_{V^\dagger}\circ Q)\circ(F\circ\mathrm{Ad}_V) = \mathrm{Ad}_{V^\dagger}\circ(Q\circ F)\circ\mathrm{Ad}_V = \mathrm{Ad}_{V^\dagger}\circ\mathrm{Ad}_V = \mathrm{id}_{\mathcal{B}}.$$

Then, applying Lemma 1 again, but to the *new* morphism on the right, gives a unitary U satisfying the conditions claimed.

Lemma 3. *Given a composable pair* $(\mathcal{C},\zeta) \xrightarrow{(G,R)} (\mathcal{B},\xi) \xrightarrow{(F,Q)} (\mathcal{A},\omega)$ *in* **NCFS**, *each in standard form, there exist* permutation *matrices* $P_x \in \mathcal{M}_{m_x}$ *such that*

$$(\mathcal{C},\zeta) \xrightarrow{(\mathrm{Ad}_{P^\dagger}\circ F\circ G, R\circ Q\circ\mathrm{Ad}_P)} (\mathcal{A},\omega\circ\mathrm{Ad}_P)$$

is also in standard form, where $P := \bigoplus_{x\in X} P_x$ *and the multiplicities* $c^{G\circ F}_{zx}$ *of* $\mathrm{Ad}_{P^\dagger}\circ G\circ F$ *are given by* $c^{G\circ F}_{zx} = \sum_{y\in Y} c^G_{zy}c^F_{yx}$.

Proof. The composite $F\circ G$ is given by

$$F(G(C)) = F\Bigg(\underbrace{\bigoplus_{y\in Y}\boxplus_{z\in Z}\big(\mathbb{1}_{c^G_{yz}}\otimes C_z\big)}_{B_y}\Bigg) = \bigoplus_{x\in X}\underbrace{\boxplus_{y\in Y}\Bigg(\mathbb{1}_{c^F_{yx}}\otimes\boxplus_{z\in Z}\big(\mathbb{1}_{c^G_{yz}}\otimes C_z\big)\Bigg)}_{A_x}.$$

The matrix A_x takes the more explicit form (with zeros in unfilled entries)

$$A_x = \mathrm{diag}\left(\mathbb{1}_{c^F_{1x}}\otimes\begin{bmatrix}\mathbb{1}_{c^G_{11}}\otimes C_1 & & \\ & \ddots & \\ & & \mathbb{1}_{c^G_{u1}}\otimes C_u\end{bmatrix}, \dots, \mathbb{1}_{c^F_{tx}}\otimes\begin{bmatrix}\mathbb{1}_{c^G_{11}}\otimes C_1 & & \\ & \ddots & \\ & & \mathbb{1}_{c^G_{u1}}\otimes C_u\end{bmatrix}\right).$$

From this, one sees that the number of times C_z appears on the diagonal is $\sum_{y\in Y} c^G_{zy}c^F_{yx}$. However, the positions of C_z are not all next to each other. Hence, a permutation matrix P_x is needed to put them into standard form.

Notation 2. *Given a composable pair* $(\mathcal{C},\zeta) \xrightarrow{(G,R)} (\mathcal{B},\xi) \xrightarrow{(F,Q)} (\mathcal{A},\omega)$ *in standard form as in Lemma 3, the states* $\zeta\circ R$ *and* $\xi\circ Q$ *will be decomposed as*

$$\zeta\circ R = \sum_{y\in Y} q^R_y\mathrm{tr}(\sigma^R_y\ \cdot\ ) \qquad \text{and} \qquad \xi\circ Q = \sum_{x\in X} p^Q_x\mathrm{tr}(\rho^Q_x\ \cdot\ ).$$

Lemma 4. *Given a morphism* (F,Q) *in standard form as in Notation 2 such that all states are faithful, there exist strictly positive matrices* $\alpha_{yx}\in\mathcal{M}_{c^F_{yx}}$ *for all* $x\in X$ *and* $y\in Y$ *such that*

$$\mathrm{tr}\left(\sum_{x\in X}\alpha_{yx}\right) = 1 \ \ \forall\, y\in Y, \qquad p^Q_x\rho^Q_x = \boxplus_{y\in Y}(\alpha_{yx}\otimes q_y\sigma_y)\ \forall\, x\in X, \quad \text{and}$$

$$Q_{yx}(A_x) = \mathrm{tr}_{\mathcal{M}_{c^F_{yx}}}\big((\alpha_{yx} \otimes \mathbb{1}_{n_y})A_{x;yy}\big) \qquad \forall\, y \in Y,\ A_x \in \mathcal{M}_{m_x},\ x \in X.$$

Proof. Because Q and R are disintegrations of $(F, \xi \circ Q, \xi)$ and $(G, \zeta \circ R, \zeta)$, respectively, the claim follows from the non-commutative disintegration theorem [9, Theorem 5.67] and the fact that F is an injective $*$-homomorphism. The α_{yx} matrices are strictly positive by the faithful assumption.

If $(\mathcal{C}, \zeta) \xrightarrow{(G,R)} (\mathcal{B}, \xi) \xrightarrow{(F,Q)} (\mathcal{A}, \omega)$ is a composable pair, a consequence of Lemma 4 is

$$\zeta \circ R \circ Q = \sum_{x \in X} \mathrm{tr}\left(\left(\boxplus_{y \in Y} \alpha_{yx} \otimes q^R_y \sigma^R_y\right) \cdot \right). \tag{2.2}$$

3 The Relative Entropy as a Functor

Definition 3. *Set* RE : **NCFS** $\to \mathbb{B}(-\infty, \infty]$ *to be the assignment that sends a morphism* $(\mathcal{B}, \xi) \xrightarrow{(F,Q)} (\mathcal{A}, \omega)$ *to* $S(\omega \parallel \xi \circ Q)$ *(the assignment is trivial on objects). Here,* $\mathbb{B}M$ *is the one object category associated to any monoid*[2] M*,* $S(\cdot \parallel \cdot)$ *is the* **relative entropy** *of two states on the same* C^**-algebra, which is defined on an ordered pair of states* (ω, ω')*, with* $\omega \preceq \omega'$ *(meaning* $\omega'(a^*a) = 0$ *implies* $\omega(a^*a) = 0$*), on* $\mathcal{A} = \bigoplus_{x \in X} \mathcal{M}_{m_x}$ *by*

$$S(\omega \parallel \omega') := \mathrm{tr}\left(\bigoplus_{x \in X} p_x \rho_x \Big(\log(p_x \rho_x) - \log(p'_x \rho'_x)\Big)\right),$$

where $0 \log 0 := 0$ *by convention. If* $\omega \not\preceq \omega'$*, then* $S(\omega \parallel \omega') := \infty$*.*

Lemma 5. *Using the notation from Definition 3, the following facts hold.*

(a) RE *factors through* $\mathbb{B}[0, \infty]$*.*
(b) RE *vanishes on the subcategory* **NCFP**.
(c) RE *is invariant with respect to changing a morphism to standard form, i.e. in terms of the notation introduced in Lemma 1,*

$$\mathrm{RE}\left((\mathcal{B}, \xi) \xrightarrow{(F,Q)} (\mathcal{A}, \omega)\right) = \mathrm{RE}\left((\mathcal{B}, \xi) \xrightarrow{(\mathrm{Ad}_{U^\dagger} \circ F, Q \circ \mathrm{Ad}_U)} (\mathcal{A}, \omega \circ \mathrm{Ad}_U)\right).$$

Proof. Left as an exercise.

Proposition 1. *For a composable pair* $(\mathcal{C}, \zeta) \xrightarrow{(G,R)} (\mathcal{B}, \xi) \xrightarrow{(F,Q)} (\mathcal{A}, \omega)$ *in* **NCFS** *(with all states and CPU maps faithful),*[3] $S(\omega \parallel \zeta \circ R \circ Q) = S(\xi \parallel \zeta \circ R) + S(\omega \parallel \xi \circ Q)$*, i.e.* $\mathrm{RE}\big((F \circ G, R \circ Q)\big) = \mathrm{RE}\big((G, R)\big) + \mathrm{RE}\big((F, Q)\big)$*.*

[2] The morphisms of $\mathbb{B}M$ from that single object to itself equals the set M and the composition is the monoid multiplication. Here, the monoid is $(-\infty, \infty]$ under addition, with the convention that $a + \infty = \infty$ for all $a \in (-\infty, \infty]$.

[3] Faithfulness guarantees the finiteness of all expressions. More generally, our proof works if the appropriate absolute continuity conditions hold. Also, note that the "conditional expectation property" in [11] is a special case of functoriality applied to a composable pair of morphisms of the form $(\mathbb{C}, \mathrm{id}_{\mathbb{C}}) \xrightarrow{(!_{\mathcal{B}}, R)} (\mathcal{B}, \xi) \xrightarrow{(F,Q)} (\mathcal{A}, \omega)$, where $!_{\mathcal{B}} : \mathbb{C} \to \mathcal{B}$ is the unique unital linear map. Indeed, Petz' $\mathcal{A}$, $\mathcal{B}$, E, $\omega_{|\mathcal{A}}$, $\varphi_{|\mathcal{A}}$, and φ, are our $\mathcal{B}$, $\mathcal{A}$, Q, ξ, R, and $R \circ Q$, respectively (ω is the same).

Proof. By Lemma 5, it suffices to assume (F,Q) and (G,R) are in standard form. To prove the claim, we expand each term. First,[4]

$$
\begin{aligned}
S(\omega \,\|\, \xi \circ Q) &\xlongequal{\text{Lem 4}} -S(\omega) - \sum_{x\in X} \operatorname{tr}\left(p_x\rho_x \log\left(\boxplus_{y\in Y} \alpha_{yx} \otimes q_y\sigma_y\right)\right) \\
&= -S(\omega) - \sum_{x\in X}\sum_{y\in Y} \operatorname{tr}\Big(p_x\rho_{x;yy}\big(\log(\alpha_{yx}) \otimes \mathbb{1}_{n_y}\big)\Big) \\
&\quad - \sum_{x\in X}\sum_{y\in Y} \operatorname{tr}\Big(p_x \operatorname{tr}_{\mathcal{M}_{c^F_{yx}}}(\rho_{x;yy}) \log(q_y\sigma_y)\Big).
\end{aligned}
\tag{3.1}
$$

The last equality follows from the properties of the trace, partial trace, and logarithms of tensor products. By similar arguments,

$$
\begin{aligned}
S(\xi \,\|\, \zeta \circ R) &\xlongequal{(2.1)} \sum_{x\in X}\sum_{y\in Y} \operatorname{tr}\Big(p_x \operatorname{tr}_{\mathcal{M}_{c^F_{yx}}}(\rho_{x;yy}) \log(q_y\sigma_y)\Big) \\
&\quad - \sum_{x\in X}\sum_{y\in Y} \operatorname{tr}\Big(p_x \operatorname{tr}_{\mathcal{M}_{c^F_{yx}}}(\rho_{x;yy}) \log(q^R_y\sigma^R_y)\Big)
\end{aligned}
\tag{3.2}
$$

and

$$
\begin{aligned}
S(\omega \,\|\, \zeta \circ R \circ Q) &\xlongequal{(2.2)} -S(\omega) - \sum_{x\in X}\sum_{y\in Y} \operatorname{tr}\Big(p_x\rho_{x;yy}\Big(\log(\alpha_{yx}) \otimes \mathbb{1}_{n_y}\Big)\Big) \\
&\quad - \sum_{x\in X}\sum_{y\in Y} \operatorname{tr}\Big(p_x \operatorname{tr}_{\mathcal{M}_{c^F_{yx}}}(\rho_{x;yy}) \log(q^R_y\sigma^R_y)\Big).
\end{aligned}
\tag{3.3}
$$

Hence, (3.1) + (3.2) = (3.3), which proves the claim.

Example 1. The usual chain rule for the quantum conditional entropy is a special case of Proposition 1. To see this, set $\mathcal{A} := \mathcal{M}_{d_A}, \mathcal{B} := \mathcal{M}_{d_B}, \mathcal{C} := \mathcal{M}_{d_C}$ with $d_A, d_B, d_C \in \mathbb{N}$. Given a density matrix ρ_{ABC} on $\mathcal{A} \otimes \mathcal{B} \otimes \mathcal{C}$, we implement subscripts to denote the associated density matrix after tracing out a subsystem. The chain rule for the conditional entropy states

$$
H(AB|C) = H(A|BC) + H(B|C),
\tag{3.4}
$$

where (for example)

$$
H(B|C) := \operatorname{tr}(\rho_{BC} \log \rho_{BC}) - \operatorname{tr}(\rho_C \log \rho_C)
$$

is the ***quantum conditional entropy*** of ρ_{BC} given ρ_C. One can show that

$$
\operatorname{RE}\big((F \circ G, R \circ Q)\big) = H(AB|C) + \log(d_A) + \log(d_B),
$$

$$
\operatorname{RE}\big((G,R)\big) = H(B|C) + \log(d_B), \quad \text{and} \quad \operatorname{RE}\big((F,Q)\big) = H(A|BC) + \log(d_A)
$$

[4] Equation (3.1) is a generalization of Equation (3.2) in [1], which plays a crucial role in proving many claims. We will also use it to prove affinity of RE.

by applying Proposition 1 to the composable pair

$$\Big(\mathcal{C}, \mathrm{tr}(\rho_C\,\cdot\,)\Big) \xrightarrow{(G,R)} \Big(\mathcal{B}\otimes\mathcal{C}, \mathrm{tr}(\rho_{BC}\,\cdot\,)\Big) \xrightarrow{(F,Q)} \Big(\mathcal{A}\otimes\mathcal{B}\otimes\mathcal{C}, \mathrm{tr}(\rho_{ABC}\,\cdot\,)\Big),$$

where G and F are the standard inclusions, $\upsilon_B := \frac{1}{d_B}\mathbb{1}_{d_B}$, $\upsilon_A := \frac{1}{d_A}\mathbb{1}_{d_A}$, and R and Q are the CPU maps given by $R := \mathrm{tr}_{\mathcal{B}}(\upsilon_B \otimes 1_{\mathcal{C}}\,\cdot\,)$, and $Q := \mathrm{tr}_{\mathcal{A}}(\upsilon_A \otimes 1_{\mathcal{B}} \otimes 1_{\mathcal{C}}\,\cdot\,)$. This reproduces (3.4).

Proposition 1 does not fully prove functoriality of RE. One still needs to check functoriality in case one of the terms is infinite (e.g. if $S(\omega \parallel \zeta\circ R\circ Q) = \infty$, then at least one of $S(\xi \parallel \zeta\circ R)$ or $S(\omega \parallel \xi\circ Q)$ must be infinite, and conversely). This will be addressed in future work. In the remainder, we prove affinity of RE.

Definition 4. *Given $\lambda \in [0,1]$, set $\overline{\lambda} := 1-\lambda$. The λ-weighted* **convex sum** *$\lambda(\mathcal{A},\omega)\oplus\overline{\lambda}(\overline{\mathcal{A}},\overline{\omega})$ of objects $(\mathcal{A},\omega)$ and $(\overline{\mathcal{A}},\overline{\omega})$ in* **NCFS** *is given by the pair $(\mathcal{A}\oplus\overline{\mathcal{A}}, \lambda\omega\oplus\overline{\lambda}\overline{\omega})$, where $(\lambda\omega\oplus\overline{\lambda}\overline{\omega})(A\oplus\overline{A}) := \lambda\omega(A)+\overline{\lambda}\overline{\omega}(\overline{A})$ whenever $A\in\mathcal{A}$, $\overline{A}\in\overline{\mathcal{A}}$. The* **convex sum** *$\lambda(F,Q)\oplus\overline{\lambda}(\overline{F},\overline{Q})$ of $(\mathcal{B},\xi)\xrightarrow{(F,Q)}(\mathcal{A},\omega)$ and $(\overline{\mathcal{B}},\overline{\xi})\xrightarrow{(\overline{F},\overline{Q})}(\overline{\mathcal{A}},\overline{\omega})$ is the morphism $(F\oplus\overline{F}, Q\oplus\overline{Q})$. A functor* **NCFS** $\xrightarrow{\mathfrak{L}}$ *$\mathbb{B}[0,\infty]$ is* **affine** *iff $\mathfrak{L}(\lambda(F,Q)\oplus\overline{\lambda}(\overline{F},\overline{Q})) = \lambda\mathfrak{L}(F,Q)+\overline{\lambda}\mathfrak{L}(\overline{F},\overline{Q})$ for all pairs of morphisms in* **NCFS** *and $\lambda\in[0,1]$.*

Proposition 2. *Let $(\mathcal{B},\xi)\xrightarrow{(F,Q)}(\mathcal{A},\omega)$ and $(\overline{\mathcal{B}},\overline{\xi})\xrightarrow{(\overline{F},\overline{Q})}(\overline{\mathcal{A}},\overline{\omega})$ be two morphisms for which $\mathrm{RE}(F,Q)$ and $\mathrm{RE}(\overline{F},\overline{Q})$ are finite. Then $\mathrm{RE}\big(\lambda(F,Q)\oplus\overline{\lambda}(\overline{F},\overline{Q})\big) = \lambda\mathrm{RE}(F,Q)+\overline{\lambda}\mathrm{RE}(\overline{F},\overline{Q})$.*

Proof. When $\lambda\in\{0,1\}$, the claim follows from the convention $0\log 0 = 0$. For $\lambda\in(0,1)$, temporarily set $\mu := \mathrm{RE}\big(\lambda(F,Q)\oplus\overline{\lambda}(\overline{F},\overline{Q})\big)$. Then

$$\begin{aligned}
\mu \overset{(3.1)}{=\!=} & \sum_{x\in X}\mathrm{tr}\Big(\lambda p_x\rho_x\log(\lambda p_x\rho_x)\Big) + \sum_{\overline{x}\in\overline{X}}\mathrm{tr}\Big(\overline{\lambda}\overline{p}_{\overline{x}}\overline{\rho}_{\overline{x}}\log\big(\overline{\lambda}\overline{p}_{\overline{x}}\overline{\rho}_{\overline{x}}\big)\Big)\\
&-\sum_{x,y}\Big[\mathrm{tr}\Big(\lambda p_x\rho_{x;yy}\big(\log(\alpha_{yx})\otimes\mathbb{1}_{n_y}\big)\Big)+\mathrm{tr}\Big(\lambda p_x\mathrm{tr}_{\mathcal{M}_{c^F_{yx}}}(\rho_{x;yy})\log(\lambda q_y\sigma_y)\Big)\Big]\\
&-\sum_{\overline{x},\overline{y}}\Big[\mathrm{tr}\Big(\overline{\lambda}\overline{p}_{\overline{x}}\rho_{\overline{x};\overline{y}\overline{y}}\big(\log(\overline{\alpha}_{\overline{y}\overline{x}})\otimes\mathbb{1}_{\overline{n}_{\overline{y}}}\big)\Big)+\mathrm{tr}\Big(\overline{\lambda}\overline{p}_{\overline{x}}\mathrm{tr}_{\mathcal{M}_{c^{\overline{F}}_{\overline{y}\overline{x}}}}(\rho_{\overline{x};\overline{y}\overline{y}})\log\big(\overline{\lambda}\overline{q}_{\overline{y}}\overline{\sigma}_{\overline{y}}\big)\Big)\Big],
\end{aligned}$$

where we have used bars to denote analogous expressions for the algebras, morphisms, and states with bars over them. From this, the property $\log(ab) = \log(a)+\log(b)$ of logarithms is used to complete the proof.

In summary, we have taken the first steps towards illustrating that the quantum relative entropy may have a functorial description along similar lines to those of the classical one in [1]. Using the recent non-commutative disintegration theorem [9], we have proved parts of affinity and functoriality of the relative entropy. The importance of functoriality comes from the connection between the quantum relative entropy and the reversibility of morphisms [10, Theorem 4]. For example, optimal hypotheses are Bayesian inverses [8, Theorem 8.3], which admit stronger compositional properties [8, Propositions 7.18 and 7.21] than alternative recovery maps in quantum information theory [13, Section 4].[5] In future work, we hope to prove functoriality (without any faithfulness assumptions), continuity, and a complete characterization.

Acknowledgements. The author thanks the reviewers of GSI'21 for their numerous helpful suggestions. This research has also received funding from the European Research Council (ERC) under the European Union's Horizon 2020 research and innovation program (QUASIFT grant agreement 677368).

References

1. Baez, J.C., Fritz, T.: A Bayesian characterization of relative entropy. Theory Appl. Categ. **29**(16), 422–457 (2014)
2. Baez, J.C., Fritz, T., Leinster, T.: A characterization of entropy in terms of information loss. Entropy **13**(11), 1945–1957 (2011). https://doi.org/10.3390/e13111945
3. Farenick, D.R.: Algebras of Linear Transformations. Universitext. Springer, New York (2001). https://doi.org/10.1007/978-1-4613-0097-7
4. Furber, R., Jacobs, B.: From Kleisli categories to commutative C^*-algebras: probabilistic Gelfand duality. Log. Methods Comput. Sci. **11**(2), 1:5, 28 (2015). https://doi.org/10.2168/LMCS-11(2:5)2015
5. Gagné, N., Panangaden, P.: A categorical characterization of relative entropy on standard Borel spaces. In: The Thirty-third Conference on the Mathematical Foundations of Programming Semantics (MFPS XXXIII). Electronic Notes in Theoretical Computer Science, vol. 336, pp. 135–153. Elsevier Sci. B. V., Amsterdam (2018). https://doi.org/10.1016/j.entcs.2018.03.020
6. Parzygnat, A.J.: Discrete probabilistic and algebraic dynamics: a stochastic Gelfand-Naimark theorem (2017). arXiv preprint: arXiv:1708.00091 [math.FA]
7. Parzygnat, A.J.: A functorial characterization of von Neumann entropy (2020). arXiv preprint: arXiv:2009.07125 [quant-ph]
8. Parzygnat, A.J.: Inverses, disintegrations, and Bayesian inversion in quantum Markov categories (2020). arXiv preprint: arXiv:2001.08375 [quant-ph]
9. Parzygnat, A.J., Russo, B.P.: Non-commutative disintegrations: existence and uniqueness in finite dimensions (2019). arXiv preprint: arXiv:1907.09689 [quant-ph]
10. Petz, D.: Sufficient subalgebras and the relative entropy of states of a von Neumann algebra. Commun. Math. Phys. **105**(1), 123–131 (1986). https://doi.org/10.1007/BF01212345
11. Petz, D.: Characterization of the relative entropy of states of matrix algebras. Acta Math. Hung. **59**(3–4), 449–455 (1992). https://doi.org/10.1007/BF00050907
12. Umegaki, H.: Conditional expectation in an operator algebra. IV. Entropy and information. Kodai Math. Sem. Rep. **14**(2), 59–85 (1962). https://doi.org/10.2996/kmj/1138844604
13. Wilde, M.M.: Recoverability in quantum information theory. Proc. R. Soc. A. **471**, 20150338 (2015). https://doi.org/10.1098/rspa.2015.0338

[5] One must assume faithfulness for some of the calculations in [13, Section 4]. The compositional properties in [8, Proposition 7.21], however, need no such assumptions.

Frobenius Statistical Manifolds and Geometric Invariants

Noemie Combe(✉), Philippe Combe, and Hanna Nencka

Max Planck Institute for Maths in the Sciences,
Inselstrasse 22, 04103 Leipzig, Germany
noemie.combe@mis.mpg.de

Abstract. Using an algebraic approach, we prove that statistical manifolds, related to exponential families and with flat structure connection have a Frobenius manifold structure. This latter object, at the interplay of beautiful interactions between topology and quantum field theory, raises natural questions, concerning the existence of Gromov–Witten invariants for those statistical manifolds. We prove that an analog of Gromov–Witten invariants for those statistical manifolds (GWS) exists. Similarly to its original version, these new invariants have a geometric interpretation concerning intersection points of para-holomorphic curves. It also plays an important role in the learning process, since it determines whether a system has succeeded in learning or failed.

Keywords: Statistical manifold · Frobenius manifold · Gromov–Witten invariants · Paracomplex geometry

Mathematics Subject Classification 53B99 · 62B10 · 60D99 · 53D45

1 Introduction

For more than 60 years statistical manifolds have been a domain of great interest in information theory [4], machine learning [2,3,5,6] and in decision theory [6].

Statistical manifolds (related to exponential families) have an F-manifold structure, as was proved in [7]. The notion of F-manifolds, developed in [10], arose in the context of mirror symmetry. It is a version of classical Frobenius manifolds, requiring less axioms.

In this paper, we restrict our attention to statistical manifolds related to **exponential families** (see [6] p. 265 for a definition), which have Markov invariant metrics and **being totally geodesic maximal submanifolds** (see [6] p. 182 and [11] p. 180 for foundations of differential geometry). The latter condition implies the existence of affine flat structures. Using a *purely algebraic framework*, we determine the necessary condition to have a Frobenius manifold structure. It is possible to encapsulate a necessary (and important) property of

This research was supported by the Max Planck Society's Minerva grant. The authors express their gratitude towards MPI MiS for excellent working conditions.

F. Nielsen and F. Barbaresco (Eds.): GSI 2021, LNCS 12829, pp. 565–573, 2021.
https://doi.org/10.1007/978-3-030-80209-7_61

Frobenius' structure within the Witten–Dijkgraaf–Verlinde–Verlinde (WDVV) highly non-linear PDE system:

$$\forall a, b, c, d : \sum_{ef} \Phi_{abe} g^{ef} \Phi_{fcd} = (-1)^{a(b+c)} \sum_{ef} \Phi_{bce} g^{ef} \Phi_{fad}. \tag{1}$$

Notice that the (WDVV) system expresses *a flatness condition of the manifold* (i.e. vanishing of the curvature), see [3,12,16].

However, we investigate the (WDVV) condition from a purely *algebraic framework*, giving a totally different insight on this problem, and thus avoiding tedious computations of differential equations. *Algebraically* the (WDVV) condition is expressed by the associativity and potentiality conditions below (see [12], pp. 19–20 for a detailed exposition):

* **Associativity:** For any (flat) local tangent fields u, v, w, we have:

$$t(u, v, w) = g(u \circ v, w) = g(u, v \circ w), \tag{2}$$

  where t is a rank 3 tensor, g is a metric and $\circ$ is a multiplication on the tangent sheaf.
* **Potentiality:** t admits locally everywhere locally a potential function Φ such that, for any local tangent fields ∂_i we have

$$t(\partial_a, \partial_b, \partial_c) = \partial_a \partial_b \partial_c \Phi. \tag{3}$$

We prove explicitly that statistical manifolds related to exponential families with a flat structure connection (i.e. $\alpha = \pm 1$, in the notation of Amari, indexing the connection $\overset{\alpha}{\nabla}$, see [2]) strictly obey to the axioms of a *Frobenius manifold.* This algebraic approach to statistical Frobenius manifolds allows us to define an analog of Gromov–Witten invariants: the statistical Gromov–Witten invariants (GWS). This plays an important role in the learning process, since it determines whether a system has succeeded in learning or failed. Also, it has a geometric interpretation (as its original version) concerning the intersection of (para-)holomorphic curves.

In this short note, we do not discuss the Cartan–Koszul bundle approach of the affine and projective connection [4] and which is related to the modern algebraic approach of the Kazan–Moscow school [13,15,17].

2 Statistical Manifolds and Frobenius Manifolds

We rapidly recall material from the previous part [8]. Let $(\Omega, \mathcal{F}, \lambda)$ be a measure space, where $\mathcal{F}$ denotes the σ-algebra of elements of Ω, and λ is a σ-finite measure. We consider the family of parametric probabilities $\mathfrak{S}$ on the measure space $(\Omega, \mathcal{F})$, absolutely continuous wrt λ. We denote by $\rho_\theta = \frac{dP_\theta}{d\lambda}$, the Radon–Nikodym derivative of $P_\theta \in \mathfrak{S}$, wrt to λ and denote by S the associated family of probability densities of the parametric probabilities. We limit ourselves to the case where S is a smooth topological manifold.

$$S = \left\{ \rho_\theta \in L^1(\Omega, \lambda),\ \theta = \{\theta_1, \dots \theta_n\};\ \rho_\theta > 0\ \lambda - a.e.,\ \int_\Omega \rho_\theta d\lambda = 1 \right\}.$$

This generates the space of probability measures absolutely continuous with respect to the measure λ, i.e. $P_\theta(A) = \int_A \rho_\theta d\lambda$ where $A \subset \mathcal{F}$.

We construct its tangent space as follows. Let $u \in L^2(\Omega, P_\theta)$ be a tangent vector to S at the point ρ_θ.

$$T_\theta = \left\{ u \in L^2(\Omega, P_\theta); \mathbb{E}_{P_\theta}[u] = 0, u = \sum_{i=1}^{d} u^i \partial_i \ell_\theta \right\},$$

where $\mathbb{E}_{P_\theta}[u]$ is the expectation value, w.r. to the probability P_θ.

The tangent space of S is isomorphic to the n-dimensional linear space generated by the centred random variables (also known as score vector) $\{\partial_i \ell_\theta\}_{i=1}^n$, where $\ell_\theta = \ln \rho_\theta$.

In 1945, Rao [14] introduced the Riemannian metric on a statistical manifold, using the Fischer information matrix. The statistical manifold forms a (pseudo)-Riemannian manifold.

In the basis, where $\{\partial_i \ell_\theta\}, \{i = 1, \dots, n\}$ where $\ell_\theta = \ln \rho_\theta$, the Fisher metric are just the *covariance matrix* of the score vector. Citing results of [5] (p. 89) we can in particular state that:

$$g_{i,j}(\theta) = \mathbb{E}_{P_\theta}[\partial_i \ell_\theta \partial_j \ell_\theta]$$

$$g^{i,j}(\theta) = \mathbb{E}_{P_\theta}[a_\theta^i a_\theta^j],$$

where $\{a^i\}$ form a dual basis to $\{\partial_j \ell_\theta\}$:

$$a_\theta^i(\partial_j \ell_\theta) = \mathbb{E}_{P_\theta}[a_\theta^i \partial_j \ell_\theta] = \delta_j^i$$

with

$$\mathbb{E}_{P_\theta}[a_\theta^i] = 0.$$

Definition 1. *A Frobenius manifold is a manifold M endowed with an affine flat structure*[1]*, a compatible metric g, and an even symmetric rank 3 tensor t. Define a symmetric bilinear multiplication on the tangent bundle:*

$$\circ : TM \otimes TM \to TM.$$

M endowed with these structures is called pre-Frobenius.

Definition 2. *A pre-Frobenius manifold is Frobenius if it verifies the associativity and potentiality properties defined in the introduction as Eqs. (2) and (3).*

---

[1] Here the affine flat structure is equivalently described as complete atlas whose transition functions are affine linear. Since the statistical manifolds are (pseudo)-Riemannian manifolds this condition is fulfilled.

Example 1. Take the particular case of (M, g, t), where $t = \nabla - \nabla^* = 0$ and $\nabla \neq 0$. So, $\nabla = \nabla^*$. In this case, the WDVV condition is not satisfied, and therefore the Frobenius manifold axioms are not fulfilled. Indeed, as stated in [3] p. 199, the (WDVV) equations holds only if the curvature vanishes. Therefore, if $t = 0$ and if the manifold has a non flat curvature i.e. $\nabla \neq 0$ then this manifold is not a Frobenius one.

We now discuss the necessary conditions to have a statistical Frobenius manifold. As was stated above, **our attention throughout this paper is restricted only to exponential families, having Markov invariant metrics.**

Let (S, g, t) be a statistical manifold equipped with the (Fischer–Rao) Riemannian metric g and a 3-covariant tensor field t called the *skewness tensor*. It is a covariant tensor of rank 3 which is fully symmetric:

$$t : TS \times TS \times TS \to \mathbb{R},$$

given by

$$t|_{p_\theta}(u, v, w) = \mathbb{E}_{P_\theta}[u_\theta v_\theta w_\theta].$$

In other words, in the score coordinates, we have:

$$t_{ijk}(\theta) = \mathbb{E}_{P_\theta}[\partial_i \ell_\theta \partial_j \ell_\theta \partial_k \ell_\theta].$$

Denote the mixed tensor by $\bar{t} = t.g^{-1}$. It is bilinear map $\bar{t} : TS \times TS \to TS$, given componentwise by:

$$\bar{t}_{ij}^k = g^{km} t_{ijm}, \tag{4}$$

where $g^{km} = \mathbb{E}_{P_\theta}[a_\theta^k a_\theta^m]$. **NB:** This is written using Einstein's convention.

Remark 1. The Einstein convention will be used throughout this paper, whenever needed.

We have:

$$\bar{t}_{ij}^k = \bar{t}|_{p_\theta}(\partial_i \ell_\theta, \partial_j \ell_\theta, a^k) = \mathbb{E}_{P_\theta}[\partial_i \ell_\theta \partial_j \ell_\theta a_\theta^k].$$

As for the connection, it is given by:

$$\overset{\alpha}{\nabla}_X Y = \overset{0}{\nabla}_X Y + \frac{\alpha}{2}\bar{t}(X, Y), \quad \alpha \in \mathbb{R}, X, Y \in T_p S$$

where $\overset{\alpha}{\nabla}_X Y$ denotes the α-covariant derivative.

Remark 2. Whenever we have a pre-Frobenius manifold (S, g, t) we call the connection $\overset{\alpha}{\nabla}$ the *structure connection.*

In fact $\overset{\alpha}{\nabla}$ is the unique torsion free connection satisfying:

$$\overset{\alpha}{\nabla} g = \alpha t,$$

i.e.

$$\overset{\alpha}{\nabla}_X g(Y,Z) = \alpha t(X,Y,Z).$$

Proposition 1. *The tensor $\bar{t}: TS \times TS \to TS$ allows to define a multiplication $\circ$ on TS, such that for all $u, v, \in T_{p_\theta}S$, we have:*

$$u \circ v = \bar{t}(u,v).$$

Proof. By construction, in local coordinates, for any $u, v, \in T_{p_\theta}S$, we have $u = \partial_i \ell_\theta$ and $v = \partial_j \ell_\theta$. In particular, $\partial_i \ell_\theta \otimes \partial_j \ell_\theta = \bar{t}_{ij}^k \partial_k \ell_\theta$, which by calculation turns to be $\mathbb{E}_{P_\theta}[\partial_i \ell_\theta \partial_j \ell_\theta a_\theta^k]$.

Lemma 1. *For any local tangent fields $u, v, w \in T_{p_\theta}S$ the associativity property holds:*

$$g(u \circ v, w) = g(u, v \circ w).$$

Proof. Let us start with the left hand side of the equation. Suppose that $u = \partial_i \ell_\theta$, $v = \partial_j \ell_\theta$, $w = \partial_l \ell_\theta$. By previous calculations: $\partial_i \ell_\theta \circ \partial_j \ell_\theta = \bar{t}_{ij}^k \partial_k \ell_\theta$. Insert this result into $g(u \circ v, w)$, which gives us $g(\partial_i \ell_\theta \circ \partial_j \ell_\theta, \partial_l \ell_\theta)$, and leads to $g(\bar{t}_{ij}^k \partial_k \ell_\theta, \partial_l \ell_\theta)$. By some calculations and formula (4) it turns out to be equal to $t(u,v,w)$.

Consider the right hand side. Let $g(u, v \circ w) = g(\partial_i \ell_\theta, \partial_j \ell_\theta \circ \partial_l \ell_\theta)$. Mimicking the previous approach, we show that this is equivalent to $g(\partial_i \ell_\theta, \bar{t}_{jl}^k \partial_l \ell_\theta)$, which is equal to $t(u,v,w)$.

Theorem 1. *The statistical manifold (S,g,t), related to exponential families, is a Frobenius manifold only for $\alpha = \pm 1$.*

Proof. The statistical manifold S comes equipped with a Riemannian metric g, and a skew symmetric tensor t. We have proved that (S,g,t) is an associative pre-Frobenius manifold (the associativity condition is fulfilled, by Lemma 1). It remains to show that it is Frobenius. We invoke the theorem 1.5 of [12], p. 20 stating that the triplet (S,g,t) is Frobenius if and only if the *structure connection* $\overset{\alpha}{\nabla}$ is flat. The pencil of connections depending on a parameter α are defined by:

$$\overset{\alpha}{\nabla}_X Y = \overset{0}{\nabla}_X Y + \frac{\alpha}{2}(X \circ Y), \quad \alpha \in \mathbb{R}, X, Y \in T_p S$$

where $\overset{\alpha}{\nabla}_X Y$ denotes the α-covariant derivative. By a direct computation, we show that only for $\alpha = \pm 1$, the structure connection is flat. Therefore, the conclusion is straightforward.

The (WDVV) PDE version expresses geometrically a flatness condition for a given manifold. We establish the following statement.

Proposition 2. *For $\alpha = \pm 1$, the (WDVV) PDE system are always (uniquely) integrable over (S,g,t).*

Proof. The WDVV equations are always integrable if and only if the curvature is null. In the context of (S, g, t) the curvature tensor of the covariant derivative is null for $\alpha = \pm 1$. Therefore, in this context the WDVV equation is always integrable (uniquely).

Corollary 1. *The Frobenius manifold (S, g, t), related to exponential families, and indexed by $\alpha = \pm 1$ verifies the potentiality condition.*[2]

3 Statistical Gromov–Witten Invariants and Learning

We introduce Gromov–Witten invariants for statistical manifolds, in short (GWS). Originally, those invariants are rational numbers that count (pseudo) holomorphic curves under some conditions on a (symplectic) manifold. The (GWS) encode deep geometric aspects of statistical manifolds, (similarly to its original version) concerning the intersection of (para-)holomorphic curves. Also, this plays an important role in the learning process, since it determines whether a system has succeeded in learning or failed.

Let us consider the (formal) Frobenius manifold (H, g). We denote k a (super) commutative $\mathbb{Q}$-algebra. Let H be a k-module of finite rank and $g : H \otimes H \to k$ an even symmetric pairing (which is non degenerate). We denote H^* the dual to H. The structure of the Formal Frobenius manifold on (H, g) is given by an even potential $\mathbf{\Phi} \in k[[H^*]]$:

$$\mathbf{\Phi} = \sum_{n \geq 3} \frac{1}{n!} Y_n,$$

where $Y_n \in (H^*)^{\otimes n}$ can also be considered as an even symmetric map $H^{\otimes n} \to k$. This system of *Abstract Correlation Functions* in (H, g) is a system of (symmetric, even) polynomials. The Gromov–Witten invariants appear in these multilinear maps.

We go back to statistical manifolds. Let us consider the discrete case of the exponential family formula:

$$\sum_{\omega \in \Omega} \exp\{-\sum \beta^j X_j(\omega)\} = \sum_{\omega \in \Omega} \sum_{m \geq 1} \frac{1}{m!} \left\{ -\sum_j \beta^j X_j(\omega) \right\}^{\otimes m}, \tag{5}$$

where $\beta = (\beta_0, ..., \beta_n) \in \mathbb{R}^{n+1}$ is a canonical affine parametrisation, $X_j(\omega)$ are directional co-vectors, belonging to a finite cardinality $n + 1$ list $\mathcal{X}_n$ of random variables. These co-vectors represent necessary and sufficient statistics of

[2] Another interpretation goes as follows. Since for $\alpha = \pm 1$, (S, g, t) is a Frobenius manifold (i.e. a flat manifold), it satisfies the condition to apply theorem 4.4 in [3]. There exist potential functions (which are convex functions) $\Psi(\theta)$ and $\Phi(\eta)$ such that the metric tensor is given by: $g_{ij} = \frac{\partial^2}{\partial \theta^i \partial \theta^j} \Phi$ and $g^{ij} = \frac{\partial^2}{\partial \eta^i \partial \eta^j} \Psi$, where there exists a pair of dual coordinate systems (θ, η) such that $\theta = \{\theta_1, \ldots, \theta_n\}$ is α-affine and $\eta = \{\eta_1, ..., \eta_n\}$ is a $-\alpha$-affine coordinate system. Convexity refers to local coordinates and not to any metric.

the exponential family. We have $X_0(\omega) \equiv 1$, and $X_1(\omega), \ldots, X_n(\omega)$ are linearly independent co-vectors. The family in (5) describes an analytical n-dimensional hypersurface in the statistical manifold. It can be uniquely determined by $n+1$ points in general position.

Definition 3. *Let k be the field of real numbers. Let S be the statistical manifold. The Gromov–Witten invariants for statistical manifolds (GWS) are given by the multi-linear maps:*

$$\tilde{Y}_n : S^{\otimes n} \to k.$$

One can also write them as follows:

$$\tilde{Y}_n \in \left(-\sum_j \beta^j X_j(\omega) \right)^{\otimes n}.$$

These invariants appear as part of the potential function $\tilde{\mathbf{\Phi}}$ which is a Kullback–Liebler entropy function.

One can write the relative entropy function:

$$\tilde{\mathbf{\Phi}} = ln \sum_{\omega \in \Omega} \exp\left(-\sum_j \beta^j X_j(\omega)\right). \tag{6}$$

Therefore, we state the following:

Proposition 3. *The entropy function $\tilde{\mathbf{\Phi}}$ of the statistical manifold relies on the (GWS).*

Proof. Indeed, since $\tilde{\mathbf{\Phi}}$, in formula (6) relies on the polylinear maps $\tilde{Y}_n \in \left(-\sum_j \beta^j X_j(\omega)\right)^{\otimes n}$, defining the (GWS), the statement follows.

Consider the tangent fiber bundle over S, the space of probability distributions, with Lie group G. We denote it by (TS, S, π, G, F), where TS is the total space of the bundle $\pi : TS \to S$ is a continuous surjective map and F the fiber. Recall that for any point ρ on S, the tangent space at ρ is isomorphic to the space of bounded, signed measures vanishing on an ideal I of the $\sigma-$algebra. The Lie group G acts (freely and transitively) on the fibers by $f \overset{h}{\mapsto} f+h$, where h is a parallel transport, and f an element of the total space (see [7] for details).

Remark 3. Consider the (local) fibre bundle $\pi^{-1}(\rho) \cong \{\rho\} \times F$. Then F can be identified to a module over the algebra of paracomplex numbers $\mathfrak{C}$ (see [7,8] for details). By a certain change of basis, this rank 2 algebra generated by $\{e_1, e_2\}$, can always be written as $\langle 1, \varepsilon | \, \varepsilon^2 = 1 \rangle$.

We call a *canonical basis* for this paracomplex algebra, the one given by: $\{e_+, e_-\}$, where $e_\pm = \frac{1}{2}(1 \pm \varepsilon)$. Moreover, any vector $X = \{x^i\}$ in the module over the algebra is written as $\{x^{ia} e_a\}$, where $a \in \{1, 2\}$.

Lemma 2. *Consider the fiber bundle (TS, S, π, G, F). Consider a path γ being a geodesic in S. Consider its fiber F_γ. Then, the fiber contains two connected compontents: (γ^+, γ^-), lying respectively in totally geodesic submanifolds E^+ and E^-.*

Proof. Consider the fiber above γ. Since for any point of S, its the tangent space is identified to module over paracomplex numbers. This space is decomposed into a pair of subspaces (i.e. eigenspaces with eigenvalues $\pm\varepsilon$) (see [8]). The geodesic curve in S is a path such that $\gamma = (\gamma^i(t)) : t \in [0,1] \to S$. In local coordinates, the fiber budle is given by $\{\gamma^{ia} e_a\}$, and $a \in \{1,2\}$. Therefore, the fiber over γ has two components (γ^+, γ^-). Taking the canonical basis for $\{e_1, e_2\}$, implies that (γ^+, γ^-) lie respectively in the subspaces E^+ and E^-. These submanifolds are totally geodesic in virtue of Lemma 3 in [8].

We define a learning process through the Ackley–Hilton–Sejnowski method [1], which consists in minimising the Kullback–Leibler divergence. By Propositions 2 and 3 in [9], we can restate it geometrically, as follows:

Proposition 4. *The learning process consists in determining if there exist intersections of the paraholomorphic curve γ^+ with the orthogonal projection of γ^- in the subspace E^+.*

In particular, a learning process succeeds whenever the distance between a geodesic γ^+ and the projected one in E^+ shrinks to become as small as possible.

More formally, as was depicted in [5] (sec. 3) let us denote by Υ the set of (centered) random variables over $(\Omega, \mathcal{F}, P_\theta)$ which admit an expansion in terms of the scores under the following form:

$$\Upsilon_P = \{X \in \mathbb{R}^\Omega \mid X - \mathbb{E}_P[X] = g^{-1}(\mathbb{E}_P[X d\ell]), d\ell\}.$$

By direct calculation, one finds that the log-likelihood $\ell = ln\rho$ of the usual (parametric) families of probability distributions belongs to Υ_p as well as the difference $\ell - \ell^*$ of log-likelihood of two probabilities of the same family. Being given a family of probability distributions such that $\ell \in \Upsilon_P$ for any P, let $\mathcal{U}_P$, let us denote P^* the set such that $\ell - \ell^* \in \Upsilon_p$. Then, for any $P^* \in \mathcal{U}_p$, we define $K(P, P^*) = \mathbb{E}_P[\ell - \ell^*]$.

Theorem 2. *Let (S, g, t) be statistical manifold. Then, the (GWS) determine the learning process.*

Proof. Since $K(P, P^*) = \mathbb{E}_P[\ell - \ell^*]$, this implies that $K(P, P^*)$ is minimised whenever there is a successful learning process. The learning process is by definition given by *deformation* of a pair of geodesics, defined respectively in the pair of totally geodesic manifolds E^+, E^-. Therefore, the (GWS), which arise in the $\tilde{Y}_n$ in the potential function $\tilde{\mathbf{\Phi}}$, which is directly related to the relative entropy function $K(P, P^*)$. Therefore, the (GWS) determine the learning process.

Similarly as in the classical (GW) case, the (GWS) count intersection numbers of certain para-holomorphic curves. In fact, we have the following statement:

Corollary 2. *Let* (TS, S, π, G, F) *be the fiber bundle above. Then, the (GWS) determine the number of intersection of the projected* γ^- *geodesic onto* E^+*, with the* $\gamma^+ \subset E^+$ *geodesic.*

References

1. Ackley, D., Hilton, G., Sejnowski, T.: Learning algorithm for Boltzmann machine. Cogn. Sci. **9**, 147–169 (1985)
2. Amari, S.: Differential-Geometrical Methods in Statistics. Lecture Notes in Statistics, vol. 28. Springer, Heidelberg (1985). https://doi.org/10.1007/978-1-4612-5056-2
3. Ay, N., Jost, J., Vân Lê, H., Schwachhöfer, L.: Information Geometry. A Series of Modern Surveys in Mathematics, vol. 64. Springer, Cham (2017). https://doi.org/10.1007/978-3-319-56478-4
4. Barbaresco, F.: Jean-Louis Koszul and the elementary structures of information geometry. In: Nielsen, F. (ed.) Geometric Structures of Information. SCT, pp. 333–392. Springer, Cham (2019). https://doi.org/10.1007/978-3-030-02520-5_12
5. Burdet, G., Combe, Ph., Nencka, H.: Statistical manifolds, self-parallel curves and learning processes. Seminar on Stochastic Analysis, Random Fields and Applications (Ascona 1996), Progr. Probab., vol. 45, pp. 87–99. Birkhäuser, Basel (1999)
6. Čensov, N.: Statistical Decision Rules and Optimal Inference. Translations of Mathematical Monographs, vol. 53. American Mathematical Society (1982)
7. Combe, N., Manin, Y.I.: F-manifolds and geometry information. Bull. Lond. Math. Soc. **52**, 777–792 (2020)
8. Combe, N.C., Combe, Ph., Nencka, H.: Statistical manifolds & Hidden symmetries. (Submitted to GSI 2021 conference)
9. Combe, Ph., Nencka, H.: Information geometry and learning in formal neural networks. In: Bourguignon, J.-P., Nencka, H. (eds.) Contemporary Mathematics, vol. 203, pp. 105–116. American Mathematical Society, Providence (1997)
10. Hertling, C., Manin, Yu.I.: Weak Frobenius manifolds. Int. Math. Res. Notices **6**, 277–286 (1999)
11. Kobayashi, S., Nomizu, K.: Foundations of Differential Geometry 1, pp. 81–89. Wiley, New York (1996). (1945)
12. Manin, Yu.I.: Frobenius Manifolds, Quantum Cohomology, and Moduli Spaces. American Mathematical Society Colloquium Publications 47 (1999)
13. Norden, A.P.: On intrinsic geometry of the second kind on hypersurfaces of an affine space. (Russian) Izv. Vysš. Učebn. Zaved. Matematika **4**(5), 172–183 (1958)
14. Rao, C.R.: Information and the accuracy attainable in the estimation of statistical parameters. C.R. Bull. Calcutta Math. Soc. **37**, 81–89 (1945)
15. Rozenfeld, B.: Geometry of Lie Groups. Springer, Dordrecht (1997). https://doi.org/10.1007/978-0-387-84794-8
16. Sikorski, R.: Differential modules. Colloquium Mathematicum Fasc. **1**(24), 45–79 (1971)
17. Shirokov, A.P.: Spaces over algebras and their applications. J. Math. Sci. **108**(2), 232–248 (2002)

Geometric Deep Learning

SU(1, 1) Equivariant Neural Networks and Application to Robust Toeplitz Hermitian Positive Definite Matrix Classification

Pierre-Yves Lagrave[1(✉)], Yann Cabanes[2,3], and Frédéric Barbaresco[2]

[1] Thales Research and Technology, Palaiseau, France
pierre-yves.lagrave@thalesgroup.com
[2] Thales Land and Air Systems, Limours, France
frederic.barbaresco@thalesgroup.com
[3] Institut de Mathématiques de Bordeaux, Bordeaux, France

Abstract. In this paper, we propose a practical approach for building SU(1, 1) equivariant neural networks to process data with support within the Poincaré disk $\mathbb{D}$. By leveraging on the transitive action of SU(1, 1) on $\mathbb{D}$, we define an equivariant convolution operator on $\mathbb{D}$ and introduce a Helgason-Fourier analysis approach for its computation, that we compare with a conditional Monte-Carlo method. Finally, we illustrate the performance of such neural networks from both accuracy and robustness standpoints through the example of Toeplitz Hermitian Positive Definite (THPD) matrix classification in the context of radar clutter identification from the corresponding Doppler signals.

Keywords: Equivariant neural networks · Hyperbolic embedding · Homogeneous spaces · Burg algorithm · Radar clutter.

1 Introduction

Group-Convolutional Neural Networks (G-CNN) are becoming more and more popular thanks to their conceptual soundness and to their ability to reach state-of-the-art accuracies for a wide range of applications [3,9,12]. In this paper, we propose a scalable approach for building SU(1, 1) equivariant neural networks and provide numerical results for their application to Toeplitz Hermitian Positive Definite (THPD) matrix classification.

Dealing with THPD matrices is indeed of particular interest for Doppler signal processing tasks, as the input data can be represented by auto-correlation matrices assuming some stationarity assumptions [5], which can themselves be embedded into a product space of Poincaré disks after proper rescaling [6] by leveraging on the Trench-Verblunsky theorem. We have instanciated a SU(1, 1) equivariant neural network in this context and have provided accuracy and robustness results by using simulated data according to some realistic radar clutter modeling [6].

F. Nielsen and F. Barbaresco (Eds.): GSI 2021, LNCS 12829, pp. 577–584, 2021.
https://doi.org/10.1007/978-3-030-80209-7_62

1.1 Related Work and Contribution

G-CNN have initially been introduced in the seminal work [8] as a generalization of usual CNN [15] by introducing group-based convolution kernels to achieve equivariance with respect to the action of finite groups and have further been generalized to more generic actions [9,14]. In particular, [10] proposed a sound theory for the case of the transitive action of a compact group on its homogeneous space by leveraging on group representation theory, a topic also covered in [7].

In [2], G-CNN on Lie groups have been introduced for homogeneous spaces by decomposing the kernel function on a B-splines basis. Anchoring in similar ideas, [12] introduced a very generic approach providing equivariance to the action of any Lie group with a surjective exponential map and which is applicable to any input data representable by a function defined on a smooth manifold and valued in a vectorial space.

Practical applications of G-CNN mainly relate to groups of transforms in Euclidean spaces such as SE(2), SO(3), $\mathbb{R}_+^*$, or the semi-product of those with the translation group. In the present paper, we work with the group of isometric transforms $SU(1,1)$ acting on the hyperbolic space represented as the Poincaré disk $\mathbb{D}$.

Although existing works provide tools for specifying $SU(1,1)$ based G-CNN on which we will anchor, working with non-compact groups is quite challenging from a numerical stanpoint. For instance, the Monte-Carlo approach of [12] faces scalability issue with the number of layers of the considered network due to the large sampling requirement for $SU(1,1)$. As a remediation for our specific $SU(1,1)$ instanciation, we propose specifying the convolution operator on the corresponding homogeneous space $\mathbb{D}$ rather than on the group itself in order to leverage on Helgason-Fourier analysis, following the ideas introduced for compact group actions in [14] and generalized in [7].

Finally, we have applied our approach to the problem of classifying THPD matrices arising in the context of Doppler signals processing. We propose here an alternative approach to [5] by using a $SU(1,1)$ equivariant neural network operating on the hyperbolic representation of the THPD matrices manifold.

1.2 Preliminaries

We denote $\mathbb{D}$ the Poincaré unit disk $\mathbb{D} = \{z = x + iy \in \mathbb{C} / |z| < 1\}$ and by $\mathbb{T}$ the unit circle $\mathbb{T} = \{z = x + iy \in \mathbb{C} / |z| = 1\}$. We further introduce the following Lie groups:

$$SU(1,1) = \left\{ g_{\alpha,\beta} = \begin{bmatrix} \alpha & \beta \\ \bar{\beta} & \bar{\alpha} \end{bmatrix}, |\alpha|^2 - |\beta|^2 = 1, \alpha, \beta \in \mathbb{C} \right\} \tag{1}$$

$$U(1) = \left\{ \begin{bmatrix} \alpha / |\alpha| & 0 \\ 0 & \bar{\alpha} / |\alpha| \end{bmatrix}, \alpha \in \mathbb{C} \right\} \tag{2}$$

We can endow $\mathbb{D}$ with a transitive action $\circ$ of $SU(1,1)$ defined as it follows

$$\forall g_{\alpha,\beta} \in SU(1,1),\ \forall z \in \mathbb{D},\ g_{\alpha,\beta} \circ z = \frac{\alpha z + \beta}{\bar{\beta} z + \bar{\alpha}} \tag{3}$$

The Cartan decomposition associated with SU(1, 1) allows building a diffeomorphism between SU(1, 1) (seen as a topological space) and $\mathbb{T}\times\mathbb{D}$ [11]. As $\mathbb{T} \simeq U(1)$ as a group, we have $\mathbb{D} \simeq \mathrm{SU}(1,1)/\mathrm{U}(1)$ and we can therefore associate $z \in \mathbb{D}$ with an element of the quotient space $\mathrm{SU}(1,1)/\mathrm{U}(1)$.

We consider in the following a supervised learning set-up in which the input data space $\mathscr{X}$ corresponds to functionals defined on $\mathbb{D}$ and valued in a vectorial space V. Hence, a data point $x_i \in \mathscr{X}$ will be represented by the graph of a function $f_{x_i} : \mathbb{D} \to V$, so that $x_i = \{(z, f_{x_i}(z)),\ z \in \mathbb{D}\}$. We then extend the action $\circ$ of SU(1, 1) on $\mathbb{D}$ to the considered input data space by writing, $\forall g \in \mathrm{SU}(1,1)$ and $\forall x \in \mathscr{X}$, $g\circ x = \{(z, f_x(g \circ z)),\ z \in \mathbb{D}\}$.

2 Equivariant Convolution Operator

We introduce in this section a convolution operator defined on the Poincaré disk $\mathbb{D}$ and which is equivariant with respect to the action of SU(1, 1). Corresponding feature maps can then be embedded within usual neural network topologies. After introducing local convolution kernels, we discuss the numerical aspects associated with practical computations.

2.1 Convolution on $\mathbb{D}$

Following [13], we will denote by μ^G the Haar measure of $G = \mathrm{SU}(1,1)$ that is normalized according to $\int_G F(g \circ 0_{\mathbb{D}})\, d\mu^G(g) = \int_{\mathbb{D}} F(z)\, dm(z)$ for all $F \in L^1(\mathbb{D}, dm)$, where $L^1(X, d\mu)$ refers to the set of functions from X to $\mathbb{C}$ which are integrable with respect to the measure μ, with $0_{\mathbb{D}}$ the center of $\mathbb{D}$ and where the measure dm is given for $z = z_1 + iz_2$ by $dm(z) = \frac{dz_1 dz_2}{(1-|z|^2)^2}$. We can then build feature maps $\psi : \mathbb{D} \to \mathbb{C}$ by leveraging on the convolution operator on $\mathbb{D}$.

More precisely, for a kernel function $k_\theta : \mathbb{D} \to \mathbb{C}$ parameterized by $\theta \in \mathbb{R}^\ell$, an input feature map $f : \mathbb{D} \to \mathbb{C}$ and an element $z \in \mathbb{D}$, we will consider the following operator ψ^θ

$$\psi^\theta(z) = (k_\theta \star_{\mathbb{D}} f)(z) = \int_G k_\theta(g^{-1} \circ z)\, f(g \circ 0_{\mathbb{D}})\, d\mu^G(g) \tag{4}$$

as long as the right handside integral is well defined. The operator ψ^θ can actually be seen as a specific group-convolution operator which is constant over each coset so that it only requires to be evaluated on $\mathbb{D}$.

Denoting L_g the left shift operator such that $\forall z \in \mathbb{D}$, $\forall g \in G$ and $\forall \phi : \mathbb{D} \to \mathbb{C}$, $L_g\phi(z) = \phi(g^{-1} \circ z)$, we show that ψ^θ is equivariant with respect to the action of G by writing:

$$\begin{aligned}
L_h\psi^\theta(z) &= \int_G k_\theta\left(g^{-1} \circ (h^{-1} \circ z)\right) f(g \circ 0_{\mathbb{D}})\, d\mu^G(g) \\
&\overset{\text{action}}{=} \int_G k_\theta\left((hg)^{-1} \circ z\right) f(g \circ 0_{\mathbb{D}})\, d\mu^G(g) \\
&\overset{\text{Haar}}{=} \int_G k_\theta(g^{-1} \circ z)\, L_h f(g \circ 0_{\mathbb{D}})\, d\mu^G(g) = k_\theta \star_{\mathbb{D}} (L_h f)
\end{aligned}$$

2.2 Kernel Modeling and Locality

To accomodate with the Euclidean nature of a wide range of kernel functionals, we propose projecting the input data to k_θ to a vectorial space with the logarithm map $\log_{\mathbb{D}}$ from $\mathbb{D}$ to its tangent space $(\mathcal{T}\mathbb{D})_{0_{\mathbb{D}}}$ at its center. Following [12], we will generically model the kernel functional as a simple neural network. However, for the sake of computational efficiency, specific parameterizations can also be envisioned such as the use of a B-splines expansion as in [2], or more generically, the use of a well suited functional expansion in $(\mathcal{T}\mathbb{D})_{0_{\mathbb{D}}}$.

To localize the kernel action, we leverage on the geodesic distance in the Poincaré disk $\rho_{\mathbb{D}}$ and we then restrict the integration in the convolution operator to the points "close" to z according to the metric $\rho_{\mathbb{D}}$, leading to

$$\psi^{\theta,M}(z) = \int_{\rho_{\mathbb{D}}(z, g \circ 0_{\mathbb{D}}) \leq M} k_\theta\left(\log_{\mathbb{D}}\left(g^{-1} \circ z\right)\right) f\left(g \circ 0_{\mathbb{D}}\right) d\mu^G(g) \tag{5}$$

where M is a threshold to be considered as a model hyperparameter and which could be assimilated as the filter bandwidth in standard CNN. It should also be noted that the localization of the kernel function ensures the definition of the integral in the right handside of (5).

2.3 Numerical Aspects

In practice, the integral (5) cannot be computed exactly and a numerical scheme is therefore required for evaluating the convolution maps on a discrete grid of points in $\mathbb{D}$. In this section, we tackle the scalability challenges associated with $\mathrm{SU}(1,1)$ equivariant networks operating on the hyperbolic space $\mathbb{D}$.

Monte-Carlo Method. Following [12], a first approach to the estimation of (5) is to compute, $\forall z \in \mathbb{D}$, the following Monte-Carlo estimator,

$$\psi_N^{\theta,M}(z) = \frac{1}{N} \sum_{i=1}^{N} k_\theta\left(\log_{\mathbb{D}}\left(g_i^{-1} \circ z\right)\right) f\left(g_i \circ 0_{\mathbb{D}}\right) \tag{6}$$

where the $g_i \in \mathrm{SU}(1,1)$ are drawn according to the conditional Haar measure $\mu^G\big|_{\mathcal{B}(z,M)}$, where $\mathcal{B}(z,M)$ is the ball centered in z and of radius M measured according to the metric $\rho_{\mathbb{D}}$. To do so efficiently, we first sample elements in $\mathcal{B}(0_{\mathbb{D}}, M)$ and then move to $\mathcal{B}(z,M)$ through the group action. The obtained coset elements are then lifted to $\mathrm{SU}(1,1)$ by sampling random elements of $\mathrm{U}(1)$.

Although appealing from a an implementation standpoint, this approach is severely limited by its exponential complexity in the number of layers of the considered neural network. To illustrate this point, let's consider a simple architecture with ℓ convolutional layers of one kernel each and let's assume that we are targeting an evaluation of the last convolution map of the layer ℓ on m points. By denoting n_i the number of evaluation points required for the evaluation of $\psi_i^{\theta_i}$ and by using a simple induction, we show that $n_i = mN^{\ell-i+1}$, for $1 \leq i \leq \ell$.

Approaches such as [17], further generalized in [12], have been introduced to reduce the computational cost by using a clever reordering of the computations. However, the corresponding complexity remains exponential and therefore prevents from building deep networks with this Monte-Carlo approach, especially when working with groups such as $\mathrm{SU}(1,1)$ which require significant sampling.

Helgason-Fourier Analysis. The use of Fourier analysis within G-CNN architectures to evaluate convolution operators defined on Euclidean spaces has been previously suggested [9,14]. By leveraging on similar ideas, we here propose an alternative approach to the Monte-Carlo sampling on $\mathrm{SU}(1,1)$ by using Helgason-Fourier (HF) transforms [13] of functions defined on the hyperbolic space $\mathbb{D}$.

For points $b \in \partial\mathbb{D}$ and $z \in \mathbb{D}$, we denote $\langle z, b\rangle$ the algebraic distance to the center of the horocyle based at b through z. Let's then consider $f : \mathbb{D} \to \mathbb{C}$ such that $z \to f(z)e^{(1-i\rho)\langle z,b\rangle} \in L^1(\mathbb{D}, dm)$, $\forall\rho \in \mathbb{R}$. The HF transform of f is then defined on $\mathbb{R} \times \partial\mathbb{D}$ by

$$\mathcal{HF}(f)(\rho, b) = \int_{\mathbb{D}} f(z)\, e^{(1-i\rho)\langle z,b\rangle} dm(z) \tag{7}$$

[13] gives the formal expression for the horocyclic wave $e^{\nu\langle z,b\rangle}$, for $\nu \in \mathbb{C}$. Assuming corresponding integrability conditions, the HF transform of the convolution of two functions $F_1, F_2 : \mathbb{D} \to \mathbb{C}$ is given by

$$\mathcal{HF}(F_1 \star_{\mathbb{D}} F_2)(\rho, b) = \mathcal{HF}(F_1)(\rho, b)\,\mathcal{HF}(F_2)(\rho, b) \tag{8}$$

As for the Euclidean case, an inversion formula also exists and is given by

$$f(z) = \frac{1}{4\pi}\int_{\mathbb{R}}\int_{\partial\mathbb{D}} \mathcal{HF}(f)(\rho, b)\, e^{(1+i\rho)\langle z,b\rangle}\rho \tanh\left(\frac{\pi\rho}{2}\right) d\rho db \tag{9}$$

Using HF transforms allows working with independent discretization grids for the convolution maps computation across the network structure, but the pointwise evaluation of each map requires the computation of three integrals, instead of one in the Monte-Carlo approach. Also, the numerical estimation of the integrals (7) and (9) is quite challenging and leads to convergence issues with classical schemes. We have obtained satisfactory results by using the Quasi-Monte-Carlo approach [4] and the corresponding GPU-compatible implementation provided by the authors. A differentiable estimation of $\mathcal{HF}\left(z \to k_{\theta_i}(z)\,\mathbf{1}_{\rho_{\mathbb{D}}(0_{\mathbb{D}},z)\leq M}\right)$ can therefore be obtained efficiently by using a well chosen functional expansion of the kernel (e.g., basis of the Fock-Bargmann Hilbert space [11], B-splines [3], eigenfunctions of the Laplacian on $\mathbb{D}$ [13]) and by pre-computing the HF transforms of the corresponding basis elements.

3 Application to THPD Matrix Classification

Let's denote by $\mathcal{T}_n^+$ the set of Toeplitz Hermitian Positive Definite (THPD) matrices of size n. As described in [6], the regularized Burg algorithm allows transforming a given matrix $\Gamma \in \mathcal{T}_n^+$ into a power factor in $\mathbb{R}_+^*$ and reflection coefficients μ_i in a $(n-1)$ dimensional product of Poincaré spaces $\mathbb{D}^{n-1} = \mathbb{D} \times ... \times \mathbb{D}$. Combining this with the formalism introduced in Sect. 1.2, we can therefore see an element $\Gamma \in \mathcal{T}_n^+$ as a power factor in $\mathbb{R}_+^*$ and $n-1$ coset elements in $\mathrm{SU}(1,1)/\mathrm{U}(1)$.

In the following, we will focus on radar clutter classification and consider in this context rescaled THPD matrices which can be represented by the reflection coefficients only. We further assume that the ordering of these coefficients is not material for our considered classification task, so that we will represent a given element $\Gamma \in \mathcal{T}_n^+$ by folding into $\mathbb{D}$ with the function $f_\Gamma : \mathbb{D} \to \mathbb{C}$, with $f_\Gamma(\mu) = \sum_{i=1}^{n-1} \mu^{\mathbb{C}} \mathbf{1}_{\{\mu=\mu_i\}}$ for $\mu \in \mathbb{D}$ and where $\mu \to \mu^{\mathbb{C}}$ is the canonical embedding from $\mathbb{D}$ to $\mathbb{C}$.

3.1 Classification Problem

We have considered the problem of pathological radar clutter classification as formulated in [6]. More precisely, we represent each cell by its THPD auto-correlation matrix Γ, our goal being to predict the corresponding clutter $c \in \{1, ..., n_c\}$ from the observation of Γ. Within our formalism, the training samples are of the form (f_{Γ_i}, c_i) and the input data have been obtained by simulating the auto-correlation matrix of a given cell according to

$$Z = \sqrt{\tau} R^{1/2} x + b_{radar} \tag{10}$$

where τ is a positive random variable corresponding to the clutter texture, R a THPD matrix associated with a given clutter, $x \sim \mathcal{N}_{\mathbb{C}}(0, \sigma^x)$ and $b_{radar} \sim \mathcal{N}_{\mathbb{C}}(0, \sigma)$, with $\mathcal{N}_{\mathbb{C}}(0, t)$ referring to the complex gaussian distribution with mean 0 and standard deviation t. In the following, b_{radar} will be considered as a source of thermal noise inherent to the sensor.

3.2 Numerical Results

We have instanciated a simple neural network constituted of one $\mathrm{SU}(1,1)$ convolutional layer with two filters and ReLu activation functions, followed by one fully connected layer and one softmax layer operating on the complex numbers represented as 2-dimensional tensors. The kernel functions are modeled as a neural networks with one layer of 16 neurons with swish activation functions. The two convolution maps have been evaluated on the same grid constituted of 100 elements of $\mathbb{D}$ sampled according to the corresponding volume measure.

To appreciate the improvement provided by our approach, we will compare the obtained results with those corresponding to the use of a conventional neural network with roughly the same numbers of trainable parameters and operating on the complex reflection coefficients. In the following, we will denote $\mathcal{N}_\sigma^G$ (resp. $\mathcal{N}_\sigma^{FC}$) the neural network with $\mathrm{SU}(1,1)$ equivariant convolutional (resp. fully connected) layers and trained on 400 THPD matrices of dimension 10 corresponding to 4 different classes (100 samples in each class) which have been simulated according to (10) with a thermal noise standard deviation σ.

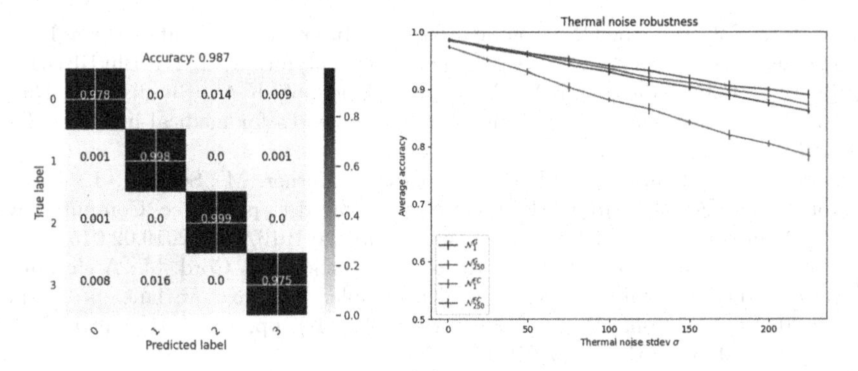

Fig. 1. Left handside: confusion matrix corresponding to the evaluation of $\mathcal{N}_1^G$ on the testing set T_1, averaged over 10 realizations. Right handside: average accuracy results of the algorithms $\mathcal{N}_1^G$, $\mathcal{N}_1^{FC}$, $\mathcal{N}_{250}^G$ and $\mathcal{N}_{250}^{FC}$ on the testing sets T_σ shown as a function of σ, together with the corresponding standard deviation as error bars

In order to evaluate the algorithms $\mathcal{N}_\sigma^G$ and $\mathcal{N}_\sigma^{FC}$, we have considered several testing sets T_σ consisting in 2000 THPD matrices of dimension 10 (500 samples in each of the 4 classes) simulated according to (10) with a thermal noise standard deviation σ. The obtained results are shown on Fig. 1 where it can in particular be seen that $\mathcal{N}_1^G$ reaches similar performances as $\mathcal{N}_{250}^G$ and $\mathcal{N}_{250}^{FC}$ while significantly outperforming $\mathcal{N}_1^{FC}$ as σ increases, meaning that our approach allows to achieve some degree of robustness with respect to the variation of σ through the use of equivariant layers.

4 Conclusions and Further Work

Following recent work on G-CNN specification, we have introduced an approach for building $\mathrm{SU}(1,1)$ equivariant neural networks and have discussed how to deal with corresponding numerical challenges by leveraging on Helgason-Fourier analysis. We have instanciated the proposed algorithm for the problem of THPD matrix classification arising in the context of Doppler signals processing and have obtained satisfactory accuracy and robustness results. Further work will include studying $\mathrm{SU}(n,n)$ equivariant neural networks and their applicability to the problem of classifying Positive Definite Block-Toeplitz matrices by leveraging on their embedding into Siegel spaces [1]. We will also investigate the link with the coadjoint representation theory which may allow generalizing our approach while transferring convolution computation techniques to coadjoint orbits [16].

References

1. Barbaresco, F.: Radar micro-doppler signal encoding in Siegel unit poly-disk for machine learning in Fisher metric space. In: 2018 19th International Radar Symposium (IRS), pp. 1–10 (2018). https://doi.org/10.23919/IRS.2018.8448021

2. Bekkers, E.J.: B-spline CNNs on lie groups. In: International Conference on Learning Representations (2020). https://openreview.net/forum?id=H1gBhkBFDH
3. Bekkers, E.J., Lafarge, M.W., Veta, M., Eppenhof, K.A., Pluim, J.P., Duits, R.: Roto-translation covariant convolutional networks for medical image analysis (2018)
4. Borowka, S., Heinrich, G., Jahn, S., Jones, S., Kerner, M., Schlenk, J.: A GPU compatible Quasi-Monte Carlo integrator interfaced to pySecDec. Comput. Phys. Commun. **240**, 120–137 (2019). https://doi.org/10.1016/j.cpc.2019.02.015
5. Brooks, D., Schwander, O., Barbaresco, F., Schneider, J., Cord, M.: A Hermitian positive definite neural network for micro-doppler complex covariance processing. In: 2019 International Radar Conference (RADAR), pp. 1–6 (2019). https://doi.org/10.1109/RADAR41533.2019.171277
6. Cabanes, Y., Barbaresco, F., Arnaudon, M., Bigot, J.: Toeplitz Hermitian positive definite matrix machine learning based on Fisher metric. In: Nielsen, F., Barbaresco, F. (eds.) GSI 2019. LNCS, vol. 11712, pp. 261–270. Springer, Cham (2019). https://doi.org/10.1007/978-3-030-26980-7_27
7. Chakraborty, R., Banerjee, M., Vemuri, B.C.: H-CNNs: convolutional neural networks for Riemannian homogeneous spaces. arXiv preprint arXiv:1805.05487 1 (2018)
8. Cohen, T., Welling, M.: Group equivariant convolutional networks. In: Balcan, M.F., Weinberger, K.Q. (eds.) Proceedings of The 33rd International Conference on Machine Learning. Proceedings of Machine Learning Research, vol. 48, pp. 2990–2999. PMLR, New York, 20–22 June 2016. http://proceedings.mlr.press/v48/cohenc16.html
9. Cohen, T.S., Geiger, M., Köhler, J., Welling, M.: Spherical CNNs. CoRR abs/1801.10130 (2018)
10. Cohen, T.S., Geiger, M., Weiler, M.: A general theory of equivariant CNNs on homogeneous spaces. In: Wallach, H., Larochelle, H., Beygelzimer, A., Alché-Buc, F., Fox, E., Garnett, R. (eds.) Advances in Neural Information Processing Systems, vol. 32, pp. 9145–9156. Curran Associates, Inc. (2019)
11. del Olmo, M.A., Gazeau, J.P.: Covariant integral quantization of the unit disk. J. Math. Phys. **61**(2) (2020). https://doi.org/10.1063/1.5128066
12. Finzi, M., Stanton, S., Izmailov, P., Wilson, A.G.: Generalizing convolutional neural networks for equivariance to lie groups on arbitrary continuous data (2020)
13. Helgason, S.: Groups and geometric analysis (1984)
14. Kondor, R., Trivedi, S.: On the generalization of equivariance and convolution in neural networks to the action of compact groups. In: Dy, J., Krause, A. (eds.) Proceedings of the 35th International Conference on Machine Learning. Proceedings of Machine Learning Research, vol. 80, pp. 2747–2755. PMLR, Stockholmsmässan, 10–15 July 2018. http://proceedings.mlr.press/v80/kondor18a.html
15. Lecun, Y., Bottou, L., Bengio, Y., Haffner, P.: Gradient-based learning applied to document recognition. Proc. IEEE **86**(11), 2278–2324 (1998). https://doi.org/10.1109/5.726791
16. Rieffel, M.A.: Lie group convolution algebras as deformation quantizations of linear poisson structures. Am. J. Math. **112**(4), 657–685 (1990)
17. Wu, W., Qi, Z., Fuxin, L.: PointConv: deep convolutional networks on 3d point clouds. In: Proceedings of the IEEE/CVF Conference on Computer Vision and Pattern Recognition (CVPR), June 2019

Iterative SE(3)-Transformers

Fabian B. Fuchs[1], Edward Wagstaff[1(✉)], Justas Dauparas[2], and Ingmar Posner[1]

[1] Department of Engineering Science, University of Oxford, Oxford, UK
{fabian,ed}@robots.ox.ac.uk
[2] Institute for Protein Design, University of Washington, Seattle, WA, USA

Abstract. When manipulating three-dimensional data, it is possible to ensure that rotational and translational symmetries are respected by applying so-called *SE(3)-equivariant* models. Protein structure prediction is a prominent example of a task which displays these symmetries. Recent work in this area has successfully made use of an SE(3)-equivariant model, applying an iterative SE(3)-equivariant attention mechanism. Motivated by this application, we implement an iterative version of the SE(3)-Transformer, an SE(3)-equivariant attention-based model for graph data. We address the additional complications which arise when applying the SE(3)-Transformer in an iterative fashion, compare the iterative and single-pass versions on a toy problem, and consider why an iterative model may be beneficial in some problem settings. We make the code for our implementation available to the community (https://github.com/FabianFuchsML/se3-transformer-public).

Keywords: Deep learning · Equivariance · Graphs · Proteins

1 Introduction

Tasks involving manipulation of three-dimensional (3D) data often exhibit rotational and translational symmetry, such that the overall orientation or position of the data is not relevant to solving the task. One prominent example of such a task is protein structure refinement [1]. The goal is to improve on the initial 3D structure – the position and orientation of the structure, i.e. the frame of reference, is not important to the goal. We would like to find a mapping from the initial structure to the final structure such that if the initial structure is rotated and translated then the predicted final structure is rotated and translated in the same way. This symmetry between input and output is known as *equivariance*. More specifically, the group of translations and rotations in 3D is called the special Euclidean group and is denoted by SE(3). The relevant symmetry is known as SE(3) equivariance.

In the latest Community-Wide Experiment on the Critical Assessment of Techniques for Protein Structure Prediction (CASP14) structure-prediction challenge, DeepMind's AlphaFold 2 team [2] successfully applied machine learning

F. B. Fuchs and E. Wagstaff—Contributed equally.

F. Nielsen and F. Barbaresco (Eds.): GSI 2021, LNCS 12829, pp. 585–595, 2021.
https://doi.org/10.1007/978-3-030-80209-7_63

techniques to win the categories "regular targets" and "interdomain prediction" by a wide margin. This is a important achievement as it opens up new routes for understanding diseases and drug discovery in cases where protein structures cannot be experimentally determined [3]. At this time, the full implementation details of the AlphaFold 2 algorithm are not public. Consequently, there is a lot of interest in understanding and reimplementing AlphaFold 2 [4–8].

One of the core aspects that enabled AlphaFold 2 to produce very high quality structures is the end-to-end iterative refinement of protein structures [2]. Concretely, the inputs for the refinement task are the estimated coordinates of the protein, and the outputs are updates to these coordinates. This task is equivariant: when rotating the input, the update vectors are rotated identically. To leverage this symmetry, AlphaFold 2 uses an SE(3)-equivariant attention network. The first such SE(3)-equivariant attention network described in the literature is the SE(3)-Transformer [9]. However, in the original paper, the SE(3)-Transformer is only described as a single-pass predictor, and its use in an iterative fashion is not considered.

In this paper, we present an implementation of an iterative version of the SE(3)-Transformer, with a discussion of the additional complications which arise in this iterative setting. In particular, the backward pass is altered, as the gradient of the loss with respect to the model parameters could flow through basis functions. We conduct toy experiments to compare the iterative and single-pass versions of the architecture, draw conclusions about why this architecture choice has been made in the context of protein structure prediction, and consider in which other scenarios it may be a useful choice. The code will be made publicly available[1].

2 Background

In this section we provide a brief overview of SE(3) equivariance, with a description of some prominent SE(3)-equivariant machine learning models. To situate this discussion in a concrete setting, we take motivation from the task of protein structure prediction and refinement, which is subject to SE(3) symmetry.

2.1 Protein Structure Prediction and Refinement

In protein structure prediction, we are given a target sequence of amino acids, and our task is to return 3D coordinates of all the atoms in the encoded protein. Additional information is often needed to solve this problem – the target sequence may be used to find similar sequences and related structures in protein databases first [10,11]. Such coevolutionary data can be used to predict likely interresidue distances using deep learning – an approach that has been dominating protein structure prediction in recent years [12–14]. Coevolutionary information is encoded in a multiple sequence alignment (MSA) [15], which can be used to learn pairwise features such as residue distances and orientations [14].

[1] https://github.com/FabianFuchsML/se3-transformer-public.

These pairwise features are constraints on the structure of the protein, which inform the prediction of the output structure. One could start with random 3D coordinates for the protein chain and use constraints from learnt pairwise features to find the best structure according to those constraints. The problem can then be approached in an iterative way, by feeding the new coordinates back in as inputs and further improving the structure. This iterative approach can help to improve predictions [16].

Importantly, MSA and pairwise features do not include a global orientation of the protein – in other words, they are invariant under rotation of the protein. This allows for the application of SE(3)-equivariant networks, which respect this symmetry by design, as done in AlphaFold 2 [2]. The predictions of an SE(3)-equivariant network, in this case predicted shifts to backbone and side chain atoms, are always relative to the arbitrary input frame of reference, without the need for data augmentation. SE(3)-equivariant networks may also be applied in an iterative fashion, and when doing so it is possible to propagate gradients through the whole structure prediction pipeline. This full gradient propagation contrasts with the disconnected structure refinement pipeline of the first version of AlphaFold [12].

2.2 Equivariance and the SE(3)-Transformer

A function, task, feature[2], or neural network is *equivariant* if transforming the input results in an equivalent transformation of the output. Using rotations $\mathbf{R}$ as an example, this condition reads:

$$f(\mathbf{R} \cdot \vec{x}) = \mathbf{R} \cdot f(\vec{x}) \tag{1}$$

In the following, we will focus on 3D rotations only – this group of rotations is denoted SO(3). Adding translation equivariance, i.e. going from SO(3) to SE(3), is easily achieved by considering relative positions or by subtracting the center of mass from all coordinates.

A set of data relating to points in 3D may be represented as a graph. Each node has spatial coordinates, as well as an associated feature vector which encodes further relevant data. A node could represent an atom, with a feature vector describing its momentum. Each edge of the graph also has a feature vector, which encodes data about interactions between pairs of nodes. In an equivariant problem, it is crucial that equivariance applies not only to the positions of the nodes, but also to all feature vectors – for SO(3) equivariance, the feature vectors must rotate to match any rotation of the input. To distinguish such equivariant features from ordinary neural network features, we refer to them as *filaments*.

A filament could, for example, encode momentum and mass as a 4 dimensional vector, formed by concatenating the two components. A momentum vector would typically be 3 dimensional (also called *type-1*), and is rotated by a 3×3 matrix. The mass is scalar information (also called *type-0*), and is invariant to rotation.

[2] A *feature* of a neural network is an input or output of any layer of the network.

The concept of *types* is used in the representation theory of SO(3), where a type-ℓ feature is rotated by a $(2\ell + 1) \times (2\ell + 1)$ Wigner-D matrix.[3] Because a filament is a concatenation of features of different types, the entire filament is rotated by a block diagonal matrix, where each block is a Wigner-D matrix [17–19].

At the input and output layers, the filament structure is determined by the task at hand. For the intermediate layers, arbitrary filament structures can be chosen. In the following, we denote the structure of a filament as a dictionary. E.g. a filament with 3 scalar values (e.g. RGB colour channels) and one velocity vector has 3 type-0 and 1 type-1 feature: {`0:3, 1:1`}. This filament is a feature vector of length 6.

There is a large and active literature on machine learning methods for graph data, most importantly graph neural networks [20–24]. The SE(3)-Transformer [9] in particular is a graph neural network explicitly designed for SE(3)-equivariant tasks, making use of the filament structure and Wigner-D matrices discussed above to enforce equivariance of all features at every layer of the network.

Alternative Approaches to Equivariance: The Wigner-D matrix approach[4] at the core of the SE(3)-Transformer is based on closely related earlier works [17–19]. In contrast, Cohen et al. [25] introduced rotation equivariance by storing copies corresponding to each element of the group in the hidden layers – an approach called regular representations. This was constrained to 90 degree rotations of images. Recent regular representation approaches [26–28] extend this to continuous data by sampling the (infinite) group elements and map the group elements to the corresponding Lie algebra to achieve a smoother representation.

2.3 Equivariant Attention

The second core aspect of an SE(3)-Transformer layer is the self-attention [29] mechanism. This is widely used in machine learning [21,30–35] and based on the principle of keys, queries and values – where each is a learned embedding of the input. The word 'self' describes the fact that keys, queries and values are derived from the same context. In graph neural networks, this mechanism can be used to have nodes attend to their neighbours [21,29,36–38]. Each node serves as a focus point and queries information from the surrounding points. That is, the feature vector f_i of node i is transformed via an equivariant mapping into a query q_i. The feature vectors f_j of the surrounding points j are mapped to equivariant keys k_{ij} and values v_{ij}[5]. A scalar product between key and query – together with a softmax normalisation – gives the attention weight. The scalar product of two rotation equivariant features of the same type gives an invariant feature. Multiplying the invariant weights w_{ij} with the equivariant values v_{ij} gives an equivariant output.

[3] We can think of type-0 features as rotating by the 1×1 rotation matrix (1).

[4] This approach rests on the theory of *irreducible representations* [17,18].

[5] Note that often, the keys and values do not depend on the query node, i.e. $k_{ij} = k_j$. However, in the SE(3)-Transformer, keys and values depend on the relative position between query i and neighbour j as well as on the feature vector f_j.

3 Implementation of an Iterative SE(3)-Transformer

Here, we describe the implementation of the iterative SE(3)-Transformer covering multiple aspects such as gradient flow, equivariance, weight sharing and avoidance of information bottlenecks.

3.1 Gradient Flow in Single-Pass vs. Iterative SE(3)-Transformers

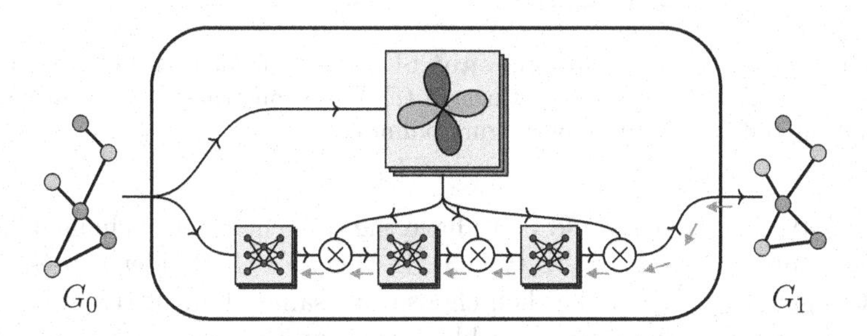

Fig. 1. Gradient flow (orange) in a conventional single-pass SE(3)-Transformer mapping from a graph (left) to an updated graph (right). The equivariant basis kernels (top) need not be differentiated as gradients do not flow through them. (Color figure online)

At the core of the SE(3)-Transformer are the kernel matrices $\mathbf{W}(\vec{x}_j - \vec{x}_i)$ which form the equivariant linear mappings used to obtain keys, queries and values in the attention mechanism. These matrices are a linear combination of basis matrices. The weights for this linear combination are computed by trainable neural networks which take scalar edge features as input. The basis matrices are *not* learnable. They are defined by spherical harmonics and Clebsch-Gordan coefficients, and depend only on the relative positions in the input graph G_0 [9].

Typically, equivariant networks have been applied to tasks where 3D coordinates in the input are mapped to an invariant or equivariant output in a single pass. The relative positions of nodes do not change until the final output, and the basis matrices therefore remain constant. Gradients do not flow through them, and the spherical harmonics need not be differentiated, as can be seen in Fig. 1.

When applying an iterative SE(3)-Transformer (Fig. 2), each block i outputs a graph G_i. This allows for re-evaluating the interactions between pairs of nodes. The relative positions, and thus the basis matrices, change before the final output, and gradients thus flow through the basis. The spherical harmonics used to construct the basis are smooth functions, and therefore backpropagating through them is possible. We provide code implementing this backpropagation.

3.2 Hidden Representations Between Blocks

In the simplest case, each SE(3)-Transformer block outputs a single type-1 feature per point, which is then used as a relative update to the coordinates before applying the second block. This, however, introduces a bottleneck in the information flow. Instead, we choose to maintain the dimensionality of the hidden

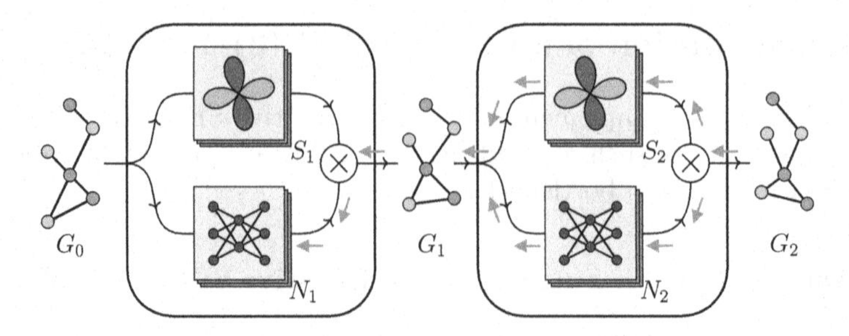

Fig. 2. Gradient flow (orange) in an iterative SE(3)-Transformer with multiple position updates. Now the gradients do flow through the basis construction (S_2) meaning this part has to be differentiated. (Color figure online)

features. A typical choice would be to have the same number of channels (e.g. 4) for each feature type (e.g., up to type 3). In this case the hidden representation reads {`0:4,1:4,2:4,3:4`}. We then choose this same filament structure for the outputs of each SE(3)-Transformer block (except the last) and the inputs to the blocks (except the first). This way, the amount of information saved in the hidden representations is constant throughout the entire network.

3.3 Weight Sharing

If each iteration is expected to solve the same sub-task, weight sharing can make sense in order to reduce overfitting. The effect on memory consumption and speed should, however, be negligible during training, as the basis functions still have to be evaluated and activations have to evaluated and stored separately for each iteration. In our implementation, we chose not to share weights between different SE(3)-Transformer blocks. The number of trainable parameters hence scales linearly with the number of iterations. In theory, this enables the network to leverage information gained in previous steps for deciding on the next update. One benefit of this choice is that it facilitates using larger filaments between the blocks as described in Sect. 3.2 to avoid information bottlenecks. A downside is that it fixes the number of iterations of the SE(3)-Transformer.

3.4 Gradient Descent

In this paper, we will apply the SE(3)-Transformer to an energy optimisation problem, loosely inspired by the protein structure prediction problem. For convex optimsation problems, gradient descent is a simple yet effective algorithm. However, long amino acid chains with a range of different interactions can be assumed to create many local minima. Our hypothesis is that the SE(3)-Transformer is better at escaping the local minimum of the starting configuration and likely to find a better global minimum. We add an optional post-processing step of gradient descent, which optimises the configuration within the minimum that the SE(3)-Transformer found. In real-world protein folding, these two steps do not have to use the same potential. Gradient descent needs a differentiable potential, whereas the SE(3)-Transformer is not subject to that constraint.

4 Experiments

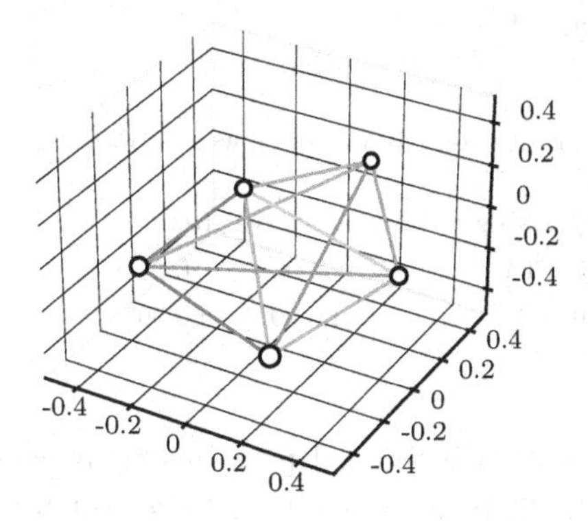

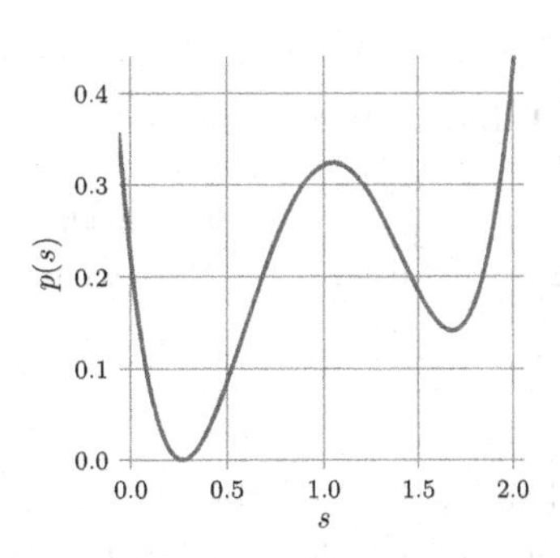

Fig. 3. The left plot shows a configuration of nodes, with a potential between each pair. The gradient of each potential is represented by the colour of the edges. Blue edges indicate repulsion and orange edges indicate attraction. Stronger colour represents stronger interaction. Our experiment uses ten nodes, but we show only five here for visual clarity. The right plot shows the double-minimum, pairwise potential $p(s)$, with parameter $a = 0$. (Color figure online)

We study a physical toy problem as a proof-of-concept and to get insights into what type of tasks could benefit from iterative predictions. We consider an energy minimisation problem with 10 particles, which we will refer to as nodes. Each pair (n_i, n_j) of nodes in the graph interacts according to a potential $p_{ij}(r_{ij})$, where r_{ij} is the distance between the nodes. The goal is to minimise the total value of the potential across all pairs of nodes in the graph (Fig. 3).

We choose a pairwise potential with two local minima. This creates a complex global potential landscape that is not trivially solvable for gradient descent:

$$s_{ij} = r_{ij} - a_{ij} - 1 \tag{2}$$

$$p_{ij}(s_{ij}) = s_{ij}^4 - s_{ij}^2 + \frac{s_{ij}}{10} + p_{\min} \tag{3}$$

Here $p_{\min} \approx 0.32$ ensures that $p_{ij}(s_{ij})$ attains a minimum value of zero. The parameter $a_{ij} = a_{ji}$ is a random number between 0.1 and 1.0 – this stochasticity is necessary to avoid the existence of a single optimal solution for all examples, which the network could simply memorise.

We consider three models for solving this problem: (i) the **single-pass** SE(3)-Transformer (12 layers); (ii) a three-block **iterative** SE(3)-Transformer (4 × 3 layers); (iii) **gradient descent** (GD) on the positions of the nodes. We also evaluate a combination of first applying an SE(3)-Transformer and then running GD on the output. We evaluate the iterative SE(3)-Transformer both with and without propagation of basis function gradients as described in Sect. 3.1 (the latter case corresponds to removing the gradient flow from the upper half of Fig. 2). We run GD until the update changes the potential by less than a fixed tolerance value. We ran each model between 15 and 25 times – the σ values in Tables 1 and 2 represent 1-σ confidence intervals (computed using Student's t-distribution) for the mean performance achieved by each model. All models except GD have the same overall number of parameters.

Table 1. Average energy of the system after optimisation (lower is better).

	Gradient descent	Single-pass	No basis gradients	Iterative	Iterative + GD
Energy	0.0619	0.0942	0.0704	0.0592	0.0410
σ	±0.0001	±0.0002	±0.0025	±0.0011	±0.0003

The results in Table 1 show that the iterative model performs significantly better than the single-pass model, approximately matching the performance of gradient descent. The performance of the iterative model is significantly degraded if gradients are not propagated through the basis functions. The best performing method was the SE(3)-Transformer followed by gradient descent. The fact that the combination of SE(3)-Transformer and gradient descent outperforms pure gradient descent demonstrates that the SE(3)-Transformer finds a genuinely different solution to the one found by gradient descent, moving the problem into a better gradient descent basin.

Table 2. Performance comparison of single-pass and iterative SE(3)-Transformers for different neighbourhood sizes K and a fully connected version (FC).

	FC single (K9)	FC iterative (K9)	K5 single	K5 iterative	K3 single	K3 iterative
Energy	0.0942	0.0592	0.1321	0.0759	0.1527	0.0922
σ	±0.0002	±0.0011	±0.0003	±0.0050	±0.0001	±0.0036

So far, every node attended to all $N-1$ other nodes in each layer, hence creating a fully connected graph. In Table 2, we analyse how limited neighborhood sizes [9] affect the performance, where a neighborhood size of $K = 5$ means that every node attends to 5 other nodes. This reduces complexity from $\mathcal{O}(N^2)$ to $\mathcal{O}(NK)$, which is important in many practical applications with large graphs, such as proteins. We choose the neighbors of each node by selecting the nodes with which it interacts most strongly. In the iterative SE(3)-Transformer, we update the neighbourhoods in each step as the interactions change.

Table 2 shows that the iterative version consistently outperforms the single-pass version across multiple neighborhood sizes, with the absolute difference being the biggest for the smallest K. In particular, it is worth noting that the iterative version with $K = 3$ outperforms the fully connected single-pass network.

In summary, the iterative version consistently outperforms the single-pass version in finding low energy configurations. We emphasise that this experiment is no more than a proof-of-concept. However, we expect that when moving to larger graphs (e.g. proteins often have more than 10^3 atoms), being able to explore different configurations iteratively will only become more important for building an effective optimiser.

Acknowledgements. We thank Georgy Derevyanko, Oiwi Parker Jones and Rob Weston for helpful discussions and feedback. This research was funded by the EPSRC AIMS Centre for Doctoral Training at the University of Oxford and The Open Philanthropy Project Improving Protein Design.

A Network Architecture and Training Details

For the single-pass SE(3)-Transformer we use 12 layers. For the iterative version we use 3 iterations with 4 layers in each iteration. In both cases, for all hidden layers, we use type-0, type-1, and type-2 representations with 4 channels each. The attention uses a single head. The model was trained using an Adam optimizer with a cosine annealing learning rate decay starting at 10^{-3} and ending at 10^{-4}. Our gradient descent implementation uses a step size of 0.02 and stops optimizing when the norm of every position update is below 0.001. We train the SE(3)-Transformers for 100 epochs with 5000 examples per epoch, which takes about 90 to 120 min on an Nvidia Titan V GPU. We did not measure the runtime increase of backpropagating through the basis functions in a controlled environment, but an informal analysis suggested a single digit percentage increase in wall-clock time.

References

1. Adiyaman, R., McGuffin, L.J.: Methods for the refinement of protein structure 3D models. Int. J. Mol. Sci. **20**(9), 2301 (2019)
2. Jumper, J., et al.: High accuracy protein structure prediction using deep learning. In: Fourteenth Critical Assessment of Techniques for Protein Structure Prediction (Abstract Book), 22:24 (2020)
3. Kuhlman, B., Bradley, P.: Advances in protein structure prediction and design. Nat. Rev. Mol. Cell Biol. **20**(11), 681–697 (2019)
4. Rubiera, C.O.: Casp14: what google deepmind's alphafold 2 really achieved, and what it means for protein folding, biology and bioinformatics (2020). https://www.blopig.com/blog/2020/12/casp14-what-google-deepminds-alphafold-2-really-achieved-and-what-it-means-for-protein-folding-biology-and-bioinformatics/
5. Lupoglaz. Openfold2 (2021). https://github.com/lupoglaz/OpenFold2/tree/toy_se3
6. Wang, P.: Se3 transformer - pytorch (2021). https://github.com/lucidrains/se3-transformer-pytorch
7. Markowitz, D.: AlphaFold 2 explained: a semi-deep dive (2020). https://towardsdatascience.com/alphafold-2-explained-a-semi-deep-dive-fa7618c1a7f6
8. AlQuraishi, M.: Alphafold2 @ casp14: "it feels like one's child has left home" (2020). https://moalquraishi.wordpress.com/2020/12/08/alphafold2-casp14-it-feels-like-ones-child-has-left-home/

9. Fuchs, F.B., Worrall, D.E., Fischer, V., Welling, M.: Se(3)-transformers: 3D roto-translation equivariant attention networks. In: Advances in Neural Information Processing System (NeurIPS) (2020)
10. UniProt Consortium: UniProt: a worldwide hub of protein knowledge. Nucleic Acids Res. **47**(D1), D506–D515 (2019)
11. Protein data bank: The single global archive for 3D macromolecular structure data. Nucleic Acids Res. **47**(D1), D520–D528 (2019)
12. Senior, A.W., et al.: Improved protein structure prediction using potentials from deep learning. Nature **577**(7792), 706–710 (2020)
13. Jinbo, X.: Distance-based protein folding powered by deep learning. Proc. Nat. Acad. Sci. **116**(34), 16856–16865 (2019)
14. Yang, J., Anishchenko, I., Park, H., Peng, Z., Ovchinnikov, S., Baker, D.: Improved protein structure prediction using predicted interresidue orientations. Proc. Nat. Acad. Sci. **117**(3), 1496–1503 (2020)
15. Steinegger, M., Meier, M., Mirdita, M., Vöhringer, H., Haunsberger, S.J., Söding, J.: HH-suite3 for fast remote homology detection and deep protein annotation. BMC Bioinformatics **20**(1), 1–15 (2019)
16. Greener, J.G., Kandathil, S.M., Jones, D.T.: Deep learning extends de novo protein modelling coverage of genomes using iteratively predicted structural constraints. Nat. Commun. **10**(1), 1–13 (2019)
17. Thomas, N., et al.: Tensor field networks: Rotation- and translation-equivariant neural networks for 3D point clouds. ArXiv Preprint (2018)
18. Weiler, M., Geiger, M., Welling, M., Boomsma, W., Cohen, T.: 3D steerable CNNs: learning rotationally equivariant features in volumetric data. In: Advances in Neural Information Processing Systems (NeurIPS) (2018)
19. Kondor, R.: N-body networks: a covariant hierarchical neural network architecture for learning atomic potentials. ArXiv preprint (2018)
20. Kipf, T.N., Fetaya, E., Wang, K.-C., Welling, M., Zemel, R.S.: Neural relational inference for interacting systems. In: Proceedings of the International Conference on Machine Learning, ICML (2018)
21. Veličković, P., Cucurull, G., Casanova, A., Romero, A., Lio, P., Bengio, Y.: Graph attention networks. In: International Conference on Learning Representations (ICLR) (2018)
22. Wang, X., Girshick, R., Gupta, A., He, K.: Non-local neural networks. In: IEEE Conference on Computer Vision and Pattern Recognition (CVPR) (2017)
23. Battaglia, P., et al.: Relational inductive biases, deep learning, and graph networks. arXiv (2018)
24. Wu, Z., Pan, S., Chen, F., Long, G., Zhang, C., Philip, S.Y.: A comprehensive survey on graph neural networks. IEEE Trans. Neural Netw. Learn. Syst. **32**(1), 4–24 (2020)
25. Cohen, T., Welling, M.: Group equivariant convolutional networks. In: Proceedings of the International Conference on Machine Learning, ICML (2016)
26. Finzi, M., Stanton, S., Izmailov, P., Wilson, A.: Generalizing convolutional neural networks for equivariance to lie groups on arbitrary continuous data. In: Proceedings of the International Conference on Machine Learning, ICML (2020)
27. Hutchinson, M., Lan, C.L., Zaidi, S., Dupont, E., Teh, Y.W., Kim, H.: Lietransformer: equivariant self-attention for lie groups. ArXiv Preprint (2020)
28. Bekkers, E.J.: B-spline CNNs on lie groups. In: International Conference on Learning Representations (2019)
29. Vaswani, A., et al.: Attention is all you need. In: Advances in Neural Information Processing Systems (NeurIPS) (2017)

30. Lee, J., Lee, Y., Kim, J., Kosiorek, A.R., Choi, S., Teh, Y.W.: Set transformer: a framework for attention-based permutation-invariant neural networks. In: Proceedings of the International Conference on Machine Learning, ICML (2019)
31. Ramachandran, P., Parmar, N., Vaswani, A., Bello, I., Levskaya, A., Shlens, J.: Stand-alone self-attention in vision models. In: Advances in Neural Information Processing System (NeurIPS) (2019)
32. van Steenkiste, S., Chang, M., Gre, K., Schmidhuber, J.: Relational neural expectation maximization: unsupervised discovery of objects and their interactions. In: International Conference on Learning Representations (ICLR) (2018)
33. Fuchs, F.B., Kosiorek, A.R., Sun, L., Jones, O.P., Posner, I.: End-to-end recurrent multi-object tracking and prediction with relational reasoning. arXiv preprint (2020)
34. Yang, J., Zhang, Q., Ni, B.: Modeling point clouds with self-attention and gumbel subset sampling. In: IEEE Conference on Computer Vision and Pattern Recognition (CVPR) (2019)
35. Xie, S., Liu, S., Tu, Z.C.Z.: Attentional shapecontextnet for point cloud recognition. In: IEEE Conference on Computer Vision and Pattern Recognition (CVPR) (2018)
36. Lin, Z., et al.: A structured self-attentive sentence embedding. In: International Conference on Learning Representations (ICLR) (2017)
37. Hoshen, Y.: VAIN: attentional multi-agent predictive modeling. In: Advances in Neural Information Processing Systems (NeurIPS) (2017)
38. Shaw, P., Uszkoreit, J., Vaswani, A.: Self-attention with relative position representations. In: Annual Conference of the North American Chapter of the Association for Computational Linguistics (NAACL-HLT) (2018)

End-to-End Similarity Learning and Hierarchical Clustering for Unfixed Size Datasets

Leonardo Gigli(✉), Beatriz Marcotegui, and Santiago Velasco-Forero

Centre for Mathematical Morphology - Mines ParisTech - PSL University, Fontainebleau, France
leonardo.gigli@mines-paristech.fr

Abstract. Hierarchical clustering (HC) is a powerful tool in data analysis since it allows discovering patterns in the observed data at different scales. Similarity-based HC methods take as input a fixed number of points and the matrix of pairwise similarities and output the dendrogram representing the nested partition. However, in some cases, the entire dataset cannot be known in advance and thus neither the relations between the points. In this paper, we consider the case in which we have a collection of realizations of a random distribution, and we want to extract a hierarchical clustering for each sample. The number of elements varies at each draw. Based on a continuous relaxation of Dasgupta's cost function, we propose to integrate a triplet loss function to Chami's formulation in order to learn an optimal similarity function between the points to use to compute the optimal hierarchy. Two architectures are tested on four datasets as approximators of the similarity function. The results obtained are promising and the proposed method showed in many cases good robustness to noise and higher adaptability to different datasets compared with the classical approaches.

1 Introduction

Similarity-based HC is a classical unsupervised learning problem and different solutions have been proposed during the years. Classical HC methods such as Single Linkage, Complete Linkage, Ward's Method are conceivable solutions to the problem. Dasgupta [1] firstly formulated this problem as a discrete optimization problem. Successively, several continuous approximations of the Dasgupta's cost function have been proposed in recent years. However, in the formulation of the problem, the input set and the similarities between points are fixed elements, i.e., the number of samples to compute similarities is fixed. Thus, any change in the input set entails a reinitialization of the problem and a new solution must be found. In addition to this, we underline that the similarity function employed is a key component for the quality of the solution. For these reasons, we are interested in an extended formulation of the problem in which we assume as input a family of point sets, all sampled from a fixed distribution. Our goal is

F. Nielsen and F. Barbaresco (Eds.): GSI 2021, LNCS 12829, pp. 596–604, 2021.
https://doi.org/10.1007/978-3-030-80209-7_64

to find at the same time a "good" similarity function on the input space and optimal hierarchical clustering for the point sets. The rest of this paper is organized as follows. In Sect. 2, we review related works on hierarchical clustering, especially the hyperbolic hierarchical clustering. In Sect. 3, the proposed method is described. In Sect. 4, the experimental design is described. And finally, Sect. 5 concludes this paper.

2 Related Works

Our work is inspired by [2] and [3] where for the first time continuous frameworks for hierarchical clustering have been proposed. Both papers assume that a given weighted-graph $\mathcal{G} = (V, E, W)$ is given as input. [2] aims to find the best ultrametric that optimizes a given cost function. Basically, they exploit the fact that the set $\mathcal{W} = \{w : E \rightarrow \mathbb{R}_+\}$ of all possible functions over the edges of $\mathcal{G}$ is isomorphic to the Euclidean subset $\mathbb{R}_+^{|E|}$ and thus it makes sense to "*take a derivative according to a given weight function*". Along with this they show that the min-max operator function $\Phi_\mathcal{G} : \mathcal{W} \rightarrow \mathcal{W}$, defined as $(\forall \tilde{w} \in \mathcal{W}, \forall e_{xy} \in E) \quad \Phi_\mathcal{G}(\tilde{w})(e_{xy}) = \min_{\pi \in \Pi_{xy}} \max_{e' \in \pi} \tilde{w}(e')$, where Π_{xy} is the set of all paths from vertex x to vertex y, is sub-differentiable. As insight, the min-max operator maps any function $w \in \mathcal{W}$ to its associated subdominant ultrametric on $\mathcal{G}$. These two key elements are combined to define the following minimization problem over $\mathcal{W}$, $w^* = \arg\min_{\tilde{w} \in \mathcal{W}} J(\Phi_\mathcal{G}(\tilde{w}), w)$, where the function J is a differentiable loss function to optimize. In particular, since the metrics $\tilde{w}$ are indeed vectors of $\mathbb{R}^{|E|}$, the authors propose to use the L^2 distance as a natural loss function. Furthermore, they come up with other regularization functions, such as a cluster-size regularization, a triplet loss, or a new differentiable relaxation of the famous Dasgupta's cost function [1]. The contribution of [3] is twofold. On the one hand, inspired by the work of [4], they propose to find an optimal embedding of the graph nodes into the Poincaré Disk, observing that the internal structure of the hierarchical tree can be inferred from leaves' hyperbolic embeddings. On the other hand, they propose a direct differentiable relaxation of the Dasgupta's cost and prove theoretical guarantees in terms of clustering quality of the optimal solution for their proposed function compared with the optimal hierarchy for the Dasgupta's function. As said before, both approaches assume a dataset $\mathcal{D}$ containing n datapoints and pairwise similarities $(w_{ij})_{i,j \in [n]}$ between points are known in advance. Even though this is a very general hypothesis, unfortunately, it does not include cases where part of the data is unknown yet or the number of points cannot be estimated in advance, for example point-cloud scans. In this paper, we investigate a formalism to extend previous works above to the case of n varies between examples. To be specific, we cannot assume anymore that a given graph is known in advance and thus we cannot work on the earlier defined set of functions $\mathcal{W}$, but we would rather look for optimal embeddings of the node features. Before describing the problem, let us review hyperbolic geometry and hyperbolic hierarchical clustering.

2.1 Hyperbolic Hierarchical Clustering

The Poincaré Ball Model $(\mathbb{B}^n, g^{\mathbb{B}})$ is a particular hyperbolic space, defined by the manifold $\mathbb{B}^n = \{x \in \mathbb{R}^n \mid \|x\| < 1\}$ equipped with the following Riemaniann metric $g_x^{\mathbb{B}} = \lambda_x^2 g^E$, where $\lambda_x := \frac{2}{1-\|x\|^2}$, where $g^E = \mathbf{I}_n$ is the Euclidean metric tensor. The distance between two points in the $x, y \in \mathbb{B}^n$ is given by $d_{\mathbb{B}}(x,y) = \cosh^{-1}\left(1 + 2\frac{\|x-y\|_2^2}{(1-\|x\|_2^2)(1-\|y\|_2^2)}\right)$. It is thus straightforward to prove that the distance of a point to the origin is $d_o(x) := d(o,x) = 2\tanh^{-1}(\|x\|_2)$. Finally, we remark that $g_x^{\mathbb{B}}$ defines the same angles as the Euclidean metric. The angle between two vectors $u, v \in T_x\mathbb{B}^n \setminus \{\mathbf{0}\}$ is defined as $\cos(\angle(u,v)) = \frac{g_x^{\mathbb{B}}(u,v)}{\sqrt{g_x^{\mathbb{B}}(u,u)}\sqrt{g_x^{\mathbb{B}}(v,v)}} = \frac{\langle u,v\rangle}{\|u\|\|v\|}$, and $g^{\mathbb{B}}$ is said to be *conformal* to the Euclidean metric. In our case, we are going to work on the Poincaré Disk that is $n = 2$. The interested reader may refer to [5] for a wider discussion on Hyperbolic Geometry. Please remark that the geodesic between two points in this metric is either the segment of the circle orthogonal to the boundary of the ball or the straight line that goes through the origin in case the two points are antipodal. The intuition behind the choice of this particular space is motivated by the fact that the curvature of the space is negative and geodesic coming out from a point has a *"tree-like"* shape. Moreover, [3] proposed an analog of the Least Common Ancestor (LCA) in the hyperbolic space. Given two leaf nodes i, j of a hierarchical T, the LCA $i \vee j$ is the closest node to the root r of T on the shortest path π_{ij} connecting i and j. In other words $i \vee j = \arg\min_{k \in \pi_{ij}} d_T(r,k)$, where $d_T(r,k)$ measures the length of the path from the root node r to the node k. Similarly, the *hyperbolic lowest common ancestor* between two points z_i and z_j in the hyperbolic space is defined as the closest point to the origin in the geodesic path, denoted $z_i \rightsquigarrow z_j$, connecting the two points: $z_i \vee z_j := \arg\min_{z \in z_i \rightsquigarrow z_j} d(o,z)$. Thanks to this definition, it is possible to decode a hierarchical tree starting from leaf nodes embedding into the hyperbolic space. The decoding algorithm uses a union-find paradigm, iteratively merging the most similar pairs of nodes based on their hyperbolic LCA distance to the origin. Finally [3] also proposed a continuous version of Dasgupta's cost function. Let $Z = \{z_1, \ldots, z_n\} \subset \mathbb{B}^2$ be an embedding of a tree T with n leaves, they define their cost function as:

$$C_{\text{HYPHC}}(Z; w, \tau) = \sum_{ijk}(w_{ij} + w_{ik} + w_{jk} - w_{\text{HYPHC},ijk}(Z; w, \tau)) + \sum_{ij} w_{ij}, \quad (1)$$

where $w_{\text{HYPHC},ijk}(Z; w, \tau) = (w_{ij}, w_{ik}, w_{jk}) \cdot \sigma_\tau(d_o(z_i \vee z_j), d_o(z_i \vee z_k), d_o(z_j \vee z_k))^\top$, and $\sigma_\tau(\cdot)$ is the scaled softmax function $\sigma_\tau(w)_i = e^{w_i/\tau}/\sum_j e^{w_j/\tau}$. We recall that w_{ij} are the pair-wise similarities, which in [3] are assumed to be known, but in this work are learned.

3 End-to-End Similarity Learning and HC

Let us consider the example of k continuous random variables that take values over an open set $\Omega \subset \mathbb{R}^d$. Let $\mathcal{X}_t = \{x_1^{(t)}, \ldots, x_{n_t}^{(t)}\}$ a set of points obtained as realization of the k random variables at step t. Moreover, we assume to be in a semi-supervised setting. Without loss of generality, we expect to know the associated labels of the first l points in $\mathcal{X}_t$, for each t. Each label takes value in $[k] = \{1, \ldots, k\}$, and indicates from which distribution the point has been sampled. In our work, we aim to obtain at the same time a good similarity function $\delta : \Omega \times \Omega \to \mathbb{R}_+$ that permits us to discriminate the points according to the distribution they have been drawn and an optimal hierarchical clustering for each set $\mathcal{X}_t$. Our idea to achieve this goal is to combine the continuous optimization framework proposed by Chami [3] along with deep metric learning to learn the similarities between points. Hence, we look for a function $\delta_\theta : \Omega \times \Omega \to \mathbb{R}_+$ such that

$$\min_{\theta, Z \in \mathcal{Z}} C_{\text{HYPHC}}(Z, \delta_\theta, \tau) + \mathcal{L}_{\text{triplet}}(\delta_\theta; \alpha). \tag{2}$$

The second term of the equation above is the sum over the set $\mathcal{T}$ of triplets:

$$\mathcal{L}_{\text{triplet}}(\delta_\theta; \alpha) = \sum_{(a_i, p_i, n_i) \in \mathcal{T}} \max(\delta_\theta(a_i, p_i) - \delta_\theta(a_i, n_i) + \alpha, 0), \tag{3}$$

where a_i is the anchor input, p_i is the positive input of the same class as a_i, n_i is the negative input of a different class from a_i and $\alpha > 0$ is the margin between positive and negative values. One advantage of our formalism is that it allows us to use deep learning approach, i.e., backpropagation and gradient descend optimization to optimize the model's parameters. As explained before, we aim to learn a similarity function and at the same time find an optimal embedding for a family of point sets into the hyperbolic space which implicitly encodes a hierarchical structure. To achieve this, our idea is to model the function δ_θ using a neural network whose parameters we fit to optimize the loss function defined in (2). Our implementation consists of a neural network NN_θ that carries out a mapping $\text{NN}_\theta : \Omega \to \mathbb{R}^2$. The function δ_θ is thus written as:

$$\delta_\theta(x, y) = \cos(\angle(\text{NN}_\theta(x), \text{NN}_\theta(y))), \tag{4}$$

We use the cosine similarity for two reasons. The first comes from the intuition that points belonging to the same cluster will be forced to have small angles between them. As a consequence, they will be merged earlier in the hierarchy. The second reason regards the optimization process. Since the hyperbolic metric is *conformal* to the Euclidean metric, the cosine similarity allows us to use the same the Riemannian Adam optimizer [6] in (2). Once computed the similarities, the points are all normalized at the same length to embed them into the Hyperbolic space. The normalization length is also a trainable parameter of the model. Accordingly, we have selected two architectures. The first is a Multi-Layer-Perceptron (MLP) composed of four hidden layers, and the second is a model composed of three layers of Dynamic Graph Egde Convolution (DGCNN) [7].

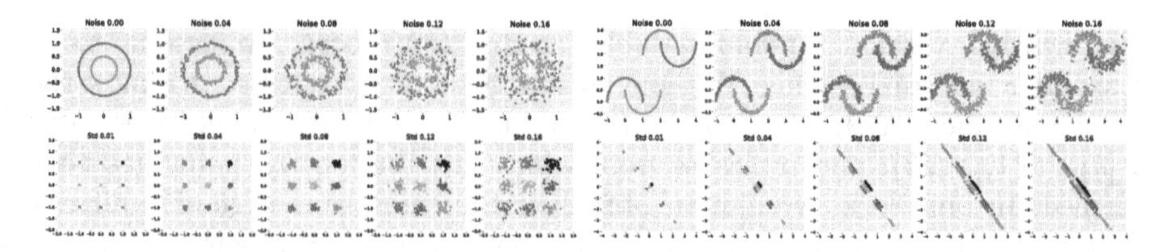

Fig. 1. Different examples of *circles*, *moons*, *blobs* and *anisotropics* that have been generated varying noise value to evaluate robustness and performance of proposed method.

4 Experiments

Experiments Setup: Inspired by the work of [2] we took into account four sample generators in Scikit-Learn to produce four datasets as it is illustrated in Fig. 1. For each dataset we generate, the training set is made of 100 samples and the validation set of 20 samples. The test set contains 200 samples. In addition, each sample in the datasets contains a random number of points. In the experiments we sample from 200 to 300 points each time and the labels are known only for the 30% of them. In *circles* and *moons* datasets increasing the value of noise makes the clusters mix each other and thus the task of detection becomes more difficult. Similarly, in *blobs* and *anisotropics*, we can increase the value of standard deviation to make the problem harder. Our goal is to explore how the models embed the points and separate the clusters. Moreover, we want to investigate the robustness of the models to noise. For this reason, in these datasets, we set up two different levels of difficulty according to the noise/standard deviation used to generate the sample. In *circles* and *moons* datasets, the easiest level is represented by samples without noise, while the harder level contains samples whose noise value varies up to 0.16. In Blobs and Anisotropic datasets, we chose two different values of Gaussian standard deviation to generate the sample. In the easiest level, the standard deviation value is fixed at 0.08, while in the harder level is at 0.16. For each level of difficulty, we trained the models and compared the architectures. In addition, we used the harder level of difficulty to test all the models.

Architectures: The architectures we test are MLP and DGCNN. The dimension of hidden layers is 64. After each Linear layer we apply a LeakyReLU defined as $\text{LeakyReLU}(x) = \max\{x, \eta x\}$ with a negative slope $\eta = 0.2$. In addition, we use Batch Normalization [8] to speed up and stabilize convergence.

Metrics: Let k the number of clusters that we want to determine. For the evaluation, we consider the partition of k clusters obtained from the hierarchy and we measure the quality of the predictions using Average Rand Index (ARI), Purity, and Normalized Mutual Information Score (NMI). Our goal is to test the ability of the two types of architectures selected to approximate function in (4).

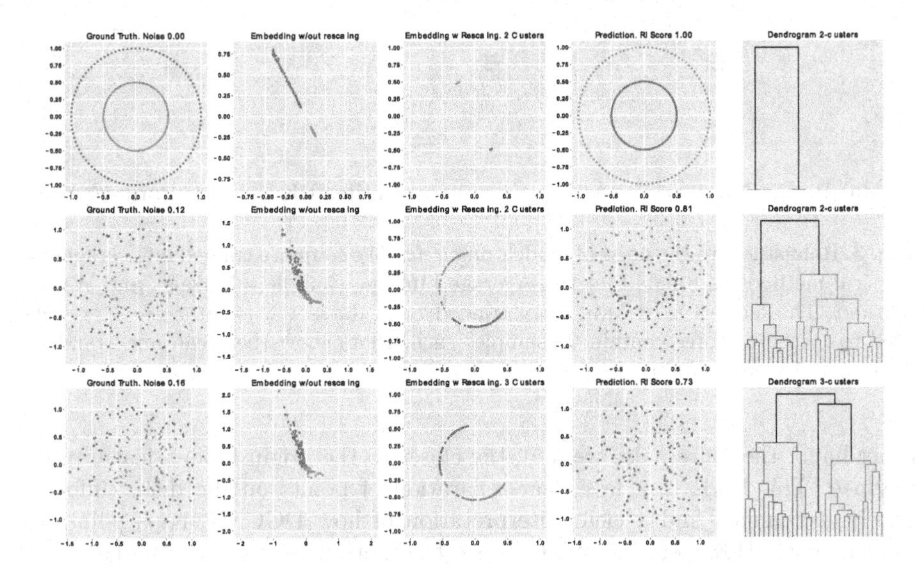

Fig. 2. Effect of noise on predictions in the *circles* database. The model used for prediction is an MLP trained without noise. From top to bottom, each row is a case with an increasing level of noise. In the first column input points, while in the second column we illustrate hidden features. Points are colored according to ground truth. The third column illustrates hidden features after projection to Poincaré Disk. The fourth column shows predicted labels, while the fifth column shows associated dendrograms.

Visualize the Embeddings: Let first discuss the results obtained on *circles* and *moons*. In order to understand and visualize how similarities are learned, we first trained the architectures at the easiest level. Figure 2 illustrates the predictions carried out by models trained using samples without noise. Each row in the figures illustrates the model's prediction on a sample generated with a specific noise value. The second column from the left of sub-figures depicts hidden features in the feature space $\mathcal{H} \subset \mathbb{R}^2$. The color assigned to hidden features depends on points' labels in the ground truth. The embeddings in the Poincaré Disk (third column from the left) are obtained by normalizing the features to a learned scale. Furthermore, the fourth column of sub-figures shows the prediction obtained by extracting flat clustering from the hierarchy decoded from leaves embedding in the Poincaré Disk. Here, colors assigned to points come from predicted labels. The number of clusters is chosen in order to maximize the average rand index score. It is interesting to remark how differently the two architectures extract features. Looking at the samples without noise, it is straightforward that hidden features obtained with MLP are aligned along lines passing through the origin. Especially in the case of *circles* (Fig. 2), hidden features belonging to different clusters are mapped to opposite sides with respect to the origin, and after rescaling hidden features are clearly separated in the hyperbolic space. Indeed, picking cosine similarity in (4) we were expecting this kind of solution. On the

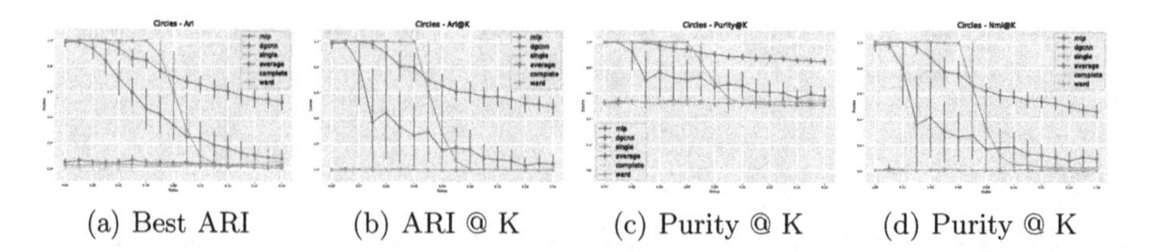

(a) Best ARI (b) ARI @ K (c) Purity @ K (d) Purity @ K

Fig. 3. Robustness to the noise of models on *circles*. We compare trained models against classical methods as Single Linkage, Average Linkage, Complete Linkage, and Ward's Method. The models used have been trained on a dataset without noise. Test sets used to measure scores contain 20 samples each. Plots show the mean and standard deviation of scores obtained.

other hand, the more noise we add, the closer to the origin hidden features are mapped. This leads to a less clear separation of points on the disk. Unfortunately, we cannot find a clear interpretation of how DGCNN maps points to hidden space. However, also in this case, the more noise we add, the harder is to discriminate between points of different clusters[1].

Table 1. Scores obtained by MLP and DGCNN on four datasets: *circle*, *moons*, *blobs*, and *anisotropics*. In each dataset the models have been tested on the same test set containing 200 samples.

Dataset	k	Noise/Cluster std	Model	Hidden	Temp	Margin	Ari@k $\pm$ s.d	Purity@k $\pm$ s.d	Nmi@k $\pm$ s.d	Ari $\pm$ s.d
circle	2	0.0	MLP	64	0.1	1.0	0.871 ± 0.153	0.965 ± 0.0427	0.846 ± 0.167	0.896 ± 0.123
circle	2	[0.0–0.16]	MLP	64	0.1	1.0	0.919 ± 0.18	0.972 ± 0.0848	0.895 ± 0.187	0.948 ± 0.0755
circle	2	0.0	DGCNN	64	0.1	1.0	0.296 ± 0.388	0.699 ± 0.188	0.327 ± 0.356	0.408 ± 0.362
circle	2	[0.0–0.16]	DGCNN	64	0.1	1.0	0.852 ± 0.243	0.947 ± 0.116	0.826 ± 0.247	0.9 ± 0.115
moons	4	0.0	MLP	64	0.1	1.0	0.895 ± 0.137	0.927 ± 0.108	0.934 ± 0.0805	0.955 ± 0.0656
moons	4	[0.0–0.16]	MLP	64	0.1	1.0	0.96 ± 0.0901	0.971 ± 0.0751	0.972 ± 0.049	0.989 ± 0.017
moons	4	0.0	DGCNN	64	0.1	1.0	0.718 ± 0.247	0.807 ± 0.187	0.786 ± 0.191	0.807 ± 0.172
moons	4	[0.0–0.16]	DGCNN	64	0.1	1.0	0.917 ± 0.123	0.942 ± 0.0992	0.941 ± 0.0726	0.966 ± 0.0455
blobs	9	0.08	MLP	64	0.1	0.2	0.911 ± 0.069	0.939 ± 0.057	0.953 ± 0.025	0.958 ± 0.022
blobs	9	0.16	MLP	64	0.1	0.2	0.985 ± 0.0246	0.992 ± 0.0198	0.99 ± 0.0115	0.992 ± 0.00821
blobs	9	0.08	DGCNN	64	0.1	0.2	0.856 ± 0.0634	0.891 ± 0.0583	0.931 ± 0.025	0.921 ± 0.0401
blobs	9	0.16	DGCNN	64	0.1	0.2	0.894 ± 0.0694	0.92 ± 0.0604	0.95 ± 0.0255	0.948 ± 0.0336
aniso	9	0.08	MLP	64	0.1	0.2	0.86 ± 0.0696	0.904 ± 0.0631	0.922 ± 0.0291	0.925 ± 0.0287
aniso	9	0.16	MLP	64	0.1	0.2	0.952 ± 0.0503	0.968 ± 0.044	0.972 ± 0.0189	0.976 ± 0.0133
aniso	9	0.08	DGCNN	64	0.1	0.2	0.713 ± 0.0835	0.793 ± 0.0727	0.844 ± 0.0401	0.795 ± 0.0652
aniso	9	0.16	DGCNN	64	0.1	0.2	0.84 ± 0.0666	0.879 ± 0.0595	0.922 ± 0.0274	0.914 ± 0.0436

Comparison with Classical HC Methods: In Fig. 3 we compare models trained at easier level of difficulty against classical methods such as Single, Complete, Average and Ward's method Linkage on *circles*, *moons*, *blobs* and *anisotropics* respectively. The plots show the degradation of the performance of models

[1] Supplementary Figures are available at https://github.com/liubigli/similarity-learning/blob/main/GSI2021_Appendix.pdf.

as we add noise to samples. Results on *circles* say that Single Linkage is the method that performs the best for small values of noise. However, MLP shows better robustness to noise. For high levels of noise, MLP is the best method. On the other hand, DGCNN exhibits a low efficacy also on low levels of noise. Other classical methods do not achieve good scores on this dataset. A similar trend can also be observed in the *moons* dataset. Note that, in this case, MLP is comparable with Single Linkage also on small values of noise, and its scores remain good also on higher levels of noise. DGCNN and other classical methods perform worse even in this data set. Results obtained by MLP and DGCNN on *blobs* dataset are comparable with the classical methods, even though the performances of models are slightly worse compared to classical methods for higher values of noise. On the contrary, MLP and DGCNN achieve better scores on the *anisotropics* dataset compared to all classical models. Overall, MLP models seem to act better than DCGNN ones in all the datasets.

Benchmark of the Models: Table 1 reports the scores obtained by the trained models on each dataset. Each line corresponds to a model trained either at an easier or harder level of difficulty. The test set used to evaluate the results contains 200 samples generated using the harder level of difficulty. Scores obtained demonstrate that models trained at the harder levels of difficulty are more robust to noise and achieve better results. As before, also in this case MLP is, in general, better than DGCNN in all the datasets considered.

5 Conclusion

In this paper, we have studied the metric learning problem to perform hierarchical clustering where the number of nodes per graph in the training set can vary. We have trained MLP and DGCNN architectures on five datasets by using our proposed protocol. The quantitative results show that overall MLP performs better than DGCNN. The comparison with the classic methods proves the flexibility of the solution proposed to the different cases analyzed, and the results obtained confirm higher robustness to noise. Finally, inspecting the hidden features, we have perceived how MLP tends to project points along lines coming out from the origin. To conclude, the results obtained are promising and we believe that it is worth testing this solution also on other types of datasets such as 3D point clouds.

References

1. Dasgupta, S.: A cost function for similarity-based hierarchical clustering. In: Proceedings of the 48 Annual ACM Symposium on Theory of Computing, pp. 118–127 (2016)
2. Chierchia, G., Perret, B.: Ultrametric fitting by gradient descent. In: NIPS, pp. 3181–3192 (2019)
3. Chami, I., Gu, A., Chatziafratis, V., Ré, C.: From trees to continuous embeddings and back: Hyperbolic hierarchical clustering. In: NIPS vol. 33 (2020)

4. Monath, N., Zaheer, M., Silva, D., McCallum, A., Ahmed, A.: Gradient-based hierarchical clustering using continuous representations of trees in hyperbolic space. In: 25th ACM SIGKDD Conference on Discovery & Data Mining, pp. 714–722 (2019)
5. Brannan, D.A., Esplen, M.F., Gray, J.: Geometry. Cambridge University Press, Cambridge (2011)
6. Bécigneul, G., Ganea, O.E.: Riemannian adaptive optimization methods. arXiv preprint arXiv:1810.00760 (2018)
7. Wang, Y., Sun, Y., Liu, Z., Sarma, S.E., Bronstein, M.M., Solomon, J.M.: Dynamic graph CNN for learning on point clouds. ACM Trans. Graph. (TOG) **38**(5), 1–12 (2019)
8. Ioffe, S., Szegedy, C.: Batch normalization: Accelerating deep network training by reducing internal covariate shift. arXiv preprint arXiv:1502.03167 (2015)

Information Theory and the Embedding Problem for Riemannian Manifolds

Govind Menon(✉)

Division of Applied Mathematics, Brown University, Providence, RI 02912, USA
govind_menon@brown.edu

Abstract. This paper provides an introduction to an information theoretic formulation of the embedding problem for Riemannian manifolds developed by the author. The main new construct is a stochastic relaxation scheme for embedding problems and hard constraint systems. This scheme is introduced with examples and context.

Keywords: Geometric information theory · Embedding theorems

1 The Embedding Problem for Riemannian Manifolds

The purpose of this paper is to outline an information theoretic formulation of the embedding problem for Riemannian manifolds developed by the author. The main new contribution is a stochastic relaxation scheme that unifies many hard constraint problems. This paper is an informal introduction to this method.

The modern definition of a manifold was formalized by Whitney in 1936 [11]. He defined manifolds as abstract spaces covered by locally compatible charts, thus providing a rigorous description of the intuitive idea that a manifold is a topological space that 'locally looks like Euclidean space'. Whitney's definition should be contrasted with the 19th century idea of manifolds as hypersurfaces in Euclidean space. The embedding problem for a differentiable manifold checks the compatibility of these notions. Given an (abstractly defined) n-dimensional differentiable manifold $\mathcal{M}^n$, an embedding of $\mathcal{M}^n$ is a smooth map $u : \mathcal{M}^n \to \mathbb{R}^q$ that is one-to-one and whose derivative $Du(x)$ has full rank at each $x \in \mathcal{M}^n$.

An embedded manifold carries a pullback metric, denoted $u^\sharp e$, where e denotes the identity metric on $\mathbb{R}^q$. Assume that $\mathcal{M}^n$ is a Riemannian manifold equipped with a metric g. We say that an embedding $u : \mathcal{M}^n \to \mathbb{R}^q$ is *isometric* if $u^\sharp e = g$. In any local chart U, this is the nonlinear PDE

$$\sum_{\alpha=1}^{q} \frac{\partial u^\alpha}{\partial x^i}\frac{\partial u^\alpha}{\partial x^j}(x) = g_{ij}(x), \quad x \in U, \quad 1 \leq i, j \leq n. \tag{1}$$

Let us contrast Eq. (1) with isometric embedding of finite spaces.

Partial support for this work was provided by the National Science Foundation (DMS 1714187), the Simons Foundation (Award 561041), the Charles Simonyi Foundation and the School of Mathematics at the Institute for Advanced Study. The author is grateful to the anonymous referees of this paper for many valuable comments.

F. Nielsen and F. Barbaresco (Eds.): GSI 2021, LNCS 12829, pp. 605–612, 2021.
https://doi.org/10.1007/978-3-030-80209-7_65

(a) Embedding metric spaces: We are given a finite set K with a distance function $\rho(x,y)$ and we seek a map $u : K \to \mathbb{R}^q$ such that

$$|u(x) - u(y)| = \rho(x,y), \quad x,y \in K. \tag{2}$$

(b) Graph embedding: We are given a graph $G = (V,E)$ with a distance function $\rho : E \to \mathbb{R}_+$ that associates a length to each edge. The graph embedding problem is to find $u : V \to \mathbb{R}^q$ such that

$$|u(e_+) - u(e_-)| = \rho(e), \quad e \in E, \tag{3}$$

where $e_\pm$ denote the vertices at the two ends of an edge $e \in E$.

In each of these problems an abstractly defined metric space is being mapped into the reference space $\mathbb{R}^q$. The LHS is the length measured in $\mathbb{R}^q$. The RHS is the intrinsic distance on the given space. In Eq. (1), the equality of length is expressed infinitesimally, which is why we obtain a PDE.

Modern understanding of (1) begins with the pioneering work of Nash in the 1950s [8,9]. His work led to the following results: for $q = n+1$ and $g \in C^0$ there are infinitely many C^1 isometric embeddings (assuming no topological obstructions); when $q \geq n + n(n+1)/2 + 5$ and $g \in C^\infty$ there are infinitely many C^∞ isometric embeddings. These results are improvements of Nash's original work, but follow his ideas closely. Nash's work has been systematized in two distinct ways: as Gromov's h-principle in geometry and as hard implicit function theorems in analysis. However, several fundamental questions remain unresolved [5].

Our interest in the area was stimulated by an unexpected link with turbulence [2]. A long-standing goal in turbulence is to construct Gibbs measures for the Euler equations of ideal incompressible fluids whose statistical behavior is in accordance with experiments. This connection suggests the application of statistical mechanics to embeddings. The construction of Gibbs measures for embeddings allows us to formalize the question 'What does a typical isometric embedding look like?'. This is in contrast with the questions 'Does an isometric embedding exist? If so, how smooth is it?' resolved by Nash and Gromov.

The Gibbs measures have a natural information theoretic construction. We model embedding as a stochastic process in which an observer in $\mathbb{R}^q$ makes a copy of a given Riemannian geometry by measurement of distances at finer and finer scales. This is a Bayesian interpretation suited to the interplay between geometry and information theory at this conference. However, prior to the author's work there was no attempt to study (1) with probabilistic methods or to treat Eqs. (1)–(3) through a common framework. Further, the devil lies in the details, since any new attempt must be consistent with past work and must be nailed down with complete rigor. This paper discusses only the evolution equations. Analysis of these equations will be reported in forthcoming work.

2 The Role of Information Theory

The embedding theorems are interesting both for their conceptual and technical depth. They arise in apparently unrelated fields and they have been studied by disparate techniques. Within mathematics, Nash's three papers on manifolds appear unrelated on first sight. Mathematical techniques for graph embedding, mainly stimulated by computer science, appear to have little relation to the embedding problem for manifolds [7]. The embedding problem also appears under the guise of the nonlinear sigma models in quantum field theory. In his breakthrough work, Friedan showed that the renormalization of the nonlinear sigma model is the Ricci flow [3,4]. However, Friedan's technique, renormalization by expansion in dimension, is notoriously hard to pin down mathematically.

Such a diversity of methods and applications is bewildering until one recognizes that it offers a route to a radical conceptual simplification. In order to obtain a unified treatment, it is necessary to insist on a minimalistic formulation of embedding that does not rely in a fundamental manner on the structure of the space being embedded (e.g. whether it is a graph, manifold, or metric space). Such a formulation must be consistent with both Nash and Friedan's approach to the problem, as well as applications in computer science. This line of reasoning suggests that the appropriate foundation must be information theory – it is the only common thread in the above applications.

The underlying perspective is as follows. We view embedding as a form of information transfer between a source and an observer. The process of information transfer is complete when all measurements of distances by the observer agree with those at the source. This viewpoint shifts the emphasis from the structure of the space to an investigation of the process by which length is measured. In the Bayesian interpretation, the world is random and both the source and the observer are stochastic processes with well-defined parameters (we construct these processes on a Gaussian space to be concrete). Thus, embedding is simply 'replication' and the process of replication is complete when all measurements by the observer and the source agree on a common set of questions (here it is the question: 'what is the distance between points x and y?'). From this standpoint, there is no fundamental obstruction to embedding, except that implied by Shannon's channel coding theorem.

The challenge then is to implement this viewpoint with mathematical rigor. In order to explain our method, we must briefly review Nash's techniques. The main idea in [8] is that when $q = n + 2$ one can relax the PDE $u^\sharp g = e$ to a space of subsolutions and then introduce highly structured corrugations in the normal directions at increasingly fine scales. This iteration 'bumps up' smooth subsolutions towards a solution. In [9], Nash introduces a geometric flow, which evolves an immersion and a smoothing operator simultaneously. Unlike [8], which is brief and intuitive, the paper [9] is lengthy and technical, introducing what is now known as the Nash-Moser technique. A central insight in our approach is that one can unify these methods by introducing a reproducing kernel that evolves stochastically with the subsolution.

A rigid adherence to an information theoretic approach to embedding contradicts Nash's results in the following sense. If all that matters is agreement in the measurement of distances between different copies of a manifold, the historical emphasis in mathematics on the role of codimension and regularity cannot be a fundamental feature of the problem. All embeddings are just copies of the same object, an abstractly defined manifold. Typical embeddings of the manifold $(\mathcal{M}^n, g)$ into $\mathbb{R}^p$ and $\mathbb{R}^q$ for $p, q \geq n$ should have different regularity: a C^∞ metric g may yield a crumpled embedding in $\mathbb{R}^p$ and a smooth embedding in $\mathbb{R}^q$ for $p < q$, but since it is only the measurement of length that matters, there is no preferred embedding. Thus, existence and regularity of solutions to (1) must be treated separately. This mathematical distinction acquires salience from its physical meaning. While Wilson renormalization is often seen as a technique for integrating out frequencies scale-by-scale, it reflects the role of gauge invariance in the construction of a physical theory. The only true measurements are those that are independent of the observer. Therefore, in order to ensure consistency between mathematical and physical approaches to the isometric embedding problem it is necessary to develop a unified theory of embeddings that does not rely substantially on codimension. Nash's methods do not meet this criterion.

Finally, embedding theorems may be used effectively in engineering only if the rigorous formulation can be supported by fast numerical methods. Here too Nash's techniques fail the test. The first numerical computations of isometric embeddings are relatively recent [1]. Despite the inspiring beauty of these images, they require a more sophisticated computational effort than is appropriate for a problem of such a fundamental nature. The numerical scheme in [1] is ultimately based on [8]. Thus, it requires the composition of functions, which is delicate to implement accurately. The models proposed below require only semidefinite programming and the use of Markov Chain Monte Carlo, both of which are standard techniques, supported by excellent software.

These remarks appear to be deeply critical of Nash's work, but the truth is more mysterious. The imposition of stringent constraints – consistency with applications, physics and numerical methods – has the opposite effect. It allows us to strip Nash's techniques down to their essence, revealing the robustness of his fundamental insights. By using information theory to mediate between these perspectives, we obtain a new method in the statistical theory of fields.

3 Renormalization Group (RG) Flows

3.1 General Principles

The structure of our method is as follows. Many hard constraint systems and nonlinear PDE such as $u^\sharp e = g$ admit relaxations to subsolutions. We will begin with a subsolution and improve it to a solution by adding fluctuations in a bandlimited manner. These ideas originate in Nash's work [8,9]. We sharpen his procedure as follows:

1. The space of subsolutions is augmented with a Gaussian filter. More precisely, our unknown is a subsolution u_t and a reproducing kernel L_t, $t \in [0, \infty)$. In physical terms, the unknown is a thermal system.
2. We introduce a stochastic flow for (u_t, L_t). This allows us to interpolate between the discrete time iteration in [8] and C^1 time evolution in [9], replacing Nash's feedback control method with stochastic control theory.

3. We apply the modeling principles of continuum mechanics – a separation between kinematics and energetics – to obtain a semidefinite program (SDP) for the covariance of the Itô SDE.
4. A principled resolution of the SDP is the most subtle part of the problem. We illustrate two approaches: low-rank matrix completion and Gibbs measures for the SDP. In physical terms, this is a choice of an equation of state.

The use of reproducing kernels provides other insights too. The Aronszajn-Moore theorem asserts that the reproducing kernel L_t is in one-to-one correspondence with a Hilbert space $\mathcal{H}_{L_t}$. Thus, (u_t, L_t) describes a stochastically evolving affine Hilbert space $(u_t, \mathcal{H}_{L_t})$ much like subspace tracking in machine learning.

3.2 An Example: Random Lipschitz Functions

Let us illustrate the structure of the RG flows on a model problem. We construct random Lipschitz functions as solutions to the Hamilton-Jacobi equation

$$|\nabla u(x)|^2 = 1, \quad x \in \mathbb{T}^n, \quad u : \mathbb{T}^n \to \mathbb{R}. \tag{4}$$

Equation (4) is simpler than Eq. (1) because the unknown u is a scalar. We say that $v : \mathbb{T}^n \to \mathbb{R}$ is a smooth subsolution if $v \in C^\infty$ and $|\nabla v(x)| < 1$ for $x \in \mathbb{T}^n$. Define the residual $r(x; v)$, trace l, and the density matrix P by

$$r(x;v) = (1-|\nabla v(x)|^2)_+^{1/2}, \quad l = \int_{\mathbb{T}^n} L(x,x)\,dx, \quad P(x,y) = \frac{1}{l}L(x,y), \; x,y\in\mathbb{T}^n. \tag{5}$$

The simplest RG flow associated to Eq. (4) is the stochastic evolution

$$du_t(x)du_t(y) = (PSP)(x,y)\,dt, \quad \dot{P} = PSP - \mathrm{Tr}(PSP)P, \quad \frac{\dot{l}}{l} = \mathrm{Tr}(PSP), \tag{6}$$

where S is a covariance kernel constructed from $r(x; u_t)$ as follows

$$S(x,y) = \nabla r(x) \cdot \nabla r(y) = \sum_{i=1}^{n} \partial_{x_i} r(x;u_t) \partial_{y_i} r(y;u_t), \quad x, y \in \mathbb{T}^n. \tag{7}$$

Both S and P are integral operators on $L^2(\mathbb{T}^n)$ and PSP denotes the natural composition of such operators.

The first equation in (6) reflects *stochastic kinematics*. As in Nash's work, we are bumping up a subsolution, but now by stochastic fluctuations with covariance tensor PSP. The density matrix P smoothes the correction S, so that fluctuations are band-limited. The last equation shows that l_t is slaved to (u_t, P_t). The study of (u_t, P_t) and (u_t, L_t) is equivalent for this reason and both choices offer different insights. The equation for (u_t, L_t) (after a change of time-scale) is

$$du_t(x)du_t(y) = \dot{L}(x,y)\,dt, \quad \dot{L} = LSL. \tag{8}$$

Observe that this equation is invariant under reparametrization of time.

The specific relation between S and r in Eq. (7) emerges from an explicit rank-one solution to a matrix completion problem. More generally, all RG flows for Eq. (4) require the resolution of an SDP that provides a covariance kernel S, given a residual r. An alternate resolution of this question, involving Gibbs measures for the SDP, is described in the next section.

3.3 Interpretation

The intuition here is as follows. Assume given initial conditions (u_0, L_0) where u_0 is a smooth subsolution and L_0 is a band-limited reproducing kernel. Standard SDE theory implies the existence of strong solutions to Eqs. (6) and (8). Equation (8) tells us that L_t is increasing in the Loewner order, so that the Hilbert spaces $\mathcal{H}_{L_t}$ are ordered by inclusion. This corresponds to the subsolutions getting rougher and rougher, while staying band-limited. On the other hand, Eq. (6) tells us that u_t and P_t are bounded martingales, so that $(u_\infty, P_\infty) := \lim_{t\to\infty}(u_t, P_t)$ exists by the martingale convergence theorem. The limit u_∞ is always a random subsolution to (4). Our task is to find the smallest space L_0 so that u_∞ is a solution to (4), thus providing random Lipschitz functions.

The analogous evolution for the isometric embedding problem (1) is obtained from similar reasoning. A subsolution is a map $v : \mathcal{M}^n \to \mathbb{R}^q$ such that $v^\sharp e$ satisfies the matrix inequality $v^\sharp e(x) < g(x)$ at each $x \in \mathcal{M}^n$. The residual $r(x; v)$ is the matrix square-root of the metric defect $g - v^\sharp e(x)$. The covariance kernel L_t is also now a matrix valued kernel. Thus, the generalization reflects the tensorial nature of (1) and does not change the essence of Eq. (6). The associated flow makes precise the idea that embedding is a process of estimation of the metric g by estimators $u_t^\sharp e$. At each scale t, we choose the best correction to u_t given the Gaussian prior P_t and a principled resolution of an SDP.

Finally, Eq. (8) has a simple physical interpretation. The RG flows model quasistatic equilibration of the thermal system (u_t, L_t). This is perhaps the most traditional thermodynamic picture of the flow of heat, dating back to Clausius, Gibbs and Maxwell. What is new is the mathematical structure. The mean and covariance evolve on different time-scales, so that the system is always in local equilibrium. This insight originates in Nelson's derivation of the heat equation [10]. Like Nelson, we stress the foundational role of stochastic kinematics and time-reversibility. However, unlike Nelson, we rely on information theory as the foundation for heat flow, not *a priori* assumptions about a background field. The flows are designed so that $\dot{L}_t$ is always the 'most symmetric' fluctuation field with respect to the prior. This offers a rigorous route to the construction of Gibbs measures by renormalization, using different techniques from Friedan's work. This is why we term our model an RG flow.

4 Isometric Embedding of Finite Metric Spaces into $\mathbb{R}^q$

In this section we show that RG flows for Eqs. (1)–(3) may be derived from common principles. This goes roughly as follows: the discrete embeddings (2)–(3) have subsolutions and we use a stochastic flow analogous to (8) to push these up to solutions. The main insights in this section are the role of an underlying SDP and the use of low-rank kernels L_t as finite-dimensional analogs of smoothing operators. Formally, we expect embeddings of the manifold to be the continuum limit of discrete embeddings of geodesic triangulations of the manifold. However, this has not yet been established rigorously.

Assume given a finite metric space (K, ρ). Equation (2) is a hard constraint system that may not have a solution. For example, an equilateral triangle cannot be isometrically embedded into $\mathbb{R}$. It is necessary to relax the problem. Following Nash [8], let us say that a map $v : K \to \mathbb{R}^q$ is *short* if $|v(x) - v(y)| < \rho(x, y)$ for each pair of distinct points $x, y \in K$. These are our subsolutions.

Let $\mathbb{P}(n, q)$ denote the space of covariance tensors for $\mathbb{R}^q$-valued centered Gaussian processes on K. Our state space is $\mathcal{S}_q = \{(u, L) \in \mathbb{R}^{nq} \times \mathbb{P}(n, q)\}$. The RG flow analogous to Eq. (8) is the Itô SDE

$$du_t^i(x) du_t^j(y) = \dot{L}(x, y)^{ij}\, dt, \quad \dot{L}_t = C(u_t, L_t), \quad x, y \in K, \quad 1 \le i, j \le q. \tag{9}$$

The rest of this section describes the use of SDP to determine C. First, we set $C(u, L) \equiv 0$ when u is not short. When u is short its *metric defect* is

$$r^2(x, y; u) = \left(\rho^2(x, y) - |u(x) - u(y)|^2\right)_+, \quad x, y \in K. \tag{10}$$

We'd like to choose $C(u, L)$ to correct a solution by $r^2(x, y; u)$ on average. To this end, assume $du^i(x) du^j(y) = Q^{ij}(x, y)\, dt$ and use Itô's formula to compute

$$d\,|u(x) - u(y)|^2 = 2\,(u(x) - u(y)) \cdot (du(x) - du(y)) + \Diamond Q\, dt. \tag{11}$$

The Itô correction, captured by the $\Diamond$ operator defined below, provides the expected bump up in lengths

$$(\Diamond Q)\,(x, y) := \sum_{j=1}^{q} \left(Q^{jj}(x, x) + Q^{jj}(y, y) - 2Q^{jj}(x, y)\right). \tag{12}$$

In order to correct by the metric defect, Q must satisfy the linear constraints

$$(\Diamond Q)\,(x, y) = r^2(x, y; u), \quad x, y \in K. \tag{13}$$

A second set of constraints is imposed by the Cameron-Martin theorem: the Gaussian measure associated to $\dot{L}$ must be absolutely continuous with respect to that of L. Explicitly, this means that $Q = ALA^T$ where A is a linear transformation, given in coordinates by $A^{ij}(x, y)$. This restriction is trivial when L has full-rank; but when L is rank-deficient, it provides $\mathbb{P}(n, q)$ with a sub-Riemannian geometry. We use the notation $Q \in T_L\mathbb{P}(n, q)$ to recognize this constraint.

These constraints describe a *matrix completion problem*: choose $Q \in T_L\mathbb{P}(n, q)$ that satisfies (13). This may not have a solution, so we introduce the convex set

$$\mathcal{P} = \{Q \in T_L\mathbb{P}(n, q) \,|\, \Diamond Q(x, y) \le r^2(x, y),\ x, y \in K\}. \tag{14}$$

Our model design task is to make a principled choice of a point $C(u, L)$ in $\mathcal{P}$.

Equation (7) is obtained by choosing a rank-one solution to the analogous operator completion problem for (4). But one may also use the theory of SDP to provide other resolutions of the above matrix completion problem. Interior point methods for SDP associate a barrier $F_{\mathcal{P}}$ to $\mathcal{P}$ and it is natural to choose

$$C(u,L) = \text{argmin}_{Q\in\mathcal{P}} F_{\mathcal{P}}(Q). \tag{15}$$

When $F_{\mathcal{P}}$ is the *canonical barrier* associated to $\mathcal{P}$ we find that C is the *analytic center* of $\mathcal{P}$. This choice is similar in its minimalism to (7). The barrier $F_{\mathcal{P}}$ is a convex function on $\mathcal{P}$ whose Hessian $D^2F_{\mathcal{P}}$ provides a fundamental Riemannian metric on $\mathcal{P}$ [6]. It provides a natural microcanonical ensemble for embedding.

We may also introduce Gibbs measures on $T_L\mathbb{P}(n,q)$ that have the density

$$p_\beta(Q) = \frac{1}{Z_\beta}e^{-\beta E_r(Q)},\; Z_\beta = \int_{T_L\mathbb{P}(n,q)} e^{-\beta E_r(Q)}\,dQ,\; C = \int_{T_L\mathbb{P}(n,q)} Q\,p_\beta(Q)\,dQ, \tag{16}$$

where the energies E_r replace the constraints $\lozenge Q \leq r^2$ with suitable penalties. In these models, the covariance $C(u,L)$ is the most symmetric choice at scale t, with respect to the Gibbs measure p_β. As noted in Sect. 3.3, this is why these models may be termed renormalization group flows.

References

1. Borrelli, V., Jabrane, S., Lazarus, F., Thibert, B.: Flat tori in three-dimensional space and convex integration. Proc. Natl. Acad. Sci. U.S.A. **109**(19), 7218–7223 (2012). https://doi.org/10.1073/pnas.1118478109
2. De Lellis, C., Székelyhidi Jr., L.: Dissipative continuous Euler flows. Invent. Math. **193**(2), 377–407 (2013). https://doi.org/10.1007/s00222-012-0429-9
3. Friedan, D.: Nonlinear models in $2+\varepsilon$ dimensions. Ph.D. thesis, Lawrence Berkeley Laboratory, University of California (1980)
4. Friedan, D.: Nonlinear models in $2+\varepsilon$ dimensions. Phys. Rev. Lett. **45**(13), 1057–1060 (1980). https://doi.org/10.1103/PhysRevLett.45.1057
5. Gromov, M.: Geometric, algebraic, and analytic descendants of Nash isometric embedding theorems. Bull. Amer. Math. Soc. (N.S.) **54**(2), 173–245 (2017). https://doi.org/10.1090/bull/1551
6. Hildebrand, R.: Canonical barriers on convex cones. Math. Oper. Res. **39**(3), 841–850 (2014)
7. Indyk, P., Matoušek, J., Sidiropoulos, A.: Low-distortion embeddings of finite metric spaces. In: Handbook of Discrete and Computational Geometry, vol. 37, p. 46 (2004)
8. Nash, J.: C^1 isometric imbeddings. Ann. Math. **2**(60), 383–396 (1954). https://doi.org/10.2307/1969840
9. Nash, J.: The imbedding problem for Riemannian manifolds. Ann. Math. **2**(63), 20–63 (1956). https://doi.org/10.2307/1969989
10. Nelson, E.: Dynamical Theories of Brownian Motion. Princeton University Press, Princeton (1967)
11. Whitney, H.: Differentiable manifolds. Ann. Math. (2) **37**(3), 645–680 (1936). https://doi.org/10.2307/1968482

cCorrGAN: Conditional Correlation GAN for Learning Empirical Conditional Distributions in the Elliptope

Gautier Marti[1(✉)], Victor Goubet[2], and Frank Nielsen[3]

[1] Hong Kong SAR, China
gautier.marti@polytechnique.edu
[2] ESILV - École Supérieure d'Ingénieurs Léonard de Vinci, Paris, France
[3] Sony Computer Science Laboratories Inc., Tokyo, Japan

Abstract. We propose a methodology to approximate conditional distributions in the elliptope of correlation matrices based on conditional generative adversarial networks. We illustrate the methodology with an application from quantitative finance: Monte Carlo simulations of correlated returns to compare risk-based portfolio construction methods. Finally, we discuss about current limitations and advocate for further exploration of the elliptope geometry to improve results.

Keywords: Generative adversarial networks · Correlation matrices · Elliptope geometry · Empirical distributions · Quantitative finance · Monte Carlo simulations

1 Introduction

Since the seminal Generative Adversarial Networks (GANs) paper [10], adversarial training has been successful in several areas which are typically explored by machine learning researchers (e.g. image generation in computer vision, voice and music generation in audio signal processing). To a smaller extent, GANs were also applied for generating social and research citations networks. We are not aware of existing generative adversarial networks for sampling matrices from an empirical distribution defined in the elliptope, the set of correlation matrices, except our previous work [18]. In this work, we propose extensions of the original CorrGAN model keeping in mind matrix information geometry [20].

This body of work can be motivated by applications in quantitative finance, and possibly in other fields relying on correlation and covariance matrices. Having access to a generative model of realistic correlation matrices can enable large scale numerical experiments (e.g. Monte Carlo simulations) when current mathematical guidance falls short. In the context of quantitative finance, it can improve the falsifiability of statements and more objective comparison of empirical methods. For example, paper A and paper B concurrently claim their novel portfolio

G. Marti—Independent Researcher

F. Nielsen and F. Barbaresco (Eds.): GSI 2021, LNCS 12829, pp. 613–620, 2021.
https://doi.org/10.1007/978-3-030-80209-7_66

allocation methods are the state of the art, out-performing the rest of the literature on well-chosen period and universe of assets. Can we verify and obtain better insights than these claims? We will briefly illustrate that the generative models presented in this paper can help perform a more reliable comparison between portfolio allocation methods. Moreover, by analyzing the simulations, we are able to extract insights on which market conditions are more favorable to such or such portfolio allocation methods.

2 Related Work

Research related to our work can be found in three distinct areas: generation of random correlation matrices, generative adversarial networks (in finance), information geometry of the elliptope in machine learning.

2.1 Sampling of Random Correlation Matrices

Seminal methods detailed in the literature are able to generate random correlation matrices without any structure. These methods are sampling uniformly (in a precise sense) in the set of correlation matrices, e.g. the onion method [9]. Extended onion methods and vines can be used to generate correlation matrices with large off-diagonal values [17], but they are not providing any other particular structure by design. Other methods allow to simulate correlation matrices with given eigenvalues [5], or with the Perron–Frobenius property [14]. CorrGAN [18] is the first and so far only method which attempts to learn a generative model from a given empirical distribution of correlation matrices. Preliminary experiments have shown that the synthetic correlation matrices sampled from the generative model have similar properties than the original ones.

2.2 GANs in Finance

Generative Adversarial Networks (GANs) in Finance are just starting to be explored. Most of the research papers (and many are still at the preprint stage) were written after 2019. The main focus and motivation behind the use of GANs in Finance is *data anonymization* [1], followed by univariate time series (e.g. stock returns) modeling [8,15,24,26]. Also under investigation: modeling and generating tabular datasets which are often the format under which the so-called 'alternative data' hedge funds are using is structured; An example of such 'alternative data' can be credit card (and other) transactions which can also be modeled by GANs [27]. To the best of our knowledge, only our previous work [18] tackles the problem of modeling the joint behaviour of a large number of assets.

2.3 Geometry of the Elliptope in Machine Learning

Except a recent doctoral thesis [4], the core focus of the research community is on the Riemannian geometry of the SPD cone, e.g. [2,19], SPDNet [13] and its extension SPDNet with Riemannian batch normalization [3]. Concerning the elliptope, a non-Riemannian geometry (Hilbert's projective geometry) has also been considered for clustering [21]. The Hilbert elliptope distance is an example of non-separable distance which satisfies the information monotonicity property [21].

3 Our Contributions

We suggest the use of GANs to learn empirical conditional distributions in the correlation elliptope; We highlight that the correlation elliptope has received less attention from the information geometry community in comparison to the symmetric positive definite cone of covariance matrices; We highlight where information geometric contributions could help improve on the proposed models (the Euclidean Deep Convolutional Generative Adversarial Network (DCGAN) is already competitive if not fully satisfying from a geometric point of view); Finally, we illustrate the use of such generative models with Monte Carlo simulations to understand the empirical properties of portfolio allocation methods in quantitative finance.

4 The Set of Correlation Matrices

4.1 The Elliptope and its Geometrical Properties

Let $\mathcal{E} = \{C \in \mathbb{R}^{n \times n} \mid C = C^\top, \forall i \in \{1, \ldots, n\}, C_{ii} = 1, \forall x \in \mathbb{R}^n, x^\top C x \geq 0\}$ be the set of correlation matrices, also known as the correlation elliptope.

The convex compact set of correlation matrices $\mathcal{E}$ is a strict subspace of the set of covariance matrices whose Riemannian geometry has been well studied in the information geometry literature, e.g. Fisher-Rao distance [25], Fréchet mean [19], PCA [12], clustering [16,23], Riemannian-based kernel in SVM classification [2]. However, $\mathcal{E}$ endowed with the Riemannian Fisher-Rao metric is not totally geodesic (cf. Figs. 1, 2), and this has strong implications on the principled use of the tools developed for the covariance matrices. For example, consider SPDNet [13] (with an eventual Riemannian batch normalization layer [3]). Despite this neural network is fed with correlation matrices as input, it will generate covariance matrices in subsequent layers. How can we stay in the elliptope in a principled and computationally efficient fashion?

In Fig. 1, each 2×2 covariance matrix is represented by a 3D point (x, y, z). The blue segment $(x = z = 1)$ is the set of 2×2 correlation matrices. In green, the geodesic (using the Riemannian Fisher-Rao metric for covariances) between correlation matrix $\gamma(0) = (1, -0.75, 1)$ and correlation matrix $\gamma(1) = (1, 0.75, 1)$. The geodesic $t \in [0, 1] \to \gamma(t)$ (and in particular the Riemannian mean $\gamma(0.5)$) is not included in the blue segment representing the correlation matrix space.

In Fig. 2 we compare several means:

- (M1) Euclidean mean $\frac{A+B}{2}$,
- (M2) Riemannian barycenter (in general, a covariance matrix $\Sigma^\star$),

- (M3) $C^\star = \text{diag}(\Sigma^\star)^{-\frac{1}{2}} \Sigma^\star \text{diag}(\Sigma^\star)^{-\frac{1}{2}}$ ($\Sigma^\star$ normalized by variance),
- (M4) Fréchet mean constrained to the space of correlation matrices,
- (M5) projection of $\Sigma^\star$ onto the elliptope using the Riemannian distance.

We observe that (M3) is close but not equal to (M4) and (M5), which seems to be equivalent. We have not found yet how to compute (M4) or (M5) efficiently. Alternatively, a recent doctoral dissertation [4] on Riemannian quotient structure for correlation matrices might help progress in this direction.

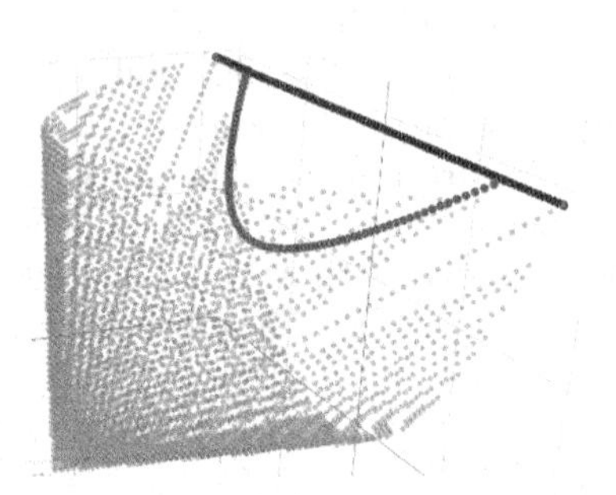

Fig. 1. 2×2 correlation matrices (blue); Fisher-Rao geodesic (green) between two matrices. The geodesic does not stay inside the set of correlation matrices. (Color figure online)

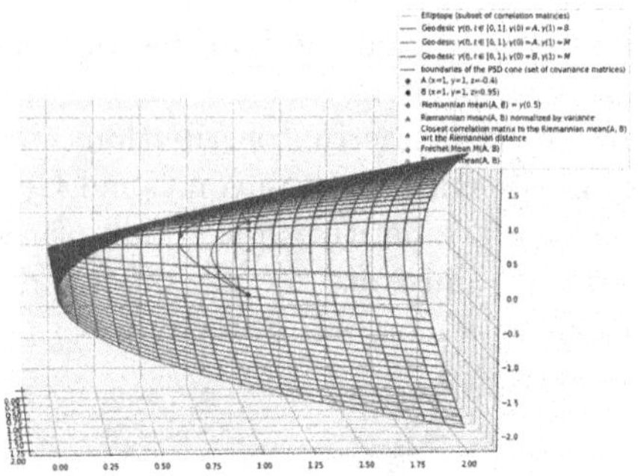

Fig. 2. Visualizing various means.

With a better understanding on how to define and efficiently compute a Riemannian mean for correlation matrices, we can think of adapting SPDNet to correlation matrices, and eventually use it as a component of the generative adversarial network architecture.

4.2 Financial Correlations: Stylized Facts

Since we aim at learning empirical (conditional) distributions, different domains may have different type of typical correlation matrices whose set describes a subspace of the full correlation elliptope. In the present work, we will showcase the proposed model on correlation matrices estimated on stocks returns. We briefly describe the known properties, also known as stylized facts, of these correlation matrices. We leave for future work or research collaborations to explore the empirical space of correlation matrices in other domains.

The properties of financial correlation matrices are well-known and documented in the literature. However, until CorrGAN [18], no single model was able to generate correlation matrices verifying a handful of them at the same time: (SF 1) Distribution of pairwise correlations is significantly shifted to the positive; (SF 2) Eigenvalues follow the Marchenko–Pastur distribution, but for a very large first eigenvalue; (SF 3) Eigenvalues follow the Marchenko–Pastur distribution, but for ≈5% of other large eigenvalues; (SF 4) Perron-Frobenius property (first eigenvector has positive entries); (SF 5) Hierarchical structure of correlation clusters; (SF 6) Scale-free property of the corresponding Minimum Spanning Tree.

When markets are under stress, it is widely known among practitioners that the average correlation is high. More precisely, the whole correlation structure changes and becomes different than the ones typical of rallying or steady markets. These characteristics tied to the market regime would be even harder to capture with a fully specified mathematical model.

5 Learning Empirical Distributions in the Elliptope Using Generative Adversarial Networks

We suggest that conditional generative adversarial networks are promising models to generate random correlation matrices verifying all these not-so-well-specified properties.

5.1 CorrGAN: Sampling Realistic Financial Correlation Matrices

The original CorrGAN [18] model is essentially a DCGAN (architecture based on CNNs). CNNs have 3 relevant properties which align well with stylized facts of financial correlation matrices: shift-invariance, locality, and compositionality (aka hierarchy). Based on our experiments, this architecture yields the best off-the-shelf results most consistently. However, outputs (synthetic matrices) need to be post-processed by a projection algorithm (e.g. [11]) as they are not strictly valid (but close to) correlation matrices. This goes against the end-to-end optimization of deep learning systems. Despite CorrGAN yields correlation matrices which verify the stylized facts, we can observe with PCA (or MDS, t-SNE) that the projected empirical and synthetic distributions do not match perfectly. We also noticed some instability from one trained model to another (well-known GAN defect). For example, in our experiments, the average Wasserstein distance (using the POT package [7]) between any two (PCA-projected) training sets is $\mu_E := 6.7 \pm \sigma_E := 6.8$ (max distance $:= 15$) whereas the average distance between a training set and a generated synthetic set is $\mu_G := 18.8 \pm \sigma_G := 8$ (min distance $:= 8$). Ideally, we would obtain $\mu_E \approx \mu_G$ instead of $\mu_E \ll \mu_G$. We observed that CorrGAN was not good at capturing the different modes of the distribution. This motivated the design of a conditional extension of the model.

5.2 Conditional CorrGAN: Learning Empirical Conditional Distributions in the Elliptope

We describe in Fig. 3 the architecture of cCorrGAN, the conditional extension of the CorrGAN model [18]. To train this model, we used the following experimental protocol: 1. We built a training set of correlation matrices labeled with one of the three classes {`stressed`, `normal`, `rally`} depending on the performance of an equally weighted portfolio over the period of the correlation estimation; 2. We trained several cCorrGAN models; 3. We evaluated them qualitatively (PCA projections, reproductions of stylized facts), and quantitatively (confusion matrix of a SPDNet classifier [3], Wasserstein distance between projected empirical and synthetic distributions) to select the best model. Samples obtained from the best model are shown in Fig. 4.

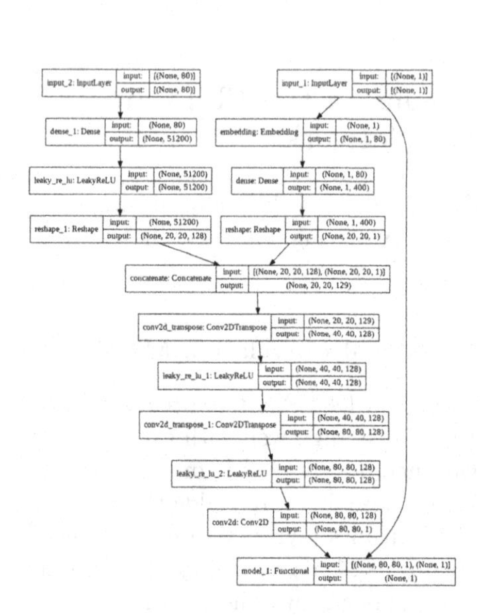

Fig. 3. Architecture of the conditional GAN (cCorrGAN) which generates 80×80 correlation matrices corresponding to a given market regime {`stressed`, `normal`, `rally`}

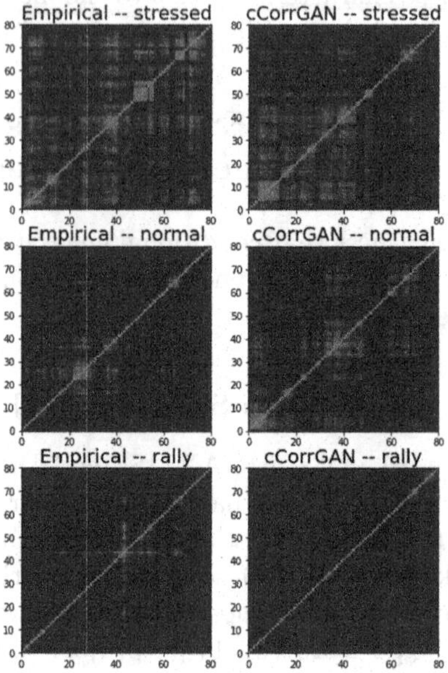

Fig. 4. Empirical (left) and cCorrGAN generated (right) correlation matrices for three market regimes

6 Application: Monte Carlo Simulations of Risk-Based Portfolio Allocation Methods

Monte Carlo simulations to test the robustness of investment strategies and portfolio construction methods were the original motivation to develop CorrGAN, and its extension conditional CorrGAN. The methodology is summarized in Fig. 5: 1. Generate a random correlation matrix from a given market regime using cCorrGAN, 2. Extract features from the correlation matrix, 3. Estimate the in-sample and out-of-sample risk on synthetic returns which verify the generated correlation structure, 4. Predict and explain (with Shapley values)

a) performance decay from in-sample to out-of-sample, b) out-performance of a portfolio allocation method over another. The idea of using Shapley values to explain portfolio construction methods can be found in [22].

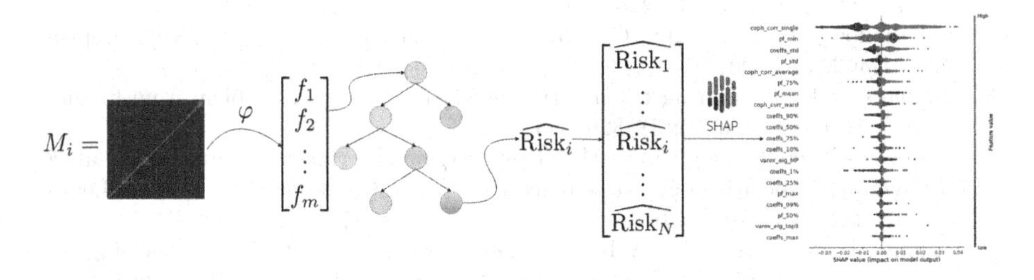

Fig. 5. Methodology for analysing the results of the Monte Carlo simulations

Using such methodology, enabled by the sampling of realistic random correlation matrices conditioned on a market regime, we can find, for example, that the Hierarchicaly Risk Parity (HRP) [6] outperforms the naive risk parity when there is a strong hierarchy in the correlation structure, well-separated clusters, and a high dispersion in the first eigenvector entries, typical of a normal or rallying market regime. In the case of a stressed market where stocks plunge altogether, HRP performs on-par (or slightly worse because of the risk of identifying a spurious hierarchy) than the simpler naive risk parity.

Acknowledgments. We thank ESILV students for helping us running experiments for hundreds of GPU-days: Chloé Daems, Thomas Graff, Quentin Bourgue, Davide Coldebella, Riccardo Rossetto, Mahdieh Aminian Shahrokhabadi, Marco Menotti, Johan Bonifaci, Clément Roux, Sarah Tailee, Ekoue Edem Ekoue-Kouvahey, Max Shamiri.

References

1. Assefa, S.: Generating synthetic data in finance: opportunities, challenges and pitfalls. Challenges and Pitfalls (2020)
2. Barachant, A., Bonnet, S., Congedo, M., Jutten, C.: Classification of covariance matrices using a Riemannian-based kernel for BCI applications. Neurocomputing **112**, 172–178 (2013)
3. Brooks, D., Schwander, O., Barbaresco, F., Schneider, J.Y., Cord, M.: Riemannian batch normalization for SPD neural networks. Adv. Neural Inf. Process. Syst. **32**, 15489–15500 (2019)
4. David, P.: A Riemannian Quotient Structure for Correlation Matrices with Applications to Data Science. Ph.D. thesis, The Claremont Graduate University (2019)
5. Davies, P.I., Higham, N.J.: Numerically stable generation of correlation matrices and their factors. BIT Numer. Math. **40**(4), 640–651 (2000)
6. De Prado, M.L.: Building diversified portfolios that outperform out of sample. J. Portfolio Manage. **42**(4), 59–69 (2016)
7. Flamary, R., Courty, N.: POT Python Optimal Transport library (2017). https://pythonot.github.io/

8. Fu, R., Chen, J., Zeng, S., Zhuang, Y., Sudjianto, A.: Time series simulation by conditional generative adversarial net. arXiv preprint arXiv:1904.11419 (2019)
9. Ghosh, S., Henderson, S.G.: Behavior of the NORTA method for correlated random vector generation as the dimension increases. ACM Trans. Modeling Comput. Simul. (TOMACS) **13**(3), 276–294 (2003)
10. Goodfellow, I.J., et al.: Generative adversarial networks. arXiv preprint arXiv:1406.2661 (2014)
11. Higham, N.J.: Computing the nearest correlation matrix-a problem from finance. IMA J. Numer. Anal. **22**(3), 329–343 (2002)
12. Horev, I., Yger, F., Sugiyama, M.: Geometry-aware principal component analysis for symmetric positive definite matrices. In: Asian Conference on Machine Learning. PMLR, pp. 1–16 (2016)
13. Huang, Z., Van Gool, L.: A Riemannian network for SPD matrix learning. In: Proceedings of the AAAI Conference on Artificial Intelligence, vol. 31 (2017)
14. Hüttner, A., Mai, J.F.: Simulating realistic correlation matrices for financial applications: correlation matrices with the Perron-Frobenius property. J. Stat. Comput. Simul. **89**(2), 315–336 (2019)
15. Koshiyama, A., Firoozye, N., Treleaven, P.: Generative adversarial networks for financial trading strategies fine-tuning and combination. In: Quantitative Finance, pp. 1–17 (2020)
16. Lee, H., Ahn, H.J., Kim, K.R., Kim, P.T., Koo, J.Y.: Geodesic clustering for covariance matrices. Commun. Stat. Appl. Methods **22**(4), 321–331 (2015)
17. Lewandowski, D., Kurowicka, D., Joe, H.: Generating random correlation matrices based on vines and extended onion method. J. Multivariate Anal. **100**(9), 1989–2001 (2009)
18. Marti, G.: CorrGAN: sampling realistic financial correlation matrices using generative adversarial networks. In: ICASSP 2020–2020 IEEE International Conference on Acoustics, Speech and Signal Processing (ICASSP), pp. 8459–8463. IEEE (2020)
19. Moakher, M.: A differential geometric approach to the geometric mean of symmetric positive-definite matrices. SIAM J. Matrix Anal. Appl. **26**(3), 735–747 (2005)
20. Nielsen, F., Bhatia, R.: Matrix Information Geometry. Springer, New York (2013). https://doi.org/10.1007/978-3-642-30232-9
21. Nielsen, F., Sun, K.: Clustering in Hilbert's projective geometry: the case studies of the probability simplex and the elliptope of correlation matrices. In: Nielsen, F. (ed.) Geometric Structures of Information. SCT, pp. 297–331. Springer, Cham (2019). https://doi.org/10.1007/978-3-030-02520-5_11
22. Papenbrock, J., Schwendner, P., Jaeger, M., Krügel, S.: Matrix evolutions: Synthetic correlations and explainable machine learning for constructing robust investment portfolios (2020)
23. Shinohara, Y., Masuko, T., Akamine, M.: Covariance clustering on Riemannian manifolds for acoustic model compression. In: 2010 IEEE International Conference on Acoustics, Speech and Signal Processing, pp. 4326–4329. IEEE (2010)
24. Takahashi, S., Chen, Y., Tanaka-Ishii, K.: Modeling financial time-series with generative adversarial networks. Physica A Stat. Mech. Appl. **527**, 121261 (2019)
25. Wells, J., Cook, M., Pine, K., Robinson, B.D.: Fisher-Rao distance on the covariance cone. arXiv preprint arXiv:2010.15861 (2020)
26. Wiese, M., Knobloch, R., Korn, R., Kretschmer, P.: Quant GANs: deep generation of financial time series. Quant. Financ. **20**(9), 1419–1440 (2020)
27. Zheng, Y.J., Zhou, X.H., Sheng, W.G., Xue, Y., Chen, S.Y.: Generative adversarial network based telecom fraud detection at the receiving bank. Neural Networks **102**, 78–86 (2018)

Topological and Geometrical Structures in Neurosciences

Topological Model of Neural Information Networks

Matilde Marcolli(✉)

California Institute of Technology, Pasadena, CA 91125, USA
matilde@caltech.edu

Abstract. This is a brief overview of an ongoing research project, involving topological models of neural information networks and the development of new versions of associated information measures that can be seen as possible alternatives to integrated information. Among the goals are a geometric modeling of a "space of qualia" and an associated mechanism that constructs and transforms representations from neural codes topologically. The more mathematical aspects of this project stem from the recent joint work of the author and Yuri Manin [18], while the neuroscience modeling aspects are part of an ongoing collaboration of the author with Doris Tsao.

1 Motivation

Before describing the main aspects of this project, it is useful to present some of the main motivations behind it.

1.1 Homology as Functional to Processing Stimuli

In recent experiments on rhesus monkey, Tevin Rouse at Marlene Cohen's lab at Carnegie Mellon showed that visual attention increases firing rates while decreasing correlation of fluctuations of pairs of neurons, and also showed that topological analysis of the activated networks of V4 neurons reveals a peak of non-trivial homology generators in response to stimulus processing during visual attention, [25].

A similar phenomenon, with the formation of a high number of non-trivial homology generators in response to stimuli, was produced in the analysis directed by the topologist Katrin Hess [23] of simulations of the neocortical microcircuitry. The analysis of [23] proposes the interpretation that these topological structures are necessary for the processing of stimuli in the brain cortex.

These findings are very intriguing for two reasons: (1) they link topological structures in the activated neural circuitry to phenomena like attention; (2) they

Supported by Foundational Questions Institute and Fetzer Franklin Fund FFF grant number FQXi-RFP-1804, SVCF grant 2018-190467 and FQXi-RFP-CPW-2014, SVCF grant 2020-224047.

F. Nielsen and F. Barbaresco (Eds.): GSI 2021, LNCS 12829, pp. 623–633, 2021.
https://doi.org/10.1007/978-3-030-80209-7_67

suggest that a sufficient amount of topological complexity serves a functional computational purpose.

The first point will be discussed more at length below, in Sect. 3. The second point is relevant because it suggests a better mathematical setting for modeling neural information networks architecture in the brain. Indeed, although the work of [23] does not offer a theoretical explanation of why topology is needed for stimulus processing, there is a well known context in the theory of computation where a similar situation occurs, which may provide the key for the correct interpretation, namely the theory of concurrent and distributed computing [11, 15]. Note that in some case, like the toroidal topology arising from grid cell activity [14], a neuronal functional explanation for the topology is in fact known, though not a distributed computing interpretation.

In the mathematical theory of concurrent and distributed computing, one considers a collection of sequential computing entities (processes) that cooperate to solve a problem (task). The processes communicate by applying operations to objects in a shared memory, and they are asynchronous, in the sense that they run at arbitrary varying speeds. Concurrent and distributed algorithms and protocols decide how and when each process communicates and shares with others. The main questions are how to design such algorithms that are efficient in the presence of noise, failures of communication, and delays, and how to understand when a an algorithm exists to solve a particular task.

Protocols for concurrent and distributed computing are modeled using (directed) simplicial sets. For example, in distributed algorithms an initial or final state of a process is a vertex, any $d+1$ mutually compatible initial or final states are a d-dimensional simplex, and each vertex is labelled by a different process. The complete set of all possible initial and final states is then a simplicial set. A decision task consists of two simplicial sets of initial and final states and a simplicial map (or more generally correspondence) between them. The typical structure consists of an input complex, a protocol complex, and an output complex, with a certain number of topology changes along the execution of the protocol [15].

There are very interesting topological obstruction results in the theory of distributed computing [15], which show that a sufficient amount of non-trivial homology in the protocol complex is necessary for a decision task problem to be solvable. Thus, the theory of distributed computing shows explicitly a setting where a sufficient amount of topological complexity (measured by non-trivial homology) is necessary for computation.

The working hypothesis we would like to consider here is that the brain requires a sufficient amount of non-trivial homology *because* it is carrying out a concurrent/distributed computing task that can only run successfully in the presence of enough nontrivial homology. This suggests that the mathematical modeling of architectures of neural information networks should be formulated in such a way as to incorporate additional structure keeping track of associated concurrent/distributed computational systems.

An important aspect of the topological analysis of brain networks is the presence of directed structures on the simplicial sets involved. A directed structure is not an orientation, but rather a kind of "flow of information" structure, as in concurrent and distributed computing [11]. A directed n-cube models the concurrent execution of n simultaneous transitions between states in a concurrent machine. In directed algebraic topology one considers topological (simplicial) spaces where paths and homotopies have a preferred direction and cannot be reversed, and directed simplicial sets can be regarded as a higher dimensional generalization of the usual notion of directed graphs. The categorical viewpoint is natural when one thinks of paths in a directed graph as forming a free category generated by the edges. A similar notion for higher dimensional directed complexes exists in the form of ω-categories [16,29].

1.2 Neural Code and Homotopical Representations

It is known from the work of Curto and collaborators (see [8] and references therein) that the geometry of the stimulus space can be reconstructed *up to homotopy* from the binary structure of the neural code. The key observation behind this reconstruction result is a simple topological property: under the reasonable assumption that the place field of a neuron (the preferred region of the stimulus space that causes the neuron to respond with a high firing rate) is a convex open set, the binary code words in the neural code represent the overlaps between these regions, which determine a simplicial complex (the simplicial nerve of the open covering of the stimulus space). Under the convexity hypothesis the homotopy type of this simplicial complex is the same as the homotopy type of the stimulus space. Thus, the fact that the binary neural code captures the complete information on the intersections between the place fields of the individual neurons is sufficient to reconstruct the stimulus space, but only up to homotopy.

This suggests another working hypothesis. One can read the result about the neural code and the stimulus space reconstruction as suggesting that the brain *represents* the stimulus space through a *homotopy type*. We intend to use this idea as the basic starting point and propose that the experience of *qualia* should be understood in terms of the processing of external stimuli via the construction of associated homotopy types.

The homotopy equivalence relation in topology is weaker but also more flexible than the notion of homeomorphism. The most significant topological invariants, such as homotopy and homology groups, are homotopy invariants. Heuristically, homotopy describes the possibility of deforming a topological space in a one parameter family. In particular, a *homotopy type* is an equivalence class of topological spaces up to (weak) homotopy equivalence, which roughly means that only the information about the space that is captured by its homotopy groups is retained. There is a direct connection between the formulation of topology at the level of homotopy types and "higher categorical structures". For our purposes here it is also worth mentioning that simplicial sets provide a good combinatorial models for the (weak) homotopy type of topological spaces.

Thus, these results suggest that a good mathematical modeling of network architectures in the brain should also include a mechanism that generates homotopy types, through the information carried by the network via neural codes.

1.3 Consciousness and Informational Complexity

The idea of computational models of consciousness based on some measure of informational complexity is appealing in as it proposes quantifiable measures of consciousness tied up to an appropriate notion of complexity and interrelatedness of the underlying system.

Tononi's *integrated information* theory (or Φ function) is a proposal for such a quantitative correlate of consciousness, which roughly measures of the least amount of effective information in a whole system that is not accounted for by the effective information of its separate parts. More precisely, effective information is computed by a Kullback–Leibler divergence (relative entropy) between an observed output probability distribution and a hypothetical uniformly distributed input. The integrated information considers all possible partitions of a system into subsystems (two disjoint sets of nodes in a network for instance) and assigns to each partition a symmetrized and normalized measure of effective information with inputs and outputs on the two sides of the partition, and minimizes over all partitions. The main idea is therefore that integrated information is a measure of informational complexity and interconnectedness of a system [30]. It is proposed as a measure of consciousness because it appears to capture quantitatively several aspects of conscious experience (such as its compositional and integrated qualities, see [30]).

An attractive feature of the integrated information model is an idea of consciousness that is not a binary state (an entity either does or does not have consciousness) but that varies by continuous degrees and is tied up to the intuitive understanding that sufficient complexity and interconnectedness is required for the acquisition of a greater amount of consciousness. The significant ethical implications of these assumptions are evident. On the other hand, one of the main objections to using integrated information as a measure of consciousness is the fact that, while it is easy to believe that sufficient informational complexity may be a *necessary* condition for consciousness, the assumption that it would also be a *sufficient* condition appears less justifiable.

Indeed, this approach to a mathematical modeling of consciousness has been criticized on the ground that it is easy to construct simple mathematical models exhibiting a very high value of the Φ function (Aaronson's Vandermonde matrix example), that can hardly be claimed to possess consciousness. Generally, one can resort to the setting of coding theory to generate many examples of sufficiently good codes (for example the algebro-geometric Reed-Solomon error-correcting codes) that indeed exhibit precisely the typical from of high interconnectedness that leads to large values of integrated information.

Another valid objection to the use of integrated information lies in its computational complexity, given that it requires a computation and minimization over the whole set of partitions of a given set (which in a realistic example like a biological brain is already in itself huge). Other computational difficulties with integrated information and some variants of the same model are discussed in detail in [20]. A possible approach to reducing the computational complexity may be derived from the free energy principle of [12] and its informational applications, though we will not discuss it here.

Our goal is not to improve the mathematical theory of integrated information, but to construct a different kind of mathematical model based on topology and in particular homotopy theory. However, assuming that high values of (some version of) integrated information may indeed provide a valid necessary condition for consciousness, it is natural to ask how a topological model of the kind we will describe in the next section would constrain the values of integrated information.

In our setting, the best approach to infer the existence of lower bounds for some integrated information is to reformulate integrated information itself from a topological viewpoint as discussed in [18], along the lines of the cohomological formulations of information of [4] and [31,32].

Treating integrated information as cohomological can have advantages in terms of computability: it is well known in the context of simplicial sets that computational complexity at the level of chains can be much higher than at the level of (co)homology.

2 Homotopical Models of Neural Information Networks

We summarize here the main mathematical aspects of this project, with special emphasis on the author's joint work with Manin in [18] on a mathematical framework for a homotopy theoretic modeling of neural information networks in the brain, subject to informational and metabolic constraints, performing some concurrent/distributed computing tasks, and generating associated homotopy types from neural codes structures.

2.1 Gamma Spaces and Information Networks

The starting point of our approach is the notion of Gamma spaces, introduced by Graeme Segal in the 1970s to model homotopy theoretic spectra. A Gamma space is a functor from finite (pointed) sets to simplicial sets. The Segal construction associated such a functor to a given category $\mathcal{C}$ with some basic properties (categorical sum and zero object), by assigning to a finite set X the nerve of a category of $\mathcal{C}$-valued summing functors that map subsets of a given finite set and inclusions between them to objects of the category $\mathcal{C}$ and morphisms in a way that is additive over disjoint subsets. One should think of summing functors as categorically valued measures that can be used to consistently assign a structure (computational, information-theoretic, etc.) to all subsystems of a given system. The enrichment of the notion of Gamma space with probabilistic data was developed in [19]. The nerve of the category of summing functors is a topological model that parameterizes all the possible ways of assigning structures described by the target category $\mathcal{C}$ to the system so that they are consistent over subsystems.

A general mathematical setting for a theory of resources and conversion of resources was developed in [6] and [13] in terms of symmetric monoidal categories. These can be taken as targets of the summing functors in the Gamma space construction. Thus, the language of Gamma spaces can be understood as describing, through the category of summing functors, all possible consistent assignments of resources to subsystems of a given system in such a way that resources of independent subsystems behave additively. The fact that Gamma spaces are built using data of subsystems and relations between subsystems is reminiscent of the ideas underlying the notion of integrated information, but it is also richer and more flexible in its applicability. One of the main advantages of the categorical framework is that the categorical structure ensures that many fundamental properties carry over naturally between different categories. As applications of category theory in other contexts (such as theoretical computer science and physics) have shown, this often renders the resulting constructions much more manageable.

The target monoidal category of resources can describe computational resources (concurrent and distributed computing structures that can be implementable on a given network) by using a category of transition systems in the sense of [33]. One can similarly describe other kinds of metabolic or informational resources. Different kinds of resources and constraints on resources assigned to the same underlying network can be analyzed through functorial relations between these categories. Gamma spaces, via spectra, have associated homotopy groups providing invariants detecting changes of homotopy types.

2.2 Neural Codes and Coding Theory

In the theory of error-correcting codes, binary codes are evaluated in terms of their effectiveness at encoding and decoding. This results in a two-parameter space (transmission rate and relative minimum distance). There are bounds in the space of code parameters, such as the Gilbert-Varshamov bound (related to the Shannon entropy and expressing the typical behavior of random codes) or the asymptotic bound (related to Kolmogorov complexity and marking a boundary above which only isolated very good codes exist). Examples of good codes above both the Gilbert-Varshamov and the asymptotic bound typically come from algebro-geometric constructions. As mentioned above, these are precisely the type of codes that can generate high values of integrated information.

It is known, however, that neural codes are typically not good codes according to these parameters: this is essentially because these codes are related to the combinatorics of intersections of sets in open coverings and this can make the Hamming distance between code words very small. Manin suggests in [17] the possible existence of a secondary auxiliary level of neural codes involved in the formation of place maps consisting of very good codes that lie near or above the asymptotic bound. This idea that the bad neural codes are combined in the brain into some higher structure that determines good codes can now be made precise through the recent work of Rishidev Chaudhuri and Ila Fiete [5], on bipartite expander Hopfield networks. Their model is based on Hopfield networks with stable states determined by sparse constraints with expander structure, which have good error correcting properties and are good models of memory, where the network dynamics extracts higher order correlations from the neural codes.

This suggests that a version of the model of Chaudhuri and Fiete, together with a categorical form of the Hopfield network dynamics discussed below, may be used to address the step from the (poor code) neural codes to the (good codes) bipartite expander Hopfield networks, and from these to a lower bound on the values of an appropriate integrated information measure.

2.3 Dynamics of Networks in a Categorical Setting

The setting described in the previous subsection is just a kinematic scaffolding, in the sense that so far we have discussed the underlying geometry of the relevant configuration space, but not yet how to introduce dynamics. Hopfield network dynamics is a good model for how brain networks evolve over time with excitatory/inhibitory synaptic connections. Recent work of Curto and collaborators has revealed beautiful topological structures in the dynamics of Hopfield networks [9,10].

In [18] we developed a version of the Hopfield network dynamics where the evolution equations, viewed as finite difference equations, take place in a category of summing functors, in the context of the Gamma space models described above. These induce the usual non-linear Hopfield dynamics when summing functors take values in a suitable category of weighted codes. It also induces a dynamical system on the topological space given by the realization of the nerve of the category of summing functors. This formulation has the advantage that all the associated structures, resources with constraints, concurrent/distributed computing architectures, informational measures and constraints, are all evolving dynamically according to a common dynamics and functorial relations between the respective categories. Phenomena such as excitatory-inhibitory balance can also be investigated in this context. One aspect of the dynamics that is interesting in this context is whether it involves changes in homotopy types, which can be especially relevant to the interpretation discussed in §3 below.

This leads to a related question, which is a mechanism in this categorical setting of dynamical information networks, for operations altering stored homotopy types, formulated in terms of a suitable "calculus" of homotopy types. This points to the important connection between homotopy types and logic (in the form of Church's type theory), which is provided by the developed in Voevodsky's "homotopy type theory", see for instance [21], a setting designed to replace set-theory based mathematical logic with a model of the logic theory of types based on simplicial sets.

2.4 Combined Roles of Topology

Topology in neuroscience, as mentioned above, has appeared in three different roles: clique complexes of subnetworks activated in response to stimuli and their persistent homology, topological internal reconstruction of external inputs and output through the nerve complex of the code generated by the activation of cells given their receptive fields, and information topology through information structures and their cohomology in the sense of [31]. These roles seem apparently unrelated. However, in the setting we are describing in [18] we see that all of these are simultaneously accounted for. There is a functorial mapping from a category of summing functors associated to a network to the category of information structures that generated a functorial assignment of a cohomological version of integrated information. Moreover, clique complexes of subnetworks and nerve complexes of codes are also recovered from the spectra obtained through the Gamma space formalism form categories of summing functors on networks.

3 Toward a Model of Qualia and Experience

In this final section we describe some tentative steps in the direction of using the topological setting described above to model qualia and experience.

A homotopy type is a (simplicial) space viewed up to continuous deformations that preserve homotopy invariants. One can imagine a time series of successive representations of an external stimulus space that either remain within the same homotopy type or undergo sudden transitions that causes detectable jumps in some of the homotopy groups. This collection of discrete topological invariants given by the homotopy groups (or homology as a simpler invariant) as detectors triggered by sudden changes of the stimulus space that does not preserve the homotopy type of the representation. We propose that such "jumps" in the homotopy type are related to experience.

In "Theater of Consciousness" models like [1], it is observed how the mind carries out a large number of tasks that are only "intermittently conscious", where a number of processes continue to be carried out unconsciously with "limited access" to consciousness, with one process coming to the forefront of consciousness (like actors on the stage in the theater metaphor) in response to an unexpected change. In the setting we are proposing, changes that do not alter the stored homotopy type are like the processes running in the unconscious background in the theater of consciousness, while the jumps in the discrete homotopy invariants affect the foregrounding to the stage of consciousness. This suggests that one should not simply think of the stored homotopy type as the reconstruction of a stimulus space at a given time instant. It is more likely that a whole time sequence should be stored as a homotopy type, changes to which will trigger an associated conscious experience. Some filtering of the stored information is also involved, that can be modeled through a form of persistent topology. The Gamma spaces formalism can be adapted to this persistent topology setting, as shown in [18].

In this setting we correspondingly think of "qualia space" as a space of homotopy types. This can be compared, for example, with the proposal of a qualia space in the theory of integrated information [2], based on probabilities on the set of states of the system, with a conscious experience represented as a shape in this qualia space. In [2], a "quale" is described as the set of all probability measures on all subsystems and mappings between them corresponding to inclusions. Notice how this setting is very closely related to the notion of probabilistic Gamma spaces developed in [19].

One can fit within this framework the observed role of homology as a detector of experience. The neuroscience experiments of [22], for example, conducted measurement of persistent homology of activated brain networks, comparing resting-state functional brain activity after intravenous infusion of either a placebo or psilocybin. The results they obtained show a significant change in homology in the case of psilocybin, with many transient structures of low stability and a small number of more significant persistent structures that are not observed in the case of placebo. These persistent homology structure are a correlate of the psychedelic experience. This evidence supports the idea that changes in topological invariants are related to experience and different states of consciousness.

The approach described here has possible points of contacts with other recent ideas about consciousness, beyond the integrated information proposal, including [3,24,27], for which a comparative analysis would be interesting.

3.1 Neural Correlates of Consciousness and the Role of Topology

In recent work [28], Ann Sizemore, Danielle Bassett and collaborators conducted a detailed topological analysis of the human connectome and identified explicit generators for the most prominently persistent non-trivial elements of both the first and second persistent homology groups. The persistence is computed with respect to a filtering of the connectivity matrix. The resulting generators identify which networks of connections between different cortical and subcortical regions involve non-trivial homology. Moreover, their analysis shows that the subcortical regions act as cone vertices for many homology generators formed by cortical regions. Based on the fact that electrostimulation of the posterior, but not of the anterior cortex, elicit various forms of somatosensory experience, the posterior cortex was proposed as a possible region involved in a neural correlate of consciousness (see [7]). A question to investigate within the mathematical framework outlined in the previous section is whether nontrivial topological structures in the connectome can be linked to candidate substrates in the wiring of brain regions for possible neural correlate of consciousness.

References

1. Baars, B.J.: In the Theater of Consciousness. Oxford University Press, Oxford (2001)
2. Balduzzi, D., Tononi, G.Q.: The geometry of integrated information. PLoS Comput. Biol. **5**(8), e1000462 (2009)

3. Baudot, P.: Elements of Consciousness and Cognition. Biology, Mathematic, Physics and Panpsychism: an Information Topology Perspective. arXiv:1807.04520
4. Baudot, P., Bennequin, D.: The homological nature of entropy. Entropy **17**(5), 3253–3318 (2015)
5. Chaudhuri, R., Fiete, I.: Bipartite expander Hopfield networks as self-decoding high-capacity error correcting codes. In: 33rd Conference on Neural Information Processing Systems (NeurIPS 2019), Vancouver, Canada (2019)
6. Coecke, B., Fritz, T., Spekkens, R.W.: A mathematical theory of resources. Inf. Comput. **250**, 59–86 (2016). [arXiv:1409.5531]
7. Crick, F.C., Koch, C.: Towards a neurobiological theory of consciousness. Semin. Neurosci. **2**, 263–275 (1990)
8. Curto, C.: What can topology tell us about the neural code? Bull. Amer. Math. Soc. (N.S.) **54**(1), 63–78 (2017)
9. Curto, C., Geneson, J., Morrison, K.: Stable fixed points of combinatorial threshold-linear networks. arXiv:1909.02947
10. Curto, C., Langdon, C., Morrison, C.: Robust motifs of threshold-linear networks. arXiv:1902.10270
11. Fajstrup, L., Goubault, E., Haucourt, E., Mimram, S., Raussen, M.: Perspectives. In: Raussen, M. (ed.) Directed Algebraic Topology and Concurrency, pp. 151–153. Springer, Cham (2016). https://doi.org/10.1007/978-3-319-15398-8_8
12. Friston, K.: A free energy principle for biological systems. Entropy **14**, 2100–2121 (2012)
13. Fritz, T.: Resource convertibility and ordered commutative monoids, Math. Str. Comp. Sci. 27(6), 850–938 arXiv:1504.03661
14. Gardner, R.J., et al.: Toroidal topology of population activity in grid cells. bioRxiv 2021.02.25.432776
15. Herlihy, M., Kozlov, D., Rajsbaum, S.: Distributed Computing Through Combinatorial Topology. Elsevier (2014)
16. Kapranov, M., Voevodsky, V.: Combinatorial-geometric aspects of polycategory theory: pasting schemes and higher Bruhat orders (list of results). Cab. Top. Géom. Diff. Catégoriques **32**, 11–27 (1991)
17. Manin, Yu.I.: Error-correcting codes and neural networks, Selecta Math. (N.S.) **24**(1), 521–530 (2018)
18. Manin, Yu.I., Marcolli, M.: Homotopy Theoretic and Categorical Models of Neural Information Networks. arXiv:2006.15136
19. Marcolli, M.: Gamma spaces and Information. J. Geom. Phys. **140**, 26–55 (2019)
20. Mediano, P.AM., Seth, A.K., Barrett, A.B.: Measuring integrated information: comparison of candidate measures in theory and simulation, Entropy **21**, 17 (2019)
21. Pelayo, A., Warren, M.A.: Homotopy type theory and Voevodsky's univalent foundations. Bull. Amer. Math. Soc. (N.S.) **51**(4), 597–648 (2014)
22. Petri, G., et al.: Homological scaffolds of brain functional networks. J. R. Soc. Interface **11**(101), 20140873 (2014)
23. Reimann, M.W., et al.: Cliques of Neurons bound into cavities provide a missing link between structure and function. Front. Comput. Neurosci. **11**, 48 (2017)
24. Ondobaka, S., Kilner, J., Friston, K.: The role of interoceptive inference in theory of mind. Brain Cogn. **112**, 64–68 (2017)
25. Rouse, T.: Topological Analysis of Attention, talk at the Fields Institute Focus Program workshop New Mathematical Methods for Neuroscience, Toronto (2020). http://gfs.fields.utoronto.ca/talks/Topological-Analysis-Attention
26. Segal, G.: Categories and cohomology theories. Topology **13**, 293–312 (1974)

27. Seth, A.K.: Interoceptive inference, emotion, and the embodied self. Trends Cogn. Sci. **17**(11), 565–573 (2013)
28. Sizemore, A.E., Giusti, C., Kahn, A., Vettel, J.M., Betzel, R.F., Bassett, D.S.: Cliques and cavities in the human connectome. J. Comput. Neurosci. **44**(1), 115–145 (2017). https://doi.org/10.1007/s10827-017-0672-6
29. Steiner, R.: The Algebra of directed complexes. Appl. Categor. Struct. **1**, 247–284 (1993)
30. Tononi, G.: An information integration theory of consciousness. BMC Neurosci. **5**, 42 (2004)
31. Vigneaux, J.P.: Generalized information structures and their cohomology. arXiv:1709.07807
32. Vigneaux, J.P.: Topology of statistical systems. A cohomological approach to information theory, Ph.D. Thesis, Institut de mathématiques de Jussieu, Université de Paris Diderot (2019)
33. Winskel, G., Nielsen, M.: Categories in concurrency in Semantics and logics of computation (Cambridge, 1995), vol. 14, pp. 299–354, Publications Newton Institute, Cambridge University Press (1997)

On Information Links

Pierre Baudot(✉)

Median Technologies, 1800 Route des Crêtes Bâtiment B, 06560 Valbonne, France
pierre.baudot@mediantechnologies.com,
https://sites.google.com/view/pierrebaudot/

Abstract. In a joint work with D. Bennequin [8], we suggested that the (negative) minima of the 3-way multivariate mutual information correspond to Borromean links, paving the way for providing probabilistic analogs of linking numbers. This short note generalizes the correspondence of the minima of k multivariate interaction information with k Brunnian links in the binary variable case. Following [16], the negativity of the associated K-L divergence of the joint probability law with its Kirkwood approximation implies an obstruction to local decomposition into lower order interactions than k, defining a local decomposition inconsistency that reverses Abramsky's contextuality local-global relation [1]. Those negative k-links provide a straightforward definition of collective emergence in complex k-body interacting systems or dataset.

Keywords: Mutual information · Homotopy links · Contextuality · Synergy · Emergence

1 Introduction

Previous works established that Gibbs-Shannon entropy function H_k can be characterized (uniquely up to the multiplicative constant of the logarithm basis) as the first class of a Topos cohomology defined on random variables complexes (realized as the poset of partitions of atomic probabilities), where marginalization corresponds to localization (allowing to construct Topos of information) and where the coboundary operator is an Hochschild's coboundary with a left action of conditioning ([7,30], see also the related results found independently by Baez, Fritz and Leinster [4,5]). Vigneaux could notably underline the correspondence of the formalism with the theory of contextuality developed by Abramsky [1,30]. Multivariate mutual informations I_k appear in this context as coboundaries [6,7], and quantify refined and local statistical dependences in the sens that n variables are mutually independent if and only if all the I_k vanish (with $1 < k < n$, giving $(2^n - n - 1)$ functions), whereas the Total Correlations G_k quantify global or total dependences, in the sens that n variables are mutually independent if and only if $G_n = 0$ (theorem 2 [8]). As preliminary uncovered by several related studies, information functions and statistical structures not only present some co-homological but also homotopical features that are finer invariants [6,9,21]. Notably, proposition 9 in [8], underlines a correspondence of the minima $I_3 = -1$

F. Nielsen and F. Barbaresco (Eds.): GSI 2021, LNCS 12829, pp. 634–644, 2021.
https://doi.org/10.1007/978-3-030-80209-7_68

of the mutual information between 3 binary variables with Borromean link. For $k \geq 3$, I_k can be negative [15], a phenomenon called synergy and first encountered in neural coding [11] and frustrated systems [22] (cf. [8] for a review and examples of data and gene expression interpretation). However, the possible negativity of the I_k has posed problems of interpretation, motivating a series of study to focus on non-negative decomposition, "unique information" [10,25,31]. Rosas et al. used such a positive decomposition to define emergence in multivariate dataset [28]. Following [6,8], the present work promotes the correspondence of emergence phenomena with homotopy links and information negativity. The chain rule of mutual-information goes together with the following inequalities discovered by Matsuda [22]. For all random variables $X_1; ..; X_k$ with associated joint probability distribution P we have: $I(X_1; ..; X_{k-1}|X_k; P) \geq 0 \Leftrightarrow I_{k-1} \geq I_k$ and $I(X_1; ..; X_{k-1}|X_k; P) < 0 \Leftrightarrow I_{k-1} < I_k$ that characterize the phenomenon of information negativity as an increasing or decreasing sequence of mutual information. It means that positive conditional mutual informations imply that the introduction of the new variable decreased the amount of dependence, while negative conditional mutual informations imply that the introduction of the new variable increase the amount of dependence. The meaning of conditioning, notably in terms of causality, as been studied at length in Bayesian network study, DAGs (cf. Pearl's book [26]), although not in terms of information at the nice exception of Jakulin and Bratko [16]. Following notably their work and the work of Galas and Sakhanenko [13] and Peltre [27] on Möbius functions, we adopt the convention of Interaction Information functions $J_k = (-1)^{k+1} I_k$, which consists in changing the sign of even multivariate I_k (remark 10 [8]). This sign trick, as called in topology, makes that even and odd J_k are both superharmonic, a kind of pseudo-concavity in the sens of theorem D [7]. In terms of Bayesian networks, interaction information (respectively conditional interaction information) negativity generalize the independence relation (conditional independence resp.), and identifies common consequence scheme (or multiple cause). Interaction information (Conditional resp.) negativity can be considered as an extended or "super" independence (Conditional resp.) property, and hence $J(X_1; ..; X_n|X_k; P) < 0$ means that $(X_1; ..; X_n)$ are n "super" conditionally independent given X_k. Moreover such negativity captures common causes in DAGs and Bayesian networks. Notably, the cases where 3 variables are pairwise independent but yet present a (minimally negative) J_3 what is called the Xor problem in Bayesian network [16]. In [8], we proposed that those minima correspond to Borromean links. k-links are the simplest example of rich and complex families of link or knots (link and knot can be described equivalently by their braiding or braid word). k-links are prototypical homotopical invariants, formalized as link groups by Milnor [24]. A Brunnian link is a nontrivial link that becomes a set of trivial unlinked circles if any one component is removed. They are peculiarly attracting because of their beauty and apparent simplicity, and provide a clear-cut illustration of what is an emergent or purely collective property. They can only appear in 3-dimensional geometry and above (not in 1 or 2 dimension), and their complement in the 3-sphere were coined as "the mystery of three-manifolds" by Bill Thurston in the first slide of his presentation for Perelman proof of Poincaré conjecture (cf. Fig. 1). In what follows, we will first generalize the correspondence of J_k minima with k-links for arbitrary k and then establish that negativity of interaction information detects incompatible probability, which allows an interpretation with respect to Abramsky's formalism as contextual. The information links are clearly in line with the principles of higher structures developed by [2,3] that uses links to account for group interactions beyond pair interaction to catch the essence of many multi-agent interactions. Unravelling the formal relation of information links with the work of Baas, or with Khovanov homology [17] are open questions.

2 Information Functions - Definitions

Entropy. The joint-entropy is defined by [29] for any joint-product of k random variables $(X_1,..,X_k)$ with $\forall i \in [1,..,k], X_i \leq \Omega$ and for a probability joint-distribution $\mathbb{P}_{(X_1,..,X_k)}$:

$$H_k = H(X_1,..,X_k;P) = c \sum_{x_1 \in [N_1],..,x_k \in [N_k]}^{N_1 \times .. \times N_k} p(x_1,..,x_k) \ln p(x_1,..,x_k) \quad (1)$$

where $[N_1 \times ... \times N_k]$ denotes the "alphabet" of $(X_1,...,X_k)$. More precisely, H_k depends on 4 arguments: first, the sample space: a finite set N_Ω; second a probability law P on N_Ω; third, a set of random variable on N_Ω, which is a surjective map $X_j : N_\Omega \to N_j$ and provides a partition of N_Ω, indexed by the elements x_{j_i} of N_j. X_j is less fine than Ω, and write $X_j \leq \Omega$, or $\Omega \to X_j$, and the joint-variable (X_i, X_j) is the less fine partition, which is finer than X_i and X_j; fourth, the arbitrary constant c taken here as $c = -1/\ln 2$. Adopting this more exhaustive notation, the entropy of X_j for P at Ω becomes $H_\Omega(X_j;P) = H(X_j;P_{X_j}) = H(X_{j*}(P))$, where $X_{j*}(P)$ is the *marginal* of P by X_j in Ω.

Kullback-Liebler Divergence and Cross Entropy. Kullback-Liebler divergence [20], is defined for two probability laws P and Q having included support ($support(P) \subseteq support(Q)$) on a same probability space $(\Omega, \mathcal{F}, P)$, noted $p(x)$ and $q(x)$ by:

$$D_\Omega(X_j;P,Q) = D(X_j;p(x)||q(x)) = c \sum_{x \in \mathscr{X}} p(x) \ln \frac{q(x)}{p(x)} \quad (2)$$

Multivariate Mutual Informations. The k-mutual-information (also called co-information) are defined by [15,23]:

$$I_k = I(X_1;...;X_k;P) = c \sum_{x_1,...,x_k \in [N_1 \times ... \times N_k]}^{N_1 \times ... \times N_k} p(x_1.....x_k) \ln \frac{\prod_{I \subset [k]; card(I)=i; i \text{ odd}} p_I}{\prod_{I \subset [k]; card(I)=i; i \text{ even}} p_I} \quad (3)$$

For example, $I_2 = c\sum p(x_1,x_2)\ln\frac{p(x_1)p(x_2)}{p(x_1,x_2)}$ and the 3-mutual information is the function $I_3 = c\sum p(x_1,x_2,x_3)\ln\frac{p(x_1)p(x_2)p(x_3)p(x_1,x_2,x_3)}{p(x_1,x_2)p(x_1,x_3)p(x_2,x_3)}$. We have the alternated sums or inclusion-exclusion rules [7,15,22]:

$$I_n = I(X_1;...;X_n;P) = \sum_{i=1}^{n}(-1)^{i-1}\sum_{I\subset[n];card(I)=i} H_i(X_I;P) \tag{4}$$

And the dual inclusion-exclusion relation [6]:

$$H_n = H(X_1,...,X_n;P) = \sum_{i=1}^{n}(-1)^{i-1}\sum_{I\subset[n];card(I)=i} I_i(X_I;P) \tag{5}$$

As noted by Matsuda [22], it is related to the general Kirkwood superposition approximation, derived in order to approximate the statistical physic quantities and the distribution law by only considering the k first k-body interactions terms, with $k < n$ [18]:

$$\hat{p}(x_1,...,x_n) = \frac{\prod_{I\subset[n];card(I)=n-1} p(x_I)}{\frac{\prod_{I\subset[n];card(I)=n-2} p(x_I)}{\frac{\vdots}{\prod_{i=1}^{n} p(x_i)}}} = \prod_{k=1}^{n-1}(-1)^{k-1}\prod_{I\subset[n];card(I)=n-k} p(x_I) \tag{6}$$

For example:

$$\hat{p}(x_1,x_2,x_3) = \frac{p(x_1,x_2)p(x_1,x_3)p(x_2,x_3)}{p(x_1)p(x_2)p(x_3)} \tag{7}$$

We have directly that $I_3 = D(p(x_1,...,x_n)||\hat{p}(x_1,...,x_n))$ and $-I_4 = D(p(x_1,...,x_n)||\hat{p}(x_1,...,x_n))$. It is hence helpful to introduce the "twin functions" of k-mutual Information called the **k-interaction information** [8,16] (multiplied by minus one compared to [16]), noted J_n:

$$J_n = J(X_1;...;X_n;P) = (-1)^{i-1} I_n(X_1;...;X_n;P) \tag{8}$$

Then we have $J_n = D(p(x_1,...,x_n)||\hat{p}(x_1,...,x_n))$. Hence, J_n can be used to quantify how much the Kirkwood approximation of the probability distribution is "good" (in the sense that $p = \hat{p}$ if and only if $J_n = D(p||\hat{p}) = 0$). The alternated sums or inclusion-exclusion rules becomes [16]:

$$J_n = J(X_1;...;X_n;P) = \sum_{i=1}^{n}(-1)^{n-i}\sum_{I\subset[n];card(I)=i} H_i(X_I;P) \tag{9}$$

And now, their direct sums give the multivariate entropy, for example: $H_3 = H(X_1,X_2,X_3) = J(X_1)+J(X_2)+J(X_3)+J(X_1;X_2)+J(X_1;X_3)+J(X_2;X_3)+J(X_1;X_2;X_3)$ Depending on the context ([8] (p. 15)) and the properties one wishes to use, one should use for convenience either I_k or J_k functions.

Conditional Mutual Informations. The conditional mutual information of two variables $X_1; X_2$ knowing X_3, also noted $X_3.I(X_1; X_2)$, is defined as [29]:

$$I(X_1; X_2|X_3; P) = c \sum_{x_1,x_2,x_3 \in [N_1 \times N_2 \times N_3]}^{N_1 \times N_2 \times N_3} p(x_1, x_2, x_3) \ln \frac{p(x_1, x_3)p(x_2, x_3)}{p(x_3)p(x_1, x_2, x_3)} \quad (10)$$

3 Information k-Links

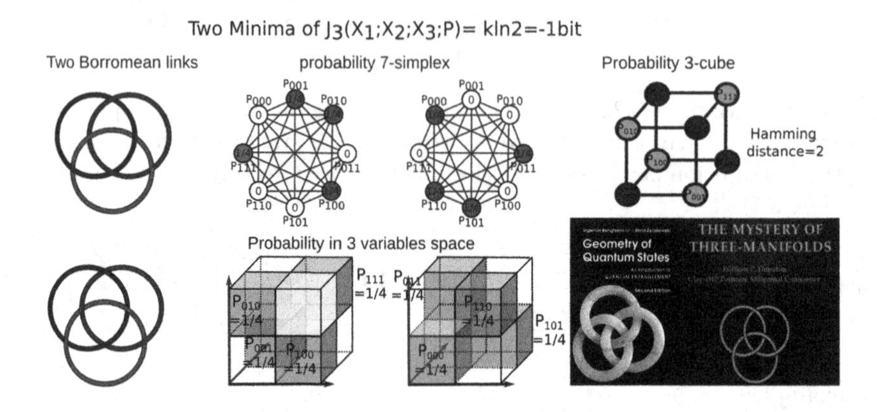

Fig. 1. Borromean 3-links of information (adapted from [8]) from left to right. The two Borromean links (mirror images). The corresponding two probability laws in the probability 7-simplex for 3 binary random variables, and below the representation of the corresponding configuration in 3 dimensional data space. The variables are individually maximally entropic ($J_1 = 1$), fully pairwise independent ($J_2 = 0$ for all pairs), but minimally linked by a negative information interaction ($J_3 = -1$). The corresponding graph covering in the 3-cube of the two configurations. The Hamming distance between two vertex having the same probability value is 2. The cover of Bengtsson and Życzkowski's book on "the geometry of quantum states" and the first slide of the conference on "The Mystery of 3-Manifolds" by Bill Thurston at the Clay-IHP Milenial conference.

Consider 3 binary random variables, then we have:

Theorem 1 (Borromean links of information [8]**).** *The absolute minimum of J_3, equal to -1, is attained only in the two cases of three two by two independent unbiased variables satisfying $p_{000} = 1/4, p_{001} = 0, p_{010} = 0, p_{011} = 1/4, p_{100} = 0, p_{101} = 1/4, p_{110} = 1/4, p_{111} = 0$, or $p_{000} = 0, p_{001} = 1/4, p_{010} = 1/4, p_{011} = 0, p_{100} = 1/4, p_{101} = 0, p_{110} = 0, p_{111} = 1/4$. These cases correspond to the two borromean 3-links, the right one and the left one (cf. Fig. 1, see [8] p. 18).*

For those 2 minima, we have $H_1 = 1$, $H_2 = 2$, $H_3 = 2$, and $I_1 = 1$, $I_2 = 0$, $I_3 = -1$, and $G_1 = 1$, $G_2 = 0$, $G_3 = 1$, and $J_1 = 1$, $J_2 = 0$, $J_3 = -1$. The same can be shown for arbitrary n-ary variables, which opens the question of the possibility or not to classify more generaly others links. For example, for ternary variables, a Borromean link is achieved for the state $p_{000} = 1/9, p_{110} = 1/9, p_{220} = 1/9, p_{011} = 1/9, p_{121} = 1/9, p_{201} = 1/9, p_{022} = 1/9, p_{102} = 1/9, p_{212} = 1/9$, and all others atomic probabilities are 0. The values of information functions are the same but in logarithmic basis 3 (trits) instead of 2 (bits). In other word, the probabilistic framework may open some new views on the classification of links.

We now show the same result for the 4-Brunnian link, considering 4 binary random variables:

Theorem 2 (4-links of information[1]). *The absolute minimum of J_4, equal to -1, is attained only in the two cases of four two by two and three by three independent unbiased variables satisfying* $p_{0000} = 0, p_{0001} = 1/8, p_{0010} = 1/8, p_{0011} = 0, p_{0101} = 0, p_{1001} = 0, p_{0111} = 1/8, p_{1011} = 1/8, p_{1111} = 0, p_{1101} = 1/8, p_{1110} = 1/8, p_{0110} = 0, p_{1010} = 0, p_{1100} = 0, p_{1000} = 1/8, p_{0100} = 1/8$, *or* $p_{0000} = 1/8, p_{0001} = 0, p_{0010} = 0, p_{0011} = 1/8, p_{0101} = 1/8, p_{1001} = 1/8, p_{0111} = 0, p_{1011} = 0, p_{1111} = 1/8, p_{1101} = 0, p_{1110} = 0, p_{0110} = 1/8, p_{1010} = 1/8, p_{1100} = 1/8, p_{1000} = 0, p_{0100} = 0$. *These cases correspond to the two 4-Brunnian links, the right one and the left one (cf. Fig. 2).*

Proof. The minima of J_4 are the maxima of I_4. We have easily $I_4 \geq -\min(H(X_1), H(X_2), H(X_3), H(X_4))$ and $I_4 \leq \min(H(X_1), H(X_2), H(X_3), H(X_4))$ [22]. Consider the case where all the variables are k-independent for all $k < 4$ and all H_1 are maximal, then a simple combinatorial argument shows that $I_4 = \binom{1}{4}.1 - \binom{2}{4}.2 + \binom{3}{4}.3 + \binom{4}{4}.H_4$, which gives $I_4 = 4 - H_4$. Now, since $I_4 \leq \min(H(X_1), H(X_2), H(X_3), H(X_4)) \leq \max(H(X_i) = 1$, we have $H_4 = 4 - 1 = 3$ and $I_4 = 1$ or $J_4 = -1$, and it is a maxima of I_4 because it achieves the bound $I_4 \leq \min(H(X_1), H(X_2), H(X_3), H(X_4)) \leq \max(H(X_i)) = 1$. To obtain the atomic probability values and see that there are two such maxima, let's consider all the constraint imposed by independence. We note the 16 unknown:

$$a{=}p_{0000},\quad b{=}p_{0011},\quad c{=}p_{0101},\quad d{=}p_{0111},\quad e{=}p_{1001},\quad f{=}p_{1011},\quad g{=}p_{1101},\quad h{=}p_{1110},$$
$$i{=}p_{0001},\quad j{=}p_{0010},\quad k{=}p_{0100},\quad l{=}p_{0110},\quad m{=}p_{1000},\quad n{=}p_{1010},\quad o{=}p_{1100},\quad p{=}p_{1111}.$$

The maximum entropy (or 1-independence) of single variable gives 8 equations:

$$a{+}b{+}c{+}d{+}i{+}j{+}k{+}l{=}1/2,\quad e{+}f{+}g{+}h{+}m{+}n{+}o{+}p{=}1/2,\quad a{+}b{+}e{+}f{+}i{+}j{+}m{+}n{=}1/2,$$
$$c{+}d{+}g{+}h{+}k{+}l{+}o{+}p{=}1/2,\quad a{+}c{+}e{+}g{+}i{+}k{+}m{+}o{=}1/2,\quad b{+}d{+}f{+}h{+}j{+}l{+}n{+}p{=}1/2,$$
$$a{+}h{+}j{+}k{+}l{+}m{+}n{+}o{=}1/2,\quad b{+}c{+}d{+}e{+}f{+}g{+}i{+}p{=}1/2.$$

[1] As the proof relies on a weak concavity theorem D [7] which proof has not been provided yet, this theorem shall be considered as a conjecture as long as the proof of theorem D [7] is not given.

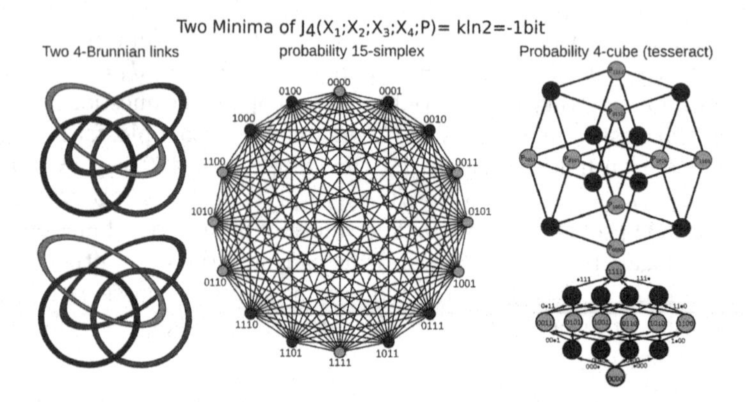

Fig. 2. 4-links of information from left to right. The two 4-links (mirror images). The corresponding two probability laws in the probability 15-simplex for 4 binary random variables. The variables are individually maximally entropic ($J_1 = 1$), fully pairwise and tripletwise independent ($J_2 = 0$, $J_3 = 0$, the link is said Brunnian), but minimally linked by a negative 4-information interaction ($J_4 = -1$). The corresponding graph covering in the 4-cube of the two configurations, called the tesseract. The Hamming distance between two vertex having the same probability value is 2. The corresponding lattice representation is illustrated below.

The 2-independence of (pair of) variables gives the 24 equations:

$$a+b+i+j=1/4,\quad c+d+k+l=1/4,\quad e+f+m+n=1/4,\quad g+h+o+p=1/4,\quad a+c+i+k=1/4,$$
$$b+d+j+l=1/4,\quad e+g+m+o=1/4,\quad f+h+n+p=1/4,\quad a+j+k+l=1/4,\quad b+c+d+i=1/4,$$
$$h+m+n+o=1/4,\quad e+f+g+p=1/4,\quad a+e+i+m=1/4,\quad b+f+j+n=1/4,\quad c+g+k+o=1/4,$$
$$d+h+l+p=1/4,\quad a+j+m+n=1/4,\quad b+e+f+i=1/4,\quad h+k+l+o=1/4,\quad c+d+g+p=1/4,$$
$$a+k+m+o=1/4,\quad c+e+g+i=1/4,\quad h+j+l+n=1/4,\quad b+d+f+p=1/4.$$

The 3-independence of (triplet of) variables gives the 32 equations:

$$a+j=1/8, b+i=1/8, k+l=1/8, c+d=1/8, m+n=1/8, e+f=1/8, h+o=1/8, g+p=1/8,$$
$$a+i=1/8, b+j=1/8, c+h=1/8, d+l=1/8, e+m=1/8, f+n=1/8, vg+o=1/8, h+p=1/8,$$
$$a+m=1/8, e+i=1/8, n+j=1/8, b+f=1/8, k+o=1/8, c+g=1/8, h+l=1/8, d+p=1/8,$$
$$a+k=1/8, c+i=1/8, j+l=1/8, b+d=1/8, m+o=1/8, e+g=1/8, k+n=1/8, f+p=1/8.$$

Since J_4 is super-harmonic (weakly concave, theorem D [7])(See Footnote 1) which implies that minima of J happen on the boundary of the probability simplex, we have one additional constraint that for example a is either 0 or $1/8$. Solving this system of 64 equations with a computer with $a = 0$ or $a = 1/(2^3)$ gives the two announced solutions. Alternatively, one can remark that out of the 64 equations only $\sum_{k=1}^{4} \binom{k}{4} - 2 + 1 = 2^4 - 2 + 1 = 15$ are independent with $a = 0$ or $a = 1/(2^3)$, the 2 systems are hence fully determined and we have 2 solutions. Alternatively, it could be possible to derive a geometric proof using the 4-cube as covering graph (called tesseract), of probability simplex, establishing that 0 probabilities only connects 1/8 probabilities as illustrated in Fig. 2.

For those 2 minima, we have $H_1 = 1$, $H_2 = 2$, $H_3 = 3$, $H_4 = 3$, and $I_1 = 1$, $I_2 = 0$, $I_3 = 0$, $I_4 = 1$, and $G_1 = 1$, $G_2 = 0$, $G_3 = 0$, $G_4 = 1$, $J_1 = 1$,$J_2 = 0$, $J_3 = 0$, $J_3 = -1$.

The preceding results generalizes to k-Brunnian link: consider k binary random variables then we have the theorem:

Theorem 3 (k-links of information *(See Footnote 1))*. *The absolute minimum of J_k, equal to -1, is attained only in the two cases j-independent j-uplets of unbiased variables for all $1 < j < k$ with atomic probabilities $p(x_1, ..., x_k) = 1/2^{k-1}$ or $p(x_1, ..., x_k) = 0$ such that the associated vertex of the associated k-hypercube covering graph of $p(x_1, ..., x_k) = 1/2^{k-1}$ connects a vertex of $p(x_1, ..., x_k) = 0$ and conversely. These cases correspond to the two k-Brunnian links, the right one and the left one.*

Proof. Following the same line as previously. We have $J_k \geq -\min(H(X_1)...,H(X_k))$ and $I_k \leq \min(H(X_1), ..., H(X_k))$. Consider the case where all the variables are i-independent for all $i < k$ and all H_1 are maximal, then a simple combinatorial argument shows that $I_k = \sum_{i=1}^{k-1}(-1)^{i-1}\binom{i}{k}.i + (-1)^{k-1}H(X_1, ..., X_k)$ that is $I_k = k - H(X_1, ..., X_k)$ now since $I(X_1; ...; X_k) \leq \min(H(X_1), ..., H(X_k)) \leq \max(H(X_i) = 1$, we have $H(X_1, ..., X_k) = k - 1$ and $I_k = (-1)^{k-1}$ or $J_4 = -1$, and it is a minima of J_k because it saturates the bound $J(X_1; ...; X_k) \geq \max(H(X_i)) = -1$. i-independent for all $i < k$ imposes a system of $2^k - 2 + 1 = 15$ independent equations (the +1 is for $\sum p_i = 1$). Since J_k is weakly concave (theorem D [7])(See Footnote 1) which is equivalent to say that minima of J happen on the boundary of the probability simplex, we have one additional constraint that for example a is either 0 or $1/2^{k-1}$. It gives 2 systems of equations that are hence fully determined and we have 2 solutions. The probability configurations corresponding to those two solutions can be found by considering the k-cube as covering graph (well known to be bipartite: it can be colored with only two colors) of the probability $2^k - 1$-simplex, establishing that 0 probabilities only connects $1/2^{k-1}$ probabilities.

For those 2 minima, we have: for $-1 < i < k$ $H_i = i$ and $H_k = k - 1$. We have $I_1 = 1$, for $1 < i < k$ $I_i = 0$, and $I_k = -1^{k-1}$. We have $G_1 = 1$, for $1 < i < k$ $G_i = 0$, and $G_k = 1$. We have $J_1 = 1$,for $1 < i < k$ $J_i = 0$,and $J_k = -1$.

4 Negativity and Kirkwood Decomposition Inconsistency

The negativity of K-L divergence can happen in certain cases, in an extended context of measure theory. A measure space is a probability space that does not necessarily realize the axiom of probability of a total probability equal to 1 [19], e.g. $P(\Omega) = \sum q_i = k$, where k is an arbitrary real number (for real measure). In the seminal work of Abramsky and Brandenburger [1], two probability laws P and Q are contextual whenever there does not exist any joint probability that would correspond to such marginals (but only non-positive measure). In such cases, measures are said incompatible, leading to some obstruction to the existence of a global section probability joint-distribution: P and Q are locally consistent but globally inconsistent. This section underlines that interaction information negativity displays a kind of "dual" phenomena, that we call inconsistent decomposition (or indecomposability), whenever interaction information is negative on a consistent global probability law then no local Kirkwood decomposition can be consistent, leading to an obstruction to decomposability (e.g. globally consistent but locally inconsistent decomposition measure).

Definition 1 (Inconsistent decomposition). *A measure space P is consistent whenever $P(\Omega) = \sum p_i = 1$ (and hence a probability space), and inconsistent otherwise.*

We first show that given a probability law P, if $D(P,Q) < 0$ then Q is inconsistent and cannot be a probability law.

Theorem 4 (K-L divergence negativity and inconsistency). *Consider a probability space with probability P and a measure space with measure Q, the negativity of the Kullback-Leibler divergence $D_\Omega(X_j; P, Q)$ implies that Q is inconsistent.*

Proof. The proof essentially relies on basic argument of the proof of convexity of K-L divergence or on Gibbs inequalities. Consider the K-L divergence between P and Q: $D_\Omega(X_j; P, Q) = D(X_j; p(x)||q(x)) = c\sum_{i\in\mathscr{X}} p_i \ln \frac{q_i}{p_i}$. Since $\forall x > 0$, $\ln x \leqslant x - 1$ with equality if and only if $x = 1$, hence we have with $c = -1/\ln 2$; $k\sum_{i\in\mathscr{X}} p_i \ln \frac{q_i}{p_i} \geq c\sum_{i\in\mathscr{X}} p_i(\frac{q_i}{p_i} - 1)$. We have $c\sum_{i\in\mathscr{X}} p_i(\frac{q_i}{p_i} - 1) = c\left(\sum_{i\in\mathscr{X}} q_i - \sum_{i\in\mathscr{X}} p_i\right)$. P is a probability law, then by the axiom 4 of probability [19], we have $\sum q_i \neq 1$ and hence $c\left(\sum_{i\in\mathscr{X}} q_i - \sum_{i\in\mathscr{X}} p_i\right) = c\left(\sum_{i\in\mathscr{X}} q_i - 1\right)$. Hence if $D_\Omega(X_j; P, Q) < 0$ then $\sum_{i\in\mathscr{X}} q_i > 1$, and since $\sum_{i\in\mathscr{X}} q_i > 1$, by definition Q is inconsistent.

Theorem 5 (Interaction negativity and inconsistent Kirkwood decomposition, adapted from [16] p. 13). *for $n > 2$, if $J_n < 0$ then no Kirkwood probability decomposition subspace P_{X_K} defined by the variable products of (X_K) variables with $K \subset [n]$ is consistent.*

Proof. The theorem is proved by remarking, that interaction negativity corresponds precisely to cases where the Kirkwood approximation is not possible (fails) and would imply a probability space with $\sum p_i > 1$ which contradicts the axioms of probability theory. We will use a proof by contradiction. Let's assume that the probability law follows the Kirkwood approximation which can be obtained from $n-1$ products of variables and we have 6:

$$\hat{p}(x_1, ..., x_n) = \frac{\prod_{I\subset[n];card(I)=n-1} p(x_I)}{\frac{\prod_{I\subset[n];card(I)=n-2} p(x_I)}{\vdots \atop \prod_{i=1}^{n} p(x_i)}} \tag{11}$$

Now consider that $J_n < 0$ then, since $-\log x < 0$ if and only if $x > 1$, we have $\left(\frac{\prod_{I\subset[n];card(I)=i;i \text{ odd}} p_I}{\prod_{I\subset[n];card(I)=i;i \text{ even}} p_I}\right)^{(-1)^{n-1}} > 1$, which is the same as $\frac{\hat{p}(x_1,...,x_n)}{p(x_1,...,x_n)} > 1$ and hence $\hat{p}(x_1,...,x_n) > p(x_1,...,x_n)$. Then summing over all atomic probabilities, we obtain $\sum_{x_1\in[N_1],...,x_n\in[N_n]}^{N_1\times...\times N_n} \hat{p}(x_1,...,x_n) > \sum_{x_1\in[N_1],...,x_n\in[N_n]}^{N_1\times...\times N_n} p(x_1,...,x_n)$ and hence $\sum_{x_1\in[N_1],...,x_n\in[N_n]}^{N_1\times...\times N_n} \hat{p}(x_1,...,x_n) > 1$ which contradicts axiom 4 of probability [19]. Hence there does not exist probability law on only $n-1$ products of variables satisfying $J_n < 0$. If there does not exist probability law on only $n-1$ products of variables, then by marginalization on the $n-1$ products, there does not exist such probability law for all $2 < k < n$ marginal distributions on k variable product.

Remark: Negativity of interaction information provides intuitive insight into contextual interactions as obstruction to decomposition-factorization into lower order interactions, which is classical here in the sense that it does not rely on quantum formalism: quantum information extends this phenomenon to self or pairwise interactions: Information negativity happens also for $n = 2$ or $n = 1$ precisely for the states that violate Bell's inequalities [12,14].

Acknowledgments. I thank warmly anonymous reviewer for helpful remarks improving the manuscript and Daniel Bennequin whose ideas are at the origin of this work.

References

1. Abramsky, S., Brandenburger, A.: The sheaf-theoretic structure of non-locality and contextuality. New J. Phys. **13**, 1–40 (2011)
2. Baas, N.: New states of matter suggested by new topological structures. Int. J. Gen. Syst. **42**(2), 137–169 (2012)
3. Baas, N.: On the mathematics of higher structures. Int. J. Gen. Syst. **48**(6), 603–624 (2019)
4. Baez, J.C., Fritz, T.: A bayesian characterization of relative entropy. Theory Appl. Categories **29**(16), 422–456 (2014)
5. Baez, J., Fritz, T., Leinster, T.: A characterization of entropy in terms of information loss. Entropy **13**, 1945–1957 (2011)
6. Baudot, P.: The Poincare-Shannon machine: statistical physics and machine learning aspects of information cohomology. Entropy **21**(9), 881 (2019)
7. Baudot, P., Bennequin, D.: The homological nature of entropy. Entropy **17**(5), 3253–3318 (2015)
8. Baudot, P., Tapia, M., Bennequin, D., Goaillard, J.: Topological information data analysis. Entropy 21(9), 869 (2019)
9. Bennequin, D., Peltre, O., Sergeant-Perthuis, G., Vigneaux, J.: Extra-fine sheaves and interaction decompositions. arXiv:2009.12646 (2020)
10. Bertschinger, N., Rauh, J., Olbrich, E., Jost, J., Ay, N.: Quantifying unique information. Entropy **16**, 2161–2183 (2014)
11. Brenner, N., Strong, S., Koberle, R., Bialek, W.: Synergy in a neural code. Neural Comput. **12**, 1531–1552 (2000)
12. Cerf, N., Adami, C.: Negative entropy and information in quantum mechanic. Phys. Rev. Lett. **79**, 5194 (1997)

13. Galas, D., Sakhanenko, N.: Symmetries among multivariate information measures explored using möbius operators. Entropy **21**(1(88)), 1–17 (2019)
14. Horodecki, M., Oppenheim, J., Winter, A.: Quantum state merging and negative information. Comm. Math. Phys. **1**, 107–136 (2007)
15. Hu, K.T.: On the amount of information. Theory Probab. Appl. **7**(4), 439–447 (1962)
16. Jakulin, A., Bratko, I.: Quantifying and visualizing attribute interactions. arXiv:cs/0308002 (2004)
17. Khovanov, M.: Link homology and categorification. In: International Congress of Mathematicians, vol. II, pp. 989–999. European Mathematical Society, Zürich (2006)
18. Kirkwood, J.: Statistical mechanics of fluid mixtures. J. Chem. Phys. **3**, 300 (1935)
19. Kolmogorov, A.N.: Grundbegriffe der Wahrscheinlichkeitsrechnung. (English translation (1950): Foundations of the Theory of Probability.). Springer, Berlin (Chelsea, New York) (1933)
20. Kullback, S., Leibler, R.: On information and sufficiency. Ann. Math. Stat. **22**, 79–86 (1951)
21. Manin, Y., Marcolli, M.: Homotopy theoretic and categorical models of neural information networks. arXiv:2006.15136 (2020)
22. Matsuda, H.: Physical nature of higher-order mutual information: intrinsic correlations and frustration. Phys. Rev. E **62**(3), 3096–3102 (2000)
23. McGill, W.: Multivariate information transmission. Psychometrika **19**, 97–116 (1954)
24. Milnor, J.: Link groups. Ann. Math. **59**(2), 177–195 (1954)
25. Olbrich, E., Bertschinger, N., Rauh, J.: Information decomposition and synergy. Entropy **17**(5), 3501–3517 (2015)
26. Pearl, J.: Probabilistic Reasoning in Intelligent Systems: Networks of Plausible Inference, 2nd edn. Morgan Kaufmann, San Francisco (1988)
27. Peltre, O.: Message-passing algorithms and homology. From thermodynamics to statistical learning. PhD thesis of Institut Mathematique de Jussieu arXiv:2009.11631 (2020)
28. Rosas, F.E., et al.: Reconciling emergences: an information-theoretic approach to identify causal emergence in multivariate data. PLoS Comput. Biol. **16**(12), p1-23 (2020)
29. Shannon, C.E.: A mathematical theory of communication. Bell Syst. Tech. J. **27**, 379–423 (1948)
30. Vigneaux, J.: Topology of statistical systems. A cohomological approach to information theory. Ph.D. thesis, Paris 7 Diderot University (2019)
31. Williams, P., Beer, R.: Nonnegative decomposition of multivariate information. arXiv:1004.2515v1 (2010)

Betti Curves of Rank One Symmetric Matrices

Carina Curto[1(✉)], Joshua Paik[1], and Igor Rivin[2]

[1] The Pennsylvania State University, State College, USA
ccurto@psu.edu

[2] Edgestream Partners LP and Temple University, Philadelphia, USA

Abstract. Betti curves of symmetric matrices were introduced in [3] as a new class of matrix invariants that depend only on the relative ordering of matrix entries. These invariants are computed using persistent homology, and can be used to detect underlying structure in biological data that may otherwise be obscured by monotone nonlinearities. Here we prove three theorems that characterize the Betti curves of rank 1 symmetric matrices. We then illustrate how these Betti curve signatures arise in natural data obtained from calcium imaging of neural activity in zebrafish.

Keywords: Betti curves · Topological data analysis · Calcium imaging

1 Introduction

Measurements in biology are often related to the underlying variables in a nonlinear fashion. For example, a brighter calcium imaging signal indicates higher neural activity, but a neuron with twice the activity of another does not produce twice the brightness. This is because the measurement is a monotone nonlinear function of the desired quantity. How can one detect meaningful structure in matrices derived from such data? One solution is to try to estimate the monotone nonlinearity, and invert it. A different approach, introduced in [3], is to compute new matrix invariants that depend only on the relative ordering of matrix entries, and are thus invariant to the effects of monotone nonlinearities.

Figure 1a illustrates the pitfalls of trying to use traditional linear algebra methods to estimate the underlying rank of a matrix in the presence of a monotone nonlinearity. The original 100×100 matrix A is symmetric of rank 5 (top left cartoon), and this is reflected in the singular values (bottom left). In contrast, the matrix B with entries $B_{ij} = f(A_{ij})$ appears to be full rank, despite having exactly the same *ordering* of matrix entries: $B_{ij} > B_{k\ell}$ if and only if $A_{ij} > A_{k\ell}$. The apparently high rank of B is purely an artifact of the monotone

This work was supported by NIH R01 NS120581 to CC. We thank Enrique Hansen and Germán Sumbre of École Normale Supérieure for providing us with the calcium imaging data that is used in our Applications section.

F. Nielsen and F. Barbaresco (Eds.): GSI 2021, LNCS 12829, pp. 645–655, 2021.
https://doi.org/10.1007/978-3-030-80209-7_69

nonlinearity f. This motivates the need for matrix invariants, like Betti curves, that will give the same answer for A and B. Such invariants depend only on the ordering of matrix entries and do not "see" the nonlinearity [3].

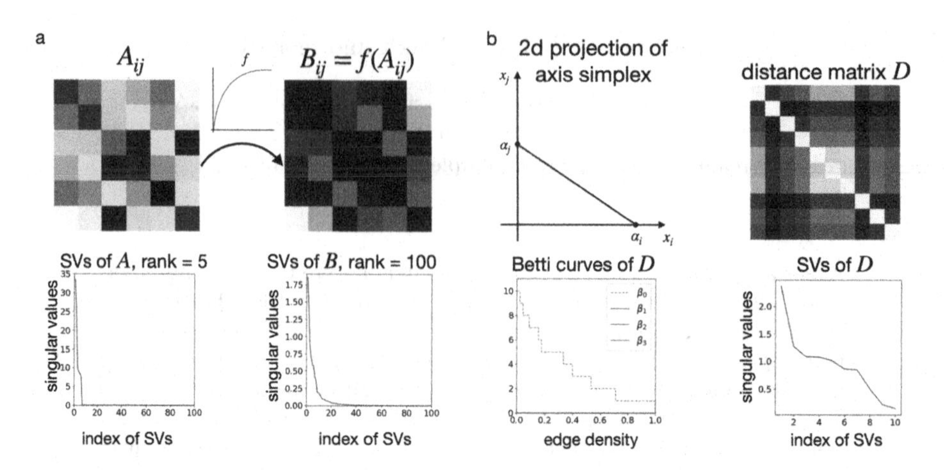

Fig. 1. Low rank structure is obscured by monotone nonlinearities. The matrix A is a 100×100 symmetric matrix of rank 5. B is obtained from A by applying the monotone nonlinearity $f(x) = 1 - e^{-5x}$ entrywise. This alters the singular values so that B appears to be full rank, but this is purely an artifact of the nonlinearity. (b) An axis simplex in $\mathbb{R}^{10}$ generates a distance matrix that is full rank, and this is evident in the singular values. The Betti curves, however, are consistent with an underlying rank of 1.

In this paper we will characterize all Betti curves that can arise from rank 1 matrices. This provides necessary conditions that must be satisfied by any matrix whose underlying rank is 1 – that is, whose ordering is the same as that of a rank 1 matrix. We then apply these results to calcium imaging data of neural activity in zebrafish and find that correlation matrices for cell assemblies have Betti curve signatures of rank 1.

It turns out that many interesting matrices have an underlying rank of 1. For example, consider the distance matrix induced by an *axis simplex*, meaning a simplex in $\mathbb{R}^n$ whose vertices are $\alpha_1\mathbf{e_1}, \alpha_2\mathbf{e_2}, \ldots, \alpha_n\mathbf{e_n}$, where $\mathbf{e_1}, \mathbf{e_2}, \ldots, \mathbf{e_n}$ are the standard basis vectors. The Euclidean distance matrix D for these points has off-diagonal entries $D_{ij} = \sqrt{\alpha_i^2 + \alpha_j^2}$, and is typically full rank (Fig. 1b, top). But this matrix has underlying rank 1. To see this, observe that the matrix with entries $\sqrt{\alpha_i^2 + \alpha_j^2}$ has the same ordering as the matrix with entries $\alpha_i^2 + \alpha_j^2$, and this in turn has the same ordering as the matrix with entries $e^{\alpha_i^2 + \alpha_j^2} = e^{\alpha_i^2} e^{\alpha_j^2}$, which is clearly rank 1. Although the singular values do not reflect this rank 1 structure, the Betti curves do (Fig. 1b, bottom).

Adding another vertex at the origin to an axis simplex yields a *simplex with an orthogonal corner.* The distance matrix is now $(n + 1) \times (n + 1)$, and is again given by $D_{ij} = \sqrt{\alpha_i^2 + \alpha_j^2}$ but with $i, j = 0, \ldots, n$ and $\alpha_0 = 0$. The same argument as above shows that this matrix also has underlying rank 1.

The organization of this paper is as follows. In Sect. 2 we provide background on Betti curves and prove three theorems characterizing the Betti curves of symmetric rank 1 matrices. In Sect. 3 we illustrate our results by computing Betti curves for pairwise correlation matrices obtained from calcium imaging data of neural activity in zebrafish. All Betti curves were computed using the well-known persistent homology package Ripser [1].

2 Betti Curves of Symmetric Rank 1 Matrices

Betti Curves. Given a real symmetric $n \times n$ matrix M, the $\binom{n}{2}$ off-diagonal entries M_{ij} for $i < j$ can be sorted in increasing order. We denote by $\widehat{M}$ be the corresponding ordering matrix, where $\widehat{M}_{ij} = k$ if M_{ij} is the k-th smallest entry. From the ordering, we can construct an increasing sequence of graphs $\{G_t(M)\}$ for $t \in [0, 1]$ as follows: for each graph, the vertex set is $1, \ldots, n$ and the edge set is

$$E(G_t(M)) = \left\{(i, j) \mid \widehat{M}_{ij} \leq t\binom{n}{2}\right\}.$$

Note that for $t = 0$, $G_t(M)$ has no edges, while at $t = 1$ it is the complete graph with all edges. Although t is a continuous parameter, it is clear that there are only a finite number of distinct graphs in the family $\{G_t(M)\}_{t \in [0,1]}$. Each new graph differs from the previous one by the addition of an edge[1]. When it is clear from the context, we will denote $G_t(M)$ as simply G_t.

For each graph G_t, we build a *clique complex* $X(G_t)$, where every k-clique in G_t is filled in by a k-simplex. We thus obtain a filtration of clique complexes $\{X(G_t) \mid t \in [0, 1]\}$. The i-th Betti curve of M is defined as

$$\beta_i(t) = \beta_i(X(G_t)) = \text{rank} H_i(X(G_t), \mathbf{k}),$$

where H_i is the i-th homology group with coefficients in the field $\mathbf{k}$. The Betti curves clearly depend only on the ordering of the entries of M, and are thus invariant to (increasing) monotone nonlinearities like the one shown in Fig. 1a. They are also naturally invariant to permutations of the indices $1, \ldots, n$, provided rows and columns are permuted in the same way. See [3, 4] for more details.

Figure 2a shows the ordering matrix $\widehat{M}$ of a small matrix M. The complete sequence of clique complexes is depicted in panel c, and the corresponding Betti curves in panel b. Note that $\beta_2 = \beta_3 = 0$ for all values of the edge density t. On the other hand, $\beta_1(t) = 1$ for some intermediate values of t where a 1-dimensional hole arises in the clique complex $X(G_t)$. Note that $\beta_0(t)$ counts the number of connected components for each clique complex, and is thus monotonically decreasing for any matrix.

[1] In the non-generic case of equal entries, multiple edges may be added at once.

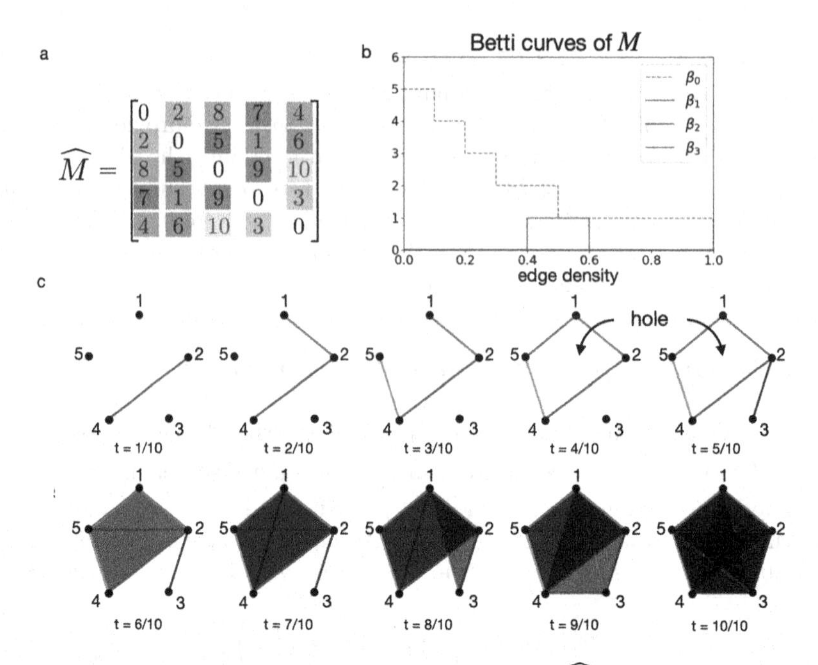

Fig. 2. Betti curves of real symmetric matrices. (a) $\widehat{M}$ is the ordering matrix for a symmetric 5×5 matrix M of rank 4. (b) We plot the Betti curves $\beta_0, \ldots, \beta_3$ induced by M. (c) The filtration of clique complexes $X(G_t)$ for $\widehat{M}$ from which the Betti curves are computed. Note that at times $t = 4/10$ and $t = 5/10$, a one-dimensional hole appears, contributing to nonzero β_1 values in (b). At time $t = 6/10$, this hole is filled by the cliques created after adding edge $(2, 5)$, and thus $\beta_1(t)$ goes back to zero.

Our main results characterize Betti curves of symmetric rank 1 matrices. We say that a vector $\mathbf{x} = (x_1, \ldots, x_n)$ *generates* a rank 1 matrix M if $M = \mathbf{x}^\mathbf{T}\mathbf{x}$. Perhaps surprisingly, there are significant differences in the ordering matrices $\widehat{M}$ depending on the sign pattern of the x_i. The simplest case is when M is generated by a vector $\mathbf{x}$ with all positive or all negative entries. When $\mathbf{x}$ has a mix of positive and negative entries, then M has a block structure with two diagonal blocks of positive entries and two off-diagonal blocks of negative entries. This produces qualitatively distinct $\widehat{M}$. Even more surprising, taking $M = -\mathbf{x}^\mathbf{T}\mathbf{x}$ qualitatively changes the structure of the ordering in a way that Betti curves can detect. This corresponds to building clique complexes from M by adding edges in reverse order, from largest to smallest. Here we consider all four cases: $M = \mathbf{x}^\mathbf{T}\mathbf{x}$ and $M = -\mathbf{x}^\mathbf{T}\mathbf{x}$ for $\mathbf{x}$ a vector whose entries are either (i) all the same sign or (ii) have a mix of positive and negative signs.

When $\mathbf{x}$ has all entries the same sign, we say the matrix $M = \mathbf{x}^\mathbf{T}\mathbf{x}$ is *positive rank one* and $M = -\mathbf{x}^\mathbf{T}\mathbf{x}$ is *negative rank one*. Observe in Fig. 3a that for a positive rank one matrix, $\beta_k(t)$ for $k > 0$ is identically zero. In Fig. 3d we see that the same is true for negative rank one matrices, though the $\beta_0(t)$ curve

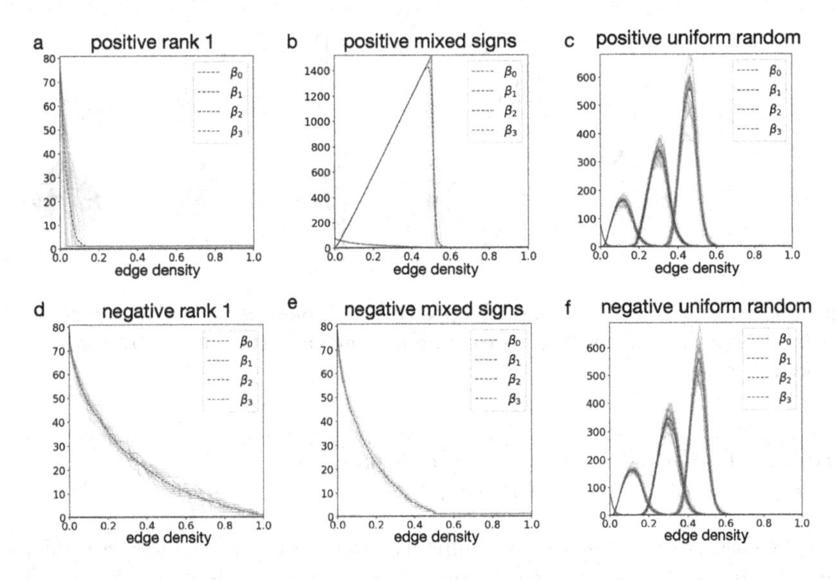

Fig. 3. Betti curves for different classes of symmetric rank 1 matrices. For each of the six figures, we generate Betti curves $\beta_0, \beta_1, \beta_2$, and β_3 of twenty five randomly generated 80×80 symmetric matrices. (a) The Betti curve of a positive rank one matrix $M = \mathbf{x}^T\mathbf{x}$ where all entries of row vector $\mathbf{x}$ are chosen uniformly from $[0, 1]$. (d) The Betti curve of a negative rank one matrix, where the only difference from the positive case is that $M = -\mathbf{x}^T\mathbf{x}$. (b, e) follow similar suit, except the entries of $\mathbf{x}$ are chosen uniformly from the interval $[-1, 1]$ and thus has mixed signs. (c, f) are random symmetric matrices constructed by making a matrix X with i.i.d. entries in $[0, 1]$, and setting $M_\pm = \pm(X + X^T)/2$. For random matrices, there is no difference in the Betti curves between the positive and negative versions.

decreases more slowly than in the positive rank one case. The vanishing of the higher Betti curves in both cases can be proven.

Theorem 1. *Let M be a positive rank one matrix or a negative rank one matrix. The k-th Betti curve $\beta_k(t)$ is identically zero for all $k > 0$ and $t \in [0, 1]$.*

Proof. Without loss of generality, let $\mathbf{x} = (x_1, x_2, \ldots, x_n)$ be a vector such that $0 \leq x_1 \leq x_2 \leq \cdots \leq x_n$. Let $M = \mathbf{x}^T\mathbf{x}$ be a rank one matrix generated by $\mathbf{x}$ and for a given $t \in [0, 1]$, consider the graph G_t. It is clear that, because x_1 is minimal, if (i, j) is an edge of G_t, then $(1, i)$ and $(1, j)$ are also edges of G_t, since $x_1 x_i \leq x_i x_j$ and $x_1 x_j \leq x_i x_j$. It follows that $X(G_t)$ is the union of a cone and a collection of isolated vertices, and hence is homotopy equivalent to a set of points. It follows that $\beta_k(t) = 0$ for all $k > 0$ and all $t \in [0, 1]$, while $\beta_0(t)$ is the number of isolated vertices of G_t plus one (for the cone component). See Fig. 4 for an example. The same argument works for $-M$, where the cone vertex corresponds to x_n instead of x_1. □

The axis simplices we described in the Introduction are all positive rank one. We thus have the following corollary.

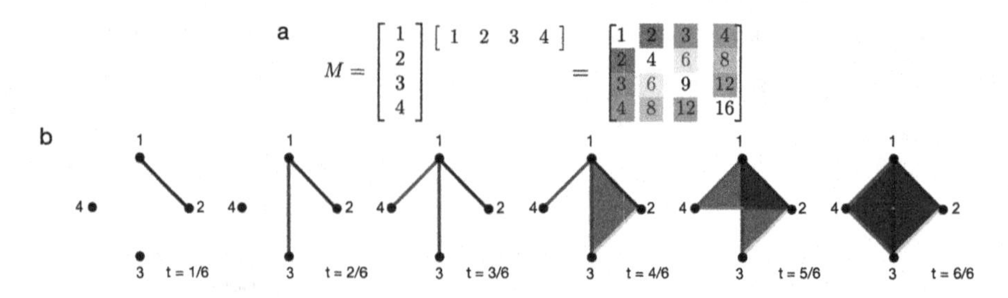

Fig. 4. Illustration for the proof of the positive rank one case. Note that vertex 1 has minimal value in the generating vector $\mathbf{x}$, and hence all other nodes are connected to 1 before any other node.

Corollary 1. *The k-th Betti curve of a distance matrix induced by an axis simplex or by a simplex with an orthogonal corner is identically zero for $k > 0$.*

In the cases where $\mathbf{x}$ has mixed signs, the situation is a bit more complicated. Figure 3b shows that for *positive mixed sign* matrices $M = \mathbf{x}^T\mathbf{x}$, the first Betti curve $\beta_1(t)$ ramps up linearly to a high value and then quickly crashes down to 1. In contrast, Fig. 3e shows that the *negative mixed sign* matrices $M = -\mathbf{x}^T\mathbf{x}$ have vanishing $\beta_1(t)$ and have a similar profile to the positive and negative rank one Betti curves in Fig. 3a, d. Note, however, that β_0 decreases more quickly than the negative rank one case, but more slowly than positive rank one.

Figure 5 provides some intuition for the mixed sign cases. The matrix $M = \mathbf{x}^T\mathbf{x}$ splits into blocks, with the green edges added first. Because these edges belong to a bipartite graph, the number of 1-cycles increases until G_t is a complete bipartite graph with all the green edges (see Fig. 5b). This is what allows $\beta_1(t)$ to increase approximately linearly. Once the edges corresponding to the diagonal blocks are added, we obtain coning behavior on each side similar to what we saw in the positive rank one case. The 1-cycles created by the bipartite graph quickly disappear and the higher-order Betti curves all vanish. On the other hand, for the negative matrix $-M = -\mathbf{x}^T\mathbf{x}$, the green edges will be added last. This changes the $\beta_1(t)$ behavior dramatically, as all 1-cycles in the bipartite graph are automatically filled in with cliques because both sides of the bipartite graph are complete graphs by the time the first green edge is added.

Lemma 1. *Let M be an $n \times n$ positive mixed sign rank one matrix, generated by a vector $\mathbf{x}$ with precisely ℓ negative entries. Then $G_t(M)$ is bipartite for $t \leq t_0 = \ell(n-\ell)/\binom{n}{2}$ and is the complete bipartite graph, $K_{\ell,n-\ell}$, at $t = t_0$.*

The next two theorems characterize the Betti curves for the positive and negative mixed sign cases. We assume $\mathbf{x} = (x_1, x_2, \ldots, x_\ell, x_{\ell+1}, \ldots, x_n)$, where $x_1 \leq x_2 \leq \cdots \leq x_\ell < 0 \leq x_{\ell+1} \leq \cdots \leq x_n$.

Theorem 2. *Let $M = \mathbf{x}^T\mathbf{x}$. Then $\beta_1(t) \leq (\ell-1)(n-\ell-1)$, with equality when G_t is the complete bipartite graph $K_{\ell,n-\ell}$. The higher Betti curves all vanish: $\beta_k(t) = 0$ for $k > 1$ and all $t \in [0,1]$.*

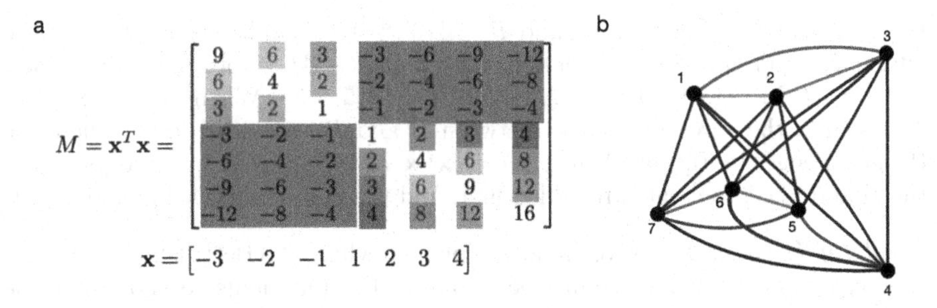

Fig. 5. (a) A mixed sign rank one matrix $M = \mathbf{x}^T\mathbf{x}$ with negative entries in green. (b) The graph G_t induced by M at $t = 1$, with edge colors corresponding to matrix colorings. The negative entries (green) get added first and form a complete bipartite graph. In the negative version, $-\mathbf{x}^T\mathbf{x}$, the green edges are added last. This leads to qualitative differences in the Betti curves. (Color figure online)

Theorem 3. *Let $M = -\mathbf{x}^T\mathbf{x}$. Then $\beta_k(t) = 0$ for $k > 0$ and all $t \in [0, 1]$.*

To prove these theorems, we need a bit more algebraic topology. Let A and B be simplicial complexes with disjoint vertex sets. Then *the join* of A and B, denoted $A * B$, is defined as

$$A * B = \{\sigma_A \cup \sigma_B \mid \sigma_A \in A \text{ and } \sigma_B \in B\}.$$

The homology of the join of two simplicial complexes, was computed by J. W. Milnor [5, Lemma 2.1]. We give the simpler version with field coefficients:

$$\widetilde{H}_{r+1}(A * B) = \bigoplus_{i+j=r} \widetilde{H}_i(A) \otimes \widetilde{H}_j(B),$$

where $\widetilde{H}$ denotes the reduced homology groups[2]. Recall that a simplicial complex A is called *acyclic* if $\beta_i(A) = \text{rank}(H_i(A)) = 0$ for all $i > 0$.

Corollary 2. *If A and B are acyclic, then $H_i(A * B) = 0$ for all $i > 1$. Also $\beta_1(A * B) = \text{rank}H_1(A * B) = \text{rank}(\widetilde{H}_0(A) * \widetilde{H}_0(B)) = (\beta_0(A) - 1)(\beta_0(B) - 1)$.*

Example 1. The complete bipartite graph, $K_{\ell,m}$, is the join of two zero dimensional complexes, S_ℓ and S_m. We see that $\beta_1(K_{\ell,m}) = (\ell - 1)(m - 1)$. This immediately gives us the upper bound on $\beta_1(t)$ from Theorem 2.

Example 2. Let G be a graph such that $V(G) = B \cup R$. Let G_B be the subgraph induced by B and G_R the subgraph induced by R. Assume that every vertex in B is connected by an edge to every vertex in R. Then $X(G) = X(G_B) * X(G_R)$.

We are now ready to prove Theorems 2 and 3. Recall that in both theorems, M is generated by a vector $\mathbf{x}$ with positive and negative entries. We will use

[2] Note that these are the same as the usual homology groups for $i > 0$.

the notation $B = \{i \mid x_i < 0\}$ and $R = \{i \mid x_i \geq 0\}$, and refer to these as the "negative" and "positive" vertices of the graphs $G_t(M)$. Note that $|B| = \ell$ and $|R| = n - \ell$. In Fig. 5b, $B = \{1, 2, 3\}$ and $R = \{4, 5, 6, 7\}$. When $M = \mathbf{x}^T\mathbf{x}$, as in Theorem 2, the "crossing" edges between B and R are added first. (These are the green edges in Fig. 5b.) When $M = -\mathbf{x}^T\mathbf{x}$, as in Theorem 3, the edges within the B and R components are added first, and the crossing edges are added last.

Proof (of Theorem 2). We've already explained where the Betti 1 bound $\beta_1(t) \leq (\ell - 1)(n - \ell - 1)$ comes from (see Example 1). It remains to show that the higher Betti curves all vanish for $t \in [0, 1]$. Let t_0 be the value at which $G_{t_0}(M)$ is the complete bipartite graph with parts B and R. For $t \leq t_0$, when edges corresponding to negative entries of M are being added, $G_t(M)$ is always a bipartite graph (see Lemma 1). It follows that $G_t(M)$ has no cliques of size greater than two, and so $X(G_t(M)) = G_t(M)$ is a one-dimensional simplicial complex. Thus, $\beta_k(t) = 0$ for all $k > 1$.

For $t > t_0$, denote by M_B and M_R the principal submatrices induced by the indices in B and R, respectively. Clearly, M_B and M_R are positive rank one matrices, and hence by Theorem 1 the clique complexes $X(G_t(M_B))$ and $X(G_t(M_R))$ are acyclic for all t. Moreover, for $t > t_0$ we have that $X(G_t(M)) = X(G_t(M_B)) * X(G_t(M_R))$, where $*$ is the join (see Example 2). It follows from Corollary 2 that $\beta_k(t) = 0$ for all $k > 1$. □

The proof of Theorem 3 is a bit more subtle. In this case, the edges within each part B and R are added first, and the "crossing" edges come last. The conclusion is also different, as we prove that the $X(G_t(M))$ are all acyclic.

Proof (of Theorem 3). Recall that $M = -\mathbf{x}^T\mathbf{x}$, and let t_0 be the value at which all edges for the complete graphs K_B and K_R have been added, so that $G_{t_0}(M) = K_B \sqcup K_R$ (note that this is a disjoint union). Let M_B and M_R denote the principal submatrices induced by the indices in B and R, respectively. Note that for $t \leq t_0$, $X(G_t(M)) = X(G_t(M_B)) \sqcup X(G_t(M_R))$. Since M_B and M_R are both positive rank one matrices, and the clique complexes $X(G_t(M_B))$ and $X(G_t(M_R))$ are disjoint. It follows that $X(G_t(M))$ is acyclic for $t \leq t_0$.

To see that $X(G_t(M))$ is acyclic for $t > t_0$, we will show that $X(G_t(M))$ is the union of two contractible simplicial complexes whose intersection is also contractible. Let M^* be the positive rank one matrix generated by the vector of absolute values, $y = (|x_1|, \ldots, |x_n|)$, so that $M^* = y^T y$. Now observe that for $t > t_0$, the added edges in $G_t(M)$ correspond to entries of the matrix where $M_{ij} = -x_i x_j = |x_i||x_j| = M^*_{ij}$. This means that the "crossing" edges have the same relative ordering as those of the analogous graph filtration $F_t(M^*)$. Let s be the value at which the last edge added to $F_s(M^*)$ is the same as the last edge added to $G_t(M)$. (In general, $s \neq t$ and $F_s(M^*) \neq G_t(M)$.)

Since M^* is positive rank one, the clique complex $X(F_s(M^*))$ is the union of a cone and some isolated vertices (see the proof of Theorem 1). Let $\widetilde{F}_s$ be the largest connected component of $F_s(M^*)$ – that is, the graph obtained by removing isolated vertices from $F_s(M^*)$. The clique complex $X(\widetilde{F}_s)$ is a cone, and hence contractible (not merely acyclic). Now define two simplicial complexes, Δ_1 and Δ_2, as follows:

$$\Delta_1 = X(\widetilde{F}_s \cup K_B) \text{ and } \Delta_2 = X(\widetilde{F}_s \cup K_R).$$

It is easy to see that both Δ_1 and Δ_2 are contractible. Moreover, $X(G_t) = \Delta_1 \cup \Delta_2$. Since the intersection $\Delta_1 \cap \Delta_2 = X(\widetilde{F}_s)$ is contractible, using the Mayer-Vietoris sequence we can conclude that $X(G_t)$ is contractible for all $t > t_0$. □

3 Application to Calcium Imaging Data in Zebrafish Larvae

Calcium imaging data for neural activity in the optic tectum of zebrafish larvae was collected by the Sumbre lab at École Normale Superieure. The individual time series of calcium activation for each neuron were preprocessed to produce neural activity rasters. This in turn was used to compute cross correlograms (CCGs) that capture the time-lagged pairwise correlations between neurons. We then integrated the CCGs over a time window of $\pm\tau_{\max}$ for $\tau_{\max} = 1$ s, to obtain pairwise correlation values:

$$C_{ij} = \frac{1}{2\tau_{\max} r_i r_j} \frac{1}{T} \int_{-\tau_{\max}}^{\tau_{\max}} \int_{-T}^{T} f_i(t) f_j(t+\tau) dt d\tau.$$

Here T is the total time of the recording, $f_i(t)$ is the raster data for neuron i taking values of 0 or 1 at each (discrete) time point t. The "firing rate" r_i is the proportion of 1s for neuron i over the entire recording.

The Sumbre lab independently identified cell assemblies: subsets of neurons that were co-active spontaneously and in response to visual stimuli, and which are believed to underlie functional units of neural computation. Figure 6 shows the Betti curves of principal submatrices of C for two pre-identified cell assemblies. The Betti curves for the assemblies display the same signature we saw for rank 1 matrices, and are thus consistent with an underlying low rank. In contrast, the singular values of these same correlation matrices (centered to have zero mean) appear random and full rank, so the structure detected by the Betti curves could not have been seen with singular values. To check that this low rank Betti curve signature is not an artifact of the way the correlation matrix C was computed, we compared the Betti curves to a randomly chosen assembly of the same size that includes the highest firing neuron of the original assembly. (We included the highest firing neuron because this could potentially act as a coning vertex, yielding the rank 1 Betti signature for trivial reasons.) In summary, we found that the real assemblies exhibit rank 1 Betti curves, while those of randomly-selected assemblies do not.

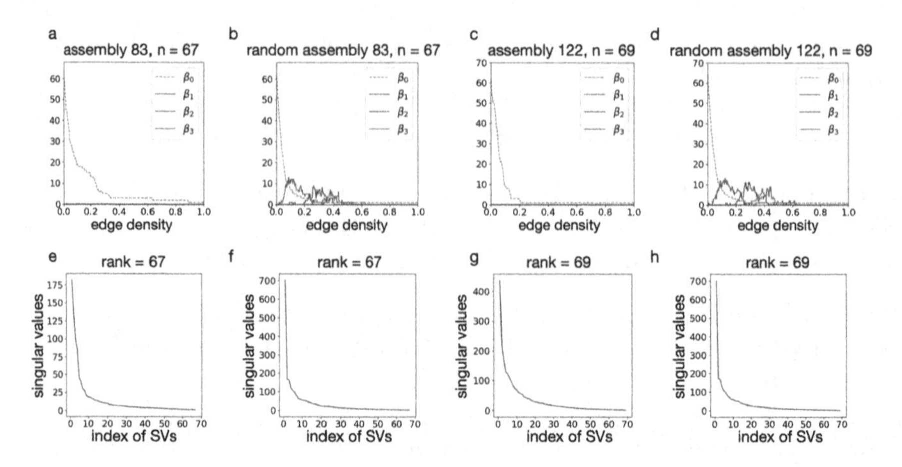

Fig. 6. We plot Betti curves and singular values of two submatrices induced by preidentified cell assemblies 83 and 122 (data courtesy of the Sumbre lab at Ecole Normale Superieure). The Betti curves are consistent with a rank one matrix. In contrast, Betti curves for random assembly whose pairwise correlations come from the same CCG matrix of the same order. Below each Betti curve, we plot the normalized singular values and visually see that they are full rank and appear close to the singular values of a random assembly of the same size.

4 Conclusion

In this paper, we proved three theorems that characterized the Betti curves of rank 1 symmetric matrices. We also showed these rank 1 signatures are present in some cell assembly correlation matrices for zebrafish. A limitation of these theorems, however, is that the converse is not in general true. Identically zero Betti curves do not imply rank 1, though our computational experiments suggest they are indicative of low rank, similar to the case of low-dimensional distance matrices [3]. So while we can use these results to rule out an underlying rank 1 structure, as in the random assemblies, we can not conclude from Betti curves alone that a matrix has underlying rank 1.

Code. All code used for the construction of the figures, except for Fig. 6, is available on Github. Code and data to produce Fig. 6 is available upon request. https://github.com/joshp112358/rank-one-betti-curves.

References

1. Bauer, U.: Ripser: efficient computation of Vietoris-Rips persistence barcodes. arxiv:1908.02518 (2019)
2. Fors, G.: On the foundation of algebraic topology. arxiv:1904.12552 (2004)
3. Giusti, C., Pastalkova, E., Curto, C., Itskov, V.: Clique topology reveals intrinsic geometric structure in neural correlations. Proc. Nat. Acad. Sci. **112**(44), 13455–13460 (2015)

4. Kahle, M.: Topology of random clique complexes. Discrete Math. **309**(6), 1658–1671 (2009)
5. Milnor, J.: Construction of universal bundles. II. Ann. Math. **2**(63), 430–436 (1956)
6. Zomorodian, A.: Topology for Computing. Cambridge University Press, Cambridge (2018)

Algebraic Homotopy Interleaving Distance

Nicolas Berkouk(✉)

EPFL, Laboratory for Topology and Neuroscience, Lausanne, Switzerland
nicolas.berkouk@epfl.ch

Abstract. The theory of persistence, which arises from topological data analysis, has been intensively studied in the one-parameter case both theoretically and in its applications. However, its extension to the multi-parameter case raises numerous difficulties, where it has been proven that no barcode-like decomposition exists. To tackle this problem, algebraic invariants have been proposed to summarize multi-parameter persistence modules, adapting classical ideas from commutative algebra and algebraic geometry to this context. Nevertheless, the crucial question of their stability has raised little attention so far, and many of the proposed invariants do not satisfy a naive form of stability.

In this paper, we equip the homotopy and the derived category of multi-parameter persistence modules with an appropriate interleaving distance. We prove that resolution functors are always isometric with respect to this distance. As an application, this explains why the graded-Betti numbers of a persistence module do not satisfy a naive form of stability. This opens the door to performing homological algebra operations while keeping track of stability. We believe this approach can lead to the definition of new stable invariants for multi-parameter persistence, and to new computable lower bounds for the interleaving distance (which has been recently shown to be NP-hard to compute in [2]).

1 Introduction

Persistence theory appeared in the early 2000s as a theoretical framework to define topological descriptors of real-world datasets. At its roots, persistence studies filtrations of topological spaces built out of datasets, and defines algebraic invariants of these filtrations (such as the now famous *barcodes*) together with a notion of metric between these invariants. It has since been widely developed and applied. We refer the reader to [14] for an extended exposition of the classical theory and of its applications.

The need for studying persistence modules obtained from functions valued in higher-dimensional partially ordered sets naturally arises from the context of data analysis, see for example [11]. However, as shown in [4], the category $\text{Pers}(\mathbb{R}^d)$ of functors $(\mathbb{R}^d, \leq) \to \text{Mod}(\mathbf{k})$ seems to be too general at first sight

Supported by Innosuisse grant 45665.1 IP-ICT.

F. Nielsen and F. Barbaresco (Eds.): GSI 2021, LNCS 12829, pp. 656–664, 2021.
https://doi.org/10.1007/978-3-030-80209-7_70

for $d \geq 2$ to allow for computer friendly analysis. Indeed, it contains a full sub-category isomorphic to the one of $\mathbb{Z}^n$-graded-$\mathbf{k}[x_1, ..., x_n]$-modules.

Ideas from algebraic geometry [7] and combinatorial commutative algebra [5,13] have since been developed in the context of persistence to study the category of persistence modules with multiple parameters from a computational point of view. These works propose new invariants for summing up the algebraic structure of a given persistence module, arising from homological algebra methods. While informative about the algebraic structure of a given module, the -crucial- question of the stability of these invariants is not studied yet. Indeed, data arising from the real world comes with noise, and one cannot avoid a theoretical study of the dependency of topological descriptors to small perturbations in view of practical applications.

In this paper, we show that simple homological operations –such as considering graded-Betti numbers of a persistence module– are not stable in a naive way with respect to the interleaving distance. To overcome this problem, we equip the homotopy (resp. derived) category of persistence modules of an interleaving-like distance and prove that the projective resolution functors (resp. localization functor) are distance-preserving with respect to the distances we introduce. These distance comparison theorems open the door to perform homological algebra operations on free resolutions while keeping track of stability, and has already drawn attention in later works [8,12]. The future directions of research we shall undertake consist in studying precisely which meaningful stable invariants one can derive from our distance comparison theorems, which we explain at the end of the paper.

We acknowledge that some work has been undertaken to study interleaving distances on the homotopy category of topological spaces [3]. It should be noticed that these results are of a different nature than ours for that they are purely topological.

2 Multiparameter Persistence

Throughout the paper, we let $\mathbf{k}$ be a field, and $d \in \mathbb{Z}_{\geq 0}$. We equip $\mathbb{R}^d$ with the product order $\leq$ defined by coordinate-wise comparisons. We denote by $(\mathbb{R}^d, \leq)$ the associated partially ordered set (poset) category. We refer the reader to [10] for a detailed exposition of the content of this section.

A *persistence module* over $\mathbb{R}^d$ is a functor from $(\mathbb{R}^d, \leq)$ to the category of $\mathbf{k}$-vector spaces $\mathrm{Mod}(\mathbf{k})$. We denote by $\mathrm{Pers}(\mathbb{R}^d)$ the category of persistence modules over $\mathbb{R}^d$ with natural transformations of functors as homomorphisms. It is easily shown to be a Grothendieck category. The main example of persistence modules over $\mathbb{R}^d$ is as follows. Given a function $u : X \to \mathbb{R}^d$ from the topological space X (not necessarily continuous), its associated *sub-level sets filtration* is the functor $\mathcal{S}(u)$ from the poset category $(\mathbb{R}^d, \leq)$ to the category of topological spaces $\mathbf{Top}$, defined by $\mathcal{S}(u)(s) = u^{-1}\{x \in \mathbb{R}^d \mid x \leq s\}$, and $\mathcal{S}(u)(s \leq t)$ is the inclusion of $u^{-1}\{x \in \mathbb{R}^d \mid x \leq s\}$ into $u^{-1}\{x \in \mathbb{R}^d \mid x \leq t\}$. For $n \geq 0$, the *n-th sublevel set persistence modules* associated to u is the functor $\mathcal{S}_n(u) := \mathrm{H}_n^{\mathrm{sing}} \circ \mathcal{S}(u)$, where $\mathrm{H}_n^{\mathrm{sing}}$ denotes the n-th singular homology functor with coefficients in $\mathbf{k}$.

Another example is given by the following. A subset $\mathcal{I} \subset \mathbb{R}^d$ is an *interval*, if it satisfies: (i) for any $x, z \in \mathcal{I}$, and $y \in \mathbb{R}^d$ satisfying $x \leq y \leq z$, then $y \in \mathcal{I}$; (ii) for $x, y \in \mathcal{I}$ there exists a finite sequence $(r_i)_{i=0..n}$ in $\mathcal{P}$ such that $r_0 = x$, $r_n = y$ and for any $i < n$, $r_i \leq r_{i+1}$ or $r_{i+1} \leq r_i$.

Definition 1. *Given $\mathcal{I}$ an interval of $(\mathbb{R}^d, \leq)$, one defines $\mathbf{k}^{\mathcal{I}}$ the* interval persistence module *associated to $\mathcal{I}$ by:*

$$\begin{cases} \mathbf{k}^{\mathcal{I}}(x) = \mathbf{k} \text{ if } x \in \mathcal{I},\ 0 \text{ otherwise}, \\ \mathbf{k}^{\mathcal{I}}(x \leq y) = id_{\mathbf{k}} \text{ if } x \text{ and } y \in \mathcal{I},\ 0 \text{ otherwise}. \end{cases}$$

One can easily prove that $\mathbf{k}^{\mathcal{I}}$ is an indecomposable persistence module.

Given $\varepsilon \in \mathbb{R}$, the *$\varepsilon$-shift* is the functor $\cdot[\varepsilon] : \mathrm{Pers}(\mathbb{R}^d) \to \mathrm{Pers}(\mathbb{R}^d)$ defined by $M[\varepsilon](s) = M\big(s + (\varepsilon, ..., \varepsilon)\big)$ and $M[\varepsilon](s \leq t) = M\big(s + (\varepsilon, ..., \varepsilon) \leq t + (\varepsilon, ..., \varepsilon)\big)$ for $s \leq t$ in $\mathbb{R}^d$. For $\varphi \in \mathrm{Hom}_{\mathrm{Pers}(\mathbb{R}^d)}(M, N)$, we set $\varphi[\varepsilon](s) = \varphi(s + (\varepsilon, ..., \varepsilon))$. It is immediate to verify that $\cdot[\varepsilon] \circ \cdot[\varepsilon'] = \cdot[\varepsilon'] \circ \cdot[\varepsilon] = \cdot[\varepsilon + \varepsilon']$. When $\varepsilon \geq 0$, one also defines the *ε-smoothing morphism of M* as the natural transformation $M \Rightarrow M[\varepsilon]$ defined by $\tau_\varepsilon^M(t) = M(t \leq t + (\varepsilon, ..., \varepsilon))$, $t \in \mathbb{R}^d$.

Given $M, N \in \mathrm{Pers}(\mathbb{R}^d)$ and $\varepsilon \geq 0$, an *ε-interleaving* between M and N is the data of two morphisms $f : M \to N[\varepsilon]$ and $g : N \to M[\varepsilon]$ such that $g[\varepsilon] \circ f = \tau_{2\varepsilon}^M$ and $f[\varepsilon] \circ g = \tau_{2\varepsilon}^N$. If such morphisms exist, we write $M \sim_\varepsilon N$. One shall observe that $M \sim_0 N$ if and only if M and N are isomorphic. Hence, ε-interleavings can be thought of as a weaker form of isomorphisms.

Definition 2. *Let $M, N \in Pers(\mathbb{R}^d)$. One defines their* interleaving distance *as the possibly infinite number:*

$$d_I(M, N) = \inf\{\varepsilon \in \mathbb{R}_{\geq 0} \mid M \sim_\varepsilon N\}.$$

The interleaving distance is an extended pseudo-distance on $\mathrm{Pers}(\mathbb{R}^d)$ (a pseudo-distance that is not always finite). It allows to express the stability properties of the sublevel-sets persistence functor as celebrated by the algebraic stability theorem [10, Theorem 5.3]: let $u, v : X \to \mathbb{R}^d$ be two maps from a topological space X to $\mathbb{R}^d$, and $n \geq 0$, one has

$$d_I(\mathcal{S}_n(u), \mathcal{S}_n(v)) \leq \sup_{x \in X} |u(x) - v(x)|.$$

One defines B^d to be the algebra of generalized polynomials with coefficients in $\mathbf{k}$ and real positive exponents, that is polynomials that expresses as a finite sum $\sum_i \alpha_i x_1^{e_i^1} ... x_d^{e_i^d}$, with $\alpha_i \in \mathbf{k}$ and $e_i^j \in \mathbb{R}_{\geq 0}$. Given $e = (e_1, ..., e_d) \in (\mathbb{R}_{\geq 0})^d$, one shall denote the monomial $x_1^{e_1} ... x_d^{e_d}$ by x^e. The polynomial $\mathbf{k}$-algebra B^d is naturally endowed with a $\mathbb{R}^d$-grading, given by the decomposition of $\mathbf{k}$-vector spaces: $B^d = \bigoplus_{e \in \mathbb{R}^d_{\geq 0}} \mathbf{k} \cdot x^e$. We denote by $\mathbb{R}^d$-grad-B^d-mod the category of $\mathbb{R}^d$ graded modules over B^d.

Theorem 1 [4]. *There exists an equivalence of abelian categories α : $Pers(\mathbb{R}^d) \to \mathbb{R}^d$-grad-$B^d$-mod.*

This equivalence of categories explains the complexity and impossibility to give a combinatorial classification of $\mathrm{Pers}(\mathbb{R}^d)$ when $d \geq 2$ as the existence of barcodes in the case $n = 1$.

3 Homotopy and Derived Interleavings

Let $\mathscr{C}$ be an additive category. We shall denote by $C(\mathscr{C})$ (resp. $K(\mathscr{C})$, resp. $D(\mathscr{C})$) the category of chain complexes of $\mathscr{C}$ (resp. the homotopy category of $\mathscr{C}$, resp. the derived category of $\mathscr{C}$). We denote $C^-(\mathscr{C})$, $K^-(\mathscr{C})$, $D^-(\mathscr{C})$ their full subcategories consisting of chain complexes X such that there exists $N \in \mathbb{Z}$ such that $X^n = 0$ for $n \geq N$. We also write Q for the localization functor $K(\mathscr{C}) \to D(\mathscr{C})$. We refer the reader to [9, Chapter 1], for the definitions of these constructions. For $\varepsilon \in \mathbb{R}$, since the functor $\cdot[\varepsilon]$ is an additive auto-equivalence of $\mathrm{Pers}(\mathbb{R}^d)$, it is exact and preserves projective objects. Consequently, $\cdot[\varepsilon]$ preserves quasi-isomorphisms and induces a well-defined triangulated endofunctor of $\mathrm{D}(\mathrm{Pers}(\mathbb{R}^d))$, by applying $\cdot[\varepsilon]$ degree-wise to a chain complex. We will still denote $\cdot[\varepsilon]$ the induced functors on $C(\mathrm{Pers}(\mathbb{R}^d))$, $K(\mathrm{Pers}(\mathbb{R}^d))$ and $\mathrm{D}(\mathrm{Pers}(\mathbb{R}^d))$. Let X be an object of $C(\mathrm{Pers}(\mathbb{R}^d))$. It is immediate to verify that the collection $(\tau_\varepsilon^{X^i})_{i\in\mathbb{Z}}$ defines a homomorphism of chain complexes

$$\tau_\varepsilon^X \in \mathrm{Hom}_{C(\mathrm{Pers}(\mathbb{R}^d))}(X, X[\varepsilon]).$$

We shall denote by $[\tau_\varepsilon^X]$ (resp. $\{\tau_\varepsilon^X\}$) the image of τ_ε^X in $K(\mathrm{Pers}(\mathbb{R}^d))$ (resp. $\mathrm{D}(\mathrm{Pers}(\mathbb{R}^d))$). Given $\varepsilon, \varepsilon' \geq 0$, the equality $\tau_{\varepsilon'}^{X[\varepsilon]} \circ \tau_\varepsilon^X = \tau_{\varepsilon+\varepsilon'}^X$ implies:

$$[\tau_{\varepsilon'}^{X[\varepsilon]}] \circ [\tau_\varepsilon^X] = [\tau_{\varepsilon+\varepsilon'}^X] \quad \text{and} \quad \{\tau_{\varepsilon'}^{X[\varepsilon]}\} \circ \{\tau_\varepsilon^X\} = \{\tau_{\varepsilon+\varepsilon'}^X\}. \tag{1}$$

Definition 3. *Let X and Y be two objects of $C(Pers(\mathbb{R}^d))$, and $\varepsilon \geq 0$. An ε-*homotopy-interleaving *(resp. an ε-*derived-interleaving*) between X and Y is the data of two morphisms of $K(Pers(\mathbb{R}^d))$, $f : X \to Y[\varepsilon]$ and $g : X \to Y[\varepsilon]$ (resp. two morphisms of $D(Pers(\mathbb{R}^d))$, $\tilde{f} : X \to Y[\varepsilon]$ and $\tilde{g} : X \to Y[\varepsilon]$) such that:*

$$g[\varepsilon] \circ f = [\tau_{2\varepsilon}^X] \quad (\textit{resp.} \quad \tilde{g}[\varepsilon] \circ \tilde{f} = \{\tau_{2\varepsilon}^X\}),$$
$$f[\varepsilon] \circ g = [\tau_{2\varepsilon}^Y] \quad (\textit{resp.} \quad \tilde{f}[\varepsilon] \circ \tilde{g} = \{\tau_{2\varepsilon}^Y\}).$$

*In this situation, we say that X and Y are ε-*homotopy-interleaved *(resp. ε-*derived-interleaved*) and write $X \sim_\varepsilon^K Y$ (resp. $X \sim_\varepsilon^D Y$).*

Definition 4. *The homotopy interleaving distance (resp. derived interleaving distance) between X and Y in $C(Pers(\mathbb{R}^d))$ is the possibly infinite number:*

$$d_I^K(X,Y) = \inf\{\varepsilon \in \mathbb{R}_{\geq 0} \mid X \sim_\varepsilon^K Y\} \ (\textit{resp.}\ d_I^D(X,Y) = \inf\{\varepsilon \in \mathbb{R}_{\geq 0} \mid X \sim_\varepsilon^D Y\}).$$

Proposition 1. *The map d_I^K (resp. d_I^D) is an extended pseudo-distance on $Obj(K(Pers(\mathbb{R}^d)))$ (resp. $Obj(D(Pers(\mathbb{R}^d)))$).*

Proof. For X an object of $C(\text{Pers}(\mathbb{R}^d))$, we have $d_I^K(X, X) = d_I^D(X, X) = 0$ since isomorphisms are 0-interleavings. The triangle inequality is an easy consequence of Eq. 1.

Definition 5. *A projective resolution functor of an abelian category $\mathscr{C}$ is the following data: for all objects X in $K^-(\mathscr{C})$, (i) a projective resolution $j(X)$, (ii) a quasi-isomorphism $i_X : j(X) \to X$.*

Note that Lemma 13.23.3 in [16] states that the data of a resolution functor is equivalent to the data of a unique functor $P : K^-(\mathscr{C}) \to K^-(\mathscr{P})$, where $\mathscr{P}$ is the full sub-category of projective objects of X, together with a unique 2-isomorphism $Q \circ P \overset{\sim}{\Rightarrow} Q$. From Theorem 1, the category $\text{Pers}(\mathbb{R}^d)$ is equivalent to a category of modules over a commutative ring. As a consequence, it is abelian and has enough projectives. We recall the following important proposition about projective resolution functors.

Proposition 2. (13.23.4, [16]). *Assume $\mathscr{C}$ is an abelian category with enough projectives. Then there exists a projective resolution functor of $\mathscr{C}$. Moreover, for any two such functors P, P' there exists a unique isomorphism of functors $P \overset{\sim}{\Rightarrow} P'$ such that $Q \circ P \overset{\sim}{\Rightarrow} Q$ equals the composition $Q \circ P \overset{\sim}{\Rightarrow} Q \circ P' \overset{\sim}{\Rightarrow} Q$.*

Consequently, $\text{Pers}(\mathbb{R}^d)$ admits a projective resolution functor $(j, i_{\text{Pers}(\mathbb{R}^d)})$ with associated functor P. We will keep writing P for its restriction to $\text{Pers}(\mathbb{R}^d)$, identified as the full subcategory of $K^-(\text{Pers}(\mathbb{R}^d))$ with objects complexes concentrated in degree 0. We can now state the distance comparison theorem, that we will prove in this section.

Theorem 2. *The following diagram of categories commutes up to isomorphism of functors, and all functors are object-wise isometric.*

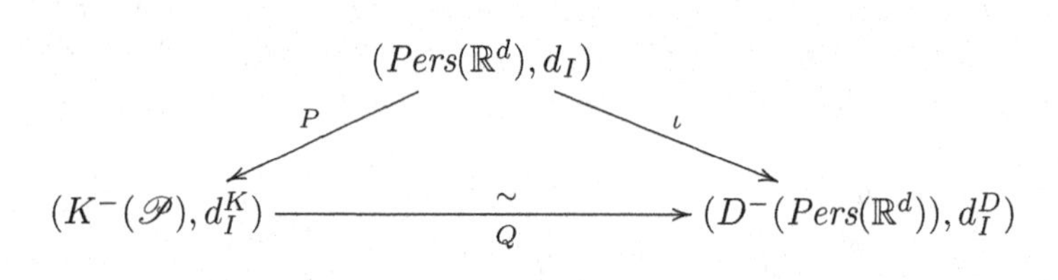

The functor $P(- \cdot [\varepsilon])[-\varepsilon]$ clearly is a resolution functor. Let $\eta_\varepsilon : P(- \cdot [\varepsilon])[-\varepsilon] \overset{\sim}{\Rightarrow} P$ be the unique isomorphism of functors given by Proposition 2, and set $\chi_\varepsilon := \eta_\varepsilon[\varepsilon] : P(- \cdot [\varepsilon]) \overset{\sim}{\Rightarrow} P(-)[\varepsilon]$.

Lemma 1. *Let $M \in Pers(\mathbb{R}^d)$ and $\varepsilon \geq 0$, then $P(\tau_\varepsilon^M) = \chi_\varepsilon(M)[\varepsilon] \circ \left[\tau_\varepsilon^{P(M)}\right]$.*

Consequently: $H^0\left(\left[\tau_\varepsilon^{P(M)}\right]\right) = \tau_\varepsilon^M$.

Proof. Consider a chain morphism $\varphi : P(M)[\varepsilon] \to P(M[\varepsilon])$ such that $[\varphi] = \chi_\varepsilon(M)[\varepsilon]$. Observe the commutativity of the following diagram in $C(\text{Pers}(\mathbb{R}^d))$:

$$
\begin{array}{ccccccccc}
\cdots \longrightarrow & P^i(M) & \longrightarrow \cdots \longrightarrow & P^1(M) & \longrightarrow & P^0(M) & \longrightarrow & M & \longrightarrow 0 \\
& \downarrow \tau_\varepsilon^{P^i(M)} & & \downarrow \tau_\varepsilon^{P^1(M)} & & \downarrow \tau_\varepsilon^{P^0(M)} & & \downarrow \tau_\varepsilon^{M} & \\
\cdots \longrightarrow & P^i(M)[\varepsilon] & \longrightarrow \cdots \longrightarrow & P^1(M)[\varepsilon] & \longrightarrow & P^0(M)[\varepsilon] & \longrightarrow & M[\varepsilon] & \longrightarrow 0 \\
& \downarrow \varphi^i & & \downarrow \varphi^1 & & \downarrow \varphi^0 & & \downarrow \mathrm{id}_{M[\varepsilon]} & \\
\cdots \longrightarrow & P^i(M[\varepsilon]) & \longrightarrow \cdots \longrightarrow & P^1(M[\varepsilon]) & \longrightarrow & P^0(M[\varepsilon]) & \longrightarrow & M[\varepsilon] & \longrightarrow 0
\end{array}
$$

Therefore, $\varphi \circ \tau_\varepsilon^{P(M)}$ is one lift of τ_ε^M, which by characterization of lifts of morphism to projective resolutions [15, Lemma 6.7] proves that

$$P(\tau_\varepsilon^M) = \left[\varphi \circ \tau_\varepsilon^{P(M)}\right] = [\varphi] \circ \left[\tau_\varepsilon^{P(M)}\right] = \chi_\varepsilon(M)[\varepsilon] \circ \left[\tau_\varepsilon^{P(M)}\right].$$

Lemma 2. *Let M and N be two persistence modules in Pers($\mathbb{R}^d$) and $\varepsilon \geq 0$.*

1. *If M and N are ε-interleaved with respect to $f : M \to N[\varepsilon]$ and $g : N \to M[\varepsilon]$ in Pers($\mathbb{R}^d$), then $P(M)$ and $P(N)$ are ε-homotopy-interleaved with respect to $\chi_\varepsilon(N)[\varepsilon] \circ P(f)$ and $\chi_\varepsilon(M)[\varepsilon] \circ P(g)$ in K(Pers($\mathbb{R}^d$)).*
2. *Conversely, if $P(M)$ and $P(N)$ are ε-homotopy-interleaved with respect to $\alpha : P(M) \to P(N)[\varepsilon]$ and $\beta : P(N) \to P(M)[\varepsilon]$ in K(Pers($\mathbb{R}^d$)), then M and N are ε-interleaved with respect to $H^0(\alpha)$ and $H^0(\beta)$ in Pers($\mathbb{R}^d$).*

Proof. 1. Applying the functor P to the ε-interleavings gives: $P(\tau_{2\varepsilon}^M) = P(g[\varepsilon]) \circ P(f)$ and $P(\tau_{2\varepsilon}^N) = P(f[\varepsilon]) \circ P(g)$. Therefore we have a commutative diagram in K(Pers($\mathbb{R}^d$)):

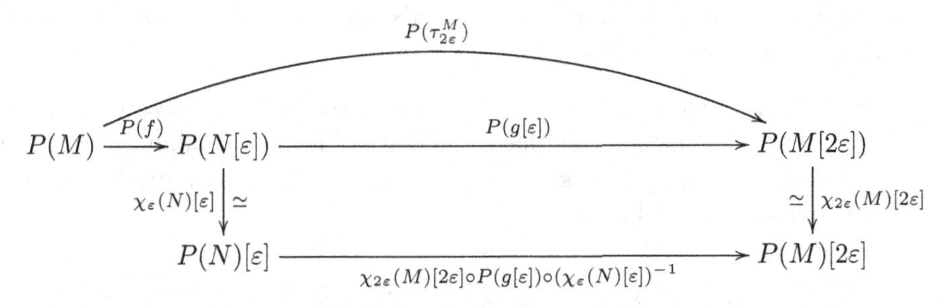

We introduce the morphisms $\tilde{f} = \chi_\varepsilon(N)[\varepsilon] \circ P(f)$ and $\tilde{g} = \chi_\varepsilon(M)[\varepsilon] \circ P(g)$. By the unicity of χ_ε, we obtain the following:

$$\tilde{g}[\varepsilon] = \chi_{2\varepsilon}(M)[2\varepsilon] \circ P(g[\varepsilon]) \circ (\chi_\varepsilon(N)[\varepsilon])^{-1}, \quad \mathrm{H}^0\left(\tilde{g}[\varepsilon] \circ \tilde{f}\right) = \tau_{2\varepsilon}^M. \qquad (2)$$

From Eq. 2 and Lemma 1, we deduce: $\tilde{g}[\varepsilon] \circ \tilde{f} = [\tau_{2\varepsilon}^{P(M)}]$. The above equations also hold when intertwining f and g, M and N, $\tilde{f}$ and $\tilde{g}$. Therefore, $\tilde{f}$ and $\tilde{g}$ define a homotopy ε-homotopy-interleaving between $P(N)$ and $P(M)$.

2. The converse is obtained by applying H^0 to the ε-homotopy-interleaving morphisms, and Lemma 1 according to which $\mathrm{H}^0([\tau_\varepsilon^{P(M)}]) = \tau_\varepsilon^M$.

Proof. (of Theorem 2)

There only remains to prove that ι and Q are distance preserving. Since by definition, $\iota(\tau_\varepsilon^M) = \{\tau_\varepsilon^M\}$, it is clear that ι sends ε-interleavings to ε-derived-interleavings. Conversely, assume that $\iota(M)$ and $\iota(N)$ are ε-derived-interleaved. Then since ι is fully-faithful, applying H^0 to the ε-derived-interleaving morphisms in the derived category leads to a ε-interleaving in $\mathrm{Pers}(\mathbb{R}^d)$.

By definition, for X an object of $K^-(\mathscr{P})$, $Q([\tau_\varepsilon^X]) = \{\tau_\varepsilon^X\}$, which shows that Q preserves interleavings. Now assume that $Q(X)$ and $Q(Y)$ are ε-interleaved. Since Q is fully-faithful and $Q([\tau_\varepsilon^X]) = \{\tau_\varepsilon^X\}$, one deduces the existence of a ε-interleaving between X and Y in $K^-(\mathscr{P})$.

4 Application: Explaining the Instability of the Graded-Betti Numbers of Persistence Modules

Given $a \in \mathbb{R}^d$, let $U_a = \{x \in \mathbb{R}^d \mid a \leq x\}$ and define $F_a := \mathbf{k}^{U_a}$ (see Definition 1). A finite free persistence module over $\mathbb{R}^d$ is a persistence module F such that there exists a function $\xi(F) : \mathbb{R}^d \to \mathbb{Z}_{\geq 0}$ with finite support such that $F \simeq \bigoplus_{a \in \mathbb{R}^d} F_a^{\oplus \xi(F)(a)}$[1]. We will say that a projective resolution of $M \in \mathrm{Pers}(\mathbb{R}^d)$ is finite free if each step of the resolution is a finite free persistence module. A minimal finite free resolution of M, if it exists, is a finite free resolution $X \in C(\mathrm{Pers}(\mathbb{R}^d))$ such that any other finite free resolution $Y \in C(\mathrm{Pers}(\mathbb{R}^d))$ of M splits as $Y \simeq X \oplus Y'$, for some $Y' \in C(\mathrm{Pers}(\mathbb{R}^d))$ [15, Theorem 7.5].

Definition 6 (12.1, [15]). *Let $M \in \mathrm{Pers}(\mathbb{R}^d)$ and assume that M admits a minimal finite free resolution $F \in C(\mathrm{Pers}(\mathbb{R}^d))$. The n-th graded-Betti number of M is the function $\beta_n(M) := \xi(F^{-n})$.*

We can find two persistence modules which are arbitrary close with respect to the interleaving distance and have different graded-Betti numbers. For $\varepsilon > 0$, consider $M := 0$ and $N_\varepsilon := \mathbf{k}^{[0,\varepsilon)^2}$ in $\mathrm{Pers}(\mathbb{R}^2)$. Then N_ε admits the following minimal finite free resolution:

$$C_\varepsilon = 0 \longrightarrow F_{(\varepsilon,\varepsilon)} \xrightarrow{\binom{1}{1}} F_{(\varepsilon,0)} \oplus F_{(0,\varepsilon)} \xrightarrow{(1\ -1)} F_{(0,0)} \longrightarrow 0$$

Therefore $\beta^0(N_\varepsilon)(x) = 1$ for $x = (0,0)$ and $\beta^0(M)(x) = 0$ otherwise, and $\beta^0(M)$ is the zero function. Observe that the interleaving distance between M and N_ε is $\frac{\varepsilon}{2}$, consequently:

$$d_I(M, N_\varepsilon) = \frac{\varepsilon}{2} \quad \text{and} \quad \sup_{x \in \mathbb{R}^d} |\beta^0(M)(x) - \beta^0(N_\varepsilon)(x)| = 1.$$

[1] Where $F_a^{\oplus \xi(F)(a)}$ is the direct sum of $\xi(F)(a)$ copies of F_a.

This very simple example shows that graded-Betti numbers are extremely sensitive to noise. An arbitrary small perturbation of the input persistence modules can lead to an arbitrary large change in the graded-Betti numbers. The fundamental reason for this, is that interleavings only lift to minimal free resolution up to homotopy, and we know from Theorem 2 that $d_I^K(C_\varepsilon, 0) = d_I(N_\varepsilon, 0) = \frac{\varepsilon}{2}$. In our example, it is not possible to construct an $\frac{\varepsilon}{2}$-interleaving between C_ε and 0 without taking homotopies into account, which explains what we call the graded-Betti numbers instability.

5 Future Directions of Work

In [1], the author bounds the bottleneck distance between two finite free persistence modules by a factor times their interleaving distance. One could ask whether it is possible to define a bottleneck distance between minimal free resolutions of two persistence modules (that would allow to match free indecomposable modules across degrees), and to bound this distance by a multiple of the homotopy interleaving distance. This could lead to a computable lower bound to the interleaving distance (which has been shown in [2] to be NP-hard to compute for $d \geq 2$), without relying on any kind of decomposition theorems.

Another interesting direction of research would be to understand the implications of [6] to our setting. Indeed, in this work, the authors classify all rank functions on the derived category of vector spaces. This naturally leads to a classification of rank functions for the derived category of persistence modules. Whether these rank functions satisfy a form of stability with respect to our derived interleaving distance should be investigated in the future.

References

1. Bjerkevik, H.B.: Stability of higher-dimensional interval decomposable persistence modules (2016)
2. Bjerkevik, H.B., Botnan, M.B., Kerber, M.: Computing the interleaving distance is NP-hard. In: Foundations of Computational Mathematics (2018)
3. Blumberg, A.J., Lesnick, M.: Universality of the homotopy interleaving distance (2017)
4. Carlsson, G.: Zomorodian. The theory of multidimensional persistence. Discrete and Computational Geometry, Afra (2009)
5. Chacholski, W., Scolamiero, M., Vaccarino, F.: Combinatorial presentation of multidimensional persistent homology. J. Pure Appl. Algebra **221**, 1055–1075 (2017)
6. Chuang, J., Lazarev, A.: Rank functions on triangulated categories (2021)
7. Heather, A.H., Schenck, H., Tillmann, U.: Stratifying multiparameter persistent homology, Nina Otter (2017)
8. Hiraoka, Y., Ike, Y., Yoshiwaki, M.: Algebraic stability theorem for derived categories of zigzag persistence modules (2020)
9. Kashiwara, M., Schapira, P.: Sheaves on Manifolds. Springer, Berlin (1990). https://doi.org/10.1007/978-3-662-02661-8
10. Lesnick, M.: The theory of the interleaving distance on multidimensional persistence modules. Foundations of Computational Mathematics (2015)

11. Lesnick, M., Wright, M.: Interactive visualization of 2-D persistence modules. arXiv:1512.00180
12. Milicevic, N.: Convolution of persistence modules (2020)
13. Miller, E.: Modules over posets: commutative and homological algebra (2019)
14. Steve, Y.: Oudot. American Mathematical Society, Persistence theory, From quiver representations to data analysis (2015)
15. Peeva, I.: Graded Syzygies. AA, vol. 16, 1st edn. Springer-Verlag, London (2011). https://doi.org/10.1007/978-0-85729-177-6
16. Stacks Project, Resolution functors. https://stacks.math.columbia.edu/tag/013U

A Python Hands-on Tutorial on Network and Topological Neuroscience

Eduarda Gervini Zampieri Centeno[1,2(✉)], Giulia Moreni[1], Chris Vriend[1,3], Linda Douw[1], and Fernando Antônio Nóbrega Santos[1,4]

[1] Amsterdam UMC, Vrije Universiteit Amsterdam, Anatomy and Neurosciences, Amsterdam Neuroscience, De Boelelaan 1117, Amsterdam, Netherlands
e.centeno@amsterdamumc.nl

[2] Bordeaux Neurocampus, Université de Bordeaux, Institut des Maladies Neurodégénératives, CNRS UMR 5293, 146 Rue Léo Saignat, Bordeaux, France

[3] Amsterdam UMC, Vrije Universiteit Amsterdam, Psychiatry, Amsterdam Neuroscience, De Boelelaan 1117, Amsterdam, Netherlands

[4] Institute of Advanced Studies, University of Amsterdam, Oude Turfmarkt 147, 1012 Amsterdam, GC, Netherlands

Abstract. Network neuroscience investigates brain functioning through the prism of connectivity, and graph theory has been the main framework to understand brain networks. Recently, an alternative framework has gained attention: topological data analysis. It provides a set of metrics that go beyond pairwise connections and offer improved robustness against noise. Here, our goal is to provide an easy-to-grasp theoretical and computational tutorial to explore neuroimaging data using these frameworks, facilitating their accessibility, data visualisation, and comprehension for newcomers to the field. We provide a concise (and by no means complete) theoretical overview of the two frameworks and a computational guide on the computation of both well-established and newer metrics using a publicly available resting-state functional magnetic resonance imaging dataset. Moreover, we have developed a pipeline for three-dimensional (3-D) visualisation of high order interactions in brain networks.

Keywords: Network neuroscience · Data visualisation · Topological data analysis

1 Introduction

Network neuroscience sees the brain through an integrative lens by mapping and modelling its elements and interactions [3,21]. The main theoretical framework from complex network science used to model, estimate, and simulate brain networks is graph theory [9,24]. A graph is comprised of a set of interconnected elements, also known as vertices and edges. Vertices (also known as nodes) in a network can be, for example, brain areas, while edges (also known as links) are a representation of the functional connectivity between pairs of

F. Nielsen and F. Barbaresco (Eds.): GSI 2021, LNCS 12829, pp. 665–673, 2021.
https://doi.org/10.1007/978-3-030-80209-7_71

vertices [42]. Several descriptive graph metrics[1] [16] can then be calculated to describe the brain network's characteristic [21,26], and they have consistently allowed researchers to identify non-random features of brain networks. An example is the ground-breaking discovery that the brain (like most other real-world networks) follows a 'small-world network' architecture [2,43], indicating a compromise between wiring cost and optimal efficiency. Using graph theory, many insights have been gathered on the healthy and diseased brain neurobiology [19,26]. Algebraic topological data analysis (TDA) provides another prism on brain connectivity investigation beyond the 'simple' pairwise connections (*i.e.*, higher-order interactions). With TDA, one can identify a network's shape and its invariant properties (*i.e.*, coordinate and deformation invariances [47]). Moreover, TDA often provides more robustness against noise than graph theoretical analysis [41], a significant neuroimaging data issue [30]. Although TDA has only recently been adopted to network neuroscience [15,39], it has already shown exciting results on brain network data [14,18]. However, clinical scientists' comprehension and application can be hindered by TDA's complexity and mathematical abstraction. Here, we want to facilitate the use of network neuroscience and its constituents graph theory and TDA by the general neuroscientific community by providing both computational and theoretical explanation of the primary metrics, strongly inspired by [21]. The work is divided into a longer manuscript [13] containing several resources (see Table in [13]) and more theoretical explanations, and a publicly available Jupyter Notebook online (https://github.com/multinetlab-amsterdam/network_TDA_tutorial). In these notebooks, we use third-party Python packages for part of the computations (*e.g.*, *networkx* [25] for the graph theory metrics and *gudhi* [32] for persistent homology) and provide practical scripts for some TDA metrics and 3-D visualisations of simplicial complexes (a new addition to the field). Our tutorial focuses on resting-state functional magnetic resonance imaging (rsfMRI) data; however, the main concepts and tools discussed in this paper can be extrapolated to other imaging modalities, biological or complex networks. The extended version [13] covers the most commonly used graph metrics in network neuroscience, also in line with reference [21], and TDA. However, due to the size constraint, here we prioritize the latter.

1.1 Starting Point: The Adjacency Matrix

The basic unit on which graph theory and TDA are applied in the context of rsfMRI is the adjacency or functional connectivity matrix [3,21]. Typically, rsfMRI matrices are symmetric and do not specify the direction of connectivity (*i.e.*, activity in area A drives activity in area B), therefore yielding undirected networks. To further analyse the sfMRI connectivity matrix, one has to decide

[1] Notice that the notion of metric in mathematics defines distance between two points in a set [16], which is distinct from what we are using in this work. We denote as metric any quantity that can be computed, *i.e.*, "measured", in a brain network or simplicial complex.

whether to keep or not edges' weights (*e.g.*, correlation values in rsfMRI connectivity) or to absolutise negative weights (or anticorrelations) [21,27]. These decisions influence the computation of the different metrics described in the tutorial and matter for the biological interpretation of the results [21]. In this tutorial we use an undirected, weighted, and absolutised connectivity matrix.

1.2 Topological Data Analysis

TDA uses topology and geometry methods to study the shape of the data [11] and can identify a network's different characteristics by addressing a network's high-order structure [4,10,28]. A core success of TDA is the ability to provide robust results when compared with alternative methods, even if the data are noisy [18,41]. One of the benefits of using TDA in network neuroscience is the possibility of finding global properties of a network that are preserved regardless of the way we represent the network [35], as we illustrate below. Those properties are the so-called topological invariants. Here, we cover some fundamental TDA concepts: filtration, simplicial complexes, Euler characteristic, phase-transitions, Betti numbers, curvature, and persistent homology.

Simplicial Complexes. In TDA, we consider that the network as a multidimensional structure called the simplicial complex. Such a network is not only made up of the set of vertices (0-simplex) and edges (1-simplex) but also of triangles (2-simplex), tetrahedrons (3-simplex), and higher k-dimensional structures. In short, a k-simplex is an object in k-dimensions and, in our work, is formed by a subset of $k+1$ vertices of the network.

Filtration. Consists of a nested sequence of simplicial complexes. Here, a filtration is defined by changing the density d of the network, from $0 \leq d \leq 1$. This yields a nested sequence of networks, in which increasing d leads to a more densely connected network. In neuroscience, It can be used to avoid arbitrary threshold/density choices, which are usually made in the field.

We can encode a network into a simplicial complex in several ways [17,29,31]. Here, we focus on building a simplicial complex only from the brain network's cliques, *i.e.*, we create the so-called clique complex of a brain network. In a network, a k-clique is a subset of the network with k all-to-all connected nodes. 0-clique corresponds to the empty set, 1-cliques correspond to nodes, 2-cliques to links, 3-cliques to triangles, *etc.*. In the clique complex, each $k+1$ clique is associated with a k-simplex. This choice for creating simplexes from cliques has the advantage that we can still use pairwise signal processing to create a simplicial complex from brain networks, such as in [23]. It is essential to mention that other strategies to build simplicial complexes beyond pairwise signal processing are still under development, such as applications using multivariate information theory together with tools from algebraic topology [1,5–7,22,36]. In our Jupyter Notebook [12], we provide the code to visualise the clique complex developed in [38] (Fig. 1).

The Euler Characteristic. The Euler characteristic is one example of topological invariants: the network properties that do not depend on a specific graph representation. We first introduce the Euler characteristic for polyhredra. Later, we translate this concept to brain networks. In 3-D convex polyhedra, the Euler characteristic is defined as the numbers of vertices minus edges plus faces. For convex polyhedra without cavities (holes in its shape), which are isomorphous to the sphere, the Euler characteristic is always two. If we take the cube and make a cavity, the Euler drops to zero as it is in the torus. If we make two cavities in a polyhedral (as in the bitorus), the Euler drops to minus two. We can understand that the Euler characteristic tells us something about a polyhedron's topology and its analogous surface. In other words, if we have a surface and we make a discrete representation of it (*e.g.*, a surface triangulation), its Euler characteristic is always the same, regardless of the way we do it. We can now generalise the definition of Euler characteristic to a simplicial complex in any dimension. Thus, the high dimensional version of the Euler characteristic is expressed by the alternate sum of the numbers $Cl_k(d)$ of the k-cliques (which are $(k-1)$-simplexes) present in the network's simplicial complex for a given value of the density threshold d:

$$\chi(d) = Cl_1 - Cl_2 + ... + Cl_n = \sum_{k=1}^{n}(-1)^k Cl_(k)(d).$$

Betti Numbers. Another set of topological invariants are the Betti numbers (β). Given that a simplicial complex is a high-dimensional structure, β_k counts the number of k-dimensional holes in the simplicial complex. These are topological invariants that correspond, for each $k \geq 0$, to the number of k-dimensional holes in the simplicial complex [47]. In a simplicial complex, there can be many of these k-holes and counting them provide the Betti number β, e.g., if β_2 is equal to five, there are 5 two-dimensional holes. The Euler characteristics of a simplicial complex can also be computed using the β via the following formula [17]:

$$\chi = \beta_0 - \beta_1 + \beta_2 - \ldots (-1)^{k_{\max}} \beta_{k_{\max}} = \sum_{k=0}^{k_{max}} (-1)^k \beta_k ,$$

where $k_{\max}$ the maximum dimension that we are computing the cycles.

Curvature. Curvature is a TDA metric that can link the global network properties described above to local features [20,38,44]. It allows us to compute topological invariants for the whole-brain set of vertices and understand the contribution of specific individual nodal, or subnetwork, geometric proprieties to the brain network's global properties. Several approaches to defining a curvature for networks are available [33,44], including some already used in neuroscientific investigations [38]. We illustrate the curvature approach linked to topological phase transitions, previously introduced for complex systems [20,33,45]. To compute the curvature, filtration is used to calculate the clique participation rank (*i.e.*, the number of k-cliques in which a vertex i participates for density d) [40], which we denote here by $Cl_{ik}(d)$. The curvature of the vertex based on the participation rank is then defined as:

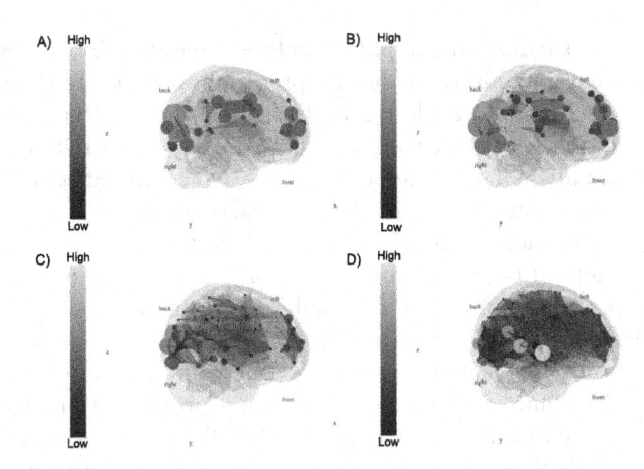

Fig. 1. Simplex 3-D visualisation. Here we visualise the number of 3-cliques (triangles) in a functional brain network as we increase the edge density d (0.01, 0.015, 0.02, and 0.025, from A to D). For higher densities, we have a more significant number of 3-cliques compared to smaller densities. The vertex colour indicates the clique participation rank. We used fMRI data from the 1000 Functional Connectome Project [8].

$$\kappa_i = \sum_{k=1}^{k_{\max}} (-1)^{k+1}, \frac{Cl_{ik}\ (d)}{k}$$

where $Cl_{ik} = 1$ since each vertex i participates in a single 1-clique (the vertex itself), and $k_{\max}$ the maximum number of vertices that are all-to-all connected in the network (see in Fig. 1 the participation in 3-cliques). We use the Gauss-Bonnet theorem for networks to link the local (nodal) curvature to the network's global properties (its Euler characteristic). Conversely, by summing up all the curvatures of the network across different thresholds, one can reach the alternate sum of the numbers Cl_k of k-cliques (a subgraph with k all-to-all connected vertices) present in the network's simplicial complex for a given density threshold $d \in [0, 1]$. By doing so, we also write the Euler characteristics as a sum of the curvature of all network vertices, *i.e.*,

$$\chi(d) = \sum_{i=1}^{N} k_i(d).$$

1.3 Discussion

This tutorial explains some of the primary metrics related to two network neuroscience branches - graph theory and TDA -, providing short theoretical backgrounds and code examples accompanied by a publicly available Jupyter Notebook, with a special section on visualisations of simplicial complexes and curvature computation in brain data. Here, we did not aim to provide a use-case report but rather a hands on computational resource. Finally, we would like to mention some relevant limitations in interpretation when using these metrics in connectivity-based data. Considering that rsfMRI data is often calculated as a temporal correlation between time series using Pearson's correlation coefficient, a bias on the number of triangles can emerge [46]. This affects TDA (where the impact depends on how high-order interactions are defined) and graph-theoretical metrics (such as the clustering coefficient), with networks based on this statistical method being automatically more clustered than random models. The proper way to determine and infer high-order interactions in the brain is an ongoing challenge in network neuroscience [1,5–7,22,36]. Moreover, it is essential to think about the computational cost. The computation of cliques falls in the clique problem, an NP (nonpolynomial time) problem; thus, listing cliques may require exponential time as the size of the cliques or networks grows [34]. What we can do for practical applications is to limit the clique size that can be reached by the algorithm, which determines the dimension of the simplicial complex in which the brain network is represented. This arbitrary constraint implies a theoretical simplification, limiting the space or the dimensionality in which we would analyse brain data. Another issue is that, to finish TDA computations in a realistic time frame, the researcher might need to establish a maximal threshold/density for convergence even after reducing the maximal clique size. Even though TDA approaches lead to substantial improvements in network science, apart from applications using the Mapper algorithm [37], the limitations mentioned above contribute to losing information on the data's shape. In conclusion, graph theory has been widely used in network neuroscience, but newer methods such as TDA are gaining momentum. To further improve the field, especially for users in the domain of clinical network neuroscience, it is imperative to make the computation of the developed metrics accessible and easy to comprehend and visualise. We hope to have facilitated the comprehension of some aspects of network and topological neuroscience, the computation and visualisation of some of its metrics.

References

1. Barbarossa, S., Sardellitti, S.: Topological signal processing over simplicial complexes. IEEE Trans. Sig. Process. **68**, 2992–3007 (2020). https://doi.org/10.1109/TSP.2020.2981920
2. Bassett, D.S., Bullmore, E.T.: Small-world brain networks revisited. Neuroscientist **23**(5), 499–516 (2017). https://doi.org/10.1177/1073858416667720
3. Bassett, D.S., Sporns, O.: Network neuroscience. Nat. Neurosci. **20**(3), 353 (2017). https://doi.org/10.1038/nn.4502
4. Battiston, F., et al.: Networks beyond pairwise interactions: structure and dynamics. Phys. Rep. **874**, 1–92 (2020). https://doi.org/10.1016/j.physrep.2020.05.004
5. Baudot, P.: The poincare-shannon machine: statistical physics and machine learning aspects of information cohomology. Entropy **21**(9), 881 (2019). https://doi.org/10.3390/e21090881
6. Baudot, P., Bennequin, D.: The homological nature of entropy. Entropy **17**(5), 3253–3318 (2015). https://doi.org/10.3390/e17053253

7. Baudot, P., et al.: Topological information data analysis. Entropy **21**(9), 869 (2019). https://doi.org/10.3390/e21090869
8. Biswal, B.B., et al.: Toward discovery science of human brain function. Proc. Nat. Acad. Sci. USA **107**(10), 4734–4739 (2010). https://doi.org/10.1073/pnas.0911855107
9. Bullmore, E., Sporns, O.: Complex brain networks: graph theoretical analysis of structural and functional systems. Nat. Rev. Neurosci. **10**(3), 186–198 (2009). https://doi.org/10.1038/nrn2575
10. Carlsson, G.: Topological methods for data modelling. Nat. Rev. Phys. **2**(12), 697–708 (2019). https://doi.org/10.1038/s42254-020-00249-3
11. Carlsson, G.: Topology and data. Bull. Am. Math. Soc. **46**(2), 255–308 (2009). https://doi.org/10.1090/S0273-0979-09-01249-X
12. Centeno, E., Santos, F.N., MultinetLAB: Notebook for Network and Topological Analysis in Neuroscience (2021). https://doi.org/10.5281/zenodo.4483651
13. Centeno, E.G.Z., et al.: A hands-on tutorial on network and topological neuroscience. bioRxiv. p. 2021.02.15.431255. https://doi.org/10.1101/2021.02.15.431255
14. Curto, C.: What can topology tell us about the neural code? Bull. Am. Math. Soc. **54**(1), 63–78 (2017). https://doi.org/10.1090/bull/1554
15. Curto, C., Itskov, V.: Cell groups reveal structure of stimulus space. PLOS Comput. Biol. **4**(10), e1000205 (2008). https://doi.org/10.1371/journal.pcbi.1000205
16. Do Carmo, M.P.: Differential Geometry of Curves and Surfaces: Revised and Updated Second Edition. Courier Dover Publications, New York (2016)
17. Edelsbrunner, H., Harer, J.: Computational topology: an introduction, vol. 69. 1st edn. American Mathematical Society, Providence, USA (2010). https://doi.org/10.1090/mbk/069
18. Expert, P., et al.: Editorial: topological neuroscience. Network Neurosci. **3**(3), 653–655 (2019). https://doi.org/10.1162/netne00096
19. Farahani, F.V., Karwowski, W., Lighthall, N.R.: Application of graph theory for identifying connectivity patterns in human brain networks: a systematic review. Front. Neurosci. **13**, 585 (2019). https://doi.org/10.3389/fnins.2019.00585
20. Farooq, H., et al.: Network curvature as a hallmark of brain structural connectivity. Nat. Commun. **10**(1), 4937 (2019). https://doi.org/10.1038/s41467-019-12915-x
21. Fornito, A., Zalesky, A., Bullmore, E.: Fundamentals of Brain Network Analysis, 1st edn. Academic Press, San Diego (2016)
22. Gatica, M., et al.: High-order interdependencies in the aging brain. bioRxiv (2020)
23. Giusti, C., Pastalkova, E., Curto, C., Itskov, V.: Clique topology reveals intrinsic geometric structure in neural correlations. Proc. Nat. Acad. Sci. USA **112**(44), 13455–13460 (2015). https://doi.org/10.1073/pnas.1506407112
24. Gross, J.L., Yellen, J.: Handbook of Graph Theory, 1st edn. CRC Press, Boca Raton (2003)
25. Hagberg, A., Swart, P., S Chult, D.: Exploring network structure, dynamics, and function using networkx. In: Varoquaux, G.T., Vaught, J.M. (ed.) Proceedings of the 7th Python in Science Conference (SciPy 2008), pp. 11–15 (2008)
26. Hallquist, M.N., Hillary, F.G.: Graph theory approaches to functional network organization in brain disorders: a critique for a brave new small-world. Network Neurosci. **3**(1), 1–26 (2018). https://doi.org/10.1162/netna00054
27. Jalili, M.: Functional brain networks: does the choice of dependency estimator and binarization method matter? Sci. Rep. **6**, 29780 (2016). https://doi.org/10.1038/srep29780

28. Kartun-Giles, A.P., Bianconi, G.: Beyond the clustering coefficient: a topological analysis of node neighbourhoods in complex networks. Chaos Solitons Fract. X **1**, 100004 (2019). https://doi.org/10.1016/j.csfx.2019.100004
29. Lambiotte, R., Rosvall, M., Scholtes, I.: From networks to optimal higher-order models of complex systems. Nat. Phys. **15**(4), 313–320 (2019). https://doi.org/10.1038/s41567-019-0459-y
30. Liu, T.T.: Noise contributions to the FMRI signal: an overview. Neuroimage **143**, 141–151 (2016). https://doi.org/10.1016/j.neuroimage.2016.09.008
31. Maletić, S., Rajković, M., Vasiljević, D.: Simplicial complexes of networks and their statistical properties. In: Bubak, M., van Albada, G.D., Dongarra, J., Sloot, P.M.A. (eds.) ICCS 2008. LNCS, vol. 5102, pp. 568–575. Springer, Heidelberg (2008). https://doi.org/10.1007/978-3-540-69387-1_65
32. Maria, C., Boissonnat, J.-D., Glisse, M., Yvinec, M.: The Gudhi library: simplicial complexes and persistent homology. In: Hong, H., Yap, C. (eds.) ICMS 2014. LNCS, vol. 8592, pp. 167–174. Springer, Heidelberg (2014). https://doi.org/10.1007/978-3-662-44199-2_28
33. Najman, L., Romon, P. (eds.): Modern Approaches to Discrete Curvature. LNM, vol. 2184. Springer, Cham (2017). https://doi.org/10.1007/978-3-319-58002-9
34. Pardalos, P.M., Xue, J.: The maximum clique problem. J. Glob. Optim. **4**(3), 301–328 (1994)
35. Petri, G., et al.: Homological scaffolds of brain functional networks. J. R. Soc. Interface **11**(101), 20140873 (2014). https://doi.org/10.1098/rsif.2014.0873
36. Rosas, F.E., et al.: Quantifying high-order interdependencies via multivariate extensions of the mutual information. Phys. Rev. E **100**(3), 032305 (2019). https://doi.org/10.1103/PhysRevE.100.032305
37. Saggar, M., et al.: Towards a new approach to reveal dynamical organization of the brain using topological data analysis. Nat. Commun. **9**(1), 1399 (2018). https://doi.org/10.1038/s41467-018-03664-4
38. Santos, F.A.N., et al.: Topological phase transitions in functional brain networks. Phys. Rev. E **100**(3–1), 032414 (2019). https://doi.org/10.1103/PhysRevE.100.032414
39. Singh, G., et al.: Topological analysis of population activity in visual cortex. J. Vis. **8**(8), 11 (2008). https://doi.org/10.1167/8.8.11
40. Sizemore, A.E., Giusti, C., Kahn, A., Vettel, J.M., Betzel, R.F., Bassett, D.S.: Cliques and cavities in the human connectome. J. Comput. Neurosci. **44**(1), 115–145 (2017). https://doi.org/10.1007/s10827-017-0672-6
41. Sizemore Blevins, A., Bassett, D.S.: Reorderability of node-filtered order complexes. Phys. Rev. E **101**(5–1), 052311 (2020). https://doi.org/10.1103/PhysRevE.101.052311
42. Sporns, O.: Graph theory methods: applications in brain networks. Dialogues Clin. Neurosci. **20**(2), 111–121 (2018)
43. Watts, D.J., Strogatz, S.H.: Collective dynamics of 'small-world' networks. Nature **393**(6684), 440 (1998). https://doi.org/10.1038/30918
44. Weber, M., et al.: Curvature-based methods for brain network analysis. arXiv preprint arXiv:1707.00180 (2017)
45. Wu, Z., et al.: Emergent complex network geometry. Sci. Rep. **5**, 10073 (2015). https://doi.org/10.1038/srep10073

46. Zalesky, A., Fornito, A., Bullmore, E.: On the use of correlation as a measure of network connectivity. Neuroimage **60**(4), 2096–2106 (2012). https://doi.org/10.1016/j.neuroimage.2012.02.001
47. Zomorodian, A.J.: Topology for Computing, vol. 16, 1st edn. Cambridge University Press, New York (2005)

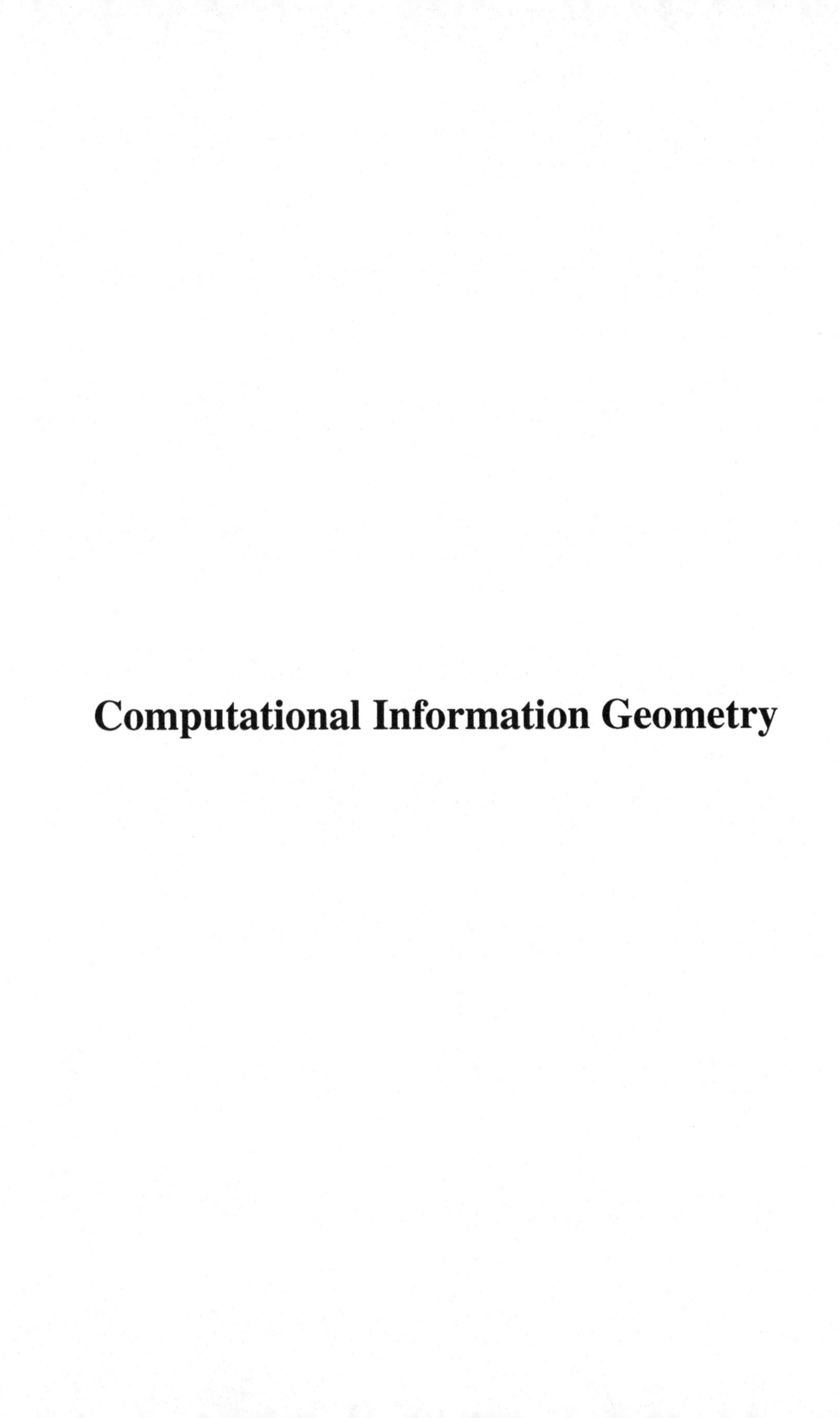

Computational Information Geometry

Computing Statistical Divergences with Sigma Points

Frank Nielsen[1(✉)] and Richard Nock[2]

[1] Sony Computer Science Laboratories Inc., Tokyo, Japan
Frank.Nielsen@acm.org
[2] Data61, CSIRO, Sydney, Australia
Richard.Nock@data61.csiro.au
https://FrankNielsen.github.io/

Abstract. We show how to bypass the integral calculation and express the Kullback-Leibler divergence between any two densities of an exponential family using a finite sum of logarithms of density ratio evaluated at sigma points. This result allows us to characterize the exact error of Monte Carlo estimators. We then extend the sigma point method to the calculations of q-divergences between densities of a deformed q-exponential family.

Keywords: Exponential family · Kullback-Leibler divergence · Deformed exponential family · q-divergence · Unscented transform

1 Introduction

Let $(\mathcal{X}, \mathcal{F}, \mu)$ be a measure space [6] with sample space $\mathcal{X}$, σ-algebra of events $\mathcal{F}$, and positive measure μ (e.g., Lebesgue measure or counting measure). The Kullback-Leibler divergence [15] (KLD), also called the relative entropy [9], between two probability measures P and Q dominated by μ with respective Radon-Nikodym densities $p = \frac{\mathrm{d}P}{\mathrm{d}\mu}$ and $q = \frac{\mathrm{d}Q}{\mathrm{d}\mu}$ is defined by:

$$D_{\mathrm{KL}}[p:q] := \int_{\mathcal{X}} p(x) \log \frac{p(x)}{q(x)} \mathrm{d}\mu(x) = E_p\left[\log\left(\frac{p(x)}{q(x)}\right)\right] \neq D_{\mathrm{KL}}[q:p].$$

The definite integral may be infinite (e.g., the KLD between a standard Cauchy density and a standard normal density) or finite (e.g., the KLD between a standard normal density and a standard Cauchy density). The KLD may admit a closed-form formula or not depending on its density arguments (see Risch pseudo-algorithm [23] in symbolic computing which depends on an oracle). When the KLD is not available in closed-form, it is usually estimated by Monte Carlo sampling using a set $\mathcal{S}_m = \{x_1, \ldots, x_m\}$ of m i.i.d. variates from $X \sim p(x)$ as follows $\hat{D}_{\mathrm{KL}}^{\mathcal{S}_m}[p:q] = \frac{1}{m}\sum_{i=1}^m \log\left(\frac{p(x_i)}{q(x_i)}\right)$. Under mild conditions [16], this Monte Carlo estimator $\hat{D}_{\mathrm{KL}}[p:q]$ is consistent: $\lim_{m\to\infty} \hat{D}_{\mathrm{KL}}^{\mathcal{S}_m}[p:q] = D_{\mathrm{KL}}[p:q]$.

F. Nielsen and F. Barbaresco (Eds.): GSI 2021, LNCS 12829, pp. 677–684, 2021.
https://doi.org/10.1007/978-3-030-80209-7_72

When the densities p and q belong to a common parametric family $\mathcal{E} = \{p_\theta : \theta \in \Theta\}$ of densities, i.e., $p = p_{\theta_1}$ and $q = p_{\theta_2}$, the KLD yields an *equivalent* parameter divergence: $D_{\mathrm{KL}}^{\mathcal{E}}(\theta_1 : \theta_2) := D_{\mathrm{KL}}(p_{\theta_1} : p_{\theta_2})$. In the reminder, we use the notation $[\cdot : \cdot]$ to enclose the arguments of a statistical divergence and the notation $(\cdot : \cdot)$ to enclose the parameters of a parameter divergence. The ':' notation indicates that the distance may be asymmetric.

A statistical divergence is said separable when it can be written as the definite integral of a scalar divergence. For example, the f-divergences [10] $I_f[p : q] := \int_{\mathcal{X}} p(x) f\left(\frac{q(x)}{p(x)}\right) \mathrm{d}\mu(x)$, induced by a convex function f (strictly convex at 1 with $f(1) = 0$) are separable but the Wasserstein optimal transport distances [22] are not separable. We say that a separable statistical divergence $D[p_{\theta_1} : p_{\theta_2}]$ between two densities p_{θ_1} and p_{θ_2} of a parametric family can be written equivalently using *sigma points* if there exists a finite set $\{\omega_1, \ldots, \omega_s\}$ of s points (called the sigma points) and a bivariate scalar function h so that

$$D[p_{\theta_1} : p_{\theta_2}] := \int_{\mathcal{X}} d(p_{\theta_1}(x) : p_{\theta_2}(x)) \mathrm{d}\mu(x) = \frac{1}{s} \sum_{i=1}^{s} h(p_{\theta_1}(\omega_i), p_{\theta_2}(\omega_i)).$$

The term sigma points is borrowed from the *unscented transform* [14], and conveys the idea that the integral can be calculated exactly using a discrete sum $\sum$ (Sigma), hence the terminology. Sigma points have also been used to approximate the KLD between mixture of Gaussians [13].

The paper is organized as follows: In Sect. 2, we recall the basics of exponential families and show how the KLD between two densities of an exponential family can be calculated equivalently from a reverse Bregman divergence induced by the cumulant function of the family. In Sect. 3, we show how to express the KLD between densities of an exponential family using sigma points (Theorem 1), and discuss on how this results allows one to quantify the exact error of Monte Carlo estimators (Theorem 2). Section 4 extends the method of sigma points to q-divergences between densities of deformed q-exponential families (Theorem 3).

2 Exponential Families and Kullback-Leibler Divergence

A natural exponential family [5] (NEF) is a set of parametric densities

$$\mathcal{E} = \left\{p_\theta(x) = \exp\left(\theta^\top t(x) - F(\theta) + k(x)\right) \ : \ \theta \in \Theta\right\}, \tag{1}$$

admitting sufficient statistics $t(x)$, where $k(x)$ an auxiliary measure carrier term. The normalizing function

$$F(\theta) := \log\left(\int_{\mathcal{X}} \exp(\theta^\top t(x)) \mathrm{d}\mu(x)\right), \tag{2}$$

is called the cumulant function [17] (or free energy, or log-partition function) and is a strictly smooth convex real analytic function [5] defined on the open convex natural parameter space Θ. Let D denote the dimension of the parameter space Θ (i.e., the order of the exponential family) and d the dimension of the sample space $\mathcal{X}$. We consider full *regular* exponential families [5] so that Θ is a non-empty open convex domain and $t(x)$ is a minimal sufficient statistic [11].

Many common families of distributions [19] $\{p_\lambda(x)\ \lambda \in \Lambda\}$ (e.g., Gaussian family, Beta family, etc.) are exponential families in disguise after reparameterization: $p_\lambda(x) = p_{\theta(\lambda)}(x)$. We call parameter λ the ordinary parameter and get the equivalent natural parameter as $\theta(\lambda) \in \Theta$. A dual parameterization of exponential families is the *moment parameter* [5]: $\eta = E_{p_\theta}[t(x)] = \nabla F(\theta)$.

When densities $p = p_{\theta_1}$ and $q = p_{\theta_2}$ both belong to the same exponential family, we have the following well-known closed-form expressions [3] for the KLD:

$$D_{\mathrm{KL}}[p_{\theta_1} : p_{\theta_2}] = {B_F}^*(\theta_1 : \theta_2) = B_F(\theta_2 : \theta_1), \tag{3}$$

where D^* indicates the reverse divergence $D^*(\theta_1 : \theta_2) := D(\theta_2 : \theta_1)$, and B_F is the Bregman divergence [7] induced by a strictly convex and differentiable real-valued function F:

$$B_F(\theta_1 : \theta_2) := F(\theta_1) - F(\theta_2) - (\theta_1 - \theta_2)^\top \nabla F(\theta_2). \tag{4}$$

There exists a bijection between regular exponential families and a class of Bregman divergences called regular Bregman divergences [4] highlighting the role of the Legendre transform $F^*(\eta) := \sup_{\theta\in\Theta} \sum_{i=1}^D \theta_i \eta_i - F(\theta)$ of F. Notice that although Eq. 3 expresses the KLD as an equivalent (reverse) Bregman divergence thus bypassing the definite integral calculation (or infinite sums, e.g., Poisson case), the cumulant function F may not always be available in closed-form (e.g., polynomial exponential families [20]), nor computationally intractable (e.g., restricted Boltzmann machines [1] on discrete sample spaces).

3 Sigma Points for the Kullback-Leibler Divergence

Let us observe that the generator F of a Bregman divergence B_F (Eq. 4) is defined up to an affine term. That is, generator $F(\theta)$ and generator $G(\theta) = F(\theta) + a^\top\theta + b$ for $a \in \mathbb{R}^D$ and $b \in \mathbb{R}$ yields the same Bregman divergence $B_F = B_G$. Thus we may express the KLD equivalent Bregman divergence of Eq. 3 with the following generator

$$F_\omega(\theta) := -\log(p_\theta(\omega)) = F(\theta) - \underbrace{(\theta^\top t(\omega) + k(\omega))}_{\text{affine term}} \tag{5}$$

for *any* prescribed $\omega \in \mathcal{X}$ instead of the cumulant function F of Eq. 2. We get

$$D_{\mathrm{KL}}[p_{\lambda_1} : p_{\lambda_2}] = B_{F_\omega}(\theta(\lambda_2) : \theta(\lambda_1)), \tag{6}$$

$$= \log\left(\frac{p_{\lambda_1}(\omega)}{p_{\lambda_2}(\omega)}\right) - (\theta(\lambda_2) - \theta(\lambda_1))^\top \nabla_\theta F_\omega(\theta(\lambda_1)). \tag{7}$$

Since we have $\nabla F_\omega(\theta(\lambda_1)) = -(t(\omega) - \nabla F(\theta(\lambda_1)))$, it follows that:

$$D_{\mathrm{KL}}[p_{\lambda_1} : p_{\lambda_2}] = \log\left(\frac{p_{\lambda_1}(\omega)}{p_{\lambda_2}(\omega)}\right) + (\theta(\lambda_2) - \theta(\lambda_1))^\top (t(\omega) - \nabla F(\theta(\lambda_1))), \quad \forall \omega \in \mathcal{X}. \tag{8}$$

When the exponential family has order $D = 1$, we can choose a single sigma point ω such that $t(\omega) = F'(\theta(\lambda_1))$ as illustrated in the following example:

Example 1. Consider the exponential family of Rayleigh distributions $\mathcal{E} = \left\{p_\lambda(x) = \frac{x}{\lambda^2}\exp\left(-\frac{x^2}{2\lambda^2}\right)\right\}$, for $\mathcal{X} = [0, \infty)$ with $\theta = -\frac{1}{2\lambda^2}$, $t(x) = x^2$ and $\nabla F(\theta(\lambda)) = 2\lambda^2$. Let us choose $\omega^2 = 2\lambda_1^2$ (i.e., $\omega = \lambda_1\sqrt{2}$). such that $(\theta(\lambda_2) - \theta(\lambda_1))^\top (t(\omega) - \nabla F(\theta(\lambda_1))) = 0$. We get $D_{\mathrm{KL}}[p_{\lambda_1} : p_{\lambda_2}] = \log\left(\frac{p_{\lambda_1}\left(\frac{1}{\lambda_1}\right)}{p_{\lambda_2}\left(\frac{1}{\lambda_1}\right)}\right) = 2\log\left(\frac{\lambda_2}{\lambda_1}\right) + \frac{\lambda_1^2}{\lambda_2^2} - 1$.

Let us notice that ω may not belong to the support but the closed convex hull of the support (e.g., Poisson family). Consider now exponential families of order $D > 1$. Since Eq. 8 holds for any $\omega \in \Omega$, let us choose s values for ω (i.e., $\omega_1, \ldots, \omega_s$), and *average* Eq. 8 for these s values. We obtain

$$D_{\mathrm{KL}}[p_{\lambda_1} : p_{\lambda_2}] = \frac{1}{s}\sum_{i=1}^{s} \log\left(\frac{p_{\lambda_1}(\omega_i)}{p_{\lambda_2}(\omega_i)}\right) + (\theta(\lambda_2) - \theta(\lambda_1))^\top \left(\frac{1}{s}\sum_{i=1}^{s} t(\omega_i) - \nabla F(\theta(\lambda_1))\right). \tag{9}$$

Thus we get the following theorem:

Theorem 1. *The Kullback-Leibler divergence between two densities p_{λ_1} and p_{λ_2} belonging to a full regular exponential family $\mathcal{E}$ of order D can be expressed as the average sum of logarithms of density ratios:*

$$D_{\mathrm{KL}}[p_{\lambda_1} : p_{\lambda_2}] = \frac{1}{s}\sum_{i=1}^{s} \log\left(\frac{p_{\lambda_1}(\omega_i)}{p_{\lambda_2}(\omega_i)}\right),$$

where $\omega_1, \ldots, \omega_s$ are $s \leq D + 1$ distinct points of $\mathcal{X}$ chosen such that $\frac{1}{s}\sum_{i=1}^{s} t(\omega_i) = E_{p_{\lambda_1}}[t(x)]$.

The bound $s \leq D + 1$ follows from Carathéodory's theorem [8]. Notice that the sigma points depend on λ_1: $\omega_i = \omega_i(\lambda_1)$. Here, we inverted the role of the "observations" $\omega_1, \ldots, \omega_s$ and parameter λ (or θ) in the maximum likelihood estimation's normal equation by requiring $\frac{1}{s}\sum_{i=1}^{s} t(\omega_i) = E_{p_{\lambda_1}}[t(x)] = \nabla F(\theta_1)$.

Example 2. Consider the full family of d-dimensional normal densities. We shall use $2d$ vectors ω_i to express $D_{\mathrm{KL}}[p_{\mu_1,\Sigma_1} : p_{\mu_2,\Sigma_2}]$ as follows:

$$\omega_i = \mu_1 - \sqrt{d\lambda_i}e_i, \omega_{i+d} = \mu_1 + \sqrt{d\lambda_i}e_i, \quad i \in \{1, \ldots, d\},$$

where the λ_i's are the eigenvalues of Σ_1 with the e_i's the corresponding eigenvectors. It can be checked that $\frac{1}{2d}\sum_{i=1}^{2d} \omega_i = E_{p_{\lambda_1}}[x] = \mu_1$ and $\frac{1}{2d}\sum_{i=1}^{2d} \omega_i\omega_i^\top =$

$\mu_1\mu_1^\top + \Sigma_1$. Moreover, we have $\sqrt{\lambda_i}e_i = [\sqrt{\Sigma_1}]_{\cdot,i}$, the i-th column of the square root of the covariance matrix of Σ_1. Thus it follows that

$$D_{\mathrm{KL}}[p_{\mu_1,\Sigma_1} : p_{\mu_2,\Sigma_2}]$$
$$= \frac{1}{2d}\sum_{i=1}^{d}\left(\log\left(\frac{p_{\mu_1,\Sigma_1}(\mu_1 - [\sqrt{d\Sigma_1}]_{\cdot,i})}{p_{\mu_2,\Sigma_2}(\mu_1 - [\sqrt{d\Sigma_1}]_{\cdot,i})}\right) + \log\left(\frac{p_{\mu_1,\Sigma_1}(\mu_1 + [\sqrt{d\Sigma_1}]_{\cdot,i})}{p_{\mu_2,\Sigma_2}(\mu_1 + [\sqrt{d\Sigma_1}]_{\cdot,i})}\right)\right),$$

where $[\sqrt{d\Sigma_1}]_{\cdot,i} = \sqrt{\lambda_i}e_i$ denotes the vector extracted from the i-th column of the square root matrix of $d\Sigma_1$. This formula matches the ordinary formula for the KLD between the two Gaussian densities p_{μ_1,Σ_1} and p_{μ_2,Σ_2}:

$$D_{\mathrm{KL}}[p_{\mu_1,\Sigma_1} : p_{\mu_2,\Sigma_2}] = \frac{1}{2}\left((\mu_2 - \mu_1)^\top \Sigma_2^{-1}(\mu_2 - \mu_1) + \operatorname{tr}\left(\Sigma_2\Sigma_1^{-1}\right) + \log\left(\frac{|\Sigma_2|}{|\Sigma_1|}\right) - d\right).$$

See [21] for other illustrating examples.

3.1 Assessing the Errors of Monte Carlo Estimators

The Monte Carlo (MC) stochastic approximation [16] of the KLD is given by:

$$\widetilde{D}_{\mathrm{KL}}^{\mathcal{S}_m}[p_{\lambda_1} : p_{\lambda_2}] = \frac{1}{m}\sum_{i=1}^{m}\log\frac{p_{\lambda_1}(x_i)}{p_{\lambda_2}(x_i)}, \tag{10}$$

where $\mathcal{S}_m = \{x_1, \ldots, x_m\}$ is a set of m independent and identically distributed variates from p_{λ_1}. Notice that the MC estimators are also average sums of logarithms of density ratios. Thus we can quantify exactly the error when we perform MC stochastic integration of the KLD between densities of an exponential family from Eq. 9. We have:

Theorem 2. *The error* $\left|\widetilde{D}_{\mathrm{KL}}^{\mathcal{S}_m}[p_{\lambda_1} : p_{\lambda_2}] - D_{\mathrm{KL}}[p_{\lambda_1} : p_{\lambda_2}]\right|$ *of the Monte Carlo estimation of the KLD between two densities* p_{λ_1} *and* p_{λ_2} *of an exponential family is*

$$\left|(\theta(\lambda_2) - \theta(\lambda_1))^\top\left(\frac{1}{m}\sum_{i=1}^{m}t(x_i) - E_{p_{\lambda_1}}[t(x)]\right)\right|.$$

When $m \to \infty$, we have $\frac{1}{m}\sum_{i=1}^{m}t(x_i) \to E_{p_{\lambda_1}}[t(x)]$, and the MC error tends to zero, hence proving the efficiency of the MC estimator (under mild conditions [16]). If we know $\eta_1 = \nabla F(\theta_1) = E_{p_{\lambda_1}}[t(x)]$ and the log density ratios, then we can calculate both the MC error and the KL divergence using Eq. 9.

4 Sigma Points for the q-divergences

We extend the sigma point method to the computation of q-divergences between densities of a deformed q-exponential family [1,2,12]. For $q > 0$ and $q \neq 1$, let

$\log_q(u) = \frac{1}{1-q}(u^{1-q} - 1)$ (for $u > 0$) be the deformed logarithm and $\exp_q(u) = (1 + (1-q)u)^{\frac{1}{1-q}}$ (for $u > -\frac{1}{1-q}$) be the reciprocal deformed exponential. When $q \to 1$, we have $\log_q(u) \to \log(u)$ and $\exp_q(u) \to \exp(u)$. The q-exponential family is defined as the set of densities such that $\log_q p_\theta(x) := \sum_{i=1}^D \theta_i t_i(x) - F_q(\theta)$, where $F_q(\theta)$ is the normalizer function (q-free energy) which is proven to be strictly convex and differentiable [2]. Thus we can define a corresponding Bregman divergence [1] $B_{F_q}(\theta_1 : \theta_2)$ between parameters θ_1 and θ_2 corresponding to two densities p_{θ_1} and p_{θ_2}. Then we can reconstruct the corresponding statistical divergence D_q, called the q-divergence, so that $D_q[p_{\theta_1} : p_{\theta_2}] = B_{F_q}(\theta_2 : \theta_1)$ where:

$$D_q[p : r] := \frac{1}{(1-q)h_q(p)}\left(1 - \int p^q(x) r^{1-q}(x) \mathrm{d}\mu(x)\right) = B_{F_q}(\theta_1 : \theta_2),$$

where $h_q(p) := \int p^q(x)\mathrm{d}\mu(x)$. We can also rewrite the q-divergence as $D_q[p : r] = {}^qE_p[\log_q p(x) - \log_q r(x)]$, where ${}^qE_p$ denotes the q-expectation:

$${}^qE_p[f(x)] := \frac{1}{h_q(p)} \int p(x)^q f(x) \mathrm{d}\mu(x) = \nabla F_q(\theta) = \eta.$$

Let us use the equivalent Bregman generator $F_{q,\omega}(\theta) = -\log_q p_\theta(\omega) = F_q(\theta) - \theta^\top t(\omega) - k(\omega)$ instead of F_q (with $\nabla F_{q,\omega}(\theta) = \nabla F_q(\theta) - t(\omega)$), and choose at most $s \leq D+1$ sigma points ω_i's so that $\frac{1}{s}\sum_{i=1}^s t(\omega_i) = \nabla F_q(\theta_1)$. We get the following theorem:

Theorem 3. *The q-divergence for $q \neq 1$ between two densities p_{λ_1} and p_{λ_2} belonging to a q-exponential family $\mathcal{E}$ of order D can be expressed as the average sum of difference of q-logarithms:*

$$D_q[p_{\lambda_1} : p_{\lambda_2}] = \frac{1}{s}\sum_{i=1}^s \left(\log_q p_{\lambda_1}(\omega_i) - \log_q p_{\lambda_2}(\omega_i)\right)$$

where $\omega_1, \ldots, \omega_s$ are $s \leq D+1$ distinct samples of $\mathcal{X}$ chosen such that $\frac{1}{s}\sum_{i=1}^s t(\omega_i) = \nabla F_q(\theta_1) = {}^qE_{p_{\lambda_1}}[t(x)]$.

Example 3. Consider the family $\mathcal{C}$ of Cauchy distributions (a q-Gaussian family [1] for $q = 2$): $\mathcal{C} := \left\{p_\lambda(x) := \frac{s}{\pi(s^2+(x-l)^2)}, \quad \lambda := (l, s) \in \mathbb{R} \times \mathbb{R}_+\right\}$, where l is the location parameter (median) and s is the scale parameter (probable error). The deformed 2-logarithm is $\log_{\mathcal{C}}(u) := 1 - \frac{1}{u}$ (for $u \neq 0$) and its reciprocal 2-exponential function is $\exp_{\mathcal{C}}(u) := \frac{1}{1-u}$ (for $u \neq 1$). The Cauchy probability density can be factorized [18] as a 2-exponential family $p_\theta(x) = \exp_{\mathcal{C}}(\theta^\top x - F_2(\theta))$ where θ denotes the 2D natural parameters:

$$\log_{\mathcal{C}}(p_\theta(x)) = \underbrace{\left(2\pi\frac{l}{s}\right)x + \left(-\frac{\pi}{s}\right)x^2}_{\theta^\top t(x)} - \underbrace{\left(\pi s + \pi\frac{l^2}{s} - 1\right)}_{F(\theta)}. \tag{11}$$

Thus the natural parameter is $\theta(l,s) = (\theta_1,\theta_2) = \left(2\pi\frac{l}{s}, -\frac{\pi}{s}\right) \in \Theta = \mathbb{R}\times\mathbb{R}_-$ for the sufficient statistics $t(x) = (t_1(x) = x, t_2(x) = x^2)$, and the deformed log-normalizer (free energy) is $F_C(\theta) = -\frac{\pi^2}{\theta_2} - \frac{\theta_1^2}{4\theta_2} - 1$. The gradient of the free energy is $\eta = \nabla F_C(\theta) = \begin{bmatrix} -\frac{\theta_1}{2\theta_2} \\ \frac{\pi^2}{\theta_2^2} + \frac{\theta_1^2}{4\theta_2^2} \end{bmatrix} = \begin{bmatrix} s \\ s^2 + l^2 \end{bmatrix}$. The 2-divergence D_C between two Cauchy densities p_{λ_1} and p_{λ_2} is

$$D_C[p_{\lambda_1} : p_{\lambda_2}] := \frac{1}{\int p_{\lambda_1}^2(x)\mathrm{d}x}\left(\int \frac{p_{\lambda_1}^2(x)}{p_{\lambda_2}(x)}\mathrm{d}x - 1\right), \tag{12}$$

$$= \frac{\pi}{s_2}\|\lambda_1 - \lambda_2\|^2 = B_{F_C}(\theta(\lambda_2) : \theta(\lambda_1)), \tag{13}$$

and amounts to an equivalent Bregman divergence on swapped parameters.

Using a computer algebra system, we calculate the definite integrals $h_2(p_\lambda) = \int_{\mathbb{R}} p_\lambda^2(x)\mathrm{d}x = \frac{1}{2\pi s}$, $\int_{\mathbb{R}} p_\lambda^2(x)x\mathrm{d}x = \frac{1}{2\pi}\frac{l}{s}$ and $\int_{\mathbb{R}} p_\lambda^2(x)x^2\mathrm{d}x = \frac{l^2+s^2}{2\pi s}$. The 2-escort distributions are $\hat{p}_\lambda := \frac{1}{h_2(p_\lambda)}p_\lambda^2(x)$. Thus we have the following 2-expectations of the sufficient statistics ${}^2E_{p_\lambda}[x] = E_{\hat{p}_\lambda}[x] = l$ and ${}^2E_{p_\lambda}[x^2] = E_{\hat{p}_\lambda}[x^2] = l^2 + s^2$. Notice that the first and second moments for the Cauchy distribution are undefined but that the first 2-expectation moments for the Cauchy densities are well-defined and finite. Let us choose two sigma points ω_1 and ω_2 by solving the system $\frac{1}{2}(\omega_1 + \omega_2) = a = l$ and $\frac{1}{2}(\omega_1^2 + \omega_2^2) = b = l^2 + s^2$. We find $\omega_1 = a - \sqrt{b - a^2} = l - s$ and $\omega_2 = a + \sqrt{b - a^2} = l + s$. For q-Gaussians [24] (or q-normals) with $\lambda = (\mu, \sigma)$ (including the Cauchy family for $q = 2$), we have ${}^qE_{p_\lambda}[x] = l$ and ${}^qE_{p_\lambda}[x^2] = l^2 + s^2$. Thus we check that we have

$$D_q[p_{\mu_1,\sigma_1} : p_{\mu_2,\sigma_2}] = \frac{1}{2}\left(\log_q p_{\mu_1,\sigma_1}(\mu_1 - \sigma_1) - \log_q p_{\mu_2,\sigma_2}(\mu_1 - \sigma_1)\right.$$
$$\left. + \log_q p_{\mu_1,\sigma_1}(\mu_1 + \sigma_1) - \log_q p_{\mu_2,\sigma_2}(\mu_1 + \sigma_1)\right).$$

Similarly, the χ-divergence between densities of a χ-deformed exponential family [1] can be calculated using sigma points (χ-deformed exponential families include the q-exponential families). See [21] for additional results.

References

1. Amari, S.I.: Information Geometry and Its Applications. Applied Mathematical Sciences. Springer, Japan (2016). https://doi.org/10.1007/978-4-431-55978-8
2. Amari, S.I., Ohara, A.: Geometry of q-exponential family of probability distributions. Entropy **13**(6), 1170–1185 (2011)
3. Azoury, K.S., Warmuth, M.K.: Relative loss bounds for on-line density estimation with the exponential family of distributions. Mach. Learn. **43**(3), 211–246 (2001)
4. Banerjee, A., Merugu, S., Dhillon, I.S., Ghosh, J.: Clustering with Bregman divergences. J. Mach. Learn. Res. **6**, 1705–1749 (2005)
5. Barndorff-Nielsen, O.: Information and Exponential Families. John Wiley & Sons, Hoboken (2014)

6. Billingsley, P.: Probability and Measure, 3rd edn. Wiley, New York (1995)
7. Bregman, L.M.: The relaxation method of finding the common point of convex sets and its application to the solution of problems in convex programming. USSR Comput. Math. Math. Phys. **7**(3), 200–217 (1967)
8. Carathéodory, C.: Über den Variabilitätsbereich der Koeffizienten von Potenzreihen, die gegebene Werte nicht annehmen. Mathematische Annalen **64**(1), 95–115 (1907)
9. Cover, T.M., Thomas, J.A.: Elements of Information Theory. John Wiley & Sons, Hoboken (2012)
10. Csiszár, I.: Information-type measures of difference of probability distributions and indirect observation. Studia Scientiarum Mathematicarum Hungarica **2**, 229–318 (1967)
11. Del Castillo, J.: The singly truncated normal distribution: a non-steep exponential family. Ann. Inst. Stat. Math. **46**(1), 57–66 (1994)
12. Gayen, A., Kumar, M.A.: Projection theorems and estimating equations for power-law models. J. Multivariate Anal. **184**, 104734 (2021)
13. Goldberger, J., Greenspan, H.K., Dreyfuss, J.: Simplifying mixture models using the unscented transform. IEEE Trans. Pattern Anal. Mach. Intell. **30**(8), 1496–1502 (2008)
14. Julier, S., Uhlmann, J., Durrant-Whyte, H.F.: A new method for the nonlinear transformation of means and covariances in filters and estimators. IEEE Trans. Autom. Control **45**(3), 477–482 (2000)
15. Kullback, S., Leibler, R.A.: On information and sufficiency. Ann. Math. Stat. **22**(1), 79–86 (1951)
16. Lemieux, C.: Monte Carlo and quasi-Monte Carlo sampling. Springer Science & Business Media, New York (2009). https://doi.org/10.1007/978-0-387-78165-5
17. Nielsen, F.: An elementary introduction to information geometry. Entropy **22**(10), 1100 (2020)
18. Nielsen, F.: On Voronoi diagrams on the information-geometric Cauchy manifolds. Entropy **22**(7), 713 (2020)
19. Nielsen, F., Garcia, V.: Statistical exponential families: A digest with flash cards. Technical Report. arXiv:0911.4863 (2009)
20. Nielsen, F., Nock, R.: Patch matching with polynomial exponential families and projective divergences. In: Amsaleg, L., Houle, M.E., Schubert, E. (eds.) SISAP 2016. LNCS, vol. 9939, pp. 109–116. Springer, Cham (2016). https://doi.org/10.1007/978-3-319-46759-7_8
21. Nielsen, F., Nock, R.: Cumulant-free closed-form formulas for some common (dis)similarities between densities of an exponential family. arXiv:2003.02469 (2020)
22. Peyré, G., Cuturi, M.: Computational optimal transport: with applications to data science. Found. Trends® Mach. Learn. **11**(5–6), 355–607 (2019)
23. Risch, R.H.: The solution of the problem of integration in finite terms. Bull. Am. Math. Soc. **76**(3), 605–608 (1970)
24. Tanaya, D., Tanaka, M., Matsuzoe, H.: Notes on geometry of q-normal distributions. In: Recent Progress in Differential Geometry and Its Related Fields, pp. 137–149. World Scientific (2012)

Remarks on Laplacian of Graphical Models in Various Graphs

Tomasz Skalski[1,2(✉)]

[1] Wrocław University of Science and Technology, Wybrzeże Wyspiańskiego 27, 50-370 Wrocław, Poland
tomasz.skalski@pwr.edu.pl
[2] LAREMA, Université d'Angers, Angers, France

Abstract. This paper discusses the connection between inverses of graph Laplacians with the diagonal increased by a positive vector and the covariance matrices of associated Gaussian graphical models.

Afterwards we study the way of modification of a graph Laplacian on the case of daisy graphs.

Keywords: Graphical models · Laplacian of a graph · Daisy graphs

1 Introduction

Let $G = (V, E, C)$ be a simple undirected graph, where $V = \{1, 2, \ldots, n\}$ is a set of vertices, $E \subset \binom{n}{2}$ is a set of edges and $C \subset V$ is a set of source vertices, which will be called later as a "root set" or a "root". As a degree of a vertex deg(v) we treat a number of its neighbours.

In our convention a graph Laplacian is defined as $L(G) = \{l_{i,j}\}_{1 \leq i,j \leq n}$ with $l_{i,i} = deg(i)$, $l_{i,j} = -1$ if $\{i, j\}$ is an edge in G and zeroes otherwise, cf. e.g. [2].

Note that the graph Laplacian is a singular matrix, since its entries in each row (and each column) sum up to 0. Therefore we introduce an augmented graph Laplacian by adding 1 to every entry $l_{c,c}$ corresponding to $c \in C$. In other words,

$$L^*(G) := L(G) + E_c$$

with E_c being a square matrix with 1 in (c, c) and zero outside. We define $L^*(G) = \{l^*_{i,j}\}_{1 \leq i,j \leq n}$ with $l^*_{i,j} = l_{i,j} + \mathbb{1}\{i = j \in C\}$

2 Trees

Let $T = (V, E)$ be an undirected tree. We may orient it in a following way: Choose one root vertex $C = \{c\}$. Then we orient every edge in a direction

The author is grateful to Hélène Massam (1949–2020) for her comments and fruitful discussion in scope of this work and he dedicates this work to her. The author owes thanks to the reviewers of the GSI conference and to Bartosz Kołodziejek for their comments and remarks. This research was supported by a French Government Scholarship.

F. Nielsen and F. Barbaresco (Eds.): GSI 2021, LNCS 12829, pp. 685–692, 2021.
https://doi.org/10.1007/978-3-030-80209-7_73

from c. Following this method we may induce a partial order $\leq$ on the set of vertices such that $v \leq v'$ if and only if there exists a directed path from v' to v.

For every vertex $v \in V$ we define its ancestry $\mathrm{AN}(v) := \{w \in V : v \leq w\}$ as a set of vertices in a unique path from v to c. Note that both the partial order and AN depend strictly on the choice of c.

Now consider an n-dimensional Gaussian random variable $(X_1, \ldots, X_n)$ with a covariance matrix $\Sigma = \{\sigma_{i,j}\}_{1\leq i,j\leq n}$ such that

$$\sigma_{i,j} = |\mathrm{AN}(i) \cap \mathrm{AN}(j)|$$

for every $1 \leq i, j \leq n$.

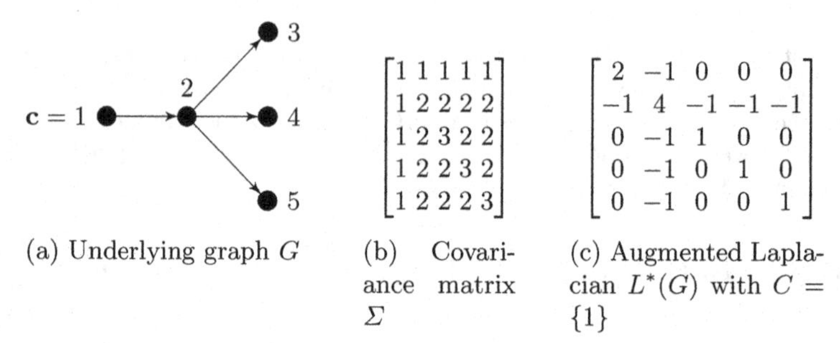

(a) Underlying graph G (b) Covariance matrix Σ (c) Augmented Laplacian $L^*(G)$ with $C = \{1\}$

Theorem 1. *Let $(X_1, \ldots, X_n)$ be a Gaussian graphical model with an underlying graph G being a tree rooted in $C = \{c\}$. Assume that its covariance matrix $\Sigma = (\sigma_{i,j})$ satisfies*

$$\sigma_{i,j} = |\mathrm{AN}(i) \cap \mathrm{AN}(j)|.$$

Then the precision matrix $K = \Sigma^{-1}$ of $(X_1, \ldots, X_n)$ is equal to $L_c^* = L_c^*(G)$.

Proof. It suffices to prove that $L_c^*(G)\cdot\Sigma = I_n$. At first consider the case $i = j = c$. Observe that for any $1 \leq k \leq n$ we have $\sigma_{c,k} = 1$. Therefore:

$$(L_c^*\Sigma)_{c,c} = \sum_{k=1}^{n} l_{c,k}^*\sigma_{k,c} = l_{c,c}^*\sigma_{c,j} + \sum_{k\sim c} l_{c,k}^*\sigma_{k,c} = (\deg(c)+1) + \deg(c)\cdot(-1)\cdot 1 = 1.$$

Now consider $i = j \neq c$:

$$\begin{aligned}(L_c^*\Sigma)_{i,i} &= \sum_{k=1}^{n} l_{i,k}^*\sigma_{k,i} = l_{i,i}^*\sigma_{i,i} + \sum_{\substack{k\sim i\\ k\in\mathrm{AN}(i)}} l_{i,k}^*\sigma_{k,i} + \sum_{\substack{k\sim i\\ i\in\mathrm{AN}(k)}} l_{i,k}^*\sigma_{k,i}\\ &= \deg(i)\sigma_{i,i} - (\sigma_{i,i} - 1) - (\deg(i) - 1)\sigma_{i,i} = 1.\end{aligned}$$

Now we will prove that the outside of the diagonal $(L^*\Sigma)$ consists only of zeroes. Observe that for $i = c, j \neq c$ we have:

$$(L_c^*\Sigma)_{i,j} = \sum_{k=1}^{n} l_{i,k}^*\sigma_{k,j} = l_{i,i}^*\sigma_{i,j} + \sum_{\substack{k\sim i\\ k\notin \mathrm{AN}(j)}} l_{i,k}^*\sigma_{k,j} + \sum_{\substack{k\sim i\\ k\in \mathrm{AN}(j)}} l_{i,k}^*\sigma_{k,j}$$
$$= (\deg(i)+1)\cdot 1 - (\deg(i)-1)\cdot 1 - 1\cdot 2 = 0.$$

On the other hand, if $j = c, i \neq c$, then:

$$(L_c^*\Sigma)_{i,j} = \sum_{k=1}^{n} l_{i,k}^*\sigma_{k,j} = \sum_{k=1}^{n} l_{i,k}\cdot 1 = 0,$$

because each row of the graph Laplacian sums up to 0. Now we let $i, j \neq c, i \neq j$. If $i \in \mathrm{AN}(j)$, then:

$$(L_c^*\Sigma)_{i,j} = \sum_{k=1}^{n} l_{i,k}^*\sigma_{k,j} = \sum_{k=1}^{n} l_{i,k}\cdot\sigma_{k,j}$$
$$= \deg(i)\cdot\sigma_{i,j} - (\deg(i)-2)\sigma_{i,j} - 1\cdot(\sigma_{i,j}-1) - 1\cdot(\sigma_{i,j}+1) = 0.$$

Analogously, if $j \in \mathrm{AN}(i)$, then $\sigma_{k,j} = \sigma_{i,j}$ for every $k \sim i$. Therefore:

$$(L_c^*\Sigma)_{i,j} = \sum_{k=1}^{n} l_{i,k}^*\sigma_{k,j} = \sum_{k=1}^{n} l_{i,k}\cdot\sigma_{k,j} = \deg(i)\cdot\sigma_{i,j} - \deg(i)\cdot\sigma_{i,j} = 0.$$

The same argument may be applied in the only case left, when $i \neq j; i, j \neq c, i \notin \mathrm{AN}(j)$ and $j \notin \mathrm{AN}(i)$. Therefore here also $(L_c^*\Sigma)_{i,j} = 0$.

3 Discussion and Non-tree Graphs

3.1 Non-tree Graphs

Cycles and Complete Graphs. The description of the inverse of $L^*(G)$ for general G is much harder for G not being a tree. So far we are not able to present a general formula, thus we consider examples of such inverse for specific classes of graphs G. Below we show some examples of $n \times \Sigma = [L^*(G)]^{-1}$ for cycles C_n and complete graphs K_n with the root $C = \{1\}$:

Theorem 2. *The inverse matrix $\Sigma = (\sigma_{i,j})$ of $L^*(C_n)$ is a symmetric matrix satisfying*

$$\sigma_{i,j} = 1 + \frac{(i-1)(n-j+1)}{n}, \quad \textit{for} \quad i \leq j.$$

Proof. Again, we show $L^*\Sigma = I_n$, we assume that the cycle is $1 \to 2 \to 3 \to \ldots \to n \to 1$ and take $C = \{1\}$. Then

$$(L^*\Sigma)_{c,c} = \sum_{k=1}^{n} l_{c,k}\sigma_{k,c} = \sum_{k=1}^{n} l_{c,k} = 1.$$

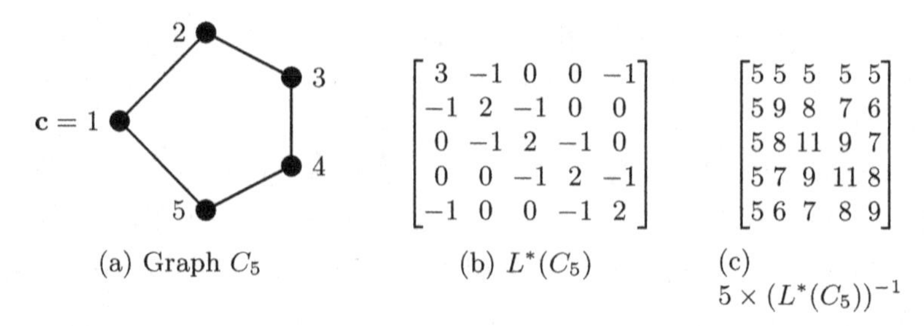

(a) Graph C_5 (b) $L^*(C_5)$ (c) $5 \times (L^*(C_5))^{-1}$

If $j \neq c$, then we have

$$
\begin{aligned}
(L^*\Sigma)_{c,j} &= \sum_{k=1}^{n} l_{c,k}\sigma_{k,j} = l_{c,c}\sigma_{c,j} + l_{c,2}\sigma_{2,j} + l_{c,n}\sigma_{n,j} \\
&= 3 \cdot 1 + (-1) \cdot \left(1 + \frac{1 \cdot (n-j+1)}{n}\right) + (-1) \cdot \left(1 + \frac{(j-1) \cdot 1}{n}\right) \\
&= 3 - 1 - \frac{n-j+1}{n} - 1 - \frac{j-1}{n} = 0
\end{aligned}
$$

and, analogously,

$$
(L^*\Sigma)_{i,c} = \sum_{k=1}^{n} l_{i,k}\sigma_{k,c} = \sum_{k=1}^{n} l_{i,k} = 0.
$$

If $i, j \neq c$, then on the main diagonal we have

$$
\begin{aligned}
(L^*\Sigma)_{i,i} &= \sum_{k=1}^{n} l_{i,k}\sigma_{k,i} l_{i,i-1}\sigma_{i-1,i} + l_{i,i}\sigma_{i,i} + l_{i,i+1}\sigma_{i+1,i} \\
&= -\left(1 + \frac{(i-2)(n-i+1)}{n}\right) + 2\left(1 + \frac{(i-1)(n-i+1)}{n}\right) - \left(1 + \frac{(i-1)(n-i)}{n}\right) \\
&= \frac{-(i-2)(n-i+1) + 2(i-1)(n-i+1) - (i-1)(n-i)}{n} \\
&= \frac{(n-i+2) + (i-1)}{n} = 1.
\end{aligned}
$$

Note that the above calculations are also true for $i = n$ and the $(n+1)^{st}$ row/column of L and Σ being treated as c^{th} row/column.

Now we only need to consider the outside of the main diagonal. Note that $\sigma_{i,n} = 1 + \frac{i-1}{n}$. Thus

$$
(L^*\Sigma)_{i,n} = \sum_{k=1}^{n} l_{i,k}\sigma_{k,n} = -\sigma_{i-1,n} + 2\sigma_{i,n} - \sigma_{i+1,n} = 0.
$$

Also,

$$
\begin{aligned}
(L^*\Sigma)_{n,j} &= \sum_{k=1}^{n} l_{n,k}\sigma_{k,j} = -\sigma_{n-1,j} + 2\sigma_{n,j} - \sigma_{c,j} \\
&= -1 - \frac{(j-1) \cdot 2}{n} + 2 + \frac{(j-1) \cdot 1}{n} \cdot 2 - 1 = 0.
\end{aligned}
$$

Finally, for $i, j \neq 1, n$, $i \neq j$ we have

$$(L^* \Sigma)_{i,j} = \sum_{k=1}^{n} l_{i,k}\sigma_{k,j} = -\sigma_{i-1,j} + 2\sigma_{i,j} - \sigma_{i+1,j} = (*)$$

Observe that as i, j, n are pairwise distinct, either $i+1 \leq j$ or $i-1 \geq j$. Therefore

$$\begin{cases} i+1 \leq j \Rightarrow (*) = -1 - \frac{(i-2)(n-j+1)}{n} + 2 + 2 \cdot \frac{(i-1)(n-j+1)}{n} - 1 - \frac{(i)(n-j+1)}{n} = 0, \\ i-1 \geq j \Rightarrow (*) = -1 - \frac{(j-1)(n-i+2)}{n} + 2 + 2 \cdot \frac{(j-1)(n-i+1)}{n} - 1 - \frac{(j-1)(n-i)}{n} = 0, \end{cases}$$

which ends the proof.

Theorem 3

$$[L^*(K_n)]^{-1}_{i,j} = \begin{cases} 1 & \textit{if } i = 1 \textit{ or } j = 1, \\ 1 + \frac{2}{n} & \textit{if } 1 < i = j, \\ 1 + \frac{1}{n} & \textit{else.} \end{cases}$$

Proof. For proving that claim we observe that $L^*(K_n) = L(K_n) + E_1 = nI_n - J_n + E_1$. Thus $L^*(K_n)$ belongs to an associative algebra being the 5-dimensional matrix space spanned by $I_n, E_1, J_n, E_1 J_n$ and $J_n E_1$. Therefore $(L^*(K_n))^{-1}$ has to be looked for under the form

$$\frac{1}{n} I_n + aE_1 + bJ_n + cE_1 J_n + dJ_n E_1.$$

Clearly $(L^*(K_n))^{-1}$ is a symmetric matrix, thus $d = c$. Solving the linear system in a, b, c gives $a = 0, b = (n+1)/n$ and $c = -1/n$.

Daisy Graphs. The daisy graphs (cf. e.g. Nakashima and Graczyk [1]) may be understood as a notion between a complete bipartite graph and a complete graph. To be more specific, the daisy graph $D_{a,b}$ is built as a sum of the complete bipartite graph $K_{a,b}$ and a complete subgraph K_a, i.e.

$$D_{a,b} := (V = V_A \uplus V_B, E), \qquad V = V_A \uplus V_B, \qquad (x, y) \in E \iff \{x, y\} \cap V_A \neq \emptyset.$$

We may interpret V_A and B_B as the internal and the external part of a daisy, respectively. To give an intuition, graphical models based on daisy graphs may be useful for analysis of a internal features of data (without knowledge of any independence among them) and b mutually conditionally independent external factors, which can influence the internal environment.

Note that for $b = 1$ we have $D_{a,1}$ being a complete graph K_{a+1} and for $a = 1$ we have $D_{1,a}$ being a star graph, which is a tree (cf. Sect. 2).

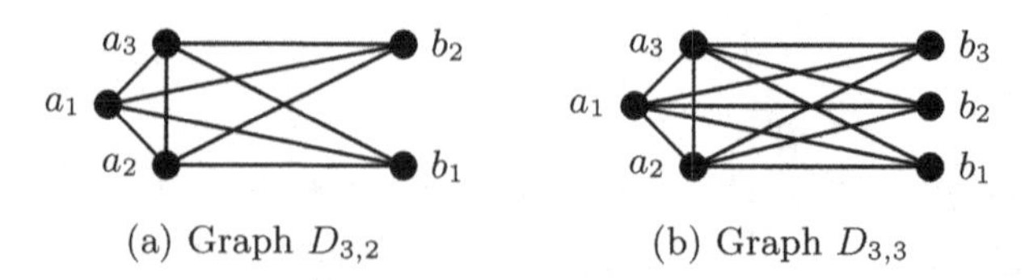

(a) Graph $D_{3,2}$ (b) Graph $D_{3,3}$

The augmented Laplacian (and its inverse) depends on the choice of a root set $C \subset V$. Below we consider four choices of a root in $D_{a,b}$ and their augmented Laplacians:

$$L^*_{\text{in}} := L(D_{a,b}) + E_c \qquad \text{for } C = \{c\}, \quad c \in V_A,$$

$$L^{**}_{\text{in}} := L(D_{a,b}) + \sum_{c \in A} E_c \qquad \text{for } C = V_A,$$

$$L^*_{\text{ex}} := L(D_{a,b}) + E_c \qquad \text{for } C = \{c\}, \quad c \in V_B,$$

$$L^{**}_{\text{ex}} := L(D_{a,b}) + \sum_{c \in A} E_c \qquad \text{for } C = V_B.$$

Without loss of generality we imply such ordering on vertices that the internal vertices precede the external ones. Moreover, concerning cases of one rooted internal (external) vertex we label it as the first (last) vertex.

The exact formulas for $(L^*)^{-1}(D_{a,b})$ are presented below:

Theorem 4: Inverses of augmented Laplacians of daisy graphs

$[\mathrm{L}^*_{\text{in}}(D_{a,b})]^{-1}_{i,j} = \begin{cases} 1 & \text{if } i=1 \text{ or } j=1, \\ 1+\frac{1}{n} & \text{if } 2 \le i,j \le a, i \ne j, \\ 1+\frac{2}{n} & \text{if } 2 \le i = j \le a, \\ 1+\frac{1}{n} & \text{if } 2 \le i \le a < j, \\ 1+\frac{a-1}{an} & \text{if } a < i,j, i \ne j, \\ 1+\frac{a+n-1}{an} & \text{if } a < i = j. \end{cases}$	$[\mathrm{L}^{**}_{\text{in}}(D_{a,b})]^{-1}_{i,j} = \begin{cases} \frac{n}{a(n+1)} & \text{if } i,j \le a, i \ne j, \\ \frac{n+a}{a(n+1)} & \text{if } i = j \le a, \\ \frac{1}{a} & \text{if } i \le a < j, \\ \frac{1}{a} & \text{if } a < i,j, i \ne j, \\ \frac{2}{a} & \text{if } a < i = j. \end{cases}$
$[\mathrm{L}^*_{\text{ex}}(D_{a,b})]^{-1}_{i,j} = \begin{cases} 1 & \text{if } i=n \text{ or } j=n, \\ 1+\frac{1}{a} & \text{if } a < i,j < n, i \ne j, \\ 1+\frac{2}{a} & \text{if } a < i = j < n, \\ 1+\frac{1}{a} & \text{if } i \le a < j < n, \\ 1+\frac{n-1}{an} & \text{if } i,j \le a, i \ne j, \\ 1+\frac{a+n-1}{an} & \text{if } i = j \le a. \end{cases}$	$[\mathrm{L}^{**}_{\text{ex}}(D_{a,b})]^{-1}_{i,j} = \begin{cases} \frac{n+1}{bn} & \text{if } i,j \le a, i \ne j, \\ \frac{n+b+1}{bn} & \text{if } i = j \le a, \\ \frac{1}{b} & \text{if } i \le a < j, \\ \frac{a}{(a+1)b} & \text{if } a < i,j, i \ne j, \\ \frac{n}{(a+1)b} & \text{if } a < i = j. \end{cases}$

Proof. Similarly to 3, the proof relies on the associative algebras being 6-dimensional (for L^*_{ex}) or 10-dimensional (for L^*_{in}) matrix spaces. The resulting inverses of augmented Laplacians follow from solving the corresponding systems of equations.

3.2 Eigenvalues of L^*

Below we show that if the root consists of only one vertex, then the determinant of $L^*(G)$ does not depend on the choice of the root vertex. Moreover, it can be proved that it is equal to the number of spanning trees of G.

Remark 1. *Let $\lambda_1 \leq \lambda_2 \leq \ldots \leq \lambda_n$ be the eigenvalues of L^*. Then*

$$\prod_{i=1}^{n} \lambda_i = \det(L^*) = \#\{\textit{spanning trees } (G)\}.$$

Proof. Recall that $L^*(G)$ differs from $L(G)$ only at $l_{c,c}$, where c is a root vertex of G. Let M be a matrix obtained from L by replacing the cth column with a column with 1 at the cth row and zeroes elsewhere. Then

$$\det(L^*(G)) = \det(L(G)) + \det(M).$$

Clearly, the Laplacian matrix $L(G)$ is singular. We can observe that $det(M)$ is a $(n-1) \times (n-1)$ cofactor of $L(G)$. By the Kirchhoff's matrix-tree Theorem any $(n-1) \times (n-1)$ cofactor of $L(G)$ is equal to the number of spanning trees of G.

3.3 Discussion

Interpretations of the Inverse of the Augmented Laplacian. Below we couple the obtained inverses of L^* matrices with the covariance matrices of some classical examples of random walks. At first let us remind that the covariance function of the Wiener process $(W_t)_{t \geq 0}$ is $\mathrm{Cov}(W_s, W_t) = \min\{s, t\}$. For example, $(W_1, W_2, \ldots, W_n)$ has covariance matrix equal to $(\Sigma)_{i,j} = \min\{i, j\}$. At 1 we proved that this is exactly the inverse of the augmented Laplacian of a path graph with an initial vertex in one of its endpoints. This observation stays consistent

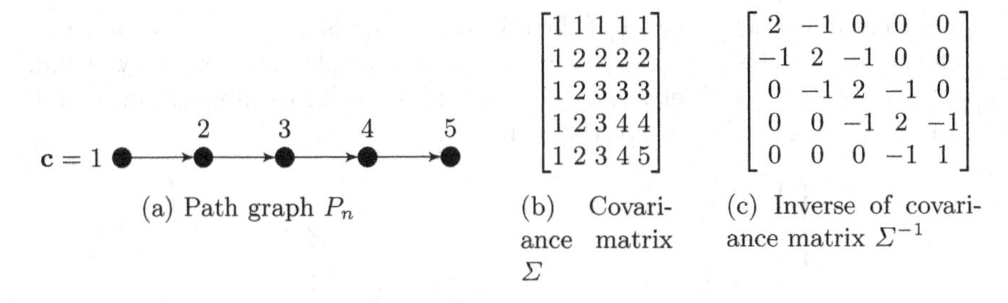

$$\begin{bmatrix} 1&1&1&1&1 \\ 1&2&2&2&2 \\ 1&2&3&3&3 \\ 1&2&3&4&4 \\ 1&2&3&4&5 \end{bmatrix} \qquad \begin{bmatrix} 2&-1&0&0&0 \\ -1&2&-1&0&0 \\ 0&-1&2&-1&0 \\ 0&0&-1&2&-1 \\ 0&0&0&-1&1 \end{bmatrix}$$

(a) Path graph P_n (b) Covariance matrix Σ (c) Inverse of covariance matrix Σ^{-1}

with the conditional independence of W_{t_1} and W_{t_2} under W_t for any $t_1 < t < t_2$. Similarly, the broader class of trees and their augmented Laplacians can be connected with sums of the standard Gaussian random variables 'branched' according to the underlying tree graph.

To observe the analogy of the previous examples for the cycle graph C_n, we may note that replacing the edge $(n, 1)$ with $(n, n+1)$ (with $n+1 \notin V$) gives a path graph $([c=1] - [2] - \ldots - [n+1])$.

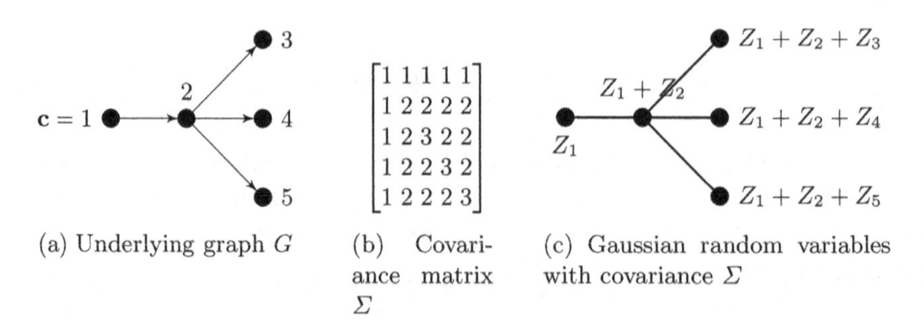

(a) Underlying graph G (b) Covariance matrix Σ (c) Gaussian random variables with covariance Σ

Therefore we may consider the model corresponding to the cycle graph as the Wiener model $(W_1, \ldots, W_n, W_{n+1})$ conditioned by $W_{n+1} = W_1$. This gives a sum of a random variable $W_1 \sim \mathcal{N}(0,1)$ and a Brownian bridge "tied down" at 1 and $(n+1)$. Therefore the covariance matrix of $(W_1, \ldots, W_n)$ is equal to

$$\sigma_{i,j} = 1 + \frac{(i-1)(n-j+1)}{n}, \quad i \le j,$$

cf. Theorem 2.
In order to find a model with a covariance matrix equal to

$$[L^*(K_n)]^{-1}_{i,j} = \begin{cases} 1 & \text{if } i = 1 \text{ or } j = 1, \\ 1 + \frac{2}{n} & \text{if } 1 < i = j, \\ 1 + \frac{1}{n} & \text{else,} \end{cases}$$

observe that all vertices (except of the initial one) are isomorphic and connected, therefore each of their correspondent random variables are mutually equally dependent Let $c = 1$ and let $Z_1, \ldots, Z_n$ be i.i.d. random variables from $\mathcal{N}(0,1)$. Therefore $X_1, X_2, \ldots, X_n$ are of the form

$$\begin{cases} X_1 = Z_1 \\ X_2 = Z_1 + \alpha Z_2 + \beta(Z_3 + Z_4 + \ldots + Z_n) \\ X_3 = Z_1 + \alpha Z_3 + \beta(Z_2 + Z_4 + \ldots + Z_n) \\ \ldots \\ X_n = Z_1 + \alpha Z_n + \beta(Z_2 + Z_3 + \ldots + Z_{n-1}). \end{cases}$$

The restriction on the covariance matrix of $(X_1, \ldots, X_n)$ induces a system of equations, which is satisfied only for $\beta = \frac{1-\sqrt{\frac{1}{n}}}{n-1}$ and $\alpha = \beta + \sqrt{\frac{1}{n}}$.

References

1. Nakashima, H., Graczyk, P.: Wigner and Wishart Ensembles for graphical models. arXiv:2008.10446 [math.ST] (2020)
2. Brouwer, A.E., Haemers, W.H.: Spectra of Graphs. Springer, New York (2011). https://doi.org/10.1007/978-1-4614-1939-6

Classification in the Siegel Space for Vectorial Autoregressive Data

Yann Cabanes[1,2(✉)] and Frank Nielsen[3,4]

[1] Thales LAS Limours, Limours, France
[2] Institute of Mathematics of Bordeaux, Bordeaux, France
[3] Sony Computer Science Laboratories Inc., Tokyo, Japan
Frank.Nielsen@acm.org
[4] École Polytechnique, LIX, Palaiseau, France

Abstract. We introduce new geometrical tools to cluster data in the Siegel space. We give the expression of the Riemannian logarithm and exponential maps in the Siegel disk. These new tools help us to perform classification algorithms in the Siegel disk. We also give the expression of the sectional curvature in the Siegel disk. The sectional curvatures are negative or equal to zero, and therefore the curvature of the Siegel disk is non-positive. This result proves the convergence of the gradient descent performed when computing the mean of a set of matrix points in the Siegel disk.

Keywords: Siegel space · Riemannian manifold · Riemannian exponential map · Riemannian logarithm map · Sectional curvature · Machine learning · Information geometry · Var model

1 Complex Vectorial Autoregressive Gaussian Models

We present here the multidimensional linear autoregressive model which generalize the one-dimensional model presented in [3].

1.1 The Multidimensional Linear Autoregressive Model

We assume that the multidimensional signal can be modeled as a centered stationary autoregressive multidimensional Gaussian process of order $p-1$:

$$U(k) + \sum_{j=1}^{p-1} A_j^{p-1}\, U(k-j) = W(k) \tag{1}$$

where W is the prediction error vector which we assume to be a standard Gaussian random vector and the prediction coefficients A_j^{p-1} are square matrices.

This work was initiated while the author was at École Polytechnique, France. This work was further developed during a thesis with Thales LAS France and the Institute of Mathematics of Bordeaux. We thank the French MoD, DGA/AID for funding (convention CIFRE AID $N°$2017.0008 & ANRT $N°$2017.60.0062).

F. Nielsen and F. Barbaresco (Eds.): GSI 2021, LNCS 12829, pp. 693–700, 2021.
https://doi.org/10.1007/978-3-030-80209-7_74

1.2 Three Equivalent Representation Spaces

There are at least three equivalent spaces to represent our model parameter, they are described in detail in [6]. The first one is the set of Hermitian Positive Definite Block-Toeplitz matrices corresponding to the covariance matrix of the large column vector U defined by:

$$U \stackrel{\text{def}}{=} \left[U(0)^T, \ldots, U(p-1)^T\right]^T \tag{2}$$

The second one is a product space: a HPD matrix (which characterizes the average correlation matrix) and the coefficients $\left(A_i^i\right)_{i=1,\ldots,p-1}$ (which characterize the multidimensional autoregressive model). The third representation space looks like the second one: the coefficients A_i^i are slightly modified to belong to the Siegel disk which metric has been studied in [6,8].

1.3 A Natural Metric Coming from Information Geometry

In [6], the three equivalent representation spaces presented in Sect. 1.2 are endowed with a natural metric coming from information geometry. Indeed, the model assumptions done in Sect. 1.1 are equivalent to the assumption that the large vector U described in Eq. (2) is the realisation of a Gaussian process with zero mean and an Hermitian Positive Definite Block-Toeplitz covariance matrix. We define a metric on this space as the restriction of the information geometry metric on the space of Gaussian processes with zero mean and an Hermitian Positive Definite covariance matrices. We finally transpose this metric to the equivalent representation space constituted of a HPD matrix and coefficients in the Siegel disks. Luckily, the metric in this space is a product metric. The metric on the HPD manifold is the information geometry metric [6]. The metric on the Siegel disk is described in detail in Sect. 2.

2 The Siegel Disk

In this section we present a Riemannian manifold named the Siegel disk and introduce the Riemannian logarithm and exponential maps. These tools will be very useful to classify data in the Siegel space, as shown in Sect. 3. We also introduce the formula of the Siegel sectional curvature and prove it to be non-positive which proves the convergence of classification algorithms based on mean computations, such as the k-means algorithm. The Siegel disk generalizes the Poincaré disk described in [3,4].

2.1 The Space Definition

Definition 1. *The Siegel disk is defined as the set of complex matrices M of shape $N \times N$ with singular values lower than one, which can also be written:*

$$SD_N = \left\{M \in \mathbb{C}^{N \times N}, I - MM^H > 0\right\} \tag{3}$$

or equally:

$$SD_N = \left\{ M \in \mathbb{C}^{N \times N}, I - M^H M > 0 \right\}. \tag{4}$$

We use the partial ordering of the set of complex matrices: we note $A > B$ when the difference $A - B$ is a positive definite matrix. As the Siegel disk is here defined as an open subset of the complex matrices $\mathbb{C}^{N \times N}$, its tangent space at each point can also be considered as $\mathbb{C}^{N \times N}$.

Note that another definition of the Siegel disk also exists in other papers [8], imposing an additional symmetry condition on the matrix M: $M = M^T$. We will not require the symmetry condition in our work.

Property 1. The Siegel disk can also be defined as the set of complex matrices M with a linear operator norm lower than one: $SD_N = \left\{ M \in \mathbb{C}^{N \times N}, |||M||| < 1 \right\}$, where $|||M||| = \sup_{X \in \mathbb{C}^{N \times N}, \|X\|=1} (\|MX\|)$.

2.2 The Metric

Square of the Line Element ds

$$ds^2 = \operatorname{trace}\left(\left(I - \Omega\Omega^H\right)^{-1} d\Omega \left(I - \Omega^H\Omega\right)^{-1} d\Omega^H \right) \tag{5}$$

The Scalar Product. $\forall \Omega \in SD_N, \forall v, w \in \mathbb{C}^{N \times N}$:

$$\langle v, w \rangle_\Omega = \frac{1}{2}\operatorname{trace}\left(\left(I - \Omega\Omega^H\right)^{-1} v \left(I - \Omega^H\Omega\right)^{-1} w^H \right) \tag{6}$$

$$+ \frac{1}{2}\operatorname{trace}\left(\left(I - \Omega\Omega^H\right)^{-1} w \left(I - \Omega^H\Omega\right)^{-1} v^H \right) \tag{7}$$

The norm of a vector belonging to the tangent space is therefore:

$$\|v\|_\Omega^2 = \operatorname{trace}\left(\left(I - \Omega\Omega^H\right)^{-1} v \left(I - \Omega^H\Omega\right)^{-1} v^H \right) \tag{8}$$

The Distance

$$d_{SD_N}^2(\Omega, \Psi) = \frac{1}{4}\operatorname{trace}\left(\log^2\left(\frac{I + C^{1/2}}{I - C^{1/2}} \right) \right) \tag{9}$$

$$= \operatorname{trace}\left(\operatorname{arctanh}^2\left(C^{1/2} \right) \right) \tag{10}$$

with $C = (\Psi - \Omega)\left(I - \Omega^H\Psi\right)^{-1}\left(\Psi^H - \Omega^H\right)\left(I - \Omega\Psi^H\right)^{-1}$.

2.3 The Isometry

In [6], the following function is said to be an isometry for the Siegel distance described in Eq. (9).

$$\Phi_{\Omega}(\Psi) = \left(I - \Omega\Omega^H\right)^{-1/2} (\Psi - \Omega) \left(I - \Omega^H \Psi\right)^{-1} \left(I - \Omega^H \Omega\right)^{1/2} \tag{11}$$

Property 2. The differential of the isometry Φ has the following expression:

$$D\Phi_{\Omega}(\Psi)[h] = \left(I - \Omega\Omega^H\right)^{1/2} \left(I - \Psi\Omega^H\right)^{-1} h \left(I - \Omega^H \Psi\right)^{-1} \left(I - \Omega^H \Omega\right)^{1/2} \tag{12}$$

Property 3. The inverse of the function Φ described in Eq. (11) is:

$$\Phi_{\Omega}^{-1}(\Psi) = \Phi_{\Omega}(-\Psi) \tag{13}$$

2.4 The Riemannian Logarithm Map

Riemannian Logarithm Map at 0. In [6], the logarithm map at 0 is given by the formula :

$$\log_0(\Omega) = \mathcal{V}\ \Omega \tag{14}$$

with:

$$\mathcal{V} = \mathscr{L}\left(\left(\Omega\Omega^H\right)^{1/2}, \log\left(\frac{I + \left(\Omega\Omega^H\right)^{1/2}}{I - \left(\Omega\Omega^H\right)^{1/2}}\right)\right) \tag{15}$$

where $\mathscr{L}(A, Q)$ is defined as the solution of:

$$AZ + ZA^H = Q \tag{16}$$

However, the expression of the logarithm map at zero given in [6] can be greatly simplified.

Property 4. The Riemannian logarithm map of the Siegel disk at zero has the following expression:

$$\log_0(\Omega) = \operatorname{arctanh}(X)\, X^{-1}\Omega \quad \text{where } X = \left(\Omega\Omega^H\right)^{1/2} \tag{17}$$

Note that for $|||X||| < 1$, we can develop the function arctanh in whole series:

$$\operatorname{arctanh}(X) = \sum_{n=0}^{+\infty} \frac{X^{2n+1}}{2n+1} \tag{18}$$

Hence for all X in the Siegel disk, we can write the product $\operatorname{arctanh}(X)\, X^{-1}$ the following way:

$$\operatorname{arctanh}(X)\, X^{-1} = \sum_{n=0}^{+\infty} \frac{X^{2n}}{2n+1} \tag{19}$$

This new expression is also valid when the matrix X is not invertible.

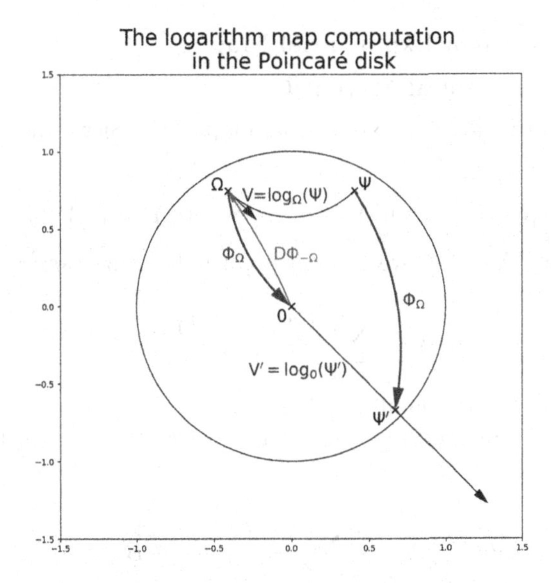

Fig. 1. The Poincaré disk logarithm map computation

2.5 The Riemannian Logarithm Map at Any Point

To compute the Riemannian logarithm map at a point Ω the key idea here is to transport the problem at zero, compute a certain logarithm at zero and transport the result back to Ω. If we want to compute the logartithm map: $\log_{\Omega}(\Psi)$, we first transport both Ω and Ψ using the isometry Φ_{Ω} given in Eq. (11). The point Ω is sent to zero, and we denote $\Psi^{'}$ the image of Ψ by Φ_{Ω} :

$$\Psi^{'} \stackrel{\text{def}}{=} \Phi_{\Omega}(\Psi) = \left(I - \Omega\Omega^{H}\right)^{-1/2}(\Psi - \Omega)\left(I - \Omega^{H}\Psi\right)^{-1}\left(I - \Omega^{H}\Omega\right)^{1/2} \tag{20}$$

Then we compute the logarithm map at zero $\log_0\left(\Psi^{'}\right)$:

$$V^{'} \stackrel{\text{def}}{=} \log_0\left(\Psi^{'}\right) = \operatorname{arctanh}(X)\, X^{-1}\Psi^{'} \quad \text{where } X = \left(\Psi^{'}\Psi^{'\,H}\right)^{1/2} \tag{21}$$

And finally, we transport back the logarithm to the point Ω using the differential of the isometry Φ given in Eq. (12):

$$V \stackrel{\text{def}}{=} \log_{\Omega}(\Psi) = D\Phi_{-\Omega}(0)\left[V^{'}\right] = \left(I - \Omega\Omega^{H}\right)^{1/2} V^{'} \left(I - \Omega^{H}\Omega\right)^{1/2} \tag{22}$$

2.6 The Riemannian Exponential Map

Riemannian Exponential Map at 0

Property 5. The Riemannian exponential map of the Siegel disk at zero has the following expression:

$$\exp_0(V) = \tanh(Y)\, Y^{-1} V \quad \text{where } Y = \left(V V^H\right)^{1/2} \tag{23}$$

Note that for $|||X||| < \frac{\pi}{2}$, we can develop the function tanh in whole series:

$$\tanh(X) = \sum_{n=1}^{+\infty} \frac{2^{2n}\left(2^{2n}-1\right)}{(2n)!} B_{2n} X^{2n-1} \tag{24}$$

where B_{2n} are the Bernoulli numbers.

Hence for all X in the Siegel disk, we can write the product $\tanh(X)\, X^{-1}$ the following way:

$$\tanh(X)\, X^{-1} = \sum_{n=1}^{+\infty} \frac{2^{2n}\left(2^{2n}-1\right)}{(2n)!} B_{2n} X^{2n-2} \tag{25}$$

This new expression is also valid when the matrix X is not invertible.

2.7 The Riemannian Exponential Map at Any Point

To compute the Riemannian exponential map at a point Ω the key idea here is to transport the problem at zero (as for the logarithm), compute a certain exponential at zero and transport the result back to Ω. If we want to compute the exponential map: $\exp_\Omega(V)$, we first transport the vector V at zero using the differential of the isometry Φ given in Eq. (12):

$$V' \stackrel{\text{def}}{=} D\Phi_\Omega(\Omega)[V] \tag{26}$$

$$= \left(I - \Omega\Omega^H\right)^{1/2} \left(I - \Omega\Omega^H\right)^{-1} V \left(I - \Omega^H\Omega\right)^{-1} \left(I - \Omega^H\Omega\right)^{1/2} \tag{27}$$

$$= \left(I - \Omega\Omega^H\right)^{-1/2} V \left(I - \Omega^H\Omega\right)^{-1/2} \tag{28}$$

Then we compute the exponential map at zero $\exp_0\left(V'\right)$:

$$\Psi' \stackrel{\text{def}}{=} \exp_0\left(V'\right) = \tanh(Y)\, Y^{-1} V' \quad \text{where } Y = \left(V' V'^{\,H}\right)^{1/2} \tag{29}$$

And finally, we transport back the exponential to the point Ω using the isometry $\Phi_{-\Omega}$ which is the inverse of isometry Φ_Ω (see Property 3) and transport the point 0 back to Ω and the point Ψ' back to $\exp_\Omega(V)$:

$$\Psi \stackrel{\text{def}}{=} \exp_\Omega(V) \tag{30}$$

$$= \Phi_{-\Omega}\left(\Psi'\right) \tag{31}$$

$$= \left(I - \Omega\Omega^H\right)^{-1/2} \left(\Psi' + \Omega\right) \left(I + \Omega^H \Psi'\right)^{-1} \left(I - \Omega^H\Omega\right)^{1/2} \tag{32}$$

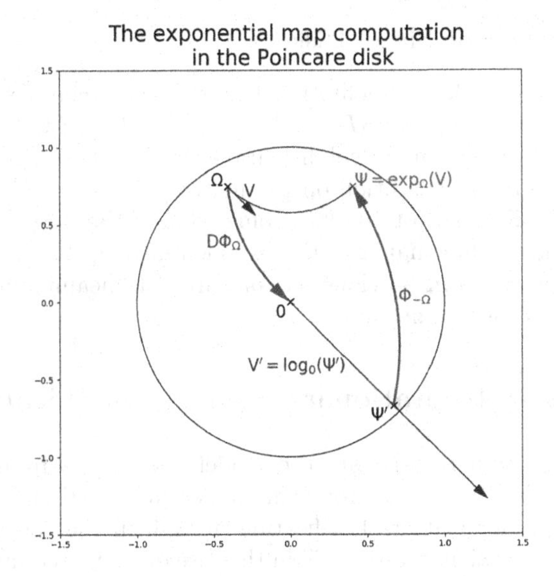

Fig. 2. The Poincaré disk exponential map computation

2.8 The Geodesics

The expression the geodesics can be obtained using the exponential map: the geodesic starting from Ω with velocity V is given by the application:

$$\zeta(t) : t \mapsto \exp_{\Omega}(tV) \tag{33}$$

2.9 Sectional Curvature of the Siegel Space

We first focus on the sectional curvature at 0. We can then obtain the sectional curvature at any point using the isometry Φ defined in Eq. (11).

Let σ be a section defined by the two first vectors of an orthogonal basis $(E_1, \ldots, E_n)$ of the tangent space of the Siegel disk at the point $\Omega = 0$.

Theorem 1. *The sectional curvature at zero of the plan σ defined by E_1 and E_2 has the following expression:*

$$K(\sigma) = -\frac{1}{2}\left(\left\|E_1 E_2^H - E_2 E_1^H\right\|^2 + \left\|E_1^H E_2 - E_2^H E_1\right\|^2\right) \tag{34}$$

As a consequence, we have:

$$-4 \leqslant K(\sigma) \leqslant 0 \quad \forall \sigma \tag{35}$$

The Siegel disk is therefore a Hadamard manifold. The bounds of the sectional curvature provides a proof of the convergence of certain algorithms calculating the Riemannian p-mean [1] or the circumcenter [2] of a set of points on a manifold.

2.10 The Symmetric Siegel Disk

We defined the Siegel disk in definition 1 as the set of complex matrices with singular values lower than one: $SD_N = \{M \in \mathbb{C}^{N \times N}, I - MM^H > 0\}$. We recall that another definition of the Siegel disk also exists in other papers [8], imposing an additional symmetry condition on the matrix M: $M = M^T$. Note that the symmetric Siegel disk is a totally flat submanifold of the Siegel disk. Hence the formula of the logarithm map 2.5, the exponential map 2.7 and the sectional curvature 2.9 computed in previous sections are still meaningful when working in the submanifold of symmetric matrices.

3 Application to Stationary Signals Classification

An effective algorithm to estimate the model parameters from the raw data is described in [5,7]. The classification model presented in this article has been applied to radar clutter classification in [3,4] in the special case of one-dimensional complex signals and in [5] in the case of simulated multidimensional radar signals. This model will be applied in future work to stereo audio signals classification in the case of two-dimensional real signals.

References

1. Arnaudon, M., Barbaresco, F., Yang, L.: Riemannian medians and means with applications to radar signal processing. IEEE J. Sig. Process.**7**, 595 –604 (2013)
2. Arnaudon, M., Nielsen, F.: On Approximating the Riemannian 1-Center. arXiv:1101.4718v3. https://arxiv.org/pdf/1101.4718.pdf
3. Cabanes, Y., Barbaresco, F., Arnaudon, M., Bigot, J.: Toeplitz Hermitian positive definite matrix machine learning based on fisher metric. In: Nielsen, F., Barbaresco, F. (eds.) GSI 2019. LNCS, vol. 11712, pp. 261–270. Springer, Cham (2019). https://doi.org/10.1007/978-3-030-26980-7_27
4. Cabanes, Y., Barbaresco, F., Arnaudon, M., Bigot, J.: Unsupervised machine learning for pathological radar clutter clustering: the P-Mean-shift algorithm. In: C&ESAR (2019). IEEE, Rennes, France (2019)
5. Cabanes, Y., Barbaresco, F., Arnaudon, M., Bigot, J.: Matrix extension for Pathological Radar Clutter Machine Learning. hal-02875440 (2020)
6. Jeuris, B., Vandrebril, R.: The Kähler mean of Block-Toeplitz matrices with Toeplitz structured blocks. SIAM J. Matrix Anal. Appl. **37**(3), 1151–1175 (2016)
7. Morf, M., Augusto Vieira, D.T.L.L., Kailath, T.: Recursive multichannel maximum entropy spectral estimation. IEEE Trans. Geosci. Electr. **GE-16**(2), 85–94 (1978)
8. Nielsen, F.: The Siegel-Klein disk: Hilbert geometry of the Siegel disk domain. Entropy **22**(9), 1019 (2020)

Information Metrics for Phylogenetic Trees via Distributions of Discrete and Continuous Characters

Maryam K. Garba[1], Tom M. W. Nye[2(✉)], Jonas Lueg[3], and Stephan F. Huckemann[3]

[1] Department of Mathematical Sciences, Bayero University, Kano, Nigeria
[2] School of Mathematics, Statistics and Physics, Newcastle University, Newcastle upon Tyne, UK
tom.nye@ncl.ac.uk
[3] Felix-Bernstein-Institute for Mathematical Statistics in the Biosciences, Georg-August-Universität, Göttingen, Germany

Abstract. A wide range of metrics between phylogenetic trees are used in evolutionary molecular biology. They are typically based directly on the branching patterns and edge lengths represented by the trees. Metrics have recently been proposed which are based on the information content of distributions of genetic characters induced by the trees. We first show how these metrics lead to a change to the topology of the underlying tree space. Next we show via computational methods that the metrics are stable under changes to the Markov process used to generate characters, at least in the case of 5 taxa. As a result, a Gaussian process defined over the edges of trees can be used to compute the metrics, leading to a substantial computational efficiency over DNA nucleotide-valued Markov process models.

1 Introduction

Phylogenetic trees or phylogenies represent evolutionary relationships between species, and are a fundamental tool in evolutionary molecular biology. Phylogenies are usually inferred from genetic sequence data of extant species, and phylogenetic inference has become a mature statistical field [4]. A large number of different metrics between phylogenetic trees have been developed, and these play a key role in posterior analysis of samples of trees and in hypothesis testing. Some distances measure differences in the branching structure alone [1] while others also take edge lengths into account [2]. However, the definitions of most metrics are divorced from the models of genetic sequence evolution used to infer trees.

In contrast, Garba et al. [6] developed *probabilistic* or *information* metrics, which are defined by comparing the distributions on genetic sequences induced by phylogenetic trees. The idea is to take a metric such as the Hellinger or

Acknowledging DFG HU 1575/7, DFG GK 2088, DFG SFB 1465 and the Niedersachsen Vorab of the Volkswagen Foundation.

F. Nielsen and F. Barbaresco (Eds.): GSI 2021, LNCS 12829, pp. 701–709, 2021.
https://doi.org/10.1007/978-3-030-80209-7_75

Jensen-Shannon metric between distributions, and pull this back onto the space of trees. Garba et al. [6] focused on computational aspects of calculating these metrics, and also showed how they depart in behaviour from existing metrics. More recently, novel geometries in a space of phylogenetic trees have been constructed by defining distances infinitesimally using the Fisher information metric, and globally as the infimum of the lengths of paths between trees [5]. Papers [6] and [5] are related, since infinitesimally the Fisher information metric is proportional to the information metrics considered in [6] (see [5], Theorem 3.1). The motivation for [5] was to develop a geodesic metric space of phylogenies. A geodesic metric space is a metric space in which every pair of points is joined by at least one path whose length is equal to the distance between the points. The well-known BHV tree space of Billera, Holmes and Vogtmann [2] is a geodesic metric space, for which the metric is locally the Euclidean distance between the vector of edge lengths for trees with the same branching pattern. The existence of geodesics in the BHV tree space makes it amenable for a variety of statistical methods, such as the calculation of Fréchet means [9].

Two geometries for phylogenetic trees are developed in [5]. One is the information geometry of a two-state discrete-valued Markov process defined over the edges of trees, while the second is the information geometry of Gaussian process defined over the edges of trees. Geometries using Markov processes with state space $\{A, C, G, T\}$ can also be defined. The Gaussian process is an approximation to the two-state discrete-valued Markov process, which in turn is an approximation to a simple Markov model of DNA evolution over each tree [7].

Our aim in this short paper is to show, at least for trees with 5 taxa, that the geometries obtained via these three Markov processes are similar. The implication is that the information geometry based on the Gaussian process model is well-motivated biologically, while having the advantage of computational efficiency over the DNA-based model. Along the way, we show how use of information-based metrics motivates adding a boundary at infinity to BHV tree space, thereby changing its topology. The new space is actually a space of forests, called *wald space* [5,8].

2 Background

2.1 Phylogenetic Trees

Let $L = \{1, \dots, N\}$ be a fixed set of *species* or *taxa*. An unrooted phylogenetic tree on L, is a pair (T, ℓ) which satisfies the following.

(i) T is a tree for which the degree 1 vertices (or *leaves*) are bijectively labelled by the elements of L.
(ii) ℓ is a mapping $\ell : E \to (0, \infty)$, where E is the edge set of T.
(iii) There are no vertices with degree 2.

For any edge e in a tree, $\ell(e)$ is called the *length* of e and can be interpreted as a degree of evolutionary divergence along that edge. Although we focus on unrooted trees, since the character distribution associated to each unrooted tree is easier to explain in this context, it is straightforward instead to work with rooted trees: an additional taxon, labelled 0 can be added to the taxon set L to mark the position of the root.

From now on we will just use the term *tree* to refer to an unrooted phylogenetic tree with taxon set L. Given $(T, \ell) \in \mathcal{U}_N$, we call T the *topology* of the tree. Every tree contains N *pendant edges*, which are the edges containing the leaves. The remaining edges are called *internal edges*. Every tree contains between 0 and $N-3$ internal edges. Trees with no internal edges are called *star trees* and consist of a single degree N vertex, connected via the N pendant edges to the leaves. Trees with $N-3$ internal edges are called *resolved*, and every vertex other than the leaves has degree 3.

Let $\mathcal{U}_N$ denote the set of all unrooted phylogenetic trees on $L = \{1, \ldots, N\}$, including both resolved and unresolved trees. Although they worked with rooted trees, the work of Billera et al. [2] can be used to embed $\mathcal{U}_N$ in $\mathbb{R}^M$ where M increases exponentially with N. The embedded space is a product of $(0, \infty)^N$ (which parametrizes the pendant edge lengths) with a union of copies of $[0, \infty)^{N-3}$ which are glued along their boundaries. Each copy of $[0, \infty)^{N-3}$ is called an orthant, and its interior parametrizes the trees with some fixed resolved topology. The boundary of each $[0, \infty)^{N-3}$ corresponds to certain unresolved trees obtained by collapsing one or more internal edges. Billera et al. [2] proved the existence of a metric d_{BHV} on $\mathcal{U}_N$: it is the product metric of the Euclidean metric on the pendant edge lengths and a metric which is locally Euclidean within each orthant. More details are given by [2].

Later, we will consider the case of $N = 5$ taxa so each resolved topology contains $N-3 = 2$ internal edges. The corresponding orthants are 2-dimensional, and each point in an orthant determines the lengths of the two internal edges. We will draw plots using three positive orthogonal axes in $\mathbb{R}^2$ and 3 associated orthants.

2.2 Substitution Models and Character Distributions

Phylogenetic trees are inferred from genetic sequence data or sometimes physical traits. A Markov process X defined over the edges of each tree and taking values in some set Ω is used to model the evolution of genetic sequences or traits given the tree. A *character* is a map $L \to \Omega$ or equivalently, an element of Ω^N.

If the Markov process X is stationary, each phylogenetic tree (T, ℓ) uniquely determines the distribution of X at the leaves and hence a distribution of characters. For example, Ω is often the set of DNA letters $\{A, C, G, T\}$, and each phylogenetic tree determines a distribution on $\{A, C, G, T\}^N$. The Markov process X is called a *substitution model* since it represents how locations in DNA sequences mutate. Given a set of characters which are assumed to be independent observations, maximum likelihood or Bayesian inference methods are used to infer the underlying phylogeny from the data.

We will consider three different Markov process models defined on trees. The first two are commonly used in the phylogenetic literature. The third is constructed as a continuous-valued approximation to the first model.

1. The symmetric two-state model. Under the symmetric two-state model, $\Omega = \{0,1\}$ and the transition probabilities of the Markov process are defined in terms of the path length $\ell_{t_1 t_2}$ between any two points $t_1, t_2 \in T$:

$$\Pr(X(t_2) = j \,|\, X(t_1) = i) = \begin{cases} \frac{1}{2}(1 + e^{-\ell_{t_1 t_2}}) & \text{when } i = j \text{ and} \\ \frac{1}{2}(1 - e^{-\ell_{t_1 t_2}}) & \text{when } i \neq j \end{cases}$$

for $i, j \in \Omega$. The stationary distribution of this Markov process is the uniform distribution on Ω.

2. The Jukes-Cantor model. The Jukes-Cantor model [7] is a model of DNA substitution so $\Omega = \{A, C, G, T\}$. It is the 4-state analog of the symmetric two-state model, in that all transitions between different states have equal rates. We will denote the Markov process by Y, and the transition probabilities are

$$\Pr(Y(t_2) = j \,|\, Y(t_1) = i) = \begin{cases} \frac{1}{4}(1 + 3e^{-4\ell_{t_1 t_2}/3}) & \text{when } i = j \text{ and} \\ \frac{1}{4}(1 - e^{-4\ell_{t_1 t_2}/3}) & \text{when } i \neq j \end{cases}$$

for $i, j \in \Omega$. The constant 4/3 appears in the exponent to ensure that, like the two-state symmetric model, on average one transition occurs per unit edge length on T. The stationary distribution is the uniform distribution on Ω.

3. A Gaussian process model. In this case the Markov process Z is continuous-valued with $\Omega = \mathbb{R}$. The transition kernel for Z between $t_1, t_2 \in T$ is specified by

$$\{Z(t_2) \,|\, Z(t_1) = z\} \sim N(ze^{-\ell_{t_1 t_2}}, 1 - e^{-2\ell_{t_1 t_2}}).$$

The stationary distribution is the standard normal distribution $N(0,1)$. The Markov process Z was defined specifically to act as a continuous-valued analog of the two-state symmetric process X [5]. In particular, if $X_1, \ldots, X_N$ and $Z_1, \ldots, Z_N$ represent the values of the two Markov processes at the leaves L of a tree (T, ℓ), then it can be shown that

$$4\mathrm{Cov}(X_i, X_j) = 1 - \exp(-2\ell_{ij}) = \mathrm{Cov}(Z_i, Z_j).$$

(It emerges that the factor of 4 makes no difference to the geometry obtained.) The process Z can be obtained as an approximation to binomial random variables corresponding to many independent copies of the two-state symmetric process X.

Lemma 1. *Each of the three models determines an injective map from $\mathcal{U}_N$ to the set of distributions of characters Ω^N.*

For details of the proof see [5,10].

3 Information Metrics from Distributions of Discrete Characters

3.1 Definition

Suppose we fix one of the Markov processes defined above, and let $p_{(T,\ell)}$ denote the distribution of characters Ω^N induced by a tree (T, ℓ). Any metric d on distributions then pulls back to the tree space $\mathcal{U}_N$ via

$$d^*((T_1, \ell_1), (T_2, \ell_2)) = d(p_{(T_1,\ell_1)}, p_{(T_2,\ell_2)}).$$

For example, d can be the Hellinger metric or Jensen-Shannon metric. We call the pull back metrics *information metrics* or *probabilistic metrics.* [6] considered such metrics in the case of discrete-valued Markov substitution models. Computation of these metrics in the discrete case involves a sum over Ω^N. We will only work with $N = 4$ or 5, in which cases the computation is tractable.

3.2 Properties

A key property of the information metrics is that they behave differently from d_{BHV} as edge lengths are scaled. If we take two distinct trees (T_1, ℓ_1) and (T_2, ℓ_2) and simultaneously scale all edge lengths in the trees by a factor s, then in the limit $s \to \infty$ the information metrics between the trees decreases to zero [6]. This is because in the limit the distribution induced on Ω^N by both trees is the same: it is the product of N independent copies of the stationary distribution on Ω. In evolutionary terms, as edge lengths approach infinity the number of mutations on each edge becomes large, so the genetic sequences at the leaves of the tree are independent from one another.

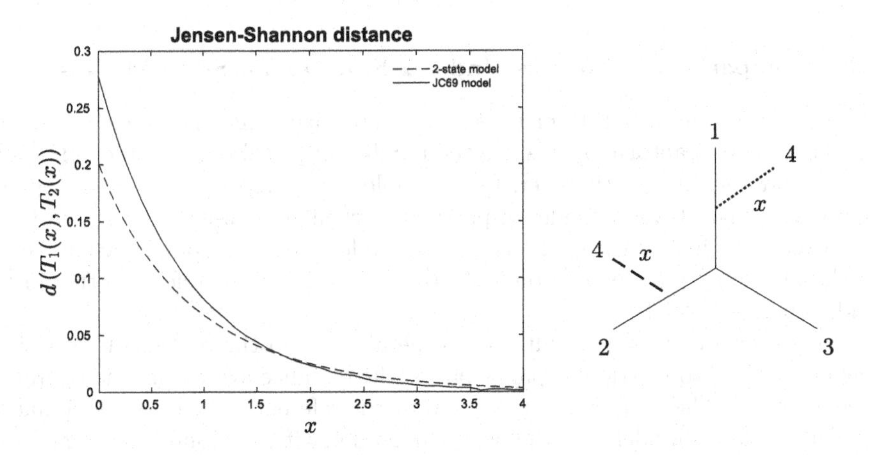

Fig. 1. Jensen-Shannon distance between two 4-taxon trees for the two-state symmetric and Jukes-Cantor Markov models. Right: the trees are identical apart from the placement of taxon 4 (dotted line for T_1, dashed for T_2). The distance is plotted as a function of the length x of the edge containing taxon 4.

This is not the only way the boundary at infinity can be approached. Consider the star tree on $L = \{1, 2, 3\}$ for which all edge lengths are 0.4 (for example). For $i = 1, 2$, let (T_i, ℓ_i) be the tree obtained by attaching a fourth taxon by an edge with length x midway along the edge containing taxon i. As shown in Fig. 1, as x increases the Jensen-Shannon metric between the trees (T_1, ℓ_1) and (T_2, ℓ_2) decreases to zero. (The Hellinger metric behaves very similarly.) In the limit, the value of the Markov process at taxon 4 is independent of the values at the other taxa. The limit can be thought of as the forest consisting of the original star tree on taxa 1, 2, 3 together with the isolated vertex $4 \in L$. In fact this forest is obtained as the limit as $x \to \infty$ independently of the point at which taxon 4 is attached to the star tree. Although the information metrics are well-defined on $\mathcal{U}_N$, it is natural to augment the space by adding on the boundary at infinity, especially since the boundary is finitely far from the rest of the space:

Lemma 2. *Fix a tree $(T, \ell) \in \mathcal{U}_N$. Setting the edge lengths of a subset S of edges of T to infinity determines a distribution p_S on Ω^N, such that $d(p_{(T,\ell)}, p_S) < \infty$ where d denotes the Hellinger or Jensen-Shannon metric.*

The proof follows from Theorem 4.1 and Theorem 5.1 in [5], respectively for the Markov processes X and Z. However, it is apparent that the proof also holds for the Jukes-Cantor model Y.

The wald space described by [5] and [8] includes this boundary and certain other points (such as trees with taxon labels on internal vertices) which are excluded from BHV tree space. The boundary at infinity contains forests obtained by expanding subsets of edges to infinite length, which is the same as removing these edges from trees. In particular, it contains the forest of N isolated vertices which is obtained, for example, when the pendant edges become infinitely long. Wald space is constructed in such a way that the pull back of information metrics are true metrics as opposed to pseudometrics.

3.3 Comparison of Metrics Under 2-State and 4-State Models

Computing the information metrics is less intensive using the two-state model than the Jukes Cantor model, since the number of possible characters ($|\Omega|^N$) is smaller for the former. However, from a biological perspective we would prefer to use the Jukes-Cantor model in particular when the trees were inferred from DNA data. To justify use of the two-state model, we investigated the similarity of the metric by plotting contours of distance from a fixed reference tree in $\mathcal{U}_5$ under the two models.

Figure 2 shows typical results. The general shape of the contours is very similar under the two metrics, in particular as the distance away from the reference tree increases. The minimum represented by the reference tree is less steep under the Jukes-Cantor model than the two-state model, with the contours more widely spaced. Overall, however, the two-state model provides a good approximation to

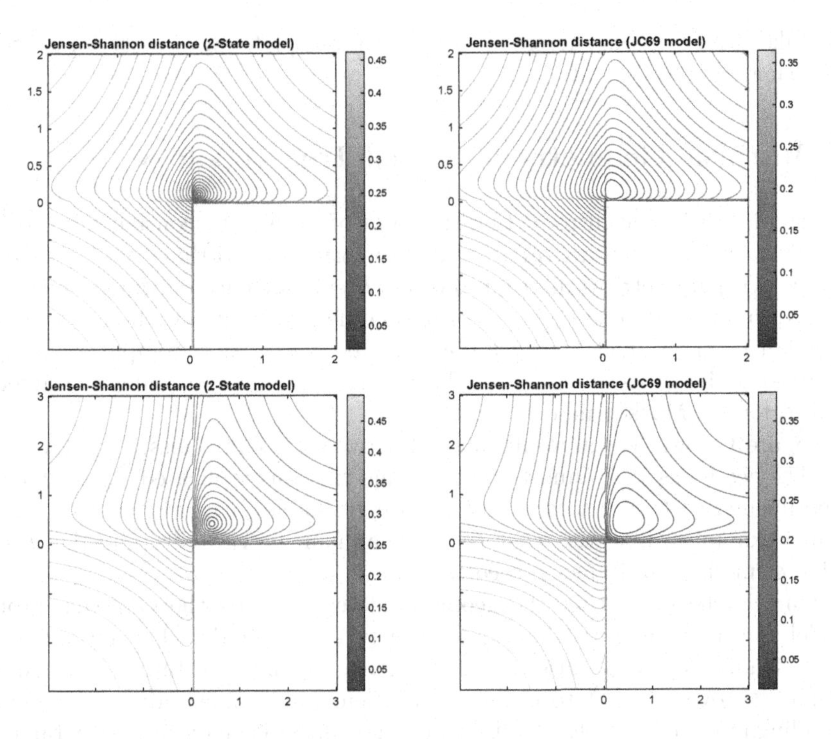

Fig. 2. Contour plots of the Jensen-Shannon distance from a fixed reference tree in $\mathcal{U}_5$. In each plot, three neighbouring orthants in $\mathcal{U}_5$ are shown (see the last paragraph of Sect. 2.1). The contours show the distance from the reference tree positioned in the centre of concentric contours in the top right orthant. Top row: both internal edges on the reference tree have length 0.1. Bottom row: both internal edges on the reference tree have length 0.4. Left column: two-state model. Right: Jukes-Cantor model.

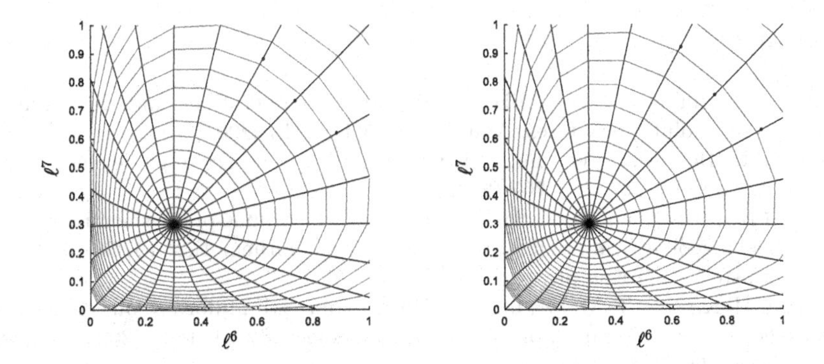

Fig. 3. Contours of distance and a selection of geodesics fired from a point in $\mathcal{U}_5$. The axes give the lengths of the two internal edges. Left: Fisher information metric computed under the two-state model. Right: metric computed using the Gaussian process model.

the Jukes-Cantor model across $\mathcal{U}_5$. Other contour plots for a variety of reference trees show similar results.

4 Intrinsic Geometry for Fixed Tree Topology

As mentioned in the introduction, the aim of [5] was to construct a geodesic metric space of trees using information geometry. Each orthant of $\mathcal{U}_N$ is a manifold which parametrizes distributions on Ω^N via a choice of Markov model, as defined in Sect. 2.2. The Fisher information metric [3] gives each orthant a Riemannian inner product, and geodesics can be constructed computationally by integrating the geodesic equation. These can be compared under the different choices for the Markov model.

Figure 3 shows typical results under the two-state and Gaussian process models. The geodesics look almost identical, and this was replicated across different experiments, with the plots almost indistinguishable. It can be shown on some simple examples, however, that the Riemannian metrics determined with the different models are in fact not equal [5].

Our conclusion is that the geometry determined by the Gaussian process model is very similar to that under the two-state model. The former model has a significant computational advantage: while the two-state model requires summation over $\{0, 1\}^N$ to evaluate the Riemannian inner product, the corresponding integral under the Gaussian process model is analytically tractable. As shown by [5], the Fisher information metric under the Gaussian model is the same as the affine invariant Riemannian metric on positive definite matrices, where each tree is associated to the covariance matrix between the random variables $Z_1, \ldots, Z_N$ at the leaves. This raises the possibility of using geodesics in the ambient space of positive definite matrices, which are easily computed, to help construct information geodesics in wald space [8].

References

1. Allen, Benjamin L., Steel, Mike: Subtree transfer operations and their induced metrics on evolutionary trees. Ann. Comb. **5**(1), 1–15 (2001)
2. Billera, L., Holmes, S., Vogtmann, K.: Geometry of the space of phylogenetic trees. Adv. Appl. Math. **27**, 733–767 (2001)
3. Sueli I.R. Costa, Sandra A. Santos, and Joao E. Strapasson. Fisher information distance: a geometrical reading. Discrete Applied Mathematics, 197:59–69, 2015
4. J. Felsenstein. Inferring phylogenies. Sinauer, 2004
5. Garba, M. K., Nye, T. M. W., Lueg, J., Huckemann, S. F.: Information geometry for phylogenetic trees. Journal of Mathematical Biology **82**(3), 1–39 (2021). https://doi.org/10.1007/s00285-021-01553-x
6. Maryam K. Garba, Tom M. W. Nye, and Richard J. Boys. Probabilistic distances between trees. Syst. Bio., 67(2):320–327, 2018
7. T. H. Jukes and C. R. Cantor. Evolution of protein molecules. In Mammalian protein metabolism, pages 21–132. Elsevier, 1969

8. Lueg, J., Nye, T.M.W., Garba, M.K., Huckemann, S.F.: Phylogenetic wald spaces. In: Nielsen, F., Barbaresco, F. (eds.) GSI 2021, LNCS, vol. 12829, pp. 710–717. Springer, Heidelberg (2021)
9. Miller, E., Owen, M., Provan, J.S.: Polyhedral computational geometry for averaging metric phylogenetic trees. Adv. Appl. Math. **68**, 51–91 (2015)
10. Rogers, J.S.: On the consistency of maximum likelihood estimation of phylogenetic trees from nucleotide sequences. Syst. Bio. **46**(2), 354–357 (1997)

Wald Space for Phylogenetic Trees

Jonas Lueg[1(✉)], Maryam K. Garba[2], Tom M. W. Nye[3], and Stephan F. Huckemann[1]

[1] Felix-Bernstein-Institute for Mathematical Statistics in the Biosciences, Georg-August-Universität, Göttingen, Germany
jonas.Lueg@uni-goettingen.de

[2] Department of Mathematical Sciences, Bayero University, Kano, Nigeria

[3] School of Mathematics, Statistics and Physics, Newcastle University, Newcastle upon Tyne, UK

Abstract. The recently introduced wald space models phylogenetic trees from an evolutionary perspective. We show that it is a stratified space and propose algorithms to compute geodesics. In application we compute a Fréchet mean of three trees of different topologies that is fully resolved, unlike in BHV-space. Both, preliminary results on geodesics and on means suggest that wald space features less stickiness than BHV-space, making it an alternative model for statistical investigations.

1 Introduction

Phylogenetic trees reflect biological species' evolution. They are built from genetic variation over a set of taxa. Curiously, building them for the same set of taxa, from different genes, however, often result in fundamentally different trees, e.g. Rokas et al. (2003). This generates a call for statistics, for instance averaging over different trees while controlling their uncertainty. Also, this is a call for geometry, designing suitable spaces of trees that are both, biologically meaningful and numerically tractable.

A seminal model has been proposed twenty years ago by Billera et al. (2001), abbreviated as the BHV-model. It has the favorable property of being a Riemann stratified space of globally nonpositive curvature, thus admitting unique geodesics and unique Fréchet means. Additionally, since it is locally flat, an abundance of successful algorithms have been developed for their computation that suffer only from inherent combinatorial complexity, e.g. Owen (2011); Bačák (2014); Miller et al. (2015); Brown and Owen (2018).

While this model is mathematically intriguing, more recently new models have been developed with geometries more closely reflecting stochastic biological fundamentals of gene mutations, e.g. Moulton and Steel (2004); Shiers et al. (2016); Garba et al. (2020). In Garba et al. (2018), metrics for phylogenetic trees based on the information geometry of the two-state and four-state model

Acknowledging DFG HU 1575/7, DFG GK 2088, DFG SFB 1465 and the Niedersachsen Vorab of the Volkswagen Foundation.

F. Nielsen and F. Barbaresco (Eds.): GSI 2021, LNCS 12829, pp. 710–717, 2021.
https://doi.org/10.1007/978-3-030-80209-7_76

were proposed (four states because gene entries are taken from one of the four nucleotide bases). This study was continued in Garba et al. (2020, 2021) and - as a further simplification - a continuous model has been proposed with moments matching those of the two-state model.

In this contribution, we briefly review the definition of our new *wald space* (cf. Garba et al. (2020)) and propose algorithms to compute geodesics and Fréchet means. On the one hand, the wald space is geometrically more challenging. It is a stratified space that is isometrically embedded in the space of positive symmetric $N \times N$ matrices $\mathcal{P}$ (where $N \in \mathbb{N}$ is the number of taxa) equipped with the well known affine invariant geometry of globally nonpositive curvature – hence the need for algorithms as sophisticated as those of the BHV-space. On the other hand, we believe it is biologically more meaningful than the BHV-space. For example, in BHV-space the distance of two different trees with edge lengths becoming arbitrary large diverges to infinity. In wald space, such trees converge to the completely disconnected forest, a member of the wald space, along with other forests. Hence these two trees become more and more similar. Simulations and data analyses reveal advantage of wald space: degenerate trees seem to be less *sticky* (sticky means have degenerate limiting distributions) in wald space than in BHV-space, cf. Hotz et al. (2013); Huckemann et al. (2015); Barden et al. (2013, 2018), thus more easily allowing for statistical inference.

Wald space was first proposed at the Oberwolfach workshop 1804 (2018) in the black forest which is the *Schwarzwald* in German.

2 Wald Space

Let $N \in \mathbb{N}$ denote the number of taxa. A *phylogenetic forest* (F, ℓ) is

(i) a forest $F = (V, E)$ with a finite number of *vertices* V, undirected *edges* E such that any two vertices $u, v \in V$ are connected by at most one edge denoted by $\{u, v\}$ and *labeled* vertices $L = \{1, \ldots, N\} \subseteq V$, where $v \in V \setminus L$ implies that $\deg(v) \geq 3$,
(ii) with a mapping $\ell\colon E \to (0, \infty)$.

Two phylogenetic forests are equivalent, $(F_1, \ell_1) \sim (F_2, \ell_2)$, if their label sets agree $L_1 = L = L_2$ and if there is a graph isomorphism $f\colon V_1 \to V_2$ such that

(i) $f(u) = u$ for all $u \in L$, and
(ii) $\ell_1(\{u, v\}) = \ell_2(\{f(u), f(v)\})$ for all $\{u, v\} \in E_1$.

Definition 1. *Every equivalence class $W = [F, \ell]$ is called a* wald *and all equivalence classes form the* wald space $\mathcal{W}$*, its geometric structure is defined further below. Disregarding the edge lengths map ℓ, every equivalence class of forests F with regards to (i) above, is a* wald topology. *For a given wald $W = [F, \ell]$, the* grove *of W is $\mathcal{W}_W$ which comprises all $W' = [F', \ell'] \in \mathcal{W}$ where F' and F have the same wald topology.*

In the following, for any connected $u, v \in V$, $E(u,v)$ is the set of edges along the unique path connecting u and v. For $u = v$, we set any sum over $E(u,u)$ equal zero.

With this notation, the map ϕ sending $W = [F, \ell]$ to the $N \times N$ matrix with coordinate entry at $u, v \in L$,

$$\big(\phi(W)\big)_{uv} = \big(\phi([F,\ell])\big)_{uv} := \begin{cases} \exp\Big(-\sum_{e \in E(u,v)} \ell(e)\Big), & \text{if } u \text{ and } v \text{ are connected,} \\ 0, & \text{else,} \end{cases} \tag{1}$$

is well defined and maps $\mathcal{W}$ injectively into the set of symmetric positive $N \times N$ matrices $\mathcal{P}$, cf. Garba et al. (2020).

Recall from Garba et al. (2020, 2021) that the affine invariant Riemannian metric on $\mathcal{P}$ corresponds to the Fisher information geometry for zero-mean nondegenerate N-dimensional Gaussians induced by tree-indexed Gaussian processes, a continuous generalisation of the two-state model. This metric has the advantage of turning $\mathcal{P}$ into a Riemannian manifold of global nonpositive curvature (e.g. Lang (1999)), guaranteeing unique geodesics and unique Fréchet means (e.g. Sturm (2003)). The squared distance induced on $\mathcal{P}$ is given by

$$d_{\mathcal{P}}^2(P,Q) = \operatorname{Tr}\left[\log\left(\sqrt{P}^{-1} Q \sqrt{P}^{-1}\right)^2\right] = \sum_{i=1}^{N} \log(\mu_i)^2,$$

where $\sqrt{P}$ is the unique positive definite square root of P and μ_i are the eigenvalues of $P^{-1}Q$.

Definition 2. *The metric $d_{\mathcal{W}}$ of the* wald space *is the pullback of $d_{\mathcal{P}}$ under ϕ, which is given for $W_1, W_2 \in \mathcal{W}$ by*

$$d_{\mathcal{W}}(W_1, W_2) = \inf_{\substack{\gamma\colon [0,1] \to \mathcal{W} \\ \phi\circ\gamma \text{ cont. path,} \\ \gamma(0)=W_1, \gamma(1)=W_2}} L_{d_{\mathcal{P}}}(\phi \circ \gamma),$$

where $L_{d_{\mathcal{P}}}(\gamma)$ is the length of the path γ measured in $d_{\mathcal{P}}$. If no such path exists, we set $d_{\mathcal{W}}(W_1, W_2) = \infty$.

As previously noted, trees with edge lengths ℓ tending to infinity move infinitively far apart in the BHV geometry. In the wald geometry the distance between these trees goes to zero. This is reflected in the following reparametrization $W = [F, \lambda]$ with $\lambda := 1 - \exp(-\ell)$, recasting (1) as

$$\big(\phi(W)\big)_{uv} = \big(\phi([F,\lambda])\big)_{uv} := \begin{cases} \prod_{e \in E(u,v)} \big(1 - \lambda(e)\big), & \text{if } u \text{ and } v \text{ are connected,} \\ 0, & \text{else.} \end{cases}$$

In particular, if $W = [F, \lambda]$, $F = (V, E)$, has $|E|$ edges, vectorizing $\lambda \in (0,1)^{|E|}$, we have the following identification for the grove of W:

$$\mathcal{W}_W \cong (0,1)^{|E|}.$$

Theorem 1. *1. For every wald* $W = [F, \lambda]$, $F = (V, E)$ *with grove* $\mathcal{W}_W$, *the mapping* $(0,1)^{|E|} \cong \mathcal{W}_W \xrightarrow{\phi} \mathcal{P}$ *is an embedding.*
2. If $W = [F, \lambda]$ *with a fully resolved (i.e. binary) tree* F *then* $\mathcal{W}_W$ *is an open subset of* $\mathcal{W}$.

Proof. cf. Lueg et al. (2021).

In consequence, $\mathcal{W}$ is a stratified space with strata given by groves. As BHV-space can be viewed as a subset of wald space, cf. Garba et al. (2020), BHV-orthants are subsets of groves. In contrast to BHV-space, groves are not only connected to the star stratum (trees without interior edges), they are also connected to forest strata including the completely disconnected forest (consisting of N isolated vertices, no edges), which lies on the boundary of the star stratum.

3 Geodesics in Wald Space

We propose different algorithms to compute geodesics between two fully resolved trees W_1 and W_2, where Algorithm 4 is only applicable if W_1 and W_2 lie in a common grove $\mathcal{W}_W$. Dropping the embedding map ϕ, we consider wald space $\mathcal{W}$ as a subset of the ambient space $\mathcal{P}$. To this end, for $P, Q \in \mathcal{P}$, denote the unique geodesic between P and Q by $\gamma_{P,Q} \colon [0,1] \to \mathcal{P}$, the Riemann exponential and logarithm by $\mathrm{Exp}_P^{(\mathcal{P})} \colon T_P\mathcal{P} \to \mathcal{P}$ and $\mathrm{Log}_P^{(\mathcal{P})} \colon \mathcal{P} \to T_P\mathcal{P}$, respectively, the orthogonal tangent space projection by $\pi_W \colon T_P\mathcal{P} \to T_W\mathcal{W}$ and define the projection $\pi \colon \mathcal{P} \to \mathcal{W}, P \mapsto \pi(P) := \mathrm{argmin}_{W \in \mathcal{W}}\, d_{\mathcal{P}}(P, W)$, where π is only well-defined for $P \in \mathcal{P}$ close enough to $\mathcal{W}$. The following is a very simple but naive algorithm.

Algorithm 1 (Naive Projection (NP)). Given $3 \le n \in \mathbb{N}$, $W_1, W_2 \in \mathcal{W}$, for $i = 1, \ldots, n$ compute

(1) $X_i = \pi\big(\gamma_{W_1,W_2}(\frac{i-1}{n-1})\big)$.

Return $(X_1, \ldots, X_n)$.

The next algorithm makes small (approximately geodesic) steps and successively takes the geodesic from the newest point to the destination (note the X_{i-1} and Y_{i-1} in the subscript in the update step).

Algorithm 2 (Successive Projection (SP)). Given $3 \le n \in \mathbb{N}$, $W_1, W_2 \in \mathcal{W}$, set $X_1 := W_1$ and $Y_1 := W_2$. For $i = 2, \ldots, \lfloor \frac{n}{2} \rfloor$, do

(1) $X_i := \pi\big(\gamma_{X_{i-1},Y_{i-1}}(\frac{1}{n-i+1})\big)$ and
(2) $Y_i := \pi\big(\gamma_{Y_{i-1},X_{i-1}}(\frac{1}{n-i+1})\big)$.

If n is even, return $(X_1, \ldots, X_{\lfloor \frac{n}{2} \rfloor}, Y_{\lfloor \frac{n}{2} \rfloor}, \ldots, Y_1)$.
If n is odd, set $Z := \pi\big(\gamma_{X_{\lfloor \frac{n}{2} \rfloor}, Y_{\lfloor \frac{n}{2} \rfloor}}(\frac{1}{2})\big)$ and return $(X_1, \ldots, X_{\lfloor \frac{n}{2} \rfloor}, Z, Y_{\lfloor \frac{n}{2} \rfloor}, \ldots, Y_1)$.

The following two algorithms are inspired by Schmidt et al. (2006). They update a given path iteratively and perform a straightening of the path, eventually leading to a geodesic (cf. Figs. 1–4).

Algorithm 3 (Extrinsic Path Straightening (EPS)). Let $3 \leq n \in \mathbb{N}$, $m \in \mathbb{N}$, $W_1, W_2 \in \mathcal{W}$ and suppose $(X_1, \dots, X_n)$ is a path in $\mathcal{W}$ from W_1 to W_2. For $j = 1, \dots, m$, do

(1) for $i = 2, \dots, n-1$ compute $V_i = \frac{1}{2}\big(\mathrm{Log}_{X_i}(X_{i-1}) + \mathrm{Log}_{X_i}(X_{i+1})\big)$ and
(2) update $(X_2, \dots, X_{n-1})$: for $i = 2, \dots, n-1$ compute $X_i := \pi\big(\mathrm{Exp}^{(\mathcal{P})}_{X_i}(V_i)\big)$.

Return $(X_1, \dots, X_n)$.

Exploiting the manifold structure of groves, for two walds $W_1, W_2 \in \mathcal{W}_{[F]}$ with the same fully resolved tree F, we change Algorithm 3 slightly and thus avoid using the projection.

Algorithm 4 (Intrinsic Path Straightening (IPS)). Let $3 \leq n \in \mathbb{N}$, $m \in \mathbb{N}$, $W_1, W_2 \in \mathcal{W}_W$ and suppose $(X_1, \dots, X_n)$ is a path in $\mathcal{W}_W$ from $X_1 := W_1$ to $X_n := W_2$. For $j = 1, \dots, m$, do

(1) for $i = 2, \dots, n-1$ compute $V_i = \frac{1}{2}\pi_{X_i}\Big(\big(\mathrm{Log}_{X_i}(X_{i-1}) + \mathrm{Log}_{X_i}(X_{i+1})\big)\Big)$ and
(2) update $(X_2, \dots, X_{n-1})$: for $i = 2, \dots, n-1$ compute $X_i := \mathrm{Exp}^{(\mathcal{W}_W)}_{X_i}(V_i)$.

Return $(X_1, \dots, X_n)$.

We measure the quality of a proposal $(X_1, \dots, X_n)$, $3 \leq n \in \mathbb{N}$ by its length,

$$L(X_1, \dots, X_n) = \sum_{i=1}^{n-1} d_{\mathcal{P}}(X_i, X_{i+1})$$

and its energy,

$$E(X_1, \dots, X_n) = \frac{1}{2}\sum_{i=1}^{n-1} d_{\mathcal{P}}(X_i, X_{i+1})^2.$$

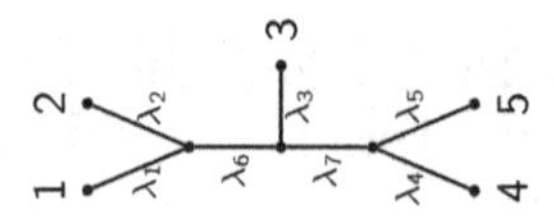

Fig. 1. Tree with edge weights $\lambda^{(1)} = (0.5, \dots, 0.5, 0.1, 0.8)$ and $\lambda^{(2)} = (0.5, \dots, 0.5, 0.9, 0.1)$ for computation of geodesics in Fig. 2.

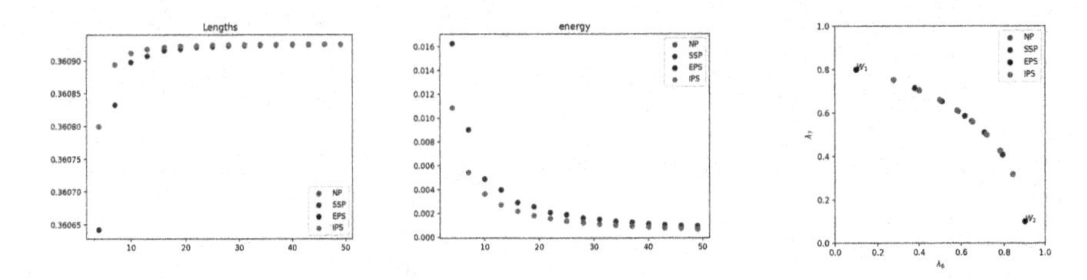

Fig. 2. Length (left) and energy (center) of paths between the two trees from Fig. 1 obtained from the four algorithms for $n = 4, 7, \ldots, 46, 49$. Right: coordinates λ_6, λ_7 of the paths obtained from the four algorithms for $n = 10$. Note that (NP), (IPS), (EPS) almost coincide.

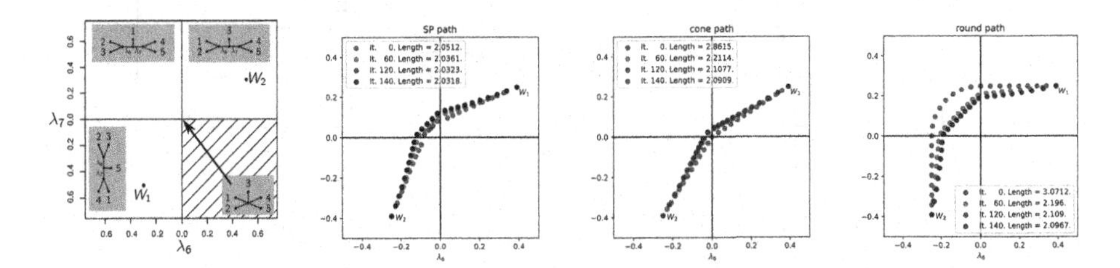

Fig. 3. Left: the coordinate representation (only interior edges) of different neighbouring groves and two walds $W_1, W_2 \in \mathcal{W}$. Second to left to right: Selected iterations of the (EPS) algorithm for different starting paths: the output of the (SP) algorithm, the cone path and a round path, respectively. All paths have $n = 25$ points.

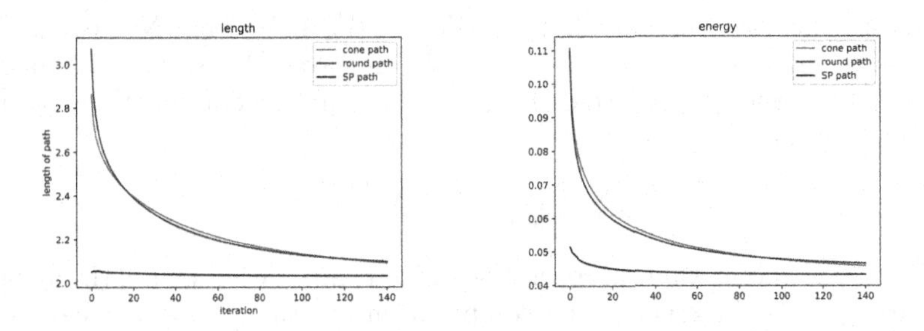

Fig. 4. Left: length of the paths for the iterations of the (EPS) algorithm for different starting paths. Right: energy of the paths for the iterations of the (EPS) algorithm for different starting paths.

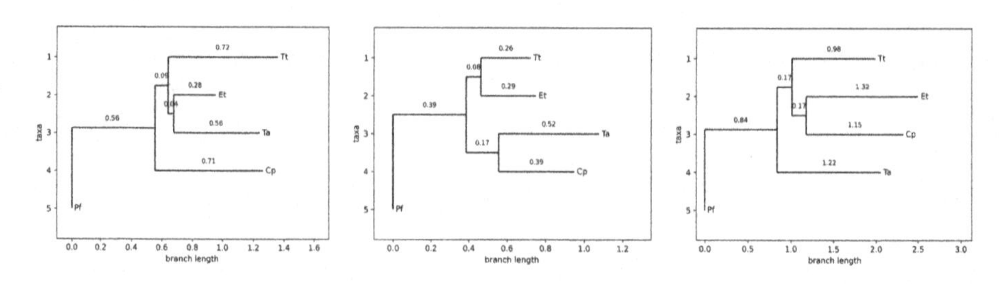

Fig. 5. Three trees $W_1, W_2, W_3 \in \mathcal{W}$ from Nye et al. (2016). Their Fréchet means are depicted in Fig. 6.

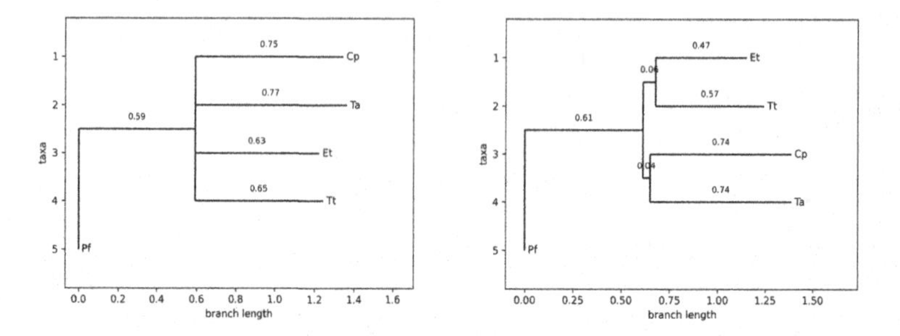

Fig. 6. Fréchet means from the three trees from Fig. 5. Left: in BHV-space, right: in wald space.

4 Comparing Fréchet Means

For illustration, we take $n = 3$ trees $W_1, \ldots, W_n \in \mathcal{W}$ from Nye et al. (2016) depicted in Fig. 5, each of which having $N = 5$ leaves (3 taxa and the root were removed from the original trees for computational tractability). We compute their Fréchet means

$$W^* \in \underset{W \in \mathcal{W}}{\operatorname{argmin}} \sum_{k=1}^{n} d_{\mathcal{W}}(W_k, W)^2$$

in BHV-space and in wald space, cf. Fig. 6. For computation we use the algorithm of Sturm (2003). In general, the computation of other types of means is also possible (e.g. the Riemannian 1-center, cf. Arnaudon et al. (2013)).

While in BHV-space, the Fréchet mean is unique, in wald space its uniqueness is dubious. For both spaces we have performed 15 iterations after which the final subsequent iterates were less than 0.05 apart, respectively. Remarkably, the mean tree in BHV-space is a star tree. In wald space, however, it is a fully resolved tree.

References

Arnaudon, M., Nielsen, F.: On approximating the Riemannian 1-center. Computational Geometry **46**(1), 93–104 (2013)

Barden, D., Le, H., Owen, M.: Central limit theorems for Fréchet means in the space of phylogenetic trees. Electron. J. Probab **18**(25), 1–25 (2013)

Barden, D., Le, H., Owen, M.: Limiting behaviour of Fréchet means in the space of phylogenetic trees. Annals of the Institute of Statistical Mathematics **70**(1), 99–129 (2016). https://doi.org/10.1007/s10463-016-0582-9

Bačák, M.: Computing Medians and Means in Hadamard Spaces. SIAM Journal on Optimization **24**(3), 1542–1566 (2014)

Billera, L., Holmes, S., Vogtmann, K.: Geometry of the space of phylogenetic trees. Advances in Applied Mathematics **27**(4), 733–767 (2001)

Brown, D. G. and M. Owen (2018, May). Mean and Variance of Phylogenetic Trees. arXiv:1708.00294 [math, q-bio, stat]. arXiv: 1708.00294

Garba, M.K., Nye, T.M., Boys, R.J.: Probabilistic Distances Between Trees. Systematic Biology **67**(2), 320–327 (2018)

Garba, M. K., Nye, T. M. W., Lueg, J., Huckemann, S. F.: Information geometry for phylogenetic trees. Journal of Mathematical Biology **82**(3), 1–39 (2021). https://doi.org/10.1007/s00285-021-01553-x

Garba, M. K., T. M. W. Nye, J. Lueg, and S. F. Huckemann (2021). Information metrics for phylogenetic trees via distributions of discrete and continuous characters. In: Nielsen, F., Barbaresco, F. (Eds.) GSI 2021, LNCS 12829, pp. 701–709 (2021). https://doi.org/10.1007/978-3-030-80209-7_75

Hotz, T., Huckemann, S., Le, H., Marron, J.S., Mattingly, J., Miller, E., Nolen, J., Owen, M., Patrangenaru, V., Skwerer, S.: Sticky central limit theorems on open books. Annals of Applied Probability **23**(6), 2238–2258 (2013)

Huckemann, S., Mattingly, J.C., Miller, E., Nolen, J.: Sticky central limit theorems at isolated hyperbolic planar singularities. Electronic Journal of Probability **20**(78), 1–34 (2015)

Lang, S.: Fundamentals of Differential Geometry. Graduate Texts in Mathematics. Springer-Verlag, New York (1999)

Lueg, J., T. Nye, M. Garba, and S. F. Huckemann (2021). Phylogenetic wald spaces. manuscript

Miller, E., Owen, M., Provan, J.S.: July). Polyhedral computational geometry for averaging metric phylogenetic trees. Advances in Applied Mathematics **68**, 51–91 (2015)

Moulton, V., Steel, M.: Peeling phylogenetic 'oranges'. Advances in Applied Mathematics **33**(4), 710–727 (2004)

Nye, T. M., X. Tang, G. Weyenberg, and Y. Yoshida (2016). Principal component analysis and the locus of the Fréchet mean in the space of phylogenetic trees. arXiv preprint arXiv:1609.03045

Owen, M.: Computing geodesic distances in tree space. SIAM Journal on Discrete Mathematics **25**(4), 1506–1529 (2011)

Rokas, A., B. L. Williams, N. King, and S. B. Carroll (2003, October). Genome-scale approaches to resolving incongruence in molecular phylogenies. Nature 425(6960), 798–804

Schmidt, Frank R., Clausen, Michael, Cremers, Daniel: Shape Matching by Variational Computation of Geodesics on a Manifold. In: Franke, Katrin, Müller, Klaus-Robert., Nickolay, Bertram, Schäfer, Ralf (eds.) DAGM 2006. LNCS, vol. 4174, pp. 142–151. Springer, Heidelberg (2006). https://doi.org/10.1007/11861898_15

Shiers, N., Zwiernik, P., Aston, J.A., Smith, J.Q.: The correlation space of gaussian latent tree models and model selection without fitting. Biometrika **103**(3), 531–545 (2016)

Sturm, K.: Probability measures on metric spaces of nonpositive curvature. Contemporary mathematics **338**, 357–390 (2003)

A Necessary Condition for Semiparametric Efficiency of Experimental Designs

Hisatoshi Tanaka(✉)

School of Political Science and Economics, Waseda University, Shinjuku, Tokyo 169-8050, Japan
hstnk@waseda.jp

Abstract. The efficiency of estimation depends not only on the method of estimation but also on the distribution of data. In statistical experiments, statisticians can at least partially design the data-generating process to obtain high estimation performance. This paper proposes a necessary condition for a semiparametrically efficient experimental design. We derived a formula to determine the efficient distribution of the input variables. The paper also presents an application to the optimal bid design problem of contingent valuation survey experiments.

Keywords: Optimal design · Semiparametric efficiency · Binary response model · Contingent valuation survey experiments

1 Introduction

This paper investigates a class of simple statistical experiments described by a 4-tuple,

$$\mathcal{E} = \{(\mu, \nu, \rho, \varphi) \mid \mu \in \mathcal{M}, \nu \in \mathcal{N}\}. \tag{1}$$

Here, $\mathcal{M}$ is a set of probability measures on $(\mathcal{W}, \mathcal{A})$, $\mathcal{N}$ is a set of probability measures on $(\mathcal{X}, \mathcal{B})$, ρ is a measurable map from $\mathcal{W} \times \mathcal{X}$ to $(\mathcal{Y}, \mathcal{C})$, and φ is a functional on $\mathcal{M}$. In every experiment $(\mu, \nu, \rho, \varphi) \in \mathcal{E}$, input x is drawn from ν, output $y = \rho(\omega, x)$ with $\omega \sim \mu$ is observed, and the value of $\varphi(\mu)$ is estimated from independent copies of (x, y).

For example, imagine that n lightning bulbs exist, whose lifetime hours $\omega_1, \ldots, \omega_n$ are i.i.d. random variables distributed according to μ. To estimate the expected lifetime hours $\varphi(\mu) = E_\mu \omega$, the following experiment was conducted. First, all n bulbs are turned on at time 0. Second, one bulb was sampled without replacement at time x, and its status was observed. If the sampled bulb was alive, y was set to 1. Otherwise, y was set to 0. The procedure was repeated n times until all bulbs were sampled. Finally, data of independent pairs (x_1, y_1), ..., (x_n, y_n) are obtained, and $E_\mu \omega$ will be consistently estimated by using existing estimation methods, such as the nonparametric maximum likelihood estimation [6].

Note that the efficiency of the estimation depends not only on the estimation method but also on the distribution ν of $x_1, \ldots, x_n$. In an extreme case where

F. Nielsen and F. Barbaresco (Eds.): GSI 2021, LNCS 12829, pp. 718–725, 2021.
https://doi.org/10.1007/978-3-030-80209-7_77

$x_1 = \cdots = x_n = 0$, trivial outcomes $y_1 = \cdots = y_n = 1$ will be obtained unless some bulbs have an initial failure. In the opposite extreme case, in which $x_1 = \cdots = x_n = +\infty$, $y_1 = \cdots = y_n = 0$ will occur with probability one. In both cases, the data are quite insufficient in that a consistent estimation of $E_\mu \omega$ is not possible. Finding the best distribution ν of x, with which the experiment produces the most informative data, is therefore an interesting problem.

The remainder of this paper is organized as follows: Sect. 2 formally states the problem of the paper. The efficient design is formulated as a minimizer of the Fisher information norm of the gradient of a functional defined on a statistical model set. Section 3 proposes a necessary condition for the efficient design. Section 4 presents application examples of the main theorem. Specifically, the optimal bid design problem of contingent valuation survey experiments is solved.

2 The Model

2.1 Tangent Space of a Statistical Model

This section introduces a theory of semiparametric estimation to formulate an efficient design problem. Terms and definitions given in the following are according to [9]. Closely related approaches were also found in [1] and [8]. See [7] for the generalization and comparison of different approaches.

Let $\mathcal{M}$ be a set of probability measures on $(\mathcal{W}, \mathcal{A})$, and let $\mu \in \mathcal{M}$. A map $t \mapsto \mu_t$ from $(-\epsilon, \epsilon) \subset \mathbb{R}$ to $\mathcal{M}$ such that $\mu_0 = \mu$ is *differentiable in quadratic mean* at $t = 0$ if $\alpha \in L_2(\mu)$ exists so that

$$\lim_{t \to 0} \int \left(\frac{\sqrt{d\mu_t} - \sqrt{d\mu}}{t} - \frac{1}{2} \alpha \sqrt{d\mu} \right)^2 = 0. \tag{2}$$

Note that α is a quadratic-mean version of the "score function" $(d/dt)_{t=0} \log d\mu_t$. A collection of those differentiable maps $t \mapsto \mu_t$ is denoted by $\mathcal{M}(\mu)$. A *tangent set* $T_\mu \mathcal{M}$ of $\mathcal{M}$ at μ is a set of *tangent vectors* α as in (2). In the following, assume that $T_\mu \mathcal{M}$ is a linear space. Further, assume that $t \mapsto \mu_{ht} \in \mathcal{M}(\mu)$ for every $h \in \mathbb{R}$ and $t \mapsto \mu_t \in \mathcal{M}(\mu)$. On $T_\mu \mathcal{M}$, the *Fisher information metric* $\langle \cdot, \cdot \rangle_\mu$ is given by

$$\langle \alpha, \alpha' \rangle_\mu = \int_{\mathcal{W}} \alpha \alpha' \, d\mu \tag{3}$$

for every α and α' in $T_\mu \mathcal{M}$. The *Fisher information norm* $\| \cdot \|_\mu$ is given by $\|\alpha\|_\mu = \langle \alpha, \alpha \rangle_\mu^{1/2}$.

Proposition 1 [6]. *Let $\overline{T_\mu \mathcal{M}}$ be the closure of a tangent set $T_\mu \mathcal{M}$ with respect to $\| \cdot \|_\mu$. Then, $\overline{T_\mu \mathcal{M}} = L_2^0(\mu) := \{ \alpha \in L_2(\mu) \mid \int \alpha \, d\mu = 0 \}$.*

2.2 Score Operator

Let $\mathcal{N}$ be a class of probability measures on $(\mathcal{X}, \mathcal{B})$, and let $\mathcal{P}$ be a class of probability measures on $(\mathcal{X}\times\mathcal{Y}, \mathcal{B}\otimes\mathcal{C})$. For every $P \in \mathcal{P}$, let $\mathcal{P}(P)$ be a collection of differentiable-in-quadratic-mean maps $t \in (-\epsilon, \epsilon) \mapsto P_t \in \mathcal{P}$ so that $P_0 = P$ and

$$\lim_{t\to 0} \int \left(\frac{\sqrt{dP_t} - \sqrt{dP}}{t} - \frac{1}{2}\beta\sqrt{dP} \right)^2 = 0 \tag{4}$$

with $\beta \in L_2(P)$. The tangent set $T_P\mathcal{P}$ of $\mathcal{P}$ at P is a set of β as in (4). In the following, assume that $T_P\mathcal{P}$ is a linear space. The Fisher information norm on $T_P\mathcal{P}$ is given by $\|\beta\|_P = \left(\int \beta(x,y)\, dP(x,y)\right)^{1/2}$ for every $\beta \in T_P\mathcal{P}$.

Let $\rho_\nu(\mu)(\in \mathcal{P})$ be the joint distribution of (x, y) given by

$$\rho_\nu(\mu)(D) = \int \{(x, \rho(\omega, x)) \in D\}\, \mu(d\omega)\nu(dx), \quad D \in \mathcal{B}\otimes\mathcal{C}, \tag{5}$$

where $\omega \sim \mu$, $x \sim \nu$, and $y = \rho(\omega, x)$. The *score operator* $(d\rho_\nu)_\mu : T_\mu\mathcal{M} \mapsto L_2(\rho_\nu(\mu))$ associated with the map $\rho_\nu : \mathcal{M} \mapsto \mathcal{P}$, $\mu \mapsto \rho_\nu(\mu)$, is given by

$$((d\rho_\nu)_\mu\alpha)(x, y) = E_{\mu\times\nu}(\alpha(\omega) \mid x, y), \quad (x, y) \in \mathcal{X}\times\mathcal{Y}, \tag{6}$$

for every $\alpha \in T_\mu\mathcal{M}$ [7]. A tangent set of a subclass $\rho_\nu(\mathcal{M}) := \{\rho_\nu(\mu) \in \mathcal{P} \mid \mu \in \mathcal{M}\}$, which is a set of statistical models to be estimated in experiment $(\mu, \nu, \rho, \varphi)$, is the range of the score operator as follows:

$$T_{\rho_\nu(\mu)}\rho_\nu(\mathcal{M}) = R((d\rho_\nu)_\mu) = (d\rho_\nu)_\mu(T_\mu\mathcal{M}). \tag{7}$$

Here, $R(\cdot)$ denotes the range of given operators. The score operator is linear and continuous according to the Fisher information metrics. The conjugate operator $(d\rho_\nu)^*_\mu : L_2(\rho_\nu(\mu)) \mapsto L_2(\mu)$, which satisfies $\langle (d\rho_\nu)_\mu\alpha, \beta\rangle_{\rho_\nu(\mu)} = \langle \alpha, (d\rho_\nu)^*_\mu\beta\rangle_\mu$ for every $\alpha \in L_2(\mu)$ and $\beta \in L_2(\rho_\nu(\mu))$, is given by

$$((d\rho_\nu)^*_\mu\beta)(\omega) = E_{\mu\times\nu}(\beta(x, y) \mid \omega), \quad \omega \in \mathcal{W}. \tag{8}$$

2.3 Efficiency Bound

Assume that

(A1) there exists a linear, continuous operator $\varphi'_\mu : T_\mu\mathcal{M} \mapsto \mathbb{R}$ such as

$$\lim_{t\to 0} \frac{\varphi(\mu_t) - \varphi(\mu)}{t} = \varphi'_\mu\alpha \tag{9}$$

for every differentiable-in-quadratic-mean path $t \mapsto \mu_t \in \mathcal{M}(\mu)$ with a tangent vector $\alpha \in T_\mu\mathcal{M}$.

According to Riesz's representation theorem, the *gradient function* $\partial\varphi_\mu \in L_2^0(\mu)$ uniquely exists so that $\varphi'_\mu\alpha \equiv \langle \partial\varphi_\mu, \alpha\rangle_\mu$. Assume also that

(A2) there exists a functional $\kappa : \mathcal{P} \mapsto \mathbb{R}$ such that $\kappa(\rho_\nu(\mu)) \equiv \varphi(\mu)$ for all $\mu \in \mathcal{M}$.

The functional κ is said to be *differentiable at* $P \in \mathcal{P}$ *relative to* $\mathcal{P}(P)$ if a linear, continuous operator $\kappa'_P : T_P\mathcal{P} \mapsto \mathbb{R}$ such as

$$\lim_{t\to 0} \frac{\kappa(P_t) - \kappa(P)}{t} = \kappa'_P \beta \tag{10}$$

for every differentiable-in-quadratic-mean path $t \mapsto P_t \in \mathcal{P}(P)$ exists. The *efficient influence function* $\partial\kappa_P$ is the Riesz representation of κ'_P on $\overline{T_P\mathcal{P}}$. According to van der Vaart's differentiability theorem [9], κ is differentiable at $\rho_\nu(\mu)$ relative to $\rho_\nu(\mathcal{M}(\mu)) := \{t \mapsto \rho_\nu(\mu_t) \mid t \mapsto \mu_t \in \mathcal{M}(\mu)\}$ if and only if

$$\partial\varphi_\mu \in R((d\rho_\nu)^*_\mu). \tag{11}$$

When (11) is satisfied, $\partial\kappa_{\rho_\nu(\mu)}$ is related to $\partial\varphi_\mu$ by using the *score equation*:

$$\partial\varphi_\mu = (d\rho_\nu)^*_\mu(\partial\kappa_{\rho_\nu(\mu)}), \quad \partial\kappa_{\rho_\nu(\mu)} \in \overline{R((d\rho_\nu)_\mu)}. \tag{12}$$

[9] also shows that the Fisher information norm $\|\partial\kappa_{\rho_\nu(\mu)}\|^2_{\rho_\nu(\mu)}$ provides the lower bound of the asymptotic variances of all regular estimators of $\varphi(\mu)$. Let $\mathcal{N}^*$ be a subclass of $\mathcal{N}$ so that $\mathcal{N}^* = \{\nu \in \mathcal{N} \mid \partial\varphi_\mu \in R((d\rho_\nu)^*_\mu)\}$, and then an efficiency criterion of experimental designs is given by

$$\text{l.b.}(\varphi(\mu) \mid \nu) := \begin{cases} \|\partial\kappa_{\rho_\nu(\mu)}\|^2_{\rho_\nu(\mu)} & \text{if } \nu \in \mathcal{N}^* \\ +\infty & \text{if } \nu \notin \mathcal{N}^*. \end{cases} \tag{13}$$

Definition 1. *The probability measure* ν^* *is efficient for experiment* $\mathcal{E}$ *at* μ *if*

$$\text{l.b.}(\varphi(\mu) \mid \nu^*) \le \text{l.b.}(\varphi(\mu) \mid \nu) \tag{14}$$

for every $\nu \in \mathcal{N}$.

3 Main Results

A main theorem of the paper is given as follows:

Theorem 1. *If* $\nu^* \in \mathcal{N}^*$ *is efficient for experiment* $\mathcal{E}$ *at* μ, *then*

$$x \mapsto E_{\mu\times\nu^*}(\partial\kappa_{\rho_{\nu^*}(\mu)}(x, y)^2 \mid x) \tag{15}$$

is ν^**-a.s. constant on* $\mathcal{X}$.

The intuition behind the condition is obtained from the following expression:

$$\text{l.b.}(\varphi(\mu) \mid \nu^*) = \int E_{\mu\times\nu^*}(\partial\kappa_{\rho_{\nu^*}(\mu)}(x, y)^2 \mid x)\, \nu^*(dx). \tag{16}$$

If the lower bound is minimized at $\nu = \nu^*$, any small perturbations added to ν^* would not significantly change the value of $\text{l.b.}(\varphi(\mu) \mid \nu^*)$. This is possible only if the integrand $E_{\mu\times\nu^*}(\partial\kappa_{\rho_{\nu^*}(\mu)}(x, y)^2 \mid x)$ of (16) is independent of x.

Lemma 1. *For every $\nu \in \mathcal{N}^*$, $E_\nu\big[E_{\mu\times\nu}(\partial\kappa_{\rho_\nu(\mu)}(x,y)^2 \mid x)\big]^{1/2} < \infty$.*

Proof. Let $\nu \in \mathcal{N}^*$. There exists $\beta^* \in L_2(\rho_\nu(\mu))$ so that $\partial\varphi_\mu = (d\rho_\nu)^*_\mu\beta^*$. Let Π_ν be the orthogonal projection from $L_2(\rho_\nu(\mu))$ to $\overline{R((d\rho_\nu)_\mu)}$; then,

$$\langle\beta^* - \Pi_\nu\beta^*, (d\rho_\nu)_\mu\alpha\rangle_{\rho_\nu(\mu)} = \langle(d\rho_\nu)^*_\mu(\beta^* - \Pi_\nu\beta^*), \alpha\rangle_\mu = 0 \tag{17}$$

for every $\alpha \in L_2(\mu)$. Hence, there exists $\delta_\nu \in Ker((d\rho_\nu)^*_\mu)$ so that $\Pi_\nu\beta^* = \beta^* - \delta_\nu \in \overline{R((d\rho_\nu)_\mu)}$ and $(d\rho_\nu)^*_\mu\Pi_\nu\beta^* = (d\rho_\nu)^*_\mu\beta^* = \partial\varphi_\mu$, which shows that $\partial\kappa_{\rho_\nu(\mu)} = \Pi_\nu\beta^*$ is the efficient influence function of κ at $\rho_\nu(\mu)$. By Jensen's inequality and the property of projections,

$$\begin{aligned}\int\big[E_{\mu\times\nu}(\partial\kappa^2_{\rho_\nu(\mu)}(x,y) \mid x)\big]^{1/2}\,\nu(dx) &\le \left(\int E_{\mu\times\nu}(\partial\kappa^2_{\rho_\nu(\mu)}(x,y) \mid x)\,\nu(dx)\right)^{1/2}\\ &= \|\Pi_\nu\beta^*\|_{\rho_\nu(\mu)} \le \|\beta^*\|_{\rho_\nu(\mu)} < \infty.\end{aligned}$$

Lemma 2. *Define a map $\Gamma : \mathcal{N}^* \mapsto \mathcal{N}$ by*

$$\frac{d(\Gamma\nu)}{d\nu}(x) = C^{-1}\big[E_{\mu\times\nu}(\partial\kappa_{\rho_\nu(\mu)}(x,y)^2 \mid x)\big]^{1/2}, \tag{18}$$

where $C = E_\nu\big[E_{\mu\times\nu}(\partial\kappa_{\rho_\nu(\mu)}(x,y)^2 \mid x)\big]^{1/2}$. Then, $\Gamma\nu \in \mathcal{N}^*$ for every $\nu \in \mathcal{N}^*$.

Proof. For the simplicity of description, let $\nu' = \Gamma\nu$. For every $\nu \in \mathcal{N}^*$, there exists $\beta^* \in L_2(\rho_\nu(\mu))$ such that $\partial\kappa_{\rho_\nu(\mu)} = \Pi_\nu\beta^*$. We define $B_0 \subset \mathcal{X}$ as

$$B_0 = \big\{x \in \mathcal{X} \mid E_{\mu\times\nu}(\partial\kappa_{\rho_\nu(\mu)}(x,y)^2|x) = 0\big\}. \tag{19}$$

Let $\nu_0(B) := \nu(B\backslash B_0)$ and $\nu^\perp(B) := \nu(B \cap B_0)$ for $B \in \mathcal{B}$ so that $\nu' \sim \nu_0$, $\nu' \perp \nu^\perp$, $\nu = \nu_0 + \nu^\perp$, and $(d\nu'/d\nu)(x) = (d\nu'/d\nu_0)(x) = C^{-1}\big[E_{\mu\times\nu}(\partial\kappa^2_{\rho_\nu(\mu)}|x)\big]^{1/2}$. Note that $\partial\kappa_{\rho_\nu(\mu)}(x,\rho(\omega,x))\{x \in B_0\} \equiv 0$ μ-a.s. and that $(d\nu_0/d\nu')\partial\kappa_{\rho_\nu(\mu)} \in L_2(\rho_{\nu'}(\mu))$. Let $\Pi_{\nu'} : L_2(\rho_{\nu'}(\mu)) \mapsto \overline{R((d\rho_{\nu'})_\mu)}$ be the orthogonal projection, then

$$\begin{aligned}(d\rho_{\nu'})^*_\mu\left(\Pi_{\nu'}\frac{d\nu_0}{d\nu'}\partial\kappa_{\rho_\nu(\mu)}\right)(\omega) &= \int\left(\frac{d\nu_0}{d\nu'}(x)\right)\partial\kappa_{\rho_\nu(\mu)}(x,\rho(\omega,x))\,\nu'(dx)\\ &= \int\partial\kappa_{\rho_\nu(\mu)}(x,\rho(\omega,x))\,(\nu-\nu^\perp)(dx) = \left((d\rho_\nu)^*_\mu\,\partial\kappa_{\rho_\nu(\mu)}\right)(\omega) = \partial\varphi_\mu(\omega).\end{aligned}$$

Therefore, $\partial\varphi_\mu \in R((d\rho_{\nu'})^*_\mu)$ holds at $\nu' = \Gamma\nu$.

Lemma 3. *For every $\nu \in \mathcal{N}^*$, $\mathrm{l.b.}(\varphi(\mu)\,|\,\Gamma\nu) \le \mathrm{l.b.}(\varphi(\mu)\,|\,\nu)$.*

Proof. Let $\nu' = \Gamma\nu$. Because the efficient influence function of κ at $\rho_{\nu'}(\mu)$ is given by $\partial\kappa_{\rho_{\nu'}(\mu)} = \Pi_{\nu'}(d\nu_0/d\nu')\partial\kappa_{\rho_\nu(\mu)}$,

$$\mathrm{l.b.}(\varphi(\mu)|\nu') \le \left\|\left(\frac{d\nu_0}{d\nu'}\right)\partial\kappa_{\rho_\nu(\mu)}\right\|^2_{\rho_{\nu'}(\mu)} = C^2 \le \big\|\partial\kappa_{\rho_\nu(\mu)}\big\|^2_{\rho_\nu(\mu)} = \mathrm{l.b.}(\varphi(\mu)|\nu).$$

Proof of Theorem 1. Let $\zeta^* = E_{\mu\times\nu^*}(\partial\kappa_{\rho_{\nu^*}(\mu)}(x,y)^2 \mid x)$, and then

$$\text{l.b.}(\varphi(\mu)|\Gamma\nu^*) \leq (E_{\nu^*}\sqrt{\zeta^*})^2 \leq E_{\nu^*}\zeta^* = \text{l.b.}(\varphi(\mu)|\nu^*) \leq \text{l.b.}(\varphi(\mu)|\Gamma\nu^*) \quad (20)$$

according to the previous lemma. Therefore, $E_{\nu^*}\zeta^* = (E_{\nu^*}\sqrt{\zeta^*})^2$, which implies that $\text{Var}_{\nu^*}\zeta^* = 0$. □

4 Examples

4.1 Model Without Information Loss

Consider an experiment $\mathcal{E}$ with $\mathcal{W} = \mathcal{X} = \mathcal{Y} = \mathbb{R}$ and $\rho(\omega, x) = \omega + x$. Let $\mathcal{M}$ be a set of probability measures on $\mathbb{R}$, and let $\mathcal{N} \subseteq \mathcal{M}$. The model set is $\rho_\nu(\mathcal{M}) = \{\rho_\nu(\mu) \mid \mu \in \mathcal{M}\}$, in which $\rho_\nu(\mu)(D) = \int\{(x, x+\omega) \in D\}\,\mu(d\omega)\nu(dx)$. The score operator $(d\rho_\nu)_\mu$ maps each $\alpha \in L_2^0(\mu)$ to $((d\rho_\nu)_\mu\alpha)(x,y) = \alpha(y-x)$. The adjoint operator is $((d\rho_\nu)_\mu^*\beta)(\omega) = \int \beta(x, \omega + x)\,\nu(dx)$. The efficient influence function of $\kappa(\rho_\nu(\mu)) = \varphi(\mu)$ is $\partial\kappa_{\rho_\nu(\mu)}(x,y) = \partial\varphi_\mu(y-x)$, and the efficiency bound is provided by

$$\text{l.b.}(\varphi(\mu)|\nu) = \int \partial\varphi_\mu(y-x)^2\rho_\nu(\mu)(dx,dy) = \int \partial\varphi_\mu(\omega)^2\mu(d\omega), \quad (21)$$

which is independent of ν. Therefore, an arbitrary $\nu \in \mathcal{N}$ is efficient for $\mathcal{E}$, and $E_{\mu\times\nu}(\partial\kappa^2_{\rho_\nu(\mu)}(x,y) \mid x) = \int \partial\varphi_\mu(\omega)^2\mu(d\omega)$ is a constant that is independent of $\nu \in \mathcal{N}$.

In this example, ω is always observable since $\omega = y - x$. Therefore, the distribution of x has no impact on efficiency of the estimation. The example suggests that the choice of ν becomes significant only when a model with information loss is estimated.

4.2 Parametric Models

Let $\mathcal{M} = \{\mu_\theta \mid \theta \in \Theta\}$ be a collection of statistical models parameterized by $\theta \in \Theta \subset \mathbb{R}^l$. Assume that every μ_θ is absolutely continuous with respect to a reference measure π. Let $\dot{\ell}_\theta = (\partial/\partial\theta)(d\mu_\theta/d\pi)$, and then $T_{\mu_\theta}\mathcal{M} = \{\tau'\dot{\ell}_\theta : \tau \in \mathbb{R}^l\}$, and $\overline{R((d\rho_\nu)_{\mu_\theta})} = R((d\rho_\nu)_{\mu_\theta}) = \{\tau'(d\rho_\nu)_{\mu_\theta}(\dot{\ell}_\theta) : \tau \in \mathbb{R}^l\}$. Let $\varphi(\mu_\theta)$ be the target of estimation and let $\varphi_0(\theta) := \varphi(\mu_\theta)$. Then, $\kappa(\rho_\nu(\mu_\theta)) = \varphi_0(\theta)$ is differentiable if and only if $\tau \in \mathbb{R}^l$ so that

$$(\nabla_\theta\varphi_0)'\dot{\ell}_\theta(\omega) = \tau' E_\theta\left[E_\theta(\dot{\ell}_\theta(\omega)|x,y)\,\middle|\,\omega\right]. \quad (22)$$

Let $\mathcal{N}^*$ be a collection of ν with which (22) is satisfied by some τ. For every $\nu \in \mathcal{N}^*$, the efficient influence function is solved as $\partial\kappa_{\rho_\nu(\mu_\theta)} = (\tau^*)'\dot{\ell}_\theta$, where $\tau^* = (E_\theta[E_\theta(\dot{\ell}_\theta|x,y)E_\theta(\dot{\ell}_\theta|x,y)'])^{-1}E_\theta(\dot{\ell}_\theta\dot{\ell}_\theta')\nabla_\theta\varphi_0$. The efficiency bound of $\varphi_0(\theta)$ is now given by $\text{l.b.}(\varphi_0(\theta)|\nu) = (\tau^*)'E_\theta(\dot{\ell}_\theta\dot{\ell}_\theta')(\tau^*)$, which is minimized with respect to ν only when $x \mapsto E_\theta(\partial\kappa_{\rho_\nu(\mu_\theta)}(x,y)^2 \mid x) = (\tau^*)'E_\theta(\dot{\ell}_\theta\dot{\ell}_\theta'|x)(\tau^*)$ is a constant map. Because $\omega \perp\!\!\!\perp x$, the condition is satisfied by an arbitrary $\nu \in \mathcal{N}^*$.

The example suggests that, when a parametric model of $\mathcal{M}$ is assumed, choice of ν is irrelevant to estimation efficiency, while it is still relevant to identification and differentiability of the parameter. In other words, the efficient design problem becomes degenerated when the model is parametric.

4.3 Dichotomous Choice Contingent Valuation Experiment

Consider an experiment $\mathcal{E}$ with $\mathcal{W} = \mathcal{X} = [0,\infty)$, $\mathcal{Y} = \{0,1\}$, and $\rho(\omega,x) = \{\omega < x\}$. This *dichotomous choice contingent valuation* (DC-CV) experiment is one of the most widely used experimental methods in environmental economics [2]. The results of Monte Carlo simulations of the DC–CV experiments, showing the sensitivity of the estimates to the choice of ν, are reported in [3]. An efficient design for the DC–CV experiment to estimate the mean $E_\mu\omega$ was proposed by [4] and [5]. In the following, their results are generalized to estimate arbitrary smooth functionals $\varphi(\mu)$.

Let $\mathcal{M}$ be a set of probability measures on $[0,\infty)$, where every $\mu \in \mathcal{M}$ is equivalent to the Lebesgue measure $\lambda = \lambda_{[0,\infty)}$ on $[0,\infty)$. Let $\mathcal{N} = \mathcal{M}$, and then the model set $\rho_\nu(\mathcal{M}) = \{\rho_\nu(\mu)|\mu \in \mathcal{M}\}$ is provided by

$$\rho_\nu(\mu)(D) = \int_D \mu[0,x)^y \mu[x,\infty)^{1-y}\,\nu(dx)\delta_{\mathcal{Y}}(dy) \tag{23}$$

for every Borel set $D \subset \mathbb{R}^2$, where $\delta_{\mathcal{Y}}$ is the Dirac measure on $\mathcal{Y}$. Note that $E_{\mu\times\nu}(y \mid x) = \mu[0,x)$ holds. The score operator $(d\rho_\nu)_\mu$ maps each $\alpha \in L_2^0(\mu)$ to

$$((d\rho_\nu)_\mu\alpha)(x,y) = \frac{y - \mu[0,x)}{\mu[0,x)\mu[x,\infty)}\int_0^x \alpha\,d\mu, \tag{24}$$

and the adjoint operator is provided by

$$((d\rho_\nu)_\mu^*\beta)(\omega) = \int_0^\omega \beta(x,0)\,\nu(dx) + \int_\omega^\infty \beta(x,1)\,\nu(dx). \tag{25}$$

Assume that the gradient $\omega \mapsto \partial\varphi_\mu(\omega)$ is differentiable with respect to ω with derivative $(\partial\varphi_\mu)'(\omega) = (d/d\omega)(\partial\varphi_\mu)(\omega)$ and that $\lim_{M\uparrow\infty}\partial\varphi_\mu(M)\mu[M,\infty) = 0$. If $\partial\varphi_\mu \in R((d\rho_\nu)_\mu)$ at some $\nu \sim \lambda$,

$$\beta(x,y) := -\frac{d\lambda}{d\nu}(x)(\partial\varphi_\mu)'(x)(y - \mu[0,x)) \tag{26}$$

solves $\partial\varphi_\nu = (d\rho_\nu)_\mu^*\beta$ because

$$\begin{aligned}((d\rho_\nu)_\mu^*\beta)(\omega) &= \int_0^\omega (\partial\varphi_\mu)'(x)\,dx - \int_0^\infty (\partial\varphi_\mu)'(x)\,\mu[x,\infty)\,dx \\ &= \partial\varphi_\mu(\omega) - \lim_{M\uparrow\infty}\partial\varphi_\mu(M)\mu[M,\infty) = \partial\varphi_\mu(\omega).\end{aligned}$$

If $\beta \in \overline{R((d\rho_\nu)_\mu)}$ is satisfied, $\partial\kappa_{\rho_\nu(\mu)} = \beta$ is the efficient influence function to estimate $\varphi(\mu)$. Let ν^* be efficient for $\mathcal{E}$ at μ. Then, a positive constant C exists so that, for any $x \in [0, \infty)$,

$$C = E_{\mu\times\nu^*}\left[\left(-\frac{d\lambda}{d\nu^*}(x)\ (\partial\varphi_\mu)'(x)\ (y - \mu[0,x))\right)^2 \middle|\ x\right],$$

which implies

$$\frac{d\nu^*}{d\lambda}(x) = \frac{|(\partial\varphi_\mu)'(x)|\ \sqrt{\mu[0,x)\mu[x,\infty)}}{\int_0^\infty |(\partial\varphi_\mu)'(\xi)|\ \sqrt{\mu[0,\xi)\mu[\xi,\infty)}\, d\xi}. \tag{27}$$

Particularly when $\varphi(\mu) = E_\mu\omega = \int_0^\infty \omega\,\mu(d\omega)$, $\partial\varphi_\mu = \omega - E_\mu\omega$ and $(\partial\varphi_\mu)' \equiv 1$, and the formula (27) becomes equivalent to the result of [4] and [5]. It is also confirmed that $\partial\varphi_\mu \in R((d\rho_\nu)_\mu)$ and $\beta \in \overline{R((d\rho_\nu)_\mu)}$ are satisfied at $\nu = \nu^*$.

References

1. Ay, N., Jost, J., Lê, H.V., Schwachhöfer, L.: Information geometry. Ergebnisse der Mathematik und ihrer Grenzgebiete **3**, 64 (2017). Springer
2. Carson, R., Hanemann, M.: Contingent valuation (Chap. 17). In: Maler, K.G., Vincent, J.R. (eds.) Handbook of Environmental Economics, 1st edn, vol. 2, pp. 821–936. Elsevier, Amsterdam (2006)
3. Cooper, J., Loomis, J.: Sensitivity of willingness-to-pay estimates to bid design in dichotomous choice contingent valuation models. Land Econ. **68**(2), 211–224 (1992)
4. Cooper, J.C.: Optimal bid selection for dichotomous choice contingent valuation surveys. J. Environ. Econ. Manag. **24**(1), 25–40 (1993)
5. Duffield, J.W., Patterson, D.A.: Inference and optimal design for a welfare measure in dichotomous choice contingent valuation. Land Econ. **67**(2), 225–239 (1991)
6. Groeneboom, P., Wellner, J.A.: Information Bounds and Nonparametric Maximum Likelihood Estimation. Birkhäuser, Basel (1992)
7. Lê, H.V.: Diffeological statistical models, the Fisher metric and probabilistic mappings. Mathematics **8**(2) (2020)
8. Pistone, G., Sempi, C.: An infinite-dimensional geometric structure on the space of all the probability measures equivalent to a given one. Ann. Stat. **23**(5), 1543–1561 (1995)
9. van der Vaart, A.W.: On differentiable functionals. Ann. Stat. **19**(1), 178–204 (1991)

Parametrisation Independence of the Natural Gradient in Overparametrised Systems

Jesse van Oostrum[1(✉)] and Nihat Ay[1,2,3]

[1] Hamburg University of Technology, 21073 Hamburg, Germany
{jesse.van,nihat.ay}@tuhh.de
[2] Leipzig University, 04109 Leipzig, Germany
[3] Santa Fe Institute, Santa Fe, NM 87501, USA

Abstract. In this paper we study the natural gradient method for overparametrised systems. This method is based on the natural gradient field which is invariant with respect to coordinate transformations. One calculates the natural gradient of a function on the manifold by multiplying the ordinary gradient of the function by the inverse of the Fisher Information Matrix (FIM). In overparametrised models, the FIM is degenerate and therefore one needs to use a generalised inverse. We show explicitly that using a generalised inverse, and in particular the Moore-Penrose inverse, does not affect the parametrisation independence of the natural gradient. Furthermore, we show that for singular points on the manifold the parametrisation independence is not even guaranteed for non-overparametrised models.

Keywords: Natural gradient · Riemannian metric · Deep learning · Information geometry

1 Introduction

Within the field of deep learning, gradient methods have become ubiquitous tools for parameter optimisation. The natural gradient method, first proposed by Amari [1], is an efficient method for performing gradient descent. It is an active field of study within information geometry [2] and has been shown to be extremely effective in many applications [3,8,9]. The natural gradient is defined independently of a specific parametrisation. Although it is an open problem, there is work supporting the idea that the efficiency of learning of the method is due to this invariance [11]. The natural gradient is calculated by multiplying the ordinary gradient by the inverse of the Fisher Information Matrix (FIM) associated with the statistical manifold. For high-dimensional parameter spaces, this inversion is computationally expensive but different solutions have been proposed [2,5,7]. In many practical applications of machine learning, and in

F. Nielsen and F. Barbaresco (Eds.): GSI 2021, LNCS 12829, pp. 726–735, 2021.
https://doi.org/10.1007/978-3-030-80209-7_78

particular deep learning, one deals with overparametrised models, in which different configurations of the parameters correspond to the same output distribution. This causes the FIM to be degenerate. In this case, one uses a generalised inverse of the FIM to calculate the natural gradient [4]. The Moore-Penrose (MP) inverse is the canonical choice for this. The definition of the MP inverse is based on the Euclidean inner product defined on the parameter space. Using the MP inverse is therefore thought to affect the parametrisation independence of the natural gradient [7], and thus potentially the performance of the natural gradient method.

In this paper we show explicitly that for overparametrised models parametrisation independence is not affected when using a generalised, and in particular MP, inverse. We will show that for non-singular points (see Sect. 2) on the manifold, a generalised inverse does not introduce parametrisation dependence. Furthermore, we will show that for singular points parametrisation independence is not even guaranteed for non-overparametrised models.

2 Parametrisation (In)dependence of the Natural Gradient

Let $(\mathcal{Z}, g)$ be a Riemannian manifold, $\Xi \subset \mathbb{R}^d$ a smooth manifold of parameters, $\phi : \Xi \to \mathcal{Z}$ a smooth map (taking the role of a parametrisation), $\mathcal{M} \equiv \phi(\Xi) \subset \mathcal{Z}$ a model, and $\mathcal{L} : \mathcal{M} \to \mathbb{R}$ a smooth (loss) function. We call $p \in \mathcal{M}$ *non-singular* if $\mathcal{M}$ is locally an embedded submanifold of $\mathcal{Z}$ around p and we denote the set of non-singular points with $\text{Smooth}(\mathcal{M})$. A point p is called *singular* if it is not non-singular. The gradient of $\mathcal{L}$ at $p \in \text{Smooth}(\mathcal{M})$ is defined implicitly as follows:

$$g_p\left(\text{grad}_p\mathcal{L}, \cdot\right) = d\mathcal{L}_p(\cdot). \tag{1}$$

By the Riesz representation theorem, this defines the gradient uniquely. We define the pushforward of the tangent vector on the parameter space through the parametrisation as $\partial_i(\xi) \equiv d\phi_\xi\left(\frac{\partial}{\partial \xi^i}|_\xi\right)$, and the matrix[1] $G(\xi)$ be such that $G_{ij}(\xi) = g_{\phi(\xi)}\left(\partial_i(\xi), \partial_j(\xi)\right)$. We denote the vector of coordinate derivatives with $\nabla_{\partial(\xi)}\mathcal{L} \equiv (\partial_1(\xi)\mathcal{L}, ..., \partial_d(\xi)\mathcal{L}) \in \mathbb{R}^d$. Furthermore, following the Einstein summation convention, we write $a^i b_i$ for the sum $\sum_i a^i b_i$.

2.1 Gradient in Non-singular Points, $p \in \text{Smooth}(\mathcal{M})$

We assume that ϕ is a *proper parametrisation* such that for all ξ for which $\phi(\xi) \in \text{Smooth}(\mathcal{M})$ we have: $\text{span}(\{\partial_1(\xi), ..., \partial_d(\xi)\}) = T_{\phi(\xi)}\mathcal{M}$.

Definition 1. *A generalised inverse of a matrix A, denoted A^+, is a matrix satisfying the following property:*

$$AA^+A = A. \tag{2}$$

[1] Note that this becomes the FIM when g is the Fisher metric.

Note that this definition implies that for w in the image of A, i.e. $w = Av$, we have:

$$AA^+w = AA^+Av = Av = w. \tag{3}$$

This shows that AA^+ is the identity operator on the image of A.

Theorem 1. *For $\xi \in \Xi$ such that $\phi(\xi) = p \in \text{Smooth}(\mathcal{M})$, we have:*

$$\text{grad}_p\mathcal{L} = \left(G^+(\xi)\nabla_{\partial(\xi)}\mathcal{L}\right)^i \partial_i(\xi). \tag{4}$$

From this theorem we can conclude that the natural gradient of a loss function $\mathcal{L}$, which is itself defined independently of any parametrisation in Eq. (1), can be obtained by applying a generalised inverse of the FIM to the vector of coordinate derivatives $\nabla_{\partial(\xi)}\mathcal{L}$. In particular this proves that the MP inverse preserves the parameter independence in overparametrised models.

The proof of Theorem 1 will be based on the following:

Lemma 1. *Let $(V, \langle\cdot,\cdot\rangle)$ be a finite-dimensional inner product space and V^* its dual space. Let $\{e_i\}_{i\in\{1,...,d\}} \subset V$ be such that $\text{span}(\{e_i\}_{i\in\{1,...,d\}}) = V$. We denote $A_{ij} = \langle e_i, e_j\rangle$. For $v' \in V^*$ and $(v^1, ..., v^d) \in \mathbb{R}^d$ we have:*

$$v'(\cdot) = \langle v^i e_i, \cdot\rangle \iff v'(e_j) = A_{ij}v^i \;\forall j \in \{1, ..., d\}. \tag{5}$$

Moreover, the vector $(v'(e_1), ..., v'(e_d)) \in \mathbb{R}^d$ is in the image of A.

Proof. ($\Longrightarrow$) filling in gives:

$$v'(e_j) = \langle v, e_j\rangle = v^i\langle e_i, e_j\rangle = A_{ij}v^i \tag{6}$$

($\Longleftarrow$) Let $w = w^j e_j \in V$ be arbitrary. We have:

$$v'(w) = w^j v^i A_{ij} = w^j v^i\langle e_i, e_j\rangle = \langle v^i e_i, w_j e_j\rangle = \langle v^i e_i, w\rangle \tag{7}$$

From the Riesz representation theorem it follows that there exists a $v \in V$ such that $v'(\cdot) = \langle v, \cdot\rangle$, which can be written as $v = v^i e_i$. From the above it follows that multiplying A with $(v^1, ..., v^d)$ gives $(v'(e_1), ..., v'(e_d))$ and is therefore in the image of A. □

Proof (of Theorem 1). We now let $T_p\mathcal{M}$ take the role of V, $d\mathcal{L}_p$ the role of v', $\partial_i(\xi)$ the role of e_i, and $G(\xi)$ the role of A. From the lemma we know that for all $x = (x^1, ..., x^d) \in \mathbb{R}^d$ such that:

$$\partial_j(\xi)\mathcal{L} = G_{ij}(\xi)x^i \quad \forall j \in \{1, ..., d\}, \tag{8}$$

we have:

$$\text{grad}_p\mathcal{L} = x^i\partial_i(\xi). \tag{9}$$

Equation (8) can be written in vector form as follows:

$$\nabla_{\partial(\xi)}\mathcal{L} = G(\xi)x. \tag{10}$$

Suppose there exists a generalised inverse of $G(\xi)$, denoted $G^+(\xi)$. From the lemma we know that $\nabla_{\partial(\xi)}\mathcal{L}$ is in the image of $G(\xi)$. We saw above that $G(\xi)G^+(\xi)$ is the identity operator on the image of $G(\xi)$ and therefore we have that $x = G^+(\xi)\nabla_{\partial(\xi)}\mathcal{L}$ satisfies (10). Plugging this into (9) gives:

$$\mathrm{grad}_p\mathcal{L} = \left(G^+(\xi)\nabla_{\partial(\xi)}\mathcal{L}\right)^i \partial_i(\xi), \tag{11}$$

which is what we wanted to show. □

Remark 1. Note that although the calculation of the gradient vector from the perspective of the manifold is independent of the parametrisation, the final result of the gradient step might still depend on this. This is because the direction of the gradient step is only at the point p parallel to the gradient vector, but will in general not follow the gradient flow exactly after leaving point p. This is however not an issue specific to overparametrised models but with the natural gradient in general. See Section 12 of [6] for exact bounds on the invariance.

Parametrisation Dependence Within the Parameter Space
Let us denote the gradient of $\mathcal{L}$ on the parameter space Ξ at a point ξ as follows:

$$\mathrm{grad}^{\Xi}_{\xi}\mathcal{L} = \left(G^+(\xi)\nabla_{\partial(\xi)}\mathcal{L}\right)^i \frac{\partial}{\partial\xi^i}|_{\xi} \in T_{\xi}\Xi. \tag{12}$$

In this example, we will show explicitly that $\mathrm{grad}^{\Xi}_{\xi}\mathcal{L}$ *does* depend on the choice of parametrisation and therefore on the inner product of the parameter space, when G^+ is the MP inverse of G. However, we show as well that this dependence disappears when $\mathrm{grad}^{\Xi}_{\xi}\mathcal{L}$ is mapped to $T_{\phi(\xi)}\mathcal{M}$ through $d\phi$.

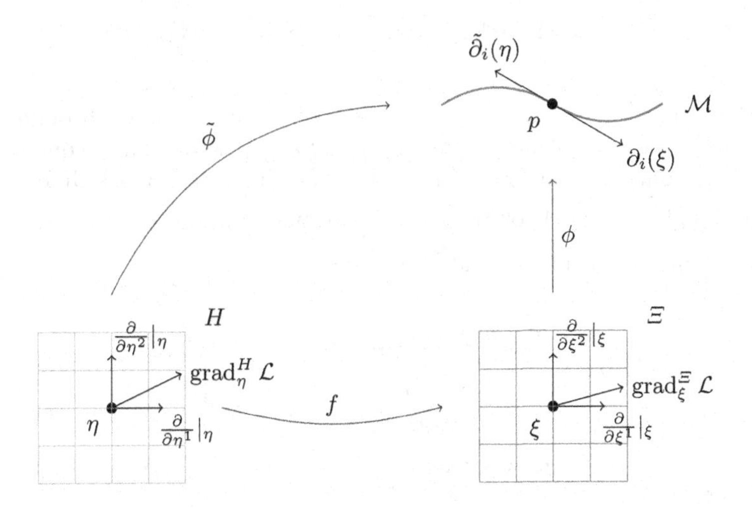

Fig. 1. Two parameterisations of $\mathcal{M}$ with different gradient vectors on the parameter space

Let us consider an alternative parametrisation $\tilde{\phi} : H \ni \eta \mapsto \tilde{\phi}(\eta) \in \mathcal{M}$ such that $\tilde{\phi} = \phi \circ f$ for a diffeomorphism $f : H \to \Xi$ (see Fig. 1). We define correspondingly: $\tilde{\partial}_i(\eta) = d\tilde{\phi}_\eta\left(\frac{\partial}{\partial \eta^i}|_\eta\right)$, and $\widetilde{G}_{ij}(\eta) = g_{\tilde{\phi}(\eta)}\left(\tilde{\partial}_i(\eta), \tilde{\partial}_j(\eta)\right)$, and $F(\eta)$ the matrix of partial derivatives of f at η, that is: $F_i^j(\eta) = \frac{\partial f^j}{\partial \eta^i}(\eta)$. For $\xi = f(\eta)$ we get the following relations:

$$\tilde{\partial}_i(\eta) = F_i^j(\eta)\, \partial_j(\xi), \tag{13}$$

$$\nabla_{\tilde{\partial}(\eta)} \mathcal{L} = F(\eta)\, \nabla_{\partial(\xi)} \mathcal{L}, \tag{14}$$

$$\widetilde{G}(\eta) = F(\eta)\, G(\xi)\, F^T(\eta). \tag{15}$$

We map $\mathrm{grad}_\eta^H \mathcal{L}$ to $T_\xi \Xi$ through df_η and get:

$$df_\eta \mathrm{grad}_\eta^H \mathcal{L} = df_\eta \left(\left(\widetilde{G}^+(\eta) \nabla_{\tilde{\partial}(\eta)} \mathcal{L} \right)^i \frac{\partial}{\partial \eta^i}|_\eta \right) \tag{16}$$

$$= \left(\left(F(\eta) G(\xi) F^T(\eta) \right)^+ F(\eta)\, \nabla_{\partial(\xi)} \mathcal{L} \right)^i F_i^j(\eta) \frac{\partial}{\partial \xi^j}|_\xi \tag{17}$$

$$= \left(F^T(\eta) \left(F(\eta) G(\xi) F^T(\eta) \right)^+ F(\eta)\, \nabla_{\partial(\xi)} \mathcal{L} \right)^j \frac{\partial}{\partial \xi^j}|_\xi. \tag{18}$$

From Theorem 1 and the fact that F is of full rank we know that $F\, \nabla_{\partial(\xi)} \mathcal{L}$ lies in the image of FGF^T. Therefore, by the definition of the MP inverse, we have that $\left(FGF^T\right)^+ F\, \nabla_{\partial(\xi)} \mathcal{L} = \arg\min_x \{||x|| : FGF^T x = F\, \nabla_{\partial(\xi)} \mathcal{L}\}$, where $||\cdot||$ is the Euclidean norm on $\mathbb{R}^d$. By substituting $y = F^T x$, the coefficients in (18) become:

$$y_H \equiv F^T(\eta) \left(F(\eta) G(\xi) F^T(\eta) \right)^+ F(\eta)\, \nabla_{\partial(\xi)} \mathcal{L} \tag{19}$$

$$= \arg\min_y \{|| \left(F^T(\eta) \right)^{-1} y|| : G(\xi) y = \nabla_{\partial(\xi)} \mathcal{L}\}. \tag{20}$$

Note that $|| \left(F^T\right)^{-1} (\cdot)||$ is the pushforward of the norm on H through f. This shows nicely the equivalence of the gradient for on the one hand constructing a different parametrisation ($\tilde{\phi}$), and on the other hand defining a different inner product $\left(|| \left(F^T\right)^{-1} (\cdot)||\right)$ for the existing parametrisation (ϕ).

Comparing the result to $\mathrm{grad}_\xi^\Xi \mathcal{L}$ gives:

$$\mathrm{grad}_\xi^\Xi \mathcal{L} = \left(G^+(\xi) \nabla_{\partial(\xi)} \mathcal{L} \right)^i \frac{\partial}{\partial \xi^i}|_\xi \tag{21}$$

$$= (y_\Xi)^i \frac{\partial}{\partial \xi^i}|_\xi, \tag{22}$$

$$y_\Xi = \arg\min_y \{||y|| : G(\xi) y = \nabla_{\partial(\xi)} \mathcal{L}\}. \tag{23}$$

Because the norms in (20) and (23) are different, generally $y_H \neq y_\Xi$. However, both satisfy $Gy = \nabla_{\partial(\xi)}\mathcal{L}$ and therefore $G(y_H - y_\Xi) = 0$. This implies:

$$d\phi_\xi \left(df_\eta \text{grad}_\eta^H \mathcal{L} - \text{grad}_\xi^\Xi \mathcal{L} \right) = (y_H - y_\Xi)^i \, \partial_i(\xi) \tag{24}$$

$$= 0 \tag{25}$$

where the last equality follows from non-degeneracy of the norm on $T_{\phi(\xi)}\mathcal{M}$:

$$|| \, (y_H - y_\Xi)^i \, \partial_i(\xi) ||_g^2 = (y_H - y_\Xi)^T \, G(\xi) \, (y_H - y_\Xi) = 0. \tag{26}$$

This shows that the dependency on the inner product disappears when the gradient is mapped to the manifold $\mathcal{M}$. See Example 1 in the appendix for a worked-out example of the above discussion.

2.2 Gradient in Singular Points, $p \notin$ Smooth($\mathcal{M}$)

In this section we will show that for $p \notin \text{Smooth}(\mathcal{M})$, even for a non-degenerate FIM the procedure for calculating the gradient described in Theorem 1 gives a result that is parametrisation-dependent.

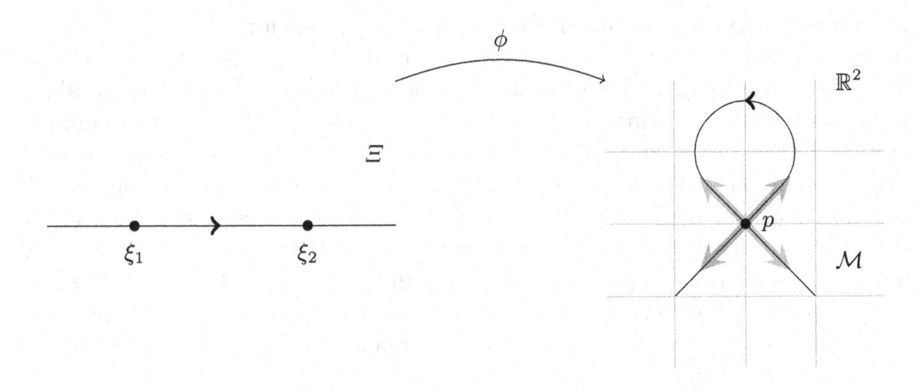

Fig. 2. Example parametrisation that contains a singular point

Let us consider the case in which ϕ is a smooth map from an interval on the real line to $\mathbb{R}^2$ as depicted in Fig. 2. We have that ξ_1 and ξ_2 are both mapped to the same point p in $\mathbb{R}^2$. Note that $\mathcal{M}$ is in this case not a locally embedded submanifold around p and thus p is a singular point. For $p \notin \text{Smooth}(\mathcal{M})$, we will denote the RHS of (4) with $\overline{\text{grad}}_p^{\partial(\xi)}\mathcal{L}$, to distinguish it from $\text{grad}_p\mathcal{L}$ which is only defined for non-singular points. Note that $G(\xi_1)$ is a real number different from zero and therefore non-degenerate. Applying the same procedure as above for finding the gradient at p gives:

$$\overline{\text{grad}}_p^{\partial(\xi_1)}\mathcal{L} = G^+(\xi_1)\nabla_{\partial(\xi_1)}\mathcal{L} \; \partial(\xi_1) \tag{27}$$

$$= G^{-1}(\xi_1)\frac{\partial \mathcal{L} \circ \phi}{\partial \xi}(\xi_1)\partial(\xi_1). \tag{28}$$

Note that $G^{-1}(\xi_1)\frac{\partial \mathcal{L}\circ\phi}{\partial \xi}(\xi_1)$ is a scalar. The resulting vector will therefore lie in the span of $\partial(\xi_1)$ illustrated by the blue arrows in the figure.

Now let $f : \Xi \to \Xi$ be a diffeomorphism such that $f(\xi_1) = \xi_2$. An alternative parametrisation of $\mathcal{M}$ is given by:

$$\tilde{\phi} = \phi \circ f. \tag{29}$$

We define equivalently for this alternative parametrisation $\tilde{\partial}_i(\xi) \equiv d\tilde{\phi}_\xi(\frac{\partial}{\partial \xi^i}|_\xi)$, and $\widetilde{G}(\xi)$ such that $\widetilde{G}_{ij}(\xi) = g_p\left(\tilde{\partial}_i(\xi), \tilde{\partial}_j(\xi)\right)$. Calculating the gradient for this parametrisation gives:

$$\overline{\mathrm{grad}}_p^{\tilde{\partial}(\xi_1)}\mathcal{L} = \widetilde{G}^{-1}(\xi_1)\frac{\partial \mathcal{L}\circ\tilde{\phi}}{\partial \xi}(\xi_1)\tilde{\partial}(\xi_1). \tag{30}$$

Note that this vector is in the span of $\tilde{\partial}(\xi_1)$ denoted by the red arrows in the figure and therefore in general different from (28). This shows that for $p \notin \mathrm{Smooth}(\mathcal{M})$, this procedure is not guaranteed to be parametrisation-independent. See Example 2 in the appendix for a worked-out example of this.

Necessity of Assumption of Proper Parametrisation

In the beginning of this section we assumed that ϕ is a proper parametrisation. We will now investigate the necessity of this assumption. We start by recalling some basic facts from smooth manifold theory: Let M, N be smooth manifolds and $F : M \to N$ a smooth map. We call a point $p \in M$ a *regular point* if $dF_p : T_pM \to T_{F(p)}N$ is surjective and a *critical point* otherwise. A point $q \in N$ is called a *regular value* if all the elements in $F^{-1}(q)$ are regular points, and a *critical value* otherwise. If M is n-dimensional, we say that a subset $S \subset M$ has measure zero in M, if for every smooth chart (U, ψ) for M, the subset $\psi(S \cap U) \subset \mathbb{R}^n$ has n-dimensional measure zero. That is: $\forall \delta > 0$, there exists a countable cover of $\psi(S \cap U)$ consisting of open rectangles, the sum of whose volumes is less than δ.

Proposition 1. *If* Smooth($\mathcal{M}$) *is a manifold, then the image of the set of points for which ϕ is not proper has measure zero in* Smooth($\mathcal{M}$).

Proof. From the definition of Smooth($\mathcal{M}$) we know that for every $p \in \mathrm{Smooth}(\mathcal{M})$ there exists a U_p open in $\mathcal{Z}$ such that $U_p \cap \mathcal{M}$ is an embedded submanifold of $\mathcal{Z}$. Let $U \equiv \bigcup_{p\in \mathrm{Smooth}(\mathcal{M})} U_p$. Note that: $U \cap \mathcal{M} = \mathrm{Smooth}(\mathcal{M})$ and therefore $\phi^{-1}(U) = \phi^{-1}(\mathrm{Smooth}(\mathcal{M}))$. Since U is open in $\mathcal{Z}$, $\phi^{-1}(\mathrm{Smooth}(\mathcal{M}))$ is an open subset of Ξ and thus an embedded submanifold. Therefore we can consider the map:

$$\phi|_{\phi^{-1}(\mathrm{Smooth}(\mathcal{M}))} : \phi^{-1}(\mathrm{Smooth}(\mathcal{M})) \to \mathrm{Smooth}(\mathcal{M}) \tag{31}$$

and note that the image of the set of points for which ϕ is not proper is equal to the set of critical values of $\phi|_{\phi^{-1}(\mathrm{Smooth}(\mathcal{M}))}$ in Smooth($\mathcal{M}$). A simple application of Sard's theorem gives the result. □

3 Conclusion

We have shown that for non-singular points, the natural gradient of a loss function on an overparametrised model is parametrisation-independent. Subsequently, we have shown that in singular points, there is no guarantee for the natural gradient to be parametrisation-independent, even in non-overparametrised models. In this paper, we have only looked at the case for which the FIM is known. In practice, one often has to use the empirical FIM based on the available data. When inverting this matrix it can be beneficial to apply Tikhonov regularisation [6]. Note that by letting the regularisation parameter go to zero, the MP inverse is obtained. Also only one type of singularity has been discussed. A further direction of investigation is the behaviour of the natural gradient on different types of singular points on the manifold, see also [10].

Acknowledgements. The authors acknowledge the support of the Deutsche Forschungsgemeinschaft Priority Programme "The Active Self" (SPP 2134).

Appendix

Example 1. Let us consider the specific example where:

$$Z = \Xi = H = \mathbb{R}^2, \tag{32}$$

$$\phi(\xi^1, \xi^2) = (\xi^1 + \xi^2, 0) \tag{33}$$

$$f(\eta^1, \eta^2) = (2\eta^1, \eta^2) \tag{34}$$

$$\mathcal{L}(x, y) = x^2 \tag{35}$$

Plugging this into the expressions derived above gives:

$$\tilde{\phi}(\eta^1, \eta^2) = (2\eta^1 + \eta^2, 0) \tag{36}$$

$$\partial_1(\xi) = \partial_2(\xi) = \frac{\partial}{\partial x}|_{\phi(\xi)} \qquad F(\eta) = \begin{bmatrix} 2 & 0 \\ 0 & 1 \end{bmatrix} \tag{37}$$

$$\tilde{\partial}_1(\eta) = 2\frac{\partial}{\partial x}|_{\tilde{\phi}(\eta)} \qquad G(\xi) = \begin{bmatrix} 1 & 1 \\ 1 & 1 \end{bmatrix} \tag{38}$$

$$\tilde{\partial}_2(\eta) = \frac{\partial}{\partial x}|_{\tilde{\phi}(\eta)} \qquad \widetilde{G}(\eta) = \begin{bmatrix} 4 & 2 \\ 2 & 1 \end{bmatrix} \tag{39}$$

$$\nabla_{\partial(\xi)}\mathcal{L} = (2(\xi^1 + \xi^2), 2(\xi^1 + \xi^2)) \tag{40}$$

$$\nabla_{\tilde{\partial}(\eta)}\mathcal{L} = (4(2\eta^1 + \eta^2), 2(2\eta^1 + \eta^2)) \tag{41}$$

Now we fix $\eta = (1, 1)$ and $\xi = f(\eta) = (2, 1)$. We start by computing y_Ξ. From the above we know that:

$$y_\Xi = \arg\min_y\{||y|| : G(\xi)y = \nabla_{\partial(\xi)}\mathcal{L}\}. \tag{42}$$

It can be easily verified that this gives $y_\Xi = (3,3)$. For y_H we get:

$$y_H = \arg\min_y \{|| \left(F^T\right)^{-1} y|| : G(\xi)y = \nabla_{\partial(\xi)}\mathcal{L}\} \tag{43}$$

which gives: $y_H = (4\frac{4}{5}, 1\frac{1}{5})$. Evidently we have $y_\Xi \neq y_H$. Note however that when we map the difference of the two gradient vectors from $T_{(2,1)}\Xi$ to $T_{(3,0)}\mathcal{M}$ through $d\phi_{(2,1)}$ we get:

$$d\phi_\xi \left(df_\eta \mathrm{grad}^H_\eta \mathcal{L} - \mathrm{grad}^\Xi_\xi \mathcal{L} \right) = (y_H - y_\Xi)^i \, \partial_i(\xi) \tag{44}$$

$$= (4\frac{4}{5} - 3)\frac{\partial}{\partial x}|_{(3,0)} + (1\frac{1}{5} - 3)\frac{\partial}{\partial x}|_{(3,0)} \tag{45}$$

$$= 0 \tag{46}$$

which shows that although the gradient vectors can be dependent on the parametrisation or inner product on the parameter space, when mapped to the manifold $\mathcal{M}$ they are invariant.

Example 2. We illustrate the discussion in Sect. 2.2 with a specific example. Let us consider the following parametrisation:

$$\phi : (-1/8\pi, 5/8\pi) \to (\mathbb{R}^2, \bar{g}) \tag{47}$$

$$t \mapsto (x, y) = (\sin(2t), \sin(t)) \tag{48}$$

This gives $\xi_1 = 0, \xi_2 = \frac{1}{2}\pi$ in the above discussion. We get the following calculation for $\overline{\mathrm{grad}}_p^{\partial(0)}\mathcal{L}$:

$$\partial^{(0)} = d\phi_t \left(\frac{\partial}{\partial t}|_t \right) \tag{49}$$

$$= 2\cos(2t)\frac{\partial}{\partial x}|_{\phi(t)} + \cos(t))\frac{\partial}{\partial y}|_{\phi(t)} \tag{50}$$

$$\partial^{(0)} = 2\frac{\partial}{\partial x}|_{(0,0)} + \frac{\partial}{\partial y}|_{(0,0)} \tag{51}$$

$$G(0) = 2^2 + 1^1 = 5 \tag{52}$$

$$\overline{\mathrm{grad}}_p^{\partial(0)}\mathcal{L} = \frac{1}{5}\frac{\partial}{\partial t}|_{t=0}\Big(\mathcal{L}(\sin(2t), \sin(t)) \Big) \left(2\frac{\partial}{\partial x}|_{(0,0)} + \frac{\partial}{\partial y}|_{(0,0)} \right) \tag{53}$$

Now let:

$$f : (-1/8\pi, 5/8\pi) \to (-1/8\pi, 5/8\pi) \tag{54}$$

$$t \mapsto -(t - \frac{1}{4}\pi). \tag{55}$$

We define the alternative parametrisation $\tilde{\phi} = \phi \circ f$. Note that we have $\tilde{\phi}(t) = (-\sin(2t), \sin(t))$ and thus $\tilde{\phi}(0) = \phi(0) = (0, 0)$. A similar calculation as before gives:

$$\tilde{\partial}^{(0)} = -2\frac{\partial}{\partial x}|_{(0,0)} + \frac{\partial}{\partial y}|_{(0,0)} \tag{56}$$

$$\overline{\text{grad}}_p^{\tilde{\partial}(0)} \mathcal{L} = \frac{1}{5}\frac{\partial}{\partial t}|_{t=0}\Big(\mathcal{L}(-\sin(2t), \sin(t))\Big)\left(-2\frac{\partial}{\partial x}|_{(0,0)} + \frac{\partial}{\partial y}|_{(0,0)}\right). \tag{57}$$

Note that because $\partial^{(0)} \neq \tilde{\partial}^{(0)}$ (53) and (57) are not equal to each other. We can therefore conclude that in this case the expression for the gradient is parametrisation-dependent.

References

1. Amari, S.I.: Natural gradient works efficiently in learning. Neural Comput. **10**(2), 251–276 (1998)
2. Ay, N.: On the locality of the natural gradient for learning in deep Bayesian networks. Inf. Geom. 1–49 (2020). https://doi.org/10.1007/s41884-020-00038-y
3. Ay, N., Montúfar, G., Rauh, J.: Selection criteria for neuromanifolds of stochastic dynamics. In: Yamaguchi, Y. (ed.) Advances in Cognitive Neurodynamics (III), pp. 147–154. Springer, Dordrecht (2013). https://doi.org/10.1007/978-94-007-4792-0_20
4. Bernacchia, A., Lengyel, M., Hennequin, G.: Exact natural gradient in deep linear networks and application to the nonlinear case. In: NIPS (2019)
5. Grosse, R., Martens, J.: A Kronecker-factored approximate fisher matrix for convolution layers. In: International Conference on Machine Learning, pp. 573–582. PMLR (2016)
6. Martens, J.: New insights and perspectives on the natural gradient method. arXiv preprint arXiv:1412.1193 (2014)
7. Ollivier, Y.: Riemannian metrics for neural networks I: feedforward networks. Inf. Infer. J. IMA **4**(2), 108–153 (2015)
8. Van Hasselt, H.: Reinforcement learning in continuous state and action spaces. In: Wiering, M., van Otterlo, M. (eds.) Reinforcement Learning. ALO, vol. 12, pp. 207–251. Springer, Heidelberg (2012). https://doi.org/10.1007/978-3-642-27645-3_7
9. Várady, C., Volpi, R., Malagò, L., Ay, N.: Natural wake-sleep algorithm. arXiv preprint arXiv:2008.06687 (2020)
10. Watanabe, S.: Algebraic geometry and statistical learning theory. Cambridge University Press, Cambridge (2009)
11. Zhang, G., Martens, J., Grosse, R.: Fast convergence of natural gradient descent for overparameterized neural networks. arXiv preprint arXiv:1905.10961 (2019)

Properties of Nonlinear Diffusion Equations on Networks and Their Geometric Aspects

Atsumi Ohara(✉) and Xiaoyan Zhang

University of Fukui, Fukui 910-8507, Japan
ohara@fuee.u-fukui.ac.jp

Abstract. We consider a fairly wide class of nonlinear diffusion equations on networks, and derive several common and basic behaviors of solutions to them. Further, we demonstrate that the Legendre structure can be naturally introduced for such a class of dynamical systems, and discuss their information geometric aspects.

Keywords: Nonlinear diffusion on networks · Graph Laplacian · Information geometry

1 Introduction

In the last two decades increasing attention to diffusion phenomena on huge and complex networks has widely spread in not only statistical physics but also various research fields and their applications such as computer-communication networks, machine learning, random walks, biological networks, epidemics, commercial or traffic networks and so on (e.g., [1,2], just to name a few).

Presently many research results and applications are reported by use of linear diffusion equations. On the other hand, important and theoretically interesting aspects to be developed more would be an understanding of nonlinear diffusion phenomena and their relevance with network topology.

Among them anomalous diffusion on disordered medium, such as fractal graphs, has been well-known and recent results can be found in e.g., [3]. Synchronization phenomena of coupled oscillators, such as the Kuramoto model [4], have been extensively explored, e.g., [5], which are closely related to nonlinear diffusion. See also [6] for another type of synchronization.

In this paper we first derive the common and basic properties of network diffusion equation composed by strictly monotone-increasing functions, which is considered to be a physically natural assumption to describe various phenomena. In addition, we define the Legendre structure for such a class of dynamical systems and discuss information geometric aspects behind them.

Supported by JSPS Grant278 in-Aid (C) 19K03633.

F. Nielsen and F. Barbaresco (Eds.): GSI 2021, LNCS 12829, pp. 736–743, 2021.
https://doi.org/10.1007/978-3-030-80209-7_79

Notation: In this paper the set of n-dimensional real column vectors and n by n real square matrices are respectively denoted by $\mathbf{R}^n$ and $\mathbf{R}^{n\times n}$. We express the transpose of a vector x and a matrix A by x^T and A^T, respectively. A positive (resp. nonnegative) vector x is denoted by $x > 0$ (resp. $x \geq 0$). When a real symmetric matrix A is positive definite (resp. positive semidefinite), we write $A \succ 0$ (resp. $A \succeq 0$). Along the context 0 is interpreted as a vector or a matrix all the entries of which are zeros. The real column vector $\mathbf{1} := (1\ 1\ \cdots\ 1)^T$ is often used. The unit column vector of which the i-th entry is one and the others are all zeros is denoted by e_i.

2 Preliminaries

Consider an undirected finite graph $\mathcal{G} = (\mathcal{V}, \mathcal{E})$, where $\mathcal{V} := \{1, 2, \cdots, n\}$ is the set of n numbered *vertices* ($n \geq 2$) and the set of *edges* $\mathcal{E}$ is a subset of unordered pairs (i, j) in $\mathcal{V} \times \mathcal{V}$. We assume that m is the number of edges, i.e., $|\mathcal{E}| = m$.

Let $(i_0, i_1, \cdots, i_l)$ be a sequence of $l+1$ distinct vertices. We call the sequence a *path* connecting i_0 and i_l if $l \geq 1$ and $(i_{k-1}, i_k) \in \mathcal{E}$ for all $k = 1, \cdots, l$. The edge (i, i) is said to be a *self-loop.* A graph is called *connected* if there exists a path connecting two arbitrary distinct vertices. A *tree* is a connected graph with no self-loops for which any two distinct vertices are connected by exactly one path. A *spanning tree* of $\mathcal{G}$ is a tree $\mathcal{G}' = (\mathcal{V}', \mathcal{E}')$ satisfying $\mathcal{V}' = \mathcal{V}$ and $\mathcal{E}' \subset \mathcal{E}$.

For any subset $\mathcal{U} \subset \mathcal{V}$ we define the interior of $\mathcal{U}$ by $\overset{\circ}{\mathcal{U}} := \{i \in \mathcal{U} \mid (i, j) \in \mathcal{E} \Rightarrow j \in \mathcal{U}\}$. The boundary of $\mathcal{U}$ is denoted by $\partial\mathcal{U} := \mathcal{U} \backslash \overset{\circ}{\mathcal{U}} = \{i \in \mathcal{U} \mid \exists j \notin \mathcal{U},\ (i, j) \in \mathcal{E}\}$.

The (weighted) *graph Laplacian* [7,8] $L \in \mathbf{R}^{n\times n}$ for an undirected graph $\mathcal{G}$ with no self-loops is a real symmetric matrix defined by

$$L := D_{\mathrm{w}} - W, \tag{1}$$

where the entries w_{ij} of W satisfy $w_{ij} = w_{ji} > 0$ for $(i, j) \in \mathcal{E}$ and $w_{ij} = 0$ for $(i, j) \notin \mathcal{E}$, and D_{w} is a diagonal matrix with entries $d_i := \sum_{j=1}^{n} w_{ij}$. Note that L is expressed as

$$L = \sum_{(i,j)\in\mathcal{E}} w_{ij}(e_i - e_j)(e_i - e_j)^T = B_0 \widehat{W} B_0^T, \tag{2}$$

where $B_0 \in \mathbf{R}^{n\times m}$ is a matrix consisting of column vectors $e_i - e_j$, $(i, j) \in \mathcal{E}$ and $\widehat{W} \in \mathbf{R}^{m\times m}$ is a diagonal matrix with the corresponding entries w_{ij}.

Hence, L is positive semidefinite, but not positive definite because $L\mathbf{1} = 0$. Further, recalling a spanning tree of $\mathcal{G}$, we see that the following three statements are equivalent: i) $\mathcal{G}$ is connected, ii) $\mathrm{rank} B_0 = n - 1$, iii) the eigenvalue zero of L is simple (i.e., the algebraic multiplicity is one).

Given a real vector $x \in \mathbf{R}^n$, the i-th entry of Lx for the unweighted Laplacian ($w_{ij} = w_{ji} = 1$, $\forall(i, j) \in \mathcal{E}$) is

$$(Lx)_i = \sum_{j=1}^{n} w_{ij}(x_i - x_j) = \sum_{(i,j)\in\mathcal{E}} x_i - x_j.$$

Hence, the negative of the unweighted Laplacian $-L$ can be interpreted as a discrete analogue of the usual Laplacian in continuous spaces.

3 Nonlinear Diffusion Equations on Finite Graphs

We will denote the nonnegative orthant by

$$\mathbf{R}^n_+ := \{x = (x_i) \in \mathbf{R}^n \mid x_i \geq 0,\ i = 1, \cdots, n\},$$

and its interior and boundary, respectively, by

$$\mathbf{R}^n_{++} := \mathrm{int}\mathbf{R}^n_+ = \{x = (x_i) \in \mathbf{R}^n \mid x_i > 0,\ i = 1, \cdots, n\},$$
$$\partial\mathbf{R}^n_+ := \{x = (x_i) \in \mathbf{R}^n_+ \mid x_i = 0,\ \exists i = 1, \cdots, n\}.$$

Let $\mathcal{G} = (\mathcal{V}, \mathcal{E})$ for $|\mathcal{V}| = n \geq 2$ and $|\mathcal{E}| = m$ be a connected undirected graph with no self-loops. Consider a function $x_i(t)$ of vertices $i \in \mathcal{V}$ and the time $t \in \mathbf{R}$.

Let f be a function that is continuously differentiable and strictly monotone increasing on $\mathbf{R}_{++}$. We denote the class of such functions by $\mathcal{F}$. For a vector $x = (x_i) \in \mathbf{R}^n$ and a diagonal mapping $F : \mathbf{R}^n \to \mathbf{R}^n$ defined by

$$F(x) := (f(x_1)\ f(x_2)\ \cdots\ f(x_n))^T, \quad f \in \mathcal{F},$$

we consider the following nonlinear ODE on $t \geq t_0$ and $\mathbf{R}^n$ with an initial vector $x(t_0) \in \mathbf{R}^n_{++}$:

$$\dot{x}(t) := \frac{d}{dt}x(t) = -LF(x(t)), \tag{3}$$

where L is the Laplacian matrix of the graph $(\mathcal{V}, \mathcal{E})$. When $f(s) = s$, the ODE is referred to the *linear diffusion equation* on a network (e.g., [9]) and its behavior is well-understood because the solution is explicitly expressed by $x(t) = \exp\{-L(t - t_0)\}x(t_0)$. The Eq. (3) is formally regarded as a nonlinear generalization. Physically the equation $\dot{x}_i = \sum_{(i,j)\in\mathcal{E}} w_{ij}(f(x_j) - f(x_i))$ implies the Fick's law where each flux on an edge (i, j) is proportional to the difference $f(x_j) - f(x_i)$ and a diffusion coefficient w_{ij}.

The following results in this section demonstrate some of basic properties of the ODE (3), which commonly share with the linear diffusion equation.

For the sake of brevity we sometimes use y_i instead of $f(x_i)$, i.e., $y = (y_i) := F(x) \in \mathbf{R}^n$. Further, we respectively denote by $\underline{k}$ and $\overline{k}$ the arbitrary indices satisfying $x_{\underline{k}}(t) := \min_i\{x_i(t)\}$ and $x_{\overline{k}}(t) := \max_i\{x_i(t)\}$ at each $t \geq t_0$.

Proposition 1. *Let $x(t_0) \in \mathbf{R}^n_{++}$ be an initial vector for the ODE (3). Then the following statements hold for $f \in \mathcal{F}$:*

i) The solution for (3) uniquely exists and stays in $\mathbf{R}^n_{++}$ for all $t \geq t_0$.
ii) The sum of entries of the initial vector $\sigma := \sum_{i=1}^n x_i(t_0)$ is conserved and the unique equilibrium x_e expressed by

$$x_\mathrm{e} := \frac{\sigma}{n}\mathbf{1} \tag{4}$$

is globally asymptotically stable.

iii) Two quantities $\min_k\{x_k(t)\}$ *and* $\max_k\{x_k(t)\}$ *are, respectively, non-decreasing and non-increasing for all* $t \geq t_0$*, i.e.,*

$$\frac{d}{dt}\min_i\{x_i(t)\} = \frac{d}{dt}x_{\underline{k}}(t) \geq 0, \quad \frac{d}{dt}\max_i\{x_i(t)\} = \frac{d}{dt}x_{\overline{k}}(t) \leq 0. \tag{5}$$

Proof) Omitted.
Since the class $\mathcal{F}$ admits functions with a property $\lim_{s\searrow 0} f(s) = -\infty$, such as the logarithmic function, the proposition restricts an initial vector in $\mathbf{R}^n_{++}$.

When we add to $\mathcal{F}$ an assumption that $f(0) := \lim_{s\searrow 0} f(s)$ is finite to be defined as a value at $s = 0$, every function $f \in \mathcal{F}$ with the assumption is continuous on $\mathbf{R}_+$. Note that we still cannot expect that F is Lipschitz on $\mathbf{R}^n_+$ because it might hold that $\lim_{s\searrow 0} f'(s) = \infty$. One of such examples would be $f(s) = s^q$ for $0 < q < 1$.

However, by only this additional assumption we can extend the region of initial vectors to $\mathbf{R}^n_+$ as is shown in the following corollary.

Corollary 1. *Let* f *be in* $\mathcal{F}$*. If* $f(0)$ *is finitely defined and satisfies* $f(0) = \lim_{s\searrow 0} f(s)$*, then the statements i), ii) and iii) in Proposition 1 hold for* $x(t_0) \in \mathbf{R}^n_+$ *and* $t \geq t_0$*.*

Proof) Omitted.

Proposition 2. *Let* $\mathcal{U}$ *be a connected subset of* $\mathcal{V}$*. For* $T > t_0$ *consider the solution to (3),* $x(t) = (x_i(t))$*, on* $[t_0, T]$*. Then* $x_i(t)$ *attains its maximum and minimum on the parabolic boundary*

$$\partial_{\mathrm{P}}([t_0, T] \times \mathcal{U}) := (\{t_0\} \times \mathcal{U}) \cup ((t_0, T] \times \partial\mathcal{U})).$$

Proof) Omitted.

Remark 1. The statement iii) in Proposition 1 also follows from Proposition 2 because $\partial\mathcal{U} = \emptyset$ in case of $\mathcal{U} = \mathcal{V}$.

We can understand that the statements ii) and iii) in Proposition 1 show averaging property, positivity and volume-preserving properties of solutions to (3). Further, we can interpret Proposition 2 as the *maximum and minimum principle.* Thus, we confirm that the nonlinear ODE (3) still shares such fundamental properties with the linear diffusion equation.

4 Convergence Rate

For the statement ii) in the Proposition 1, the Lyapunov function

$$V(x) := y^T L y, \quad y := F(x)$$

plays an important role. We derive one of simple convergence rates of the ODE (3) using $V(x)$.

For $x \in \mathbf{R}^n_+$ let $x^\perp$ be the orthogonal projection of x to the hyperplane $\mathcal{H}_0 := \{x | \langle \mathbf{1}, x\rangle = 0\}$, i.e. we can uniquely decompose x such as

$$x = c\mathbf{1} + x^\perp, \quad \langle \mathbf{1}, x^\perp \rangle = 0. \quad \exists c > 0.$$

For $y = F(x)$, it follows that

$$x^\perp(t) = 0 \Leftrightarrow y^\perp(t) = 0 \Leftrightarrow x(t) = x_e.$$

We shall give an upperbound of the convergence rate in terms of $y^\perp$. Let $\lambda_1 = 0 < \lambda_2 \leq \cdots \leq \lambda_n$ and v_i, $i = 1, \cdots, n$ be, respectively, eigenvalues and the corresponding normalized and mutually orthogonal eigenvectors of L, where $v_1 = \mathbf{1}/n$. Note that we can represent as $L = \sum_{i=2}^n \lambda_i v_i v_i^T$ and $y^\perp = \sum_{i=2}^n a_i v_i$ for some reals a_i. For the Lyapunov function $V(x) = y^T Ly/2$ we have

$$\frac{1}{2}\lambda_2 \|y^\perp\|_2^2 \leq V(x) = \frac{1}{2}(y^\perp)^T L y^\perp \leq \frac{1}{2}\lambda_n \|y^\perp\|_2^2, \tag{6}$$

$$\dot{V}(x) = \frac{\partial V}{\partial x}\dot{x} = -y^T L^T G(x) L y \leq -\lambda_2^2 \min_i \{f'(x_i)\} \|y^\perp\|_2^2, \tag{7}$$

$$\text{where} \quad G(x) = (f'(x_i)\delta_{ij}),$$

along the trajectory of (3). For the simplicity, we write $\gamma(t) := \min_i\{f'(x_i(t))\}$. If an initial vector $x(t_0)$ is in $\mathbf{R}^n_{++}$, it follows from i) of Proposition 1 that $\gamma(t) > 0$ for all $t \geq t_0$.

Thus, by (6) and (7) it holds that

$$\dot{V}(x(t)) \leq -\frac{2\lambda_2^2}{\lambda_n}\gamma(t) V(x(t)), \quad \forall t \geq t_0,$$

and by the principle of comparison we have

$$V(x(t)) \leq V(x(t_0)) \exp\left\{-\frac{2\lambda_2^2}{\lambda_n}\int_{t_0}^t \gamma(\tau)d\tau\right\}, \quad \forall t \geq t_0.$$

Recall Proposition 1 and define the set $\mathcal{C} := \{x \in \mathbf{R}^n \mid \forall i, \ x_i \geq x_{\underline{k}}(t_0) > 0\} \cap \{x \in \mathbf{R}^n \mid \sum_{i=1}^n x_i = \sigma\}$. Since $\mathcal{C}$ is a compact and invariant set for (3), there exists a minimum value γ^* of $\min_i\{f'(x_i)\}$ for $x = (x_i)$ in $\mathcal{C}$, i.e.,

$$\gamma^* := \min_{x \in \mathcal{C}} f'(x_i) = \min_{s \in \mathcal{I}} f'(s),$$

where the interval $\mathcal{I}$ is defined by $[x_{\underline{k}}(t_0), \sigma - (n-1)x_{\underline{k}}(t_0)]$. From (6) we have

$$\begin{aligned}\|y^\perp(t)\|_2 &\leq \|y^\perp(t_0)\|_2 \sqrt{\frac{\lambda_n}{\lambda_2}} \exp\left\{-\frac{\lambda_2^2}{\lambda_n}\int_{t_0}^t \gamma(\tau)d\tau\right\} \\ &\leq \|y^\perp(t_0)\|_2 \sqrt{\frac{\lambda_n}{\lambda_2}} \exp\left\{-\frac{\lambda_2^2\gamma^*}{\lambda_n}(t - t_0)\right\}.\end{aligned}$$

5 Information Geometric Viewpoints

We discuss the relation with the ODE (3) and the Legendre structure.

Since f is invertible, we can consider the inverse function h, i.e., $f \circ h = h \circ f = \mathrm{id}$, where $\circ$ is the composition of functions. Using their primitive functions $\widehat{f}(s) := \int f(s)ds$ and $\widehat{h}(s) := \int h(s)ds$, define the following functions:

$$\psi(y) := \sum_{i=1}^{n} \widehat{h}(y_i) + c_1, \quad \varphi(x) := \sum_{i=1}^{n} \widehat{f}(x_i) + c_2,$$

where c_1 and c_2 are constants. Note that ψ and φ are convex functions with positive definite Hessian matrices G and G^{-1}, respectively. Further the *Legendre conjugate* [10] of φ is obtained as

$$\varphi^*(y) := \max_x \{x^T y - \varphi(x)\} = \sum_{i=1}^{n} h(y_i)y_i - (\widehat{f} \circ h)(y_i) - c_2$$

from the extreme value conditions: $y_i = (\partial\varphi/\partial x_i)(x) = f(x_i)$, i.e., $x_i = h(y_i)$. Since it holds that

$$\frac{\partial \varphi^*}{\partial y_i}(y) = h'(y_i)y_i + h(y_i) - (f \circ h)(y_i)h'(y_i) = h(y_i) = \frac{\partial \psi}{\partial y_i}(y),$$

we have $\psi(y) = \varphi^*(y) + c_1 + c_2$. Thus, ψ and φ are the Legendre conjugate of each other up to constant, and the coordinate system $\{y_i\}_{i=1}^n$ can be interpreted as the Legendre transform of $\{x_i\}_{i=1}^n$.

By regarding the Hessian matrix G of $\varphi(x)$, i.e.,

$$G(x) = \frac{\partial^2}{\partial x_i \partial x_j}\varphi(x) = \left(f'(x_i)\delta_{ij}\right).$$

as a Riemannian metric, we can consider Hessian geometric structure $(\mathbf{R}_{++}, G)$.

For two points $x = (x_i)$ and $\tilde{x} = (\tilde{x}_i)$ on $\mathbf{R}^n_+$, let us represent the point x as $y = (y_i) = F(x)$ using the coordinates $\{y_i\}_{i=1}^n$ transformed from $\{x_i\}_{i=1}^n$. Consider the following function on $\mathbf{R}^n_{++} \times \mathbf{R}^n_{++}$, which is sometimes called *divergence* [11]:

$$\begin{aligned} D(x||\tilde{x}) &:= \psi(y) + \varphi(\tilde{x}) - \sum_{i=1}^{n} y_i \tilde{x}_i \\ &= \sum_{i=1}^{n} \left\{ (\widehat{h} \circ f)(x_i) + \widehat{f}(\tilde{x}_i) - f(x_i)\tilde{x}_i \right\} + c_1 + c_2. \end{aligned}$$

The quantity $D(x||x)$ is constant because

$$\frac{\partial}{\partial x_i} \sum_{i=1}^{n} \left\{ (\widehat{h} \circ f)(x_i) + \widehat{f}(x_i) - f(x_i)x_i \right\} = 0, \quad i = 1, \cdots, n.$$

Hence, by defining c_1 and c_2 to satisfy $c_1 + c_2 = -\sum_{i=1}^{n}\left\{(\widehat{h} \circ f)(x_i) + \widehat{f}(x_i) - f(x_i)x_i\right\}$ for some $x = (x_i) \in \mathbf{R}^n_{++}$, we have, by the Legendre conjugacy,

$$D(x||\tilde{x}) \geq 0$$

and the equality holds if and only if $x = \tilde{x}$.

Without loss of generality we can take $\sigma = 1$ and consider the relative interior of the probability simplex $\mathcal{S}^{n-1} := \{x \in \mathbf{R}^n_{++} \mid \sum_{i=1}^{n} x_i = 1\}$ as a state space of the ODE (3) by the properties of Proposition 1. For the equilibrium $x_\mathrm{e} = \mathbf{1}/n$ we have

$$\begin{aligned} \frac{d}{dt}D(x_\mathrm{e} \parallel x(t)) &= \sum_{i=1}^{n}\{f(x_i) - f(1/n)\}\dot{x}_i \\ &= (y - F(\mathbf{1}/n))^T\dot{x} = -y^T Ly + f(1/n)\mathbf{1}^T Ly \leq 0, \end{aligned} \tag{8}$$

and on $\mathcal{S}^{n-1}$ the equality holds if and only if $x = x_\mathrm{e}$. By the similar argument, we also find that $d\varphi(x)/dt \leq 0$ on $\mathcal{S}^{n-1}$ and the equality holds at x_e. Thus, $D(x_\mathrm{e} \parallel x)$ and $\varphi(x) - \varphi(x_\mathrm{e})$ are also Lyapunov functions for the equilibrium x_e in addition to $V(x)$.

These facts are closely related with the increase of entropy or what is called the *H-theorem* in statistical physics. Actually the following example, in the case when $f(s) = \log(s)$ and an initial vector on $\mathcal{S}^{n-1}$, exhibits a typical one.

Example

$$f(s) = \log s, \quad h(s) = \exp s, \qquad \hat{f}(s) = s\log s - s, \quad \hat{h}(s) = \exp s$$

$$\psi(y) = \sum_{i=1}^{n} \exp y_i, \quad \varphi(x) = \sum_{i=1}^{n}(x_i \log x_i - x_i), \quad G(x) = (\delta_{ij}/x_i).$$

$$D(x||\tilde{x}) = \sum_{i=1}^{n}\{\tilde{x}_i \log(\tilde{x}_i/x_i) + x_i - \tilde{x}_i\}$$

Note that $-\varphi(x)$ and $G(x)$ on $\mathcal{S}^{n-1}$ are respectively called the Shannon entropy and the *Fisher information*. Further, $KL(x||\tilde{x}) := D(\tilde{x}||x)$ on $\mathcal{S}^{n-1} \times \mathcal{S}^{n-1}$ is called the *Kullback-Leibler divergence* or relative entropy [11].

From the viewpoint of the Hessian geometric structure $(\mathbf{R}^n_{++}, G)$, the following property of the ODE (3) is straightforward:

Proposition 3. *The ODE (3) can be regarded as a gradient system with respect to a Riemannian metric $G(x)$ for the Lyapunov function $V(x) = F(x)^T LF(x)/2$, i.e.,*

$$\dot{x} = -G(x)^{-1}\frac{\partial V(x)}{\partial x}$$

6 Concluding Remarks

We have derived several properties a class of nonlinear network diffusion equations with the graph Laplacian. The nonlinearity we have assumed is considered to be physically natural and cover many important examples governed by Fick's law.

The basic and common behaviors of solutions are summarized in the Sect. 3. We have analyzed the convergence rate of the solution in the Sect. 4. It is seen that the rate is roughly evaluated by $\min_i\{f'(x_i)\}$ on a certain compact set. Finally, in the Sect. 5 we have derived the Legendre structure of the class of equations. Using the structure, we have shown that each equation can be regarded as a gradient flow of a certain Lyapunov function via a Hessian metric.

Complete version with omitted proofs can be found in the near future.

References

1. Barrat, A., Barthelemy, M., Vespignani, A.: Dynamical Processes on Complex Networks. Cambridge University Press, Cambridge (2008)
2. Thurner, S., Hanel, R., Klimek, P.: Introduction to the Theory of Complex Systems. Oxford University Press, Oxford (2018)
3. Kumagai, T.: Random Walks on Disordered Media and Their Scaling Limits. Lecture Notes in Mathematics, vol. 2101. Springer, Cham (2014). https://doi.org/10.1007/978-3-319-03152-1
4. Kuramoto, Y.: Self-entrainment of a population of coupled non-linear oscillators. In: Araki, H. (ed.) International Symposium on Mathematical Problems in Theoretical Physics. Lecture Notes in Physics, vol. 39, pp. 420–422. Springer, Heidelberg (1975). https://doi.org/10.1007/BFb0013365
5. Arenas, A., Díaz-Guilera, A., Kurths, J., Moreno, Y., Zhou, C.: Phys. Rep. **469**, 93–153 (2008)
6. Zhang, X., Senyo, T., Sakai, H., Ohara, A.: Behaviors of solutions to network diffusion equation with power-nonlinearity. Eur. Phys. J. Spec. Top. **229**(5), 729–741 (2020). https://doi.org/10.1140/epjst/e2020-900202-5
7. Chung, F.R.K.: Spectral Graph Theory. AMS, Providence (1997)
8. Merris, R.: Laplacian matrices of graphs: a survey. Linear Algebra Appl. **197–198**, 143–176 (1994)
9. Newman, M.: Networks: An Introduction. Oxford University Press Inc., New York (2018)
10. Rockafellar, R.T.: Convex Analysis. Princeton University Press, Princeton (1970)
11. Amari, S., Nagaoka, H.: Methods of Information Geometry. AMS and Oxford (2000)

Rényi Relative Entropy from Homogeneous Kullback-Leibler Divergence Lagrangian

Goffredo Chirco(✉)

Istituto Nazionale di Fisica Nucleare - Sezione di Napoli, Complesso Universitario di Monte S. Angelo, ed. 6, Via Cintia, 80126 Napoli, Italy
goffredo.chirco@na.infn.it

Abstract. We study the *homogeneous* extension of the Kullback-Leibler divergence associated to a *covariant* variational problem on the statistical bundle. We assume a finite sample space. We show how such a divergence can be interpreted as a Finsler metric on an extended statistical bundle, where the time and the time score are understood as extra random functions defining the model—. We find a relation between the homogeneous generalisation of the Kullback-Leibler divergence and the Rényi relative entropy, the Rényi parameter being related to the time-reparametrization lapse of the Lagrangian model. We investigate such intriguing relation with an eye to applications in physics and quantum information theory.

Keywords: Kullback-Leibler divergence · General covariance · Rényi relative entropy · Non-parametric information geometry · Statistical bundle

1 Information Geometry on the Statistical Bundle

The probability simplex on a finite sample space Ω, with $\#\Omega = N$, is denoted by $\Delta(\Omega)$, and $\Delta^\circ(\Omega)$ its interior. The uniform probability function is μ, $\mu(x) = \frac{1}{N}$, $x \in \Omega$. We regard $\Delta^\circ(\Omega)$ as the *maximal exponential family* $\mathcal{E}(\mu)$, in the sense that each strictly positive density q can be written as $q \propto \mathrm{e}^v$, where v is defined up to a constant. The expected value of v with respect to the density q is $\mathbb{E}_q[v]$.

A non metric, non-parametric presentation of Information Geometry (IG) [1] is realised via a *joint* geometrical structure given by the probability simplex together with the set of q-integrable functions $v \in L(q)$: the couple (q, v) forming a *statistical vector bundle*

$$S\mathcal{E}(\mu) = \{ (q, v) \mid q \in \mathcal{E}(\mu), v \in L_0^2(q) \}, \tag{1}$$

with base $\Delta^\circ(\Omega)$. For each $q \in \Delta^\circ(\Omega)$, $L^2(q)$ is the vector space of real functions of Ω endowed with the inner product $\langle u, v \rangle_q = \mathbb{E}_q[u\, v]$, and it holds $L^2(q) = \mathbb{R} \oplus L_0^2(p)$ [15,17]. A (differential) geometry for the statistical bundle $S\mathcal{E}(\mu)$ is naturally provided by an *exponential atlas* of charts given for each $p \in \mathcal{E}(\mu)$ by

F. Nielsen and F. Barbaresco (Eds.): GSI 2021, LNCS 12829, pp. 744–751, 2021.
https://doi.org/10.1007/978-3-030-80209-7_80

$$s_p \colon S\Delta^\circ(\Omega) \ni (q,v) \mapsto \left(\log\frac{q}{p} - \mathbb{E}_p\left[\log\frac{q}{p}\right], {}^{\mathrm{e}}\mathbb{U}_q^p v\right) \in S_p\Delta^\circ(\Omega) \times S_p\Delta^\circ(\Omega)$$

where ${}^{\mathrm{e}}\mathbb{U}_q^p$ denotes the *exponential transport*, defined for each $p, q \in \Delta^\circ(\Omega)$ by

$${}^{\mathrm{e}}\mathbb{U}_q^p : S_q\Delta^\circ(\Omega) \ni v \mapsto v - \mathbb{E}_p\left[v\right] \in S_p\Delta^\circ(\Omega)$$

As $s_p(p,v) = (0,v)$, we say that s_p is the chart *centered at* p.

We can write then

$$q = \exp\left(v - K_p(v)\right) \cdot p = e_p(v), \tag{2}$$

and see the mapping $s_p(q) = e_p(v)^{-1}$ as a section of the bundle.

The *cumulant function* $K_p(v) = \mathbb{E}_p\left[\log\frac{p}{q}\right] = \mathrm{D}\left(p \,\|\, q\right)$ is the expression in chart of the Kullback-Leibler divergence $p \mapsto \mathrm{D}\left(p \,\|\, q\right)$. The other divergence $\mathrm{D}\left(q \,\|\, p\right) = \mathbb{E}_q\left[\log\frac{q}{p}\right] = \mathbb{E}_q\left[v\right] - K_p(v)$ is the convex conjugate of the cumulant in the chart centered at p.

With respect to any of the convex functions $K_p(v)$ the maximal exponential family is a Hessian manifold. The given exponential atlas then provides the statistical bundle with an *affine* geometry, with a dual covariant structure induced by the inner product on the fibers given by the duality pairing between $S_q\mathcal{E}\left(\mu\right)$ and its dual ${}^*S_q\mathcal{E}\left(\mu\right)$, and the associated dual affine transports [10].

2 From Divergences on $\mathcal{E}\left(\mu\right) \times \mathcal{E}\left(\mu\right)$ to Lagrangians on $S\mathcal{E}\left(\mu\right)$

A divergence is a smooth mapping $D \colon \mathcal{E}\left(\mu\right) \times \mathcal{E}\left(\mu\right) \to \mathbb{R}$, such that for all $q, r \in \mathcal{E}\left(\mu\right)$ it holds $D(q,r) \geq 0$ and $D(q,r) = 0$ iff $q = r$. Every divergence can be associated to a Lagrangian function on the statistical bundle via the *canonical mapping* [5]

$$\mathcal{E}\left(\mu\right)^2 \ni (q,r) \mapsto (q, s_q(r)) = (q,w) \in S\mathcal{E}\left(\mu\right), \tag{3}$$

where $r = \mathrm{e}^{w - K_q(w)} \cdot q$, that is, $w = s_q(r) = e_r^{-1}(w)$.

Remark 1. Such a mapping appears to be an instance of a general integral relation between the $\mathcal{G} = \mathcal{E}\left(\mu\right) \times \mathcal{E}\left(\mu\right)$ intended as a *pair groupoid* and the associated Lie algebroid $\mathfrak{Lie}(\mathcal{G})$ corresponding to the tangent bundle $S\mathcal{E}\left(\mu\right)$, see e.g. [11].

The inverse mapping is the *retraction* mapping

$$S\mathcal{E}\left(\mu\right) \ni (q,w) \mapsto (q, e_q(w)) = (q,r) \in \mathcal{E}\left(\mu\right)^2. \tag{4}$$

As the curve $t \mapsto e_q(tw)$ has null exponential acceleration [16], one could say that Eq. (4) defines the *exponential* mapping of the exponential connection, while Eq. (3) defines the so-called *logarithmic* mapping.

The expression in a chart centered at p of the mapping of Eq. (4) is affine:

$$\begin{aligned} &S_p\mathcal{E}\left(\mu\right) \times S_p\mathcal{E}\left(\mu\right) \to \mathcal{E}\left(\mu\right) \times \mathcal{E}\left(\mu\right) \to S\mathcal{E}\left(\mu\right) \to S_p\mathcal{E}\left(\mu\right) \times S_p\mathcal{E}\left(\mu\right) \\ &(p,u,v) \mapsto (e_p(u), e_p(v)) \mapsto (e_p(u), s_{e_p(u)}(e_p(v))) \mapsto (p, u, (v-u)). \end{aligned}$$

The correspondence above maps every divergence D into a *divergence Lagrangian*, and conversely,

$$L(q, w) = D(q, e_q(w)), \quad D(q, r) = L(q, s_q(r)). \tag{5}$$

Notice that, according to our assumptions on the divergence, the divergence Lagrangian defined in Eq. (5) is non-negative and zero if, and only if, $w = 0$.

3 Kullback-Leibler Dissimilarity Functional

Let us consider, as a canonical example, a Lagrangian given by the Kullback-Leibler divergence $\mathrm{D}(q \,\|\, r)$, which, by means of the canonical mapping in (3), corresponds to the cumulant function $K_q(w)$ on $S_q\mathcal{E}(\mu)$. This is a case of high regularity as we assume the densities q and r positive and connected by an open exponential arc. The Hessian structure of the exponential manifold is reflected in the hyper-regularity of the cumulant Lagrangian.

Let $\mathbb{R} \ni t \mapsto q(t) \in \mathcal{E}(\mu)$ be a smooth curve on the exponential manifold, intended as a one-dimensional parametric model. In the exponential chart centered at p, the velocity of the curve $q(t)$ is computed as

$$\frac{d}{dt} s_p(q(t)) = \frac{d}{dt}\left(\log \frac{q(t)}{p} - \mathbb{E}_p\left[\log \frac{q(t)}{p}\right]\right) = \frac{\dot{q}(t)}{q(t)} - \mathbb{E}_p\left[\frac{\dot{q}(t)}{q(t)}\right] = \\ {}^{\mathrm{e}}\mathbb{U}^p_{q(t)} \frac{\dot{q}(t)}{q(t)} = {}^{\mathrm{e}}\mathbb{U}^p_{q(t)} \frac{d}{dt} \log q(t). \tag{6}$$

where partial derivatives are defined in the trivialisations given by the affine charts. By expressing the tangent at each time t in the moving frame at $q(t)$ along the curve, we define the *velocity* of the curve as the *score function* of the one-dimensional parametric model (see e.g. [8, §4.2])

$$\overset{\star}{q}(t) = {}^{\mathrm{e}}\mathbb{U}^{q(t)}_p \frac{d}{dt} s_p(q(t)) = \dot{u}(t) - \mathbb{E}_{q(t)}\left[\dot{u}(t)\right] = \frac{d}{dt} \log q(t) = \frac{\dot{q}(t)}{q(t)}. \tag{7}$$

The mapping $q \mapsto (q, \overset{\star}{q})$ is a lift of the curve to the statistical bundle whose expression in the chart centered at p is $t \mapsto (u(t), \dot{u}(t))$.

We shall now understand the cumulant Lagrangian as a divergence between q and r at any time, which compares the two probabilities at times infinitesimally apart. This amounts to consider the two probabilities as two one-dimensional parametric models constrained via the canonical mapping defined with respect to the score velocity vector, namely $r(t) = e_{q(t)}(\overset{\star}{q}(t))$.

By summing these divergences in time, we define a *dissimilarity* functional as the integral of the Kullback-Leibler cumulant Lagrangian $L\colon S\mathcal{E}(\mu) \times \mathbb{R} \to \mathbb{R}$ along the model

$$\begin{aligned}(q, r) \mapsto A[q] &= \int L(q(t), \overset{\star}{q}(t), t)\, dt \\ &= \int \mathrm{D}\left(q(t) \,\|\, e_{q(t)}(\overset{\star}{q}(t)), t\right)\, dt = \int K_{q(t)}(\overset{\star}{q}(t))\, dt.\end{aligned} \tag{8}$$

As shown in [5,16], via $A[q]$, one can consistently define a *variational principle* on the statistical bundle, leading to a *non-parametric* expression of the Euler-Lagrange equations in the statistical bundle. Indeed, if q is an extremal of the action integral, one gets

$$\frac{D}{dt}\text{grad}_{\text{e}} L(q(t), \overset{\star}{q}(t), t) = \text{grad} L(q(t), \overset{\star}{q}(t), t), \tag{9}$$

where grad indicates the natural gradient at $q(t)$ on the manifold, grad_{e} the natural fiber-derivative with respect to the score, and $\frac{D}{dt}$ the mixture covariant derivative, as defined in [5].

The variational problem on the statistical bundle provides a natural setting for *accelerated* optimization on the probability simplex, with the divergence working as a kinetic energy *regulariser* for the scores, leading to faster converging and more stable optimization algorithms (see e.g. [9]).

More generally, the cumulant Lagrangian in (8) can be used to formulate a generalised geodesic principle on the statistical bundle, in terms of a class of local non-symmetric, non-quadratic generalizations of the Riemannian metrics. To this aim, beside smoothness and convexity, we need in first place the divergence Lagrangian to be *positive homogenous* of the first order in the scores. This brings new structures into play.

4 Re-parametrization Invariance of the Dissimilarity Action

Homogeneous Lagrangians (more precisely, positively homogeneous of degree one) lead to actions that are invariant under time re-parametrizations. Consider the action in Eq. (8) and introduce a formal time parameter τ, such that $t = f(\tau)$ and $q(t) \to q(f(\tau)) = q(\tau)$. This elevates the time integration variable t to the rank of an independent dynamical variable, with f an *arbitrary* function, for which we can assume $\dot{f}(\tau) > 0$.

As a consequence, the re-parametrised action reads

$$\begin{aligned} A[(q,t)] = \int d\tau\, L(q(\tau), \overset{\star}{q}(\tau)/\dot{f}(\tau)))\, \dot{f}(\tau) \\ = \int d\tau\, \mathrm{D}\left(q(\tau) \,\|\, e_q(\overset{\star}{q}(\tau)/\dot{f}(\tau))\right)\, \dot{f}(\tau), \end{aligned}$$

where $dt \to df(\tau) = \frac{d}{d\tau} f\, d\tau = \dot{f}\, d\tau$.

The new Lagrangian

$$\tilde{L}(q, f, v, \dot{f}) = \mathrm{D}\left(q \,\|\, e_q(v/\dot{f})\right)\, \dot{f} = K_q(\overset{\star}{q}/\dot{f})\, \dot{f} \tag{10}$$

is defined on the *extended* statistical bundle $\tilde{S\mathcal{E}}(\mu) = S\mathcal{E}(\mu) \times \mathbb{R}$, with base $\tilde{\mathcal{E}}(\mu) = \mathcal{E}(\mu) \times \mathbb{R}$.

In particular, $\tilde{L}(q, f, v, \dot{f})$ is *homogeneous* of degree one in the velocity $(v, \dot{f}) \in \tilde{S}_{(q,f)}\mathcal{E}(\mu)$, that is

$$\tilde{L}(q, f, \lambda v, \lambda \dot{f}) = \lambda \tilde{L}(q, f, v, \dot{f}). \tag{11}$$

In terms of $\tilde{L}$, the dissimilarity action becomes covariant to time reparametrisation, providing a generalisation of the notion of *information length* (see e.g.[6]), which allows for a generalised geodesic principle on the (extended) statistical bundle.

We shall focus on the very notion of homogeneous cumulant function. Our interest in this sense is twofold. On the one hand, as already noticed in [12], the homogenised cumulant function $\tilde{L}(q, f, v, \dot{f}) = K_q(\overset{\star}{q}/\dot{f})\,\dot{f}$ generalises the Hessian geometry of $\mathcal{E}(\mu)$ to a *Finsler* geomety for the extended exponential family $\tilde{\mathcal{E}}(\mu)$, with $\tilde{L}$ playing the role of the Finsler metric. On the other hand, as it will be shown in the last section, the reparametrization invariance *symmetry* property of the homogeneous cumulant can be used to newly motivate the definition of Rényi relative entropy, with the Rényi parameter appearing as associated to the time-reparametrization lapse of the model.

5 Finsler Structure on the Extended Statistical Bundle

A Finsler metric on a differentiable manifold M is a continuous non-negative function $F : TM \to [0, +\infty)$ defined on the tangent bundle, so that for each point $x \in M$, (see e.g. [7])

$$F(v + w) \leq F(v) + F(w) \quad \text{for every } v, w \text{ tangent to } M \text{ at } x(\textit{subadditivity})$$
$$F(\lambda v) = \lambda F(v)\ \forall\, \lambda \geq 0 \quad (\textit{positivehomogeneity})$$
$$F(v) > 0 \text{ unless } v = 0 \quad (\textit{positivedefiniteness})$$

In our setting, by definition, for each point $\tilde{q} = (q, f) \in \tilde{\mathcal{E}}(\mu)$, and for $\dot{f}(\tau) > 0$, the Lagrangian $\tilde{L}$ is positively homogeneous, continuous, and non-negative on the extended statistical bundle $\tilde{S}\mathcal{E}(\mu)$. By labelling $\tilde{v} = (v, \dot{f}) \in \tilde{S}_{(q,f)}\mathcal{E}(\mu)$, subadditivity requires

$$\tilde{L}(\tilde{q}, \tilde{v} + \tilde{w}) \leq \tilde{L}(\tilde{q}, \tilde{v}) + \tilde{L}(\tilde{q}, \tilde{w}) \tag{12}$$

for every $\tilde{v}, \tilde{w} \in \tilde{S}_{(q,f)}\mathcal{E}(\mu)$. By noticing that $\tilde{v} + \tilde{w} = (v, \dot{f}) + (w, \dot{f}) = (v + w, 2\dot{f})$, the subadditivity of $\tilde{L}$ is easily proved via Cauchy-Schwartz inequality.

We have

$$\begin{aligned}
\tilde{L}(\tilde{q}, \tilde{v} + \tilde{w}) &= \tilde{L}(q, f, v + w, 2\dot{f}) = 2\dot{f} \log \mathbb{E}_q \left[\exp\left(\frac{v + w}{2\dot{f}}\right)\right] \\
&= 2\dot{f} \log \mathbb{E}_q \left[\exp\left(\frac{v}{2\dot{f}}\right) \exp\left(\frac{w}{2\dot{f}}\right)\right] \\
&\leq 2\dot{f} \log \left(\mathbb{E}_q \left[\left(\exp\left(\frac{v}{2\dot{f}}\right)\right)^2\right]^{1/2} \mathbb{E}_q \left[\left(\exp\left(\frac{w}{2\dot{f}}\right)\right)^2\right]^{1/2}\right) \\
&= \dot{f} \log \mathbb{E}_q \left[\exp\left(\frac{v}{v_f}\right)\right] + \dot{f} \log \mathbb{E}_q \left[\exp\left(\frac{w}{\dot{f}}\right)\right] \\
&= \tilde{L}(\tilde{q}, \tilde{v}) + \tilde{L}(\tilde{q}, \tilde{w}).
\end{aligned}$$

Together with the positive homogeneity expressed in (11) and the positive definiteness inherited by the definition of the KL divergence, we get that for each f, the homogeneous cumulant function $\tilde{L}$ satisfies the properties of a Finsler metric.

6 Re-parametrization Invariance Gives $(1/\dot{f})$-Rényi Divergence

The homogeneous Lagrangian $(q, \overset{\star}{q}, f, \dot{f}) \mapsto \dot{f} K_q(\overset{\star}{q}/\dot{f})$ describes a family of scaled Kullback-Leibler divergence measure, $\dot{f}(\tau)\,\mathrm{D}\,(q(\tau)\,\|\,\tilde{r}(\tau))$, between the probabilities $q(\tau)$ and $\tilde{r}(\tau)$ infinitesimally apart in (sample) space and time, such that $\tilde{r}(\tau)$ can be expressed via retraction (exponential) mapping $\tilde{r}(\tau) = e_{q(\tau)}(\overset{\star}{q}(\tau)/\dot{f}(\tau))$ at any time. The same expression can be easily rewritten in terms of unscaled distributions q and r, by virtue of the canonical (log) mapping $\overset{\star}{q}(t) = s_{q(t)}(r(t))$. Indeed, we have

$$\begin{aligned}\dot{f}\,K_q(\overset{\star}{q}/\dot{f}) &= \dot{f}\log\mathbb{E}_q\left[\exp\left(\frac{\overset{\star}{q}}{\dot{f}}\right)\right] = \dot{f}\log\mathbb{E}_q\left[\exp\left(\frac{1}{\dot{f}}\left(\log\frac{r}{q} - \mathbb{E}_q\left[\log\frac{r}{q}\right]\right)\right)\right]\\ &= \dot{f}\log\mathbb{E}_q\left[\left(\frac{r}{q}\right)^{\frac{1}{\dot{f}}}\right] + \mathrm{D}\,(q\,\|\,r) = \dot{f}\log\mathbb{E}_\mu\left[r^{\frac{1}{\dot{f}}}q^{1-\frac{1}{\dot{f}}}\right] + \mathrm{D}\,(q\,\|\,r)\,. \quad (13)\end{aligned}$$

where we use the definition $K_p(v) = \mathbb{E}_p\left[\log\frac{p}{q}\right] = \mathrm{D}\,(p\,\|\,q)$ in the second line. We removed the dependence on time in (13) in order to ease the notation.

Therefore, we see that for $\dot{f} = 1/1-\alpha$, the term $\dot{f}\log\mathbb{E}_\mu\left[r^{1/\dot{f}}\,q^{1-1/\dot{f}}\right]$ corresponds to the definition of the α-Rényi divergence [18] with a minus sign

$$D_\alpha(q\|r) = \frac{1}{\alpha-1}\log\mathbb{E}_\mu\left[r^{1-\alpha}\,q^{\alpha}\right],$$

while the second term, giving the KL for the unscaled distributions, derives from the centering contribution in the canonical map.

Remark 2. The Rényi parameter can be put in direct relation with the *lapse* factor of the reparametrization symmetry. Since the α parameter in Rènyi's entropy is constant, in this sense relating the lapse $\dot{f}$ to α amounts to restrict to reparametrization function f which are *linear* in time.

Remark 3. The two terms together in expression Eq. (13) define a family of free energies, typically espressed as $(1/KT)\,F_\alpha(q,q') = D_\alpha(q\|q') - \log Z$ (see e.g. [3]), and the action in Eq. (8) can be understood as an integrated free energy associated with the transition from $q(\tau)$ to $r(\tau)$ along a curve in the bundle. Similar structures appear both in physics and quantum information theory in the study of out-of-equilibrium systems, where they provide extra constraints on thermodynamic evolution, beyond ordinary Second Law [2].

Remark 4. While the use of the canonical mapping in our affine setting somehow naturally leads to the Rényi formula for the divergence, along with the *exponential mean* generalization of the entropy formula [13,14], the interpretation of the relation of Rènyi index and time lapse induced by reparametrization symmetry is open [19]. In our approach, setting $\dot{f} > 0$ just fixes $\alpha < 1$. A detailed thermodynamic analysis of these expressions is necessary for a deeper understanding of the map between α and $\dot{f}$.

Remark 5. The proposed result is quite intriguing when considered together with the Finsler characterization of the statistical manifold induced by the homogenized divergence, and its relation with *contact geometry* on the projectivised tangent bundle of the Finsler manifold (see e.g. [4]).

Acknowledgements. The author would like to thank G. Pistone and the anonymous referees for the careful read of the manuscript and the interesting points raised.

References

1. Amari, S., Nagaoka, H.: Methods of information geometry. American Mathematical Society (2000). (Translated from the 1993 Japanese Original by D. Harada)
2. Bernamonti, A., Galli, F., Myers, R.C., Oppenheim, J.: Holographic second laws of black hole thermodynamics. J. High Energy Phys. **2018**(7), 111 (2018). https://doi.org/10.1007/JHEP07(2018)111
3. Brandão, F., Horodecki, M., Ng, N., Oppenheim, J., Wehner, S.: The second laws of quantum thermodynamics. Proc. Natl. Acad. Sci. **112**(11), 3275–3279 (2015). https://doi.org/10.1073/pnas.1411728112
4. Chern, S.-S., Chen, W., Lam, W.K.: Lectures on Differential Geometry. World Scientific, Singapore (1999)
5. Chirco, G., Malagò, L., Pistone, G.: Lagrangian and Hamiltonian mechanics for probabilities on the statistical manifold. arXiv:2009.09431 [math.ST]
6. Crooks, G.E.: Measuring thermodynamic length. Phys. Rev. Lett. **99**, 100602 (2007). https://doi.org/10.1103/PhysRevLett.99.100602
7. Bao, D., Chern, S.-S., Shen, Z.: An Introduction to Riemann-Finsler Geometry. Graduate Texts in Mathematics, Springer, New York (2000). https://doi.org/10.1007/978-1-4612-1268-3
8. Efron, B., Hastie, T.: Computer Age Statistical Inference, Institute of Mathematical Statistics (IMS) Monographs, vol. 5. Cambridge University Press, New York (2016). https://doi.org/10.1017/CBO9781316576533
9. França, G., Jordan, M.I., Vidal, R.: On dissipative symplectic integration with applications to gradient-based optimization. arXiv:2004.06840 [math.OC]
10. Gibilisco, P., Pistone, G.: Connections on non-parametric statistical manifolds by Orlicz space geometry. IDAQP **1**(2), 325–347 (1998)
11. Grabowska, K., Grabowski, J., Kuś, M., Marmo, G.: Lie groupoids in information geometry. J. Phys. A Math. Theor. **52**(50), 505202 (2019). https://doi.org/10.1088/1751-8121/ab542e
12. Ingarden, R.S.: Information geometry of thermodynamics. In: Višek, J.Á. (ed.) Transactions of the Tenth Prague Conference: Information Theory, Statistical Decision Functions, Random Processes held at Prague, vol. A, pp. 421–428. Springer, Dordrecht (1988). https://doi.org/10.1007/978-94-009-3859-5_44

13. Kolmogorov, A.: Sur la notion de la moyenne. Atti. R. Accad. Naz. Lincei **12**, 388 (1930)
14. Nagumo, M.: Über eine klasse der mittelwerte. Japan. J. Math. Trans. Abstracts **7**, 71–79 (1930)
15. Pistone, G.: Nonparametric information geometry. In: Nielsen, F., Barbaresco, F. (eds.) GSI 2013. LNCS, vol. 8085, pp. 5–36. Springer, Heidelberg (2013). https://doi.org/10.1007/978-3-642-40020-9_3
16. Pistone, G.: Lagrangian function on the finite state space statistical bundle. Entropy **20**(2), 139 (2018). https://doi.org/10.3390/e20020139
17. Pistone, G.: Information geometry of the probability simplex. A short course. Nonlinear Phenomena Compl. Syst. **23**(2), 221–242 (2020)
18. Rényi, A.: On measures of entropy and information. In: Proceedings of the Fourth Berkeley Symposium on Mathematical Statistics and Probability: Contributions to the Theory of Statistics, vol. 1, pp. 547–561. University of California Press (1961)
19. Chirco, G.: In preparation

Statistical Bundle of the Transport Model

Giovanni Pistone(✉)

de Castro Statistics, Collegio Carlo Alberto, Piazza Arbarello 8, 10122 Turin, Italy
giovanni.pistone@carloalberto.org
https://www.giannidiorestino.it

Abstract. We discuss the statistical bundle of the manifold of two-variate stricly positive probability functions with given marginals. The fiber associated to each coupling turns out to be the vector space of interacions in the ANOVA decomposition with respect to the given weight. In this setting, we derive the form of the gradient flow equation for the Kantorovich optimal transport problem.

Keywords: Statistical bundle · Kantorovich transport problem · Gradient flow

1 Introduction

Consider a product sample space $\Omega = \Omega_1 \times \Omega_2$ and the marginal mappings $X\colon \Omega \to \Omega_1$, $Y\colon \Omega \to \Omega_2$. Both factors are finite sets, $n_i = \#\Omega_i$. If γ is a positive probability function on Ω, we denote by γ_1 and γ_2 the two margins. We denote by $L^2(\gamma)$, $L^2(\gamma_1)$ and $L^2(\gamma_2)$ the Euclidean spaces of real random variables with the given weight, and by $L_0^2(\gamma)$, $L_0^2(\gamma_1)$ and $L_0^2(\gamma_2)$ the spaces of random variables which are centered for the given probability function.

We aim studying evolution equations on the vector bundle defined in the non-parametric Information Geometry (IG) of the open probability simplex of the sample space Ω. See the tutorial [6] and the applications of this formalism in [2,5,8]. A very short review follows below.

If $\Delta^\circ(\Omega)$ is the open probability simplex, we define the *statistical bundle* $S\Delta^\circ(\Omega)$ to be the set of all couples (γ, u) with $\gamma \in \Delta^\circ(\Omega)$ and u a random variable such that $\mathbb{E}_\gamma[u] = 0$. The fiber at γ is $S_\gamma\Delta^\circ(\Omega) = L_0^2(\gamma)$. The IG dualistic structure of [1] is taken into account by saying that the dual statistical bundle $S^*\Delta^\circ(\Omega)$ is the bundle whose fibers are the duals $L_0^2(\gamma)^* = L_0^2(\gamma)$. In the following of the present note, we will make the abuse of notations of identifying the statistical bundle and its dual. Two dual affine connections are defined via the parallel transports

$$
\begin{aligned}
{}^m\mathbb{U}_\gamma^\mu &\colon S_\gamma\Delta^\circ(\Omega) \ni v \mapsto \frac{\gamma}{\mu} v \in S_\mu\Delta^\circ(\Omega), \\
{}^e\mathbb{U}_\mu^\gamma &\colon S_\mu\Delta^\circ(\Omega) \ni w \mapsto w - \mathbb{E}_\gamma[w] \in S_\gamma\Delta^\circ(\Omega),
\end{aligned}
$$

The author is supported by de Castro Statistics, Collegio Carlo Alberto. He is a member of INdAM-GNAMPA.

F. Nielsen and F. Barbaresco (Eds.): GSI 2021, LNCS 12829, pp. 752–759, 2021.
https://doi.org/10.1007/978-3-030-80209-7_81

so that $\left\langle {}^m\mathbb{U}_\gamma^\mu v, w\right\rangle_\mu = \left\langle v, {}^e\mathbb{U}_\mu^\gamma w\right\rangle_\gamma$. The inner product is trasported between fibers by $\langle u, v\rangle_\nu = \left\langle {}^m\mathbb{U}_\mu^\gamma u, {}^e\mathbb{U}_\mu^\gamma v\right\rangle_\gamma$.

For each smooth curve $t \mapsto \gamma(t) \in \Delta^\circ(\Omega)$ the *velocity* (or *score*) is $\overset{\star}{\gamma}(t) = \frac{d}{dt}\log\gamma(t)$. We look at $t \mapsto (\gamma(t), \overset{\star}{\gamma}(t))$ as a curve in the statistical bundle. In the duality we have

$$\frac{d}{dt}\mathbb{E}_{\gamma(t)}[u] = \left\langle u - \mathbb{E}_{\gamma(t)}[u], \overset{\star}{\gamma}(t)\right\rangle_{\gamma(t)}. \tag{1}$$

The mapping $\gamma \mapsto u - \mathbb{E}_\gamma[u]$ is the *gradient mapping* of $\gamma \mapsto \mathbb{E}_\gamma[u]$. It is a section of the statistical bundle. For each section F we define the *evolution equation* $\overset{\star}{\gamma} = F(\gamma)$. By writing $\overset{\star}{\gamma} = \dot{\gamma}/\gamma$, we see that the evolution equation in our sense is equivalent to the Ordinary Differential Equation (ODE) $\dot{\gamma} = \gamma F(\gamma)$.

Given a sub-manifold of $\Delta^\circ(\Omega)$, each fiber S_γ of the statistical bundle splits to define the proper sub-statistical bundle. We are interested in the sub-manifold associated to transport plans, see for example [9]. Let be given $\mu_1 \in \Delta^\circ(\Omega_1)$ and $\mu_2 \in \Delta^\circ(\Omega_2)$. The *transport model* with margins μ_1 and μ_2 is the statistical model

$$\Gamma(\mu_1, \mu_2) = \left\{\gamma \in \Delta(\Omega) \,|\, \gamma(\cdot, +) = \mu_1, \gamma(+, \cdot) = \mu_2\right\}.$$

Without restriction of generality, we assume the two margins μ_1 and μ_2 to be strictly positive. Our sub-manifold is the *open transport model*

$$\Gamma^\circ(\mu_1, \mu_2) = \left\{\gamma \in \Delta^\circ(\Omega) \,|\, \gamma(\cdot, +) = \mu_1, \gamma(+, \cdot) = \mu_2\right\}.$$

Marginalization acts on velocities as a conditional expectation. If $t \mapsto \gamma(t)$ is a smooth curve in the open transport model, then Eq. (1) with $u = f \circ X$ gives

$$0 = \frac{d}{dt}\mathbb{E}_{\mu_1}[f] = \frac{d}{dt}\mathbb{E}_{\gamma(t)}[f(X)] = \left\langle f(X), \overset{\star}{\gamma}(t)\right\rangle_{\gamma(t)} = \left\langle f, \overset{\star}{\gamma}(t)_1\right\rangle_{\mu_1},$$

with $\overset{\star}{\gamma}(t)_1(X) = \mathbb{E}_{\gamma(t)}\left[\overset{\star}{\gamma}(t)\middle| X\right]$. Similarly on the other projection. It follows that $\mathbb{E}_{\gamma(t)}\left[\overset{\star}{\gamma}(t)\middle| X\right] = 0$ and $\mathbb{E}_{\gamma(t)}\left[\overset{\star}{\gamma}(t)\middle| Y\right] = 0$. This suggests the characterization of the velocity bundle of the open transport model we develop in the following section.

2 ANOVA

Let us discuss more in detail the relevant splittings. We use the statistical language of Analysis of Variance (ANOVA). Recall that $\gamma \in \Delta^\circ(\Omega)$. The linear sub-spaces of $L^2(\gamma)$ which , respectively, express the *γ-grand-mean*, the two *γ-simple effects*, and the *γ-interactions*, are

$$\begin{aligned}
B_0(\gamma) &\sim \mathbb{R},\\
B_1(\gamma) &= \left\{f \circ X \,|\, f \in L_0^2(\gamma_1)\right\},\\
B_2(\gamma) &= \left\{f \circ Y \,|\, f \in L_0^2(\gamma_2)\right\},\\
B_{12}(\gamma) &= (B_0(\gamma) + B_1(\gamma) + B_2(\gamma))^\perp,
\end{aligned} \tag{2}$$

where the orthogonality is computed in the γ weight, that is in the inner product of $L^2(\gamma)$, $\langle u, v\rangle_\gamma = \mathbb{E}_\gamma[uv]$. Each element of the space $B_0(\gamma) + B_1(\gamma) + B_2(\gamma)$ has the form $u = u_0 + f_1(X) + f_2(Y)$, where $u_0 = \mathbb{E}_\gamma[u]$ and f_1, f_2 are uniquely defined.

Proposition 1. *For each $\gamma \in \Delta(\Omega)$ there exist a unique orthogonal splitting*

$$L^2(\gamma) = \mathbb{R} \oplus (B_1(\gamma) + B_2(\gamma)) \oplus B_{12}(\gamma).$$

Namely, each $u \in L^2(\gamma)$ can be written uniquely as

$$u = u_0 + (u_1 + u_2) + u_{12}, \tag{3}$$

where $u_0 = \mathbb{E}_\gamma[u]$ and $(u_1 + u_2)$ is the γ-orthogonal projection of $u - u_0$ unto $(B_1(\gamma) + B_2(\gamma))$.

Proof. Each couple of spaces in Eq. (2) has trivial intersection $\{0\}$, hence the splitting in Eq. (3) is unique. As $\mathbb{E}_\gamma[u] = u_0$ and $B_0(\gamma)$ is orthogonal to $B_1(\gamma) + B_2(\gamma)$, then u_{12} is the orthogonal projection of $u - \mathbb{E}_\gamma[u]$ unto $(B_1(\gamma) + B_2(\gamma))^\perp$, while $u_1 + u_2$ is the orthogonal projection of u onto $(B_1(\gamma) + B_2(\gamma))$. $\square$

Let us derive a system of equations for the simple effects. By definition, $u_{12} = (u - u_0) - (u_1 + u_2)$ is γ-orthogonal to each $g_1 \in B_1(\gamma)$ and each $g_2 \in B_2(\gamma)$. The orthogonal projections on $B_1(\gamma)$ and $B_2(\gamma)$ are the conditional expectation with respect to X and Y, respectively,

$$0 = \mathbb{E}_\gamma[(u - u_0) - (u_1 + u_2)| X],$$
$$0 = \mathbb{E}_\gamma[(u - u_0) - (u_1 + u_2)| Y].$$

Assume zero mean, $u_0 = 0$. We have the system of equations

$$\mathbb{E}_\gamma[u| X] = u_1 + \mathbb{E}_\gamma[u_2| X]$$
$$\mathbb{E}_\gamma[u| Y] = \mathbb{E}_\gamma[u_1| Y] + u_2$$

The ANOVA decomposition proves the existence of u_1 and u_2. The following $(n_1 + n_2) \times (n_1 + n_2)$ linear system holds:

$$\begin{cases} \sum_{y \in \Omega_2} \gamma_{2|1}(y|x) u(x, y) & = u_1(x) + \sum_{y \in \Omega_2} \gamma_{2|1}(y|x) u_2(y), \quad x \in \Omega_1 \\ \sum_{x \in \Omega_1} \gamma_{1|2}(x|y) u(x, y) & = \sum_{x \in \Omega_1} \gamma_{1|2}(x|y) u_1(x) + u_2(y), \quad y \in \Omega_2 \end{cases}$$

where we use the conditional probabilities $\gamma_{2|1}(y|x)\gamma_1(x) = \gamma_{1|2}(x|y)\gamma_2(y) = \gamma(x, y)$, $x \in \Omega_1$ and $y \in \Omega_2$.

We write the previous system in matrix form with blocks

$$\begin{aligned}
\Gamma_{2|1} &= [\gamma_{2|1}(y|x)]_{x\in\Omega_1,y\in\Omega_2} \in \mathbb{R}^{n_1\times n_2} \\
\Gamma_{1|2} &= [\gamma_{1|2}(x|y)]_{x\in\Omega_1,y\in\Omega_2} \in \mathbb{R}^{n_1\times n_2} \\
\boldsymbol{u}_1 &= (u_1(x)|x\in\Omega_1) \\
\boldsymbol{u}_2 &= (u_2(x)|x\in\Omega_2) \\
\boldsymbol{u}_{2|1} &= (\sum_{y\in\Omega_2} u(x,y)\gamma_{2|1}(y|x)|x\in\Omega_1) \\
\boldsymbol{u}_{1|2} &= (\sum_{x\in\Omega_1} u(x,y)\gamma_{1|2}(x|y)|x\in\Omega_2),
\end{aligned}$$

as

$$\begin{bmatrix} I_{n_1} & \Gamma_{2|1} \\ \Gamma_{1|2}^T & I_{n_2} \end{bmatrix} \begin{bmatrix} \boldsymbol{u}_1 \\ \boldsymbol{u}_2 \end{bmatrix} = \begin{bmatrix} \boldsymbol{u}_{2|1} \\ \boldsymbol{u}_{1|2} \end{bmatrix} \tag{4}$$

The block matrix is not globally invertible because the kernel contains the vector $\mathbf{1}_{n_1} \oplus \mathbf{1}_{n_2}$, hence we look for a generalised inverse of the block matrix which exists uniquely because of the interpretation as the orthogonal projection of $f \in L_0^2(\gamma)$ onto $(B_1(\gamma)+B_2(\gamma))$. Generalised inverses of the Shur complements $(I_{n_1}-\Gamma_{2|1}\Gamma_{1|2}^T)$ and $(I_{n_2}-\Gamma_{1|2}^T\Gamma_{2|1})$, if available, suggests the following generalised block inversion formula:

$$\begin{bmatrix} I_{n_1} & \Gamma_{2|1} \\ \Gamma_{1|2}^T & I_{n_2} \end{bmatrix}^+ = \begin{bmatrix} (I_{n_1} - \Gamma_{2|1}\Gamma_{1|2}^T)^+ & -\Gamma_{2|1}(I_{n_2} - \Gamma_{1|2}^T\Gamma_{2|1})^+ \\ -\Gamma_{1|2}^T(I_{n_1} - \Gamma_{2|1}\Gamma_{1|2}^T)^+ & (I_{n_2} - \Gamma_{1|2}^T\Gamma_{2|1})^+ \end{bmatrix} \tag{5}$$

Proposition 2. *The matrices $P = \Gamma_{2|1}\Gamma_{1|2}^T$ and $Q = \Gamma_{1|2}^T\Gamma_{2|1}$ are Markov, with invariant probability γ_1 and γ_2 respectively, and strictly positive, hence ergodic. It holds*

$$(I-P)^+ = \sum_{k=0}^{\infty} P^k \ , \quad (I-Q)^+ = \sum_{k=0}^{\infty} Q^k.$$

Proof. The element $P_{x_1x_2}$, $x_1, x_2 \in \Omega_1$, of $P = \Gamma_{2|1}\Gamma_{1|2}^T \in \mathbb{R}^{\Omega_1\times\Omega_2}$ is

$$P_{x_1x_2} = \sum_{y\in\Omega_2} \gamma_{2|1}(y|x_1)\gamma_{1|2}(x_2|y),$$

so that P is a Markov matrix with strictly positive entries and stationary probability γ_1. Because of the ergodic theorem, $P^n \to P^\infty = \gamma_1^T\mathbf{1}$, $n\to\infty$, so that

$$\lim_{n\to\infty}\left(\sum_{k=0}^{n} P^k\right)(I-P) = \lim_{n\to\infty}\left(I - P^{n+1}\right) = I - P^\infty,$$

so that for each $\boldsymbol{f}_1 \in L(\Omega_1)$ it holds

$$\lim_{n\to\infty}\left(\sum_{k=0}^{n} P^k\right)(I-P)\boldsymbol{f}_1 = \boldsymbol{f}_1 - \mathbb{E}_{\gamma_1}[\boldsymbol{f}_1],$$

where the last term is the orthogonal projection Π_{γ_1} of $\boldsymbol{f}_1$ onto $L_0(\gamma_1)$. □

Proposition 3. *The generalised inverses* $(I-P)^+$ *and* $(I-Q)^+$ *are defined and*

$$\begin{bmatrix} \boldsymbol{f}_1 \\ \boldsymbol{f}_2 \end{bmatrix} = \begin{bmatrix} (I-P)^+ & -\Gamma_{2|1}(I-Q)^+ \\ -\Gamma_{1|2}(I-P)^+ & (I-Q)^+ \end{bmatrix} \begin{bmatrix} \boldsymbol{f}_{2|1} \\ \boldsymbol{f}_{1|2} \end{bmatrix}$$

solves Eq. (4).

Proof. Let us write

$$(I_{n_1} - \Gamma_{2|1}\Gamma_{1|2}^T)^+ = \lim_{n\to\infty} \left(\sum_{k=0}^{n} (\Gamma_{2|1}\Gamma_{1|2}^T)^k \right) \Pi_{\gamma_1}$$

so that

$$(I_{n_1} - \Gamma_{2|1}\Gamma_{1|2}^T)^+(I_{n_1} - \Gamma_{2|1}\Gamma_{1|2}^T) = \Pi_{\gamma_1},$$

because $\Pi_{\gamma_1}(I_{n_1} - \Gamma_{2|1}\Gamma_{1|2}^T) = (I_{n_1} - \Gamma_{2|1}\Gamma_{1|2}^T)$.

Similarly, for $Q = \Gamma_{2|1}^T\Gamma_{1|2}$, we have

$$(I_{n_1} - \Gamma_{2|1}^T\Gamma_{1|2})^+(I_{n_1} - \Gamma_{2|1}^T\Gamma_{1|2}) = \Pi_{\gamma_2},$$

It follows that Eq. (4) is solved with Eq. (5). □

We now apply the ANOVA decomposition to the study of the open transport model geometry as a sub-bundle of the statistical bundle $S\Delta^\circ(\Omega)$. Let us write the ANOVA decomposition of the statistical bundle as

$$S_\gamma\Delta^\circ(\Omega) = (B_1(\gamma) + B_2(\gamma)) \oplus B_{12}(\gamma).$$

Proposition 4. *1. Let* $t \mapsto \gamma(t) \in \Gamma^\circ(\mu_1, \mu_2)$ *be a smooth curve with* $\gamma(0) = \gamma$. *Then the velocity at* γ *belongs to the interactions,* $\overset{\star}{\gamma}(0) \in B_{12}(\gamma)$.
2. Given any interaction $v \in B_{12}(\gamma)$, *the curve* $t \mapsto \gamma(t) = (1+tv)\gamma$ *stays in* $\Gamma^\circ(\mu_1, \mu_2)$ *for* t *in a neighborhood of 0 and* $v = \overset{\star}{\gamma}(0)$.

Proof. We already know that $\mathbb{E}_\gamma\left[\overset{\star}{\gamma}(0)\right] = 0$. For each $f \in L_0^2(\mu_1)$, we have $f \circ X \in B_1(\gamma)$, so that

$$\left\langle f\circ X, \overset{\star}{\gamma}(0) \right\rangle_\gamma = \left. \frac{d}{dt}\mathbb{E}_{\gamma(t)}\left[f\circ X\right] \right|_{t=0} = \left. \frac{d}{dt}\mathbb{E}_{\gamma_1(t)}\left[f\right] \right|_{t=0} = \left. \frac{d}{dt}\mathbb{E}_{\mu_1}\left[f\right] \right|_{t=0} = 0.$$

Same argument in the other margin.

Given $v \in B_{12}(\gamma)$, we show there exist an open interval I around 0 such that $I \ni t \mapsto \gamma(t) = (1+tv)\gamma$ is a regular curve in $\Gamma^\circ(\mu_1,\mu_2) \subset \Delta^\circ(\Omega)$. It is indeed a curve in $\Delta^\circ(\Omega)$ because $\sum_{x,y}\gamma(x,y)(1+tv(x,y)) = \mathbb{E}_\gamma\left[1+tv\right] = 1$ and because $\gamma(x,y)(1+tV(x,y)) > 0$ for all x, y, provided $t \in]-(\max v)^{-1}, -(\min v)^{-1}[$. The velocity exists,

$$\overset{\star}{\gamma}(0) = \left. \frac{d}{dt}\log\left((1+tv)\gamma\right) \right|_{t=0} = \left. \frac{v}{1+tv} \right|_{t=0} = v.$$

Finally, let us compute the margins of $(x, y) \mapsto \gamma(x, y; t) = (1 + tv(x, y))\gamma(x, y)$. The argument is the same for both margins. For all $x \in \Omega_1$,

$$\sum_y (1 + tv(x, y))\gamma(x, y) = \gamma_1(x) + t \sum_y v(x, y)\gamma(x, y)$$
$$= \mu_1(x) + t\mathbb{E}_\gamma \left[(X = x)v\right] = \mu_1(x),$$

where $(X = x)$ is the indicator function of $\{X = x\}$. □

In conclusion, the IG structure of the open transport model is the following.

Proposition 5. *The* transport model bundle *with margins μ_1 and μ_2 is the sub-statistical bundle*

$$S\Gamma^\circ(\mu_1, \mu_2) = \left\{ (\gamma, v) \middle| \gamma \in \Gamma^\circ(\mu_1, \mu_2), \mathbb{E}_\gamma \left[v \middle| X \right] = \mathbb{E}_\gamma \left[v \middle| Y \right] = 0 \right\}. \tag{6}$$

The transport ${}^m\mathbb{U}_\gamma^{\bar\gamma}$ maps the fiber at γ to the fiber at $\bar\gamma$.

Proof. The first statement has been already proved. If $\gamma, \bar\gamma \in \Gamma^\circ(\mu_1, \mu_2)$ and $v \in S_\gamma\Gamma^\circ(\mu_1, \mu_2)$, we have

$$\mathbb{E}_{\bar\gamma} \left[{}^m\mathbb{U}_\gamma^{\bar\gamma} v \middle| X \right] \propto \mathbb{E}_\gamma \left[\frac{\bar\gamma}{\gamma} {}^m\mathbb{U}_\gamma^{\bar\gamma} v \middle| X \right] = \mathbb{E}_\gamma \left[v \middle| X \right] = 0.$$

Remark 1. In the statistical bundle setup, at each $p \in \Delta^\circ(\Omega)$ there is a chart s_p derived from the exponential representation, namely

$$s_p \colon \Delta^\circ(\Omega) \ni q \mapsto \log \frac{q}{p} - \mathbb{E}_\gamma \left[\log \frac{q}{p} \right] = u \in S_p\Delta^\circ(\Omega),$$

so that

$$u \mapsto \mathrm{e}^{u - K_p(u)} \cdot p\,, \quad K_p(u) = \log \mathbb{E}_p \left[\mathrm{e}^u \right].$$

The sub-manifold of the transport model is flat in the mixture geometry and there is no simple expression of the exponential coordinate. However, the splitting of the dual statistical bundle suggests a mixed parameterization of $\Delta^\circ(\Omega)$. If $u \in S_\gamma\Delta^\circ(\Omega)$, consider the splitting $u = u_1 + u_2 + u_{1,2}$. If $1 + u_{12} > 0$, then

$$u \mapsto \gamma(u) \propto \mathrm{e}^{u_1 + u_2}(1 + u_{12}) \cdot \gamma \in \Delta^\circ(\Omega),$$

where $(1 + u_{12}) \cdot \gamma \in \Gamma^\circ(\gamma_1, \gamma_2)$ and, for each given u_{12}, we have an exponential family orthogonal to $\Gamma^\circ(\gamma_1, \gamma_2)$.

3 Gradient Flow of the Transport Problem

Let us discuss the optimal transport problem in the framework of the transport model bundle of Eq. (6). Let be given a cost function $c \colon \Omega_1 \times \Omega_2 = \Omega \to \mathbb{R}$ and define the expected cost function $C \colon \Delta(\Omega) \ni \gamma \mapsto \mathbb{E}_\gamma \left[c\right]$.

Proposition 6. *The function $\gamma \mapsto C(\gamma)$ restricted to the open transport model $\Gamma^\circ(\mu_1, \mu_2)$ has statistical gradient in $S\Gamma^\circ(\mu_1, \mu_2)$ given by*

$$\operatorname{grad} C \colon \gamma \mapsto c_{12,\gamma} = c - c_{0,\gamma} - (c_{1,\gamma} + c_{2,\gamma}) \in S_\gamma \Gamma^\circ(\mu_1, \mu_2). \tag{7}$$

Proof. For each smooth curve $t \mapsto \gamma(t)$, $\gamma(0) = \gamma$, we have

$$\frac{d}{dt} C(\gamma(t)) = \frac{d}{dt} \mathbb{E}_{\gamma(t)}[c] = \mathbb{E}_{\gamma(t)}\left[c \overset{\star}{\gamma}(t)\right] = \left\langle c_{12,\gamma(t)}, \overset{\star}{\gamma}(t) \right\rangle_{\gamma(t)} .$$

where we have used the splitting at $\gamma(t)$,

$$c = c_{0,\gamma(t)} + \left(c_{1,\gamma(t)} + c_{2,\gamma(t)}\right) + c_{12,\gamma(t)}$$

together with the fact that the velocity $\overset{\star}{\gamma}(t)$ is an interaction. □

It follows that the equation of the *gradient flow of C* is

$$\overset{\star}{\gamma} = -\left(c - c_{0,\gamma} - (c_{1,\gamma} + c_{2,\gamma})\right).$$

Remark 2. As the gradient mapping $\operatorname{grad} C(\gamma)$ is defined to be the orthogonal projection of the cost c onto the space of γ-interactions $B_{12}(\gamma)$, one could consider its extension to all $\hat{\gamma} \in \Gamma(\mu_1, \mu_2)$. If $\hat{\gamma}$ is a zero of the extended gradient map, $\operatorname{grad} C(\hat{\gamma}) = 0$, then it holds

$$c(x, y) = c_{0,\gamma} + c_{1,\gamma}(x) + c_{2,\gamma}(y) , \quad (x, y) \in \operatorname{supp} \hat{\gamma}.$$

We expect any solution $t \mapsto \gamma(t)$ of the gradient flow to converge to a coupling $\bar{\gamma} = \lim_{t \to \infty} \gamma(t) \in \Delta(\Omega)$ such that $\mathbb{E}_{\bar{\gamma}}[c]$ is the value of the Kantorovich optimal transport problem, $\mathbb{E}_{\bar{\gamma}}[c] = \min\{\mathbb{E}_\gamma[c] | \gamma \in \Gamma(\mu_1, \mu_2)\}$. These informal arguments should be compared with the support properties of the optimal solutions of the Kantorovich problem, see [4]. For the finite state space case, see also [3,7].

Remark 3. The splitting of the statistical bundle suggests that each $\gamma \in \Delta(\Omega_1 \times \Omega_2)$ stays at the intersection of two orthogonal manifolds, namely, the model with the margins of γ and the additive exponential model $e^{u_1 + u_2 - K_\gamma(u_1 + u_2)} \cdot \gamma$, $u_1 \in B_1(\gamma)$ and $u_2 \in B_2(\gamma)$.

Remark 4. The simplest non-trivial case is the toy example with $n_1 = n_2 = 2$. Let us use the coding $\Omega_1 = \Omega_2 = \{+1, -1\}$. Any function u on Ω has the pseudo-Boolean form $u(x, y) = a_0 + a_1 x + a_2 y + a_{12} xy$. Notice that the monomials $1, x, y, xy$ are orthogonal in the counting measure. In particular, a probability has the form $\gamma(x, y) = \frac{1}{4}(1 + b_1 x + b_2 y + b_{12} xy)$ with margins $\mu_1(x) = \frac{1}{2}(1 + b_1 x)$, $\mu_2(y) = \frac{1}{2}(1 + b_2 y)$. The coefficients are the values of the moments. Given $b_1, b_2 \in]-1, +1[$ to fix positive margins, the transport model is the 1-parameter family

$$\gamma(x, y; \theta) = \frac{1}{4}(1 + b_1 x + b_2 y + \theta xy) , \quad -1 + |b_1 + b_2| \leq \theta \leq 1 - |b_1 - b_2| .$$

The conditional probability functions are

$$\gamma_{1|2}(x|y) = \frac{1}{2}\left(1 + \frac{b_1 - b_2\theta}{1 - b_2^2}x + \frac{-b_1b_2 + \theta}{1 - b_2^2}xy\right)$$
$$\gamma_{2|1}(y|x) = \frac{1}{2}\left(1 + \frac{b_2 - b_1\theta}{1 - b_1^2}y + \frac{-b_1b_2 + \theta}{1 - b_1^2}xy\right)$$

Let us compute the simple effects of a generic $u = \alpha_0 + \alpha_1 x + \alpha_2 y + \alpha_{12} xy$. We have $u_0 = \alpha_0 + \alpha_1 b_1 + \alpha_2 b_2 + \alpha_{12}\theta$, so that

$$u(x, y) - u_0 = \alpha_1(x - b_1) + \alpha_2(y - b_2) + \alpha_{12}(xy - \theta).$$

The simple effects are $u_1(x)+u_2(y) = \beta_1(x-b_1)+\beta_2(y-b_2)$ and the orthogonality conditions $0 = \mathbb{E}_\gamma\left[u_{12}(x - b_1)\right] = \mathbb{E}_\gamma\left[u_{12}(y - b_2)\right]$ become

$$(\alpha_1 - \beta_1)(1 - b_1^2) + (\alpha_2 - \beta_2)(\theta - b_1b_2) = -\alpha_{12}(b_2 - \theta b_1),$$
$$(\alpha_1 - \beta_1)(\theta - b_1b_2) + (\alpha_2 - \beta_2)(1 - b_2^2) = -\alpha_{12}(b_1 - \theta b_2).$$

The system has a unique closed-form solution.

Acknowledgments. The author thanks three anonymous referees whose comments suggested a complete revision of the first version. The author thanks L. Malagò and L. Montrucchio for valuable suggestions. We first presented the gradient flow argument of Sect. 3 to the "Computational information geometry for image and signal processing" Conference, Sep 21–25 2015 ICMS, Edinburgh.

References

1. Amari, S., Nagaoka, H.: Methods of Information Geometry. American Mathematical Society (2000). Translated from the 1993 Japanese original by Daishi Harada
2. Chirco, G., Malagò, L., Pistone, G.: Lagrangian and Hamiltonian dynamics for probabilities on the statistical bundle, arXiv:2009.09431 (2020)
3. Montrucchio, L., Pistone, G.: Kantorovich distance on a weighted graph (2019). arXiv:1905.07547 [math.PR]
4. Peyré, G., Cuturi, M.: Computational optimal transport. Found. Trends Mach. Learn. **11**(5–6), 355–607 (2019). arXiv:1803.00567v2
5. Pistone, G.: Lagrangian function on the finite state space statistical bundle. Entropy **20**(2), 139 (2018). https://doi.org/10.3390/e20020139. http://www.mdpi.com/1099-4300/20/2/139
6. Pistone, G.: Information geometry of the probability simplex. A short course. Nonlinear Phenomena Complex Syst. **23**(2), 221–242 (2020)
7. Pistone, G., Rapallo, F., Rogantin, M.P.: Finite space Kantorovich problem with an MCMC of table moves. Electron. J. Stat. **15**(1), 880–907 (2021). https://doi.org/10.1214/21-EJS1804
8. Pistone, G., Rogantin, M.P.: The gradient flow of the polarization measure. With an appendix (2015). arXiv:1502.06718
9. Santambrogio, F.: Optimal Transport for Applied Mathematicians: Calculus of Variations, PDEs, and Modeling. Birkhäuser (2015)

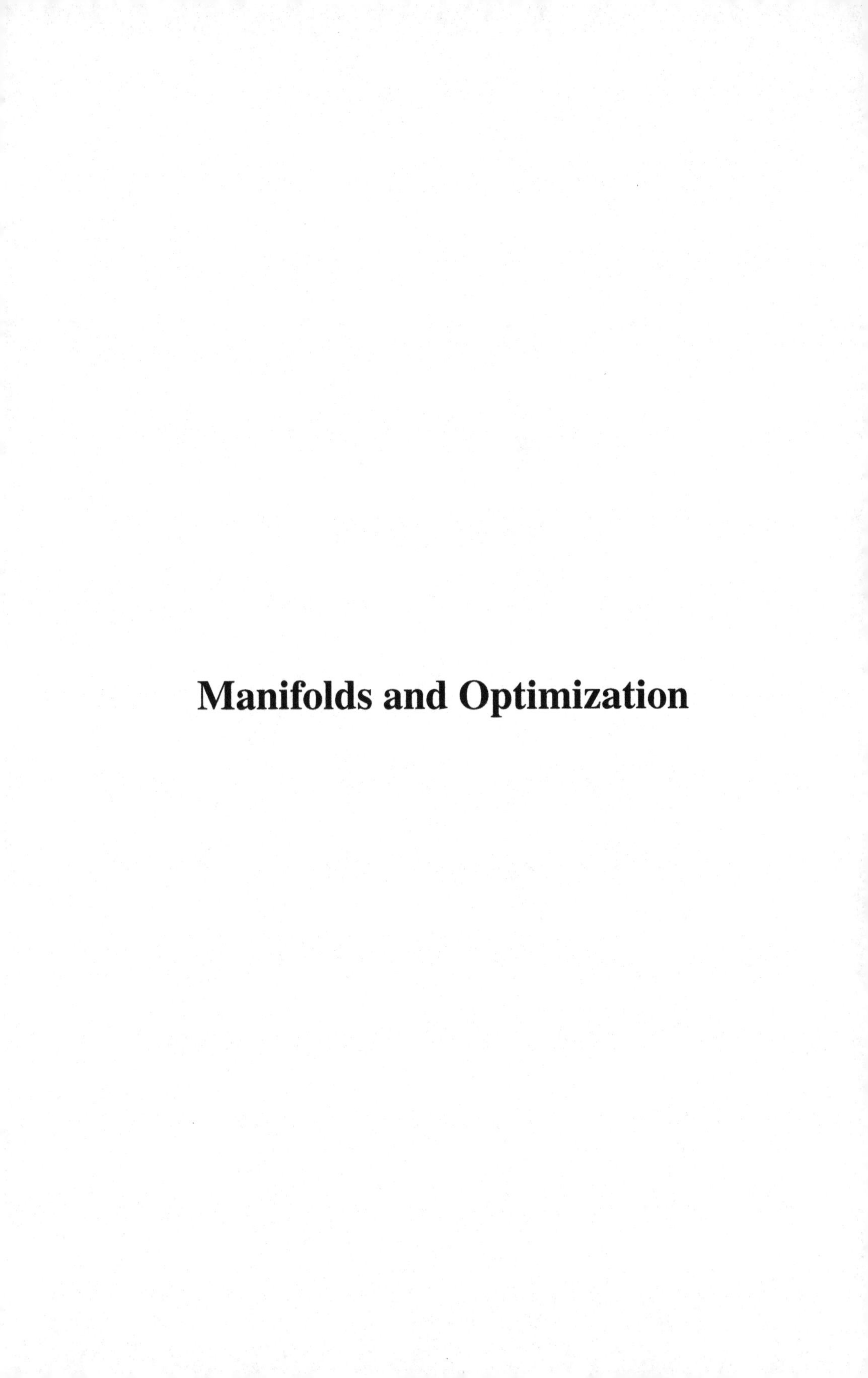

Manifolds and Optimization

Efficient Quasi-Geodesics on the Stiefel Manifold

Thomas Bendokat(✉) and Ralf Zimmermann

Department of Mathematics and Computer Science, University of Southern Denmark (SDU), Odense, Denmark
{bendokat,zimmermann}@imada.sdu.dk

Abstract. Solving the so-called geodesic endpoint problem, i.e., finding a geodesic that connects two given points on a manifold, is at the basis of virtually all data processing operations, including averaging, clustering, interpolation and optimization. On the Stiefel manifold of orthonormal frames, this problem is computationally involved. A remedy is to use quasi-geodesics as a replacement for the Riemannian geodesics. Quasi-geodesics feature constant speed and covariant acceleration with constant (but possibly non-zero) norm. For a well-known type of quasi-geodesics, we derive a new representation that is suited for large-scale computations. Moreover, we introduce a new kind of quasi-geodesics that turns out to be much closer to the Riemannian geodesics.

Keywords: Stiefel manifold · Geodesic · Quasi-geodesic · Geodesic endpoint problem

1 Introduction

Connecting two points on the Stiefel manifold with a geodesic requires the use of an iterative algorithm [10], which raises issues such as convergence and computational costs. An alternative is to use *quasi-geodesics* [2,5–8]. The term is used inconsistently. Here, we mean *curves with constant speed and covariant acceleration with constant (but possibly non-zero) norm.* The term quasi-geodesics is motivated by the fact that actual geodesics feature a constant-zero covariant acceleration. Such quasi-geodesics have been considered in [2,5–7], where a representation of the Stiefel manifold with square matrices was used.

We introduce an economic way to compute these quasi-geodesics at considerably reduced computational costs. Furthermore, we propose a new kind of quasi-geodesics, which turn out to be closer to the true Riemannian geodesic but come at a slightly higher computational cost than the aforementioned economic quasi-geodesics. Both kinds of quasi-geodesics can be used for a wide range of problems, including optimization and interpolation of a set of points.

F. Nielsen and F. Barbaresco (Eds.): GSI 2021, LNCS 12829, pp. 763–771, 2021.
https://doi.org/10.1007/978-3-030-80209-7_82

2 The Stiefel Manifold

This introductory exposition follows mainly [4]. The manifold of orthonormal frames in $\mathbb{R}^{n\times p}$, i.e. the *Stiefel manifold*, is $\mathrm{St}(n,p) := \left\{U \in \mathbb{R}^{n\times p} \mid U^T U = \mathrm{I}_p\right\}$. The *tangent space* at any point $U \in \mathrm{St}(n,p)$ can be parameterized as

$$T_U\mathrm{St}(n,p) := \left\{UA + U_\perp B \mid A \in \mathfrak{so}(p),\ B \in \mathbb{R}^{(n-p)\times p}\right\},$$

where $\mathfrak{so}(p)$ denotes the real $p \times p$ skew-symmetric matrices and $U_\perp$ denotes an arbitrary but fixed orthonormal completion of U such that $\begin{pmatrix}U\ U_\perp\end{pmatrix} \in \mathrm{O}(n)$. A Riemannian metric on $T_U\mathrm{St}(n,p)$ is induced by the *canonical* inner product

$$g_U\colon T_U\mathrm{St}(n,p) \times T_U\mathrm{St}(n,p) \to \mathbb{R},\ g_U(\Delta_1,\Delta_2) := \mathrm{tr}\big(\Delta_1^T(\mathrm{I}_n - \frac{1}{2}UU^T)\Delta_2\big).$$

The metric defines geodesics, i.e. locally shortest curves. The *Riemannian exponential* gives the geodesic from a point $U \in \mathrm{St}(n,p)$ in direction $\Delta = UA + U_\perp B \in T_U\mathrm{St}(n,p)$. Via the QR-decomposition $(\mathrm{I}_n - UU^T)\Delta = QR$, with $Q \in \mathrm{St}(n,p)$ and $R \in \mathbb{R}^{p\times p}$, it can be calculated as

$$\mathrm{Exp}_U(t\Delta) := \begin{pmatrix}U\ U_\perp\end{pmatrix}\mathrm{exp_m}t\begin{pmatrix}A & -B^T\\ B & 0\end{pmatrix}\begin{pmatrix}\mathrm{I}_p\\ 0\end{pmatrix} = \begin{pmatrix}U\ Q\end{pmatrix}\mathrm{exp_m}t\begin{pmatrix}A & -R^T\\ R & 0\end{pmatrix}\begin{pmatrix}\mathrm{I}_p\\ 0\end{pmatrix},$$

where $\mathrm{exp_m}$ denotes the matrix exponential. The inverse problem, i.e. given $U, \tilde{U} \in \mathrm{St}(n,p)$, find $\Delta \in T_U\mathrm{St}(n,p)$ with $\tilde{U} = \mathrm{Exp}_U(\Delta)$, is called the *geodesic endpoint problem* and is associated with computing the Riemannian logarithm [10]. There is no known closed formula. Yet, as suggested in [2,6], one can exploit the quotient relation between $\mathrm{St}(n,p)$ and the *Grassmann manifold* [3,4] of p-dimensional subspaces of $\mathbb{R}^n$: Let $U, \tilde{U} \in \mathrm{St}(n,p)$. Then the columns of U and $\tilde{U}$ span subspaces, i.e. points on the Grassmannian. For the Grassmannian, the Riemannian logarithm is known [3], which means that we know how to find $U_\perp B \in T_U\mathrm{St}(n,p)$ and $R \in \mathrm{O}(p)$ such that $\tilde{U}R = \begin{pmatrix}U\ U_\perp\end{pmatrix}\mathrm{exp_m}\left(\begin{smallmatrix}0 & -B^T\\ B & 0\end{smallmatrix}\right)\left(\begin{smallmatrix}\mathrm{I}_p\\ 0\end{smallmatrix}\right)$. Denote $A := \mathrm{log_m}(R^T)$, where $\mathrm{log_m}$ is the principle matrix logarithm. Then

$$\tilde{U} = \begin{pmatrix}U\ U_\perp\end{pmatrix}\mathrm{exp_m}\begin{pmatrix}0 & -B^T\\ B & 0\end{pmatrix}\begin{pmatrix}\mathrm{I}_p\\ 0\end{pmatrix}\mathrm{exp_m}(A). \tag{1}$$

If there is a Stiefel geodesic from U to $\tilde{U}$, then there is also $U\tilde{A} + U_\perp\tilde{B} \in T_U\mathrm{St}(n,p)$ such that

$$\tilde{U} = \begin{pmatrix}U\ U_\perp\end{pmatrix}\mathrm{exp_m}\begin{pmatrix}\tilde{A} & -\tilde{B}^T\\ \tilde{B} & 0\end{pmatrix}\begin{pmatrix}\mathrm{I}_p\\ 0\end{pmatrix}. \tag{2}$$

Given $U, \tilde{U}$, we cannot find $U\tilde{A} + U_\perp\tilde{B} \in T_U\mathrm{St}(n,p)$ directly for (2), but we can find $UA + U_\perp B \in T_U\mathrm{St}(n,p)$ for (1). On the other hand, given $U \in \mathrm{St}(n,p)$ and $\Delta = UA + U_\perp B \in T_U\mathrm{St}(n,p)$, we can define a $\tilde{U} \in \mathrm{St}(n,p)$ via (1).

3 Quasi-Geodesics on the Stiefel Manifold

We first reduce the computational effort associated with the quasi-geodesics of [2,5,6]. Then, we introduce a new technique to construct quasi-geodesics.

3.1 Economy-Size Quasi-Geodesics

Similarly to [2,5,6], we use the notion of a retraction [1] as a starting point of the construction. Let M be a smooth manifold with tangent bundle TM. A *retraction* is a smooth mapping $R\colon TM \to M$ with the following properties:

1.) $R_x(0) = x$, i.e. R maps the zero tangent vector at $x \in M$ to x.
2.) The derivative at 0, $\mathrm{d}R_x(0)$ is the identity mapping on $T_0T_xM \simeq T_xM$.

Here, R_x denotes the restriction of R to T_xM. An example of a retraction is the Riemannian exponential mapping. On the Stiefel manifold, we can for any retraction $R\colon T\mathrm{St}(n,p) \to \mathrm{St}(n,p)$ and any tangent vector $\Delta \in T_U\mathrm{St}(n,p)$ define a smooth curve $\gamma_\Delta\colon t \mapsto R_U(t\Delta)$, which fulfills $\gamma_\Delta(0) = U$ and $\dot\gamma_\Delta(0) = \Delta$. The essential difference to [2,5,6] is that we work mainly with $n \times p$ matrices instead of $n \times n$ representatives, which entails a considerable cost reduction, when $p \leq \frac{n}{2}$.

The idea is to connect the subspaces spanned by the Stiefel manifold points with the associated Grassmann geodesic, while concurrently moving along the equivalence classes to start and end at the correct Stiefel representatives. This principle is visualized in [2, Fig. 1]. We define the economy-size quasi-geodesics similarly to [6, Prop. 6 and Thm. 7].

Proposition 1. *Let $U \in \mathrm{St}(n,p)$ and $\Delta = UA + U_\perp B \in T_U\mathrm{St}(n,p)$ with compact SVD $(\mathrm{I}_n - UU^T)\Delta = U_\perp B \overset{SVD}{=} Q\Sigma V^T$. The mapping $\mathcal{RS}\colon T\mathrm{St}(n,p) \to \mathrm{St}(n,p)$, defined by Δ maps to $\mathcal{RS}_U(\Delta) := (UV\cos(\Sigma) + Q\sin(\Sigma))V^T\exp_{\mathrm{m}}(A)$, is a retraction with corresponding quasi-geodesic*

$$\gamma(t) = \mathcal{RS}_U(t\Delta) = (UV\cos(t\Sigma) + Q\sin(t\Sigma))V^T\exp_{\mathrm{m}}(tA). \tag{3}$$

An orthogonal completion of $\gamma(t)$ is $\gamma_\perp(t) = \begin{pmatrix} U & U_\perp \end{pmatrix} \exp_{\mathrm{m}}\left(t\begin{pmatrix} 0 & -B^T \\ B & 0 \end{pmatrix}\right)\begin{pmatrix} 0 \\ \mathrm{I}_{n-p} \end{pmatrix}$. The quasi-geodesic γ has the following properties:

1. $\gamma(0) = U$
2. $\dot\gamma(t) = \gamma(t)A + \gamma_\perp(t)B\exp_{\mathrm{m}}(tA)$
3. $\|\dot\gamma(t)\|^2 = \frac{1}{2}\mathrm{tr}(A^TA) + \mathrm{tr}(B^TB)$ (constant speed)
4. $\ddot\gamma(t) = \gamma(t)(A^2 - \exp_{\mathrm{m}}(tA^T)B^TB\exp_{\mathrm{m}}(tA)) + 2\gamma_\perp(t)BA\exp_{\mathrm{m}}(tA)$
5. $D_t\dot\gamma(t) = \gamma_\perp(t)BA\exp_{\mathrm{m}}(tA)$
6. $\|D_t\dot\gamma(t)\|^2 = \|BA\|_F^2$ (constant-norm covariant acceleration)

Furthermore, γ is a geodesic if and only if $BA = 0$.

Proof. The fact that $\mathcal{RS}_U(0) = U$ for all $U \in \mathrm{St}(n,p)$ is obvious. Furthermore

$$\begin{aligned} \mathrm{d}\mathcal{RS}_U(0)(\Delta) &= (-UV\sin(\varepsilon\Sigma)\Sigma + Q\cos(\varepsilon\Sigma)\Sigma)V^T \exp_m(\varepsilon A)\big|_{\varepsilon=0} \\ &\quad + (UV\cos(\varepsilon\Sigma) + Q\sin(\varepsilon\Sigma))V^T \exp_{\mathrm{m}}(\varepsilon A)A\big|_{\varepsilon=0} \\ &= Q\Sigma V^T + UVV^T A = \Delta, \end{aligned}$$

so $\mathcal{RS}$ is a retraction. Note that γ can also be written as

$$\gamma(t) = \begin{pmatrix} U & U_\perp \end{pmatrix} \exp_{\mathrm{m}}\left(t\begin{pmatrix} 0 & -B^T \\ B & 0 \end{pmatrix}\right)\begin{pmatrix} \exp_{\mathrm{m}}(tA) \\ 0 \end{pmatrix},$$

by comparison with the Grassmann geodesics in [4, Thm. 2.3]. Therefore one possible orthogonal completion is given by the stated formula. The formulas for $\dot\gamma(t)$ and $\ddot\gamma(t)$ can be calculated by taking the derivative of

$$t \mapsto \begin{pmatrix} \gamma(t) & \gamma_\perp(t) \end{pmatrix} = \begin{pmatrix} U & U_\perp \end{pmatrix} \exp_{\mathrm{m}}\left(t\begin{pmatrix} 0 & -B^T \\ B & 0 \end{pmatrix}\right)\begin{pmatrix} \exp_{\mathrm{m}}(tA) & 0 \\ 0 & \mathrm{I}_{n-p} \end{pmatrix}.$$

It follows that $\|\dot\gamma(t)\|^2 = \mathrm{tr}(\dot\gamma(t)^T(\mathrm{I}_n - \frac{1}{2}\gamma(t)\gamma(t)^T)\dot\gamma(t)) = \frac{1}{2}\mathrm{tr}(A^TA) + \mathrm{tr}(B^TB)$. To calculate the covariant derivative $D_t\dot\gamma(t)$, we use $\gamma(t)^T\gamma_\perp(t) = 0$ and the formula for the covariant derivative of $\dot\gamma$ along γ from [4, eq. (2.41), (2.48)],

$$D_t\dot\gamma(t) = \ddot\gamma(t) + \dot\gamma(t)\dot\gamma(t)^T\gamma(t) + \gamma(t)\left((\gamma(t)^T\dot\gamma(t))^2 + \dot\gamma(t)^T\dot\gamma(t)\right), \tag{4}$$

cf. [6]. Since it features constant speed and constant-norm covariant acceleration, γ is a quasi-geodesic. It becomes a true geodesic if and only if $BA = 0$. □

Connecting $U, \tilde U \in \mathrm{St}(n,p)$ with a quasi-geodesic from Proposition 1 requires the inverse of $\mathcal{RS}_U$. Since $\mathcal{RS}_U$ is the Grassmann exponential – lifted to the Stiefel manifold – followed by a change of basis, we can make use of the modified algorithm from [3] for the Grassmann logarithm. The procedure is stated in Algorithm 1. Proposition 2 confirms that it yields a quasi-geodesic.

Algorithm 1. Economy-size quasi-geodesic between two given points

Input: $U, \tilde U \in \mathrm{St}(n,p)$
1: $\widetilde Q\widetilde S\widetilde R^T \overset{\mathrm{SVD}}{:=} \tilde U^T U$ ▷ SVD
2: $R := \widetilde Q\widetilde R^T$
3: $\tilde U_* := \tilde U R$ ▷ Change of basis in the subspace spanned by $\tilde U$
4: $A := \log_{\mathrm{m}}(R^T)$
5: $QSV^T \overset{\mathrm{SVD}}{:=} (\mathrm{I}_n - UU^T)\tilde U_*$ ▷ compact SVD
6: $\Sigma := \arcsin(S)$ ▷ element-wise on the diagonal
Output: $\gamma(t) = (UV\cos(t\Sigma) + Q\sin(t\Sigma))V^T\exp_{\mathrm{m}}(tA)$

Proposition 2. *Let $U, \tilde U \in \mathrm{St}(n,p)$. Then Algorithm 1 returns a quasi-geodesic γ connecting U and $\tilde U$, i.e. $\gamma(0) = U$ and $\gamma(1) = \tilde U$, in direction $\dot\gamma(0) = UA + Q\Sigma V^T$ and of length $L(\gamma) = (\frac{1}{2}\mathrm{tr}(A^TA) + \mathrm{tr}(\Sigma^2))^{\frac{1}{2}}$.*

Proof. Follows from [3, Algorithm 1], Proposition 1 and a straightforward calculation.

3.2 Short Economy-Size Quasi-Geodesics

To construct an alternative type of quasi-geodesics, we make the following observation: Denote B in (1) by $\hat{B}$ and calculate the SVD $U_\perp \hat{B} = Q\Sigma V^T$. Furthermore, compute $R \in \mathrm{O}(p)$ as in Algorithm 1 and denote $a := \log_\mathrm{m}(R^T) \in \mathfrak{so}(p)$ and $b := \Sigma V^T \in \mathbb{R}^{p\times p}$. Then we can rewrite (1) as

$$\begin{aligned}\tilde{U} &= \begin{pmatrix} U & Q \end{pmatrix} \exp_\mathrm{m}\begin{pmatrix} 0 & -b^T \\ b & 0 \end{pmatrix} \begin{pmatrix} \mathrm{I}_p \\ 0_{p\times p} \end{pmatrix} \exp_\mathrm{m}(a) \\ &= \begin{pmatrix} U & Q \end{pmatrix} \exp_\mathrm{m}\begin{pmatrix} 0 & -b^T \\ b & 0 \end{pmatrix} \exp_\mathrm{m}\begin{pmatrix} a & 0 \\ 0 & c \end{pmatrix} \begin{pmatrix} \mathrm{I}_p \\ 0 \end{pmatrix} \qquad \text{for any } c \in \mathfrak{so}(p).\end{aligned}$$

Without the factor c, this is exactly what lead to the quasi-geodesics (3). There are however also matrices $A \in \mathfrak{so}(p), B \in \mathbb{R}^{p\times p}$ and $C \in \mathfrak{so}(p)$ satisfying

$$\begin{pmatrix} A & -B^T \\ B & C \end{pmatrix} = \log_\mathrm{m}\left(\exp_\mathrm{m}\begin{pmatrix} 0 & -b^T \\ b & 0 \end{pmatrix} \exp_\mathrm{m}\begin{pmatrix} a & 0 \\ 0 & c \end{pmatrix}\right). \tag{5}$$

This implies that

$$\rho(t) = \begin{pmatrix} U & Q \end{pmatrix} \exp_\mathrm{m}\left(t\begin{pmatrix} A & -B^T \\ B & C \end{pmatrix}\right) \begin{pmatrix} \mathrm{I}_p \\ 0 \end{pmatrix} \tag{6}$$

is a curve from $\rho(0) = U$ to $\rho(1) = \tilde{U}$. It is indeed the projection of the geodesic in $\mathrm{O}(n)$ from $\begin{pmatrix} U & U_\perp \end{pmatrix}$ to $\begin{pmatrix} \tilde{U} & \tilde{U}_\perp \end{pmatrix}$ for some orthogonal completion $\tilde{U}_\perp$ of $\tilde{U}$. If $C = 0$, then ρ is exactly the Stiefel geodesic. For $x := \left(\begin{smallmatrix} 0 & -b^T \\ b & 0 \end{smallmatrix}\right)$ and $y := \left(\begin{smallmatrix} a & 0 \\ 0 & c \end{smallmatrix}\right)$ with $\|x\| + \|y\| \leq \ln(\sqrt{2})$, we can express C with help of the (Dynkin-)Baker-Campbell-Hausdorff (BCH) series $Z(x,y) = \log_\mathrm{m}(\exp_\mathrm{m}(x)\exp_\mathrm{m}(y))$, [9, Sect. 1.3, p. 22]. To get close to the Riemannian geodesics, we want to find a c such that C becomes small. Three facts are now helpful for the solution:

1. The series $Z(x,y)$ depends only on iterated commutators $[\cdot,\cdot]$ of x and y.
2. Since the Grassmannian is symmetric, the Lie algebra $\mathfrak{so}(n)$ has a Cartan decomposition $\mathfrak{so}(n) = \mathfrak{v} \oplus \mathfrak{h}$ with $[\mathfrak{v},\mathfrak{v}] \subseteq \mathfrak{v}$, $[\mathfrak{v},\mathfrak{h}] \subseteq \mathfrak{h}$, $[\mathfrak{h},\mathfrak{h}] \subseteq \mathfrak{v}$, [4, (2.38)].
3. For x and y defined as above, we have $x \in \mathfrak{h}$ and $y \in \mathfrak{v} \subset \mathfrak{so}(2p)$.

Considering terms up to combined order 4 in x and y in the Dynkin formula and denoting the anti-commutator by $\{x,y\} = xy + yx$, the matrix C is given by

$$C = C(c) = c + \frac{1}{12}\left(2bab^T - \{bb^T, c\}\right) - \frac{1}{24}\left(2[c, bab^T] - [c, \{bb^T, c\}]\right) + \text{h.o.t.}$$

Ignoring the higher-order terms, we can consider this as a fixed point problem $0 = C(c) \Leftrightarrow c = -\frac{1}{12}(\ldots) + \frac{1}{24}(\ldots)$. Performing a single iteration starting from $c_0 = 0$ yields

$$c_1 = c_1(a,b) = -\frac{1}{6}bab^T. \tag{7}$$

With $\delta := \max\{\|a\|, \|b\|\} \leq \ln(\sqrt{2})$, where $\|\cdot\|$ denotes the 2-norm, one can show that this choice of $c(a,b)$ produces a C-block with $\|C(c)\| \leq (\frac{7}{216} + \frac{1}{1-\delta})\delta^5$.

A closer look at curves of the form (6) shows the following Proposition.

Proposition 3. *Let $U, Q \in \mathrm{St}(n,p)$ with $U^T Q = 0$ and $A \in \mathfrak{so}(p)$, $B \in \mathbb{R}^{p\times p}$, $C \in \mathfrak{so}(p)$. Then the curve*

$$\rho(t) = (U\ Q)\,\mathrm{exp_m}\left(t\begin{pmatrix} A & -B^T \\ B & C\end{pmatrix}\right)\begin{pmatrix} \mathrm{I}_p \\ 0 \end{pmatrix} \tag{8}$$

has the following properties, where $\rho_\perp(t) := (U\ Q)\,\mathrm{exp_m}\left(t\left(\begin{smallmatrix} A & -B^T \\ B & C\end{smallmatrix}\right)\right)\left(\begin{smallmatrix} 0 \\ \mathrm{I}_p \end{smallmatrix}\right)$:

1. $\rho(0) = U$
2. $\dot\rho(0) = UA + QB \in T_U\mathrm{St}(n,p)$
3. $\dot\rho(t) = (U\ Q)\,\mathrm{exp_m}\left(t\begin{pmatrix} A & -B^T \\ B & C\end{pmatrix}\right)\begin{pmatrix} A \\ B \end{pmatrix} = \rho(t)A + \rho_\perp(t)B \in T_{\rho(t)}\mathrm{St}(n,p)$
4. $\|\dot\rho(t)\|^2 = \frac{1}{2}\mathrm{tr}(A^TA) + \mathrm{tr}(B^TB)$ *(constant speed)*
5. $\ddot\rho(t) = (U\ Q)\,\mathrm{exp_m}\left(t\begin{pmatrix} A & -B^T \\ B & C\end{pmatrix}\right)\begin{pmatrix} A^2 - B^TB \\ BA + CB \end{pmatrix}$
   $= \rho(t)(A^2 - B^TB) + \rho_\perp(t)(BA + CB)$
6. $D_t\dot\rho(t) = \rho_\perp(t)CB$
7. $\|D_t\dot\rho(t)\|^2 = \|CB\|_F^2$ *(constant-norm covariant acceleration)*

Proof. This can directly be checked by calculation and making use of the formula (4) for the covariant derivative $D_t\dot\rho(t)$.

Note that the property $U^TQ = 0$ can only be fulfilled if $p \leq \frac{n}{2}$. Since they feature constant speed and constant-norm covariant acceleration, curves of the form (8) are *quasi-geodesics*. Now we can connect two points on the Stiefel manifold with Algorithm 2, making use of $\mathrm{exp_m}\left(\begin{smallmatrix} 0 & -\Sigma \\ \Sigma & 0\end{smallmatrix}\right) = \left(\begin{smallmatrix} \cos\Sigma & -\sin\Sigma \\ \sin\Sigma & \cos\Sigma\end{smallmatrix}\right)$ for diagonal Σ. Curves produced by Algorithm 2 are of the form (8). We numerically verified

---

Algorithm 2. Short economy-size quasi-geodesic between two given points

Input: $U, \tilde U \in \mathrm{St}(n,p)$
1: $\widetilde Q \widetilde S \widetilde R^T \overset{\mathrm{SVD}}{:=} \tilde U^T U$ ▷ SVD
2: $R := \widetilde Q \widetilde R^T$
3: $a := \log_{\mathrm{m}}(R^T)$
4: $QSV^T \overset{\mathrm{SVD}}{:=} (\mathrm{I}_n - UU^T)\tilde U R$ ▷ compact SVD
5: $\Sigma := \arcsin(S)$ ▷ element-wise on the diagonal
6: $b := \Sigma V^T$
7: $c := c(a,b) = -\frac{1}{6}bab^T$
8: $\begin{pmatrix} A & -B^T \\ B & C\end{pmatrix} = \log_{\mathrm{m}}\begin{pmatrix} V\cos(\Sigma)V^TR^T & -V\sin(\Sigma)\mathrm{exp_m}(c) \\ \sin(\Sigma)V^TR^T & \cos(\Sigma)\mathrm{exp_m}(c)\end{pmatrix}$
Output: $\rho(t) = (U\ Q)\,\mathrm{exp_m}\left(t\begin{pmatrix} A & -B^T \\ B & C\end{pmatrix}\right)\begin{pmatrix} \mathrm{I}_p \\ 0 \end{pmatrix}$

---

that they are closer to the Riemannian geodesic than the economy-size quasi-geodesics of Algorithm 1 in the cases we considered. As geodesics are locally shortest curves, this motivates the term *short economy-size quasi-geodesics* for the curves produced by Algorithm 2.

Note that Algorithm 2 allows to compute the quasi-geodesic $\rho(t)$ with initial velocity $\dot{\rho}(0) = UA + QB$ between two given points. The opposite problem, namely finding the quasi-geodesic $\rho(t)$ given a point $U \in \mathrm{St}(n,p)$ with tangent vector $UA + QB \in T_U\mathrm{St}(n,p)$, is however not solved, since the correct $C \in \mathfrak{so}(p)$ is missing. Nevertheless, since $\rho(t)$ is an approximation of a geodesic, the Riemannian exponential can be used to generate an endpoint.

4 Numerical Comparison

To compare the behaviour of the economy-size and the short economy-size quasi-geodesics, two random points on the Stiefel manifold $\mathrm{St}(200, 30)$ with a distance of $d \in \{0.1\pi, 0.5\pi, 1.3\pi\}$ from each other are created. Then the quasi-geodesics according to Algorithms 1 and 2 are calculated. The Riemannian distance, i.e., the norm of the Riemannian logarithm, between the quasi-geodesics and the actual Riemannian geodesic is plotted in Fig. 1. In all cases considered, the short quasi-geodesics turn out to be two to five orders of magnitude closer to the Riemannian geodesic than the economy-size quasi-geodesics.

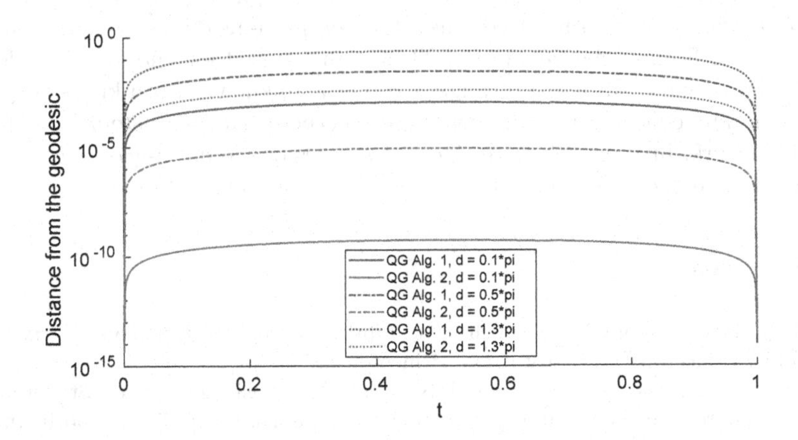

Fig. 1. Comparison of the distance of the quasi-geodesics (QG) to the true geodesic between two random points. The distance between the two points is denoted by d.

In Table 1, we display the relative deviation of the length of the quasi-geodesics from the Riemannian distance between two randomly generated points at a distance of $\frac{\pi}{2}$ on $\mathrm{St}(200, p)$, where p varies between 10 and 100. The outcome justifies the name *short* quasi-geodesics.

Table 1. Comparison of the relative deviation in length of the quasi-geodesics compared to the true geodesic between two random points on $\mathrm{St}(200, p)$ at a distance of $\frac{\pi}{2}$ for different values of p. The observable p-dependence suggests further investigations.

p	Algorithm 2	Algorithm 1	p	Algorithm 2	Algorithm 1
10	1.1997e-08	9.9308e-04	60	4.0840e-12	7.2356e-04
20	3.6416e-10	8.6301e-04	70	1.9134e-12	6.8109e-04
30	8.4590e-11	8.9039e-04	80	9.9672e-13	6.1304e-04
40	1.9998e-11	7.7265e-04	90	5.7957e-13	5.7272e-04
50	8.1613e-12	7.7672e-04	100	2.7409e-13	4.9691e-04

The essential difference in terms of the computational costs between the quasi-geodesics of Algorithm 1, those of Algorithm 2, and the approach in [6, Thm. 7] is that the they require matrix exp- and log-function evaluations of $(p \times p)$-, $(2p \times 2p)$- and $(n \times n)$-matrices, respectively.

5 Conclusion and Outlook

We have proposed a new efficient representation for a well-known type of quasi-geodesics on the Stiefel manifold, which is suitable for large-scale computations and has an exact inverse to the endpoint problem for a given tangent vector. Furthermore, we have introduced a new kind of quasi-geodesics, which are much closer to the Riemannian geodesics. These can be used for endpoint problems, but the exact curve for a given tangent vector is unknown. Both kinds of quasi-geodesics can be used, e.g., for interpolation methods like De Casteljau etc. [2,6]. In future work, further and more rigorous studies of the quasi-geodesics' length properties and the p-dependence displayed in Table 1 are of interest.

References

1. Absil, P.A., Mahony, R., Sepulchre, R.: Optimization Algorithms on Matrix Manifolds. Princeton University Press, Princeton (2008)
2. Batista, J., Krakowski, K., Leite, F.S.: Exploring quasi-geodesics on Stiefel manifolds in order to smooth interpolate between domains. In: 2017 IEEE 56th Annual Conference on Decision and Control (CDC), pp. 6395–6402 (2017)
3. Bendokat, T., Zimmermann, R., Absil, P.A.: A Grassmann Manifold Handbook: Basic Geometry and Computational Aspects. arXiv:2011.13699v2 (2020)
4. Edelman, A., Arias, T.A., Smith, S.T.: The geometry of algorithms with orthogonality constraints. SIAM J. Matrix. Anal. Appl. **20**(2), 303–353 (1998)
5. Jurdjevic, V., Markina, I., Silva Leite, F.: Extremal curves on Stiefel and Grassmann manifolds. J. Geom. Anal. **30**, 3948–3978 (2019)
6. Krakowski, K.A., Machado, L., Silva Leite, F., Batista, J.: A modified Casteljau algorithm to solve interpolation problems on Stiefel manifolds. J. Comput. Appl. Math. **311**, 84–99 (2017)

7. Machado, L., Leite, F.S., Batzies, E.: Geometric algorithm to generate interpolating splines on Grassmann and Stiefel manifolds. In: Gonçalves, J.A., Braz-César, M., Coelho, J.P. (eds.) CONTROLO 2020. LNEE, vol. 695, pp. 180–189. Springer, Cham (2021). https://doi.org/10.1007/978-3-030-58653-9_17
8. Nishimori, Y., Akaho, S.: Learning algorithms utilizing quasi-geodesic flows on the Stiefel manifold. Neurocomputing **67**, 106–135 (2005)
9. Rossmann, W.: Lie Groups: An Introduction Through Linear Groups. Oxford Graduate Texts in Mathematics. Oxford University Press, Oxford (2006)
10. Zimmermann, R.: A matrix-algebraic algorithm for the Riemannian logarithm on the Stiefel manifold under the canonical metric. SIAM J. Matrix. Anal. Appl. **38**(2), 322–342 (2017)

Optimization of a Shape Metric Based on Information Theory Applied to Segmentation Fusion and Evaluation in Multimodal MRI for DIPG Tumor Analysis

Stéphanie Jehan-Besson[1(✉)], Régis Clouard[2], Nathalie Boddaert[3,4], Jacques Grill[5,6], and Frédérique Frouin[1]

[1] Institut Curie, Université PSL, Inserm U1288, Laboratoire d'Imagerie Translationnelle en Oncologie LITO, 91400 Orsay, France
stephanie.jehan-besson@cnrs.fr
[2] Normandie Univ, UNICAEN, ENSICAEN, CNRS, GREYC, Caen, France
[3] Pediatric Radiology Department, Hôpital Necker Enfants Malades, Université de Paris, Paris, France
[4] AP-HP, IMAGINE Institute, INSERM UMR 1163, Paris, France
[5] Department of Pediatric and Adolescent Oncology, Gustave Roussy, Université Paris Saclay, Villejuif, France
[6] INSERM U981 "Molecular Predictors and New Targets in Oncology", team "Genomics and Oncogenesis of Pediatric Brain Tumors", Gustave Roussy, Université Paris Saclay, Villejuif, France

Abstract. In medical imaging, the construction of a reference shape from a set of segmentation results from different algorithms or image modalities is an important issue when dealing with the evaluation of segmentation without knowing the gold standard or when an evaluation of the inter or intra expert variability is needed. It is also interesting to build this consensus shape to merge the results obtained for the same target object from automatic or semi-automatic segmentation methods. In this paper, to deal with both segmentation fusion and evaluation, we propose to define such a "mutual shape" as the optimum of a criterion using both the mutual information and the joint entropy of the segmentation methods. This energy criterion is justified using the similarities between quantities of information theory and area measures and is presented in a continuous variational framework. We investigate the applicability of our framework for the fusion and evaluation of segmentation methods in multimodal MR images of diffuse intrinsic pontine glioma (DIPG).

Keywords: Mutual shape · Segmentation fusion · Segmentation evaluation · Shape optimization · Information theory · Neuro-oncology · Multimodal MR imaging

1 Introduction

In medical imaging, the construction of a reference shape from a set of segmentation results from different image modalities or algorithms is an important issue when eval-

F. Nielsen and F. Barbaresco (Eds.): GSI 2021, LNCS 12829, pp. 772–780, 2021.
https://doi.org/10.1007/978-3-030-80209-7_83

uating segmentation results without knowing the gold standard or comparing different ground truths. The STAPLE algorithm (Simultaneous Truth and Performance Level Estimation) proposed by Warfield et al. [19] is classically used for segmentation evaluation without gold standard. This algorithm consists in one instance of the EM (Expectation Maximization) algorithm, where the true segmentation is estimated by maximizing the likelihood of the complete data. Their approach leads to the estimation of a reference shape simultaneously with the sensitivity and specificity parameters. From these measurements, the performance level of each segmentation can be estimated and a classification of all the segmentation entries can be performed. The MAP-STAPLE [3] is semi-local and takes benefit of a small window around the pixel. On the other hand, the estimation of such a consensus shape is also an important issue for merging different segmentation results from different algorithms or image modalities. Such a fusion must take full advantage of all segmentation methods while being robust to outliers. It then requires the optimization of suitable shape metrics. We note that shape optimization algorithms have already been proposed in order to compute shape averages [2,22] or median shapes [1] by minimizing different (pseudo-)metrics on the set of measurable geometric shapes such as the Hausdorff distance in [2] or the symmetric difference between shapes in [22].

In this paper, inspired by the widely used STAPLE approach, we focus instead on the estimation of *a mutual shape* from several segmentation methods and study this problem under the umbrella of shape optimization and information theory. We propose to model the estimation of a mutual shape by introducing a continuous criterion that maximizes the mutual information between the segmentation entries and the mutual shape while minimizing the joint entropy. Our framework can be used for segmentation evaluation and/or segmentation fusion. The optimization is solved in a continuous framework using level sets and shape derivation tools as detailed in [10]. The proposed criterion can be interpreted as a measure of the symmetric difference and has interesting properties, which leads us to use the term of shape metric as detailed in Sect. 2. In the research report [9], the whole optimization framework is detailed as well as the computation of all shape derivatives. We also compared this variational mutual shape to the STAPLE method and to the minimization of a symmetric difference [22].

In terms of application, we extend our previous work [12] to the fusion of segmentation methods in the context of multimodal MR imaging in Diffuse Intrinsic Pontine Glioma (DIPG), a pediatric brain tumor for which new therapies need to be assessed. Our study concerns the fusion of tumor segmentation results obtained by different methods separately on the given modalities (T1 weighted, T2 weighted, FLAIR, T1 weighted acquired after contrast injection). Despite the known localization of the tumor inside the pons, it is difficult to segment it accurately due to the complexity of its 3D structure. We propose to test some 3D segmentation methods as segmentation entries to our fusion process. The methods proposed as entries of the segmentation fusion algorithm are based on the optimization of appropriate criteria [11,14] implemented using the level set method. Our aim was not to design a new segmentation algorithm but to investigate whether the mutual shape can lead to a better and more robust segmentation. A ground truth is available from an expert, which allows an objective evaluation of our results using the Dice Coefficient (DC) for both segmentation and fusion.

2 Proposition of a Criterion Based on Information Theory

Let $\mathcal{U}$ be a class of domains (open regular bounded sets, i.e. $\mathcal{C}^2$) of $\mathbb{R}^d$ (with $d = 2$ or 3). We denote by Ω_i an element of $\mathcal{U}$ of boundary $\partial\Omega_i$. We consider $\{\Omega_1, ..., \Omega_n\}$ a family of n shapes where each shape corresponds to the segmentation of the same unknown object O in a given image. The image domain is denoted by $\Omega \in \mathbb{R}^d$. Our aim is to compute a reference shape μ that can closely represent the true object O. We propose to define the problem using information theory and a statistical representation of the shapes. We denote by $\overline{\Omega_i}$ the complementary shape of Ω_i in Ω with $\Omega_i \cup \overline{\Omega_i} = \Omega$. The shapes can be represented using their characteristic function : $d_i(\mathbf{x}) = 1$ if $\mathbf{x} \in \Omega_i$ and $d_i(\mathbf{x}) = 0$ if $\mathbf{x} \in \overline{\Omega_i}$ where $\mathbf{x} \in \Omega$ is the location of the pixel within the image. In this paper, we consider that the shape Ω_i is represented through a random variable D_i whose observation is the characteristic function d_i. The reference shape μ is also represented through an unknown random variable T whose associated characteristic function $t(\mathbf{x}) = 1$ if $\mathbf{x} \in \mu$ and $t(\mathbf{x}) = 0$ if $\mathbf{x} \in \overline{\mu}$.

Our goal is here to mutualize the information given by each segmentation to define a consensus or reference shape. In this context, we propose to take advantage of the analogies between information measures (mutual information, joint entropy) and area measures. As previously mentioned, D_i represents the random variable associated with the characteristic function d_i of the shape Ω_i and T the random variable associated with the characteristic function t of the reference shape μ. Using these notations, $H(D_i, T)$ represents the joint entropy between the variables D_i and T, and $I(D_i, T)$ the mutual information. We then propose to minimize the following criterion:

$$E(T) = \sum_{i=1}^{n} (H(D_i, T) - I(D_i, T)) = JH(T) + MI(T), \tag{1}$$

where the sum of joint entropies is denoted by $JH(T) = \sum_{i=1}^{n} H(D_i, T)$ and the sum of mutual information by $MI(T) = -\sum_{i=1}^{n} I(D_i, T)$.

2.1 Justification of the Criterion

First of all, the introduction of this criterion can be justified by the fact that $\varphi(D_i, T) = (H(D_i, T) - I(D_i, T))$ is a metric which satisfies the following properties:

1. $\varphi(X, Y) \geq 0$
2. $\varphi(X, Y) = \varphi(Y, X)$
3. $\varphi(X, Y) = 0$ if and only if $X = Y$
4. $\varphi(X, Y) + \varphi(Y, Z) \geq \varphi(X, Z)$

Indeed, we can show that $\varphi(D_i, T) = H(T|D_i) + H(D_i|T)$ using the following classical relations between the joint entropy and the conditional entropy $H(D_i, T) = H(D_i) + H(T|D_i)$ and the relation between the mutual information and the conditional entropy $I(D_i, T) = H(D_i) - H(D_i|T)$. In [5], $H(T|D_i) + H(D_i|T)$ is shown to be a metric that satisfies the four properties mentioned above.

On the other hand, we can give another interesting geometrical interpretation of the proposed criterion. We propose to take advantage of the analogies between information measures (mutual information, joint entropy) and area measures. In [17,21], it is shown that Shannon's information measures can be interpreted in terms of area measures as follows: $H(D_i, T) = \text{mes}(\tilde{D}_i \cup \tilde{T})$ and $I(D_i, T) = \text{mes}(\tilde{D}_i \cap \tilde{T})$ with $\tilde{X}$ the abstract set associated with the random variable X and the term mes corresponds to a signed measure defined on an algebra of sets with values in $]-\infty, +\infty[$. These properties allow us to better understand the role of each term in the criterion to be optimized. Using the relation: $\text{mes}(A \setminus B) = \text{mes}(A) - \text{mes}(B) + \text{mes}(B \setminus A)$ applied to the sets $A = \tilde{D}_i \cup \tilde{T}$ and $B = \tilde{D}_i \cap \tilde{T}$, the proposed criterion may then also be interpreted as the estimation of a mean shape using a measure of the symmetric difference since $B \setminus A = \emptyset$. According to [7], a measure of the symmetric difference can be considered as a metric on the set of measurable shapes under certain properties.

2.2 Modelization of the Criterion in a Continuous Variational Setting

In order to take advantage of the previous statistical criterion (1) within a continuous shape optimization framework, and in order to optimize the criterion, we propose to express the joint and conditional probability density functions according to the reference shape μ. This difficult but essential step is detailed below for both the mutual information and the joint entropy.

We first propose to express $MI(T) = -\sum_{i=1}^{n} I(D_i, T)$ in a continuous setting according to the unknown shape μ. In order to simplify the criterion, we use the classic relation between mutual information and conditional entropy mentioned above. Since $H(D_i|T) \geq 0$ and $H(D_i)$ is independent of T, we will instead minimize $\sum_{i=1}^{n} H(D_i|T)$. Denoting by t and d_i the observations of the random variables T and D_i, the conditional entropy of D_i knowing T can be written as follows:

$$H(D_i|T) = -\sum_{t \in \{0,1\}} \left[p(t) \sum_{d_i \in \{0,1\}} p(d_i/t) \log(p(d_i/t)) \right], \tag{2}$$

with $p(T = t) = p(t)$ and $p(D_i = d_i/T = t) = p(d_i/t)$. The conditional probability $p(d_i = 1/t = 1)$ corresponds to the sensitivity parameter p_i (true positive fraction): $p_i(\mu) = p(d_i = 1/t = 1) = \frac{1}{|\mu|} \int_\mu K(d_i(\mathbf{x}) - 1) d\mathbf{x}$. The function under the integral must be chosen as a C^1 function to correctly compute the shape derivative of p_i as described in [6]. Among the available functions, we can simply choose K as a Gaussian Kernel of 0-mean and variance h where h is chosen to obtain the best estimate of p_i. We can also choose other functions such as for example $K(\mathbf{x}) = \frac{1}{1+\epsilon} * \frac{1}{(1+(\mathbf{x}^2/\epsilon^2))}$ with ϵ chosen small. Such smoothed functions simply avoid instabilities in the numerical implementation. The derivation is given for any kernel K that does not depend on μ in [9]. Some methods for estimating pdfs may involve the size of the region to estimate the kernel parameters. In this case, the derivation is not the same and must be recomputed.

The conditional probability $p(d_i = 0/t = 0)$ corresponds to the specificity parameter q_i (true negatives fraction) $q_i(\mu) = p(d_i = 0/t = 0) = \frac{1}{|\bar{\mu}|} \int_{\bar{\mu}} K(d_i(\mathbf{x})) d\mathbf{x}$. In the following, $p_i(\mu)$ is replaced by p_i and $q_i(\mu)$ by q_i. The random variable T takes the value 1 with a probability $p(t = 1) = |\mu|/|\Omega|$ and 0 with a probability

$p(t=0) = |\overline{\mu}|/|\Omega|$. The criterion (2) can then be expressed according to μ (denoted by $E_{H_C}(\mu)$):

$$E_{H_C}(\mu) = -\sum_{i=1}^{n} \Big[\frac{|\mu|}{|\Omega|}\left((1-p_i)\log(1-p_i) + p_i \log p_i\right) + \frac{|\overline{\mu}|}{|\Omega|}\left(q_i \log q_i + (1-q_i)\log(1-q_i)\right)\Big] \quad (3)$$

The parameters p_i and q_i depend explicitly on μ, which must be taken into account in the optimization process.

Let us now express, according to μ and in a continuous setting, the sum of the joint entropies $\sum_{i=1}^{n} H(D_i, T)$. The following expression of the joint entropy is considered:

$$H(D_i, T) = -\sum_{t\in\{0,1\}} \sum_{d_i\in\{0,1\}} p(d_i, t)\log\left(p(d_i, t)\right), \quad (4)$$

with $p(D_i = d_i, T = t) = p(d_i, t)$.
The following estimates for the joint probabilities are then used ($a = 0$ or $a = 1$):

$$p(d_i = a, t = 1) = \frac{1}{|\Omega|}\int_{\mu} \left(K(d_i(\mathbf{x}) - a)\right) d\mathbf{x}, \;\; p(d_i = a, t = 0) = \frac{1}{|\Omega|}\int_{\overline{\mu}} \left(K(d_i(\mathbf{x}) - a)\right) d\mathbf{x}.$$

The criterion (4) can now be expressed according to μ (denoted $E_{H_J}(\mu)$):

$$\begin{aligned} E_{H_J}(\mu) = -\sum_{i=1}^{n} \Big[& p(d_i = 1, t = 1)\log(p(d_i = 1, t = 1)) + p(d_i = 1, t = 0)\log(p(d_i = 1, t = 0)) \\ & + p(d_i = 0, t = 1)\log(p(d_i = 0, t = 1)) + p(d_i = 0, t = 0)\log(p(d_i = 0, t = 0))\Big] \end{aligned} \quad (5)$$

where $p(d_i = a, t = 1)$ and $p(d_i = a, t = 0)$ depends on μ as expressed above.

The minimization of the criterion (1) can now be expressed according to the unknown shape μ:

$$E(\mu) = E_{H_C}(\mu) + E_{H_J}(\mu) + E_H, \quad (6)$$

where the expressions of E_{H_C} and E_{H_J} are given respectively in Eqs. (3) and (5). The last term $E_H = -\sum_{i=1}^{n} H(D_i)$ does not depend on μ. Note that we also add a standard regularization term based on the minimization of $\int_{\partial\mu} ds$ where s is the arc length. This term allows to favor smooth shapes and is balanced using a positive parameter λ chosen small. In this paper $\lambda = 5$ for all the experiments.

In this given form, the minimization of such a criterion can be considered using different optimization frameworks. In this paper, we deform an initial curve (or surface), represented using a level set function [15], towards the boundaries of the region of interest according to shape derivatives whose computation is given in [9].

3 Multimodality Brain Tumors Segmentation Fusion and Evaluation

In patients with DIPG, multimodal MR images are essential for the diagnosis and the follow-up. For the present work, MR images issued from four structural modalities: T1 weighted (T1w), post-contrast T1 weighted after Gadolinium injection (T1Gd),

T2 weighted (T2w) and FLAIR images were considered. First, a pre-processing step, including spatial re-sampling, rigid co-registration and MR intensity standardization was applied to MR volumes, as detailed in [8].

Three segmentation methods were applied in order to illustrate the interest of the mutual shape estimation framework. The first segmentation method is based on the minimization of the anti-likelihood of a Gaussian distribution, called ML (Maximum likelihood) [23]. The second method, named VA, is based on the minimization of a function of the variance [10]. The third method, named KL, takes benefit of the maximization of the divergence of Kullback-Leibler as proposed in [14]. In order to optimize these three criteria, we modelled the surfaces in 3D using the level set method. Each segmentation was applied to each MR modality independently to obtain the 3D segmentation of the tumor. Other methods can be used as entries to our algorithm such as the methods proposed in [13,18]. See for example [13] for a review on supervised and unsupervised methods for multimodal brain tumors segmentation.

To mimic the RAPNO (Response Assessment in Pediatric Neuro-Oncology) guidelines for the clinical follow-up of DIPG [4], the 2D slice corresponding to the largest area tumor was selected. Our segmentation results are illustrated in Fig. 1 for such a slice of interest. On this slice, an expert delineated a contour which will be further considered as the "gold standard" and is displayed in Fig. 2(d). This expert segmentation was a compromise between the different modalities.

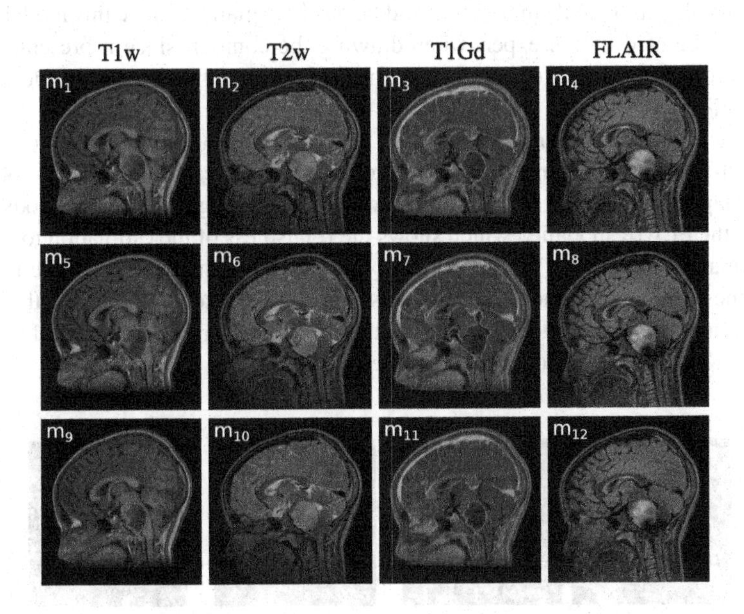

Fig. 1. Segmentation results m_1 to m_{12}, performed using 3 different segmentation methods: Maximum Likelihood ML (first row), Variance VA (second row) and Kullback-Leibler KL (third row) on the given modalities T1w, T2w, T1Gd and FLAIR displayed in columns.

We then estimated the mutual shape for both segmentation evaluation and segmentation fusion. In a first case, the 12 segmentation results (3 methods, 4 modalities) were used as entries, the resulting mutual shape being called MS1. In a second case, all the segmentation results except those obtained on the FLAIR results were used, providing a mutual shape without FLAIR, called MS2. The two mutual shapes are presented in Fig. 2, in addition with the expert segmentation.

In order to compare two shapes Ω_1 and Ω_2, the Dice Coefficient is classically used whose formula is $DC(\Omega_1, \Omega_2) = \frac{2|\Omega_1| \bigcap |\Omega_2|}{|\Omega_1|+|\Omega_2|}$. The parameter DC is close to 1 when the two shapes are similar. All segmentation methods have been compared to the expert contour in Table 1. For MS1 and MS2, the DC parameters were greater than 0.91, which is among the best results of the 12 (or 9 for MS2) segmentation results. Note that DC of MS2 slightly outperformed the DC coefficients of all the 12 segmentation methods. Thus the mutual shape enables us not to choose the optimal segmentation method, which may vary according to the context. Moreover, the mutual shape is robust to segmentation entries of lower quality (outliers). Removing FLAIR images (images for which the segmentation results are not in full agreement with expert) to compute MS2 can be useful, but is not an absolute prerequisite.

As far as the segmentation evaluation task is considered, the parameters DC obtained for the 12 segmentation results and MS1 (thus serving as a reference, without a "gold standard") were computed (Table 2). Globally the ranking of the 12 segmentation results was very similar to the ranking obtained with the expert segmentation. The same best methods (m_1,m_3) and the same outliers (m_4,m_7,m_8,m_{12}) were highlighted. The mismatch with FLAIR modality could be easily explained, since this modality was not fully integrated by the expert when drawing the contour, since it presented some semantic differences with tumor delineation, reflecting inflammation inside tumoral processes [4].

In conclusion, the mutual shape allows us to classify and to combine the different segmentation methods in an elegant and rigorous way. We can notice that the obtained mutual shape does not vary significantly if we remove the segmentation methods resulting from the FLAIR modality, which shows the robustness of the estimation to outliers. The mismatch between the FLAIR modality and the other modalities can be independent of the segmentation methods and be used as a biomarker [20]. From a theoretical point of view, we may further investigate the link between our work and the current works on Riemannian shape analysis detailed in [16].

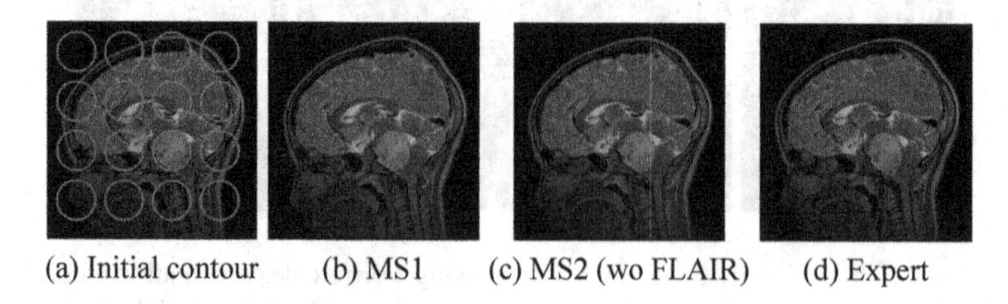

(a) Initial contour (b) MS1 (c) MS2 (wo FLAIR) (d) Expert

Fig. 2. Estimation of a mutual shape MS1 computed from the 12 segmentation methods, and the mutual shape MS2 computed without including the FLAIR modality.

Table 1. Dice Coefficients $DC(\Omega_i, \Omega_{ref})$ between the expert segmentation Ω_{ref} and Ω_i corresponding to the segmentation region given by method m_i or by the mutual shape MS1 or MS2

i	m_1	m_2	m_3	m_4	m_5	m_6	m_7	m_8	m_9	m_{10}	m_{11}	m_{12}	MS1	MS2
DC	0.920	0.887	0.915	0.731	0.865	0. 881	0.740	0.740	0.881	0.841	0. 911	0.594	0.913	0.921

Table 2. Dice Coefficients $DC(\Omega_i, MS1)$ between the MS1 mutual shape and Ω_i corresponding to the segmentation region given by the method m_i

i	m_1	m_2	m_3	m_4	m_5	m_6	m_7	m_8	m_9	m_{10}	m_{11}	m_{12}
DC	0.971	0.956	0.960	0.799	0.875	0.951	0.767	0.809	0.919	0.887	0.935	0.621

Acknowledgements. This work takes place in the BIOMEDE-IA project carried out in collaboration with Gustave-Roussy, Necker Hospital and Neurospin and supported by the association Imagine4Margo.

References

1. Berkels, B., Linkmann, G., Rumpf, M.: An SL(2) invariant shape median. JMIV **37**(2), 85–97 (2010). https://doi.org/10.1007/s10851-010-0194-6
2. Charpiat, G., Faugeras, O., Keriven, R.: Approximations of shape metrics and application to shape warping and empirical shape statistics. FCM **5**(1) (2005). https://doi.org/10.1007/s10208-003-0094-x
3. Commowick, O., Akhondi-Asl, A., Warfield, S.K.: Estimating a reference standard segmentation with spatially varying performance parameters. TMI **31**(8), 1593–1606 (2012). https://doi.org/10.1109/TMI.2012.2197406
4. Cooney, T.M., Cohen, K.J., Guimaraes, C.V., et al.: Response assessment in diffuse intrinsic Pontine Glioma: recommendations from the response assessment in pediatric neuro-oncology (RAPNO) working group. Lancet Oncol. **21**(6), e330–e336 (2020). https://doi.org/10.1016/S1470-2045(20)30166-2
5. Cover, T.M., Thomas, J.A.: Elements of Information Theory (Wiley Series in Telecommunications and Signal Processing. Wiley-Interscience, New York (2006)
6. Delfour, M.C., Zolésio, J.P.: Shapes and geometries: metrics, analysis, differential calculus, and optimization. In: Advances in Design and Control. SIAM (2001). https://doi.org/10.1137/1.9780898719826
7. Deza, M.M., Deza, E.: Encyclopedia of Distances. Springer, Berlin, Heidelberg (2016). https://doi.org/10.1007/978-3-642-30958-8
8. Goya-Outi, J., Orlhac, F., Calmon, R., et al.: Computation of reliable textural indices from multimodal brain MRI: suggestions based on a study of patients with diffuse intrinsic Pontine Glioma. Phys. Med. Biol. **63**(10), 105003 (2018). https://doi.org/10.1088/1361-6560/aabd21
9. Jehan-Besson, S., Clouard, R., Tilmant, C., et al.: A mutual reference shape for segmentation fusion and evaluation. arXiv:2102.08939 (2021)
10. Jehan-Besson, S., Herbulot, A., Barlaud, M., Aubert, G.: Shape gradient for image and video segmentation. In: Paragios, N., Chen, Y., Faugeras, O., (eds.) Handbook of Mathematical Models in Computer Vision, pp. 309–323. Springer, Boston (2006). https://doi.org/10.1007/0-387-28831-7_19

11. Jehan-Besson, S., Lecellier, F., Fadili, J., et al.: Medical image segmentation and tracking through the maximisation or the minimisation of divergence between pdfs. In: Advances in Medical Technologies and Clinical Practice, IGI Global (2011). https://doi.org/10.4018/978-1-60566-280-0.ch002
12. Jehan-Besson, S., Tilmant, C., de Cesare, A., et al.: A mutual reference shape based on information theory. In: ICIP (2014). https://doi.org/10.1109/ICIP.2014.7025178
13. Lapuyade-Lahorgue, J., Xue, J., Ruan, S.: Segmenting multi-source images using Hidden Markov fields with copula-based multivariate statistical distributions. TIP **26**(7), 3187–3197 (2017). https://doi.org/10.1109/TIP.2017.2685345
14. Lecellier, F., Jehan-Besson, S., Fadili, J.: Statistical region-based active contours for segmentation: an overview. IRBM **35**(1), 3–10 (2014). https://doi.org/10.1016/j.irbm.2013.12.002
15. Osher, S., Paragios, N.: Geometric Level Set Methods in Imaging, Vision, and Graphics. Springer, New York (2016). https://doi.org/10.1007/b97541
16. Pennec, X., Sommer, S., Fletcher, T.: Riemannian Geometric Statistics in Medical Image Analysis. Elsevier, New York (2020). https://doi.org/10.1016/C2017-0-01561-6
17. Reza, F.M.: An Introduction to Information Theory. McGraw-Hill, New York (1994)
18. Urien, H., Buvat, I., Rougon, N., Boughdad, S., Bloch, I.: 3D PET-driven multi-phase segmentation of meningiomas in MRI. In: ISBI (2016). https://doi.org/10.1109/ISBI.2016.7493294
19. Warfield, S.K., Zou, K.H., Wells III, W.M.: Simultaneous truth and performance level estimation (STAPLE): an algorithm for the validation of image segmentation. IEEE TMI **23**(7), 903–921 (2004). https://doi.org/10.1109/TMI.2004.828354
20. Yamasaki, F., Nishibuchi, I., Karakawa, S., et al.: T2-FLAIR mismatch sign and response to radiotherapy in diffuse intrinsic Pontine Glioma. Pediatr. Neurosurg. **56**(1) (2021). https://doi.org/10.1159/000513360
21. Yeung, R.W.: A new outlook on Shannon's information measures. IEEE Trans. Inf. Theory **37**(3), 466–474 (1991). https://doi.org/10.1109/18.79902
22. Yezzi, A.J., Soatto, S.: Deformotion: deforming motion, shape average and the joint registration and approximation of structures in images. IJCV **53**(2), 153–167 (2003). https://doi.org/10.1023/A:1023048024042
23. Zhu, S., Yuille, A.: Region competition: unifying snakes, region growing, and Bayes/MDL for multiband image segmentation. In: PAMI, vol. 18, pp. 884–900 (1996). https://doi.org/10.1109/34.537343

Metamorphic Image Registration Using a Semi-lagrangian Scheme

Anton François[1,2(✉)], Pietro Gori[2], and Joan Glaunès[1]

[1] MAP5, Université de Paris, Paris, France
{anton.francois,alexis.glaunes}@parisdescartes.fr
[2] LTCI, Télécom Paris, IPParis, Paris, France
pietro.gori@telecom-paris.fr
https://math-info.u-paris.fr/en/map5-laboratory/

Abstract. In this paper, we propose an implementation of both *Large Deformation Diffeomorphic Metric Mapping* (LDDMM) and Metamorphosis image registration using a semi-Lagrangian scheme for geodesic shooting. We propose to solve both problems as an inexact matching providing a single and unifying cost function. We demonstrate that for image registration the use of a semi-Lagrangian scheme is more stable than a standard Eulerian scheme. Our GPU implementation is based on `PyTorch`, which greatly simplifies and accelerates the computations thanks to its powerful automatic differentiation engine. It will be freely available at https://github.com/antonfrancois/Demeter_metamorphosis.

Keywords: Image diffeomorphic registration · LDDMM · Metamorphosis · Semi-Lagrangian scheme

1 Introduction

Diffeomophic image matching is a key component in computational anatomy for assessing morphological changes in a variety of cases. Since it does not modify the spatial organization of the image (i.e. no tearing, shearing, holes), it produces anatomically plausible transformations. Possible applications are: alignment of multi-modal images, longitudinal image registration (images of the same subject at different time points) or alignment of images of the same modality across subjects, for statistical analysis such as atlas construction.

Extensive work have been conduced to compute diffeomorphic transformations for registering medical images. One method is to use flows of time-dependent vector fields, as in LDDMM [3,5,17]. This allows us to define a right-invariant Riemannian metric on the group of diffeomorphisms. This metric can be projected onto the space of images, which thus becomes a shape space, providing useful notions of geodesics, shortest paths and distances between images [6,18]. A shortest path represents the registration between two images.

Due to the high computational cost of LDDMM, some authors proposed to use stationary vector fields, instead than time-varying, using the Lie algebra vector field exponential [1,2].

F. Nielsen and F. Barbaresco (Eds.): GSI 2021, LNCS 12829, pp. 781–788, 2021.
https://doi.org/10.1007/978-3-030-80209-7_84

Diffeomorphic maps are by definition one-to-one, which means that they are suited for matching only images characterized by the same topology. However, many clinical or morphometric studies often include an alignment step between a healthy template (or atlas) and images with lesions, alterations or pathologies, like white matter multiple sclerosis or brain tumors. Three main strategies have been proposed in the literature. Cost function masking is used in order not to take into account the lesion or tumor (by masking it) during registration [14]. This method is quite simple and easy to implement but it does not give good results when working with big lesions or tumors. The other two strategies consist in modifying either the healthy template or the pathological image in order to make them look like a pathological or healthy image respectively. For instance, in GLISTR [8], authors first make growing a tumor into an healthy image and then they register it to an image with tumor. This strategy is quite slow and computationally heavy. On the other hand, in [15], authors try to fill the lesions using inpainting in order to make the image looks like an healthy image. This strategy seems to work well only with small lesions. Similarly, in [11], authors proposed to estimate the healthy version of an image as its low-rank component, which seems to work correctly only when the sample size is quite large. Here, we propose to use a natural extension of LDDMM: the Metamorphosis framework as introduced in [10,16,18]. It was designed to jointly estimate a diffeomorphic deformation and a variation in image appearance, modeling, for instance, the apparition of a tumor. Intensity variations are also deformed along the images during the process. Metamorphosis is related to morphing in computer graphics. Metamorphosis, like LDDMM, generates a distance between objects. This time the integration follows two curves, one for the deformation and one for the intensity changes (see definition in [9,10,18]). By comparing the works of Beg and Vialard [5,17] for LDDMM and the Metamorphosis framework, one can notice that they are closely related as they end up using a very similar set of variational equations for their geodesics.

We decided to use geodesic shooting [12] as it is the only method that can theoretically ensure to get optimal paths (geodesics), even if not performing optimization until convergence. Once defined the Euler-Lagrange equations associated to the functional registration, one can integrate them. The estimated optimal paths, which are geodesics, are usually computed using the shooting algorithm and are completely encoded by the initial conditions of the system. The minimizing initial conditions, subject to the geodesic equations, are usually estimated using a gradient descent scheme (which needs the computation of the adjoint equations).

In this paper, we make the following contributions. To the best of out knowledge, we provide the first implementation of both LDDMM and Metamorphosis joined in one optimisation problem.

We also propose a full semi-Lagrangian scheme for the geodesic shooting equations and we give access to our easy to use GPU implementation fully developed with `PyTorch`.

2 Methods

LDDMM and Metamorphoses Geodesics. Let $\Omega \subset \mathbb{R}^d$ be a fixed bounded domain, where $d = \{2, 3\}$. We define a gray-scale image $I \in \mathcal{I}$ (or a gray-scale volume) as a square-integrable function defined on Ω (i.e.: $\mathcal{I} \doteq L^2(\Omega, \mathbb{R})$).

Let V be a fixed Hilbert space of l-times continuously differentiable vector fields supported on Ω and l times continuously differentiable (i.e.: $V \subset C_0^l(\Omega, \mathbb{R}^d)$). We consider that the time varying vector fields v_t are elements of $L^2([0,1]\,; V), t \in [0,1]$, [5,17,18]. A pure deformation flow can be deduced solving the ODE $\dot{\Phi}_t = v_t \cdot \Phi_t := v_t \circ \Phi_t$, $v_t \in V, \forall t \in [0,1]$. Hence, the flow at a given time t is written $\Phi_t = \int_0^t v_s \circ \Phi_s ds$ with $\Phi_0 = \mathrm{Id}$. As shown in [6], the elements of V need to be sufficiently smooth to produce a flow of diffeomorphisms. The traditional optimisation problem for LDDMM is an inexact matching. Let $I, J \in \mathcal{I}$ be a source and a target image, it aims at minimizing a cost composed of a data term (e.g. the L_2-norm, known as the sum of images squared difference (SSD)), and a regularisation term on v, usually defined as the total kinetic energy $\int_0^1 \|v_t\|_V^2 \; dt$. The goal is thus to find the "simplest" deformation to correctly match I to J.

Metamorphoses join additive intensity changes to the deformations. The goal of Metamorphosis is to register an image I to J using variational methods with an intensity additive term $z_t \in L^1([0,1]\,; V)$. The image evolution can be defined as:

$$\partial_t I_t = v_t \cdot I_t + \mu z_t = -\langle \nabla I_t, v_t \rangle + \mu z_t, \quad \text{s.t. } I_0 = I \quad \mu \in \mathbb{R}^+. \tag{1}$$

One can control the amount of deformation vs photometric changes by varying the hyperparameter $\mu \in \mathbb{R}^+$. The dot notation is used for infinitesimal step composition, here writing $v \cdot I_t$ implies that I_t is deformed by an infinitesimal vector field v. As described by Trouvé & Younes in [16,18], the $\{z_t\}$ have to be the 'leftovers' of the transport of I_t by v_t toward the exact registration, as it can be seen by rewriting Eq. 1 as $z_t = \frac{1}{\mu^2}(\partial_t I_t - v_t \cdot I_t)$. This is usually called the Metamorphic residual image or the momentum.

In order to find the optimal $(v_t)_{t \in [0,1]}$ and $(z_t)_{t \in [0,1]}$, one can minimize the exact matching functional [10,13,18] using Eq. 1:

$$E_M(v, z) = \int_0^1 \|v_t\|_V^2 + \rho \|z_t\|_{L_2}^2 dt, \quad \text{s.t. } I_1 = J, I_0 = I; \quad \rho \in \mathbb{R}^+, \tag{2}$$

As shown in [10,18], the geodesic equations for Metamorphosis are:

$$\begin{cases} v_t &= -\frac{\rho}{\mu} K \star (z_t \nabla I_t) \\ \partial_t z_t &= - \quad \nabla \cdot (z_t v_t) \\ \partial_t I_t &= -\langle \nabla I_t, v_t \rangle + \mu z_t \end{cases} \tag{3}$$

By setting $\rho = \mu$ and letting $\mu \to 0$, one recovers the geodesic equations for LDDMM as pointed out in [17,18]. In Eq. 3, $\nabla \cdot (zv) = \mathrm{div}(zv)$ is the divergence of the field v times z at each pixel, K is the chosen translation invariant reproducing kernel (of the RKHS) and $\star$ is the convolution. In practice K is often a Gaussian blurring kernel [12,17]. The last line of Eq. 3 is the advection term, simulating the movement of non diffusive material. The second (continuity) equation is a conservative form which ensures that the amount of deformation is preserved on the whole domain over time. Thus, given the initial conditions of the system, $I = I_0$ and z_0, one can integrate in time the system of Eq. 3 to obtain I_1. Note that v_0 can be computed from z_0, making z the only unknown. Furthermore, one can notice that the energy in Eq. 2 is conserved (i.e.: constant along the geodesic paths) and therefore the time integrals may be replaced by the norms at time 0.

Here, we propose to solve Metamorphosis as an inexact matching problem. This allows us to have a unifying cost function (i.e.: Hamiltonian) for both LDDMM and Metamorphosis:

$$H(z_0) = \frac{1}{2}\|I_1 - J\|_{L_2}^2 + \lambda\Big[\|v_0\|_V^2 + \rho\|z_0\|_{L_2}^2\Big] \tag{4}$$

with $\|v_0\|_V^2 = \langle z_0 \nabla I, K \star (z_0 \nabla I)\rangle$. The hyperparameters λ and μ define the amount of total regularization and intensity changes respectively.

Geodesic Shooting Integration. Integration of the geodesics is a crucial computational point for both LDDMM and Metamorphosis. In the case of image registration using LDDMM, Beg et al. [5] initially described a method based on gradient descent which could not retrieve exact geodesics, as shown in [17]. An actual shooting method was then proposed in [17] for LDDMM based registration of images. To the best of our knowledge, the only shooting method proposed in the literature for image Metamorphosis is the one proposed in [13]. It is based on a Lagrangian frame of reference and therefore it is not well suited for large images showing complicated deformations, as it could be the case when registering healthy templates to patients with large tumors. Here, we propose to use a semi-Lagrangian scheme.

From Eulerian to Semi-lagrangian Formulation. When analysing flows from ODE and PDE, two concurrent points of view are often discussed. Lagrangian schemes where one follows the stream of a set of points from initialisation, and Eulerian schemes where one consider some fixed point in space (often a grid) and evaluate the changes occurred. Eulerian schemes seem to be the most natural candidate for flow integration over an image. In fact, in the Lagrangian schemes the streams of pixels we follow can go far apart during integration, making impossible the image re-interpolation. However, the Lagrangian scheme is more numerically stable than the Eulerian one thus allowing a better convergence [4]. Indeed a necessary condition for proper convergence for Eulerian schemes is the Courant–Friedrichs–Lewy (CFL) condition, which defines a maximum value for the time step. If the time step is badly chosen, ripples may appear or, worse, the integration may fail (see Fig. 1). The minimal number of time steps required increases with the size of the image, thus increasing the number of iterations and making it very slow to use on real imaging data. For these reasons, the so-called semi-Lagrangian scheme seems to be a good compromise.

The idea is to compute the deformation of a grid corresponding to a small displacement $\mathrm{Id} - \delta t\, v_t$, and then interpolate the values of the image I_t on the grid. This can be summarized by $I_{t+\delta t} \approx I_t \circ (\mathrm{Id} - \delta t\, v_t)$. Semi-Lagrangian schemes are more stable and don't need as many iterations. Too many iterations would blur the images due to the successive bilinear or trilinear (in 3D) interpolations.

Let's reformulate Eq. 3 in a semi-Lagrangian formulation, starting by the advection part [7]. From the Eulerian formulation for the closed domain $[0,1] \times \Omega$, we can write:

$$\partial_t I_t + \langle \nabla I_t, v_t \rangle - \mu z_t = \partial_t I_t + \sum_{i=1}^{d} \partial_{x_i} I(t,x) v_{x_i}(t,x) - \mu z(t,x) = 0. \tag{5}$$

where we use for convenience the notations $v_t(x) = v(t,x) = (v_{x_1}(t,x), \cdots, v_{x_d}(t,x)), t \in [0,1], x \in \Omega$. We deduce the characteristics defined by the system of differential equations :

$$x_i'(t) = v_{x_i}(t,x) + \mu z_t, \qquad i \leq d \in \mathbb{N}^*, t \in [0,1]. \tag{6}$$

Then, we can also rewrite the continuity equation as:

$$\partial_t z_t + \nabla \cdot (z_t v_t) = \partial_t z_t + \sum_{i=1}^{d} \partial_{x_i} (z(t,x) \times v_{x_i}(t,x)) = 0 \tag{7}$$

$$= \partial_t z_t + < \nabla z_t, v_t > + (\nabla \cdot v_t) z_t = 0 \tag{8}$$

In the same way, we also extract the characteristics defined by the system of differential equations:

$$x_i'(t) = v_{x_i}(t,x) + [\nabla \cdot v(t,x)] z(t,x), \qquad i \leq d \in \mathbb{N}^*, t \in [0,1]. \tag{9}$$

Note that by using a semi-Lagrangian scheme for this part we avoid to compute a discrete approximation of $\nabla \cdot (z_t v_t)$, but still need to compute an approximation of $\nabla \cdot v_t$. However, the momentum z_t, similarly to the image I_t, is potentially non smooth, while v_t is smooth due to its expression through the convolution operator K.

In the following Section, we will compare three computational options to integrate over the geodesics: 1- the Eulerian scheme, 2- the semi-Lagrangian approach and 3- a combination of the two, where we use the semi-Lagragian scheme for the advection (Eq. 5) and the Eulerian scheme for the residuals (Eq. 9).

3 Results and Conclusions

In Fig. 1, we can observe the lack of stability of Eulerian methods compared to the semi-Lagrangian ones. Even if the chosen time step is rather small, the Eulerian scheme produces ripples (in purple in the residuals) and the integration fails (see the estimated deformation). On the contrary, semi-Lagrangian schemes converge to a better deformation with an higher time step. It should also be noted

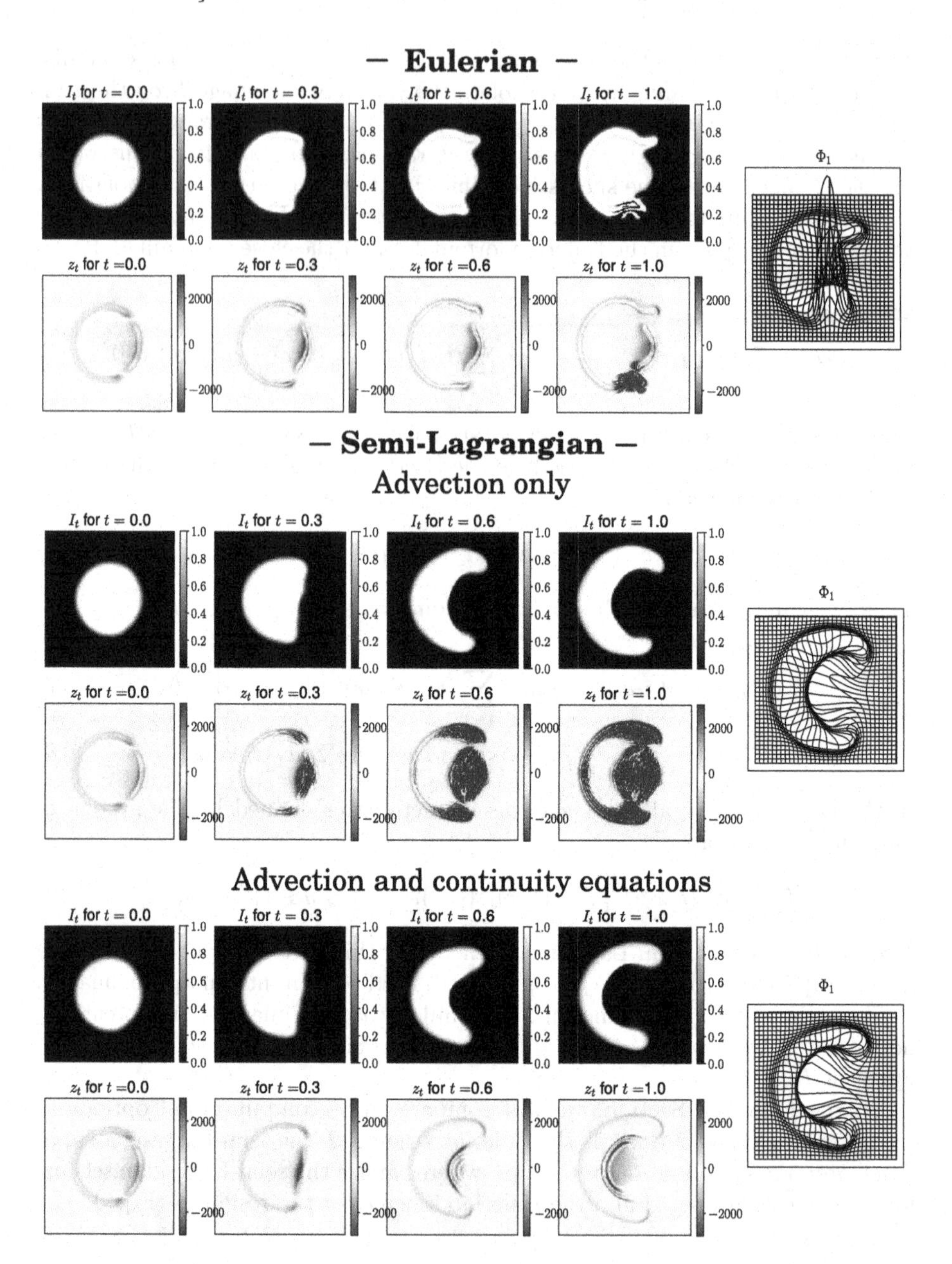

Fig. 1. Comparison between the stability of the 3 geodesic shootings schemes proposed for LDDMM. In each row, we show four intermediary shooting steps with the same initial z_0 and RKHS for v. The black and white pictures are the images, with below the corresponding z. The deformations grids on the right are obtained by integrating over all $v_t, \forall t \in [0, 1]$. Shooting was performed using a z_0 obtained from LDDMM optimisation towards a 'C' picture ($\mu = 0$). The Eulerian and semi-Lagrangian schemes have a time step of $1/38$ and $1/20$ respectively.

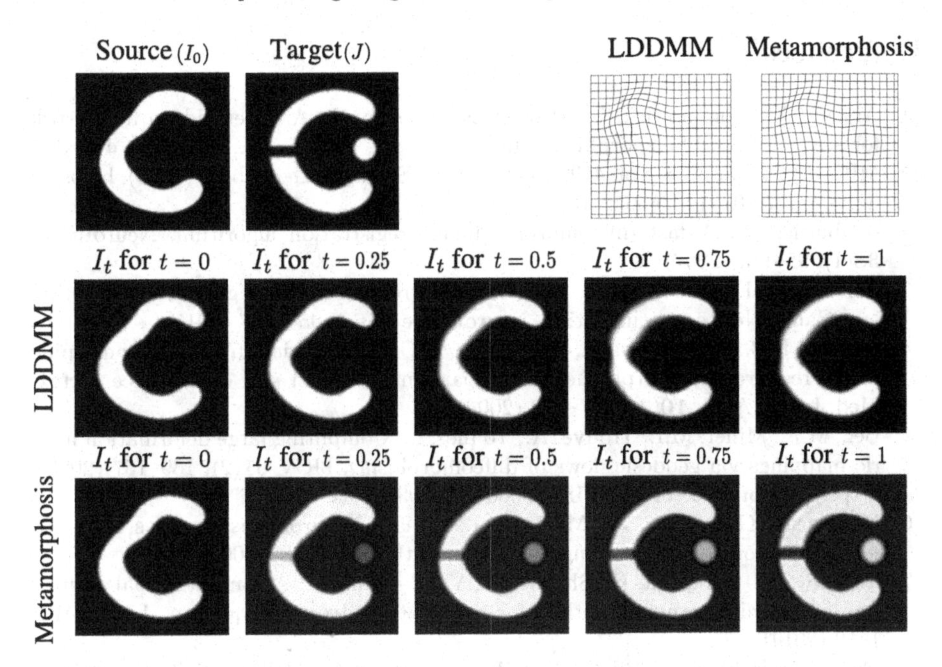

Fig. 2. Comparison of LDDMM vs Metamorphoses registration. *Top-right* final deformation grids obtained by integrating over all vector fields $(v_t)_{t\in[0,1]}$. *Bottom Rows* Image evolution during the geodesic shootings of the respective method after optimisation with Eq. 4.

that the full semi-Lagrangian scheme (advection and continuity equations) is perfectly stable without showing ripples as it is the case for the advection-only semi-Lagrangian scheme.

In Fig. 2 we can see that Metamorphosis and LDDMM describe deformations in a similar way. As we use the SSD (i.e. L^2 norm) as data term, an object in the source image is matched to another object in the target image only if they have some pixels in common. For this reason, the C form is not pushed to match the small disk on the right of the example. In the Metamorphosis deformation grid we can see that the small disk is growing, as it is less costly to create a small disk and make it grow. However, the Metamorphic registration, thanks to the intensities changes modeled by z, manages to correctly take into account the topological differences between the source and target images.

With the use of automatic differentiation, we bypass the extensive and delicate work of deriving the backward adjoint equations and implementing a discrete scheme to solve them. This allowed us to merge LDDMM and Metamorphosis into a single framework and to easily test different configurations of the problem. For this study, we optimized all costs using gradient descent. We also provide alternative optimization methods, such as L-BFGS, in our library Demeter, which will be regularly updated.

References

1. Arsigny, V., Commowick, O., Pennec, X., Ayache, N.: A log-euclidean framework for statistics on diffeomorphisms. In: Larsen, R., Nielsen, M., Sporring, J. (eds.) MICCAI 2006. LNCS, vol. 4190, pp. 924–931. Springer, Heidelberg (2006). https://doi.org/10.1007/11866565_113
2. Ashburner, J.: A fast diffeomorphic image registration algorithm. NeuroImage **38**(1), 95–113 (2007)
3. Ashburner, J., Friston, K.J.: Diffeomorphic registration using geodesic shooting and Gauss–Newton optimisation. NeuroImage **55**(3), 954–967 (2011)
4. Avants, B.B., Schoenemann, P.T., Gee, J.C.: Lagrangian frame diffeomorphic image registration: morphometric comparison of human and chimpanzee cortex. Med. Image Anal. **10**(3), 397–412 (2006)
5. Beg, M.F., Miller, M.I., Trouvé, A., Younes, L.: Computing large deformation metric mappings via geodesic flows of diffeomorphisms. IJCV **61**(2), 139–157 (2005). https://doi.org/10.1023/B:VISI.0000043755.93987.aa
6. Dupuis, P., Grenander, U., Miller, M.I.: Variational problems on flows of diffeomorphisms for image matching. Q. Appl. Math. **56**(3), 587–600 (1998)
7. Efremov, A., Karepova, E., Shaydurov, V., Vyatkin, A.: A computational realization of a semi-lagrangian method for solving the advection equation. JAM **2014**, 1–12 (2014)
8. Gooya, A., Pohl, K.M., Bilello, M., Cirillo, L., Biros, G., Melhem, E.R., Davatzikos, C.: GLISTR: Glioma image segmentation and registration. IEEE Trans. Med. Imaging **31**(10), 1941–1954 (2012)
9. Gris, B., Chen, C., Öktem, O.: Image reconstruction through metamorphosis. Inverse Prob. **36**(2), 025001 (2020)
10. Holm, D.D., Trouvé, A., Younes, L.: The Euler-Poincaré theory of metamorphosis. QAL **67**(4), 661–685 (2009)
11. Liu, X., Niethammer, M., Kwitt, R., Singh, N., McCormick, M., Aylward, S.: Low-rank atlas image analyses in the presence of pathologies. IEEE Trans. Med. Imaging **34**(12), 2583–2591 (2015)
12. Miller, M.I., Trouvé, A., Younes, L.: Geodesic shooting for computational anatomy. J. Math. Imaging Vis. **24**(2), 209–228 (2006)
13. Richardson, C.L., Younes, L.: Metamorphosis of images in reproducing kernel Hilbert spaces. Adv. Comput. Math. **42**(3), 573–603 (2015). https://doi.org/10.1007/s10444-015-9435-y
14. Ripollés, P., et al.: Analysis of automated methods for spatial normalization of lesioned brains. NeuroImage **60**(2), 1296–1306 (2012)
15. Sdika, M., Pelletier, D.: Nonrigid registration of multiple sclerosis brain images using lesion inpainting for morphometry or lesion mapping. Hum. Brain Mapp. **30**(4), 1060–1067 (2009)
16. Trouvé, A., Younes, L.: Local geometry of deformable templates. SIAM **37**(1), 17–59 (2005)
17. Vialard, F.X., Risser, L., Rueckert, D., Cotter, C.J.: Diffeomorphic 3D image registration via geodesic shooting using an efficient adjoint calculation. IJCV **97**(2), 229–241 (2011). https://doi.org/10.1007/s11263-011-0481-8
18. Younes, L.: Shapes and Diffeomorphisms. Springer, Heidelberg (2010). https://doi.org/10.1007/978-3-642-12055-8

Geometry of the Symplectic Stiefel Manifold Endowed with the Euclidean Metric

Bin Gao[1(✉)], Nguyen Thanh Son[2], P.-A. Absil[1], and Tatjana Stykel[3]

[1] ICTEAM Institute, UCLouvain, 1348 Louvain-la-Neuve, Belgium
gaobin@lsec.cc.ac.cn
[2] Department of Mathematics and Informatics, Thai Nguyen University of Sciences, Thai Nguyen, Vietnam
[3] Institute of Mathematics, University of Augsburg, Augsburg, Germany

Abstract. The symplectic Stiefel manifold, denoted by $\mathrm{Sp}(2p, 2n)$, is the set of linear symplectic maps between the standard symplectic spaces $\mathbb{R}^{2p}$ and $\mathbb{R}^{2n}$. When $p = n$, it reduces to the well-known set of $2n \times 2n$ symplectic matrices. We study the Riemannian geometry of this manifold viewed as a Riemannian submanifold of the Euclidean space $\mathbb{R}^{2n\times 2p}$. The corresponding normal space and projections onto the tangent and normal spaces are investigated. Moreover, we consider optimization problems on the symplectic Stiefel manifold. We obtain the expression of the Riemannian gradient with respect to the Euclidean metric, which is then used in optimization algorithms. Numerical experiments on the nearest symplectic matrix problem and the symplectic eigenvalue problem illustrate the effectiveness of Euclidean-based algorithms.

Keywords: Symplectic matrix · symplectic Stiefel manifold · Euclidean metric · Optimization

1 Introduction

Let J_{2m} denote the nonsingular and skew-symmetric matrix $\left[\begin{smallmatrix} 0 & I_m \\ -I_m & 0 \end{smallmatrix}\right]$, where I_m is the $m \times m$ identity matrix and m is any positive integer. The *symplectic Stiefel manifold*, denoted by

$$\mathrm{Sp}(2p, 2n) := \left\{X \in \mathbb{R}^{2n\times 2p} : X^\top J_{2n} X = J_{2p}\right\},$$

is a smooth embedded submanifold of the Euclidean space $\mathbb{R}^{2n\times 2p}$ $(p \leq n)$ [11, Proposition 3.1]. We remove the subscript of J_{2m} and I_m for simplicity if there is no confusion. This manifold was studied in [11]: it is closed and unbounded; it has dimension $4np - p(2p-1)$; when $p = n$, it reduces to the *symplectic group*, denoted by $\mathrm{Sp}(2n)$. When $X \in \mathrm{Sp}(2p, 2n)$, it is termed as a *symplectic* matrix.

Symplectic matrices are employed in many fields. They are indispensable for finding eigenvalues of (skew-)Hamiltonian matrices [4,5] and for model order reduction of Hamiltonian systems [8,15]. They appear in Williamson's theorem and the formulation

This work was supported by the Fonds de la Recherche Scientifique – FNRS and the Fonds Wetenschappelijk Onderzoek – Vlaanderen under EOS Project no. 30468160.

F. Nielsen and F. Barbaresco (Eds.): GSI 2021, LNCS 12829, pp. 789–796, 2021.
https://doi.org/10.1007/978-3-030-80209-7_85

of symplectic eigenvalues of symmetric and positive-definite matrices [6,13,16,18]. Moreover, symplectic matrices can be found in the study of optical systems [10] and the optimal control of quantum symplectic gates [20]. Specifically, some applications [6,10,15,16,20] can be reformulated as optimization problems on the set of symplectic matrices where the constraint is indeed the symplectic Stiefel manifold.

In recent decades, most of studies on the symplectic topic focused on the symplectic group ($p = n$) including geodesics of the symplectic group [9], optimality conditions for optimization problems on the symplectic group [7,14,19], and optimization algorithms on the symplectic group [10,17]. However, there was less attention to the geometry of the symplectic Stiefel manifold $\mathrm{Sp}(2p, 2n)$. More recently, the Riemannian structure of $\mathrm{Sp}(2p, 2n)$ was investigated in [11] by endowing it with a new class of metrics called *canonical-like*. This canonical-like metric is different from the standard *Euclidean* metric (the Frobenius inner product in the ambient space $\mathbb{R}^{2n\times 2p}$)

$$\langle X, Y\rangle := \mathrm{tr}(X^\top Y) \quad \text{for } X, Y \in \mathbb{R}^{2n\times 2p},$$

where $\mathrm{tr}(\,\cdot\,)$ is the trace operator. A priori reasons to investigate the Euclidean metric on $\mathrm{Sp}(2p, 2n)$ are that it is a natural choice, and that there are specific applications with close links to the Euclidean metric, e.g., the projection onto $\mathrm{Sp}(2p, 2n)$ with respect to the Frobenius norm (also known as the nearest symplectic matrix problem)

$$\min_{X\in\mathrm{Sp}(2p,2n)} \|X - A\|_{\mathrm{F}}^2 . \tag{1}$$

Note that this problem does not admit a known closed-form solution for general $A \in \mathbb{R}^{2n\times 2p}$.

In this paper, we consider the symplectic Stiefel manifold $\mathrm{Sp}(2p, 2n)$ as a Riemannian submanifold of the Euclidean space $\mathbb{R}^{2n\times 2p}$. Specifically, the normal space and projections onto the tangent and normal spaces are derived. As an application, we obtain the Riemannian gradient of any function on $\mathrm{Sp}(2p, 2n)$ in the sense of the Euclidean metric. Numerical experiments on the nearest symplectic matrix problem and the symplectic eigenvalue problem are reported. In addition, numerical comparisons with the canonical-like metric are also presented. We observe that the Euclidean-based optimization methods need fewer iterations than the methods with the canonical-like metric on the nearest symplectic problem, and Cayley-based methods perform best among all the choices.

The rest of paper is organized as follows. In Sect. 2, we study the Riemannian geometry of the symplectic Stiefel manifold endowed with the Euclidean metric. This geometry is further applied to optimization problems on the manifold in Sect. 3. Numerical results are presented in Sect. 4.

2 Geometry of the Riemannian Submanifold $\mathrm{Sp}(2p, 2n)$

In this section, we study the Riemannian geometry of $\mathrm{Sp}(2p, 2n)$ equipped with the Euclidean metric.

Given $X \in \mathrm{Sp}(2p, 2n)$, let $X_\perp \in \mathbb{R}^{2n\times(2n-2p)}$ be a full-rank matrix such that $\mathrm{span}(X_\perp)$ is the orthogonal complement of $\mathrm{span}(X)$. Then the matrix $[XJ \;\; JX_\perp]$ is nonsingular, and every matrix $Y \in \mathbb{R}^{2n\times 2p}$ can be represented as $Y = XJW + JX_{\perp}K$, where $W \in \mathbb{R}^{2p\times 2p}$ and $K \in \mathbb{R}^{(2n-2p)\times 2p}$; see [11, Lemma 3.2]. The tangent space of $\mathrm{Sp}(2p, 2n)$ at X, denoted by $\mathrm{T}_X\mathrm{Sp}(2p, 2n)$, is given by [11, Proposition 3.3]

$$\begin{aligned}\mathrm{T}_X\mathrm{Sp}(2p, 2n) &= \{XJW + JX_{\perp}K : W \in \mathcal{S}_{\mathrm{sym}}(2p), K \in \mathbb{R}^{(2n-2p)\times 2p}\} &\text{(2a)}\\ &= \{SJX : S \in \mathcal{S}_{\mathrm{sym}}(2n)\}, &\text{(2b)}\end{aligned}$$

where $\mathcal{S}_{\mathrm{sym}}(2p)$ denotes the set of all $2p \times 2p$ real symmetric matrices. These two expressions can be regarded as different parameterizations of the tangent space.

Now we consider the Euclidean metric. Given any tangent vectors $Z_i = XJW_i + JX_{\perp}K_i$ with $W_i \in \mathcal{S}_{\mathrm{sym}}(2p)$ and $K_i \in \mathbb{R}^{(2n-2p)\times 2p}$ for $i = 1, 2$, the standard Euclidean metric is defined as

$$\begin{aligned}g_{\mathrm{e}}(Z_1, Z_2) &:= \langle Z_1, Z_2\rangle = \mathrm{tr}(Z_1^\top Z_2)\\ &= \mathrm{tr}(W_1^\top J^\top X^\top XJW_2) + \mathrm{tr}(K_1^\top X_{\perp}^\top X_{\perp}K_2)\\ &\quad + \mathrm{tr}(W_1^\top J^\top X^\top JX_{\perp}K_2) + \mathrm{tr}(K_1^\top X_{\perp}^\top J^\top XJW_2).\end{aligned}$$

In contrast with the canonical-like metric proposed in [11]

$$g_{\rho,X_{\perp}}(Z_1, Z_2) := \frac{1}{\rho}\,\mathrm{tr}(W_1^\top W_2) + \mathrm{tr}(K_1^\top K_2) \quad \text{with } \rho > 0,$$

g_{e} has cross terms between W and K. Note that g_{e} is also well-defined when it is extended to $\mathbb{R}^{2n\times 2p}$. Then the normal space of $\mathrm{Sp}(2p, 2n)$ with respect to g_{e} can be defined as

$$(\mathrm{T}_X\mathrm{Sp}(2p, 2n))_{\mathrm{e}}^{\perp} := \left\{N \in \mathbb{R}^{2n\times 2p} : g_{\mathrm{e}}(N, Z) = 0 \text{ for all } Z \in \mathrm{T}_X\mathrm{Sp}(2p, 2n)\right\}.$$

We obtain the following expression of the normal space.

Proposition 1. *Given $X \in \mathrm{Sp}(2p, 2n)$, we have*

$$(\mathrm{T}_X\mathrm{Sp}(2p, 2n))_{\mathrm{e}}^{\perp} = \{JX\Omega : \Omega \in \mathcal{S}_{\mathrm{skew}}(2p)\}, \tag{3}$$

where $\mathcal{S}_{\mathrm{skew}}(2p)$ denotes the set of all $2p \times 2p$ real skew-symmetric matrices.

Proof. Given any $N = JX\Omega$ with $\Omega \in \mathcal{S}_{\mathrm{skew}}(2p)$, and $Z = XJW + JX_{\perp}K \in \mathrm{T}_X\mathrm{Sp}(2p, 2n)$ with $W \in \mathcal{S}_{\mathrm{sym}}(2p)$, we have $g_{\mathrm{e}}(N, Z) = \mathrm{tr}(N^\top Z) = \mathrm{tr}(\Omega^\top W) = 0$, where the last equality follows from $\Omega^\top = -\Omega$ and $W^\top = W$. Therefore, it yields $N \in (\mathrm{T}_X\mathrm{Sp}(2p, 2n))_{\mathrm{e}}^{\perp}$. Counting dimensions of $\mathrm{T}_X\mathrm{Sp}(2p, 2n)$ and the subspace $\{JX\Omega : \Omega \in \mathcal{S}_{\mathrm{skew}}(2p)\}$, i.e., $4np - p(2p-1)$ and $p(2p-1)$, respectively, the expression (3) holds. □

Notice that $(\mathrm{T}_X\mathrm{Sp}(2p, 2n))_{\mathrm{e}}^{\perp}$ is different from the normal space with respect to the canonical-like metric $g_{\rho,X_{\perp}}$, denoted by $(\mathrm{T}_X\mathrm{Sp}(2p, 2n))^{\perp}$, which has the expression $\{XJ\Omega : \Omega \in \mathcal{S}_{\mathrm{skew}}(2p)\}$, obtained in [11].

The following proposition provides explicit expressions for the orthogonal projection onto the tangent and normal spaces with respect to the metric g_{e}, denoted by $(\mathcal{P}_X)_{\mathrm{e}}$ and $(\mathcal{P}_X)_{\mathrm{e}}^{\perp}$, respectively.

Proposition 2. *Given $X \in \mathrm{Sp}(2p, 2n)$ and $Y \in \mathbb{R}^{2n \times 2p}$, we have*

$$(\mathcal{P}_X)_{\mathrm{e}}(Y) = Y - JX\Omega_{X,Y}, \tag{4}$$

$$(\mathcal{P}_X)_{\mathrm{e}}^{\perp}(Y) = JX\Omega_{X,Y}, \tag{5}$$

where $\Omega_{X,Y} \in \mathcal{S}_{\mathrm{skew}}(2p)$ is the unique solution of the Lyapunov equation with unknown Ω

$$X^\top X\Omega + \Omega X^\top X = 2\mathrm{skew}(X^\top J^\top Y) \tag{6}$$

and $\mathrm{skew}(A) := \frac{1}{2}(A - A^\top)$ *denotes the skew-symmetric part of A.*

Proof. For any $Y \in \mathbb{R}^{2n \times 2p}$, in view of (2a) and (3), it follows that

$$(\mathcal{P}_X)_{\mathrm{e}}(Y) = XJW_Y + JX_\perp K_Y, \qquad (\mathcal{P}_X)_{\mathrm{e}}^{\perp}(Y) = JX\Omega,$$

with $W_Y \in \mathcal{S}_{\mathrm{sym}}(2p)$, $K_Y \in \mathbb{R}^{(2n-2p)\times 2p}$ and $\Omega \in \mathcal{S}_{\mathrm{skew}}(2p)$. Further, Y can be represented as

$$Y = (\mathcal{P}_X)_{\mathrm{e}}(Y) + (\mathcal{P}_X)_{\mathrm{e}}^{\perp}(Y) = XJW_Y + JX_\perp K_Y + JX\Omega.$$

Multiplying this equation from the left with $X^\top J^\top$, it follows that

$$X^\top J^\top Y = W_Y + X^\top X\Omega.$$

Subtracting from this equation its transpose and taking into account that $W^\top = W$ and $\Omega^\top = -\Omega$, we get the Lyapunov equation (6) with unknown Ω. Since $X^\top X$ is symmetric positive definite, all its eigenvalues are positive, and, hence, Eq. (6) has a unique solution $\Omega_{X,Y}$; see [12, Lemma 7.1.5]. Therefore, the relation (5) holds. Finally, (4) follows from $(\mathcal{P}_X)_{\mathrm{e}}(Y) = Y - (\mathcal{P}_X)_{\mathrm{e}}^{\perp}(Y)$. □

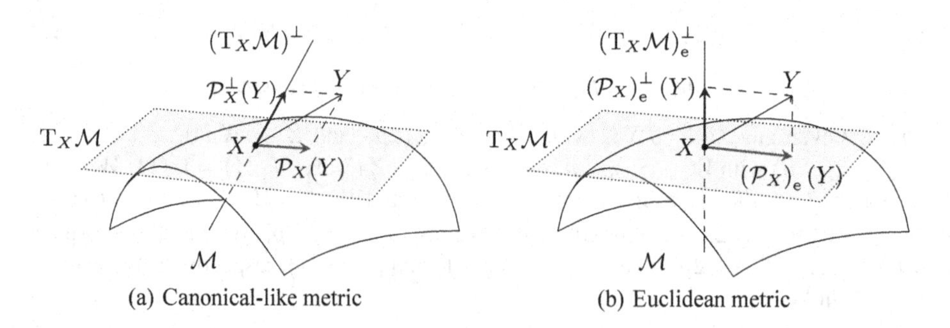

(a) Canonical-like metric (b) Euclidean metric

Fig. 1. Normal spaces and projections associated with different metrics on $\mathcal{M} = \mathrm{Sp}(2p, 2n)$

Figure 1 illustrates the difference of the normal spaces and projections for the canonical-like metric $g_{\rho,X_\perp}$ and the Euclidean metric g_{e}. Note that projections with respect to the canonical-like metric only require matrix additions and multiplications (see [11, Proposition 4.3]) while one has to solve the Lyapunov equation (6) in the Euclidean case.

The Lyapunov equation (6) can be solved using the Bartels–Stewart method [3]. Observe that the coefficient matrix $X^\top X$ is symmetric positive definite, and, hence, it has an eigenvalue decomposition $X^\top X = Q\Lambda Q^\top$, where $Q \in \mathbb{R}^{2p\times 2p}$ is orthogonal and $\Lambda = \mathrm{diag}(\lambda_1, \ldots, \lambda_{2p})$ is diagonal with $\lambda_i > 0$ for $i = 1, \ldots, 2p$. Inserting this decomposition into (6) and multiplying it from the left and right with $Q^\top$ and Q, respectively, we obtain the equation

$$\Lambda U + U\Lambda = R$$

with $R = 2Q^\top \mathrm{skew}(X^\top JY)Q$ and unknown $U = Q^\top \Omega Q$. The entries of U can then be computed as

$$u_{ij} = \frac{r_{ij}}{\lambda_i + \lambda_j}, \quad i, j = 1, \ldots, 2p.$$

Finally, we find $\Omega = QUQ^\top$. The computational cost for matrix-matrix multiplications involved to generate (6) is $O(np^2)$, and $O(p^3)$ for solving this equation.

3 Application to Optimization

In this section, we consider a continuously differentiable real-valued function f on $\mathrm{Sp}(2p, 2n)$ and optimization problems on the manifold.

The Riemannian gradient of f at $X \in \mathrm{Sp}(2p, 2n)$ with respect to the metric g_{e}, denoted by $\mathrm{grad}_{\mathrm{e}} f(X)$, is defined as the unique element of $\mathrm{T}_X\mathrm{Sp}(2p, 2n)$ that satisfies the condition $g_{\mathrm{e}}\left(\mathrm{grad}_{\mathrm{e}} f(X), Z\right) = \mathrm{D}\bar{f}(X)[Z]$ for all $Z \in \mathrm{T}_X\mathrm{Sp}(2p, 2n)$, where $\bar{f}$ is a smooth extension of f around X in $\mathbb{R}^{2n\times 2p}$, and $\mathrm{D}\bar{f}(X)$ denotes the Fréchet derivative of $\bar{f}$ at X. Since $\mathrm{Sp}(2p, 2n)$ is endowed with the Euclidean metric, the Riemannian gradient can be readily computed by using [1, Sect. 3.6] as follows.

Proposition 3. *The Riemannian gradient of a function* $f : \mathrm{Sp}(2p, 2n) \to \mathbb{R}$ *with respect to the Euclidean metric* g_{e} *has the following form*

$$\mathrm{grad}_{\mathrm{e}} f(X) = (\mathcal{P}_X)_{\mathrm{e}}\left(\nabla \bar{f}(X)\right) = \nabla \bar{f}(X) - JX\Omega_X, \tag{7}$$

where $\Omega_X \in \mathcal{S}_{\mathrm{skew}}(2p)$ *is the unique solution of the Lyapunov equation with unknown* Ω

$$X^\top X\Omega + \Omega X^\top X = 2\mathrm{skew}\left(X^\top J^\top \nabla \bar{f}(X)\right),$$

and $\nabla \bar{f}(X)$ *denotes the (Euclidean, i.e., classical) gradient of* $\bar{f}$ *at* X.

In the case of the symplectic group $\mathrm{Sp}(2n)$, the Riemannian gradient (7) is equivalent to the formulation in [7], where the minimization problem was treated as a constrained optimization problem in the Euclidean space. We notice that Ω_X in (7) is actually the Lagrangian multiplier of the symplectic constraints; see [7].

Expression (7) can be rewritten in the parameterization (2a): it follows from [11, Lemma 3.2] that

$$\mathrm{grad}_{\mathrm{e}} f(X) = XJW_X + JX_\perp K_X$$

with $W_X = X^\top J^\top \mathrm{grad}_\mathrm{e} f(X)$ and $K_X = \left(X_\perp^\top J X_\perp\right)^{-1} X_\perp^\top \mathrm{grad}_\mathrm{e} f(X)$. This suffices to compute the update $X^{k+1} = \mathcal{R}_{X^k}\left(-t_k \mathrm{grad}_\mathrm{e} f(X^k)\right)$ required in [11, Algorithm 1, line 6] in the case that $\mathcal{R}$ is the *quasi-geodesic* retraction [11, Sect. 5.1]

$$\mathcal{R}_X^{\mathrm{qgeo}}(Z) := [X, Z] \exp\left(\begin{bmatrix} -JW & JZ^\top JZ \\ I_{2p} & -JW \end{bmatrix}\right) \begin{bmatrix} I_{2p} \\ 0 \end{bmatrix} e^{JW},$$

where $Z \in \mathrm{T}_X\mathrm{Sp}(2p, 2n)$, $W = X^\top JZ$ and $\exp(\cdot)$ denotes the matrix exponential. Observe that the quasi-geodesics differ from the Euclidean geodesics; to obtain the latter, in view of (3), the term $Y(t)J\Omega$ in [11, (4.12)] has to be replaced by $JY(t)\Omega$. If $\mathcal{R}$ is instead chosen as the *Cayley* retraction [11, Definition 5.2]

$$\mathcal{R}_X^{\mathrm{cay}}(Z) := \left(I - \frac{1}{2} S_{X,Z} J\right)^{-1} \left(I + \frac{1}{2} S_{X,Z} J\right) X,$$

where $S_{X,Z} = G_X Z (XJ)^\top + XJ(G_X Z)^\top$ and $G_X = I - \frac{1}{2} XJX^\top J^\top$, then it is essential to rewrite (7) in the parameterization (2b) with S in a factorized form as in [11, Proposition 5.4]. To this end, observe that (7) is its own tangent projection and use the tangent projection formula of [11, Proposition 4.3] to obtain

$$\mathrm{grad}_\mathrm{e} f(X) = S_X JX$$

with $S_X = G_X \mathrm{grad}_\mathrm{e} f(X)(XJ)^\top + XJ(G_X \mathrm{grad}_\mathrm{e} f(X))^\top$ and $G_X = I - \frac{1}{2} XJX^\top J^\top$.

4 Numerical Experiments

In this section, we adopt the Riemannian gradient (7) and numerically compare the performance of optimization algorithms with respect to the Euclidean metric. All experiments are performed on a laptop with 2.7 GHz Dual-Core Intel i5 processor and 8GB of RAM running MATLAB R2016b under macOS 10.15.2. The code that produces the result is available from https://github.com/opt-gaobin/spopt.

First, we consider the optimization problem (1). We compare gradient-descent algorithms proposed in [11] with different metrics (Euclidean and canonical-like, denoted by "-E" and "-C") and retractions (quasi-geodesics and Cayley transform, denoted by "Geo" and "Cay"). The canonical-like metric has two formulations, denoted by "-I" and "-II", based on different choices of $X_\perp$. Hence, there are six methods involved. The problem generation and parameter settings are in parallel with ones in [11]. The numerical results are presented in Fig. 2. Notice that the algorithms that use the Euclidean metric are considerably superior in the sense of the number of iterations. Hence, in this problem the Euclidean metric may be more suitable than other metrics. However, due to their lower computational cost per iteration, algorithms with canonical-like-based Cayley retraction perform best with respect to time among all tested methods, and Cayley-based methods always outperform quasi-geodesics in each setting.

The second example is the symplectic eigenvalue problem. We compute the smallest symplectic eigenvalues and eigenvectors of symmetric positive-definite matrices in

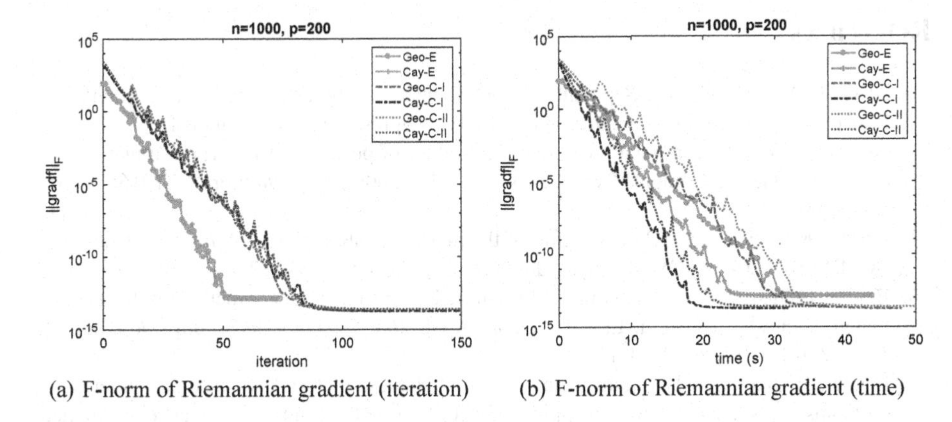

(a) F-norm of Riemannian gradient (iteration) (b) F-norm of Riemannian gradient (time)

Fig. 2. A comparison of gradient-descent algorithms with different metrics and retractions. Recall that grad$f$ differs between the "E" and "C" methods, which explains why they do not have the same initial value.

the sense of Williamson's theorem; see [16]. According to the performance in Fig. 2, we consider "Cay-E" and "Cay-C-I" as representative methods. The problem generation and default settings can be found in [16]. Note that the synthetic data matrix has five smallest symplectic eigenvalues $1, 2, 3, 4, 5$. In Table 1, we list the computed symplectic eigenvalues $d_1, \ldots, d_5$ and 1-norm errors defined as $\sum_{i=1}^{5} |d_i - i|$. The results illustrate that our methods are comparable with the structure-preserving eigensolver "symplLanczos" based on a Lanczos procedure [2].

Table 1. Five smallest symplectic eigenvalues of a 1000×1000 matrix computed by different methods

	symplLanczos	Cay-E	Cay-C-I
	0.999999999999997	1.000000000000000	0.999999999999992
	2.000000000000010	2.000000000000010	2.000000000000010
	3.000000000000014	2.999999999999995	3.000000000000008
	4.000000000000004	3.999999999999988	3.999999999999993
	5.000000000000016	4.999999999999996	4.999999999999996
Errors	4.75e−14	3.11e−14	3.70e−14

References

1. Absil, P.-A., Mahony, R., Sepulchre, R.: Optimization Algorithms on Matrix Manifolds. Princeton University Press, Princeton (2008). https://press.princeton.edu/absil
2. Amodio, P.: On the computation of few eigenvalues of positive definite Hamiltonian matrices. Future Gener. Comput. Syst. **22**(4), 403–411 (2006). https://doi.org/10.1016/j.future.2004.11.027
3. Bartels, R.H., Stewart, G.W.: Solution of the matrix equation $AX + XB = C$. Commun. ACM **15**(9), 820–826 (1972). https://doi.org/10.1145/361573.361582
4. Benner, P., Fassbender, H.: An implicitly restarted symplectic Lanczos method for the Hamiltonian eigenvalue problem. Linear Algebra Appl. **263**, 75–111 (1997). https://doi.org/10.1016/S0024-3795(96)00524-1
5. Benner, P., Kressner, D., Mehrmann, V.: Skew-Hamiltonian and Hamiltonian eigenvalue problems: theory, algorithms and applications. In: Proceedings of the Conference on Applied Mathematics and Scientific Computing, pp. 3–39 (2005). https://doi.org/10.1007/1-4020-3197-1_1
6. Bhatia, R., Jain, T.: On symplectic eigenvalues of positive definite matrices. J. Math. Phys. **56**(11), 112201 (2015). https://doi.org/10.1063/1.4935852
7. Birtea, P., Caşu, I., Comănescu, D.: Optimization on the real symplectic group. Monatshefte für Math. **191**(3), 465–485 (2020). https://doi.org/10.1007/s00605-020-01369-9
8. Buchfink, P., Bhatt, A., Haasdonk, B.: Symplectic model order reduction with non-orthonormal bases. Math. Comput. Appl. 24(2) (2019). https://doi.org/10.3390/mca24020043
9. Fiori, S.: Solving minimal-distance problems over the manifold of real-symplectic matrices. SIAM J. Matrix Anal. Appl. **32**(3), 938–968 (2011). https://doi.org/10.1137/100817115
10. Fiori, S.: A Riemannian steepest descent approach over the inhomogeneous symplectic group: application to the averaging of linear optical systems. Appl. Math. Comput. **283**, 251–264 (2016). https://doi.org/10.1016/j.amc.2016.02.018
11. Gao, B., Son, N.T., Absil, P.-A., Stykel, T.: Riemannian optimization on the symplectic Stiefel manifold. arXiv preprint arXiv:2006.15226 (2020)
12. Golub, G.H., Van Loan, C.F.: Matrix Computations, 4th edn. Johns Hopkins University Press, Baltimore (2013)
13. Jain, T., Mishra, H.K.: Derivatives of symplectic eigenvalues and a Lidskii type theorem. Can. J. Math. 1–29 (2020). https://doi.org/10.4153/S0008414X2000084X
14. Machado, L.M., Leite, F.S.: Optimization on quadratic matrix Lie groups (2002). http://hdl.handle.net/10316/11446
15. Peng, L., Mohseni, K.: Symplectic model reduction of Hamiltonian systems. SIAM J. Sci. Comput. **38**(1), A1–A27 (2016). https://doi.org/10.1137/140978922
16. Son, N.T., Absil, P.-A., Gao, B., Stykel, T.: Symplectic eigenvalue problem via trace minimization and Riemannian optimization. arXiv preprint arXiv:2101.02618 (2021)
17. Wang, J., Sun, H., Fiori, S.: A Riemannian-steepest-descent approach for optimization on the real symplectic group. Math. Meth. Appl. Sci. **41**(11), 4273–4286 (2018). https://doi.org/10.1002/mma.4890
18. Williamson, J.: On the algebraic problem concerning the normal forms of linear dynamical systems. Amer. J. Math. **58**(1), 141–163 (1936). https://doi.org/10.2307/2371062
19. Wu, R.B., Chakrabarti, R., Rabitz, H.: Critical landscape topology for optimization on the symplectic group. J. Optim. Theory Appl. **145**(2), 387–406 (2010). https://doi.org/10.1007/s10957-009-9641-1
20. Wu, R., Chakrabarti, R., Rabitz, H.: Optimal control theory for continuous-variable quantum gates. Phys. Rev. A **77**(5) (2008). https://doi.org/10.1103/PhysRevA.77.052303

Divergence Statistics

On f-divergences Between Cauchy Distributions

Frank Nielsen[1(✉)] and Kazuki Okamura[2]

[1] Sony Computer Science Laboratories Inc., Tokyo, Japan
Frank.Nielsen@acm.org
[2] Department of Mathematics, Faculty of Science, Shizuoka University, Shizuoka, Japan
okamura.kazuki@shizuoka.ac.jp

Abstract. We prove that all f-divergences between univariate Cauchy distributions are symmetric, and can be expressed as functions of the chi-squared divergence. This property does not hold anymore for multivariate Cauchy distributions.

Keywords: Cauchy distributions · Complex analysis · Maximal invariant · Information geometry

1 Introduction

Let $\mathbb{R}$, $\mathbb{R}_+$ and $\mathbb{R}_+^*$ denote the sets of real numbers, non-negative real numbers, and positive real numbers, respectively. The probability density function of a Cauchy distribution with location parameter $l \in \mathbb{R}$ and scale parameter $s \in \mathbb{R}_{++}$ is $p_{l,s}(x) := \frac{1}{\pi s\left(1+\left(\frac{x-l}{s}\right)^2\right)}, x \in \mathbb{R}$ The space of Cauchy distributions form a location-scale family: $\mathcal{C} = \left\{p_{l,s}(x) := \frac{1}{s}p\left(\frac{x-l}{s}\right) : (l,s) \in \mathbb{R}\times\mathbb{R}_{++}\right\}$, with standard density $p(x) := p_{0,1}(x) = \frac{1}{\pi(1+x^2)}$. To measure the dissimilarity between two continuous probability distributions P and Q, we consider the class of statistical f-divergences [2,10] between their corresponding probability densities functions $p(x)$ and $q(x)$ assumed to be strictly positive on $\mathbb{R}$: $I_f(p:q) := \int_{\mathbb{R}} p(x) f\left(\frac{q(x)}{p(x)}\right) \mathrm{d}x$, where $f(u)$ is a convex function on $(0,\infty)$, strictly convex at $u=1$, and satisfying $f(1)=0$ so that by Jensen's inequality, we have $I_f(p:q) \geq f(1) = 0$. The Kullback-Leibler divergence (KLD also called relative entropy) is a f-divergence obtained for $f_{\mathrm{KL}}(u) = -\log u$. In general, the f-divergences are oriented dissimilarities: $I_f(p:q) \neq I_f(q:p)$ (e.g., the KLD). The reverse f-divergence $I_f(q:p)$ can be obtained as a forward f-divergence for the conjugate function $f^*(u) := uf\left(\frac{1}{u}\right)$ (convex with $f^*(1)=0$): $I_f(q:p) = I_{f^*}(p:q)$. In general, calculating the definite integrals

The second author was supported by JSPS KAKENHI 19K14549. Further results are presented in the technical report [11].

F. Nielsen and F. Barbaresco (Eds.): GSI 2021, LNCS 12829, pp. 799–807, 2021.
https://doi.org/10.1007/978-3-030-80209-7_86

of f-divergences is non trivial: For example, the formula for the KLD between Cauchy densities was only recently obtained [1]:

$$D_{\mathrm{KL}}(p_{l_1,s_1} : p_{l_2,s_2}) := I_{f_{\mathrm{KL}}}(p : q) = \int p_{l_1,s_1}(x) \log \frac{p_{l_1,s_1}(x)}{p_{l_2,s_2}(x)} \mathrm{d}x,$$
$$= \log \left(\frac{(s_1 + s_2)^2 + (l_1 - l_2)^2}{4 s_1 s_2} \right).$$

Let $\lambda = (\lambda_1 = l, \lambda_2 = s)$. We can rewrite the KLD formula as

$$D_{\mathrm{KL}}(p_{\lambda_1} : p_{\lambda_2}) = \log \left(1 + \frac{1}{2} \chi(\lambda_1, \lambda_2) \right), \tag{1}$$

where $\chi(\lambda, \lambda') := \frac{(\lambda_1 - \lambda_1')^2 + (\lambda_2 - \lambda_2')^2}{2\lambda_2 \lambda_2'}$. We observe that the KLD between Cauchy distributions is symmetric: $D_{\mathrm{KL}}(p_{l_1,s_1} : p_{l_2,s_2}) = D_{\mathrm{KL}}(p_{l_2,s_2} : p_{l_1,s_1})$. Let $D_\chi^N(p : q) := \int \frac{(p(x)-q(x))^2}{q(x)} \mathrm{d}x$ and $D_\chi^P(p : q) := \int \frac{(p(x)-q(x))^2}{p(x)} \mathrm{d}x$ denote the Neyman and Pearson chi-squared divergences between densities $p(x)$ and $q(x)$, respectively. These divergences are f-divergences [10] for the generators $f_\chi^P(u) = (u-1)^2$ and $f_\chi^N(u) = \frac{1}{u}(u-1)^2$, respectively. Both the Neyman and Pearson χ^2-divergences between Cauchy densities are symmetric [8]: $D_\chi(p_{\lambda_1} : p_{\lambda_2}) := D_\chi^N(p_{\lambda_1} : p_{\lambda_2}) = D_\chi^P(p_{\lambda_1} : p_{\lambda_2}) = \chi(\lambda_1, \lambda_2)$, hence the naming of the function $\chi(\cdot, \cdot)$. Notice that $\chi(\cdot, \cdot)$ is a conformal squared Euclidean divergence [12].

In this work, we first give a simple proof of the KLD formula of Eq. 1 based on complex analysis in Sect. 3.1. Then we prove that all f-divergences between univariate Cauchy distributions are symmetric (Theorem 1) and can be expressed as scalar functions of the chi-squared divergence (Theorem 2) in Sect. 3. This property holds only for the univariate case as we report in Sect. 4 an example of bivariate Cauchy distributions for which the KLD is asymmetric. Let us start by recalling the hyperbolic Fisher-Rao geometry of location-scale families.

2 Information Geometry of Location-Scale Families

The Fisher information matrix [6,8] of a location-scale family with continuously differentiable standard density $p(x)$ on full support $\mathbb{R}$ is $I(\lambda) = \frac{1}{s^2} \begin{bmatrix} a^2 & c \\ c & b^2 \end{bmatrix}$, where $a^2 = E_p \left[\left(\frac{p'(x)}{p(x)} \right)^2 \right]$, $b^2 = E_p \left[\left(1 + x \frac{p'(x)}{p(x)} \right)^2 \right]$ and $c = E_p \left[\frac{p'(x)}{p(x)} \left(1 + x \frac{p'(x)}{p(x)} \right) \right]$. When the standard density is even (i.e., $p(x) = p(-x)$), we get a diagonal Fisher matrix that can be reparameterized with $\theta(\lambda) = \left(\frac{a}{b} \lambda_1, \lambda_2 \right)$ so that the Fisher matrix with respect to θ becomes $\frac{b^2}{\theta_2^2} \begin{bmatrix} 1 & 0 \\ 0 & 1 \end{bmatrix}$. It follows that the Fisher-Rao geometry is hyperbolic with curvature $\kappa = -\frac{1}{b^2} < 0$, and that the Fisher-Rao distance is $\rho_p(\lambda_1, \lambda_2) = b\, \rho_U \left(\left(\frac{a}{b} l_1, s_1 \right), \left(\frac{a}{b} l_2, s_2 \right) \right)$ where $\rho_U(\theta_1, \theta_2) = \operatorname{arccosh}(1 + \chi(\theta_1, \theta_2))$ where $\operatorname{arccosh}(u) = \log(u + \sqrt{u^2 - 1})$ for $u > 1$. For the Cauchy family, we have $a^2 = b^2 = \frac{1}{2}$ ($\kappa = -\frac{1}{b^2} = -2$), and the Fisher-Rao distance is $\rho_{\mathrm{FR}}(p_{\lambda_1} : p_{\lambda_2}) = \frac{1}{\sqrt{2}}\mathrm{arccosh}(1 + \chi(\lambda_1, \lambda_2))$. Notice that if we let $\theta = l + is \in \mathbb{C}$ then the metric in the complex upper plane $\mathbb{H}$ is $\frac{|\mathrm{d}\theta|^2}{\mathrm{Im}(\theta)^2}$ where $|x+iy| = \sqrt{x^2+y^2}$ denotes the complex modulus, and $\theta \in \mathbb{H} := \{x + iy \;:\; x \in \mathbb{R}, y \in \mathbb{R}_{++}\}$.

3 f-divergences Between Univariate Cauchy Distributions

Consider the location-scale non-abelian group LS(2) which can be represented as a matrix group [9]. A group element $g_{l,s}$ is represented by a matrix element $M_{l,s} = \begin{bmatrix} s & l \\ 0 & 1 \end{bmatrix}$ for $(l, s) \in \mathbb{R} \times \mathbb{R}_{++}$. The group operation $g_{l_{12},s_{12}} = g_{l_1,s_1} \times g_{l_2,s_2}$ corresponds to a matrix multiplication $M_{l_{12},s_{12}} = M_{l_1,s_1} \times M_{l_2,s_2}$ (with the group identity element $g_{0,1}$ being the matrix identity $I = \begin{bmatrix} 1 & 0 \\ 0 & 1 \end{bmatrix}$). A location-scale family is defined by the action of the location-group on a standard density $p(x) = p_{0,1}(x)$. That is, density $p_{l,s}(x) = g_{l,s}.p(x)$, where '.' denotes the group action. We have the following invariance for the f-divergences between any two densities of a location-scale family [9] (including the Cauchy family): $I_f(g.p_{l_1,s_1} : g.p_{l_2,s_2}) = I_f(p_{l_1,s_1} : p_{l_2,s_2}), \forall g \in \mathrm{LS}(2)$. Thus we have $I_f(p_{l_1,s_1} : p_{l_2,s_2}) = I_f\left(p : p_{\frac{l_2-l_1}{s_1},\frac{s_2}{s_1}}\right) = I_f\left(p_{\frac{l_1-l_2}{s_2},\frac{s_1}{s_2}} : p\right)$. Therefore, we may always consider the calculation of the f-divergence between the standard density and another density of the location-scale family. For example, we check that $\chi((l_1, s_1), (l_2, s_2)) = \chi\left((0, 1), \left(\frac{l_2-l_1}{s_1}, \frac{s_2}{s_1}\right)\right)$ since $\chi((0, 1), (l, s)) = \frac{(s-1)^2+l^2}{2s}$. If we assume that the standard density p is such that $E_p[X] = \int xp(x)\mathrm{d}x = 0$ and $E_p[X^2] = \int x^2p(x)\mathrm{d}x = 1$ (hence, X of unit variance), then the random variable $Y = \mu + \sigma X$ has mean $E[Y] = \mu$ and standard deviation $\sigma(Y) = \sqrt{E[(Y - \mu)^2]} = \sigma$. However, the expectation and variance of Cauchy distributions are not defined, hence we preferred the (l, s) parameterization over the (μ, σ^2) parameterization. For the Cauchy location-scale family [5], parameter l is also called the median, and parameter s the probable error.

3.1 Revisiting the KLD Between Cauchy Densities

Let us give a simple proof of Eq. 1 (first proven in [1])

Proof.

$$D_{\mathrm{KL}}(p_{l_1,s_1} : p_{l_2,s_2}) = \frac{s_1}{\pi}\int_{\mathbb{R}} \frac{\log((z-l_2)^2 + s_2^2)}{(z-l_1)^2 + s_1^2}\mathrm{d}z$$
$$-\frac{s_1}{\pi}\int_{\mathbb{R}} \frac{\log((z-l_1)^2 + s_1^2)}{(z-l_1)^2 + s_1^2}\mathrm{d}z + \log\frac{s_1}{s_2}. \tag{2}$$

As a function of z, $\frac{\log(z-l_2+is_2)}{z-l_1+is_1}$ is holomorphic on the upper-half plane $\{x+yi : y > 0\}$. By the Cauchy integral formula [7], we have that for sufficiently large R,

$$\frac{1}{2\pi i}\int_{C_R^+} \frac{\log(z-l_2+is_2)}{(z-l_1)^2+s_1^2}\mathrm{d}z = \frac{\log(l_1-l_2+i(s_2+s_1))}{2s_1 i},$$

where $C_R^+ := \{z : |z| = R, \mathrm{Im}(z) > 0\} \cup \{z : \mathrm{Im}(z) = 0, |\mathrm{Re}(z)| \leq R\}$.

Hence, by $R \to +\infty$, we get

$$\frac{s_1}{\pi}\int_{\mathbb{R}} \frac{\log(z-l_2+is_2)}{(z-l_1)^2+s_1^2}\mathrm{d}z = \log(l_1-l_2+i(s_2+s_1)). \tag{3}$$

As a function of z, $\frac{\log(z-l_2-is_2)}{z-l_1-is_1}$ is holomorphic on the lower-half plane $\{x+yi : y < 0\}$. By the Cauchy integral formula again, we have that for sufficiently large R,

$$\frac{1}{2\pi i}\int_{C_R^-} \frac{\log(z-l_2-is_2)}{(z-l_1)^2+s_1^2}\mathrm{d}z = \frac{\log(l_1-l_2-i(s_2+s_1))}{-2s_1 i},$$

where $C_R^- := \{z : |z| = R, \mathrm{Im}(z) < 0\} \cup \{z : \mathrm{Im}(z) = 0, |\mathrm{Re}(z)| \leq R\}$.

Hence, by $R \to +\infty$, we get

$$\frac{s_1}{\pi}\int_{\mathbb{R}} \frac{\log(z-l_2-is_2)}{(z-l_1)^2+s_1^2}\mathrm{d}z = \log(l_1-l_2-i(s_2+s_1)). \tag{4}$$

By Eq. 3 and Eq. 4, we have that

$$\frac{s_1}{\pi}\int_{\mathbb{R}} \frac{\log((z-l_2)^2+s_2^2)}{(z-l_1)^2+s_1^2}\mathrm{d}z = \log\left((l_1-l_2)^2+(s_1+s_2)^2\right). \tag{5}$$

In the same manner, we have that

$$\frac{s_1}{\pi}\int_{\mathbb{R}} \frac{\log((z-l_1)^2+s_1^2)}{(z-l_1)^2+s_1^2}\mathrm{d}z = \log(4s_1^2). \tag{6}$$

By substituting Eq. 5 and Eq. 6 into Eq. 2, we obtain the formula Eq. 1.

3.2 f-divergences Between Cauchy Distributions are Symmetric

Let $\|\lambda\| = \sqrt{\lambda_1^2+\lambda_2^2}$ denote the Euclidean norm of a 2D vector $\lambda \in \mathbb{R}^2$. We state the main theorem:

Theorem 1. *All f-divergences between univariate Cauchy distributions p_λ and $p_{\lambda'}$ with $\lambda = (l, s)$ and $\lambda' = (l', s')$ are symmetric and can be expressed as $I_f(p_\lambda : p_{\lambda'}) = h_f(\chi(\lambda, \lambda'))$ where $\chi(\lambda, \lambda') := \frac{\|\lambda-\lambda'\|^2}{2\lambda_2\lambda_2'}$ and $h_f : \mathbb{R}_+ \to \mathbb{R}_+$ is a scalar function.*

The proof does not always yield explicit closed-form formula for the f-divergences as it can be in general difficult to obtain h_f (e.g., α-divergences), and relies on McCullagh's complex parametrization [5] p_θ of the parameter of the Cauchy density $p_{l,s}$ with $\theta = l + is$:

$$p_\theta(x) = \frac{\mathrm{Im}(\theta)}{\pi|x-\theta|^2}.$$

We make use of the special linear group $\mathrm{SL}(2,\mathbb{R})$ for θ the complex parameter: $\mathrm{SL}(2,\mathbb{R}) := \left\{ \begin{bmatrix} a & b \\ c & d \end{bmatrix} \ : \ a,b,c,d \in \mathbb{R}, ad - bc = 1 \right\}$. Let $A.\theta := \frac{a\theta+b}{c\theta+d}$ (real linear fractional transformations) be the action of $A = \begin{bmatrix} a & b \\ c & d \end{bmatrix} \in \mathrm{SL}(2,\mathbb{R})$. McCullagh proved that if $X \sim \mathrm{Cauchy}(\theta)$ then $A.X \sim \mathrm{Cauchy}(A.\theta)$. We can also define an action of $\mathrm{SL}(2,\mathbb{R})$ to the real line $\mathbb{R}$ by $x \mapsto \frac{ax+b}{cx+d}$, $\quad x \in \mathbb{R}$, where we interpret $-\frac{d}{c} \mapsto \frac{a}{c}$ if $c \neq 0$. We remark that $d \neq 0$ if $c = 0$. This map is bijective on $\mathbb{R}$. We have the following invariance:

Lemma 1 (Invariance of Cauchy f-divergence under $\mathrm{SL}(2,\mathbb{R})$). *For any $A \in \mathrm{SL}(2,\mathbb{R})$ and $\theta \in \mathbb{H}$, we have $I_f(p_{A.\theta_1} : p_{A.\theta_2}) = I_f(p_{\theta_1} : p_{\theta_2})$.*

Proof. We prove the invariance by the change of variable in the integral. Let $D(\theta_1 : \theta_2) := I_f(p_{\theta_1} : p_{\theta_2})$. We have

$$D(A.\theta_1 : A.\theta_2) = \int_{\mathbb{R}} \frac{\mathrm{Im}(A.\theta_1)}{|x - A.\theta_1|^2} f\left(\frac{\mathrm{Im}(A.\theta_2)|x - A.\theta_1|^2}{\mathrm{Im}(A.\theta_1)|x - A.\theta_2|^2}\right) \mathrm{d}x.$$

Since $A \in \mathrm{SL}(2,\mathbb{R})$, we have $\mathrm{Im}(A.\theta_i) = \frac{\mathrm{Im}(\theta_i)}{|c\theta_i+d|^2}$, $\quad i \in \{1,2\}$.

If $x = A.y$, then, $\mathrm{d}x = \frac{\mathrm{d}y}{|cy+d|^2}$, and, $|A.y - A.\theta_i|^2 = \frac{|y-\theta_i|^2}{|cy+d|^2|c\theta_i+d|^2}$, $\quad i \in \{1,2\}$. Hence we get:

$$\int_{\mathbb{R}} f\left(\frac{\mathrm{Im}(A.\theta_2)|x - A.\theta_1|^2}{\mathrm{Im}(A.\theta_1)|x - A.\theta_2|^2}\right) \frac{\mathrm{Im}(A.\theta_2)}{|x - A.\theta_1|^1} \mathrm{d}x = \int_{\mathbb{R}} f\left(\frac{\mathrm{Im}(\theta_2)|y - \theta_1|^2}{\mathrm{Im}(\theta_1)|y - \theta_2|^2}\right) \frac{\mathrm{Im}(\theta_2)}{|y - \theta_2|^2} \mathrm{d}y.$$

We now prove Theorem 1 using the notion of *maximal invariants* of Eaton [3] (Chapter 2) that will be discussed in Sect. 3.3.

Let us rewrite the function χ with complex arguments as:

$$\chi(z,w) := \frac{|z-w|^2}{2\mathrm{Im}(z)\mathrm{Im}(w)}, \quad z,w \in \mathbb{C}. \tag{7}$$

Proposition 1 (McCullagh [5]). *The function χ defined in Eq. 7 is a* maximal invariant *for the action of the special linear group $\mathrm{SL}(2,\mathbb{R})$ to $\mathbb{H}\times\mathbb{H}$ defined by*

$$A.(z,w) := \left(\frac{az+b}{cz+d}, \frac{aw+b}{cw+d}\right), \quad A = \begin{bmatrix} a & b \\ c & d \end{bmatrix} \in \mathrm{SL}(2,\mathbb{R}), \ z,w \in \mathbb{H}.$$

That is, we have $\chi(A.z, A.w) = \chi(z,w)$, $\quad A \in \mathrm{SL}(2,\mathbb{R})$, $z,w \in \mathbb{H}$, and it holds that for every $z,w,z',w' \in \mathbb{H}$ satisfying that $\chi(z',w') = \chi(z,w)$, there exists $A \in \mathrm{SL}(2,\mathbb{R})$ such that $(A.z, A.w) = (z',w')$.

By Lemma 1 and Theorem 2.3 of [3], there exists a unique function $h_f : [0, \infty) \to [0, \infty)$ such that $h_f(\chi(z, w)) = D(z, w)$ for all $z, w \in \mathbb{H}$.

Theorem 2. *Any f-divergence between two univariate Cauchy densities is symmetric and can be expressed as a function of the chi-squared divergence:*

$$I_f(p_{\theta_1} : p_{\theta_2}) = I_f(p_{\theta_2} : p_{\theta_1}) = h_f(\chi(\theta_1, \theta_1)), \quad \theta_1, \theta_2 \in \mathbb{H}. \tag{8}$$

Thus all f-divergences between univariate Cauchy densities are symmetric. For example, we have $h_{\mathrm{KL}}(u) = \log(1 + \frac{1}{2}u)$. When the function h_f is not explicitly known, we may estimate the f-divergences using Monte Carlo importance samplings [9].

3.3 Maximal Invariants (Proof of Proposition 1)

Proof. First, let us show that

Lemma 2. *For every $(z, w) \in \mathbb{H}^2$, there exist $\lambda \geq 1$ and $A \in \mathrm{SL}(2, \mathbb{R})$ such that $(A.z, A.w) = (\lambda i, i)$.*

Proof. Since the special orthogonal group $\mathrm{SO}(2, \mathbb{R})$ is the isotropy subgroup of $\mathrm{SL}(2, \mathbb{R})$ for i and the action is transitive, it suffices to show that for every $z \in \mathbb{H}$ there exist $\lambda \geq 1$ and $A \in \mathrm{SO}(2, \mathbb{R})$ such that $\lambda i = A.z$.

Since we have that for every $\lambda > 0$, $\begin{bmatrix} 0 & -1 \\ 1 & 0 \end{bmatrix} .\lambda i = \frac{i}{\lambda}$, it suffices to show that for every $z \in \mathbb{H}$ there exist $\lambda > 0$ and $A \in \mathrm{SO}(2, \mathbb{R})$ such that $\lambda i = A.z$.

We have that

$$\begin{bmatrix} \cos\theta & -\sin\theta \\ \sin\theta & \cos\theta \end{bmatrix} .z = \frac{\frac{|z|^2-1}{2} \sin 2\theta + \mathrm{Re}(z) \cos 2\theta + i\mathrm{Im}(z)}{|z \sin\theta + \cos\theta|^2},$$

Therefore for some θ, we have $\frac{|z|^2-1}{2} \sin 2\theta + \mathrm{Re}(z) \cos 2\theta = 0$.

By this lemma, we have that for some $\lambda, \lambda' \geq 1$ and $A, A' \in \mathrm{SL}(2, \mathbb{R})$,

$$(\lambda i, i) = (A.z, A.w), \ (\lambda' i, i) = (A'.z', A'.w'),$$

$F(z, w) = F(\lambda i, i) = \frac{(\lambda-1)^2}{4\lambda} = \frac{1}{4}\left(\lambda + \frac{1}{\lambda} - 2\right)$, and $F(z', w') = F(\lambda' i, i) = \frac{(\lambda'-1)^2}{4\lambda'} = \frac{1}{4}(\lambda' + \frac{1}{\lambda'} - 2)$.

If $F(z', w') = F(z, w)$, then, $\lambda = \lambda'$ and hence $(A.z, A.w) = (A'.z', A'.w')$.

4 Asymmetric KLD Between Multivariate Cauchy Distributions

This section bears a natural question about minimal conditions for symmetry. The probability density function of a d-dimensional Cauchy distribution with parameters $\mu \in \mathbb{R}^d$ and Σ be a $d \times d$ positive-definite symmetric matrix is defined by:

$$p_{\mu,\Sigma}(x) := \frac{C_d}{(\det \Sigma)^{1/2}} \left(1 + (x-\mu)' \, \Sigma^{-1} \, (x-\mu)\right)^{-(d+1)/2}, \; x \in \mathbb{R}^d,$$

where C_d is a normalizing constant. Contrary to the univariate Cauchy distribution, we have the following:

Proposition 2. *There exist two bivariate Cauchy densities* p_{μ_1,Σ_1} *and* p_{μ_2,Σ_2} *such that* $D_{\mathrm{KL}}(p_{\mu_1,\Sigma_1} : p_{\mu_2,\Sigma_2}) \neq D_{\mathrm{KL}}(p_{\mu_2,\Sigma_2} : p_{\mu_1,\Sigma_1})$.

Proof. We let $d = 2$. By the change of variable in the integral [9], we have

$$D_{\mathrm{KL}}(p_{\mu_1,\Sigma_1} : p_{\mu_2,\Sigma_2}) = D_{\mathrm{KL}}\left(p_{0,I_2} : p_{\Sigma_1^{-1/2}(\mu_2-\mu_1),\Sigma_1^{-1/2}\Sigma_2\Sigma_1^{-1/2}}\right),$$

where I_2 denotes the unit 2×2 matrix.

Let $\mu_1 = 0, \Sigma_1 = I_2,\ \mu_2 = (0,1)^\top, \Sigma_2 = \begin{bmatrix} n & 0 \\ 0 & \frac{1}{n} \end{bmatrix}$, where n is a natural number. We will show that $D_{\mathrm{KL}}(p_{\mu_1,\Sigma_1} : p_{\mu_2,\Sigma_2}) \neq D_{\mathrm{KL}}(p_{\mu_2,\Sigma_2} : p_{\mu_1,\Sigma_1})$ for sufficiently large n. Then,

$$D_{\mathrm{KL}}(p_{\mu_1,\Sigma_1} : p_{\mu_2,\Sigma_2}) = \frac{3C_2}{2} \int_{\mathbb{R}^2} \frac{\log(1 + x_1^2/n + nx_2^2) - \log(1 + x_1^2 + x_2^2)}{(1 + x_1^2 + x_2^2)^{3/2}} \mathrm{d}x_1 \mathrm{d}x_2$$

and

$$\begin{aligned} D_{\mathrm{KL}}(p_{\mu_2,\Sigma_2} : p_{\mu_1,\Sigma_1}) &= D_{\mathrm{KL}}\left(p_{0,I_2} : p_{-\Sigma_1^{-1/2}\mu_1,\Sigma_1^{-1}}\right), \\ &= \frac{3C_2}{2} \int_{\mathbb{R}^2} \frac{\log(1 + x_1^2/n + n(x_2 + \sqrt{n})^2) - \log(1 + x_1^2 + x_2^2)}{(1 + x_1^2 + x_2^2)^{3/2}} \mathrm{d}x_1 \mathrm{d}x_2. \end{aligned}$$

Hence it suffices to show that

$$\int_{\mathbb{R}^2} \frac{\log(1 + x_1^2/n + n(x_2 + \sqrt{n})^2) - \log(1 + x_1^2/n + nx_2^2)}{(1 + x_1^2 + x_2^2)^{3/2}} \mathrm{d}x_1 \mathrm{d}x_2 \neq 0$$

for some n. Since $\log(1 + x) \leq x$, it suffices to show that

$$\int_{\mathbb{R}^2} \frac{-n^2 + n - 2x_2(n + \sqrt{n})}{(1 + x_1^2 + x_2^2)^{3/2}(1 + x_1^2/n + n(x_2 + \sqrt{n})^2)} \mathrm{d}x_1 \mathrm{d}x_2 < 0$$

for some n. We see that

$$\begin{aligned} &\int_{\mathbb{R}^2} \frac{2|x_2|(n + \sqrt{n})}{(1 + x_1^2 + x_2^2)^{3/2}(1 + x_1^2/n + n(x_2 + \sqrt{n})^2)} \mathrm{d}x_1 \mathrm{d}x_2, \\ &\leq 4n \int \frac{|x_2|}{(1 + x_1^2 + x_2^2)^{3/2}(1 + x_1^2/n + n(x_2 + \sqrt{n})^2)} \mathrm{d}x_1 \mathrm{d}x_2. \end{aligned}$$

We have that

$$n \int_{|x_2+\sqrt{n}|>n^{1/3}} \frac{|x_2|}{(1+x_1^2+x_2^2)^{3/2}(1+x_1^2/n+n(x_2+\sqrt{n})^2)} \mathrm{d}x_1 \mathrm{d}x_2,$$
$$\leq \int_{|x_2+\sqrt{n}|>n^{1/3}} \frac{|x_2|}{(1+x_1^2+x_2^2)^{3/2}(x_2+\sqrt{n})^2} dx_1 dx_2.$$

We have that $\limsup_{n\to\infty} \sup_{x;|x+\sqrt{n}|>n^{1/3}} \frac{|x|}{(x+\sqrt{n})^2} = 0$.

Hence,

$$\limsup_{n\to\infty} n \int_{|x_2+\sqrt{n}|>n^{1/3}} \frac{|x_2|}{(1+x_1^2+x_2^2)^{3/2}(1+x_1^2/n+n(x_2+\sqrt{n})^2)} \mathrm{d}x_1 \mathrm{d}x_2 = 0.$$

It holds that

$$\liminf_{n\to\infty} \int_{|x_2+\sqrt{n}|>n^{1/3}} \frac{n^2-n}{(1+x_1^2+x_2^2)^{3/2}(1+x_1^2/n+n(x_2+\sqrt{n})^2)} \mathrm{d}x_1 \mathrm{d}x_2$$

$$\geq \liminf_{n\to\infty} \int_{x_1^2+x_2^2\leq 1} \frac{n^2-n}{(1+x_1^2+x_2^2)^{3/2}(1+x_1^2+2n(x_2^2+n))} \mathrm{d}x_1 \mathrm{d}x_2$$
$$\geq \frac{1}{2} \int_{x_1^2+x_2^2\leq 1} \frac{1}{(1+x_1^2+x_2^2)^{3/2}} \mathrm{d}x_1 \mathrm{d}x_2 > 0,$$

where we have used the fact that $\{x_1^2+x_2^2 \leq 1\} \subset \{|x_2+\sqrt{n}| > n^{1/3}\}$ for large n in the first inequality, and Fatou's lemma [4] (p. 93) in the second inequality.

Hence,

$$\liminf_{n\to\infty} \int_{|x_2+\sqrt{n}|>n^{1/3}} \frac{n^2-n-4|x_2|n}{(1+x_1^2+x_2^2)^{3/2}(1+x_1^2/n+n(x_2+\sqrt{n})^2)} dx_1 dx_2$$
$$\geq \frac{1}{2} \int_{x_1^2+x_2^2\leq 1} \frac{1}{(1+x_1^2+x_2^2)^{3/2}} \mathrm{d}x_1 \mathrm{d}x_2 > 0.$$

We have that for large n,

$$\int_{|x_2+\sqrt{n}|\leq n^{1/3}} \frac{n^2-n-4|x_2|n}{(1+x_1^2+x_2^2)^{3/2}(1+x_1^2/n+n(x_2+\sqrt{n})^2)} \mathrm{d}x_1 \mathrm{d}x_2 \geq 0.$$

Thus we have that

$$\liminf_{n\to\infty} \int_{\mathbb{R}^2} \frac{n^2-n-4|x_2|n}{(1+x_1^2+x_2^2)^{3/2}(1+x_1^2/n+n(x_2+\sqrt{n})^2)} \mathrm{d}x_1 \mathrm{d}x_2 > 0.$$

References

1. Chyzak, F., Nielsen, F.: A closed-form formula for the Kullback-Leibler divergence between Cauchy distributions. arXiv preprint arXiv:1905.10965 (2019)
2. Csiszár, I.: Information-type measures of difference of probability distributions and indirect observation. Studia Scientiarum Mathematicarum Hungarica **2**, 229–318 (1967)
3. Eaton, M.L.: Group invariance applications in statistics. Institute of Mathematical Statistics Hayward, California (1989)
4. Kesavan (emeritus), S.: Measure and Integration. TRM, Springer, Singapore (2019). https://doi.org/10.1007/978-981-13-6678-9
5. McCullagh, P.: On the distribution of the Cauchy maximum-likelihood estimator. Proc. R. Soc. Lond. Ser. A Math. Phys. Sci. **440**(1909), 475–479 (1993)
6. Mitchell, A.F.S.: Statistical manifolds of univariate elliptic distributions. In: International Statistical Review/Revue Internationale de Statistique, pp. 1–16 (1988)
7. Needham, T.: Visual Complex Analysis. Oxford University Press, Oxford (1998)
8. Nielsen, F.: On Voronoi diagrams on the information-geometric Cauchy manifolds. Entropy **22**(7), 713 (2020)
9. Nielsen, F.: On information projections between multivariate elliptical and location-scale families. Technical Report. arXiv:2101.03839 (2021)
10. Nielsen, F., Nock, R.: On the chi square and higher-order chi distances for approximating f-divergences. IEEE Sign. Process. Lett. **21**(1), 10–13 (2013)
11. Nielsen, F., Okamura, K.: On f-divergences between Cauchy distributions. Technical Report (2021)
12. Nock, R., Nielsen, F., Amari, S.I.: On conformal divergences and their population minimizers. IEEE Trans. Inf. Theory. **62**(1), 527–538 (2015)

Transport Information Hessian Distances

Wuchen Li(✉)

University of South Carolina, Columbia, USA
wuchen@mailbox.sc.edu

Abstract. We formulate closed-form Hessian distances of information entropies in one-dimensional probability density space embedded with the L^2-Wasserstein metric. Some analytical examples are provided.

Keywords: Optimal transport · Information geometry · Hessian distances

1 Introduction

Hessian distances of information entropies in probability density space play crucial roles in information theory with applications in signal and image processing, inverse problems, and AI [1,6,7]. One important example is the Hellinger distance [1,2,13,16], known as a Hessian distance of negative Boltzmann-Shannon entropy in L^2 (Euclidean) space. Nowadays, the Hellinger distance has shown various useful properties in statistical and AI inference problems.

Recently, optimal transport distance, a.k.a. Wasserstein distance[1], provides another type of distance functions in probability density space [3,15]. Unlike the L^2 distance, the Wasserstein distance compares probability densities by their pushforward mapping functions. More importantly, it introduces a metric space under which the information entropies present convexity properties in term of mapping functions. These properties nowadays have been shown useful in fluid dynamics, inverse problems and AI [5,8,14].

Nowadays, one can show that the optimal transport distance itself is a Hessian distance of a second moment functional in Wasserstein space [11]. Natural questions arise. *Can we construct Hessian distances of information entropies in Wasserstein space? And what is the Hessian distance of negative Boltzmann-Shannon entropy in Wasserstein space?*

In this paper, we derive closed-form Hessian distances of information entropies in Wasserstein space supported on a one dimensional sample space. In details, given a closed and bounded set $\Omega \subset \mathbb{R}^1$, consider a (negative) f-entropy by

$$\mathcal{F}(p) = \int_\Omega f(p(x))dx,$$

[1] There are various generalizations of optimal transport distances using different ground costs defined on a sample space. For the simplicity of discussion, we focus on the L^2-Wasserstein distance.

F. Nielsen and F. Barbaresco (Eds.): GSI 2021, LNCS 12829, pp. 808–817, 2021.
https://doi.org/10.1007/978-3-030-80209-7_87

where $f\colon \mathbb{R} \to \mathbb{R}$ is a second differentiable convex function and p is a given probability density function. We show that the Hessian distance of f-entropy in Wasserstein space between probability density functions p and q satisfies

$$\mathrm{Dist_H}(p,q) = \sqrt{\int_0^1 h(\nabla_y F_p^{-1}(y)) - h(\nabla_y F_q^{-1}(y))|^2 dy},$$

where $h(y) = \int_1^y \sqrt{f''(\frac{1}{z})\frac{1}{z^{\frac{3}{2}}}}dz$, F_p, F_q are cumulative density functions (CDFs) of p, q respectively, and F_p^{-1}, F_q^{-1} are their inverse CDFs. We call $\mathrm{Dist_H}(p,q)$ *transport information Hessian distances.* Shortly, we show that the proposed distances are constructed by the Jacobian operators (derivatives) of mapping functions between density functions p and q.

This paper is organized as follows. In Sect. 2, we briefly review the Wasserstein space and its Hessian operators for information entropies. In Sect. 3, we derive closed-form solutions of Hessian distances in Wasserstein space. Several analytical examples are presented in Sect. 4.

2 Review of Transport Information Hessian Metric

In this section, we briefly review the Wasserstein space and its induced Hessian metrics for information entropies.

2.1 Wasserstein Space

We recall the definition of a one dimensional Wasserstein distance [3]. Denote a spatial domain by $\Omega = [0,1] \subset \mathbb{R}^1$. We remark that one dimensional sample space has an independent interest since it applies to univariate (one-variable) statistical problems.

Consider the space of smooth positive probability densities by

$$\mathcal{P}(\Omega) = \Big\{p(x) \in C^\infty(\Omega)\colon \int_\Omega p(x)dx = 1, \quad p(x) > 0\Big\}.$$

Given any two probability densities p, $q \in \mathcal{P}(\Omega)$, the squared Wasserstein distance in Ω is defined by

$$\mathrm{Dist_T}(p,q)^2 = \int_\Omega |T(x) - x|^2 q(x)dx,$$

where $|\cdot|$ represents a Euclidean norm and T is a monotone transport mapping function such that $T_\# q(x) = p(x)$, i.e.

$$p(T(x))\nabla_x T(x) = q(x).$$

Since $\Omega \subset \mathbb{R}^1$, then the mapping function T can be solved analytically. Concretely,

$$T(x) = F_p^{-1}(F_q(x)),$$

where F_p^{-1}, F_q^{-1} are inverse CDFs of probability densities p, q, respectively. Equivalently,

$$\mathrm{Dist}_{\mathrm{T}}(p,q)^2 = \int_\Omega |F_p^{-1}(F_q(x)) - x|^2 q(x)dx$$
$$= \int_0^1 F_p^{-1}(y) - F_q^{-1}(y)|^2 dy,$$

where we apply the change of variable $y = F_q(x) \in [0,1]$ in the second equality.

There is also a metric formulation for the L^2-Wasserstein distance. Denote the tangent space of $\mathcal{P}(\Omega)$ at a probability density p by

$$T_p\mathcal{P}(\Omega) = \Big\{\sigma \in C^\infty(\Omega) \colon \int_\Omega \sigma(x)dx = 0\Big\}.$$

The L^2-Wasserstein metric refers to the following bilinear form:

$$g_{\mathrm{T}}(p)(\sigma,\sigma) = \int_\Omega (\nabla_x\Phi(x), \nabla_x\Phi(x))p(x)dx,$$

where $\Phi, \sigma \in C^\infty(\Omega)$ satisfy an elliptical equation

$$\sigma(x) = -\nabla_x \cdot (p(x)\nabla_x\Phi(x)), \tag{1}$$

with either Neumann or periodic boundary conditions on Ω. Here, the above mentioned boundary conditions ensure the fact that σ stays in the tangent space of probability density space, i.e. $\int_\Omega \sigma(x)dx = 0$. As a known fact, the Wasserstein metric can be derived from a Taylor expansion of the Wasserstein distance. i.e.,

$$\mathrm{Dist}_{\mathrm{T}}(p, p+\sigma)^2 = g_{\mathrm{T}}(\sigma,\sigma) + o(|\sigma|_{L^2}^2),$$

where $\sigma \in T_p\mathcal{P}(\Omega)$ and $|\cdot|_{L_2}$ represents the L^2 norm. In literature, $(\mathcal{P}(\Omega), g_{\mathrm{T}})$ is often called the *Wasserstein space*.

2.2 Hessian Metrics in Wasserstein Space

We next review the Hessian operator of f-entropies in Wasserstein space [15]. Consider a (negative) f-entropy by

$$\mathcal{F}(p) = \int_\Omega f(p(x))dx,$$

where $f\colon \mathbb{R} \to \mathbb{R}$ is a one dimensional second order differentiable convex function. The Hessian operator of f-entropy in one dimensional Wasserstein space is a bilinear form satisfying

$$\mathrm{Hess}_{\mathrm{T}}\mathcal{F}(p)(\sigma,\sigma) = \int_\Omega |\nabla_x^2\Phi(x)|^2 f''(p(x))p(x)^2 dx.$$

where f'' represents the second derivative of function f and (Φ, σ) satisfies an elliptic equation (1).

In this paper, we seek closed-form solutions for transport Hessian (pseudo) distances of $\mathcal{F}$. Here, the Hessian distances of entropies are of interests in mathematical physics equations and AI inference problems [4,11]. They generalize the optimal transport distances by using the Hessian operator of entropies. Here we aim to solve an action functional in $(\mathcal{P}(\Omega), \mathrm{Hess}_{\mathrm{T}}\mathcal{F})$.

Definition 1 (Transport information Hessian distance [11]). *Define a distance function* $\mathrm{Dist}_{\mathrm{H}} : \mathcal{P}(\Omega) \times \mathcal{P}(\Omega) \to \mathbb{R}$ *by*

$$\mathrm{Dist}_{\mathrm{H}}(p,q)^2 = \inf_{p\colon [0,1]\times\Omega\to\mathbb{R}} \Big\{ \int_0^1 \mathrm{Hess}_{\mathrm{T}}\mathcal{F}(p)(\partial_t p, \partial_t p)dt \colon p(0,x) = q(x),\ p(1,x) = p(x) \Big\}. \tag{2}$$

Here the infimum is taken among all smooth density paths $p\colon [0,1] \times \Omega \to \mathbb{R}$*, which connects both initial and terminal time probability density functions* q*,* $p \in \mathcal{P}(\Omega)$.

From now on, we call $\mathrm{Dist}_{\mathrm{H}}$ the *transport information Hessian distance.* Here the terminology of "transport" corresponds to the application of Wasserstein space, while the name of "information" refers to the usage of f-entropies.

3 Transport Information Hessian Distances

In this section, we present the main result of this paper.

3.1 Formulations

We first derive closed-form solutions for transport information Hessian distances defined by (2).

Theorem 1. *Denote a one dimensional function* $h\colon \Omega \to \mathbb{R}$ *by*

$$h(y) = \int_1^y \sqrt{f''(\frac{1}{z})\frac{1}{z^{\frac{3}{2}}}}dz.$$

Then the squared transport Hessian distance of f*-entropy has the following formulations.*

(i) Inverse CDF formulation:

$$\mathrm{Dist}_{\mathrm{H}}(p,q)^2 = \int_0^1 |h(\nabla_y F_p^{-1}(y)) - h(\nabla_y F_q^{-1}(y))|^2 dy.$$

(ii) Mapping formulation:

$$\mathrm{Dist}_{\mathrm{H}}(p,q)^2 = \int_\Omega |h(\frac{\nabla_x T(x)}{q(x)}) - h(\frac{1}{q(x)})|^2 q(x)dx,$$

where T *is a mapping function, such that* $T_\# q = p$ *and* $T(x) = F_p^{-1}(F_q(x))$*. Equivalently,*

$$\mathrm{Dist}_{\mathrm{H}}(p,q)^2 = \int_\Omega |h(\frac{\nabla_x T^{-1}(x)}{p(x)}) - h(\frac{1}{p(x)})|^2 p(x)dx,$$

where T^{-1} is the inverse function of mapping function T, such that $(T^{-1})_{\#}p = q$ and $T^{-1}(x) = F_q^{-1}(F_p(x))$.

Proof. We first derive formulation (i). To do so, we consider the following two set of change of variables. First, denote $p^0(x) = q(x)$ and $p^1(x) = p(x)$. Denote the variational problem (2) by

$$\mathrm{Dist}_{\mathrm{H}}(p^0,p^1)^2 = \inf_{\Phi,p\colon [0,1]\times\Omega\to\mathbb{R}} \Big\{ \int_0^1\int_\Omega |\nabla_y^2\Phi(t,y)|^2 f''(p(t,y))p(t,y)^2 dydt\colon \\ \partial_t p(t,y) + \nabla\cdot(p(t,y)\nabla_y\Phi(t,y)) = 0,\ \text{fixed } p^0,\ p^1\Big\},$$

where the infimum is among all smooth density paths $p\colon [0,1]\times\Omega\to\mathbb{R}$ satisfying the continuity equation with gradient drift vector fields $\nabla\Phi\colon [0,1]\times\Omega\to\mathbb{R}$. Denote a monotone mapping function $T\colon [0,1]\times\Omega\to\Omega$ by

$$y = T(t,x).$$

Denote the vector field function $v\colon [0,1]\times\Omega\to\Omega$ by

$$\partial_t T(t,x) = v(t,T(t,x)) = \nabla_y\Phi(t,y).$$

Hence

$$\mathrm{Dist}_{\mathrm{H}}(p^0,p^1)^2 = \inf_{T\colon [0,1]\times\Omega\to\Omega} \Big\{ \int_0^1\int_\Omega |\nabla_y v(t,T(t,x))|^2 f''(p(t,T(t,x)))p(t,T(t,x))^2 dT(t,x)dt\colon \\ T(t,\cdot)_{\#}p(0,x) = p(t,x)\Big\},$$

where the infimum is taken among all smooth monotone transport mapping functions $T\colon [0,1]\times\Omega\to\Omega$ with $T(0,x) = x$ and $T(1,x) = T(x)$. We observe that the above variation problem leads to

$$\begin{aligned}
&\int_0^1\int_\Omega |\nabla_y\partial_t T(t,x)|^2 f''(p(t,T(t,x)))p(t,T(t,x))^2\nabla_x T(t,x)dxdt \\
=&\int_0^1\int_\Omega |\nabla_x\partial_t T(t,x)\frac{dx}{dy}|^2 f''(p(t,T(t,x)))p(t,T(t,x))^2\nabla_x T(t,x)dxdt \\
=&\int_0^1\int_\Omega |\nabla_x\partial_t T(t,x)\frac{1}{\nabla_x T(t,x)}|^2 f''(\frac{p(0,x)}{\nabla_x T(t,x)})\frac{p(0,x)^2}{\nabla_x T(t,x)}dxdt \\
=&\int_0^1\int_\Omega |\partial_t\nabla_x T(t,x)\frac{1}{(\nabla_x T(t,x))^{3/2}}\sqrt{f''(\frac{q(x)}{\nabla_x T(t,x)})}|^2 q(x)^2 dxdt.
\end{aligned} \tag{3}$$

Second, denote $y = F_q(x)$, where $y\in[0,1]$. By using a chain rule for $T(t,\cdot)_{\#}q(x) = p_t$ with $p_t := p(t,x)$, we have

$$q(x) = \frac{dy}{dx} = \frac{1}{\frac{dx}{dy}} = \frac{1}{\nabla_y F_q^{-1}(y)},$$

and

$$\nabla_x T(t,x) = \nabla_x F_{p_t}^{-1}(F_q(x)) = \nabla_y F_{p_t}^{-1}(y)\frac{dy}{dx} = \frac{\nabla_y F_{p_t}^{-1}(y)}{\nabla_y F_q^{-1}(y)}.$$

Under the above change of variables, variation problem (3) leads to

$$\begin{aligned}\mathrm{Dist_H}(p^0,p^1)^2 &= \inf_{\nabla_y F_{p_t}^{-1}\colon [0,1]^2\to\mathbb{R}} \Big\{\int_0^1\int_0^1 |\partial_t \nabla_y F_{p_t}^{-1}(y)\frac{1}{(\nabla_y F_{p_t}^{-1}(y))^{\frac{3}{2}}}\sqrt{f''(\frac{1}{\nabla_y F_{p_t}^{-1}(y)})}|^2 dydt\Big\} \\ &= \inf_{\nabla_y F_{p_t}^{-1}\colon [0,1]^2\to\mathbb{R}} \Big\{\int_0^1\int_0^1 |\partial_t h(\nabla_y F_{p_t}^{-1}(y))|^2 dydt\Big\},\end{aligned} \tag{4}$$

where the infimum is taken among all paths $\nabla_y F_{p_t}^{-1}\colon [0,1]^2 \to \mathbb{R}$ with fixed initial and terminal time conditions. Here we apply the fact that

$$\nabla_y h(y) = \sqrt{f''(\frac{1}{y})}\frac{1}{y^{\frac{3}{2}}},$$

and treat $\nabla_y F_{p_t}^{-1}$ as an individual variable. By using the Euler-Lagrange equation for $\nabla_y F_{p_t}^{-1}$, we show that the geodesic equation in transport Hessian metric satisfies

$$\partial_{tt} h(\nabla_y F_{p_t}^{-1}(y)) = 0.$$

In details, we have

$$h(\nabla_y F_{p_t}^{-1}(y)) = th(\nabla_y F_p^{-1}(y)) + (1-t)h(\nabla_y F_q^{-1}(y)),$$

and

$$\partial_t h(\nabla_y F_{p_t}^{-1}(y)) = h(\nabla_y F_p^{-1}(y)) - h(\nabla_y F_q^{-1}(y)).$$

Therefore, variational problem (4) leads to the formulation (i).

We next derive formulation (ii). Denote $y = F_q(x)$ and $T(x) = F_p^{-1}(F_q(x))$. By using formulation (i) and the change of variable formula in integration, we have

$$\begin{aligned}\mathrm{Dist_H}(p^0,p^1)^2 &= \int_0^1 |h(\nabla_y F_p^{-1}(y)) - h(\nabla_y F_q^{-1}(y))|^2 dy \\ &= \int_\Omega |h(\nabla_y F_p^{-1}(F_q(x)) - h(\nabla_y x)|^2 dF_q(x) \\ &= \int_\Omega |h(\frac{\nabla_x T(x)}{\frac{dy}{dx}}) - h(\frac{1}{\frac{dy}{dx}})|^2 q(x)dx \\ &= \int_\Omega |h(\frac{\nabla_x T(x)}{q(x)}) - h(\frac{1}{q(x)})|^2 q(x)dx.\end{aligned}$$

This finishes the first part of proof. Similarly, we can derive the transport information Hessian distance in term of the inverse mapping function T^{-1}.

Remark 1. We notice that $\mathrm{Dist_H}$ forms a class of distance functions in probability density space. Compared to the classical optimal transport distance, it emphasizes on the differences by the Jacobian (derivative) operators of mapping operators. The nondecreasing function h plays the role of "ground metric" defined on the Jacobian operators of mapping functions. In this sense, we call the transport information Hessian distance the *"Optimal Jacobian transport distance"*.

Remark 2. Transport information Hessian distances share similarities with transport Bregman divergences defined in [12]. Here we remark that transport Hessian distances are symmetric to p, q, i.e. $\mathrm{Dist}_{\mathrm{H}}(p,q) = \mathrm{Dist}_{\mathrm{H}}(q,p)$, while the transport Bregman divergences are often asymmetric to p, q; see examples in [12].

3.2 Properties

We next demonstrate that transport Hessian distances have several basic properties.

Proposition 1. *The transport Hessian distance has the following properties.*

(i) Nonnegativity:

$$\mathrm{Dist}_{\mathrm{H}}(p,q) \geq 0.$$

In addition,

$$\mathrm{Dist}_{\mathrm{H}}(p,q) = 0 \quad \textit{iff} \quad p(x+c) = q(x),$$

where $c \in \mathbb{R}$ is a constant.

(ii) Symmetry:

$$\mathrm{Dist}_{\mathrm{H}}(p,q) = \mathrm{Dist}_{\mathrm{H}}(q,p).$$

(iii) Triangle inequality: For any probability densities p, q, $r \in \mathcal{P}(\Omega)$, we have

$$\mathrm{Dist}_{\mathrm{H}}(p,r) \leq \mathrm{Dist}_{\mathrm{H}}(p,q) + \mathrm{Dist}_{\mathrm{H}}(q,r).$$

(v) Hessian metric: Consider a Taylor expansion by

$$\mathrm{Dist}_{\mathrm{H}}(p,p+\sigma)^2 = \mathrm{Hess}_{\mathrm{T}}\mathcal{F}(p)(\sigma,\sigma) + o(|\sigma|^2_{L^2}),$$

where $\sigma \in T_p\mathcal{P}(\Omega)$.

Proof. From the construction of transport Hessian distance, the above properties are satisfied. Here we only need to show (i). $\mathrm{Dist}_{\mathrm{H}}(p,q) = 0$ implies that $|h(\nabla_x T(x)) - h(1)| = |h(\nabla_x T(x))| = 0$ under the support of density q. Notice that h is a monotone function. Thus $\nabla_x T(x) = 1$. Hence $T(x) = x + c$, for some constant $c \in \mathbb{R}$. From the fact that $T_\# q = p$, we derive $p(T(x))\nabla_x T(x) = q(x)$. This implies $p(x+c) = q(x)$.

4 Closed-Form Distances

In this section, we provide several closed-form examples of transport information Hessian distances. From now on, we always denote

$$y = F_q(x) \quad \text{and} \quad T(x) = F_p^{-1}(F_q(x)).$$

Example 1 (Boltzmann-Shanon entropy). Let $f(p) = p\log p$, i.e.

$$\mathcal{F}(p) = -\mathcal{H}(p) = \int_\Omega p(x)\log p(x)dx.$$

Then $h(y) = \log y$. Hence

$$\begin{aligned} \mathrm{Dist_H}(p,q) =& \sqrt{\int_\Omega |\log \nabla_x T(x)|^2 q(x)dx} \\ =& \sqrt{\int_0^1 |\log \nabla_y F_p^{-1}(y) - \log \nabla_y F_q^{-1}(y)|^2 dy}. \end{aligned} \tag{5}$$

Remark 3 (Comparisons with Hellinger distances). We compare the Hessian distance of Boltzmann-Shannon entropy defined in either L^2 space or Wasserstein space. In L^2 space, this Hessian distance is known as the Hellinger distance, where $\mathrm{Hellinger}(p,q)^2 = \frac{1}{2}\int_\Omega |\sqrt{p(x)} - \sqrt{q(x)}|^2 dx$. Here distance (5) is an analog of the Hellinger distance in Wasserstein space.

Example 2 (Quadratic entropy). Let $f(p) = \frac{1}{2}p^2$, then $h(y) = -2(y^{-\frac{1}{2}} - 1)$. Hence

$$\begin{aligned} \mathrm{Dist_H}(p,q) =& 2\sqrt{\int_\Omega |(\nabla_x T(x))^{-\frac{1}{2}} - 1|^2 q(x)^2 dx} \\ =& 2\sqrt{\int_0^1 |(\nabla_y F_p^{-1}(y))^{-\frac{1}{2}} - (\nabla_y F_q^{-1}(y))^{-\frac{1}{2}}|^2 dy}. \end{aligned}$$

Example 3 (Cross entropy). Let $f(p) = -\log p$, then $h(y) = 2(y^{\frac{1}{2}} - 1)$. Hence

$$\begin{aligned} \mathrm{Dist_H}(p,q) =& 2\sqrt{\int_\Omega |(\nabla_x T(x))^{\frac{1}{2}} - 1|^2 dx} \\ =& 2\sqrt{\int_0^1 |(\nabla_y F_p^{-1}(y))^{\frac{1}{2}} - (\nabla_y F_q^{-1}(y))^{\frac{1}{2}}|^2 dy}. \end{aligned}$$

Example 4. Let $f(p) = \frac{1}{2p}$, then $h(y) = y - 1$. Hence

$$\begin{aligned}\mathrm{Dist_H}(p,q) =& \sqrt{\int_\Omega |\nabla_x T(x) - 1|^2 q(x)^{-1} dx} \\ =& 2\sqrt{\int_0^1 |\nabla_y F_p^{-1}(y) - \nabla_y F_q^{-1}(y)|^2 dy}.\end{aligned}$$

Example 5 (γ-entropy). Let $f(p) = \frac{1}{(1-\gamma)(2-\gamma)} p^{2-\gamma}$, $\gamma \neq 1, 2$, then $h(y) = \frac{2}{\gamma-1}(y^{\frac{\gamma-1}{2}} - 1)$. Hence

$$\begin{aligned}\mathrm{Dist_H}(p,q) =& \frac{2}{|\gamma-1|}\sqrt{\int_\Omega |(\nabla_x T(x))^{\frac{\gamma-1}{2}} - 1|^2 q(x)^{2-\gamma} dx} \\ =& \frac{2}{|\gamma-1|}\sqrt{\int_0^1 |(\nabla_y F_p^{-1}(y))^{\frac{\gamma-1}{2}} - (\nabla_y F_q^{-1}(y))^{\frac{\gamma-1}{2}}|^2 dy}.\end{aligned}$$

Acknowledgement. Wuchen Li is supported by a start-up funding in University of South Carolina.

References

1. Amari, S.: Information Geometry and Its Applications. AMS, vol. 194. Springer, Tokyo (2016). https://doi.org/10.1007/978-4-431-55978-8
2. Ay, N., Jost, J., Lê, H.V., Schwachhöfer, L.: Information Geometry. EMG-FASMSM, vol. 64. Springer, Cham (2017). https://doi.org/10.1007/978-3-319-56478-4
3. Ambrosio, L., Gigli, N., Savare, G.: Gradient Flows in Metric Spaces and in the Space of Probability Measures (2008)
4. Bauer, M., Modin, K.: Semi-invariant Riemannian metrics in hydrodynamics. Calc. Var. Partial Differ. Equ. **59**(2), 1–25 (2020). https://doi.org/10.1007/s00526-020-1722-x
5. Benamou, J., Brenier, Y.: A computational fluid mechanics solution to the Monge-Kantorovich mass transfer problem. Numerische Mathematik **84**(3), 375–393 (2000)
6. Cheng, S., Yau, S.T.: The real Monge-Ampére equation and affine flat structures. In: Proceedings 1980 Beijing Symposium Differential Geometry and Differential Equations, vol. 1, pp. 339–370 (1982)
7. Cover, T.M., Thomas, J.A.: Elements of Information Theory. Wiley Series in Telecommunications, Wiley, New York (1991)
8. Engquist, B., Froese, B.D., Yang, Y.: Optimal transport for seismic full waveform inversion. Commun. Math. Sci. **14**(8), 2309–2330 (2016)
9. Lafferty, J.D.: The density manifold and configuration space quantization. Trans. Am. Math. Soc. **305**(2), 699–741 (1988)
10. Li, W.: Transport information geometry: Riemannian calculus in probability simplex. arXiv:1803.06360 (2018)

11. Li, W.: Hessian metric via transport information geometry. J. Math. Phys. **62**, 033301 (2021)
12. Li, W.: Transport information Bregman divergences. arXiv:2101.01162 (2021)
13. Nielsen, F. (ed.): ETVC 2008. LNCS, vol. 5416. Springer, Heidelberg (2009). https://doi.org/10.1007/978-3-642-00826-9
14. Peyré, G., Cuturi, M.: Computational optimal transport. Found. Trends Mach. Learn. **11**(5–6), 355–607 (2019)
15. Villani, C.: Optimal Transport: Old and New. GL, vol. 338. Springer, Heidelberg (2009). https://doi.org/10.1007/978-3-540-71050-9
16. Zhang, J.: Divergence function, duality, and convex analysis. Neural Comput. **16**(1), 159–195 (2004)

Minimization with Respect to Divergences and Applications

Pierre Bertrand[1(✉)], Michel Broniatowski[1,2], and Jean-François Marcotorchino[3]

[1] Laboratoire de Probabilités, Statistique et Modélisation, Sorbonne Université, Paris, France
pierre.bertrand@ens-paris-saclay.fr
[2] CNRS UMR 8001, Sorbonne Université, Paris, France
[3] Institut de Statistique, Sorbonne Université, Paris, France

Abstract. We apply divergences to project a prior guess discrete probability law on pq elements towards a subspace defined by fixed margins constraints μ and ν on p and q elements respectively. We justify why the Kullback-Leibler and the Chi-square divergences are two canonical choices based on a 1991 work of Imre Csiszár. Besides we interpret the so called *indetermination* resulting from the second divergence as a construction to reduce couple matchings. Eventually, we demonstrate how both resulting probabilities arise in two information theory applications: guessing problem and task partitioning where some optimization remains to minimize a divergence projection.

Keywords: Mathematical relational analysis · Logical *indetermination* · Coupling functions · Task partitioning problem · Guessing problem

1 Introduction

The present document is interested in the poorly studied *indetermination* (also called *indeterminacy*) as a natural second canonical coupling, the first one being the usual independence. We extract both notions from a projection of a prior uniform guess $\mathbb{U}$ to a subset of the set of probabilities defined by fixed margins. The problem is closely linked with contingency table adjustment exposed by Maurice Fréchet in [9] or [10] as well as by Edwards Deming and Frederick Stephan [8] or by Sir Alan G. Wilson [18] or [19]. Although we will limit ourselves to the discrete case it can be transposed in a continuous domain (see for example [16] in the same proceeding as the present paper) leading to the optimal transport theory for which an overview and recent breakthroughs are presented in [17].

When limited to the discrete case, a work of Csiszár [7] shows that two projection functions set apart from all the others as they are the only ones to convey "natural" properties. As we shall see, the first one leads to independence while

F. Nielsen and F. Barbaresco (Eds.): GSI 2021, LNCS 12829, pp. 818–828, 2021.
https://doi.org/10.1007/978-3-030-80209-7_88

the second one, since canonic, motivates our interest for the *indetermination* coupling it generates.

We gather here a unified construction for both couplings and show *indetermination* comes with a stunning property of couple matchings minimization that one could have expected to happen under independence. Eventually this property is applied to two concrete problems: guessing and task partitioning.

2 Projection Using Divergences

Consider two given discrete probabilities $\mu = \mu_0, \ldots, \mu_p$ on p elements and $\nu = \nu_0, \ldots, \nu_q$ on q elements with non zero values (it amounts to reduce p and q). We associate to them the subspace $\mathcal{L}_{\mu,\nu}$ of probability laws π on pq elements where:

$$\pi_{u,\cdot} = \sum_{v=1}^{q} \pi_{u,v} = \mu_u, \forall 1 \leq u \leq p,$$
$$\pi_{\cdot,v} = \sum_{u=1}^{p} \pi_{u,v} = \nu_v, \forall 1 \leq v \leq q,$$

namely, μ and ν are the margins of π. $\mathcal{L}_{\mu,\nu}$ is thus defined by a set of linear constraints on the components of π where π is a probability law living inside the simplex S_{pq}.

As spotted in [3], we can interpret discrete transport problems related to contingency table adjustments, as projections of a prior guess, typically the uniform law $\mathbb{U} = \frac{1}{pq}, \ldots, \frac{1}{pq}$, to our just defined subspace $\mathcal{L}_{\mu,\nu}$.

It happens that the projections of a prior guess that we fix equal to $\mathbb{U}$ to a subspace $\mathcal{L}$ of S_{pq} defined by a set of linear constraints was deeply studied in [7]. Formally, we are minimizing a divergence (see Definition 1) between $\mathbb{U}$ and any element of $\mathcal{L}$ as exposed in Problem 1.

Definition 1 (Divergence with functional parameter ϕ also called f-divergence).
Given, a positive function $\phi : \mathbb{R}^+ \to \mathbb{R}^+$, strictly convex at 1 with $\phi(1) = 0$, and given two discrete probabilities m and n on pq elements, we define the divergence D_ϕ as:

$$D_\phi(m|n) = \sum_{u=1}^{p}\sum_{v=1}^{q} n_{u,v}\phi\left(\frac{m_{u,v}}{n_{u,v}}\right) = \mathbb{E}_n\left[\phi\left(\frac{m_{u,v}}{n_{u,v}}\right)\right].$$

Additionally we set $0 * \phi(x) = 0$ for all $x \in \mathbb{R}^+$.

Problem 1 (Generic Projection Problem).
Given a subspace $\mathcal{L}$ of S_{pq} defined by a set of linear constraints, the projection problem of the uniform law $\mathbb{U}$ to $\mathcal{L}$ using a divergence D_ϕ as defined in Definition 1, states as follows:

$$\min_{\pi \in \mathcal{L}} D_\phi\left(\pi \,|\mathbb{U}\right).$$

Remark 1.
As mentioned by an anonymous reviewer, since $\mathcal{L}$ is a $m-$affine subspace and D_ϕ is a $f-$divergence, the unicity of the optimal is stated from information geometry theory (see [14]).

A key result of the paper [7] is that, provided we add some hypotheses on the cost function we can limit ourselves to two divergences. We do not gather here those hypotheses, but one can find them obviously in the quoted paper as well as summarized in [3]. Eventually, we obtain the two eligible divergences by setting $\phi : x \mapsto x\log(x) - x + 1$ leading to the usual Kullback-Leibler divergence:

$$D_{KL}(\pi \,|\mathbb{U}) = \sum_{u=1}^{p}\sum_{v=1}^{q} \pi_{u,v} \log(pq\pi_{u,v}); \tag{1}$$

and thus to the transport problem using entropy exposed in [18] or with $\phi : x \mapsto (x-1)^2$ leading to the so-called Pearson Chi-square divergence function:

$$D_2(\pi \,|\mathbb{U}) = pq \sum_{u=1}^{p}\sum_{v=1}^{q} \left(\pi_{u,v} - \frac{1}{pq}\right)^2 \tag{2}$$

hence to the minimal transport problem exposed in [15].

Typical computations, using Karush–Kuhn–Tucker, provide for the two divergences, a closed-form expression of the projections that we summarize in the Fig. 1 below where $\pi^\times$ is the *independence* coupling of the two margins:

$$\pi^\times(u,v) = \mu_u\nu_v; \tag{3}$$

and π^+ is the so-called *indetermination* (or *indeterminacy*) coupling:

$$\pi^+(u,v) = \frac{\mu_u}{q} + \frac{\nu_v}{p} - \frac{1}{pq}. \tag{4}$$

At this point we must notice that π^+ is not always defined since it may contain negative values. We add a hypothesis on the margins μ and ν to get rid of this problem:

$$p \min_u \mu_u + q \min_v \nu_v \geq 1. \tag{5}$$

Regarding Eq. 5, a typical modification of any couple of margins to deduce eligible margins can be found in [3] while the probability to draw adapted margins if μ and ν are uniformly drawn on S_p and S_q respectively is provided in [1].

Using Mathematical Relational Analysis notations, one can show that π^+ leads to an equality between a "for" vs "against" notion which justifies the name *indetermination* historically coined in [12]. Besides, the + notation we chose in [2] comes from the Full-Monge property (see for instance the survey [6]) that the associated contingency matrix respects. Eventually, we can also extend the notion of *indeterminacy* in a continuous space and estimate an associated copula as exposed in [4]. In the present paper we shall not detail further the properties π^+ conveys and refer the interested reader to the aforementioned articles.

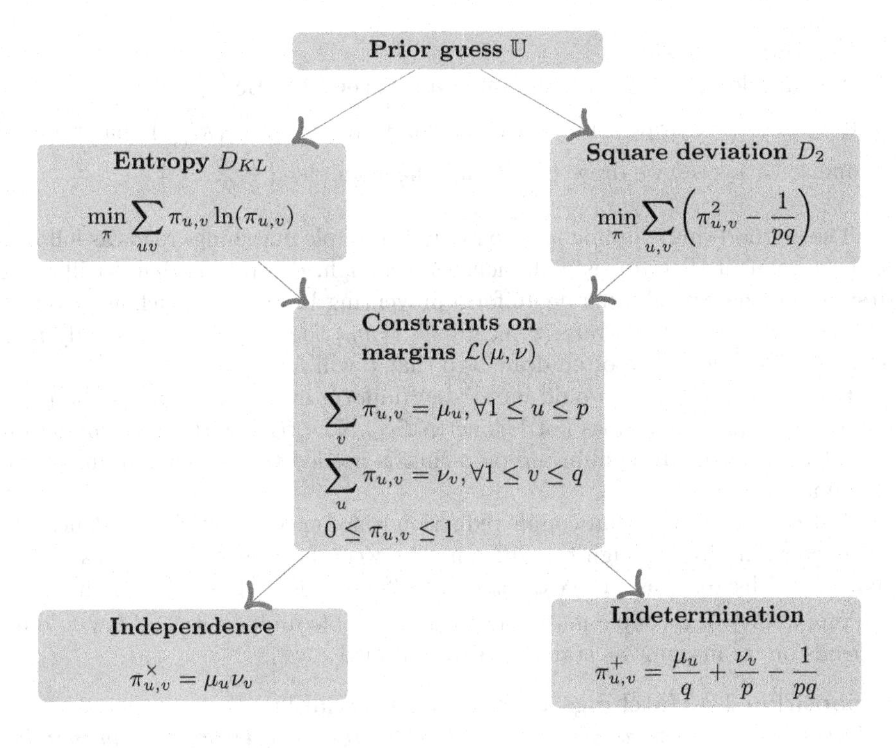

Fig. 1. Summary of a symmetric construction

3 Mimisation of Couple Matchings

We insist on a property expressed at the very beginning of the *indetermination* construction. Let us say π is a probability law on pq elements. If we independently draw two occurrences of $W = (U, V) \sim \pi$, the probability of getting a matching for a couple, that is to say $U_1 = U_2$ and $V_1 = V_2$ simultaneously, is given by $\pi_{U_1,V_1} \times \pi_{U_2,V_2} = \pi_{U_1,V_1}^2$.

Besides, it turns out that

$$D_2(\pi\,|\mathbb{U}) = pq \sum_{u=1}^{p} \sum_{v=1}^{q} \left(\pi_{u,v} - \frac{1}{pq}\right)^2 \tag{6}$$

$$= pq \sum_{u=1}^{p} \sum_{v=1}^{q} (\pi_{u,v})^2 - 1 \tag{7}$$

so that the shift introduced by $\mathbb{U}$ vanishes and minimizing $D_2(\pi\,|\mathbb{U})$ amounts to minimize the probability of a couple matching.

This property is illustrated by a decomposition of π^+ whose proof is available in [3] and which leads to a method for drawing under π^+. We first set u_0 such that it corresponds to the smallest line meaning μ_{u0} is minimal. Then, quoting $\forall u,\ \delta_u = \mu_u - \mu_{u_0}$, we can show that the following drawing realizes *indetermination*:

1. We draw u through μ.
2. We roll a loaded dice with the Bernoulli's skew: $I = \text{Be}\left(\frac{\delta_u}{\mu_u}\right)$.
3. If we get 0, we draw v under the distribution $\left(\pi^+_{u_0,1}, \ldots, \pi^+_{u_0,q}\right)$ that we normalize to 1, else, we draw v uniformly hence under $\left(\frac{1}{q}, \ldots, \frac{1}{q}\right)$.

The method this coupling uses to minimize couple matchings reads as follows. If U_1 is frequent then μ_{U_1} is high, hence δ_{U_1} is high, this means that I will often draw 1 and that v will often be uniform preventing $V_1 = V_2$ as much as possible. On the opposite, if U_1 is rare (typically $U_1 = u_0$) then $U_1 = U_2$ is already rare and I will be allowed to often draw 0 so that v will follow $\left(\pi^+_{u_0,v}\right)_v$.

Best option of course would be to stay uniform every time, this is the prior guess $\mathbb{U}$ solution but it does not belong to $\mathcal{L}_{\mu,\nu}$. The trick with indetermination is that it hides the disequilibrium on v that is needed to respect margins inside rare lines.

A stunning result is that couple reduction is not ensured by independence but by indetermination; though the difference between both is small. One can show that in the discrete case, the expected $\mathbb{L}^2$-difference is $\frac{1}{pq}$ (see [2]). Furthermore the probability of a couple matching with a variable under a second law π' only depends on its margins as stated in Proposition 1.

Proposition 1 (Matchings with another variable). *Let us suppose that π' is a second probability law on pq elements with margins μ' and ν'. We give ourselves a first couple $W = (U_1, U_2)$ drawn under π^+ and a second one $W' = (U_1', U_2')$ independently drawn under π'. Then the probability of couple matching only depends on the margins and is given by:*

$$\mathbb{P}_{\pi^+ \times \pi'}(W = W') = \sum_{u=1}^{p} \left(\frac{\mu_u \mu'_u}{q}\right) + \sum_{v=1}^{q} \left(\frac{\nu_v \nu'_v}{p}\right) - \frac{1}{pq}. \tag{8}$$

Proof. The result is due to the structure of π^+. Formally:

$$\begin{aligned}
\mathbb{P}_{\pi^+ \times \pi'}(W = W') &= \sum_{(u,v)} \pi^+_{u,v} \pi'_{u,v} \\
&= \sum_{(u,v)} \left(\frac{\mu_u}{q} + \frac{\nu_v}{p} - \frac{1}{pq}\right) \pi'_{u,v} \\
&= \sum_{u=1}^{p} \left(\frac{\mu_u \mu'_u}{q}\right) + \sum_{v=1}^{q} \left(\frac{\nu_v \nu'_v}{p}\right) - \frac{1}{pq}
\end{aligned}$$

In particular, the last Proposition 1 shows that any $\pi' \in \mathcal{L}_{\mu,\nu}$ has the same probability of a couple matching with π^+ than π^+ itself. The next section leverages on that ability to reduce couple matching in two applications.

4 Applications

4.1 Guessing Problem

In cryptography, a message u in a finite alphabet $\mathcal{U}$ of size p is typically sent from Alice to Bob while a spy whose name is Charlie tries to decode it. A common strategy for Alice to communicate efficiently and secretly with Bob consists in encoding the message using a couple of keys (public, private) for each character or a symmetric encryption which only requires one shared key between Alice and Bob. The literature concerned with the encryption method to choose according to the situation is diverse, the most-used standard is Advanced Encryption Standard described in various articles. Possibly, Charlie observes an encrypted message V in a second finite alphabet $\mathcal{V}$ of size q which is a function of the message u. We choose to gather in V any additional data Charlie may observe (the channel used, the sender's location, ...). In the end U and V are correlated but not always in a deterministic link.

Related to the cryptography situation, the guessing problem quoted hereafter as Problem 2 was first introduced in the article [13] and extended in [11] to take V into account. While in cryptography Charlie tries to decode a sequence of messages, the guessing problem focuses on decoding a unique message.

Let us formalize the situation: Charlie chooses its strategy (see Definition 2) according to the value taken by the observed second variable V: he typically adapts himself to the conveyed encryption. The probability law of the couple (U, V) is quoted $\mathbb{P}_{U,V} = \pi$ while its margins are $\mathbb{P}_U = \mu$ and $\mathbb{P}_V = \nu$ with no zero values.

Problem 2 (Original Guessing Problem or Spy Problem).
When Alice sends a message $U = u$ to Bob, the spy Charlie observes a variable $V = v$ and must find out the value u of the realization. He has access to a sequence of formatted questions for any guess $\tilde{u}$ he may have: "Does u equal $\tilde{u}$?" for which the binary answer is limited to "yes/no".

Definition 2 (Strategy).
For any value v, a strategy $\sigma_v = S|v$ of Charlie is defined by an order on $\mathcal{U}$ representing the first try, the second and so on until number p. It can be deterministic or random: we quote $\mathbb{P}_{S|v}$ its probability law. The global strategy quoted σ or S gathers all the marginal strategies S_v and is a random variable dependent only on V (not U) whose probability law is quoted $\mathbb{P}_S$.

Besides, for a given position $i \in [1, p]$, and an order σ, $\sigma[i]$ is the element in $\mathcal{U}$ corresponding to the i-th try.

The usual performance $G(\sigma, u)$ of a spy consists in counting the number of questions required to eventually obtain a "yes" in Problem 2 when Charlie proposed the order $S = \sigma$ and Alice generated the message $U = u$. For instance, with an alphabet $\mathcal{U} = \{a, b, c, d\}$, if the message is $u = c$ and the strategy σ of the spy consists in the order (b, c, a, d) (meaning he first proposes message b then c,...) we have:

$$
\begin{aligned}
G(\sigma, u) &= \sum_{i=1}^{p} i \mathbf{1}_{u=\sigma[i]} \\
&= 1 \cdot \mathbf{1}_{u=b} + 2 \cdot \mathbf{1}_{u=c} + 3 \cdot \mathbf{1}_{u=a} + 4 \cdot \mathbf{1}_{u=d} \\
&= 2 \cdot \mathbf{1}_{u=c} \\
&= 2
\end{aligned}
$$

Eventually in [13], they compute the $\rho-$moment of G as an estimation of the efficiency of a strategy S:

$$||G(S,U)||_\rho = \left[\mathbb{E}_{(S,U,V)\sim\mathbb{P}_{S,U,V}} \left(G(S,U)^\rho\right)\right]. \tag{9}$$

We notice that G actually takes as granted that Charlie has up to p trials which is not adapted in case Alice sends a sequence of messages due to a lack of time. Let us move away from the literature and measure Charlie's performance by its probability to find out after one trial the message u Alice sent. It is a reasonable measure as, if a sequence of messages is sent, he may have to jump from one to the following after only one trial. What actually matters (as mentioned by an anonymous reviewer) is the number of attempts k Charlie can try between two messages from Alice. It could be captured with the one-shot performance M using a trick and supposing that Alice repeats each message k times.

Definition 3 (one-shot performance).
For a given strategy S (which can depend on V with a non deterministic link), we define the following performance measure as the probability to find out the value u after one trial, formally:

$$M(S,U,V) = M(S,\pi) = \mathbb{P}_{(S,U,V)}\left(S[1] = U\right).$$

We suppose as for the original optimal strategy that the spy has access to the distribution $\pi = \mathbb{P}_{U,V}$. We can imagine he previously observed the non-encrypted messages in a preliminary step.

Two strategies immediately stand out:

1. S_{max}: systematically returns at v fixed (observed by hypothesis), the u associated with the maximal probability on the margin $\mathbb{P}_{U|V=v}$;
2. S_{margin}: returns at v fixed a random realization in $\mathcal{U}$ under the law $\mathbb{P}_{U|V=v}$.

Similarly as in see [13] where they prove S_{max} is the best strategy in case the performance measure is G, we can show it also maximizes the one-shot performance. Actually, M rewrites:

$$
\begin{aligned}
M(S,U,V) &= \sum_{v=1}^{q} \nu_v \mathbb{P}_{(S_v,U)\sim\mathbb{P}_{S_v}\otimes\mathbb{P}_{U|V=v}} \left(S_v[1] = U\right) && (10) \\
&= \sum_{u=1}^{p}\sum_{v=1}^{q} \pi_{u,v} \mathbb{P}_{S_v}\left(S_v[1] = u\right). && (11)
\end{aligned}
$$

Since $\mathbb{P}_{S_v}$ only depends on v, M is maximal under S_{max}, when $\mathbb{P}_{S_v} = \delta_{u_v}$ where $u_v = \text{argmax}_u \pi_{u,v}$ so that:

$$M(S,U,V) \geq M(S_{max},U,V) = \sum_{v=1}^{q} \pi_{u_v,v}. \tag{12}$$

Eventually, we quote $u_1 = \text{argmax}_u \mu_u$ and notice that $\sum_{v=1}^{q} \pi_{u_v,v} \geq \mu_{u_1}$ leading to the Proposition 2.

Proposition 2 (Charlie's best performance).
We suppose that the margins μ and ν are fixed. Then, for any coupling probability π between messages U and ciphers V, the best one-shot performance Charlie can perform always happens under S_{max}. Furthermore it admits a fixed lower-bound μ_{u_1} independent on π; to summarize:

$$M(S,U,V) \geq M(S_{max},U,V) = \sum_{v=1}^{q} \pi_{u_v,v} \geq \mu_{u_1}. \tag{13}$$

Let us suppose, commendable task if any, that Alice wants to minimize Charlie's best one-shot performance. We also suppose that the margins μ on U and ν on V are fixed. It is a common hypothesis: the alphabet $\mathcal{U}$ in which the messages are composed typically respects a distribution on letters; variable V on its own, if it represents frequencies for instance may have to satisfy occupation weights on each channel. Eventually, Alice can only leverage on the coupling between U and V.

Precisely, let us now compute the corresponding value for our two couplings, both are optimal (for Alice).

$$M^{\times} = M(S_{max}, \pi^{\times}) = \mu_{u_1} \quad M^{+} = M(S_{max}, \pi^{+}) = \mu_{u_1}$$

Regarding the second strategy S_{margin} we know it is less efficient for Charlie in term of one-shot performance. Yet, it is by far harder to cope with for the sender who cannot easily prevent random conflicts. Consequently we come back to the reduction of couple matchings (here a success for Charlie), whose "*indetermination* coupling", we know, prevents us against. Let us unfold this remark hereafter.

Replacing $\mathbb{P}_{S_v}$ by its value under the second strategy in Eq. 11 allows us to estimate the one-shot performance of S_{margin} which is given by:

$$M(S_{margin},U,V) = \sum_{u\in\mathcal{U}} \sum_{v\in\mathcal{V}} \nu_v (\pi_{u|V=v})^2. \tag{14}$$

Concerning the strategy S_{margin} we have the two bounds:

$$\frac{||\pi||_2^2}{\min_{v\in\mathcal{V}} \nu_v} \geq M(S_{margin},U,V) \geq \frac{||\pi||_2^2}{\max_{v\in\mathcal{V}} \nu_v}, \tag{15}$$

with

$$||\pi||_2 = \sqrt{\sum_{u \in \mathcal{U}} \sum_{v \in \mathcal{V}} (\pi_{u,v})^2}.$$

Depending on ν the bounds are more or less tight. If ν_v goes to the uniform, the situation becomes similar to S_{max}: both couplings are equal and optimal.

In any case, Eq. 15 shows that studying the guessing problem brings us back to Problem 1 associated with cost D_2 whose solution is given by the *indetermination* coupling of the margins. It guarantees an efficient reduction of conflicts (see Sect. 3) and eventually a weak one-shot performance under S_{margin} as expressed in Eq. (15) and an optimal one under S_{max}.

4.2 Task Partitioning

Task partitioning problem is originally introduced in [5] where the authors provide a lower bound on the moment of the number of tasks to perform. Let us follow the gathering work of [11] where they also coin a generalized task partitioning problem basically adapting it as a special case of the guessing problem.

Formally, we begin with the original problem of tasks partitioning: a finite set $\mathcal{U}$ of p tasks is given together with an integer $q \leq p$. The problem consists in creating a partition $\mathcal{A} = (\mathcal{A}_1, \ldots, \mathcal{A}_q)$ of $\mathcal{U}$ in q classes to minimize the number of tasks to perform, knowing that if one needs to perform a task $u \in \mathcal{A}_i$, it is mandatory to launch simultaneously the whole subset of tasks included within $\mathcal{A}_i$.

Practically, a task $U = u$ to perform is randomly drawn from $\mathcal{U}$ under a probability distribution $\mathbb{P}_U = \mu$ representing the tasks frequencies. As any task, the task u to perform is assigned to a unique class $\mathcal{A}_{i(u)}$ of the arbitrary partition. Hence, $A(u) = |\mathcal{A}_{i(u)}|$ counts the number of tasks to perform. Precisely, one plays on the partition knowledge to perform, in average, as few tasks as possible.

Similarly to the guessing problem, the performance of a partition $\mathcal{A}$ is estimated through the $\rho-$moment of the number of tasks $A(U)$ to perform:

$$\mathbb{E}_{U \sim \mathbb{P}_U} \left[A(U)^\rho \right]. \tag{16}$$

Inspired by the general guessing problem, paper [11] extends the task partitioning problem. Let us introduce here this generalized version, in which we are no longer interested in minimizing the number of tasks to perform but rather in reducing the number of tasks before a selected (or a chosen) task u.

Indeed, in the first version, as soon as u is drawn, an arbitrary rule imposes to perform the whole subset $\mathcal{A}_{i(u)}$ leading to realize $A(u)$ tasks. In the new version, tasks are performed sequentially in $\mathcal{A}_{i(u)}$ according to a global strategy S that can be deterministic or random. Typically, tasks may consist in a flow of signatures which an administration requires while q would be the number of workers dedicated to perform those signatures on incoming documents. A worker can be given the entitlement to perform several signatures, assistants usually do. In that case, the partition encodes the assignments of tasks to workers.

When a worker $V = v$ is assigned a document, the depositor waits until the signature. At the same time the worker follows his own strategy S_v to sign his assigned documents, meaning he can always follow the same order leading to a deterministic strategy or change every day leading to a random strategy.

With a global strategy S which gathers the workers' strategies S_v, $\forall 1 \leq v \leq q$ and for a task u to perform, the performance of a partition $\mathcal{A}$ is measured using

$$N_{S,\mathcal{A}}(u) = j \ / \ S_{i(u)}[j] = u,$$

which represents the number of tasks performed before the intended task u (u included) by the worker $i[u]$ it is affected to according to partition $\mathcal{A}$. A lower bound is provided in the paper [11].

Let us now suppose that the keys $1 \leq v \leq q$ are associated with offices that must perform a proportion ν_v of the incoming tasks which still follow a distribution μ. It actually appears as a sensible problem where a manager would have to distribute in advance tasks among teams according to the usual observed distribution of tasks and a list of available teams with their capacities.

Besides, we suppose that each team uses the strategy S_{margin} to perform tasks, meaning they randomly perform one according to their margin theoretical distribution; for a document signing, they randomly sign one.

From now on, we can rewrite our task partitioning problem under the form of a guessing problem:

- $V = v$, formerly corresponds to a worker, now it represents the information the spy has access to;
- $U = u$, formerly represents a task to perform, now it represents a sent message;
- $S = \sigma$, formerly represents the order in which tasks are performed, now it represents the order in which Charlie proposes his guesses.

Under this formalism, we are interested in measuring the probability $M(S, U, V)$ of executing u first as an extended application of the one-shot performance of Definition 3 and we have:

$$M(S, U, V) \geq \frac{||\pi||_2^2}{\max_{1 \leq v \leq q} \nu_v} \geq \frac{||\pi^+||_2^2}{\max_{1 \leq v \leq q} \nu_v}. \tag{17}$$

This inequality provides a lower bound for any distribution of the tasks among the team, no distribution can generate a worst "one-shot probability" of satisfying the intended task. In task partitioning actually, each task u is uniquely associated to a worker $v = i(u)$ so that the random variable representing the worker is deterministic conditionally to U: $\pi_{u,v} = \delta_{v=i(u)}$.

References

1. Bertrand, P.: Transport Optimal, Matrices de Monge et Pont relationnel. Ph.D. thesis, Paris 6 (2021)
2. Bertrand, P., Broniatowski, M., Marcotorchino, J.F.: Independence versus Indetermination: basis of two canonical clustering criteria. Working paper or preprint (2020). https://hal.archives-ouvertes.fr/hal-02901167
3. Bertrand, P., Broniatowski, M., Marcotorchino, J.F.: Logical indetermination coupling:a method to minimize drawing matches and its applications. Working paper or preprint (2020). https://hal.archives-ouvertes.fr/hal-03086553
4. Bertrand, P., Broniatowski, M., Marcotorchino, J.F.: Continuous indetermination and average likelihood minimization. Working paper or preprint (2021). https://hal.archives-ouvertes.fr/hal-03215096
5. Bunte, C., Lapidoth, A.: Encoding tasks and Rényi entropy. IEEE Trans. Inf. Theory **60**(9), 5065–5076 (2014). https://doi.org/10.1109/tit.2014.2329490
6. Burkard, R.E., Klinz, B., Rudolf, R.: Perspectives of Monge properties in optimization. Discrete Appl. Math. **70**, 95–161 (1996)
7. Csiszár, I., et al.: Why least squares and maximum entropy? an axiomatic approach to inference for linear inverse problems. Ann. Stat. **19**(4), 2032–2066 (1991)
8. Deming, W.E., Stephan, F.F.: On the least squares adjustment of a sampled frequency table when the expected marginal totals are known. Ann. Math. Stat. **11**, 427–444 (1940)
9. Fréchet, M.: Sur les tableaux de corrélations dont les marges sont données. Annales de l'Université de Lyon Sec. A. **14**, 53–77 (1951)
10. Fréchet, M.: Sur les tableaux de corrélations dont les marges et les bornes sont données. Revue de l'Institut de Statistique **28**, 10–32 (1960)
11. Kumar, M.A., Sunny, A., Thakre, A., Kumar, A.: A unified framework for problems on guessing, source coding and task partitioning (2019)
12. Marcotorchino, J.F.: Utilisation des comparaisons par paires en statistique des contingences. Publication du Centre Scientifique IBM de Paris et Cahiers du Séminaire Analyse des Données et Processus Stochastiques Université Libre de Bruxelles, pp. 1–57 (1984)
13. Massey, J.L.: Guessing and entropy. In: IEEE International Symposium on Information Thesis, p. 204 (1994)
14. Nielsen, F.: What is an information projection. Notices AMS **65**(3), 321–324 (2018)
15. Stemmelen, E.: Tableaux d'échanges, description et prévision. Cahiers du Bureau Universitaire de Recherche Opérationnelle **28** (1977)
16. Stummer, W.: Optimal transport with some directed distances. In: International Conference on Geometric Science of Information. Springer (2021)
17. Villani, C.: Optimal Transport. Old and New. Springer-Verlag, Berlin (2000). https://doi.org/10.1007/978-3-540-71050-9
18. Wilson, A.G.: A statistical theory of spatial distribution models. Transp. Res. **1**, 253–269 (1967)
19. Wilson, A.G.: Entropy in Urban and Regional Modelling, vol. 1. Routledge, Cambridge (2011)

Optimal Transport with Some Directed Distances

Wolfgang Stummer[1,2(✉)]

[1] Department of Mathematics, University of Erlangen–Nürnberg, Cauerstrasse 11, 91058 Erlangen, Germany
stummer@math.fau.de

[2] School of Business, Economics and Society, University of Erlangen–Nürnberg, Lange Gasse 20, 90403 Nürnberg, Germany

Abstract. We present a toolkit of directed distances between quantile functions. By employing this, we solve some new optimal transport (OT) problems which e.g. considerably flexibilize some prominent OTs expressed through Wasserstein distances.

Keywords: Scaled Bregman divergences · ϕ-divergences · Mass transport

1 Introduction

Widely used tools for various different tasks in statistics and probability (and thus, in the strongly connected fields of information theory, artificial intelligence and machine learning) are density-based "directed" (i.e. generally asymmetric) distances – called divergences – which measure the dissimilarity/proximity between two probability distributions; some comprehensive overviews can be found in the books of e.g. Liese and Vajda [19], Read and Cressie [30], Vajda [40], Csiszár and Shields [10], Stummer [36], Pardo [24], Liese and Miescke [18], Basu et al. [4]. Amongst others, some important density-based directed-distance classes are:

(1) the *Csiszar-Ali-Silvey-Morimoto ϕ-divergences* (CASM divergences) [1,8, 22]: this includes e.g. the total variation distance, exponentials of Renyi cross-entropies, and the power divergences; the latter cover e.g. the Kullback-Leibler information divergence (relative entropy), the (squared) Hellinger distance, the Pearson chi-square divergence;
(2) the *"classical" Bregman distances* (CB distances) (see e.g. Csiszar [9], Pardo and Vajda [25,26]): this includes e.g. the density-power divergences (cf. [3]) with the squared L_2-norm as special case.

More generally, Stummer [37] and Stummer and Vajda [38] introduced the concept of *scaled Bregman distances*, which enlarges and flexibilizes both the above-mentioned CASM and CB divergence/distance classes; their use for robustness

F. Nielsen and F. Barbaresco (Eds.): GSI 2021, LNCS 12829, pp. 829–840, 2021.
https://doi.org/10.1007/978-3-030-80209-7_89

of minimum-distance parameter estimation, testing as well as applications can be found e.g. in Kißlinger and Stummer [13–16], Roensch and Stummer [31–33], Krömer and Stummer [17]. An even *much wider framework of directed distances/divergences* (BS distances) was introduced in the recent comprehensive paper of Broniatowski and Stummer [6].

Another important omnipresent scientific concept is *optimal transport*; comprehensive general insights into this (and directly/closely related topics such as e.g. mass transportation problems, Wasserstein distances, optimal coupling problems, extremal copula problems, optimal assignment problems) can be found in the books of e.g. Rachev and Rüschendorf [29], Villani [41,42], Santambrogio [35], Peyre and Cuturi [27], and the survey paper of e.g. Ambrosio and Gigli [2].

In the light of the above explanations, the main goals of this paper are:

(i) To apply (for the sake of brevity, only some subset of) the BS distances to the context of *quantile functions* (see Sect. 2).
(ii) To establish a link between (i) and a new class of optimal transport problems where the cost functions are *pointwise* BS distances (see Sect. 3).

2 A Toolkit of Divergences Between Quantile Functions

By adapting and widening the concept of *scaled Bregman distances* of Stummer [37] and Stummer and Vajda [38], the recent paper of Broniatowski and Stummer [6] introduced a fairly universal, flexible, multi-component system of "directed distances" (which we abbreviate as *BS distances*) *between two arbitrary functions*; in the following, we apply (for the sake of brevity, only parts of) this to the important widely used context of quantile functions. We employ the following ingredients:

2.1 Quantile Functions

(I1) Let $\mathscr{X} :=]0,1[$ (open unit interval). For two probability distributions $\mathbb{P}$ and $\mathbb{Q}$ on the one-dimensional Euclidean space $\mathbb{R}$, we denote their cumulative distribution functions (cdf) as $F_{\mathbb{P}}$ and $F_{\mathbb{Q}}$, and write their quantile functions as

$$F_{\mathbb{P}}^{\leftarrow} := \{F_{\mathbb{P}}^{\leftarrow}(x)\}_{x\in\mathscr{X}} := \{\inf\{z \in \mathbb{R} : F_{\mathbb{P}}(z) \geq x\}\}_{x\in\mathscr{X}} ,$$
$$F_{\mathbb{Q}}^{\leftarrow} := \{F_{\mathbb{Q}}^{\leftarrow}(x)\}_{x\in\mathscr{X}} := \{\inf\{z \in \mathbb{R} : F_{\mathbb{Q}}(z) \geq x\}\}_{x\in\mathscr{X}} ;$$

if (say) $\mathbb{P}$ is concentrated on $[0,\infty[$ (i.e., the support of $\mathbb{P}$ is a subset of $[0,\infty[$), then we alternatively take (for our purposes, without loss of generality)

$$F_{\mathbb{P}}^{\leftarrow} := \{F_{\mathbb{P}}^{\leftarrow}(x)\}_{x\in\mathscr{X}} := \{\inf\{z \in [0,\infty[: F_{\mathbb{P}}(z) \geq x\}\}_{x\in\mathscr{X}}$$

which leads to the nonnegativity $F_{\mathbb{Q}}^{\leftarrow}(x) \geq 0$ for all $x \in \mathscr{X}$.

Of course, if the underlying cdf $z \to F_{\mathbb{P}}(z)$ is strictly increasing, then $x \to F_{\mathbb{P}}^{\leftarrow}(x)$ is nothing but its "classical" inverse function. Let us also mention that in quantitative finance and insurance, the quantile $F_{\mathbb{P}}^{\leftarrow}(x)$ (e.g. quoted in US dollars units) is called the value-at-risk for confidence level $x \cdot 100\%$. A detailed discussion on properties and pitfalls of univariate quantile functions can be found e.g. in Embrechts and Hofert [11].

2.2 Directed Distances—Basic Concept

We quantify the *dissimilarity* between the two quantile functions $F_{\mathbb{P}}^{\leftarrow}, F_{\mathbb{Q}}^{\leftarrow}$ in terms of *BS distances* $D_{\beta}^{c}(F_{\mathbb{P}}^{\leftarrow}, F_{\mathbb{Q}}^{\leftarrow})$ with $\beta = (\phi, M_1, M_2, M_3)$, defined by

$$
\begin{aligned}
&0 \le D^{c}_{\phi,M_1,M_2,M_3}(F_{\mathbb{P}}^{\leftarrow}, F_{\mathbb{Q}}^{\leftarrow}) \\
&:= \overline{\int}_{\mathscr{X}} \Big[\phi\Big(\tfrac{F_{\mathbb{P}}^{\leftarrow}(x)}{M_1(x)}\Big) - \phi\Big(\tfrac{F_{\mathbb{Q}}^{\leftarrow}(x)}{M_2(x)}\Big) - \phi'_{+,c}\Big(\tfrac{F_{\mathbb{Q}}^{\leftarrow}(x)}{M_2(x)}\Big) \cdot \Big(\tfrac{F_{\mathbb{P}}^{\leftarrow}(x)}{M_1(x)} - \tfrac{F_{\mathbb{Q}}^{\leftarrow}(x)}{M_2(x)}\Big)\Big] \cdot M_3(x)\,\mathrm{d}\lambda(x); \quad (1)
\end{aligned}
$$

the meaning of the integral symbol $\overline{\int}$ will become clear in (3) below. Here, in accordance with Broniatowski and Stummer [6] we use:

(I2) the Lebesgue measure λ (it is well known that in general an integral $\int \ldots \mathrm{d}\lambda(x)$ turns—except for rare cases—into a classical Riemann integral $\int \ldots \mathrm{d}x$).

(I3) (measurable) *scaling functions* $M_1 : \mathscr{X} \to [-\infty, \infty]$ and $M_2 : \mathscr{X} \to [-\infty, \infty]$ as well as a nonnegative (measurable) *aggregating function* $M_3 : \mathscr{X} \to [0, \infty]$ such that $M_1(x) \in]-\infty, \infty[$, $M_2(x) \in]-\infty, \infty[$, $M_3(x) \in [0, \infty[$ for λ-a.a. $x \in \mathscr{X}$. In analogy with the above notation, we use the symbols $M_i := \{M_i(x)\}_{x\in\mathscr{X}}$ to refer to the whole functions. In the following, $\mathscr{R}(G)$ denotes the range (image) of a function $G := \{G(x)\}_{x\in\mathscr{X}}$.

(I4) the so-called "divergence-generator" ϕ which is a continuous, convex (finite) function $\phi : E \to]-\infty, \infty[$ on some appropriately chosen open interval $E =]a, b[$ such that $[a, b]$ covers (at least) the union $\mathscr{R}(\frac{F_{\mathbb{P}}^{\leftarrow}}{M_1}) \cup \mathscr{R}(\frac{F_{\mathbb{Q}}^{\leftarrow}}{M_2})$ of both ranges $\mathscr{R}(\frac{F_{\mathbb{P}}^{\leftarrow}}{M_1})$ of $\{\frac{F_{\mathbb{P}}^{\leftarrow}(x)}{M_1(x)}\}_{x\in\mathscr{X}}$ and $\mathscr{R}(\frac{F_{\mathbb{Q}}^{\leftarrow}}{M_2})$ of $\{\frac{F_{\mathbb{Q}}^{\leftarrow}(x)}{M_2(x)}\}_{x\in\mathscr{X}}$; for instance, $E =]0,1[$, $E =]0, \infty[$ or $E =]-\infty, \infty[$; the class of all such functions will be denoted by $\Phi(]a, b[)$. Furthermore, we assume that ϕ is continuously extended to $\overline{\phi} : [a, b] \to [-\infty, \infty]$ by setting $\overline{\phi}(t) := \phi(t)$ for $t \in]a, b[$ as well as $\overline{\phi}(a) := \lim_{t\downarrow a} \phi(t)$, $\overline{\phi}(b) := \lim_{t\uparrow b} \phi(t)$ on the two boundary points $t = a$ and $t = b$. The latter two are the only points at which infinite values may appear. Moreover, for any fixed $c \in [0, 1]$ the (finite) function $\phi'_{+,c} :]a, b[\to]-\infty, \infty[$ is well-defined by $\phi'_{+,c}(t) := c \cdot \phi'_{+}(t) + (1 - c) \cdot \phi'_{-}(t)$, where $\phi'_{+}(t)$ denotes the (always finite) right-hand derivative of ϕ at the point $t \in]a, b[$ and $\phi'_{-}(t)$ the (always finite) left-hand derivative of ϕ at $t \in]a, b[$. If $\phi \in \Phi(]a, b[)$ is also continuously differentiable – which we denote by $\phi \in \Phi_{C_1}(]a, b[)$ – then for all $c \in [0, 1]$ one gets $\phi'_{+,c}(t) = \phi'(t)$ $(t \in]a, b[)$ and in such a situation we always suppress the obsolete indices $+$, c in the corresponding expressions. We also employ the continuous continuation $\overline{\phi'_{+,c}} : [a,b] \rightarrow [-\infty, \infty]$ given by $\overline{\phi'_{+,c}}(t) := \phi'_{+,c}(t)$ $(t \in]a,b[)$, $\overline{\phi'_{+,c}}(a) := \lim_{t\downarrow a} \phi'_{+,c}(t)$, $\overline{\phi'_{+,c}}(b) := \lim_{t\uparrow b} \phi'_{+,c}(t)$. To explain the precise meaning of (1), we also make use of the (finite, nonnegative) function $\psi_{\phi,c} :]a,b[\times]a,b[\rightarrow [0,\infty[$ given by $\psi_{\phi,c}(s,t) := \phi(s) - \phi(t) - \phi'_{+,c}(t) \cdot (s-t) \geq 0$ $(s,t \in]a,b[)$. To extend this to a lower semi-continuous function $\overline{\psi_{\phi,c}} : [a,b] \times [a,b] \rightarrow [0,\infty]$ we proceed as follows: firstly, we set $\overline{\psi_{\phi,c}}(s,t) := \psi_{\phi,c}(s,t)$ for all $s,t \in]a,b[$. Moreover, since for fixed $t \in]a,b[$, the function $s \rightarrow \psi_{\phi,c}(s,t)$ is convex and continuous, the limit $\overline{\psi_{\phi,c}}(a,t) := \lim_{s\rightarrow a} \psi_{\phi,c}(s,t)$ always exists and (in order to avoid overlines in (1)) will be interpreted/abbreviated as $\phi(a) - \phi(t) - \phi'_{+,c}(t) \cdot (a-t)$. Analogously, for fixed $t \in]a,b[$ we set $\overline{\psi_{\phi,c}}(b,t) := \lim_{s\rightarrow b} \psi_{\phi,c}(s,t)$ with corresponding short-hand notation $\phi(b) - \phi(t) - \phi'_{+,c}(t) \cdot (b-t)$. Furthermore, for fixed $s \in]a,b[$ we interpret $\phi(s) - \phi(a) - \phi'_{+,c}(a) \cdot (s-a)$ as

$$\overline{\psi_{\phi,c}}(s,a) := \left\{\phi(s) - \overline{\phi'_{+,c}}(a) \cdot s + \lim_{t\rightarrow a} \left(t \cdot \overline{\phi'_{+,c}}(a) - \phi(t)\right)\right\} \cdot \mathbf{1}_{]-\infty,\infty[}\left(\overline{\phi'_{+,c}}(a)\right) \\ + \infty \cdot \mathbf{1}_{\{-\infty\}}\left(\overline{\phi'_{+,c}}(a)\right),$$

where the involved limit always exists but may be infinite. Analogously, for fixed $s \in]a,b[$ we interpret $\phi(s) - \phi(b) - \phi'_{+,c}(b) \cdot (s-b)$ as

$$\overline{\psi_{\phi,c}}(s,b) := \left\{\phi(s) - \overline{\phi'_{+,c}}(b) \cdot s + \lim_{t\rightarrow b} \left(t \cdot \overline{\phi'_{+,c}}(b) - \phi(t)\right)\right\} \cdot \mathbf{1}_{]-\infty,\infty[}\left(\overline{\phi'_{+,c}}(b)\right) \\ + \infty \cdot \mathbf{1}_{\{+\infty\}}\left(\overline{\phi'_{+,c}}(b)\right),$$

where again the involved limit always exists but may be infinite. Finally, we always set $\overline{\psi_{\phi,c}}(a,a) := 0$, $\overline{\psi_{\phi,c}}(b,b) := 0$, and $\overline{\psi_{\phi,c}}(a,b) := \lim_{s\rightarrow a} \overline{\psi_{\phi,c}}(s,b)$, $\overline{\psi_{\phi,c}}(b,a) := \lim_{s\rightarrow b} \overline{\psi_{\phi,c}}(s,a)$. Notice that $\overline{\psi_{\phi,c}}(\cdot,\cdot)$ is lower semi-continuous but not necessarily continuous. Since ratios are ultimately involved, we also consistently take $\overline{\psi_{\phi,c}}\left(\frac{0}{0}, \frac{0}{0}\right) := 0$.

With (I1) to (I4), we define the *BS distance (BS divergence)* of (1) precisely as

$$0 \leq D^{c}_{\phi,M_1,M_2,M_3}\left(F^{\leftarrow}_{\mathbb{P}}, F^{\leftarrow}_{\mathbb{Q}}\right) = \overline{\int}_{\mathscr{X}} \psi_{\phi,c}\left(\frac{F^{\leftarrow}_{\mathbb{P}}(x)}{M_1(x)}, \frac{F^{\leftarrow}_{\mathbb{Q}}(x)}{M_2(x)}\right) \cdot M_3(x) \,\mathrm{d}\lambda(x) \tag{2}$$

$$:= \int_{\mathscr{X}} \overline{\psi_{\phi,c}}\left(\frac{F^{\leftarrow}_{\mathbb{P}}(x)}{M_1(x)}, \frac{F^{\leftarrow}_{\mathbb{Q}}(x)}{M_2(x)}\right) \cdot M_3(x) \,\mathrm{d}\lambda(x), \tag{3}$$

but mostly use the less clumsy notation with $\overline{\int}$ given in (1), (2) henceforth, as a shortcut for the implicitly involved boundary behaviour.

Notice that generally (with some exceptions) one has the asymmetry $D^{c}_{\phi,M_1,M_2,M_3}\left(F^{\leftarrow}_{\mathbb{P}}, F^{\leftarrow}_{\mathbb{Q}}\right) \neq D^{c}_{\phi,M_1,M_2,M_3}\left(F^{\leftarrow}_{\mathbb{Q}}, F^{\leftarrow}_{\mathbb{P}}\right)$

leading—together with the nonnegativity—to the (already used above) interpretation of $D^c_{\phi,M_1,M_2,M_3}(F_{\mathbb{P}}^{\leftarrow}, F_{\mathbb{Q}}^{\leftarrow})$ as a *candidate* of a "directed" distance/divergence. To make this *proper*, one needs to verify

(NNg) $D^c_{\phi,M_1,M_2,M_3}(F_{\mathbb{P}}^{\leftarrow}, F_{\mathbb{Q}}^{\leftarrow}) \geq 0$.
(REg) $D^c_{\phi,M_1,M_2,M_3}(F_{\mathbb{P}}^{\leftarrow}, F_{\mathbb{Q}}^{\leftarrow}) = 0$ if and only if $F_{\mathbb{P}}^{\leftarrow}(x) = F_{\mathbb{Q}}^{\leftarrow}(x)$ for λ-a.a. $x \in \mathscr{X}$.

As already indicated above, the nonnegativity (NNg) holds per construction. For the reflexivity (REg) one needs further assumptions. Indeed, in a more general context beyond quantile functions and the Lebesgue measure, Broniatowski and Stummer [6] gave conditions such that objects as in (2) and (3) satisfy (REg). We shall adapt this to the current special context, where for the sake of brevity, for the rest of this paper we shall always concentrate on the important adaptive subcase $M_1(x) := W(F_{\mathbb{P}}^{\leftarrow}(x), F_{\mathbb{Q}}^{\leftarrow}(x))$, $M_2(x) := M_1(x)$, $M_3(x) := W_3(F_{\mathbb{P}}^{\leftarrow}(x), F_{\mathbb{Q}}^{\leftarrow}(x))$, for some (measurable) functions $W : \mathscr{R}(F_{\mathbb{P}}^{\leftarrow}) \times \mathscr{R}(F_{\mathbb{Q}}^{\leftarrow}) \to [-\infty, \infty]$ and $W_3 : \mathscr{R}(F_{\mathbb{P}}^{\leftarrow}) \times \mathscr{R}(F_{\mathbb{Q}}^{\leftarrow}) \to [0, \infty]$. Accordingly, (1), (2) and (3) simplify to

$$\begin{aligned}
0 \leq\ & D^c_{\phi,W,W_3}(F_{\mathbb{P}}^{\leftarrow}, F_{\mathbb{Q}}^{\leftarrow}) := D^c_{\phi,W(F_{\mathbb{P}}^{\leftarrow},F_{\mathbb{Q}}^{\leftarrow}),W(F_{\mathbb{P}}^{\leftarrow},F_{\mathbb{Q}}^{\leftarrow}),W_3(F_{\mathbb{P}}^{\leftarrow},F_{\mathbb{Q}}^{\leftarrow})}(F_{\mathbb{P}}^{\leftarrow}, F_{\mathbb{Q}}^{\leftarrow}) \\
&= \overline{\int}_{\mathscr{X}} \Big[\phi\Big(\frac{F_{\mathbb{P}}^{\leftarrow}(x)}{W(F_{\mathbb{P}}^{\leftarrow}(x),F_{\mathbb{Q}}^{\leftarrow}(x))}\Big) - \phi\Big(\frac{F_{\mathbb{Q}}^{\leftarrow}(x)}{W(F_{\mathbb{P}}^{\leftarrow}(x),F_{\mathbb{Q}}^{\leftarrow}(x))}\Big) \\
&\quad -\phi'_{+,c}\Big(\frac{F_{\mathbb{Q}}^{\leftarrow}(x)}{W(F_{\mathbb{P}}^{\leftarrow}(x),F_{\mathbb{Q}}^{\leftarrow}(x))}\Big) \cdot \Big(\frac{F_{\mathbb{P}}^{\leftarrow}(x)}{W(F_{\mathbb{P}}^{\leftarrow}(x),F_{\mathbb{Q}}^{\leftarrow}(x))} - \frac{F_{\mathbb{Q}}^{\leftarrow}(x)}{W(F_{\mathbb{P}}^{\leftarrow}(x),F_{\mathbb{Q}}^{\leftarrow}(x))}\Big)\Big] \\
&\quad \cdot W_3(F_{\mathbb{P}}^{\leftarrow}(x), F_{\mathbb{Q}}^{\leftarrow}(x))\, \mathrm{d}\lambda(x) \qquad (4) \\
&=: \int_{\mathscr{X}} \overline{\Upsilon}_{\phi,c,W,W_3}(F_{\mathbb{P}}^{\leftarrow}(x), F_{\mathbb{Q}}^{\leftarrow}(x)) \cdot \mathrm{d}\lambda(x), \qquad (5)
\end{aligned}$$

where we employ $\overline{\Upsilon}_{\phi,c,W,W_3} : \mathscr{R}(F_{\mathbb{P}}^{\leftarrow}) \times \mathscr{R}(F_{\mathbb{Q}}^{\leftarrow}) \mapsto [0, \infty]$ defined by

$$\overline{\Upsilon}_{\phi,c,W,W_3}(u,v) := W_3(u,v) \cdot \overline{\psi_{\phi,c}}\Big(\frac{u}{W(u,v)}, \frac{v}{W(u,v)}\Big) \geq 0 \quad \text{with} \qquad (6)$$

$$\psi_{\phi,c}\Big(\frac{u}{W(u,v)}, \frac{v}{W(u,v)}\Big) := \Big[\phi\big(\tfrac{u}{W(u,v)}\big) - \phi\big(\tfrac{v}{W(u,v)}\big) - \phi'_{+,c}\big(\tfrac{v}{W(u,v)}\big) \cdot \big(\tfrac{u}{W(u,v)} - \tfrac{v}{W(u,v)}\big)\Big]. \qquad (7)$$

We give conditions for the validity of the crucial reflexivity in the following subsection; this may be skipped by the non-specialist (and the divergence expert).

2.3 Justification of Distance Properties

By construction, one gets for all $\phi \in \Phi(]a,b[)$ and all $c \in [0,1]$ the important assertion $D^c_{\phi,W,W_3}(F_{\mathbb{P}}^{\leftarrow}, F_{\mathbb{Q}}^{\leftarrow}) \geq 0$ with equality if $F_{\mathbb{P}}^{\leftarrow}(x) = F_{\mathbb{Q}}^{\leftarrow}(x)$ for λ-almost all $x \in \mathscr{X}$. As investigated in Broniatowski and Stummer [6], in order to get "sharp identifiability" (i.e. reflexivity) one needs further assumptions on $\phi \in \Phi(]a,b[)$, $c \in [0,1]$; for instance, if $\phi \in \Phi(]a,b[)$ is affine linear on the whole interval $]a,b[$ and W_3 is constant (say, 1), then $\overline{\Upsilon}_{\phi,c,W,W_3}$ takes the constant value 0, and hence $D^c_{\phi,W,W_3}(F_{\mathbb{P}}^{\leftarrow}, F_{\mathbb{Q}}^{\leftarrow}) = 0$ even in cases where $F_{\mathbb{P}}^{\leftarrow}(x) \neq F_{\mathbb{Q}}^{\leftarrow}(x)$ for λ-a.a. $x \in \mathscr{X}$. In order to avoid such and similar phenomena, we use the following

Assumption 1. *Let* $c \in [0,1]$, $\phi \in \Phi(]a,b[)$ *and*

$$\mathscr{R}\Big(\frac{F_{\mathbb{P}}^{\leftarrow}}{W\big(F_{\mathbb{P}}^{\leftarrow}, F_{\mathbb{Q}}^{\leftarrow}\big)}\Big) \cup \mathscr{R}\Big(\frac{F_{\mathbb{Q}}^{\leftarrow}}{W\big(F_{\mathbb{P}}^{\leftarrow}, F_{\mathbb{Q}}^{\leftarrow}\big)}\Big) \subset [a,b].$$

Moreover, for all $s \in \mathscr{R}\Big(\frac{F_{\mathbb{P}}^{\leftarrow}}{W\big(F_{\mathbb{P}}^{\leftarrow}, F_{\mathbb{Q}}^{\leftarrow}\big)}\Big)$, *all* $t \in \mathscr{R}\Big(\frac{F_{\mathbb{Q}}^{\leftarrow}}{W\big(F_{\mathbb{P}}^{\leftarrow}, F_{\mathbb{Q}}^{\leftarrow}\big)}\Big)$, *all* $u \in \mathscr{R}(F_{\mathbb{P}}^{\leftarrow})$ *and all* $v \in \mathscr{R}(F_{\mathbb{Q}}^{\leftarrow})$ *let the following conditions hold:*

(a) ϕ *is strictly convex at* t;
(b) if ϕ *is differentiable at* t *and* $s \neq t$, *then* ϕ *is not affine-linear on the interval* $[\min(s,t), \max(s,t)]$ *(i.e. between* t *and* s*);*
(c) if ϕ *is not differentiable at* t, $s > t$ *and* ϕ *is affine linear on* $[t,s]$, *then we exclude* $c = 1$ *for the ("globally/universally chosen") subderivative* $\phi'_{+,c}(\cdot) = c \cdot \phi'_{+}(\cdot) + (1-c) \cdot \phi'_{-}(\cdot)$;
(d) if ϕ *is not differentiable at* t, $s < t$ *and* ϕ *is affine linear on* $[s,t]$, *then we exclude* $c = 0$ *for* $\phi'_{+,c}(\cdot)$;
(e) $W_3(u,v) < \infty$;
(f) $W_3(u,v) > 0$ *if* $u \neq v$;
(g) by employing (with a slight abuse of notation) the function $\Upsilon(u,v) := W_3(u,v) \cdot \psi_{\phi,c}\Big(\frac{u}{W(u,v)}, \frac{v}{W(u,v)}\Big)$, *we set by convention* $\Upsilon(u,v) := 0$ *if* $\frac{u}{W(u,v)} = \frac{v}{W(u,v)} = a$;
(h) by convention, $\Upsilon(u,v) := 0$ *if* $\frac{u}{W(u,v)} = \frac{v}{W(u,v)} = b$;
(i) $\Upsilon(u,v) > 0$ *if* $\frac{u}{W(u,v)} = a$ *and* $\frac{v}{W(u,v)} = t \notin \{a,b\}$; *this is understood in the following way: for* $\frac{v}{W(u,v)} = t \notin \{a,b\}$, *we require* $\lim_{\frac{u}{W(u,v)} \to a} \Upsilon(u,v) > 0$ *if this limit exists, or otherwise we set by convention* $\Upsilon(u,v) := 1$ *(or any other strictly positive constant) if* $\frac{u}{W(u,v)} = a$; *the following boundary-behaviour conditions have to be interpreted analogously;*
(j) $\Upsilon(u,v) > 0$ *if* $\frac{u}{W(u,v)} = b$ *and* $\frac{v}{W(u,v)} = t \notin \{a,b\}$;
(k) $\Upsilon(u,v) > 0$ *if* $\frac{v}{W(u,v)} = a$ *and* $\frac{u}{W(u,v)} = s \notin \{a,b\}$;
(ℓ) $\Upsilon(u,v) > 0$ *if* $\frac{v}{W(u,v)} = b$ *and* $\frac{u}{W(u,v)} = s \notin \{a,b\}$;
(m) $\Upsilon(u,v) > 0$ *if* $\frac{v}{W(u,v)} = b$ *and* $\frac{u}{W(u,v)} = a$ *(as limit from (ℓ) or by convention);*
(n) $\Upsilon(u,v) > 0$ *if* $\frac{v}{W(u,v)} = a$ *and* $\frac{u}{W(u,v)} = b$ *(as limit from (k) or by convention).*

Remark 1. We could even work with a weaker assumption obtained by replacing s with $\frac{F_{\mathbb{P}}^{\leftarrow}(x)}{W\big(F_{\mathbb{P}}^{\leftarrow}(x), F_{\mathbb{Q}}^{\leftarrow}(x)\big)}$, t with $\frac{F_{\mathbb{Q}}^{\leftarrow}(x)}{W\big(F_{\mathbb{P}}^{\leftarrow}(x), F_{\mathbb{Q}}^{\leftarrow}(x)\big)}$, u with $F_{\mathbb{P}}^{\leftarrow}(x)$, v with $F_{\mathbb{Q}}^{\leftarrow}(x)$, and by requiring that then the correspondingly plugged-in conditions (a) to (n) hold for λ-a.a. $x \in \mathscr{X}$.

The following requirement is stronger than the "model-individual/dependent" Assumption 1 but is more "universally applicable":

Assumption 2. *Let* $c \in [0,1]$, $\phi \in \Phi(]a,b[)$ *on some fixed* $]a,b[\in]-\infty,+\infty[$

$$\text{such that} \quad \mathscr{R}\Big(\frac{F_{\mathbb{P}}^{\leftarrow}}{W\big(F_{\mathbb{P}}^{\leftarrow},F_{\mathbb{Q}}^{\leftarrow}\big)}\Big) \cup \mathscr{R}\Big(\frac{F_{\mathbb{Q}}^{\leftarrow}}{W\big(F_{\mathbb{P}}^{\leftarrow},F_{\mathbb{Q}}^{\leftarrow}\big)}\Big) \subset]a,b[.$$

Moreover, for all $s \in]a,b[$, *all* $t \in]a,b[$, *all* $u \in \mathscr{R}(F_{\mathbb{P}}^{\leftarrow})$ *and all* $v \in \mathscr{R}(F_{\mathbb{Q}}^{\leftarrow})$ *the conditions (a) to (n) of Assumption 1 hold.*

By adapting Theorem 4 and Corollary 1 of Broniatwoski and Stummer [6], under Assumption 1 (and hence, under Assumption 2) we obtain

(NN) $D^{c}_{\phi,W,W_3}\big(F_{\mathbb{P}}^{\leftarrow},F_{\mathbb{Q}}^{\leftarrow}\big) \geq 0.$

(RE) $D^{c}_{\phi,W,W_3}\big(F_{\mathbb{P}}^{\leftarrow},F_{\mathbb{Q}}^{\leftarrow}\big) = 0$ if and only if $F_{\mathbb{P}}^{\leftarrow}(x) = F_{\mathbb{Q}}^{\leftarrow}(x)$ for λ-a.a. $x \in \mathscr{X}$.

The non-negativity (NN) and the reflexivity (RE) say that $D^{c}_{\phi,W,W_3}\big(F_{\mathbb{P}}^{\leftarrow},F_{\mathbb{Q}}^{\leftarrow}\big)$ is indeed a "proper" divergence under Assumption 1 (and hence, under Assumption 2). Thus, the latter will be assumed for the rest of the paper.

3 New Optimal Transport Problems

For our applications to optimal transport, we impose henceforth the additional requirement that the *nonnegative* (extended) function $\overline{\Upsilon}_{\phi,c,W,W_3}$ is continuous and quasi-antitone[1] in the sense

$$\overline{\Upsilon}_{\phi,c,W,W_3}(u_1,v_1) + \overline{\Upsilon}_{\phi,c,W,W_3}(u_2,v_2) \leq \overline{\Upsilon}_{\phi,c,W,W_3}(u_2,v_1) + \overline{\Upsilon}_{\phi,c,W,W_3}(u_1,v_2)$$
$$\text{for all } u_1 \leq u_2,\ v_1 \leq v_2; \qquad (8)$$

in other words, $-\overline{\Upsilon}_{\phi,c,W,W_3}(\cdot,\cdot)$ is assumed to be continuous and quasi-monotone[2,3]. For such a setup, we consider the *novel* Kantorovich transportation problem (KTP) with the *pointwise-BS-distance-type* (pBS-type) cost function $\overline{\Upsilon}_{\phi,c,W,W_3}(u,v)$; indeed, we obtain the following

Theorem 3. *Let* $\widetilde{\Gamma}(\mathbb{P},\mathbb{Q})$ *be the family of all probability distributions* $\mathfrak{P}$ *on* $\mathbb{R} \times \mathbb{R}$ *which have marginal distributions* $\mathfrak{P}[\cdot \times \mathbb{R}] = \mathbb{P}[\cdot]$ *and* $\mathfrak{P}[\mathbb{R} \times \cdot] = \mathbb{Q}[\cdot]$. *Moreover, we denote the corresponding upper Hoeffding-Fréchet bound (cf. e.g.*

[1] Other names are: submodular, Lattice-subadditive, 2-antitone, 2-negative, Δ-antitone, supernegative, "satisfying the (continuous) Monge property/condition".

[2] Other names are: supermodular, Lattice-superadditive, 2-increasing, 2-positive, Δ-monotone, 2-monotone, "fulfilling the moderate growth property", "satisfying the measure property", "satisfying the twist condition".

[3] A comprehensive discussion on general quasi-monotone functions can be found e.g. in Chap. 6.C of Marshall et al. [21].

Theorem 3.1.1 of Rachev and Rüschendorf [29]) by $\mathfrak{P}^{com}$ *having "comonotonic" distribution function* $F_{\mathfrak{P}^{com}}(u,v) := \min\{F_{\mathbb{P}}(u), F_{\mathbb{Q}}(v)\}$ $(u,v \in \mathbb{R})$*. Then*

$$\min_{\{X\sim\mathbb{P},\,Y\sim\mathbb{Q}\}} \mathbb{E}\big[\,\overline{\Upsilon}_{\phi,c,W,W_3}(X,Y)\big] \tag{9}$$

$$= \min_{\{\mathfrak{P}\in\widetilde{\Gamma}(\mathbb{P},\mathbb{Q})\}} \int_{\mathbb{R}\times\mathbb{R}} \overline{\Upsilon}_{\phi,c,W,W_3}(u,v)\,\mathrm{d}\mathfrak{P}(u,v) \tag{10}$$

$$= \int_{\mathbb{R}\times\mathbb{R}} \overline{\Upsilon}_{\phi,c,W,W_3}(u,v)\,\mathrm{d}\mathfrak{P}^{com}(u,v) \tag{11}$$

$$= \int_{[0,1]} \overline{\Upsilon}_{\phi,c,W,W_3}(F_{\mathbb{P}}^{\leftarrow}(x), F_{\mathbb{Q}}^{\leftarrow}(x))\,\mathrm{d}\lambda(x) \tag{12}$$

$$= \overline{\int}_{]0,1[}\Big[\phi\Big(\tfrac{F_{\mathbb{P}}^{\leftarrow}(x)}{W(F_{\mathbb{P}}^{\leftarrow}(x),F_{\mathbb{Q}}^{\leftarrow}(x))}\Big) - \phi\Big(\tfrac{F_{\mathbb{Q}}^{\leftarrow}(x)}{W(F_{\mathbb{P}}^{\leftarrow}(x),F_{\mathbb{Q}}^{\leftarrow}(x))}\Big)$$

$$-\phi'_{+,c}\Big(\tfrac{F_{\mathbb{Q}}^{\leftarrow}(x)}{W(F_{\mathbb{P}}^{\leftarrow}(x),F_{\mathbb{Q}}^{\leftarrow}(x))}\Big)\cdot\Big(\tfrac{F_{\mathbb{P}}^{\leftarrow}(x)}{W(F_{\mathbb{P}}^{\leftarrow}(x),F_{\mathbb{Q}}^{\leftarrow}(x))} - \tfrac{F_{\mathbb{Q}}^{\leftarrow}(x)}{W(F_{\mathbb{P}}^{\leftarrow}(x),F_{\mathbb{Q}}^{\leftarrow}(x))}\Big)\Big]$$

$$\cdot W_3(F_{\mathbb{P}}^{\leftarrow}(x), F_{\mathbb{Q}}^{\leftarrow}(x))\,\mathrm{d}\lambda(x) \tag{13}$$

$$= D^{c}_{\phi,W,W_3}(F_{\mathbb{P}}^{\leftarrow}, F_{\mathbb{Q}}^{\leftarrow}) \geq 0, \tag{14}$$

where the minimum in (9) *is taken over all* $\mathbb{R}$*-valued random variables* X, Y *(on an arbitrary probability space* $(\Omega,\mathscr{A},\mathfrak{S})$*) such that* $\mathfrak{P}[X\in\cdot\,] = \mathbb{P}[\,\cdot\,]$, $\mathfrak{P}[Y\in\cdot\,] = \mathbb{Q}[\,\cdot\,]$. *As usual,* $\mathbb{E}$ *denotes the expectation with respect to* $\mathfrak{P}$.

The assertion (11) follows by applying Corollary 2.2a of Tchen [39] (see also – with different regularity conditions and generalizations – Cambanis et al. [7], Rüschendorf [34], Theorem 3.1.2 of Rachev and Rüschendorf [29], Theorem 3.8.2 of Müller and Stoyan [23], Theorem 2.5 of Puccetti and Scarsini [28], Theorem 2.5 of Ambrosio and Gigli [2]).

Remark 2. (i) Notice that $\mathfrak{P}^{com}$ is $\overline{\Upsilon}_{\phi,c,W,W_3}$-independent, and may not be the unique minimizer in (10). As a (not necessarily unique) minimizer in (9), one can take $X := F_{\mathbb{P}}^{\leftarrow}(U)$, $Y := F_{\mathbb{Q}}^{\leftarrow}(U)$ for some uniform random variable U on $[0,1]$.
(ii) In Theorem 3 we have shown that $\mathfrak{P}^{com}$ (cf. (11)) is an optimal transport plan of the KTP (10) with the *pointwise-BS-distance-type* (pBS-type) cost function $\overline{\Upsilon}_{\phi,c,W,W_3}(u,v)$. The outcoming minimal value is equal to $D^{c}_{\phi,W,W_3}(F_{\mathbb{P}}^{\leftarrow}, F_{\mathbb{Q}}^{\leftarrow})$ which is typically straightforward to compute (resp. approximate).

Remark 2(ii) generally contrasts to those prominently used KTP whose cost function is a power $d(u,v)^p$ of a metric $d(u,v)$ (denoted as POM-type cost function) which leads to the well-known Wasserstein distances. (Apart from technicalities) There are some overlaps, though:

Example 1. (i) Take the *non-smooth* $\phi(t) := \phi_{TV}(t) := |t-1|$ $(t\in\mathbb{R})$, $c = \frac{1}{2}$, $W(u,v) := v$, $W_3(u,v) := |v|$ to obtain $\overline{\Upsilon}_{\phi_{TV},1/2,W,W_3}(u,v) = |u-v| =: d(u,v)$.
(ii) Take $\phi(t) := \phi_2(t) := \frac{(t-1)^2}{2}$ $(t\in\mathbb{R}$, with obsolete $c)$, $W(u,v) := 1$ and $W_3(u,v) := 1$ to end up with $\overline{\Upsilon}_{\phi_2,c,W,W_3}(u,v) = \frac{(u-v)^2}{2} = \frac{d(u,v)^2}{2}$.
(iii) The *symmetric* distances $d(u,v)$ and $\frac{d(u,v)^2}{2}$ are convex functions of $u-v$ and thus continuous quasi-antitone functions on $\mathbb{R}\times\mathbb{R}$. The correspondingly

outcoming Wasserstein distances are thus considerably flexibilized by our new much more general distance $D^c_{\phi,W,W_3}(F_{\mathbb{P}}^{\leftarrow}, F_{\mathbb{Q}}^{\leftarrow})$ of (14).

Depending on the chosen divergence, one may have to restrict the support of $\mathbb{P}$ respectively $\mathbb{Q}$, for instance to $[0,\infty[$. We give some further special cases of pBS-type cost functions, which are continuous and quasi-antitone, but which are generally not symmetric and thus not of POM-type:

Example 2. "smooth" pointwise *Csiszar-Ali-Silvey-Morimoto divergences* (CASM divergences): take $\phi : [0,\infty[\mapsto \mathbb{R}$ to be a strictly convex, twice continuously differentiable function on $]0,\infty[$ with continuous extension on $t = 0$, together with $W(u,v) := v$, $W_3(u,v) := v$ ($v \in]0,\infty[$) and c is obsolete. Accordingly, $\Upsilon_{\phi,c,W,W_3}(u,v) := v \cdot \phi(\frac{u}{v}) - v \cdot \phi(1) - \phi'(1) \cdot (u-v)$, and hence the second mixed derivative satisfies $\frac{\partial^2 \Upsilon_{\phi,c,W,W_3}(u,v)}{\partial u \partial v} = -\frac{u}{v^2}\phi''(\frac{u}{v}) < 0$ ($u, v \in]0,\infty[$); thus, Υ_{ϕ,c,W,W_3} is quasi-antitone on $]0,\infty[\times]0,\infty[$. Accordingly, (9) to (13) applies to (such kind of) CASM divergences concerning $\mathbb{P},\mathbb{Q}$ having support in $[0,\infty[$. As an example, take e.g. the power function $\phi(t) := \frac{t^\gamma - \gamma \cdot t + \gamma - 1}{\gamma \cdot (\gamma - 1)}$ ($\gamma \in \mathbb{R}\backslash\{0,1\}$). A different connection between optimal transport and other kind of CASM divergences can be found in Bertrand et al. [5] in the current GSI2021 volume.

Example 3. "smooth" pointwise *classical (i.e. unscaled) Bregman divergences* (CBD): take $\phi : \mathbb{R} \mapsto \mathbb{R}$ to be a strictly convex, twice continuously differentiable function $W(u,v) := 1$, $W_3(u,v) := 1$, and c is obsolete. Accordingly, $\Upsilon_{\phi,c,W,W_3}(u,v) := \phi(u) - \phi(v) - \phi'(v) \cdot (u-v)$ and hence $\frac{\partial^2 \Upsilon_{\phi,c,W,W_3}(u,v)}{\partial u \partial v} = -\phi''(v) < 0$ ($u, v \in \mathbb{R}$); thus, Υ_{ϕ,c,W,W_3} is quasi-antitone on $\mathbb{R} \times \mathbb{R}$. Accordingly, the representation (9) to (13) applies to (such kind of) CBD. The corresponding special case of (10) is called "a relaxed Wasserstein distance (parameterized by ϕ) between $\mathbb{P}$ and $\mathbb{Q}$" in the recent papers of Lin et al. [20] and Guo et al. [12] for a *restrictive* setup where $\mathbb{P}$ and $\mathbb{Q}$ are supposed to have *compact* support; the latter two references do not give connections to divergences of quantile functions, but substantially concentrate on applications to topic sparsity for analyzing user-generated web content and social media, respectively, to Generative Adversarial Networks (GANs).

Example 4. "smooth" pointwise *Scaled Bregman Distances*: for instance, consider $\mathbb{P}$ and $\mathbb{Q}$ with support in $[0,\infty[$. Under $W = W_3$ one gets that $\Upsilon_{\phi,c,W,W}$ is quasi-antitone on $]0,\infty[\times]0,\infty[$ if the generator function ϕ is strictly convex and thrice continuously differentiable on $]0,\infty[$ (and hence, c is obsolete) and the so-called scale connector W is twice continuously differentiable such that – on $]0,\infty[\times]0,\infty[$ – $\Upsilon_{\phi,c,W,W}$ is twice continuously differentiable and $\frac{\partial^2 \Upsilon_{\phi,c,W,W}(u,v)}{\partial u \partial v} \leq 0$ (an explicit formula of the latter is given in the appendix of Kißlinger and Stummer [16]). Illustrative examples of suitable ϕ and W can be found e.g. in Kißlinger and Stummer [15].

Returning to the general context, it is straightforward to see that if $\mathbb{P}$ does not give mass to points (i.e. it has continuous distribution function $F_{\mathbb{P}}$) then there exists even a deterministic optimal transportation plan: indeed, for the map $T^{com} := F_{\mathbb{Q}}^{\leftarrow} \circ F_{\mathbb{P}}$ one has $\mathfrak{P}^{com}[\,\cdot\,] = \mathbb{P}[(id, T^{com}) \in \cdot\,]$ and thus (11) is equal to

$$\int_{\mathbb{R}} \overline{\Upsilon}_{\phi,c,W,W_3}(u, T^{com}(u))\, \mathrm{d}\mathbb{P}(u) \tag{15}$$

$$= \min_{\{T \in \widehat{\Gamma}(\mathbb{P},\mathbb{Q})\}} \int_{\mathbb{R}} \overline{\Upsilon}_{\phi,c,W,W_3}(u, T(u))\, \mathrm{d}\mathbb{P}(u) \tag{16}$$

$$= \min_{\{X \sim \mathbb{P},\, T(X) \sim \mathbb{Q}\}} \mathbb{E}\big[\, \overline{\Upsilon}_{\phi,c,W,W_3}(X, T(X))\big] \tag{17}$$

where (16) is called Monge transportation problem (MTP). Here, $\widehat{\Gamma}(\mathbb{P}, \mathbb{Q})$ denotes the family of all measurable maps $T : \mathbb{R} \mapsto \mathbb{R}$ such that $\mathbb{P}[T \in \cdot\,] = \mathbb{Q}[\,\cdot\,]$.

Acknowledgement. I am grateful to the four referees for their comments and suggestions on readability improvements.

References

1. Ali, M.S., Silvey, D.: A general class of coefficients of divergence of one distribution from another. J. Roy. Stat. Soc. B **28**, 131–140 (1966)
2. Ambrosio, L., Gigli, N.: A user's guide to optimal transport. In: Ambrosio, L., et al. (eds.) Modeling and Optimisation of Flows on Networks. LNM, vol. 2062, pp. 1–155. Springer, Heidelberg (2013). https://doi.org/10.1007/978-3-642-32160-3_1
3. Basu, A., Harris, I.R., Hjort, N.L., Jones, M.C.: Robust and efficient estimation by minimizing a density power divergence. Biometrika **85**(3), 549–559 (1998)
4. Basu, A., Shioya, H., Park, C.: Statistical Inference: The Minimum Distance Approach. CRC Press, Boca Raton (2011)
5. Bertrand, P., Broniatowski, M., Marcotorchino, J.-F.: Minimization with respect to divergences and applications. In: Nielsen, F., Barbaresco, F. (eds.) GSI 2021. LNCS, vol. 12829, pp. 818–828. Springer, Cham (2021)
6. Broniatowski, M., Stummer, W.: Some universal insights on divergences for statistics, machine learning and artificial intelligence. In: Nielsen, F. (ed.) Geometric Structures of Information. SCT, pp. 149–211. Springer, Cham (2019). https://doi.org/10.1007/978-3-030-02520-5_8
7. Cambanis, S., Simons, G., Stout, W.: Inequalities for Ek(X, Y) when the marginals are fixed. Probab. Theory Rel. Fields **36**, 285–294 (1976)
8. Csiszar, I.: Eine informationstheoretische Ungleichung und ihre Anwendung auf den Beweis der Ergodizität von Markoffschen Ketten. Publ. Math. Inst. Hungar. Acad. Sci. A **8**, 85–108 (1963)
9. Csiszar, I.: Why least squares and maximum entropy? An axiomatic approach to inference for linear inverse problems. Ann. Stat. **19**(4), 2032–2066 (1991)
10. Csiszar, I., Shields, P.C.: Information Theory and Statistics: A Tutorial. Now Publishers, Hanover (2004)
11. Embrechts, P., Hofert, M.: A note on generalized inverses. Math. Meth. Oper. Res. **77**, 423–432 (2013)
12. Guo, X., Hong, J., Lin, T., Yang, N.: Relaxed Wasserstein with application to GANs. arXiv:1705.07164v7, February 2021

13. Kißlinger, A.-L., Stummer, W.: Some decision procedures based on scaled Bregman distance surfaces. In: Nielsen, F., Barbaresco, F. (eds.) GSI 2013. LNCS, vol. 8085, pp. 479–486. Springer, Heidelberg (2013). https://doi.org/10.1007/978-3-642-40020-9_52
14. Kißlinger, A.-L., Stummer, W.: New model search for nonlinear recursive models, regressions and autoregressions. In: Nielsen, F., Barbaresco, F. (eds.) GSI 2015. LNCS, vol. 9389, pp. 693–701. Springer, Cham (2015). https://doi.org/10.1007/978-3-319-25040-3_74
15. Kißlinger, A.-L., Stummer, W.: Robust statistical engineering by means of scaled Bregman distances. In: Agostinelli, C., Basu, A., Filzmoser, P., Mukherjee, D. (eds.) Recent Advances in Robust Statistics: Theory and Applications, pp. 81–113. Springer, New Delhi (2016). https://doi.org/10.1007/978-81-322-3643-6_5
16. Kißlinger, A.-L., Stummer, W.: A new toolkit for robust distributional change detection. Appl. Stochastic Models Bus. Ind. **34**, 682–699 (2018)
17. Krömer, S., Stummer, W.: A new toolkit for mortality data analytics. In: Steland, A., Rafajłowicz, E., Okhrin, O. (eds.) SMSA 2019. SPMS, vol. 294, pp. 393–407. Springer, Cham (2019). https://doi.org/10.1007/978-3-030-28665-1_30
18. Liese, F., Miescke, K.J.: Statistical Decision Theory: Estimation, Testing, and Selection. Springer, New York (2008). https://doi.org/10.1007/978-0-387-73194-0
19. Liese, F., Vajda, I.: Convex Statistical Distances. Teubner, Leipzig (1987)
20. Lin, T., Hu, Z., Guo, X.: Sparsemax and relaxed Wasserstein for topic sparsity. In: The Twelfth ACM International Conference on Web Search and Data Mining (WSDM 2019), pp. 141–149. ACM, New York (2019). https://doi.org/10.1145/3289600.3290957
21. Marshall, A.W., Olkin, I., Arnold, B.C.: Inequalities: Theory of Majorization and Its Applications, 2nd edn. SSS, Springer, New York (2011). https://doi.org/10.1007/978-0-387-68276-1
22. Morimoto, T.: Markov processes and the H-theorem. J. Phys. Soc. Jpn. **18**(3), 328–331 (1963)
23. Müller, A., Stoyan, D.: Comparison Methods for Stochastic Models and Risks. Wiley, Chichester (2002)
24. Pardo, L.: Statistical Inference Based on Divergence Measures. Chapman & Hall/CRC, Boca Raton (2006)
25. Pardo, M.C., Vajda, I.: About distances of discrete distributions satisfying the data processing theorem of information theory. IEEE Trans. Inf. Theor. **43**(4), 1288–1293 (1997)
26. Pardo, M.C., Vajda, I.: On asymptotic properties of information-theoretic divergences. IEEE Trans. Inf. Theor. **49**(7), 1860–1868 (2003)
27. Peyre, G., Cuturi, M.: Computational optimal transport: with applications to data science. Found. Trends Mach. Learn. **11**(5–6), 355–607 (2019). Also appeared in book form at now Publishers, Hanover (2019)
28. Puccetti, G., Scarsini, M.: Multivariate comonotonicity. J. Multiv. Anal. **101**, 291–304 (2010)
29. Rachev, S.T., Rüschendorf, L.: Mass Transportation Problems Vol.I,II. Springer, New York (1998). https://doi.org/10.1007/b98893, https://doi.org/10.1007/b98894
30. Read, T.R.C., Cressie, N.A.C.: Goodness-of-Fit Statistics for Discrete Multivariate Data. Springer, New York (1988). https://doi.org/10.1007/978-1-4612-4578-0

31. Roensch, B., Stummer, W.: 3D insights to some divergences for robust statistics and machine learning. In: Nielsen, F., Barbaresco, F. (eds.) GSI 2017. LNCS, vol. 10589, pp. 460–469. Springer, Cham (2017). https://doi.org/10.1007/978-3-319-68445-1_54
32. Roensch, B., Stummer, W.: Robust estimation by means of scaled Bregman power distances. Part I. non-homogeneous data. In: Nielsen, F., Barbaresco, F. (eds.) GSI 2019. LNCS, vol. 11712, pp. 319–330. Springer, Cham (2019). https://doi.org/10.1007/978-3-030-26980-7_33
33. Roensch, B., Stummer, W.: Robust estimation by means of scaled Bregman power distances. Part II. extreme values. In: Nielsen, F., Barbaresco, F. (eds.) GSI 2019. LNCS, vol. 11712, pp. 331–340. Springer, Cham (2019). https://doi.org/10.1007/978-3-030-26980-7_34
34. Rüschendorf, L.: Solution of a statistical optimization problem by rearrangement methods. Metrika **30**(1), 55–61 (1983). https://doi.org/10.1007/BF02056901
35. Santambrogio, F.: Optimal Transport for Applied Mathematicians. Birkhäuser, Cham (2015)
36. Stummer, W.: Exponentials, Diffusions, Finance, Entropy and Information. Shaker, Aachen (2004)
37. Stummer, W.: Some Bregman distances between financial diffusion processes. Proc. Appl. Math. Mech. **7**(1), 1050503–1050504 (2007)
38. Stummer, W., Vajda, I.: On Bregman distances and divergences of probability measures. IEEE Trans. Inform. Theory **58**(3), 1277–1288 (2012)
39. Tchen, A.H.: Inequalities for distributions with given marginals. Ann. Probab. **8**(4), 814–827 (1980)
40. Vajda, I.: Theory of Statistical Inference and Information. Kluwer, Dordrecht (1989)
41. Villani, C.: Topics in Optimal Transportation. American Mathematical Society, Providence (2003)
42. Villani, C.: Optimal Transport, Old and New. Springer, Heidelberg (2009). https://doi.org/10.1007/978-3-540-71050-9

Robust Empirical Likelihood

Amor Keziou[1(✉)] and Aida Toma[2,3]

[1] Université de Reims Champagne-Ardenne, Laboratoire de Mathématiques de Reims (LMR) - CNRS UMR 9008, U.F.R. Sciences Exactes et Naturelles Moulin de la Housse, BP 1039, 51687 Reims Cedex 2, France
amor.keziou@univ-reims.fr

[2] Department of Applied Mathematics, Bucharest University of Economic Studies, Piaţa Romană No. 6, Bucharest, Romania
aida.toma@csie.ase.ro

[3] Gheorghe Mihoc-Caius Iacob Institute of Mathematical Statistics and Applied Mathematics of the Romanian Academy, Bucharest, Romania
http://www.univ-reims.fr/labo-maths

Abstract. In this paper, we present a robust version of the empirical likelihood estimator for semiparametric moment condition models. This estimator is obtained by minimizing the modified Kullback-Leibler divergence, in its dual form, using truncated orthogonality functions. Some asymptotic properties regarding the limit laws of the estimators are stated.

Keywords: Moment condition models · Robustness · Divergence measures

1 Introduction

We consider a moment condition model, namely a family $\mathcal{M}$ of probability measures Q, all defined on the same measurable space $(\mathbb{R}^m, \mathcal{B}(\mathbb{R}^m))$, such that $\int_{\mathbb{R}^m} g(x,\theta)\, dQ(x) = 0$. The unknown parameter θ belongs to the interior of a compact set $\Theta \subset \mathbb{R}^d$, and the function $g := (g_1, \ldots, g_\ell)^\top$, with $\ell \geq d$, is defined on the set $\mathbb{R}^m \times \Theta$, each g_i being a real valued function. Denoting by M the set of all probability measures on $(\mathbb{R}^m, \mathcal{B}(\mathbb{R}^m))$ and defining the sets

$$\mathcal{M}_\theta := \left\{ Q \in M \text{ s.t. } \int_{\mathbb{R}^m} g(x,\theta)\, dQ(x) = 0 \right\},\ \theta \in \Theta,$$

then the moment condition model $\mathcal{M}$ can be written under the form

$$\mathcal{M} = \bigcup_{\theta \in \Theta} \mathcal{M}_\theta. \tag{1}$$

This work was supported by a grant of the Ministry of Research, Innovation and Digitization, CNCS/CCCDI – UEFISCDI, project number PN-III-P4-ID-PCE-2020-1112, within PNCDI III.

F. Nielsen and F. Barbaresco (Eds.): GSI 2021, LNCS 12829, pp. 841–848, 2021.
https://doi.org/10.1007/978-3-030-80209-7_90

Let $X_1, \ldots, X_n$ be an i.i.d. sample with unknown probability measure P_0. We assume that the equation $\int_{\mathbb{R}^m} g(x,\theta)\, dP_0(x) = 0$ has a unique solution (in θ) which will be denoted θ_0. We consider the estimation problem of the true unknown value θ_0.

Among the most known methods for estimating the parameter θ_0, we recall the Generalized Method of Moments (GMM) of [6], the Continuous Updating (CU) estimator of [7], the Empirical Likelihood (EL) estimator of [8,14,15], the Exponential Tilting (ET) of [11], as well as the Generalized Empirical Likelihood (GEL) class of estimators of [13] that contains the EL, ET and CU estimators in particular. Some alternative methods have been proposed in order to improve the finite sample accuracy or the robustness under misspecification of the model, for example in [4,8,11,12,16].

The authors in [3] have developed a general methodology for estimation and testing in moment condition models. Their approach is based on minimizing divergences in dual form and allows the asymptotic study of the estimators (called minimum empirical divergence estimators) and of the associated test statistics, both under the model and under misspecification of the model. Using the approach based on the influence function, [18] studied robustness properties for these classes of estimators and test statistics, showing that the minimum empirical divergence estimators of the parameter θ_0 of the model are generally not robust. This approach based on divergences and duality was initially used in the case of parametric models, the results being published in the articles, [2,19,20].

The classical EL estimator represents a particular case of the class of estimators from [3], namely, when using the modified Kullback-Leibler divergence. Although the EL estimator is superior to other above mentioned estimators in what regards higher-order asymptotic efficiency, this property is valid only in the case of the correct specification of the moment conditions. It is a known fact that the EL estimator and the EL ratio test for moment condition models are not robust with respect to the presence of outliers in the sample. Also, [17] showed that, when the support of the p.m. corresponding to the model and the orthogonality functions are not bounded, the EL estimator is not root n consistent under misspecification.

In this paper, we present a robust version of the EL estimator for moment condition models. This estimator is defined by minimizing an empirical version of the modified Kullback-Leibler divergence in dual form, using truncated orthogonality functions. For this estimator, we present some asymptotic properties regarding both consistency and limit laws. The robust EL estimator is root n consistent, even under misspecification, which gives a solution to the problem noticed by [17] for the EL estimator.

2 A Robust Version of the Empirical Likelihood Estimator

Let $\{P_\theta; \theta \in \Theta\}$ be a reference identifiable model, containing probability measures such that, for each $\theta \in \Theta$, $P_\theta \in \mathcal{M}_\theta$, meaning that $\int_{\mathbb{R}^m} g(x,\theta)\, dP_\theta(x) = 0$, and θ is the unique solution of the equation. We assume that the p.m. P_0 of the data, corresponding to the true unknown value θ_0 of the parameter to be estimated, belongs to this reference model. The reference model will be associated to the truncated orthogonality function g_c, defined hereafter, that will be used in the definition of the robust version of the EL estimator of the parameter θ_0. We use the notation $\|\cdot\|$ for the Euclidean norm. Similarly as in [16], using the reference model $\{P_\theta;\, \theta \in \Theta\}$, define the function $g_c : \mathbb{R}^m \times \Theta \to \mathbb{R}^\ell$,

$$g_c(x,\theta) := H_c\left(A_\theta\left[g(x,\theta) - \tau_\theta\right]\right), \tag{2}$$

where $H_c : \mathbb{R}^\ell \to \mathbb{R}^\ell$ is the Huber's function

$$H_c(y) := \begin{cases} y \cdot \min\left(1, \frac{c}{\|y\|}\right) & \text{if } y \neq 0, \\ 0 \text{ if } y = 0, \end{cases} \tag{3}$$

and A_θ, τ_θ are determined by the solutions of the system of implicit equations

$$\begin{cases} \int g_c(x,\theta)\, dP_\theta(x) = 0, \\ \int g_c(x,\theta)\, g_c(x,\theta)^\top\, dP_\theta(x) = I_\ell, \end{cases} \tag{4}$$

where I_ℓ is the $\ell \times \ell$ identity matrix and $c > 0$ is a given positive constant. Therefore, we have $\|g_c(x,\theta)\| \leq c$, for all x and θ. We also use the function

$$h_c(x,\theta,A,\tau) := H_c\left(A\left[g(x,\theta) - \tau\right]\right), \tag{5}$$

when needed to work with the dependence on the $\ell \times \ell$ matrix A and on the ℓ-dimensional vector τ. Then,

$$g_c(x,\theta) = h_c(x,\theta,A_\theta,\tau_\theta), \tag{6}$$

where A_θ and τ_θ are the solution of (4). For given P_θ from the reference model, the triplet $(\theta, A_\theta, \tau_\theta)$ is the unique solution of the system

$$\begin{cases} \int g(x,\theta)\, dP_\theta(x) = 0, \\ \int g_c(x,\theta)\, dP_\theta(x) = 0, \\ \int g_c(x,\theta)\, g_c(x,\theta)^\top\, dP_\theta(x) = I_\ell. \end{cases} \tag{7}$$

The uniqueness is justified in [16], p. 48.

In what follows, we will use the so-called modified Kullback-Leibler divergence between probability measures, say Q and P, defined by

$$KL_m(Q,P) := \int_{\mathbb{R}^m} \varphi\left(\frac{dQ}{dP}(x)\right) dP(x), \tag{8}$$

if Q is absolutely continuous with respect to P, and $KL_m(Q,P) := +\infty$, elsewhere. The strictly convex function φ is defined by $\varphi(x) := -\log x + x - 1$, if $x > 0$, respectively $\varphi(x) := +\infty$, if $x \leq 0$. Straightforward calculus show that the convex conjugate[1] of the convex function φ is $\psi(u) = -\log(1-u)$ if $u < 1$, respectively $\psi(u) = +\infty$, if $u \geq 1$. Recall that the Kullback-Leibler divergence, between any probability measures Q and P, is defined by

$$KL(Q,P) := \int_{\mathbb{R}^m} \varphi\left(\frac{dQ}{dP}(x)\right) dP(x),$$

if Q is absolutely continuous with respect to P, and $KL_m(Q,P) := +\infty$, elsewhere. Here, the strictly convex function φ is defined by $\varphi(x) = x\log x - x + 1$, if $x \geq 0$, and $\varphi(x) = +\infty$, if $x < 0$. Notice also that $KL_m(Q,P) = KL(P,Q)$, for all probability measures Q and P. For any subset Ω of M, we define the KL_m-divergence, between Ω and any probability measure P, by

$$KL_m(\Omega,P) := \inf_{Q\in\Omega} KL_m(Q,P).$$

Define the moment condition model

$$\mathcal{M}_c := \bigcup_{\theta\in\Theta} \mathcal{M}_{c,\theta} := \bigcup_{\theta\in\Theta} \left\{Q \in M \text{ s.t. } \int_{\mathbb{R}^m} g_c(x,\theta)\, dQ(x) = 0\right\}. \tag{9}$$

For any $\theta \in \Theta$, define the set

$$\Lambda_{c,\theta} := \Lambda_{c,\theta}(P_0) := \left\{t \in \mathbb{R}^\ell \text{ s.t. } \int_{\mathbb{R}^m} \left|\psi\left(t^\top g_c(x,\theta)\right)\right| dP_0(x) < \infty\right\}.$$

Since $g_c(x,\theta)$ is bounded (with respect to x), then on the basis of Theorem 1.1 in [1] and Proposition 4.2 in [3], the following dual representation of divergence holds

$$KL_m(\mathcal{M}_{c,\theta}, P_0) = \sup_{t\in\Lambda_{c,\theta}} \int_{\mathbb{R}^m} m_c(x,\theta,t)\, dP_0(x), \tag{10}$$

where

$$m_c(x,\theta,t) := -\psi(t^\top g_c(x,\theta)) = \log(1 - t^\top g_c(x,\theta)), \tag{11}$$

and the supremum in (10) is reached, provided that $KL_m(\mathcal{M}_{c,\theta}, P_0)$ is finite. Moreover, the supremum in (10) is unique under the following assumption

$$P_0\left(\{x \in \mathbb{R}^m \text{ s.t. } \overline{t}^\top \overline{g}_c(x,\theta) \neq 0\}\right) > 0, \text{ for all } \overline{t} \in \mathbb{R}^{1+\ell}\backslash\{0\}, \tag{12}$$

where $\overline{t} := (t_0, t_1, \ldots, t_\ell)^\top$ and $\overline{g} := (g_0, g_{c,1}, \ldots, g_{c,\ell})^\top$. This last condition is satisfied if the functions $g_0(\cdot) := \mathbf{1}_{\mathbb{R}^m}(\cdot), g_{c,1}(\cdot,\theta), \ldots, g_{c,\ell}(\cdot,\theta)$ are linearly independent and P_0 is not degenerate. The empirical measure, associated to the sample $X_1, \ldots, X_n$, is defined by

$$P_n(\cdot) := \frac{1}{n}\sum_{i=1}^n \delta_{X_i}(\cdot),$$

[1] The convex conjugate, called also Fenchel-Legendre transform, of φ, is the function defined on $\mathbb{R}$ by $\psi(u) := \sup_{x\in\mathbb{R}}\{ux - \varphi(x)\}$, $\forall u \in \mathbb{R}$.

$\delta_x(\cdot)$ being the Dirac measure at the point x. Denote

$$\Lambda_{c,\theta,n} := \Lambda_{c,\theta}(P_n) = \left\{ t \in \mathbb{R}^\ell \text{ s.t. } \int_{\mathbb{R}^m} \left| \psi\left(t^\top g_c(x,\theta)\right) \right| dP_n(x) < \infty \right\} \tag{13}$$
$$= \left\{ t \in \mathbb{R}^\ell \text{ s.t. } \frac{1}{n}\sum_{i=1}^n \left| \log(1 - \sum_{j=1}^\ell t_j\, g_{c,j}(X_i,\theta)) \right| < \infty \right\}.$$

In view of relation (10), for given $\theta \in \Theta$, a natural estimator of

$$t_{c,\theta} := \arg\sup_{t \in \Lambda_{c,\theta}} \int m_c(x,\theta,t)\, dP_0(x), \tag{14}$$

can be defined by "plug-in" as follows

$$\widehat{t}_{c,\theta} := \arg\sup_{t \in \Lambda_{c,\theta,n}} \int m_c(x,\theta,t)\, dP_n(x). \tag{15}$$

A "dual" plug-in estimator of the modified Kullback-Leibler divergence, between $\mathcal{M}_{c,\theta}$ and P_0, can then be defined by

$$\widehat{KL_m}(\mathcal{M}_{c,\theta}, P_0) := \sup_{t \in \Lambda_{c,\theta,n}} \int m_c(x,\theta,t)\, dP_n(x) \tag{16}$$
$$= \sup_{(t_1,\ldots,t_\ell) \in \mathbb{R}^\ell} \left\{ \frac{1}{n}\sum_{i=1}^n \overline{\log}\left(1 - \sum_{j=1}^\ell t_j\, g_{c,j}(X_i,\theta)\right) \right\},$$

where $\overline{\log}(\cdot)$ is the extended logarithm function, i.e., the function defined by $\overline{\log}(u) = \log(u)$ if $u > 0$, and $\overline{\log}(u) = -\infty$ if $u \leq 0$. Hence,

$$KL_m(\mathcal{M}_c, P_0) := \inf_{\theta \in \Theta} KL_m(\mathcal{M}_{c,\theta}, P_0) \tag{17}$$

can be estimated by

$$\widehat{KL_m}(\mathcal{M}_c, P_0) := \inf_{\theta \in \Theta} \widehat{KL_m}(\mathcal{M}_{c,\theta}, P_0)$$
$$= \inf_{\theta \in \Theta} \sup_{(t_1,\ldots,t_\ell) \in \mathbb{R}^\ell} \left\{ \frac{1}{n}\sum_{i=1}^n \overline{\log}\left(1 - \sum_{j=1}^\ell t_j\, g_{c,j}(X_i,\theta)\right) \right\}. \tag{18}$$

Since $\theta_0 = \arg\inf_{\theta \in \Theta} KL_m(\mathcal{M}_{c,\theta}, P_0)$, where the infimum is unique, we propose then to estimate θ_0 by

$$\widehat{\theta}_c := \arg\inf_{\theta \in \Theta} \sup_{t \in \Lambda_{c,\theta,n}} \int m_c(x,\theta,t)\, dP_n(x) \tag{19}$$
$$= \arg\inf_{\theta \in \Theta} \sup_{(t_1,\ldots,t_\ell) \in \mathbb{R}^\ell} \left\{ \frac{1}{n}\sum_{i=1}^n \overline{\log}\left(1 - \sum_{j=1}^\ell t_j\, g_{c,j}(X_i,\theta)\right) \right\},$$

which can be seen as a "robust" version of the classical EL estimator. Recall that the EL estimator, see e.g. [14], can be written as

$$\widehat{\theta} = \arg\inf_{\theta \in \Theta} \sup_{(t_1,\ldots,t_\ell)\in\mathbb{R}^\ell} \left\{ \frac{1}{n}\sum_{i=1}^{n} \overline{\log}\left(1 - \sum_{j=1}^{\ell} t_j\, g_j(X_i,\theta)\right)\right\}.$$

A slightly different definition of an estimator for the parameter θ_0 was introduced in [10], where robustness and consistency properties are also stated. However, the limiting distribution of the estimator in [10] is not standard, and not easy to be obtained, due to the fact that the used bounded orthogonality functions depend on both θ and the data. The present version is simpler and does not present this difficulty. We give in the following sections, the influence function of the estimator (19), and state both consistency and the limiting distributions of all the proposed estimators (15), (16), (19) and (18).

2.1 Robustness Property

The classical EL estimator of the parameter θ_0 of a moment condition model can be obtained as a particular case of the class of minimum empirical divergence estimators introduced by [3]. [18] showed that the influence functions for the estimators from this class, so particularly the influence function of the EL estimator, are each proportional to the orthogonality function $g(x,\theta_0)$ of the model. These influence functions also coincide with the influence function of the GMM estimator obtained by [16]. Therefore, when $g(x,\theta)$ is not bounded in x, all these estimators, and particularly the EL estimator of θ_0, are not robust.

Denote $T_c(\cdot)$ the statistical functional associated to the estimator $\widehat{\theta}_c$, so that $\widehat{\theta}_c = T_c(P_n)$. The influence function $\mathrm{IF}(x;T_c,P_0)$ of T_c at P_0 is defined by

$$\mathrm{IF}(x;T_c,P_0) := \left.\frac{\partial}{\partial\varepsilon} T_c(P_{\varepsilon,x})\right|_{\varepsilon=0},$$

where $P_{\varepsilon,x}(\cdot) = (1-\varepsilon)\,P_0(\cdot) + \varepsilon\,\delta_x(\cdot)$, $\varepsilon \in]0,1[$; see e.g. [5]. The influence function $\mathrm{IF}(x;T_c,P_0)$ of the estimator $\widehat{\theta}_c$ presented in this paper is linearly related to the bounded function $g_c(x,\theta)$, more precisely, the following result holds

$$\mathrm{IF}(x;T_c,P_0) = \left\{ \left[\int \frac{\partial}{\partial\theta} g_c(y,\theta_0)\,dP_0(y)\right]^{\top} \int \frac{\partial}{\partial\theta} g_c(y,\theta_0)\,dP_0(y)\right\}^{-1} \cdot \left[\int \frac{\partial}{\partial\theta} g_c(y,\theta_0)\,dP_0(y)\right]^{\top} g_c(x,\theta_0),$$

which implies the robustness of the estimator $\widehat{\theta}_c$ of the parameter θ_0. The proof of this result is similar to the one presented in [10].

2.2 Asymptotic Properties

In this subsection, we give the limiting distributions of the proposed estimators, under some regularity assumptions similar to those used by [3].

Proposition 1. *For any fixed $\theta \in \Theta$, under some regularity assumptions, we have*

(1) $\sqrt{n}(\widehat{t}_{c,\theta} - t_{c,\theta})$ *converges in distribution to a centered normal random vector;*
(2) If $P_0 \notin \mathcal{M}_{c,\theta}$*, then* $\sqrt{n}(\widehat{KL_m}(\mathcal{M}_{c,\theta}, P_0) - KL_m(\mathcal{M}_{c,\theta}, P_0))$ *converges in distribution to a centered normal random variable;*
(3) If $P_0 \in \mathcal{M}_{c,\theta}$*, then* $2n\,\widehat{KL_m}(\mathcal{M}_{c,\theta}, P_0)$ *convergences in distribution to a* $\chi^2(\ell)$ *random variable.*

Proposition 2. *Under some regularity assumptions, we have*

(1) $\sqrt{n}\left(\widehat{\theta}_c - \theta_0\right)$ *converges in distribution to a centered normal random vector;*
(2) If $\ell > d$*, then* $2n\,\widehat{KL_m}(\mathcal{M}_c, P_0)$ *converges in distribution to a* $\chi^2(\ell - d)$ *random variable.*

References

1. Broniatowski, M., Keziou, A.: Minimization of ϕ-divergences on sets of signed measures. Stud. Sci. Math. Hung. **43**(4), 403–442 (2006)
2. Broniatowski, M., Keziou, A.: Parametric estimation and tests through divergences and the duality technique. J. Multivar. Anal. **100**(1), 16–36 (2009)
3. Broniatowski, M., Keziou, A.: Divergences and duality for estimation and test under moment condition models. J. Stat. Plann. Infer. **142**(9), 2554–2573 (2012)
4. Felipe, A., Martin, N., Miranda, P., Pardo, L.: Testing with exponentially tilted empirical likelihood. Methodol. Comput. Appl. Probab. **20**, 1–40 (2018)
5. Hampel, F.R., Ronchetti, E.M., Rousseeuw, P.J., Stahel, W.A.: Robust Statistics: The Approach Based on Influence Functions. Wiley Series in Probability and Mathematical Statistics: Probability and Mathematical Statistics. Wiley, New York (1986)
6. Hansen, L.P.: Large sample properties of generalized method of moments estimators. Econometrica **50**(4), 1029–1054 (1982)
7. Hansen, L.P., Heaton, J., Yaron, A.: Finite-sample properties of some alternative generalized method of moments estimators. J. Bus. Econ. Stat. **14**, 262–280 (1996)
8. Imbens, G.W.: One-step estimators for over-identified generalized method of moments models. Rev. Econ. Stud. **64**(3), 359–383 (1997)
9. Imbens, G.W.: Generalized method of moments and empirical likelihood. J. Bus. Econ. Stat. **20**(4), 493–506 (2002)
10. Keziou, A., Toma, A.: A robust version of the empirical likelihood estimator. Mathematics **9**(8), 829 (2021)
11. Kitamura, Y., Stutzer, M.: An information-theoretic alternative to generalized method of moments estimation. Econometrica **65**(4), 861–874 (1997)
12. Lô, S.N., Ronchetti, E.: Robust small sample accurate inference in moment condition models. Comput. Stat. Data Anal. **56**(11), 3182–3197 (2012)

13. Newey, W.K., Smith, R.J.: Higher order properties of GMM and generalized empirical likelihood estimators. Econometrica **72**(1), 219–255 (2004)
14. Owen, A.: Empirical likelihood. Chapman and Hall, New York (2001)
15. Qin, J., Lawless, J.: Empirical likelihood and general estimating equations. Ann. Stat. **22**(1), 300–325 (1994)
16. Ronchetti, E., Trojani, F.: Robust inference with GMM estimators. J. Econ. **101**(1), 37–69 (2001)
17. Schennach, S.M.: Point estimation with exponentially tilted empirical likelihood. Ann. Stat. **35**(2), 634–672 (2007)
18. Toma, A.: Robustness of dual divergence estimators for models satisfying linear constraints. Comptes Rendus Mathématiques **351**(7–8), 311–316 (2013)
19. Toma, A., Leoni-Aubin, S.: Robust tests based on dual divergence estimators and saddlepoint approximations. J. Multivar. Anal. **101**(5), 1143–1155 (2010)
20. Toma, A., Broniatowski, M.: Dual divergence estimators and tests: robustness results. J. Multivar. Anal. **102**(1), 20–36 (2011)

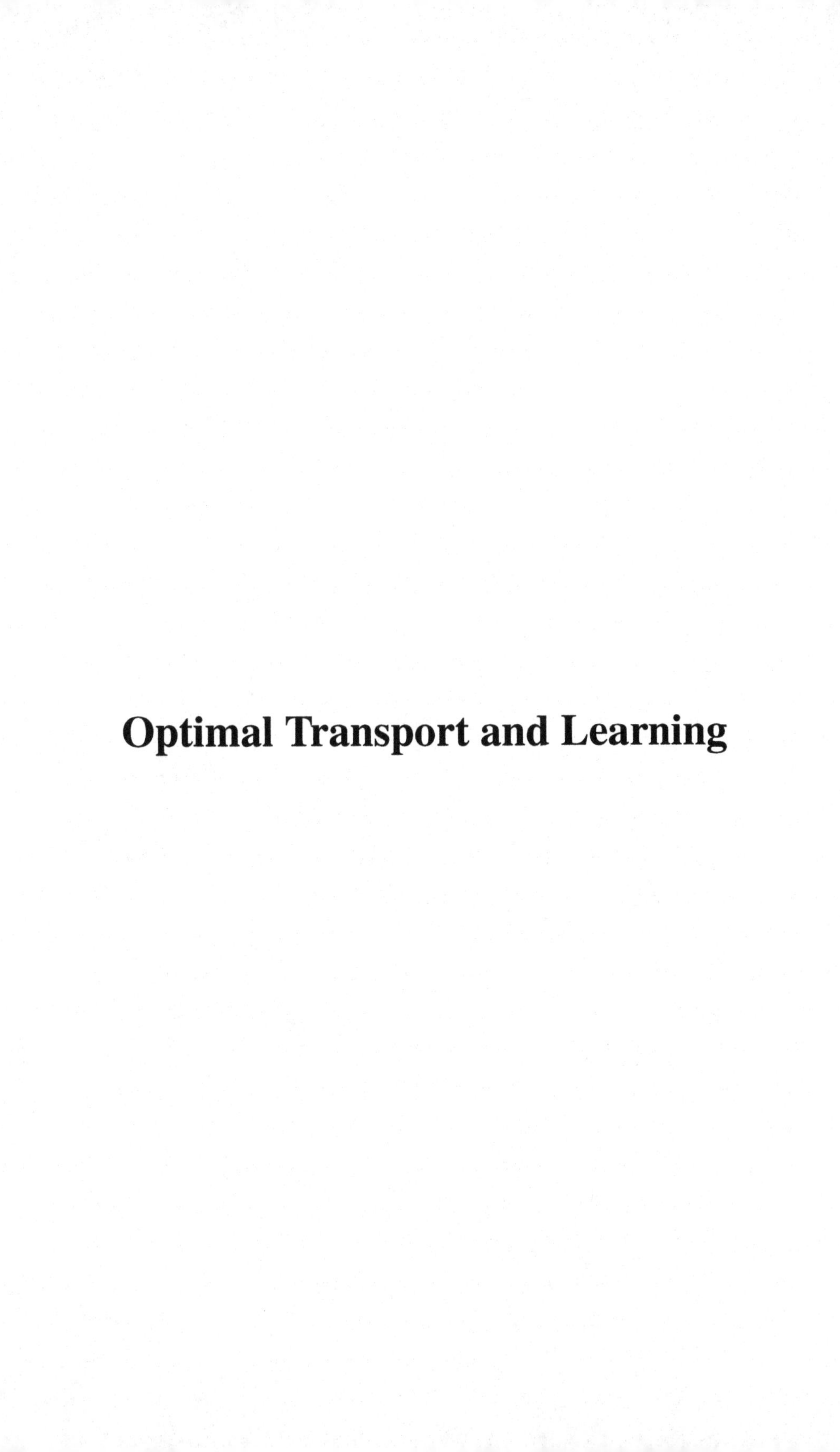

Optimal Transport and Learning

Mind2Mind: Transfer Learning for GANs

Yael Fregier and Jean-Baptiste Gouray(✉)

Univ. Artois, UR2462, Laboratoire de Mathématiques de Lens (LML), 62300 Lens, France
{yael.fregier,jbaptiste.gouray}@univ-artois.fr

Abstract. Training generative adversarial networks (GANs) on high quality (HQ) images involves important computing resources. This requirement represents a bottleneck for the development of applications of GANs. We propose a transfer learning technique for GANs that significantly reduces training time. Our approach consists of freezing the low-level layers of both the critic and generator of the original GAN. We assume an auto-encoder constraint to ensure the compatibility of the internal representations of the critic and the generator. This assumption explains the gain in training time as it enables us to bypass the low-level layers during the forward and backward passes. We compare our method to baselines and observe a significant acceleration of the training. It can reach two orders of magnitude on HQ datasets when compared with StyleGAN. We provide a theorem, rigorously proven within the framework of optimal transport, ensuring the convergence of the learning of the transferred GAN. We moreover provide a precise bound for the convergence of the training in terms of the distance between the source and target dataset.

1 Introduction

The recent rise of deep learning as a leading paradigm in AI mostly relies on computing power (with generalized use of GPUs) and massive datasets. These requirements represent bottlenecks for most practitioners outside of big labs in industry or academia and are the main obstacles to the generalization of the use of deep learning. Therefore, methods that can bypass such bottlenecks are in strong demand. Transfer learning is one of them and various methods of transfer learning (for classification tasks) specific to deep neural networks have been developed [20].

A *generative problem* is a situation in which one wants to be able to produce elements that could belong to a given data set $\mathcal{D}$. Generative Adversarial Networks (GANs) were introduced in 2014 [8,18] to tackle generative tasks with deep learning architectures and improved in [1,9] (Wasserstein GANs). They

This work was granted access to the HPC resources of IDRIS under the allocation 20XX-AD011011199R1 made by GENCI.

Y. Fregier and J.-B. Gouray—Contributed equally.

F. Nielsen and F. Barbaresco (Eds.): GSI 2021, LNCS 12829, pp. 851–859, 2021.
https://doi.org/10.1007/978-3-030-80209-7_91

immediately took a leading position in the world of generative models, comforted by progress with HQ images [11,12]. The goal of our work (see Sect. 3) is to develop in a generative setting, i.e., for GAN architectures, the analog of the *cut-and-paste* approach.

The main idea of our method is to reuse, for the training of a GAN, the weights of an autoencoder already trained on a source dataset $\mathcal{D}$. The weights of the low level layers of the generator (resp. critic) will be given by those of the decoder (resp. encoder). By analogy with humans (Sect. 3), we call *MindGAN* the high level layers of the generator. It is a subnetwork that is trained as a GAN on the encoded features of the target dataset $\mathcal{D}'$.

We provide in Sect. 4 a theorem that controls the convergence of the transferred GAN in terms of the quality of the autoencoder, the domain shift between $\mathcal{D}$ and $\mathcal{D}'$ and the quality of the MindGAN. As a consequence, our experimental results in Sect. 5 rely heavily on the choice of the autoencoder. ALAE autoencoders [17] are extremely good autoencoders that turned out to be crucial in our experiments with HQ images. Their use, in conjunction with our transfer technique, enables an **acceleration** of the training **by a factor of 656**, while keeping a good quality.

2 Preliminaries

A GAN consists of two networks trained adversarially. The *generator* $g : Z \to \chi$ associates to a vector z sampled from a latent vector space Z a vector $g(z)$ in another vector space χ while the *discriminator* $c : \chi \to \mathbb{R}$ learns to associate a value close to 1 if the vector $g(z)$ belongs to $\mathcal{D}$ and zero otherwise. Their respective loss functions, L_g and L_c are recalled in Sect. 3.

One can assume that elements of $\mathcal{D}$ can be sampled from an underlying probability distribution $\mathbb{P}_{\mathcal{D}}$ on a space χ and try to approximate it by $\mathbb{P}_\theta$, another distribution on χ that depends on some learnable parameters θ. *Generating* then means sampling from the distribution $\mathbb{P}_\theta$. Among the distributions on χ, some can be obtained from a prior distribution $\mathbb{P}_Z$ on an auxiliary latent space Z and a map $g : Z \to \chi$ as follows. The *push-forward* of $\mathbb{P}_Z$ under g is defined so that a sample is given by $g(z)$ the image through g of a sample z from the distribution $\mathbb{P}_Z$. We will denote this pushforward by $g_\sharp\mathbb{P}_Z$ and when g depends on parameters θ use instead the notation $\mathbb{P}_\theta := g_\sharp\mathbb{P}_Z$. In practice, one can choose for $\mathbb{P}_Z$ a uniform or Gaussian distribution, and for g a (de)convolution deep neural network. In our applications, we will consider for instance $Z := \mathbb{R}^{128}$ equipped with a multivariate gaussian distribution and $\chi = [-1, 1]^{28\times28}$, the space of gray level images of resolution 28×28. Hence, sampling from $\mathbb{P}_\theta$ will produce images.

The main idea behind a Wasserstein GAN is to use the *Wasserstein distance* (see Definition 1) to define by $W(\mathbb{P}_{\mathcal{D}}, \mathbb{P}_\theta)$ the loss function for this optimisation problem. More precisely, the Wasserstein distance is a distance on Borel probability measures on χ (when compact metric space). In particular, the quantity $W(\mathbb{P}_{\mathcal{D}}, \mathbb{P}_\theta)$ gives a number which depends on the parameter θ since $\mathbb{P}_\theta$ depends itself on θ. The main result of [1] asserts that if g is of class $\mathcal{C}^k$ almost everywhere as a function of θ, $W(\mathbb{P}_D, \mathbb{P}_\theta)$ is also of class $\mathcal{C}^k$ with respect to θ. As a consequence, one can solve this optimization problem by doing gradient descent for the parameters θ (using a concrete gradient formula provided by the same theorem) until the two probability distributions coincide.

In order to minimize the function $W(\mathbb{P}_{\mathcal{D}}, \mathbb{P}_\theta)$, one needs a good estimate of the Wasserstein distance. The *Rubinstein-Kantorovich duality* states that $W(\mathbb{P}, \mathbb{P}') = \max_{|c|_L \leq 1} \mathbb{E}_{x\sim\mathbb{P}}\ c(x) - \mathbb{E}_{x\sim\mathbb{P}'}\ c(x)$, where $\mathbb{E}_{x\sim\mathbb{P}} f(x)$ denotes the expected value of the function f for the probability measure $\mathbb{P}$, while the max is taken on the unit ball for the Lipschitz semi-norm. Concretely, this max is obtained by gradient ascent on a function c_θ encoded by a deep convolution neural network.

In our case, when $\mathbb{P}' := \mathbb{P}_\theta$, the term $\mathbb{E}_{x\sim\mathbb{P}'}\ c(x)$ takes the form $\mathbb{E}_{z\sim\mathbb{P}_Z}\ c_\theta(g_\theta(z))$. One recovers the diagram

$$Z \xrightarrow{g_\theta} \chi \xrightarrow{c_\theta} \mathbb{R} \tag{1}$$

familiar in the adversarial interpretation of GANs. With this observation, one understands that one of the drawbacks of GANs is that there are two networks to train. They involve many parameters during the training, and the error needs to backpropagate through all the layers of the two networks combined, i.e., through $c_\theta \circ g_\theta$. This process is computationally expensive and can trigger the vanishing of the gradient. Therefore, specific techniques need to be introduced to deal with very deep GANs, such as in [12]. The approach we present in Sect. 3 can circumvent these two problems.

Transfer learning is a general approach in machine learning to overcome the constraints of the volume of data and computing power needed for training models (see [5]). It leverages a priori knowledge from a learned task $\mathcal{T}$ on a source data set $\mathcal{D}$ in order to learn more efficiently a task $\mathcal{T}'$ on a target data set $\mathcal{D}'$. It applies in deep learning in at least two ways: *Cut and Paste* and *Fine tuning.*

Cut and Paste takes advantage of the difference between high and low-level layers of the network c. It assumes that the network c is composed of two networks c_0 and c_1 stacked one on each other, i.e., mathematically that $c = c_0 \circ c_1$ (one understands the networks as maps). While the low-level layers c_1 process low-level features of the data, usually common for similar datasets, the high-level layers c_0 are in charge of the high-level features which are very specific to each dataset. This approach consists in the following steps :

1. identify a dataset $\mathcal{D}$ similar to the data set $\mathcal{D}'$ we are interested in, both in the same space χ,
2. import a network $c : \chi \xrightarrow{c_1} M \xrightarrow{c_0} C$, already trained on the dataset $\mathcal{D}$, where C is the space of classes, M the space of features at that cut level,
3. pass the new dataset $\mathcal{D}'$ through c_1 and train c_0' on $c_1(\mathcal{D}')$, and
4. use $c' := c_0' \circ c_1$ as the new classifier on $\mathcal{D}'$.

This approach requires much fewer parameters to train, passes the data $\mathcal{D}'$ only once through c_1, and does not need to backpropagate the error through c_1.

Fine tuning is based on the same first two steps than *Cut and Paste*, but instead of steps 3 and 4, uses the weights of c to initialise the training of the new network c' on $\mathcal{D}'$. It is assumed that c and c' share the same architecture.

3 Mind to Mind Algorithm

We now adapt to GANs the cut-and-paste procedure described in Sect. 2. Let us consider factorisations of the form $g : Z \xrightarrow{g_0} M' \xrightarrow{g_1} \chi$, $c : \chi \xrightarrow{c_1} M \xrightarrow{c_0} C$.

Algorithm 1. Mind2Mind transfer learning.

Require: (c_1, g_1), an autoencoder trained on a source dataset $\mathcal{D}$, α, the learning rate, b, the batch size, n, the number of iterations of the critic per generator iteration, $\mathcal{D}'$, a target dataset, φ' and θ' the initial parameters of the critic c_0' and of the generator g_0'. We denote by L_c and L_g the losses of the c and g.
Compute $c_1(\mathcal{D}')$.
while θ' has not converged **do**
  for $t = 0, ..., n$ **do**
    Sample $\{m^{(i)}\}_{i=1}^{b} \sim c_{1\sharp}\mathbb{P}_{\mathcal{D}'}$ a batch from $c_1(\mathcal{D}')$.
    Sample $\{z^{(i)}\}_{i=1}^{b} \sim \mathbb{P}_Z$ a batch of prior samples.
    Update c_0' by descending L_c.
  end for
  Sample $\{z^{(i)}\}_{i=1}^{b} \sim \mathbb{P}_Z$ a batch of prior samples.
  Update g_0' by descending $-L_g$.
end while
return $g_1 \circ g_0'$. (Note that c_0 and g_0 were never used).

From now on, as in [9]: $L_g := \mathbb{E}_{z \sim \mathbb{P}_Z} c_0'(g_0'(z))$, $L_c := -\mathbb{E}_{m \sim c_{1\sharp}\mathbb{P}_{\mathcal{D}'}} c_0'(m) + L_g + \lambda \mathbb{E}_{m \sim c_{1\sharp}\mathbb{P}_{\mathcal{D}'}, z \sim \mathbb{P}_Z, \alpha \sim (0,1)} \{(\|\nabla c_0'(\alpha m + (1-\alpha) g_0'(z))\|_2 - 1)^2\}$.

Motivation for the Approach. The architectures of a generator and a critic of a GAN are symmetric to one another. The high-level features appear in g_0, the closest to the prior vector (resp. c_0, the closest to the prediction), while the low-level features are in g_1, the closest to the generated sample (resp c_1, the closest to the input image). Therefore, the analogy with cut and paste is to keep g_1 (the low level features of $\mathcal{D}$) and only learn the high level features g_0' of the target data set $\mathcal{D}'$. However, the only way a generator can access to information from $\mathcal{D}'$ is through the critic c via the value of $c \circ g = c_0' \circ c_1 \circ g_1 \circ g_0'$ [9]. Hence, the information needs to back-propagate through $c_1 \circ g_1$ to reach the weights of g_0'. Our main idea is to bypass this computational burden and train directly g_0' and c_0' on $c_1(\mathcal{D}')$. But this requires that the source of c_0' coincides with the target of g_0'. Therefore, we assume that $M = M'$, a first hint that autoencoders are relevant for us.

A second hint comes from an analogy with humans learning a task, like playing tennis, for instance. One can model a player as a function $Z \xrightarrow{g} \chi$, where

χ is the space of physical actions of the player. His/her coach can be understood as a function $\chi \xrightarrow{c} \mathbb{R}$, giving $c(g(z))$ as a feedback for an action $g(z)$ of g. The objective of the player can be understood as to be able to generate instances of the distribution $\mathcal{D}'$ on χ corresponding to the "tennis moves". However, in practice, a coach rarely gives his/her feedback as a score. He instead describes what the player has done and should do instead. We can model this description as a vector $c_1(g(z))$ in M, the mind of $c = c_0 \circ c_1$. In this analogy, c_1 corresponds to the coach analyzing the action, while c_0 corresponds to the coach giving a score based on this analysis. One can also decompose the player itself as $g = g_1 \circ g_0$. Here g_0 corresponds to the player conceiving the set of movements he/she wants to perform and g_1 to the execution of these actions. Therefore, two conditions are needed for the coach to help efficiently his/her student:

1. they must speak the same language in order to understand one each other,
2. the player must already have a good command of his/her motor system g_1.

In particular, the first constraint implies that they must share the same feature space, i.e., $M = M'$. A way to ensure that both constraints are satisfied is to check whether the player can reproduce a task described by the coach, i.e., that

$$g_1(c_1(x)) = x \tag{2}$$

holds. One recognizes in (2) the expression of an autoencoder. It is important to remark that usually, based on previous learning, a player already has a good motor control and he/she and his/her coach know how to communicate together. In other words g_1 and c_1 satisfy (2) before the training starts. Then the training consists only in learning g_0' and c_0' on the high level feature interpretations of the possible tennis movements, i.e., on $c_1(\mathcal{D}')$.

4 Theoretical Guarantee for Convergence

Let us start by recalling the relevant definitions.

Definition 1. *Let* $(X, \|.\|)$ *be a metric space. For two probability measure* $\mathbb{P}_1$, $\mathbb{P}_2$ *on* X, *a* transference plan with marginals $\mathbb{P}_1$ and $\mathbb{P}_2$ *is a measure* γ *on* $X \times X$ *such that:*

$$\gamma(A \times X) = \mathbb{P}_1(A) \, and \, \gamma(X \times B) = \mathbb{P}_2(B).$$

The Wasserstein distance *between* $\mathbb{P}_1$ *and* $\mathbb{P}_2$ *is defined by*

$$W(\mathbb{P}_1, \mathbb{P}_2) = \inf_{\gamma \in \Pi(\mathbb{P}_1, \mathbb{P}_2)} \mathbb{E}_{(x,y)\sim\gamma} \|x - y\|, \tag{3}$$

where $\Pi(\mathbb{P}_1, \mathbb{P}_2)$ *denotes the set of transference plans with marginals* $\mathbb{P}_1$ *and* $\mathbb{P}_2$.

Our main theorem controls the convergence of the generated distribution towards the true target distribution. Very concretely, Theorem 1 tells us that in order to control the convergence error $er_{conv} = W(\mathbb{P}_{\mathcal{D}'}, \mathbb{P}_\theta')$ of the transferred GAN towards the distribution of the target dataset $\mathcal{D}'$, we need the exact analogues of steps 1–3 of 2:

1. choose two datasets $\mathcal{D}'$ and $\mathcal{D}$ very similar, i.e., $er_{shift} = W(\mathbb{P}_{\mathcal{D}}, \mathbb{P}_{\mathcal{D}'})$ small,
2. choose a good autoencoder (c_1, g_1) so that $er_{AE} = W(\mathbb{P}_{\mathcal{D}}, AE(\mathbb{P}_{\mathcal{D}}))$ is small,
3. train well the MindGAN (g'_0, c'_0) on $c_1(\mathcal{D}')$, i.e., $er_{mind} = W(c_{1\sharp}\mathbb{P}_{\mathcal{D}'}, \mathbb{P}'^0_\theta)$ small, where $\mathbb{P}'^0_\theta = g'_{0\sharp}\mathbb{P}_Z$.

Theorem 1. *There exist two positive constants* a *and* b *such that*

$$er_{conv} \leq a \cdot er_{shift} + er_{AE} + b \cdot er_{mind}. \tag{4}$$

Proof. From the triangle inequality property of the Wasserstein metric and the definition of $\mathbb{P}'_\theta$, one has:

$$W(\mathbb{P}_{\mathcal{D}'}, \mathbb{P}'_\theta) \leq W(\mathbb{P}_{\mathcal{D}'}, AE(\mathbb{P}_{\mathcal{D}'})) + W(AE(\mathbb{P}_{\mathcal{D}'}), g_{1\sharp}\mathbb{P}'^0_\theta).$$

One concludes with Lemma 2 and Lemma 1 with $\phi = g_1$.

Lemma 1. *Let* $\phi : X \to Y$ *be a C-lipschitz map, then:*

$$W_Y(\phi_\sharp\mu, \phi_\sharp\nu) \leq CW_X(\mu, \nu).$$

Proof. Let γ be a transference plan realising $W_X(\mu, \nu)$. Define $\gamma' := (\phi \times \phi)_\sharp\gamma$. One can check that γ' defines a transference plan between $\phi_\sharp\mu$ and $\phi_\sharp\nu$. Therefore, one has the following relation

$$\begin{aligned} W_Y(\phi_\sharp\mu, \phi_\sharp\nu) &\leq \int \|x - y\| d\gamma'(x, y) \\ &= \int \|\phi(x) - \phi(y)\| d\gamma(x, y) \\ &\leq \int C\|x - y\| d\gamma(x, y) \\ &= CW_X(\mu, \nu), \end{aligned}$$

where the first inequality comes from the fact that γ' is a transference plan, the first equality from the definition of the push-forward of a measure by a map (recalled in Sect. 2), and the last equality from the choice of γ.

Lemma 2. *There exist a positive constant* a *such that*

$$W(\mathbb{P}_{\mathcal{D}'}, AE(\mathbb{P}_{\mathcal{D}'})) \leq aW(\mathbb{P}_{\mathcal{D}}, \mathbb{P}_{\mathcal{D}'}) + W(\mathbb{P}_{\mathcal{D}}, AE(\mathbb{P}_{\mathcal{D}})).$$

Proof. Applying twice the triangle inequality, one has:

$$\begin{aligned} W(\mathbb{P}_{\mathcal{D}'}, AE(\mathbb{P}_{\mathcal{D}'})) \leq\ & W(\mathbb{P}_{\mathcal{D}'}, \mathbb{P}_{\mathcal{D}}) + W(\mathbb{P}_{\mathcal{D}}, AE(\mathbb{P}_{\mathcal{D}})) \\ &+ W(AE(\mathbb{P}_{\mathcal{D}}), AE(\mathbb{P}_{\mathcal{D}'})). \end{aligned}$$

One concludes with Lemma 1 with $\phi = g_1 \circ c_1$.

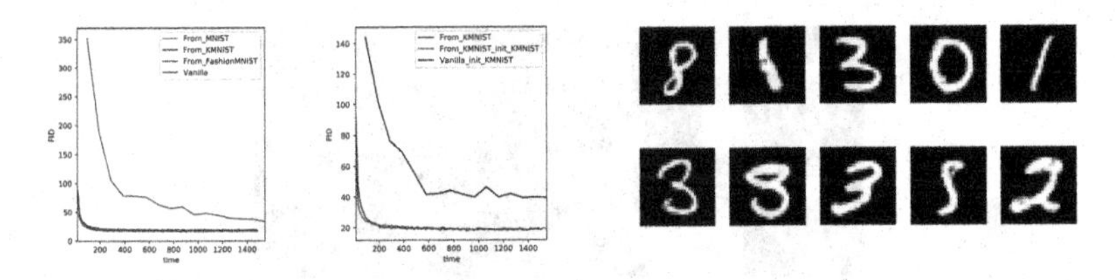

Fig. 1. Mind2Mind training and samples in 28×28.

5 Evaluation

Datasets. We have tested our algorithm at resolution 28×28 in grey levels on MNIST [13], KMNIST [4], FashionMNIST [23] and at resolution 1024×1024 in color on CelabaHQ [12] from models trained on the 60 000 first images of FFHQ [11], using the **library** Pytorch. Despite its limitations [3], we have used *FID* (Frechet Inception Distance) as **metric**. It is the current standard for evaluations of GANs. Wasserstein distance can be faithfully estimated in low dimensions. However, in these regimes our method is not efficient (cf discussion at end of Sect. 5). Our **code**, description of **hardware** and **models** are at [15].

At Resolution 28 × 28. We report the results with $\mathcal{D}' =$ MNIST for c_1 trained on each dataset. We compare our results to a Vanilla WGAN with architecture $(g_1 \circ g_0, c_0 \circ c_1)$, so that the number of parameters agrees, for fair comparison.

Baseline 1. Vanilla WGAN with gradient penalty trained on MNIST. We have used for the Vanilla WGAN a model similar to the one used in [9]. However, this model would not converge properly, due to a problem of "magnitude race." We have therefore added an ϵ-term [12]. Our results appear in the first graph on the *l.h.s* of Fig. 1, with time in seconds in abscissa. One can observe extremely fast convergence of the mindGAN to good scores, in comparison with the Vanilla WGAN. Note that we have not smoothed any of the curves. This observation suggests that our approach, beyond the gain in training time, provides a regularising mechanism. The stability of the training confirms this hypothesis. Indeed, **statistics over 10 runs** demonstrate a very small standard deviation. In particular, this regularization enabled us to use a much bigger learning rate (10^{-3} instead of 10^{-4}), adding to the speed of convergence. In terms of epochs, the MindGAN and the Vanilla WGAN learn similarly.

Baseline 2. *Fine tuning* studied in [22]. We have trained a Vanilla WGAN with gradient penalty on $\mathcal{D} =$ KMNIST, the dataset the closest to $\mathcal{D}' =$ MNIST. We have then *fine-tuned* it on $\mathcal{D}'$, i.e., trained a new network on $\mathcal{D}'$, initialized with the weights of this previously trained Vanilla WGAN. We display it on the *l.h.s* of Fig. 1 under the name *Vanilla init Kmnist*, together with our best result, namely a Mind2Mind transfer on $\mathcal{D}' =$ MNIST from $\mathcal{D} =$ KMNIST. One can observe that the Mind2Mind approach achieves significantly better performances in FID.

Fig. 2. Mind2Mind on CelebaHQ transfered from FFHQ.

The *r.h.s* of Fig. 1 displays samples of images produced by a MindGAN trained on MNIST images encoded using a KMNIST autoencoder.

At Resolution 1024 × 1024. We train a mindGAN on CelebaHQ [12] encoded with the ALAE model [17] pre-trained on FFHQ. We reach (over 5 runs) an average FID of 15.18 with an average standard deviation of 0.8. This is a better result than the FID score (19.21) of an ALAE directly trained from scratch (see table 5 from [17]). Samples are displayed in Fig. 2. Our training (1.5 h) on a GPU V100 is **224** (resp. **656**) **times faster** than the training of a **proGAN** (resp. **StyleGAN**) (14, resp. 41 days) [11,12]. On a GTX 1060 it is **112** (resp. **328**) **times faster** than the training of a proGAN (resp. StyleGAN) both on a V100. Note that a GTX 1060 costs around 200 USD while a V100 is around 8000 USD. We see two factors that can explain the acceleration of the training. The first one is that there are much less parameters to train. Indeed, our mindGAN in HD has around 870K parameters, while the ALAE model (based on a StyleGAN architecture) has 51M parameters. So this already represents a difference of almost two orders of magnitude. One can suspect that the rest of the difference comes from the fact that we bypass 18 layers in the computation of the backpropagation. We believe that the validation of this hypothesis deserves a careful experimental study.

6 Discussion

Previous works considered adversarial learning with autoencoders ([2,6,7,10,14, 17,21,24]), without considering transfer learning. It was addressed by [19] and [22] with fine-tuning (see Sect. 5), which can be also applied in our setting, see Fig. 1. Transferring layers enables faster learning at the price of possibly worse quality. However, in practice, we obtained a better asymptote with a FID of 15.18 (resp. 19.21) for the transferred model (resp. from scratch) on CelebaHQ thanks to the focus of the training on a more significant part of the network (the MindGAN). Our primary objective was not to improve training on limited data: the targeted users are practitioners without significant computing capacity. Their need is to reach a reasonable quality in a short time. In terms of wall clock time, we learn roughly 224 (resp. 656) times faster than a ProGAN (resp. StyleGAN) on CelebaHQ, at a price of a worse performance in terms of quality (FID of 15.18 compared to 8.03 for ProGAN and 4.40 for StyleGAN, see table 5 and

4 from [17] and [11]). So merits of our algorithm will depend on the tradeoff between the needs of the users in terms of quality and their constraints in terms of computing capacity.

References

1. Arjovsky, M., Chintala, S., Bottou, L.: Wasserstein generative adversarial networks. CoRR abs/1701.07875 (2017)
2. Belghazi, M.I., et al.: Hierarchical adversarially learned inference. CoRR abs/1802.01071 (2018)
3. Borji, A.: Pros and cons of GAN evaluation measures. CoRR abs/1802.03446 (2018)
4. Clanuwat, T., Bober-Irizar, M., Kitamoto, A., Lamb, A., Yamamoto, K., Ha, D.: Deep learning for classical Japanese literature. CoRR abs/1812.01718 (2018)
5. Donahue, J., et al.: A deep convolutional activation feature for generic visual recognition. In: International Conference on Machine Learning (2014)
6. Donahue, J., Krähenbühl, P., Darrell, T.: Adversarial feature learning. CoRR abs/1605.09782 (2017)
7. Dumoulin, V., et al.: Adversarially learned inference. CoRR abs/1606.00704 (2017)
8. Goodfellow, I., et al.: Generative adversarial nets. CoRR abs/1406.2661
9. Gulrajani, I., Ahmed, F., Arjovsky, M., Dumoulin, V., Courville, A.C.: Improved training of Wasserstein GANs. In: NIPS (2017)
10. Haidar, M.A., Rezagholizadeh, M.: TextKD-GAN: text generation using knowledge distillation and generative adversarial networks (2019)
11. Karras, T., Laine, S., Aila, T.: A style-based generator architecture for generative adversarial networks. CVPR (2019). https://github.com/NVlabs/stylegan
12. Karras, T., et al.: Progressive growing of GANs for improved quality, stability, and variation. https://github.com/tkarras/progressive_growing_of_gans
13. LeCun, Y., Cortes, C.: MNIST handwritten digit database (2010). http://yann.lecun.com/exdb/mnist/
14. Makhzani, A., Shlens, J., Jaitly, N., Goodfellow, I.J.: Adversarial autoencoders. CoRR abs/1511.05644 (2015)
15. (2020). https://github.com/JeanBaptisteGouray/mindgan
16. Patrini, G., et al.: Sinkhorn autoencoders. In: UAI (2019)
17. Pidhorskyi, S., et al.: Adversarial latent autoencoders. In: CVPR (2020)
18. Salimans, T., Goodfellow, I.J., Zaremba, W., Cheung, V., Radford, A., Chen, X.: Improved techniques for training GANs. In: NIPS (2016)
19. Shan, H., et al.: 3-d convolutional encoder-decoder network for low-dose CT via transfer learning from a 2-d trained network. Trans. Med. Imaging **37**, 1522–1534 (2018)
20. Tan, C., Sun, F., Kong, T., Zhang, W., Yang, C., Liu, C.: A survey on deep transfer learning. In: ICANN (2018)
21. Tolstikhin, I.O., Bousquet, O., Gelly, S., Schölkopf, B.: Wasserstein auto-encoders. CoRR abs/1711.01558 (2017)
22. Wang, Y., Wu, C., Herranz, L., van de Weijer, J., Gonzalez-Garcia, A., Raducanu, B.: Transferring GANs: generating images from limited data. In: Ferrari, V., Hebert, M., Sminchisescu, C., Weiss, Y. (eds.) ECCV 2018. LNCS, vol. 11210, pp. 220–236. Springer, Cham (2018). https://doi.org/10.1007/978-3-030-01231-1_14
23. Xiao, H., Rasul, K., Vollgraf, R.: Fashion-MNIST: a novel image dataset for benchmarking machine learning algorithms. CoRR abs/1708.07747 (2017)
24. Zhao, J.J., Kim, Y., Zhang, K., Rush, A.M., LeCun, Y.: Adversarially regularized autoencoders. In: ICML (2018)

Fast and Asymptotic Steering to a Steady State for Networks Flows

Yongxin Chen[1], Tryphon Georgiou[2(✉)], and Michele Pavon[3]

[1] School of Aerospace Engineering, Georgia Institute of Technology, Atlanta, GA 30332, USA
yongchen@gatech.edu

[2] Department of Mechanical and Aerospace Engineering, University of California, Irvine, CA 92697, USA
tryphon@uci.edu

[3] Department of Mathematics "Tullio Levi-Civita", University of Padova, 35121 Padova, Italy
pavon@math.unipd.it

Abstract. We study the problem of optimally steering a network flow to a desired steady state, such as the Boltzmann distribution with a lower temperature, both in finite time and asymptotically. In the infinite horizon case, the problem is formulated as constrained minimization of the relative entropy rate. In such a case, we find that, if the prior is reversible, so is the solution.

Keywords: Markov Decision Process · Schrödinger Bridge · Reversibility · Relative entropy rate · Regularized optimal mass transport

1 Introduction

We consider in this paper optimal steering for networks flows over a finite or infinite time horizon. The goal is to steer a Markovian evolution to a steady state with desirable properties while minimizing relative entropy or relative entropy rate for distributions on paths. For the case of infinite time horizon, we derive discrete counterparts of our *cooling* results in [3]. In particular, we characterize the optimal Markov decision policy and show that, if the prior evolution is reversible, so is the solution.

The problem relates to a special Markov Decision Process problem, cf. [6, Section 6] which is referred to as a *Schrödinger Bridge Problem* or as regularized Optimal Mass Transport. This field was born with two impressive contributions by Erwin Schrödinger in 1931/32 [31,32] dealing with a large deviation problem for the Wiener measure.

Supported in part by the NSF under grant 1807664, 1839441, 1901599, 1942523, the AFOSR under grants FA9550-20-1-0029, and by the University of Padova Research Project CPDA 140897.

F. Nielsen and F. Barbaresco (Eds.): GSI 2021, LNCS 12829, pp. 860–868, 2021.
https://doi.org/10.1007/978-3-030-80209-7_92

The delicate existence question, left open by Schrödinger, was solved in various degrees of generality by Fortet, Beurling, Jamison and Föllmer [1,17,18,22]. The connection between Schrödinger Bridges and Optimal Mass Transport was discovered more recently in [27], see also [23] and references therein. It is important to stress that Fortet's proof is algorithmic, establishing convergence of a (rather complex) iterative scheme. This predates by more than twenty years the contribution of Sinkhorn [33], who established convergence in a special case of the discrete problem where the proof is much simpler. Thus, these algorithms should be called Fortet-IPF-Sinkhorn, where IPF stands for the Iterative Proportional Fitting algorithm proposed in 1940 without proof of convergence in [15].

These "bridge" problems may be formulated in a "static" or "dynamic" form. The static, discrete space version was studied by many after Sinkhorn, see [14,30] and references therein. The dynamic, discrete problem was considered in [9–11,19,29]. It should be remarked that the latter papers feature a general "prior" measure on paths, differently from the static problem which is viewed as a regularized optimal mass transport problem. Moreover, the static formulation does not readily provide by-product information on the new transition probabilities nor does it suggest the paths where the optimal mass flows. It is therefore less suited for typical network routing applications. Convergence of the iterative scheme in a suitable projective metric was also established in [19] for the discrete case and in [4] for the continuous case.

This general topic lies nowadays at the crossroads of many fields of science such as probability, statistical physics, optimal mass transport, machine learning, computer graphics, statistics, stochastic control, image processing, etc. Several survey papers have appeared over time emphasizing different aspects of the subject, see [5–7,23,30,36].

We finally mention, for the benefit of the reader, that, starting from [37], several attempts have been made to connect Schrödinger Bridges with various versions of stochastic mechanics, the most notable of the latter being due to Fényes, Bohm, Nelson, Levy-Krener [2,16,24,28]. In a series of recent papers [12,13,34], the flows of one time marginals (*displacement interpolation*) associated to the Optimal Mass Transport (OMT), to the Schrödinger Bridge (SB) and to the Fényes-Bohm-Nelson Stochastic Mechanics (SM) are characterized as extremal curves in Wasserstein space [35]. The actions in SB and in SM feature, besides the kinetic energy term of OMT, a *Fisher Information Functional* with the plus sign for SB and the minus sign for SM, see [13, Section VI]. From the latter viewpoint, the misterious "quantum potential" appearing in the *Madelung fluid* [25] associated to the Schrödinger Equation simply originates from the Wasserstein gradient of the Fisher Information Functional.

The outline of the paper is as follows. In Sect. 2, we collect a few basic facts on the discrete Schrödinger bridge problem. In Sect. 3, we introduce the infinite-horizon steering problem by minimizing the entropy rate with respect to the prior measure. We then discuss existence for the one-step Schrödinger system leading to existence for the steering problem. Section 4 is devoted to an interesting result linking optimality in the steering problem to reversibility of the solution.

2 Discrete Schrödinger Bridges

Let $M = (m_{ij})$ be a matrix with nonnegative elements compatible with the topology of the directed, strongly connected, aperiodic graph $\mathbf{G} = (\mathcal{X}, \mathcal{E})$. Here $\mathcal{X} = \{1, 2, \ldots, n\}$ is the vertex set and $\mathcal{E} \subseteq \mathcal{X} \times \mathcal{X}$ the edge set. Compatibility means that $m_{ij} = 0$ whenever $(i, j) \notin \mathcal{E}$. Let time vary in $\mathcal{T} = \{0, 1, \ldots, N\}$. Let μ_0 be a probability distribution supported on all of $\mathcal{X}$. Consider the Markovian evolution

$$\mu_{t+1}(x_{t+1}) = \sum_{x_t \in \mathcal{X}} \mu_t(x_t) m_{x_t x_{t+1}}, \quad t = 0, 1, \ldots, N-1. \tag{1}$$

As we do not assume that the rows of M sum to one, the total mass is not necessarily preserved. We can then assign a measure to each feasible[1] path $x = (x_0, x_1, \ldots, x_N) \in \mathcal{X}^{N+1}$,

$$\mathfrak{M}(x_0, x_1, \ldots, x_N) = \mu_0(x_0) m_{x_0 x_1} \cdots m_{x_{N-1} x_N}. \tag{2}$$

One-time marginals are then recovered through

$$\mu_t(x_t) = \sum_{x_{\ell \neq t}} \mathfrak{M}(x_0, x_1, \ldots, x_N), \quad t \in \mathcal{T}.$$

We seek a probability distribution $\mathfrak{P}$ on the feasible paths, with prescribed initial and final marginals $\nu_0(\cdot)$ and $\nu_N(\cdot)$, respectively, which is closest to the "prior" measure $\mathfrak{M}$ in relative entropy[2]. Let $\mathcal{P}(\nu_0, \nu_N)$ denote the family of such probability distributions. This brings us to the statement of the *Schrödinger Bridge Problem* (SBP):

Problem 1. Determine

$$\mathfrak{P}^*[\nu_0, \nu_N] := \operatorname{argmin}\{\mathbb{D}(\mathfrak{P} \| \mathfrak{M}) \mid \mathfrak{P} \in \mathcal{P}(\nu_0, \nu_N)\}. \tag{3}$$

---

[1] $x = (x_0, x_1, \ldots, x_N)$ is feasible if $(x_i, x_{i+1}) \in \mathcal{E}, \forall i$.

[2] *Relative entropy* (divergence, Kullback-Leibler index) is defined by

$$\mathbb{D}(\mathfrak{P} \| \mathfrak{M}) := \begin{cases} \sum_x \mathfrak{P}(x) \log \frac{\mathfrak{P}(x)}{\mathfrak{M}(x)}, & \text{Supp}(\mathfrak{P}) \subseteq \text{Supp}(\mathfrak{M}), \\ +\infty, & \text{Supp}(\mathfrak{P}) \not\subseteq \text{Supp}(\mathfrak{M}), \end{cases}$$

Here, by definition, $0 \cdot \log 0 = 0$. The value of $\mathbb{D}(\mathfrak{P} \| \mathfrak{M})$ may turn out to be negative due to the different total masses in the case when $\mathfrak{M}$ is not a probability measure. The optimization problem, however, poses no challenge as the relative entropy is (jointly) convex over this larger domain and bounded below.

Theorem 1 [9,19,29]. *Assume that M^N has all positive elements. Then there exists a pair of nonnegative functions $(\varphi, \hat{\varphi})$ defined on $\mathcal{T} \times \mathcal{X}$ and satisfying the system*

$$\varphi(t, x_t) = \sum_{x_{t+1}} m_{x_t x_{t+1}} \varphi(t+1, x_{t+1}), \quad t = 0, 1, \ldots, N-1, \tag{4}$$

$$\hat{\varphi}(t+1, x_{t+1}) = \sum_{x_t} m_{x_t x_{t+1}} \hat{\varphi}(t, x_t), \quad t = 0, 1, \ldots, N-1, \tag{5}$$

as well as the boundary conditions

$$\varphi(0, x_0) \cdot \hat{\varphi}(0, x_0) := \nu_0(x_0), \quad \varphi(N, x_N) \cdot \hat{\varphi}(N, x_N) := \nu_N(x_N), \quad \forall x_0, x_N \in \mathcal{X}. \tag{6}$$

Suppose moreover that $\varphi(t, i) > 0$, $\forall t \in \mathcal{T}, \forall x \in \mathcal{X}$. Then, the Markov distribution $\mathfrak{P}^*$ in $\mathcal{P}(\nu_0, \nu_N)$ having transition probabilities

$$\pi^*_{x_t x_{t+1}}(t) = m_{x_t x_{t+1}} \frac{\varphi(t+1, x_{t+1})}{\varphi(t, x_t)} \tag{7}$$

solves Problem 1.

For an extension of this result to a time-varying M see, e.g. [9]. In Theorem 1, if $(\varphi, \hat{\varphi})$ satisfies (4)–(5)–(6), so does the pair $(c\varphi, \frac{1}{c}\hat{\varphi})$ for all $c > 0$. Hence, uniqueness for the Schrödinger system is always intended as uniqueness of rays. The solution can be computed through the Fortet-IPF-Sinkhorn[3] iteration [14, 15,18,19,33].

Consider now an "energy function" E_x on the vertex set and the Boltzmann distribution

$$\pi_T(x) = Z(T)^{-1} \exp\left[-\frac{E_x}{kT}\right], \text{ where } Z(T) = \sum_x \exp\left[-\frac{E_x}{kT}\right], \tag{8}$$

for a suitable value of the "temperature" parameter T, and k a (Boltzmann) constant in units of energy. This distribution tends to the uniform for $T \to \infty$. For $T \searrow 0$, it has the remarkable property of concentrating on the set of *minimal energy states*. This is exploited in the search for absolute minima of non convex functions or even NP-hard problems such as the traveling salesman problem. Indeed, in Metropolis-like algorithms, it is more likely at low temperatures to sample minimal energy states [20]. Simulated annealing is then used to avoid getting stuck in local minima. In the case when $M = P(T)$ is an irreducible stochastic matrix such that

$$P(T)'\pi_T = \pi_T,$$

Theorem 1 can be used to optimally steer the Markov chain to the desired $\pi_{T_{\text{eff}}}, T_{\text{eff}} < T$, at time $t = N$. To keep the chain in $\pi_{T_{\text{eff}}}$ after time N, however, one has to switch to the time-invariant transition mechanism derived in the next section.

[3] Please see the Introduction for the historical justification of this name.

3 Optimal Steering to a Steady State

Suppose π is a desired probability distribution over the state space $\mathcal{X}$ of our resources, goods, packets, vehicles, etc. We tackle below the following key question: *How should we modify M so that the new evolution stays as close as possible to the prior but admits π as invariant distribution?*

3.1 Minimizing the Relative Entropy Rate

Let $\mathfrak{M}$ be the prior measure on the feasible paths as in Sect. 2. Our goal is to find a stochastic matrix Π such that $\Pi'\pi = \pi$ and the corresponding time-invariant measure $\mathfrak{P}$ on $\mathcal{X} \times \mathcal{X} \times \cdots$ is closest to $\mathfrak{M}$. Let $\mathcal{P}$ be the family of Markovian distributions on the feasible paths. More precisely, we consider the following problem:

Problem 2.

$$\begin{aligned} &\min_{\mathfrak{P}\in\mathcal{P}} \lim_{N\to\infty} \frac{1}{N}\mathbb{D}\left(\mathfrak{P}_{[0,N]}\|\mathfrak{M}_{[0,N]}\right) \\ \text{subject to}\quad &\Pi'\pi = \pi, \\ &\Pi\mathbb{1} = \mathbb{1}, \end{aligned}$$

where $\mathbb{1}$ is the vector of all ones.

By formula (18) in [29]

$$\mathbb{D}\left(\mathfrak{P}_{[0,N]}\|\mathfrak{M}_{[0,N]}\right) = \mathbb{D}(\pi\|m_0) + \sum_{k=0}^{N-1}\sum_{i_k}\mathbb{D}(\pi_{i_k i_{k+1}}\|m_{i_k i_{k+1}})\pi(i_k). \tag{9}$$

Thus, Problem 2 becomes the *stationary Schrödinger bridge problem:*

Problem 3.

$$\begin{aligned} &\min_{(\pi_{ij})}\sum_i \mathbb{D}(\pi_{ij}\|m_{ij})\pi(i), \\ \text{subject to}\quad &\sum_i \pi_{ij}\pi(i) = \pi(j), \quad j = 1,2,\ldots,n, \\ &\sum_j \pi_{ij} = 1, \quad i = 1,2,\ldots,n, \end{aligned}$$

where we have ignored the nonnegativity of the π_{ij}. This problem is readily seen to be equivalent to a standard *one-step* Schrödinger bridge problem for the joint distributions $p(i,j)$ and $m(i,j)$ at times $t = 0,1$ with the two marginals equal to π. Following Theorem 1, suppose there exist vectors with nonnegative components $(\varphi(t,\cdot),\hat{\varphi}(t,\cdot))$, $t = 0,1$ satisfying the Schrödinger system

$$\varphi(0,i) = \sum_j m_{ij}\varphi(1,j), \tag{10a}$$

$$\hat{\varphi}(1,j) = \sum_i m_{ij}\hat{\varphi}(0,i), \tag{10b}$$

$$\varphi(0,i)\cdot\hat{\varphi}(0,i) = \pi(i), \tag{10c}$$

$$\varphi(1,j)\cdot\hat{\varphi}(1,j) = \pi(j). \tag{10d}$$

Then, we get

$$\pi^*_{ij} = \frac{\varphi(1,j)}{\varphi(0,i)} m_{ij}, \tag{11}$$

which satisfies both constraints of Problem 3. The corresponding measure $\mathfrak{P}^* \in \mathcal{P}$ solves Problem 2. There is, however, a difficulty. The assumption in Theorem 1 which guarantees existence for the Schrödinger system takes here the form that M must have all positive elements: This is typically not satisfied. A more reasonable condition is provided below.

3.2 Existence for the One-Step Schrödinger System

Definition 1 [26]. *A square matrix A is called* fully indecomposable *if there exist no pair of permutation matrices P and Q such that*

$$A = P \begin{bmatrix} A_{11} & 0 \\ A_{21} & A_{22} \end{bmatrix} Q \tag{12}$$

where A_{11} and A_{22} are nonvacuous square matrices.

The following result may be proven along the lines of [26, Theorem 5].

Proposition 1. *Suppose M is fully indecomposable and π has all positive components. Then there exists a solution to (10) with $\varphi(0,\cdot)$ and $\varphi(1,\cdot)$ with positive components which is unique in the sense of the corresponding rays.*

A question that naturally arises in this context is the following: Given the graph $\mathbf{G}$, what distributions π admit at least one stochastic matrix compatible with the topology of the graph for which they are invariant? Clearly, if all self loops are present ($(i,i) \in \mathcal{E}, \forall i \in \mathcal{X}$), any distribution is invariant with respect to the identity matrix which is compatible. Without such a strong assumption, a partial answer is provided by the following result which follows from Theorem 1 and Proposition 1 taking as M the adjacency matrix, see [8].

Proposition 2. *Let π be a probability distribution supported on all of $\mathcal{X}$, i.e. $\pi(i) > 0, \forall i \in \mathcal{X}$. Assume that the adjacency matrix A of $\mathbf{G} = (\mathcal{X}, \mathcal{E})$ is fully indecomposable. Then, there exist stochastic matrices Π compatible with the topology of the graph $\mathbf{G}$ such that*

$$\Pi'\pi = \pi.$$

4 Reversibility

In [3, Corollary 2], we have shown that, in the case of a reversible prior (Boltzmann-Gibbs density) for a stochastic oscillator, the solution of the continuous countepart of Problem 2 (minimizing the *expected input power*) is reversible. We prove next that the same remarkable property holds here in the discrete setting.

Theorem 2 [8]. *Assume that M is reversible with respect to μ, i.e., that*

$$\mathrm{diag}(\mu)M = M'\,\mathrm{diag}(\mu). \tag{13}$$

Assume that M is fully indecomposable and that μ has all positive components. Then, the solution Π^* of Problem 3 is also reversible with respect to π.

Remark 1. Notice that reversibility in (13) does imply invariance as the rows of M do not necessarily sum to one.

Corollary 1. *Let $M = A$ the adjacency matrix. Suppose A is symmetric and fully indecomposable. Suppose π has all positive components. Then, the solution to Problem 3 is reversible with respect to π. Moreover, the corresponding path measure maximizes the entropy rate.*

Proof. The first part follows from Theorem 2 observing that A being symmetric is equivalent to A being reversible with respect to $\mathbb{1}$. For the second part, recall that entropy (rate) is the opposite of relative entropy (rate) with respect to the uniform. Finally, observe that, up to a positive scalar, the adjacency matrix is the uniform on the set of edges $\mathcal{E}$.

References

1. Beurling, A.: An automorphism of product measures. Ann. Math. **72**, 189–200 (1960)
2. Bohm, D.: A suggested interpretation of the quantum theory in terms of "hidden" variables I. Phys. Rev. **85**, 166 (1952)
3. Chen, Y., Georgiou, T.T., Pavon, M.: Fast cooling for a system of stochastic oscillators. J. Math. Phys. **56**(11), 113302 (2015)
4. Chen, Y., Georgiou, T.T., Pavon, M.: Entropic and displacement interpolation: a computational approach using the Hilbert metric. SIAM J. Appl. Math. **76**(6), 2375–2396 (2016)
5. Chen, Y., Georgiou, T.T., Pavon, M.: Stochastic control liaisons: Richard Sinkhorn meets Gaspard Monge on a Schrödinger bridge, ArXiv e-prints, arXiv: 2005.10963. SIAM Review, **63**, 249–313 (2021)
6. Chen, Y., Georgiou, T.T., Pavon, M.: Optimal transport in systems and control. Ann. Rev. Contr. Robot. Auton. Syst. **4**, 89–113 (2021)

7. Chen, Y., Georgiou, T.T., Pavon, M.: Controlling uncertainty: Schrödinger's inference method and the optimal steering of probability distributions. IEEE Control Syst. Mag, (2020, to appear)
8. Chen, Y., Georgiou, T.T., Pavon, M.: Optimal steering to invariant distributions for networks flows. http://arxiv.org/abs/2102.12628
9. Chen, Y., Georgiou, T.T., Pavon, M., Tannenbaum, A.: Robust transport over networks. IEEE Trans. Aut. Control **62**(9), 4675–4682 (2017)
10. Chen, Y., Georgiou, T.T., Pavon, M., Tannenbaum, A.: Efficient-robust routing for single-commodity network flows. IEEE Trans. Aut. Control **63**, 2287–2294 (2018)
11. Chen, Y., Georgiou, T.T., Pavon, M., Tannenbaum, A.: Relaxed Schroödinger bridges and robust network routing. IEEE Trans. Control Netw. Syst. **7**(2), 923–931 (2020)
12. Conforti, G.: A second order equation for Schrödinger bridges with applications to the hot gas experiment and entropic transportation cost. Prob. Th. Rel. Fields **174**(1), 1–47 (2019)
13. Conforti, G., Pavon, M.: Extremal flows in Wasserstein space. J. Math. Phys. **59**(6), 063502 (2018)
14. Cuturi, M.: Sinkhorn distances: lightspeed computation of optimal transport. In: Advances in Neural Information Processing Systems, pp. 2292–2300 (2013)
15. Deming, W.E., Stephan, F.F.: On a least squares adjustment of a sampled frequency table when the expected marginal totals are known. Ann. Math. Statist. **11**(4), 427–444 (1940)
16. Fényes, I.: Eine wahrscheinlichkeitstheoretische Begründung und Interpretation der Quantenmechanik. Z. Phys. **132**, 81–106 (1952)
17. Föllmer, H.: Random fields and diffusion processes. In: Hennequin, P.-L. (ed.) École d'Été de Probabilités de Saint-Flour XV–XVII, 1985–87. LNM, vol. 1362, pp. 101–203. Springer, Heidelberg (1988). https://doi.org/10.1007/BFb0086180
18. Fortet, R.: Résolution d'un système d'equations de M. Schrödinger. J. Math. Pure Appl. **IX**, 83–105 (1940)
19. Georgiou, T.T., Pavon, M.: Positive contraction mappings for classical and quantum Schrödinger systems. J. Math. Phys. **56**, 033301 (2015)
20. Häggström, O.: Finite Markov Chains and Algorithmic Applications, London Mathematical Society Student Texts 52. Cambridge University Press, Cambridge (2002)
21. Hernandez, D.B., Pavon, M.: Equilibrium description of a particle system in a heat bath. Acta Appl. Math. **14**, 239–256 (1989)
22. Jamison, B.: The Markov processes of Schrödinger. Z. Wahrscheinlichkeitstheorie verw. Gebiete **32**, 323–331 (1975)
23. Léonard, C.: A survey of the Schroedinger problem and some of its connections with optimal transport. Discrete Contin. Dyn. Syst. A **34**(4), 1533–1574 (2014)
24. Levy, B.C., Krener, A.J.: Stochastic mechanics of reciprocal diffusions. J. Math. Phys. **37**, 769 (1996)
25. Madelung, E.: Quantentheorie in hydrodyn. Form Z. Physik **40**, 322–326 (1926)
26. Marshall, A.W., Olkin, I.: Scaling of matrices to achieve specified row and column sums. Numer. Math. **12**, 83–90 (1968)
27. Mikami, T.: Monge's problem with a quadratic cost by the zero-noise limit of h-path processes. Probab. Theory Relat. Fields **129**, 245–260 (2004)
28. Nelson, E.: Dynamical Theories of Brownian Motion. Princeton University Press, Princeton (1967)
29. Pavon, M., Ticozzi, F.: Discrete-time classical and quantum Markovian evolutions: maximum entropy problems on path space. J. Math. Phys. **51**, 042104 (2010)

30. Peyré, G., Cuturi, M.: Computational optimal transport. Found. Trends Machine Learn. **11**(5–6), 1–257 (2019)
31. Schrödinger, E.: Über die Umkehrung der Naturgesetze, Sitzungsberichte der Preuss Akad. Wissen. Berlin Phys. Math. Klasse **10**, 144–153 (1931)
32. Schrödinger, E.: Sur la théorie relativiste de l'électron et l'interpretation de la mécanique quantique. Ann. Inst. H. Poincaré **2**, 269 (1932)
33. Sinkhorn, R.: A relationship between arbitrary positive matrices and doubly stochastic matrices. Ann. Math. Statist. **35**(2), 876–879 (1964)
34. von Renesse, M.K.: An optimal transport view on Schrödinger's equation. Canad. Math. Bull **55**(4), 858–869 (2012)
35. Villani, C.: Topics in optimal transportation, vol. 58. AMS (2003)
36. Wakolbinger, A.: Schroedinger bridges from 1931 to 1991. Contribuciones en probabilidad y estadistica matematica **3**, 61–79 (1992)
37. Zambrini, J.C.: Stochastic mechanics according to E. Schrödinger. Phys. Rev. A **33**, 1532–1548 (1986)

Geometry of Outdoor Virus Avoidance in Cities

Philippe Jacquet[1(✉)] and Liubov Tupikina[2]

[1] Inria, Rocquencourt, France
philippe.jacquet@inria.fr
[2] Nokia Bell-Labs, Murray Hill, USA
liubov.tupikina@nokia-bell-labs.com

Abstract. We present low complexity and efficient path selection algorithms for limiting the exposure to pandemic virus during and between lock-downs. The average outdoor exposure rate reduction is around 3.

Keywords: Virus exposure · Random walks · City maps

1 Introduction

The Covid-19 pandemic has caused repetitive lock-downs for most of mankind in order to limit the virus circulation. In many countries outdoor excursion were permitted during lock-downs under strict restrictions; in France, no more than 1 h within 1 km from home. All efforts done in order to reduce the daily exposure time to virus, the lock-down's aim is to "flatten the curve" in order to limit the congestion in hospitals and must be repeated every time the curve becomes critical and before vaccines and herd-immunity take over. But social distancing needs to be applied during inter lock-down periods.

This paper is divided in two parts. In the first part we introduce an application, called Ariadne Covid. It selects outdoor excursion path with limited exposure to virus during lock-downs based on its geometrical properties. The second part is dedicated to inter lock-downs periods, and we introduce a class of algorithms to select commuting paths.

The fundamental tool in Ariadne Covid is the use of random walks which naturally leads to to a better load balance between streets while avoiding crowded areas. The application is shown to give a benefit to the first user, although to have an impact on the pandemic, it must be used by a majority of users.

During inter lock-downs periods people are allowed to commute from home to work. The path selection presented in the second part is based on drifted random walks in order to reach the destinations without too much elongations while offering balanced loads on city streets. Both applications take benefit from the geometry of city maps which leads to naturally homogeneous street networks.

F. Nielsen and F. Barbaresco (Eds.): GSI 2021, LNCS 12829, pp. 869–877, 2021.
https://doi.org/10.1007/978-3-030-80209-7_93

2 Covid Exposure During Outdoor Excursion in Cities

The virus of Covid-19 is supposed to have some probability to be transferred when an infected person is within 1 m of another person during a certain time (10 m when running or biking). The larger is the exposure time the higher is the contamination probability (the average contamination delay is around 30 min). We call exposure time the cumulated time a given person is within 1 m of any other person during an outdoor excursions.

There are 2,900 km of cumulative pedestrian path length in Paris for a population of 2.2 million inhabitants [1]. Therefore there is ample room for all of the inhabitants to walk 24/24 h within safe distance. Since only a fraction of the inhabitant are expected to be walking at the same time, it should be easy to get the safe spacing.

The problem is that people show a *social* trend to gather in the same streets, even during lock-downs, rising the exposure to virus. Data sampled via Google Street View show how negative is the impact of the social trends. In the paper we will call *social walkers* (or social joggers) the walkers who follow the social trends during outdoor excursions in the city.

We have sampled 55 km of Paris street via Google Street View. We have made 11 independent random walks starting in different *arrondissements*. A segment (30 m) of pedestrian path is sampled every 100 m. Figure 1 displays the histogram of segment densities sampled in Paris. The Street View pictures date before the lockdown, and mix several periods. The histogram shows a profile close to a Zipf distribution of coefficient close to 0.75. We have no yet records for the period during the lockdown, but we expect a more concentrated distribution with a larger Zipf coefficient, since the most of the commercial areas are closed and therefore attraction streets are reduced.

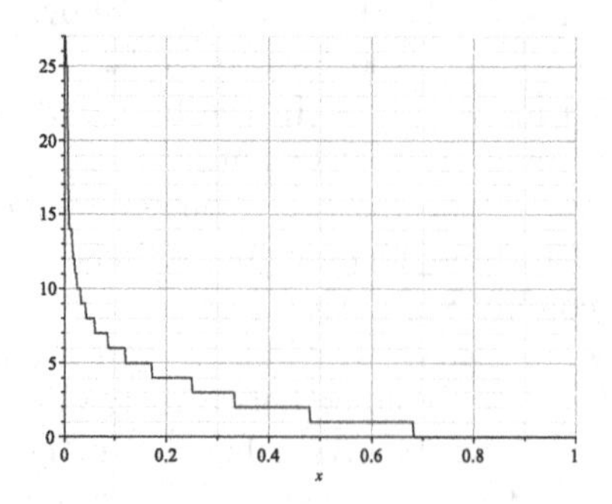

Fig. 1. Histogram of Pedestrian densities in Paris

We represent the city by an abstract graph, with the street intersection (vertices) linked by street segments (edge). We denote V the set of intersections and E the set of street segments.

3 Path Selection During Lock-Downs

3.1 Ariadne String Protocol Description

The application consists into delivering on request, a random path for the excursion and a time slot in the day. This path will reduce the exposure to the virus by avoiding the dangerous areas with an abnormal density of social walkers. In its most basic setting the path will be a pure random walk, since it shows to actually reduce the exposure rate even if the user faces only social walkers. Indeed it is shown that the outdoor exposure to virus is even more reduced compared to the situation where all inhabitants use the application.

During the excursion the user will have his/her own position displayed on her/his mobile phone. At half of the excursion time the walker will receive a notification to reverse the path in order to come back home by keeping path uniformization. The path may contain loops, because they are necessary in order to get the uniform density. If it is the case the walker will proceed with the loops. The walker may skip the loop in the way back but this will shorten the delay of the excursion. The application terminates (at least its embedded mobile part) when the user is back home. There are several options in the randomized path selection.

Time Slot Determination. The time slot is selected uniformly on the time schedule proposed by the user in his/her request. In France, the slot has a maximum duration of one hour.

Path Initialisation. The first step is to randomly select the initial direction: left or right with probability 1/2, 1/2 on the street/edge of the home address.

Non-backtracking Algorithm: at each intersection J the algorithm randomly selects the exit road excluding the entry street. If the intersection has four roads, thus three eligible exits, the weight of each exit is 1/3, 1/3, 1/3. In case of dead-end, the algorithm selects the U turn. This guarantees the uniformisation of the densities. If the cartesian distance to home address exceeds the authorized distance (1 km in France), the path should make a U-turn.

Non-backtracking Algorithm with Reduced Path Diversions: if the intersection has an even number of streets (thus an odd number of exits) more weight will be given to the exit street facing the entry street, *i.e.* the median exit. For example, the median exit has 1/2 probability to be selected, and the other exits are uniformly selected. This minor trick still maintain the uniformization of the densities.

3.2 Estimation of Exposure Reduction During Covid-19 Lock-Downs

Let E be the street set and for $s \in E$, let $\ell(s)$ its length and $\lambda(s,t)$ its linear density at time t. The total length of the pedestrian network is $L = \sum_{s \in E} \ell(s)$. The total pedestrian population at time t is $N(t) = \sum_{s \in E} \lambda(s,t)\ell(s)$. The probability that a pedestrian walk on segment s at time t is $\rho(s,t) = \frac{\ell(s)\lambda(s,t)}{N(t)}$. If we assume that the distribution on every segment is uniform (and ignoring side effects) the probability that a given pedestrian on segment s is within distance r to another given pedestrian on the same segment is $\frac{2r}{\ell(s)}$ which could be reduced to $\frac{r}{\ell(s)}$ if the dangerous zone stands forward to the pedestrian.

Thus if a total population of N pedestrians are present on the streets, each independently following the density pattern of the $\lambda(s,t)$, the probability that a pedestrian on segment s is not on situation of contact at time t is $\left(1 - \rho(s,t)\frac{2r}{\ell(s)}\right)^N$ or equivalently $\left(1 - 2r\frac{\lambda(s,t)}{N(t)}\right)^N$.

Let N_T the total population of the city. If the proportion of inhabitants walking outdoor is ν, the average exposure rate $E_G(\nu)$ satisfies

$$E_G(\nu) = 1 - \sum_{s \in E} \rho(s) \left(1 - \rho(s)\frac{2r}{\ell(s)}\right)^{\nu N_T}. \tag{1}$$

assuming that $\rho(s) := \rho(s,t)$ does not vary (too much) during the permitted hours. If $\rho(s,t)$ varies during the day and $\rho(s)$ is its average on segment s, then by concavity the above expression is an upper bound of the exposure rate.

Theorem 1. *The stationary distribution of the non-backtracking random walk is the uniform distribution density.*

By uniform distribution we mean that the average densities in every street are identical.

Proof. This is mainly a state of the art result.

But the random walk may be long to converge to its steady state. In our case it is faster because the density of inhabitant in Paris is almost uniform along each street, thanks to Hausmannian construction rules dating from the XIXth century. From now we take the simplified assumption that walkers using the random walk are indeed in uniform density in the streets.

Let $E_U(\nu)$ be the average exposure rate when all walkers are random walkers. It satisfies:

$$E_U(\nu) = 1 - \left(1 - \frac{2r}{L}\right)^{\nu N_T}. \tag{2}$$

Indeed it suffices to consider the whole pedestrian network as a single segment.

More interestingly we have the case when a single random walker proceeds among νN_T walkers following their social trends. Let's call $E_{UG}(\nu)$ her/his exposure rate:

$$E_{UG}(\nu) = 1 - \sum_{s \in E} \frac{\ell(s)}{L} \left(1 - \rho(s)\frac{2r}{\ell(s)}\right)^{\nu N_T}. \tag{3}$$

We assume that Parisians have a daily time span (10 h) when excursions are permitted.

The Fig. 2 shows the cumulative time spent within unsafe distance to any another individual. Notice that with one hour of excursion time with all social walkers, each walker cumulates nearly half of this time at less than 1 m of another individual. This duration drops to 8 min with random walking. If the excursions were extended to 5 h, the social walkers would cumulate 4 h 30 min within unsafe distance, the random walkers around 2 h 30 min.

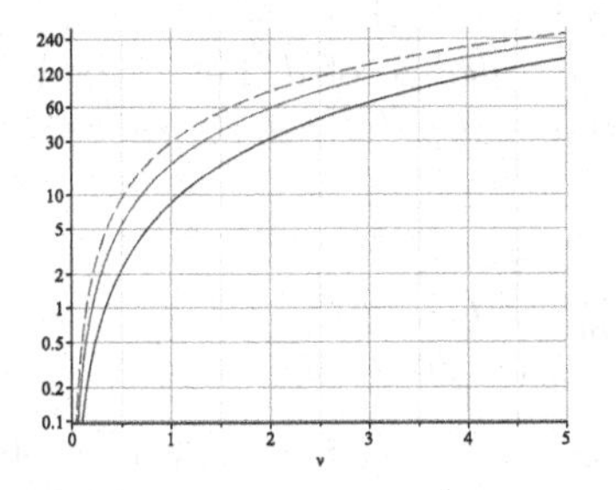

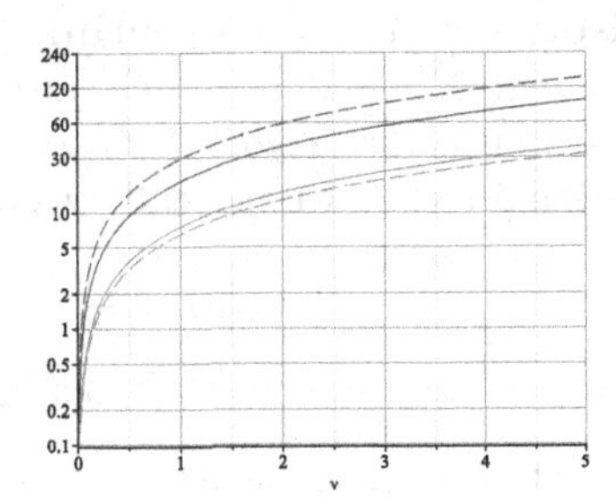

Fig. 2. Virus exposure time (min) as function of outdoor excursion duration (in hours). (Left) red: all walkers are social but randomly dispatched on 10 h (solid) on 5 h (dashed); blue: when walkers are all random. (Right) red: for a single social walker, when all other walkers are social walkers during 1 h, green: the single walker is random, dashed: social walkers select slot on an interval of 5 h instead of 20 h. (Color figure online)

4 Path Selection During Lock-Downs Periods

Contrary to lock-down outdoor excursions which are mostly aimless (walking, jogging), the excursions during inter lock-downs periods are now aimed (commuting between home and work place). The public transportation will be mostly discarded because of poor social distancing. The city of Paris has increased the number of biking lanes. However the biking lanes will be very busy because people commute around the same hours. Our work is to extend Ariadne Covid application in order to help the user to find biking routes from home to work, which limit the virus exposure.

4.1 Path Selection Algorithm

An initial position of a user is on a segment $(I_U, I_D) \in E$. The path will take the intersection among I_U and I_D which lies ahead of the destination z_f, *i.e.* if $\langle z_f - z_0 | I_U - I_D \rangle \geq 0$ then I_U is selected: $z_1 = I_U$, otherwise $z_1 = I_D$.

At step number k of the algorithm, let z_k be the intersection, where the path is currently ending. Let d be the degree of the intersection z_k. If $d = 1$, then the path backtracks. If $d > 1$ then with probability ϵ the path proceeds as a non-backtracking random walk, *i.e.* the path takes any other neighbor edge distinct of the entrance street with equal probability. Otherwise, with probability $1 - \epsilon$, the path takes one of the two best sectored streets, defined as follows. Assuming the heading of the segment (z_k, z_{k-1}) is θ_0, and we enumerate $\theta_1, \theta_2, \ldots \theta_{d-1}$ the headings in increasing order of the other exit segments of the intersection z_k. Let θ be the heading of the vector (z_k, z_f) toward the destination.

The Isotropic Walk Algorithm. Let j be such that $j\frac{2\pi}{d} < \theta \leq (j+1)\frac{2\pi}{d}$. Note here that $j+1$ must be considered by *modulo* of d. Let variables $\alpha_d(\theta)$ and $\beta_d(\theta)$ such that $\alpha_d(\theta) + \beta_d(\theta) = 1$ and $\alpha_d(\theta)e^{2ij\pi/d} + \beta_d(\theta)e^{2i(j+1)\pi/d}$ be proportional to $e^{i\theta}$. Numerically we have

$$\begin{cases} \alpha_d(\theta) = \frac{\sin(2(j+1)\pi/d-\theta)}{\sin(\theta-2j\pi/d)+\sin(2(j+1)\pi/d-\theta)} \\ \beta_d(\theta) = \frac{\sin(\theta-2j\pi/d)}{\sin(\theta-2j\pi/d)+\sin(2(j+1)\pi/d-\theta)} \end{cases}. \tag{4}$$

The selection of the next intersection z_{k+1} is done as follows: with probability $\alpha_d(\theta)$ it selects the edge corresponding to angle θ_j, and with probability $\beta_d(\theta)$ the edge corresponding to the angle θ_{j+1}. The algorithm is much less complex than the Dijkstra shortest path algorithm and contrary to the later provides path diversity. Indeed the shortest path algorithm leads to a complexity larger than the number of edges, while the randomized algorithm is at most linear, in the number of vertices, in fact proportional to the diameter of the graph.

Definition 1 (Isotropic walk condition). *A walk is isotropic, when reaching any given intersection I, the difference of an angle θ toward a destination with an angle θ_0 to an arriving edge, satisfies a certain distribution $P_I(\theta - \theta_0)$.*

Theorem 2. *Under the isotropic walk condition, and assuming the same constant speed v of all travellers in each street, the aggregation of paths leads to uniform densities of travellers on streets.*

Proof. We show that uniform rate situation is stationary. Let's consider that a walker enters an intersection I of degree d. In the epsilon mode the walker uniformly exits on the other streets. In the sectored mode, enumerating the edges in counter clockwise way from the entrance street, the isotropic walk conditions leads to the probability $p_d(j)$ that the edge j is selected:

$$p_d(j) = \int_{2j\pi/d}^{2(j+1)\pi/d} P_I(\theta)\alpha_d(\theta) + \int_{2(j-1)\pi/d}^{2j\pi/d} P_I(\theta)\beta_d(\theta). \tag{5}$$

Let us assume that before time t all streets have same exit rate ρ. Therefore the entrance flow on any of the streets is $\epsilon\rho + (1-\epsilon)\sum_{j=0}^{j=d} p_d(j)\rho = \rho$ and consequently the exit rates are uniform beyond t. Assuming a uniform speed v on each street, the density of travellers on each street is ρ/v per unit length.

4.2 The Geo-Routing Algorithm

We introduce the geo-routing algorithm ispired from [2], which is very close to the isotropic walk with the difference that we replace the $2j\pi/d$ by the actual angles of the streets. The consequences are twofold: (i) the average exit path is heading toward the destination, (ii) we lose the pure isotropic property, the densities in the exit streets will now vary with the angles between the streets.

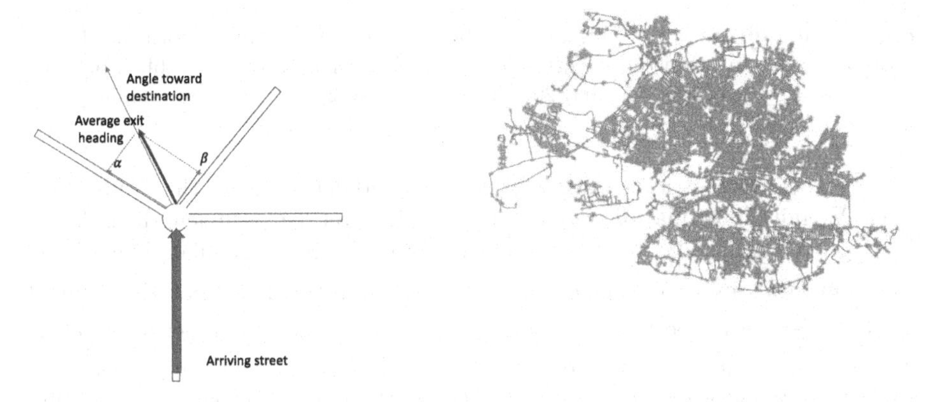

Fig. 3. Left: Illustration of steps of the randomized geo-routing algorithm. The traveller enters the intersection via the large blue arrow. Right: the map of Kaliningrad. (Color figure online)

4.3 Simulation of the Algorithms

We have simulated the algorithms on the map of the city of Kaliningrad, famous for its network of bridges, river and marshes (Fig. 3). Instead of the classic grid maps of modern cities, the natural obstacles, blocking direct headings, hamper the performances of the algorithms. Indeed we get some blocking loops. In order to cope with looping problems we used the concept of the so-called accessibility graph, see [3]. Let $G = G(V, E)$ be the street graph. The second degree accessibility graph of the graph $G * G$ or G^{*2} is the graph whose vertex set is V and the edge set is E_2 such that for $(x, y) \in V^2$, $(xy) \in E_2$ if there exists $z \in V$ such that (xz) and (zy) belong to E. We have computed G^{*2}, G^{*4} and G^{*8}. If the algorithm on a path determination fails on G because of a permanent loop, we run the algorithm on G^{*2}. Then, if it again fails on G^{*2^k} we run the algorithm on the accessibility graph of the next degree $G^{*2^{k+1}}$ until the accessibility graph becomes fully connected graph (we never get to this). When the path is determined one unfolds the path on the original graph G. We have simulated the application of the isotropic walk and the geo-routing algorithm over 1,000 requests between the same pair initial point and destination point. We set $\epsilon = 0$. The Fig. 4 shows the various path provided by the algorithms.

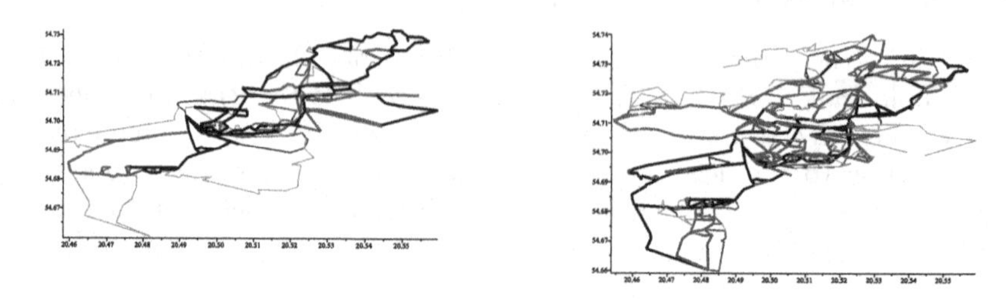

Fig. 4. Path diversity in central Königsberg. Left: geo-routing algorithm. Right: isotropic walks. In green the streets belonging to less than 25 paths, in blue, between 25 and 50, in red, between 50 and 200. In black, above 200. (Color figure online)

Let $s \in E$, $\ell(s)$ is the segment length, and let $\lambda(s)$ be the traffic load after N_0 initial-destination random pairs. The average path length is $L_G = \frac{1}{N_0}\sum_{s\in E}\lambda(s)\ell(s)$. The average travel time is L_G/v. If the simulated time is T, the segment s sees an average entrance rate $\frac{\lambda(s)}{T}$ and the density on the segment length is $\frac{\lambda(s)}{vT}$, assumed to be Poisson. If we consider a total number of travellers N, then the density on a segment is $\frac{\lambda(s)}{vT}\frac{N}{N_0}$, where N_0 is a number of travellers needed to simulate the estimate of $\lambda(s)$. Given the Poisson density, a probability that a random traveller at a random time on the segment is not within distance R_0 of another traveller is $\exp\left(-\lambda(s)\frac{2R_0}{vT}\frac{N}{N_0}\right)$, R_0 is the safe distance against the virus (for biking 10 m). We write the expression of an average cumulative exposure time $E(N)$ for a random walk (traveller):

$$E(N) = \sum_{s\in E}\frac{\lambda(s)\ell(s)}{vN_0}\left(1-\exp\left(-\lambda(s)\frac{2R_0}{vT}\frac{N}{N_0}\right)\right). \tag{6}$$

Figure 5 shows the simulation of the path length and exposure times with both algorithms after $N_0 = 5,000$ random initial-destination pairs. The geo-routing algorithm shows a slightly better path length and cumulated exposure.

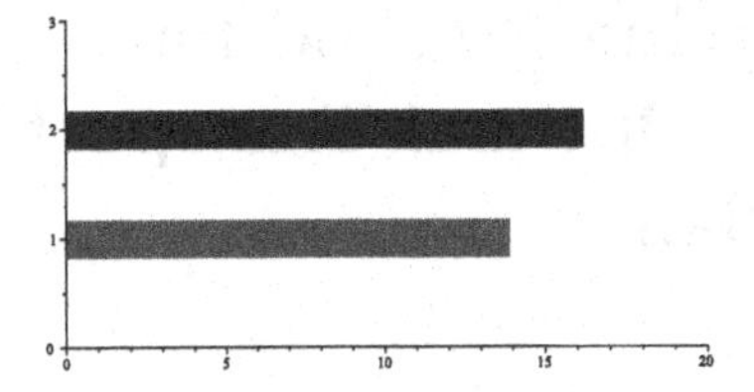

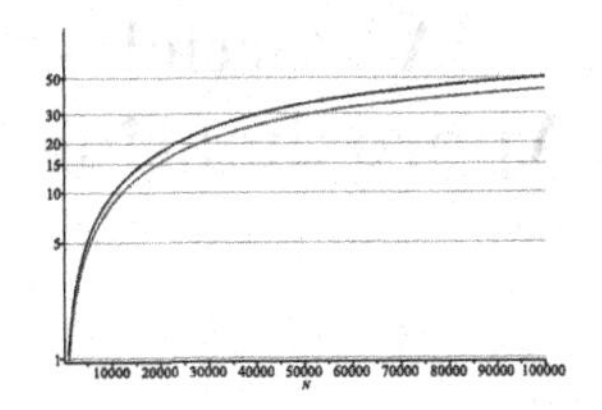

Fig. 5. In blue the isotropic routing, in brown the geo-routing in Kaliningrad (Königsberg). (Left) average path length in km. (Right) the exposure time in minute as function of travellers number N, peak time duration: 2 h. (Color figure online)

5 Conclusions

We have presented the path selection algorithms and efficient path selection, which limits outdoor virus exposure during or between lock-down periods by path diversity and exploiting the geometry of street networks in cities. The proposed solution is based on low complexity algorithms. For practical applications one needs to take into account that success rate of the algorithm implementation strongly depends on users acceptance of those rules, which is left as an outlook.

References

1. https://www.pariszigzag.fr/secret/histoire-insolite-paris/combien-y-a-t-il-de-kilometres-de-trottoirs-a-paris
2. Jacquet, P., Mans, B., Muhlethaler, P., Rodolakis, G.: Opportunistic routing in wireless ad hoc networks: Upper bounds for the packet propagation speed. IEEE J. Select. Areas Commun. **27**(7), 1192–1202 (2009)
3. Lentz, H.H.K., Selhorst, T., Sokolov, I.M.: Unfolding accessibility provides a macroscopic approach to temporal networks. Phys. Rev. Lett. **110**(11) 118701 (2013)

A Particle-Evolving Method for Approximating the Optimal Transport Plan

Shu Liu[1(✉)], Haodong Sun[1(✉)], and Hongyuan Zha[2]

[1] Georgia Institute of Technology, Atlanta, USA
{sliu459,hsun310}@gatech.edu
[2] The Chinese University of Hong Kong, Shenzhen, China

Abstract. We propose an innovative algorithm that iteratively evolves a particle system to approximate the sample-wised Optimal Transport plan for given continuous probability densities. Our algorithm is proposed via the gradient flow of certain functional derived from the Entropy Transport Problem constrained on probability space, which can be understood as a relaxed Optimal Transport problem. We realize our computation by designing and evolving the corresponding interacting particle system. We present theoretical analysis as well as numerical verifications to our method.

Keywords: Optimal Transport · Entropy Transport · Wasserstein gradient flow · Kernel Density Estimation · Interacting particle systems

1 Introduction

Optimal transport (OT) provides powerful tools for comparing probability measures in various types. The optimal Transport problem was initially formalized by Gaspard Monge [19] and later generalized by Leonid Kantorovich [12]. Later a series of significant contribution in transportation theory leads to deep connections with more mathematical branches including partial differential equations, geometry, probability theory and statistics [5,12]. Researchers in applied science fields also discover the importance of Optimal Transport. In spite of elegant theoretical results, generally computing Wasserstein distance is not an easy task, especially for the continuous case.

In this paper, instead of solving the standard Optimal Transport (OT) problem, we start with the so-called Entropy Transport (ET) problem, which can be treated as a relaxed OT problem with soft marginal constraints. Recently, the importance of Entropy Transport problem has drawn researchers' attention due to its rich theoretical properties [16]. By restricting ET problem to probability manifold and formulating the gradient flow of the target functional of the Entropy Transport problem, we derive a time-evolution Partial Differential Equation (PDE) that can be then realized by evolving an interacting particle system via Kernel Density Estimation techniques [22].

F. Nielsen and F. Barbaresco (Eds.): GSI 2021, LNCS 12829, pp. 878–887, 2021.
https://doi.org/10.1007/978-3-030-80209-7_94

Our method directly computes for the sample approximation of the optimal coupling to the OT problem between two density functions. This is very different from traditional methods like Linear Programming [20,24,27], Sinkhorn iteration [8], Monge-Ampère Equation [4] or dynamical scheme [3,15,23]; or methods involving neural network optimizations [13,18,25].

Our main contribution is to analyze the theoretical properties of the Entropy Transport problem constrained on probability space and derive its Wasserstein gradient flow. To be specific, we study the existence and uniqueness of the solution to ET problem and further study its Γ-convergence property to the classical OT problem. Then based on the gradient flow we derive, we propose an innovative particle-evolving algorithm for obtaining the sample approximation of the optimal transport plan. Our method can deal with optimal transport problem between two known densities. As far as we know, despite the classical discretization methods [3,4,15] there is no scalable way to solve this type of problem. We also demonstrate the efficiency of our method by numerical experiments.

2 Constrained Entropy Transport as a Regularized Optimal Transport Problem

2.1 Optimal Transport Problem and Its Relaxation

In our research, we will mainly focus on Euclidean Space $\mathbb{R}^d$. We denote $\mathcal{P}(E)$ as the probability space defined on the given measurable set E. The **Optimal Transport** problem is usually formulated as

$$\inf_{\substack{\gamma\in\mathcal{P}(\mathbb{R}^d\times\mathbb{R}^d),\\ \gamma_1=\mu,\gamma_2=\nu}} \iint c(x,y)\, d\gamma(x,y). \tag{1}$$

Here $\mu,\nu\in\mathcal{P}(\mathbb{R}^d)$, we denote γ_1 as the marginal distribution of γ w.r.t. component x and γ_2 as the marginal distribution w.r.t. y. We call the optimizer of (1)[1] as **Optimal Transport plan** and we denote it as γ_{OT}.

We can also reformulate (1) as $\min_{\gamma\in\mathcal{P}(\mathbb{R}^d\times\mathbb{R}^d)}\left\{\mathcal{E}_\iota(\gamma|\mu,\nu)\right\}$ where

$$\mathcal{E}_\iota(\gamma|\mu,\nu)=\iint c(x,y)d\gamma(x,y)+\int \iota\left(\frac{d\gamma_1}{d\mu}\right)d\mu+\int \iota\left(\frac{d\gamma_2}{d\nu}\right)d\nu \tag{2}$$

Here ι is defined as $\iota(1)=0$ and $\iota(s)=+\infty$ when $s\neq 1$. We now derive a relaxed version of (1) by replacing $\iota(\cdot)$ with $\Lambda F(\cdot)$, where $\Lambda>0$ is a tunable positive parameter and F is a smooth convex function with $F(1)=0$ and 1 is the unique minimizer. In our research, we mainly focus on $F(s)=s\log s-s+1$ [16]. It is worth mentioning that after such relaxation, the constraint term $\int F(\frac{d\gamma_1}{d\mu})\,d\mu$ is usually called Kullback-Leibler (KL) divergence [14] and is denoted as $D_{\mathrm{KL}}(\gamma_1\|\mu)$.

[1] When μ,ν are absolute continuous with respect to the Lebesgue measure on $\mathbb{R}^d$, the optimizer of (1) is guaranteed to be unique.

From now on, we should focus on the following functional involving cost $c(x,y) = h(x-y)$ with h as a strictly convex function, and enforcing the marginal constraints by using KL-divergence:

$$\mathcal{E}_{\Lambda,\mathrm{KL}}(\gamma|\mu,\nu) = \iint_{\mathbb{R}^d\times\mathbb{R}^d} c(x,y)\, d\gamma(x,y) + \Lambda D_{\mathrm{KL}}(\gamma_1\|\mu) + \Lambda D_{\mathrm{KL}}(\gamma_2\|\nu). \quad (3)$$

2.2 Constrained Entropy Transport Problem and Its Properties

For the following discussions, we always assume that $\mu, \nu \in \mathcal{P}(\mathbb{R}^d)$ are **absolute continuous** with respect to the Lebesgue measure on $\mathbb{R}^d$. We now focus on solving the following problem

$$\min_{\gamma\in\mathcal{P}(\mathbb{R}^d\times\mathbb{R}^d)} \{\mathcal{E}_{\Lambda,\mathrm{KL}}(\gamma|\mu,\nu)\}. \quad (4)$$

A similar problem

$$\min_{\gamma\in\mathcal{M}(\mathbb{R}^d\times\mathbb{R}^d)} \{\mathcal{E}_{\Lambda,\mathrm{KL}}(\gamma|\mu,\nu)\} \quad (5)$$

has been studied in [16] with $\mathcal{P}(\mathbb{R}^d \times \mathbb{R}^d)$ being replaced by the space of positive measures $\mathcal{M}(\mathbb{R}^d \times \mathbb{R}^d)$ and is named as **Entropy Transport** problem therein. In our case, since we restrict γ to probability space, we call (4) **constrained Entropy Transport problem** and call $\mathcal{E}_{\Lambda,\mathrm{KL}}$ the Entropy Transport functional.

Let us denote $\mathcal{E}_{\min} = \inf_{\gamma\in\mathcal{P}(\mathbb{R}^d\times\mathbb{R}^d)}\{\mathcal{E}_{\Lambda,\mathrm{KL}}(\gamma|\mu,\nu)\}$. The following theorem shows the existence of the optimal solution to problem (4). It also describes the relationship between the solution to the constrained ET problem (4) and the solution to the general ET problem (5):

Theorem 1. *Suppose $\tilde{\gamma}$ is the solution to original Entropy Transport problem* (5). *Then we have $\tilde{\gamma} = Z\gamma$, here $Z = e^{-\frac{\mathcal{E}_{\min}}{2\Lambda}}$ and $\gamma \in \mathcal{P}(\mathbb{R}^d \times \mathbb{R}^d)$ is the solution to constrained Entropy Transport problem* (4).

The proof is provided in [17]. The following corollary guarantees the uniqueness of optimal solution to (4):

Corollary 1. *The constrained ET problem* (4) *admits a unique optimal solution.*

Despite the discussions on the constrained problem (4) with fixed Λ, we also establish asymptotic results for (4) with quadratic cost $c(x,y) = |x-y|^2$ as $\Lambda \to +\infty$. For the rest of this section, we define:

$$\mathcal{P}_2(E) = \left\{\gamma \middle|\ \gamma \in \mathcal{P}(E), \gamma \ll \mathscr{L}^d,\ \int_E |x|^2 d\gamma(x) < +\infty\right\} \quad E \text{ measurable.}$$

Here we denote $\mathscr{L}^d$ as the Lebesgue measure on $\mathbb{R}^d$.

Let us now consider $\mathcal{P}_2(\mathbb{R}^d \times \mathbb{R}^d)$ and assume it is equipped with the topology of weak convergence. We are able to establish the following Γ-convergence result.

Theorem 2. *Suppose* $c(x,y) = |x-y|^2$, *let us assume* $\mu, \nu \in \mathcal{P}_2(\mathbb{R}^d)$, $\mu, \nu \ll \mathscr{L}^d$ *and both* μ, ν *satisfy the Logarithmic Sobolev inequality with constants* $K_\mu, K_\nu > 0$. *Let* $\{\Lambda_n\}$ *be a positive increasing sequence with* $\lim_{n\to\infty} \Lambda_n = +\infty$.

We consider the sequence of functionals $\{\mathcal{E}_{\Lambda_n,\mathrm{KL}}(\cdot|\mu,\nu)\}$. *Recall the functional* $\mathcal{E}_\iota(\cdot|\mu,\nu)$ *defined in* (2). *Then* $\{\mathcal{E}_{\Lambda_n,\mathrm{KL}}(\cdot|\mu,\nu)\}$ Γ*- converges to* $\mathcal{E}_\iota(\cdot|\mu,\nu)$ *on* $\mathcal{P}_2(\mathbb{R}^d \times \mathbb{R}^d)$.

Furthermore, (4) *with functional* $\mathcal{E}_{\Lambda_n,\mathrm{KL}}(\cdot|\mu,\nu)$ *admits a unique optimal solution, let us denote it as* γ_n. *At the same time, the Optimal Transport problem* (1) *also admits a unique optimal solution, we denote it as* γ_{OT}. *Then* $\lim_{n\to\infty} \gamma_n = \gamma_{OT}$ *in* $\mathcal{P}_2(\mathbb{R}^d \times \mathbb{R}^d)$.

Remark 1. We say a distribution μ satisfies the Log-Sobolev inequality with $K > 0$ if for any $\tilde{\mu} \ll \mu$, $D_{\mathrm{KL}}(\tilde{\mu}|\mu) \leq \frac{1}{2K}\mathcal{I}(\tilde{\mu}|\mu)$ always holds. Here $\mathcal{I}(\tilde{\mu}|\mu) = \int |\nabla \log\left(\frac{d\tilde{\mu}}{d\mu}\right)|^2 \, d\tilde{\mu}$.

Theorem 2 justifies the asymptotic convergence property of the approximation solutions $\{\gamma_n\}$ to the desired Optimal Transport plan γ_{OT}. The proof of Theorem 2 is provided in [17].

3 Wasserstein Gradient Flow Approach for Solving the Regularized Problem

3.1 Wasserstein Gradient Flow

There are already numerous references [2,11,21] regarding Wasserstein gradient flows of functionals defined on the Wasserstein manifold-like structure $(\mathcal{P}_2(\mathbb{R}^d), g^W)$. The Wasserstein manifold-like structure is the manifold $\mathcal{P}_2(\mathbb{R}^d)$ equipped with a special metric g^W compatible to the 2-Wasserstein distance [2,26]. Under this setting, the Wasserstein gradient flow of a certain functional $\mathcal{F}$ can thus be formulated as:

$$\frac{\partial \gamma_t}{\partial t} = -\mathrm{grad}_W \mathcal{F}(\gamma_t) \tag{6}$$

3.2 Wasserstein Gradient Flow of Entropy Transport Functional

We now come back to our constrained Entropy Transport problem (4). There are mainly two reasons why we choose to compute the Wasserstein gradient flow of functional $\mathcal{E}_{\Lambda,\mathrm{KL}}(\cdot|\mu,\nu)$:

- Computing the Wasserstein gradient flow is equivalent to applying gradient descent method to determine the minimizer of the ET functional (3);
- In most of the cases, Wasserstein gradient flows can be viewed as a time evolution PDE describing the density evolution of a stochastic process. As a result, once we derived the gradient flow, there will be a natural particle version associated to the gradient flow, which makes the computation of gradient flow tractable.

Now let us compute the Wasserstein gradient flow of $\mathcal{E}_{\Lambda,\mathrm{KL}}(\cdot|\mu,\nu)$:

$$\frac{\partial \gamma_t}{\partial t} = -\mathrm{grad}_W \mathcal{E}_{\Lambda,\mathrm{KL}}(\gamma_t|\mu,\nu), \quad \gamma_t|_{t=0} = \gamma_0 \tag{7}$$

To keep our notations concise, we denote $\rho(\cdot,t) = \frac{d\gamma_t}{d\mathscr{L}^{2d}}$, $\varrho_1 = \frac{d\mu}{d\mathscr{L}^d}$, $\varrho_2 = \frac{d\nu}{d\mathscr{L}^d}$, we can show that the previous Eq. (7) can be written as:

$$\frac{\partial \rho}{\partial t} = \nabla \cdot (\rho\, \nabla(c(x,y) + \Lambda \log(\frac{\rho_1(x,t)}{\varrho_1(x)}) + \Lambda \log(\frac{\rho_2(y,t)}{\varrho_2(y)}))) \tag{8}$$

Here $\rho_1(\cdot,t) = \int \rho(\cdot,y,t)\, dy$ and $\rho_2(\cdot,t) = \int \rho(x,\cdot,t)\, dx$ are density functions of marginals of γ_t.

3.3 Relating the Wasserstein Gradient Flow to a Particle System

Let us treat (8) as certain kind of continuity equation, i.e. we treat $\rho(\cdot,t)$ as the density of the time-evolving random particles. Then the vector field that drives the random particles at time t should be $-\nabla(c(x,y) + \Lambda \log\left(\frac{\rho_1(x,t)}{\varrho_1(x)}\right) + \Lambda \log\left(\frac{\rho_2(y,t)}{\varrho_2(y)}\right))$. This helps us design the following dynamics $\{(X_t, Y_t)\}_{t\geq 0}$: (here $\dot{X}_t$ denotes the time derivative $\frac{dX_t}{dt}$)

$$\begin{cases} \dot{X}_t = -\nabla_x c(X_t, Y_t) + \Lambda(\nabla \log \varrho_1(X_t) - \nabla \log \rho_1(X_t, t)); \\ \dot{Y}_t = -\nabla_y c(X_t, Y_t) + \Lambda(\nabla \log \varrho_2(Y_t) - \nabla \log \rho_2(Y_t, t)); \end{cases} \tag{9}$$

Here $\rho_1(\cdot,t)$ denotes the probability density of random variable X_t and $\rho_2(\cdot,t)$ denotes the density of Y_t. If we assume (9) is well-posed, then the density $\rho_t(x,y)$ of (X_t, Y_t) solves the PDE (8).

When we take a closer look at (9), the movement of the particle (X_t, Y_t) at certain time t depends on the probability density $\rho(X_t, Y_t, t)$, which can be approximated by the distribution of the surrounding particles near (X_t, Y_t). Generally speaking, we plan to evolve (9) as a particle aggregation model in order to converge to a sample-wised approximation of the Optimal Transport plan γ_{OT} for OT problem (1).

4 Algorithmic Development

To simulate the stochastic process (9) with the Euler scheme, we apply the Kernel Density Estimation [22] here to approximate the gradient log function $\nabla \log \rho(x)$ by convolving it with kernel $K(x,\xi)$[2]:

$$\nabla \log \rho(x) \approx \nabla \log(K * \rho)(x) = \frac{(\nabla_x K) * \rho(x)}{K * \rho(x)} \tag{10}$$

[2] In this paper, we choose the Radial Basis Function (RBF) as the kernel: $K(x,\xi) = \exp(-\frac{|x-\xi|^2}{2\tau^2})$.

Algorithm 1. Random Batch Particle Evolution Algorithm

Input: The density functions of the marginals ϱ_1, ϱ_2, timestep Δt, total number of iterations T, parameters of the chosen kernel K
Initialize: The initial locations of all particles $X_i(0)$ and $Y_i(0)$ where $i = 1, 2, \cdots, n$,
for t = 1,2,$\cdots$,T **do**
  Shuffle the particles and divide them into m batches: $\mathcal{C}_1, \cdots, \mathcal{C}_m$
  for each batch $\mathcal{C}_q$ **do**
    Update the location of each particle (X_i, Y_i) $(i \in \mathcal{C}_q)$ according to (11)
  end for
end for
Output: A sample approximation of the optimal coupling: $X_i(T), Y_i(T)$ for $i = 1, 2, \cdots, n$

Here $K * \rho(x) = \int K(x,\xi)\rho(\xi)d\xi$, $(\nabla_x K) * \rho(x) = \int \nabla_x K(x,\xi)\rho(\xi)d\xi$[3]. Such technique is also known as blobing method, which was first studied in [6] and has already been applied to Bayesian sampling [7]. With the blobing method, $\nabla \log \rho(x)$ is evaluated based on the locations of the particles:

$$\frac{\mathbb{E}_{\xi\sim\rho}\nabla_x K(x,\xi)}{\mathbb{E}_{\xi\sim\rho}K(x,\xi)} \approx \frac{\sum_{k=1}^{N}\nabla_x K(x,\xi_k)}{\sum_{k=1}^{N}K(x,\xi_k)} \qquad \xi_1, ..., \xi_N, \text{i.i.d.} \sim \rho$$

Now we are able to simulate (9) with the following interacting particle system involving N particles $\{(X_i, Y_i)\}_{i=1,...,N}$. For the i-th particle, we have:

$$\begin{cases} \dot{X}_i(t) = -\nabla_x c(X_i(t), Y_i(t)) - \Lambda\left(\nabla V_1(X_i(t)) + \dfrac{\sum_{k=1}^{N}\nabla_x K(X_i(t), X_k(t))}{\sum_{k=1}^{N}K(X_i(t), X_k(t))}\right) \\ \dot{Y}_i(t) = -\nabla_y c(X_i(t), Y_i(t)) - \Lambda\left(\nabla V_2(Y_i(t)) + \dfrac{\sum_{k=1}^{N}\nabla_x K(Y_i(t), Y_k(t))}{\sum_{k=1}^{N}K(Y_i(t), Y_k(t))}\right) \end{cases} \tag{11}$$

Here we denote $V_1 = -\log \varrho_1$, $V_2 = -\log \varrho_2$. Since we only need the gradients of V_1, V_2 , our algorithm can deal with unnormalized probability measures. We numerically verify that when $t \to \infty$, the empirical distribution $\frac{1}{N}\sum_{i=1}^{N}\delta_{(X_i(t),Y_i(t))}$ will converge to the optimal distribution γ_{cET} that solves (4) with sufficient large N and Λ, while the rigorous proof is reserved for our future work. The algorithm scheme is summarized in the algorithm 1.

Remark 2. Inspired by [10], we apply the Random Batch Methods (RBM) here to reduce the computational effort required to approximate $\nabla \log \rho(x)$ with blobing method: We divide all N particles into m batches equally, and we only consider the particles in the same batch as the particle X_i when we evaluate the $\nabla \log \rho(X_i)$. Now in each time step, the computational cost is reduced from $\mathcal{O}(n^2)$ to $\mathcal{O}(n^2/m)$.

[3] Notice that we always use $\nabla_x K$ to denote the partial derivative of K with respect to the first components.

5 Numerical Experiments

In this section, we test our algorithm on several toy examples.

Gaussian. We first apply the algorithm to compute the sample approximation of the Optimal Transport plan between two 1D Gaussian distributions $\mathcal{N}(-4,1), \mathcal{N}(6,1)$[4]. We set $\lambda = 200, \Delta t = 0.001, c(x,y) = |x-y|^2$ and run it with 1000 particles (X_i, Y_i)'s for 1000 iterations. We initialize the particles by drawing 1000 i.i.d. sample points from $\mathcal{N}(-20,4)$ as X_i's and 1000 i.i.d. sample points from $\mathcal{N}(20,2)$ as Y_i's. The empirical results are shown in Fig. 1 and Fig. 2.

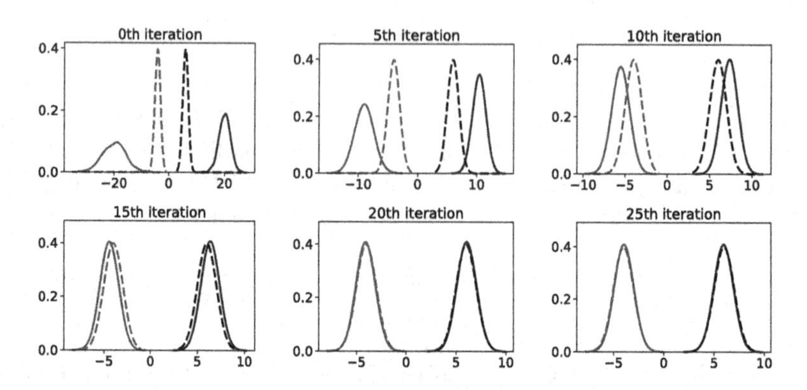

Fig. 1. Marginal plot for 1D Gaussian example. The red and black dashed lines correspond to two marginal distribution respectively and the solid blue and green lines are the kernel estimated density functions of particles at certain iterations. After first 25 iterations, the particles have matched the marginal distributions very well. (Color figure online)

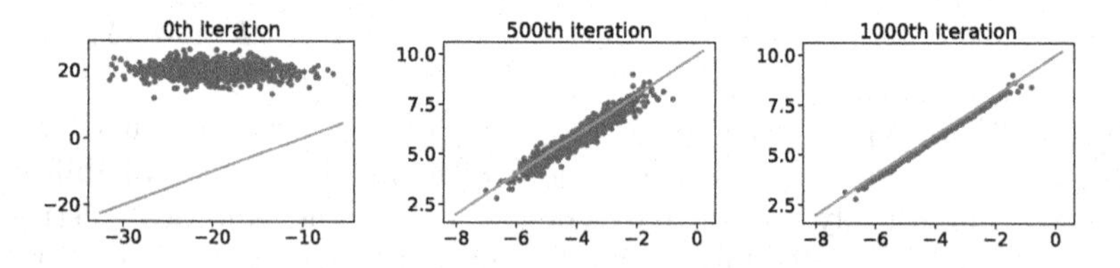

Fig. 2. The sample approximation for 1D Gaussian example. The orange dash line corresponds to the Optimal Transport map $T(x) = x + 10$. (Color figure online)

Gaussian Mixture. Then we apply the algorithm to two 1D Gaussian mixture $\varrho_1 = \frac{1}{2}\mathcal{N}(-1,1) + \frac{1}{2}\mathcal{N}(1,1), \varrho_2 = \frac{1}{2}\mathcal{N}(-2,1) + \frac{1}{2}\mathcal{N}(2,1)$. For experiment, we set $\lambda = 60, \Delta t = 0.0004, c(x,y) = |x-y|^2$ and run it with 1000 particles (X_i, Y_i)'s for 5000 iterations. We initialize the particles by drawing 2000 i.i.d. sample points from $\mathcal{N}(0,2)$ as X_i's and Y_i's. Figure 3 gives a visualization of the marginal distributions and the Optimal Transport map.

[4] Here $\mathcal{N}(\mu, \sigma^2)$ denotes the Gaussian distribution with mean value μ and variace σ^2.

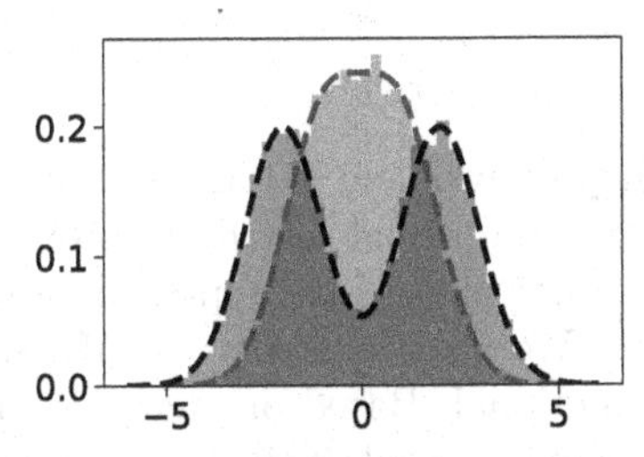

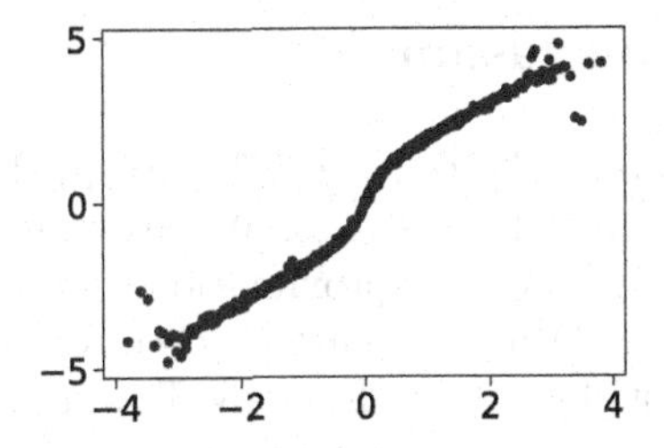

Fig. 3. 1D Gaussian mixture. Left. Marginal plot. The dash lines correspond to two marginal distributions. The histogram indicates the distribution of particles after 5000 iterations. Right. Sample approximation for the optimal coupling.

Wasserstein Barycenters. Our framework can be easily extended to solve the Wasserstein barycenter problem [1,9]

$$\min_{\mu \in \mathcal{P}_2(\mathbb{R}^d)} \sum_{i=1}^{m} \lambda_i W_2^2(\mu, \mu_i) \tag{12}$$

where $\lambda_i > 0$ are the weights. Similar to our previous treatment on OT problem, we can relax the marginal constraints in (12) and consider

$$\min_{\gamma \in \mathcal{P}(\mathbb{R}^{(m+1)d})} \int_{\mathbb{R}^{(m+1)d}} \sum_{j=1}^{m} \lambda_j |x - x_j|^2 \, d\gamma(x, x_1, ..., x_m) + \sum_{j=1}^{m} \Lambda_j D_{\mathrm{KL}}(\gamma_j \| \mu_j) \tag{13}$$

Then we can apply similar particle-evolving method to compute for problem (13), which can be treated as an approximation of the original barycenter problem (12). Here is an illustrative example: Given two Gaussian distributions $\rho_1 = \mathcal{N}(-10, 1), \rho_2 = \mathcal{N}(10, 1)$, and cost function $c(x, x_1, x_2) = w_1|x - x_1|^2 + w_2|x - x_2|^2$, we can compute sample approximation of the barycenter $\bar{\rho}$ of ρ_1, ρ_2. We try different weights $[w_1, w_2] = [0.25, 0.75], [0.5, 0.5], [0.75, 0.25]$ to test our algorithm. The experimental results are shown in Fig. 4. The distribution of the particles corresponding to the barycenter random variable X_0 converges to $\mathcal{N}(5, 1), \mathcal{N}(0, 1), \mathcal{N}(-5, 1)$ successfully after 2000 iterations.

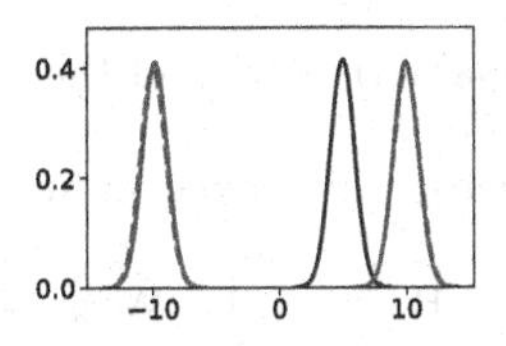

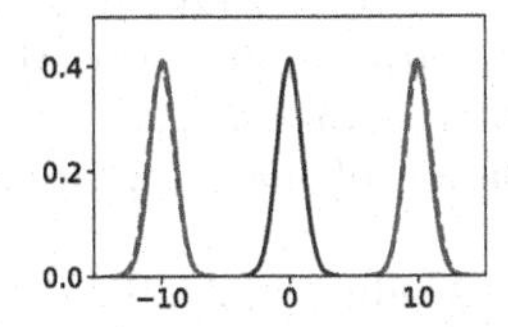

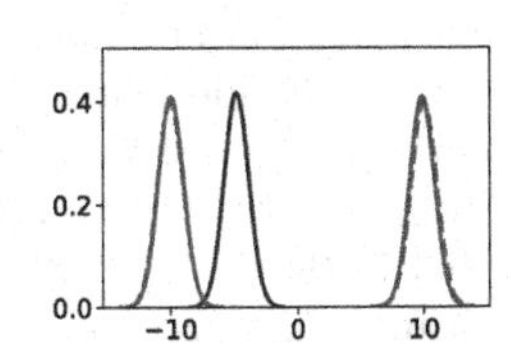

Fig. 4. Density plots for 1D Wasserstein barycenter example. The red dashed lines correspond to two marginal distributions respectively and the solid green lines are the kernel estimated density functions of the particles X_1's and X_2's. The solid blue line represents the kernel estimated density function of the particles corresponding to the barycenter. $[w1, w2] = [0.25, 0.75], [0.5, 0.5], [0.75, 0.25]$ from left to right. (Color figure online)

6 Conclusion

We propose the constrained Entropy Transport problem (4) and study some of its theoretical properties. We discover that the optimal distribution of (4) can be treated as an approximation to the optimal plan of the Optimal Transport problem (1) in the sense of Γ-convergence. We also construct the Wasserstein gradient flow of the Entropy Transport functional. Based on that, we propose a novel algorithm that computes for the sample-wised optimal transport plan by evolving an interacting particle system. We demonstrate the effectiveness of our method by several illustrative examples. More theoretical analysis and numerical experiments on higher dimensional cases including comparisons with other methods will be included in our future work.

References

1. Agueh, M., Carlier, G.: Barycenters in the Wasserstein space. SIAM J. Math. Anal. **43**(2), 904–924 (2011)
2. Ambrosio, L., Gigli, N., Savaré, G.: Gradient Flows: In Metric Spaces and in the Space of Probability Measures. Springer, Heidelberg (2008). https://doi.org/10.1007/978-3-7643-8722-8
3. Benamou, J.D., Brenier, Y.: A computational fluid mechanics solution to the Monge-Kantorovich mass transfer problem. Numer. Math. **84**(3), 375–393 (2000)
4. Benamou, J.D., Froese, B.D., Oberman, A.M.: Numerical solution of the optimal transportation problem using the Monge-Ampère equation. J. Comput. Phys. **260**, 107–126 (2014)
5. Brenier, Y.: Polar factorization and monotone rearrangement of vector-valued functions. Commun. Pure Appl. Math. **44**(4), 375–417 (1991)
6. Carrillo, J.A., Craig, K., Patacchini, F.S.: A blob method for diffusion. Calc. Var. Partial. Differ. Equ. **58**(2), 53 (2019)
7. Chen, C., Zhang, R., Wang, W., Li, B., Chen, L.: A unified particle-optimization framework for scalable Bayesian sampling. arXiv preprint arXiv:1805.11659 (2018)
8. Cuturi, M.: Sinkhorn distances: lightspeed computation of optimal transport. In: Advances in Neural Information Processing Systems, pp. 2292–2300 (2013)
9. Cuturi, M., Doucet, A.: Fast computation of Wasserstein barycenters. In: International Conference on Machine Learning, pp. 685–693. PMLR (2014)
10. Jin, S., Li, L., Liu, J.G.: Random batch methods (RBM) for interacting particle systems. J. Comput. Phys. **400**, 108877 (2020)
11. Jordan, R., Kinderlehrer, D., Otto, F.: The variational formulation of the Fokker-Planck equation. SIAM J. Math. Anal. **29**(1), 1–17 (1998)
12. Kantorovich, L.: On translation of mass (in Russian), c r. Doklady. Acad. Sci. USSR **37**, 199–201 (1942)
13. Korotin, A., Egiazarian, V., Asadulaev, A., Safin, A., Burnaev, E.: Wasserstein-2 generative networks. arXiv preprint arXiv:1909.13082 (2019)
14. Kullback, S., Leibler, R.A.: On information and sufficiency. Ann. Math. Stat. **22**(1), 79–86 (1951)
15. Li, W., Ryu, E.K., Osher, S., Yin, W., Gangbo, W.: A parallel method for earth mover's distance. J. Sci. Comput. **75**(1), 182–197 (2018)

16. Liero, M., Mielke, A., Savaré, G.: Optimal entropy-transport problems and a new Hellinger-Kantorovich distance between positive measures. Invent. Math. **211**(3), 969–1117 (2018)
17. Liu, S., Sun, H., Zha, H.: Approximating the optimal transport plan via particle-evolving method. arXiv preprint arXiv:2105.06088 (2021)
18. Makkuva, A., Taghvaei, A., Oh, S., Lee, J.: Optimal transport mapping via input convex neural networks. In: International Conference on Machine Learning, pp. 6672–6681. PMLR (2020)
19. Monge, G.: Mémoire sur la théorie des déblais et des remblais. Histoire de l'Académie Royale des Sciences de Paris (1781)
20. Oberman, A.M., Ruan, Y.: An efficient linear programming method for optimal transportation. arXiv preprint arXiv:1509.03668 (2015)
21. Otto, F.: The geometry of dissipative evolution equations: the porous medium equation. Comm. Partial Differ. Equ. **26**(1–2), 101–174 (2001)
22. Parzen, E.: On estimation of a probability density function and mode. Ann. Math. Stat. **33**(3), 1065–1076 (1962)
23. Ruthotto, L., Osher, S.J., Li, W., Nurbekyan, L., Fung, S.W.: A machine learning framework for solving high-dimensional mean field game and mean field control problems. Proc. Natl. Acad. Sci. **117**(17), 9183–9193 (2020)
24. Schmitzer, B.: A sparse multiscale algorithm for dense optimal transport. J. Math. Imaging Vis. **56**(2), 238–259 (2016)
25. Seguy, V., Damodaran, B.B., Flamary, R., Courty, N., Rolet, A., Blondel, M.: Large-scale optimal transport and mapping estimation. arXiv preprint arXiv:1711.02283 (2017)
26. Villani, C.: Optimal Transport: Old and New, vol. 338. Springer, Heidelberg (2008). https://doi.org/10.1007/978-3-540-71050-9
27. Walsh III, J.D., Dieci, L.: General auction method for real-valued optimal transport. arXiv preprint arXiv:1705.06379 (2017)

Geometric Structures in Thermodynamics and Statistical Physics

Schrödinger Problem for Lattice Gases: A Heuristic Point of View

Alberto Chiarini[2], Giovanni Conforti[1(✉)], and Luca Tamanini[3]

[1] IP Paris, Palaiseau, France
giovanni.conforti@polytechnique.edu
[2] TU Eindhoven, Eindhoven, The Netherlands
alberto.chiarini@tue.nl
[3] Université Paris Dauphine PSL, Paris, France
tamanini@ceremade.dauphine.fr

Abstract. Aim of this paper is to take the first steps in the study of the Schrödinger problem for lattice gases (SPLG), which we formulate relying on classical results in large deviations theory. Our main contributions are a dynamical characterization of optimizers through a coupled system of PDEs and a precise study of the evolution and convexity of the quasi-potential along Schrödinger bridges. In particular, our computations show that, although SPLG does not admit a variational interpretation through Otto calculus, the fundamental geometric properties of the classical Schrödinger problem for independent particles still admit a natural generalization. These observations motivate the development of a Riemannian calculus on the space of probability measures associated with the class of geodesic distances studied in [3]. All our computations are *formal*, further efforts are needed to turn them into rigorous results.

Keywords: Large deviations · Schrödinger problem · Optimal transport · Non-linear mobility · Displacement convexity · Gamma calculus

1 Introduction

In its original formulation, the Schrödinger problem (henceforth SP) is the problem of finding the most likely evolution of a cloud of independent Brownian particles conditionally to the observation of its initial (at $t = 0$) and final (at $t = T$) configurations. From a physical standpoint, it is very natural to address the same question for a system of *interacting* particles. This has led to the so-called mean field Schrödinger problem [1], where the underlying particle system is made of diffusions interacting through their empirical distribution. Without relying on a large deviations interpretation, other generalizations of the Schrödinger problem have been proposed, see e.g. [7,9]. All the versions of the Schrödinger problem considered in these works can be interpreted as the problem of controlling a gradient flow in such a way that a target distribution is attained and the control

F. Nielsen and F. Barbaresco (Eds.): GSI 2021, LNCS 12829, pp. 891–899, 2021.
https://doi.org/10.1007/978-3-030-80209-7_95

energy is minimized, once the space of probability measures is endowed with the Riemannian metric of optimal transport, also called Otto metric [13]. In the recent work [12] the Schrödinger problem is formulated in a purely metric setting, thus without relying on a Riemannian structure. In this general context, the authors obtain a Γ-convergence result of SP towards the corresponding geodesic problem. However, there are interesting results available for the classical SP that do rely on geometric, rather than metric considerations and are therefore unlikely to hold under the weak hypotheses of [12]. These include the convexity of the entropy along Schrödinger bridges, the symmetric form of optimality conditions for the dual variables, also called Schrödinger-Kantorovich potentials, and the interpretation of optimizers as solutions to Newton's laws. In this note we consider the Schrödinger problem for lattice gases (henceforth SPLG) and formally establish the analogous of some of the aforementioned geometric results. This task is not trivial since SPLG does not admit an interpretation in terms of Otto calculus and Wasserstein geometry. The metric aspects and small-noise limit of this problem have been studied in [12, Sec 6.4], although without noticing the connection with large deviations theory. Leaving precise statements to the main body of the article, we summarize our main contributions.

- We derive necessary optimality conditions in the form of (non-linear) coupled Hamilton-Jacobi-Bellman and Fokker-Planck equations. We also show that through a change of variable inspired by time-reversal considerations we can write optimality conditions in a form akin to the celebrated Schrödinger system, that is to say, two HJB equations having the same form except for the terms containing the time derivative, which have opposite signs.
- We prove representation formulas for the first and second derivative of the entropy along Schrödinger bridges that formally resemble those for the classical Schrödinger problem. Under the generalized McCann condition put forward in [3], we show that the entropy is convex along Schrödinger bridges.

We stress that all our calculations are formal and there are serious technical issues to be overcome in order to turn them into rigorous results. We plan to address this non-trivial task in future work(s).

2 Schrödinger Problem for Lattice Gases (SPLG)

Stochastic lattice gases may be described as a large collection of interacting particles performing random walks on the macroscopic blow-up $\Lambda_N = N\Lambda \cap \mathbb{Z}^d$ of a set $\Lambda \subseteq \mathbb{R}^d$. Particles are indistinguishable and the effect of the interaction is that the jump rates depend on the empirical particle distribution. Thermal reservoirs are modeled by adding creation/annihilation of particles at the boundary $\partial\Lambda_N$ and the influence of an external field is modeled by perturbing the rates giving a net drift toward a specified direction. Notable examples include the simple exclusion process, zero-range processes and the Glauber-Kawasaki dynamics. In the diffusive limit $N \to +\infty$, the empirical distribution of these systems converges to the hydrodynamic limit equation

$$\partial_t \rho + \nabla \cdot (J(t,\rho)) = 0, \quad \text{where } J(t,\rho) = -D(\rho)\nabla\rho + \chi(\rho)E(t).$$

The vector field $J(t,\rho)$ is called the hydrodynamic current, $\chi(\rho) \geq 0$ is the mobility, $D(\rho) \geq 0$ the diffusion coefficient and $E(t)$ the external field. D and χ are connected through the Einstein relation

$$D(\rho) = f''(\rho)\chi(\rho), \tag{1}$$

where f is the free energy of the model. The above equations are supplemented by the boundary conditions

$$f'(\rho(t,x)) = \lambda(t,x), \quad x \in \partial\Lambda.$$

All coefficients χ, D, E, λ are macroscopic quantities reflecting the microscopic description of the particle system: for example, the simple exclusion process corresponds to $\chi(\rho) = \rho(1-\rho)$, $J(t,\rho) = -\nabla\rho$ and its hydrodynamic limit was studied in [8]; the zero-range dynamics corresponds to $\chi(\rho) = \varphi(\rho)$, $J(t,\rho) = -\varphi'(\rho)\nabla\rho$, see [2]. The rate function quantifying the large deviations from the hydrodynamic limit is given by the following "fundamental formula" in the context of macroscopic fluctuation theory [2]

$$\mathcal{I}_{[0,T]}(\rho, j) = \frac{1}{4}\int_0^T\int_\Lambda (j - J(t,\rho)) \cdot \chi(\rho)^{-1} \cdot (j - J(t,\rho))\,\mathrm{d}x\mathrm{d}t, \tag{2}$$

where ρ is the local density of particles and j is connected to ρ by the continuity equation. It is worth mentioning that the rate function (2) captures the large deviations behavior of other relevant interacting systems beyond lattice gases such as Ginzburg-Landau models [6]. The Schrödinger problem for lattice gases is therefore given by

$$\mathcal{C}_T(\mu,\nu) = \min\{\mathcal{I}_{[0,T]}(\rho, j) \,:\, \partial_t\rho + \nabla \cdot j = 0,\ \rho_0 = \mu, \rho_T = \nu\}. \tag{\textbf{SPLG}}$$

3 Optimality Conditions

From now on we consider the simpler situation of periodic boundary conditions, where $\Lambda = \mathbb{T}^d$ is the d-dimensional torus and there is no external field, that is $E = 0$. To simplify computations we also assume that the mobility χ is a scalar quantity. Note that the optimal current and density for (**SPLG**) need to satisfy $\chi(\rho)^{-1}(j - J(t,\rho)) = \nabla H$ for some *corrector* H. Indeed, if we let $u = \chi(\rho)^{-1}(j - J(t,\rho))$ and w is any smooth vector field satisfying $\nabla \cdot (\chi(\rho)w) = 0$, then $\partial_t\rho + \nabla \cdot (j + \varepsilon\chi(\rho)w) = 0$ for all ε, and by the optimality of (ρ, j)

$$0 = \frac{\mathrm{d}}{\mathrm{d}\varepsilon}\bigg|_{\varepsilon=0} \mathcal{I}_{[0,T]}(\rho, j + \varepsilon w\chi(\rho)) = \frac{1}{2}\int_0^T\int_\Lambda u \cdot \chi(\rho) w\,\mathrm{d}x\mathrm{d}t,$$

which implies that u belongs to the closure of smooth gradients in $L^2(\chi(\rho))$.

We can then rewrite (**SPLG**) as the problem of minimizing

$$\bar{\mathcal{I}}_{0,T}(\rho, H) = \frac{1}{4}\int_0^T\int_\Lambda |\nabla H|^2 \chi(\rho)\, \mathrm{d}x\mathrm{d}t \tag{3}$$

among all curves (ρ, H) satisfying the continuity equation $\partial_t \rho + \nabla\cdot(\chi(\rho)(\nabla H - f''(\rho)\nabla\rho)) = 0$ and such that $\rho(0,\cdot) = \mu$, $\rho(T,\cdot) = \nu$.

Theorem 1. *Let* (ρ, H) *be an optimal solution. Then*

$$\begin{cases} \partial_t\rho - \nabla\cdot(D(\rho)\nabla\rho) + \nabla\cdot(\chi(\rho)\nabla H) = 0, \\ \partial_t H + D(\rho)\Delta H + \frac{1}{2}|\nabla H|^2\chi'(\rho) = 0, \\ \rho(0,\cdot) = \mu, \;\; \rho(T,\cdot) = \nu. \end{cases} \tag{4}$$

Proof. The equation for ρ is simply a rewriting of the continuity equation with the Einstein relation $D(\rho) = f''(\rho)\chi(\rho)$. The equation for H requires more care. We call *variation* a curve $\tilde{\rho} : (-\varepsilon, \varepsilon)\times[0,T]\times\Lambda \to \mathbb{R}_{\geq 0}$ such that

- $\tilde{\rho}(u, 0, \cdot) = \mu$, $\tilde{\rho}(u, T, \cdot) = \nu$ for all $u \in (-\epsilon, \epsilon)$,
- $\tilde{\rho}(0, t, \cdot) = \mu$, $\tilde{\rho}(0, t, \cdot) = \rho(t, \cdot)$ for all $t \in [0, T]$.

Let $\psi : (-\varepsilon, \varepsilon)\times[0,T]\times\Lambda \to \mathbb{R}$ be a potential such that $\forall u \in (-\varepsilon, \varepsilon)$ we have $\nabla\psi(u, 0, \cdot) \equiv \nabla\psi(u, T, \cdot) \equiv 0$, and consider the variation which satisfies

$$\partial_u\tilde{\rho} + \nabla\cdot(\chi(\tilde{\rho})\nabla\psi) = 0, \tag{5}$$

then there is a potential $\tilde{H} : (-\varepsilon, \varepsilon)\times[0,T]\times\Lambda \to \mathbb{R}$ such that $\forall t \in [0,T]$

$$\nabla\tilde{H}(0, t, \cdot) = \nabla H(t, \cdot), \quad \partial_t\tilde{\rho} - \nabla\cdot(D(\tilde{\rho})\nabla\tilde{\rho}) + \nabla\cdot(\chi(\tilde{\rho})\nabla\tilde{H}) = 0. \tag{6}$$

Imposing the condition

$$\iint \partial_u\big(\nabla\tilde{H}\chi(\tilde{\rho})\nabla\tilde{H}\big)\big|_{u=0}\mathrm{d}x\mathrm{d}t = 0,$$

we are led to analyze at $u = 0$ the two terms

$$A = \iint 2\nabla\partial_u\tilde{H}\cdot\nabla\tilde{H}\chi(\tilde{\rho})\,\mathrm{d}x\mathrm{d}t, \qquad B = \iint |\nabla\tilde{H}|^2\chi'(\tilde{\rho})\,\partial_u\tilde{\rho}\,\mathrm{d}x\mathrm{d}t.$$

We have, using Eqs. (5), (6) and a series of integration by parts

$$
\begin{aligned}
A &= \iint 2\tilde{H}\,\partial_u \nabla\cdot\big(\chi(\tilde{\rho})\nabla\tilde{H}\big)\,\mathrm{d}x\mathrm{d}t\\
&= \iint 2\tilde{H}\,\partial_u\big(-\partial_t\tilde{\rho}+\nabla\cdot(D(\tilde{\rho})\nabla\tilde{\rho})\big)\,\mathrm{d}x\mathrm{d}t\\
&= \iint 2\partial_t\tilde{H}\partial_u\tilde{\rho} - 2\nabla\tilde{H}\cdot\big(D'(\tilde{\rho})\partial_u\tilde{\rho}\nabla\tilde{\rho}+D(\tilde{\rho})\nabla\partial_u\tilde{\rho}\big)\,\mathrm{d}x\mathrm{d}t\\
&= \iint 2[\partial_t H - D'(\rho)\nabla H\cdot\nabla\rho + \nabla\cdot(D(\rho)\nabla H)]\partial_u\tilde{\rho}\,\mathrm{d}x\mathrm{d}t\\
&= \iint 2[\partial_t H + D(\rho)\Delta H]\partial_u\tilde{\rho}\,\mathrm{d}x\mathrm{d}t\\
&= \iint \nabla\Big(2[\partial_t H + D(\rho)\Delta H]\Big)\cdot\nabla\psi\,\chi(\rho)\,\mathrm{d}x\mathrm{d}t.
\end{aligned}
$$

Likewise, using (5) and integration by parts

$$B = \iint \nabla\big(\nabla H\chi'(\rho)\nabla H\big)\cdot\nabla\psi\,\chi(\rho)\,\mathrm{d}x\mathrm{d}t.$$

Imposing criticality, i.e. $A+B=0$ we find that along any variation,

$$\iint \nabla\Big([\partial_t H + D(\rho)\Delta H) + \frac{1}{2}|\nabla H|^2\chi'(\rho)]\Big)\cdot\nabla\psi\,\chi(\rho)\,\mathrm{d}x\mathrm{d}t = 0.$$

Since $\nabla\psi$ can be chosen arbitrarily, the conclusion follows.

The Schrödinger system

$$
\begin{cases}
\partial_t H + \frac{1}{2}|\nabla H|^2 + \frac{1}{2}\Delta H = 0,\\
-\partial_t \hat{H} + \frac{1}{2}|\nabla \hat{H}|^2 + \frac{1}{2}\Delta \hat{H} = 0.
\end{cases}
$$

expresses the optimality conditions for the classical Schrödinger problem. It has a peculiar structure in the sense that H and $\hat{H}$ evolve according to two HJB equations which are identical except for the ∂_t term, which changes sign from one equation to the other. The relation between the Schrödinger system and the HJB-FP system is given in the classical SP through the Hopf-Cole transform. It is natural to wonder whether such a structure is conserved in SPLG. In the next Theorem we answer affirmatively and find a change of variables, inspired by considerations on time reversal for diffusion processes, that turns (4) into a symmetric system of non-linear HJB equations.

Theorem 2. *Let (ρ, H) be optimal and define $\hat{H}$ by*

$$\forall (t,x)\in[0,T]\times\Lambda,\quad \hat{H}(t,x)+H(t,x) = 2f'(\rho(t,x)). \tag{7}$$

Then we have

$$\begin{cases} \partial_t H + D(\rho)\Delta H + \frac{1}{2}|\nabla H|^2\chi'(\rho) = 0, \\ -\partial_t \hat{H} + D(\rho)\Delta \hat{H} + \frac{1}{2}|\nabla \hat{H}|^2\chi'(\rho) = 0. \end{cases}$$

Proof. We have

$$\partial_t \hat{H} = 2f''(\rho)\partial_t\rho - \partial_t H. \tag{8}$$

We analyze the first term using the continuity equation and $f''(\rho)\nabla\rho = \nabla(f'(\rho))$.

$$\begin{aligned} 2f''(\rho)\partial_t\rho &= 2f''(\rho)\nabla\cdot(\chi(\rho)(\nabla(f'(\rho)) - \nabla H)) \\ &= 2D(\rho)(\Delta(f'(\rho)) - \Delta H) + \chi'(\rho)\big[2|\nabla(f'(\rho))|^2 - 2\nabla(f'(\rho))\cdot\nabla H\big]. \end{aligned}$$

Plugging this relation into (8) and using (4) to handle $\partial_t H$ we finally arrive at

$$\begin{aligned} \partial_t \hat{H}_t &= D(\rho)\big[\Delta(2f'(\rho) - H)\big] \\ &\quad + \frac{1}{2}\chi'(\rho)\big[4|\nabla(f'(\rho))|^2 - 4\nabla(f'(\rho))\cdot\nabla H + |\nabla H|^2\big] \\ &= D(\rho)\big[\Delta(2f'(\rho) - H)\big] + \frac{1}{2}\chi'(\rho)|\nabla(2f'(\rho) - H)|^2. \end{aligned}$$

The conclusion follows from the definition of $\hat{H}$, see (7).

As a corollary of the optimality conditions we deduce that the function

$$t \mapsto \int_\Lambda \nabla H \cdot \nabla \hat{H} \mathrm{d}x, \qquad t \in [0,T],$$

is conserved along Schrödinger bridges. We denote $\mathcal{E}_T(\mu,\nu)$ its constant value. Beside having the interpretation of a conserved total energy, it is very natural to conjecture that it expresses the derivative of the cost, viewed as a function of the time horizon T, i.e.

$$\frac{\mathrm{d}}{\mathrm{d}T}\mathcal{C}_T(\mu,\nu) = -\mathcal{E}_T(\mu,\nu).$$

Rigorous results in this direction have been obtained in [5] for the classical SP.

4 Convexity Along Optimal Flow

We consider the energy functional

$$\mathcal{F}(\rho) = \int_\Lambda f(\rho)\,\mathrm{d}x. \tag{9}$$

This energy functional is naturally associated to the dynamical rate function and in large deviations theory it is often called the quasi-potential, [2]. A remarkable fact with many interesting consequences is that the relative entropy is convex along classical Schrödinger bridges. It is natural to wonder if this remains true in the non-linear setting. We show two formulas for the first and second derivative of the entropy that have to be seen as the non-linear analogue of [4,10] and the stochastic counterpart of those in [3].

Proposition 1. *We have*

$$\frac{\mathrm{d}}{\mathrm{d}t}\mathcal{F}(\rho) = \frac{1}{4}\int |\nabla H|^2 \chi(\rho)\,\mathrm{d}x - \frac{1}{4}\int |\nabla \hat{H}|^2 \chi(\rho)\,\mathrm{d}x.$$

Proof. Using the formula (7) we find $\partial_t \rho_t + \frac{1}{2}\nabla\cdot((\nabla H_t - \nabla \hat{H}_t)\chi(\rho)) = 0$, whence, after an integration by parts we obtain

$$\frac{\mathrm{d}}{\mathrm{d}t}\mathcal{F}(\rho) = \frac{1}{2}\int \nabla(f'(\rho))(\nabla H - \nabla \hat{H})\chi(\rho)\,\mathrm{d}x.$$

The desired conclusion follows again from (7).

Theorem 3. *Let g be the primitive of $\chi'(\cdot)D(\cdot)$ such that $g(0) = 0$. Along any optimal flow (ρ, H) we have*

$$\begin{aligned}\frac{\mathrm{d}}{\mathrm{d}t}\frac{1}{4}\int |\nabla H|^2\chi(\rho)\,\mathrm{d}x &= \frac{1}{2}\int (D(\rho)\chi(\rho) - g(\rho))(\Delta H)^2\,\mathrm{d}x \\ &+ \frac{1}{2}\int g(\rho)\Big(\frac{1}{2}\Delta|\nabla H|^2 - \nabla\Delta H\cdot\nabla H\Big) - \frac{1}{2}|\nabla H|^2|\nabla\rho|^2 D(\rho)\chi''(\rho)\,\mathrm{d}x,\end{aligned} \tag{10}$$

$$\begin{aligned}-\frac{\mathrm{d}}{\mathrm{d}t}\frac{1}{4}\int |\nabla \hat{H}|^2\chi(\rho)\,\mathrm{d}x &= \frac{1}{2}\int (D(\rho)\chi(\rho) - g(\rho))(\Delta \hat{H})^2\,\mathrm{d}x \\ &+ \frac{1}{2}\int g(\rho)\Big(\frac{1}{2}\Delta|\nabla \hat{H}|^2 - \nabla\Delta \hat{H}\cdot\nabla \hat{H}\Big) - \frac{1}{2}|\nabla \hat{H}|^2|\nabla\rho|^2 D(\rho)\chi''(\rho)\,\mathrm{d}x.\end{aligned} \tag{11}$$

In particular,

$$\begin{aligned}\frac{\mathrm{d}^2}{\mathrm{d}t^2}\mathcal{F}(\rho) &= \frac{1}{2}\int (D(\rho)\chi(\rho) - g(\rho))\big((\Delta H)^2 + (\Delta\hat{H})^2\big)\,\mathrm{d}x \\ &+ \frac{1}{2}\int g(\rho)\Big(\frac{1}{2}\Delta|\nabla H|^2 - \nabla\Delta H\cdot\nabla H + \frac{1}{2}\Delta|\nabla\hat{H}|^2 - \nabla\Delta\hat{H}\cdot\nabla\hat{H}\Big)\,\mathrm{d}x \\ &- \frac{1}{4}\int \Big(|\nabla H|^2 + |\nabla\hat{H}|^2\Big)|\nabla\rho|^2 D(\rho)\chi''(\rho)\,\mathrm{d}x.\end{aligned} \tag{12}$$

Proof. We only prove (10), since (11) is completely analogous and (12) follows from Proposition 1. A direct calculation gives

$$\frac{\mathrm{d}}{\mathrm{d}t}\frac{1}{2}\int |\nabla H|^2\chi(\rho)\,\mathrm{d}x = A + B$$

with

$$A = \int \nabla(\partial_t H)\cdot\nabla H\chi(\rho)\,\mathrm{d}x, \quad B = \frac{1}{2}\int |\nabla H|^2\chi'(\rho)\partial_t\rho\,\mathrm{d}x.$$

Using the optimality conditions (4) we write $A = A_1 + A_2$ with

$$A_1 = -\int \nabla(D(\rho)\Delta H)\cdot\nabla H\chi(\rho)\,\mathrm{d}x, \quad A_2 = -\frac{1}{2}\int \nabla\big(|\nabla H|^2\chi'(\rho)\big)\cdot\nabla H\chi(\rho)\,\mathrm{d}x.$$

Also we have $B = B_1 + B_2$ with

$$B_1 = \frac{1}{2}\int |\nabla H|^2 \chi'(\rho)\nabla\cdot(-\nabla H \chi(\rho))\,\mathrm{d}x, \quad B_2 = \frac{1}{2}\int |\nabla H|^2 \chi'(\rho)\nabla\cdot(D(\rho)\nabla\rho)\,\mathrm{d}x.$$

Integrating by parts yields $B_1 + A_2 = 0$. Integrating twice by parts, and applying the definition of g we also get

$$\begin{aligned} A_1 &= \int D(\rho)\chi(\rho)(\Delta H)^2 \mathrm{d}x + \int \Delta H\, \nabla H \cdot \nabla(g(\rho))\,\mathrm{d}x \\ &= \int \big[D(\rho)\chi(\rho) - g(\rho)\big](\Delta H)^2\,\mathrm{d}x - \int \nabla\Delta H \cdot \nabla H\, g(\rho)\,\mathrm{d}x. \end{aligned} \tag{13}$$

To analyze B_2 we integrate twice more by parts and get

$$\begin{aligned} B_2 &= -\frac{1}{2}\int |\nabla H|^2 |\nabla\rho|^2 \chi''(\rho) D(\rho)\,\mathrm{d}x - \frac{1}{2}\int \nabla\big(|\nabla H|^2\big)\cdot\nabla(g(\rho))\,\mathrm{d}x \\ &= -\frac{1}{2}\int |\nabla H|^2 |\nabla\rho|^2 \chi''(\rho) D(\rho)\,\mathrm{d}x + \frac{1}{2}\int \Delta\big(|\nabla H|^2\big) g(\rho)\,\mathrm{d}x. \end{aligned}$$

The conclusion follows from this last identity, $A_2 + B_1 = 0$ and (13).

4.1 Convexity Under the Generalized McCann Condition

In [3] the authors proposed a generalization of McCann's condition [11] which implies the convexity of an energy functional along the geodesic problem built setting $D = 0$ in (**SPLG**). For the functional $\mathcal{F}$ we consider, the condition reads

$$\chi'' \leq 0, \quad D(\cdot)\chi(\cdot) \geq (1 - 1/d)\, g(\cdot), \tag{14}$$

where either $\chi' \geq 0$ or χ is defined on some interval $[0, M)$ where it is non-negative and concave. With the help of Bochner's formula

$$\frac{1}{2}\Delta|\nabla H|^2 - \nabla\Delta H \cdot \nabla H \geq \frac{1}{d}(\Delta H)^2,$$

we immediately derive from Theorem 3 the following

Corollary 1. *Under the generalized McCann's condition* (14) *it follows that the quasi-potential* (9) *is convex along Schrödinger bridges.*

It is straightforward to check that this condition is verified in the setting of the simple exclusion process.

References

1. Backhoff, J., Conforti, G., Gentil, I., Léonard, C.: The mean field Schrödinger problem: ergodic behavior, entropy estimates and functional inequalities. Prob. Theory Relat. Fields **178**, 475–530 (2020). https://doi.org/10.1007/s00440-020-00977-8

2. Bertini, L., De Sole, A., Gabrielli, D., Jona-Lasinio, G., Landim, C.: Macroscopic fluctuation theory. Rev. Mod. Phys. **87**(2), 593 (2015)
3. Carrillo, J.A., Lisini, S., Savaré, G., Slepčev, D.: Nonlinear mobility continuity equations and generalized displacement convexity. J. Funct. Anal. **258**(4), 1273–1309 (2010)
4. Conforti, G.: A second order equation for Schrödinger bridges with applications to the hot gas experiment and entropic transportation cost. Prob. Theory Relat. Fields **174**, 1–47 (2018). https://doi.org/10.1007/s00440-018-0856-7
5. Conforti, G., Tamanini, L.: A formula for the time derivative of the entropic cost and applications. J. Funct. Anal. **280**(11), 108964 (2021)
6. Donsker, M.D., Varadhan, S.R.S.: Large deviations from a hydrodynamic scaling limit. Commun. Pure Appl. Math. **42**(3), 243–270 (1989)
7. Gentil, I., Léonard, C., Ripani, L.: Dynamical aspects of generalized Schrödinger problem via otto calculus-a heuristic point of view (2018). To appear in Rev. Mat. Iberoam
8. Kipnis, C., Olla, S., Varadhan, S.R.S.: Hydrodynamics and large deviation for simple exclusion processes. Commun. Pure Appl. Math. **42**(2), 115–137 (1989)
9. Léger, F., Li, W.: Hopf Cole transformation via generalized Schrödinger bridge problem. J. Differ. Equ. **274**, 788–827 (2021)
10. Léonard, C.: On the convexity of the entropy along entropic interpolations. In: Measure Theory in Non-Smooth Spaces, pp. 194–242. Sciendo Migration (2017)
11. McCann, R.: A convexity principle for interacting gases. Adv. Math. **128**(1), 153–179 (1997)
12. Monsaingeon, L., Tamanini, L., Vorotnikov, D.: The dynamical Schrödinger problem in abstract metric spaces. Preprint at arXiv:2012.12005 (2020)
13. Otto, F., Villani, C.: Generalization of an inequality by Talagrand and links with the logarithmic Sobolev inequality. J. Funct. Anal. **173**(2), 361–400 (2000)

A Variational Perspective on the Thermodynamics of Non-isothermal Reacting Open Systems

François Gay-Balmaz[1(✉)] and Hiroaki Yoshimura[2]

[1] CNRS & École Normale Supérieure, Paris, France
francois.gay-balmaz@lmd.ens.fr
[2] School of Science and Engineering, Waseda University, Okubo, Shinjuku, Tokyo 169-8555, Japan
yoshimura@waseda.jp

Abstract. We review the variational formulation of nonequilibrium thermodynamics as an extension of the Hamilton principle and the Lagrange-d'Alembert principle of classical mechanics. We focus on the case of open systems that include the power exchange due to heat and matter transfer, with special emphasis on reacting systems which are very important in biological science.

Keywords: Variational formulation · Thermodynamics · Open systems · Chemical reactions

1 Variational Formulation in Nonequilibrium Thermodynamics

In this section we review the variational formulations of *isolated* and *open* thermodynamic systems as extensions of the variational formulation of classical mechanics [2–5]. In a similar way to the Lagrange-d'Alembert principle of nonholonomic mechanics, the variational formulation consists of a critical action principle subject to two types of constraints: a *kinematic constraints* on the critical curve and a *variational constraint* on the variations to be considered in the critical action principle. For nonequilibrium thermodynamics, these two constraints are related in a specific way, which are referred to as *constraints of thermodynamic type*. The kinematic constraints are phenomeneological constraints since they are constructed from the thermodynamic fluxes J_α associated to the irreversible processes α of the system, which are given by phenomenological expressions [1]. An important ingredient in the variational formulation is the

H.Y. is partially supported by JSPS Grant-in-Aid for Scientific Research (17H01097), JST CREST Grant Number JPMJCR1914, the MEXT Top Global University Project, Waseda University (SR 2021C-134) and the Organization for University Research Initiatives (Evolution and application of energy conversion theory in collaboration with modern mathematics).

F. Nielsen and F. Barbaresco (Eds.): GSI 2021, LNCS 12829, pp. 900–908, 2021.
https://doi.org/10.1007/978-3-030-80209-7_96

concept of *thermodynamic displacements* Λ^α, such that $\dot{\Lambda}^\alpha = X^\alpha$, with X^α the thermodynamic force of the process α.

1.1 Variational Formulation in Mechanics

Consider a mechanical system with configuration manifold Q and Lagrangian $L : TQ \to \mathbb{R}$, defined on the tangent bundle of Q. In absence of nonholonomic constraints and irreversible processes, the equations of motion are given by the Euler-Lagrange equations which arise from the *Hamilton principle*

$$\delta \int_{t_0}^{t_1} L(q, \dot{q}) \mathrm{d}t = 0 \tag{1}$$

for arbitrary variations δq of the curve $q(t)$ with $\delta q_{t=t_0,t_1} = 0$. Assume now that the system is subject to linear nonholonomic constraints on velocities given by a vector subbundle $\Delta \subset TQ$. The equations of motion for the nonholonomic system follow from the *Lagrange-d'Alembert principle* given as follows:

$$\delta \int_{t_0}^{t_1} L(q, \dot{q}) \mathrm{d}t = 0 \tag{2}$$

$$\text{with} \quad \dot{q} \in \Delta(q) \quad \text{and} \quad \delta q \in \Delta(q). \tag{3}$$

The first condition in (3) is the constraint on the critical curve $q(t)$ of (2), called *the kinematic constraint*, while the second condition is the constraint on the variation δq to be considered in (2), called *the variational constraint.* In local coordinates, the constraints (3) take the form

$$A_i^l(q)\dot{q}^i = 0 \quad \text{and} \quad A_i^l(q)\delta q^i = 0, \quad l = 1, ..., N. \tag{4}$$

1.2 Variational Formulation for Isolated Thermodynamic Systems

The variational formulation of *isolated* systems [2,3] is an extension of the Hamilton principle (1) and the Lagrange-d'Alembert principle (2)–(3) which falls into the following abstract formulation. Let $\mathcal{Q}$ be a manifold and $\mathcal{L} : T\mathcal{Q} \to \mathbb{R}$ a Lagrangian. Let C_V be a variational constraint, i.e., a submanifold $C_V \subset T\mathcal{Q} \times_{\mathcal{Q}} T\mathcal{Q}$ such that the set $C_V(x, v) := C_V \cap (\{(x, v)\} \times T_x\mathcal{Q})$ is a vector subspace of $T_x\mathcal{Q}$ for every $(x, v) \in T\mathcal{Q}$. We consider the variational formulation:

$$\delta \int_{t_0}^{t_1} \mathcal{L}(x, \dot{x}) \mathrm{d}t = 0 \tag{5}$$

$$\text{with} \qquad \dot{x} \in C_V(x, \dot{x}) \qquad \text{and} \qquad \delta x \in C_V(x, \dot{x}). \tag{6}$$

In a similar way to (3), the first condition in (6) is a *kinematic constraint* on the critical curve $x(t)$ while the second condition in (6) is a *variational constraint.* Variational and kinematic constraints related as in (6) are called *constraints of the thermodynamic type.* In local coordinates the constraints (6) take the form

$$A_i^l(x, \dot{x})\dot{x}^i = 0 \qquad \text{and} \qquad A_i^l(x, \dot{x})\delta x^i = 0, \qquad l = 1, ..., N.$$

Example 1: A Thermomechanical System. One of the simplest example of a thermodynamic system is the case of a mechanical system with only one entropy variable. The Lagrangian of this system is a function $L : TQ \times \mathbb{R} \to \mathbb{R}$ depending on the position and velocity of the mechanical part of the system and on the entropy. A standard expression is $L(q, \dot{q}, S) = \frac{1}{2}m|\dot{q}|^2 - U(q, S)$ where U is an internal energy depending on both the position q and the entropy S of the system. We assume that the irreversible process is described by a friction force $F^{\rm fr} : TQ \times \mathbb{R} \to T^*Q$. The variational formulation is

$$\delta \int_{t_0}^{t_1} L(q, \dot{q}, S) \mathrm{d}t = 0, \qquad \text{subject to} \tag{7}$$

$$\frac{\partial L}{\partial S}\dot{S} = \left\langle F^{\rm fr}, \dot{q}\right\rangle \quad \text{and} \quad \frac{\partial L}{\partial S}\delta S = \left\langle F^{\rm fr}, \delta q\right\rangle . \tag{8}$$

It yields the equations of motion

$$\frac{d}{dt}\frac{\partial L}{\partial \dot{q}} - \frac{\partial L}{\partial q} = F^{\rm fr}, \qquad \frac{\partial L}{\partial S}\dot{S} = \left\langle F^{\rm fr}, \dot{q}\right\rangle . \tag{9}$$

One checks that the total energy $E_{\rm tot} = \langle \frac{\partial L}{\partial \dot{q}}, \dot{q}\rangle - L$ is preserved by the equations of motion (9) while the entropy equation is $\dot{S} = -\frac{1}{T}\left\langle F^{\rm fr}, \dot{q}\right\rangle$, where $T = -\frac{\partial L}{\partial S}$ is the temperature. It is assumed that L is such that T is strictly positive. Hence, from the second law, $F^{\rm fr}$ must be a dissipative force.

One easily checks that (7)–(8) is a particular instance of (5)–(6) with $x = (q, S)$, $\mathfrak{L}(x, \dot{x}) = L(q, \dot{q}, S)$, and C_V given from (8).

Example 2: The Heat and Matter Exchanger. We consider a thermodynamic system made from several compartments exchanging heat and matter. We assume that there is a single species and we denote by N_A and S_A the number of moles and the entropy of the species in the compartment A, $A = 1, ..., K$. The internal energies are given as $U_A(S_A, N_A)$ where we assume that the volume of each compartment is constant. The variational formulation is based on the concept of *thermodynamic displacement* associated to an irreversible process, defined such that its time rate of change equals the thermodynamic force of the process. In our case, the thermodynamic forces are the temperatures $T^A = \frac{\partial U}{\partial S_A}$ and the chemical potentials $\mu^A = \frac{\partial U}{\partial N_A}$, so the thermodynamic displacements are variables Γ^A and W^A with $\dot{\Gamma}^A = T^A$ and $\dot{W}^A = \mu^A$. We denote by $\mathcal{J}^{A \to B}$ the molar flow rate from compartment A to compartment B due to diffusion of the species. We also introduce the heat fluxes J_{AB}, $A \neq B$ associated to heat transfer and define $J_{AA} := -\sum_{A \neq B} J_{AB}$. The variational formulation for this class of systems is

$$\delta \int_{t_0}^{t_1} \Big[L(S_1, ..., S_K, N_1, ..., N_K) + \sum_A \dot{W}^A N_A + \sum_A \dot{\Gamma}^A (S_A - \Sigma_A) \Big] \mathrm{d}t = 0 \tag{10}$$

$$\begin{aligned} \frac{\partial L}{\partial S_A}\dot{\Sigma}_A &= \sum_B J_{AB}\dot{\Gamma}^B + \sum_B \mathcal{J}^{B \to A}\dot{W}^A, \quad A = 1, ..., K \\ \frac{\partial L}{\partial S_A}\delta\Sigma_A &= \sum_B J_{AB}\delta\Gamma^B + \sum_B \mathcal{J}^{B \to A}\delta W^A, \quad A = 1, ..., K. \end{aligned} \tag{11}$$

In our case, the Lagrangian is $L(S_1, ..., S_K, N_1, ..., N_K) = -\sum_A U_A(S_A, N_A)$ and from (10)–(11) one gets the system of equations

$$\dot{N}_A = \sum_B \mathcal{J}^{B\to A}, \quad T^A \dot{S}_A = -\sum_B J_{AB}(T^B - T^A) - \sum_B \mathcal{J}^{B\to A}\mu^A, \quad A = 1, ..., K,$$

together with $\dot{\Gamma}^A = T^A$, $\dot{W}^A = \mu^A$, and $\dot{\Sigma}_A = \dot{S}_A$. One checks that the total energy and number of moles are preserved while the total entropy satisfies

$$\dot{S} = \sum_{A<B} \left(\frac{1}{T^B} - \frac{1}{T^A}\right) J_{AB}(T^B - T^A) + \sum_{A<B} \left(\frac{\mu^A}{T^A} - \frac{\mu^B}{T^B}\right) \mathcal{J}^{B\to A},$$

which dictates the phenomenological expressions for J_{AB} and $\mathcal{J}^{B\to A}$, see [5].

One easily checks that (10)–(11) is a particular case of (5)–(6) with $x = (S_1, N_1, \Gamma^1, W^1, \Sigma_1, ...)$, with $\mathfrak{L}(x, \dot{x})$ given by the integrand in (10), and with C_V given by (11).

1.3 Variational Formulation for Open Thermodynamics Systems

The variational formulation of *open* systems [4] is an extension of the Hamilton principle (1), the Lagrange-d'Alembert principle (2)–(3), and the variational formulation (5)–(6) with constraints of thermodynamic type. It falls into the following abstract formulation.

Let $\mathcal{Q}$ be a manifold and $\mathfrak{L} : \mathbb{R} \times T\mathcal{Q} \to \mathbb{R}$ be a Lagrangian, possibly time dependent. Let $\mathfrak{F}^{\rm ext} : \mathbb{R} \times T\mathcal{Q} \to T^*\mathcal{Q}$ be a given exterior force. We consider a time dependent variational constraint C_V given by a submanifold $C_V \subset (\mathbb{R} \times T\mathcal{Q}) \times_{\mathbb{R}\times\mathcal{Q}} T(\mathcal{Q}\times\mathbb{R})$ such that the set $C_V(t, x, v) := C_V \cap (\{(t, x, v)\} \times T_{(t,x)}(\mathbb{R}\times\mathcal{Q}))$ is a vector subspace of $T_{(t,x)}(\mathbb{R} \times \mathcal{Q})$ for every $(t, x, v) \in \mathbb{R} \times T\mathcal{Q}$. It is thus locally defined with the help of functions $A^l_i(t, x, \dot{x})$ and $B^l(t, x, \dot{x})$. We consider the variational formulation:

$$\delta \int_{t_0}^{t_1} \mathfrak{L}(t, x, \dot{x}) dt + \int_{t_0}^{t_1} \left\langle \mathfrak{F}^{\rm ext}(t, x, \dot{x}), \delta x \right\rangle dt = 0, \qquad \text{subject to} \tag{12}$$

$$A^l_i(t, x, \dot{x})\dot{x}^i + B^l(t, x, \dot{x}) = 0 \quad \text{and} \quad A^l_i(t, x, \dot{x})\delta x^i = 0, \qquad l = 1, ..., N. \tag{13}$$

In a similar way to the previous cases, the first condition in (13) is a *kinematic constraint* on the critical curve $x(t)$ while the second condition in (13) is a *variational constraint*. Variational and kinematic constraints related as in (6) in the time-dependent setting are also called *constraints of the thermodynamic type*.

Remark 1. The constraints in (13) are explicitly time dependent. Allowing this time dependence is important for the applications to open thermodynamic systems. Consistently with this, we have considered that both $\mathfrak{L}$ and $\mathfrak{F}^{\rm ext}$ may be explicitly time dependent in (12), by defining them on $\mathbb{R} \times T\mathcal{Q} \ni (t, x, \dot{x})$. It turns out that this time dependence is very natural if one considers the geometric setting underlying (12)–(13), see Remark 2.

Remark 2. The appropriate geometric setting underlying (12)–(13) is that of time dependent mechanics, seen as a special instance of the geometric setting of classical field theories. The basic object is the configuration bundle $\mathcal{Y} \to \mathcal{X}$, here given by $\mathcal{Y} = \mathbb{R} \times \mathcal{Q} \to \mathbb{R}$ where $\mathcal{X} = \mathbb{R}$. Then, the Lagrangian in (12) is defined on the first jet bundle $J^1\mathcal{Y} = \mathbb{R} \times T\mathcal{Q}$ of $\mathbb{R} \times \mathcal{Q} \to \mathbb{R}$ and the variational constraint C_V mentioned above is $C_V \subset J^1\mathcal{Y} \times_{\mathcal{Y}} T\mathcal{Y}$, see [6].

2 Open Reacting Systems

In this section we show how the variational formulation (12)–(13) for open thermodynamic systems is used to model the dynamics of an open system exchanging heat and mass with the exterior and involving chemical reactions.

2.1 General Setting, Thermodynamic Forces and Displacements

We assume that the system involves R chemical species $I = 1, ..., R$ and r chemical reactions $a = 1, ..., r$. Chemical reactions may be represented by

$$\sum_I \nu'^a_I I \underset{a_{(2)}}{\overset{a_{(1)}}{\rightleftarrows}} \sum_I \nu''^a_I I, \quad a = 1, ..., r,$$

where $a_{(1)}$ and $a_{(2)}$ are the forward and backward reactions associated to the reaction a, and ν''^a_I, ν'^a_I are the forward and backward stoichiometric coefficients. Mass conservation during each reaction is given by

$$\sum_I m_I \nu^a_I = 0 \quad \text{for } a = 1, ..., r \ \ \text{(Lavoisier law)},$$

where $\nu^a_I := \nu''^a_I - \nu'^a_I$ and m_I is the molecular mass of species I.

We denote by $U = U(S, V, N_1, ..., N_R)$ the internal energy of the system, written in terms of the entropy S, the volume V, and the number of moles N_I of each species $I = 1, ..., R$.

We have already seen above the thermodynamic forces, displacements, and fluxes associated to heat and matter transfer. For chemical reactions, the thermodynamic forces are the affinities of the reactions defined by

$$\mathcal{A}^a = -\sum_I \nu^a_I \mu^I, \quad a = 1, ..., r.$$

Following our definition, the corresponding thermodynamic displacement, denoted ν^a, satisfies

$$\dot{\nu}^a = -\mathcal{A}^a, \quad a = 1, ..., r.$$

The thermodynamic fluxes are the *rates of extent* denoted J_a given by phenomenological laws, see Sect. 2.4.

We assume that the system has several ports through which species can flow into or out of the system. To fix the ideas, we assume that for each species there is one inlet (i) and one outlet (o) with corresponding molar flow rates into the system denoted $\mathcal{J}_I^i > 0$ and $\mathcal{J}_I^o < 0$, $I = 1, ..., R$. The associated entropy flow is $\mathcal{J}_{S,I}^k = \mathcal{J}_I^k \mathsf{S}_I^k$, where S_I^k is the molar entropy of species I at $k = i, o$. We also assume that there is a heat source with entropy flow rate $\mathcal{J}_{S,h}$. We denote by μ_k^I, T_k^I, $k = i, o$, the chemical potentials and temperatures at the ports and by T^h the temperature of the heat source.

2.2 Variational Setting

Following the general approach mentioned above, the variational formulation is:

$$\delta \int_{t_0}^{t_1} \Big[L(q, \dot{q}, S, N_1, ..., N_R) + \sum_I \dot{W}^I N_I + \dot{\Gamma}(S - \Sigma) \Big] \mathrm{d}t + \int_{t_0}^{t_1} \langle F^{\mathrm{ext}}, \delta q \rangle \, \mathrm{d}t = 0 \tag{14}$$

with kinematic and variational constraints

$$\begin{aligned} \frac{\partial L}{\partial S} \dot{\Sigma} &= \left\langle F^{\mathrm{fr}}, \dot{q} \right\rangle + \sum_a J_a \dot{\nu}^a + \sum_{I,k} \left(\mathcal{J}_I^k (\dot{W}^I - \mu_k^I) + \mathcal{J}_{S,I}^k (\dot{\Gamma} - T_k^I) \right) + \mathcal{J}_{S,h} (\dot{\Gamma} - T^h) \\ \dot{\nu}^a &= \sum_I \nu_I^a \dot{W}^I \end{aligned} \tag{15}$$

$$\begin{aligned} \frac{\partial L}{\partial S} \delta \Sigma &= \left\langle F^{\mathrm{fr}}, \delta q \right\rangle + \sum_a J_a \delta \nu^a + \sum_I^k \left(\mathcal{J}_I^k \delta W^I + \mathcal{J}_{S,I}^k \delta \Gamma \right) + \mathcal{J}_{S,h} \delta \Gamma \\ \delta \nu^a &= \sum_I \nu_I^a \delta W^I . \end{aligned} \tag{16}$$

We note that the variational constraint (16) follows from the phenomenological constraint (15) by formally replacing the time derivatives $\dot{\Sigma}$, $\dot{q}$, $\dot{\nu}^a$, $\dot{W}^I$, $\dot{\Gamma}$ by the corresponding virtual displacements $\delta\Sigma$, δq, $\delta\nu^a$, δW^I, $\delta\Gamma$, and by removing all the terms that depend uniquely on the exterior, i.e., the terms $\mathcal{J}_I^k \mu_k^I$, $\mathcal{J}_{S,I}^k T_k^I$, and $\mathcal{J}_{S,h} T^h$. This is consistent with the general setting in (13).

Taking variations of the integral in (14), integrating by parts, and using $\delta q(t_0) = \delta q(t_1) = 0$, $\delta W^I(t_0) = \delta W^I(t_1) = 0$, and $\delta\Gamma(t_0) = \delta\Gamma(t_1) = 0$ and using the variational constraint (16), we get the following conditions:

$$\begin{aligned} &\delta q : \frac{d}{dt} \frac{\partial L}{\partial \dot{q}} - \frac{\partial L}{\partial q} = F^{\mathrm{fr}} + F^{\mathrm{ext}}, \qquad \delta \Gamma : \dot{S} = \dot{\Sigma} + \sum_{I,k} \mathcal{J}_{S,I}^k + \mathcal{J}_{S,h}, \\ &\delta N_I : \dot{W}^I = -\frac{\partial L}{\partial N_I}, \quad \delta S : \dot{\Gamma} = -\frac{\partial L}{\partial S}, \quad \delta W^I : \dot{N}_I = \sum_k \mathcal{J}_I^k + \sum_a J_a \nu_I^a . \end{aligned} \tag{17}$$

By the third and fourth equations the variables Γ and W^I are thermodynamic displacements. Using (15), we get the following system of evolution equations for the curves $q(t)$, $S(t)$, $N_I(t)$:

$$\begin{cases} \dfrac{d}{dt}\dfrac{\partial L}{\partial \dot q} - \dfrac{\partial L}{\partial q} = F^{\rm fr} + F^{\rm ext}, \qquad \dfrac{d}{dt}N_I = \sum_k \mathcal{J}_I^k + \sum_a J_a \nu_I^a, \\ \dfrac{\partial L}{\partial S}\Big(\dot S - \sum_{I,k} \mathcal{J}_{S,I}^k - \mathcal{J}_{S,h}\Big) = \langle F^{\rm fr}, \dot q\rangle - \sum_{a,I} J_a \nu_I^a \dfrac{\partial L}{\partial N_I} \\ \qquad - \sum_{I,k}\Big[\mathcal{J}_I^k\Big(\dfrac{\partial L}{\partial N_I} + \mu_k^I\Big) + \mathcal{J}_{I,S}^k\Big(\dfrac{\partial L}{\partial S} + T_k^I\Big)\Big] - \mathcal{J}_{S,h}\Big(\dfrac{\partial L}{\partial S} + T^h\Big). \end{cases} \tag{18}$$

2.3 The First and Second Law of Thermodynamics

The energy balance for this system is computed as

$$\frac{d}{dt}E = \langle F^{\rm ext}, \dot q\rangle + \mathcal{J}_{S,h}T^h + \sum_{I,k}(\mathcal{J}_I^k \mu_k^I + \mathcal{J}_{S,I}^k T_k^I) =: P_W^{\rm ext} + P_H^{\rm ext} + P_M^{\rm ext},$$

with the three terms representing the power associated to the transfer of mechanical work, heat, and matter into the system, respectively. From the last equation in (18), the entropy equation reads

$$\dot S = I + \sum_{I,k} \mathcal{J}_{S,I}^k + \mathcal{J}_{S,h}, \tag{19}$$

where I is the rate of internal entropy production given by

$$\begin{aligned} I = \;& \underbrace{-\frac{1}{T}\langle F^{\rm fr}, \dot q\rangle}_{\text{mechanical friction}} \; \underbrace{+\frac{1}{T}\sum_a J_a \mathcal{A}^a}_{\text{chemistry}} \\ & \underbrace{+\frac{1}{T}\sum_{I,k}\Big[\mathcal{J}_I^k\big(\mu_k^I - \mu^I\big) + \mathcal{J}_{S,I}^k\big(T_k^I - T\big)\Big]}_{\text{mixing of matter flowing into the system}} \underbrace{+\frac{1}{T}\mathcal{J}_{S,h}\big(T^h - T\big)}_{\text{heating}}. \end{aligned} \tag{20}$$

From the second equation in (17) and from (19) we obtain the interpretation of the variable Σ, namely, $\dot\Sigma = I$ is the rate of internal entropy production. The second and third terms in (19) represent the entropy flow rate into the system associated to the ports and the heat sources. The second law requires $I \geq 0$, whereas the sign of the rate of entropy flow into the system is arbitrary. Of course, at the outlet in which $\mu_o^I = \mu^I$ and $T_o^I = T^I$, the corresponding entropy production term vanishes. The second term in (19) can be written in terms of the molar enthalpy H_I^k of species I at $k = i, o$, as $\frac{1}{T}\sum_I^k \mathcal{J}_I^k(\mathsf{H}_k^I - T\mathsf{S}_k^I - \mu^I)$.

2.4 Entropy Production Associated to Chemical Reactions

We give here the expression of the entropy production for elementary chemical reactions in a form general enough to cover the case of real gas mixtures and non-isothermal chemical reactions.

For a mixture of gas, the affinity of a reaction can be written in the form

$$\mathcal{A}^a(T, p, N_1, ..., N_R) = RT \ln \Big(K_a^\circ(T) \prod_A (f_A/p_0)^{-\nu_A^a} \Big),$$

where the state function f_A is the fugacity of component A, p_0 is a standard or reference pressure, and

$$K_a^\circ(T) := \exp \Big(-\frac{1}{RT} \sum_A \nu_A^a \mu_0^A(T, p_0) \Big)$$

is the thermodynamic equilibrium constant of reaction a, with μ_0^A the chemical potential of component A seen as a perfect gas. The quotient $\mathsf{a}_A = f_A/p_0$ is the activity of component A, defined with respect to the prefect gas reference state. The rates of extend J_a have the general expression

$$\frac{J_a}{V} = R_f^a - R_r^a = k_f^a(T) \prod_{A=1}^R \mathsf{a}_A^{\kappa'^a_A} - k_r^a(T) \prod_{A=1}^R \mathsf{a}_A^{\kappa''^a_A}$$

with R_f^a, R_r^a the forward and reverse reaction rates, k_f^a, k_r^a the forward and reverse rate constants, and κ'^a_A, κ''^a_A the order of the reactants and of the products [7]. The quantities k_f^a, k_r^a are given by phenomenological expressions, the most widely used being the *Arrhenius equation*, and satisfy the condition $K_a^\circ(T) = k_f^a(T)/k_r^a(T)$. For elementary reactions, we have $\kappa'^a_A = \nu'^a_A$, $\kappa'^a_A = \nu'^a_A$, hence the entropy production term in (20) associated to chemical reactions is

$$\frac{1}{T} \sum_a J_a \mathcal{A}^a = \sum_a VR \left(R_f^a - R_r^a \right) \ln \frac{R_f^a}{R_r^a} \geq 0.$$

Remark 3. The Lagrangian variational formulation presented above can be transformed into the Hamiltonian setting when the given Lagrangian is nondegenerate with respect to the mechanical variable. We will show the Hamiltonian variational formulation for thermodynamics as a future work.

References

1. de Groot, S.R., Mazur, P.: Nonequilibrium Thermodynamics. North-Holland (1969)
2. Gay-Balmaz, F., Yoshimura, H.: A Lagrangian variational formulation for nonequilibrium thermodynamics. Part I: discrete systems. J. Geom. Phys. **111**, 169–193 (2016)
3. Gay-Balmaz, F., Yoshimura, H.: A Lagrangian variational formulation for nonequilibrium thermodynamics. Part II: continuum systems. J. Geom. Phys. **111**, 194–212 (2016)
4. Gay-Balmaz, F., Yoshimura, H.: A variational formulation of nonequilibrium thermodynamics for discrete open systems with mass and heat transfer. Entropy **20**(3), 163 (2018)

5. Gay-Balmaz, F., Yoshimura, H.: From Lagrangian mechanics to nonequilibrium thermodynamics: a variational perspective. Entropy **21**(1), 8–39 (2019)
6. Gay-Balmaz, F., Yoshimura, H.: Dirac structures in nonequilibrium thermodynamics for simple open systems. J. Math. Phys. **61**(9), 092701 (2021)
7. Kondepudi, D., Prigogine, I.: Modern Thermodynamics. John Wiley & Sons, New York (1998)

On the Thermodynamic Interpretation of Deep Learning Systems

Rita Fioresi[1](✉), Francesco Faglioni[2], Francesco Morri[3], and Lorenzo Squadrani[1]

[1] University of Bologna, Bologna, Italy
rita.fioresi@unibo.it, lorenzo.squadrani@studio.unibo.it
[2] University of Modena, Modena, Italy
francesco.faglioni@unimore.it
[3] Politecnico di Torino, Torino, Italy
francesco.morri@studenti.polito.it

Abstract. In the study of time evolution of the parameters in Deep Learning systems, subject to optimization via SGD (stochastic gradient descent), temperature, entropy and other thermodynamic notions are commonly employed to exploit the Boltzmann formalism. We show that, in simulations on popular databases (CIFAR10, MNIST), such simplified models appear inadequate: different regions in the parameter space exhibit significantly different temperatures and no elementary function expresses the temperature in terms of learning rate and batch size, as commonly assumed. This suggests a more conceptual approach involving contact dynamics and Lie Group Thermodynamics.

Keywords: Deep Learning · Statistical mechanics · Lie groups machine learning

1 Introduction

In the study of artificial neural networks, thermodynamics and statistical mechanics modeling proved to be a driving force leading to the development of new algorithms, starting from the pioneering work by Jaynes [8], going from Hopfield neural networks [7] to Boltzmann machines [1] and their newer restricted and deep versions [12]. As the new successful family of *Deep Learning* algorithms emerged, the language of thermodynamics and statistical mechanics is commonly employed to draw analogies and boost intuition on the functioning of optimizers based on SGD (Stochastic Gradient Descent), see [5,6] and refs. therein.

Our purpose is to establish a dictionary connecting neural networks notions commonly used in such algorithms (e.g. loss, parameters, learning rate, minibatch, etc.) and statistical mechanics concepts (e.g. particles, masses, energy, temperature, etc.), so that the analogies may be exploited with a deeper understanding and go beyond a qualitative analysis.

F. Nielsen and F. Barbaresco (Eds.): GSI 2021, LNCS 12829, pp. 909–917, 2021.
https://doi.org/10.1007/978-3-030-80209-7_97

Our paper is organized as follows. In Sect. 2 we establish the above mentioned correspondence, relating thermodynamics concepts with neural networks ones. We then validate our model in Sect. 3 through some experiments and in Sect. 4 we suggest a more conceptual approach based on the formalism of contact dynamics and Lie Groups Thermodynamics [3,4,13].

2 Thermodynamics and Deep Learning

Let $\Sigma = \{z_i \,|1 \leq i \ \leq N\} \subset \mathbb{R}^D$ represent a dataset of size N, i. e. $|\Sigma| = N$ and let $f = \frac{1}{N}\sum_{i=1}^{N} f_i$ be the loss function, f_i being the loss of the i-th datum z_i. A popular choice for f, for example, is the Kullback-Leibler divergence of the Amari loss [2] (Softmax). We assume the training to take place through Stochastic Gradient Descent (SGD) with minibatch $\mathcal{B}$, $|\mathcal{B}| << N$. We call $\overline{x} \in \mathbb{R}^d$ the vector consisting of the learning parameters of the model. The value of the kth parameter is thus x_k. Since parameters evolve in time during training, we write $\overline{x}(t)$, or $x_k(t)$ to emphasize this. In practical implementations, with optimization obtained via Gradient Descent (GD), t is a discrete variable indicating the timestep:

$$\overline{x}(t+1) = \overline{x}(t) - \eta \nabla f \tag{1}$$

η denoting the learning rate. Equation (1) is often expressed in its continuous version as:

$$\frac{d\overline{x}}{dt} = -\eta \nabla f \tag{2}$$

Notice that, since Σ is fixed, the loss function $f(t)$ at time t is determined by the parameters $\overline{x}(t)$. If SGD is used for optimization, Eq. (2) could be substituted with (see [6]):

$$\frac{d\overline{x}}{dt} = -\eta \nabla f_{\mathcal{B}} \tag{3}$$

where the full loss function is replaced with $f_{\mathcal{B}} = \frac{1}{|\mathcal{B}|}\sum_{i=1}^{|\mathcal{B}|} f_i$ and at each time step $\mathcal{B}$ is chosen in Σ. Our purpose is to show that if (2) is properly interpreted in a thermodynamics context, we can analyze effectively the dynamics of SGD, without introducing stochastic variables as in (3) studied in [6], besides the ones intrinsic to Boltzmann statistical mechanics and its far reaching generalizations (see [3,4,11] and refs. therein).

We now proceed with our thermodynamic interpretation. Let $\overline{x}(t)$ represent the position (or geometry) of a mechanical system at time t and consider the loss $f(t)$ as the potential energy associated with the geometry of the system.

We first look at the system as conservative, i.e., the force acting on each particle according to (2) is the negative gradient of the potential:

$$F_k = -\frac{\partial f}{\partial x_k} \qquad \text{or} \qquad \overline{F} = -\nabla f$$

The velocity at each optimization step is:

$$v_k(t) = \frac{x_k(t) - x_k(t-1)}{\Delta t} \qquad \text{or} \qquad \overline{v}(t) = \frac{\overline{x}(t) - \overline{x}(t-1)}{\Delta t}$$

We assign masses m_k to each particle. i.e. to each parameter x_k. If the mechanical system were ideal and isolated, it would evolve following Newton's law:

$$F_k = m_k \frac{dv_k}{dt}$$

For finite time steps Δt, the corresponding position update would be

$$\Delta x_k = v_k \Delta t + \frac{1}{2}\frac{F_k}{m_k}(\Delta t)^2 \tag{4}$$

In this case, the total energy would be conserved (up to numerical integration errors) and the system would convert potential into kinetic energy and vice versa.

In the language of atomistic simulations, this is referred to as *Constant Energy* dynamics, but it is *not* what occurs during neural network training. In fact, typical optimization algorithms use the gradient to update coordinates, not velocities. The equation ruling the dynamics is in fact (1): $\Delta\overline{x} = -\eta\nabla f\,\Delta t$. Let us rewrite (4), taking the force term as the gradient of the potential:

$$\Delta x_k = v_k \Delta t - \frac{\Delta t}{2m_k}\frac{\partial f}{\partial x_k}\,\Delta t \tag{5}$$

This is the same as $\Delta x = -\eta\nabla f \Delta t$, if the velocities are set to zero before taking each step and $\eta = \Delta t/(2\,m_k)$. Notice that the higher the masses, the lower is the learning rate, since the parameters are "harder to move". The variations of the learning rate can be seen equivalently as altering the time step: a smaller learning rate means a slower simulation, that is a smaller time step. In Sect. 3 we will study the dependence of the key hyperparameters η and $|\mathcal{B}|$ from the temperature, (see [6]), but this relation will be more elusive.

We define the *instantaneous temperature* $\mathcal{T}(t)$ of the system as the kinetic energy $\mathcal{K}(t)$ divided by the number of degrees of freedom d and a constant k_B to obtain the desired units:

$$\mathcal{T}(t) = \frac{\mathcal{K}(t)}{k_B\,d} = \frac{1}{k_B\,d}\sum_{k=1}^{d}\frac{1}{2}m_k\,v_k(t)^2$$

The *thermodynamics temperature* is then the time average of $\mathcal{T}(t)$:

$$T = \frac{1}{\tau}\int_0^{\tau}\mathcal{T}(t)\,dt = \frac{1}{\tau k_B\,d}\int_0^{\tau}\mathcal{K}(t) = \frac{K}{k_B\,d}$$

where K is the average kinetic energy and τ is long enough to yield small fluctuations in T and depends on the time scale of the individual particle motions. In practice, to perform mechanical simulations at constant (or regularly varying) temperature, coupling with a thermal reservoir is introduced by rescaling the velocities every fixed number of steps to match the desired temperature. These are called *constant temperature* simulations; we set the temperature equal to zero every step (instead every few steps as usual). The mechanical equivalent, is a dynamic simulation where heat is extracted from the system at each step. Performing such an optimization with the gradient descent (GD), has a clear mechanical interpretation: it leads to a (local) minimum in the potential or, equivalently, a minimum of the total energy at zero temperature, when the kinetic energy vanishes.

We now turn to examine the effect of SGD, Stochastic Gradient Descent. With SGD, the gradient for a given geometry changes each time it is computed, because of the random choice of the minibatch $\mathcal{B}$. This amounts to a residual velocity associated to each particle even after equilibrium is reached. Hence, our system does not evolve according to Newton dynamics and in particular the mechanical energy is not constant.

Once the equilibrium is reached, the macroscopic parameters (the loss and temperature) will no longer change, i.e. they will have small fluctuations only. In particular $\langle v \rangle = 0$, that is, we expect the average value of velocity to vanish. Notice that here we have a key difference between GD (gradient descent) and SGD (stochastic gradient descent): at equilibrium

$$\sigma^2 = \langle v^2 \rangle - \langle v \rangle^2$$

$\sigma^2 = 0$ for GD but $\sigma^2 = \langle v^2 \rangle$ for SGD. This corresponds to the physical fact that GD reaches the equivalent of the zero Kelvin (no temperature), while with SGD we maintain a residual finite temperature. With a constant temperature simulation we will achieve the minimum *free energy* and not a minimum of the potential energy, that is our loss function.

We now recall the *principle of equipartition of energy*: *at thermodynamic equilibrium, all accessible degrees of freedom have, on a sufficiently long time average, the same kinetic energy.*

Let K_{av} be the time average kinetic energy of particle (i.e. parameter) k. By the principle of equipartition of energy, this is $1/d$ the total kinetic energy:

$$K_{av} = K/d = k_B T$$

Hence knowing the average value of v_k^2 at equilibrium, i.e., the variance of the gradient, this equation allows us to compute the temperature (if we set all masses equal to 1).

In the next section we will perform experiments to test our thermodynamic model and the relation between some of the notions we introduced. In Sect. 4, we shall interpret the continuous version of the time evolution of our system as the dynamics of a mechanical system with Hamiltonian H consisting of the sum of a conservative term $H^{\mathrm{mech}} = K + V$ and a dissipative term.

3 Experiments

Our experiments were performed on both the MNIST and CIFAR10 datasets (see [9,10]) obtaining similar results. We report the experiments on the MNIST dataset only. We are using a LeNet modified architecture in colab platform (Fig. 1):

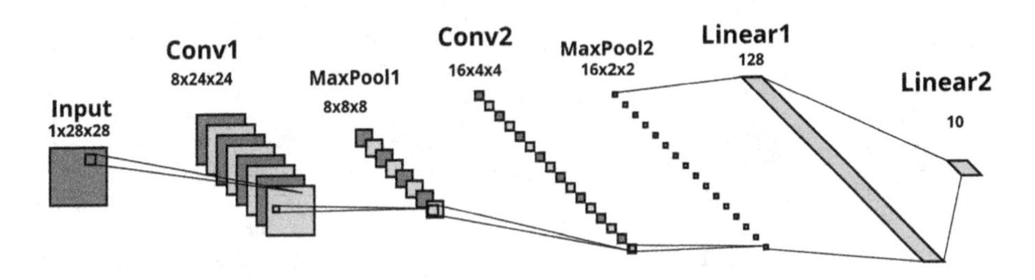

Fig. 1. Modified LeNet

This is an accurate, yet simple, network consisting of two convolutional layers (Conv1, Conv2) followed by two linear ones (Linear1, Linear2), with a low number of parameters. Batchnormalization and maxpool are also used.

We use SGD to optimize the network, with a constant for the regularization penalty $\lambda = 4 \cdot 10^{-2}$ and minibatch size $\beta = 32$. We start our training with learning rate $\eta = 10^{-2}$, then decrease it to 10^{-3} after 300 epochs and finally to 10^{-4} at 600 epochs. The loss function $f_{\mathcal{B}}$ during training and average temperature at equilibrium in layers are expressed in Fig. 2 and 3.

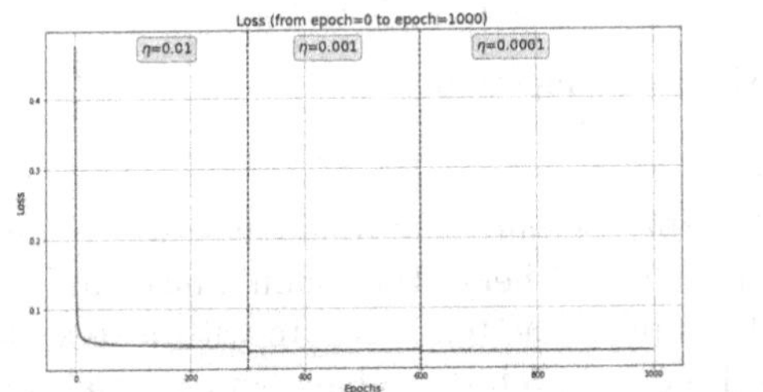

Fig. 2. Loss function

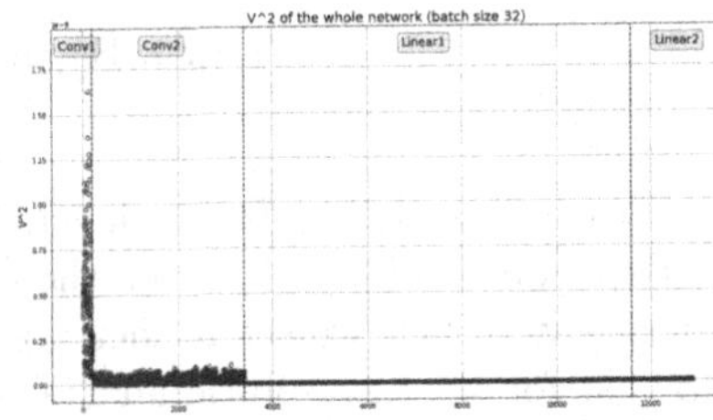

Fig. 3. Temperature in layers

Notice that different layers exhibit significantly different temperatures.

In Fig. 4 and 5 we describe the behaviour of the temperature T, as defined in our previous section, depending on the inverse $1/\beta$ of the minibatch size and the learning rate η. Despite in the literature ([6] and refs therein) T is commonly believed to behave proportionally to such parameters, we discover in practice quite a different behaviour.

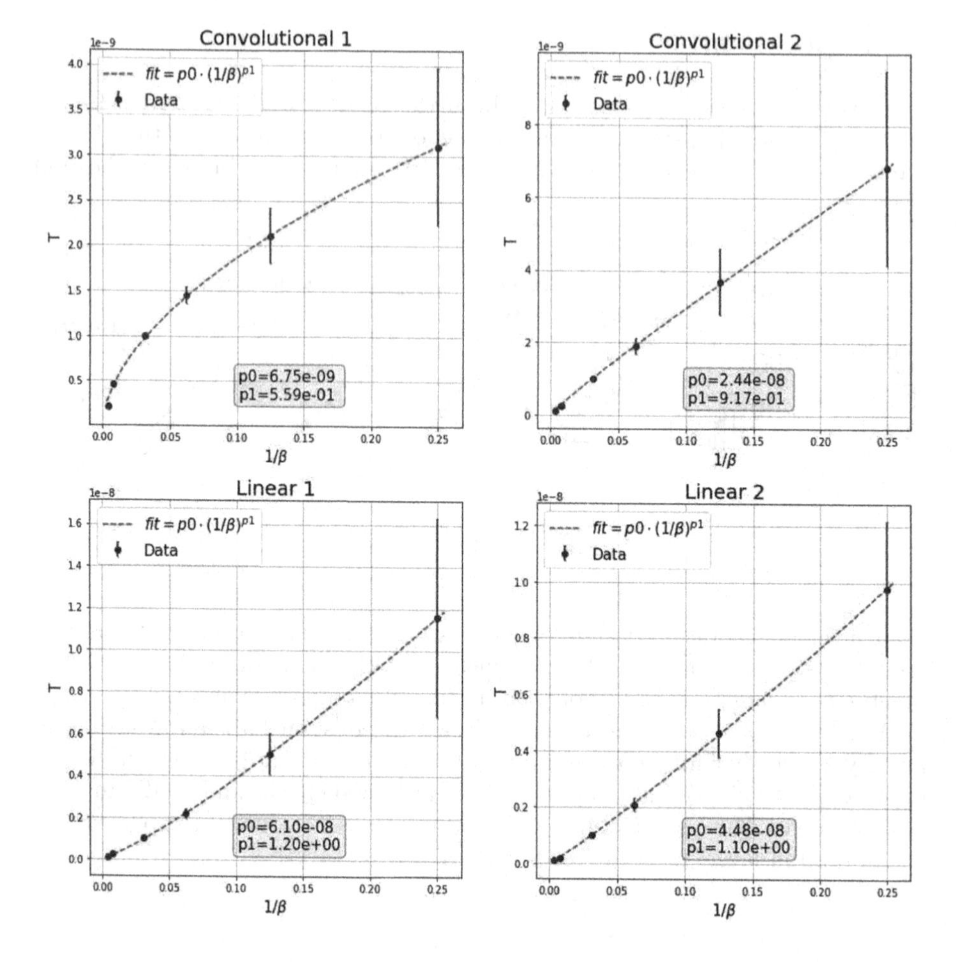

Fig. 4. Batch size and temperature

Notice first that different layers exhibit significantly different behaviour in the dependence of η and β from the temperature, hence they should be examined separately. In fact we see a linear behaviour of the temperature with respect to $1/\beta$ just for the Convolutional 2 and Linear 2. While we have an essential non linearity for the others. Similar considerations hold for the quadratic behaviour with respect to the learning rate. We now look at the temperature of the filters of the first convolutional layer at equilibrium (see the values of parameters in Conv 1 in Fig. 6).

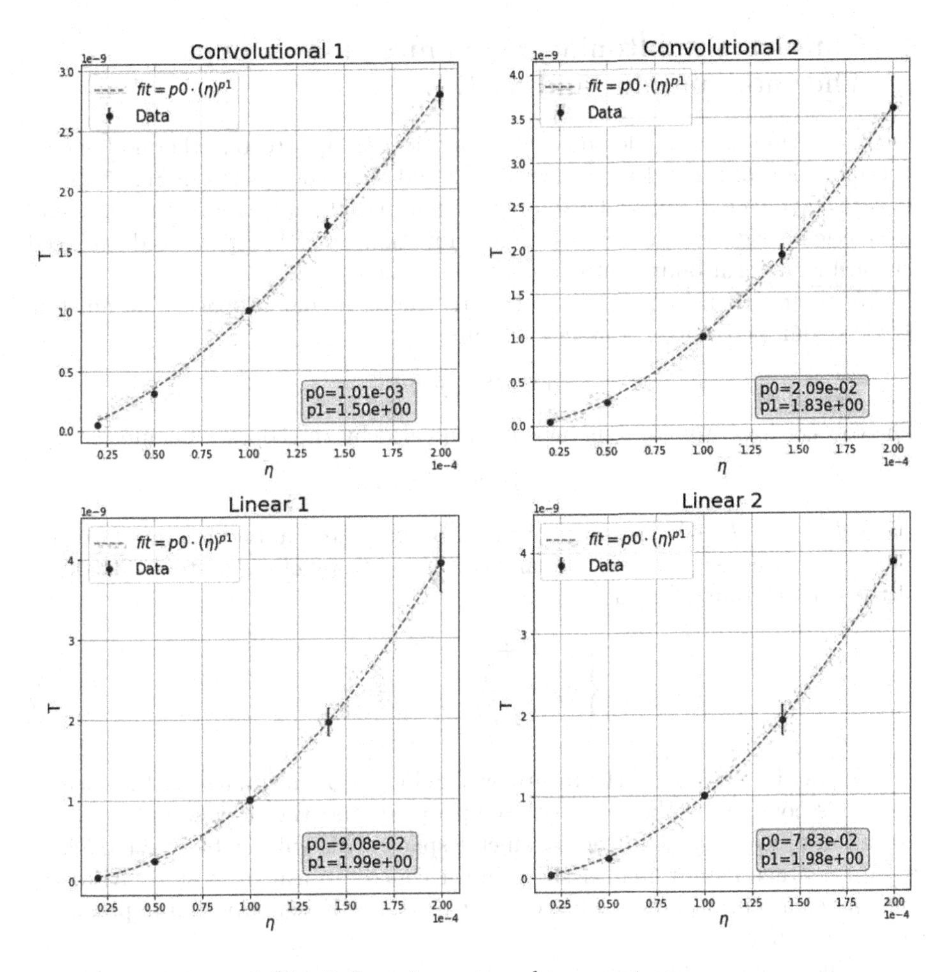

Fig. 5. Learning rate and temperature

Clearly the parameters of different filters have different temperature behaviours at equilibrium: some filters tend to stay stable while others keep changing. A possible interpretation of this is that some filters are more effective than others, so once learnt the system will not forget them. Vice versa, non effective filters in image recognition still change since they do not contribute to loss reduction.

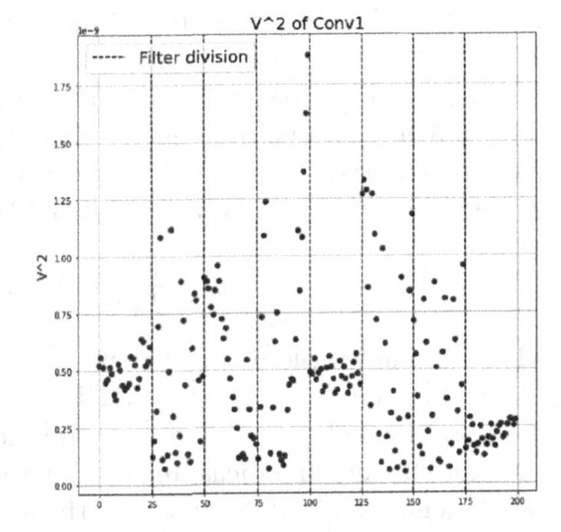

Fig. 6. Temperature of filters in the first convolutional layer

4 Contact Hamiltonian Dynamics, Lie Groups Thermodynamics and SGD

In this section we provide some mathematical insight to our thermodynamic interpretation of SGD described in Sec. 2 and the experiments in Sec. 3. Since at each step we extract heat from our system, we cannot assume that the sum of kinetic energy and potential (the loss function) H^{mech} is preserved; we need to consider a term taking into account dissipation.

We assume then (see [4], Sec. 5) that the thermodynamic space of parameters $\mathbb{R}^{d+1}$ is equipped with a contact structure:

$$\alpha = dS - p_a dq^a$$

The contact hamiltonian dynamics is then ruled by the contact hamiltonian:

$$H = H^{\mathrm{mech}} + \mathcal{V}(S)$$

where $H^{\mathrm{mech}} = K + V$ as above. We could take as first approximation (see [4]), $\mathcal{V}(S) = \gamma S$, where γ is a constant and S is the entropy of the system. This leads to the contact Hamilton equations:

$$\begin{cases} \dot{q}^a = \frac{\partial H}{\partial p_a} \\ \dot{p}_a = -\frac{\partial H}{\partial q^a} - p_a \frac{\partial H}{\partial S} \\ \dot{S} = p_a \frac{\partial H}{\partial p_a} \end{cases} \tag{6}$$

We plan to measure in the future, with a long enough simulation the entropy S in this context, by knowing the temperature and the area sampled by the evolution of the system in the parameter space. This will enable the modelling with a contact hamiltonian system and a concrete mean to test it. Also, the entropy comes into play in other contexts, like the Koszul-Souriau approach to thermodynamics.

We now make some considerations on Lie group thermodynamics and suggest possibly future mathematically interesting directions. Consider the action of G the Galilei group on space time. As Souriau proves, this action is hamiltonian, so it makes sense to speak of its moment map. We cannot however expect generalized Gibbs states to exist for the full Galilean group, but, as specified in [11] 7.3.3 only for one-parameter subgroups. If we are able to run our experiment long enough, the system would explore a large portion of the parameter space, so to test the partition function predicted by the probability function ([11] 7.1.1):

$$\rho_b = \frac{1}{P(b)} e^{-\langle J,b\rangle}, \qquad b \in \mathrm{Lie}(G) \tag{7}$$

where J denotes the moment map.

We also believe that if would be mathematically interesting to fit contact dynamics for a thermodynamics system into the framework of Souriau Lie group thermodynamics and measurements on this simple model could be an experimental evidence that these these two theories are effective and equivalent ways to describe popular machine learning systems.

5 Conclusions

We described a parallelism between SGD dynamics in Deep Learning and a thermodynamics system. Experiments show that the temperatures of each layer behaves independently, hence it is necessary to treat layers as independent systems. Furthermore, the temperature of each layer does not depend in a consistent and simple way on the size of the minibatch and the learning rate: extra care must then be exerted when defining the relation of the temperature with such key hyperparameters. Insight from Lie group thermodynamics and its generalization to contact hamiltonian dynamics suggests to push this analogy further to obtain quantitative experimental results.

Acknowledgements. We are indebted with Prof. P. Chaudhari, Dr. A. Achille and Prof. S. Soatto for many helpful discussions. We also thank our Referees for valuable suggestions.

References

1. Ackley, D.H., Hinton, G.E., Sejnowski, T.J.: A learning algorithm for Boltzmann machines. Cogn. Sci. **9**(1), 147–169 (1985)
2. Amari, S.-I.: Natural gradient works efficiently in learning. Neural Comput. **10**(2), 251–276 (1998)
3. Barbaresco, F.: Lie group statistics and Lie group machine learning based on Souriau Lie groups thermodynamics and Koszul-Souriau-Fisher metric: new entropy definition as generalized Casimir invariant function in coadjoint representation. Entropy **22**(6), 642 (2020)
4. Bravetti, A.: Contact Hamiltonian dynamics: the concept and its use. Entropy **19**(10), 535 (2017)
5. Fioresi, R., Chaudhari, P., Soatto, S.: A geometric interpretation of stochastic gradient descent using diffusion metrics. Entropy **22**(1), 101 (2020)
6. Chaudhari, P., Soatto, S.: Stochastic gradient descent performs variational inference, converges to limit cycles for deep networks. In: 2018 Information Theory and Applications Workshop (ITA), pp. 1–10. IEEE (2018)
7. Hopfield, J.J.: Neurons with graded response have collective computational properties like those of two-state neurons. PNAS **81**(10), 3088–3092 (1984)
8. Jaynes, E.T.: Information theory and statistical mechanics. Phys. Rev. **106**(4), 620 (1957)
9. Krizhevsky, A., Sutskever, I., Hinton, G.E.; ImageNet classification with deep convolutional neural networks. In: Advances in Neural Information Processing Systems, vol. 25, pp. 1097–1105 (2012)
10. LeCun, Y., Bengio, Y., Hinton, G.: Deep learning. Nature **521**(7553), 436–444 (2015)
11. Marle, C.-M.: From tools in symplectic and Poisson geometry to J. M. Souriau's theories of statistical mechanics and thermodynamics. Entropy **18**(10), 370 (2016)
12. Salakhutdinov, R., Hinton, G.: Deep Boltzmann machines. In: Artificial Intelligence and Statistics, pp. 448–455. PMLR (2009)
13. Simoes, A.A., De Leon, M., Valcazar, M.L. and De Diego, D.M.: Contact geometry for simple thermodynamical systems with friction. Proc. Roy. Soc. A **476**(2241), 20200244 (2020)

Dirac Structures in Thermodynamics of Non-simple Systems

Hiroaki Yoshimura[1(✉)] and François Gay-Balmaz[2]

[1] Waseda University, Okubo, Shinjuku, Tokyo 169-8555, Japan
yoshimura@waseda.jp
[2] Ecole Normale Supérieure, Paris, France
francois.gay-balmaz@lmd.ens.fr

Abstract. We present the Dirac structures and the associated Dirac system formulations for *non-simple* thermodynamic systems by focusing upon the cases that include irreversible processes due to friction and heat conduction. These systems are called non-simple since they involve several entropy variables. We review the variational formulation of the evolution equations of such non-simple systems. Then, based on this, we clarify that there exists a Dirac structure on the Pontryagin bundle over a thermodynamic configuration space and we develop the Dirac dynamical formulation of such non-simple systems. The approach is illustrated with the example of an adiabatic piston.

Keywords: Dirac structures · Non-simple systems · Thermodynamics · Adiabatic piston

1 Variational Formulation of Non-simple Systems

Before exploring Dirac structures underlying the thermodynamics of non-simple systems, we review the variational setting of such non-simple systems by focusing on the internal irreversible processes associated with friction and heat conduction.

1.1 Setting for Thermodynamics of Non-simple Systems

Non-simple Systems with Friction and Heat Conduction. Consider an adiabatically closed system $\boldsymbol{\Sigma} = \cup_{A=1}^{P} \boldsymbol{\Sigma}_A$ which consists of P simple thermodynamic systems $\boldsymbol{\Sigma}_A$, in which we include the irreversible processes due to friction and heat conduction between subsystems. Here a *simple* thermodynamic system

H.Yoshimura is partially supported by JSPS Grant-in-Aid for Scientific Research (17H01097), JST CREST Grant Number JPMJCR1914, the MEXT Top Global University Project, Waseda University (SR 2021C-134) and the Organization for University Research Initiatives (Evolution and application of energy conversion theory in collaboration with modern mathematics).

F. Nielsen and F. Barbaresco (Eds.): GSI 2021, LNCS 12829, pp. 918–925, 2021.
https://doi.org/10.1007/978-3-030-80209-7_98

denotes a system that has only one variable to represent the thermodynamic state, usually denoted by entropy. Since $\mathbf{\Sigma}$ is an interconnected system of simple subsystems $\mathbf{\Sigma}_1, ..., \mathbf{\Sigma}_P$, it becomes a "non-simple" system that has several entropy (or temperature) variables (see [7]) and we note that all the irreversible processes are *internal.* For each simple subsystem $\mathbf{\Sigma}_A$, $A = 1, ..., P$, $S_A \in \mathbb{R}$ indicates its entropy variable. Here, we assume that the mechanical configuration of $\mathbf{\Sigma}$ is given by independent mechanical variables $q = (q^1, ..., q^n) \in Q$, where Q is the mechanical configuration manifold of $\mathbf{\Sigma}$.

Friction, Heat Conduction and External Forces. Let $F^{\mathrm{ext}\to A} : T^*Q \times \mathbb{R}^P \to T^*Q$ be an external force that acts on $\mathbf{\Sigma}_A$ and hence the total exterior force is $F^{\mathrm{ext}} = \sum_{A=1}^{P} F^{\mathrm{ext}\to A}$. Let $F^{\mathrm{fr}(A)} : T^*Q \times \mathbb{R}^P \to T^*Q$ be the friction forces associated with the irreversible processes of each subsystem $\mathbf{\Sigma}_A$, which yield an entropy production for subsystem $\mathbf{\Sigma}_A$. Associated with the heat exchange between $\mathbf{\Sigma}_A$ and $\mathbf{\Sigma}_B$, let J_{AB} be the fluxes such that for $A \neq B$, $J_{AB} = J_{BA}$ and for $A = B$, $J_{AA} := -\sum_{B \neq A} J_{AB}$, where $\sum_{A=1}^{P} J_{AB} = 0$ for all B.

Thermodynamic Displacements. In our formulation, we introduce the concept of thermodynamic displacements, see [3,4]. For the case of heat exchange, we define the *thermal displacements* Γ^A, $A = 1, ..., P$ such that its time rate $\dot{\Gamma}^A$ becomes the temperature of $\mathbf{\Sigma}_A$. We also introduce a new variable Σ_A associated with the internal entropy production.

1.2 Variational Formulation of Non-simple Systems

The Lagrange-d'Alembert Principle for Non-simple Systems. Now we consider a variational formulation of Lagrange-d'Alembert type for non-simple systems with friction and heat conduction, which is a natural extension of Hamilton's principle in mechanics (see [4]).

Given a Lagrangian $L : TQ \times \mathbb{R}^P \to \mathbb{R}$ and an external force $F^{\mathrm{ext}} : TQ \times \mathbb{R}^P \to T^*Q$, find the curves $q(t)$, $S_A(t)$, $\Gamma^A(t)$, $\Sigma_A(t)$ which are critical for the *variational condition*

$$\delta \int_{t_1}^{t_2} \Big[L(q, \dot{q}, S_A) + \dot{\Gamma}^A (S_A - \Sigma_A) \Big] \mathrm{d}t + \int_{t_1}^{t_2} \left\langle F^{\mathrm{ext}}, \delta q \right\rangle \mathrm{d}t = 0,$$

subject to the *phenomenological constraint*

$$\frac{\partial L}{\partial S_A} \dot{\Sigma}_A = \left\langle F^{\mathrm{fr}(A)}, \dot{q} \right\rangle + J_{AB} \dot{\Gamma}^B, \quad \text{for } A = 1, ..., P, \tag{1}$$

and for variations subject to the *variational constraint*

$$\frac{\partial L}{\partial S_A} \delta \Sigma_A = \left\langle F^{\mathrm{fr}(A)}, \delta q \right\rangle + J_{AB} \delta \Gamma^B, \quad \text{for } A = 1, ..., P, \tag{2}$$

with $\delta q(t_1) = \delta q(t_2) = 0$ and $\delta \Gamma^A(t_1) = \delta \Gamma^A(t_2) = 0$, $A = 1, ..., P$.

By direct computations, we obtain the following evolution equations:

$$\begin{cases} \dfrac{d}{dt}\dfrac{\partial L}{\partial \dot{q}} = \dfrac{\partial L}{\partial q} - \displaystyle\sum_{A=1}^{P} \dfrac{\dot{\Gamma}^A}{\frac{\partial L}{\partial S_A}} F^{\mathrm{fr}(A)} + F^{\mathrm{ext}}, \\ \dfrac{\partial L}{\partial S_A} + \dot{\Gamma}^A = 0, \quad A = 1, ..., P, \\ \dot{S}_A - \dot{\Sigma}_A + \displaystyle\sum_{B=1}^{P} \dfrac{\dot{\Gamma}^A}{\frac{\partial L}{\partial S_A}} J_{BA} = 0, \quad A = 1, ..., P. \end{cases} \tag{3}$$

From the second equation in (3), the temperature of the subsystem $\mathbf{\Sigma}_A$, i.e., T^A can be obtained as $\dot{\Gamma}^A = -\frac{\partial L}{\partial S_A} =: T^A$. Because $\sum_{A=1}^{P} J_{AB} = 0$ for all B, the last equation in (3) yields $\dot{S}_A = \dot{\Sigma}_A$. Hence, together with (1), we obtain the following Lagrange-d'Alembert equations for the curves $q(t)$ and $S_A(t)$:

$$\begin{cases} \dfrac{d}{dt}\dfrac{\partial L}{\partial \dot{q}} - \dfrac{\partial L}{\partial q} = \displaystyle\sum_{A=1}^{P} F^{\mathrm{fr}(A)} + F^{\mathrm{ext}}, \\ \dfrac{\partial L}{\partial S_A}\dot{S}_A = \left\langle F^{\mathrm{fr}(A)}, \dot{q} \right\rangle - \displaystyle\sum_{B=1}^{P} J_{AB}\left(\dfrac{\partial L}{\partial S_B} - \dfrac{\partial L}{\partial S_A}\right), \quad A = 1, ..., P. \end{cases} \tag{4}$$

The First Law of Energy Balance. For the total energy $E : TQ \times \mathbb{R}^P \to \mathbb{R}$ given by $E(q, v_q, S_A) = \left\langle \frac{\partial L}{\partial v_q}(q, v_q, S_A), v_q \right\rangle - L(q, v_q, S_A)$, we have $\frac{d}{dt}E = \langle F^{\mathrm{ext}}, \dot{q} \rangle = P_W^{\mathrm{ext}}$ along the solution curve of (4). If the Lagrangian is given by $L(q, v, S_1, ..., S_P) = \sum_{A=1}^{P} L_A(q, v, S_A)$, the evolution equations for $\mathbf{\Sigma}_A$ are

$$\frac{d}{dt}\frac{\partial L_A}{\partial \dot{q}} - \frac{\partial L_A}{\partial q} = F^{\mathrm{fr}(A)} + F^{\mathrm{ext} \to A} + \sum_{B=1}^{P} F^{B \to A}, \quad A = 1, ..., P,$$

where $F^{B \to A}$ is the internal force exerted by $\mathbf{\Sigma}_B$ on $\mathbf{\Sigma}_A$. From Newton's third law, we have $F^{B \to A} = -F^{A \to B}$. Denoting by E_A the total energy of $\mathbf{\Sigma}_A$, we have

$$\frac{d}{dt}E_A = P_W^{\mathrm{ext} \to A} + \sum_{B=1}^{P} P_W^{B \to A} + \sum_{B=1}^{P} P_H^{B \to A}, \tag{5}$$

where $P_W^{\mathrm{ext} \to A} = \left\langle F^{\mathrm{ext} \to A}, \dot{q} \right\rangle$ is the mechanical power that flows from the exterior into $\mathbf{\Sigma}_A$, $P_W^{B \to A} = \sum_{B=1}^{P} \left\langle F^{B \to A}, \dot{q} \right\rangle$ is the internal mechanical power that flows from $\mathbf{\Sigma}_B$ into $\mathbf{\Sigma}_A$, and $P_H^{B \to A} = \sum_{B=1}^{P} J_{AB}\left(\frac{\partial L}{\partial S_B} - \frac{\partial L}{\partial S_A}\right)$ is the internal heat power from $\mathbf{\Sigma}_B$ to $\mathbf{\Sigma}_A$. It follows that the power exchange can be written as $P_H^{B \to A} = J_{AB}(T^A - T^B)$.

The Second Law and Internal Entropy Production. The total entropy of the system is $S = \sum_{A=1}^{P} S_A$. Therefore, it follows from (4) that the rate of total entropy production of the system is given by

$$\dot{S} = -\sum_{A=1}^{P} \frac{1}{T^A} \left\langle F^{\mathrm{fr}(A)}, \dot{q} \right\rangle + \sum_{A<B}^{P} J_{AB} \left(\frac{1}{T^B} - \frac{1}{T^A} \right) (T^B - T^A),$$

which becomes always positive because of the second law. This is consistent with the phenomenological relations of the form

$$F_i^{\mathrm{fr}(A)} = -\lambda_{ij}^A \dot{q}^j \quad \text{and} \quad J_{AB} \frac{T^A - T^B}{T^A T^B} = \mathscr{L}_{AB}(T^B - T^A). \tag{6}$$

In the above, λ_{ij}^A and $\mathscr{L}_{AB}$ are functions of the state variables, where the symmetric part of λ_{ij}^A are positive semi-definite and with $\mathscr{L}_{AB} \geq 0$ for all A, B. From the second relation, we get $J_{AB} = -\mathscr{L}_{AB} T^A T^B = -\kappa_{AB}$, with $\kappa_{AB} = \kappa_{AB}(q, S_A, S_B)$ the heat conduction coefficients between $\mathbf{\Sigma}_A$ and $\mathbf{\Sigma}_B$.

2 Dirac Formulation of Non-simple Systems

In this section, we develop the Dirac formulation for the dynamics of non-simple systems by means of an induced Dirac structure the Pontryagin bundle; for the details, see [1,5,6].

2.1 Dirac Structures in Thermodynamics

Thermodynamic Configuration Space. For our class of non-simple systems, let $\mathscr{Q} = Q \times V$ be a *thermodynamic configuration space*, where Q denotes the mechanical configuration space with mechanical variables $q \in Q$ as before and $V = \mathbb{R}^P \times \mathbb{R}^P \times \mathbb{R}^P$ is the thermodynamic space with thermodynamic variables $(S_A, \Gamma^A, \Sigma_A) \in V$. We denote by $x = (q, S_A, \Gamma^A, \Sigma_A)$ an element of $\mathscr{Q}$, by (x, v) an element in the tangent bundle $T\mathscr{Q}$ where $v = (v_q, v_{S_A}, v_{\Gamma^A}, v_{\Sigma_A}) \in T_x\mathscr{Q}$, and by (x, p) an element of the cotangent bundle $T^*\mathscr{Q}$, where $p = (p_q, p_{S_A}, p_{\Gamma^A}, p_{\Sigma_A}) \in T_x^*\mathscr{Q}$.

Nonlinear Constraints of Thermodynamic Type. Let $C_V \subset T\mathscr{Q} \times_{\mathscr{Q}} T\mathscr{Q}$ be the *variational constraint* locally given as

$$C_V = \left\{ (x, v, \delta x) \in T\mathscr{Q} \times_{\mathscr{Q}} T\mathscr{Q} \;\middle|\; \frac{\partial L}{\partial S_A} \delta\Sigma_A = \left\langle F^{\mathrm{fr}(A)}, \delta q \right\rangle + \sum_{B=1}^{P} J_{AB} \delta\Gamma^B, A = 1, ..., P \right\}. \tag{7}$$

For every $(x, v) \in T\mathscr{Q}$, we consider the subspace of $T_x\mathscr{D}$ given by

$$C_V(x, v) := C_V \cap \left(\{(x, v)\} \times T_x\mathscr{Q} \right) \subset T_x\mathscr{Q}.$$

The *kinematic constraint* associated to C_V is defined by

$$C_K = \{(x,v) \in T\mathscr{Q} \mid (x,v) \in C_V(x,v)\}, \tag{8}$$

which is given locally as

$$C_K = \left\{ (x,v) \in T\mathscr{Q} \;\middle|\; \frac{\partial L}{\partial S_A} v_{\Sigma_A} = \left\langle F^{\mathrm{fr}(A)}, v_q \right\rangle + \sum_{B=1}^{P} J_{AB} v_{\Gamma^B}, A = 1, ..., P \right\}. \tag{9}$$

Variational and kinematic constraints C_V and C_K are called *nonlinear constraints of thermodynamic type* if they are related as in (8).

Note also that the annihilator of $C_V(x,v) \subset T_x\mathscr{Q}$, defined by

$$C_V(x,v)^\circ = \left\{ (x,\zeta) \in T_x^*\mathscr{Q} \mid \langle \zeta, \delta x \rangle = 0, \ \ \forall \delta x \in C_V(x,v) \right\} \subset T_x^*\mathscr{Q},$$

is given, in coordinates $\zeta = (\zeta_q, \zeta_{S_A}, \zeta_{\Gamma^A}, \zeta_{\Sigma_A}) \in T_x^*\mathscr{Q}$, by

$$C_V(x,v)^\circ = \left\{ (x,\zeta) \in T_x^*\mathscr{Q} \;\middle|\; \zeta_q + \frac{\zeta_{\Sigma_A}}{\frac{\partial L}{\partial S_A}} F^{\mathrm{fr}(A)} = 0, \quad \zeta_{S_A} = 0, \right.$$

$$\left. \zeta_{\Gamma_A} = \frac{\zeta_{\Sigma_A}}{\frac{\partial L}{\partial S_A}} \sum_{B=1}^{P} J_{AB}, \ \ A = 1, ..., P \right\}.$$

Dirac Structures on the Pontryagin Bundle. $\mathscr{P} = T\mathscr{Q} \oplus T^*\mathscr{Q}$. The Pontryagin bundle $\mathscr{P}$ is defined as the Whitney sum bundle of $\mathscr{P} = T\mathscr{Q} \oplus T^*\mathscr{Q}$, with vector bundle projection, $\pi_{(\mathscr{P},\mathscr{Q})} : \mathscr{P} = T\mathscr{Q} \oplus T^*\mathscr{Q} \to \mathscr{Q}$, $\mathrm{x} = (x,v,p) \mapsto x$. Given a variational constraint $C_V \subset T\mathscr{Q} \times_{\mathscr{Q}} T\mathscr{Q}$ as in (7), we define the *induced distribution* $\Delta_{\mathscr{P}}$ on $\mathscr{P}$ by

$$\Delta_{\mathscr{P}}(x,v,p) := \left(T_{(x,v,p)} \pi_{(\mathscr{P},\mathscr{Q})} \right)^{-1} (C_V(x,v)) \subset T_{(x,v,p)}\mathscr{P},$$

for each $(x,v,p) \in \mathscr{P}$. Locally, this distribution reads

$$\Delta_{\mathscr{P}}(x,v,p) = \left\{ (x,v,p,\delta x, \delta v, \delta p) \in T_{(x,v,p)}\mathscr{P} \mid (x,\delta x) \in C_V(x,v) \right\}.$$

Further, the presymplectic form on $\mathscr{P}$ is defined from the canonical symplectic form $\Omega_{T^*\mathscr{Q}}$ on $T^*\mathscr{Q}$ as $\Omega_{\mathscr{P}} := \pi^*_{(\mathscr{P},T^*\mathscr{Q})} \Omega_{T^*\mathscr{Q}}$, which is locally given by using local coordinates $(x,v,p) = (q, S_A, \Gamma^A, \Sigma_A, v_q, v_{S_A}, v_{\Gamma^A}, v_{\Sigma_A}, p_q, p_{S_A}, p_{\Gamma^A}, p_{\Sigma_A})$ for each $\mathrm{x} = (x,v,p) \in \mathscr{P}$ as

$$\Omega_{\mathscr{P}} = dq \wedge dp_q + dS_A \wedge dp_{S_A} + d\Gamma_A \wedge dp_\Gamma + d\Sigma_A \wedge dp_{\Sigma_A}.$$

Definition 1. *The Dirac structure $D_{\Delta_{\mathscr{P}}}$ induced on $\mathscr{P}$ from $\Delta_{\mathscr{P}}$ and $\omega_{\mathscr{P}}$ is defined by, for each $\mathrm{x} \in \mathscr{P}$,*

$$D_{\Delta_{\mathscr{P}}}(\mathrm{x}) := \left\{ (u_{\mathrm{x}}, \alpha_{\mathrm{x}}) \in T_{\mathrm{x}}\mathscr{P} \times T_{\mathrm{x}}^*\mathscr{P} \mid u_{\mathrm{x}} \in \Delta_{\mathscr{P}}(\mathrm{x}) \text{ and } \right.$$

$$\left. \langle \alpha_{\mathrm{x}}, w_{\mathrm{x}} \rangle = \Omega_{\mathscr{P}}(\mathrm{x})(u_{\mathrm{x}}, w_{\mathrm{x}}) \text{ for all } w_{\mathrm{x}} \in \Delta_{\mathscr{P}}(\mathrm{x}) \right\}.$$

Proposition 1. *The local expression of the Dirac condition, for each* $\mathrm{x} = (x, v, p)$,

$$\big((x, v, p, \dot{x}, \dot{v}, \dot{p}), (x, v, p, \alpha, \beta, \gamma)\big) \in D_{\Delta_{\mathscr{P}}}(x, v, p)$$

is equivalent to

$$(x, \dot{x}) \in C_V(x, v), \quad \beta = 0, \quad \gamma = \dot{x}, \quad \dot{p} + \alpha \in C_V(x, v)^{\circ}.$$

In coordinates $(\alpha, \beta, \gamma) = (\alpha_q, \alpha_{S_A}, \alpha_{\Gamma^A}, \alpha_{\Sigma_A}, \beta_q, \beta_{S_A}, \beta_{\Gamma^A}, \beta_{\Sigma_A}, \gamma_q, \gamma_{S_A}, \gamma_{\Gamma^A}, \gamma_{\Sigma_A})$, *this condition reads as*

$$\begin{cases} \dot{p}_q + \alpha_q + (\dot{p}_{\Sigma_A} + \alpha_{\Sigma_A}) \dfrac{1}{\frac{\partial L}{\partial S_A}} F^{\mathrm{fr}(A)} = 0, \ \ \dot{p}_{S_A} + \alpha_{S_A} = 0, \\ \dot{p}_{\Gamma_A} + \alpha_{\Gamma_A} = \dot{p}_{\Sigma_A} + \alpha_{\Sigma_A} \dfrac{1}{\frac{\partial L}{\partial S_A}} \displaystyle\sum_{B=1}^{P} J_{AB}, \\ \beta_q = 0, \ \ \beta_{S_A} = 0, \ \ \beta_{\Gamma^A} = 0, \ \ \beta_{\Sigma_A} = 0, \\ \dot{q} = \gamma_q, \ \ \dot{S}_A = \gamma_{S_A}, \ \ \dot{\Gamma}_A = \gamma_{\Gamma^A}, \ \ \dot{\Sigma}_A = \gamma_{\Sigma_A}, \\ \dfrac{\partial L}{\partial S_A} \gamma_{\Sigma_A} = \left\langle F^{\mathrm{fr}(A)}, \gamma_q \right\rangle + \displaystyle\sum_{B=1}^{P} J_{AB} \gamma_{\Sigma_B}. \end{cases}$$

2.2 Dirac Formulation for Thermodynamics of Non-simple Systems

Dirac Dynamical Systems on $\mathscr{P} = T\mathscr{Q} \oplus T^*\mathscr{Q}$. For a given Lagrangian $L(q, v_q, S_A)$ on $TQ \times \mathbb{R}^P$, we introduce an *augmented Lagrangian* by $\mathscr{L}(q, S_A, \Gamma^A, \Sigma_A, v_q, v_{\Gamma^A}) := L(q, v_q, S_A) + v_{\Gamma^A}(S_A - \Sigma_A)$.

In the above, note that the augmented Lagrangian may be regarded as a (degenerate) Lagrangian function $\mathscr{L}(x, v)$ on $T\mathscr{Q}$. Further, define the *generalized energy* $\mathscr{E}$ on $\mathscr{P} = T\mathscr{Q} \oplus T^*\mathscr{Q}$ as

$$\mathscr{E}(x, v, p) := \langle p, v \rangle - \mathscr{L}(x, v).$$

Given an external force $F^{\mathrm{ext}}(q, v_q, S_A)$, which may be regarded as a map $F^{\mathrm{ext}} : T\mathscr{Q} \to T^*\mathscr{Q}$, a horizontal one-form $\widetilde{F}^{\mathrm{ext}} : \mathscr{P} \to T^*\mathscr{P}$ is induced by

$$\left\langle \widetilde{F}^{\mathrm{ext}}(x, v, p), u \right\rangle = \left\langle F^{\mathrm{ext}}(x, v), T\pi_{(\mathscr{P}, \mathscr{Q})}(u) \right\rangle, \quad \text{for all } u \in T_{(x,v,p)}\mathscr{P}.$$

Theorem 1. *Given* C_V *and* C_K *as in* (7) *and* (9), *the solution curve* $\mathrm{x} = (x(t), v(t), p(t))$ *of the Dirac system*

$$\Big((x, v, p, \dot{x}, \dot{v}, \dot{p}), \mathbf{d}\mathscr{E}(x, v, p) - \widetilde{F}^{\mathrm{ext}}(x, v, p)\Big) \in D_{\Delta_{\mathscr{P}}}(x, v, p), \tag{10}$$

satisfies the equations of motion

$$(x,\dot{x}) \in C_V(x,v), \quad p - \frac{\partial \mathscr{L}}{\partial v} = 0, \quad \dot{x} = v, \quad \dot{p} - \frac{\partial \mathscr{L}}{\partial x} - F^{\text{ext}}(x,v) \in C_V(x,v)^{\circ}.$$

In coordinates, we obtain the system

$$\begin{cases} \dot{p}_q = \dfrac{\partial L}{\partial q} - \displaystyle\sum_{A=1}^{P} \dfrac{\dot{\Gamma}^A}{\frac{\partial L}{\partial S_A}} F^{\text{fr}(A)} + F^{\text{ext}} = 0, \; \dot{p}_{\Gamma^A} + \displaystyle\sum_{B=1}^{P} \dfrac{\dot{\Gamma}^A}{\frac{\partial L}{\partial S_A}} J_{BA} = 0, \\ \dot{q} = v_q, \; \dot{\Gamma}_A = v_{\Gamma^A}, \; \dot{\Sigma}_A = v_{\Sigma_A}, \\ p_q = \dfrac{\partial L}{\partial v_q}, \; p_{\Gamma^A} = S_A - \Sigma_A, \; \dfrac{\partial L}{\partial S_A} + \dot{\Gamma}^A = 0, \\ \dfrac{\partial L}{\partial S_A} v_{\Sigma_A} = \left\langle F^{\text{fr}(A)}, v_q \right\rangle + \displaystyle\sum_{B=1}^{P} J_{AB} v_{\Gamma^B}. \end{cases} \tag{11}$$

This system yields the Lagrange-d'Alembert evolution Eq. (4) for non-simple thermodynamic systems with friction and heat conduction.,

Along the solution curve $(x(t), v(t), p(t)) \in \mathscr{P}$ of the Dirac dynamical system in (10), the energy balance equation holds as

$$\frac{d}{dt}\mathscr{E}(x,v,p) = \left\langle F^{\text{ext}}(x,v), \dot{x} \right\rangle.$$

2.3 Example of the Adiabatic Piston

The Adiabatic Piston. Now we consider a piston-cylinder system that is consisted of two cylinders connected by a rod, each of which contains a fluid (or an ideal gas) and is separated by a movable piston, as in Fig. 1 (see [2]).

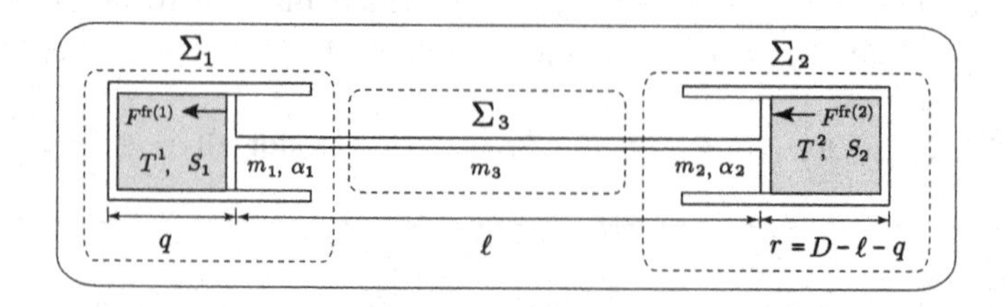

Fig. 1. The adiabatic piston problem

The system $\boldsymbol{\Sigma}$ is an interconnected system that is composed of three simple systems; the two pistons $\boldsymbol{\Sigma}_1, \boldsymbol{\Sigma}_2$ with mass m_1, m_2 and the connecting rod $\boldsymbol{\Sigma}_3$

with mass m_3. As in Fig. 1, q and $r = D - \ell - q$ denote the distances between the bottom and the top in each piston where $D = \text{const}$. Choose the state variables (q, v_q, S_1, S_2) (the entropy of $\mathbf{\Sigma}_3$ is constant), and the Lagrangian is

$$L(q, v_q, S_1, S_2) = \frac{1}{2} M v_q^2 - U_1(q, S_1) - U_2(q, S_2),$$

where $M := m_1 + m_2 + m_3$, $U_1(q, S_1) := \mathsf{U}_1(S_1, V_1 = \alpha_1 q, N_1)$, and $U_2(q, S_2) := \mathsf{U}_2(S_2, V_2 = \alpha_2 r, N_2)$, with $\mathsf{U}_i(S_i, V_i, N_i)$ the internal energies of the fluids, N_i the constant numbers of moles, and α_i the constant areas of the cylinders, $i = 1, 2$. As in (6), we have $F^{\mathrm{fr}(A)}(q, \dot{q}, S_A) = -\lambda^A \dot{q}$, with $\lambda^A = \lambda^A(q, S^A) \geq 0$, $A = 1, 2$ and $J_{AB} = -\kappa_{AB} =: -\kappa$, where $\kappa = \kappa(S_1, S_2, q) \geq 0$ is the heat conductivity of the connecting rod.

From the Dirac system formulation (11), we obtain the evolution equations as

$$\begin{cases} \dot{p}_q = \Pi_1(q, S_1)\alpha_1 - \Pi_2(q, S_2)\alpha_2 - (\lambda^1 + \lambda^2)\dot{q}, \\ \dot{q} = v_q, \quad p_q = M v_q, \\ T^1(q, S_1)\dot{S}_1 = \lambda^1 \dot{q}^2 + \kappa \left(T^2(q, S_2) - T^1(q, S_1) \right), \\ T^2(q, S_2)\dot{S}_2 = \lambda^2 \dot{q}^2 + \kappa \left(T^1(q, S_1) - T^2(q, S_2) \right), \end{cases}$$

where $T^i(q, S_i) = \frac{\partial U_i}{\partial S_i}(q, S_i)$, $\frac{\partial U_1}{\partial q} = -\Pi_1(q, S_1)\alpha_1$, and $\frac{\partial U_2}{\partial q} = \Pi_2(q, S_2)\alpha_2$.

Since the system is isolated, we recover the first law $\frac{d}{dt}E = 0$, where $E = \frac{1}{2} M \dot{q}^2 + U_1(q, S_1) + U(q, S_2)$. The second law is also recovered as

$$\frac{d}{dt} S = \left(\frac{\lambda^1}{T^1} + \frac{\lambda^2}{T^2} \right) \dot{q}^2 + \kappa \frac{(T^2 - T^1)^2}{T^1 T^2} \geq 0.$$

References

1. Courant, T.J.: Dirac manifolds. Trans. Amer. Math. Soc. **319**, 631–661 (1990)
2. Gruber, C.: Thermodynamics of systems with internal adiabatic constraints: time evolution of the adiabatic piston. Eur. J. Phys. **20**, 259–266 (1999)
3. Gay-Balmaz, F., Yoshimura, H.: A Lagrangian variational formulation for nonequilibrium thermodynamics. Part I: discrete systems. J. Geom. Phys. **111**, 169–193 (2016)
4. Gay-Balmaz, F., Yoshimur, H.: From Lagrangian mechanics to nonequilibrium thermodynamics: a variational perspective. Entropy **21**(1), 8 (2019)
5. Gay-Balmaz, F., Yoshimura, H.: Dirac structures in nonequilibrium thermodynamics for simple open systems. J. Math. Phys. **61**(9), 092701 (2020)
6. Gay-Balmaz, F., Yoshimura, H.: Dirac structures and variational formulation of port-Dirac systems in nonequilibrium thermodynamics. IMA J. Math. Control Inf. **37**(4), 1298–1347 (2020)
7. Stueckelberg, E.C.G., Scheurer, P.B.: Thermocinétique Phénoménologique Galiléenne. Birkhäuser (1974)

Author Index